COMPARATIVE PATHOPHYSIOLOGY AND TOXICOLOGY OF CYCLOOXYGENASES

COMPARATIVE PATHOPHYSIOLOGY AND TOXICOLOGY OF CYCLOOXYGENASES

ZAHER A. RADI

Pfizer Worldwide Research and Development
Drug Safety Research and Development
Cambridge, Massachusetts

WILEY

A JOHN WILEY & SONS, INC., PUBLICATION

Published by John Wiley & Sons, Inc., Hoboken, New Jersey.
Published simultaneously in Canada.

Library of Congress Cataloging-in-Publication Data:

Radi, Zaher A.
Comparative pathophysiology and toxicology of cyclooxygenases / Zaher A. Radi.
p. cm.
Includes Bibliographical references and index.
ISBN 978-0-470-57754-7 (cloth)
1. Cyclooxygenases – Pathophysiology. 2. Cyclooxygenases–Toxicology. I. Title.
RM405.5.R33 2012
615.9′5137–dc23

2012002908

10 9 8 7 6 5 4 3 2 1

This book is dedicated with love and appreciation to my father, Dr. Ahmad Khalil Radi; my mother; my brothers; my sisters; my wife, Rawan; and my kids, Ahmad, Waleed, Laith, and Lamees.

CONTENTS

PREFACE

This book covers the hot topic of cyclooxygenases (COXs) and fills the gap in the discipline of pathophysiology and toxicology of prostanoids. Information on the comparative aspects in the field of prostanoids and COX inhibitors is relatively scant. COXs catalyze the conversion of arachidonic acid into prostaglandins, which play a significant role in the health and disease of many organ systems, including the gastrointestinal tract, skeletal, renal, cardiovascular, ocular, and respiratory systems. It is known that COX-1 is expressed constitutively and is found in most normal tissues, while COX-2 can be expressed at low levels in normal tissues and is induced strongly by proinflammatory mediators. Inhibitors of COX activity include (1) conventional nonselective nonsteroidal anti-inflammatory drugs and (2) COX-2 selective nonsteroidal anti inflammatory drugs (COX-2 s-NSAIDs).

This comprehensive book covers the toxicology and pathophysiological role of COXs and the mechanisms of their toxicity after their inhibition in various body systems. The book will be of interest to toxicologic pathologists (e.g., (Society of Toxicologic Pathologists, European Society of Toxicologic Pathologists, and American College of Veterinary Pathologists), toxicologists (Society of Toxicology and Academy of Toxicological Sciences), pharmacologists (American Society for Pharmacology and Experimental Therapeutics), and scientists and researchers (Inflammation Research Association) in both academia and the pharmaceutical industry working in the field of prostanoid research and drug development of COXs; or related inhibitors as therapeutic agents. There are at present no books that address the comparative aspects of COXs; hopefully, this book will fill that gap.

Inhibition of COX-1 often elicits gastrointestinal (GI) toxicity in animals and humans. Therefore, COX-2 s-NSAIDs medicines were developed to provide a selective COX-2 agent while minimizing the attendant COX-1-mediated GI toxicities. Various nonclinical species differ in their response to COX inhibition, and COX expression varies across various body systems and within each body system. For example, rats and dogs overpredict COX inhibition for renal effects such as renal handling of electrolytes in humans. COX inhibitors in various animal models are shown to have both beneficial and detrimental effects on the skeletal system, such as on healing of ligament or tendon tears in animal models; however, large clinical trials are lacking. There is a large gap in the literature regarding these comparative aspects of prostanoid research. The book addresses such differences in expression, pathophysiology, and toxicology across species.

The books begins with an introduction that details the discovery of COXs, includes a historical background, and explains how COXs were ultimately used as therapeutic agents. The book is designed to focus on the comparative aspects of COXs based on body systems. The first chapter covers the effects and pathophysiology of COX inhibitors on the GI tract and mechanisms of toxicity. Subsequent chapters cover the pathophysiology of COX inhibition on bone, tendon, and ligament healing, COX inhibitors and renal system pathophysiology and mechanisms of toxicity, the pathophysiological role of COX inhibition in the ocular system, and COX inhibition in the respiratory and cardiovascular systems. The final chapter focuses on current research efforts to design COX inhibitors that are devoid of adverse side effects.

Zaher A. Radi

INTRODUCTION: DISCOVERY OF CYCLOOXYGENASES AND HISTORICAL PERSPECTIVE

ASPIRIN

The anti-inflammatory, antipyretic, and analgesic effects of medicinal plants have been recognized since ancient times in Egypt and Greece. Various herbal medicines, such as extracts of willow bark and leaves of myrtle, have been used for the relief of pain and swelling caused by inflammatory conditions. The ancient Greeks chewed the bark of willow trees to alleviate pain and fever. In 1543 B.C., about 3500 years ago, the Ebers Papyrus, an ancient Egyptian medical text, recommended the application of a decoction of the dried leaves of myrtle to the abdomen and back to expel rheumatic pains from the womb. Hippocrates, about 2400 years ago, recommended the juices of the poplar tree for treating eye diseases and those of the white willow tree (*Salix abla*) bark to relieve the pain of childbirth and to reduce fever. By the middle of the nineteenth century the active ingredient in these medicinal remedies was found to contain salicylates (Vane and Botting, 1998). Around A.D. 30, Aulus Cornelius Celsus described the four cardinal signs of inflammation as rubor (redness), calor (heat), dolor (pain), and tumor (swelling), and used extracts of willow leaves to relieve these signs of inflammation. In the second century A.D. and throughout Roman times, Dioscorides and Galen, Greek physicians, recommended willow bark to treat mild to moderate pain. In China and other parts of Asia, salicylate-containing plants were being applied therapeutically. The *Spiraea* plant species (called meadowsweet) had a long history of medicinal use by Native Americans. This plant contains methyl salicylate and other salicylates. The anti-inflammatory and analgesic effects of *Salix* and *Spiraea* species were known to the early inhabitants of North America and South Africa (Vane and Botting, 1998).

The era of salicylates started with a letter sent in 1758 by Reverend Edward Stone to the Royal Society in London in which he described "an account of the success of the bark of willow in the cure of agues" (Stone, 1763). Investigation of the therapeutically active substance in willow extract began early in the nineteenth century. In 1803, Wilkinson extracted the active substance from *Cortex salicis*

Comparative Pathophysiology and Toxicology of Cyclooxygenases, First Edition. Zaher A. Radi.

latifoliae, which was later found to contain high levels of tannin. In the early 1820s, Brugnateli, Fontana, and Buchner obtained therapeutically active substances of a less contaminated nature, and further purification was achieved by Buchner in 1828 at the University of Munich. Buchner removed the tannins and obtained a yellowish substance, which he called *salicin*, the Latin name for willow (Mahdi et al., 2006). In 1829, the pure crystalline form of salicin was obtained by Henri Leroux, a French pharmacist (Leroux, 1830; Vane and Botting, 1998). In 1833, Lowig prepared salicin from oil of wintergreen (*Gaultheria*) and meadowsweet (*Spiraea ulmaria*). In 1874, the commercial organic synthesis of salicylic acid was formulated by Kolbe and his colleagues and led to the founding of the Heyden Chemical Company (Vane and Botting, 2003). In 1876, Maclagan, a physician from Scotland, used salicin, the bitter principle of the common white willow, to reduce the fever, pain, and inflammation of rheumatic fever (Maclagan, 1876). By the middle of the nineteenth century, the active ingredient of these herbal medicines was found to be acetylsalicylic acid (aspirin). The name *aspirin* is derived from A = *acetyl* and *Spirsäure* = an old (German) name for salicylic acid.

Acetylsalicylic acid was synthesized by a chemistry professor, Charles Frederick von Gerhard, at Strasbourg, Germany in 1853 and the quest for the pharmaceutical development of similar remedies began. In 1895, Arthur Eichengru, director of the chemical research laboratories at the Bayer Company, assigned this task to Felix Hoffman one of his chemists. Hoffman also had personal reasons for wanting a more acceptable salicylic acid derivative. Hoffman's father had been taking salicylic acid for many years to treat his arthritis and had recently discovered that he could no longer take the drug without vomiting. Therefore, there was a need for a better tolerated salicylic acid derivative. Hoffman searched through the scientific literature and found a way of acetylating the hydroxyl group on the benzene ring of salicylic acid to form acetylsalicylic acid. After initial laboratory tests, Hoffman's father was given the drug; it was pronounced effective and later confirmed as such by a more impartial clinical trial. By 1899, the pharmaceutical manufacturing house of Frederick Bayer in Germany had named and released palatable acetylsalicylic acid (aspirin) onto the market (Vane and Botting, 1998). Thus, by the early twentieth century the antipyretic and anti-inflammatory therapeutic actions of aspirin were recognized. Similar several other antipyretic and anti-inflammatory drugs were discovered later [e.g., phenacetin, acetaminophen (paracetamol), phenylbutazone]. Because these drugs were clearly distinct from the glucocorticoids, all of them except acetaminophen were called *nonsteroidal anti-inflammatory drugs* (NSAIDs). However, endoscopic studies in 1938 found that long-term aspirin use was associated with gastric mucosa hyperemia, congestion, erosions, and ulcers (Douthwaite and Lintott, 1938).

PROSTAGLANDINS

In 1982, the presence of two distinct cyclooxygenase (COX) isoforms in brain tissue with differing sensitivities to indomethacin was suggested (Lysz and Needleman, 1982). However, COX enzymes were not purified and sequenced until 1988 (Merlie et al., 1988; DeWitt et al., 1989). The early discovery of prostaglandins (PGs) took place before cyclooxygenases were discovered. Initially, a fatty acid with potent vasoactive properties was identified in human seminal fluid (Goldblatt, 1933; von Euler, 1935). In the 1930s, Ulf von Euler of Sweden named this fatty acid *prosta glandin*, thinking it originated from the prostate gland, and he suggested that PGs play a role in sperm transport (von Euler, 1935). In the 1960s, the structure of the first two PGs (named PGE and PGF because of their partition into ether and phosphate buffer, respectively) was established (Bergström et al., 1964). Furthermore, others also discovered in the 1970s a group of lipid mediators called PGs and their relationship to inflammation (Piper and Vane, 1969; Vane, 1971,2000). Hamberg et al. (1975) discovered rabbit aorta contracting substance (RCS), which provided the first clue to the relationship between aspirin and the PGs when it was found that aspirin blocked RCS released during anaphylaxis from isolated guinea pig lungs (Palmer et al., 1970). Shortly thereafter, it was found, using guinea pig lung homogenate, that acetylsalicylic acid and indomethacin (both nonselective nonsteroidel anti-inflammatory drugs), but not morphine or hydrocortisone, also blocked PG synthesis (Vane, 1971). Smith and Willis found that aspirin prevented the release of PGs from aggregating human platelets (Smith and Willis, 1971). Additionally, it was demonstrated that aspirin and indomethacin blocked PG release from the isolated perfused spleen of dogs (Ferreira et al., 1971). Collier and Flower also demonstrated that aspirin inhibited human seminal PGs (Collier and Flower, 1971). Today, 10 specific molecular groups of PGs exist, designated by the letters A through J. Among the PGs, the most widely distributed and best characterized are those derived from arachidonic acid (i.e., PGE_2, PGD_2, PGI_2 and $PGF_{2\alpha}$). Eight types and subtypes of membrane PG receptors are conserved in mammals from mice to humans: the PGD receptor, DP; four subtypes of the PGE receptor, EP_1, EP_2, EP_3, and EP_4; the PGF receptor, FP; the PGI receptor, IP; and the thromboxane A receptor, TPA. They have different cell- and tissue-specific functions (Narumiya and Fitzgerald, 2001). PGs are known to have anti-inflammatory and antipyretic properties. For example, PGE_2 is a potent vasodilator that increases blood flow, increases vascular permeability, contributes to edema formation in inflammation, and sensitizes peripheral sensory nerve endings located at the site of inflammation (Williams and Peck, 1977; Bley et al., 1998). $PGF_{2\alpha}$ and PGE_2 play a role in the ovulation process, luteolysis, and pregnancy (Franczak et al., 2006). In 1999, Jakobsson et al. identified and characterized a novel human enzyme called microsomal PGE synthase-1 (mPGES-1), a member of the membrane-associated proteins involved in the eicosanoid and glutathione metabolism (MAPEG) superfamily with the ability to catalyze the conversion of PGH_2 to PGE_2. In addition, a cytosolic form of PGE synthase, termed cPGES-1, which also isomerizes PGH_2 to PGE_2, was also cloned (Tanioka et al., 2000). In 2002, another isoform of membrane-associated PGE synthase, termed mPGES-2, was identified (Tanikawa et al., 2002). These PG discoveries led to the identification of the COX enzyme, which led to the generation of PGs (Vane, 1971; Vane and Botting, 1998). Among the three PGE synthase enzymes, mPGES-1 has received much attention since it is inducible and linked functionally with COX-2 (Mancini et al., 2001).

CYCLOOXYGENASES

The most important lipid mediators are known as eicosanoids (Greek *eicosa* = 20, for 20-carbon fatty acid derivatives). The polyunsaturated fatty acid arachidonic acid (AA) is the precursor for the biosynthesis of eicosanoids. In mammalian cells, AA cannot be synthesized de novo and must be obtained either through diet or through conversion of linoleic acid by chain elongation to dihomo-γ-linolenic acid (Funk, 2001). The discovery of the importance of fatty acids and AA to health came from experiments in laboratory animals, where it was found that feeding a fat-free diet to rats led to growth retardation, reproductive disturbances, scaly skin, kidney lesions, and excessive water consumption (Burr and Burr, 1930). Arachidonic acid is found in mammalian cell membrane phospholipids. When the cell receives a stimulius (e.g., inflammation, trauma), the phospholipase $A_2(PLA_2)$ enzyme is activated to release AA from membrane phospholipids. Prostaglandin H (PGH) synthases, also called cyclooxygenases (COXs), are membrane-bound enzymes whose main role is to catalyze the first two steps (cyclooxygenation and peroxidation) to transform AA into the intermediate cyclic endoperoxidase prostaglandin $H_2(PGH_2)$ and prostaglandin $G_2(PGG_2)$. In the first step, COX cyclizes and adds two molecules of O_2 to AA to form the cyclic hydroperoxide PGG_2. Therefore, the term *cyclooxygenase* originated from this first reaction involving the transformation of AA to PG via cyclization and oxygenation. In the second step, COX reduces PGG_2 to PGH_2. PGH_2 is highly unstable and serves as an intermediate precursor for a variety of prostanoids [e.g., thromboxane $A_2(TXA_2)$, prostacyclin $(PGI_2), PGD_2, PGE_2, PGF_2$] with diverse biological actions (Funk, 2001) (Fig. I-1) (Radi, 2009). COX is present in the endoplasmic reticulum and nuclear membrane. The primary source of COX isolation and characterization originated from ovine seminal vesicles (Hemler et al., 1976; Miyamoto et al., 1976).

COX isoforms are called COX-1, COX-2, and COX-3. The COX-1 (constitutive) gene encodes a 2.8-kb message and is localized to the endoplasmic reticulum. It is expressed constitutively in most normal tissues throughout the body and is responsible for the synthesis of PGs in many cell types. High levels of COX-1 are present in monocytes, endothelial cells, platelets, and seminal vesicles (Smith and DeWitt, 1996). Therefore, COX-1 is considered as a "housekeeping" enzyme involved in normal physiological processes. COX-1 is inhibited by aspirin, resulting in irreversible acetylation of PGHS-1 and inhibition of platelet TXA_2 formation (Shimokawa and Smith, 1992). TXA_2 is involved in platelet aggregation and PGE_2 is involved in protecting the gastric mucosa. COX-2 (inducible) is localized to the endoplasmic reticulum and the nuclear membrane and is found in low amounts in healthy tissue, but is induced rapidly and strongly in inflamed and pathological tissues by proinflammatory mediators and cytokines (Radi, 2009). Cytokines (e.g., IL-1, IL-6, IL-8), immunomodulatory proteins [tumor necrosis factor alpha (TNFα), gamma interferon (IFNγ)], bacterial lipopolysaccarides (LPs), paramyxoviruses, and growth factors [epidermal growth factor (EGF), platelet-derived growth factor (PDGF)] can induce COX-2 expression in various cell types, including fibroblasts;

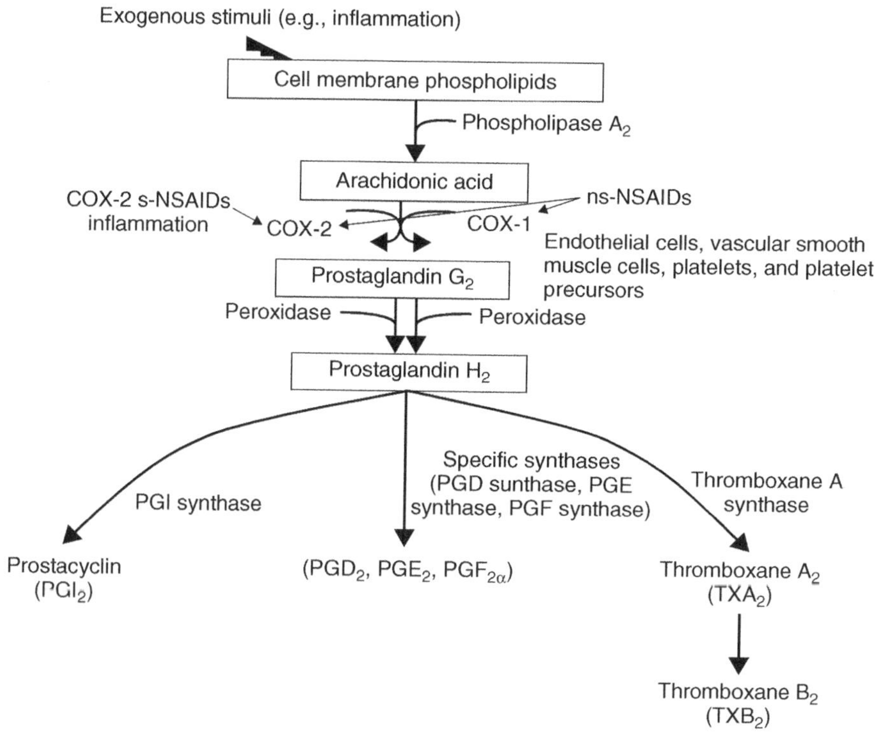

FIGURE I-1 Prostaglandin (PG) metabolism pathway. Arachidonic acid (AA) is found in mammalian cell membrane phospholipids. When the cell receives a stimuli (e.g., inflammation, trauma), the phospholipase $A_2(PLA_2)$ enzyme is activated to release AA from membrane phospholipids. Prostaglandin H (PGH) synthases, also called cyclooxygenases (COXs), are membrane-bound enzymes whose main role is to catalyze the first two steps (cyclooxygenation and peroxidation) to transform AA into the intermediate cyclic endoperoxidase prostaglandin $H_2(PGH_2)$ and prostaglandin $G_2(PGG_2)$. In the first step, COX cyclizes and adds two molecules of O_2 to AA to form the cyclic hydroperoxide PGG_2. In the second step, COX reduces PGG_2 to PGH_2. PGH_2 is highly unstable and serves as the precursor intermediate for a variety of prostanoids [e.g., thromboxane $A_2(TXA_2)$, prostacyclin $(PGI_2), PGD_2, PGE_2, PGF_2$]. [The final, definitive version of this figure was published in Z. A. Radi, *Toxicologic Pathology*, 37(1), 2009, pp. 34–46.]

macrophages; epithelial, endothelial, and smooth muscle cells; synoviocytes; and osteoblasts (Morita, 2002; Radi et al., 2010).

Polyclonal antiserum against sheep seminal vesicle PGH synthase, which cross-reacts with human cyclooxygenase, was used to determine PGH synthase modulation by interleukin-1 (IL-1) cytokine in cultured human dermal fibroblast cells (Raz et al., 1988). It was demonstrated that IL-1 treatment resulted in dose–response curves for stimulation of PE_2 formation, increased cellular COX activity, and the increased synthetic rate of newly formed COX. The authors concluded that the IL-1 effect is mediated primarily, if not solely, via induction of

COX synthesis (Raz et al., 1988). Another experiment found that LPS stimulation of human monocytes resulted in increased COX enzyme activity without affecting phospholipase and thromboxane synthase activities. Additionally, a PG inhibitor, dexamethasone, completely blocked LPS-induced prostanoid release by inhibiting the activity of COX (Fu et al., 1990).

Both COX isoforms map to distinct chromosomes. In humans, COX-1 and COX-2 are localized on chromosomes 9 and 1, respectively (Funk et al., 1991; Tay et al., 1994). The COX isoforms differ in length, the COX-2 gene being smaller. The COX-1 gene is approximately 22 kb in length, contains 11 exons, and is transcribed as a 2.8-kb mRNA (Yokoyama and Tanabe, 1989; Funk et al., 1991). The COX-2 gene is approximately 8 kb in length, contains 10 exons and 9 introns, and is transcribed as 4.6-, 4-, and 2.8-kb mRNAs variants (Hla and Neilson, 1992). The amino acid sequences of COX-1 and COX-2 are about 80% identical. There are amino acid carboxyl (C) and N terminus differences between the two COX isoforms. COX-1 has a 17-amino acid hydrophobic leader sequence near its N terminus; COX-2, on the other hand, has an 18-amino acid residue located near its C terminus (Funk et al., 1991; Jones et al., 1993; Tay et al., 1994). COX-1 has 576 amino acids with a molecular mass of 70 kDa (Yokoyama et al., 1989; Funk et al., 1991). The human COX-2 cDNA encodes a polypeptide that contains 604 amino acids with a molecular mass of 70 kDa (Hla and Neilson, 1992; Jones et al., 1993). The COX-2 cDNA of several other species (e.g., rat, mouse, chicken, dog, sheep, rabbit, guinea pig, horse, mink, cow) was also cloned later. A number of sequence elements control the half-life of an mRNA, either by stimulating or inhibiting degradation. In mammalian cells, sequence elements rich in adenosine and uridine, called AU-rich elements (AREs), were identified by their ability to target host mRNAs toward rapid degradation (Barreau et al., 2006). COX-2 mRNA has multiple AUUUA instability sequences that mediate transcript rapid degradation (Barreau et al., 2006).

In 2002, a third COX enzyme, COX-3, although still controversial, was discovered. COX-3 is expressed primarily in the canine cerebral cortex (Chandrasekharan et al., 2002). COX-3 was suggested to represent a potential primary central mechanism for the action of acetaminophen. Functional studies have demonstrated that the dog COX-3 possesses glycosylation-dependent cyclooxygenase activity and is potently inhibited by some, but not all, NSAIDs. COX-3 in dogs is selectively inhibited by drugs such as acetaminophen, phenacetin, antipyrine, and dipyrone (Chandrasekharan et al., 2002). The existence of human COX-3 and its relationship to analgesic drugs such as acetaminophen is yet to be determined. Database analysis of human COX-1 showed a frameshift induced by intron 1, possibly revealing COX-3 to be a virtual protein in humans. Western blot analysis of human aorta tissue using polyclonal antibodies directed against the first 13 amino acids of the predicted human and dog/mouse COX-3 detected a 65-kDa protein postulated to be human COX-3 (Schwab et al., 2003).

However, Schneider et al. (2005) confirmed that the spliced variant COX-1 protein is not detectable in human tissue and does not possess any catalytic activity in expression studies. Interestingly, the expression of intron 1 retaining COX-1 splice variants was studied in the human colon cancer cell line Caco-2 and in human colonic tissue samples. The investigators observed an increase of the COX-3 transcript, but not of protein, in human colonic tissue. In addition, COX-3 was up-regulated in hypertonic conditions in a human colon cancer cell line, and transcript-specific degradation by RNA interference increases COX-1 and COX-2 mRNA. Thus, although the transcript is not translated, it may play a regulatory role in COX-mediated epithelial osmoregulation (Nurmi et al., 2005). COX-3 has also been cloned in the mouse and rat (Snipes et al., 2005; Kis et al., 2006).

Inhibitors of COX activity include (1) conventional nonselective nonsteroidal anti-inflammatory drugs (ns-NSAIDs) and (2) COX-2 selective nonsteroidal anti-inflammatory drugs (COX-2 s-NSAIDs). The antipyretic, analgesic, and anti-inflammatory properties of NSAIDs are due to blocking the transformation of AA into PGH_2 and subsequent PGs. Examples of NSAIDs include, but are not limited to, aspirin, ibuprofen, ketoprofen, piroxicam, naproxen, sulindac, fenoprofen, indomethacin, meclofenamate, and phenylbutazone (Radi and Khan, 2006). After the early introduction of aspirin, the 1940s saw the introduction of phenylbutazone, the fenamates appeared in the 1950s, indomethacin in the 1960s, the proprionates in the 1970s, and the oxicams in the 1980s (Flower, 2003). The development of the first category of ns-NSAIDs began with phenylbutazone in 1946. Epidemiological studies revealed that inhibition of COX-1 often elicits gastrointestinal (GI) toxicity in animals and humans. Therefore, COX-2 s-NSAIDs were developed to provide a selective COX-2 agent while minimizing the attendant COX-1-mediated GI toxicities.

COX-2 SELECTIVE NSAIDs

In 1991, before the discovery of the COX-2 selective inhibitors, the Dupont Company reported the development of DuP697, a potent inhibitor of paw swelling in unestablished and established adjuvant arthritis in rats (Gans et al., 1990). DuP697 was a novel orally effective PG synthesis inhibitor with potent anti-inflammatory effects and reduced ulcerogenic properties (Gans et al., 1990). DuP697 showed activity in in vitro assays of COX using seminal vesicle or rat kidney preparations (known to contain predominantly COX-1), but was more effective against rat brain prostanoid synthesis (Flower, 2003). DuP697 served as a building block for the synthesis of COX-2 selective inhibitors and as the basic chemical model for the coxibs. In 1993, another COX-2 s-NSAID, NS-398 [*N*-(2-cyclohexyloxy-4-nitrophenyl)methane sulfonamide], was found to be associated with fewer gastric side effects (Futaki et al., 1993). The crystal structures of COX-1 and COX-2 were identified in 1994 and 1996, respectively (Picot et al., 1994). For the COX-1 crystal structure, attention was drawn to the importance of arginine 120 (which interacted with the carboxyl group of both substrate and inhibitors) and tyrosine 355 in determining the stereospecificity of NSAID binding (Picot et al., 1994; Kurumbail et al., 1996). For the COX-2 crystal structure, the structure of murine COX-2 and complexes with flurbiprofen, indomethacin, and SC-558 were determined at resolutions of 3 to 2.5 Å (Kurumbail et al., 1996). The identification of COX-2 s-NSAIDs was based on (1) cell lines with COX-1 and COX-2 activities, (2) isolated recombinant

COX enzymes, (3) cellular sources of COX-1 (platelets) and COX-2 (stimulated macrophages), (4) cell lines such as Chinese hamster ovary cells transfected with recombinant COX-1 or COX-2, and (5) whole blood assays in which PG levels were measured.

Thus, celecoxib (Celebrex) and rofecoxib (Vioxx), the first COX-2 s-NSAIDs inhibitors to reach the market, were based on DuP697 (Flower, 2003). Chemically, rofecoxib and celecoxib belonged to a diaryl-substituted class of compounds. Celecoxib is a sulfonamide-substituted 1,5-diaryl pyrazole compound, whereas rofoecoxib is a methylsulfonylphenyl compound. Celecoxib and rofecoxib were approved in 1998 and 1999, respectively. Celecoxib exhibits anti-inflammatory, analgesic and antipyretic activities in animal models and was approved in the United States in December 1998 and in Europe (Sweden) in December 2000 for symptomatic relief in the treatment of osteoarthritis and rheumatoid arthritis in humans. In subsequent years, the U.S. Food and Drug Administration placed a boxed warning on all prescription NSAIDs.

REFERENCES

Barreau C, Paillard L, Osborne HB. AU-rich elements and associated factors: Are there unifying principles? Nucleic Acids Res 2006;33:7138–7150.

Bergström S, Danielsson H, Samuelsson B. The enzymatic formation of prostaglandin E2 from arachidonic acid. Biochim Biophys Acta 1964;90:207–210.

Bley KR, Hunter JC, Eglen RM, Smith JA. The role of IP prostanoid receptors in inflammatory pain. Trends Pharmacol Sci 1998;19:141–147.

Burr GO, Burr MM. On the nature and role of the fatty acids essential in nutrition. J Biol Chem 1930;86:587–621.

Chandrasekharan NV, Dai H, Roos KL, Evanson NK, et al. COX-3, a cyclooxygenase-1 variant inhibited by acetaminophen and other analgesic/antipyretic drugs: cloning, structure, and expression. Proc Natl Acad Sci USA 2002;99:13926–13931.

Collier JG, Flower RJ. Effect of aspirin on human seminal prostaglandins. Lancet 1971;2:852–853.

DeWitt DL, el-Harith EA, Smith WL. Molecular cloning of prostaglandin G/H synthase. Adv Prostaglandin Thromboxane Leukot Res 1989;19:454–457.

Douthwaite A, Lintott CA. Gastroscopic observations of the effect of aspirin and certain other substances on the stomach. Lancet 1938;2:1222.

Ferreira SH, Moncada S, Vane Jr. Indomethacin and aspirin abolish prostaglandin release from the spleen. Nat New Biol 1971;231:237–239.

Flower RJ. The development of COX-2 inhibitors. Nat Rev Drug Discov 2003;2:179–191.

Franczak A, Kotwica G, Kurowicka B, Oponowicz A, et al. Expression of enzymes of cyclooxygenase pathway and secretion of prostaglandin E2 and F2alpha by porcine myometrium during luteolysis and early pregnancy. Theriogenology 2006;66:1049–1056.

Fu JY, Masferrer JL, Seibert K, Raz A, et al. The induction and suppression of prostaglandin H2 synthase (cyclooxygenase) in human monocytes. J Biol Chem 1990;265:16737–16740.

Funk CD. Prostaglandins and leukotrienes: advances in eicosanoid biology. Science 2001;294: 1871–1875.

Funk CD, Funk LB, Kennedy ME, Pong AS, et al. Human platelet/erythroleukemia cell prostaglandin G/H synthase: cDNA cloning, expression, and gene chromosomal assignment. FASEB J 1991; 5:2304–2312.

Futaki N, Arai I, Hamasaka Y, Takahashi S, et al. Selective inhibition of NS-398 on prostanoid production in inflamed tissue in rat carrageenan-air-pouch inflammation. J Pharm Pharmacol 1993;45: 753–755.

Gans KR, Galbraith W, Roman RJ, Haber SB, et al. Anti-inflammatory and safety profile of DuP 697, a novel orally effective prostaglandin synthesis inhibitor. J Pharmacol Exp Ther 1990;254:180–187.

Goldblatt MW. A depressor substance in seminal fluid. J Soc Chem Ind (Lond) 1933;52:1056–1057.

Hamberg M, Svensson J, Samuelsson B. Thromboxanes: a new group of biologically active compounds derived from prostaglandin endoperoxides. Proc Natl Acad Sci USA 1975;72:2994–2998.

Hemler M, Lands WE. Purification of the cyclooxygenase that forms prostaglandins: demonstration of two forms of iron in the holoenzyme. J Biol Chem 1976;251:5575–5579.

Hla T, Neilson K. Human cyclooxygenase-2 cDNA. Proc Natl Acad Sci USA 1992;89:7384–7388.

Jakobsson PJ, Thoren S, Morgenstern R, Samuelsson B. Identification of human prostaglandin E synthase: a microsomal, glutathione-dependent, inducible enzyme, constituting a potential novel drug target. Proc Natl Acad Sci USA 1999;96:7220–7225.

Jones DA, Carlton DP, McIntyre TM, Zimmerman GA, et al. Molecular cloning of human prostaglandin endoperoxide synthase type II and demonstration of expression in response to cytokines. J Biol Chem 1993;268:9049–9054.

Kis B, Snipes JA, Gaspar T, Lenzser G, et al. Cloning of cyclooxygenase-1b (putative COX-3) in mouse. Inflamm Res 2006;55:274–278.

Kurumbail RG, Stevens AM, Gierse JK, McDonald JJ, et al. Structural basis for selective inhibition of cyclooxygenase-2 by anti-inflammatory agents. Nature 1996;384:644–648.

Leroux H. Discovery of salicine. J Chim Med 1830;6:340–432.

Lysz TW, Needleman P. Evidence for two distinct forms of fatty acid cyclooxygenase in brain. J Neurochem 1982;38:1111–1117.

Maclagan TJ. The treatment of acute rheumatism by salicin. Lancet 1876;2:601–604.

Mahdi JG, Mahdi AJ, Bowen ID. The historical analysis of aspirin discovery, its relation to the willow tree and antiproliferative and anticancer potential. Cell Prolif 2006;39:147–155.

Mancini JA, Blood K, Guay J, Gordon R, et al. Cloning, expression, and up-regulation of inducible rat prostaglandin E synthase during lipopolysaccharide-induced pyresis and adjuvant-induced arthritis. J Biol Chem 2001;276:4469–4475.

Merlie JP, Fagan D, Mudd J, Needleman P. Isolation and characterization of the complementary DNA for sheep seminal vesicle prostaglandin endoperoxide synthase (cyclooxygenase). J Biol Chem 1988;263:3550–3553.

Miyamoto T, Ogino N, Yamamoto S, Hayaishi O. Purification of prostaglandin endoperoxide synthetase from bovine vesicular gland microsomes. J Biol Chem 1976;251:2629–2636.

Morita I. Distinct functions of COX-1 and COX-2. Prostaglandins Other Lipid Mediat 2002;68–69: 165–175.

Narumiya, S, Fitzgerald, GA. Genetic and pharmacological analysis of prostanoid receptor function. J Clin Invest 2001;108:25–30.

Nurmi JT, Puolakkainen PA, Rautonen NE. Intron 1 retaining cyclooxygenase 1 splice variant is induced by osmotic stress in human intestinal epithelial cells. Prostaglandins Leukot Essent Fatty Acids 2005;73:343–350.

Palmer MA, Piper PJ, Vane Jr. The release of rabbit aorta contracting substance (RCS) from chopped lung and its antagonism by anti-inflammatory drugs. Br J Pharmacol 1970;40:581P–582P.

Picot D, Loll PJ, Garavito RM. The x-ray crystal structure of the membrane protein prostaglandin H2 synthase-1. Nature 1994;367:243–249.

Piper PJ, Vane Jr. Release of additional factors in anaphylaxis and its antagonism by anti-inflammatory drugs. Nature 1969;223:29–35.

Radi ZA. Pathophysiology of cyclooxygenase inhibition in animal models. Toxicol Pathol 2009;37: 34–46.

Radi ZA, Khan NK. Effects of cyclooxygenase inhibition on the gastrointestinal tract. Exp Toxicol Pathol 2006;58:163–173.

Radi ZA, Meyerholz DK, Ackermann MR. Pulmonary cyclooxygenase-1 (COX-1) and COX-2 cellular expression and distribution after respiratory syncytial virus and parainfluenza virus infection. Viral Immunol 2010;23:43–48.

Raz A, Wyche A, Siegel N, Needleman P. Regulation of fibroblast cyclooxygenase synthesis by interleukin-1. J Biol Chem 1988;263:3022–3028.

Schneider C, Boeglin WE, Brash AR. Human cyclo-oxygenase-1 and an alternative splice variant: contrasts in expression of mRNA, protein and catalytic activities. Biochem J 2005;385:57–64.

Schwab JM, Beiter T, Linder JU, Laufer S, et al. COX-3: a virtual pain target in humans? FASEB J 2003;17:2174–2175.

Shimokawa T, Smith WL. Prostaglandin endoperoxide synthase: the aspirin acetylation region. J Biol Chem 1992;267:12387–12392.

Smith JB, Willis AL. Aspirin selectively inhibits prostaglandin production in human platelets. Nat New Biol 1971;231:235–237.

Smith WL, Dewitt DL. Prostaglandin endoperoxide H synthases-1 and -2. Adv Immunol 1996;62: 167–215.

Snipes JA, Kis B, Shelness GS, Hewett JA, et al. Cloning and characterization of cyclooxygenase-1b (putative cyclooxygenase-3) in rat. J Pharmacol Exp Ther 2005;313:668–676.

Stone E, Rev. An account of the success of the bark of the willow in the cure of agues. Philos Trans 1763;53:195–200.

Tanikawa N, Ohmiya Y, Ohkubo H, Hashimoto K, et al. Identification and characterization of a novel type of membrane-associated prostaglandin E synthase. Biochem Biophys Res Commun 2002;291:884–889.

Tanioka T, Nakatani Y, Semmyo N, et al. Molecular identification of cytosolic prostaglandin E2 synthase that is functionally coupled with cyclooxygenase-1 in immediate prostaglandin E2 biosynthesis. J Biol Chem 2000;275:32775–32782.

Tay A, Squire JA, Goldberg H, Skorecki K. Assignment of the human prostaglandin- endoperoxide synthase 2 (PTGS2) gene to 1q25 by fluorescence in situ hybridization. Genomics 1994;23:718–719.

Vane Jr. Inhibition of prostaglandin synthesis as a mechanism of action for aspirin-like drugs. Nat New Biol 1971;231:232–235.

Vane JR. The fight against rheumatism: from willow bark to COX-1 sparing drugs. J Physiol Pharmacol 2000;51:573–586.

Vane JR, Botting RM. Anti-inflammatory drugs and their mechanism of action. Inflamm Res 1998;47 Suppl 2:S78–S87.

Vane JR, Botting RM. The mechanism of action of aspirin. Thromb Res 2003;110:255–258.

von Euler US. A depressor substance in the vesicular gland. J Physiol 1935;84:21P.

Williams TJ, Peck MJ. Role of prostaglandin-mediated vasodilatation in inflammation. Nature 1977; 270:530–532.

Yokoyama C, Tanabe T. Cloning of human gene encoding prostaglandin endoperoxide synthase and primary structure of the enzyme. Biochem Biophys Res Commun 1989;165:888–894.

CHAPTER 1

GASTROINTESTINAL TRACT

INTRODUCTION

Rheumatoid arthritis (RA) is a chronic systemic inflammatory autoimmune disorder characterized by inflammation of synovial joints leading to progressive erosion of cartilage and bone. RA is associated with swelling, pain, and stiffness of multiple synovial joints, with an annual incidence of 31 per 100,000 women and 13 per 100,000 men. RA is more common in women than in men by a ratio of approximately 3 : 1 (Doan and Massarotti, 2005). Osteoarthritis (OA) is a common, age-related disorder of synovial joints that is pathologically characterized by irregularly distributed loss of cartilage more frequently in areas of increased load, sclerosis of subchondral bone, subchondral cysts, marginal osteophytes, increased metaphyseal blood flow, and variable synovial inflammation. Cyclooxygenase (COX) has two distinct membrane-anchored isoenzymes: a constitutively expressed (COX-1) and a highly induced (COX-2) isoenzyme. Following exogenous stimuli (i.e., inflammation), arachidonic acid is liberated by phospholipases. COX-1 and COX-2 are rate-limiting enzymes with COX and peroxidase activities that catalyze the conversion of arachidonic acid to prostaglandin (PG) endoperoxide (PGG_2) and prostanoids, which are then reduced to PGH_2 (Eling et al., 1990). PGH_2 is further metabolized to thromboxane A_2 (TXA_2), prostacyclin (PGI_2), PGD_2, $PGF_{2\alpha}$, and PGE_2 (Fig. 1-1) (Dannenberg et al., 2001; Radi and Khan, 2006b; Radi, 2009). Prostanoids, including TXA_2 and PGI_2, help regulate vascular tone and thrombosis via COX activity. TXA_2 is a vasoconstrictor that is largely platelet derived and COX-1 dependent, and it promotes platelet adhesion and aggregation and smooth muscle cell proliferation. PGI_2 is an endothelial-derived vasodilator with antiaggregatory platelet functions but is both COX-1 and COX-2 dependent (Kearney et al., 2004).

The inhibitors of COX activity include nonselective nonsteroidal anti-inflammatory drugs (ns-NSAIDs) and COX-2 selective nonsteroidal anti-inflammatory drugs (s-NSAIDs). Nonselective NSAIDs, at therapeutic doses, inhibit both COX-1 and COX-2 (Fig. 1-1) (Dannenberg et al., 2001; Radi and Khan, 2006b; Radi, 2009). The analgesic and anti-inflammatory properties of NSAIDs are linked to COX-2 inhibition, while many of the gastrointestinal tract (GI) toxicities and side effects have been linked variably to COX-1 and/or COX-2 inhibition and, in some cases, directly to the secondary pharmacologic properties of

Comparative Pathophysiology and Toxicology of Cyclooxygenases, First Edition. Zaher A. Radi.

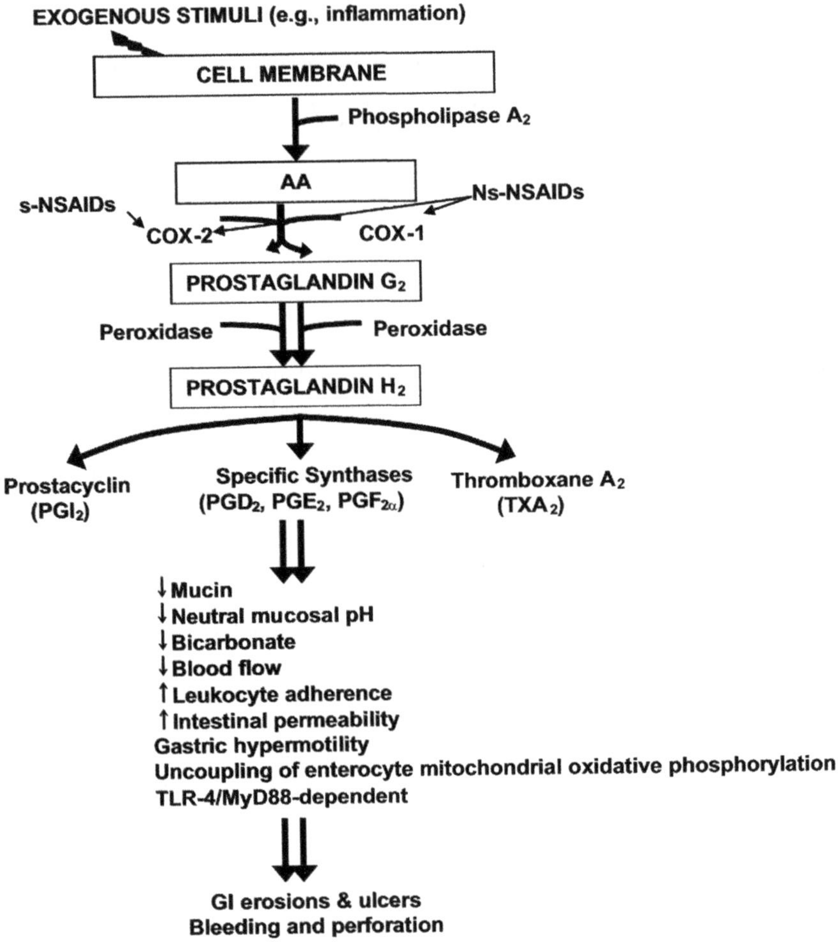

FIGURE 1-1 Pathophysiological role of prostaglandins (PGs) in the gastrointestinal (GI) tract and effects of ns-NSAIDs and COX-2 s-NSAIDs. Following an exogenous stimulus (e.g., inflammation), cell membrane phospholipid is liberated to arachidonic acid (AA) by phospholipase A_2. Both COX-1 and COX-2 catalyze the conversion of AA into various PGs. COX-1 is the predominant isoform in the normal GI tract (gastric fundus, corpus, antrum and/or pylorus, duodenum, jejunum, ileum, cecum, and colon), while COX-2 expression is up-regulated during inflammatory or neoplastic conditions. Nonselective NSAIDs (e.g., carprofen, etodolac, flunixin meglumine, ketoprofen, indomethacin, phenylbutazone) inhibit COX-1 and COX-2, while selective s-NSAIDs (e.g., celecoxib, firocoxib, rofecoxib, lumiracoxib, valdecoxib) spare COX-1 and inhibit only COX-2. Potential mechanisms of ns-NSAID-mediated GI toxicity include (1) increased intestinal epithelial permeability, (2) uncoupling of mitochondrial oxidative phosphorylation, (3) gastric hypermotility, (4) decreased epithelial cell secretion of bicarbonates, (5) decreased mucin secretion, (6) decreased blood flow, (7) decreased neutral pH of mucosa, (8) leukocyte infiltration, and (9) TLR-4/MyD88-dependent into the GI mucosa after injury. Loss of these GI protective mechanisms can lead to GI erosion, ulcers, bleeding, and perforation. (Reprinted from Z. A. Radi and N. K. Khan, Effects of cyclooxygenase inhibition on the gastrointestinal tract, *Experimental and Toxicologic Pathology*, 58, pp. 163–173. Copyright © 2006, with permission from Elsevier.)

the select drugs. COX-2 selective NSAIDs (i.e., celecoxib, deracoxib, etoricoxib, firocoxib, lumiracoxib, parecoxib, robenacoxib, rofecoxib, and valdecoxib) were developed to provide a drug that is selective for COX-2, which at therapeutic doses demonstrated therapeutic benefits comparable to those of conventional ns-NSAIDs without the attendant COX-1-mediated toxicities (Radi and Khan, 2006b). The first human-use COX-2 s-NSAIDs were celecoxib and rofecoxib, approved for the treatment of OA and RA. Later drug developments would produce deracoxib, etoricoxib, firocoxib, parecoxib, lumiracoxib, robenacoxib, and valdecoxib. The focus of this chapter is a detailed examination of the comparative expression of COX-1 and COX-2, the effects of COX-2 selective and nonselective NSAID inhibition on the GI system, and the pathophysiological mechanisms of such GI effects and toxicities.

COMPARATIVE COX-1 AND COX-2 EXPRESSION IN THE GI TRACT

GI expression of COX-1 and COX-2 in various species is summarized in Table 1-1 (Radi and Khan, 2006b; Radi, 2009). COX-1 (and not COX-2) is the predominant isoform in the normal GI tract (i.e., gastric fundus, corpus, antrum and/or pylorus, duodenum, jejunum, ileum, cecum, and colon) and is expressed normally in canine, humans, and nonhuman primates. The COX-2 isoform is nearly absent in these species, except in rats and for low levels in the large intestine (Kargman et al., 1996; Seibert et al., 1997; Koki et al., 2002a; Maziasz et al., 2003). Both COX-1 and COX-2 are present in the normal human gastric mucosa and colon (Jackson et al., 2000; Fornai et al., 2006).

COX-1 is found in the mucosal epithelium, vascular endothelium, neurones of myenteric ganglia, and in smooth muscle cells of the tunica muscularis. However, expression levels of COX-1 in the GI tract show wide intra-anatomical and interspecies variability. For example, both the gastric antrum and pyloric region of dogs contain 10-fold more COX-1 protein than is contained in the small intestine

TABLE 1-1 Comparative COX-1 and COX-2 Expression in the Gastrointestinal Tract

	COX-1					COX-2				
Location	Dog	Horse	Human	Monkey	Rat	Dog	Horse	Human	Monkey	Rat
Stomach fundus	×		×	×	×			×		×
Pyloric antrum	×		×	×	×			×		×
Duodenum	×		×	×	×					×
Jejunum	×	×	×	×	×		×			×
Ileum	×		×	×	×					×
Cecum	×		×	×	×					×
Colon	×		×	×	×	×		×	×	×

Source: Reprinted from Z. A. Radi and N. K. Khan, Effects of cyclooxygenase inhibition on the gastrointestinal tract, *Experimental and Toxicologic Pathology*, 58, pp. 163–173. Copyright © 2006, with permission from Elsevier.

(Seibert et al., 1997). In comparison, the nonhuman primate small intestine has fivefold more COX-1 protein than that in rodent or canine small intestine tissues, and rodents express less COX-1 in the GI tract than do nonhuman primates or humans (Kargman et al., 1996). In rats, weak COX-1 immunostaining is found in the stomach, small intestine, and colon (Burdan et al., 2008). In a rat model of colitis, both COX-1 and COX-2 were expressed in the normal colon and present in neurons of myenteric ganglia, while COX-2 was up-regulated in rats with colitis (Fornai et al., 2006). In rabbits, only COX-1 was detected in parietal cells, while both COX-1 and COX-2 were expressed in gastric glands, with the relative protein density of COX-1 being sixfold higher than that of COX-2 (Nandi et al., 2009). In humans, the highest and lowest areas of COX-1 expression are in the small intestine and gastric fundus/antrum, respectively (Kargman et al., 1996). Strong gastric parietal cell COX-1 and COX-2 immunoreactivity has been observed in the normal human gastric mucosa (Jackson et al., 2000). COX-2 up-regulation has been described within the mucosa in the presence of inflammation or ulcers. COX-1 and COX-2 immunostaining was increased at the rim of ulcers and in *Helicobacter pylori* gastritis, particularly at the mid-glandular zone and lamina propria inflammatory cells (Jackson et al., 2000). Some studies suggest that the predominant source of increased gastric PGE_2 in *H. pylori* infection in humans is probably COX-1 derived (Scheiman et al., 2003). In inflammatory bowel disease (IBD), COX-1 was localized in the crypt epithelium of the normal ileum and colon and its expression was unchanged. COX-2 expression, on the other hand, was undetectable in normal ileum or colon but was induced in apical epithelial cells of inflamed foci in IBD (Singer et al., 1998). In another study, COX-2 expression up-regulation occurred in neural cells of the myenteric plexus in patients with active IBD (Roberts et al., 2001).

COX-2 is normally absent (except in the colonic mucosa) in the intestinal tract in dogs, nonhuman primates, and humans (Koki et al., 2002a; Maziasz et al., 2003). In horses, COX-1 and COX-2 were expressed in nonischemic- and ischemic-injured jejunal mucosa tissues obtained 18 h after recovery, with ischemia causing significant up-regulation of both COX isoforms (Tomlinson et al., 2004). In rats, COX-2 is present at low levels in close association with macrophages in the region of gut-associated lymphoid tissue (Kargman et al., 1996). COX-2 expression was observed in the rat fundus and pylorus regions of the stomach, intestinal tract (jejunum, ileum, duodenum, cecum, colon, and rectum), and intestinal tract parasympathetic ganglia of the submucosa and muscularis (Haworth et al., 2005). The highest level of COX-2 expression was noted at the ileocecal junction in rats (Haworth et al., 2005). This ileal-side high level of COX-2 expression may explain the spontaneous ulceration and perforation of the distal ileum in COX-2 knockout ($COX\text{-}2^{-/-}$) rodents (Sigthorsson et al., 2002). There is site-dependent susceptibility to intestinal injury that is related to local prostanoid homeostasis. For example, the rat cecum is particularly sensitive to long-term, low-dose indomethacin administration (NygAArd et al., 1995). COX-2 immunostaining was observed in the small intestine lamina propria in mice (Hull et al., 1999).

COX-2 can be induced in pathological conditions and in the inflamed GI mucosa, and its inhibition by NSAIDs has been hypothesized to delay the resolution of GI injury (Kishimoto et al., 1998). Increased COX-2 expression, observed

TABLE 1-2 Comparative Susceptibility to Toxicity and Location[a] of Lesions in the Gastrointestinal Tract After COX Inhibition

GI injury	Interspecies differences
Relative susceptibility at therapeutic exposures	Rat > dog > monkey > human
Upper GI most common site	Human > monkey > dog > rat
Lower small intestine most common site	Rat > dog > monkey[b] > human

Source: Reprinted from Z. A. Radi and N. K. Khan, Effects of cyclooxygenase inhibition on the gastrointestinal tract, *Experimental and Toxicologic Pathology*, 58, pp. 163–173. Copyright © 2006, with permission from Elsevier.

[a] Location of injury similar for both ns-NSAIDs and s-NSAIDs, with the exception that injuries with s-NSAIDs occur at high exposure multiples.

[b] Injury to lower GI tract uncommon in monkeys.

maximally at 24 h, has been observed in a rat model of ischemia/reperfusion–induced acute gastric mucosal injury (Kishimoto et al., 1998). COX-2 expression was increased in cultured rat gastric mucosal cells in vitro and after acid-induced gastric injury to rats in vivo (Sawaoka et al., 1997; Erickson et al., 1999; Sun et al., 2000). COX-2 may play a role in rodent postoperative ileus since intestinal manipulation induced COX-2 within resident muscularis macrophages, a discrete subpopulation of myenteric neurons and recruited monocytes (Schwarz et al., 2001). Additionally, COX-2 has been shown to be induced in various hyperplastic and neoplastic lesions of the GI tract, such as colon cancer, familial adenomatous polyposus (FAP), and sporadic adenomatous polyps in the colon (Soslow et al., 2000; Khan et al., 2001 Koki et al., 2002b). Up-regulation of COX-2 has been demonstrated within adenomas of the small and large intestine of multiple intestinal neoplasia (Min) mice (Hull et al., 1999). These observations support the use of NSAIDs in the treatment of epithelial cancer. In fact, a COX-2 s-NSAID, celecoxib, has been approved as adjunct therapy to the usual care (e.g., endoscopic surveillance, surgery) to reduce the number of adenomatous colorectal polyps in humans. In humans, gastric mucosal expression of COX-2 is increased in gastritis and gastric ulceration (Tatsuguchi et al., 2000; Bhandari et al., 2005).

In summary, COX-1 (and not COX-2) is the predominant isoform in the normal GI tract. There appears to be significant interspecies differences in both the level of COX-1 expression and the ratio of COX-1 and COX-2 expression in the GI tract. Both COX-1 expression and relative COX-1/COX-2 expression are highest in some animal species, including the dog and rat, compared with humans and nonhuman primates, which may partly explain the overt sensitivity of these species to subtherapeutic doses of ns-NSAIDs (Table 1-2) (Radi and Khan, 2006b; Radi, 2009).

EFFECTS OF ns-NSAIDS ON THE GI TRACT

Several ns-NSAID classes (Table 1-3) are used in human and veterinary medicine for their anti-inflammatory, analgesic, and antipyretic effects. In veterinary medicine, phenylbutazone, meclofenamic acid, meloxicam, carprofen, and

TABLE 1-3 Nonselective NSAID Major Classes

ns-NSAID class	Generic name	Trade name
Arylpropionic acid	Ibuprofen	Advil, Motrin, Nuprin
	Naproxen	Aleve, Naprosyn
	Ketoprofen	Orudis, Oruvail
	Carprofen	Rimadyl, Zenecarp, Novox
	Fenoprofen	Nalfon
	Flurbiprofen	Flurofen, Ansaid
Enolic acids	Piroxicam	Feldene
	Tenoxicam	Tilcotil
	Lornoxicam	Xefo
Acetic acids	Etodolac	Etogesic
	Indomethacin	Indocin
	Diclofenac	Cataflam, Voltarin
	Sulindac	Clinoril
	Nabumetne	Relfen
Aminonicotinic acids	Flunixin meglumine	Banamine
Pyrazoles	Phenylbutazone	Zolandin
Salicylic acids	Acetylsalicylic acid	Aspirin
Anthranilic acids	Meclofenamate	Ponstel, Arquel, Meclofen

etodolac are approved for use in dogs in the United States (Fox and Johnston, 1997; Budsberg et al., 1999); flunixin meglumine, meclofenamic acid, naproxen, and phenylbutazone are approved for horses (Kopcha and Ahl, 1989). Although the broad range of applications for ns-NSAID therapy makes it an attractive prescriptive choice, it has been partnered adversely with GI toxicity. Implications to humans have included nonulcer dyspepsia and serious GI-related side effects, such as gastric and duodenal ulcers, erosions, bleeding, perforation, esophagitis, and esophageal strictures (Lanza et al., 1983; Bjorkman, 1996; Mason, 1999; Schoenfeld et al., 1999; Scheiman, 2003). Similar to use in humans, chronic ns-NSAID use in dogs has also been associated with serious GI side effects, manifested as bleeding, ulceration, erosions, perforations, peritonitis, melena, anemia, anorexia, and abdominal pain (Ewing, 1972; Roudebush and Morse, 1981; Cosenza, 1984; Daehler, 1986; Stanton and Bright, 1989; Wallace et al., 1990; Ricketts et al., 1998; Reed, 2002). The incidence of GI ulceration is greatly increased in animals receiving ns-NSAIDs in combination with steroids; therefore, this combination should be avoided (Johnston and Budsberg, 1997; Reed, 2002).

To limit the potential for serious complications associated with NSAID use in veterinary medicine, clinicians should take the following precautions before prescribing NSAIDs: (1) verify that corticosteroids and other NSAIDs are not being given concurrently with the NSAID prescribed, (2) adhere to the dosages recommended, (3) advise clients of potential safety risks and their clinical signs, and (4) avoid use in at-risk cases (Lascelles et al., 2005b). At-risk indications are (1) a history of GI ulceration, (2) geriatric patients (older animals have reduced clearance capacity and are more susceptible to NSAID GI toxicity), (3) the use of aspirin, (4) GI comorbidities (e.g., preexisting GI ulcer, *H. pylori* colonization, liver disease), and (5) clinical chemistry (e.g., indications of impaired hepatic function, hypoproteinemia) (Lascelles et al., 2005b). The comparative GI effects of various classes of ns-NSAIDs are detailed below.

Effects of Arylpropionic Acid ns-NSAIDs on the GI Tract

Arylpropionic acids represent the largest class of widely prescribed group of ns-NSAIDs, which include ibuprofen, naproxen, ketoprofen, carprofen, fenoprofen, and flurbiprofen (Table 1-3). These drugs are approved for use in the treatment of RA, OA, and ankylosing spondylitis.

Ibuprofen (Advil, Motrin, and Nuprin) is one of the most commonly used ns-NSAIDs and is supplied as tablets. It probably ranks after aspirin and paracetamol in nonprescription over-the-counter (OTC) drugs used in humans for the relief of symptoms of pain, inflammation, and fever (Rainsford, 2009). In dogs, ibuprofen has been used as an anti-inflammatory agent. However, dogs are much more sensitive than humans to the development of GI toxicity from ibuprofen administration. At therapeutic doses, adverse GI effects observed in dogs include vomiting, diarrhea, anorexia, abdominal pain, nausea, and GI bleeding. Dogs given 8 or 16 mg/kg per day of ibuprofen orally for 30 days showed no clinical signs of toxicity. However, postmortem examination revealed the presence of gastric ulcers or erosions, usually in the antrum or pylorus, less often in the fundic or cardiac regions of the stomach, and intestinal inflammation (Adams et al., 1969). No GI lesions were noted in a 4-mg/kg per day dose in this study in dogs. In another study, ibuprofen oral repeated dosing at 8 and 16 mg/kg per day (0.4- and 0.9-fold multiples of the human dose, respectively) for one month in dogs caused GI pathology comprised of bloody or discolored stools and intestinal ulceration and/or perforation (Hallesy et al., 1973). During 26-week repeat-dose toxicity in dogs given ibuprofen in a 16-mg/kg per day dose, clinical signs of GI toxicity characterized by frequent vomiting, diarrhea with occasional passage of fresh blood, and loss of weight were noted in week 8 of dosing (Adams et al., 1969). Dogs given 4- and 2-mg/kg per day doses had no evidence of GI toxicity. On the other hand, similar GI pathology was observed in rats only after six months of repeated dosing and at a higher oral dose of 180 mg/kg per day (9.7-fold multiples of the human dose) (Hallesy et al., 1973). The approximate LD_{50} values for ibuprofen are 800 mg/kg orally and 320 mg/kg intraperitoneally in the mouse and 1600 mg/kg orally and 1300 mg/kg subcutaneously in the rat (Adams et al., 1969). The proportions of the dose recovered from the ligated stomach and ligated intestine of rats at various times after introduction of ^{14}C-labeled ibuprofen were measured. Only 73% of the dose was recovered from the stomach and its contents 3 min after dosing, while no radioactivity was detected in plasma (Adams et al., 1969). Intestinal absorption of ibuprofen in rats was so rapid that by 3 min the plasma concentration of radioactivity was maximal (Adams et al., 1969). Thus, although some absorption occurs in the stomach, the main site of ibuprofen absorption, at least in rats, is the intestine. Rats given ibuprofen for 26 weeks orally at 180 mg/kg per day grew normally but were anemic (had low erythrocyte counts, hemoglobin concentration, and hematocrits) by the final week of dosing, and a few rats had intestinal ulcers (Adams et al., 1969). Therefore, due to its narrow margin of safety, ibuprofen is generally not recommended for use or should be used with caution in dogs and also in cats or ferrets (Cathers et al., 2000). Ibuprofen toxicosis can occur in dogs and ferrets after accidental ingestion. When ibuprofen toxicosis is suspected, serum, urine, and liver samples can be used for toxicological analyses for ibuprofen by gas chromatography and mass spectrophotometry (Cathers et al., 2000). In laboratory animals, ibuprofen administration has been linked to vomiting, gastric irritation, and ulceration and duodenal and jejunal ulceration in rats, dogs, rabbits, and monkeys (Adams et al., 1969; Scherkl and Frey, 1987; Elliott et al., 1988; Godshalk et al., 1992; Arai et al., 1993). Pregnant female rabbits given 60 or 20 mg/kg per day of ibuprofen on days 1 to 29 of pregnancy grew less than controls and had stomach ulcers (Adams et al., 1969). Female rabbits receiving 7.5 mg/kg per day grew normally, but some had gastric ulcers or erosions (Adams et al., 1969). Thus, pregnant rabbits are highly sensitive to ibuprofen, and during pregnancy the intestinal tract, at least in rabbits, is more sensitive than that of nonpregnant animals (Adams et al., 1969). The route of ibuprofen administration affects the GI pathology observed. Little or no GI damage occurred when ibuprofen was given daily by oral administration to nonhuman primates at doses up to 300 mg/kg for 90 days. However, intravenous (IV) administration of ibuprofen to nonhuman primates over a 24-h period in four equal doses of 75 mg/kg at 6-h intervals resulted in gastric erosions or ulcers (Elliott et al., 1988). When given IV for 14 days at 100 and 200 mg/kg per day using the same 24-h dosing conditions described above, nonhuman primates showed gastric and/or duodenal ulcers (Elliott et al., 1988). In rats, acute single oral doses under 500 mg/kg of ibuprofen were free of GI pathological changes (Elliott et al., 1988). However, acute IV administration of ibuprofen at a dose of 270 mg/kg given in four equal doses of 67.5 mg/kg at 6-h intervals over a 24-h period resulted in gastric and intestinal (ileum and jejunum) ulcerations (Elliott et al., 1988). Therefore, the acute (24 h) IV route of ibuprofen administration in rats is more ulcerogenic than the oral route.

One factor that has been correlated with GI events with ns-NSAID use is the drug plasma elimination half-life ($t_{1/2}$). There is less gastric mucosal adaptation with NSAIDs that have long half-lives. For example, due to the short plasma $t_{1/2}$ of elimination of ibuprofen, approximately 2 h, and its rapid absorption, ibuprofen at doses of 200 and 800 mg/kg has low possibilities of serious GI events and complications (i.e., epigastric or abdominal pain, dyspepsia, flatulence, nausea, heartburn, diarrhea, constipation, vomiting) in humans (Rainsford, 2009). Another GI event that can be associated with ns-NSAID intake is chronic anemia. For example, daily treatment (800 mg three times daily) with ibuprofen has also been associated with significant fecal blood loss in healthy volunteers (Bowen et al., 2005). Several ns-NSAIDs, including ibuprofen, were compared for GI events in a large two-year epidemiological safety study involving 30,000 to 40,000 rheumatic patients in centers in Germany, Switzerland, and Austria, known as the Safety Profile of

Antirheumatics in Long-Term Administration (SPALA) (Rainsford, 2009). This SPALA study found ibuprofen to be associated with the lowest numbers of GI events. In another large-scale study, fewer GI events were observed in patients taking ibuprofen at doses up to 1200 mg daily for 7 days compared with aspirin and paracetamol (Rampal et al., 2002).

Similarly, naproxen (Aleve, Naprosyn), which has a variably longer half-life across species ($t_{1/2}$ is approximately 14 h in humans, 2 h in nonhuman primates, 35 h in dogs, 9 h in guinea pigs, and 5 h in rats), is used in humans and dogs for its anti-inflammatory, analgesic, and antipyretic properties (Hallesy et al., 1973; Rainsford, 2009). The long half-life of naproxen in dogs appears to be due to its extensive enterohepatic recirculation. With the exception of the dog, all species excreted naproxen and its metabolic transformation products predominantly in the urine. In dogs, naproxen is eliminated primarily through the bile and feces, whereas in other species, the primary route of elimination is through the kidneys (Runkel et al., 1972). Once in the blood after oral administration, naproxen is absorbed fully and rapidly in all species. Naproxen is indicated for temporary relief of fever and minor aches and pains due to backache, headache, and toothache in humans (Runkel et al., 1972; Rainsford, 2009). In dogs, naproxen has been shown to cause gastric ulceration and hemorrhage, melena, vomiting, abdominal pain, weakness, and hemorrhagic gastroenteropathy, including transmural pyloric perforation (Daehler, 1986; Stanton and Bright, 1989). Repeated oral administration of naproxen to dogs for one month at 15 mg/kg per day (1.3-fold multiple of human dose) and for three months at 5 mg/kg per day (0.4-fold multiples of human dose) caused GI pathology of bloody or discolored stools and intestinal ulceration and/or perforation (Hallesy et al., 1973). Additionally, mice and rats administered an acute oral dose of naproxen displayed bloody or discolored stools and intestinal ulceration and perforation (Rainsford et al., 2003). A single oral naproxen dose of 250 mg/kg to rats caused death in 6 to 8 days and abdominal adhesions, and small intestine necrotic foci were observed at necropsy (Elliott et al., 1988). Repeated oral administration of naproxen to rats for six months at 30 mg/kg (2.6-fold multiple of human dose) and for 22 months at 2, 10, and 30 mg/kg per day (0.2-, 0.9-, and 2.6-fold multiples of human dose, respectively) caused GI pathology of bloody or discolored stools and intestinal ulceration and/or perforation (Hallesy et al., 1973). Repeated oral administration of naproxen in nonhuman primates for six months at doses up to 120 mg/kg per day (10.4-fold multiples of human dose) was well tolerated with no adverse GI pathological toxicity (Hallesy et al., 1973). The pig closely resembles humans in respect to anatomy, physiological functions in the GI tract, and the histological and pathophysiological changes in the development of gastric ulcers induced by NSAIDs (Rainsford et al., 2003). Daily oral administration of naproxen to pigs at doses up to 45 mg/kg (3.9-fold multiples of human dose) for one year was well tolerated with no adverse GI pathology (Hallesy et al., 1973). In an experimental pig model using healthy Landrace males, naproxen induced gastroduodenal ulcers and erosions when given orally for 10 days at a dose of 100 or 150 mg/kg per day (Rainsford et al., 2003). In horses, naproxen has been used for the treatment of inflammatory conditions and pain from myositis and soft tissue injuries, and it has a reasonable margin of safety. However, adverse GI ulceration has been reported in horses (Lees and Higgins, 1985). In humans, GI events associated with naproxen include GI erosions and ulcers, dyspepsia, upper abdominal pain, nausea, diarrhea, constipation, abdominal distension, and flatulence (Lohmander et al., 2005).

Ketoprofen (Orudis, Oruvail) has a $t_{1/2}$ of approximately 8.5 h and is used to treat RA in humans (Rainsford, 2009). It is used in dogs, cats, and horses to treat postsurgical and musculoskeletal pain, colic, synovitis, and OA. Its $t_{1/2}$ in dogs and cats is approximately 2 to 3 h and 2 h in horses. Ketoprofen was ulcerogenic when used in laboratory animal models (Rainsford, 1977) but to a lesser degree than other ns-NSAIDs (phenylbutazone and flunixin meglumine). Similarly, in horses, ketoprofen was less toxic than phenylbutazone and flunixin meglumine (MacAllister et al., 1993). Gastric-duodenal erosion and/or hemorrhage and ulceration of the glandular and nonglandular portions of the stomach were noted in dogs and horses, respectively (MacAllister et al., 1993; Forsyth et al., 1998). Similar ulceration was seen in a one-month oral toxicity study with ketoprofen in rats and dogs (Julou et al., 1976). In a study of Sprague–Dawley rats given a 10-mg/kg dose of ketoprofen subcutaneously that had undergone ovariectomy, many rats died or were euthanized within 3 to 7 days after surgery, due to clinical illness that was related to GI ulceration (Lamon et al., 2008). The safety profile of a reduced dosage of ketoprofen (0.25 mg/kg per day) was evaluated in a 30-day oral study in healthy beagle dogs. Mild to moderate gastric mucosal injuries, especially in the pyloric antrum, were observed in this study (Narita et al., 2006). Gastric lesions were observed in a long-term (up to 90 days) study after oral administration of various ns-NSAIDs (i.e., carprofen, etodolac, flunixin meglumine, and ketoprofen) in dogs (Luna et al., 2007). In addition, the bleeding time was significantly longer by day 7 in dogs treated with meloxicam, ketoprofen, and flunixin meglumine (Luna et al., 2007). Scaring in the pyloric antrum suggestive of ulceration healing was present in one of 12 monkeys following 12 months of ketoprofen treatment (Julou et al., 1976).

Interestingly, a study in hamster cheek pouch microcirculation showed that topically applied ketoprofen lysine salt significantly inhibited both the leukocyte adhesion and microvascular leakage induced by bradykinin (Daffonchio et al., 2002). A kallikrein-kinn cascade such as bradykinin has been shown to be involved in gastric ulcers (Sawant et al., 2001). Therefore, this study by Daffonchio et al. suggests that in addition to COX inhibition, ketoprofen may have an antagonistic effect on bradykinin, which may contribute to its ulcerogenic potential. In humans, ketoprofen is often used in a once-daily 200-mg sustained-release formulation to treat rheumatic diseases, especially in elderly patients. Long-term safety and prospective studies on 20,000 patients showed that ketoprofen is associated with a 28% rate of such GI events as peptic ulcers, bleeding, melena, and black stools (Le Loet 1989; Schattenkirchner, 1991). Most of these serious GI side effects occurred during the first three months of treatment.

Carprofen (Rimadyl, Zenecarp, Novox) is approved for use in dogs to treat pain and inflammation associated with OA and pain associated with soft tissue or orthopedic surgery. The $t_{1/2}$ in dogs is approximately 8 h and is highly variable in cats (20 ± 16 h). In dogs, biliary secretion predominates, and 70% of an IV dose of carprofen is excreted in the feces, while 8 to 15% of the dose is excreted in the urine. In rats, fecal excretion due to biliary secretion varies from 60 to 75%, and urinary excretion accounts for 20 to 30% of an IV dose (Rubio et al., 1980). Therefore, excretion in dogs, rats, and cattle is mainly fecal after biliary secretion, whereas it is primarily urinary in horses. In dogs, most carprofen is metabolized by direct conjugation to an ester glucuronide followed by oxidation to phenol and further conjugation. These conjugated phenols are eliminated in the feces. Carprofen has produced GI lesions that are mild but of no clinical relevance or significance compared with placebos (Reimer et al., 1999). Typical adverse GI effects of this drug include vomiting, diarrhea, and change in appetite (Raekallio et al., 2006). A transient decrease in serum protein and albumin concentrations (concentrations were lower in treated dogs than in those that received placebo at 4 weeks, but not at 8 weeks) was observed after daily administration of carprofen in a two-month study in dogs (Raekallio et al., 2006). When administered orally daily in a 4-mg/kg dose, carprofen induced the lowest frequency of adverse GI effects compared with etodolac, flunixin meglumine, ketoprofen, and meloxicam in a 90-day study in dogs (Luna et al., 2007).

GI ulceration and bleeding are sometimes accompanied secondarily by anemia and hypoproteinemia, due to blood and protein loss (Adams et al., 1969; Lanas et al., 2003). In a 14-day safety study (according to the Rimadyl package insert) involving oral administration of 10 mg/lb twice daily (10 times the recommended total daily dose), two of eight dogs exhibited hypoproteinemia (hypoalbuminemia). Three incidents of black or bloody stool were observed in one dog. Five of eight dogs exhibited reddened areas of duodenal mucosa on gross pathological examination. Histological examination of these areas revealed no evidence of ulceration but did show minimal congestion of the lamina propria in two of the five dogs. In separate safety studies lasting 13 and 52 weeks, respectively, dogs were administered orally up to 11.4 mg/lb per day (5.7 times the recommended total daily dose of 2 mg/lb) of carprofen. In both studies the drug was well tolerated clinically by all the animals. No gross or histological changes were seen in any of the animals treated.

In cats, carprofen is an effective analgesic for soft tissue and orthopedic procedures and is approved in several countries (Australia, France, Germany, United Kingdom) for use at 4 mg/kg for daily subcutaneous or intravenous administration (Steagall et al., 2009). Carprofen was well tolerated, and no clinical or endoscopic adverse GI effects were seen in cats after its administration in clinical trails for up to 5 days (Möllenhoff et al., 2005; Steagall et al., 2009). Although caprofen is not used routinely in nonhuman primates for postoperative analgesia, a dose of 2.2 mg/kg carprofen intramuscularly, or a combination of 0.01 mg/kg buprenorphine and 2.2 mg/kg carprofen intramuscularly provided more reliable postoperative analgesia than did buprenorphine alone (Allison et al., 2007). Although carprofen has been used to treat mastitis in cattle, it is not generally recommended for use in large animals, due to its long $t_{1/2}$ (30 to 40 h).

Fenoprofen (Nalfon) has a relatively short, 3-h half-life. GI events are similar to those with naproxen or ibuprofen. In rats, single fenprofen oral doses of 1000 to 1600 mg/kg resulted in death and small intestine necrosis and abdominal adhesions (Elliott et al., 1988). Flurbiprofen (Flurofen, Ansaid) caused abdominal adhesions and small intestinal necrosis or ulceration in rats after either acute oral (at 80 and 125 mg/kg) or intraperiptoneal (at 125, 320, and 500 mg/kg) administration (Elliott et al., 1988). Chronic administration in a three-month study in rats caused ulcerative gastritis in 4- and 8-mg/kg doses and 0.5-, 2-, and 4-mg/kg doses in another two-year study (Elliott et al., 1988).

Effects of Enolic Acid (Oxicam) ns-NSAIDs on the GI Tract

Piroxicam (Feldene) is one of few enolic acid derivatives (Table 1-3) that is absorbed completely after oral administration and that undergoes enterohepatic recirculation. Due to its antitumor activity, it is used in dogs and cats to treat some cancers, such as transitional cell carcinoma (TCC) and oral squamous cell carcinoma. The average estimated $t_{1/2}$ is approximately 40 h in dogs and 12 h in cats. Due to this long $t_{1/2}$ in dogs, the steady state is typically not reached for 7 to 12 days. GI irritation was seen in some dogs after bladder TCC treatment with piroxicam orally in a 0.3-mg/kg dose (Knapp et al., 1994). Gastric ulcers occurred in rats after once-daily piroxicam administration in doses of 2.7, 5.3, and 6.7 mg/kg, which are equieffective for indomethacin (10, 20, and 25 mg/kg) (Aguwa, 1985). Gastric ulcers were induced in rats after two oxicam oral dosing. However, the incidence of such lesions was higher for tenoxicam (Tilcotil) (10.2 mg/kg) than for diclofenac sodium (34 mg/kg, equivalent to 6.8 mg/kg tenoxicam) or piroxicam (6.2 mg/kg) (al-Ghamdi et al., 1991). Other subchronic 14- and 28-day studies in rats assessed the GI effects of equipotent doses of meloxicam (3.75 and 7.5 mg/kg) and piroxicam (5 and 10 mg/kg) in rats. Both drugs dose-dependently caused multiple gastric erosions and hemorrhage. Meloxicam led to greater gastric damage than with piroxicam on day 14, although these results were not significant (Villegas et al., 2002). In a dose-escalation study of piroxicam with oral doses ranging from 0.5 mg/kg every 48 h to 1.5 mg/kg every 48 h in dogs, a dose-limiting GI irritation or ulceration occurred in dogs that received 1.5 mg/kg, with a maximum tolerated dose of 1 mg/kg (Knapp et al., 1992). Lornoxicam (Xefo), a novel ns-NSAID compound in the same chemical class as piroxicam and tenoxicam, caused GI lesions in monkeys (Atzpodien et al., 1997). In the dose-range-finding study, animals were dosed orally for 6 weeks with 0.25, 0.5, 1, or 2 mg lornoxicam/kg per day. GI toxicity was observed in the 1- and 2-mg/kg per day dose groups only. Toxicity included mortality, diarrhea, prostration, decreased body weight gain and food consumption, fecal occult blood, anemia, hypoalbuminemia, GI erosions, and ulcerations (Atzpodien et al., 1997). A follow-up chronic study was conducted using dose levels of 0.125, 0.25, or 0.5 mg/kg per day for 52 weeks. The high-dose level was increased to 0.6 mg/kg/day from week 39 to week 52. Histopathological examination of the GI tract revealed erosions, ulcerations, and inflammation in both males and females at 0.5 or 0.6 mg/kg per day. Cinicopathological findings included decreased hematocrit and hypoproteinemia and hypoalbuminemia (Atzpodien et al., 1997).

In a clinical study in elderly patients with knee OA, piroxicam at a dose of 20 mg/day for 3 weeks resulted in elevation of the gastric mucosa endoscopic score in 78% of the subjects compared to the beginning of the study, and 22% of the subjects developed ulcers. Mild dyspepsia symptoms after piroxicam administration were positive in 67% of subjects (Girawan et al., 2004). Significantly higher bleeding was found in a 28-day study in healthy male volunteers using a 20-mg piroxicam dose compared with a placebo. In addition, endoscopy scores were significantly higher with piroxicam than in the meloxicam group at a dose of 7.5 mg (Patoia et al., 1996). In another 28-day study in healthy volunteers, significant macroscopic gastric mucosal damage occurred within 24 h of 20-mg piroxicam administration; however, such GI damage resolved in most subjects by day 28 (Lipscomb et al., 1998).

Effects of Acetic Acid Derivative ns-NSAIDs on the GI Tract

Acetic acid derivates include etodolac, indomethacin, diclofenac, sulindac, and nabumetone (Table 1-3). Etodolac (Etogesic) is approved for use in dogs with OA and has been studied in horses. Adverse reactions to etodolac in dogs include vomiting, soft or dark brown stool, and diarrhea with blood, as reported in a three-month oral toxicity study at a dose of 25 mg/kg, as well as gastric and small intestinal ulceration with associated weight loss, anorexia, anemia, and hypoproteinemia in a one year chronic toxicity study at doses of 40 and 80 mg/kg (Budsberg et al., 1999). In a 28-day study in healthy dogs, etodolac was given orally once a day at an average dose of 12.8 mg/kg and gastroduodenal endoscopy was performed. Only minor gastric lesions were observed (Reimer et al., 1999). In an experimental study of the GI effects of etodolac in horses, jejunum was exposed to 2 h of ischemia during anesthesia, and then horses received etodolac at 23 mg/kg IV every 12 h. Tissue specimens were obtained from ischemic-injured and nonischemic jejunum immediately after ischemia and 18 h after recovery from ischemia. The investigators found that ischemic-injured tissue from horses treated with etodolac had significantly lower transepithelial electric resistance and retarded recovery of the jejunal mucosa barrier after 18 h of reperfusion (Tomlinson et al., 2004).

In rats, indomethacin (Indocin) caused gastric mucosal bleeding, cecal ulceration, and small intestine (jejunum and ileum) ulcers, perforations, and adhesions (Kent et al., 1969; Brodie et al., 1970; Schriver et al., 1975; Fang et al., 1977; Arai et al., 1993; Anthony et al., 1994; Sigthorsson et al., 1998; Campbell and Blikslager, 2000; Altinkaynak et al., 2003; Takeuchi et al., 2004), with gastric damage being significantly greater in arthritic rats than in normal rats (McCafferty et al., 1995). The half-life of indomethacin in plasma ranges from hours in rats to minutes in dogs and monkeys. There are significant species differences in the distribution and excretion of indomethacin (Yesair et al., 1970). In rats, plasma clearance of indomethacin by liver, although low, is 30 times the clearance rate by kidney, and the reabsorption of indomethacin from the intestine is extensive. Desmethylindomethacin, the major metabolite, is cleared from plasma equally by liver and kidney and is not reabsorbed from the intestine of rats. In dogs, indomethacin is secreted in bile extensively and rapidly, and is eventually excreted in their feces as an unchanged drug and minimally metabolized to eschlorobenzoylindomethacin, which is excreted in urine. Nonhuman primates are similar to dogs in that the liver was more than 10 times as effective as the kidneys in clearing total radioactivity from plasma. However, the primates differed from dogs in that the drug was maximally reabsorbed from the intestine (Yesair et al., 1970). In a six-month repeat oral daily dosing study in rats, indomethacin at doses of 2 and 4 mg/kg (0.7- and 1.3-fold multiples over human dose, respectively) caused GI pathology characterized by bloody or discolored stools and intestinal ulceration and/or perforation (Hallesy et al., 1973). Indomethacin increased the incidence and ulcer index of duodenal ulcers in arthritic rats on days 14 and 28 of arthritis (DiPasquale and Welaj, 1973). Additionally, the intestinal ulcerogenic response to indomethacin was markedly aggravated in arthritic rats, and the onset of the ulceration was much earlier in arthritic rats than in normal rats (Kato et al., 2007). GI hemorrhage and ulceration, potentially attributed to the extensive enterohepatic recirculation of indomethacin, were seen in dogs (Duggan et al., 1975) (Figs. 1-2 and 1-3). In a one-month repeat oral dosing study in dogs, indomethacin at 6 and 18 mg/kg per day (1.9- and 5.8-fold multiples over human dose) caused GI pathology of bloody or discolored stools and intestinal ulceration and/or perforation (Hallesy et al., 1973). In an experimental pig model for human GI disease, indomethacin was given orally for 10 days at a

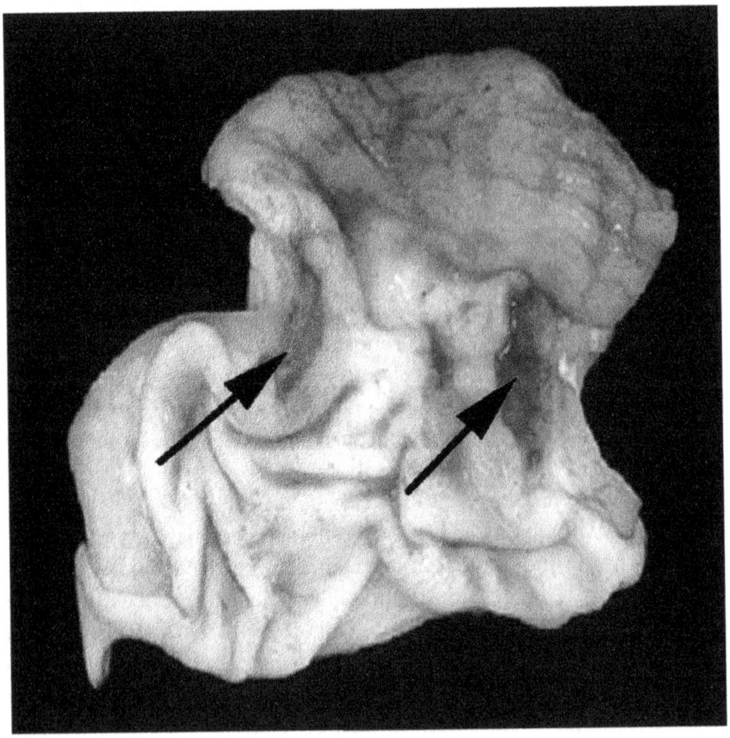

FIGURE 1-2 Severe indomethacin-induced gastric mucosal hemorrhage and ulceration at the gastroduodenal junction (arrows) in a dog. (Reprinted from Z. A. Radi and N. K. Khan, Effects of cyclooxygenase inhibition on the gastrointestinal tract, *Experimental and Toxicologic Pathology*, 58, pp. 163–173. Copyright © 2006, with permission from Elsevier.)

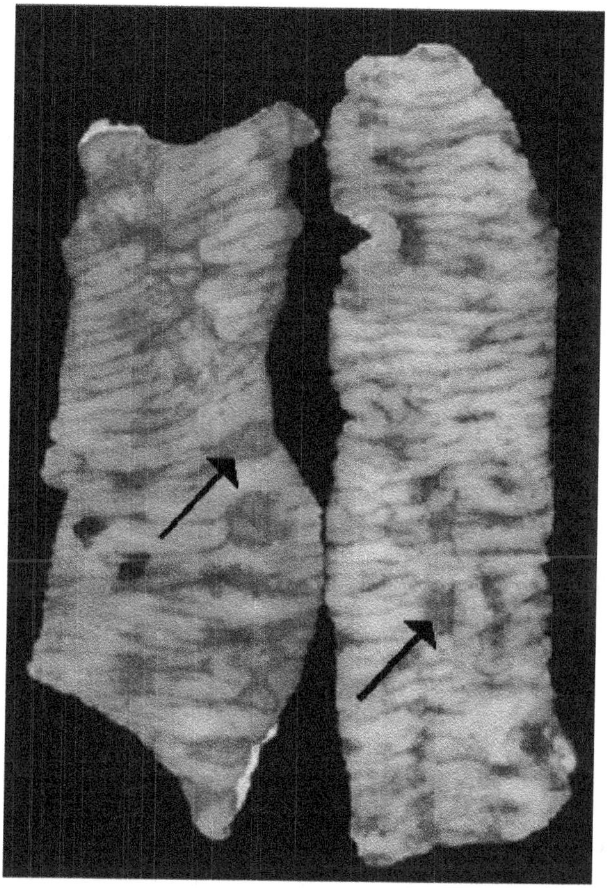

FIGURE 1-3 Indomethacin-induced small intestine mucosal ulceration (arrows) in a dog. Note the skip ulcerations typical of ns-NSAID-induced lesions in the small intestine in dogs. (Reprinted from Z. A. Radi and N. K. Khan, Effects of cyclooxygenase inhibition on the gastrointestinal tract, *Experimental and Toxicologic Pathology*, 58, pp. 163–173. Copyright © 2006, with permission from Elsevier.)

does of 10 or 20 mg/kg per day and GI effects were evaluated. Gastroduodenal ulcers and lesions occurred with indomethacin treatment at both doses. Additionally, indometacin produced focal ulcers in the cecum. The mucosal concentrations of indometacin in the gastric and intestinal mucosa correlated with mucosal injury (Rainsford et al., 2003). A 4-week toxicity study of indomethacin was conducted in nonhuman primates (marmoset) in which indomethacin was administered by oral route at dose levels of 2, 6, and 12 mg/kg per day. All animals given the daily 12-mg/kg dose and one animal given 6 mg/kg per day died during the dosing period and within 20 days. At 12 mg/kg per day, indomethacin induced severe GI toxicity, characterized by hemorrhage, ulcers, and necrosis with peritonitis (Oberto et al., 1990).

Diclofenac (Cataflam, Voltarin) has a rapid absorption, a short-half life of approximately 2 h, and is metabolized in the liver by CYP2C in humans. It is the most widely used NSAID in the world to treat RA, OA, and ankylosing spondylitis. The GI adverse effects of twice-daily administration of 75 mg of diclofenac were evaluated in one of the largest and longest individual-outcome randomized double-blind clinical studies of NSAID use in RA and OA patients. A total of 23,504 patients were randomized with mean treatment duration from 19.4 to 20.8 months (Combe et al., 2009). Significantly higher upper GI events (perforation, bleeding, obstruction, and ulcer) occurred with diclofenac than with 90- or 60-mg once-daily administration of etoricoxib (Combe et al., 2009). In rats, diclofenac acute (5 h) oral administration at 3.5, 7, and 15 mg/kg caused gastric ulcers. In rats treated with diclofenac at 15 mg/kg, pathological changes included longitudinal and diffuse gastric ulcers, particularly along the mucosal pleats, and thinning and inflammation of the intestinal wall with poor elasticity (Conforti et al., 1993).

Sulindac (Clinoril) is a prodrug whose anti-inflammatory activity (used to treat rheumatoid arthritis, osteoarthritis, ankylosing spondylitis, and acute gouty arthritis) resides in its sulfide metabolite. Sulindac is available in 200-mg tables and undergoes two major biotransformations after oral administration. It is oxidized to the sulfone and then reversibly reduced to the sulfide. The sulfide is formed largely by the action of gut microflora on sulindac excreted in the bile. The half-life of the active sulfide is approximately 18 h. Sulindac can cause serious adverse GI events in humans, including inflammation, bleeding, ulceration, and perforation of the stomach, small intestine, or large intestine (according to a sulindac package insert).

Nabumetone (Relfen) exerts its pharmacological effects via its metabolite 6-methoxy-2-naphthylacetic acid (6-MNA) and is used to treat RA and OA. 6-MNA is not a biliary secretion and is inactivated in the liver, then conjugated before excretion. Because it is a nonacidic, prodrug formulation, fewer GI events were observed after nabumentone treatment than after treatment with other NSAIDs. GI events in humans included perforations, ulcers and bleeding, nausea, abdominal pain, and dyspepsia (Bannwarth, 2008). No gastric damage was observed in a 3-day study in rats in which nabumetone was orally dosed at 79 mg/kg and 6-MNA was given IV at 34 mg/kg (Melarange et al., 1992). The GI tolerability and pathology of nabumetone and etodolac were evaluated in an extensive nonclinical acute and chronic safety study (Spangler, 1993). In a single-dose study, etodolac caused a significant increase in both gastric and intestinal damage at 6, 24, 48, and 144 h after dosing. In contrast, no significant GI damage was noted with nabumetone. In the 28-day study, a significant increase in GI damage was noted with etodolac, but not with nabumetone, despite the higher dose employed [the nabumetone dose was five times the ID_{25} (the dose that reduces inflammation by 25% in 50% of animals)] (Spangler, 1993).

Effects of Aminonicotinic Acid Derivative ns-NSAIDs on the GI Tract

Although not approved for use in cats, dogs, or food animals, flunixin meglumine (Banamine), an aminonicotinic acid derivative, is approved for use in nonhuman primates and horses to control pain, colic, and endotoxic shock (Moore et al., 1981; Lees and Higgins, 1985; Kopcha and Ahl, 1989; Kallings et al., 1999). Renal excretion contributes significantly to flunixin meglumine in horses (Lees and Higgins, 1985). Flunixin meglumine can be administered to horses by IV, intramuscular, or oral routes at the recommended therapeutic dose of 1.1 mg/kg once a day for up to 5 days. Flunixin meglumine GI toxicity at recommended doses appears to be rare. Oral administration at three times the recommended dose for 10 days failed to elicit toxicity (Lees and Higgins, 1985). However, GI ulceration and erosion occurred in horses dosed 1.1 mg/kg IV every 8 h for 12 days (MacAllister et al., 1993). Tissue specimens were obtained from ischemic-injured and nonischemic jejunum immediately after ischemia and 18 h after recovery from ischemia. The investigators found that ischemic-injured tissue from horses treated with flunixin meglumine had significantly lower transepithelial electric resistance and retarded recovery of the jejunal mucosa barrier after 18 h of reperfusion (Campbell and Blikslager, 2000; Tomlinson et al., 2004). Additionally, this ns-NSAID was linked with GI ulceration and diarrhea in horses and dogs (Traub-Dargatz et al., 1988; Carrick et al., 1989; Vonderhaar and Salisbury, 1993; Luna et al., 2007). Flunixin meglumine has a significantly longer $t_{1/2}$ in cows (approximately 8 h) compared with horses (approximately 2 h) or dogs (approximately 4 h) and is used to treat bovine pneumonia at a dose of 2 mg/kg once a day for 3 to 5 days, as well as acute mastitis. No effects on the GI tract were noted when flunixin meglumin was given experimentally to calves (Kopcha and Ahl, 1989). In dogs, flunixin at 1 mg/kg for 3 days with 4-day intervals resulted in a significantly longer bleeding time and gastric lesions (Luna et al., 2007).

Effects of Pyrazolone Derivative ns-NSAIDs on the GI Tract

Phenylbutazone (Butazolidin), a pyrazolone derivative, is a widely studied pyrazolone ns-NSAID approved for use in dogs and horses to treat OA, osteoporotic conditions, and laminitis and studied experimentally in rats, cats, and food animals. Phenylbutazone metabolite is oxyphenybutazone. When used in dogs, phenylbutazone was less toxic to this species than to humans but induced blood dyscrasia and GI injury (Watson et al., 1980; Conlon, 1988; Johnston and Budsberg, 1997). In horses, phenylbutazone has a $t_{1/2}$ that ranges from 3 to 10 h and has a narrow therapeutic index that may be related to lower plasma protein binding (Tobin et al., 1986). Absorption of phenylbutazone from the GI is influenced by the dose administered and the relationship of dosing to feeding. Access to hay can delay the time of peak plasma concentration to 18 h or longer (Tobin et al., 1986). GI-associated toxicity in hoses includes gastric ulcers and erosions, edema of the small intestine, mucosal atrophy, duodenal erosions, erosions and ulcers of the large colon, and ulcerative colitis (Mackay et al., 1983; Traub et al., 1983; Collins and Tyler, 1985; Karcher et al., 1990; Meschter et al., 1990a,b). Phenylutazone resulted in more severe GI toxicity in horses than did ketoprofen and flunixin meglumine, causing edema in the small intestine, erosions and ulcerations in the large intestine, and gastric ulceration at a dose of 4.4 mg/kg IV every 8 h for 12 days (MacAllister et al., 1993). In addition, hypoproteinemia and hypoalbuminemia secondary to protein-losing enteropathy was seen in these horses. A 10-mg/kg dose of phenylbutazone once daily for 14 days is considered toxic and caused weight loss, diarrhea, and GI erosions and ulcerations (MacAllister, 1983). In laboratory animals such as dogs and rats, phenylbutazone caused GI pathology of blood or discolored stools and intestinal ulceration and/or perforation. In dogs, daily oral dosing at 200 mg/kg per day (32.3-fold multiples of human dose) for three months caused GI lesions. In rats, repeated oral dosing at 50, 100, and 200 mg/kg per day (8.1-, 16.1-, and 32.3-fold multiples of human dose) for six months caused GI lesions (Hallesy et al., 1973).

In ruminants, phenylbutazone is used to control arthritis and laminitis and is absorbed slowly following oral administration and cleared more slowly than that in horses and dogs. Although it protected calves against local dermal inflammation and systemic shock, it partially blocked rumen stasis in goats (Van Miert et al., 1977; Eyre et al., 1981). In rats, it caused gastric mucosal ulceration, bleeding, and hemorrhage and small intestine perforation and adhesions (Shriver et al., 1977; Mersereau and Hinchey, 1981; Takeuchi et al., 2004). These lesions are attributed to increased gastric contractions induced by the drug (Mersereau and Hinchey, 1981).

Effects of Salicylic Acid Derivative ns-NSAIDs on the GI Tract

Acetylsalicylic acid (aspirin) is still used widely due to its analgesic, antipyretic, and anti-inflammatory properties. Some salicylates, such as sulfasalazine, olsalazine, and mesalamine, are used to reduce inflammation associated with inflammatory bowel disease (IBD), such as Crohn's disease and ulcerative colitis [UC] (Radi et al., 2011). These IBD drugs cause splitting of the diazo bond by colonic bacteria to give sulfapyridine and 5-aminosalicylic acid (5-ASA), which is considered to be the active moiety that is delivered to the GI mucosa (Robinson, 1989). Emerging data suggest that 5-ASA treatment reverses an imbalance between the angiogenic factor VEGF and the antiangiogenic factors endostatin and angiostatin in an experimental UC rat model (Deng et al., 2009). The authors conclude that the effect of 5-ASA in UC may be caused by the down-regulation of expression of endostatin and angiostatin by modulation of matrix metalloproteinases-2 (MMP2) and MMP9 via inhibition of TNFα (Deng et al., 2009). Acetylsalicylic acid is rapidly absorbed mostly from the upper small intestine and undergoes rapid metabolism to the hydrolyzed active product, salicylic acid. Acetylsalicylic acid is the only salicylate that irreversibly inhibits cyclooxygenase by covalent acetylation of the enzyme. Salicylic acid is eliminated by hepatic conjugation with glucuronide and glycine and by renal excretion through glomerular filtration. The safety margin of aspirin is generally wide. The elimination half-life of salicylate varies significantly across species. The $t_{1/2}$ in cats is 27 to 45 h, 4.5 to 8.5 h in dogs, 1 h in horses, and 0.5 h in cows. Therefore, aspirin dosages range from 10 to 20 mg/kg orally every 2 to 3 days in cats, 10 to 20 mg/kg orally every 12 h in dogs, and 100 mg/kg orally every 12 h in cows (Langston and Clarke, 2002). Comparison of NSAID glucuronidation between several species indicated that it was most potent in monkeys, dogs, and humans. Cats were efficient in that respect because cats tend to be deficient in some glucuronyl transferases enzymes that are important for glucuronidation (Magdalou et al., 1990). As a result, drugs that are excreted as glucuronide conjugates in other species, such as aspirin and paracetamol (acetaminophen), may have a prolonged half-life in cats, therefore increasing the risk of toxicity due to drug accumulation. Oral bioavailability of aspirin may vary due to differences in stomach content and pH (Conlon, 1988). A rise in pH increases the solubility of salicylates, and salicylate excretion depends on urinary pH; therefore, the short $t_{1/2}$ in horses is related to the basic urinary pH. Acetylsalicylic acid is poorly absorbed from the GI tract of horses after oral administration and disappears rapidly from the plasma. Although not approved for use in small animals, aspirin is most commonly used in dogs, but with associated GI complications, including mucosal erosions and hemorrhage in the pyloric antrum, cardia, and lesser curvature of the stomach (Conlon, 1988). These findings are not unusual, considering that aspirin and sodium salicylate are readily absorbed from the stomach and intestine of dogs and cats. Aspirin is rapidly deacetylated to salicylate, which is toxic to cells, affects mucosal barrier function, reduces cytosolic adenosine triphosphate, stimulates sodium transport and permeability, and increases proton dissipation from surface epithelial cells, resulting in microvascular damage, inflammation, hemorrhage, and gastric ulceration (Kauffman, 1989). In humans, aspirin use is associated with upper GI bleeding related to gastric hemorrhage and erosions distributed throughout the antrum of the stomach, especially more proximate to the body of the stomach with GI clinical manifestations of stomach upset, nausea, constipation, and diarrhea (Cryer, 2002).

Aspirin is available in a variety of preparations, such as plain, buffered, time-release, and enteric-coated. A potential strategy to combat the adverse GI effects from aspirin is administration of buffered or enteric-coated aspirin, which may prove less irritating to the dog stomach (Kauffman, 1989). Doses of 25 mg/kg of plain aspirin given at 8-h intervals for seven treatments resulted in gastric mucosal erosions in dogs, whereas there was minimal damage in dogs receiving buffered and enteric-coated preparations (Lipowitz et al., 1986). Misoprostol, a synthetic PGE_1 analog, has also been effective at decreasing endoscopically detectable mucosal gastric lesions (submucosal hemorrhage, erosion, or ulceration) in dogs given aspirin (Gullikson et al., 1987; Murtaugh et al., 1993; Johnston et al., 1995). In a dog repeat oral dosing study of aspirin at 60 mg/kg per day (onefold multiple of human dose) for three months, GI pathology of bloody or discolored stools, intestinal ulceration and/or perforation was seen (Hallesy et al., 1973). Aspirin given at 300 and 600 mg/kg per day for 4 days to Lewis rats with adjuvant-induced arthritis caused gastric mucosal bleeding and submucosal hemorrhage, which was seen at various time points of 24, 48, 72, and 96 h postdosing at necropsy (Schriver et al., 1975; Shriver et al., 1977). There are differences in the susceptibility of normal and arthritic rats to the gastric lesion-inducing properties of aspirin, with arthritic rats being more sensitive than normal rats. Oral administration of aspirin at doses of 10, 20 and 40 mg/kg caused dose-related increases in both the percentage of rats with gastric lesions and the severity of gastric lesion formation in both arthritic and nonarthritic rats. However, arthritic rats were less able to cope with the aspirin-induced insult to the gastric mucosal barrier (Katz et al., 1987).

Effects of Anthranilic Acid Derivative ns-NSAIDs on the GI Tract

Anthranilic acid derivatives include meclofenamate and mefenamic acid (Table 1-3). Meclofenamate (Ponstel, Arquel, Meclofen) is highly protein bound, metabolized by the liver, excreted in the urine and feces, and indicated to treat RA and OA in humans and musculoskeletal pain and inflammation in veterinary medicine (horses, dogs, cows). In a metaanalysis clinical study of dyspepsia, increased risk of dyspepsia was observed in meclofenamate users and NSAIDs (indomethacin, piroxicam) (Ofman et al., 2003). Dyspepsia in this study included any outcome terms, such as epigastric or upper abdominal pain/discomfort, but did not include nausea, vomiting, or heartburn. In humans, however, the most common GI event with meclofenamate is diarrhea. In cattle, oral administration of meclofenamic acid results in a biphasic pattern of absorption. Peak plasma concentration occurs at approximately 30 min and this is followed by a second peak at 4 to 6 h after dosing. The second peak is presumed to be due to enterohepatic recirculation (Aitken and Sanford, 1975). In horses, meclofenamic acid is absorbed rapidly and is effective in treating acute and chronic laminitis. It has a narrow therapeutic window and the onset of action is slow, requiring 2 to 4 days of dosing for clinical efficacy (Lees and Higgins, 1985). In rats, repeated oral dosing of mefenamic acid at 50 and 100 mg/kg per day (3.3- and 6.55-fold multiples of human dose) for 18 months caused GI pathology of bloody or discolored stool and intestinal ulceration and/or perforation (Hallesy et al., 1973). In nonhuman primates, repeated oral dosing of mefenamic acid at 400 and 600 mg/kg per day (26- and 39-fold multiples of human dose) for two years caused GI pathology similar to that described above in rats (Hallesy et al., 1973).

EFFECTS OF COX-1 INHIBITORS ON THE GI TRACT

COX-1 deficiency or inhibition is compatible with normal small intestinal integrity (Sigthorsson et al., 2002). COX-1 knockout mice do not spontaneously develop GI lesions, demonstrating that the absence of COX-1 alone is not sufficient to induce GI pathology (Langenbach et al., 1995). COX-1-deficient (COX-$1^{-/-}$) mice are normal except for a decrease in intestinal PGE_2 levels (Sigthorsson et al., 2002). COX-1 inhibition alone does not cause GI injury. A selective COX-1 inhibitor (SC-560) did not cause intestinal damage in rats (Tanaka et al., 2002a). In an in vitro study of small intestine motility, SC-560 was devoid of significant effects on horse ileal motility (Menozzi et al., 2009). Another in vitro study utilized a pig ileal intestine ischemia model and found that exposure to SC-560 recovered injured tissue to control levels as assessed by transepithelial electrical resistance (Blikslager et al., 2002). The effects of SC-560, rofecoxib, and indomethacin on the healing of colon lesions induced by dextran sulfate sodium (DSS) in the rat were investigated. The investigators found that daily administration of indomethacin and rofecoxib significantly delayed the healing of colitis, with deleterious influences on histological restitution as well as mucosal inflammation, whereas SC-560 had no effect (Tsubouchi et al., 2006). Okayama et al. (2001) found that SC-560 significantly worsened the severity of colonic damage in DSS-induced colitis in rats. Another study in a DSS-induced colitis mouse model found that rofecoxib ameliorated severe colitis and reduced the degree of inflammation by reducing neutrophil infiltration and IL-1β levels (Martin et al., 2005). In healthy rats, neither the s-NSAID rofecoxib nor the COX-1 inhibitor SC-560 when given alone at 20 mg/kg induced gastric mucosal injury. However, when rats received concurrent treatment with both SC-560 and rofecoxib, severe gastric lesions developed (Gretzer et al., 2001). Recent data suggest that COX-1 inhibition via SC-560, but not COX-2-derived PGE_2 synthesis, is involved in augmentation of NSAID-induced gastric acid secretion in isolated rabbit stomach parietal cells by enhancing expression and activation of the proton pump (Nandi et al., 2009).

Studies in animal models suggest that inhibition of COX-1 and COX-2 is required for induction of gastric ulcerogenic action of ns-NSAIDs (Wallace et al., 2000; Tanaka et al., 2001, 2002a,b; Takeuchi et al., 2004). It is thought that NSAID-induced ulcerogenesis, at least in rats, is dependent on the amount of gastric acid secretion derived from increased proton pump expression and requires inhibition of both COX-1 and COX-2 (Zinkievich et al., 2010). For example, in an experimental mouse model, small intestinal ulcers were observed when celecoxib and SC-560 were administered concurrently, but no GI damage was observed when either compound was administered independently (Sigthorsson et al., 2002). The importance of COX-1 and COX-2 simultaneous inhibition to cause GI effects is further supported by findings by Wallace et al. and Tanaka et al. Wallace et al. (2001) reported that COX-1 inhibition in rats reduced gastric mucosal blood flow but did not increase leukocyte adherence to the mesenteric vessel wall. On the other hand, COX-2 inhibition increased leukocyte adherence but did not reduce gastric mucosal blood flow (Tanaka et al., 2001). Thus, in the normal gastric mucosa, at least in rats, increased leukocyte adherence and vasoconstriction act in concert to facilitate gastric mucosal damage and lesions develop only when mucosal microcirculation and leukocyte function are impaired simultaneously. Furthermore, no spontaneous GI lesions occurred in COX-1-knockout mice, although gastric PE_2 levels were $<1\%$ of those in wild-type animals (Langenbach et al., 1995). Similarly, no GI pathology was found in COX-2-deficient mice (Morham et al., 1995).

EFFECTS OF COX-2 S-NSAIDS ON THE GI TRACT

Several COX-2 s-NSAIDs are approved for use in human and veterinary medicine (Table 1-4). Numerous nonclinical studies have demonstrated and supported the reduced GI events of COX-2 s-NSAIDs. In rats, rofercoxib did not cause damage to the stomach or small intestine (Yokota et al., 2005). When administered either orally or subcutaneously in rats, rofecoxib did not produce pathological changes in the GI mucosa, which showed normal histology (Laudanno et al., 2001). Neither rofecoxib nor celecoxib (Celebrex) caused gastric damage in normal rats after oral administration; however, both drugs caused hemorrhagic gastric lesions in arthritic rats (Kato et al., 2002). However, another study investigated the effects of celecoxib

TABLE 1-4 Major COX-2 Selective NSAIDs

Generic name	Trade name
Rofecoxib	Vioxx
Celecoxib	Celebrex
Valdecoxib	Bextra
Lumiracoxib	Prexige
Etoricoxib	Arcoxia
Deracoxib	Deramaxx
Parecoxib	Dynastat
Firocoxib	Previcox
Robenacoxib	Onsior
Meloxicam	Mobic, Metacam
Mavacoxib	Trocoxil

and rofecoxib in an experimentally induced colitis rat model. Colitis was induced by intrarectal instillation of acetic acid, which caused hemorrhagic diarrhea and weight loss. Oral administration of celecoxib at 5 mg/kg or rofecoxib at 2.5 mg/kg given twice daily reduced the degree of hemorrhagic diarrhea and the weight loss and significantly reduced the degree of colonic injury (El-Medany et al., 2005).

Clinical studies have shown that unlike ns-NSAIDs (e.g., etodolac, naproxen, ibuprofen), rofecoxib (Vioxx) did not inhibit PG synthesis or cause GI mucosal injury, even at supratherapeutic doses (Laine et al., 1995, 1999; Hawkey et al., 2000; Wight et al., 2001). There was no difference in ulceration rates of rofecoxib-treated patients as compared with a placebo and a fourfold lower depression of PG synthesis than in ibuprofen-treated patients (Laine et al., 1999; Hawkey et al., 2000). The Vioxx Gastrointestinal Outcome Research Trial (VIGOR) was a large (conducted in 301 centers in 22 countries) 13-month placebo-controlled double-blind study that compared twice the recommended dose of rofecoxib (50 mg daily) with the most common dose of naproxen (1000 mg daily) in 8076 RA patients (Bombardier et al., 2000). The primary endpoint was symptomatic ulcers, including clinical upper GI events of perforation, obstruction, and bleeding. The secondary endpoint was complicated upper GI events (perforation, obstruction, and major bleeding, resulting in a drop of 2 g or more in hemoglobin, transfusion, or hypotension). The RA patient population of VIGOR was selected because RA patients use NSAIDs chronically and have a substantially higher risk of NSAID-related GI events than do patients with OA. Rofecoxib significantly decreased the incidence of all GI endpoints studied in VIGOR. The VIGOR study showed a 54% reduction in clinical ulcers and a 57% reduction in complicated upper GI events with rofecoxib as compared with naproxen (Bombardier et al., 2000). The rate of discontinuation for any GI events (including clinical endpoints) was significantly lower in the rofecoxib group than in the naproxen group (Bombardier et al., 2000). A trial of the assessment of differences between Vioxx and naproxen to ascertain gastrointestinal tolerability and effectiveness (ADVANTAGE) was a 12-week double-blind randomized prospective trial in 5597 patients with OA in the United States and Sweden who were randomized to receive rofecoxib (25 mg daily) or naproxen (500 mg twice daily). Patients using low-dose aspirin (<81 mg/day) were included in the trial. The primary endpoint of ADVANTAGE was GI tolerability as defined by the incidence of discontinuations due to GI adverse events. The secondary endpoint was use of concomitant medication to treat GI symptoms. Most patients (71%) were women, and the mean age of study participants was 63 years. Twelve percent of patients used low-dose aspirin during the trial, and baseline characteristics of the treatment groups were similar. At the study end, a significantly lower rate of adverse GI event-related discontinuations had occurred with rofecoxib. Significantly fewer patients receiving rofecoxib required concomitant GI medications than patients receiving naproxen. Concomitant use of low-dose aspirin did not significantly affect relative rates of discontinuation due to adverse events, serious adverse events, or drug-related adverse events (Lisse et al., 2003).

Both nonclinical and clinical data show that COX-2 s-NSAIDs have a superior GI safety and improved GI tolerability profile to that of ns-NSAIDs. In dogs, no evidence of GI toxicity has been observed with celecoxib at supertherapeutic doses (Khan et al., 1997). In rats, celecoxib did not induce any damage to healthy stomachs or GI mucosa (Altinkaynak et al., 2003; Li et al., 2003), did not alter the gastric mucosal barrier (Coppelli et al., 2004), did not cause intestinal ulcers, and reduced the severity of experimental colitis (Cuzzocrea et al., 2001), but exacerbated inflammation-associated colonic injury in experimental colitis and damage induced in the stomach in a separate study (Khan et al., 1997; Zhang et al., 2004). In an experimental study in a rabbit model, the effects of valdecoxib on anastomotic healing 1 week following large bowel resection were investigated. Valdecoxib did not influence anastomotic healing or new vessel formation in the anastomotic region following large bowel resection (Neuss et al., 2009).

Similar to the nonclinical data, data from several clinical studies suggest that COX-2 s-NSAIDs have a superior GI safety profile to that of ns-NSAIDs. For example, CS-706, an s-NSAID, and naproxen were administered for 7 days to healthy men and women who did not have evidence of underlying GI lesions, and posttreatment upper GI endoscopy was conducted to assess and compare the development of GI petechiae, erosions, and ulcers. The extent of upper GI mucosal injury with CS-706 was statistically and significantly less than that for naproxen (Moberly et al., 2007). Another study compared the effects of valdecoxib (Bextra) and naproxen, administered for 6.5 days, on the upper GI mucosa of healthy older subjects (aged 65 to 75 years) as assessed by GI endoscopy. Valdecoxib was associated with a significantly lower rate of gastroduodenal, gastric, and duodenal ulcers than that of naproxen (Goldstein et al., 2006). In a 26-week clinical trial, the incidence of GI ulcers in patients receiving the COX-2 s-NSAID valdecoxib was significantly lower than in those receiving diclofenac. Additionally, valdecoxib was also associated with significantly improved GI tolerability than that with diclofenac (Pavelka et al., 2003). The incidence of upper GI bleeding and dyspeptic GI adverse experiences in patients with osteoarthritis was significantly lower with rofecoxib than with ns-NSAIDs (e.g., diclofenac, ibuprofen, nabumetone) (Langman et al., 1999). The Therapeutic Arthritis Research and Gastrointestinal Event Trial (TARGET) study compared the GI safety of lumiracoxib (Prexige) with ibuprofen and naproxen. Lumiracoxib in this TARGET study showed a three- to fourfold reduction in ulcer complications (Schnitzer et al., 2004). A large body of data has been published comparing the GI safety of COX-2 s-NSAIDs, celecoxib, and varied ns-NSAIDs. The Celecoxib Long-Term Arthritis Safety Study (CLASS) was a double-blind randomized controlled trial carried out in 7968 patients from 386 centers in the United States and Canada that compared a celecoxib dose of 400 mg twice daily (which is two- and fourfold the maximum dosage for RA and OA, respectively) with two ns-NSAIDs (diclofenac at 75 mg twice daily or ibuprofen at 800 mg three times daily) (Silverstein et al., 2000). The primary endpoint in CLASS was the incidence of ulcer complications (ulcer perforation, gastric outlet obstruction, or upper GI bleeding). The secondary endpoint was complicated and symptomatic ulcer events. Celecoxib in this CLASS study was associated with a lower combined incidence of symptomatic ulcers and ulcer complications than was ibuprofen or diclofenac (Silverstein et al., 2000). When compared with naproxen, celecoxib-treated patients also had lower rates of gastric, duodenal, and gastroduodenal ulcers (Goldstein et al., 2001). The Successive Celecoxib Efficacy and Safety Study (SUCCESS) was a 12-week double-blind randomized trial in 13,274 patients from 39 countries. The SUCCESS trial compared the incidence of upper GI hospitalizations in patients with OA taking celecoxib (200 or 400 mg daily), diclofenac (100 mg daily), or naproxen (1000 mg daily) (Singh et al., 2006). The rate of hospitalization was significantly lower in the celecoxib group. In addition, there were fewer ulcer complications in the celecoxib group than in the diclofenac or naproxen group, both in patients taking concomitant aspirin and in those not taking aspirin (Singh et al., 2006). In a separate study, video capsule endoscopy in healthy volunteers showed that celecoxib induced significantly less small bowel erosion than naproxen combined with omerprazole (Goldstein et al., 2007). In the Multinational Etoricoxib and Diclofenac Arthritis Long-Term (MEDAL) trial, the effects on GI outcome of etoricoxib (Arcoxia) and diclofenac were assessed. There were significantly fewer upper GI clinical events with etoricoxib than with diclofenac (Laine et al., 2007). Another clinical study demonstrated that several COX-2 s-NSAIDs (i.e., celecoxib, rofecoxib, valdecoxib, etoricoxib, lumiracoxib) offer greater upper GI safety and are better tolerated compared with ns-NSAIDs (Rostom et al., 2007). Upper GI mucosal effects were investigated for parecoxib (Dynastat) in a two-center double-blind randomized placebo-controlled study. Healthy subjects aged 65 to 75 years who were shown at baseline endoscopy to have no gastric or duodenal lesions received either 40 mg of parecoxib sodium IV twice daily for 7 days or 15 mg of ketorolac IV once daily for 5 days. No gastric or duodenal ulcers occurred in any subjects receiving parecoxib sodium. On the other hand, 23% of the ketorolac subjects had at least one ulcer, 16% had gastric ulcers, and 6% had duodenal ulcers (Stoltz et al., 2002). In another multicenter randomized double-blind placebo-controlled design, 123 adults with endoscopically confirmed normal upper GI mucosa received parecoxib sodium 40 mg twice daily for 7 days or ketorolac 30 mg four times daily for 5 days. No subjects treated with parecoxib sodium or placebo developed GI ulcers. Additionally, parecoxib sodium was comparable to placebo with respect to the combined incidence of erosions or ulcers. Thus, parecoxib sodium has a GI safety profile superior to that of ketorolac (Harris et al., 2004). The Meloxicam Large-Scale International Study Safety Assessment (MELISSA) trial was a large-scale double-blind randomized international trial conducted over 28 days in 9323 patients with symptomatic OA. Patients received either meloxicam 7.5 mg or diclofenac 100 mg, and significantly fewer adverse events were reported by patients receiving meloxicam than by those receiving diclofenac. Of the most common GI adverse events, there was significantly less dyspepsia, nausea and vomiting, abdominal pain, and diarrhea with meloxicam than with diclofenac. Thus, meloxicam has a significantly improved GI tolerability profile than that of diclofenac (Hawkey et al., 1998). Similarly, in another study called SELECT (the Safety and Efficacy Large-Scale Evaluation of COX-Inhibiting Therapies), 4320 patients with exacerbation of OA were treated with the recommended dose of meloxicam (7.5 mg) or piroxicam (20 mg) once daily for 28 days. There was a significantly lower incidence of GI adverse events in the meloxicam than in the piroxicam group (Dequeker et al., 1998).

Deracoxib (Deramaxx), a COX-2 s-NSAID approved for use in dogs, is indicated for the control of postoperative pain and inflammation associated with orthopedic surgery and osteoarthritis. Once absorbed, deracoxib protein binding is >90% and the half-life is 3 h. In a toxicology safety study, micronized deracoxib in gelatin capsules was administered once daily to healthy young dogs at doses of 10, 25, 50, and 100 mg/kg of body weight for up to 14 consecutive days. At the high doses of 25, 50, and 100 mg/kg, reduced body weight, vomiting, and melena occurred. Necropsy revealed gross GI lesions in dogs from all dose groups. The frequency and severity of the lesions increased with escalating doses. At 10 mg/kg, moderate diffuse congestion of gut-associated lymphoid tissues (GALT) and erosions or ulcers in the jejunum occurred. At the highest dose tested, 100 mg/kg, all dogs exhibited gastric ulcers and erosions or ulcerations of the small intestines (according to a package insert). In other 21-day and six-month toxicology studies in healthy dogs, deracoxib at lower doses of 2, 4, 6, 8, and 10 mg/kg per day did not cause abnormal GI findings as assessed by clinical observations or gross or histopathological examinations at any dose level tested (Roberts et al., 2009). Postapproval experience revealed GI events (i.e., vomiting, anorexia, diarrhea, melena, inappetence, hematemesis, hematochezia, weight loss, nausea, ulceration, perforation). However, it is not clear from this postapproval experience if deracoxib was used at the recommended doses or other NSAIDs or steroids were used. Deracoxib should only be used at approved dosages. In a retrospective study in dogs treated with deracoxib, it was found that 55% of dogs have received deracoxib at a dosage higher than that approved by the U.S. Food and Drug Administraction for the particular indication being treated. In addition, it was found that 59% of dogs have received at least one other NSAID or a corticosteroid in close temporal association (within 24 h) with deracoxib administration (Lascelles et al., 2005a). Therefore, GI perforation has been observed in dogs that received deracoxib at a higher than approved dosage or had received at least one other ns-NSAID in close temporal association with deracoxib administration (Lascelles et al., 2005a). A randomized placebo-controlled trial compared gastroscopic findings in dogs given aspirin (25 mg/kg) or deracoxib (1.5 mg/kg) for 28 days. The study found no significant differences in total scores between placebo and deracoxib-treated dogs on days 6, 14, and 28 and concluded that administration of deracoxib to healthy dogs resulted in significantly lower gastric lesion scores and fewer days of vomiting than with administration of aspirin (Sennello and Leib, 2006).

Another COX-2 s-NSAID, firocoxib (Previcox), has been proven clinically to control OA pain and inflammation in dogs (Pollmeier et al., 2006; Ryan et al., 2006). No adverse GI, hematological, or serum biochemical adverse effects were seen after oral daily administration of firocoxib for 29 days in healthy dogs (Steagall et al., 2007). In a large study with more than 1000 dogs with OA, a small withdrawal rate of 2.9% due to firocoxib GI-associated effects was observed, and no serious drug-related adverse events were reported (Ryan et al., 2006). The overall clinical efficacy of firocoxib to treat OA in horses was comparable to the ns-NSAID phenylbutazone (Doucet et al., 2008). In a study in healthy dogs, the gastric and duodenal effects of COX-2 s-NSAIDs after oral administration were investigated. Each dog received deracoxib (2 mg/kg), firocoxib (5 mg/kg), or meloxicam (0.2 mg/kg) for 3 days with a 4-week interval between successive treatments. No significant differences were found among these COX-2 s-NSAIDs regarding endoscopic GI mucosal scores, histologic scores, or COX-1 or COX-2 protein expression (Wooten et al., 2009). The effects of firocoxib on ischemic-injured jejunum mucosal recovery in horses were compared to those of flunixin meglumine. Transepithelial resistance of ischemic-injured jejunum from horses treated with flunixin meglumine was significantly lower than in firocoxib-treated horses (Cook et al., 2009). In a study in dogs, the effects of firocoxib on healing of induced gastric body and pyloric lesions were examined. Dogs were treated with firocoxib [5 mg/kg orally (PO) every 24 h] or placebo for 7 days. Healing was evaluated on days 2, 4, and 7 of treatment by endoscopic lesion scoring. Eicosanoid concentrations in plasma and at the lesion margins were determined on days 2, 4, and 7. The firocoxib group had larger pyloric lesions than the placebo, but mucosal PG production did not differ significantly from that with placebo (Goodman et al., 2009). In a blinded randomized crossover study design, cats were treated with firocoxib (1 mg/kg PO per day) and meloxicam (0.05 mg/kg PO per day) for 8 days. Blood samples and gastric and duodenal mucosal biopsy specimens were collected on days 0 (baseline; immediately before treatment), 3, and 8 of each treatment period. Firocoxib and meloxicam administration resulted in a lower plasma PGE_2 concentration than at baseline on days 3 and 8 of administration. Neither firocoxib nor meloxicam administration altered pyloric or duodenal PGE_1 synthesis (Goodman et al., 2010).

A recent COX-2 s-NSAID is robenacoxib (Onsior), which is prescribed to relieve pain and inflammation in cats and dogs. It contains four fluorine atoms and a carboxylic acid group and is chemically related to diclofenac and lumiracoxib. However, in contrast to most COX-2 s-NSAIDs, robenacoxib lacks a sulfur-containing group and is therefore considered to be chemically distinct from both the sulfone-containing rofecoxib and firocoxib class and the sulfonamide-containing celecoxib and deracoxib class (King et al., 2009). Significantly less gastric ulceration and intestinal permeability were noted in rats treated with robenacoxib

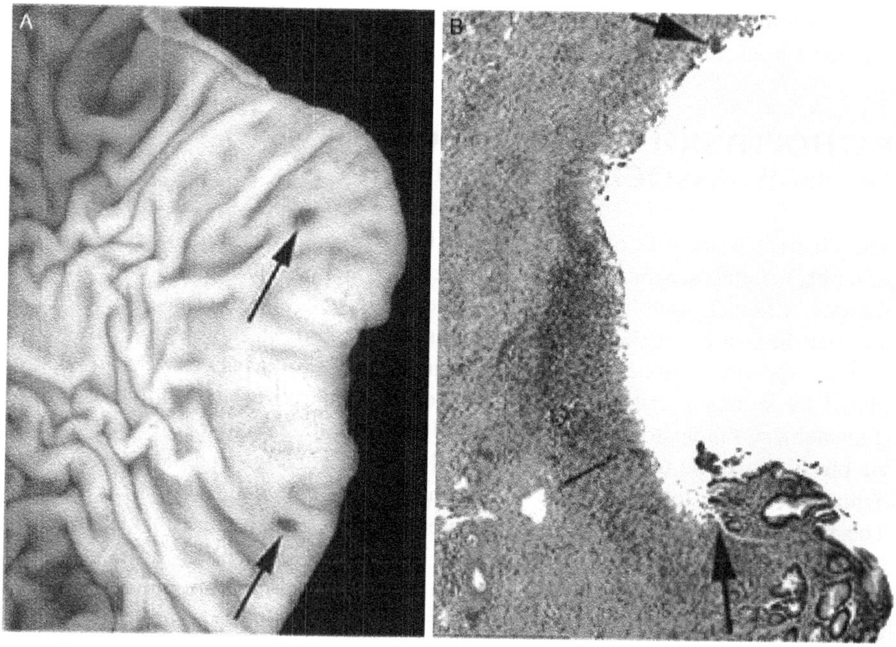

FIGURE 1-4 Meloxicam-induced gross (A) and microscopic (B) ulceration (between arrows) involving the mucosa in the pyloric region of the stomach of a dog. (Reprinted from Z. A. Radi and N. K. Khan, Effects of cyclooxygenase inhibition on the gastrointestinal tract, *Experimental and Toxicologic Pathology*, 58, pp. 163–173. Copyright © 2006, with permission from Elsevier.)

than in those treated with diclofenac (King et al., 2009). Another recent COX-2 s-NSAID is mavacoxib (Trocoxil), which is intended for the treatment of pain and inflammation associated with degenerative joint disease in dogs. No published GI safety data are currently available on this drug.

Other drugs previously compared to ns-NSAIDs for GI-related toxicity effects include meloxicam, L-745337, nimesulide, NS-398, and SC-58125. Meloxicam (Metacam, Mobic) is approved for use in dogs and cats and produced mild to moderate gastroduodenal lesions in dogs (Fig. 1-4) (Forsyth et al., 1998; Radi and Khan, 2006b; Radi, 2009); L-745,337 [5-methanesulfonamide-6-(2,4-difluorothiophenyl)-1-indanone] caused intestinal perforation in rats (Schmassmann et al., 1998); nimesulide did not cause any GI inflammation or ulcers in rats at excessive doses (Sigthorsson et al., 1998; Kataoka et al., 2000), prevented indomethacin-induced gastric ulcers in rats (Karmeli et al., 2000), and significantly decreased the extent of colitis induced by acetic acid in rats (Karmeli et al., 2000).

NS-398 produced little to no gastric ulceration in rats (Futaki et al., 1993; Masferrer et al., 1994) but had no beneficial effect on experimental colitis, whereas indomethacin did (Masferrer et al., 1994; Lesch et al., 1999). Similar to NS-398, SC-58125 yielded no GIT-related toxicity or beneficial effect in experimental colitis in rats (Seibert et al., 1994; Lesch et al., 1999). The COX-2 s-NSAID NS-398 delayed healing after acid-induced gastric injury in rats (Sun et al., 2000).

PATHOPHYSIOLOGY AND MECHANISMS OF NSAID-ASSOCIATED GI TOXICITY

The GI tract is the site of entry into the body of orally administered drugs. Due to its high metabolic and mitotic rates, the GI system is susceptible to xenobiotics toxicity. Clinical signs of NSAID-associated GI system toxicoses (i.e., vomiting and diarrhea) are generally manifestations of the body's attempt to eliminate or reduce exposure to toxicants. The medullary emetic (vomiting) center can be stimulated by stomach irritation and distension. Bright red or dark-colored vomitus (hematemesis) indicates retention of blood in the stomach due to gastric ulceration. GI bleeding can be either in the upper GI tract (occurring in the stomach or duodenum) or in the lower GI tract (occurring in jejunum, ileum, colon, or rectum). The signs of GI bleeding include hematemesis, melena, hematochezia, or occult bleeding. Hematemesis indicates bloody vomitus that is either fresh, bright red blood, or dark, digested blood. Melena is black, tarry, foul-smelling stool caused by digestion of blood in the GI system. Hematochezia indicates fresh, bright red blood passed from the rectum. Occult bleeding is seen when trace amounts of blood are detected in normal-appearing stools. An ulcer is a local defect or excavation of the GI mucosal surface that is produced by sloughing of necrotic tissues and extends into the submucosa or deeper (Figs. 1-2 to 1-4) (Radi and Khan, 2006b). One of the consequences of GI mucosa inflammatory necrosis is ulceration. Erosion is a local defect of the GI tract that is confined to the mucosa and does not penetrate the submucosa and muscularis mucosa. The prevalence of endoscopically detectable gastroduodenal erosions associated with NSAID use in humans ranges from 14 to 60%, whereas the incidence of ulcers is 14 to 31% (La Corte et al., 1999). It is estimated that the annual number of hospitalizations in the United States from NSAID-associated GI complications is at least 103,000. Also, the estimated cost of such hospitalization is \$15,000 to \$20,000 per hospitalization, and the annual direct costs of such complications exceed \$2 billion (Wolfe et al., 1999). Although NSAIDs are known to produce GI lesions in various animal species, the intraanatomical incidence of ns-NSAID-induced GI lesions in veterinary medicine is unknown, but the gross and histopathological changes in dogs treated experimentally with loxoprofen sodium were erosions and ulcerations that were limited to the pyloric gastric mucosa, jejunum, and ileum mucosa (Korytko et al., 2003). GI-associated toxicity seen with ns-NSAIDs may be attributed to direct nonspecific irritation or to various biochemical and pharmacological mechanisms linked to local COX inhibition (Hawkey and Skelly, 2002; Radi and Khan, 2006b; Radi, 2009) (Fig. 1-1). The dual-insult hypothesis by Schoen and Vender (1989) suggests that NSAIDs exert a direct toxic effect on the GI mucosa and indirect effects through active hepatic metabolites and decreases in mucosal PGs. Hepatic metabolites are excreted into the bile and subsequently into the duodenum, where they cause mucosal damage to the stomach by duodenogastric reflux and mucosal damage to the small intestine by antegrade passage through the GI tract. Detailed pathophysiological and potential mechanisms of NSAID-associated GI toxicity are discussed below.

Role of Cyclooxygenase Potency

Cyclooxygenase inhibitory potency is an important factor in NSAID GI toxicity. An indicator of the COX-1 versus COX-2 activity at therapeutic doses is the COX-1:COX-2 IC_{50} ratio in human blood. The COX-1:COX-2 IC_{50} ratio in human blood (the concentration of drug that inhibits 50% of the COX-1 activity in blood to that which inhibits COX-2 in the blood) is a good indicator of enzyme specificity; numbers greater than 1 are more COX-2 specific, and those below 1 are more COX-1 specific. There is wide variation between reports in the exact values, but the differences in COX selectivity of drugs tested have fairly good correlation between studies. Cryer and Feldman (1998) reported that the COX-1:COX-2 IC_{50} ratio of aspirin was 0.32, whereas ibuprofen was 0.6, naproxen was 1.14, and NS-398 was 24. Another report by Chan et al. (1999) indicated that the COX-1:COX-2 IC_{50} ratio in human blood for rofecoxib was 36, celecoxib was 6.6, and diclofenac was 0.4. As discussed previously, a lower incidence of adverse GI NSAID effects occurs more often with COX-2 s-NSAID. For example, substantial inhibition of platelet COX-1 will translate into an increased incidence of serious upper GI bleeding and complications (Patrono et al., 2001). Furthermore, the results of TARGET and VIGOR showed that ns-NSAIDS (naproxen and ibuprofen) cause more GI adverse events than do COX-2 s-NSAIDs (lumiracoxib and rofecoxib), which spare COX-1 activity at therapeutic dosing (Bombardier et al., 2000; Schnitzer et al., 2004).

Species Differences in NSAID-Associated Susceptibility to GI Injury

Although ns-NSAID-associated GI toxicity is prevalent throughout species, there are significant interspecies GI toxicity differences (Table 1-2) (Mahmud et al., 1996; Radi and Khan, 2006b). The degree and susceptibility of GI-associated toxicity can vary depending on the drug administered and the species tested (Radi and Khan, 2006b; Radi, 2009). For example, fenbufen has resulted in significantly fewer gastric lesions (hemorrhage and ulceration) in humans than have other ns-NSAIDs, such as naproxen and indomethacin (Lanza et al., 1983). Naproxen is well tolerated in mice, rabbits, nonhuman primates, and pigs, less well tolerated by rats, and poorly tolerated by dogs (Hallesy et al., 1973). The susceptibility of cats, dogs, and horses to the adverse effects of ns-NSAIDs is greater than that of humans (Mahmud et al., 1996). Dogs are considered a sensitive laboratory animal species for predicting the toxicity profile of NSAIDs. The no observed effect level (NOEL) dosages for GI toxicity in dogs are at or below clinical therapeutic dose levels for NSAIDs and are severalfold higher for COX-2 s-NSAIDs (Korytko et al., 2003). GI toxicity occurred at subtherapeutic ns-NSIAD lloxoprofen exposures (Korytko et al., 2003). There are significant breed differences in susceptibility to GI ulceration subsequent to ibuprofen exposure, with lower risk for the labrador breed and higher risk for the German shepherd breed (Poortinga and Hungerford, 1998). However, dogs tolerate ns-NSAIDs better than cats do. Since cats lack the glucuronyl transferase enzyme, they are delicately sensitive to the adverse effects of drugs that are glucuronidated before elimination (i.e., acetaminophen) (Court and Greenblatt, 1997). In addition, cats are at least twice as sensitive to ibuprofen as dogs are (Villar et al., 1998).

In addition, it is important not to extrapolate from studies performed in one species to another since various NSAID pharmacokinetics (PK) (a drug's absorption, distribution, metabolism, and elimination) are different across species (Lees et al., 2004). Some of the general PK properties of NSAIDs are (1) good bioavailability in monogastric species after oral dosing because of a medium to high level of lipid solubility, (2) dissolution in stomachs impaired by acidic pH, (3) possible delayed absorption by binding to digesta (e.g., horses), (4) a high degree of plasma protein binding of all drugs (except salicylate) in all species and small volume of distribution, (5) a low volume of distribution (some exceptions), and (6) marked species differences in clearance and terminal half-life (Lees et al., 2004). Such PK properties have implications in GI toxicity. For example, meloxicam has nearly 100% bioavailability after subcutaneous injection in cats, with an elimination half-life after a single dose estimated to be approximately 15 h. Therefore, the recommended dose of meloxicam (according to a package insert) in cats is 0.3 mg/kg, to be given subcutaneously only once, due to the narrow safety margin in cats. Repeated use of meloxicam in cats has been associated with diarrhea, vomiting, lethargy, and decreased appetite. Histopathological examination revealed gastrointestinal lesions ranging from GI mucosal inflammatory cell infiltration to erosions. In dogs, aspirin GI-associated side effects are dose- and preparation-related. A dose of 25 mg/kg (but not 10 mg/kg) of plain aspirin caused mucosal erosions in dogs, while no GI damage occurred in animals receiving buffered and enteric-coated preparations at similar doses of 25 mg/kg (Lipowitz et al., 1986).

GI Anatomical Differences

GI anatomical and physiological differences should be taken into consideration when NSAIDs are administered and potential GI toxicity is anticipated. For example, unlike humans, rats cannot vomit and are considered a nonvomiting species. This is due to a fold in the stomach that lies where the esophagus enters (Fox et al., 2002). Rats lack a gallbladder, and bile enters the duodenum continuously as it is made. Dogs, on the other hand, are very susceptible to vomition. Gut blood flow varies across species: 1.5, 7.5, 111, 216, 125, and 1100 mL/min in the mouse, rat, rabbit, dog, monkey, and human, respectively (Davies and Morris, 1993). Gut volume also varies across species: 1.5, 11.3, 120, 480, 230, and 1650 mL in the mouse, rat, rabbit, dog, monkey, and human, respectively (Davies and Morris, 1993). Humans, rhesus monkeys, and dogs are monogastric, and their stomachs are entirely secretory, whereas the stomachs of rodents, pigs, and horses have a glandular secretory portion and nonglandular bacterial digestion portion. Compared to humans, the GI length is shorter and the gastric emptying time is longer in dogs. Particle size can affect the rate of gastric emptying. The rate of gastric emptying in minipigs of nondigestable tablets (11 mm diameter) and granules (1 mm diameter) is slower than that in humans and dogs (Aoyagi et al., 1992). The rate of gastric emptying of both dosages in the dog tended to be faster than or similar to that in humans (Aoyagi et al., 1992). Food delayed gastric emptying in dogs, especially for tablets (Aoyagi et al., 1992). Such anatomical differences can affect the rate and extent of absorption of some NSAIDs. For example, in humans, enteric-coated aspirin was designed to reduce stomach irritation by delaying absorption until the drug reached the small intestine. In a study where dogs were administered enteric-coated aspirin, oral absorption was incomplete, gastric retention of tablets occurred, and partially digested tablets were found in the feces of some dogs (Nap et al., 1990). Ruminants have a forestomach (comprised of rumen, reticulum, and omasum) that is followed by a true stomach (abomasum). The rumen is highly permeable to volatile fatty acids released from carbohydrate metabolism and is also capable of active sodium and chloride absorption. Therefore, the large size of the rumen affects the response to oral toxicants, and initial exposure to toxicants may result in dilution and slowing of the absorption rate. Because the horse is a monogastric animal, the aspirin is absorbed rapidly following oral administration but is also removed rapidly, due to its short $t_{1/2}$. One might assume that the recommended 100-mg/kg dose of aspirin to a cow might be quite toxic; however, in this case the rumen acts as an "anatomical sustained-release device," slowly releasing the aspirin into the intestine for subsequent absorption. This slow release helps avoid potential aspirin-induced GI toxicity but allows for the maintenance of therapeutic drug concentrations (Langston and Clarke, 2002).

Enterohepatic Recirculation and NSAID Toxicity

The movement of a drug absorbed into the blood from the small intestine lumen, then carried into the liver by the hepatic portal vein, biotransformed in the liver, excreted into the bile, then through the bile duct into the lumen of duodenum, and reabsorbed into the blood via intestinal vessels is referred to as enterohepatic recirculation. Enterohepatic recirculation allows recycling of metabolized and nonmetabolized compounds and plays a role in NSAID-associated GI toxicity. Administration of NSAIDs that are not subject to enterohepatic recirculation did not produce intestinal damage in rats (Reuter et al., 1997). Additionally, ligation of the common bile duct prevented the damage normally observed after NSAID administration (Wax et al., 1970). Toxicological consequences of enteroheptic recirculation include increased drug half-life in blood, prolonged exposure, and inhibition of conjugate exports. For example, ibuprofen is particularly ulcerogenic in dogs because it undergoes enterohepatic recirculation. Indomethcin and piroxicam undergo substantial enterohepatic recirculation. Indomethacin enterohepatic recirculation is most extensive in dogs and rats and least extensive in rabbits and humans. In the dog, indomethacin remains in the enterohepatic circulation, even though it is reabsorbed from the gut, until its elimination in the feces. Biliary excretion of indomethacin and its conjugate is extensive and rapid in dogs, but slow in rats (Yesair et al., 1970). In humans and nonhuman primates, most NSAIDs undergo less enterohepatic recirculation than in the dog and rat (Beck et al., 1990). Dogs are less tolerant than rats or humans to ns-NSAIDs and more at risk for ns-NSAID-induced gastropathy (Elliott et al., 1988; Forsyth et al., 1998). Many NSAIDs, except aspirin and salicylic acid, undergo enterohepatic recirculation in rats (Beck et al., 1990). Nabumetone is formulated as a nonacidic prodrug that does not undergo enterohepatic recirculation to improve its GI safety profile.

Role of Xenobiotic Glucuronidation

Glucuronidation, catalyzed by the UDP-glucuronosyltransferase (UGT) enzyme, is a major elimination and biotransformation pathway of endogenous and exogenous xenobiotic compounds. NSAIDs are eliminated primarily through conjugation with polar sugar moieties to form glucuronides. The UGTs catalyze the conjugation of compounds that process a nucleophilic acceptor group with glucuronic acid, a relatively bulky, hydrophilic moiety, whose carboxylic acid functional group is ionized at physiological pH, thus forming metabolites with significantly different chemical and biochemical properties that in most cases have significantly decreased affinity for receptors or enzymes responsible for the biological activity of the parent compound (Siraki et al., 2005). A major biotransformation pathway for carboxylated NSAID is glucuronidation, with the resulting production of reactive acyl glucuronides (Siraki et al., 2005). Therefore, glucuronidation appears to play a role in NSAID GI-associated toxicity. Cats have remarkably low levels of UGT; therefore, they have a low capacity for hepatic glucuronidation and are more sensitive than other species to NSAID-associated GI toxicities (Court and Greenblatt, 1997; Steagall et al., 2009). Adduction, or covalent binding, of a toxicant to target macromolecules is considered a molecular mechanism of cellular injury of toxicants that damage tissues. For example, adducts detected by immunohistochemistry in the small intestine of diclofenac-treated rats represent covalent bonding of some diclofenac entity with intestinal macromolecules (Atchison et al., 2000). In addition, adduct formation after reactive metabolite generation by NSAIDs is one possible explanation for the mechanism of gut-induced toxicity. Rats given diclofenac orally at doses ranging from 10 to 100 mg/kg led to dose-dependent formation of adducts and ulcers only in the small intestine and only in animals with intact enterohepatic recirculation. Adducts formed within enterocytes by 1 h translocated to the brush border, preceded intestinal ulceration, and were intense at sites of ulceration (Atchison et al., 2000).

A major prerequisite for the enterohepatic recirculation of an NSAID is its conjugation by in the liver, which allows for hepatobiliary excretion of these highly polar metabolites. As discussed previously, in some NSAIDs (i.e., aspirin, diclofenac, ibuprofen, and indomethacin) intestinal toxicity is dependent on biliary excretion and enterohepatic circulation (Beck et al., 1990). Many carboxylic acid NSAIDs form ester glucuronides that are readily transported across the hepatic canalicular membrane by multidrug carrier systems. For example, diclofenac glucuronide is transported into bile by the canalicular conjugate export pump, Mrp2, in the rat (Seitz et al., 1998). Hepatocanalicular conjugate export pump-deficient

(TR^-) rats were used to selectively block diclofenac enterohepatic circulation without interrupting bile flow. TR^- rats were refractory to diclofenac given either intraperitoneally or perorally. However, transfer of bile containing diclofenac glucuronide significantly increased the extent of ulcer formation in both normal and TR^- rats. Moreover, induction of glucuronosyltransferase aggravated intestinal ulceration (Seitz and Boelsterli, 1998).

Aging and Stress and NSAID GI Effects

Age appears to be a contributing factor in NSAID-induced GI toxicity. The exact mechanism of this is not fully understood. It is thought that the aging gastric mucosa has impaired mucosal defense mechanisms, due to decreased mucus and bicarbonate secretion, reduced gastric emptying rate, reduced GI motility, reduced GI blood flow, increased expression on villi of enterocytes and the function of the P-glycoprotein multidrug efflux pump, and reduced GI PG production. Neonates have a poorly developed intestinal mucosa barrier that can permit absorption of various xenobiotics. Some studies in humans have demonstrated a decline in gastric and duodenal mucosal PG content with aging, associated with an increase in gastric acid secretion (Cryer et al., 1992; Goto et al., 1992). Similarly, studies in animals have yielded similar results (Uchida et al., 1990; Lee and Feldman, 1994). Uchida et al. (1990) have demonstrated a marked decrease in gastric mucosal PGI_2 level between 20 and 40 weeks of age and between 60 and 86 weeks of age in normal and ulcer-bearing rats. In another study, gastric mucosal PGs synthesis decreases with age in rats, and aged animals were more susceptible to aspirin-induced acute gastric mucosal injury (Lee and Feldman, 1994).

In a study in humans, Feldman and Cryer (1998) demonstrated that in healthy subjects with normal gastric histology, advancing age was associated with a significant decline in gastric bicarbonate secretion (HCO_3^-), Na^+, and nonparietal fluid secretion, resulting in an increase in gastric acidity, while no age-related changes in acid and parietal fluid secretion were noted. Furthermore, animal studies have shown that aging was associated with significantly lower gastric luminal pH and bicarbonate output in the rat stomach and that aging also blunted PG-mediated increases in gastric HCO_3^- secretion (Lee, 1996). Moreover, Kim et al. (1990) have shown in anaesthetized rats that although aging does not affect basal duodenal bicarbonate secretion, the duodenal bicarbonate response to a fixed load of luminal acid declines progressively with age. Finally, some NSAID clearance is altered with aging. Clearance of phenylbutazone is twice as fast in 3-year-old horses as in 8- to 10-year-old horses (Tobin et al., 1986).

A syndrome in humans called stress-related mucosal disease (SRMD) is common in critically ill patients and can result in significant morbidity. Histologically, there does not appear to be a significant inflammatory component to the gastric mucosa. The pathophysiology of this condition is multifactorial, but local mucosal ischemia and gastric acid play a critical role in disease pathogenesis. It is thought that gastric mucosal damage results from an imbalance between factors promoting mucosal injury and host defenses. A major mucosal defense factor is impaired mucosal blood flow. Aggregative factors include acid, pepsin, intramucosal acidosis, reperfusion injury, and free-radical formation (Duerksen, 2003). There does not seem to be a selective impairment of PG production in patients with SRMD. Therefore, there is no evidence that NSAIDs increase the susceptibility of critically ill patients to stress-related GI damage (Duerksen, 2003).

Disruption of GI Physiological Mucosal Defense Mechanisms

The GI system is involved primarily in breaking down food, ultimately absorbed into the body. This digestive process involves several phases: ingestion of food; secretion of mucus, water, and enzymes; fragmentation; chemical and mechanical digestion of food particles; absorption of digested food; and elimination of waste products by defecation. Fragmentation and initial digestion take place in the stomach. The stomach has gastric glands that release or secrete mucus, hydrochloric acid (HCl), gastrin, somatostatin, acetylcholine, histamine, and pepsinogen. The small intestine is a major GI site of digestion and absorption of nutrients and electrolytes. The small intestine is divided into the duodenum, jejunum, and ileum. The process of digestion is initiated in the stomach by the actions of HCl and pepsin, which break down food particles. The mucosa of the stomach and proximal duodenum are constantly exposed to gastric acid that can damage living cells. Acid secretion is stimulated by acetylcholine neurotransmitter, gastrin hormone, and histamine. PGs can inhibit acid secretion. The gastroduodenal lumen also contains bile salts and enzymes such as pepsins, lipases, proteases, and peptidases (Johnson et al., 2006). Proton pump inhibitors (PPIs) significantly decrease NSAID-induced gastric and duodenal ulcers (Lazzaroni and Porro, 2009). PPIs are a group of drugs whose main action is pronounced and long-lasting reduction of gastric acid production.

The GI mucosa has several physiological defense mechanisms that form a mucus coating called a mucosal barrier to protect itself from the damaging effects of such degradative enzymes and acidic pH. Major physical and chemical gastric mucosa defense mechanisms include a hydrophobic mucus layer, regulated intercellular tight junctions, specialized plasma membrane ion permeability, epidermal growth factors, HCO_3^- secretion, high rate of mucosal blood flow, mucosal cell hydrophobicity, and rapid epithelial turnover (Johnson et al., 2006). In addition, the gastric and duodenal mucosa are rich in PGs, which play a protective role in the GI tract via adequate perfusion of the gastroduodenal mucosa, epithelial cell secretion of bicarbonate, secretion of mucus, and maintenance of a neutral mucosa pH (Elliott et al., 1996; Scheiman, 1996).

The mucus hydrophobic layer traps secreted bicarbonate to maintain a neutral gastric mucosal pH and forms a stable water-insoluble glycoprotein gel that acts as a lubricant to prevent mechanical damage. The mucus–bicarbonate layer protects the gastric mucosa from diffusion of free hydrogen ions from the gastric lumen back into the mucosal cells (i.e., back-diffusion) (Johnson et al., 2006). If the integrity of the GI barrier is disturbed, the rate of back-diffusion of gastric acid and pepsin increases, leading to inflammation and hemorrhage. Inflammatory cells such as neutrophils and mast cells become activated and release inflammatory mediators such as histamine, leukotriens, free radicals, and proteolytic enzymes. These mediators lead subsequently to vasodilation, vasoconstriction, increased vascular permeability, and edema. Such events lead to GI mucosal ischemia, reduced mucus secretion, and reduced PG production (McConnico et al., 2008).

Inhibition of PG synthesis and abrogation of protective mechanisms lead to GI injury (Hawkey and Skelly, 2002). Misoprostol is a synthetic PGE_1 analog used to overcome NSAID-induced PG deficiency in the gastric mucosa. In the six-month randomized Misoprostol Ulcer Complication Outcomes Safety Assessment (MUCOSA) trial, the effects of concurrent administration of misoprostol on the occurrence of serious upper GI complications in patients with RA who were receiving NSAID were investigated (Agrawal and Aziz, 1998). The results of the MUCOSA study showed that misoprostol resulted in a statistically significant reduction in the incidence of serious NSAID-induced upper GI complications compared with placebo in patients with RA (Agrawal and Aziz, 1998). NSAIDs can also decrease mucosal resistance in the diseased state (Rainsford, 1982) or cause direct chemical damage to the GI mucosa (Johnston and Budsberg, 1997). When the GI mucosal integrity is compromised, a cascade of pathological events follows, leading to further mucosal barrier layer damage. COX inhibition is suggested to increase the susceptibility of the gastric mucosa to injury by inhibiting secretion of the cytoprotective mucus and bicarbonate and altering the physicochemical nature of mucus (Kauffman, 1989). Indomethacin and SC-560, but not rofecoxib, attenuated mucosal acidification in the rat stomach (Takeuchi et al., 2006). This study suggests that EP_1 receptors are essential for the increase in the secretion of HCO_3^- in response to mucosal acidification in the rat stomach. In EP_3 receptor–knockout mice, the HCO_3^- stimulatory action of PGE_2 was observed in the stomach, whereas such action was absent in the EP_1 receptor–knockout mice (Takeuchi et al., 1999). Therefore, it appears that PGE_2 receptor subtypes (EP_1 in the stomach and EP_3 in the duodenum) and COX-1 are key regulators of HCO_3^- secretion in response to gastroduodenal mucosal acidification.

GI Disequilibrium

All segments of the GI from duodenum to distal colon have mechanisms for both absorbing and secreting water and electrolytes. GI secretions play a significant role in digestion and maintenance of pH levels and acid–base balance. Disequilibrium in GI motility and secretion by NSAIDs can lead to diarrhea, dehydration, and/or systemic acidosis or alkalosis. Respiratory alkalosis may occur from stimulation of the respiratory center by phenylbutazone. Major mechanisms of diarrhea are osmosis, active secretion, exudation, and/or altered motility. The Na^+ gradient is the driving force for amino acid, oligopeptide, and sugar absorption. There are differences in Na^+ entry mechanisms, sites of HCO_3^- secretion, and sites of active K^+ transport. HCO_3^- is absorbed in the jejunum (via Na^+/H^+ exchange) and secreted in the duodenum, ileum, and colon. Na^+ crosses the small intestinal and colonic brush borders via Na^+/H^+ exchange and in the small intestine also by the Na^+ organic solute cotransport mechanism. In the distal colon, luminal Na^+ is also absorbed via an aldosterone-sensitive Na^+ channel. The intestines neither dilute nor concentrate their contents, the osmolarity of which, except in the duodenum and proximal jejunum shortly after eating, is the same as the plasma osmolarity. Osmotic diarrhea occurs when these organic solutes are absorbed, salt is absorbed with them, and water follows osmotically (i.e., transport from enterocyte to lateral intercellular space creates a local osmotic gradient that initiates water flow) (Field, 2003).

In osmotic diarrhea, agents released from inflammatory cells (PGs and leukotrienes, platelet-activating factor, histamine, serotonin) can stimulate active secretion. GI mucosal-induced inflammation interferes with GI homeostatic control. PGE_2 activates mast cells, causing histamine release, which stimulates smooth muscle activity and induces the secretion of mucus and electrolytes. Electrolytes create osmotic force for the influx of water into the lumen. Therefore, the combination of excessive fluids and increased GI motility causes diarrhea. The pathogenic importance of intestinal hypermotility in the intestinal ulcerogenic response to indomethacin has been demonstrated in rats (Takeuchi et al., 2002). Additionally, indomethacin decreased gastric mucosal PGE_2 content and produced gross pathological mucosal damage with gastric hypermotility and expression of COX-2 mRNA (Takeuchi et al., 2004). In the same study by Takeuchi et al. (2004), although SC-560 did not produce damage, it caused a decrease in the PGE_2 content and an increase in gastric motility as well as the up-regulation of COX-2 expression. Duodenal HCO_3^- secretion and luminal release of PGE_2 in rats were increased in response to mucosal acidification. This response was significantly inhibited by indomethacin but not by NS-398 or nimesulide (Hirata et al., 1997).

In addition to transporting ions, nutrients, and water, the intestinal epithelium, comprised of both enterocytes and their tight junctions (zona occludens), functions as a barrier that restricts the flow of luminal contents into the blood and lymphatics, and vice versa. In the small intestine, tight junctions are, on average, of the low-resistance type, meaning that most of the passive permeability of the epithelium to small monovalent ions and water resides in these junctional complexes (tight junctions in villi have higher resistance than do those in crypts). Colonic intercellular junctions are tighter, their resistance increasing steadily from proximal to distal portions (Field, 2003).

Although these junctions, which comprise a number of discrete proteins, are extracellular, their permeability properties are regulated by intracellular structures, especially actin filaments. Therefore, when the intestinal epithelium's barrier function is compromised by NSAID-associated erosions and ulcerations, loss of epithelial cells, and/or disruption of tight junctions, hydrostatic pressure in blood vessels and lymphatics will cause water and electrolytes, mucus, and protein to accumulate luminally, leading to exudative diarrhea.

Effects on Physiological GI Mucosal Cell Renewal Mechanisms After Mucosal Injury

The integrity of the intestinal mucosal surface barrier is generally reestablished rapidly, even after extensive damage, due to its enormous regenerative capability.

The rapid-healing mechanism of the surface mucosal epithelium is accomplished by epithelial cell migration from proliferative zones into wound, also termed epithelial restitution, epithelial cell proliferation, and differentiation. This healing mechanism is regulated by a highly complex network of factors, such as regulatory peptides within the intestinal tract mucosa, conventionally designated as growth factors and cytokines. These factors play an essential role in regulating differential epithelial cell functions to preserve normal homeostasis and integrity of the intestinal mucosa (Dignass, 2001; Sturm and Dignass, 2008). Mucosal villous contraction and the restitution mechanism represent primary repair mechanisms in the GI tract which allow resealing of the epithelial barrier within minutes or hours via reformation of tight junctions between cells (Dignass, 2001; Sturm and Dignass, 2008). Villous contraction is initiated by myofibroblasts that reside immediately beneath the epithelial basement membrane. Subsequent events include crawling of healthy epithelium adjacent to the wound, referred to as restitution. Restitution is a well-coordinated event that is dependent on epithelial cell migration but independent of cell proliferation and differentiation. The structural integrity of the mucosa is maintained by continuous cell renewal from mucosal progenitor cells. This continuous renewal is a well-coordinated process that is controlled by proliferation of progenitor cells, which enables replacement of damaged or aged surface epithelial cells. Actin filaments, focal adhesions, and focal adhesion kinase (FAK) play crucial roles in the cell motility essential for restitution (Szabó et al., 2002).

Indomethacin significantly delayed epithelial restitution in rats and reduced FAK phosphorylation and recruitment to adhesion points, as well as actin stress fiber formation in migrating surface epithelial cells (Szabó et al., 2002). NSAIDs have been found to affect intestinal restitution through decreased potassium channel [K(v)1] surface expression and trafficking (Freeman et al., 2007). In an in vitro study, intestinal epithelial cell migration in response to wounding was reduced by indomethacin, phenylbutazone, and NS-398 but not by SC-560 (Freeman et al., 2007). NSAID inhibition of intestinal cell migration was not associated with depletion of intracellular polyamines (Freeman et al., 2007). However, another study, using pig small intestinal ileal mucosa, showed that endogenous PGs, released when mucosal injury occurs, mediate local repair of small intestinal epithelium after damage by the deconjugated bile salt deoxycholate. Whereas ongoing epithelial restitution and villous contraction were prominent features of repairing mucosa, acute recovery of barrier function was uniquely dependent on PG-mediated resealing of tight junctions and lateral intercellular space. The authors conclude that failure to repair increases in paracellular pathway permeability may underlie barrier failure resulting from NSAID use in patients with underlying enteropathy (Gookin et al., 2003).

A number of other factors, such as extracellular matrix, blood clotting factors, phospholipids, short-chain fatty acids, adenine nucleotides, trace elements, and calpain, have been demonstrated to modulate intestinal epithelial repair mechanisms (Dignass, 2001; Sturm and Dignass, 2008). Calpains, cysteine proteases, are involved in numerous cellular processes, such as cell migration and invasion. Altered expression of calpain proteins contributes to NSAID effects on intestinal epithelial restitution. A multistep functional microarray genomic study was conducted using intestinal epithelial cells to identify novel signaling pathways that contribute to NSAID inhibition of GI epithelial cell migration (Raveendran et al., 2008). Raveendran et al. demonstrated that indomethacin and NS-398 decreased the expression of calpains 1, 2, and 8 proteins, whereas SC-560 had no effect on the expression of calpain proteins. Functional data were also consistent with decreased expression of calpain protein in cells treated with either NS-398 or indomethacin.

In gastric glands, a single stem cell in every gastric gland undergoes division to produce committed progenitor cells, which further differentiate into an adult epithelial cell type (Modlin et al., 2003). The stem/progenitor cell niche is made up of proliferating and differentiating epithelial cells and surrounding mesenchymal cells (Leedham et al., 2006). Intestinal subepithelial myofibroblasts (ISEMFs) are important coordinating cells that possess significant influence on their environment by virtue of their receptor profile and the signals they produce. Characteristically, ISEMFs form a protective fenestrated sheath around the stem cell compartment, creating the stem cell niche—the optimal microenvironment for stem cells to give rise to differentiated progeny (Leedham et al., 2006). The ISEMFs generate growth factors and thus promote mesenchymal-to-epithelial crosstalk and signaling to maintain the niche progenitor cell survival. Cell proliferation of progenitor cells is controlled by growth factors. The major growth factor receptor in rats during gastric ulcer healing, expressed in gastric progenitor cells, is the epidermal growth factor receptor (EGF-R) (Tarnawski et al., 1992), and the major mitogenic growth factors that activate this receptor are transforming growth factor alpha (TGFα) and insulin-like growth factor 1 (IGF-1) (Nguyen et al., 2007). EGF peptide itself is absent in normal gastric mucosa. However, it is present in the gastric lumen and can stimulate progenitor cell proliferation in case of injury. In human gastric epithelial monolayers, EGF treatment significantly stimulated cell migration and actin stress fiber formation, and increased FAK localization to focal adhesions, and phosphorylation of FAK and tensin (Szabó et al., 2002). In gastric ulcers in rats, IGF-1 promoted actin polymerization, cell proliferation, reepithelialization, and induced COX-2 in a phosphatidylinositol 3-kinase-dependent manner (Nguyen et al., 2007).

Both ns- and COX-2 s-NSAIDs may delay the healing of damaged GI mucosa. NSAIDs may delay healing after injury to the GI tract by affecting GI epithelium proliferation in response to mediators such as growth factors and nitric oxide (NO). The effects of celecoxib on normal and damaged (acid or ethanol challenged) gastric mucosa of rats were compared to those of ns-NSAIDs (Berenguer et al., 2004). In the absence of acid or ethanol challenge, only ns-NSAIDs produced appreciable gastric lesions. However, following acid or ethanol challenge, both COX-2 s- and ns-NSAIDs impaired healing of gastric mucosal damage. In rats, NS-398, a COX-2 s-NSAID, delayed gastric healing after acid-induced injury (Sun et al., 2000). Increased COX-2 expression was noted after growth stimulation by addition of serum which contains growth factors to cultured rat gastric mucosal cells in vitro and after acid-induced gastric injury to rats in vivo (Sawaoka et al., 1997; Horie-Sakata et al, 1998; Erickson et al., 1999; Sun et al., 2000). These serum growth factors were TGFα or the hepatocyte growth factor (HGF) (Horie-Sakata et al., 1998; Sawaoka et al., 1999). Celecoxib did not alter the gastric mucosal barrier or induce mucosal lesions in healthy or NO-deficient rat gastric mucosa (Copelli et al., 2004). Celecoxib appeared to worsen the flare of the colon in a rat model of inflammatory bowel disease (Singh et al., 2004). In an isolated rabbit gastric epithelial cells model, aspirin significantly retarded wound healing, but simultaneous addition of growth factors such as IGF-I and EGF significantly accelerated wound repair (Yoshizawa et al., 2000). PGE_2 and gastrin transactivate EGF-R and trigger the mitogen-activated protein kinase pathway, thereby stimulating cell proliferation and exerting a trophic action on mucosa, resulting in gastric and intestinal hypertrophy (Pai et al., 2002). Therefore, NSAIDs are contraindicated in patients with damaged GI mucosa (peptic ulceration or GI bleeding).

Effects on Leukocyte Adhesion Molecules and Trafficking

Infiltration of leukocytes into the mucosa in response to initial tissue injury has been implicated in NSAID GI injury (Reuter et al., 1997). Leukocyte–endothelial cell interactions are mediated by various cell adhesion molecules. These interactions are important for leukocyte extravasation and trafficking in many pathological conditions in several body systems, including the GI tract. There are various stages of leukocyte trafficking into sites of inflammation. An initial slowing of leukocytes on the vascular endothelium is mediated by selectins. This event is followed by (1) activation of β_2 integrins after leukocyte exposure to cytokines and proinflammatory mediators, (2) adherence of leukocyte β_2 integrins to vascular endothelial ligands [e.g., intercellular adhesion molecule-1 (ICAM-1)], (3) extravasation of leukocytes into tissues through tight junctions of endothelial cells mediated by the platelet and endothelial cell adhesion molecule 1 (PECAM-1), and (4) perivascular migration through the extracellular matrix via β_1 integrins. Inhibiting excessive leukocyte egress and subsequent free-radical-mediated damage caused by leukocyte components may attenuate or eliminate tissue damage (Radi et al., 2001). Both piroxicam and meloxicam interfered, in an in vitro experiment using flow cytometry, with neutrophil degranulation and cytokine-mediated activation changes in adhesion molecules (García-Vicu na et al., 1997). Due to their anti-inflammatory properties, NSAIDs have been used to modify leukocyte infiltration via their effects on different stages of leukocyte trafficking in various animal models (Radi et al., 2001). For example, celecoxib and indomethacin inhibited leukocyte migration induced by lipopolysaccharide injected into the cremaster muscle in a rat model. However, celecoxib was associated with reduced leukocyte rolling and adhesion, whereas indomethacin only inhibited cell adhesion (Menezes et al., 2008). A role for leukocyte trafficking, adhesion, and activation in NSAID GI-mediated toxicity has been proposed. In a study in rats, neutrophil adhesion to gastric mesenteric venules was increased by indomethacin and celecoxib but not by SC-560, whereas gastric mucosal blood flow was decreased by indomethacin and SC-560 but not by celecoxib (Wallace et al., 2000). These data suggest that the NSAID-induced decrease in gastric mucosal blood flow is COX-1-mediated, whereas NSAID-induced neutrophil vascular endothelial adhesion is COX-2-mediated. In another study in rats, neutrophil- and oxygen radical–dependent microvascular injuries were found to have a role in gastric mucosal injury induced by ns-NSAIDs. Also, reactive oxygen species (ROS) produced by activated neutrophils after indomethacin treatment in rats caused gastric mucosal injury via ROS-mediated oxidation of such macromolecules as lipids, proteins, and DNA (Naito and Yoshikawa, 2006).

Effects of GI Physiological Local pH, Gut Absorption, and Fasting

Absorption is a process whereby xenobiotics gain entrance to the body via the circulatory or lymphatic system. There are five possible processes of intestinal absorption of xenobiotics: (1) active transport, (2) passive diffusion, (3) pinocytosis, (4) filtration through "pores," and (5) lymphatic absorption. Most xenobiotics are transported across the GI mucosa by passive diffusion (Chhabra, 1979). GI physiology can have a significant impact on absorption. A number of factors, such as diet, GI motility, interference with GI flora, changes in the rate of gastric emptying, age of the animal, physical properties of a compound, and the dissolution rate of xenobiotics can influence the rate of GI absorption (Chhabra, 1979). Lipid-soluble compounds are more readily absorbed than water-soluble compounds. Additionally, xenobiotic absorption is highly dependent on local GI pH values. However, there are considerable species differences in pH along each segment of the GI tract (Table 1-5) (Smith, 1965; Dressman et al., 1990; Davies and Morris, 1993; McConnell et al., 2008).

The pH in the GI tract is a crucial factor, affecting the stability and solubility of drugs and their absorption through the mucosa (McConnell et al., 2008). The gastric acid secretion rate in the dog at the basal state is low compared to that in humans and rhesus monkeys. In an in vitro experiment, Legen and Kristl (2003) demonstrated that ketoprofen transport across the rat jejunum has pH- and energy-dependent transport mechanisms. Indomethacin-induced gastric mucosal damage in rats is markedly dependent on luminal pH (Elliott et al., 1996). Acidic compounds can potentially cause gastric toxicity. Although ibuprofen is more potent that aspirin as a cyclooxgenase inhibitor, it has less gastric toxicity in animals and humans. This is because ibuprofen is 10 times less soluble under the acidic pH conditions

TABLE 1-5 Comparative pH Values of Various Anatomical Regions of the Gastrointestinal Tract

Species	Stomach	Small intestine	Cecum	Colon
Human	1.7–6.7	5.4–7.5	6	7.5
Monkey	2.8	5.6–6	5	5.1
Dog	3.4	6.2–7.5	6.4	6.5
Rat	3.2–3.9	5–6	5.9–6.6	5.5–6.2
Mouse	3–4	4.8–5.2	4.4–4.6	4.4–5
Rabbit	1.9	6–8	6.6	7.2

in the stomach and thus is unlikely to be absorbed there (Beck et al., 1990). Unsuitable pH may cause the precipitation of acidic or basic drugs from solution or the degradation of labile compounds. The lowest pH is seen in the stomach in laboratory animals and humans. However, the fasting status affects gastric pH values. In dogs and cats, gastric acid secretion is intermittent and gastric pH during fasting can rise as high as 3 to 6.5. In rodents, gastric pH appeared higher in the fasted state (McConnell et al., 2008). These observations are unlike that seen in humans, where the gastric pH is lower in the fasting state than in the fed state (fasted pH 1.7 increases to 6.7 after meal ingestion in humans) (Dressman et al., 1990). The fasting status affects NASID GI-associated toxicity. For example, the gastric toxicity of indomethacin, diclofenac, ibuprofen, and aspirin at various doses is relatively low after oral administration to rats having access to food (Beck et al., 1990). On the other hand, fasted rats were more vulnerable to similar toxicity at all doses of these NSAIDs (Beck et al., 1990). The presence of food retards fenoprofen stomach absorption and lowers peak concentration in plasma, which is usually achieved within 2 h. In horses, food can impair the oral absorption of some NSAIDs, such as phenylbutazone. Phenylbutazone absorption from the GI tract in horses is influenced by the dose administered and the relationship of dosing to feeding. Access to hay can delay the time of peak plasma concentration to 18 h or longer (Tobin et al., 1986). The fed state of rodents, similar to the situation in humans, had no effect on intestinal pH (McConnell et al., 2008). The low intestinal pH level in rodents could have implications for the in vivo testing of oral drugs in rats and mice. For example, drugs that require a basic pH to dissolve may precipitate at the lower pH values seen in rodents. This suggests that rodent models may not be the most appropriate to use to study pH-sensitive dosage forms targeted to the human lower intestine and colon (McConnell et al., 2008).

A pH gradient has been reported at the gastric mucosal surface in several species (human, dog, rat, mouse, rabbit) (Johnson et al., 2006). This pH gradient is relatively alkaline directly at tissue surfaces and becomes more acidic at distances farther away from the surface. This alkaline layer is caused by active bicarbonate secretion and is considered a major defense mechanism. Duodenal hemorrhagic lesions were induced when pH was decreased (acid hypersecretion) (Hirata et al., 1997). In addition, such lesions were induced by histamine in rats (Hirata et al., 1997). In fact, histamine H_2-receptor antagonists have been used to reduce NSAID-induced gastric ulcers (Lazzaroni and Porro, 2009).

NSAID Topical Effect-Mediated Injury

Variation in the physicochemical properties and pharmacological profiles among the individual NSAIDs translates into interagent differences regarding the propensity to cause adverse GI effects (Bannwarth, 2008). It has been suggested that NSAID-associated GI hemorrhage and erosion are related to a topical effect of the drugs in addition to COX inhibition (Tibble et al., 2000). Enteric-coated or IV-injected drugs result in considerably less acute GI injury. In rats, the intestinal tolerability of celecoxib is thought to be related to the absence of a topical damaging effect and COX-2 selective inhibition (Tibble et al., 2000). Topical GI mucosal injury usually occurs after ingestion of the weakly acidic and lipid-soluble NSAIDs. These weak acids are not ionized in the acidic gastric environment and their lipid solubility allows them to diffuse freely across the plasma membrane into surface epithelial cells. At cellular pH, they dissociate into the ionized form, releasing hydrogen ions that are trapped within the cell (intracellular trapping hypothesis), leading to an increase in back-diffusion of gastric acids and disruption of cellular function (Wolfe et al., 1999). Topical acidic properties can cause GI mucosal damage. For example, although sulindac is administered as a non-toxic prodrug, its active metabolite, sulindac sulfide, is excreted into bile. Upon entry into the duodenum, sulindac sulfide causes duodenal topical mucosal injury by virtue of its acidic properties (Wolfe et al., 1999). After aspirin ingestion, gastric mucosal permeability is increased, as reflected by a decrease in the trans-mucosal potential difference (Baskin et al., 1976). This is caused by a direct topical effect.

Changes in GI Motility, Microcirculation, and Enterobacteria

The inner layer of the GI tract, which is in intimate contact with the contents of the lumen, is comprised of an epithelial cell layer called the mucosa. Subjacent to the mucosa is the submucosa, which is a loose connective tissue layer that contains blood vessels, lymphatics, and autonomic nerve fibers (Meissner plexus). Beneath the submucosa are circular and longitudinal muscle layers comprised of smooth muscle fibers. Contraction of these muscles is associated with mixing and propulsive movements of the intestinal contents (GI peristalsis and motility). The intestinal motility is regulated by the enteric nervous system. GI inflammation (e.g., inflammatory bowel disease) generally affects its motility. The exact pathophysiology of this dismotility is poorly understood. However, changes in myenteric neurons and smooth muscle have been proposed (Fornai et al., 2005, 2006). Emerging data suggest that cyclooxygenases may play a role in the control of GI neuromuscular functions and motility (De Backer et al., 2003; Fornai et al., 2005; 2006). Forani et al. reported that (1) in human colon, cyclooxygenases are involved in enteric circuitries exerting tonic inhibitory control on smooth muscle responses to endogenous acetylcholine; (2) both cyclooxygenase isoforms contribute to these regulatory actions; and (3) cholinergic neurons are modulated primarily by COX-1 activity, while COX-2 acts mainly at the muscular level to down-regulate muscarinic responses (Fornai et al., 2005). Gastric hypermotility occurred in indomethacin-treated rats (Takeuchi et al., 2004). Intestinal hypermotility has been implicated in NSAID-induced ulceration of the small intestine (Takeuci et al., 2002). In an experimental study in rats, proximal duodenum motility was determined in the absence and presence of different NSAIDs. Treatment with rofecoxib at 5 mg/kg or parecoxib at 0.5 mg/kg induced duodenal motility, whereas SC-560 showed no effect (Pihl and Nylander, 2006). Another study demonstrated that COX-2 activation is a critical step in diminishing bowel propulsive motility in a trinitrobenzene sulfonic acid (TNBS)–induced colitis guinea pig model (Linden et al., 2004). In this TNBS-induced colitis model, COX-2 inhibition with an s-NSAID, DFU, but not a COX-1 inhibitor, SC-560, restored to normal levels the electrical properties of myenteric neurons and the rate of propulsive motor activity (Linden et al., 2004). Thus, it is suggested that s-NSAIDs may be a possible therapeutic agent to improve bowel dysmotility.

GI mucosal and submucosal blood flow and microcirculation are important components of the gastroduodenal function and defense barrier. For example, in the stomach, the presence of luminal acid increases the delivery of vascular bicarbonate into the overlying mucus layer by mucosal microcirculation, thereby neutralizing H^+ ion invading from the lumen (Johnson et al., 2006). Indomethacin at an ulcerogenic dose of 25 mg/kg in rats enhances gastric motility and also induces microcirculatory disturbances at mucosal folds, which are caused by abnormal compression of the gastric mucosal wall and lead to increased microvascular permeability and cellular damage (Takeuchi et al., 1990). Wallace et al. (2000) reported that SC-560, but not celecoxib, produced a decrease in gastric mucosal blood flow in rats.

Enterobacterial invasion has also been implicated in NSAID-mediated GI pathophysiology. For example, the number of enterobacteria under both aerobic and anaerobic conditions is markedly increased in the intestinal mucosa following indomethacin treatment. Similarly, SC-560, with or without the coadministration of rofecoxib, increased the bacterial count in the mucosa, although rofecoxib alone did not (Takeuchi et al., 2010). The bacterial invasion in the intestinal mucosa following indomethacin treatment was blocked by prior administration of an ampicillin antibiotic, the numbers of bacteria being reduced even below control levels seen in the normal mucosa (Takeuchi et al., 2010). Enterobacteria and cytokines both play roles in the pathophysiology of NSAID-induced enteropathy. In addition, up-regulation of iNOS mRNA expression in the intestinal mucosa was observed in animals given SC-560 but not in animals given rofecoxib. Collectively, these data suggest that some NSAIDs cause GI hypermotility, followed by bacterial translocation, and GI microvascular disturbances, leading to the activation of neutrophils and expression of iNOS, and by doing so damage the intestine (Takeuchi et al., 2010).

Decreased Phosphatidylcholine Levels

A recent hypothesis related to the phosphatidylcholine (PC) role in the NSAID-associated toxicity mechanism has been proposed. Phosphatidylcholine is the major surfactant phospholipid that confers surface hydrophobic characteristics on the gastric mucosa. An improved safety profile of ibuprofen chemically associated with PC has been noted in elderly osteoarthritic patients (Lanza et al., 2008). It is possible that with age, surface phospholipid levels decrease below a critical threshold and that this reduction contributes to age-related NSAID intolerance (Lanza et al., 2008). In addition, age-associated decreases in surface hydrophobicity, PG levels, and impaired healing have been suggested to contribute to the deterioration of the barrier property of the gastric mucosa (Lanza et al., 2008). Hacklesberger et al.

(1998) observed a decrease in surface hydrophobicity in the antrum of the stomach, which is also one of the primary sites of NSAID-induced ulcers.

Impaired Drug Metabolism

Impaired drug metabolism with subsequent adverse drug effects can occur and is related to interindividual variability in drug metabolism due to polymorphisms in genes coded for drug-metabolizing enzymes such as cytochrome P450 (CYPs). Such impairment in drug metabolism would lead to increases in NSAID plasma concentrations and hence would increase the risk of developing adverse GI effects (Agúndez et al., 2009). Four NSAIDs—celecoxib, ibuprofen, lornoxicam, and piroxicam—are metabolized extensively by CYP2C9 and CYP2C8 enzymes, these enzymes being responsible for more than 90% of the primary metabolism of these drugs (Agúndez et al., 2009). Therefore, genetics has been suggested to predispose GI bleeding after NSAID use (Martinez et al., 2004). For example, inherited impairment in CYP2CP9, an enzyme responsible for the metabolism of several NSAIDs, increases the risk for severe adverse drug reactions (i.e., GI bleeding) after NSAID use (Martinez et al., 2004; Pilotto et al., 2007). This suggests that CYP2CP9 genotyping may identify subgroups of persons who are potentially at risk for NSAID-associated GI bleeding (Pilotto et al., 2007).

However, further investigation as to whether such GI bleeding is related to parent drugs or to metabolites is warranted. The impaired drug mechanism hypothesis would be relevant in long-term therapy because the drug would accumulate after multiple-dose exposure, but in many cases, patients with acute GI bleeding receive the NSAID only once, and the effect of an impaired metabolism in single-dose pharmacokinetics (PK) is less relevant, as can be expected in multiple-dose PK (Agúndez et al., 2009). An alternative mechanism for adverse GI drug effects that should be explored is whether impaired function of the main enzymes could drive the metabolism of the NSAIDs to alternative metabolic pathways and whether alternative metabolites may participate in the adverse effects (Agúndez et al., 2009).

Role of Toll-like Receptor (TLR)-4/MyD88 and Enteric Bacteria

Toll-like receptors (TLRs) comprise a family of conserved molecular structures (pathogen-associated molecular patterns) that function as sensors of microbial infection and play a central role in mucosal innate immune regulation (Rakoff-Nahoum et al., 2004). TLR activation leads to the production of cytokines and antimicrobial molecules important in the initial innate immune response. However, in addition to their function in host defense, recent findings indicate that activation of TLRs by commensal microflora is critical for protection against GI injury and associated mortality. Therefore, TLRs appear to control intestinal epithelial homeostasis and protection from injury (Rakoff-Nahoum et al., 2004). TLR-4 recognizes lipopolysaccharide (LPS), which is present in the cell wall of gram-negative bacteria. The interaction of LPS with TLR-4 and its coreceptor, MD-2, triggers signaling cascades mediated via the accessory protein MyD88-dependent pathways that lead to translocation of the transcription factor nuclear factor κ B (NFκB) and the production of proinflammatory cytokines (Medzhitov et al., 1998). TLR-4 signaling is important in the recruitment of inflammatory cells and the production of inflammatory cytokines in the intestine. In a mouse model of inflammatory bowel disease, TLR-4 was found to mediate PGE_2 production by regulation of COX-2 (Fukata et al., 2005, 2006). TLR-4 deficiency (TLR-4$^-$/$^-$) in mice results in fewer inflammatory infiltrates in the lamina propria (Fukata et al., 2005). Furthermore, a study in mice assessed the role of TLR-4 activation and signaling through MyD88 in intestinal ischemia/reperfusion (I/R)–induced damage at 2 h postischemia (Moses et al., 2009). The investigators found that a lack of TLR-4 or MyD88 attenuated intestinal damage to approximately 50% of that seen in wild-type mice. The attenuated gut injury was accompanied by decreased proinflammatory mediators, including chemokines, cytokines, and PGE_2 production. The decreased PGE_2 appeared to be mediated by COX-2 activation (Moses et al., 2009). These studies support the hypothesis that TLR-4 expression affects the extent of intestinal damage by altering COX-2-mediated PGE_2 production. However, PGE_2 alone was not sufficient to restore damage in the TLR-4-altered mice, implicating additional mechanisms of TLR-4-mediated damage. Thus, these data indicate that TLR-4 stimulation of COX-2 activation of PGE_2 production is necessary but not sufficient for intestinal I/R-induced damage and inflammation. TLR-4$^-$/$^-$ mice also had defects in mucosal repair in response to dextran sodium sulfate (DSS)–induced colitis with decreased epithelial proliferation and increased rectal bleeding (Fukata et al., 2006). Collectively, these studies suggest that TLR-4 signaling may serve a dual role in the GI tract as a mediator of both inflammation and mucosal repair (Ungaro et al., 2009).

Intestinal epithelial cells showed up-regulated COX-2 expression in a TLR-4- and MyD88-dependent fashion in the DSS colitis mouse model (Fukata et al., 2006). TLR-4 has been implicated in ns-NSAID-induced (indomethacin) small intestine damage because TLR-4$^-$/$^-$ and MyD88$^-$/$^-$ mice showed resistance to ns-NSAID-induced small intestinal damage (Watanabe et al., 2008). The physiological importance of PGs in intestinal epithelial cells and its relationship to bacterial invasion have been demonstrated. In vitro infection of human intestinal epithelial cells with invasive bacteria has been shown to induce the expression of $PGHS_2$ and the production of PGE_2 and $PGF_{2\alpha}$ (Eckmann et al., 1997). Furthermore, increased PGHS-2 expression was observed in intestinal epithelial cells in vivo after infection with invasive bacteria using a human intestinal xenograft model in SCID mice (Eckmann et al., 1997). The bacterial LPS/TLR-4 signaling pathway is also a key mechanism in NSAID-induced enteropathy. Elevation in enteric bacterial numbers and epithelial permeability of the small intestine was associated with enterohepatically recirculated ns-NSAIDs (Reuter et al., 1997). Another study investigated the effects of the *Lactobacillus casei* strain Shirota (LcS) on indomethacin-induced small intestine injury in rats. One-week treatment with viable LcS prevented indomethacin-induced intestinal injury. The investigators conclude that LcS exhibited a prophylactic effect on indomethacin-induced enteropathy by suppressing the LPS/TLR-4 signaling pathway (Watanabe et al.,

2009). Bacterial invasion in the intestinal mucosa following indomethacin treatment in rats was blocked by prior administration of the ampicillin antibiotic, the numbers of bacteria being reduced even below control levels seen in the normal mucosa (Takeuchi et al., 2010).

Role of Uncoupling of Mitochondrial Oxidative Phosphorylation

Several studies examined the role of the uncoupling of mitochondrial oxidative phosphorylation, leading to increased intestinal permeability and calcium release into the cytosol, in NSAID-mediated GI effects (Mahmud et al., 1996; Mathews, 1996; Somasundaram et al., 2000). NSAIDs increase mitochondrial respiration in vivo and in vitro without producing ATP. The higher pH inside the mitochondrial matrix deprotonates the NSAID, so protons are transported into the matrix. NSAIDs that directly uncouple or inhibit mitochondrial oxidative phosphorylation and ATP turnover include indomethacin, aspirin, diclofenac, meloxicam, and SC-236, an s-NSAID (Petrescu et al., 1997; Tibble et al., 2000; Krause et al., 2003). For example, indomethacin, but not celecoxib, uncoupled mitochondrial oxidative phosphorylation both in vitro and in vivo, caused a significant increase in small intestinal permeability, caused mucosal inflammation and a 90% decline in intestinal PGE levels, and was associated with multiple small intestinal ulcers in rats (Tibble et al., 2000). Sulindac sulfide, but not sulindac sulfone or sulindac itself, caused mitochondrial uncoupling in an isolated rat liver mitochondria (Leite et al., 2006). However, while the uncoupling of enterocyte mitochondrial oxidative phosphorylation leads to increased intestinal permeability and low-grade inflammation, concurrent decreases in mucosal prostanoids appear to be important in the development of ulcers (Somasundaram et al., 2000).

Role of Peroxisome Proliferator-Activated Receptor γ

Peroxisome proliferator-activated receptor γ (PPARγ) is a ligand-activated nuclear receptor whose activation has been linked to several pathways, including regulation of intestinal inflammation (Issemann and Green, 1990). COX-2 is elaborated during I/R injury (Sato et al., 2005). Issemann and Green examined the importance of PPARγ in inflammation and GI I/R-induced injury using a PPARγ-knockout mouse model. PPARγ-knockout mice showed exacerbated GI I/R-induced injury compared to wild-type mice. Histopathological examination of the small intestine revealed loss of villi (erosions and ulcers), hemorrhage, and inflammatory cell infiltrates. Furthermore, PPARγ activation reduced the severity of GI I/R injury and blocked up-regulation of NFκB, which is involved in the control of transcription of various inflammatory genes, such as TNFα (Issemann and Green, 1990). In a rodent model of I/R, NS-398, a COX-2 s-NSAID, reversed small intestine inflammation and injury and induced expression and nuclear translocation of PPARγ (Sato et al., 2005). In addition, several ns-NSAIDs (e.g., naproxen, ibuprofen, indomethacin, fenprofen) activate PPARγ (Jaradat et al., 2001). In a porcine I/R model, injured ileum treated with NS-398, an s-NSAID, recovered to control levels within 3 h

(Blikslager et al., 2002). Therefore, PPARγ appears to play a role in NSAID-mediated GI pathophysiological cellular mechanisms, and NSAID administration appears to have a protective effect in I/R GI conditions.

Role of Mitogen-Activated Protein Kinases

The mitogen-activated protein kinases (MAPKs) transduce a variety of extracellular signals to the transcription machinery and include three distinct mammalian types—extracellular signal-regulated kinases (ERKs), c-Jun NH_2-terminal kinases (JNKs), and p38 MAPKs (p38)—the latter having four isoforms of its own ($\alpha, \beta, \gamma, \delta$) (Radi and Khan, 2006a; Radi et al., 2009). p38 MAPKs have been shown to be crucial for COX-2 expression and PPARγ (Scherle et al., 2000; Radi et al., 2009). In fact, the p38 MAPK pathway can activate the intestinal epithelial cells *cox-2* gene promoter directly (Grishin et al., 2006). It has been found that MAPKs regulate COX-2 expression and mucosal recovery in an in vitro porcine I/R ileum model (Shifflett et al., 2004). This suggests that the MAPK pathway can have a positive regulatory effect on COX expression in I/R GI conditions, which suggests that NSAID administration would be protective under such conditions.

NSAID GI Injury-Associated Risk Factors

Epidemiological studies suggest that there are several risk factors for the development of NSAID-associated GI events in humans. These include advancing age, a high dose of NSAID, use of more than two NSAIDs, concurrent paracetamol, concurrent anticoagulants, concurrent aspirin, and prior history of peptic ulcer disease, high alcohol consumption, cigarette smoking, and *H. pylori* infection (Rainsford, 2009).

CONCLUSIONS

The comparative pathophysiologic aspects of the GI tract and interspecies COX-1 and COX-2 expression levels and the pathophysiological role of cyclooxygenases (COX-1 and COX-2) and the effects of their inhibition in the GI system are discussed in this chapter. There are significant interspecies differences in both the level of COX-1 expression and the ratio of COX-1 and COX-2 expression and susceptibility to toxicity with COX inhibition in the GI tract. Nonselective NSAIDs are used to treat a variety of inflammatory disease conditions. The ns-NSAIDs have been associated with GI toxicity in many species. Examples of GI-related toxicities are bleeding, ulceration, erosions, and perforations, distributed across the pyloric region, gastric mucosa, jejunum, ileum, duodenum, and cecum. COX-2 s-NSAIDs have a superior and improved GI tolerability profile to that of ns-NSAIDs. The analgesic and anti-inflammatory benefits of NSAIDs are linked to COX-2 inhibition, while many of the GI toxicities and side effects have variably been linked to COX-1 and/or COX-2 inhibition and, in some cases, directly to the secondary pharmacologic properties of the select drugs. Several mechanisms involved in the pathogenesis of these NSAID-associated toxicities include differences in COX enzyme potency, interspecies anatomical differences, changes in GI motility, aging, fasting status, disruption of GI physiologic mucosal defense mechanisms, effects on the physiologic GI mucosal cell renewal mechanisms, alterations in GI physiologic pH secretion regulation, inhibition of PG synthesis, impaired drug metabolism, effects of the enterohepatic recirculation, decreased phostatidylcholine levels, the role of TLR-4/MyD88, PPARs, MAPKs, and glucuronidation, neutrophil adherence, and direct chemical damage in the GI tract. COX-2 s-NSAIDs have permitted comparable therapeutic benefit to conventional ns-NSAIDs without these attendant COX-1-mediated toxicities (Radi and Khan, 2006b).

REFERENCES

Adams SS, Bough RG, Cliffe EE, Lessel B, et al. Absorption, distribution and toxicity of ibuprofen. Toxicol Appl Pharmacol 1969;15:310–330.

Agrawal NM, Aziz K. Prevention of gastrointestinal complications associated with nonsteroidal antiinflammatory drugs. J Rheumatol Suppl 1998;51:17–20.

Agúndez JA, García-Martín E, Martínez C. Genetically based impairment in CYP2C8- and CYP2C9-dependent NSAID metabolism as a risk factor for gastrointestinal bleeding: Is a combination of pharmacogenomics and metabolomics required to improve personalized medicine? Expert Opin Drug Metab Toxicol 2009;5:607–620.

Aguwa CN. Incidence of gastric ulcers by indomethacin and piroxicam in rats. Arch Toxicol 1985;56:212–213.

Aitken MM, Sanford J. Plasma levels following administration of sodium meclofenamate by various routes. Res Vet Sci 1975;19:241–244.

al-Ghamdi MS, Dissanayake AS, Cader ZA, Jain S. Tenoxicam-induced gastropathy in the rat: a comparison with piroxicam and diclofenac sodium, and the inhibitory effects of ranitidine and sucralfate. J Int Med Res 1991;19:242–248.

Allison SO, Halliday LC, French JA, Novikov DD, et al. Assessment of buprenorphine, carprofen, and their combination for postoperative analgesia in olive baboons (*Papio anubis*). J Am Assoc Lab Anim Sci 2007;46:24–231.

Altinkaynak K, Suleyman H, Akcay F. Effect of nimesulide, rofecoxib and celecoxib on gastric tissue glutathione level in rats with indomethacin-induced gastric ulcerations. Pol J Pharmacol 2003;55: 645–648.

Anthony A, Dhillon AP, Sim R, Nygard G, et al. Ulceration, fibrosis and diaphragm-like lesions in the caecum of rats treated with indomethacin. Aliment Pharmacol Ther 1994;8:417–424.

Aoyagi N, Ogata H, Kaniwa N, Uchiyama M, et al. Gastric emptying of tablets and granules in humans, dogs, pigs, and stomach-emptying-controlled rabbits. J Pharm Sci 1992;81:1170–1174.

Arai I, Hamasaka Y, Futaki N, Takahashi S, et al. Effect of NS-398, a new nonsteroidal anti-inflammatory agent, on gastric ulceration and acid secretion in rats. Res Commun Chem Pathol Pharmacol 1993;81:259–270.

Atchison CR, West AB, Balakumaran A, Hargus SJ, et al. Drug enterocyte adducts: possible causal factor for diclofenac enteropathy in rats. Gastroenterology 2000;119:1537–1547.

Atzpodien E, Mehdi N, Clarke D, Radhofer-Welte S. Subacute and chronic oral toxicity of lornoxicam in cynomolgus monkeys. Food Chem Toxicol 1997;35:465–474.

Bannwarth B. Safety of the nonselective NSAID nabumetone: focus on gastrointestinal tolerability. Drug Saf 2008;31:485–503.

Baskin WN, Ivey KJ, Krause WJ, Jeffrey GE, et al. Aspirin-induced ultrastructural changes in human gastric mucosa: correlation with potential difference. Ann Intern Med 1976;85:299–303.

Beck WS, Schneider HT, Dietzel K, Nuernberg B, et al. Gastrointestinal ulcerations induced by anti-inflammatory drugs in rats: physicochemical and biochemical factors involved. Arch Toxicol 1990;64:210–217.

Berenguer B, Alarcón de la Lastra C, Motilva V, La Casa C, et al. Effects of celecoxib on acid-challenged gastric mucosa of rats: comparison with metamizol and piroxicam. Dig Dis Sci 2004;49: 937–947.

Bhandari P, Bateman AC, Mehta RL, Patel P. Mucosal expression of cyclooxygenase isoforms 1 and 2 is increased with worsening damage to the gastric mucosa. Histopathology 2005;46:280–286.

Bjorkman DJ. Nonsteroidal anti-inflammatory drug-induced gastrointestinal injury. Am J Med 1996; 101:25S–32S.

Blikslager AT, Zimmel DN, Young KM, Campbell NB, et al. Recovery of ischaemic injured porcine ileum: evidence for a contributory role of COX-1 and COX-2. Gut 2002;50:615–623.

Bombardier C, Laine L, Reicin A, Shapiro D, et al. Comparison of upper gastrointestinal toxicity of rofecoxib and naproxen in patients with rheumatoid arthritis. VIGOR Study Group. N Engl J Med 2000;343:1520–1528.

Bowen B, Yuan Y, James C, Rashid F, et al. Time course and pattern of blood loss with ibuprofen treatment in healthy subjects. Clin Gastroenterol Hepatol 2005;3:1075–1082.

Brodie DA, Cook PG, Bauer BJ, Dagle GE. Indomethacin-induced intestinal lesions in the rat. Toxicol Appl Pharmacol 1970;17:615–624.

Budsberg SC, Johnston SA, Schwarz PD, DeCamp CE, et al. Efficacy of etodolac for the treatment of osteoarthritis of the hip joints in dogs. J Am Vet Med Assoc 1999;214:206–210.

Burdan F, Szumiło J, Gajjar B, Dudka J, et al. Immunoexpression of constitutive and inducible cyclo-oxygenase isoforms in the rat foetal and maternal digestive tract. Folia Morphol (Warsz) 2008;67:24–31.

Campbell NB, Blikslager AT. The role of cyclooxygenase inhibitors in repair of ischaemic-injured jejunal mucosa in the horse. Equine Vet J 2000;32:59–64.

Carrick JB, Papich MG, Middleton DM, Naylor JM, et al. Clinical and pathological effects of flunixin meglumine administration to neonatal foals. Can J Vet Res 1989;53:195–201.

Cathers TE, Isaza R, Oehme F. Acute ibuprofen toxicosis in a ferret. J Am Vet Med Assoc 2000;216: 1426–1428.

Chan CC, Boyce S, Brideau C, Charleson S, et al. Rofecoxib [Vioxx, MK-0966; 4-(4′-methylsulfonylphenyl)-3-phenyl-2-(5*H*)-furanone]: a potent and orally active cyclooxygenase-2 inhibitor—pharmacological and biochemical profiles. J Pharmacol Exp Ther 1999;290:551–560.

Chhabra RS. Intestinal absorption and metabolism of xenobiotics. Environ Health Perspect 1979;33:61–69.

Collins LG, Tyler DE. Experimentally induced phenylbutazone toxicosis in ponies: description of the syndrome and its prevention with synthetic prostaglandin E2. Am J Vet Res 1985;46:1605–1615.

Combe B, Swergold G, McLay J, McCarthy T, et al. Cardiovascular safety and gastrointestinal tolerability of etoricoxib vs. diclofenac in a randomized controlled clinical trial (the MEDAL study). Rheumatology (Oxford) 2009;48:425–432.

Conforti A, Donini M, Brocco G, Del Soldato P, et al. Acute anti-inflammatory activity and gastrointestinal tolerability of diclofenac and nitrofenac. Agents Actions 1993;40:176–180.

Conlon PD. Nonsteroidal drugs used in the treatment of inflammation. Vet Clin North Am Small Anim Pract 1988;18:1115–1131.

Cook VL, Meyer CT, Campbell NB, Blikslager AT. Effect of firocoxib or flunixin meglumine on recovery of ischemic-injured equine jejunum. Am J Vet Res 2009;70:992–1000.

Coppelli G, Guaita E, Spaggiari S, Coruzzi G, et al. Gastric effects of the selective cyclooxygenase-2 inhibitor, celecoxib, in the rat. Dig Liver Dis 2004;36:265–370.

Cosenza SF. Drug-induced gastroduodenal ulceration in dogs. Modern Vet Pract 1984;65:923–925.

Court MH, Greenblatt DJ. Molecular basis for deficient acetaminophen glucuronidation in cats: an interspecies comparison of enzyme kinetics in liver microsomes. Biochem Pharmacol 1997;53: 1041–1047.

Cryer B. Gastrointestinal safety of low-dose aspirin. Am J Manag Care 2002;8:S701–S708.

Cryer B, Feldman M. Cyclooxygenase-1 and cyclooxygenase-2 selectivity of widely used nonsteroidal anti-inflammatory drugs. Am J Med 1998;104:413–421.

Cryer B, Redfern JS, Goldschmiedt M, Lee E, et al. Effect of aging on gastric and duodenal mucosal prostaglandin concentrations in humans. Gastroenterology 1992;102:1118–1123.

Cuzzocrea S, Mazzon E, Serraino I, Dugo L, et al. Celecoxib, a selective cyclooxygenase-2 inhibitor reduces the severity of experimental colitis induced by dinitrobenzene sulfonic acid in rats. Eur J Pharmacol 2001;431:91–102.

Daehler MH. Transmural pyloric perforation associated with naproxen administration in a dog. J Am Vet Med Assoc 1986;189:694–695.

Daffonchio L, Novellini R, Bertuglia S. Protective effect of ketoprofen lysine salt on interleukin-1beta and bradykinin induced inflammatory changes in hamster cheek pouch microcirculation. Inflamm Res 2002;51:223–228.

Dannenberg AJ, Altorki NK, Boyle JO, Dang C, et al. Cyclo-oxygenase 2: a pharmacological target for the prevention of cancer. Lancet Oncol 2001;2:544–551.

Davies B, Morris T. Physiological parameters in laboratory animals and humans. Pharm Res 1993;10:1093–1095.

De Backer O, Leclere PG, Lefebvre RA. Pharmacological characterization of pre- and postsynaptic prostanoid receptors in pig gastric fundus. Neuropharmacology 2003;45:684–690.

Deng X, Tolstanova G, Khomenko T, Chen L, et al. Mesalamine restores angiogenic balance in experimental ulcerative colitis by reducing expression of endostatin and angiostatin: novel molecular mechanism for therapeutic action of mesalamine. J Pharmacol Exp Ther 2009;331:1071–1078.

Dequeker J, Hawkey C, Kahan A, Steinbrück K, et al. Improvement in gastrointestinal tolerability of the selective cyclooxygenase (COX)-2 inhibitor, meloxicam, compared with piroxicam: results of the safety and efficacy large-scale evaluation of COX-inhibiting therapies (SELECT) trial in osteoarthritis. Br J Rheumatol 1998;37:946–951.

Dignass AU. Mechanisms and modulation of intestinal epithelial repair. Inflamm Bowel Dis 2001;7:68–77.

DiPasquale G, Welaj P. Letter: Ulcerogenic potential of indomethacin in arthritic and non-arthritic rats. J Pharm Pharmacol 1973;25:831–832.

Doan T, Massarotti E. Rheumatoid arthritis: an overview of new and emerging therapies. J Clin Pharmacol 2005;45:751–762.

Doucet MY, Bertone AL, Hendrickson D, Hughes F, et al. Comparison of efficacy and safety of paste formulations of firocoxib and phenylbutazone in horses with naturally occurring osteoarthritis. J Am Vet Med Assoc 2008;232:91–97.

Dressman JB, Berardi RR, Dermentzoglou LC, Russell TL, et al. Upper gastrointestinal (GI) pH in young, healthy men and women. Pharm Res 1990;7:756–761.

Duerksen DR. Stress-related mucosal disease in critically ill patients. Best Pract Res Clin Gastroenterol 2003;17:327–244.

Duggan DE, Hooke KF, Noll RM, Kwan C. Enterohepatic circulation of indomethacin and its role in intestinal irritation. Biochem Pharmacol 1975;24:1749–1754.

Eckmann L, Stenson WF, Savidge TC, Lowe DC, et al. Role of intestinal epithelial cells in the host secretory response to infection by invasive bacteria: Bacterial entry induces epithelial prostaglandin H synthase-2 expression and prostaglandin E2 and F2alpha production. J Clin Invest 1997;100:296–309.

Eling TE, Thompson DC, Foureman GL, Curtis JF, et al. Prostaglandin H synthase and xenobiotic oxidation. Annu Rev Pharmacol Toxicol 1990;30:1–45.

Elliott GA, Purmalis A, Vandermeer DA, Denlinger RH. The propionic acids: gastrointestinal toxicity in various species. Toxicol Pathol 1988;16:245–250.

Elliott SL, Ferris RJ, Giraud AS, Cook GA, et al. Indomethacin damage to rat gastric mucosa is markedly dependent on luminal pH. Clin Exp Pharmacol Physiol 1996;23:432–434.

El-Medany A, Mahgoub A, Mustafa A, Arafa M, et al. The effects of selective cyclooxygenase-2 inhibitors, celecoxib and rofecoxib, on experimental colitis induced by acetic acid in rats. Eur J Pharmacol 2005;507:291–299.

Erickson BA, Longo WE, Panesar N, Mazuski JE, et al. The effect of selective cyclooxygenase inhibitors on intestinal epithelial cell mitogenesis. J Surg Res 1999;81:101–107.

Ewing GO. Indomethacin-associated gastrointestinal hemorrhage in a dog. J Am Vet Med Assoc 1972;161:1665–1668.

Eyre P, Boulard C, Deline T. Local and systemic reactions in cattle to *Hypoderma lineatum* larval toxin: protection by phenylbutazone. Am J Vet Res 1981;42:25–28.

Fang WF, Broughton A, Jacobson ED. Indomethacin-induced intestinal inflammation. Am J Dig Dis 1977;22:749–760.

Feldman M, Cryer B. Effects of age on gastric alkaline and nonparietal fluid secretion in humans. Gerontology 1998;44:222–227.

Field M. Intestinal ion transport and the pathophysiology of diarrhea. J Clin Invest 2003;111:931–943.

Fornai M, Blandizzi C, Colucci R, Antonioli L, et al. Role of cyclooxygenases 1 and 2 in the modulation of neuromuscular functions in the distal colon of humans and mice. Gut 2005;54:608–616.

Fornai M, Blandizzi C, Antonioli L, Colucci R, et al. Differential role of cyclooxygenase 1 and 2 isoforms in the modulation of colonic neuromuscular function in experimental inflammation. J Pharmacol Exp Ther 2006;317:938–945.

Forsyth SF, Guilford WG, Haslett SJ, Godfrey J. Endoscopy of the gastroduodenal mucosa after carprofen, meloxicam and ketoprofen administration in dogs. J Small Anim Pract 1998;39:421–424.

Fox J, Anderson LC, Loew FM, Quimby FW. *Laboratory Animal Medicine*, 2nd ed. San Diego, CA: Academic Press, 2002.

Fox SM, Johnston SA. Use of carprofen for the treatment of pain and inflammation in dogs. J Am Vet Med Assoc 1997;210:1493–1498.

Freeman LC, Narvaez DF, McCoy A, von Stein FB, et al. Depolarization and decreased surface expression of K channels contribute to NSAID-inhibition of intestinal restitution. Biochem Pharmacol 2007;74:74–85.

Fukata M, Michelsen KS, Eri R, Thomas LS, et al. Toll-like receptor-4 is required for intestinal response to epithelial injury and limiting bacterial translocation in a murine model of acute colitis. Am J Physiol Gastrointest Liver Physiol 2005;288:G1055–G1065.

Fukata M, Chen A, Klepper A, Krishnareddy S, et al. Cox-2 is regulated by Toll-like receptor-4 (TLR4) signaling: role in proliferation and apoptosis in the intestine. Gastroenterology 2006;131:862–877.

Futaki N, Yoshikawa K, Hamasaka Y, Arai I, et al. NS-398, a novel non-steroidal anti-inflammatory drug with potent analgesic and antipyretic effects, which causes minimal stomach lesions. Gen Pharmacol 1993;24:105–110.

García-Vicuña R, Díaz-González F, González-Alvaro I, del Pozo MA, et al. Prevention of cytokine-induced changes in leukocyte adhesion receptors by nonsteroidal antiinflammatory drugs from the oxicam family. Arthritis Rheum 1997;40:143–153.

Girawan D, Abdurachman SA, Djumhana A, Roslia J, et al. Comparison of endoscopic gastric mucosa features after administration of piroxicam to meloxicam and their correlation with dyspepsia symptoms in elderly patient with knee osteoarthritis. Acta Med Indones 2004;36:202–206.

Godshalk CP, Roush JK, Fingland RB, Sikkema D, et al. Gastric perforation associated with administration of ibuprofen in a dog. J Am Vet Med Assoc 1992;201:1734–1736.

Goldstein JL, Correa P, Zhao WW, Burr AM, et al. Reduced incidence of gastroduodenal ulcers with celecoxib, a novel cyclooxygenase-2 inhibitor, compared to naproxen in patients with arthritis. Am J Gastroenterol 2001;96:1019–1027.

Goldstein JL, Aisenberg J, Lanza F, Schwartz H, et al. A multicenter, randomized, double-blind, active-comparator, placebo-controlled, parallel-group comparison of the incidence of endoscopic gastric and duodenal ulcer rates with valdecoxib or naproxen in healthy subjects aged 65 to 75 years. Clin Ther 2006;28:340–351.

Goldstein JL, Eisen GM, Lewis B, Gralnek IM, et al. Small bowel mucosal injury is reduced in healthy subjects treated with celecoxib compared with ibuprofen plus omeprazole, as assessed by video capsule endoscopy. Aliment Pharmacol Ther 2007;25:1211–1222.

Goodman L, Torres B, Punke J, Reynolds L, et al. Effects of firocoxib and tepoxalin on healing in a canine gastric mucosal injury model: effects of firocoxib and tepoxalin on healing in a canine gastric mucosal injury model. J Vet Intern Med 2009;23:56–62.

Goodman LA, Torres BT, Reynolds LR, Budsberg SC. Effects of firocoxib, meloxicam, and tepoxalin administration on eicosanoid production in target tissues of healthy cats. Am J Vet Res 2010;71:1067–1073.

Gookin JL, Galanko JA, Blikslager AT, Argenzio RA. PG-mediated closure of paracellular pathway and not restitution is the primary determinant of barrier recovery in acutely injured porcine ileum. Am J Physiol Gastrointest Liver Physiol 2003;285:G967–G979.

Goto H, Sugiyama S, Ohara A, Hoshino H, et al. Age-associated decreases in prostaglandin contents in human gastric mucosa. Biochem Biophys Res Commun 1992;186:1443–1448.

Gretzer B, Maricic N, Respondek M, Schuligoi R, et al. Effects of specific inhibition of cyclooxygenase-1 and cyclo-oxygenase-2 in the rat stomach with normal mucosa and after acid challenge. Br J Pharmacol 2001;132:1565–1573.

Grishin AV, Wang J, Potoka DA, Hackam DJ, et al. Lipopolysaccharide induces cyclooxygenase-2 in intestinal epithelium via a noncanonical p38 MAPK pathway. J Immunol 2006;176:580–588.

Gullikson GW, Anglin CP, Kessler LK, Smeach S, et al. Misoprostol attenuates aspirin- induced changes in potential difference and associated damage in canine gastric mucosa. Clin Invest Med 1987;10:145–151.

Hackelsberger A, Platzer U, Nilius M, Schultze V, et al. Age and *Helicobacter pylori* decrease gastric mucosal surface hydrophobicity independently. Gut 1998;43:465–469.

Hallesy DW, Shott LD, Hill R. Comparative toxicology of naproxen. Scand J Rheumatol 1973;Suppl 2: 220–228.

Harris SI, Stoltz RR, LeComte D, Hubbard RC. Parecoxib sodium demonstrates gastrointestinal safety comparable to placebo in healthy subjects. J Clin Gastroenterol 2004;38:575–580.

Hawkey C, Kahan A, Steinbrück K, Alegre C, et al. Gastrointestinal tolerability of meloxicam compared to diclofenac in osteoarthritis patients. International MELISSA Study Group: Meloxicam Large-Scale International Study Safety Assessment. Br J Rheumatol 1998;37:937–945.

Hawkey C, Laine L, Simon T, Beaulieu A, et al. Comparison of the effect of rofecoxib (a cyclooxygenase 2 inhibitor), ibuprofen, and placebo on the gastroduodenal mucosa of patients with osteoarthritis: a randomized, double-blind, placebo controlled trial. The Rofecoxib Osteoarthritis Endoscopy Multinational Study Group. Arthritis Rheum 2000;43:370–377.

Hawkey CJ, Skelly MM. Gastrointestinal safety of selective COX-2 inhibitors. Curr Pharm Des 2002;8:1077–1089.

Haworth R, Oakley K, McCormack N, Pilling A. Differential expression of COX-1 and COX-2 in the gastrointestinal tract of the rat. Toxicol Pathol. 2005;33:239–245.

Hirata T, Ukawa H, Kitamura M, Takeuchi K. Effects of selective cyclooxygenase-2 inhibitors on alkaline secretory and mucosal ulcerogenic responses in rat duodenum. Life Sci 1997;61:1603–1611.

Horie-Sakata K, Shimada T, Hiraishi H, Terano A. Role of cyclooxygenase 2 in hepatocyte growth factor-mediated gastric epithelial restitution. J Clin Gastroenterol 1998;27 Suppl 1:S40–S46.

Hull MA, Booth JK, Tisbury A, Scott N, et al. Cyclooxygenase 2 is up-regulated and localized to macrophages in the intestine of Min mice. Br J Cancer 1999;79:1399–1405.

Issemann I, Green S. Activation of a member of the steroid hormone receptor superfamily by peroxisome proliferators. Nature 1990;347:645–650.

Jackson LM, Wu KC, Mahida YR, Jenkins D, et al. Cyclooxygenase (COX) 1 and 2 in normal, inflamed, and ulcerated human gastric mucosa. Gut 2000;47:762–770.

Jaradat MS, Wongsud B, Phornchirasilp S, Rangwala SM, et al. Activation of peroxisome proliferator-activated receptor isoforms and inhibition of prostaglandin H(2) synthases by ibuprofen, naproxen, and indomethacin. Biochem Pharmacol 2001;62:1587–1595.

Johnson LR, Barret KE, Gishan FK, Merchant JL, Said HM, Wood JD (eds). *Physiology of the Gastrointestinal Tract*, 4th ed. New York: Elsevier, 2006, pp. 1259–1291.

Johnston SA, Budsberg SC. Nonsteroidal anti-inflammatory drugs and corticosteroids for the management of canine osteoarthritis. Vet Clin North Am Small Anim Pract 1997;27:841–862.

Johnston SA, Leib MS, Forrester SD, Marini M. The effect of misoprostol on aspirin- induced gastroduodenal lesions in dogs. J Vet Intern Med 1995;9:32–38.

Julou L, Guyonnet JC, Ducrot R, Fournel J, et al. Ketoprofen (19.583R.P.) [2-(3- benzoylphenyl)-propionic acid]. Main pharmacological properties: line of toxicological and pharmacokinetic data. Scand J Rheumatol 1976;1976:33–44.

Kallings P, Johnston C, Drevemo S. Effects of flunixin on movement and performance of standardbred trotters on the track. Equine Vet J 1999;Suppl 30: 270–273.

Karcher LF, Dill SG, Anderson WI, King JM. Right dorsal colitis. J Vet Intern Med 1990;4:247–253.

Kargman S, Charleson S, Cartwright M, Frank J, et al. Characterization of prostaglandin G/H synthase 1 and 2 in rat, dog, monkey, and human gastrointestinal tracts. Gastroenterology 1996;111:445–454.

Karmeli F, Cohen P, Rachmilewitz D. Cyclo-oxygenase-2 inhibitors ameliorate the severity of experimental colitis in rats. Eur J Gastroenterol Hepatol 2000;12:223–231.

Kataoka H, Horie Y, Koyama R, Nakatsugi S, et al. Interaction between NSAIDs and steroid in rat stomach: safety of nimesulide as a preferential COX-2 inhibitor in the stomach. Dig Dis Sci 2000;45:1366–1375.

Kato S, Ogawa Y, Kanatsu K, Okayama M, et al. Ulcerogenic influence of selective cyclooxygenase-2 inhibitors in the rat stomach with adjuvant-induced arthritis. J Pharmacol Exp Ther 2002;303:503–509.

Kato S, Ito Y, Nishio H, Aoi Y, et al. Increased susceptibility of small intestine to NSAID- provoked ulceration in rats with adjuvant-induced arthritis: involvement of enhanced expression of TLR4. Life Sci 2007;81:1309–1316.

Katz LB, Genna T, Fuller BL, Tolman EL, et al. Altered susceptibility of arthritic rats to the gastric lesion-inducing effects of aspirin or ethanol and the antilesion effect of rioprostil. Agents Actions 1987;22:134–143.

Kauffman G. Aspirin-induced gastric mucosal injury: lessons learned from animal models. Gastroenterology 1989;96:606–614.

Kearney D, Byrne A, Crean P, Cox D, et al. Optimal suppression of thromboxane A(2) formation by aspirin during percutaneous transluminal coronary angioplasty: no additional effect of a selective cyclooxygenase-2 inhibitor. J Am Coll Cardiol 2004;43:526–531.

Kent TH, Cardelli RM, Stamler FW. Small intestinal ulcers and intestinal flora in rats given indomethacin. Am J Pathol 1969;54:237–249.

Khan KN, Paulson S, Seibert K, Maziasz T. Gastrointestinal and renal safety of a specific COX-2 inhibitor versus NSAIDs in dogs. Vet Pathol 1997;34:509.

Khan KN, Masferrer JL, Woerner BM, Soslow R, et al. Enhanced cyclooxygenase-2 expression in sporadic and familial adenomatous polyposis of the human colon. Scand J Gastroenterol 2001;36:865–869.

Kim SW, Parekh D, Townsend CM Jr, Thompson JC. Effects of aging on duodenal bicarbonate secretion. Ann Surg 1990;212:332–337.

King JN, Dawson J, Esser RE, Fujimoto R, et al. Preclinical pharmacology of robenacoxib: a novel selective inhibitor of cyclooxygenase-2. J Vet Pharmacol Ther 2009;32:1–17.

Kishimoto Y, Wada K, Nakamoto K, Kawasaki H, et al. Levels of cyclooxygenase-1 and -2 mRNA expression at various stages of acute gastric injury induced by ischemia-reperfusion in rats. Arch Biochem Biophys 1998;352:153–157.

Knapp DW, Richardson RC, Bottoms GD, Teclaw R, et al. Phase I trial of piroxicam in 62 dogs bearing naturally occurring tumors. Cancer Chemother Pharmacol 1992;29:214–218.

Knapp DW, Richardson RC, Chan TC, Bottoms GD, et al. Piroxicam therapy in 34 dogs with transitional cell carcinoma of the urinary bladder. J Vet Intern Med 1994;8:273–278.

Koki A, Khan NK, Woerner BM, Dannenberg AJ, et al. Cyclooxygenase-2 in human pathological disease. Adv Exp Med Biol 2002a;507:177–184.

Koki AT, Khan NK, Woerner BM, Seibert K, et al. Characterization of cyclooxygenase-2 (COX-2) during tumorigenesis in human epithelial cancers: evidence for potential clinical utility of COX-2 inhibitors in epithelial cancers. Prostaglandins Leukot Essent Fatty Acids 2002b;66:13–18.

Kopcha M, Ahl AS. Experimental uses of flunixin meglumine and phenylbutazone in food-producing animals. J Am Vet Med Assoc 1989;19:445–449.

Korytko PJ, Senese PB, Sellers RS, Cai W, et al. Gastrointestinal and renal toxicity associated with loxoprofen sodium in dogs. Preclinica 2003;1:107–113.

Krause MM, Brand MD, Krauss S, Meisel C, et al. Nonsteroidal antiinflammatory drugs and a selective cyclooxygenase 2 inhibitor uncouple mitochondria in intact cells. Arthritis Rheum 2003;48:1438–1444.

La Corte R, Caselli M, Castellino G, Bajocchi G, et al. Prophylaxis and treatment of NSAID-induced gastroduodenal disorders. Drug Saf 1999;20:527–543.

Laine L, Sloane R, Ferretti M, Cominelli F. A randomized double blind comparison of placebo, etodolac, and naproxen on gastrointestinal injury and prostaglandin production. Gastrointest Endosc 1995;42:428–433.

Laine L, Harper S, Simon T, Bath R, et al. A randomized trial comparing the effect of rofecoxib, a cyclooxygenase 2-specific inhibitor, with that of ibuprofen on the gastroduodenal mucosa of patients with osteoarthritis. Rofecoxib Osteoarthritis Endoscopy Study Group. Gastroenterology 1999;117:776–783.

Laine L, Curtis SP, Cryer B, Kaur A, et al. Assessment of upper gastrointestinal safety of etoricoxib and diclofenac in patients with osteoarthritis and rheumatoid arthritis in the Multinational Etoricoxib and Diclofenac Arthritis Long-Term (MEDAL) programme: a randomised comparison. Lancet 2007;369:465–473.

Lamon TK, Browder EJ, Sohrabji F, Ihrig M. Adverse effects of incorporating ketoprofen into established rodent studies. J Am Assoc Lab Anim Sci 2008;47:20–24.

Lanas A, Panes J, Pique JM. Clinical implications of COX-1 and/or COX-2 inhibition for the distal gastrointestinal tract. Curr Pharm Des 2003;9:2253–2266.

Langenbach R, Morham SG, Tiano HF, Loftin CD, et al. Prostaglandin synthase 1 gene disruption in mice reduces arachidonic acid-induced inflammation and indomethacin-induced gastric ulceration. Cell 1995;83:483–492.

Langman MJ, Jensen DM, Watson DJ, Harper SE, et al. Adverse upper gastrointestinal effects of rofecoxib compared with NSAIDs. J Am Med Assoc 1999;282:1929–1933.

Langston C, Clarke CR. The role of the clinical pharmacologist in animal health. AAPS PharmSci 2002;4:E36.

Lanza FL, Nelson RS, Greenberg BP. Effects of fenbufen, indomethacin, naproxen, and placebo on gastric mucosa of normal volunteers: a comparative endoscopic and photographic evaluation. Am J Med 1983;75:75–79.

Lanza FL, Marathi UK, Anand BS, Lichtenberger LM. Clinical trial: comparison of ibuprofen-phosphatidylcholine and ibuprofen on the gastrointestinal safety and analgesic efficacy in osteoarthritic patients. Aliment Pharmacol Ther 2008;28:431–442.

Lascelles BD, Blikslager AT, Fox SM, Reece D. Gastrointestinal tract perforation in dogs treated with a selective cyclooxygenase-2 inhibitor: 29 cases (2002–2003). J Am Vet Med Assoc 2005a;227: 1112–1117.

Lascelles BD, McFarland JM, Swann H. Guidelines for safe and effective use of NSAIDs in dogs. Vet Ther 2005b;6:237–251.

Laudanno OM, Cesolari JA, Esnarriaga J, Rista L, et al. Gastrointestinal damage induced by celecoxib and rofecoxib in rats. Dig Dis Sci 2001;46:779–784.

Lazzaroni M, Porro GB. Management of NSAID-induced gastrointestinal toxicity: focus on proton pump inhibitors. Drugs 2009;69:51–69.

Lee M, Feldman M. Age-related reductions in gastric mucosal prostaglandin levels increase susceptibility to aspirin-induced injury in rats. Gastroenterology 1994;107:1746–1750.

Lee, M. Age-related changes in gastric mucosal bicarbonate secretion in rats. Age 1996;19:55–58.

Leedham SJ, Brittan M, Preston SL, McDonald SA, et al. The stomach periglandular fibroblast sheath: all present and correct. Gut 2006;55:295–296.

Lees P, Higgins AJ. Clinical pharmacology and therapeutic uses of non-steroidal anti- inflammatory drugs ion the horse. Equine Vet J 1985;17:83–96.

Lees P, Landoni MF, Giraudel J, Toutain PL. Pharmacodynamics and pharmacokinetics of nonsteroidal anti-inflammatory drugs in species of veterinary interest. J Vet Pharmacol Ther 2004;27: 479–490.

Legen I, Kristl A. pH and energy dependent transport of ketoprofen across rat jejunum in vitro. Eur J Pharm Biopharm 2003;56:87–94.

Leite S, Martins NM, Dorta DJ, Curti C, et al. Mitochondrial uncoupling by the sulindac metabolite, sulindac sulfide. Basic Clin Pharmacol Toxicol 2006;99:294–299.

Le Loet X. Safety of ketoprofen in the elderly: a prospective study on 20,000 patients. Scand J Rheumatol Suppl 1989;83:21–27.

Lesch CA, Kraus ER, Sanchez B, Gilbertsen R, et al. Lack of beneficial effect of COX-2 inhibitors in an experimental model of colitis. Methods Find Exp Clin Pharmacol 1999;21:99–104.

Li JT, Li ZS, Zhan XB, Cui ZM, et al. Experimental study of the safety of the selective COX-2 inhibitor, celecoxib, for gastic mucosa. Chin J Dig Dis 2003;4:53–56.

Linden DR, Sharkey KA, Ho W, Mawe GM. Cyclooxygenase-2 contributes to dysmotility and enhanced excitability of myenteric AH neurones in the inflamed guinea pig distal colon. J Physiol 2004;557:191–205.

Lipowitz AJ, Boulay JP, Klausner JS. Serum salicylate concentrations and endoscopic evaluation of the gastric mucosa in dogs after oral administration of aspirin-containing products. Am J Vet Res 1986;47:1586–1589.

Lipscomb GR, Wallis N, Armstrong G, Rees WD. Gastrointestinal tolerability of meloxicam and piroxicam: a double-blind placebo-controlled study. Br J Clin Pharmacol 1998;46:133–137.

Lisse JR, Perlman M, Johansson G, Shoemaker JR, et al. Gastrointestinal tolerability and effectiveness of rofecoxib versus naproxen in the treatment of osteoarthritis: a randomized, controlled trial. Ann Intern Med 2003;139:539–546.

Lohmander LS, McKeith D, Svensson O, Malmenäs M, et al. A randomised, placebo controlled, comparative trial of the gastrointestinal safety and efficacy of AZD3582 versus naproxen in osteoarthritis. Ann Rheum Dis 2005;64:449–456.

Luna SP, Basílio AC, Steagall PV, Machado LP, et al. Evaluation of adverse effects of long-term oral administration of carprofen, etodolac, flunixin meglumine, ketoprofen, and meloxicam in dogs. Am J Vet Res 2007;68:258–264.

MacAllister CG. Effects of toxic doses of phenylbutazone in ponies. Am J Vet Res 1983;44:2277–2279.

MacAllister CG, Morgan SJ, Borne AT, Pollet RA. Comparison of adverse effects of phenylbutazone, flunixin meglumine, and ketoprofen in horses. J Am Vet Med Assoc 1993;202:71–77.

Mackay RJ, French TW, Nguyen HT, Mayhew IG. Effects of large doses of phenylbutazone administration to horses. Am J Vet Res 1983;44:774–780.

Magdalou J, Chajes V, Lafaurie C, Siest G. Glucuronidation of 2-arylpropionic acids pirprofen, flurbiprofen, and ibuprofen by liver microsomes. Drug Metab Dispos 1990;18:692–697.

Mahmud T, Rafi SS, Scott DL, Wrigglesworth JM, et al. Nonsteroidal antiinflammatory drugs and uncoupling of mitochondrial oxidative phosphorylation. Arthritis Rheum 1996;39:1998–2003.

Martín AR, Villegas I, Alarc6n de la Lastra C. The COX-2 inhibitor, rofecoxib, ameliorates dextran sulphate sodium induced colitis in mice. Inflamm Res 2005;54:145–151.

Martínez C, Blanco G, Ladero JM, García-Martín E, et al. Genetic predisposition to acute gastrointestinal bleeding after NSAIDs use. Br J Pharmacol 2004;141:205–208.

Masferrer JL, Zweifel BS, Manning PT, Hauser SD, et al. Selective inhibition of inducible cyclooxygenase 2 in vivo is antiinflammatory and nonulcerogenic. Proc Natl Acad Sci USA 1994;91:3228–3232.

Mason JC. NSAIDs and the oesophagus. Eur J Gastroenterol Hepatol 1999;11:369–373.

Mathews KA. Nonsteroidal anti-inflammatory analgesics in pain management in dogs and cats. Can Vet J 1996;37:539–545.

Maziasz TJ, Khan KNM, Talley J, Gierse J, et al. The development of drugs that target cycloxygenase-2. In: Harris RE (ed), *Cyclooxygenase 2 Blockade in Cancer Prevention and Therapy*. Clifton, UK: Humana Press, 2003, pp. 259–277.

McCafferty DM, Granger DN, Wallace JL. Indomethacin induced gastric injury and leukocyte adherence in arthritic versus healthy rats. Gastroenterology 1995;109:1173–1180.

McConnell EL, Basit AW, Murdan S. Measurements of rat and mouse gastrointestinal pH, fluid and lymphoid tissue, and implications for in-vivo experiments. J Pharm Pharmacol 2008;60:63–70.

McConnico RS, Morgan TW, Williams CC, Hubert JD, et al. Pathophysiologic effects of phenylbutazone on the right dorsal colon in horses. Am J Vet Res 2008;69:1496–1505.

Medzhitov R, Preston-Hurlburt P, Kopp E, Stadlen A, et al. MyD88 is an adaptor protein in the hToll/IL-1 receptor family signaling pathways. Mol Cell 1998;2:253–258.

Melarange R, Gentry C, O'Connell C, Blower PR, et al. Anti-inflammatory and gastrointestinal effects of nabumetone or its active metabolite, 6MNA (6-methoxy-2-naphthylacetic acid): comparison with indomethacin. Agents Actions 1992;Spec No:C82–C83.

Menezes GB, Rezende RM, Pereira-Silva PE, Klein A, et al. Differential involvement of cyclooxygenase isoforms in neutrophil migration in vivo and in vitro. Eur J Pharmacol 2008;598:118–122.

Menozzi A, Pozzoli C, Poli E, Dacasto M, et al. Effects of nonselective and selective cyclooxygenase inhibitors on small intestinal motility in the horse. Res Vet Sci 2009;86:129–135.

Mersereau WA, Hinchey EJ. Synergism between acid and gastric contractile activity in the genesis of ulceration and hemorrhage in the phenylbutazone-treated rat. Surgery 1981;90:516–522.

Meschter CL, Gilbert M, Krook L, Maylin G, et al. The effects of phenylbutazone on the intestinal mucosa of the horse: a morphological, ultrastructural and biochemical study. Equine Vet J 1990a;22:255–263.

Meschter CL, Gilbert M, Krook L, Maylin G, et al. The effects of phenylbutazone on the morphology and prostaglandin concentrations of the pyloric mucosa of the equine stomach. Vet Pathol 1990b;27:244–253.

Moberly JB, Harris SI, Riff DS, Dale JC, et al. A randomized, double-blind, one-week study comparing effects of a novel COX-2 inhibitor and naproxen on the gastric mucosa. Dig Dis Sci 2007;52:442–450.

Modlin IM, Kidd M, Lye KD, Wright NA. Gastric stem cells: an update. Keio J Med 2003;52:134–137.

Möllenhoff A, Nolte I, Kramer S. Anti-nociceptive efficacy of carprofen, levomethadone and buprenorphine for pain relief in cats following major orthopaedic surgery. J Vet Med A Physiol Pathol Clin Med 2005;52:186–198.

Moore JN, Garner HE, Shapland JE, Hatfield DG. Prevention of endotoxin-induced arterial hypoxaemia and lactic acidosis with flunixin meglumine in the conscious pony. Equine Vet J 1981;13:95–98.

Morham SG, Langenbach R, Loftin CD, Tiano HF, et al. Prostaglandin synthase 2 gene disruption causes severe renal pathology in the mouse. Cell 1995;83:473–482.

Moses T, Wagner L, Fleming SD. TLR4-mediated COX-2 expression increases intestinal ischemia/reperfusion-induced damage. J Leukoc Biol 2009;86:971–980.

Murtaugh RJ, Matz ME, Labato MA, Boudrieau RJ. Use of synthetic prostaglandin E1 (misoprostol) for prevention of aspirininduced gastroduodenal ulceration in arthritic dogs. J Am Vet Med Assoc 1993;202:251–256.

Naito Y, Yoshikawa T. Oxidative stress involvement and gene expression in indomethacin-induced gastropathy. Redox Rep 2006;11:243–253.

Nandi J, Das PK, Zinkievich JM, Baltodano JD, et al. Cyclo-oxygenase-1 inhibition increases acid secretion by modulating H^+, K^+-ATPase expression and activation in rabbit parietal cells. Clin Exp Pharmacol Physiol 2009;36:127–134.

Nap RC, Breen DJ, Lam TJ, De Bruyne JJ. Gastric retention of enteric-coated aspirin tablets in beagle dogs. J Vet Pharmacol Ther 1990;13:148–153.

Narita T, Sato R, Tomizawa N, Tani K, et al. Safety of reduced-dosage ketoprofen for long-term oral administration in healthy dogs. Am J Vet Res 2006;67:1115–1120.

Neuss H, Raue W, Müller V, Schwenk W, et al. Effects of cyclooxygenase inhibition on anastomotic healing following large bowel resection in a rabbit model: a randomized, blinded, placebo-controlled trial. Int J Colorectal Dis 2009;24:551–557.

Nguyen T, Chai J, Li A, Akahoshi T, et al. Novel roles of local insulin-like growth factor-1 activation in gastric ulcer healing: promotes actin polymerization, cell proliferation, re-epithelialization, and induces cyclooxygenase-2 in a phosphatidylinositol 3-kinase-dependent manner. Am J Pathol 2007;170:1219–1228.

NygAArd G, Anthony A, Khan K, Bounds SV, et al. Intestinal site-dependent susceptibility to chronic indomethacin in the rat: a morphological and biochemical study. Aliment Pharmacol Ther 1995;9:403–410.

Oberto G, Conz A, Giachetti C, Galli CL. Evaluation of the toxicity of indomethacin in a 4-week study, by oral route, in the marmoset (*Callithrix jacchus*). Fundam Appl Toxicol 1990;15:800–813.

Ofman JJ, Maclean CH, Straus WL, Morton SC, et al. Meta-analysis of dyspepsia and nonsteroidal antiinflammatory drugs. Arthritis Rheum 2003;49:508–518.

Okayama M, Hayashi S, Aoi Y, Nishio H, et al. Aggravation by selective COX-1 and COX-2 inhibitors of dextran sulfate sodium (DSS)–induced colon lesions in rats. Dig Dis Sci 2007;52:2095–2103.

Pai R, Soreghan B, Szabó IL, Pavelka M, et al. Prostaglandin E2 transactivates EGF receptor: a novel mechanism for promoting colon cancer growth and gastrointestinal hypertrophy. Nat Med 2002;8: 289–293.

Patoia L, Santucci L, Furno P, Dionisi MS, et al. A 4-week, double-blind, parallel-group study to compare the gastrointestinal effects of meloxicam 7.5mg, meloxicam 15mg, piroxicam 20mg and placebo by means of faecal blood loss, endoscopy and symptom evaluation in healthy volunteers. Br J Rheumatol 1996;35 Suppl 1: 61–67.

Patrono C, Patrignani P, García Rodríguez LA. Cyclooxygenase-selective inhibition of prostanoid formation: transducing biochemical selectivity into clinical read-outs. J Clin Invest 2001;108:7–13.

Pavelka K, Recker DP, Verburg KM. Valdecoxib is as effective as diclofenac in the management of rheumatoid arthritis with a lower incidence of gastroduodenal ulcers: results of a 26-week trial. Rheumatology (Oxford) 2003;42:1207–1215.

Petrescu I, Tarba C. Uncoupling effects of diclofenac and aspirin in the perfused liver and isolated hepatic mitochondria of rat. Biochim Biophys Acta 1997;1318:385–394.

Pihl L, Nylander O. Products of cyclooxygenase-2 depress duodenal function in rats subjected to abdominal surgery. Acta Physiol (Oxford) 2006;186:279–290.

Pilotto A, Seripa D, Franceschi M, Scarcelli C, et al. Genetic susceptibility to nonsteroidal anti-inflammatory drug-related gastroduodenal bleeding: role of cytochrome P450 2C9 polymorphisms. Gastroenterology 2007;133:465–471.

Pollmeier M, Toulemonde C, Fleishman C, Hanson PD. Clinical evaluation of firocoxib and carprofen for the treatment of dogs with osteoarthritis. Vet Rec 2006;159:547–451.

Poortinga EW, Hungerford LL. A case–control study of acute ibuprofen toxicity in dogs. Prev Vet Med 1998;35:115–124.

Radi ZA. Pathophysiology of cyclooxygenase inhibition in animal models. Toxicol Pathol 2009;37:34–46.

Radi ZA, Khan NK. Comparative expression and distribution of c-fos, estrogen receptor alpha (eralpha), and p38 alpha in the uterus of rats, monkeys, and humans. Toxicol Pathol 2006a;34:327–335.

Radi ZA, Khan NK. Effects of cyclooxygenase inhibition on the gastrointestinal tract. Exp Toxicol Pathol 2006b;58:163–173.

Radi ZA, Kehrli ME Jr, Ackermann MR. Cell adhesion molecules, leukocyte trafficking, and strategies to reduce leukocyte infiltration. J Vet Intern Med 2001;15:516–529.

Radi ZA, Marusak RA, Morris DL. Species comparison of the role of p38 MAP kinase in the female reproductive system. J Toxicol Pathol 2009;22:109–124.

Radi ZA, Heuvelman DM, Masferrer JL, Benson EL. Pharmacologic evaluation of sulfasalazine, FTY720, and anti-IL-12/23p40 in a TNBS-induced Crohn's disease model. Dig Dis Sci 2011;56:2283–2291.

Raekallio MR, Hielm-Björkman AK, Kejonen J, Saloned HM, et al. Evaluation of adverse effects of long-term orally administered carprofen in dogs. J Am Vet Med Assoc 2006;228:876–880.

Rainsford KD. The comparative gastric ulcerogenic activities of non-steroid anti-inflammatory drugs. Agents Actions 1977;7:573–577.

Rainsford KD. An analysis of the gastro-intestinal side-effects of non-steroidal anti- inflammatory drugs, with particular reference to comparative studies in man and laboratory species. Rheumatol Int 1982;2:1–10.

Rainsford KD. Ibuprofen: pharmacology, efficacy and safety. Inflammopharmacology 2009;17: 275–342.

Rainsford KD, Stetsko PI, Sirko SP, Debski S. Gastrointestinal mucosal injury following repeated daily oral administration of conventional formulations of indometacin and other non-steroidal anti-inflammatory drugs to pigs: a model for human gastrointestinal disease. J Pharm Pharmacol 2003;55: 661–668.

Rakoff-Nahoum S, Paglino J, Eslami-Varzaneh F, Edberg S, et al. Recognition of commensal microflora by Toll-like receptors is required for intestinal homeostasis. Cell 2004;118:229–241.

Rampal P, Moore N, Van Ganse E, Le Parc JM, et al. Gastrointestinal tolerability of ibuprofen compared with paracetamol and aspirin at over-the-counter doses. J Int Med Res 2002;30:301–308.

Raveendran NN, Silver K, Freeman LC, Narvaez D, et al. Drug-induced alterations to gene and protein expression in intestinal epithelial cell 6 cells suggest a role for calpains in the gastrointestinal toxicity of nonsteroidal anti-inflammatory agents. J Pharmacol Exp Ther 2008;325:389–399.

Reed S. Nonsteroidal anti-inflammatory drug-induced duodenal ulceration and perforation in a mature rottweiler. Can Vet J 2002;43:971–972.

Reimer ME, Johnston SA, Leib MS, Duncan RB, et al. The gastroduodenal effects of buffered aspirin, carprofen, and etodolac in healthy dogs. J Vet Intern Med 1999;13:472–477.

Reuter BK, Davies NM, Wallace JL. Nonsteroidal anti-inflammatory drug enteropathy in rats: role of permeability, bacteria, and enterohepatic circulation. Gastroenterology 1997;112:109–117.

Ricketts AP, Lundy KM, Seibel SB. Evaluation of selective inhibition of canine cyclooxygenase 1 and 2 by carprofen and other nonsteroidal anti-inflammatory drugs. Am J Vet Res 1998;59:1441–1446.

Roberts ES, Van Lare KA, Marable BR, Salminen WF. Safety and tolerability of 3-week and 6- month dosing of Deramaxx (deracoxib) chewable tablets in dogs. J Vet Pharmacol Ther 2009;32:329–337.

Roberts PJ, Morgan K, Miller R, Hunter JO, et al. Neuronal COX-2 expression in human myenteric plexus in active inflammatory bowel disease. Gut 2001;48:468–472.

Robinson MG. New oral salicylates in the therapy of chronic idiopathic inflammatory bowel disease. Gastroenterol Clin North Am 1989;18:43–50.

Rostom A, Muir K, Dubé C, Jolicoeur E, et al. Gastrointestinal safety of cyclooxygenase-2 inhibitors: a Cochrane Collaboration systematic review. Clin Gastroenterol Hepatol 2007;5:818–828.

Roudebush P, Morse GE. Naproxen toxicosis in a dog. J Am Vet Med Assoc 1981;179:805–806.

Rubio F, Seawall S, Pocelinko R, DeBarbieri B, et al. Metabolism of carprofen, a nonsteroid anti-inflammatory agent, in rats, dogs, and humans. J Pharm Sci 1980;69:1245–1253.

Runkel R, Chaplin M, Boost G, Segre E, et al. Absorption, distribution, metabolism, and excretion of naproxen in various laboratory animals and human subjects. J Pharm Sci 1972;61:703–708.

Ryan WG, Moldave K, Carithers D. Clinical effectiveness and safety of a new NSAID, firocoxib: a 1,000 dog study. Vet Ther 2006;7:119–126.

Sato N, Kozar RA, Zou L, Weatherall JM, et al. Peroxisome proliferator-activated receptor gamma mediates protection against cyclooxygenase-2-induced gut dysfunction in a rodent model of mesenteric ischemia/reperfusion. Shock 2005;24:462–469.

Sawant S, Snyman C, Bhoola K. Comparison of tissue kallikrein and kinin receptor expression in gastric ulcers and neoplasms. Int Immunopharmacol 2001;1:2063–2080.

Sawaoka H, Tsuji S, Tsuji M, Gunawan ES, et al. Expression of the cyclooxygenase-2 gene in gastric epithelium. J Clin Gastroenterol 1997; S105–S110.

Sawaoka H, Tsuji S, Tsuji M, Gunawan ES, et al. Involvement of cyclooxygenase-2 in proliferation and morphogenesis induced by transforming growth factor alpha in gastric epithelial cells. Prostaglandins Leukot Essent Fatty Acids 1999;61:315–322.

Schattenkirchner M. Long-term safety of ketoprofen in an elderly population of arthritic patients. Scand J Rheumatol Suppl 1991;91:27–36.

Scheiman JM. NSAIDs, gastrointestinal injury, and cytoprotection. Gastroenterol Clin North Am 1996;25:279–298.

Scheiman JM. Gastroduodenal safety of cyclooxygenase-2 inhibitors. Curr Pharm Des 2003;9:2197–2206.

Scheiman JM, Greenson JK, Lee J, Cryer B. Effect of cyclooxygenase-2 inhibition on human *Helicobacter pylori* gastritis: mechanisms underlying gastrointestinal safety and implications for cancer chemoprevention. Aliment Pharmacol Ther 2003;17:1535–1543.

Scherkl R, Frey HH. Pharmacokinetics of ibuprofen in the dog. J Vet Pharmacol Ther 1987;10:261–265.

Scherle PA, Ma W, Lim H, Dey SK, et al. Regulation of cyclooxygenase-2 induction in the mouse uterus during decidualization: an event of early pregnancy. J Biol Chem 2000;275:37086–37092.

Schmassmann A, Peskar BM, Stettler C, Netzer P, et al. Effects of inhibition of prostaglandin endoperoxide synthase-2 in chronic gastro-intestinal ulcer models in rats. Br J Pharmacol 1998;123:795–804.

Schnitzer TJ, Burmester GR, Mysler E, Hochberg MC, et al. Comparison of lumiracoxib with naproxen and ibuprofen in the Therapeutic Arthritis Research and Gastrointestinal Event Trial (TARGET), reduction in ulcer complications: randomised controlled trial. Lancet 2004;364:665–674.

Schoen RT, Vender RJ. Mechanisms of nonsteroidal anti-inflammatory drug-induced gastric damage. Am J Med 1989;86:449–458.

Schoenfeld P, Kimmey MB, Scheiman J, Bjorkman D, et al. Review article: nonsteroidal anti- inflammatory drug-associated gastrointestinal complications—guidelines for prevention and treatment. Aliment Pharmacol Ther 1999;13:1273–1285.

Schriver DA, White CB, Sandor A, Rosenthale ME. A profile of the rat gastrointestinal toxicity of drugs used to treat inflammatory diseases. Toxicol Appl Pharmacol 1975;32:73–83.

Schwarz NT, Kalff JC, Türler A, Engel BM, et al. Prostanoid production via COX-2 as a causative mechanism of rodent postoperative ileus. Gastroenterology 2001;121:1354–1371.

Seibert K, Zhang Y, Leahy K, Hauser S, et al. Pharmacological and biochemical demonstration of the role of cyclooxygenase 2 in inflammation and pain. Proc Natl Acad Sci USA 1994;91:12013–12017.

Seibert K, Zhang Y, Leahy K, Hauser S, et al. Distribution of COX-1 and COX-2 in normal and inflamed tissues. Adv Exp Med Biol 1997;400A:167–170.

Seitz S, Boelsterli UA. Diclofenac acyl glucuronide, a major biliary metabolite, is directly involved in small intestinal injury in rats. Gastroenterology 1998;115:1476–1482.

Seitz S, Kretz-Rommel A, Oude Elferink RP J, Boelsterli UA. Selective protein adduct formation of diclofenac glucuronide is critically dependent on the rat canalicular conjugate export pump (Mrp2). Chem Res Toxicol 1998;11:513–519.

Sennello KA, Leib MS. Effects of deracoxib or buffered aspirin on the gastric mucosa of healthy dogs. J Vet Intern Med 2006;20:1291–1296.

Shifflett DE, Jones SL, Moeser AJ, Blikslager AT. Mitogen-activated protein kinases regulate COX-2 and mucosal recovery in ischemic-injured porcine ileum. Am J Physiol Gastrointest Liver Physiol 2004;286:G906–G913.

Shriver DA, Dove PA, White CB, Sandor A, et al. A profile of the gastrointestinal toxicity of aspirin, indomethacin, oxaprozin, phenylbutazone, and fentiazac in arthritic and Lewis normal rats. Toxicol Appl Pharmacol 1977;42:75–83.

Sigthorsson G, Jacob M, Wrigglesworth J, Somasundaram S, et al. Comparison of indomethacin and nimesulide, a selective cyclooxygenase-2 inhibitor, on key pathophysiologic steps in the pathogenesis of nonsteroidal anti-inflammatory drug enteropathy in the rat. Scand J Gastroenterol 1998;33:728–735.

Sigthorsson G, Simpson RJ, Walley M, Anthony A, et al. COX-1 and 2, intestinal integrity, and pathogenesis of nonsteroidal anti-inflammatory drug enteropathy in mice. Gastroenterology 2002;122:1913–1923.

Silverstein FE, Faich G, Goldstein JL, Simon LS, et al. Gastrointestinal toxicity with celecoxib vs nonsteroidal anti-inflammatory drugs for osteoarthritis and rheumatoid arthritis: the CLASS study—a randomized controlled trial. Celecoxib Long-Term Arthritis Safety Study. J Am Med Assoc 2000;284:1247–1255.

Singer II, Kawka DW, Schloemann S, Tessner T, et al. Cyclooxygenase 2 is induced in colonic epithelial cells in inflammatory bowel disease. Gastroenterology 1998;115:297–306.

Singh G, Fort JG, Goldstein JL, Levy RA, et al. Celecoxib versus naproxen and diclofenac in osteoarthritis patients: SUCCESS-I Study. Am J Med 2006;119:255–266.

Singh VP, Patil CS, Jain NK, Kulkarni SK. Aggravation of inflammatory bowel disease by cyclooxygenase-2 inhibitors in rats. Pharmacology 2004;72:77–84.

Siraki AG, Chevaldina T, O'Brien PJ. Application of quantitative structure–toxicity relationships for acute NSAID cytotoxicity in rat hepatocytes. Chem Biol Interact 2005;151:177–191.

Smith HW. Observations on the flora of the alimentary tract of animals and factors affecting its composition. J Pathol Bacteriol 1965;89:95–122.

Somasundaram S, Sigthorsson G, Simpson RJ, Watts J, et al. Uncoupling of intestinal mitochondrial oxidative phosphorylation and inhibition of cyclooxygenase are required for the development of NSAID-enteropathy in the rat. Aliment Pharmacol Ther 2000;14:639–650.

Soslow RA, Dannenberg AJ, Rush D, Woerner BM, et al. COX-2 is expressed in human pulmonary, colonic, and mammary tumors. Cancer 2000;89:2637–2645.

Spangler RS. Gastrointestinal damage demonstrated with nabumetone or etodolac in preclinical studies. Am J Med 1993;95:35S–39S.

Stanton ME, Bright RM. Gastroduodenal ulceration in dogs: retrospective study of 43 cases and literature review. J Vet Intern Med 1989;3:238–244.

Steagall PV, Mantovani FB, Ferreira TH, Salcedo ES, et al. Evaluation of the adverse effects of oral firocoxib in healthy dogs. J Vet Pharmacol Ther 2008;30:218–223

Steagall PV, Moutinho FQ, Mantovani FB, Passarelli D, et al. Evaluation of the adverse effects of subcutaneous carprofen over six days in healthy cats. Res Vet Sci 2009;86:115–120.

Stoltz RR, Harris SI, Kuss ME, LeComte D, et al. Upper GI mucosal effects of parecoxib sodium in healthy elderly subjects. Am J Gastroenterol 2002;97:65–71.

Sturm A, Dignass AU. Epithelial restitution and wound healing in inflammatory bowel disease. World J Gastroenterol 2008;14:348–353.

Sun WH, Tsuji S, Tsuji M, Gunawan ES, et al. Cyclo-oxygenase-2 inhibitors suppress epithelial cell kinetics and delay gastric wound healing in rats. J Gastroenterol Hepatol 2000;15:752–761.

Szabó IL, Pai R, Jones MK, Ehring GR, et al. Indomethacin delays gastric restitution: association with the inhibition of focal adhesion kinase and tensin phosphorylation and reduced actin stress fibers. Exp Biol Med (Maywood) 2002;227:412–424.

Takeuchi K, Okada M, Ebara S, Osano H. Increased microvascular permeability and lesion formation during gastric hypermotility caused by indomethacin and 2- deoxy-D-glucose in the rat. J Clin Gastroenterol 1990;12 Suppl 1:S76–S84.

Takeuchi K, Ukawa H, Furukawa O, Kawauchi S, et al. Prostaglandin E receptor subtypes involved in stimulation of gastroduodenal bicarbonate secretion in rats and mice. J Physiol Pharmacol 1999;50:155–167.

Takeuchi K, Miyazawa T, Tanaka A, Kato S, et al. Pathogenic importance of intestinal hypermotility in NSAID-induced small intestinal damage in rats. Digestion 2002;66:340–341.

Takeuchi K, Tanaka A, Hayashi Y, Kubo Y. Functional mechanism underlying COX-2 expression following administration of indomethacin in rat stomachs: importance of gastric hypermotility. Dig Dis Sci 2004;49:180–187.

Takeuchi K, Aihara E, Sasaki Y, Normura Y, et al. Involvement of cyclooxygenase-1, prostaglandin E2 and EP1 receptors in acid-induced HCO_3-secretion in stomach. J Physiol Pharmacol 2006;57:661–676.

Takeuchi K, Tanaka A, Kato S, Amagase K, et al. Roles of COX inhibition in pathogenesis of NSAID-induced small intestinal damage. Clin Chim Acta 2010;411:459–466.

Tanaka A, Araki H, Komoike Y, Hase S, et al. Inhibition of both COX-1 and COX-2 is required for development of gastric damage in response to nonsteroidal anti-inflammatory drugs. J Physiol Paris 2001;95:21–27.

Tanaka A, Araki H, Hase S, Komoike Y, et al. Up-regulation of COX-2 by inhibition of COX-1 in the rat: a key to NSAID-induced gastric injury. Aliment Pharmacol Ther 2002a;16:90–101.

Tanaka A, Hase S, Miyazawa T, Ohno R, et al. Role of cyclooxygenase (COX)-1 and COX-2 inhibition in nonsteroidal anti-inflammatory drug-induced intestinal damage in rats: relation to various pathogenic events. J Pharmacol Exp Ther 2002b;303:1248–1254.

Tarnawski A, Stachura J, Durbin T, Sarfeh IJ, et al. Increased expression of epidermal growth factor receptor during gastric ulcer healing in rats. Gastroenterology 1992;102:695–698.

Tatsuguchi A, Sakamoto C, Wada K, Akamatsu T, et al. Localisation of cyclooxygenase 1 and cyclooxygenase 2 in Helicobacter pylori–related gastritis and gastric ulcer tissues in humans. Gut 2000;46:782–789.

Tibble JA, Sigthorsson G, Foster R, Bjarnason I. Comparison of the intestinal toxicity of celecoxib, a selective COX-2 inhibitor, and indomethacin in the experimental rat. Scand J Gastroenterol 2000;35:802–807.

Tobin T, Chay S, Kamerling S, Woods WE, et al. Phenylbutazone in the horse: a review. J Vet Pharmacol Ther 1986;9:1–25.

Tomlinson JE, Wilder BO, Young KM, Blikslager AT. Effects of flunixin meglumine or etodolac treatment on mucosal recovery of equine jejunum after ischemia. Am J Vet Res 2004;65:761–769.

Traub JL, Gallina AM, Grant BD, Reed SM, et al. Phenylbutazone toxicosis in the foal. Am J Vet Res 1983;44:1410–1418.

Traub-Dargatz JL, Bertone JJ, Gould DH, Wrigley RH, et al. Chronic flunixin meglumine therapy in foals. Am J Vet Res 1988;49:7–12.

Tsubouchi R, Hayashi S, Aoi Y, Nishio H, et al. Healing impairment effect of cyclooxygenase inhibitors on dextran sulfate sodium-induced colitis in rats. Digestion 2006;74:91–100.

Uchida M, Kawano O, Misaki N, Saitoh K, et al. Healing of acetic acid–induced gastric ulcer and gastric mucosal PGI2 level in rats. Dig Dis Sci 1990;35:80–85.

Ungaro R, Fukata M, Hsu D, Hernandez Y, et al. A novel Toll-like receptor 4 antagonist antibody ameliorates inflammation but impairs mucosal healing in murine colitis. Am J Physiol Gastrointest Liver Physiol 2009;296:G1167–G1179.

Van Miert AS, Van Der Wal-Komproe LE, Van Duin CT. Effects of antipyretic agents on fever and ruminal stasis induced by endotoxins in conscious goats. Arch Int Pharmacodyn Ther 1977;225:39–50.

Villar D, Buck WB, Gonzalez JM. Ibuprofen, aspirin and acetaminophen toxicosis and treatment in dogs and cats. Vet Hum Toxicol 1998;40:156–162.

Villegas I, Alarcón de la Lastra C, Martín MJ, Motilva V, et al. Gastric damage induced by subchronic administration of preferential cyclooxygenase-1 and cyclooxygenase-2 inhibitors in rats. Pharmacology 2002;66:68–75.

Vonderhaar MA, Salisbury SK. Gastroduodenal ulceration associated with flunixin meglumine administration in three dogs. J Am Vet Med Assoc 1993;203:92–95.

Wallace JL, Keenan CM, Granger DN. Gastric ulceration induced by nonsteroidal anti- inflammatory drugs is a neutrophil-dependent process. Am J Physiol 1990;259:G462–G467.

Wallace JL, McKnight W, Reuter BK, Vergnolle N. NSAID-induced gastric damage in rats: requirement for inhibition of both cyclooxygenase 1 and 2. Gastroenterology 2000;119:706–714.

Watanabe T, Higuchi K, Kobata A, Nishio H, et al. Non-steroidal anti-inflammatory drug- induced small intestinal damage is Toll-like receptor 4 dependent. Gut 2008;57:181–187.

Watanabe T, Nishio H, Tanigawa T, Yamagami H, et al. Probiotic *Lactobacillus casei* strain Shirota prevents indomethacin-induced small intestinal injury: involvement of lactic acid. Am J Physiol Gastrointest Liver Physiol 2009;297:G506–G513.

Watson AD, Wilson JT, Turner DM, Culvenor JA. Phenylbutazone induced blood dyscrasias suspected in three dogs. Vet Rec 1980;107:239–241.

Wax J, Clinger WA, Varner P, Bass P, et al. Relationship of the enterohepatic cycle to ulcerogenesis in the rat small bowel with flufenamic acid, Gastroenterology 1970;58:772–780.

Wight NJ, Gottesdiener K, Garlick NM, Atherton CT, et al. Rofecoxib, a COX-2 inhibitor, does not inhibit human gastric mucosal prostaglandin production. Gastroenterology 2001;120:867–873.

Wolfe MM, Lichtenstein DR, Singh G. Gastrointestinal toxicity of nonsteroidal antiinflammatory drugs. N Engl J Med 1999;340:1888–1899.

Wooten JG, Blikslager AT, Marks SL, Law JM, et al. Effect of nonsteroidal anti- inflammatory drugs with varied cyclooxygenase-2 selectivity on cyclooxygenase protein and prostanoid concentrations in pyloric and duodenal mucosa of dogs. Am J Vet Res 2009;70:1243–1249.

Yesair DW, Callahan M, Remington L, Kensler CJ. Role of the entero-hepatic cycle of indomethacin on its metabolism, distribution in tissues and its excretion by rats, dogs and monkeys. Biochem Pharmacol 1970;19:1579–1590.

Yokota A, Taniguchi M, Takahira Y, Tanaka A, et al. Rofecoxib produces intestinal but not gastric damage in the presence of a low dose of indomethacin in rats. J Pharmacol Exp Ther 2005;314:302–309.

Yoshizawa T, Watanabe S, Hirose M, Yamamoto J, et al. Effects of growth factors on aspirin-induced inhibition of wound repair in a rabbit gastric epithelial cell model. Aliment Pharmacol Ther 2000;Suppl 1: 176–182.

Zhang L, Lu YM, Dong XY. Effects and mechanism of the selective COX-2 inhibitor, celecoxib, on rat colitis induced by trinitrobenzene sulfonic acid. Chin J Dig Dis 2004;5:110–114.

Zinkievich JM, George S, Jha S, Nandi J, et al. Gastric acid is the key modulator in the pathogenesis of NSAID-induced ulceration in rats. Clin Exp Pharmacol Physiol 2010;37:654–661.

CHAPTER 2

BONE–TENDON–LIGAMENT SYSTEM

INTRODUCTION

The skeletal system plays an essential role in providing mechanical support for body stature and locomotion, protecting vital body organs, and maintaining mineral homeostasis. A healthy skeleton must be maintained by constant bone modeling to carry out these crucial functions throughout life (Feng and McDonald, 2011). Musculoskeletal pain treatment is essential for improving healing of traumatic injuries and surgical procedures, and for improving patients' quality of life (O'Conner and Lysz, 2008). Cyclooxygenases (COX-1 and COX-2) are rate-limiting enzymes that catalyze the conversion of arachidonic acid to prostaglandins (PGs). COX-2 is induced at sites of inflammation, injury, and pain (Fig. 2-1) (Radi and Khan, 2005; Radi, 2009). Thus, COX inhibitors such as conventional nonselective NSAIDs (ns-NSAIDs) and COX-2 selective NSAIDs (COX-2 s-NSAIDs) are used to treat musculoskeletal pain, including postsurgical orthopedic analgesia. It has been hypothesized that NSAIDs can modulate bone, ligament, or tendon healing by modulating PG production (O'Conner and Lysz, 2008). The majority of evidence to support such hypotheses has been derived from nonclinical studies in various animal models, as no well-controlled large-scale clinical or epidemiological studies have been conducted (Radi and Khan, 2005).

In this chapter we examine nonclinical and clinical literature data on PG metabolism in the skeleton and NSAID pathophysiological impact on bone, ligament, and tendon healing.

COMPARATIVE PHYSIOLOGICAL AND ANATOMICAL ASPECTS OF THE SKELETON

The skeletal system consists of the bones of the skull, thorax, and vertebral column, which form the axial skeleton, and the bones of the upper and lower extremities, which form the appendicular skeleton. The long appendicular bones and the vertebral bodies are divided anatomically into four regions: (1) epiphysis, (2) metaphyseal growth plates (physis), (3) metaphysis, and (4) diaphysis. The skeleton is

Comparative Pathophysiology and Toxicology of Cyclooxygenases, First Edition. Zaher A. Radi.

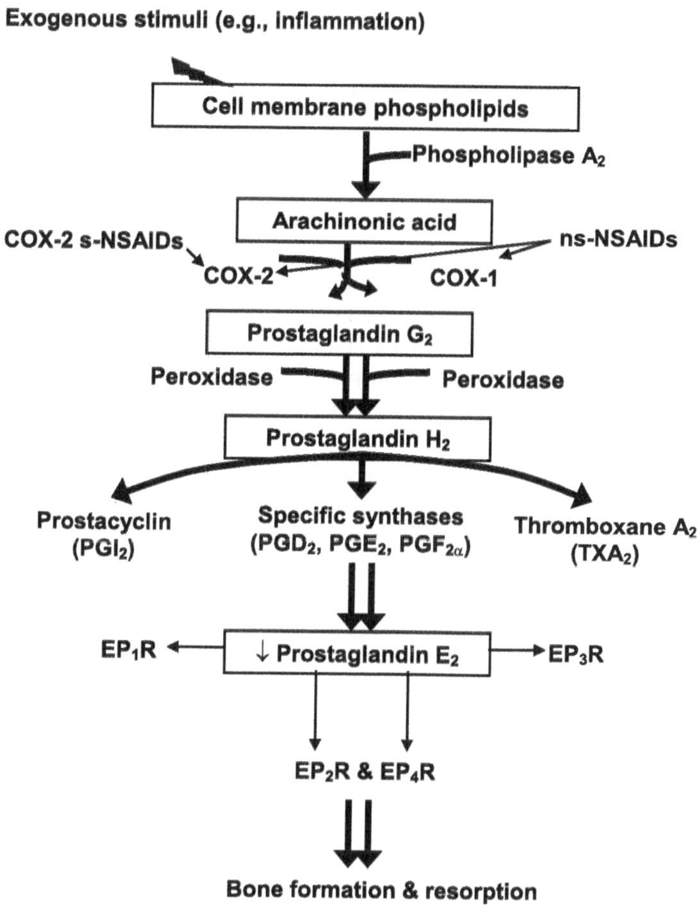

FIGURE 2-1 Role of prostaglandins (PGs) in bone metabolism in health and disease and effects of ns-NSAIDs and COX-2 s-NSAIDs. Following an exogenous stimulus (e.g., inflammation), cell membrane phospholipid is liberated to arachidonic acid (AA) by phospholipase A_2.Both COX-1 and COX-2 catalyze the conversion of AA into various PGs. PGE_2 is the most abundant PG produced by osteoblastic cells. PGE_2 interacts with four receptor types (EP_1R, EP_3R, EP_2R, and EP_4R), of which EP_2R and EP_4R stimulate bone formation and resorption. COX-1 is expressed constitutively by osteocytes and osteoblasts, while COX-2 expression is up-regulated during the initial stages of bone repair or under inflammatory or neoplastic conditions. Nonselective NSAIDs (e.g., carprofen, etodolac, flunixin meglumine, ketoprofen, indomethacin, meloxicam, phenylbutazone) inhibit COX-1 and COX-2, whereas COX-2 s-NSAIDs (e.g., celecoxib, lumiracoxib, parecoxib) spare COX-1 and inhibit only COX-2. (Reprinted with kind permission from Springer Science & Business Media: Z. A. Radi and N. K. Khan, *Inflammation Research*, 54(9), pp. 358–366. Copyright © 2005.)

composed of a dense cortical shell with a bone marrow space that contains trabecular bone and bone marrow elements. In addition, the skeleton is a metabolically active and dynamic organ that stores minerals (i.e., calcium, phosphorus, and carbonate), is a binding site for toxic ions such as lead, and undergoes continuous remodeling throughout life (Raisz, 1999a). There are two types of bones: cancellous and compact. Both are formed in layers and thus are called lamellar bone. Cancellous (spongy) bone is found in the interior of bones and is composed of trabeculae. Compact (cortical) bone has a densely packed calcified matrix. Bone is organized into osteons or haversian systems in the compact bone. Physical forces on the skeleton can lead to deformations and pathological conditions. This was recognized in the nineteenth century as Wolff's law, which is also called the "law of bone transformation" (Hert, 1990). Wolff's law, conceived in 1892 by Julius Wolff, a German anatomist, states that healthy bone will adapt to an applied force and loading stress. If loading on a particular bone increases, the bone will remodel itself over time to overcome such stress. The internal architecture of bone trabeculae undergoes adaptive changes, followed by secondary changes to the external cortical portion of the bone (Freiberg, 1902). This theory is known today as bone modeling and remodeling. Bone remodeling is the restructuring process of existing bone, which is in constant resorption and formation. Bone remodeling is vital for the structural and metabolic integrity of the skeleton. In aged rats, remodeling takes place in vertebral and tibial cancellous bone (Erben, 1996). Growth of the epiphysis contributes to the overall bone length. Growth of long bones depends on a functional metaphyseal growth plate (physis). The metaphyseal growth plate consists of hyaline cartilage and can be divided into three anatomical zones: resting or reserve, proliferative, and hypertrophic. The resting zone serves as a source of cells for the proliferating zone.

There are species differences in skeleton growth dynamics. In a nonhuman primate model (cynomolgus macaques), the status of full-thickness chondral defect repair was evaluated at different time points after microfracture (Gill et al., 2005). The chondral defects were created on the femoral condyles and trochlea of animals and evaluated by gross and histological examination at 6 and 12 weeks. At 6 weeks, there was limited chondral repair and ongoing resorption of subchondral bone. By 12 weeks, the defects were filled completely and showed more mature cartilage and bone repair. Thus, significant improvements in the extent and quality of cartilage repair were observed at the 6- to 12-week time points after microfracture in nonhuman primates (Gill et al., 2005).

The growth plate is thickest when there is rapid growth and becomes thin when growth slows. In larger mammals, growth plates close (become entirely replaced by bone) at skeletal maturity and longitudinal growth diminishes. In humans and rabbits, the growth plate is resorbed completely following puberty, resulting in fusion of the epiphysis to the metaphysis. This process of physeal closure is primarily under the control of estrogen in both males and females (Weise et al., 2001). Rats and mice, however, maintain a growth plate until old age. Bone growth rate in rats increases between 1 and 5 weeks, then declines until skeletal maturity, which is achieved at approximately 11 to 13 weeks (Kember, 1973; Hunziker and Schenk, 1989). During the period of rapid growth (up to 5 weeks), the height of the growth plate is greatest, whereas during the period when growth slows (8 to 16 weeks), the height of the growth plate decreases progressively (Roach et al., 2003). Some bones in the rat skeleton retain lifelong growth and do not fuse their epiphyses (Dawson, 1925). Many long bone epiphyseal growth plates in male rats remain open until 30 months of age (Dawson, 1925). In contrast, bone elongation at other skeleton sites, such as the proximal tibia and distal tibia, ceases at 15 and 3 months of age, respectively, in female rats (Dawson, 1925; Ke et al., 1993; Erben, 1996).

In humans, the mechanical properties of cortical bone specimens from human femora and tibiae were determined in terms of tension, torsion, and compression for a population ranging in age from 21 to 86 years (Burstein et al., 1976). No significant differences were found in the mechanical properties of males and females. Tibial bone had higher values of ultimate strength, stiffness, and ultimate strain than those for femoral bone. Consistent decreases with age of all mechanical properties except plastic modulus were found in the femoral but not in the tibial bone. No consistent significant differences in tension properties were found in specimens from normal, osteoporotic, and corticosteroid-treated persons. In addition, tensile testing to failure was carried out on 235 tensile cortical specimens (five per bone) that had been machined from 47 femora from human calvera (McCalden et al., 1993). The donors ranged in age from 20 to 102 years at the time of death. After mechanical testing, the porosity, mineralization, and microstructure were examined. Linear regression analysis showed that mechanical properties deteriorated markedly with age. Ultimate stress, strain, and energy absorption decreased by 5, 9, and 12% per decade, respectively. The porosity of bone increased significantly with age, whereas the mineral content was not affected. Microstructural analysis demonstrated that the amount of haversian bone increased with age. This demonstrates the importance of age-related changes in porosity to the decline in mechanical properties. Changes in porosity accounted for 76% of the reduction in strength (McCalden et al., 1993). Mineral content did not play a major role. Thus, the quantitative changes in aging bone tissue, rather than the qualitative changes, influence the mechanical competence of bone.

Another study analyzed iliac bone histomorphometric data from 58 healthy male and female subjects aged 1.5 to 23 years (Parfitt et al., 2000). There was a significant increase with age in core width, with corresponding increases in both cortical width and cancellous width. In cancellous bone there were increases in bone volume and trabecular thickness but not in trabecular number, wall thickness, interstitial thickness, and inferred erosion depth. Mineral apposition rates declined on the periosteal envelope and on all subdivisions of the endosteal envelope. The active osteoblast life span increased substantially. The bone formation rate was almost eight times higher on the outer than on the inner periosteum, and more than four times higher on the inner than on the outer endocortical surface. On the cancellous surface, the bone formation rate and activation frequency declined (Parfitt et al., 2000).

Age can also influence skeleton growth, and aging has been shown to diminish bone blood flow in rats and humans (Prisby et al., 2007). For example, blood flow was measured in conscious young adult (4 to 6 months old) and aged (24 to 26 months old) male Fischer 344 rats using radiolabeled microspheres. Blood flow of femurs in aged rats was 21% and 28% lower in the proximal and distal metaphyses, respectively, and 45% lower in the diaphyseal marrow (Prisby et al., 2007). Growth plates of femurs and tibiae in old Wistar rats 62 to 80 weeks old were compared to those from young rats 2 to 16 weeks old. Four categories of growth plate morphology were present in the old rats that were not present in younger rats: (1) formation of a bone band parallel to the metaphyseal edge of the growth plate; (2) extensive areas of acellularity, which were resistant to resorption and/or remodeling; (3) extensive remodeling and bone formation within cellular regions of the growth plate; and (4) direct bone formation by former growth plate chondrocytes (Roach et al., 2003). In a femoral fracture healing mouse model, healing was evaluated in nice either 7 to 9 or 52 to 56 weeks old. Aging in these mice was associated with a decreased rate of chondrogenesis, bone formation, callus vascularization, delayed remodeling, and altered expression of genes involved in bone repair and remodeling (Naik et al., 2009).

There are gender differences in skeleton dynamics. The dynamics of femoral midshaft collagen turnover was examined in adult humans. Males have a lower turnover rate than females when older but have a higher turnover rate when younger (Hedges et al., 2007). Male and female human femur cortical bone specimens were studied to determine if there are gender- and age-related differences. The degree of tissue mineralization decreased significantly with age in females but did not change in males. Total bone mineralization was higher in females than in males until 50 years of age but was lower in elderly females than in elderly males. Mineralization of osteons varied significantly with gender and age. Thus, in females, mineralization started at a higher level than in males but was lower in the sixth decade, falling below the level in males. Mineralization was far more stable throughout life in males (Bergot et al., 2009). A mutation in low-density lipoprotein receptor–related protein 5 (LRP5) has been shown to increase bone mass and density in humans and animals. Interestingly, within LRP5 mutant mice, there were significant gender-related differences in some of the trabecular bone structural parameters at various bone sites (distal femur, femoral neck, and vertebral body). However, unlike trabecular structural parameters, the gender-specific differences were not found in the trabecular strength of LRP5 transgenic mice (Dubrow et al., 2007). Gender can have an impact on NSAID elimination. For example, the elimination time for celecoxib in female rats is much longer than that in male rats (14 versus 4 h) (Paulson et al., 2000). This pharmacokinetic difference would affect drug concentration in the blood and should be taken into consideration when designing fracture-healing experimental studies in rodents.

Additionally, there are anatomical differences in the skeleton that can affect the rate of bone healing, such as vascular histology (e.g., the femur and mandible are more highly vascularized than the fibula), biomechanics (e.g., quadrupeds versus bipeds), wounding and healing histology and responses, and bone histomorphology (Auer et al., 2007). The base structure of the cortex in dogs and humans is composed of secondary osteones, whereas the sheep cortex is comprised predominantly of lamellar bone. However, during bone healing and remodeling, secondary osteones are activated in all of these species (Auer et al., 2007). Both animals and human individuals have very different base levels of bone mass. Thus, genotype may influence specific structural and material properties (Judex et al., 2002). For example, material properties may differ by as much as 200% in different laboratory mice strains relative to structural morphology (Auer et al., 2007). There is significant genetic variation in the femur cross-sectional area, density, and mechanical properties between inbred mouse strains (Wergedal et al., 2005). Different anatomical sites may also influence the pattern of repair. This is multifactorial, partly because of the unique mechanobiology of different skeletal bones and partly as a consequence of biological differences (Auer et al., 2007). For example, the metacarpal and metatarsal bones of small ruminants are fused bones with minimal soft tissue coverage. Tendons and ligaments are found on the dorsal and palmar/plantar aspects of this bone. The tibia has a bone anatomy that provides a subcutaneous surface craniomedially and muscle cover laterocaudally. By contrast, the femur is predominantly covered by muscle on all aspects (Auer et al., 2007).

Healing within specific bones may also differ at bone anatomical sites with predominantly cortical healing at diaphyseal sites, and cancellous healing at metaphyseal sites (Auer et al., 2007). The extent of cancellous bone also differs between species and is generally limited in relation to the extent seen in humans. Bony defects of a size that will not heal during the lifetime of the animal may be termed critical size defects. Experiments on critical-size bone defects demonstrated species-dependent healing (Schmitz and Hollinger, 1986). A 4-mm circular craniomandibular bone defect in mice and 8-mm defect in rats are often both nonhealing without therapeutic intervention at 12 to 24 weeks. However, an 8-mm-diameter defect in a larger animal model (e.g., rabbit, dog, and sheep) will heal spontaneously since the surrounding soft tissue and remaining perimeter bone interfacial area are better able to supply signals and regenerative milieu to this relatively smaller defect volume. Critical-size calvarial defects increase to 17- to 35-mm diameters in larger animals (guinea pigs, rabbits, dogs, and sheep). Seemingly, the linear dimensions of a critical defect in bone do not translate simply across models, nor to humans, nor across different bone defect sites (e.g., cranial to long bone) (Auer et al., 2007). Additionally, the anatomical site of a critical defect has an effect on its size and, therefore, on possible nutrient transport limitations to healing. An 8-mm-diameter circular defect in rat parietal bones will heal spontaneously, whereas an 8-mm-diameter circular defect in the human parietal bone will not. These wound-site transport scaling effects also have a profound impact on the delivery of exogenous therapeutic and biotechnology components (living cells, growth factors, drugs) to fracture or defect sites (Auer et al., 2007).

The likelihood of surrounding soft tissue prolapse into the bone wound site as a function of increasing size can pose a problem during bone healing. The amount of repair and the time required to accomplish repair of 4-cm segmental fibular transplant in adult male dogs were determined at 2 to 48 weeks after transplantation by torsional stress testing, microradiography, and tetracycline labeling. The transplanted cortical bone was greatly weakened at 6 weeks to six months but was nearly normal at one year. The strength of the transplant appeared to be related to the amount of porosity of the matrix rather than to the quality or completeness of the biological repair. The repair was ordered rather than random. The initial resorption caused increased porosity, which was slowly offset by apposition of new bone, a process that was dependent on general skeletal metabolic activity (Enneking et al., 1975). This effect extends beyond mammalian species in larger bone wounds relative to wound size. Relatively larger bone volumes will regenerate in rodents and amphibians (e.g., newt) compared to higher mammalian species (Auer et al., 2007). Animal physical activity levels should also be considered and quantified where possible, as it also modulates the bone repair process (Connolly et al., 2003). It was demonstrated that a femoral fracture mouse model that moved more had larger external calluses containing more cartilage and demonstrated lower torsional stiffness at the same time point (Connolly et al., 2003).

ROLE OF PROSTAGLANDINS IN SKELETON METABOLISM

The skeleton structural elements include bone cells (osteoclasts and osteobalsts), bone matrix [collagen fibers, bone morphogenic proteins (BMPs), and glycoproteins], and minerals. Osteoblasts express the receptor activator of nuclear factor κB ligand (RANKL), which interacts with RANK expressed on osteoclasts to stimulate the formation of osteoclasts. Osteoprotegerin (OPG) is a secreted protein that is produced and released by activated osteoblastic cells. In 1965, Marshall Urist pioneered the concept of the presence of a substance naturally present in bone that is responsible for its regeneration and repair, called BMP, which is also known as osteogenic protein (OP) (Urist, 1965). BMPs are members of the transforming growth factor beta (TGFβ) gene superfamily of growth and differentiation factors. Members of the BMP family were originally cloned and characterized by their ability to induce ectopic bone formation. The action of BMPs is mediated through receptor kinases and transcription factors called Smads (Vaibhav et al., 2007). BMPs are pleiotropic regulators mediating various sequential cellular responses: chemotaxis of cells, mitosis and proliferation of progenitor cells, differentiation into chondroblasts, cartilage calcification, vascular invasion, bone formation, remodeling and bone marrow differentiation. Sixteen different proteins (BMP-1 through BMP-16) have been identified. Of these various cloned BMPs, the bone inductive ability of BMP-7 and BMP-2 has been well characterized. Numerous studies have been conducted on large critically sized diaphyseal segmental defects in rats, rabbits, dogs, sheep, and nonhuman primates to evaluate the osteoinductive properties of BMPs. In these nonclinical studies it was observed that implantation of the BMPs with carrier matrices in bone defects led to biomechanically and biologically sound bone formation (Vaibhav et al., 2007). Furthermore, both BMP-7 and BMP-2 have been shown to have clinical utility in the healing of nonunion fractures (Govender et al., 2002; Dimitriou et al., 2005). The interaction between NSAIDs and the bone anabolic BMP-7 pathway was assessed in a study in C57BL/6 mice where BMP-7 expression and healing of fractures of the right femora were evaluated (Spiro et al., 2010). Each mouse was implanted with a 21-day-release pellet containing 2.5 mg diclofenac/pellet (5 mg diclofenac/kg body weight per day; the constant blood level of diclofenac is 0.238 μg/mL). Contact x-rays were performed on all mice 5, 10,

15, and 20 days after BMP-7 implantation. At day 20 after implantation, all mice were euthanized, and radiographic, micro computed tomographic, histologic, and histomorphometric analyses were performed on ectopic bone nodules. Immunohistochemical analyses of fracture callus sections on days 10, 15, and 20 after surgery showed high expression of COX-2 and BMP-7 in osteoblasts, chondrocytes, and fibrous tissue of the fracture callus in both groups. This coexpression of COX-2 and BMP-7 indicated a possible role of both proteins during the fracture-healing process. Analyses of the fractured femora at day 20 following surgery revealed impairment of fracture healing due to diclofenac treatment. The callus bone volume per tissue volume was decreased significantly in diclofenac-treated mice compared to controls (Spiro et al., 2010).

It has been demonstrated, using human mesenchymal stem cells, that PGE_2 increases BMP-2 expression via binding to the EP_4 receptor (Arikawa et al., 2004). BMP-2 is used clinically to stimulate bone formation and accelerate repair. In an in vitro study, the effects of BMP-2 and PGE_2 on the expression of RANKL and OPG mRNA in bone marrow cultures from mice were assessed on days 4 to 6, prior to the peak of osteoclast formation (Blackwell et al., 2009). BMP-2 alone showed no effect on RANKL mRNA. Although BMP-2 decreased OPG mRNA 32 to 54% compared with the control, the RANKL/OPG ratio was not increased significantly. Treatment with PGE_2 both increased RANKL mRNA and decreased OPG mRNA compared with the control, resulting in an increased RANKL/OPG ratio. BMP-2 in combination with PGE_2 resulted in small, nonsignificant increases in RANKL and decreases in OPG compared with PGE_2 alone. However, the RANKL/OPG ratio was increased by BMP-2 and PGE_2 compared with PGE_2 alone by $64 \pm 29\%$ on day 6 in the experiment (Blackwell et al., 2009). Therefore, BMP-2 can increase osteoclast formation in response to PGE_2 by increasing the RANKL/OPG ratio in osteoblasts. This is supported by a study in a mouse collagen-induced arthritis model in which celecoxib was given orally at doses of 16 and 75 mg/kg. Celecoxib suppressed arthritis-related increase in bone resorption at low and high doses and prevented trabecular bone mass reduction at high doses. In addition, reduction in osteoclast surface number and suppression of osteoclast development in bone marrow through inhibition of the RANKL/OPG ratio, and IL-6 mRNA expression in inflammatory synovial tissue was noted (Taketa et al., 2008).

The bone remodeling cycle is divided into four phases: activation, resorption, reversal, and formation (Raisz, 1999a). Osteoblasts and osteoclasts, the most important contributors to the bone remodeling cycle, produce PGs, which are shown to modulate their function in normal bone metabolism and in bone healing (Okada et al., 2000, 2003; Raisz et al., 2001; Li X. et al., 2002; Gajraj et al., 2003). For example, in an in vitro model using rat femora, changes in PG synthesis and concentration were correlated with changes in the quantity of trabecular regeneration and acceleration of bone healing (Sun et al., 1999). Osteoblasts and osteoclasts cooperate to maintain bone homeostasis. Osteoblasts are mature "bone-building" metabolically active cells that differentiate from mesenchymal stem cells and are responsible for new bone formation and osteoid (unmineralized organic bone matrix that subsequently undergoes mineralization) secretion. Osteoclasts are multinucleated cells that arise from hematopoietic monocyte/macrophage precursors

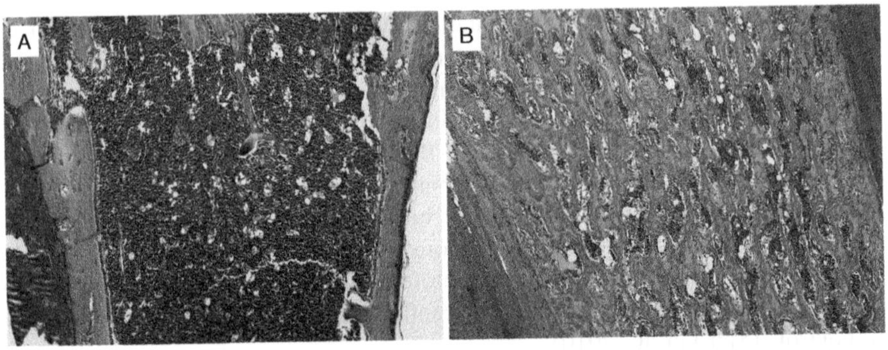

FIGURE 2-2 (A) Histopathology of a bone section from a control mouse. Hematoxylin and eosin (H&E) stain, original maginification ×10. (B) Histopathology of a bone section from an *op/op* (osteopetrosis) mouse. Note that the bone marrow cavity is effaced by interweaving thick bony trabeculae consistent with osteopetrosis, H&E, original magnification ×20.

and are the major resorptive cells of bone. Osteoclasts release hydrolytic enzymes that dissolve the bone matrix. Thus, osteoclasts are "bone-chewing" cells that function in the resorption of bone, removing the mineral and the organic matrix.

Macrophage colony-stimulating factor (M-CSF) induces monocyte/macrophage differentiation to osteoclasts and is critical to osteoclast replication and bone resorption (Radi et al., 2009, 2011). Thus, mice lacking M-CSF develop osteopetrosis, increased bone mass, in which affected bones are dense and lack a medullary cavity (Fig. 2-2) (Radi et al., 2009). Osteopetrosis also occurs in humans, dogs, horses, sheep, and cattle. The basis for osteopetrosis is the failure of osteoclasts to resorb bone and shape trabeculae. RANKL has been shown to be a marker and a mediator of bone loss in collagen-induced arthritis (CIA) models in rats (Stolina et al., 2005). RANKL-knockout mice develop severe osteopetrosis, a defect in tooth eruption, and completely lack osteoclasts as a result of an inability of osteoblasts to support osteoclastogenesis (Kong et al., 1999). RANKL was found to be expressed in activated T-cells, lymph nodes, the spleen, the thymus, intestinal lymphoid patches, and immature thymocytes (Anderson et al., 1997; Wong et al., 1997). Interestingly, RANKL-deficient mice lack all lymph nodes but have normal splenic structure and Peyer's patches (Kong et al., 1999). OPG inhibits osteoblast–osteoclast interaction, and OPG treatment reverses bone loss. OPG-deficient mice display severe osteoporosis and a decrease in total bone density characterized by severe trabecular and cortical bone porosity, marked thinning of the parietal bones of the skull, and a high incidence of fractures (Bucay et al., 1998). In a CIA model in rats, OPG effectively prevented joint destruction, even though it had no impact on the inflammatory aspects of the disease. In addition, OPG depleted osteoclast numbers by over 75% and diminished bone erosion scores by over 60% (Romas et al., 2002).

Bone cells can produce several growth factors that affect bone formation or resorption, such as insulin growth factor 1 (IGF-1), fibroblast growth factor 2 (FGF-2), platelet-derived growth factor (PDGF), vascular endothelial growth factor (VEGF), and TGFβ (Raisz, 1999a). The expression of VEGF in periodontal disease and after COX-2 inhibition, using meloxicam, was investigated in Wistar rats (Oliveira et al., 2008). VEGF expression was significantly higher at diseased sites than at healthy sites. A reduction in alveolar bone resorption was observed after meloxicam treatment with a positive correlation between COX-2 and VEGF mRNA in gingival tissues and periodontal disease. Meloxicam significantly reduced the increased mRNA VEGF expression in diseased tissues after 14 days of treatment. Therefore, VEGF plays an important role in the progression of periodontal disease, and COX-2 inhibitor therapy can modify the progression of experimentally induced periodontitis in rats by reducing VEGF expression and alveolar bone loss (Oliveira et al., 2008).

PGs play an important role in the pathophysiologic processes of bone metabolism. PGE_2 is the most abundant PG produced by osteoblastic cells, and plays many roles in bone metabolism. PGE_2 has been shown to add bone-to-bone envelopes and to induce woven bone formation (Jee and Ma, 1997). In addition, systemic administration of PGE_2 stimulated bone formation and increased bone mass in infants and animals (Ringel et al., 1982; Marks and Miller, 1988; Yang et al., 1993; Suponitzky et al., 1998). PGE_2 infusion in rabbits increased blood flow to the bone, bone marrow, and muscle of the lower limbs, except in the osteotomy area (Keller et al., 1991). The release of PGE_2 and PGF from the tibiae of rabbits and the surrounding muscle in vitro after fracture and pinning, or pinning alone, has been compared to the release from unoperated tissues. The fractured tibiae released significantly more PGE_2 and PGF than did the control tibiae 3 to 14 days after the operation (Dekel et al., 1981). Furthermore, PGE_2 caused a dose-dependent stimulation of callus formation in rabbits (Keller et al., 1993). In dogs, the administration of PGE_2 increased the extent of mineralization on the bone surface of healing ribs, osteoid formation in all three envelopes, and the resorption surface in the cortical endosteal envelope, indicating accelerated remodeling (Shih and Norrdin, 1986; Norrdin and Shis, 1988). In rats, continuous systemic PGE_2 administration elevated metaphyseal cancellous bone mass to 3.5-fold that of the control (Ke et al., 1992). In dogs, subcutaneous administration of PGE_2 resulted in cancellous and cortical bone formation and increased skeletal turnover (High et al., 1987; Li et al., 1990).

In an in vitro study using OCCM-30 cells (a mouse cementoblast cell line), PGE_2 promoted cementoblast-mediated cementoclastogenesis by regulating the expression of RANKL and OPG via the EP_4 pathway (Oka et al., 2007). The effects of COX-2 expression regulation on calvarial osteoblastic proliferation and apoptosis was investigated in vitro using primary osteoblasts (POBs) that were cultured from calvariae of COX-2 wild-type (WT) and knockout (KO) mice. PGE_2 production and ALP activity were increased in WT cultures compared to KO cultures. In contrast, cell numbers were decreased in WT compared to KO cells by day 4 of culturing. Proliferation, measured on days 3 to 7 of culture, was twofold greater in KO than in WT POBs. There was no significant effect of COX-2 genotype on apoptosis under basal culture conditions on day 5 of culturing. Cell growth was decreased in KO POBs by the addition of the PGE_2 agonist and increased in WT POBs by the addition of NS398, a COX-2 s-NSAID. In contrast, differentiation and cell growth in marrow stromal cell (MSC) cultures were increased in MSCs from WT mice compared to MSCs from KO mice, and exogenous PGE_2 increased cell growth in KO MSC cultures. Thus, PGs secondary to COX-2 expression decrease osteoblastic proliferation, at least in cultured calvarial mouse cells, but increase growth of osteoblastic precursors in MSC cultures (Xu et al., 2007). Physiologically, PGs may mediate the response to mechanical forces (Raisz et al., 1999b). In humans, impact loading was associated with an increased production of PG (Thorsen et al., 1996). Other possible roles for PGs in bone metabolism include stimulation of cAMP production and induction of COX-2, as demonstrated in mice osteoblasts (Sakuma et al., 2004). In pathological conditions, large amounts of PGE_2 have also been observed in chronic osteomyelitis and in osteoid osteoma, probably due to increased osteoblastic and osteoclastic activity (Plotquin et al., 1991; Mungo et al., 2002).

PGE_2 interacts with four receptor subtypes (EP_1R, EP_3R, EP_2R, and EP_4R), of which the latter two are shown to stimulate bone formation and bone resorption, as demonstrated in rat bone tissue and osteoblastic cells (Feyen et al., 1984; Kawaguchi et al., 1995; Lin et al., 1995; Jee and Ma, 1997; Wadleigh and Herschman, 1999; Suzawa et al., 2000; Weinreb et al., 2001). The skeletal phenotype of aged EP_4R-KO mice was studied and compared to the fracture healing in aged KO male mice versus WT mice (Li et al., 2005). There was no significant difference in body weight and femoral length between knockout KO and WT mice at 15 to 16 months of age. Radiographic bone analysis revealed lower bone mass in both axial and long bones of KO mice relative to WT mice. Microcomputed tomography images of the distal femurs showed thinner cortices, fewer trabeculae, and a deteriorated trabecular network in KO mice. Total bone content, trabecular content, and cortical content in the distal femur were lower in KO mice than in WT controls. Histomorphometric measurements showed that the trabecular bone volume and bone formation rate were significantly decreased, whereas osteoclast numbers on the trabecular, eroded, and endocortical surface were significantly increased in KO mice (Li et al., 2005). In addition, fracture healing was examined in KO and WT mice subjected to a transverse femoral fracture. Callus formation, the unorganized meshwork of woven bone that forms after a fracture, was significantly delayed in the fractured femurs of KO mice compared with those of WT mice. KO mice had significant decreases in total callus area, cartilaginous callus area, and bony callus area 2 weeks after a fracture. By 4 weeks, complete bony bridging was seen in WT mice but not in KO mice (Li et al., 2005). Some studies suggest that bone marrow stromal cells (BMSCs) play a major role in the anabolic effect of PGE_2 (Shamir et al., 2004). In a study in rats, a selective EP_4 antagonist, L-161,982, was used to investigate the effects on rat BMSCs in vitro and in vivo. In vitro, PGE_2 (100 nM) increased nodule formation and alkaline phosphatase (ALP) activity, a marker of bone formation, 1.5- to 2-fold in cultures of rat BMSCs. These effects were abolished by the EP_4 antagonist. Furthermore, PGE_2 increased the number of surviving adherent BMSCs by approximately 225%, and the EP_4 antagonist prevented this effect as well. The antagonist had no effect on basal levels of nodule formation and adherent cell number. In vivo, daily systemic administration of PGE_2 at 6 mg/kg for 2 weeks increased cancellous bone area (by approximately 50%) and increased nodule formation (measured as mineralized area) in ex vivo stromal cultures by approximately 50%. Preadministration of the EP_4 antagonist at 10 mg/kg abrogated the increase in bone mass as well as the increase in nodule formation. The investigators conclude that PGE_2 stimulates osteoblastic commitment of BMSCs via activation of the EP_4 receptor (Shamir et al., 2004).

The sternum bone-healing effects after ONO-4819, an EP_4 agonist, was administered in a 300-μg dose to diabetic Wistar rats was investigated. Sternum bone healing and incidence of sternal wound complications were evaluated 4 weeks after the operation. Sternum bone wound complications developed in five rats in the control group but in only one rat in the EP_4 group. Histological examination revealed an almost completely healed sternum filled with regenerated bone tissue only in the EP_4 group. Both bone mineral content and bone mineral density, as assessed with dual-energy x-ray absorptiometry, were higher in the EP_4 group than in the control group (Marui et al., 2006). Thus, EP_4 agonist accelerated sternal bone healing and decreased the incidences of sternal wound complications in diabetic rats. Another study investigated femur bone biomechanical properties of EP_4-KO transgenic female mice. EP_4-KO mice exhibited less structural (ultimate/yield load) and apparent material (ultimate/yield stress) strength in the femoral shaft and vertebral body than that in WT mice. Vertebral body stiffness and femoral neck ultimate load (structural strength) were marginally lower in the EP_4-KO than in the WT mice. In addition, the EP_4-KO mice had a smaller distal femur and vertebral bone volume/total volume trabecular thickness than that in the WT mice (Akhter et al., 2006).

Cultured calvarial cells from EP_4-KO mice had no increase in bone resorption when exposed to PGE_2 (Miyaura et al., 2000). Bone formation was induced in WT mice by infusion of an EP_4-selective agonist and not agonists specific for other EP subtypes (Yoshida et al., 2002). In a fracture-healing mouse model, COX-2-deficient ($COX^{-/-}$) mice showed delayed initiation and impaired endochondral bone repair, accompanied by a severe angiogenesis deficiency. Administration of an EP_4 agonist markedly improved the impaired healing, as evidenced by restoration of bony callus formation on day 14, a nearly complete reversal of bone formation, and an approximately 70% improvement in angiogenesis in the callus (Xie et al., 2009). In a study in 12-week-old rats, the effects of the administration of an EP_4 agonist, ONO-4819, on the healing of drill-hole injury in the femoral diaphysis bone was investiagted. The rats were injected subcutaneously with the EP_4-agonist compound at 10 or 30 μg/kg twice a day for 5, 7, 14, 21, and 28 days after the femoral bone injury was created (Tanaka et al., 2004). The regenerated cortical bone volume and content increased dose-dependently on days 14 and 21 following ONO-3819 treatment. In addition, the injured bone from the animals treated with a high dose of 30 μg/kg of ONO-3819 had higher values of osteoclast surface and expression levels of osteocalcin and tartrate-resistant acid phosphatase (TRAP) mRNAs, an iron-containing protein that is highly expressed by osteoclasts, on day 14 day after treatment, and increased expression levels of EP_4, BMP-2, and RANKL mRNAs on day 7 day (Tanaka et al., 2004). In another study in rabbits, bioactive titanium plates were inserted into the tibia bone and examined histologically and biomechanically at 4,

8, and 16 weeks. An EP_4 agonist, ONO-AE-724, was administrated systemically every 2 weeks after surgery. A low-dose group (10 μg/kg) and a high-dose group (100 μg/kg) were compared. The bonding strength of bioactive titanium in the EP_4-agonist groups was significantly higher than that in the control group at both 4 and 8 weeks, and enhanced bone remodeling and direct bonding around the bioactive titanium plates were observed only in the EP_4-agonist groups at 4 weeks. EP_4 agonist enhanced bone formation around the bioactive titanium plate and achieved early direct bone bonding (Onishi et al., 2008). An EP_2R agonist, CP-533,536, was delivered at doses of 0.3, 1 or 3 mg/kg into the bone marrow of the proximal tibial metaphysis of 6-week-old male rats by a single injection on day 1. The local anabolic bone effects were analyzed using peripheral quantitative computed tomography, bone histomorphometry, and biomechanical testing on day 7. The CP-533,536 dose-dependently stimulated local lamellar bone formation on trabecular, endocortical, and periosteal surfaces, and thus increased bone mineral content and bone strength at the injected site (Li et al., 2003).

The direct effects of selective EP_4 and EP_2 receptor (R) agonists with PGE_2 on the differentiation of cultured murine calvarial osteoblastic cells were investigated. Both the EP_4R agonist and PGE_2 increased alkaline phosphatase activity and osteocalcin mRNA levels in the osteoblastic cell cultures. The EP_2R agonist had a smaller effect. The investigators concluded that the anabolic effect of PGE_2 in calvarial osteoblastic cell cultures is largely mediated by activation of the EP_4R agonist, while activation of the EP_2R is less effective (Alander et al., 2006). In another study, the effects of the EP_2R agonist and PGE_2 on both calvarial primary osteoblasts (POBs) and marrow stromal cells (MSC) cultured from mice with deletion of one (Het) or both (KO) alleles of the EP_2R agonist were compared to their WT littermates. The 1-month-old mice used to provide cells in these studies did not show any significant differences in their femurs by static histomorphometry. EP_2R agonist enhanced osteoblastic differentiation as measured by ALP mRNA expression, and activity as well as osteocalcin, noncollagenous protein in bone extracellular matrix synthesized by osteoblastic cells, mRNA expression, and mineralization in the WT cell cultures from both marrow and calvariae. These effects were somewhat diminished in cultures from Het mice and abrogated in cultures from KO mice. PGE_2 effects were greater than those of the EP_2R agonist, particularly in POB cultures and were only moderately diminished in Het and KO cell cultures. Thus, activation of the EP_2R agonist is able to enhance differentiation of osteoblasts, and PGE_2 has an additional anabolic effect, probably mediated by the EP_4R (Choudhary et al., 2008a).

Prostaglandin E synthase (PGES) functions as the terminal enzyme in the biosynthesis of PGE_2 (Radi and Ostroski, 2007). The microsomal PGES-1 (mPGES-1) is known to play a critical role in the production of PGE_2 and is involved in various bone pathophysiological events (Yamakawa et al., 2008). Radiologic and histologic analyses were conducted on skeletons of mPGES-1-deficient mice and compared to WT mice. The mPGES-1-deficient mice had unaffected skeletal phenotypes or osteoarthritis under normal physiological conditions. In contrast, fracture healing was impaired by mPGES-1 deficiency, with half of the mice remaining in a bone fracture nonunion state after 21 days. Normal fracture healing was restored by the reintroduction of mPGES-1. Further analyses revealed impaired chondrocytes proliferation in cartilage in mice with PGES-1 deficiency, at an early stage of fracture healing (Yamakawa et al., 2008).

Collectively, there is abundant evidence to support a significant role of PGs in bone metabolism in health and disease, which can be modulated by exogenous administration of PGs. PGE_2 and its receptors, EP_2R and EP_4R, appear to be the main contributors to bone metabolism and pathophysiological mechanisms.

THE PROCESS OF BONE HEALING AND POTENTIAL ROLE OF PROSTAGLANDINS

Bone healing is a dynamic process that involves various stages in which PGs and many other factors play an important role in completing the repair process (Radi and Khan, 2005). PGs can stimulate bone formation by increasing replication and differentiation of osteoblasts (Raisz, 1999b). Several studies in infants and experimental animals receiving systemic administration of PGE_2 have demonstrated the role of PGs in bone formation (Ringel et al., 1982; Marks and Miller, 1988; Yang et al., 1993; Jee and Ma, 1997; Suponitzky and Weinreb, 1998). Therefore, before examining the effects of ns-NSAIDs and COX-2 s-NSAIDs on the skeleton, a clear understanding of the pathophysiological process of bone healing and PG contributions to this process is a critical starting point. The bone-healing process can be subdivided into three phases: inflammatory response, bone resorption, and bone production (Fig. 2-3) (Radi and Khan, 2005).

Inflammatory Response

When the local blood supply is disrupted following bone fracture, several immediate events take place at the fracture line: hemorrhage and hematoma formation, hypoxia-mediated osteocyte necrosis, and clotting with fibrin mesh formation (Probst and Spiegel, 1997; Radi and Khan, 2005). Hematoma formation develops within the fracture site as a result of red blood cell extravasation (bleeding) and occurs acutely during the first few hours and days. The main function of this fracture hematoma is to prevent further hemorrhage via activation of platelets and the coagulation pathway and favors bone repair. The fracture hematoma microenvironment is rich in PGs, kinins, and noncollagenous proteins. Loss of stability decreased local oxygen, soft tissue injury, and ruptured vessels, and necrosis at the fracture site leads to an inflammatory response. The fracture site is sealed off by the fibrin mesh, which serves as a framework for inflammatory cell (macrophages, neutrophils, lymphocytes, and mast cells) infiltration and its by-products, such as cytokines (IL-1, IL-6, IL-18, TNF) and growth factors [platelet-derived growth factor (PDG), transforming growth factor 1 ($TGF\beta_1$), fibroblast growth factor (FGF)] (Fig. 2-3A and B) (Einhorn, 1998; Radi and Khan, 2005). Platelets—disk-shaped anucleate cell fragments that are shed from megakaryocytes in the bone marrow into the bloodstream—are the first to arrive at the sites of bone fracture and they serve to control hemostasis (Probst and Spiegel, 1997). When vascular endothelium

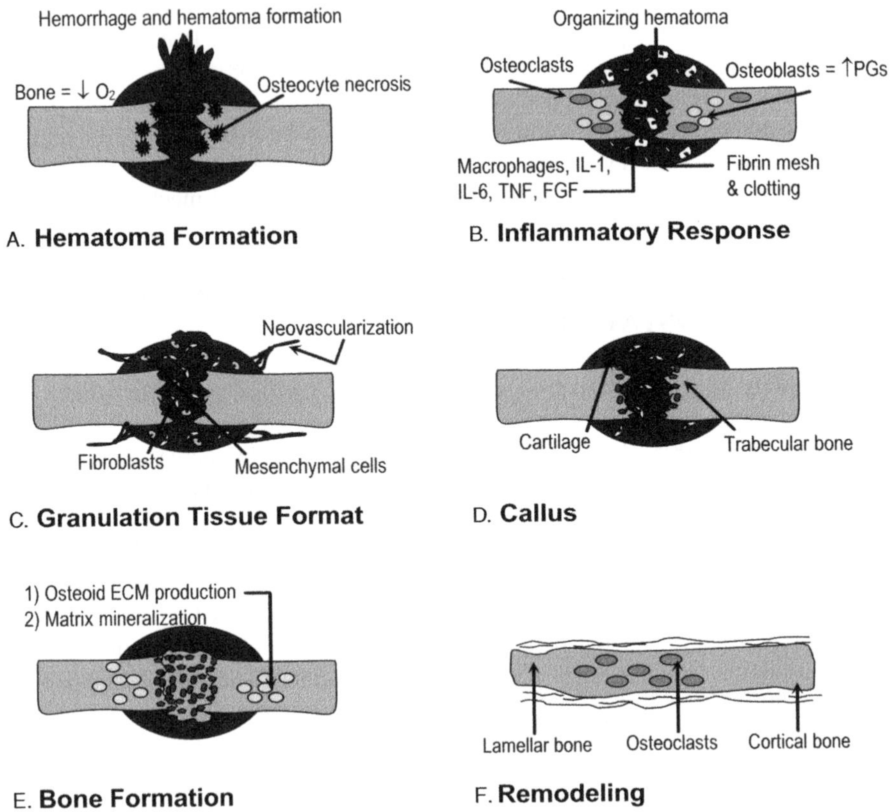

FIGURE 2-3 Sequence of events that occur during the process of bone healing. Immediately after bone fracture, there is hemorrhage with subsequent hematoma formation and osteocyte necrosis due to hypoxia (A). This is followed by an inflammatory response in which various inflammatory cells (e.g., macrophages), prostaglandins, cytokines (e.g., IL-6, TNF), and growth factors (e.g., FGF, PDGF) are released with fibrin mesh formation (B). Later, granulation tissue formation with fibroblast and mesenchymal cell migration and new capillary formation occurs (C). Bone resorption is then followed by cartilage and callus bone formation (D and E). In this process, osteoid extracellular matrix and matrix mineralization occur. Finally, bone formation and remodeling occurs (F). (Reprinted with kind permission from Springer Science & Business Media: Z. A. Radi and N. K. Khan, *Inflammation Research*, 54(9), pp. 358–366. Copyright © 2005.)

integrity is compromised and collagen in the subendothelium is exposed, platelets adhere to the exposed endothelium, a process leading to platelet activation and aggregation and release of such factors as PDGF, TGFβ, adenosine diphosphate (ADP), fibrinogen, and thromboxane A_2 (Probst and Spiegel, 1997). These factors contribute to the formation of a mature clot and stabilization of the primary hemostatic plug. In addition, this platelet aggregation process leads to increases in serum complement factors such as C5a, which possesses chemotactic activity for leukocytes leading to leukocyte infiltration into the fractured site. This process of leukocyte trafficking is mediated by various adhesion molecules on leukocytes and vascular endothelial cells (Radi et al., 2001).

The first blood leukocytes to infiltrate the fractured site are neutrophils followed by monocytes, which differentiate into macrophages, and lymphocyes (Probst and Spiegel, 1997). Macrophages release various proteolytic enzymes, extracellular matrix proteins, and growth factors such as PDGF, IGF, and FGF. FGF leads to mesenchymal cell proliferation and stimulates fibroblasts and osteoblasts. FGFR-3 is a negative regulator of bone growth (Deng et al., 1996). Genetic disruption of the murine FGFR-3 results in prolongation of endochondral bone growth with expansion of the proliferating and hypertrophic zones of the growth plate (Colvin et al., 1996). The pathogenesis of achondroplasia, the most common form of human dwarfism, is due to a mutation in the FGFR-3 gene. Furthermore, PD176067, a reversible, selective ATP-competitive inhibitor of FGF receptor tyrosine kinase, was administered to mature (11 month old) female rats by gavage for 14 days at doses of 2.5, 5, and 10 mg/kg per day. Physeal dysplasia (distal femur, proximal tibia, and sternum) occurred in all drug-treated animals and was characterized by a dose-related increased thickness of the zones of chondrocyte proliferation and hypertrophy, and marked thickening of the zone of ossification. Cartilage hyperplasia was characterized by proliferation of chondrocytes along margins of the synchondrosis and subperiosteum of sternebrae (Brown et al., 2005). The inflammatory process, which peaks by 48 h and disappears by approximately 10 days, also helps immobilize the fractured bone by eliciting pain.

Proinflammatory stimuli and cytokines and growth factors release influence PG production, especially PGE_2 (Zhang et al., 2002; Gajraj, 2003; Gerstenfeld and Einhorn, 2004). PGs are produced abundantly by osteoblasts and are increased in fracture callus during the first 2 weeks following injury (Dekel et al., 1981; Gajraj, 2003; Gerstenfeld et al., 2003a). It was demonstrated that COX-2 mRNA levels showed peak expression during the first 2 weeks of fracture healing and then returned to basal levels by 3 weeks (Gerstenfeld et al., 2003b). The gene expression and role of COX-2 and inducible nitric oxide synthase (iNOS) in growth plate injury responses were analyzed. In addition, the effects of blocking COX-2 and iNOS activities with celecoxib and aminoguanidine, respectively, were assessed 2 days prior to and until 7 days after injury in a rat tibial growth plate injury model (Arasapam et al., 2006). Up-regulation of COX-2 expression was noted on days 1 and 4 and iNOS on day 1. Histological analysis of growth plate injury sites revealed significant reductions in inflammatory cell infiltrates (particularly neutrophils) on day 1 in treated groups compared to controls. While bony tissue proportions at injury sites were unaffected by either treatment, mesenchymal tissue proportions were larger, and bone remodeling appeared delayed, with a smaller bone marrow proportion on day 14 in both treatment groups. These findings suggest that COX-2 and iNOS mediate the injury-induced inflammatory response (Arasapam et al., 2006).

Furthermore, increased expression of cartilage-related (collagen-2, collagen-10, Sox-9) and bone-related molecules (osteocalcin, cbfα-1) suggested the involvement of both endochondral and direct bone formation mechanisms during bone repair (Arasapam et al., 2006). After the initial hematoma formation, granulation tissue forms, which is characterized by neovascularization, mesenchymal cell migration, and ingrowth of fibroblasts and expression of several BMPs (Fig. 2-3C) (Radi and Khan, 2005). Hematoma and granulation tissue begin to develop into cartilaginous matrix, and chondrocyte proliferation and differentiation start to form fibrocartilage callus. In a mouse fracture-healing model, L-Sox5, Sox6, and Sox9, cartilage-related genes, were found to be involved in the activation and maintenance of chondrogenesis during fracture healing. In addition, enhanced chondrogenesis by BMP-2 was found to be mediated via an L-Sox5/Sox6/Sox9-dependent pathway (Usitalo et al., 2001).

Bone Resorption

The next phase is a proliferative and reparative phase that involves new capillary formation, a process called angiogenesis, and bony callus formation. Callus refers to an unorganized meshwork of woven bone that forms after a fracture. Several of the growth factors released by inflammatory cells contribute to the angiogenesis process. Bone is subject to continuous breakdown (resorption) by osteoclasts and rebuilding (formation) by osteoblasts in order to fulfill its functions (Radi and Khan, 2005). Osteoblasts secrete the organic matrix of bone, which is composed of approximately 90% of collagen I (Probst and Spiegel, 1997). Bone matrix contains growth factors such as IGF, PDGF, and TGFβ in high concentrations. These growth factors are likely to exert anabolic effects on growth and/or maintenance of bone and cartilage tissue (Probst and Spiegel, 1997). During bone healing, osteoclasts release proteases that dissolve bone mineral matrix and collagen, remove damaged bone, and release matrix-bound growth factors. PGs increase the number and activity of osteoclasts and subsequent bone resorption (Raisz, 1999b). PGE_2 is the most potent agonist that stimulates bone resorption and formation (Yang et al., 1993). In addition, other PGs, such as PGI_2 and TXA_2 analogs, are shown to increase the number of multinuclear osteoclasts, osteoclastic bone resorption, and rate of orthodontic tooth movement, as noted in the rat, suggesting a role for these PGs in connective tissue remodeling (Gurton et al., 2004).

Parathyroid hormone (PTH) is a potent stimulator of osteoclastic bone resorption. Osteoblasts signal to osteoclast precursors and prepare the bone surface for osteoclast attachment and initiation of the resorption process. Production of PGs in bone and subsequent action on bone cells was enhanced indirectly by the induction of COX-2 (Tai et al., 1997; Raisz, 1999b). Osteoclasts do not have receptors for PTH, but osteoblasts do. Treatment of marrow stromal cell and calvarial osteoblast cultures with PTH stimulated osteoblast differentiation when COX-2 activity was inhibited by NS-398, a COX-2 s-NSAID (Choudhary et al., 2008b). Thus, it has been suggested that COX-2 and associated PG synthesis, when stimulated by PTH, play a critical role in advancing bone resorption, while cytokines such as IL-13 and IL-4 inhibit resorption, emphasizing the role of local environment in the production of COX-2 and PGs (Onoe et al., 1996; Okada et al., 2000; Radi and Khan, 2005). In fact, the correlation between the lack of COX-2 and PTH have been investigated in COX-2-KO and WT mice. In these 10-month-old male KO mice, serum calcium and PTH, but not phosphorus, levels were increased compared with those in WT mice. 1,25-Dihydroxyvitamin D_3 levels were markedly elevated in KO mice. Skeletal analysis showed small nonsignificant decreases in cortical bone density and either an increase (distal femur, by microcomputed tomography) or no difference (distal femur, by static histomorphometry) in trabecular bone density in KO mice. There was a trend toward an increased percentage of osteoblastic and osteoclastic surfaces, and the rates of trabecular bone formation and mineral apposition were increased in KO mice relative to WT mice. However, rates of trabecular bone formation and mineral apposition were increased in 10-month-old WT females compared with males and did not increase further in female KO mice. The authors conclude that COX-2-KO mice with intact renal function have primary hyperparathyroidism, and the effects of increased PTH and 1,25-dihydroxyvitamin D_3 are to increase bone turnover, which may be compensated for by the absence of COX-2 (Xu et al., 2005). COX-2-KO mice in inbred strains can have renal dysfunction with secondary hyperparathyroidism (HPTH), making direct effects of COX-2 KO on bone difficult to assess. The effects of COX-2 KO on bone responses after 3 weeks of intermittent PTH were investigated in healthy male COX-2-KO mice. Intermittent PTH increased femoral bone mineral density and cortical bone area more in KO mice than in WT mice and increased trabecular bone volume in the distal femur in both WT and KO mice. PTH also increased serum markers of bone formation and resorption more in KO than in WT mice but increased the ratio of osteoblastic surface to osteoclastic surface only in the KO mice. In addition, PTH increased femoral mineral apposition rates and bone formation rates in KO mice more than in WT mice (Xu et al., 2010). Therefore, COX-2 deficiency leads to increased anabolic responses to PTH. Using murine MC3T3-E1 and MC-4 cells that are stably transfected with murine COX-2, a role for the calcium–calcineurin–nuclear factor of activated T-cell (NFAT) signaling pathway in the PTH induction of COX-2 has been proposed (Huang et al., 2010).

Other essential factors involved in bone resorption include the osteoclast differentiation factor, a member of the TNF ligand family that is identical to RANKL, and the previously mentioned EP_2R and EP_4R (Tsukii et al., 1998; Raisz, 1999b; Okada et al., 2003). Specifically, EP_4R is essential for osteoclast formation, as demonstrated by stimulated bone formation and prevention of bone loss upon activation (Ono et al., 2003; Yoshida et al., 2002). In addition, EP_2-and EP_4-KO mice have impaired osteoclastogenesis and bone resorption, respectively (Li et al., 2000; Miyaura et al., 2000; Sakuma et al., 2000). In a femoral fracture-healing mouse model, the effects of EP_4 agonist administration on bone healing were evaluated in either young (7 to 9 week old) or aged (52 to 56 week old) mice (Naik et al., 2009). Local administration of an EP_4 agonist to the fracture repair site in aged mice enhanced the rate of chondrogenesis and bone formation to levels observed in young mice (Naik et al., 2009). The effects of global or targeted deletion of the EP_4R on the response of osteoblasts to PG in vitro and on bone histomorphometry in aged mice were studied. The study compared WT, global heterozygote (G-HET), targeted heterozygote (T-HET), and KO mice. KO mice had one targeted and one global deletion of the EP_4R. In cultures of calvarial osteoblasts, PGE_2 increased ALP activity in cells from WT mice, and this effect was decreased significantly in cells from either G-HET or T-HET mice and decreased further in cells from

KO mice. A selective agonist for the EP_4R increased ALP activity and osteocalcin mRNA levels in cells from WT mice but not those from KO mice. A COX-2 s-NSAID, NS-398, decreased osteoblast differentiation in WT but not in KO cells. At 15 to 18 months of age there were no differences in serum creatinine, calcium, PTH, body weight, or bone mineral density among the various genotypes. Histomorphometry showed no consistent changes in bone volume or bone formation (Gao et al., 2009).

Bone Formation

Bone remodeling occurs via osteoclastic resorption of the new woven bone with subsequent osteoblastic replacement of matrix with lamellar bone (Cruess and Dumont, 1975). The steps involved in bone formation in adults include mesenchymal stem cell recruitment, proliferation, and differentiation with endochondral and intramembranous bone ossification (Bruder et al., 1994; Einhorn, 1998; Radi and Khan, 2005). Mesenchymal cells play a critical role in endochondral bone ossification. Mesenchymal cells first differentiate into chondrocytes, which subsequently undergo terminal differentiation and apoptosis, leading to calcification of the matrix. This calcified matrix lays the ground for primary bone formation by serving as a template, whereby osteoblasts deposit bone directly onto calcified cartilage (de Crombrugghe et al., 2001). In intramembranous bone ossification, the mesenchymal cells differentiate directly into osteoblasts (de Crombrugghe et al., 2001). Signaling pathway gene signatures that take place during intramembranous bone regeneration were examined in a unilateral femoral ablation model using 6-month-old Sprague–Dawley male rats (Kuroda et al., 2005). During inflammation (days 1 to 5), several genes, including COX-1 and COX-2, were down-regulated. Other proinflammatory cytokines, TNFα and IL-1β, exhibited increasing levels around day 5. During repair (days 3 to 10), growth factors, receptors, and inhibitor genes for TGFβ; FGF; BMP-2, BMP-4, and BMP-7; VEGF; and IGF-I were up-regulated. The fracture-site microenvironment is acidic and hypoxic. The pH of the fracture hematoma was shown to be acidic immediately after the trauma, becoming alkaline with ongoing fracture repair (Brighton and Krebs, 1972). Interestingly, osteoblast expression of VEGF is modulated by such an extracellular microenvironment. For example, whereas there is an increased osteoblast expression of VEGF in hypoxia (Steinbrech et al., 2001), acidic pH decreased VEGF production significantly under hypoxic conditions (Spector et al., 2000). In addition, markers of osteoblast function such as alkaline phosphatase (ALP), collagen type I, osteonectin, osteopontin, and osteocalcin exhibited peak expression at day 5 or 7. The remodeling phase (days 10 to 14) was characterized by peaks for cytokines associated with osteoclastic activity, including RANKL, cathepsin K, TNFα, IL-6, and COX-2 (Kuroda et al., 2005).

The structure of the adult skeleton is determined, in large part, by its genome (Judex et al., 2002). The differential response of trabecular bone to an anabolic (low-level mechanical vibration) and a catabolic (disuse) mechanical stimulus was evaluated in three strains of adult mice. In low-bone-mineral-density C57BL/6J mice, the low-level mechanical signal caused significantly larger bone formation rates (BFRs) in the proximal tibia, but the removal of functional weight bearing did not significantly alter the BFRs. In mid-bone-mineral-density BALB/cByJ mice, mechanical stimulation also increased BFRs, whereas disuse significantly decreased BFRs. In contrast, neither anabolic nor catabolic mechanical signals influenced any index of bone formation in high-bone-mineral-density C3H/HeJ mice (Judex et al., 2002). It is suggested that PGs release from cells at times of mechanical loading, which is a primary cellular mechanism of bone formation (Li J. et al., 2002). For example, infants with congenital heart disease who are given infusions of PGE_1 to maintain patency of the ductus arteriosus exhibit striking new periosteal bone formation (Faye-Petersen et al., 1996). Cultures of primary bone cells from the iliac crest of postmenopausal women have increased PGE_2 and PGI_2 production after fluid shear mechanical stress (Joldersma et al., 2000). As mentioned previously, fracture healing is also affected (Thorsen et al., 1996).

The role that COX-2 may play in the initiation of periosteal cortical bone healing was investigated in a mouse femoral grafting model (Xie et al., 2008). In this model, fresh femur cortical bone grafts from COX-2-KO mice were transplanted into both KO and WT mice. Similarly, grafts from WT mice were transplanted into both WT and KO mice. Histologic analyses showed that WT-to-WT transplantation resulted in normal endochondral bone healing, as evidenced by marked induction of neovascularization and periosteal bone formation on the donor graft. In contrast, transplantation of a KO graft into a KO host led to a marked, 96%, reduction in bone formation and near-elimination of donor cell–initiated periosteal bone formation. Transplantation of a WT graft into a KO host resulted in an 87% reduction of bone formation; however, when a KO graft was transplanted into a WT host, KO donor periosteal cell–initiated endochondral bone formation was restored. Histomorphometric analyses demonstrated a 10-fold increase in bone formation and three-fold increase in cartilage formation compared to KO-to-KO transplantation, suggesting that COX-2-deficient donor cells were capable of differentiating and forming bone when placed in a WT host. Thus, the authors suggested that COX-2 is critical for initiation of periosteal cortical bone healing and that elimination of COX-2 at the early stage of bone healing could lead to detrimental effects on periosteal progenitor cell–initiated cortical bone repair (Xie et al., 2008).

In summary, the process of bone formation (osteogenesis) involves three main steps: extracellular organic matrix (osteoid) production, mineralization of the matrix to form bone, and bone remodeling by resorption and reformation (Fig. 2-3D to F) (Gajraj, 2003; Radi and Khan, 2005). Bone matrix synthesis and the secretion of a collagen-rich ground substance by osteoblatsts are essential for later mineralization and are thus involved in osteogenesis (Gajraj, 2003).

COX-1 AND COX-2 EXPRESSION IN BONE, TENDON, AND LIGAMENT DURING REPAIR AND IN PATHOLOGICAL CONDITIONS

The expression of COX-1 and COX-2 in normal bone, during repair, and in pathological conditions is summarized in Table 2-1 (Radi and Khan, 2005). Studies have

TABLE 2-1 Summary of COX-1 and COX-2 Expression in Normal Bone cells, During Repair, and Under Pathological Conditions

COX-1	COX-2
Osteocytes	Canine osteosarcoma
Osteoblasts	Fracture healing
Osteoclasts	Periodontal inflammation (osteoblasts and cementoblasts)
Canine osteosarcoma	Canine pulmonary osteosarcoma
Canine pulmonary osteosarcoma	Activated human articular chondrocytes
	Pediatric human osteosarcoma

Source: Reprinted with kind permission from Springer Science & Business Media: Z. A. Radi, and N. K. Khan, *Inflammation Research* 54(9), pp. 358–366. Copyright © 2005.

shown that COX-1 is expressed constitutively by osteocytes and osteoblasts that surround alveolar bone (Mungo et al., 2002; Miyauchi et al., 2004). COX-1 was expressed constitutively in the callus and quantities remained unchanged during the healing process in a rat model of fractured femur. COX-1 and COX-2 expression were evaluated in various articular structures from 21-day-old male rat fetuses. COX-1, but not COX-2, was expressed in the anterior and posterior cruciate ligament of the knee joint and the labrum of the hip and shoulder. Both COX-1 and COX-2 were detected in the articular cartilage and joint capsule and the intraarticular disk of the temporomandibular joint and meniscus of the knee joint (Burdan et al., 2009). Oxygen plays a physiological role during bone repair and fracture sites and is generally hypoxic (Heppenstall et al., 1975). The role of hypoxia on PG release and COX expression was investigated in hypoxic osteoblasts in vitro (Lee et al., 2010). PGE_2 levels alone were significantly elevated under hypoxic conditions. COX-1 and COX-2 expression, using Western blot analysis, did not change. Treatment of cells with resveratrol, a COX-1 inhibitor, significantly reduced PGE_2 levels in media from hypoxic osteoblasts at 12 h. Furthermore, NS-398, a COX-2 s-NSAID, had no effect on PGE_2 levels from hypoxic osteoblasts at 6 h; however, treatment for 12 or 24 h significantly reduced the levels of PGE_2 released from cells. Therefore, COX may play a role in hypoxia-induced PGE_2 production (Lee et al., 2010).

COX-1 and COX-2 expression that takes place during intramembranous bone regeneration was examined in a unilateral femoral ablation model using 6-month-old Sprague–Dawley male rats (Kuroda et al., 2005). During inflammation (days 1 to 5), COX-1 and COX-2 were down-regulated (Kuroda et al., 2005). A recent study investigated COX gene expression of the healing and remodeling process initiated by fatigue loading in the rat ulna. Stress fracture healing involved direct remodeling that progressed along the fracture line as well as woven bone proliferation at the site of the fracture. At 4 days there was a significant increase in COX-1 mRNA expression (Kidd et al., 2010). At 4 h after fracture, there was a prominent peak increase in COX-2 mRNA expression (Kidd et al., 2010). The effects of metabolic acidosis on COX-1 and COX-2 expression were assessed in vitro using primary cells isolated from neonatal CD-1 mouse calvariae that were cultured in a neutral or physiologically acidic medium. The mRNA levels for COX-1 and COX-2 were measured by quantitative real-time polymerase chain reaction (RT-PCR), and the protein levels were measured by immunoblot analysis (Krieger et al., 2007). The investigators found that incubation of calvarial bone cells in an acidic medium significantly increased COX-2 mRNA and protein levels without a change in COX-1 expression levels. In addition, increased COX-2 protein levels in response to acidity were also observed in cultured calvariae bone cells (Krieger et al., 2007). Therefore, it is suggested, at least in vitro, that COX-2 expression is critical for acid-induced cell-mediated bone resorption.

COX-1, but not COX-2, was expressed in osteoblasts and osteoclasts of dog normal skeleton (Mohammed et al., 2004). COX-1 and COX-2 expression was evaluated in a naturally occurring canine model of secondary osteoarthritis (OA). Hip joint capsules (HJCs) with synovial tissue and femoral head (FH) subchondral bones were collected from normal dogs and dogs undergoing total hip replacement for coxofemoral joint OA (Lascelles et al., 2009). There was no significant difference in COX-1 expression. However, significantly more COX-1 and COX-2 were present in OA FH tissue than in normal FH tissue (Lascelles et al., 2009). Additionally, significantly more COX-2 protein was present in the OA HJCs than in normal joints (Lascelles et al., 2009). Thus, COX-2 is an appropriate target for pain management in OA. In neoplastic skeleton tumors in dogs, COX-1 was present in 8 of 13 osteosarcoma, and COX-2 was present in 3 of the 13 tumors (Mohammed et al., 2004). This suggests a role for COX-2 in skeleton neoplasms. Furthermore, COX-2 expression has been documented in other skeleton neoplastic conditions (e.g., bone sarcomas, osteoid osteomas) (Dickens et al., 2002; Mullins et al., 2004). The expression levels of COX-2 were evaluated in patients with extremity osteosarcoma. COX-2 was expressed in most of the cases. No correlation was found between COX-2 expression intensity and variables such as gender, age, anatomical site, necrosis after chemotherapy, and surgical stage. However, strong COX-2 expression was associated with low metastasis-free survival. An age older than 20 years and strong COX-2 expression independently predicted increased risk of metastasis. These data suggest that COX-2 overexpression in the primary tumor correlates with the occurrence of distant metastasis in patients with osteosarcoma (Urakawa et al., 2009). High COX-2 expression was observed in human chondrosarcomas (Schrage et al., 2010). Human osteoblastomas and osteosarcomas show high COX-2 expression (Hosono et al., 2007). Increased COX-2 expression by osteoblasts and cementoblasts has also been observed in deep periodontal tissue after application of lipopolysaccharide and in activated human articular chondrocytes (Geng et al., 1995; Miyauchi et al., 2004).

In an in vitro study using rat bone cell lines, ultrasound stimulation (US) increased COX-2 expression and PGE_2 formation, as detected by Western blotting analysis, PCR, and enzyme-linked immunosorbent assay (ELISA) (Tang et al., 2006). The signaling pathway involved in this US-induced COX-2 expression in rat osteoblasts is through up-regulation of cell membrane integrins and activation of FAK, phosphatidylinositol 3-kinase (PI3K), Akt, ERK, and NFκB. Long-term US stimulation enhanced the differentiation of osteoblasts and bone nodule formation because of the increase in COX-2 expression (Tang et al., 2006). Another study examined the effects of US on human chondrocytes and found that US stimulation transiently increased the surface expression of $\alpha2, \alpha5, \beta1$, or $\beta3$ but not $\alpha3$ or $\alpha4$ integrins in human chondrocytes (Hsu et al., 2007). US stimulation increased PGE_2 formation as well as the protein and mRNA levels of COX-2. Antibodies against either anti-integrin β1 and β3 or anti-integrin β1 and β3 integrin, using small interference RNA, attenuated the US-induced COX-2 expression. Integrin-linked kinase (ILK) inhibitor (KP-392), Akt inhibitor, NF-κB inhibitor (PDTC), or IκB protease inhibitor (TPCK) also inhibited the potentiating action of US. Ultrasound stimulation promotes kinase activity of ILK and the phosphorylation of Akt. In addition, US stimulation also induces IKKα/β phosphorylation, IκBα phosphorylation, IκBα degradation, p65 phosphorylation at Ser276, p65 and p50 translocation from the cytosol to the nucleus, and κB-luciferase activity. The binding of p65 to the NFκB element, as well as the recruitment of p300 and the enhancement of p50 acetylation on the COX-2 promoter, were enhanced by US (Hsu et al., 2007). Therefore, US stimulation increases COX-2 expression, at least in human chondrocytes in vitro, via the integrin/ILK/Akt/NF-κB, and p300 signaling pathway. COX expression after exposure of human primary bone cells to mechanical stress was investigated in vitro. Cells were obtained from the iliac crest of elderly (56 to 80 year-old) women. Increased COX-2, but not COX-1, mRNA expression was observed in this study (Joldersma et al., 2000).

In a femoral fracture-healing mouse model, COX-2 expression was evaluated in either young (7- to 9-week-old) or old (52- to 56-week-old) mice (Naik et al., 2009). COX-2 expression in young mice peaked at 5 days, coinciding with the transition of mesenchymal progenitors to cartilage and the onset of expression of early cartilage markers. COX-2 was expressed primarily in early cartilage precursors. COX-2 expression was reduced by 75% and 65% in fractures from aged mice compared with young mice on days 5 and 7, respectively (Naik et al., 2009). At the site of bone fracture, COX-2 mRNA was elevated, with peak levels expressed during the first 14 days of healing followed by return to basal level at 21 days and complete normalization by 42 days (Gerstenfeld et al., 2003a). These results are consistent with other data and suggest that COX-2 is important in early or activated stages of healing, but probably exhibits little effect on the long-term healing process (Gerstenfeld et al., 2003a).

COX-1 and COX-2 mRNA expression was evaluated in the human patellar tendon in rest and after exercise. Patellar tendon biopsy samples were taken from six people (three men and three women) before and 4 h after a bout of resistance exercise (RE) (three sets of 10 repetitions at approximately 70% of one-repetition maximum) and from a separate group of six (three men and three women) before and 24 h after RE. Samples were analyzed by RT-PCR. COX-1's were the most abundant COX mRNAs before exercise and remained unchanged after exercise. COX-2 was also expressed in tendon tissue at rest and was unchanged after exercise (Trappe et al., 2008). Application to human periodontal ligament cells of a compressive force of 2 g/cm^2 for 3 to 48 h significantly stimulated COX-2 expression in both time- and age-dependent manners. These increases in COX-2 expression were dramatically larger in periodontal ligament cells obtained from human donors over the age of 35, thus implicating aging as a factor that stimulates COX-2 expression and PGE_2 after compressive force application to such cells (Mayahara et al., 2007). Another in vitro study investigated the effects of applying mechanical force on human periodontal ligament cells COX-2 expression. Cyclic tension force application to human periodontal ligament cells resulted in PGE_2 release to the culture medium in a time-dependent manner. Although hardly detectable in controls, COX-2 mRNA expression increased dramatically on days 3 and 5 in response to tension force. In contrast, COX-1 mRNA levels detected in controls were not affected by tension force. As assessed by immunohistochemistry, COX-2 protein was increased significantly by tension force around the unstained cell nucleus in a time-dependent manner. When NS-398, a COX-2 s-NSAID, was added to the medium, PGE_2 synthesis increased by tension force was completely inhibited (Shimizu et al., 1998). In a rat supraspinatus tendon overuse injury model, mRNA expression levels of COX-2 were initially increased at 3 days, decreased by 1 week, peaked at 8 weeks, and decreased by 16 weeks of overuse (Perry et al., 2005).

In a study in rabbits, the expression of genes associated with tendon degeneration after static loads was studied. Flexor tendons were obtained from New Zealand white rabbits and secured in a tissue-loading system. A static load of 0, 2, 4, or 6 MPa was applied to tendons for 20 h, and COX-2 mRNA expression was measured by reverse transcription–polymerase chain reaction (RT-PCR). COX-2 expression was increased only at the high stress load of 6 MPa (Asundi and Rempel, 2008). COX-2 expression was evaluated in the rotator cuff tear in rabbits. A rotator cuff tear was surgically created and the expression of PGE_2 and COX-2 was analyzed. In the supernatant of the tissue culture of the torn tendon, PGE_2, the mediator of pain and the product of COX-2, was detected. The production of PGE_2 increased significantly 7 days after injury, and then decreased by day 21. RT-PCR analysis confirmed the mRNA expression of COX-2 in the torn tendon. In addition, immunohistochemical study demonstrated that cells in the tendon stump were immunopositive for COX-2. Furthermore, in the joint affected, articular chondrocytes in the remote area from the tear had strong COX-2 expression (Koshima et al., 2007). Using a rabbit model of flexor tendon injury, COX-2 mRNA levels were assessed by RT-PCR at 3, 6, 12, and 24 days after injury. Increased levels of COX-2 mRNA were detected in both tendon and tendon sheath following injury in a time-dependent manner. COX-2 was increased predominantly in the tendon (Berglund et al., 2007).

The effects of cast immobilization on COX-2 expression were studied after rat Achilles tendon rupture. RT-PCR was performed at 8 and 17 days postrupture to assess COX-1 and COX-2 in the healing region. At 8 days postinjury, tendon mRNA levels were comparable in both groups. However, by day 17, COX-1 mRNA levels in the mobilized group increased significantly. Corresponding mRNA levels in the immobilized group decreased during the same period. There were no significant differences in the expression of COX 2 among the various groups, indicating that injury-associated expression of these molecules is not overtly influenced by loading (Bring et al., 2009).

In summary and consistent with other organ systems, COX-1 expression is generally observed in normal bone, tendon, and ligament. COX-2 expression is increased during the initial stages of the bone repair process or inflammatory conditions or skeleton neoplasms.

EFFECTS OF ns-NSAIDs ON BONE HEALING

Nonselective NSAIDs are widely used in patients with bone and joint pathology and acute musculoskeletal injury (Brooks and Day, 1991; Vaile and Davis, 1998). The use of NSAIDs can have therapeutic benefits in osteoporosis and bone loss prevention in older women (Bauer et al., 1996). For example, in an in vitro study with or without basic fibroblast growth factor (b-FGF) stimulation, NSAIDs (indomethacin and etodolac) have been shown to reduce osteoclast formation and bone resorption (Bell et al., 1994; Hurley et al., 1998; Kawaguchi et al., 2000; Raisz et al., 2001). Furthermore, several studies have demonstrated that NSAIDs can be used for the prevention of heterotopic ossification following joint replacement and as a prophylactic against heterotopic bone formation in humans (Wahlstrom et al., 1991; Kjaersgaard-Andersen et al., 1993; Moed et al., 1994; Burssens et al., 1995; Persson et al., 1998; Vielpeau et al., 1999; Wurnig et al., 1999; Burd et al., 2003; Grohs et al., 2007; Saudan et al., 2007) and in nonclinical animal models (Moed et al., 1994; Radi and Khan, 2005; Rapuano et al., 2008).

Variable results were noted when NSAIDs evaluated in nonclinical models of bone healing following fracture, osteotomy, or other surgical orthopedic procedures. For example, impaired bone healing was observed in rats and rabbits after treatment with aspirin, diclofenac, ibuprofen, indomethacin, and tenoxicam (Bo et al., 1976; Sudmann and Bang, 1979; Sudmann et al., 1979; Allen et al., 1980; Tornkvist et al., 1980; Lindholm and Tornkvist, 1981; Sato et al., 1986; Keller et al., 1987; Engesaeter et al., 1992; Hogevold et al., 1992; Obeid et al., 1992; Altman et al., 1995; Glassman et al., 1998; Bichara et al., 1999; Beck et al., 2003; Giordano et al., 2003; Jaffre et al., 2003; Riew et al., 2003; Brown et al., 2004). In those nonclinical models the bone effects included inhibition of haversian (cortical bone) remodeling, alterations in spinal fusions, poor mineralization of bone matrix, decreased osteoid formation and callus bone formation, decreased bone mineral content, reduced mechanical strength of replacement tissue, decreased amounts of cortical bone, and inhibition of new bone formation (Bo et al., 1976; Lindholm et al., 1981; Keller et al., 1987; Lane et al., 1990; Obeid et al., 1992; Dimar et al., 1996; Glassman et al., 1998; Ho et al., 1998; Ankarath et al., 2003; Seidenberg and An, 2004). One of the earliest experiments of ns-NSAID effects on bone healing took place in the 1980s. Allen et al. (1980) examined the effects of indomethacin and aspirin in 45-day-old male CD rats by subjecting them under anesthesia to fractures of the right ulna and radius. Indomethacin was given at a 2- or 4-mg/kg per day dose and aspirin at a 100-, 200-, or 300-mg/kg per day dose for 21 days. After 21 days, the rats were euthanized and histological sections of the fracture sites were examined in a blinded fashion to evaluate the effects on healing. The results showed a drug- and dose-related retardation of fracture healing which was statistically significant at all dose levels of indomethacin and at the highest level of aspirin (100, 200, and 300 mg/kg per day), as compared to controls. In another study in Wistar rats, the short-term (up to 3 days) effects of indomethacin therapy at a 2-mg/kg per day dose on the healing of intramedullary pinned osteotomies of the femur were investigated. The indomethacin-treated rats showed statistically significant decreases in bone strength, absorbed energy, and bending rigidity when tested mechanically in cantilever bending (Hogevold et al., 1992). Similarly, Altman et al. (1995) studied the effects of ibuprofen at a 30-mg/kg per day dose and indomethacin at a 1-mg/kg per day dose administered orally for up to 12 weeks to rats with femur fractures. Both ibuprofen and indomethacin retarded fracture healing, with significant differences in mechanical healing found between the control and experimental groups after 10 weeks of drug administration. Both drugs also induced qualitative histological changes manifested by delayed maturation of callus, which was noticeable earlier than the difference found by mechanical testing of bone (Altman et al., 1995). In older rats (six to nine months), high doses of indomethacin were associated with histological evidence of increased fibrogenesis and decreased osteogenesis and remodeling of bone fractures (Elves et al., 1982). Indomethacin treatment was associated with increased collagen synthesis in fracture callus in rats as assessed in vitro via [^{3}H]proline incorporation (Ro et al., 1978).

The effects of ibuprofen on the healing and remodeling of bone and cartilage in the temporomandibular joint of rabbits were investigated in a 4-week study. Three groups of rabbits were operated on to create a groove and a hole in the articular surface of both the right and left mandibular condyles. Following surgery, one group was used as a control and did not receive any medication, a second group was given a daily dose of 17 mg/kg of ibuprofen, and the third group was given a daily dose of 34 mg/kg of ibuprofen. A highly significant difference in the healing of bone and cartilage as assessed clinically and histologically was observed between the control group and the ibuprofen-treated group (Obeid et al., 1992). The skeletal abnormalities were assessed after ibuprofen oral administration at doses of 180, 60, 20, and 7.5 mg/kg per day to pregnant rats from days 1 to 20 of pregnancy and from day 1 of pregnancy to parturition (Adams et al., 1969). Although different types of skeletal abnormalities were found, these generally fell within the accepted range of normal variation and were similar to those of untreated control rats (Adams et al., 1969). In another study, the prenatal skeletal developmental effect of several ns-NSAIDs (tolmetin, ibuprofen, piroxicam) and a COX-2 s-NSAID, DUF, were investigated after intragastrical administration to pregnant Wistar rats from days 7 to 21 of gestation. The initial dose was set at 8.5 mg/kg per dose for tolmetin and ibuprofen and at 0.3 and 0.2 mg/kg per dose for piroxicam and DFU, respectively. The middle dose was increased 10 times. The highest dose, except for ibuprofen, was increased 100 times. The highest dose for ibuprofen was set at 200 mg/kg. Tolmetin and ibuprofen were administered three times a day. Piroxicam and DFU were dosed once daily. The epiphysis (epiphyseal chondrocytes of long bones and menisci), as well as cells in periosteium, articular capsule, and ligaments of the 21-day-old fetuses, had normal physiological expression of COX-1 and COX-2, as well as cathepsin K; collagen types I, II, and X; osteopontin; osteocalcin; and TNFα. Increased developmental skeletal variation was noted in groups exposed to the highest dose of these ns-NSAIDs. Bone variations, including decrease of bone mineralization, limited to various parts of the axial skeleton and metacarpal bodies, were noted. Occasionally, increases in primary ossification centers were seen in fetuses obtained from two litters exposed to the lowest dose of tolmetin and DFU. Unlike nonselective NSAIDs, DFU caused a significant decrease of fetal length only. Unlike the increased number of skeletal variations observed in fetuses exposed to the highest doses of ns-NSAIDs, both COX-2 s-and ns-NSAIDs did not disturb joint formation and morphology of femoral epiphyses when administered even at high maternal toxic doses. No histological differences in joint formation were noted in the NSAID-treated group compared with controls (Burdan et al., 2005).

The postnatal developmental effects of piroxicam were assessed in rabbits and rats after oral administration at 2, 5, and 10 mg/kg per day. No adverse-effect postnatal developments were observed (Perraud et al., 1984). Etodolac was tested in a 12-week-old Wister rat fracture model by altering the period of administration from early to late. After closed fractures had been created in the middle of femoral shafts, a 20-mg/kg per day dose of etodolac was administrated intraperitoneally in three ways: Group I received it for 3 weeks, group II for just the first week after operation, and group III for just the third (final) week. Bone maturation was estimated by a radiographic scoring system and mechanically by a three-point bending test. In both the radiographic and mechanical studies, groups I and II showed lower scores, indicating that even a short period of administration of etodolac in the early phase of fracture healing creates a risk of delayed healing (Endo et al., 2005). The effects of several NSAIDs—diclofenac sodium, piroxicam, parecoxib, lornoxicam, meloxicam, ketoprofen, and ketorolac—on bone marrow mesenchymal stem cell (MSC) proliferation and osteogenic and chondrogenic differentiation were investigated in vitro using trabecular bone MSC from healthy volunteers (Pountos et al., 2010). It was found that none of the NSAIDs tested at a broad range of concentrations significantly affected MSC proliferation or osteogenic differentiation. However, NSAIDs affected chondrogenic potential with a reduction in soluble glycosaminoglycan content by 45% and 55% with diclofenac and ketorolac, respectively. Thus, these data suggest that NSAIDs may inhibit bone formation via blockage of MSC chondrogenic differentiation, which is an important intermediate phase in normal endochondral bone formation (Pountos et al., 2010).

Phenylbutazone (PBZ), an ns-NSAID, is often used in horses for the management of pain associated with conditions such as osteoarthritis. Recommendations typically include administration of PBZ at a 4.4-mg/kg dose. The effects of PBZ on bone activity and bone formation were investigated in 1- to 2-year-old healthy horses. Biopsy was conducted to obtain unicortical bone specimens from one tibia on day 0 and from the contralateral tibia on day 14. Fluorochromic markers were administered intravenously (IV) 2 days prior to and on days 0, 10, 15, and 25 after the biopsy was performed. PBZ was administered orally once at 4.4 mg/kg of body weight dose. Osteonal density and activity, the mineral apposition rate (MAR), and the percentage of mineralized tissue filling of the biopsy-induced defects in cortical bone were assessed. The MAR was decreased significantly in horses treated with PBZ. Although a regional acceleratory phenomenon was observed in cortical bone in PBZ-treated and control groups, such a phenomenon was decreased significantly in horses treated with PBZ. Osteonal activity was similar at all time points in all horses. In control horses, the percentage of mineralized tissue filling of the cortical defects was significantly greater in defects present for 30 days than in defects present for 14 days. Differences in the percentages of mineralized tissue were not detected in horses treated with PBZ. Thus, PBZ decreased the MAR in cortical bone and appeared to decrease the healing rate of cortical defects in horses (Rohde et al., 2000). In another experimental study in horses, PBZ was administered orally twice daily at a dose of 4.4 mg/kg for 3 days and subsequently at a dose of 2.2 mg/kg for 7 days to adult female horses without clinical or radiographic evidence of joint disease. Biomarkers of cartilage aggrecan synthesis (chondroitin sulfate 846) and type II collagen synthesis (procollagen type II C-propeptide) and degradation (collagen type II cleavage) were assayed. Biomarkers of bone synthesis (osteocalcin) and resorption (C-terminal telopeptide of type I collagen) were also measured. No significant differences were found between control and treatment groups or temporally for the biomarkers chondroitin sulfate 846, procollagen type II C-propeptide, collagen type II cleavage, and C-terminal telopeptide of type I collagen in serum or synovial fluid. However, a significant increase in osteocalcin concentration occurred in synovial fluid during treatment in the treated group. No treatment effect was detected for serum osteocalcin concentration. Therefore, continuous PBZ administration at recommended doses altered some biomarkers in healthy equine joints after short periods of administration (Fradette et al., 2007).

The exact mechanism of potential effects on bone healing is under debate. Studies have suggested that osteoblastic proliferation is involved, but not via inhibition of PG synthesis (Ho et al., 1998,1999). Others have demonstrated that PGE_2 is markedly reduced, as in the LPS-challenged synovial explants equine model, treated with phenylbutazone, flunixin meglumine, ketoprofen, carprofen, and meloxicam (Moses et al., 2001). Edwall et al. (1992) demonstrated in male Sprague–Dawley rats that IGF-I was increased at 8 days after tibial fracture and that this level was depressed after indomethacin administration at a dose of 3 mg/kg per day given via the intramuscular route, thus implicating IGF-I both in fracture healing and in ns-NSAID inhibitory response.

In contrast, other studies suggest that NSAIDs have no significant effect on bone repair or that their effects can be attributed only partially to the inhibition of PG synthesis (Sudmann et al., 1979, 1982; Mbugua et al., 1989; Huo et al., 1991; Reikeraas and Engebratsen, 1998; Akman et al., 2002; Seidenberg and An, 2004). For example, no effects on collagen synthesis, an important marker of bone healing, were observed in vitro in rat fracture callus (Ro et al., 1978). In a separate study, quantitative histomorphometric indices of bone formation and resorption, as well as cartilage histology, were not significantly different between rats treated with indomethacin and nontreated rats (Boiskin et al., 1988). In a rabbit distal tibia fracture model, the effects of two ns-NSAIDs, piroxicam and flunixin, on post-fracture limb swelling, joint stiffness, and torsional bone strength were examined. After 3 weeks of treatment, limb swelling in rabbits treated with low-dose piroxicam was reduced by 39%, and high-dose piroxicam reduced limb swelling by 86%. Flunixin reduced swelling by 53%. Neither piroxicam nor flunixin affected fracture healing as measured by ankle stiffness or tibia torsional strength (More et al., 1989). β-Tricalcium phosphate (β-TCP) is a biomaterial used in oral and maxillofacial surgery because it can induce a rapid proliferation of woven bone (Nyangoga et al., 2010). In a study in New Zealand rabbits, ketoprofen, a ns-NSAID, was given for 8 and 28 days to evaluate its effects on the healing of β-TCP bone graft. β-TCP induced metaplastic bone trabeculae as early as 8 days postsurgery. Ketoprofen did not affect the amount of bone formed or the number of macrophages labeled. No significant alterations in fracture biomechanics, as measured by torsion testing and fracture stage, were observed in mature (Nyangoga et al., 2010) Sprague–Dawley rats after daily oral administration of ibuprofen at a 30-mg/kg dose, starting 3 days following fracture, over a 12-week interval. Fracture histology and serum osteocalcin levels were no different in the ibuprofen-treated animals from those in controls. Furthermore, the histomorphometric parameters of bone remodeling, including bone volume and the bone formation rate in the intact tail vertebrae of these animals with unilateral femur fractures, were no different between ibuprofen-treated and control rats (Huo et al., 1991).

In Wistar rats, the left tibia were osteotomized, stabilized with a Kirschner wire, and rats were randomized into four groups (Krischak et al., 2007). Group 1 received a placebo, group 2 received a central analgesic, tramadol, at 20 mg/kg per day throughout the study; and groups 3 and 4 were treated with sodium diclofenac at 5 mg/kg per day. Group 3 received diclofenac for 7 days, followed by placebo until euthanasia (short term), and group 4 animals received diclofenac for the full period (long term). Animals were euthanized 21 days after osteotomy. All osteotomies, as assessed by histopathological examination, healed successfully and independent of the drug treatment. Delayed callus maturation, as assessed by histomorphometry, was observed in the long-term diclofenac-treated mice, with significantly higher amounts of cartilage and less bone, particularly in the outermost region of periosteal callus. Short-term diclofenac application did not significantly alter callus differentiation (Krischak et al., 2007). In a randomized placebo-controlled crossover study in mongrel dogs, indomethacin and phenylbutazone did not affect bone healing (Mbugua et al., 1989). The effect of an ns-NSAID, etodolac, on a beagle dog model of osteoarthritis in the right temporomandibular joint (TMJ) was investigated using radiology and histopathological examination 8 weeks postsurgical operation. From 2 to 6 weeks after surgery, the dogs were given 15 mg/kg of etodolac daily. The etodolac-treated dogs had fewer morphological differences between the unoperated control (left) and the operated (right) TMJ, and steoarthritic changes in the TMJ were significantly less severe in the treated group than in the untreated controls (Miyamoto et al., 2007). A middiaphyseal transverse osteotomy (stabilized with an intramedullary pin) of the right tibia was performed in each dog. The effects of carprofen administration on healing of a tibial osteotomy were investigated in dogs. Dogs received carprofen (2.2 mg/kg PO every 12 h) for 120 days. At 120 days after surgery, stiffness, elastic modulus, and flexural rigidity values in the carprofen group were significantly lower than corresponding values in the control group. Furthermore, histological evaluation revealed that the cartilage area within the callus in the carprofen group was significantly greater than that in the control group (Ochi et al., 2011). Thus, 120-day administration of carprofen appeared to inhibit bone healing in dogs that underwent tibial osteotomy. In a study of Dutch milk goats, the effects of NSAID treatment on bone graft ingrowth were evaluated. Allograft bone was obtained from the sternum of 3 donor goats; the other 27 goats were divided into three groups. The first group did not receive any NSAID, the second group received ketoprofen subcutaneously in a 2.2-mg/kg dose once daily, and the third group received meloxicam subcutaneously in a 0.5-mg/kg dose once daily. The treatment period was 6 weeks for all goats. No difference in bone in-growth in titanium bone chambers was noted, whether the goats were treated with ketoprofen or meloxicam or with medication at all. Furthermore, there was comparable in-growth of bone irrespective of whether an autograft, rinsed allograft, or irradiated rinsed allograft was used (van der Heide et al., 2008).

Clinical studies that assessed the effects of ns-NSAIDs on bone healing are few. Giannoudis et al. (2000) looked retrospectively at 377 patients treated with intramedullary nailing for femoral shaft fractures. Femur diaphysis fracture nonunion had occurred in 32 of these patients. Of those remaining, the investigators selected 67 patients of similar gender and age whose fractures had united. They then examined several differences among those patients, including postoperative ns-NSAID use, between the two groups. Significant association between nonunion and the use of ns-NSAIDs after injury and delayed healing was noted in patients who took NSAIDs and whose fractures had united. Another study looked at the association between NSAID use and the incidence of humeral shaft fracture nonunion. Of the 9995 humeral shaft fractures, 105 patients developed nonunions (1.1%), and 1032 (10.3%) were exposed to NSAIDs in the 90 days after fracture. NSAID exposure within the first 90 days was associated significantly with nonunion. Exposure to ns-NSAIDs in the period 61 to 90 days after a humeral shaft fracture was associated with nonunion (Bhattacharyya et al., 2005). However, these associations are not necessarily causal. Similarly, Butcher and Marsh (1996) retrospectively examined 94 patients who had sustained tibial fractures. The authors found that patients who received ns-NSAIDs as part of their treatment had an average length of time to union 7.6 weeks longer: from 16.7 weeks to 24.3 weeks. The Canadian Multicentre Osteoporosis Study is a longitudinal randomly selected population-based community cohort study that looked at the effects of the COX-2 s-NSAIDs, celecoxib and rofecoxib on bone mineral density (BMD). Data were from men ($n = 2004$) and postmenopausal women age 65 and older ($n = 2776$) who underwent a BMD measurement and structured interview in the fifth year of the study. To assess dose effects, rofecoxib and celecoxib doses were standardized such that starting daily doses for osteoarthritis (12.5 mg for rofecoxib and 200 mg for celecoxib) were considered equivalent. For patients taking higher doses of these medications, their dose was considered as a multiple of these starting doses. For purposes of analysis, each subject was categorized into nonuser, low-dose daily user (25 mg equivalent dose), and high-dose daily user (50 mg equivalent dose) and the relationship between dose and BMD was assessed. Persons were considered daily users of rofecoxib or celecoxib if they reported taking the medication every day. In men, daily use of COX-2 s-NSAIDs was associated with a 2.4 to 5.3% lower hip and spine BMD compared with that of nonusers. However, in postmenopausal women not using estrogen replacement therapy, daily COX-2 s-NSAID use was associated with a 0.9 to 5.7% higher BMD at most sites. COX-2 s-NSAIDs did not have an effect in women on estrogen replacement therapy (Richards et al., 2006).

This may indicate a gender difference in the effects of NSAIDs on BMD between men and women older than 65 years.

A retrospective study looked at whether patients (112 patients) with an acetabular fracture who received the ns-NSAID indomethacin for prophylaxis against heterotopic ossification (HO) were at risk of bone fractures, delayed healing, or nonunion (Burd et al., 2003). Those patients at risk for HO were randomized to receive either radiation or indomethacin; the remaining patients served as controls. The results demonstrated a higher incidence of nonunion of long bone fractures in the indomethacin group (11 of 38 patients, or 26%) versus in the nonindomethacin (radiation and control) group (5 of 74 patients, or 7%) (Burd et al., 2003). In a randomized double-blind study involving 42 postmenopausal women 52 to 79 years old with a displaced Colles' fracture, the effects of the ns-NSAID piroxicam given at 20 mg/day PO for 8 weeks on posttraumatic osteopenia reduction and recovery were investigated. The patients were treated with a below-elbow paster slab for 4 weeks after the reduction. The bone mineral content of the forearm bones was measured with a single-photon absorptiometer 8 weeks after the fracture. Piroxicam actually reduced the osteopenia caused by external fixation (Adolphson et al., 1991, 1993). No significant differences were found in the bone mineral decrease in the radius (mean of 7%) and the ulna (mean of 5%) among the patients treated with piroxicam versus 10% in the radius and 7% in the ulna in the placebo group. Furthermore, piroxicam did not decrease the rate of fracture healing. The patients who received piroxicam had significantly less pain during plaster treatment, but there was no difference in the rate of functional recovery between the groups (Adolphson et al., 1991; 1993). The effect of 1 week of treatment with naproxen on the formation of HO after cemented total hip arthroplasty was studied in a prospective trial. Twenty-seven patients received 500 mg naproxen twice daily for 7 days postoperatively and were compared with a control group of 23 patients from a previous study who had not received any type of NSAID. Three months after the operation, HO had developed in 52% patients in the control group and in 11% patients in the naproxen-treated group. One year after the operation, 17% patients in the naproxen-treated group and 52% in the control group had HO. Severe ossification developed in three patients in the control group and in none of the naproxen-treated group. Thus, naproxen given for 1 week can decrease the incidence of HO after total hip arthroplasty (Gebuhr et al., 1995). A retrospective cohort study was conducted in a general medical practice setting in the UK (using data from the General Practice Research Database). Regular NSAID users (who received three or more NSAID prescriptions) aged 18 years or older were compared with matched control patients and incidental NSAID users. The study comprised 214,577 regular NSAID users, 286,850 incidental NSAID users, and 214,577 control patients. The relative rate of nonvertebral fractures during regular NSAID treatment compared with control was 1.47 and that of hip fracture 1.08. No differences in nonvertebral fractures were found between the regular and incidental NSAID users. The rate of nonvertebral fractures among users of diclofenac and naproxen was similar to that of ibuprofen. Thus, the results of this study are not supportive of clinically significant effects of NSAIDs on bone metabolism (van Staa et al., 2000). In a case study, a patient with an ankle joint fracture-dislocation was treated preoperatively with indomethacin.

Indomethacin was found to have an inhibitory effect on bone formation (Sudmann and Hagen, 1976). The effect of 2-week flurbiprofen treatment fracture healing was investigated in patients with distal radial fracture, and no effects were noted (Davis and Ackroyd, 1988). In a retrospective observational study, ketorolac had no effect on spinal fusion when used at a dose of 30 mg IV every 6 h for 48 h in humans after spinal fusion surgery (Pradhan et al., 2008).

In summary, NSAIDs are effective anti-inflammatory drugs that have been used for the treatment of various orthopedic conditions and are not associated with adverse effects on normal bones or joints. However, the effects of these drugs on bone healing have variable findings among animal models and humans.

EFFECTS OF COX-2 s-NSAIDs ON BONE HEALING

COX-2 s-NSAIDs spare the COX-1 pathway (Fig. 2-1) (Radi and Khan, 2005). Several studies using these drugs have shown minimal effects on bone metabolism. For example, in an in vitro study using bone culture, treatment of neonatal mouse calvarial with celecoxib, a COX-2 s-NSAID, inhibited IL-1β-stimulated bone resorption completely, but not PTH-induced bone resorption (Igarashi et al., 2002). Bone formation that occurred after mechanical loading and osteoclast formation following stimulation with b-FGF was inhibited after NS-398 treatment (Forwood, 1996; Hurley et al., 1998; Kawaguchi et al., 2000). In another study, the effects of parecoxib and NS-398 on osteoclast and osteoblast differentiation and activity were tested in vitro. Cells were harvested from the tibiae and femora of 8- to 12-week-old mice. The COX-2 s-NSAIDs parecoxib and NS398 both significantly inhibited osteoclast differentiation by 93%, 94%, and 74% of the control for 100 μM indomethacin, 100 μM parecoxib, and 3 μM NS398, respectively. In addition, COX-2 inhibition reduced the resorption activity of mature osteoclasts. COX inhibition also significantly inhibited osteoblast differentiation from human mesenchymal stem cells. Simultaneously, the number of adipocytes was increased significantly. The molecular mechanism of induced adipocyte differentiation is unclear and needs further investigation (Kellinsalmi et al., 2007). A series of four inhibitors with differing selectivities for COX-1 and COX-2 each inhibited b-FGF-induced calcium release from cultured mouse calvariae with an order of potency indicating COX-2 to be the primary actor (Kawaguchi et al., 2000). However, under in vivo conditions, NS-398 did not affect new periosteal bone formation on the tibiae of rats (Forwood 1996; Li J. et al., 2002). Prostaglandin E_2 production was significantly reduced in marrow cultures from mice lacking PGHS-2 (PGHS$-2^{-/-}$). Furthermore, the decreases in PGE_2 production and osteoclastogenesis that were noted after NS-398 treatment of bone marrow cell cultures were reversible upon addition of PGE_2 to cultures (Okada et al., 2000). In female Sprague–Dawley rats, the effects of celecoxib on closed femoral fracture healing and PGs production was studied. Celecoxib treatment at a dose of 4 mg per/kg/day reduced fracture callus PGE_2 and $PGF_{2\alpha}$ levels by more than 60% (Simon and O'Conner, 2007).

Experimental data generated in COX-2-KO mice are unreplicatable in humans or other animal models, and is, therefore, questionable (Radi and Khan, 2005). Effects on bone fracture healing that were observed in COX-2-KO models included impaired fracture healing and decreased bone density, reduced bone resorption and osteoclast formation, and lack of inflammatory response (Okada et al., 2000; Simon et al., 2002). This inhibition of fracture healing was correlated with the reduced capacity of COX-2-KO mice to produce differentiated bone cells. Therefore, the relation between COX-2 gene ablation in mice and the clinical effects of pharmacological inhibition of COX-2 on bone healing remains undefined (Zhang et al., 2002). The effects noted in COX-2-KO mice included alterations of intramembranous and endochondral bone formation, suggesting a possible role of COX-2 on osteoblasts and osteoclast formation, and osteoclastogenesis, as indicated by in vitro bone marrow cell cultures (Li et al., 2000; Zhang et al., 2002; Ono et al., 2003). Bone morphology in multiple trabecular and cortical regions within the distal and diaphyseal femur of 4-month-old WT and KO (COX-$2^{-/-}$) male and female mice was investigated using microcomputed tomography. COX-$2^{-/-}$ female mice had normal bone geometry and trabecular microarchitecture at four months of age, whereas the male KO mice displayed reduced bone volume fractions within the distal femoral metaphysis. Furthermore, male COX-$2^{-/-}$ mice had a significant reduction in cortical bone mineral density within the central cortical diaphysis and distal epiphysis and metaphysis. Biomechanical testing via four-point bending showed that male COX-$2^{-/-}$ mice had a significant increase in postyield deformation, indicating a ductile bone phenotype in male COX-$2^{-/-}$ mice (Robertson et al., 2006). However, bones of COX-2-KO mice exhibited no gross or histological bone abnormalities in a separate study (Okada et al., 2000). The effect of COX-2 on bone response after placement of titanium implants in the middle of the femurs of 9-week-old male WT (COX-$2^{+/+}$) and COX-$2^{-/-}$ mice was investigated (Chikazu et al., 2007). Mice were evaluated at 1, 2, 4, 7, and 56 days after implantation. COX-2 and osteocalcin expression was induced in bone surrounding implants in COX-$2^{+/+}$ mice but not in COX-$2^{-/-}$ mice. In cortical bone, the implant surface was in direct contact with newly formed bone lamellae in COX-$2^{+/+}$ mice, and new bone formation was minimal in COX-$2^{-/-}$ mice (Chikazu et al., 2007). Thus, the authors suggested that COX-2 may interfere with osseointegration of dental implants.

Similar to studies with ns-NSAIDs, results on the effects of COX-2 on bone fracture healing with COX-2 s-NSAIDs have been variable (Tables 2-2 and 2-3) (Radi and Khan, 2005). Studies in rats and rabbits have suggested that celecoxib and rofecoxib impaired bone healing, as evidenced by delayed fracture healing, reduced intramembranous and endochondral bone formation, decreased osteoblastogenesis and osteoclastogenesis, persistence of mesenchyme and cartilage in the callus, reduction or absence of cartilage in the fracture callus, and reduced bony in-growth (Elder et al., 2001, Zhang et al., 2001, 2002; Goodman et al., 2002; Simon et al., 2002; Radi and Khan, 2005; Radi, 2009). However, celecoxib did not significantly inhibit the rate of spinal fusion in rats, rabbits, or humans, whereas indomethacin showed biomechanical and radiographic evidence of delayed healing at 4 weeks postsurgery (Long et al., 2002; Brown et al., 2004). PGs regulate bone

TABLE 2-2 Variable Nonclinical Findings on Bone Healing After Celecoxib Treatment

Animal model	Celecoxib daily dose (mg/kg)	Bone model	Post-fracture dosing duration	Post-fracture healing assessment duration	Findings vs. controls	Reference
Rat Male	3	Femur fracture	4, 8, 12 weeks	4, 8, 12 weeks	Celecoxib did not delay healing	Brown et al., 2004
Female	3, 6	Femur fracture	10 days	8 weeks	Significant ↓ in fracture healing	Bergenstock et al., 2005
Mouse (male)	10, 50	Tibia fracture	4, 8, 12 weeks	4, 8, 12 weeks	No significant differences in biomechanical, biochemical, or histological parameters	Mullis et al., 2006
Dog (female)	100, 200	Knee osteoarthritis (OA)	15 weeks after knee surgery	15 weeks after knee OA surgery	No detrimental effects on proteoglycan turnover and content of osteoarthritic cartilage	Mastbergen et al., 2006
Mouse	25	Bone allograft healing	2, 5 weeks after bone allograft	2, 5 weeks after bone allograft	↓ new bone in-growth	O'Keefe et al., 2006
Rat Female	2, 4, 8	Femur fracture	5, 15 days	8 weeks	Significant ↓ in fracture callus and ↑ nonunions at all doses	Simon and O'Conner, 2007
Male	5 to 30	Femur fracture	2, 4, 8, 12 weeks	2, 4, 8, 12 weeks	↓ biomechanical strength	Herbenick et al., 2008

TABLE 2-3 Variable Nonclinical Findings on Wound Healing After Celecoxib Treatment

Animal model	Celebrex daily dose (mg/kg)	Wound model	Post-injury dosing duration	Post-Injury healing assessment duration	Findings vs. controls	Reference
Mouse	7.5, twice a day	*ob/ob* diabetic	8 days	8 days	No significant effects	Kämpfer et al., 2005
Rat Male	10	Rotator cuff repair	2, 4, 8 weeks	2, 4, 8 weeks	Significant ↓ in tendon-to-bone healing	Cohen et al., 2006
Female	10	Patellar tendon transection	14 days	14 days	↓ failure load and healing strength	Ferry et al., 2007
Male	1500 ppm	Fasciocutaneous flap survival, revascularization and healing	12 hrs	12 h	No effects on healing	Wax et al., 2007
Mouse	20	Hemophilia B	2, 4, 6, 8, 10, 12, 15 days	2, 4, 6, 8, 10, 12, 15 days	No delay in cutaneous wound healing	Hoffman et al., 2009

formation and resorption via their direct actions on both osteoblasts and osteoclasts, and the effects of various mediators of bone cell functions are enhanced via the induction of COX-2 in these cells. PGE_2 is the most abundant PG produced by osteoblasts, which plays a critical role in bone-to-bone formation, woven bone formation, and bone resorption, primarily through the PGE_2 and PGE_4 (EP_2 and EP_4) receptor subtypes (Radi and Khan, 2005; Radi et al., 2009). The mechanism of COX effects have not been established definitively, but may be related to marked reductions in PGE_2 and osteoblastic proliferation with or without PG inhibition (Radi and Khan, 2005; Radi, 2009). Nonclinical studies have suggested that celecoxib may have an effect on bone and wound healing, as evidenced by delayed bone fracture and wound healing (Tables 2-2 and 2-3). Brown et al. (2004) examined the effects of the daily administration (beginning 1 day after fracture) of celecoxib (3 mg/kg per day) for up to 12 weeks on femur fracture healing in adult male Wistar rats. The data from this study indicated that celecoxib did not delay healing at either 4, 8, or 12 weeks following fracture. However, histological analysis of bone fractures revealed the presence of more fibrous tissue (fibrous healing) and less woven bone formation (endochondral and immature bone formation) at 4 and 8 weeks in the celecoxib-treated rats, but mechanical strength and radiographic signs of healing were similar to those in the control rats. Bergenstock et al. (2005) measured fracture healing at 8 weeks in female Sprague–Dawley rats treated once daily for 10 days after fracture with celecoxib at 3 and 6 mg/kg per day. Celecoxib treatment caused a significant decrease in fracture healing (Bergenstock et al., 2005). Mullis et al. (2006) did a bone biomechanical, biochemical, and histological assessment at 4, 8, and 12 weeks after fracture creation in the tibiae of male C57BL/6N mice given dosages of 10 and 50 mg/kg per day of celecoxib for up to 12 weeks. No differences in biomechanical, biochemical, or histological parameters were noted in the celecoxib-treated mice. The effects of low-intensity pulsed ultrasound (LIPUS) and celecoxib on ulna stress fractures were investigated in rats. Animals were treated for 5 days per week with celecoxib at 5 mg/kg. One to three hours following drug administration, all animals were treated with unilateral active-LIPUS, and the contralateral inactive-LIPUS and ulnas were evaluated histologically at 2, 4, and 8 weeks following induction of stress fractures. Neither LIPUS nor celecoxib influenced bone resorption, but each had significant and opposite effects on the intracortical bone formation rate. There was no interaction between LIPUS and celecoxb, indicating that the beneficial LIPUS effect was not mediated by the cyclooxygenase-2 pathway. LIPUS accelerated stress fracture healing, whereas celecoxib delayed repair. When used in combination, the beneficial LIPUS effect was not impaired by celecoxib (Li et al., 2007).

In a study by Simon and O'Connon (2007), the authors examined the effects of celecoxib on femur fracture following daily administration (before and after fracture) of celecoxib at 2, 4, and 8 mg/kg per day in rats for at least 8 weeks. Beginning 4 h after fracture and continuing for 5 or 15 days, the results showed that all doses of celecoxib reduced the mechanical properties of the fracture callus and caused a significant increase in the proportion of nonunion fractures. Conversely, when celecoxib treatment was given prior to fracture or 14 days (delayed) after fracture, no significant decline in the fracture-healing process was noted. In another study, structural femoral bone allograft healing was delayed in C57BL/6 mice given celecoxib at doses of 25 mg/kg per day for 2 or 5 weeks (O'Keefe et al., 2006). Delayed ulna stress fracture repair was noted in male rats following treatment 5 days per week with 5 mg/kg of celecoxib for up to 8 weeks (Li et al., 2007). Herbenick et al. (2008) found that daily supplementation to male Sprague–Dawley rats of celecoxib, at a mean dose of 3.2 mg/day, for up to 12 weeks decreased femur biomechanical strength but did not affect callus formation, and the effect on the strength of fracture callus was most pronounced in the early inflammatory phase (Herbenick et al., 2008). Celecoxib at 4 mg/kg administered daily for up to 2 weeks following femur fracture in female Sprague–Dawley rats resulted in the reduction of the callus cell proliferation rate at day 4 but not at day 2, 7, or 10 (Cottrel and O'Conner, 2009). However, celecoxib 10 mg/kg did not significantly inhibit the rate of spinal fusion in rabbits (Long et al., 2002). In addition, there were no detrimental effects of celecoxib on proteoglycan turnover and the content of osteoarthritic cartilage in a female beagle dog osteoarthritis model when celecoxib was administered at 100 or 200 mg/kg per day for at least 15 weeks after knee surgery (Mastbergen et al., 2006).

The effects of celecoxib on fracture healing were compared to acetaminophen. Mid-diaphyseal femur fractures were produced in female Sprague–Dawley rats, and the rats were given once-daily oral dosing of 60 or 300 mg/kg of acetaminophen or 3 or 6 mg/kg of celecoxib for 10 days after fracture. Fracture healing was measured after 8 weeks by radiographic examination, mechanical testing, and histology. Radiographic scoring revealed that acute celecoxib treatment impaired fracture healing significantly, whereas acetaminophen treatment had no effect. Mechanical testing supported the radiographic observations. No negative effects of celecoxib or acetaminophen treatment on the structural properties (peak torque and torsional rigidity) of the healing were detected. Celecoxib treatment, but not acetaminophen treatment, significantly reduced the material properties (maximum shear stress and shear modulus) of the healing femurs. Postmechanical testing examination of the healing femurs found that 73% of the vehicle- or acetaminophen-treated femurs had healed as unions, 27% failed as incomplete unions, and none failed as nonunions. In contrast, only 21% of the fractured femurs from the celecoxib-treated rats had healed as unions, 53% failed as incomplete unions, and 26% failed as nonunions. The proportion of nonunions among the celecoxib-treated rats was significantly higher than that of the control and acetaminophen-treated rats. Histological examination indicated that celecoxib treatment altered normal fracture callus morphology in which cartilage rather than new bone abutted the fracture site (Bergenstock et al., 2005). The effects of twice-daily oral administration of celecoxib; loxoprofen, an ns-NSAID; and SC-58560, a COX-1 s-NSAID, were compared using an adjuvant-induced arthritis (AIA) rat model. Drugs were given for 10 days beginning 15 days after adjuvant injection. Celecoxib was administered at 3 mg/kg per day, eloxoprofen at 3 mg/kg per day, and SC-58560 at 10 mg/kg per day. The therapeutic effects on three-dimensional architectural bone changes under arthritic conditions (e.g., the bone volume/total tissue volume ratio and the amount of trabecular bone pattern factor) were determined using microcomputed tomography (micro-CT). In addition, dual-energy x-ray absorptiometry two-dimensional bone analysis was performed to compare with micro-CT analysis. The AIA rats had substantial inflammatory bone destruction and erosion, which allows for significant changes in the three-dimensional architectural index. This inflammatory bone destruction was potently suppressed by celecoxib, only moderately suppressed by eloxoprofen, and not at all suppressed by SC-58560 (Noguchi et al., 2008). The clinical and joint destruction (radiographic and histopathological) effects of twice-daily oral administration of 0.01 to 3 mg/kg celecoxib and 0.013 to 3 mg/kg eloxoprofen were evaluated in the AIA model. The drugs were given by twice-daily oral administration for 10 days beginning 15 days after adjuvant injection. Celecoxib significantly inhibited paw swelling, hyperalgesic response, and joint destruction (both radiographic and histopathological findings) in these arthritic rats. These bone efficacy effects of celecoxib were superior to those of eloxoprofen (Noguchi et al., 2005). In a mouse collagen-induced arthritis model, celecoxib, deracoxib, 5-lipoxygenase (5-LO), or leukotriene A4 (LTA_4) hydrolase inhibitor, administered by oral gavage twice daily in a 0.1-mL volume, when administered alone decreased severity but had little effect on disease incidence, defined as any sign of redness and swelling in the paws. However, the combination of COX-2 s-NSAIDs with 5-LO inhibitors produced significant decreases in both incidence and severity, and normal-appearing COX-2/LT inhibitor paws in all experiments were examined histologically and found to have no signs of infiltration or cartilage joint destruction (Anderson et al., 2009). Therefore, results from these nonclinical studies suggest that celecoxib may have an effect on bone healing, especially at earlier stages, but appears to bear no significant impact on the ultimate long-term outcome.

The effects of administration of rofecoxib at 12.5 mg once-daily dosing on fibula osteotomy healing in male New Zealand white rabbits was compared to administration of ibuprofen at 150 mg once a day for 28 days after surgery. Bone healing was assessed by histomorphometry at 3 and 6 weeks after osteotomy, and at 6 and 12 weeks by torsional mechanical testing (O'Connor et al., 2009). Fracture callus morphology was abnormal in the rofecoxib-treated rabbits, and torsional mechanical testing showed that fracture healing was impaired. Ibuprofen treatment caused persistence of cartilage within the fracture callus and reduced peak torque at 6 weeks after osteotomy. In the fracture allowed to progress to possible healing, nonunion was seen in some of the fibulae from the rofecoxib- and ibuprofen-treated animals (O'Connor et al., 2009). In another study in adult male New Zealand rabbits, the short-term (for 5 days) effects of several NSAIDs (indometacin, meloxicam, or rofecoxib) on healing after middiaphyseal osteotomy of the right ulna were investigated. Radiological, biomechanical, and histomorphometric evaluation were performed at 6 weeks (Karachalios et al., 2007). The incidence of radiologically incomplete union in the rofecoxib-treated rabbits was similar to that in the control group. Statistically, all the biomechanical parameters were significantly lower in the indometacin and meloxicam groups than in the control group. Only the fracture load values were found to be statistically significantly lower in the rofecoxib group. Histomorphometric parameters were adversely affected in all groups, with the specimens of the rofecoxib group showing the least negative effect. The effects of rofecoxib dosed at 3 mg/kg once a day on femoral fractures in male rats was investigated and revealed impaired fracture healing based on histopathological and radiologic observations (Simon et al., 2002). Thus, short-term administration of low therapeutic doses of a rofecoxib had a minor negative effect on bone healing (Karachalios et al., 2007). NS-398 had no effect on periosteal bone formation in the rat tibia (Forwood, 1996; Li J. et al., 2002). In a study in mice, animals underwent surgery to induce an open transverse middiaphyseal femoral fracture, which was then fixed using a custom-made external fixator, and rofecoxib was administered orally at 5 mg/kg per day. Effects on fracture healing were investigated using radiographic assessment, histological analysis, biomechanical testing, and laser Doppler flowmetry to assess blood flow across the fracture gap. Poor healing was observed by radiography only on day 32. Histological analysis demonstrated that the control animals healed more quickly (at days 24 and 32) and had more callus and less fibrous tissue (at days 8 and 32) than those of the rofecoxib-treated animals. Biomechanical testing revealed that the control animals were stronger at day 32. The rofecoxib-treated mice exhibited a lower median flow from day 4 onward that was significant at days 4, 16, and 24 (Murnaghan and Marsh, 2006). Thus, rofecoxib reduced blood flow across the fracture gap and correlated with reduced fracture-healing outcomes in mice. In a study in male Wistar rats, a closed, nondisplaced femoral fracture was induced, and rofecoxib at 8 mg/kg or ibuprofen at 30 mg/kg was provided in food for 4 weeks. Both rofecoxib and ibuprofen diminished fracture-healing parameters during the early stages of fracture repair. Compared with ibuprofen, rofecoxib was significantly more likely to produce nonunions. Gross nonunions were noted in 64.7% of rofecoxib-treated rats, 17.6% of ibuprofen-treated rats, and 0% in placebo controls. Mean callus width was 8.9 mm for rofecoxib, 8.9 mm for ibuprofen, and 8 mm for placebo controls. Mean healing maturity was 1.6 in the rofecoxib-treated group, 1.7 in the ibuprofen-treated group, and 2.7 in the placebo controls. Mean fracture angulation was 30.8° for rofecoxib, 14.3° for ibuprofen, and 13.4° for placebo. Mean histologic healing was 5.75 for rofecoxib, 6.35 for ibuprofen, and 8.25 for placebo (Leonelli et al., 2006).

The effects of an orally administered ns-NSAID, ketorolac, and a COX-2 s-NSAID, valdecoxib, on fracture healing in rats for either 7 or 21 days were assessed with biomechanical, histological, and biochemical analyses (Gerstenfeld et al., 2007). No significant effects on the rate of fracture nonunion were noted at either 21 or 35 days. The 21-day treatment produced significantly more nonunions in valdecoxib-treated rats than in either ketorolac-treated or control animals, but these differences disappeared by 35 days. The reversibility of the effects after drug withdrawal was assessed in fracture calluses. The assessment showed that ketorolac treatment led to two- to threefold lower levels of PGE_2 than those with valdecoxib. Histological analysis showed delayed remodeling of calcified cartilage and reduced bone formation in association with valdecoxib treatment (Gerstenfeld et al., 2007).

The long-term effect of intraperitoneal administration of parecoxib sodium daily at a 1.06-mg/kg dose for 7 days on middiaphyseal bone healing in rats, when applied in a pattern similar to clinical treatment patterns, that is, in a high dose and for a short period after bone fracture, was evaluated (Akritopoulos et al., 2009). Bone union and callus formation fracture healing were evaluated by x-rays at 28 and 42 days after surgery. No statistically significant differences were noted in the parecoxib-treated rats (Akritopoulos et al., 2009). In another study in female Wistar rats, parecoxib was also administered intraperitoneally at 0.05 mg per 100 g body weight for 7 days after a closed tibial fracture. At 2, 3, and 6 weeks after surgery, bone mineral density (BMD) at the fracture site was measured using dual-energy x-ray absorptiometry (DEXA). The BMD at the fracture site was calculated as the average of the results after 2, 3, and 6 weeks. Mean BMD was lower in the parecoxib group than in the control group. However, there were no statistically significant differences in mechanical properties of the healing fractures after 6 weeks (Dimmen et al., 2008). Furthermore, parecoxib at 0.05 mg per 100 g body weight and indomethacin at 0.0625 mg per 100 g body weight were administered to female Wistar rats for 7 days after being subjected to a closed tibial fracture stabilized with an intramedullary nail. Two and three weeks after surgery, the bone density at the fracture site was measured using DEXA. Parecoxib decreased BMD at the fracture site 3 weeks after fracture and indomethacin after 2 weeks. Both parecoxib and indomethacin reduced the ultimate bending moment and the bending stiffness of the healing fractures after 3 weeks (Dimmen et al., 2009). Gregory and Forwood (2007) studied woven bone in female Wistar rats. Periosteal woven bone callus was initiated in the right tibia following a single bout of four-point bending applied as a haversine wave for 300 cycles at a frequency of 2 Hz and a magnitude of 65 N. The rats were injected daily with a COX-2 s-NSAID, 5,5-dimethyl-3-3(3 fluorophenyl)-4-(4-methylsulfonal)phenyl-2(5*H*)-furanone (DFU) at 2 mg/kg starting 7 days postloading, and the tibiae were examined 2, 3, 4, and 5 weeks postloading. No significant difference in peak woven bone area was observed in the DFU-treated rats. Treatment with DFU resulted in a temporal defect in woven bone formation, where the achievement of peak woven bone area was delayed by 1 week. However, woven bone remodeling was observed in DFU-treated rats at 21 days postloading, demonstrating that remodeling of the periosteal callus is not prevented in the presence of a COX-2 s-NSAID in the rat. The effect of meloxicam on bone healing of calvarial defects was assessed in male Wistar rats. A linear incision was made through the skin of the scalp, a full-thickness flap was reflected, and a 4-mm round defect was made with a trephine drill. The animals were injected daily via the subcutaneous route with 3 mg/kg of meloxicam for 15 or 45 days. The bone filling was measured histometrically. Intergroup comparisons demonstrated that the meloxicam groups had a significant reduction in bone healing compared to their respective controls (Gurgel et al., 2005). In a rat AIA model, etoricoxib was given orally at a 0.6-mg/kg dose, twice daily, starting on day 0 and until euthanasia on day 21, and the effects on bone were evaluated and radiographs were taken from adjuvant injected and noninjected hind paws on days 0 and 21. Etoricoxib treatment resulted in a clear reduction in adjuvant-induced radiographic changes of bone and joint structures compared with the vehicle group (Riendeau et al., 2001).

In summary, nonclinical data regarding the effects of NSIADs on bone healing have been variable and results do suggest that nonselective NSAIDs and COX-2 selective NSAIDs may have some effect on bone fracture healing. Such an effect is apparent at the earlier stages of the healing process but bears no significant impact on the overall long-term healing of the fracture (Ho et al., 1998;

Gerstenfeld et al., 2003b; Brown et al., 2004; Radi and Khan, 2005; Radi 2009; Nyangoga et al., 2010).

EFFECTS OF ns-NSAIDs ON LIGAMENT AND TENDON HEALING

In comparison to bone fracture healing, little information is available to evaluate the effects of NSAIDs on ligament and tendon healing. With the exception of some in vitro data, long-term NSAID use has not been associated with adverse effects on normal tendon or ligament healing. Nonclinical models have shown variable results on the healing of these structures, both beneficial and deleterious (Radi and Khan, 2005). In an in vitro study, diclofenac and aceclofenac had no significant effect on tendon cell proliferation or glycosaminoglycan synthesis; however, indomethacin and naproxen inhibited cell proliferation within patella tendons and inhibited glycosaminoglycan synthesis in both the digital flexor and patella tendons of rabbits (Riley et al., 2001). Treatment of nonhuman primates with oral ibuprofen significantly reduced the force required for tendon gliding following flexor tendon injury and reduced the breaking strength of completely divided and repaired extensor tendons (Kulick et al., 1986). Flurbiprofen treatment following Achilles tendon transection in rats resulted in better organization of the extracellular collagenous matrix, but decreased the cross-sectional area and failure load (Yuan et al., 2003). Indomethacin, by contrast, improved tendon healing after tendon transection in rats, as demonstrated by the reduced cross-sectional area, without change in failure load (Forslund et al., 2003). Furthermore, indomethacin significantly increased tensile strength at 16 weeks after tenotomy of plantaris longus tendons in rabbits, but did not change the total collagen content (Forslund et al., 2003). Piroxicam treatment on days 1 to 6 after injury to medial contralateral ligaments of rats also resulted in a 42% increase in the strength of ligaments at day 14 postinjury (Dahners et al., 1988). In contrast, continuous treatment with indomethacin had no significant effect on mechanical performance or histologic appearance at 2 or 6 weeks after tenotomy of rabbit Achilles tendons, and treatment for 14 days with ibuprofen had no effect on mechanical strength at 14 or 28 days after injury to medial contralateral ligaments of rabbits (Carlstedt et al., 1986; Moorman et al., 1999). Additionally, neither indomethacin- nor ibuprofen-treated rabbits regained their torsional strength after 5 or 8 weeks of treatment (Thomas et al., 1991). Diclofenac had no effect on tendon healing (Tornkvist et al., 1984). From these observations, it can be concluded that treatment with NSAIDs in experimental animal models has shown variable intra- and interspecies effects (Radi and Khan, 2005). The effects of several NSAIDs (ibuprofen, acetaminophen, naproxen, piroxicam, celecoxib, valdecoxib) on patellar tendon healing were investigated in Sprague–Dawley rats that underwent transection of the patellar tendon at the inferior pole of the patella (Ferry et al., 2007). At 14 days the repair site tissue was analyzed biochemically. The control group demonstrated greater maximum load than did the celecoxib, valdecoxib, and piroxicam groups. The acetaminophen and ibuprofen groups were also significantly stronger than the celecoxib group, but not statistically different from the control group. The investigators concluded that with the exception of ibuprofen, these NSAIDs had a detrimental effect on healing strength at the bone–tendon junction, as demonstrated by decreased failure loads and increased failures of the cerclage suture. Acetaminophen had no effect on healing strength. The biomechanical properties closely paralleled the total collagen content at the injury site, suggesting that these agents may alter healing strength by decreasing collagen content (Ferry et al., 2007).

EFFECTS OF COX-2 s-NSAIDs ON LIGAMENT AND TENDON HEALING

Similar to ns-NSAIDs, COX-2 s-NSAIDs have not been widely evaluated in nonclinical models of ligament and tendon healing (Table 2-3) (Radi and Khan, 2005). In a rat model of Achilles tendon transection, parecoxib showed an adverse effect on early tendon repair (postinjury days 1 to 5), but improved tendon remodeling when given during later stages of repair (days 6 to 14) (Virchenko et al., 2004). When given throughout the healing phase, both indomethacin and celecoxib improved the outcome after Achilles tendon transection in rats (Yuan et al., 2003). In another study, celecoxib administration for 6 days after surgical transection of the right medial collateral ligament in rats was associated with a 32% lower load to failure on day 14 compared to untreated/transected ligaments (Elder et al., 2001). In the same model, SC-560, a relatively selective COX-1 inhibitor, had no effect on the strength of injured ligaments (Bogatov et al., 2003). Because an ns-NSAID was not evaluated concurrently in the celecoxib study, direct comparison of COX-2 selective versus nonselective COX inhibition was not possible (Radi and Khan, 2005). Ekman (2002) discussed several additional shortcomings associated with this investigation in a "Letter to the Editor" and suggests alternatively that celecoxib use is actually beneficial in ankle injury in humans.

Nonclinical studies on the effects of celecoxib on wound healing are highly variable and contradictory. Significant inhibition of tendon-to-bone healing was noted after celecoxib treatment in a Sprague–Dawley rat model of acute rotator cuff repair (Cohen et al., 2006). Ferry et al. (2007) examined the effects of daily administration of celecoxib, 10 mg/kg, for 14 days on patellar tendon healing in female Sprague–Dawley rats and found that celecoxib decreased healing at the bone–tendon junction. In contrast, using a diabetic *ob/ob* mouse model, Kämpfer et al. (2005) demonstrated that celecoxib, 7.5 mg/kg, given twice daily for at least 8 days beginning at day 5 following wounding, did not impair wound healing. In an in vitro cell proliferation study, celecoxib administration was shown to result in a dose-dependent inhibition of rat Achilles tendon cell proliferation and migration; however, no effects on types I and II collagen expression were observed (Tsai et al., 2007). Celecoxib, as a supplement to food at a dose of 1500 ppm, showed no deleterious effects on fasciocutaneous flap survival, revascularization, and healing for up to 12 h postdose in a Sprague–Dawley male rat model (Wax et al., 2007). Celecoxib therapy may inhibit adhesion formation during tendon healing, with subsequent improvement in functional recovery. Another study found that celecoxib, at a higher dose, 20 mg/kg, for up to 15 days does not delay cutaneous wound healing in a hemophilia B mouse model (Hoffman et al., 2009). In a study in adult rats, bilateral transection of the knee medial collateral ligaments was performed and animals were treated 5 days per week with celecoxib at 5 mg/kg. One to three hours after drug administration, all animals were treated with unilateral active low-intensity pulsed ultrasound (US) and contralateral inactive low-intensity pulsed US. Equal numbers of animals were tested mechanically at 2, 4, and 12 weeks after injury. US and celecoxib treatment did not influence ligament mechanical properties at any time point. After 2 weeks, ligaments treated with active low-intensity pulsed US were 34.2% stronger, 27% stiffer, and could absorb 54.4% more energy before failure than could ligaments treated with inactive low-intensity pulsed US, whereas ligaments from the celecoxib-treated group could absorb 33.3% less energy than could ligaments from the control group. There were no US or celecoxib effects after 4 and 12 weeks of intervention. Thus, low-intensity pulsed US accelerated, but did not improve, ligament healing, whereas celecoxib delayed, but did not impair, healing (Warden et al., 2006). Collectively, the effects of celecoxib on wound healing in nonclinical studies appear highly variable.

CONCLUSIONS

In this chapter we described the nonclinical and clinical roles of PGs in bone metabolism and healing, data to illustrate these roles, expression of COX-1/COX-2 in normal and diseased states of bone, and an overall assessment of the effects of nonselective NSAIDs versus COX-2 s-NSAIDs on healing. PGs are known modulators of tissue metabolism, including bone, ligament, and tendons, via the action of osteoclasts and osteoblasts. PGE_2 is the most abundant PG produced by osteoblasts, and has been demonstrated to play a critical role in bone-to-bone formation, woven bone formation, and bone resorption, primarily through the EP_2 and EP_4 PG receptor subtypes. PGs are additionally demonstrated to release in response to mechanical force and stimulate cAMP. These roles of PGs in metabolism translate to bone healing. In the inflammatory stage of healing, PGs, particularly PGE_2, are produced abundantly by osteoblasts and accumulate in the fracture callus, creating a rich microenvironment of PG activity. PGE_2, PGI_2, and TXA_2 then stimulate bone resorption via increase and activity of osteoclasts, followed by bone remodeling and/formation (Radi and Khan, 2005). The production of PGs in bone has been enhanced indirectly by COX-2 induction. COX-1 is constitutively present in normal skeleton and COX-2 is up-regulated during initial stages of bone repair, inflammation, or in neoplastic skeleton conditions.

Nonclinical models revealed that NSAIDs and COX-2 s-NSAIDs could impair bone, tendon, and ligament healing with variable results. The exact mechanism of COX effects is not well defined but may be related to significant reductions in PGE_2 and osteoblastic proliferation with or without PG inhibition. Results from animal models do suggest that ns-NSAIDs and COX-2 s-NSAIDs may have a minimal effect on bone, tendon, and ligament healing, especially at earlier stages, but bear no significant impact on the ultimate long-term outcome

(Radi and Khan, 2005). Despite the abundant nonclinical evidence that PGs produce effects on bone metabolism, little relevant clinical evidence is available to adequately assess the significance of PG inhibition by nonselective or COX-2 selective inhibitors. Although inhibition of COX-2 may play a role in these effects, there is no clear convincing evidence that COX-2 s-NSAIDs are any more deleterious than conventional ns-NSAIDs.

Overall, the relevance of the available nonclinical data to humans is unknown due to (1) no replicable data of associated orthopedic problems with ns-NSAIDs or COX-2 s-NSAIDs in humans from randomized large, controlled clinical reports; (2) inconsistent results from nonclinical studies that ns-NSAIDs or COX-2 s-NSAIDs affect the long-term outcome of bone, tendon, or ligament healing; (3) intra- and interspecies dose and sensitivity differences; (4) undefined compensating systemic and local factors that may also regulate bone repair; (5) the unknown effects of underlying conditions (age, rate of bone turnover, or disease affliction) that may influence bone healing; (6) differences in dosing duration and levels between experimental animal models and clinical applications; and (7) the lack of large controlled clinical trials needed to better assess the orthopedic effects of these drugs (Radi and Khan, 2005; Radi, 2009).

REFERENCES

Adams SS, Bough RG, Cliffe EE, Lessel B, et al. Absorption, distribution and toxicity of ibuprofen. Toxicol Appl Pharmacol 1969;15:310–330.

Adolphson P, Jonsson U, Dalen N. Piroxicam-induced reduction in osteopenia after external fixation of rabbit tibia. Acta Orthop Scand 1991;62:363–366.

Adolphson P, Abbaszadegan H, Jonsson U, Dalen N, et al. No effects of piroxicam on osteopenia and recovery after Colles' fracture: a randomized, double-blind, placebo-controlled, prospective trial. Arch Orthop Trauma Surg 1993;112:127–130.

Akhter MP, Cullen DM, Pan LC. Bone biomechanical properties in EP4 knockout mice. Calcif Tissue Int 2006;78:357–362.

Akman S, Gogus A, Sener N, Bilgic B, et al. Effect of diclofenac sodium on union of tibial fractures in rats. Adv Ther 2002;19:119–125.

Akritopoulos P, Papaioannidou P, Hatzokos I, Haritanti A, et al. Parecoxib has non-significant long-term effects on bone healing in rats when administered for a short period after fracture. Arch Orthop Trauma Surg 2009;129:1427–1432.

Alander CB, Raisz LG. Effects of selective prostaglandins E2 receptor agonists on cultured calvarial murine osteoblastic cells. Prostaglandins Other Lipid Mediat 2006;81:178–183.

Allen HL, Wase A, Bear WT. Indomethacin and aspirin: effect of nonsteroidal anti-inflammatory agents on the rate of fracture repair in the rat. Acta Orthop Scand 1980;51:595–600.

Altman RD, Latta LL, Keer R, Renfree K, et al. Effect of nonsteroidal antiinflammatory drugs on fracture healing: a laboratory study in rats. J Orthop Trauma 1995;9:392–400.

Anderson DM, Maraskovsky E, Billingsley WL, et al. A homologue of the TNF receptor and its ligand enhance T-cell growth and dendritic-cell function. Nature 1997;390:175–179.

Anderson GD, Keys KL, De Ciechi PA, Masferrer JL. Combination therapies that inhibit cyclooxygenase-2 and leukotriene synthesis prevent disease in murine collagen induced arthritis. Inflamm Res 2009;58:109–117.

Ankarath S, Raman R, Giannoudis PV. Non-steroidal anti-inflammatory drugs in orthopaedic practice: an update. Curr Orthop 2003;17:144–149.

Arasapam G, Scherer M, Cool JC, Foster BK, et al. Roles of COX-2 and iNOS in the bony repair of the injured growth plate cartilage. J Cell Biochem 2006;99:450–461.

Arikawa T, Omura K, Morita I. Regulation of bone morphogenetic protein-2 expression by endogenous prostaglandin E2 in human mesenchymal stem cells. J Cell Physiol 2004;200:400–406.

Asundi KR, Rempel DM. MMP-1, IL-1beta, and COX-2 mRNA expression is modulated by static load in rabbit flexor tendons. Ann Biomed Eng 2008;36:237–243.

Auer JA, Goodship A, Arnoczky S, Pearce S, et al. Refining animal models in fracture research: seeking consensus in optimising both animal welfare and scientific validity for appropriate biomedical use. BMC Musculoskelet Disord 2007;8:72.

Bauer DC, Orwoll ES, Fox KM, Vogt TM, et al. Aspirin and NSAID use in older women: effect on bone mineral density and fracture risk. Study of Osteoporotic Fractures Research Group. J Bone Miner Res 1996;11:29–35.

Beck A, Krischak G, Sorg T, Augat P, et al. Influence of diclofenac (group of nonsteroidal anti-inflammatory drugs) on fracture healing. Arch Orthop Trauma Surg 2003;123:327–332.

Bell NH, Hollis BW, Shary JR, Eyre DR, et al. Diclofenac sodium inhibits bone resorption in post-menopausal women. Am J Med 1994;96:349–353.

Bergenstock M, Min W, Simon AM, Sabatino C, et al. A comparison between the effects of acetaminophen and celecoxib on bone fracture healing in rats. J Orthop Trauma 2005;19:717–723.

Berglund M, Hart DA, Wiig M. The inflammatory response and hyaluronan synthases in the rabbit flexor tendon and tendon sheath following injury. J Hand Surg Eur Vol 2007;32:581–587.

Bergot C, Wu Y, Jolivet E, Zhou LQ, et al. The degree and distribution of cortical bone mineralization in the human femoral shaft change with age and sex in a microradiographic study. Bone 2009;45:435–442.

Bhattacharyya T, Levin R, Vrahas MS, Solomon DH. Nonsteroidal antiinflammatory drugs and nonunion of humeral shaft fractures. Arthritis Rheum 2005;53:364–367.

Bichara J, Greenwell H, Drisko C, Wittwer JW, et al. The effect of postsurgical naproxen and a bioabsorbable membrane on osseous healing in intrabony defects. J Periodontol 1999;70:869–877.

Blackwell KA, Hortschansky P, Sanovic S, et al. Bone morphogenetic protein 2 enhances PGE(2)-stimulated osteoclast formation in murine bone marrow cultures. Prostaglandins Other Lipid Mediat 2009;90:76–80.

Bo J, Sudmann E, Marton PF. Effect of indomethacin on fracture healing in rats. Acta Orthop Scand 1976;47:588–599.

Bogatov VB, Weinhold P, Dahners LE. The influence of a cyclooxygenase-1 inhibitor on injured and uninjured ligaments in the rat. Am J Sports Med 2003;31:574–576.

Boiskin I, Epstein S, Ismail F, Fallon MD, Levy W. Long term administration of prostaglandin inhibitors in vivo fail to influence cartilage and bone mineral metabolism in the rat. Bone Miner 1988;4:27–36.

Brighton CT, Krebs AG. Oxygen tension of healing fractures in the rabbit. J Bone Joint Surg Am 1972;54:323–332.

Bring D, Reno C, Renstrom P, Salo P, Hart D, Ackermann P. Prolonged immobilization compromises up-regulation of repair genes after tendon rupture in a rat model. Scand J Med Sci Sports 2009;20: 411–417.

Brooks PM, Day RO. Nonsteroidal antiinflammatory drugs: differences and similarities. N Engl J Med 1991;324:1716–1725.

Brown AP, Courtney CL, King LM, Groom SC, Graziano MJ. Cartilage dysplasia and tissue mineralization in the rat following administration of a FGF receptor tyrosine kinase inhibitor. Toxicol Pathol 2005;33:449–455.

Brown KM, Saunders MM, Kirsch T, Donahue HJ, Reid JS. Effect of COX-2-specific inhibition on fracture-healing in the rat femur. J Bone Joint Surg Am 2004;86-A:116–123.

Bruder SP, Fink DJ, Caplan AI. Mesenchymal stem cells in bone development, bone repair, and skeletal regeneration therapy. J Cell Biochem 1994;56:283–294.

Bucay N, Sarosi I, Dunstan CR. Osteoprotegerin-deficient mice develop early onset osteoporosis and arterial calcification. Genes Dev 1998;12:1260–1268.

Burd TA, Hughes MS, Anglen JO. Heterotopic ossification prophylaxis with indomethacin increases the risk of long-bone nonunion. J Bone Joint Surg Br 2003;85:700–705.

Burdan F, Szumiło J, Marzec B, Klepacz R, et al. Skeletal developmental effects of selective and non-selective cyclooxygenase-2 inhibitors administered through organogenesis and fetogenesis in Wistar CRL:(WI)WUBR rats. Toxicology 2005;216:204–223.

Burdan F, Szumiło J, Dworzański W, Szumiło M, et al. Immunoexpression of constitutive and inducible cyclooxygenase isoforms in distinguishing and accessory structures of synovial joints in rat foetuses. Folia Morphol (Warsz) 2009;68:59–64.

Burssens A, Thiery J, Kohl P, Molderez A, et al. Prevention of heterotopic ossification with tenoxicam following total hip arthroplasty: a double-blind, placebo-controlled dose-finding study. Acta Orthop Belg 1995;61:205–211.

Burstein AH, Reilly DT, Martens M. Aging of bone tissue: mechanical properties. J Bone Joint Surg Am 1976;58:82–86.

Butcher CK, Marsh DR. Non steroidal anti-inflammatory drugs delay tibial fracture union. Injury 1996;27:375.

Carlstedt CA, Madsen K, Wredmark T. The influence of indomethacin on tendon healing: a biomechanical and biochemical study. Arch Orthop Trauma Surg 1986;105:332–336.

Chikazu D, Tomizuka K, Ogasawara T, Saijo H, et al. Cyclooxygenase-2 activity is essential for the osseointegration of dental implants. Int J Oral Maxillofac Surg 2007;36:441–446.

Choudhary S, Alander C, Zhan P, Gao Q, et al. Effect of deletion of the prostaglandin EP2 receptor on the anabolic response to prostaglandin E2 and a selective EP2 receptor agonist. Prostaglandins Other Lipid Mediat 2008a;86:35–40.

Choudhary S, Huang H, Raisz L, Pilbeam C. Anabolic effects of PTH in cyclooxygenase-2 knockout osteoblasts in vitro. Biochem Biophys Res Commun 2008b;372:536–541.

Cohen DB, Kawamura S, Ehteshami JR, Rodeo SA. Indomethacin and celecoxib impair rotator cuff tendon-to-bone healing. Am J Sports Med 2006;34:362–369.

Colvin JS, Bohne BA, Harding GW, et al. Skeletal overgrowth and deafness in mice lacking fibroblast growth factor receptor-3. Nat Genet 1996;12:390–397.

Connolly CK, Li G, Bunn JR, Mushipe M, Dickson GR, Marsh DR. A reliable externally fixated murine femoral fracture model that accounts for variation in movement between animals. J Orthop Res 2003;21:843–849.

Cottrell JA, O'Connor JP. Pharmacological inhibition of 5-lipoxygenase accelerates and enhances fracture-healing. J Bone Joint Surg Am 2009;91:2653–2665.

Cruess RL, Dumont J. Fracture healing. Can J Surg 1975;18:403–413.

Dahners LE, Gilbert JA, Lester GE, Taft TN, et al. The effect of a nonsteroidal antiinflammatory drug on the healing of ligaments. Am J Sports Med 1988;16:641–646.

Davis TR, Ackroyd CE. Non-steroidal anti-inflammatory agents in the management of Colles' fractures. Br J Clin Pract 1988;42:184–189.

Dawson AB. The age order of epiphyseal union in the long bones of the albino rat. Anat Rec 1925;31:1–17.

de Crombrugghe B, Lefebvre V, Nakashima K. Regulatory mechanisms in the pathways of cartilage and bone formation. Curr Opin Cell Biol 2001;13:721–727.

Dekel S, Lenthall G, Francis MJ. Release of prostaglandins from bone and muscle after tibial fracture: an experimental study in rabbits. J Bone Joint Surg Br 1981;63-B:185–189.

Deng C, Wynshaw-Boris A, Zhou F, et al. Fibroblast growth factor receptor 3 is a negative regulator of bone growth. Cell 1996;84:911–921.

Dickens DS, Kozielski R, Khan J, Forus A, et al. Cyclooxygenase-2 expression in pediatric sarcomas. Pediatr Dev Pathol 2002;5:356–364.

Dimar JR 2nd, Ante WA, Zhang YP, Glassman SD. The effects of nonsteroidal anti-inflammatory drugs on posterior spinal fusions in the rat. Spine 1996;21:1870–1876.

Dimitriou R, Dahabreh Z, Katsoulis E, Matthews SJ, et al. Application of recombinant BMP-7 on persistent upper and lower limb non-unions. Injury 2005;36 Suppl 4:S51–S59.

Dimmen S, Nordsletten L, Engebretsen L, Steen H, Madsen JE. Negative effect of parecoxib on bone mineral during fracture healing in rats. Acta Orthop 2008;79:438–444.

Dimmen S, Nordsletten L, Madsen JE. Parecoxib and indomethacin delay early fracture healing: a study in rats. Clin Orthop Relat Res 2009;467:1992–1999.

Dubrow SA, Hruby PM, Akhter MP. Gender specific LRP5 influences on trabecular bone structure and strength. J Musculoskelet Neuronal Interact 2007;7:166–173.

Edwall D, Prisell PT, Levinovitz A, Jennishe E, et al. Expression of insulin-like growth factor I messenger ribonucleic acid in regenerating bone after fracture: influence of indomethacin. J Bone Miner Res 1992;7:207–213.

Einhorn TA. The cell and molecular biology of fracture healing. Clin Orthop Relat Res 1998:S7–S21.
Ekman EF. A cyclooxegenase-2 inhibitor impairs ligament healing in the rat. Am J Sports Med 2002;30:457.
Elder CL, Dahners LE, Weinhold PS. A cyclooxygenase-2 inhibitor impairs ligament healing in the rat. Am J Sports Med 2001;29:801–805.
Elves MW, Bayley I, Roylance PJ. The effect of indomethacin upon experimental fractures in the rat. Acta Orthop Scand 1982;53:35–41.
Endo K, Sairyo K, Komatsubara S, Sasa T, et al. Cyclooxygenase-2 inhibitor delays fracture healing in rats. Acta Orthop 2005;76:470–474.
Engesaeter LB, Sudmann B, Sudmann E. Fracture healing in rats inhibited by locally administered indomethacin. Acta Orthop Scand 1992;63:330–333.
Enneking WF, Burchardt H, Puhl JJ, Piotrowski G. Physical and biological aspects of repair in dog cortical-bone transplants. J Bone Joint Surg Am 1975;57:237–252.
Erben RG. Trabecular and endocortical bone surfaces in the rat: modeling or remodeling? Anat Rec 1996;246:39–46.
Faye-Petersen OM, Johnson WH Jr, Carlo WA, Hedlund GL, et al. Prostaglandin E1–induced hyperostosis: clinicopathologic correlations and possible pathogenetic mechanisms. Pediatr Pathol Lab Med 1996;16:489–507.
Feng X, McDonald JM. Disorders of bone remodeling. Annu Rev Pathol 2011;6:121–145.
Ferry ST, Dahners LE, Afshari HM, Weinhold PS. The effects of common anti-inflammatory drugs on the healing rat patellar tendon. Am J Sports Med 2007;35:1326–1333.
Feyen JH, van der Wilt G, Moonen P, Di Bon A, et al. Stimulation of arachidonic acid metabolism in primary cultures of osteoblast-like cells by hormones and drugs. Prostaglandins 1984;28:769–781.
Forslund C, Bylander B, Aspenberg P. Indomethacin and celecoxib improve tendon healing in rats. Acta Orthop Scand 2003;74:465–469.
Forwood MR. Inducible cyclo-oxygenase (COX-2) mediates the induction of bone formation by mechanical loading in vivo. J Bone Miner Res 1996;11:1688–1693.
Fradette ME, Céleste C, Richard H, Beauchamp G, et al. Effects of continuous oral administration of phenylbutazone on biomarkers of cartilage and bone metabolism in horses. Am J Vet Res 2007;68: 128–133.
Freiberg AH. Wolff's law and the functional pathogenesis of deformity. J Bone Joint Surg Am 1902;s1-15:200–220.
Gajraj NM. The effect of cyclooxygenase-2 inhibitors on bone healing. Reg Anesth Pain Med 2003;28: 456–465.
Gao Q, Zhan P, Alander CB, Kream BE, et al. Effects of global or targeted deletion of the EP4 receptor on the response of osteoblasts to prostaglandin in vitro and on bone histomorphometry in aged mice. Bone 2009;45:98–103.
Gebuhr P, Wilbek H, Soelberg M. Naproxen for 8 days can prevent heterotopic ossification after hip arthroplasty. Clin Orthop Relat Res 1995;314:166–169.
Geng Y, Blanco FJ, Cornelisson M, Lotz M. Regulation of cyclooxygenase-2 expression in normal human articular chondrocytes. J Immunol 1995;155:796–801.
Gerstenfeld LC, Einhorn TA. COX inhibitors and their effects on bone healing. Expert Opin Drug Saf 2004;3:131–136.
Gerstenfeld LC, Cullinane DM, Barnes GL, Graves DT, et al. Fracture healing as a post-natal developmental process: molecular, spatial, and temporal aspects of its regulation. J Cell Biochem 2003a;88:873–884.
Gerstenfeld LC, Thiede M, Seibert K, Mielke C, et al. Differential inhibition of fracture healing by non-selective and cyclooxygenase-2 selective non-steroidal anti-inflammatory drugs. J Orthop Res 2003b;21:670–675.
Gerstenfeld LC, Al-Ghawas M, Alkhiary YM, Cullinane DM, et al. Selective and nonselective cyclooxygenase-2 inhibitors and experimental fracture-healing: reversibility of effects after short-term treatment. J Bone Joint Surg Am 2007;89:114–125.
Giannoudis PV, MacDonald DA, Matthews SJ, Smith RM, et al. Nonunion of the femoral diaphysis: the influence of reaming and non-steroidal anti-inflammatory drugs. J Bone Joint Surg Br 2000;82:655–658.

Gill TJ, McCulloch PC, Glasson SS, Blanchet T, et al. Chondral defect repair after the microfracture procedure: a nonhuman primate model. Am J Sports Med 2005;33:680–685.

Giordano V, Giordano M, Knackfuss IG, Apfel MI, Gomes RD. Effect of tenoxicam on fracture healing in rat tibiae. Injury 2003;34:85–94.

Glassman SD, Rose SM, Dimar JR, Puno RM, Campbell MJ, Johnson Jr. The effect of postoperative nonsteroidal anti-inflammatory drug administration on spinal fusion. Spine 1998;23:834–838.

Goodman S, Ma T, Trindade M, Ikenoue T, et al. COX-2 selective NSAID decreases bone ingrowth in vivo. J Orthop Res 2002;20:1164–1169.

Govender S, Csimma C, Genant HK, Valentin-Opran A, et al. Recombinant human bone morphogenetic protein-2 for treatment of open tibial fractures: a prospective, controlled, randomized study of four hundred and fifty patients. J Bone Joint Surg Am 2002;84-A:2123–2134.

Gregory LS, Forwood MR. Cyclooxygenase-2 inhibition delays the attainment of peak woven bone formation following four-point bending in the rat. Calcif Tissue Int 2007;80:176–183.

Grohs JG, Schmidt M, Wanivenhaus A. Selective COX-2 inhibitor versus indomethacin for the prevention of heterotopic ossification after hip replacement: a double-blind randomized trial of 100 patients with 1-year follow-up. Acta Orthop 2007;78:95–98.

Gurgel BC, Ribeiro FV, Silva MA, Nociti FH Jr, et al. Selective COX-2 inhibitor reduces bone healing in bone defects. Braz Oral Res 2005;19:312–316.

Gurton AU, Akin E, Sagdic D, Olmez H. Effects of PGI2 and TxA2 analogs and inhibitors in orthodontic tooth movement. Angle Orthod 2004;74:526–532.

Hedges RE, Clement JG, Thomas CD, O'Connell TC. Collagen turnover in the adult femoral mid-shaft: modeled from anthropogenic radiocarbon tracer measurements. Am J Phys Anthropol 2007;133:808–816.

Heppenstall RB, Grislis G, Hunt TK. Tissue gas tensions and oxygen consumption in healing bone defects: tissue gas tensions and oxygen consumption in healing bone defects. Clin Orthop Relat Res 1975;106:357–365.

Herbenick MA, Sprott D, Stills H, Lawless M. Effects of a cyclooxygenase 2 inhibitor on fracture healing in a rat model. Am J Orthop (Belle Mead NJ) 2008;37:E133–E137.

Hert J. Wolff's law of transformation after 100 years. Acta Chir Orthop Traumatol Cech 1990;57: 465–476.

High WB. Effects of orally administered prostaglandin E-2 on cortical bone turnover in adult dogs: a histomorphometric study. Bone 1987;8:363–373.

Ho ML, Chang JK, Wang GJ. Effects of ketorolac on bone repair: a radiographic study in modeled demineralized bone matrix grafted rabbits. Pharmacology 1998;57:148–159.

Ho ML, Chang JK, Chuang LY, Hsu HK, et al. Effects of nonsteroidal anti-inflammatory drugs and prostaglandins on osteoblastic functions. Biochem Pharmacol 1999;58:983–990.

Hoffman M, Pradhan S, Monroe DM. Celecoxib does not delay cutaneous wound healing in haemophilia B mice. Haemophilia 2009;15:615–616.

Hogevold HE, Grogaard B, Reikeras O. Effects of short-term treatment with corticosteroids and indomethacin on bone healing: a mechanical study of osteotomies in rats. Acta Orthop Scand 1992;63:607–611.

Hosono A, Yamaguchi U, Makimoto A, Endo M, et al. Utility of immunohistochemical analysis for cyclo-oxygenase 2 in the differential diagnosis of osteoblastoma and osteosarcoma. J Clin Pathol 2007;60:410–414.

Hsu HC, Fong YC, Chang CS, Hsu CJ, et al. Ultrasound induces cyclooxygenase-2 expression through integrin, integrin-linked kinase, Akt, NF-kappaB and p300 pathway in human chondrocytes. Cell Signal 2007;19:2317–2328.

Huang H, Chikazu D, Voznesensky OS, Herschman HR, et al. Parathyroid hormone induction of cyclooxygenase-2 in murine osteoblasts: role of the calcium–calcineurin–NFAT pathway. J Bone Miner Res 2010;25:819–829.

Hunziker EB, Schenk RK. Physiological mechanisms adopted by chondrocytes in regulating longitudinal bone growth in rats. J Physiol 1989;414:55–71.

Huo MH, Troiano NW, Pelker RR, Gundberg CM, et al. The influence of ibuprofen on fracture repair: biomechanical, biochemical, histologic, and histomorphometric parameters in rats. J Orthop Res 1991;9:383–390.

Hurley MM, Lee SK, Raisz LG, Bernecker P, et al. Basic fibroblast growth factor induces osteoclast formation in murine bone marrow cultures. Bone 1998;22:309–316.

Igarashi K, Woo JT, Stern PH. Effects of a selective cyclooxygenase-2 inhibitor, celecoxib, on bone resorption and osteoclastogenesis in vitro. Biochem Pharmacol 2002;63:523–532.

Jaffre B, Watrin A, Loeuille D, Gillet P, et al. Effects of antiinflammatory drugs on arthritic cartilage: a high-frequency quantitative ultrasound study in rats. Arthritis Rheum 2003;48:1594–601.

Jee WS, Ma YF. The in vivo anabolic actions of prostaglandins in bone. Bone 1997;21:297–304.

Joldersma M, Burger EH, Semeins CM, Klein-Nulend J. Mechanical stress induces COX-2 mRNA expression in bone cells from elderly women. J Biomech 2000;33:53–61.

Judex S, Donahue LR, Rubin C. Genetic predisposition to low bone mass is paralleled by an enhanced sensitivity to signals anabolic to the skeleton. FASEB J 2002;16:1280–1282.

Kämpfer H, Schmidt R, Geisslinger G, Pfeilschifter J, et al. Wound inflammation in diabetic ob/ob mice: functional coupling of prostaglandin biosynthesis to cyclooxygenase-1 activity in diabetes-impaired wound healing. Diabetes 2005;54:1543–1551.

Karachalios T, Boursinos L, Poultsides L, Khaldi L, et al. The effects of the short-term administration of low therapeutic doses of anti-COX-2 agents on the healing of fractures: an experimental study in rabbits. J Bone Joint Surg Br 2007;89:1253–1260.

Kawaguchi H, Pilbeam CC, Harrison JR, Raisz LG. The role of prostaglandins in the regulation of bone metabolism. Clin Orthop Relat Res 1995;313:36–46.

Kawaguchi H, Chikazu D, Nakamura K, Kumegawa M, et al. Direct and indirect actions of fibroblast growth factor 2 on osteoclastic bone resorption in cultures. J Bone Miner Res 2000;15:466–473.

Ke HZ, Jee WS, Mori S, Li XJ, Kimmel DB. Effects of long-term daily administration of prostaglandin-E2 on maintaining elevated proximal tibial metaphyseal cancellous bone mass in male rats. Calcif Tissue Int 1992;50:245–252.

Ke HZ, Jee WS, Ito H, Setterberg RB, et al. Greater bone formation induction occurred in aged than young cancellous bone sites. Bone 1993;14:481–485.

Keller J, Bunger C, Andreassen TT, Bak B, et al. Bone repair inhibited by indomethacin: effects on bone metabolism and strength of rabbit osteotomies. Acta Orthop Scand 1987;58:379–383.

Keller J, Hansen ES, He SZ, Kjaersgaard-Andersen P, et al. Early hemodynamic response to tibial osteotomy in rabbits: influence of indomethacin and prostaglandin E2. J Orthop Res 1991;9:539–544.

Keller J, Klamer A, Bak B, Suder P. Effect of local prostaglandin E2 on fracture callus in rabbits. Acta Orthop Scand 1993;64:59–63.

Kellinsalmi M, Parikka V, Risteli J, Hentunen T, et al. Inhibition of cyclooxygenase-2 down-regulates osteoclast and osteoblast differentiation and favours adipocyte formation in vitro. Eur J Pharmacol 2007;572:102–110.

Kember NF. Aspects of the maturation process in growth cartilage in the rat tibia. Clin Orthop 1973;95:288–294.

Kidd LJ, Stephens AS, Kuliwaba JS, Fazzalari NL, et al. Temporal pattern of gene expression and histology of stress fracture healing. Bone 2010;46:369–378.

Kjaersgaard-Andersen P, Nafei A, Teichert G, Kristensen O, et al. Indomethacin for prevention of heterotopic ossification: a randomized controlled study in 41 hip arthroplasties. Acta Orthop Scand 1993;64:639–642.

Kong YY, Yoshida H, Sarosi I, Tan HL, et al. OPGL is a key regulator of osteoclastogenesis, lymphocyte development and lymph-node organogenesis. Nature 1999;397:315–323.

Koshima H, Kondo S, Mishima S, Choi HR, et al. Expression of interleukin-1beta, cyclooxygenase-2, and prostaglandin E2 in a rotator cuff tear in rabbits. J Orthop Res 2007;25:92–97.

Krieger NS, Frick KK, LaPlante Strutz K, et al. Regulation of COX-2 mediates acid-induced bone calcium efflux in vitro. J Bone Miner Res 2007;22:907–917.

Krischak GD, Augat P, Sorg T, Blakytny R, et al. Effects of diclofenac on periosteal callus maturation in osteotomy healing in an animal model. Arch Orthop Trauma Surg 2007;127:3–9.

Kulick MI, Smith S, Hadler K. Oral ibuprofen: evaluation of its effect on peritendinous adhesions and the breaking strength of a tenorrhaphy. J Hand Surg [Am] 1986;11:110–120.

Kuroda S, Virdi AS, Dai Y, Shott S, et al. Patterns and localization of gene expression during intramembranous bone regeneration in the rat femoral marrow ablation model. Calcif Tissue Int 2005;77:212–225.

Lane N, Coble T, Kimmel DB. Effect of naproxen on cancellous bone in ovariectomized rats. J Bone Miner Res 1990;5:1029–1035.

Lascelles BD, King S, Roe S, Marcellin-Little DJ, Jones S. Expression and activity of COX-1 and 2 and 5-LOX in joint tissues from dogs with naturally occurring coxofemoral joint osteoarthritis. J Orthop Res 2009;27:1204–1208.

Lee CM, Genetos DC, Wong A, Yellowley CE. Prostaglandin expression profile in hypoxic osteoblastic cells. J Bone Miner Metab 2010;28:8–16.

Leonelli SM, Goldberg BA, Safanda J, Bagwe MR, et al. Effects of a cyclooxygenase-2 inhibitor (rofecoxib) on bone healing. Am J Orthop (Belle Mead NJ) 2006;35:79–84.

Li J, Burr DB, Turner CH. Suppression of prostaglandin synthesis with NS-398 has different effects on endocortical and periosteal bone formation induced by mechanical loading. Calcif Tissue Int 2002;70: 320–329.

Li J, Waugh LJ, Hui SL, Burr DB, et al. Low-intensity pulsed ultrasound and nonsteroidal anti-inflammatory drugs have opposing effects during stress fracture repair. J Orthop Res 2007;25:1559–1567.

Li M, Ke HZ, Qi H, Healy DR, et al. A novel, non-prostanoid EP2 receptor-selective prostaglandin E2 agonist stimulates local bone formation and enhances fracture healing. J Bone Miner Res 2003;18:2033–2042.

Li M, Healy DR, Li Y, Simmons HA, et al. Osteopenia and impaired fracture healing in aged EP4 receptor knockout mice. Bone 2005;37:46–54.

Li X, Okada Y, Pilbeam CC, Lorenzo JA, et al. Knockout of the murine prostaglandin EP2 receptor impairs osteoclastogenesis in vitro. Endocrinology 2000;141:2054–2061.

Li X, Pilbeam CC, Pan L, Breyer RM, Raisz LG. Effects of prostaglandin E2 on gene expression in primary osteoblastic cells from prostaglandin receptor knockout mice. Bone 2002;30:567–573.

Li XJ, Jee WS, Li YL, Patterson-Buckendahl P. Transient effects of subcutaneously administered prostaglandin E2 on cancellous and cortical bone in young adult dogs. Bone 1990;11:353–364.

Lin CH, Jee WS, Ma YF, Setterberg RB. Early effects of prostaglandin E2 on bone formation and resorption in different bone sites of rats. Bone 1995;17:255S–259S.

Lindholm TS, Tornkvist H. Inhibitory effect on bone formation and calcification exerted by the anti-inflammatory drug ibuprofen: an experimental study on adult rat with fracture. Scand J Rheumatol 1981;10:38–42.

Long J, Lewis S, Kuklo T, Zhu Y, et al. The effect of cyclooxygenase-2 inhibitors on spinal fusion. J Bone Joint Surg Am 2002;84-A:1763–1768.

Marks SC Jr, Miller S. Local infusion of prostaglandin E1 stimulates mandibular bone formation in vivo. J Oral Pathol 1988;17:500–505.

Marui A, Hirose K, Maruyama T, Arai Y, et al. Prostaglandin E2 EP4 receptor-selective agonist facilitates sternal healing after harvesting bilateral internal thoracic arteries in diabetic rats. J Thorac Cardiovasc Surg 2006;131:587–593.

Mastbergen SC, Marijnissen AC, Vianen ME, Zoer B, et al. Inhibition of COX-2 by celecoxib in the canine groove model of osteoarthritis. Rheumatology (Oxford) 2006;45:405–413.

Mayahara K, Kobayashi Y, Takimoto K, Suzuki N, et al. Aging stimulates cyclooxygenase-2 expression and prostaglandin E2 production in human periodontal ligament cells after the application of compressive force. J Periodont Res 2007;42:8–14.

Mbugua SW, Skoglund LA, Lokken P. Effects of phenylbutazone and indomethacin on the post-operative course following experimental orthopaedic surgery in dogs. Acta Vet Scand 1989;30:27–35.

McCalden RW, McGeough JA, Barker MB, Court-Brown CM. Age-related changes in the tensile properties of cortical bone. The relative importance of changes in porosity, mineralization, and microstructure. J Bone Joint Surg Am 1993;75:1193–1205.

Miyamoto H, Onuma H, Shigematsu H, Suzuki S, et al. The effect of etodolac on experimental temporomandibular joint osteoarthritis in dogs. J Craniomaxillofac Surg 2007;35:358–363.

Miyauchi M, Hiraoka M, Oka H, Sato S, et al. Immuno-localization of COX-1 and COX-2 in the rat molar periodontal tissue after topical application of lipopolysaccharide. Arch Oral Biol 2004;49:739–746.

Miyaura C, Inada M, Suzawa T, Sugimoto Y, et al. Impaired bone resorption to prostaglandin E2 in prostaglandin E receptor EP4-knockout mice. J Biol Chem 2000;275:19819–19823.

Moed BR, Resnick RB, Fakhouri AJ, Nallamothu B, et al. Effect of two nonsteroidal antiinflammatory drugs on heterotopic bone formation in a rabbit model. J Arthroplasty 1994;9:81–87.
Mohammed SI, Khan KN, Sellers RS, Hayek MG, et al. Expression of cyclooxygenase-1 and 2 in naturally-occurring canine cancer. Prostaglandins Leukot Essent Fatty Acids 2004;70:479–483.
Moorman CT 3rd, Kukreti U, Fenton DC, Belkoff SM. The early effect of ibuprofen on the mechanical properties of healing medial collateral ligament. Am J Sports Med 1999;27:738–741.
More RC, Kody MH, Kabo JM, Dorey FJ, et al. The effects of two nonsteroidal anti-inflammatory drugs on limb swelling, joint stiffness, and bone torsional strength following fracture in a rabbit model. Clin Orthop 1989;247:306–312.
Moses VS, Hardy J, Bertone AL, Weisbrode SE. Effects of anti-inflammatory drugs on lipopolysaccharide-challenged and -unchallenged equine synovial explants. Am J Vet Res 2001;62:54–60.
Mullins MN, Lana SE, Dernell WS, Ogilvie GK, et al. Cyclooxygenase-2 expression in canine appendicular osteosarcomas. J Vet Intern Med 2004;18:859–865.
Mullis BH, Copland ST, Weinhold PS, Miclau T, et al. Effect of COX-2 inhibitors and non-steroidal anti-inflammatory drugs on a mouse fracture model. Injury 2006;37:827–837.
Mungo DV, Zhang X, O'Keefe RJ, Rosier RN, et al. COX-1 and COX-2 expression in osteoid osteomas. J Orthop Res 2002;20:159–162.
Murnaghan M, Li G, Marsh DR. Nonsteroidal anti-inflammatory drug-induced fracture nonunion: An inhibition of angiogenesis? J Bone Joint Surg Am 2006;88 Suppl 3: 140–147.
Naik AA, Xie C, Zuscik MJ, Kingsley P, et al. Reduced COX-2 expression in aged mice is associated with impaired fracture healing. J Bone Miner Res 2009;24:251–264.
Noguchi M, Kimoto A, Kobayashi S, Yoshino T, et al. Effect of celecoxib, a cyclooxygenase-2 inhibitor, on the pathophysiology of adjuvant arthritis in rat. Eur J Pharmacol 2005;513:229–235.
Noguchi M, Kimoto A, Sasamata M, Miyata K. Micro-CT imaging analysis for the effect of celecoxib, a cyclooxygenase-2 inhibitor, on inflammatory bone destruction in adjuvant arthritis rats. J Bone Miner Metab 2008;26:461–468.
Norrdin RW, Shih MS. Systemic effects of prostaglandin E2 on vertebral trabecular remodeling in beagles used in a healing study. Calcif Tissue Int 1988;42:363–368.
Nyangoga H, Aguado E, Goyenvalle E, Baslé MF, et al. A non-steroidal anti-inflammatory drug (ketoprofen) does not delay beta-TCP bone graft healing. Acta Biomater 2010;6:3310–3317.
Obeid G, Zhang X, Wang X. Effect of ibuprofen on the healing and remodeling of bone and articular cartilage in the rabbit temporomandibular joint. J Oral Maxillofac Surg 1992;50:843–849.
Ochi H, Hara Y, Asou Y, Harada Y, et al. Effects of long-term administration of carprofen on healing of a tibial osteotomy in dogs. Am J Vet Res 2011;72:634–641.
O'Connor JP, Lysz T. Celecoxib, NSAIDs and the skeleton. Drugs Today (Barc) 2008;44:693–709.
O'Connor JP, Capo JT, Tan V, Cottrell JA, et al. A comparison of the effects of ibuprofen and rofecoxib on rabbit fibula osteotomy healing. Acta Orthop 2009;80:597–605.
Oka H, Miyauchi M, Sakamoto K, et al. PGE2 activates cementoclastogenesis by cementoblasts via EP4. J Dent Res 2007;86:974–979.
Okada Y, Lorenzo JA, Freeman AM, Tomita M, et al. Prostaglandin G/H synthase-2 is required for maximal formation of osteoclast-like cells in culture. J Clin Invest 2000;105:823–832.
Okada Y, Pilbeam C, Raisz L, Tanaka Y. Role of cyclooxygenase-2 in bone resorption. J Uoeh 2003;25:185–195.
O'Keefe RJ, Tiyapatanaputi P, Xie C, Li TF, et al. COX-2 has a critical role during incorporation of structural bone allografts. Ann NY Acad Sci 2006;1068:532–542.
Oliveira TM, Sakai VT, Machado MA, Dionísio TJ, et al. COX-2 inhibition decreases VEGF expression and alveolar bone loss during the progression of experimental periodontitis in rats. J Periodontol 2008;79:1062–1069.
Onishi E, Fujibayashi S, Takemoto M, Neo M, et al. Enhancement of bone-bonding ability of bioactive titanium by prostaglandin E2 receptor selective agonist. Biomaterials 2008;29:877–883.
Ono K, Akatsu T, Kugai N, Pilbeam CC, et al. The effect of deletion of cyclooxygenase-2, prostaglandin receptor EP2, or EP4 in bone marrow cells on osteoclasts induced by mouse mammary cancer cell lines. Bone 2003;33:798–804.
Onoe Y, Miyaura C, Kaminakayashiki T, Nagai Y, et al. IL-13 and IL-4 inhibit bone resorption by suppressing cyclooxygenase-2-dependent prostaglandin synthesis in osteoblasts. J Immunol 1996;156: 758–764.

Parfitt AM, Travers R, Rauch F, Glorieux FH. Structural and cellular changes during bone growth in healthy children. Bone 2000;27:487–494.

Paulson SK, Zhang JY, Breau AP, Hribar JD, et al. Pharmacokinetics, tissue distribution, metabolism, and excretion of celecoxib in rats. Drug Metab Dispos 2000;28:514–521.

Perraud J, Stadler J, Kessedjian MJ, Monro AM. Reproductive studies with the anti-inflammatory agent, piroxicam: modification of classical protocols. Toxicology 1984;30:59–63.

Perry SM, McIlhenny SE, Hoffman MC, Soslowsky LJ. Inflammatory and angiogenic mRNA levels are altered in a supraspinatus tendon overuse animal model. J Shoulder Elbow Surg 2005;14:79S–83S.

Persson PE, Sodemann B, Nilsson OS. Preventive effects of ibuprofen on periarticular heterotopic ossification after total hip arthroplasty: a randomized double-blind prospective study of treatment time. Acta Orthop Scand 1998;69:111–115.

Plotquin D, Dekel S, Katz S, Danon A. Prostaglandin release by normal and osteomyelitic human bones. Prostaglandins Leukot Essent Fatty Acids 1991;43:13–15.

Pountos I, Giannoudis PV, Jones E, English A, et al. NSAIDS inhibit in vitro MSC chondrogenesis but not osteogenesis. J Cell Mol Med 2010;15:525–534.

Pradhan BB, Tatsumi RL, Gallina J, Kuhns CA, et al. Ketorolac and spinal fusion: Does the perioperative use of ketorolac really inhibit spinal fusion? Spine (Phila Pa 1976) 2008;33:2079–2082.

Prisby RD, Ramsey MW, Behnke BJ, Dominguez JM 2nd, et al. Aging reduces skeletal blood flow, endothelium-dependent vasodilation, and NO bioavailability in rats. J Bone Miner Res 2007;22: 1280–1288.

Probst A, Spiegel HU. Cellular mechanisms of bone repair. J Invest Surg 1997;10:77–86.

Radi ZA. Pathophysiology of cyclooxygenase inhibition in animal models. Toxicol Pathol 2009;37:34–46.

Radi ZA, Koza-Taylor PH, Bll RR, Obert LA, et al. Increased serum enzyme levels associated with kupffer cell reduction with no signs of hepatic or skeletal muscle injury. Am J Pathol 2011;179:240–247.

Radi ZA, Khan NK. Effects of cyclooxygenase inhibition on bone, tendon, and ligament healing. Inflamm Res 2005;54(9):358–366.

Radi ZA, Ostroski R. Pulmonary and cardiorenal cyclooxygenase-1 (COX-1), -2 (COX- 2), and microsomal prostaglandin E synthase-1 (mPGES-1) and -2 (mPGES-2) expression in a hypertension model. Mediators Inflamm 2007;2007:85091.

Radi ZA, Kehrli ME Jr, Ackermann MR. Cell adhesion molecules, leukocyte trafficking, and strategies to reduce leukocyte infiltration. J Vet Intern Med 2001;15:516–529.

Radi ZA, Guzman RE, Bell RR. Increased connective tissue extracellular matrix in the op/op model of osteopetrosis. Pathobiology 2009;76:199–203.

Raisz LG. Physiology and pathophysiology of bone remodeling. Clin Chem 1999a;45:1353–1358.

Raisz LG. Prostaglandins and bone: physiology and pathophysiology. Osteoarthritis Cartilage 1999b;7:419–421.

Raisz LG. Potential impact of selective cyclooxygenase-2 inhibitors on bone metabolism in health and disease. Am J Med 2001;110 Suppl 3A:43S–45S.

Rapuano BE, Boursiquot R, Tomin E, MacDonald DE, et al. The effects of COX-1 and COX-2 inhibitors on prostaglandin synthesis and the formation of heterotopic bone in a rat model. Arch Orthop Trauma Surg 2008;128:333–344.

Reikeraas O, Engebretsen L. Effects of ketoralac tromethamine and indomethacin on primary and secondary bone healing: an experimental study in rats. Arch Orthop Trauma Surg 1998;118:50–52.

Richards JB, Joseph L, Schwartzman K, Kreiger N, et al. The effect of cyclooxygenase-2 inhibitors on bone mineral density: results from the Canadian Multicentre Osteoporosis Study. Osteoporos Int 2006;17:1410–1419.

Riendeau D, Percival MD, Brideau C, Charleson S, et al. Etoricoxib (MK-0663): preclinical profile and comparison with other agents that selectively inhibit cyclooxygenase-2. J Pharmacol Exp Ther 2001;296:558–566.

Riew KD, Long J, Rhee J, Lewis S, et al. Time-dependent inhibitory effects of indomethacin on spinal fusion. J Bone Joint Surg Am 2003;85-A:632–634.

Riley GP, Cox M, Harrall RL, Clements S, et al. Inhibition of tendon cell proliferation and matrix glycosaminoglycan synthesis by non-steroidal anti-inflammatory drugs in vitro. J Hand Surg [Br] 2001;26:224–228.

Ringel RE, Brenner JI, Haney PJ, Burns JE, et al. Prostaglandin-induced periostitis: a complication of long-term PGE1 infusion in an infant with congenital heart disease. Radiology 1982;142:657–658.
Ro J, Langeland N, Sander J. Effect of indomethacin on collagen metabolism of rat fracture callus in vitro. Acta Orthop Scand 1978;49:323–328.
Roach HI, Mehta G, Oreffo RO, Clarke NM, et al. Temporal analysis of rat growth plates: cessation of growth with age despite presence of a physis. J Histochem Cytochem 2003;51:373–383.
Robertson G, Xie C, Chen D, Awad H, et al. Alteration of femoral bone morphology and density in COX-2$^{-/-}$ mice. Bone 2006;39:767–772.
Rohde C, Anderson DE, Bertone AL, Weisbrode SE. Effects of phenylbutazone on bone activity and formation in horses. Am J Vet Res 2000;61:537–543.
Romas E, Sims NA, Hards DK, et al. Osteoprotegerin reduces osteoclast numbers and prevents bone erosion in collagen-induced arthritis. Am J Pathol 2002;161:1419–1427.
Sakuma Y, Tanaka K, Suda M, Yasoda A, et al. Crucial involvement of the EP4 subtype of prostaglandin E receptor in osteoclast formation by proinflammatory cytokines and lipopolysaccharide. J Bone Miner Res 2000;15:218–227.
Sakuma Y, Li Z, Pilbeam CC, Alander CB, et al. Stimulation of cAMP production and cyclooxygenase-2 by prostaglandin E(2) and selective prostaglandin receptor agonists in murine osteoblastic cells. Bone 2004;34:827–834.
Sato S, Kim T, Arai T, Maruyama S, et al. Comparison between the effects of dexamethasone and indomethacin on bone wound healing. Jpn J Pharmacol 1986;42:71–78.
Saudan M, Saudan P, Perneger T, Riand N, et al. Celecoxib versus ibuprofen in the prevention of heterotopic ossification following total hip replacement: a prospective randomised trial. J Bone Joint Surg Br 2007;89:155–159.
Schmitz JP, Hollinger JO. The critical size defect as an experimental model for craniomandibulofacial nonunions. Clin Orthop Relat Res 1986;205:299–308.
Schrage YM, Machado I, Meijer D, Briaire-de Bruijn I, et al. COX-2 expression in chondrosarcoma: A role for celecoxib treatment? Eur J Cancer 2010;46:616–624.
Seidenberg AB, An YH. Is there an inhibitory effect of COX-2 inhibitors on bone healing? Pharmacol Res 2004;50:151–156.
Shamir D, Keila S, Weinreb M. A selective EP4 receptor antagonist abrogates the stimulation of osteoblast recruitment from bone marrow stromal cells by prostaglandin E2 in vivo and in vitro. Bone 2004;34:157–162.
Shih MS, Norrdin RW. Effect of prostaglandin E2 on rib fracture healing in beagles: histomorphometric study on periosteum adjacent to the fracture site. Am J Vet Res 1986;47:1561–1564.
Shimizu N, Ozawa Y, Yamaguchi M, Goseki T, et al. Induction of COX-2 expression by mechanical tension force in human periodontal ligament cells. J Periodontol 1998;69:670–677.
Simon AM, Manigrasso MB, O'Connor JP. Cyclo-oxygenase 2 function is essential for bone fracture healing. J Bone Miner Res 2002;17:963–976.
Simon AM, O'Connor JP. Dose and time-dependent effects of cyclooxygenase-2 inhibition on fracture-healing. J Bone Joint Surg Am 2007;89:500–511.
Spector JA, Mehrara BJ, Greenwald JA, Saadeh PB, et al. Osteoblast expression of vascular endothelial growth factor is modulated by the extracellular microenvironment. Am J Physiol Cell Physiol 2001;280:C72–C80.
Spiro AS, Beil FT, Baranowsky A, Barvencik F, et al. BMP-7-induced ectopic bone formation and fracture healing is impaired by systemic NSAID application in C57BL/6-mice. J Orthop Res 2010;28: 785–791.
Steinbrech DS, Mehrara BJ, Saadeh PB, Greenwald JA, et al. VEGF expression in an osteoblast-like cell line is regulated by a hypoxia response mechanism. Am J Physiol Cell Physiol 2000;278:C853–C860.
Stolina M, Adamu S, Ominsky M, et al. RANKL is a marker and mediator of local and systemic bone loss in two rat models of inflammatory arthritis. J Bone Miner Res 2005;20:1756–1765.
Sudmann E, Bang G. Indomethacin-induced inhibition of haversian remodelling in rabbits. Acta Orthop Scand 1979;50:621–627.
Sudmann E, Hagen T. Indomethacin-induced delayed fracture healing. Arch Orthop Unfallchir 1976;85:151–154.

Sudmann E, Dregelid E, Bessesen A, Morland J. Inhibition of fracture healing by indomethacin in rats. Eur J Clin Invest 1979;9:333–339.

Sudmann E, Tveita T, Hald J Jr. Lack of effect of indomethacin on ordered growth of the femur in rats. Acta Orthop Scand 1982;53:43–49.

Sun JS, Tsuang YH, Lin FH, Liu HC, et al. Bone defect healing enhanced by ultrasound stimulation: an in vitro tissue culture model. J Biomed Mater Res 1999;46:253–261.

Suponitzky I, Weinreb M. Differential effects of systemic prostaglandin E2 on bone mass in rat long bones and calvariae. J Endocrinol 1998;156:51–57.

Suzawa T, Miyaura C, Inada M, Maruyama T, et al. The role of prostaglandin E receptor subtypes (EP1, EP2, EP3, and EP4) in bone resorption: an analysis using specific agonists for the respective EPs. Endocrinology 2000;141:1554–1559.

Tai H, Miyaura C, Pilbeam CC, Tamura T, et al. Transcriptional induction of cyclooxygenase-2 in osteoblasts is involved in interleukin-6-induced osteoclast formation. Endocrinology 1997;138: 2372–2379.

Taketa T, Sakai A, Tanaka S, Nakai K, et al. Selective cyclooxygenase-2 inhibitor prevents reduction of trabecular bone mass in collagen-induced arthritic mice in association with suppression of RANKL/OPG ratio and IL-6 mRNA expression in synovial tissues but not in bone marrow cells. J Bone Miner Metab 2008;26:143–151.

Tanaka M, Sakai A, Uchida S, Tanaka S, et al. Prostaglandin E2 receptor (EP4) selective agonist (ONO-4819.CD) accelerates bone repair of femoral cortex after drill-hole injury associated with local upregulation of bone turnover in mature rats. Bone 2004;34:940–948.

Tang CH, Yang RS, Huang TH, Lu DY, et al. Ultrasound stimulates cyclooxygenase-2 expression and increases bone formation through integrin, focal adhesion kinase, phosphatidylinositol 3-kinase, and Akt pathway in osteoblasts. Mol Pharmacol 2006;69:2047–2057.

Thomas J, Taylor D, Crowell R, Assor D. The effect of indomethacin on Achilles tendon healing in rabbits. Clin Orthop Relat Res 1991;308–311.

Thorsen K, Kristoffersson AO, Lerner UH, Lorentzon RP. In situ microdialysis in bone tissue: stimulation of prostaglandin E2 release by weight-bearing mechanical loading. J Clin Invest 1996;98:2446–2449.

Tornkvist H, Lindholm TS. Effect of ibuprofen on mass and composition of fracture callus and bone: an experimental study on adult rat. Scand J Rheumatol 1980;9:167–171.

Tornkvist H, Lindholm TS, Netz P, Stromberg L, et al. Effect of ibuprofen and indomethacin on bone metabolism reflected in bone strength. Clin Orthop Relat Res 1984: 255–259.

Trappe TA, Carroll CC, Jemiolo B, Trappe SW, et al. Cyclooxygenase mRNA expression in human patellar tendon at rest and after exercise. Am J Physiol Regul Integr Comp Physiol 2008;294:R192–R199.

Tsai WC, Hsu CC, Chou SW, Chung CY, et al. Effects of celecoxib on migration, proliferation and collagen expression of tendon cells. Connect Tissue Res 2007;48:46–51.

Tsukii K, Shima N, Mochizuki S, Yamaguchi K, et al. Osteoclast differentiation factor mediates an essential signal for bone resorption induced by 1 alpha,25-dihydroxyvitamin D3, prostaglandin E2, or parathyroid hormone in the microenvironment of bone. Biochem Biophys Res Commun 1998;246: 337–341.

Urakawa H, Nishida Y, Naruse T, Nakashima H, et al. Cyclooxygenase-2 overexpression predicts poor survival in patients with high-grade extremity osteosarcoma: a pilot study. Clin Orthop Relat Res 2009;467:2932–2938.

Urist M. Bone formation by osteoinduction. Science 1965;150:893–899.

Usitalo H, Hiltunen A, Ahonen M, Gao TJ, et al. Accelerated up-regulation of L-Sox5, Sox6, and Sox9 by BMP-2 gene transfer during murine fracture healing. J Bone Miner Res 2001;16:1837–1845.

Vaibhav B, Nilesh P, Vikram S, Anshul C. Bone morphogenic protein and its application in trauma cases: a current concept update. Injury 2007;38:1227–1235.

Vaile JH, Davis P. Topical NSAIDs for musculoskeletal conditions: a review of the literature. Drugs 1998;56:783–799.

van der Heide HJ, Hannink G, Buma P, Schreurs BW. No effect of ketoprofen and meloxicam on bone graft ingrowth: a bone chamber study in goats. Acta Orthop 2008;79:548–554.

van Staa TP, Leufkens HG, Cooper C. Use of nonsteroidal anti-inflammatory drugs and risk of fractures. Bone 2000;27:563–568.

Vielpeau C, Joubert JM, Hulet C. Naproxen in the prevention of heterotopic ossification after total hip replacement. Clin Orthop Relat Res 1999: 279–288.

Virchenko O, Skoglund B, Aspenberg P. Parecoxib impairs early tendon repair but improves later remodeling. Am J Sports Med 2004;32:1743–1747.

Wadleigh DJ, Herschman HR. Transcriptional regulation of the cyclooxygenase-2 gene by diverse ligands in murine osteoblasts. Biochem Biophys Res Commun 1999;264:865–870.

Wahlstrom O, Risto O, Djerf K, Hammerby S. Heterotopic bone formation prevented by diclofenac: prospective study of 100 hip arthroplasties. Acta Orthop Scand 1991;62:419–421.

Warden SJ, Avin KG, Beck EM, DeWolf ME, et al. Low-intensity pulsed ultrasound accelerates and a nonsteroidal anti-inflammatory drug delays knee ligament healing. Am J Sports Med 2006;34:1094–1102.

Wax MK, Reh DD, Levack MM. Effect of celecoxib on fasciocutaneous flap survival and revascularization. Arch Facial Plast Surg 2007;9:120–124.

Weinreb M, Machwate M, Shir N, Abramovitz M, et al. Expression of the prostaglandin E(2) [PGE(2)] receptor subtype EP(4) and its regulation by PGE(2) in osteoblastic cell lines and adult rat bone tissue. Bone 2001;28:275–281.

Weise M, De-Levi S, Barnes KM, et al. Effects of estrogen on growth plate senescence and epiphyseal fusion. Proc Natl Acad Sci USA 2001;98:6871–6876.

Wergedal JE, Sheng MH, Ackert-Bicknell CL, Beamer WG, et al. Genetic variation in femur extrinsic strength in 29 different inbred strains of mice is dependent on variations in femur cross-sectional geometry and bone density. Bone 2005;36:111–122.

Wong BR, Rho J, Arron J, Robinson E, et al. TRANCE is a novel ligand of the tumor necrosis factor receptor family that activates c-Jun N-terminal kinase in T cells. J Biol Chem 1997;272:25190–194.

Wurnig C, Schwameis E, Bitzan P, Kainberger F. Six-year results of a cementless stem with prophylaxis against heterotopic bone. Clin Orthop Relat Res 1999;361:150–158.

Xie C, Ming X, Wang Q, Schwarz EM, et al. COX-2 from the injury milieu is critical for the initiation of periosteal progenitor cell mediated bone healing. Bone 2008;43:1075–1083.

Xie C, Liang B, Xue M, Lin AS, et al. Rescue of impaired fracture healing in COX-2$^{-/-}$ mice via activation of prostaglandin E2 receptor subtype 4. Am J Pathol 2009;175:772–785.

Xu M, Choudhary S, Goltzman D, Ledgard F, et al. Do cyclooxygenase-2 knockout mice have primary hyperparathyroidism? Endocrinology 2005;146:1843–1853.

Xu M, Choudhary S, Voznesensky O, Gao Q, et al. Basal bone phenotype and increased anabolic responses to intermittent parathyroid hormone in healthy male COX-2 knockout mice. Bone 2010;47:341–352.

Xu Z, Choudhary S, Okada Y, Voznesensky O, et al. Cyclooxygenase-2 gene disruption promotes proliferation of murine calvarial osteoblasts in vitro. Bone 2007;41:68–76.

Yamakawa K, Kamekura S, Kawamura N, Saegusa M, et al. Association of microsomal prostaglandin E synthase 1 deficiency with impaired fracture healing, but not with bone loss or osteoarthritis, in mouse models of skeletal disorders. Arthritis Rheum 2008;58:172–183.

Yang RS, Liu TK, Lin-Shiau SY. Increased bone growth by local prostaglandin E2 in rats. Calcif Tissue Int 1993;52:57–61.

Yoshida K, Oida H, Kobayashi T, Maruyama T, et al. Stimulation of bone formation and prevention of bone loss by prostaglandin E EP4 receptor activation. Proc Natl Acad Sci USA 2002;99:4580–4585.

Yuan J, Murrell GA, Wei AQ, Appleyard RC, et al. Addition of nitric oxide via nitroflurbiprofen enhances the material properties of early healing of young rat Achilles tendons. Inflamm Res 2003;52:230–237.

Zhang X, Morham SG, Langenbach R, Young DA, et al. Evidence for a direct role of cyclo-oxygenase 2 in implant wear debris-induced osteolysis. J Bone Miner Res 2001;16:660–670.

Zhang X, Schwarz EM, Young DA, Puzas JE, et al. Cyclooxygenase-2 regulates mesenchymal cell differentiation into the osteoblast lineage and is critically involved in bone repair. J Clin Invest 2002;109:1405–1415.

CHAPTER 3

RENAL SYSTEM

INTRODUCTION

In the kidney, prostaglandins (PGs) are important mediators of hemodynamic regulation, salt and water homeostasis, and renin (REN) release (Harris and Breyer, 2001). The main PGs in the kidney are PGE_2 and prostacyclin (PGI_2), which is synthesized from arachidonic acid (AA) by enzymatic reactions, particularly cyclooxygenases and prostaglandin E synthases (PGESs) (Whorton et al., 1978; Radi and Ostroski, 2007). Cyclooxygenase (COX)-derived PGs have two membrane-anchored isoenzymes, COX-1 and COX-2. COX-1 is expressed constitutively and found in most normal body tissues, while COX-2 is expressed in normal tissues at a low level and is strongly induced by proinflammatory mediators (Radi and Ostroski, 2007; Radi, 2009). The membrane-associated PGES-1 (mPGES-1) is inducible and linked functionally to COX-2, while mPGES-2 is constitutive and coupled to both COX isoforms (Radi and Ostroski, 2007). These PG-derived COXs and PGEs play a role in the kidney, and their inhibition can contribute to alterations in renal homeostasis. For example, because COX-1 is expressed in the collecting ducts of the nephron, an active area in the regulation of sodium excretion in both laboratory animals and humans (Currie and Needleman, 1984), it is not surprising that blood pressure increases in COX-1-deficient mice compared with wild-type controls (Kawada et al., 2005).

Nonselective nonsteroidal anti-inflammatory drugs (ns-NSAIDs), which inhibit both COX-1 and COX-2, may aggravate REN-independent sodium-sensitive hypertension, possibly in part by inhibition of the COX-1 responsible for sodium excretion (Okumura et al., 2002). Thus, the reduction in renal blood flow and sodium and water retention contributes to the adverse effects of NSAIDs on the kidney, especially in humans or animals suffering from conditions with altered renal function, such as salt depletion, lupus nephritis, liver cirrhosis, or nephritic syndrome. The incidence of NSAID-associated renal side effects ranges from 1 to 5% (DeMaria and Weir, 2003). However, it is important to consider the interspecies renal physiological and anatomical differences that may exist and differences in the involvement of COX in the generation of renal PGs under normal and altered kidney function. The comparative aspects of the renal system

Comparative Pathophysiology and Toxicology of Cyclooxygenases, First Edition. Zaher A. Radi.

are reviewed in this chapter, with a discussion of the pathophysiological effects of COX inhibition on electrolyte handling and the renal system.

COMPARATIVE PHYSIOLOGICAL, DEVELOPMENTAL, AND ANATOMICAL ASPECTS OF THE RENAL SYSTEM

Understanding the comparative physiological, developmental, and anatomical aspects of renal function in diverse species is important for an understanding of the pathological and toxicological consequences of COX inhibition in the kidney. The kidneys are paired organs located in the posterior part of the abdomen. The mammalian kidney develops through three overlapping, successive stages: the pronephrons, the mesonephrons, and the metanephrons (the adult permanent kidney) (Dressler, 2006). The metanephric kidney undergoes three basic phases of growth. The first phase is called nephrogenesis, the development of new nephrons. Nephrogensis varies across species and is completed during the perinatal period in most mammalian species. In humans, nephrogenesis starts during week 5 of gestation and is complete at 36 weeks of fetal gestation (MacDonald and Emery, 1959). The period between gestation weeks 18 to 32 is a critical point in renal development in humans because nephrogenetic development reaches its peak (Gasser et al., 1993). Nephrogenesis continues after birth for 2 weeks in the dog (Kleinman and Reuter, 1973), 4 to 6 weeks in the rat (Solomon, 1977), and 3 weeks in the pig (Friis, 1980). The puppy kidney is immature in both structure and function at birth (Evan et al., 1979). Changes in glomerular and tubular growth in the dog are complete by approximately 2 months of age. In monkeys, nephrogenesis is complete at birth (Lumbers, 1995), and in mice, sheep, and guinea pig, nephrogenesis completes before birth (Fouser and Avner, 1993; Lumbers, 1995).

Cyclooxygenase-2 may be involved in nephrogenesis in humans, and its inhibition using NSAID inhibitors during the third trimester may be responsible for fetal renal syndromes (Khan et al., 2001b). In utero exposure to early prolonged high-dose indomethacin or ibuprofen resulted in kidney structural abnormalities in the developing human fetus and renal failure in neonates (Kaplan et al., 1994). Kidney postnatal development was evaluated in COX-2 null mice. Mice homozygous for a targeted deletion of COX-2 ($^{-/-}$) were compared with wild-type littermates ($^{+/+}$). Beginning at postnatal day 10 (PN10), kidney growth was suppressed in COX-2 ($^{-/-}$) mice. By PN10, kidneys from COX-2 ($^{-/-}$) mice had thin nephrogenic cortexes and small, crowded subcapsular glomeruli. The kidney pathological changes increased with age with progressive outer cortical dysplasia, cystic subcapsular glomeruli, loss of proximal tubules, and tubular atrophy and cyst formation. Therefore, deficiency of COX-2, at least in null mice, resulted in progressive and specific renal architectural disruption and functional deterioration beginning in the final phases of nephrogenesis (Norwood et al., 2000). Chronic suppression of PG synthesis by indomethacin, an ns-NSAID, in the pregnant rhesus monkey during late gestation resulted in fetal renal dysgenesis (Novy, 1978). In a recent study using the flatworm *Schmidtea mediterranea*, interference with the expression of epidermal growth factor receptor (EGFR5) during the regeneration process showed that protonephridia could no longer undergo branching morphogenesis. Thus, EGFR5 appears to be a crucial regulator of branching morphogenesis and maintenance (Rink et al., 2011). Such a model could be relevant for studying kidney physiology, nephrogenesis, and regeneration and the effects of NSAIDs in mammalian species.

The kidney is bean-shaped in most mammals, whereas it is heart-shaped in the horse and lobulated in the ox (Braun, 1991; Hoffmann et al., 2000). Regarding blood flow, in humans and most mammals, each kidney is supplied by a single renal artery, but it is not uncommon for one or more accessory renal arteries to be present (el-Galley and Keane, 2000). Two distinct anatomical regions can be identified on the cut surface of the kidney: an outer region called the cortex and an inner region called the medulla (Fig. 3-1). In the mammalian kidney, renal blood flow is regulated via the afferent arterioles. In avian species, however; the primary nutrient blood source to the renal cortex is the renal portal system via the afferent renal portal vein. The primary blood supply to the avian glomeruli and distal convoluted tubules is the central artery, located between the intralobular (central) vein and the cortex (Meteyer et al., 2005). Normal renal function is predicated on adequate perfusion (pressure and volume). Although the kidneys are only about 0.5% of body weight, they typically receive 20 to 25% of the cardiac output. Only about 10% of oxygen consumption occurs in the kidneys. There is heterogeneity in intrarenal flow distribution in the kidney. Approximately 85% of renal flow can be associated with cortex, 14% with outer medulla, and 1% with inner medulla (Khan et al., 2011). The very low oxygen tension in the renal medulla makes it vulnerable to NSAID-mediated hypoxic injury, which can be prevented by PGs (Brezis and Rosen, 1995).

The comparative anatomical aspects of the mammalian renal pelvis and medulla have been investigated (Pfeiffer, 1968). Simple renal medulla is present in most mammals (rabbits, guinea pigs, cats, dogs, pigs, humans). Complex renal medulla is present in animals with a high urine-concentrating ability (rat, mouse, desert sand rat). The medulla in the kidney of nonhuman primates (rhesus macaques, cynomolgus macaques) is relatively small compared to the medulla in the kidney of human. In humans, beavers, and pigs, the renal medulla is divided into striated conical masses (pyramids), and thus the kidney from these species is termed multipapillate. Cats, horses, dogs, rats, nonhuman primates, and many other laboratory animals have a single renal pyramid, and the kidney in these species is called unipapillate. The renal pelvis is lined by transitional epithelium and represents the expanded portion of the upper urinary tract. In mammals possessing a unipapillate kidney, the papilla is surrounded directly by the renal pelvis (Pfeiffer, 1968). In some species, such as horses, many mucous glands are present in the renal pelvis (Mutoh et al., 1992).

The primary functions of kidneys can be divided into four major categories: (1) control of the body's fluid and electrolytes, (2) elaboration of hormones, (3) excretion of the waste products of metabolism, and (4) selected metabolic activities. Nitrogenous waste products consisting of urea, creatinine, and ammonia

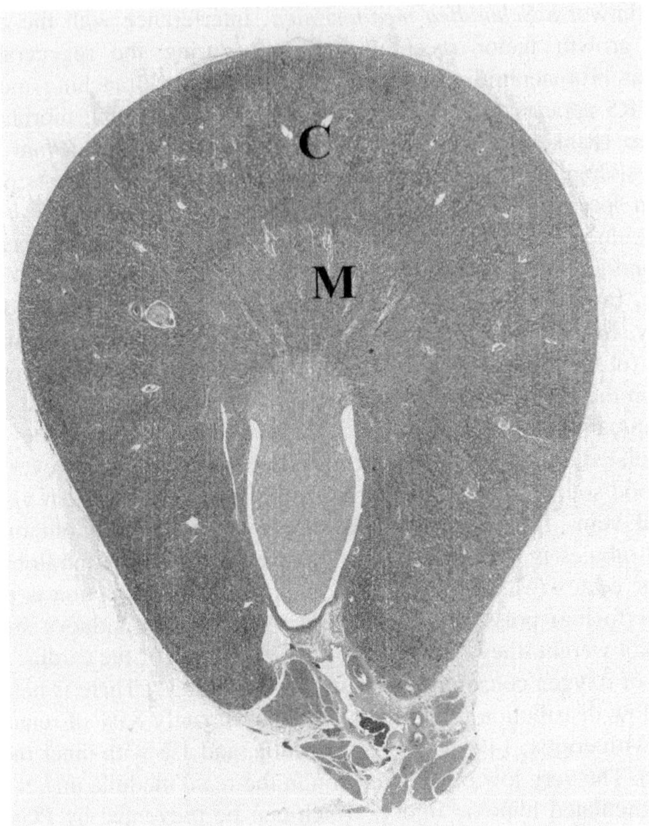

FIGURE 3-1 Two distinct anatomical regions can be identified on the cut surface of the kidney: an outer region called the cortex (C) and an inner region called the medulla (M).

ion are excreted in the urine; thus, any significant alterations in renal function are manifested in retention of these waste products in the body (Khan et al., 2011). The kidney excretes nitrogenous waste products and is central to body water and solute homeostasis. In avian species, during embryonic development of the bird within the shelled egg, nitrogenous waste products accumulate and cannot be voided. Insoluble uric acid is the main by-product rather than urea (Lierz, 2003). About 65% of uric acid is chemically bound to proteins. Thus, hyperuricemia is an indication of kidney injury in birds. In pigs and rabbits, secretion of urate commonly occurs by the proximal tubules (Chonko, 1980; Roch-Ramel et al., 1980). In humans, tubular reabsorption of filtered urate predominates, but tubular secretion also occurs (Roch-Ramel, 1979).

The nephron is the basic structural and functional unit of the kidney that performs such primary functions. The main nephron components consist of the renal or malpighian corpuscle (glomerulus and Bowman's capsule), the proximal tubule, the thin limbs, the distal tubule, and the connecting segment or connecting tubule (Khan et al., 2011). There are species differences in kidney nephron numbers. For example, each human kidney contains about 1,000,000 nephrons, adult rat kidney approximately 30,000 nephrons, dog kidney approximately 500,000 nephrons, cat kidney approximately 200,000 nephrons, and pig kidney approximately 1,250,000 nephrons. Furthermore, the interspecies and strain differences in the number of nephrons can have an impact on the development of renal pathologies. The F344 rat strain is shown to be more vulnerable to development of progressive kidney damage, due to the relatively low nephron number (Szabó et al., 2008). Spontaneously hypertensive rats (SHR) have 10 to 25% fewer nephrons than do their normotensive counterparts (Skov et al., 1994). There is evidence that fetal growth restriction is associated with impaired nephrogenesis and reduced numbers of mature nephrons (Hinchliffe et al., 1992; Bassan et al., 2000). In addition, because of the fewer nephrons, the younger animals have much less capacity to concentrate urine than do the adults. In mice with a null mutation of the gene for COX-2, there is a severe reduction in the number of functional nephrons developed in the fetal kidney (Dinchuk et al., 1995). In Sprague–Dawley rats, reduction in COX-2 activity during the nephrogenic period leads to a modest but significant reduction in nephron number. However, the consequent age-dependent deterioration of the renal structure and renal function is significantly enhanced only in male rats (Sáez et al., 2009). The comparative lengths of various nephron segments are detailed in Table 3-1.

Two main populations of nephrons are those possessing a short loop of Henle and those with a long loop of Henle (juxtamedullary). In birds, cortical (reptilian type) nephrons are most common (up to 90%) and are located in the lobule cortex. Their glomeruli are smaller and they do not possess the loop of Henle. Medullary (mammalian type) nephrons have long loops of Henle, reaching far into the medulla (Lierz, 2003). This is important because the countercurrent hypothesis for urine concentration states that the maximal urine concentration that can be achieved is related directly to the length of loops of Henle (Gottschalk and Mylle, 1958). Within the mammalian kidney, urea concentration (creating hypertonicity) within the medulla is the key factor enabling concentration of urine. As urea is not an important product of metabolism in birds and loops of Henle are absent in 90% of avian nephrons, the capacity of the kidney in birds to concentrate urine is less than in mammals. Water is conserved by resorption within the coprodeum and rectum (Lierz, 2003).

TABLE 3-1 Comparative Lengths of Various Nephron Segments

	Range of nephron segment length (mm)			
Species	Proximal tubule	Distal tubule	Loop of Henle	References
Human	7.2–23.1	1.6–4.2	0–5.6	Pai, 1935
Nonhuman primate (rhesus monkey)	3.6–9.6	1.4–6.6	0.9–1.9	Bennet et al., 1968
Dog	9–24	3–3.9	5.2–15	Watson, 1966; Bennet et al., 1968
Rat	9–13.5	2.5–2.7	5–7.3	Walker and Oliver, 1941

The glomeruli, capillary networks lined by a thin layer of endothelial cells, have a central region of mesangial cells (glomerular mesangium) with surrounding mesangial matrix material, visceral epithelial cells (podocytes) and an associated basement membrane, parietal epithelial cells of Bowman's capsule and glomerular basement membrane, and Bowman's space or urinary space (narrow cavity between the two visceral and parietal epithelial layers) (Khan et al., 2011). The four main functions of the glomeruli are (1) plasma ultrafiltration and removal of plasma solutes (e.g., glucose, urea, electrolytes) and water; (2) regulation of blood pressure through various molecules, such as angiotensin and aldosterone; (3) regulation of tubular metabolism via regulating peritubular blood flow; and (4) removal of macromolecules (e.g., soluble immune complexes) from the circulation (Robertson, 1998). Thus, the glomerular filtration rate (GFR) is the basic determinant in urine formation and a key measurement of kidney function. GFR continues to increase after birth and reaches adult levels at 1 to 2 years of age in humans (Gomez et al., 1999). In rats, GFR increases sharply during the first weeks of postnatal life (Horster, 1977). Discrimination in GFR for larger molecules occurs based on molecular size, configuration, and charge. For example, small molecules (approximately 5000 Dal) pass through the glomerular barrier, and molecules such as albumin (approximately 68,000 Dal) do not pass the barrier. Any pathophysiological process that interferes with the structural integrity and function of the glomerulus can result in abnormal loss of protein in the urine, predominantly albumin. Therefore, a hallmark of drugs that damage glomerular epithelial cells in mammals is the induction of proteinuria. However, in contrast to most other mammals, proteinuria is normal in mice, with the highest levels in sexually mature male mice (Wicks, 1941). Mice excrete only a drop or two of urine at a time, and it is highly concentrated (Fox et al., 2002). The filtering system per gram of tissue is twice that of the rat. They can concentrate urine to 4300 mOsm/L compared to 1160 mOsm/L for humans. Large amounts of protein, including creatinine, in the urine are normal, which is different from the situation in other mammals (Fox et al., 2002). To some extent, the high total solids content of urine and high specific gravity reflect this passage of protein into the urine (Parfentjev and Perlzweig, 1933). In addition, there are gender and strain differences in the mouse kidney, and sexual maturity in male mice leads to sexual dimorphism in the kidney, as evidenced by larger renal cortices, larger renal corpuscles, and cuboidal epithelium lining the parietal layer of Bowman's capsule (Ahmadizadeh et al., 1984). The parietal epithelium layer is mainly squamous in DBA female mice (Ahmadizadeh et al., 1984). Furthermore, sexual dimorphism in renal and plasma potassium content was demonstrated in adult male CD1 mice. These mice have a higher plasma potassium concentration than that of females or 40-day-old male mice (Cremades et al., 2003). Such anatomical differences can play a role in susceptibility to toxic renal insults. The glomeruli of neonatal pigs are permeable to colostral protein for a few days (Bergelin and Karlsson, 1974). Administration of indomethacin, an ns-NSAID, to women in threatened preterm labor results in abnormalities in glomerular and tubular morphology in the fetal kidney and a loss of fetal and neonatal urine production (Kaplan et al., 1994; van der Heijden et al., 1994).

Visceral epithelium is continuous with the parietal epithelium at the vascular pole, where the afferent and efferent arterioles enter and exit the glomerulus, and the parietal layer of Bowman's capsule continues into the epithelium of the proximal tubule at the urinary pole. The average diameter of the glomerulus is approximately 120, 130, and 200 μm in the rat, beagle dog, and human kidney, respectively (Khan et al., 1991). Glomerular growth during human childhood has been evaluated. It was found that the estimated mean maximum glomerular diameter, which is 112 μm for newborns, increased approximately 3.6 μm per year, reaching 167 μm at age 15. Glomerular size correlated better with age than with height, weight, or body mass index, and glomeruli in the inner cortex were significantly larger than those in the middle and outer cortex. Glomerular size was significantly greater in female children than in male children over the age range examined (Moore et al., 1993). As mentioned above, the glomerulus is responsible for the production of an ultrafiltrate of plasma. Filtration barrier between the blood and the urinary space is composed of fenestrated endothelium, peripheral glomerular basement membrane, and slit pores between the foot processes of the visceral epithelial cells. Vascular endothelial growth factor (VEGF) is produced by the visceral epithelial cells, and VEGF receptors are expressed on the surface of glomerular endothelial cells. VEGF is an important regulator of endothelial cell function and permeability, induces the mobilization of caveolae and formation of fenestrae, and is important for endothelial cell survival and repair in glomerular diseases that result in endothelial cell damage (Veron et al., 2010). The effects of an s-NSAID, celecoxib, on fetal renal growth in rabbits were evaluated. Celecoxib was given orally at a dose level of 30 mg/kg per day from 13 to 20 days of gestation (group 1) or from day 13 to day 28 of gestation (group 2). The fetuses were delivered by cesarean section at 29 days of gestation. This maternal administration of celecoxib at therapeutic doses did not affect maternal fetal renal growth in either group. However, VEGF levels were elevated in group 2 (Hartleroad et al., 2005).

Visceral epithelial cells (podocytes) are the largest cells in the glomerulus. They have long cytoplasmic processes (trabeculae) that extend from the main cell body and divide into individual foot processes called pedicels that come into direct contact with the lamina rara externa of the glomerular basement membrane (GBM). The GBM is a size-dependent barrier, and particles with radius greater than 3.5 nm do not pass through such a barrier. Podocytes are capable of endocytosis, and the heterogeneous content of their lysosomes reflects the uptake of proteins and other components from the ultrafiltrate. In addition, large numbers of microtubules, microfilaments, and intermediate filaments are present in the cytoplasm of podocytes. Filtration slit or slit pore is bridged by a thin membrane called the filtration slit membrane or slit diaphragm. Nephrin and CD2-associated protein (CD2AP) are required for normal filtration to occur. Nephrin is a transmembrane protein that belongs to the immunoglobulin family of adhesion molecules and located in the slit diaphragm. CD2AP is an adapter molecule that binds to nephrin and is believed to connect nephrin to the cytoskeleton (Pavenstädt et al., 2003). It has been demonstrated that preexisting expression of podocyte COX-2 in a mouse strain (B6/D2) that otherwise is resistant to adriamycin injury sensitized the podocytes to injury, indicated by foot process effacement and albuminuria (Cheng et al., 2007). These responses were associated with a further increase in endogenous podocyte COX-2 expression and a decrease in nephrin expression. Long-term treatment with the s-NSAID SC58236 attenuated the albuminuria and foot process effacement and restored nephrin expression. In addition, in BALB/c mice endogenous podocyte COX-2 expression also increased in association with adriamycin-induced podocyte injury (Cheng et al., 2007). Thus, overexpression of COX-2 predisposes to podocyte injury. Podocin is connected to both nephrin and CD2AP. Podocin is an integral membrane protein that is expressed in the foot process membrane at the site of insertion of the slit diaphragm (Takeda, 2003; Fukasawa et al., 2009).

Podocyte injury is a major cause of progressive glomerular damage. Prostaglandin synthesis has been demonstrated in rat glomerular epithelial cells (Kreisberg et al., 1982). Increased podocyte COX-2 expression has been reported in rat models of subtotal renal ablation (Wang et al., 1998; Kong et al., 2003) and diabetic nephropathy (Komers et al., 2001). Various membrane components have been identified on the surface of podocytes. Podocalyxin is a principal sialoprotein on the urinary surface of the podocytes. Plasma membrane of the foot processes has a negatively charged surface coat rich in sialic acid (Takeda, 2003). Podoplanin is a 43-kDa membrane protein identified in the plasma membrane, which has a role in the maintenance of podocyte shape. Anionic sites on the podocytes as well as the presence of an intact slit diaphragm are important in establishing the permi-selective properties of the filtration barrier. Recent data suggest that podocyte injury can be repaired via parietal epithelial cells of Bowman's capsule, which may represent a potential source for podocyte regeneration. An abnormal proliferative response of renal progenitors to podocyte injury can generate hyperplastic glomerular lesions, which are observed in several glomerular disorders (Smeets et al., 2009; Romagnani, 2011).

The glomerular mesangium represents the central region of the glomerulus and is comprised of mesangial cells (intercapillary arrays) and their surrounding matrix material and separated from the capillary lumen by the endothelium. In mammals, the glomerular capillaries form a complex network of anastomosing channels; while in birds single unbranched glomerular capillary coils are present around the periphery of the renal corpuscle (Casotti and Braun, 1995). Mesangial cells represent a specialized pericyte and possess many of the functional properties of smooth muscle cells (Mené, 1996). The contractile mesangial cell processes bridge the gap in the glomerular basement membrane encircling the capillary, prevent capillary wall distention secondary to elevation of intracapillary pressure, and play a role in the regulation of glomerular filtration. Contraction decreases glomerular filtration by reducing blood flow through capillary loops. Cell contraction is stimulated by a variety of vasoactive agents such as angiotensin II (Ang II), leukotrienes, and platelet-activating factor (Mené, 1996). Mesangial cells possess membrane receptors for peptidic and nonpeptidic hormones, have signaling pathways involving intracellular cAMP and cGMP accumulation, and synthesize proteins such as extracellular matrix protein (ECM) and serine proteases. Mesangial cells exhibit phagocytic properties and participate in the clearance or disposal of macromolecules from the mesangium (Mené, 1996). The mesangial matrix possesses fixed, negatively charged sites that influence the filtration of macromolecules. Thus, mesangial cells play a central role in the physiology and pathophysiology of the glomerulus.

There are significant species differences in the glomerulus mesangium. In the glomerulus of humans, only contractile resident mesangial cells (and not resident macrophages) have been identified (Sraer et al., 1993). Macrophages are detected only in pathological conditions in human kidney glomeruli, which suggests that they are not resident but rather, infiltrating cells. Mesangial receptors, most notably Ang II receptors, are present on mesangial cell membranes and are linked to PGE_2 synthesis and to cell contraction. In humans, spontaneous prostanoid synthesis is low and is increased by the induction of cyclooxygenase by sodium butyrate in the medium. The amount of PGE_2 synthesized by human mesangial cells is quantitatively low compared with that in rats (Sraer et al., 1993). In rats, the role of the renin–angiotensin system in the glomerulus is clearly demonstrated, including the presence of Ang II receptors on isolated glomeruli and mesangial cells. Mesangial cells play a role in the regulation of single-nephron GFR. In humans, Ang II also exerts control on GFR, but a direct contractile effect of Ang II on mesangial cells is more difficult to demonstrate (Sraer et al., 1993). Cyclooxygenase-2 may participate in a number of pathological processes in immune-mediated renal glomerular diseases. In human mesangial cells in vitro, progressive accumulation of collagen I in the extracellular medium induced a significant increase in COX-2 expression. Focal adhesion kinase (FAK), the phosphatidylinositol 3-kinase (PI3K)/AKT pathway, and cAMP response element (CRE) activation were demonstrated with such increased COX-2 expression. Thus, collagen type I increases COX-2 expression via the FAK/PI3K/AKT/cAMP response element binding protein signaling pathway (Alique et al., 2011). Engulfment and cell motility 1 (ELMO1), a bipartite guanine nucleotide exchange factor (GEF) for the small GTPase Rac 1, was identified as a susceptibility gene for glomerular disease. In human mesangial cells in vitro, ELMO1 interacted with COX-2 and ELMO1 was identified as a posttranslational regulator of COX-2 activity, and COX-2 activity increased fibronectin (an ECM) promoter activity (Yang and Sorokin, 2011).

Glomerular basement membrane (GBM) is 100 to 300 nm thick and is composed of a central dense layer, the lamina densa, and two thinner, more electron-lucent layers, the lamina rara externa and the lamina rara interna (Takami et al., 1991). Special stains such as methenamine silver and periodic acid–Schiff (PAS) can be used to visualize the GBM by light microscopy. Collagen IV is the major constituent of the GBM. Glomerular basement membrane is produced by the podocytes. Anionic sites exist in all three layers of the GBM, which consist of glycosaminoglycans rich in heparan sulfate (Timpl, 1986). Glomerular basement membrane plays a role in establishing the permeability to plasma proteins. Glomerular capillary wall possesses both size- and charge-selective properties and therefore restricts the passage of molecules the size of albumin or larger. Prostaglandins may play a role in GBM immunologic diseases. For example, in an experimental rat model of membranous nephropathy, a thromboxane $A_2(TXA_2)$ synthase inhibitor that was given after proteinuria is fully developed decreased the deposition of both rabbit immunoglobulin G and rat immunoglobulin G on GBM in these nephritic rats (Nagao et al., 1996). Thus, thromboxanes may exacerbate glomerular pathology.

The glomerular juxtaglomerular apparatus is located at the vascular pole of the glomerulus. It contains two distinct cell types: juxtaglomerular granular cells of the afferent arteriole that contain renin and Ang II and agranular extraglomerular mesangial cells (lacis cells or pseudomeissnerian cells of Goormaghtigh), located between the afferent and efferent arterioles. In addition, there is a specialized tubule component called macula densa (MD), which is a specialized region of the thick ascending limb adjacent to the hilus of the glomerulus (Khan et al., 2011). The MD senses change in the distal tubule in the luminal concentrations of sodium and chloride via absorption of sodium and chloride across the luminal membrane by the Na^+–K^+–Cl^- cotransporter. Signals generated by acute changes in sodium chloride concentration are transferred via the MD cells to the glomerular arterioles to control the glomerular filtration rate and to the renin-secreting cells in the afferent arteriole. The immunohistochemical expression of COX-2 has been evaluated in human fetal kidneys ranging between 15 and 23 weeks of gestational age. Strong expression of COX-2 was localized primarily in the MD and the thick ascending limb of the loop of Henle, and in rare glomerular podocytes and vascular endothelial cells. There was a progressive decrease in COX-2 immunoreactivity from the most immature nephrons adjacent to the metanephric regions to the well-developed nephrons in the middle to inner cortex (Khan et al., 2001b). Data in rats suggest that COX-2, which is expressed in the MD, regulates tubuloglomerular feedback, at least partly, via two mechanisms: (1) generation of TXA_2 or related agonists of thromboxane/PGH_2 receptor and (2) reduction of neuronal nitric oxide synthase–dependent nitric oxide (Araujo and Welch, 2009). COX-2 colocalizes with neuronal nitric oxide synthase (nNOS) in the MD. Up-regulation of COX-2 expression has been demonstrated in porcine kidney MD with chronic NOS inhibition (Kommareddy et al., 2011). The MD-derived COX-2 products could counteract renal vasoconstriction in NOS inhibition-induced hypertension and maintain renin synthesis/secretion (Kommareddy et al., 2011).

The renin–angiotensin–aldosterone system (RAAS) plays an important role in the control of cardiovascular and renal homeostasis by regulating vascular tone, blood pressure (BP), and fluid volume. Ang II is a physiologically active component of the RAAS, produced via an enzymatic cascade that begins with angiotensinogen (AGT) cleaving REN to form Ang I, which is then cleaved by the angiotensin-converting enzyme (ACE) to form Ang II. Ang II causes vasoconstriction directly by activating Ang II type 1 (AT1) receptors on vascular smooth muscle and affects fluid volume via AT1 receptor activation in the proximal tubule, resulting in renal sodium and water reabsorption. Ang II regulates fluid balance by stimulating aldosterone secretion from the adrenal glands (Radi and Ostroski, 2007). PGs are modulators of physiological functions and contribute to REN release, regulation of renal microvascular hemodynamics, salt balance, and BP control via mechanisms involving the regulation of vascular tone and renal excretory function (Radi and Ostroski, 2007). A baseline absence of REN in juxtaglomerular cells was observed in the Toll-like receptor 4–knockout mouse. This was specific to juxtaglomerular cells, as REN was easily detected in other arteriolar segments (El-Achkar et al.,

2007). Dopamine, a neurotransmitter, serves important physiological functions in the mammalian kidney and is a major regulator of proximal tubule salt reabsorption and is a modulator of REN release. There is crosstalk between intrarenal dopaminergic and COX-2 systems. In an experimental study in male Sprague–Dawley rats, acute treatment with a D1-like agonist, fenoldopam, inhibited REN release. Two mechanisms have been proposed for such intrarenal dopaminergic and COX-2 system crosstalk: (1) indirect inhibition via decreasing proximal salt reabsorption and modulating MD COX-2 expression and activity, and (2) direct stimulation of REN via activation of D1-like receptors (Zhang et al., 2009). In birds, oriental white-backed vultures, a proposed mechanism by which the ns-NSAID diclofenac induced renal failure is through the inhibition of the modulating effect of PG on Ang II-mediated adrenergic stimulation. It is proposed that renal portal valves open in response to adrenergic stimulation, redirecting portal blood to the caudal vena cava and bypassing the kidney. If diclofenac removes a modulating effect of PGs on the renal portal valves, indiscriminant activation of these valves would redirect the primary nutrient blood supply away from the renal cortex. Therefore, this leads to ischemic necrosis of the cortical proximal convoluted tubules in these birds (Meteyer et al., 2005).

The length of various nephron segments (proximal tubule, loop of Henle, distal tubule) varies across species. Proximal convoluted tubules (PCTs) begin at the urinary pole of the glomerulus. PCTs have three morphologically distinct segments (S1, S2, S3) in the rat, rabbit, mouse, and rhesus monkey (Khan et al., 2011). There are species and strains differences in PCT. For example, Sprague–Dawley rats have somewhat fewer, but longer, PCTs than those of Wistar rats (Solomon, 1977). PCTs have brush borders that form the apical or luminal surface of the tubules with microvilli. Basolateral plasma membrane contains receptors for various peptide hormones, such as insulin, growth hormone, and parathyroid hormone (Corvilain et al., 1962). Megalin is a 600-kDa glycoprotein belonging to the low-density-lipoprotein receptor gene family and represents the antigen for the rat Heymann nephritis model of membraneous glomerulonephritis. Megalin is located in the brush border, coated pits, and endocytic vesicles. It mediates the reabsorption of numerous proteins and polypeptides and serves as a receptor for numerous ligands (e.g., β_2-microglobulin, lysozyme, cytochrome *c*, epidermal growth factor, albumin, insulin, prolactin, vitamin B_{12}, vitamin D) (Christensen and Verroust, 2002).

The major function of PCTs is reabsorption of Na^+, HCO_3^-, Cl^-, K^+, Ca^{2+}, PO_4^{3-}, calcium, water, and organic solutes (glucose and amino acids). Approximately half of the ultrafiltrate is reabsorbed. Fluid reabsorption mechanism is coupled with an active transport of Na^+. The rate of fluid absorption from the proximal tubule to the peritubular capillaries is influenced by the hydraulic and oncotic pressures across the tubule and capillary wall, the main anions transported being HCO_3^- and Cl^- with HCO^3. Approximately 70% of potassium is reabsorbed in the PCTs (Khan et al., 2011). Unlike most mammals, where calcium and magnesium are excreted mainly in the bile, urine is the major route of calcium and magnesium excretion in rabbits (Cheeke and Amberg, 1973). Peritubular interstitial cells produce erythropoietin, a glycoprotein hormone that regulates survival, proliferation, and maturation of erythroid progenitor cells into mature red blood cells, whose deficiency leads to anemia (Jelkmann, 1992). This hormone is produced in peritubular interstitial cells in several experimental animals, including rats, mice, sheep, and nonhuman primates (Fisher et al., 1996). PGE_2 can affect erythropoitein activity under hypoxic conditions (Gross et al., 1976). Renal secretion of organic electrolytes of diverse chemical structures plays a critical role in limiting the body's exposure to toxic compounds of exogenous and endogenous origin. The active renal secretion of drugs occurs via a transport process in the proximal tubules, which consists of uptake from systemic circulation via the basolateral membrane and subsequent efflux into urine through the luminal membrane by efflux transporters or passive permeability (Dresser et al., 2001). Uptake of substances in the systemic circulation may subsequently lead to efflux of such substances across the brush-border membrane of proximal tubule cells by various efflux transporters, such as the multidrug resistance P-glycoprotein (P-gp) or the breast cancer resistance protein (BCRP) (Deeley et al., 2006). Other transporters, such as the multidrug and toxin extrusion proteins 1 and 2 (MATE1, MATE2-K), have been identified in the brush-border membrane of proximal tubular cells in humans and mice (Omote et al., 2006) and play an important role in the renal secretion of xenobiotics with highly diverse chemical structures (Ohta et al., 2009).

Renal secretion of organic cations (OCs) (including bases) and organic anions (OAs) (including acids) has physiological, pharmacological, and toxicological implications. Transporters that are involved in the uptake of OC and OA across the basolateral membrane into the cells of PTCs have been identified in human kidney (Hosoyamada et al., 1999). In general, substrates for pathways involved in renal OC transport include a diverse array of primary, secondary, tertiary, or quaternary amines that have a net positive charge on the amine nitrogen at physiological pH. Renal organic ion transporters [e.g., organic anion transporter (OAT1) and organic cation transporters (OCT1 and OCT2)] play an important role in the excretion of numerous commonly used drugs, including NSAIDs. For example, ns-NSAIDs such as ketoprofen and naproxen are effective in altering the pharmacokinetic behavior of acyclovir, a guanosine analog antiviral drug, by inhibiting the OAT-mediated tubular secretion of acyclovir in rats (Gwak et al., 2005). Some studies revealed that a prototypic substrate and/or inhibitor of the renal OAT1 system consists of a hydrophobic core to which at least one negatively charged group, often a carboxyl, is attached (Ullrich, 1997). PGs are secreted into the urine by OAT system in renal proximal tubules (Nagai et al., 2006). The role of human OAT1 (hOAT1) renal transporters in adefovir-mediated transport and cytotoxicity and its relationship to NSAIDs was investigated using several NSAIDs of varied structures in both the hydrophobic core and the carboxylic acid moiety (Mulato et al., 2000). In general, the inhibitory potency did not seem to be influenced significantly by the type of carboxylic acid. Diflunisal, the most potent inhibitor of adefovir transport, which contains a carboxyl moiety attached directly to the hydrophobic aromatic ring, was only slightly more potent than some of ns-NSAIDs containing a propionic acid side chain (i.e., ketoprofen, flurbiprofen, naproxen) or acetic acid side chain (i.e., indomethacin, diclofenac). However, etodolac, which is also a derivative of acetic acid, was a poor hOAT1 inhibitor. Etodalac was the only ns-NSAID with an alkyl substituent in the immediate vicinity of the carboxylic acid side chain, which presumably decreased the planarity of the molecule and may reduce its interaction with the substrate-binding site of hOAT1. Two ns-NSAIDs lacking the carboxyl moiety (phenacetin and piroxicam) clearly showed less inhibitory effect. With respect to the hydrophobic part of the molecule, two aromatic rings linked by a short spacer appear to be characteristic for ns-NSAIDs with the highest inhibitory potency (i.e., diflunisal, ketoprofen, and flurbiprofen) (Mulato et al., 2000).

In general, there was a correlation between the inhibitory potency of NSAIDs in the adefovir transport assay and their cytoprotective effect. Ketoprofen, flurbiprofen, and naproxen were potent inhibitors of adefovir transport and exhibited cytoprotection superior to that of probenecid, a uricosuric drug that increases uric acid excretion in the urine. For diclofenac and ibuprofen, both the inhibitory activities and the cytoprotective effects were comparable to those of probenecid. Diclofenac, the most potent hOAT1 inhibitor among the NSAIDs tested, still showed a distinct cytoprotective effect, despite being tested at a significantly lower concentration, due to its intrinsic cytotoxicity. Furthermore, piroxicam, one of the least potent inhibitors, showed only a minor reduction in the hOAT1-specific cytotoxicity of adefovir. In contrast, indomethacin did exhibited almost no cytoprotection in hOAT1-expressing cells despite its efficient inhibition of hOAT1 transport activity (Mulato et al., 2000). Another study, using the rat homolog of hOAT1, found that acetylsalicylate, salicylate, and indomethacin were taken up by OAT1, and the uptake rate of these NSAIDs was enhanced by the outwardly directed dicarboxylate gradient (Apiwattanakul et al., 1999). Rat OAT1 shares 88% amino acid residues with hOAT1 and showed a higher degree of homology with hOAT1 (Cihlar et al., 1999).

The distal nephron is composed of distal convoluted tubules (DCTs), thick ascending limb (TAL), and collecting ducts (Khan et al., 2011). Sodium and water reabsorption and potassium addition take place in the distal nephron. Sodium is transported passively in the TAL but actively elsewhere in the distal nephron and is responsive to aldosterone. In contrast, chloride is transported actively in the TAL (Khan et al., 2011). The effect of aldosterone on the distal nephron is an increase in sodium reabsorption in exchange for potassium. Normal water intake occurs in response to thirst, which is stimulated by hypertonicity. Hypertonicity leads to the release of vasopressin, also called antidiuretic hormone (ADH). Vasopressin binds to receptors in the distal nephron and collecting duct cells and activates adenylyl cyclase and increases intracellular cyclic AMP. Prostaglandins inhibit the response to vasopressin, and COX inhibitors can increase the action of ADH. In rats, bradykinin regulates ion transport in the TAL via a COX-2-mediated mechanism (Rodriguez et al., 2004). In a Sprague–Dawley rats sepsis model, Toll-like receptor 4 (TLR-4) was found to be necessary for COX-2 expression only in cortical and medullary thick ascending loops (cTAL and mTAL) which express and secrete Tamm–Horsfall protein (El-Achkar et al., 2007). Thus, dependence of sepsis-induced renal COX-2 expression on TLR-4 is tubule specific in the distal nephron. The collecting ducts are composed of two cell types: principal cells and intercalated cells. Principal cells resorb sodium and water and secrete potassium, and intercalated cells secrete either hydrogen or bicarbonate and reabsorb potassium. In an experiment in mouse collecting duct principal cells it was found that (1) a short-term increase of extracellular osmolarity decreases aquaporin-2 (AQP2) expression through inhibition of AQP2 gene transcription; (2) a long-term increase of extracellular tonicity, but not osmolarity, enhances AQP2 expression via stimulation of AQP2 gene transcription, and (3) long-term hypertonicity and cAMP-protein kinase increases AQP2 expression through synergistic but independent mechanisms (Hasler et al., 2005). In addition, the collecting ducts contain heat shock protein-70 (HSP-70), which is an osmotic stress protein (Khan et al., 2011). In rats, COX-2 inhibition reduced medullary HSP-70 expression and induced papillary apoptosis in dehydration (Neuhofer et al., 2004).

ROLE OF PROSTAGLANDINS IN THE RENAL SYSTEM

PGs are lipid mediators derived from the enzymatic metabolism of arachidonic acid initiated by COX. PGs commonly have a carboxylic acid group with pK_a values of approximately 5 and exist as charged molecules at physiological pH (Nagai et al., 2006). Prostanoid synthases include PGE synthase (PGES), prostacyclin synthase (PGIS), PGD synthase (PGDS), PGF synthase (PGFS), and thromboxane (TX) synthase, responsible for $PGE_2, PGI_2, PGD_2, PGF_{2\alpha}$, and TXA_2 biosynthesis, respectively (Radi, 2009). Renal COX-1 and COX-2 activity produces five primary prostanoids: $PGE_2, PGF_{2\alpha}, PGI_2, TXA_2$, and PGD_2. These prostanoids act locally via specific transmembrane G-protein-coupled receptors designated EP (for E-prostanoid receptor), FP, DP, IP, and TP, respectively (Breyer and Breyer, 2001). The kidney is capable of synthesizing all types of PGs, especially PGE_2 and PGI_2, which influence urinary sodium excretion directly through inhibition of the tubular transport function and indirectly through the regulation of RAAS system activity. A summary of various PGs and their effects on the kidney is provided in Table 3-2.

PGE_2 synthesis is predominant along the renal tubules, the main sites of synthesis being the medullary collecting tubule and, to a lesser extent, the cortical collecting tubule (CCT) and the thin limb of Henle's loop (Bonvalet et al., 1987). The NaK-ATPase maintains transmembrane electrochemical gradients in mammalian cells through the extracellular transport K^+ in exchange for intracellular Na^+, and inhibition of such Na^+ transport in CCT leads to natriuresis. PGE_2 regulates renal tubular transport processes and can be involved in natriuresis and diuresis or the regulation of renin release. PGE_2 reduced sodium reabsorption in the juxtamedullary nephrons (Stokes and Kokko, 1977) and inhibited sodium absorption in the rabbit CCT (Warden and Stokes, 1993). Indomethacin given to patients with malignant hypertension in a 25-mg dose three times daily caused sodium retention (Frolich et al., 1976). However, some studies have suggested, using EP_1-deficient mice, that PGE_2 modulates urine concentration by acting at the EP_1 receptors, not in the CCT, within the hypothalamus to promote arginine–vasopressin synthesis in response to acute water deprivation (Kennedy et al., 2007). Thus, it is thought that PGE_2 affects at least three signaling mechanisms in the CCT via (1) increased cAMP generation, (2) inhibition of vasopressin-stimulated cAMP accumulation,

TABLE 3-2 Summary of Various Prostaglandins and their Effects in the Kidney

PG	Renal synthesis location	Renal effects
PGE_2	Medullary collecting tubule, cortical collecting tubule, thin limb of Henle's	Diuresis, vasodilatation; stimulates renin release, maintains renal blood flow and GFR, increases erythropietin
PGI_2	Vasculature, macula densa, cortical thick ascending limb	Vasodilation and renal blood flow, renin release, dieresis, natriuresis, maintains GFR
PGE_1	Vasculature, cortical tubules	Vasodilation maintains renal blood flow and GFR; decreases macrophage infiltration
PGD_2	Glomeruli	Increases renin release and renal blood flow; antifibrotic
$PGF_{2\alpha}$	Glomeruli, medulla	Vasodilation, natriuresis, dieresis
TXA_2/TXB_2	Glomeruli, smooth muscle cells in renal arterioles, transitional cell epithelium of renal pelvis, distal convoluted tubules, collecting tubules, connecting tubules	Vasoconstriction, renal hemodynamics, and GFR, renovascular actions of angiotensin II, glomerular thrombosis
PGA_2	Cortical and medullary tubules	Increases renal blood flow and erythropoietin release

and (3) increased intracellular calcium (Hébert et al., 1991). Some in vitro studies have shown that EP_2 and EP_1 mediate the regulatory effects of PGE_1 and PGE_2 on the transcription of the Na^+,K^+-ATPase in Madin–Darby canine kidney (MDCK) cells (Matlhagela et al., 2006). Using primary proximal tubule cell culture system, chronic incubation with PGE_1 and PGE_2 resulted in an increase in the levels of mRNA for Na,K-ATPase as well as the total level of cellular Na,K-ATPase (Herman et al., 2010). In another in vitro study of renal cortex incubated with PGE_2, it was demonstrated that PGE_2 inhibited both phosphate and volume absorption in mouse proximal convoluted tubules perfused in vitro and reduced brush-border membrane vesicle NaPi-2a protein abundance in the renal cortex (Syal et al., 2006).

Overproduction of PGE_2 has been shown to be associated with increased GFR in diabetic nephropathy (DeRubertis and Craven, 1993). Oral administration of an antagonist selective for the PGE receptor EP_1 subtype, ONO-8713, at a dose of 1000 ppm in regular chow prevented the progression of nephropathy in streptozotocin-induced diabetic rats. ONO-8713 reduced renal and glomerular hypertrophy, decreased mesangial expansion, inhibited transcriptional activation of transforming growth factor beta (TGFβ) and fibronectin, and completely suppressed proteinuria (Makino et al., 2002). Urinary excretion of PGE_2 is significantly elevated in type 1 (insulin-dependent) diabetic women with nephropathy compared with normoalbuminuric type 1 diabetic women and nondiabetic women, and indomethacin treatment, 150 mg/day, induced a significant reduction in urinary

PGE_2 excretion (Hommel et al., 1987). These data indicate that the PGE_2–EP_1 system plays a crucial role in the development of diabetic renal injury. Antidiuretic hormone (ADH) increases renal PGE_2 synthesis (Walker et al., 1978). In hypophysectomized dogs, indomethacin increased urine osmolality in response to ADH (Anderson et al., 1975).

It is known that PGE_2 can cause either vasodilatation, via activating EP_2/EP_4 receptors, or vasoconstriction, via activating EP_1/EP_3 receptors. Thus, a role for the PGE_2/EP_1 signaling pathway in the development of renal injury in hypertension has been proposed and investigated. Oral administration of an EP_1-selective antagonist prevented the progression of renal damage in stroke-prone spontaneously hypertensive rats (SHRSP), a model of human malignant hypertension (Suganami et al., 2003). EP_1-deficient mice have a urine concentration defect due to decreased vasopressin release, resulting in hypotension (Kennedy et al., 2007). In the *db/db* mouse (with homozygote mutation in leptin receptor) model of type 2 diabetes, systolic blood pressure was significantly elevated compared with control heterozygote male mice. Exogenous application of PGE_2 or a selective agonist of the EP_1 receptor elicited arteriolar constrictions that were significantly enhanced in *db/db* mice compared with controls. Moreover, oral administration of an EP_1 receptor antagonist, AH6809, at 10 mg/kg per day for 4 days significantly reduced the systolic blood pressure in *db/db* but not in control mice. The investigators concluded that activation of EP_1 receptors increases arteriolar tone, which could contribute to the development of hypertension in the *db/db* mice (Rutkai et al., 2009). PGE_2 can maintain GFR by dilating the afferent arterioles (Edwards, 1985; Schlondorff, 1993). In addition, medullary PGE_2 promotes renal sodium excretion via the EP_2 receptor. For example, C57BL/6J mice placed on a high-NaCl diet exhibited increased medullary COX-2 and mPGES1 expression (Chen et al., 2008). In addition, COX-2 expression dramatically increased in the MD and cortical thick ascending limb in both humans and rodents with salt and volume depletion (Harris et al., 1994; Komhoff et al., 2000). Some human syndromes, such as Bartter-like syndrome, are associated with significant renal salt and water loss. In this syndrome, there is also marked by increased urinary PGE_2 excretion (Komhoff et al., 2000). The MD cells synthesize and release PGE_2 during reduced luminal salt content, and this response is important in the control of renin release and renal vascular resistance during salt deprivation (Peti-Peterdi et al., 2003). In vitro studies of glomerular podocytes have shown that PGE_2 regulates distinct cellular functions via the EP_1 and EP_4 receptors, thereby increasing Ca^{2+} and cAMP, respectively (Bek et al., 1999). PGE_2 evoked cAMP production and renin secretion by juxtaglomerular cells from salt-deprived rats was significantly higher than with cells obtained from salt-loaded animals (Jensen et al., 1999). Thus, renocortical PGE_2 stimulates renin secretion and maintains renal blood flow during low-salt states, whereas medullary PGE_2 promotes salt excretion in response to a high salt intake (Jensen et al., 1999).

PGE_1 is a potent vasodilator of all arterioles. The protective role of PGE_1 in the kidney has been demonstrated in several studies. PGE_1 suppressed macrophage-infiltration and ameliorated injury in an experimental model of macrophage-dependent glomerulonephritis (Cattell et al., 1990). In a rat model of acute macrophage-dependent glomerular injury, administration of 15(*s*)-15-methyl PGE_1 (M-PGE_1) ameliorated proteinuria by day 4, decreased glomerular hypercellularity, and reduced the numbers of infiltrating macrophages (Cattell et al., 1990). PGE_1 in pharmacological quantities retarded the development of an autoimmune, lupus-like disorder in an MRL mouse model. The effects included (1) prevention of peripheral T-lymphoid hyperplasia, (2) preservation of T-cell mitogenic responses, (3) inhibition of immune complex–mediated glomerulonephritis, and (4) reduction in the amounts of circulating immune complexes (Kelley et al., 1981). In a study of male Wistar rats, oral administration of 200 μg/kg of a PGE_1 analog, misoprostol, for 5 days was protective against toxic and ischemic renal injury induced by cisplatin (Ozer et al., 2011). In a female Sprague–Dawley rat model of acute renal microvascular injury, PGE_1 decreased the capillary loss, relieved hypoxia, and protected renal function (Sun et al., 2011). In another study in a Sprague–Dawley rat model of acute aristolochic acid nephropathy, PGE_1 significantly ameliorated the renal microvascular injury, decreased hypoxia, and protected renal function (Sun et al., 2011). In dogs, PGE_1 infusion increased renal blood flow and significantly attenuated the postischemic fall in GFR and renal concentrating ability as well as the postischemic increase of plasma creatinine and blood urea nitrogen (Johnston et al., 1967; Torsello et al., 1989). In humans it has been suggested that PGE_1 may prevent renal dysfunction after cardiopulmonary bypass (Abe et al., 1993).

TXA_2 receptors (TPs) have been found in various regions of the kidney. Thromboxane synthase has been detected in the glomeruli of rat kidneys (Vitzthum et al., 2002). In the rat kidney, TX receptor mRNA, using in situ hybridization, was localized in glomeruli, smooth muscle cells in renal arterioles, and transitional cell epithelium of renal pelvis (Abe et al., 1995). Similarly, immunostainable TX receptor was observed in the rat kidney in the glomeruli, arterial walls, luminal membranes of thick ascending limbs of Henle's loop, the luminal and basolateral membranes of either distal convoluted tubules or connecting tubules, and the basolateral membranes of collecting tubules (Takahashi et al., 1996). Another study investigated the localization of thromboxane A_2/prostaglandin H_2(TXA_2/PGH_2) receptor protein in the rat kidney. In the mouse kidney, TXA_2 receptor mRNA was expressed most abundantly in the glomerulus, followed by the distal convoluted tubule, proximal tubule, thick ascending limb of Henle, and outer medullary collecting duct (Asano et al., 1996). The investigators found that the receptor localized both in glomeruli and in tubules. In the glomeruli, the receptor was most prominent along the lumen of glomerular capillary loops. Parietal epithelial cells of the Bowman's capsule, podocytes, and mesangial cells also had receptor expression. In the tubules, the receptor was localized most prominently at the base of the brush border of proximal tubules and at the luminal surface of thick ascending limbs and distal convoluted tubules (Bresnahan et al., 1996).

TXA_2 is a potent vasoconstrictor. In fact, injection of synthetic TXA_2 in a dose of 20 ng/kg directly into the renal artery of anesthetized pigs resulted in a dose-related decrease in renal blood flow (Cirino et al., 1990). In addition, TXA_2 regulates mesangial cell contraction through elevation of intracellular cAMP (Mené and Dunn, 1988). Improved renal function was noted in a randomized clinical crossover study in which eight patients with lupus nephritis were given a TXA_2 synthetase inhibitor, DP-1904, orally at a dose of 400 mg/day twice daily for 4 days (Yoshida et al., 1996). In a study in male Sprague–Dawley rats, the effects on renal function of acute bile duct ligation were studied for 2 days. The consequences of administration of a TX receptor antagonist, BAYu3405, given at a total dose of 10 mg/kg per day via intraperitoneal injections, were then examined. It was shown that TX receptor antagonist can partially prevent renal dysfunction in experimental cholestasis (Holt et al., 1999). A role for TXA_2 in glomerular thrombosis has been suggested. For example, in a rat model of type 2 diabetes, Otsuka Long–Evans Tokushima Fatty strain, the progress of nephropathy as evidenced by an increase in proteinuria paralleled an increase in the urinary excretion of 2,3-dinor-TXB_2. An inhibitor of TXA_2 synthase, OKY-046, at 100 mg/kg in these rats suppressed the proteinuria and the formation of glomerular thrombi (Okumura et al., 2003). Thromboxane receptors play a role in the renovascular actions of angiotensin II (Ang II). For example, the selective TP receptor antagonist terutroban attenuated renal damage in a double-transgenic (harboring human renin and angiotensinogen genes) rat model of hypertension (Sebeková et al., 2008). TXA_2 plays a role in the regulation of renal hemodynamics. For example, TP-KO mice had normal basal mean arterial blood pressure (MAP) and GFR but reduced renal blood flow and increased filtration fraction (FF) and renal vascular resistance (RVR) and markers of reactive oxygen species (ROS) (Kawada et al., 2004). Ang II infusion into TP-KO mice reduced Ang I and increased aldosterone, but caused a blunted increase in MAP and failed to increase FF and ROS (Kawada et al., 2004). Thus, it is suggested that TP receptors regulate renal hemodynamics during Ang II slow pressor response. An additional role for TXA_2 in renal pathophysiology has been suggested. In TP-receptor-KO mice, the glomerular heat-aggregated bovine serum albumin content was lower than that of WT mice. Furthermore, a stable analog of TXA_2, U-46619, increased in a dose-dependent manner the uptake of heat-aggregated bovine serum albumin by mesangial cells in WT mice but not in TP-KO mice (Nagao et al., 2001). These findings suggest that TXA_2 retarded the clearance of aggregated protein in nephritic glomeruli (Nagao et al., 2001). Another study found that a thromboxane receptor antagonist, S18886, attenuated renal oxidant stress and proteinuria in a diabetic apolipoprotein E–deficient mouse model (Xu et al., 2006).

PGI plays a role in vasodilatation and is localized in vasculature in the rat kidney (Vitzthum et al., 2002). Some studies showed that PGIS is expressed in the MD and cortical thick ascending limb in 5/6-nephrectomized rats (Tomida et al., 2001). In the MD, PGI_2 mediated renin release (Ito et al., 1989). Some studies showed that PGIS deletion is associated with ischemic kidney damage and have demonstrated a role for PGI_2 in maintaining RBF (Yokoyama et al., 2002). These PGI_2-deficient mice developed progressive morphological abnormalities in the kidneys, characterized by renal atrophy, surface irregularity, fibrosis, cyst, arterial sclerosis, and hypertrophy of vessel walls. Thickening of the thoracic aortic media and adventitia was also observed (Yokoyama et al., 2002). Interestingly, the IP receptor deficiency in mice does not result in the same kidney damage as that seen in PGIS deficiency (Murata et al., 1997), suggesting that other signaling mechanisms (e.g., renin pathway) may contribute to PGI_2's effect on RBF. The systemic blood pressure and plasma renin effects after intravenous injection of PGI_2 were investigated. A dose-dependent antihypertensive effect in conscious rats with spontaneous and chronic renal hypertension was observed. In addition, PGI_2 induced an increase of plasma renin activity in anesthetized rats (Schölkens, 1978). The effects of increased renal PGI_2, using kidney-specific PGI_2 transgenic (Tg) mice, on endotoxin-induced acute kidney injury (AKI) were investigated (Wang et al., 2007). Increased sensitivity to endotoxin was noted in the PGI_2 Tg mice compared with the wild-type mice. The authors concluded that an excess of renal PGI_2 in mice may obscure beneficial effects against endotoxemia-related AKI by activating the PGI_2–cyclic AMP–renin pathway (Wang et al., 2007). PGI_2 can lead to natriuresis (Villa et al., 1997).

PGD_2 is produced localized in renal glomeruli (Hassid et al., 1979). The exact role of PGD_2 in the kidney is poorly understood. Dose–response renal effects of PGD_2 (0.125, 0.25, 0.5, and 0.75 μg/kg per minute) infused intravenously in dogs were studied. PGD_2 administration resulted in a significant dose-dependent increase in renal artery flow, urine output, creatinine clearance, plasma renin activity, sodium excretion, potassium excretion, and pulmonary artery pressure. A significant decrease occurred in renal resistance and arterial PO_2. There were no appreciable changes in mean arterial pressure, heart rate, hematocrit, platelet count, arterial pH, and PCO_2 (Rao et al., 1987). These results suggest that PGD_2 given under these experimental conditions and doses can improve renal blood flow and function, at least in dogs. In an in vitro study using rat kidney cells, it was shown that PGD_2 in the renal papilla plays a role in the maintenance of phosphatidylcholine biosynthesis (Fernandez-Tome et al., 2004). An antifibrotic role for PGD_2 has been suggested (Zhang et al., 2006). Lipocalin-type prostaglandin D synthase/β-trace (L-PGDS) is an enzyme-synthesizing prostaglandin $D_2(PGD_2)$ and a secretory protein of the lipocalin superfamily. It is synthesized in choroid plexus or leptomeninges in the brain and secreted steadily through cerebrospinal fluid into circulating blood. L-PGDS has a small molecular mass of 26,000 Da. Thus, L-PGDS can easily pass through glomerular capillary walls of the kidney, and its urinary levels may reflect a change in the permeability of glomerular capillary walls and can be used as a biomarker to predict renal injury in type 2 diabetic patients (Uehara et al., 2009). In fact, the urinary and serum concentrations of L-PGDS increased in patients with type 2 diabetes and hypertension (Ragolia et al., 2005; Ogawa et al., 2006). Atherosclerosis and diabetic nephropathy are two common pathologies associated with type 2 diabetes. Thus, the L-PGDS role in the kidney has been investigated using L-PGDS-KO mice. L-PGDS-KO mice became glucose-intolerant and insulin-resistant at an accelerated rate compared with the C57BL/6 control strain of mice. In addition, these mice developed kidney structural changes indicative of diabetic nephropathy, such as glomerular hypertrophy, fibrosis, and basement membrane thickening, and these changes were further exacerbated on the diabetogenic diet. Furthermore, RT-PCR showed that L-PGDS mRNA is expressed in kidney cortex and outer medulla in mice (Ragolio et al., 2005). Thus, urinary levels of L-PGDS have also been proposed to be a relevant biomarker for renal injury (Ragolio et al., 2005; Ogawa et al., 2006). PGA_2 is localized in the renal cortical and medullary tubules (Perez and McGuckin, 1972). PGA_2 can increase renal blood flow, lead to vasodilation, and increase erythropoietin release (Gross et al., 1976; Banks and Jacobson, 1985). $PGF_{2\alpha}$ is localized in the renal medulla (Lee et al., 1967; Gross et al., 1976). $PGF_{2\alpha}$ contributes to diuresis and natriuresis (Zook and Strandhoy, 1981) and causes vasodilation (Lebel and Grose, 1991). The local synthesis and accumulation of prostanoids in the kidney can play a role in urinary tract obstruction. Increased kidney concentrations of PGE_2, $PGF_{2\alpha}$, and TXB_2 have been demonstrated to be predominantly in the renal inner medulla, with the highest concentration 12 h after the occlusion and then largely normalized at 24 h (Nørregaard et al., 2010).

PG receptors have been localized in various microanatomic regions of the kidney. E-type prostanoid receptors (EPs) are the receptors that mediate the actions of PGE_2 and are members of the superfamily of G-protein-coupled receptors. There are four PGE_2 receptors, designated EP_1, EP_2, EP_3, and EP_4 (Morath et al., 1999). These receptors can play a role in regulation of Na^+ handling, vasodilatation, vasoconstriction, and/or blood pressure. It is well accepted that PGE_2 stimulation of the EP_1 receptor is linked to smooth muscle contraction and increased intracellular Ca^{2+}. Thus, PGE_2 stimulation of EP_1 receptors produces contractile responses. Expression of the EP_1 receptor subtype protein in renal tissue was detected in human kidney mainly in connecting segments, cortical and medullary collecting ducts, and in the media of arteries and afferent and efferent arterioles (Morath et al., 1999). In rabbit kidney cell cultures, EP_1 receptor mRNA expression predominates in the cortical collecting duct (Guan et al., 1998). It is suggested, at least in rabbit kidneys, that EP_1 receptor activation mediates PGE_2-dependent inhibition of Na^+ absorption in the collecting duct, thereby contributing to its natriuretic effects (Guan et al., 1998). EP_1-deficient mice exhibit lower systolic blood pressure and increased plasma renin levels (Stock et al., 2001). In addition, oral administration of an EP_1 receptor antagonist, SC-51322, reduced blood pressure in spontaneously hypertensive rats (Guan et al., 2007). Furthermore, some studies have shown that the EP_1 receptor facilitates Ang II-mediated vasoconstriction (Guan et al, 2007). Selective genetic disruption of the EP_1 receptor in mice blunted the acute pressor response to Ang II and reduced chronic Ang II-driven hypertension (Guan et al., 2007). In primary cultures of inner medullary collecting duct (IMCD) cells, EP_1 receptor activation prevented the aldosterone-induced epithelial sodium channel (αENaC), the rate-limiting step for Na^+ reabsorption, and up-regulation (González et al., 2009). Furthermore, infusion of Ang II to rats caused an increase in plasma aldosterone and induced a significant increase in αENaC mRNA and protein in kidney cortex and medulla. This effect of Ang II infusion on αENaC expression was blocked by coadministration of spironolactone, a mineralocorticoid receptor (MR) antagonist, indicating an MR-dependent effect (González et al., 2009). This work suggested that activation of EP_1 receptor during RAAS activation can modify the aldosterone-mediated αENaC expression in the renal medulla (González et al., 2009). However, renal concentrating and diluting ability in the EP_1 receptor–KO mice appears comparable to that of WT mice, although the KO mice may have altered vasopressin release (Kennedy et al., 2007). Arginine–vasopressin (AVP) or ADH has been established as an intrarenal vasoconstrictor, capable of decreasing perfusion, mostly of the inner medulla, with less impact in the outer medulla.

AVP facilitates water reabsorption in renal collecting duct principal cells by activation of vasopressin V2 receptors and the subsequent translocation of water channels (aquaporin-2, AQP2) from intracellular vesicles into the plasma membrane (Tamma et al., 2003). The EP_1 receptor is abundantly expressed in the principal cells of the collecting duct, where it is thought to offset the buffering influence of EP_3 receptors on the antidiuretic actions of AVP (Tamma et al., 2003). Prostaglandin E_2 antagonizes AVP-induced water reabsorption. Using primary rat inner medullary collecting duct (IMCD) cells, Tamma et al. (2003) showed that stimulation of prostaglandin EP_3 receptors induced ρ activation and actin polymerization in resting IMCD cells but did not modify the intracellular localization of AQP2. However, AVP-, dibutyryl cAMP-, and forskolin-induced AQP2 translocation was strongly inhibited. This inhibitory effect was independent of increases in cAMP and cytosolic Ca^{2+}. In addition, stimulation of EP_3 receptors inhibited the AVP-induced ρ inactivation and the AVP-induced F-actin depolymerization. Thus, the signaling pathway underlying the diuretic effects of PGE_2 includes cAMP- and Ca^{2+}-independent ρ activation and F-actin formation.

The EP_3 receptor mRNA, using in situ hybridization, was detected in medullary thick ascending limb and cortical and medullary collecting ducts with intense labeling, while no labeling of glomeruli, proximal tubules, or cortical thick ascending limbs was observed in rabbit kidney (Breyer et al., 1993). EP_3 counteracts AVP action by reducing cAMP production (Hébert et al., 1993). Activation of EP_3 leads to acute inhibition of sodium–potassium–chloride transporter 2 (NKCC2) and aquaporin 2 (AQP2) because the apical trafficking of these proteins is dependent on cAMP (Nielsen et al., 1995; Ortiz, 2006). In a study of EP_3-receptor expression in the developing rabbit kidney. The investigators found that the highest levels of EP_3 mRNA were observed at 2 weeks of age. EP_3 mRNA was observed in the medullary thick ascending limb, cortical collecting duct, and inner medullary collecting duct of adult and immature kidneys (Bonilla-Felix and Jiang, 1996). In the mouse kidney, EP_3 receptor mRNA was located in the tubules in the outer medulla and in the distal tubules in the cortex (Sugimoto et al., 1994). EP_3-deficient mice concentrated and diluted their urine normally in response to a series of physiological stimuli (Fleming et al., 1998). A study in male mice found that EP_3 receptors mediate renal vasoconstriction and are capable of buffering PGE_2-mediated renal vasodilation (Audoly et al., 2001). In human kidneys, RNA for the GI-coupled EP_3 receptor was expressed primarily in the cortical and outer medullary collecting duct as well as in the medullary thick ascending limb; however, it was absent from the inner medullary collecting duct (Breyer et al., 1996). Another study found that EP_3 receptor subtype protein, using immunohistochemistry, is strongly expressed in glomeruli, Tamm–Horsfall negative late distal convoluted tubules, connecting segments, and cortical and medullary collecting ducts, as well as in the media and the endothelial cells of arteries and arterioles of human kidney (Morath et al., 1999).

The EP_2 receptor subtype protein was detectable only in the media of arteries and arterioles of human kidney (Morath et al., 1999). Although only low levels of EP_2 receptor mRNA are detected in the kidney and its precise intrarenal localization is uncertain, mice with targeted disruption of the EP_2 receptor exhibit salt-sensitive hypertension, suggesting that this receptor may also play an important role in salt excretion. In contrast, EP_4 receptor mRNA is expressed predominantly in the glomerulus, where it may contribute to the regulation of glomerular hemodynamics and renin release. The IP receptor mRNA is highly expressed near the glomerulus, in the afferent arteriole, where it may also dilate renal arterioles and stimulate renin release. Conversely, TP receptors in the glomerulus may counteract the effects of these dilator prostanoids and increase glomerular resistance. At present there is little evidence for DP receptor expression in the kidney (Breyer and Breyer, 2001; Imig, 2006). In a mouse model of LPS-induced renal glomerular inflammation, it was demonstrated that EP_2 and EP_4 receptors modulate expression of the chemokine CCL2 (MCP-1) (Zahner et al., 2009). Some studies show that EP_4 and EP_2 receptor agonists improve renal function and/or increase survival in the mercury chloride rat model of acute renal failure, supporting a protective role of $EP_{2/4}$ in the maintenance of renal function (Vukicevic et al., 2006).

Three PGE synthases have been identified: microsomal PGE synthase-1 (mPGES-1), microsomal PGE synthase-2 (mPGES-2), and cytosolic PGE synthase (cPGES). Expression of mPGES-1 is induced by cytokines and inflammatory stimuli. In contrast, the expression of cPGES and mPGES-2 is not inducible (Radi and Ostroski, 2007). In general, mPGES-1 is located in the perinuclear compartment of the cell preferentially and couples to COX-2, whereas cPGES couples predominantly with constitutive COX-1. Also, mPGES-2 couples with both COX isoforms. The expression of mPGES was investigated in the rat and mouse kidney. It was shown that there is a parallel expression with COX-2 in the cortical thick ascending limb and macula densa, and coexpression with COX-1 in glomerular mesangial cells, distal convoluted tubule, connecting tubule, and nonintercalated collecting duct cells, where COX-2 was absent (Campean et al., 2003). Cortical fibroblasts were positive for COX-1 and PGES in mice, and medullary interstitial cells expressed all three enzymes (Campean et al., 2003). Another study have shown that mPGES-1 is highly expressed in rabbit distal convoluted tubule, medullary collecting ducts, and cortical collecting ducts, predominantly colocalizing with COX-1 and, to a lesser extent, is coexpressed with COX-2 in macula densa and medullary interstitial cells (Schneider et al., 2004). High expression of mPGES-1 is present in the collecting duct (Vitzthum et al., 2002; Campean et al., 2003; Schneider et al., 2004). Also, mPGES-1 has been detected in the thick ascending limb and MD (Campean et al., 2003). Low levels of mPGES-1 can be detected in the cortical and medullary interstitial cells (Campean et al., 2003; Schneider et al., 2004; Vitzthum et al., 2002). It is thought that mPGES-1 is coupled functionally to COX-2, due to colocalization of mPGES1 and COX-2 in the cortical thick ascending limb and medullary interstitial cells (Murakami et al., 2002). Low levels of mPGES2 and cPGES are also detected in the kidney (Yang et al., 2006; Zhang et al., 2003).

COX-1 AND COX-2 EXPRESSION IN THE KIDNEY

The early immunohistochemical localization of COX in rat, rabbit, guinea pig, sheep, and cow in the renal cortex began in 1978 (Smith and Bell, 1978). COX-1 is the most abundant isoform of COX in the kidney and is localized in the renal vasculature, collecting ducts, and papillary interstitial cells across various species (Smith and Bell, 1978; Khan et al., 1998; Sellers et al., 2004; Radi and Ostroski, 2007; Radi, 2009). Table 3-3 summarizes the comparative renal expression of COX-1 (Radi and Ostroski 2007; Radi, 2009). No differences in COX-1 renal regional localization were found among various species (Khan et al., 1998). Evidence in various species (cow, dog, guinea pig, human, monkey, mouse, rabbit, rat, and sheep) demonstrated that high levels of COX-1 expression occurred in the collecting ducts (Smith et al., 1978; Khan et al., 1998; Campean et al., 2003). In kidney cells isolated from Sprague–Dawley rats, COX-1 mRNA was found to be widely distributed and expressed at substantial levels along the entire collecting duct, in MTAL segments, in glomeruli freed of adherent MD, and in cultured renal interstitial cells, mesangial, and collecting duct-derived cell lines. Low or high dietary salt did not affect COX-1 expression (Yang et al., 1998). COX-1 expression has been investigated in various pathological conditions. Nephrogenic diabetes insipidus (NDI) is a renal disorder characterized by an inability to concentrate urine despite normal or elevated plasma concentrations of the antidiuretic hormone (ADH) arginine vasopressin (Morello and Bichet, 2001). Nephrogenic diabetes insipidus is caused either by mutation in the vasopressin type 2 receptor (*AVPR2*) or aquaporin-2 (*AQP2*) gene, hypercalcemia, or lithium treatment (Singer et al., 1972; Morello and Bichet, 2001; Faerch et al., 2009). In NDI there is a peripheral resistance to ADH and polyuria (excessive or abnormally large production and/or passage of urine), with hyposthenuria (excretion of urine of low specific gravity) and polydipsia (excessive thirst) being the cardinal clinical manifestations of the disease. NSAIDs have been used to treat patients with NDI (Usberti et al., 1980; Pattaragarn and Alon, 2003). Therefore, COX expression has been investigated in animal models of NDI. In

TABLE 3-3 Comparative Expression of COX-1 in the Kidney[a]

	Species				
Renal microanatomical location	Human	Monkey	Dog	Rat	Mouse
Vasculature	+	+	+/−	+	++
Collecting duct	+	+	+	+	+++
Interstitial cells	++	++	+++	+++	++

Source: The final, definitive version of this table was published in Z. A. Radi, *Toxicologic Pathology*, 37(1), 2009, pp. 34–46.

[a] + minimal; ++ mild; +++ moderate; ++++ marked.

an NDI model, Wistar rats were treated with lithium for 4 weeks to induce the disease. There was a marked decrease in the expression of COX-1 in the inner medulla of the kidney, and COX-1 expression was marginally reduced in the inner stripe of outer medulla in this NDI model (Kotnik et al., 2005). COX-1 expression was studied in a two-kidney/one-clip (2K1C) Goldblatt hypertensive rat model. In this 2K1C hypertension model, the defect is highly dependent on the enhanced activity of the RAAS. Decreased renal perfusion pressure in the compromised kidney maximally activates RAAS and its inability to balance the rise in blood pressure, which is partly the result of elevated Ang II levels (Theilig et al., 2006). Immunoreactive COX-1 in the 2K1C model was localized to the extraglomerular mesangium and distal convoluted tubules (DCTs), connecting tubule, and cortical collecting duct, with the exception of the intercalated cells (Theilig et al., 2006). In a hypertension transgenic mouse model that expresses the human renin and angiotensinogen genes, the highest degree of renal COX-1 expression occurred in the DCTs, cortical collecting ducts, and medullary collecting ducts, while mild-to-moderate COX-1 expression occurred in other microanatomic renal locations. PCT lacked COX-1 expression (Radi and Ostroski, 2007).

In kidney sections from CD1 mice, COX-1 expression before and after NSAID treatment was investigated (Meskell and Ettarh, 2011). Immunohistochemical analysis before NSAID treatment and in control kidneys revealed COX-1 expression in the renal cortex and medulla, with significantly more expression evident within the cortex. Immunoreactivity in the superficial cortex was more pronounced than in the rest of the cortex or the medulla. COX-1 protein was expressed in the glomeruli and parietal cells of the glomerular capsule, epithelial cells of the proximal and distal convoluted tubules, and the collecting ducts. Cytoplasmic COX-1 protein within the distal tubule was present in the juxtaglomerular region, adjacent to the glomerulus. Interstitial cells in the renal cortex did not show reactivity for COX-1. Within the medulla, pronounced expression of COX-1 was apparent within the endothelial cells of the renal vasculature, but the epithelial cells of the medullary collecting ducts and the interstitial cells in the medulla showed poor reactivity for COX-1 (Meskell and Ettarh, 2011). In kidneys treated with the ns-NSAID indomethacin, COX-1 immunostaining was present in the cortical and medullary areas. COX-1 immunoreactive protein was apparent in the outer parietal layer of the Bowman's capsule and in some cells of the glomerulus. COX-1 expression was observed within cells of the proximal and distal convoluted tubules and the cortical collecting ducts. COX-1 expression was present within the endothelial cells of the renal blood vessels, and the medullary collecting duct expression was minimal (Meskell and Ettarh, 2011). There was no staining for COX-1 immunoreactive protein in the cortical or the medullary interstitium (Meskell and Ettarh, 2011). Thus, at therapeutic doses of 10 mg/kg, indomethacin did not alter renal COX-1 expression, at least by immunohistochemistry, relative to immunoreactivity in untreated control kidney. In the same study in mice, the ns-NSAID nimesulide was used and renal COX-1 expression was examined (Meskell and Ettarh, 2011). Significant expression of COX-1 protein was present in the superficial cortical region adjacent to the renal capsule and evidence of some COX-1 immunoreactivity in the outer parietal epithelial cell layer of Bowman's capsule. Very minimal COX-1 immunoreactivity was evident in the proximal convoluted tubule. There was COX-1 protein expression within cells of the cortical distal convoluted tubules, with significant localization within the cortical collecting ducts. Focal immunostaining was seen in endothelial cells of blood vessels and collecting duct cells in the renal medulla (Meskell and Ettarh, 2011). Therefore, the investigators concluded that a therapeutic dose of 15 mg/kg of nimesulide reduced renal COX-1 expression within the first 24 h of treatment.

The ontogeny of renal COX expression has been investigated. In a study in developing rats, COX mRNA expression changes was examined in renal medulla and in cortex at various ages (Ogawa et al., 2001). COX-1 mRNA expression did not change with age in either the cortex or medulla. COX-2 mRNA expression was highest in 1-week-old rats and lowest in 4- and 8-week-old rats. Lipopolysaccharide (LPS) treatment did not alter COX-1 mRNA expression in infantile or adult rats. In adults, LPS at 1, 5, and 10 mg/kg induced COX-2 mRNA expression in renal medulla; the higher doses, 5 and 10 mg/kg, induced COX-2 expression in cortex. In infantile rats, COX-2 mRNA, which was already high in the unmanipulated state, was increased further by only 1 mg/kg LPS in both renal cortex and medulla (Ogawa et al., 2001). Another study examined the age-related changes in COX mRNA expression induced by hypoxia in the renal cortex and medulla of developing rats (Ogawa et al., 2002). Expression of COX-1 mRNA did not change in response to hypoxia in the cortex or medulla in either infantile or adult rats. In infantile rats, COX-2 mRNA expression was not induced by 1 or 4 h of hypoxia. In adults, 1- and 4-h exposures to hypoxia induced COX-2 mRNA in the renal cortex, and 1 h of exposure induced COX-2 mRNA in the medulla (Ogawa et al., 2002). Thus, age-related changes in COX-2 mRNA expression in the developing rats were observed after acute exposure to hypoxia.

Since COX-1 is expressed in the collecting ducts of the nephron, an active area in the regulation of sodium excretion in both laboratory animals and humans (Currie and Needleman, 1984), it is not surprising that BP increases in COX-1-deficient mice compared with WT controls (Kawada et al., 2005). Thus, ns-NSAIDs may aggravate renin-independent sodium-sensitive hypertension, possibly in part by inhibition of the COX-1 responsible for sodium excretion (Okumura et al., 2002). In rats infused with the COX-2 s-NSAID NS-398 or meloxicam, direct renal interstitial volume expansion significantly increased renal interstitial hydrostatic pressure and fractional excretion of sodium (Gross et al., 1999).

In humans, COX-1 expression was examined in kidney tissues obtained from patients aged 17 to 97 years old. These patients' deaths were unrelated to renal or cardiovascular pathologies (mostly car accidents). Strong immunoreactivity was detected in the collecting duct epithelium, and less intense staining was seen in the descending thin loops of Henle and interstitial cells in the papilla. Many of the arcuate arteries also expressed COX-1, which appeared to be confined to the tunica media (Nantel et al., 1999). In another study of adult normal human kidney, COX-1 expression was noted in the collecting duct cells, interstitial cells, endothelial cells, and smooth muscle cells of pre- and postglomerular vessels. In fetal human kidney, COX-1 was expressed primarily in podocytes and collecting duct cells. Expression levels of COX-1 in both cell types increased markedly from subcapsular to juxtamedullary cortex (Komhoff et al., 1997). Another study examined COX-1 expression in normal human kidney tissues using immunohistochemistry, Western blotting, and RT-PCR. Autopsy samples from healthy trauma victims and samples from biopsy surgical specimens were included in the study. COX-1 was found in blood vessels, interstitial cells, smooth muscle cells, platelets, and mesothelial cells (Zidar et al., 2009).

COX-2 is generally an inducible enzyme under settings of inflammation. In the kidney, COX-2 is expressed constitutively in macula densa and thick ascending limb in cortex and inner medulla (Ferreri et al., 1999; Campean et al., 2003) and can be found in interstitial cells (Yang et al., 1999). Enhanced COX-2 expression has been demonstrated in several renal pathologies, such as glomerulonephritis (Schneider et al., 1999), passive Heymann nephritis (Heise et al., 1998), and lupus nephritis (Tomasoni et al., 1998). COX-2-deficient mice exhibited severe disruption of renal development and function, suggesting an important role for COX-2 in renal development (Dinchuk et al., 1995; Morham et al., 1995). COX-2 is expressed in variable locations in the kidney across species (Khan et al., 1998, 2001b). COX-2 expression was studied in human fetal kidneys ranging between 15 and 23 weeks of gestational age. Strong expression of COX-2 was localized primarily in the MD and the thick ascending limb of the loop of Henle, and in rare glomerular podocytes and vascular endothelial cells. There was a progressive decrease in COX-2 immunoreactivity from the most immature nephrons adjacent to the metanephric regions to the well-developed nephrons in the middle to inner cortex. In contrast to the adult human kidney, this temporal and spatial expression of COX-2 in the fetal kidney suggests that COX-2 enzyme may be involved in nephrogenesis, and its inhibition by NSAIDs during the third trimester may be responsible for fetal renal syndromes (Khan et al., 2001b). The temporal expression of COX-2 throughout embryonic and fetal development was investigated in the rat by immunohistochemistry and in situ hybridization. COX-2 expression was not detectable in any tissues from developing embryos during gestation days 7 to 13, but was observed in the fetal growth period (gestation days 15 to 20) in several organs, including the kidney (Stanfield et al., 2003). Interspecies differences may exist for the involvement of COX-2 in the generation of renal PGs in states of normal and altered kidney function (Khan and Alden, 2002). For example, normal rodent and canine kidneys have prominent constitutive COX-2 expression in the MD and thick ascending limb of loop of Henle; whereas COX-2 is absent at these sites in the normal nonhuman primate and human kidney (Sellers et al., 2004). COX-2 levels also differ according to maturation, with high levels expressed in the macula densa and thick ascending limbs of fetal kidneys and minimal expression upon renal maturation (Khan et al., 2001b; Khan and Alden, 2002). In rat kidney cells, COX-2 mRNA was restricted to the MD, cortical thick ascending limb, and at significantly lower levels, in the inner medullary collecting duct (Yang et al., 1998). Divergent regulation of COX-2 in cortex and medulla by dietary salt has been suggested (Yang et al., 1998). In the inner medulla of rats treated with a high-salt diet, COX-2 mRNA increased 4.5-fold and protein levels increased 9.5-fold. In contrast, cortical COX-2 mRNA levels decreased 2.9-fold in rats on a high-salt diet and increased 3.3-fold in rats on a low-salt diet. A low-salt diet increased COX-2 mRNA 7.7-fold in MD and 3.3-fold in CTAL (Yang et al., 1998).

The role for MD COX-2 in the regulation of renin in renovascular hypertension has been proposed (Hartner el al., 1998). For example, the expression of COX-2 and renin was investigated in the kidneys of rats with two-kidney/one-clip renovascular hypertension. COX-2 mRNA levels were increased in clipped kidneys but remained unchanged or slightly decreased in nonclipped kidneys. COX-2 protein was expressed mainly in the MD and occasionally in distal tubular cells not associated with the MD (Hartner et al., 1998). COX-2 mRNA and protein expression, but not COX-1, is markedly decreased in the cortex and increased in the inner medulla in response to unilateral ureteral obstruction (Chou et al., 2003) and bilateral ureteral obstruction (Cheng et al., 2004). COX-2 expression was examined in kidney tissues obtained from patients aged 17 to 97 years old. These patients' deaths were unrelated to renal or cardiovascular pathologies (mostly car accidents). In the cortex, COX-2 immunoreactivity was identified in the MD, in the glomerular afferent arteriole, in podocytes in a subset of glomeruli, and in a limited number of thick loops of Henle. In the medulla, COX-2 immunoreactivity was limited to interstitial cells and the tunica media of arcuate arteries. The presence of COX-2 in the MD was more evident in kidney specimens obtained from older subjects. COX-2 immunoreactivity was detected in the MD of all kidneys from subjects aged 48 years or greater. Only two of five kidneys (40%) from subjects less than 40 years of age showed COX-2 MD immunoreactivity. COX-2-positive immunostaining was also detected within podocytes of the some glomeruli in six of 10 kidneys (60%) (Nantel et al., 1999).

There is a regulatory association between COX-2 and renin (Schnermann, 2001). It has been suggested that COX-2 mediates increased renin content induced by a low-sodium diet in mice. A low-sodium diet in mice increased renal renin content, but this increase was blocked by treatment with the COX-2 s-NSAID NS-398. In addition, treatment with NS-398 reduced renin mRNA in response to a low-sodium diet (Harding et al., 1997). Demonstration that indomethacin and SC-58236, a COX-2 s-NSAID, can decrease BP and suppress plasma renin activity in rats lends further support to this association (Jackson et al., 1981; Wang et al., 1999). In a streptozotocin-induced diabetes mouse model, increased podocyte expression of COX-2 increased susceptibility to development of diabetic nephropathy, and the podocyte injury was due in part to increased expression and activity of the pro-renin receptor. These mice exhibited significant renal changes characterized by albuminuria, foot process effacement, and GBM thickening (Cheng et al., 2011). In addition, the role that Ang II can play in modulating the regulation of renin release has been proposed. Administration of the angiotensin-converting (ACE) inhibitor captopril to rats fed a normal or sodium-deficient diet resulted in an increase in COX-2 mRNA and renal cortical COX-2 immunoreactivity, with the most pronounced expression in the macula densa (Cheng et al., 1999). Administration of an Ang I receptor antagonist, losartan, also significantly increased cortical COX-2 mRNA expression and COX-2 immunoreactivity in rats on control diet and further increased COX-2 expression in salt-deficient animals (Cheng et al., 1999). To investigate the role of an Ang I receptor in the regulation of cTALH/MD

COX-2 expression, mutant mice homozygous for both Agtr1a and Agtr1b null mutations (Agtr1a$^{-/-}$, Agtr1b$^{-/-}$) were used. These mice had large increases in immunoreactive COX-2 expression in the cTALH/MD (Cheng et al., 1999). Rats were treated with captopril for 1 week with or without the COX-2 s-NSAID SC58236. Plasma renin activity increased significantly in the captropril group, and this increase was inhibited significantly by simultaneous treatment with SC-58236. Thus, these studies indicated that Ang II inhibitors augment up-regulation of renal cortical COX-2 in states of volume depletion, suggesting that negative feedback by the renin–angiotensin system modulates renal cortical COX-2 expression and that COX-2 is a mediator of increased renin production in response to inhibition of Ang II production (Cheng et al., 1999). Conversely, COX-2-KO mice had reduced renin content and activity (Yang et al., 2000).

Endogenous glucocorticoids play a complex physiological role in modulating renal electrolyte balance and blood flow. These endogenous glucocorticoids act via glucocorticoid receptors and indirectly enhance sodium reabsorption in epithelial cells isolated from proximal tubules (Gong et al., 2008). One of the glucocorticoid metabolizing enzymes is 11β-hydroxysteroid dehydrogenase (11β-HSD1). The normal human kidney expresses functional 11β-HSD1 protein in proximal tubules and in medullary interstitial cells. Interestingly, in the normal human kidney, 11β-HSD1 colocalizes with COX-2 in proximal tubule cells (Gong et al., 2008).

EFFECTS OF ns-NSAIDs ON THE KIDNEY

Over-the-counter and prescription NSAIDs are widely used by the adult population in the United States (Paulose-Ram et al., 2003). The incidence of NSAID-associated renal side effects was investigated. In one study that used a computerized medical records system, the incidence of ibuprofen-associated renal impairment and risk factors for its development was 18% in 1908 patients treated with ibuprofen (Murray et al., 1990). In a recent survey of a total of 12,065 adult (aged 20 years or older), the use (nearly every day for 30 days or longer) of any NSAID (ibuprofen, naproxen, sulindac, piroxicam, indomethacin, tolmetin, or diclofenac) was reported by 2.5%, 2.5%, and 5% of the U.S. population with no, mild, and moderate-to-severe chronic kidney disease (CKD), respectively (Plantinga et al., 2011). NSAIDs have been associated with greater risk for development of acute kidney injury in the general population (Huerta et al., 2005) and with increased risk for rapid CKD progression in a community-based elderly population (Gooch et al., 2007). The most common renal side effects of ns-NSAIDs are functional and include interference with fluid and electrolyte homeostasis. Other less frequent side effects include acute renal failure, interstitial nephritis, nephritic syndrome, atrophy of subcapsular cortex in dogs, and renal papillary necrosis (Whelton, 1999; Rossert, 2001; Khan and Alden, 2002). Elderly persons are more predisposed to NSAID-associated renal adverse effects for several reasons: (1) age-associated changes in renal function; (2) the prevalence of comorbid conditions (congestive heart failure, hypertension, hepatic cirrhosis, renal insufficiency), and (3) the pervasive use of concomitant drugs that affect kidney function (diuretics, antihypertensives) (Ailabouni and Eknoyan, 1996).

NSAIDs contribute to changes in blood pressure and edema, resulting from renal flow and fluid retention imbalance. NSAIDs may aggravate renin-independent sodium-sensitive hypertension, possibly in part by inhibition of the COX-1 responsible for sodium excretion (Okumura et al., 2002). In addition, NSAIDs may elevate blood pressure and antagonize the blood pressure-lowering effect of antihypertensive medication to an extent that may potentially increase hypertension-related morbidity (Johnson et al., 1994). Inhibition of cortical and medullary PGs production by NSAIDs can lead to electrolyte abnormalities and fluid retention, such as sodium, potassium, and water retention. However, clinically detectable edema occurs in less than 5% of patients and is easily managed in patients and readily reversible upon discontinuation of the use of NSAID (Whelton and Hamilton, 1991). Furosemide is a loop diuretic used in the treatment of congestive heart failure and edema. Renal function and excretion of water and salt were evaluated in furosemide-treated patients with well-controlled congestive heart failure. In addition, four doses of naproxen (500 mg) were given every 12 h in a double-blind crossover design. Naproxen significantly decreased the urinary excretion of water (19%), sodium (26%), and chloride (26%), and decreased osmolal clearance (18%). Plasma renin activity, aldosterone, or free water clearance did not change significantly with treatment (Eriksson et al., 1987). The effects of ibuprofen and indomethacin on renal function and electrolytes in the presence and absence of furosemide in healthy volunteers who are on a restricted sodium diet were evaluated (Passmore et al., 1990). Neither indomethacin (50 mg) nor ibuprofen (400 and 800 mg) affected renal blood flow, GFR, or electrolyte excretion before furosemide. However, renal blood flow and GFR were increased significantly in the first 20 min after furosemide, and these changes were attenuated significantly by indomethacin compared with placebo and ibuprofen 400 mg (Passmore et al., 1990). In a placebo-controlled double-blind crossover design study, the effects of 7-day oral administration of naproxen in a 500-mg dose in the morning and a 250-mg dose in the evening in patients with polyarthritis and stable impaired renal function were investigated (Eriksson et al., 1990). Naproxen reduced GFR and renal plasma flow by 18% and 13%, respectively, Plasma renin activity decreased by 38% during naproxen treatments. No significant change in plasma aldosterone was observed during treatment, but urinary aldosterone declined significantly by 34% and albuminuria decreased by 41%. No discernible effects on base excess, on excretion of water, sodium, or potassium, or on osmolal clearance were noted. However, serum potassium increased slightly but significantly during naproxen treatment (Eriksson et al., 1990). The effect of a single 500-mg dose of naproxen on renal function in healthy volunteers was studied. No effects on GFR were noted. Over the first 4 to 8 h of the study, fractional excretion of sodium was reduced by approximately 50% (Dixey et al., 1987). Thus, naproxen should be used with caution in patients with fluid retention, hypertension, or heart failure. Another study examined renal function in healthy volunteers who received ibuprofen, naproxen, or sulindac in a randomized double-blind fashion. Renal function and plasma renin activity were not affected (Brater et al., 1985). On the average, 0.6% of a dose of ketoprofen or naproxen was found in the urine of normal male volunteers assayed immediately after its collection (Upton et al., 1980). Between approximately 60 and 85% of the dose of these drugs can be excreted in the urine as conjugates, which rapidly hydrolyze at body temperature, at room temperature, and even during frozen storage, thereby regenerating the parent drug. Thus, only a small fraction of naproxen is excreted unchanged in vivo (Upton et al., 1980). Only a small fraction of the filtered naproxen amount was excreted into the urine (Cox et al., 1990). This means that naproxen must be reabsorbed extensively. Reabsorption is probably a passive process, dependent on the urinary flow and pH (Cox et al., 1990). As a result of active uptake of naproxen across the basolateral membrane and facilitated diffusion across the brush-border membrane, naproxen can accumulate in the tubular cells (Cox et al., 1990). In isolated perfused rat kidneys naproxen accumulates considerably. Increasing the naproxen-perfused concentration results in a decrease in the K/P ratio which may be related to the saturation of active secretion of naproxen (Cox et al., 1990).

The effects of NSAIDs on urinary sodium and potassium excretion were studied in the rat. Diclofenac (30 mg/kg) and flurbiprofen (125 mg/kg) were administered orally once daily for 4 days. Urine was collected 0 to 8 h after each dose, and urinary sodium and potassium excretion and urine flow rate were compared with placebo. Both diclofenac, and flurbiprofen significantly reduced the excretion rate of sodium and potassium (Harirforoosh and Jamali, 2005). The effects of the ns-NSAID sodium meclofenamate at a 5-mg/kg dose given intravenously in pentobarbital-anesthetized and sodium-replete dogs were investigated. Sodium meclofenamate significantly reduced the urinary excretion of sodium by 47% but did not affect mean arterial blood pressure, renal blood flow, GFR, or urine volume (Blasingham and Nasjletti, 1980). In nonclinical toxicology studies, the acute 2-week oral toxicity profile of naproxen was determined in Swiss Webster albino mice, Sprague–Dawley rats, Golden Syrian hamsters, and beagle dogs with multiples of the maximum human therapeutic dose of 107, 47, 357, and 87, respectively. Subacute (1- and 3-month) repeated oral dosing studies were also conducted in mice, rats, rabbits, dogs, monkeys, and pigs with multiples of the maximum human therapeutic dose up to 1.3-fold. These acute and subacute studies showed that repeated oral dosing of naproxen did not cause significant histopathological changes in the kidney and was well tolerated by mice, rabbits, monkeys, and pigs, less well by rats, and poorly by dogs. Based on these results, rats, monkeys, and pigs were animals of choice for the chronic studies. Rats dosed for up to 2.6 times the maximum human therapeutic dose per day for 6 and 22 months showed no significant kidney histopathological changes but had increased urinary volume, decreased body weight gain, and a shorter survival time than did control rats or rats given 0.2 or 0.9 times the maximum human dose per day. No meaningful renal changes were seen for monkeys given oral doses up to 10.4 times the maximum human per day for six months or for pigs given doses up to 3.9 times the maximum human per day for one year (Hallesy et al., 1973).

NSAID-induced hyperkalemia most often occurs from impaired renal potassium excretion (Perazella, 2000). Potassium homeostasis is impaired by NSAIDs through the inhibition of renal PG (PGE_2 and PGI_2) synthesis (Garella and Matarese, 1984). These PGs stimulate renal synthesis of renin and influence the subsequent synthesis of aldosterone (Garella and Matarese, 1984). This leads to a hyporeninemic hypoaldosteronism state, which is probably the major mechanism by which NSAIDs cause hyperkalemia (Tan et al., 1979). Up to 46% of patients treated with indomethacin develop hyperkalemia (Zimran, 1985).

Nonselective NSAIDs can reduce GFR and renal blood flow (Clive and Stoff, 1984; Dunn et al., 1988) and may aggravate renin-independent sodium-sensitive hypertension, possibly in part by inhibition of the COX-1 responsible for sodium excretion (Okumura et al., 2002). However, studies in which the effect of naproxen was examined using healthy volunteers showed no reduction in GFR (Brater et al., 1985; Dixey et al., 1987). The physiological status (i.e., exercise, salt restriction, and dehydration) can affect the outcome of NSAID use. For example, the renal effects of the maximal recommended dose of ibuprofen at 1.2 g/day versus a placebo in humans subjected to progressive renal stresses (e.g., treadmill exercise, heat, dehydration, fasting) were tested. These combined stressors caused dramatic decreases in effective renal plasma flow, GFR, and sodium excretion. Baseline GFR decreased in all groups including placebo, with a significantly greater decrease in GFR in the ibuprofen-treated group. Ibuprofen treatment showed small but statistically significant effects on GFR during exercise in a sodium- and volume-depleted state (Farquhar et al., 1999). In another randomized double-blind three-way crossover study with placebo, the effects of 2 weeks of treatment with ibuprofen and etodolac on renal hemodynamics (GFR, RPF) and filtration fraction (FF), tubular function and plasma concentrations of the hormones, and renin (PRC) and arginine vasopressin (AVP) in healthy subjects were examined (Svendsen et al., 2000). In addition, the effects on the urinary excretion of albumin and α-GST as markers of renal injury were also studied. No differences were found between the three treatments (i.e., placebo, ibuprofen, and etodolac) in the effects on GFR, RPF, FF, free water clearance, urinary output, or fractional excretion of potassium and sodium. However, in contrast to etodolac, ibuprofen caused a significant decrease in both lithium clearance and the fractional excretion of lithium, suggesting an increase in the reabsorption in the proximal tubules. PRC was reduced significantly by ibuprofen but not by etodolac. None of the drugs changed AVP. Fourteen days of treatment with ibuprofen caused a significant decrease in the urinary excretion of α-GST, while no changes were seen after etodolac. None of the drugs changed the urinary excretion of albumin (Svendsen et al., 2000). Thus, a 14-day administration of etodolac or ibuprofen in therapeutic doses did not affect renal hemodynamics, the net excretion of electrolytes, or the urinary excretion of albumin in healthy subjects.

However, in contrast to etodolac, ibuprofen caused a reduction in PRC, suggesting that COX-1 may be involved in basal renin release in humans. In addition, ibuprofen decreased lithium excretion, suggesting that COX-1 is involved in the reabsorption of sodium and/or water in the proximal tubules. The reduction in the urinary excretion of α-GST by ibuprofen may be caused by an inhibition of the detoxification enzyme by ibuprofen. Overall, the study indicated that only small differences in the effects of the two drugs on renal function in healthy subjects exist during a treatment period of 2 weeks (Svendsen et al., 2000). In another clinical study, normal volunteers received placebo, ibuprofen, naproxen, or sulindac in a randomized double-blind fashion. After control periods assessing the effect of the NSAIDs alone, a 40-mg dose of furosemide was administered. Renal function, plasma renin activity, and urinary PGs were not affected during control collections, while all three ns-NSAIDs decreased TXB_2. After furosemide, all ns-NSAIDs decreased fractional excretions of Na^+ and Cl^-, renin, and TXB_2. Sulindac and ibuprofen decreased urinary PGE_2, whereas naproxen did not (Brater et al., 1985). The effects of 600 mg of ibuprofen given three times daily on renal water and sodium excretion was measured during fasting in a randomized placebo-controlled double-blind crossover study of 17 healthy humans. The subjects received a standardized diet on day 1, fasted at day 2, and received an IV infusion of 3% NaCl on day 3. Ibuprofen decreased fractional excretion of sodium and increased the urinary beta fraction of the epithelial sodium channel during a 24-h fasting period, during the first 2 h after 24 h of fasting, and during the following period with hypertonic saline infusion. In addition, ibuprofen decreased urinary excretions of aquaporin 2 and PGE_2 at all parts of the study. Ibuprofen did not change the response in AVP, urinary cAMP, urinary output, and free water clearance during any of these periods (Lauridsen et al., 2011). Thus, urinary sodium excretion decreases considerably in healthy men during fasting, so a state with an increased PGs synthesis and a possible role for epithelial sodium channels is suggested. Ibuprofen elimination is through hepatic biotransformation to inactive metabolites excreted by kidney filtration and secretion. About 70% of ibuprofen is excreted in the urine as metabolites or unchanged drug.

An acute ibuprofen overdose in dogs, cats, and ferrets is associated with renal signs of toxicosis (Villar et al., 1998). At doses greater than 175 mg/kg in dogs, there was an increased risk of acute renal failure (polyuria, polydipsia, oliguria, and uremia). The effects of anesthesia on renal function were examined in beagle dogs. Acetaminophen was given at 15 mg/kg and ibuprofen at 10 mg/kg intravenously. Both drugs decreased RBF and GFR by approximately 20 to 30% in normal anesthetized sodium-repleted dogs. Although acetaminophen produced similar changes in RBF and GFR in the low-sodium dogs, ibuprofen caused a significantly greater renal vasoconstriction (Colletti et al., 1999). In general, the healthy newborn, in particular the premature infant, have a low GFR, and drugs that may affect GFR and electrolyte handling (i.e., NSAIDs) should be given with caution to young individuals. In an experiment with New Zealand white rabbits, increasing doses of ibuprofen (0.02, 0.2, 2.0 mg/kg body weight) were given to newborn rabbits. This resulted in a significant, dose-dependent decrease in RBF and GFR. This was accompanied by a reduction in urinary volume and urinary sodium excretion, whereas mean arterial pressure remained unchanged. There was a very steep rise in renal vascular resistance (Chamaa et al., 2000). Thus, when NSAIDs are given to neonates, renal function has to be followed very carefully. The effects of clinically comparable doses of ibuprofen and indomethacin on renal vascular perfusion were investigated in newborn piglets (Speziale et al., 1999). The newborn piglets were injected with two ns-NSAIDs, ibuprofen (20 mg/kg) and indomethacin (0.3 mg/kg). Ibuprofen increased renal cortical and medullary resistance by 44 and 52% while indomethacin raised renal cortical and medullary resistance by 66 and 71% at 60 min postinjection, respectively. Indomethacin and, to a lesser extent, ibuprofen increased renal vascular resistance (Speziale et al., 1999). No significant changes in pulmonary vascular pressures of strenuously exercising thoroughbreds have been observed after phenylbutazone administration in horses (Manohar et al., 1996). In addition, NSAIDs may elevate blood pressure and antagonize the blood pressure–lowering effect of antihypertensive medication to an extent that may potentially increase hypertension-related morbidity (Johnson et al., 1994).

Analgesic nephropathy is characterized by chronic interstitial nephritis and renal papillary necrosis or calcifications. The hallmark of analgesic nephropathy is renal papillary necrosis (RPN) (Fig. 3-2). The only permanent renal complication of NSAIDs is RPN, which is very rare (Whelton and Hamilton, 1991). Studies in rats show that NSAIDs readily cause RPN; however, studies have shown that laboratory animals (i.e., rats, dogs, rabbits) are unusually susceptible to RPN, which is related to their unique renal anatomy and physiology compared to other species (Wiseman and Reinert, 1975; Khan et al., 1998). Thus, laboratory animals are not necessarily a good model for predicting RPN in humans. The effects of two papillotoxic agents, including an ns-NSAID, indomethacin, and a chemical agent, 2-bromoethanamine hydrobromide (2-BEA), on COX-1 and COX-2 in the renal papilla in female Wistar rats during the development of RPN were investigated. Indomethacin was given at 75 mg/kg as a single dose or as a daily dose of 10 mg/kg for 5 days. 2-BEA

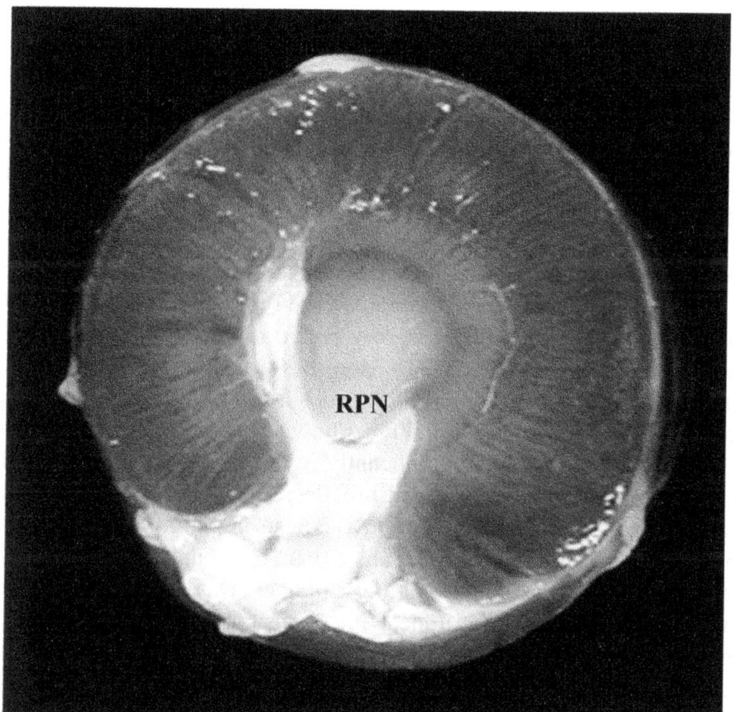

FIGURE 3-2 Analgesic nephoropathy (renal papillary necrosis). Courtesy of Dr. N. Khan. [The final, definitive version of this figure was published in Z. A. Radi, *Toxicologic Pathology*, 37(1), 2009, pp. 34–46. Copyright © Sage Publications. All rights reserved.]

was given in a daily dose of 100 mg/kg for 4 days to induce lesions of RPN. The single 75-mg/kg dose of indomethacin did not cause light microscopic changes of RPN. However, RPN was observed in rats administered indomethacin 10 mg/kg per day for 1 week or 2-BEA for 5 days. The immunohistochemical analyses of kidneys showed that both COX-1 and COX-2 were present in the renal papilla of control rats. In animals treated with indomethacin (75 mg/kg), a slight to moderate decrease in both COX isoforms was observed in essentially normal renal papillary cells within 2 h, followed by an increase in COX-2 immunoreactivity in the renal papilla, macula densa, and thick ascending limbs (at both 10 and 75 mg/kg). No changes in the expression of COX isoforms in the intact papilla occurred as a result of 2-BEA; however, cells undergoing degeneration and necrosis lost immunoreactivity to both COX isoforms (Khan et al., 1998). The earliest lesions of RPN are characterized by loss of interstitial glycosaminoglycans and focal interstitial cell degeneration near the tip of the papilla, which represents precursor lesions of RPN (Bennett et al., 1996). Papillary interstitial cells are the first cell type to undergo morphological changes in NSAID-related RPN. These cells have limited proliferating capacity; therefore, repeated injury is irreversible and may result in a local loss of vasodilatory PGs and development of secondary ischemia (Khan et al., 1998). In cases with full expression of injury, degeneration and necrosis extended to the vascular endothelium and the epithelial cells of the loops of Henle, and the collecting ducts also involved a larger portion of the papilla (Khan et al., 1998). It has been speculated that the papillae of rats and dogs are susceptible to chemical injury owing to sluggish blood flow, which predisposes to ischemia and accumulation of toxic substances (Khan and Alden, 2002). Two mechanisms of RPN have been proposed (Duggin, 1980). The first mechanism involves metabolic or functional features of the renal papilla, which makes it susceptible to ischemia resulting from decreased RBF. The second mechanism is the selective concentration of the drug within the renal papilla (Duggin, 1980).

NSAIDs are commonly used in the neonatal period to treat or prevent patent ductus arteriosus (PDA) in preterm infants. The renal effect of ibuprofen treatment for patent ductus arteriosus (PDA) in very preterm infants during the first month of life was examined. Infants aged 27 to 31 weeks of gestation were assigned to two different groups according to ibuprofen exposure during care of their PDA status assessed by echocardiography. Renal function was evaluated at baseline and weekly for one month. GFR decreased significantly in the ibuprofen group after treatment withdrawal. Adjusted analysis proved this decreased GFR to be sustained during the first month of life. Tubular function was also impaired during the first month in ibuprofen-treated infants (Vieux et al., 2010). The renal effects of ibuprofen and indomethacin on suckling and weanling Sprague–Dawley rat were investigated. Newborn rats received intraperitoneal injections of either ibuprofen (10 mg/kg) on the first day of postnatal life (P1) followed by 5 mg/kg on P2 and P3 or indomethacin (0.2 mg/kg) on P1, followed by 0.1 mg/kg indomethacin on P2 and P3. No kidney histological differences were noted among the groups. However, in suckling rats, indomethacin suppressed PGE_2 and COX-2 expression and increased $PGF_{2\alpha}$, whereas ibuprofen increased COX-2 and Ang II. Although both ns-NSAIDs suppressed 6-keto$PGF_{1\alpha}$ and TXB_2 levels in suckling rats, the effect was sustained in weanling rats with indomethacin (Hasan et al., 2008). Therefore, renal function of infants receiving ns-NSAIDs should be monitored carefully and drugs that are eliminated by glomerular filtration handled cautiously during this period (Vieux et al., 2010).

In the human, glomerulogenesis is complete at 36 weeks of gestation. The metanephron develops at 5 weeks of gestation, with vesicular glomeruli development occurring at 18 weeks of gestation. Glomerular tubular development occurs from 24 weeks of gestation and is complete at 36 weeks (Abrahamson, 1991; Horster et al., 1999; Naruse et al., 2000). The mesangial cells in almost all glomeruli at the late M stage acquire the adult phenotype before birth. In the rat, glomerulogenesis continues after birth until 14 days of life (Lasaitiene et al., 2003). At birth, the neonatal rat kidney is similar to that of a 24-week-gestation human fetus. Some ns-NSAIDs, such as indomethacin, are often effective in delaying delivery for 24 to 48 h. Indomethacin suppresses preterm labor by interfering with PG-induced myometrial contractions. However, investigators have shown that indomethacin treatment given to pregnant women as a short course (the last 2 days of pregnancy) can lead to a significant functional impairment of the kidneys in their offspring, manifested as suppression of both GFR and urine production (van der Heijden et al., 1988). Some studies have shown that maternal indomethacin exposure immediately before delivery was shown not to be associated with increased neonatal complications for infants delivered between 24 and 32 weeks of gestation (Vermillion and Newman, 1999). The effects of several ns-NSAIDs on glomerulogenesis were investigated in Sprague–Dawley pregnant dams. Rat pups were given indomethacin, intraperitoneally at a dose of 0.2 mg/kg, ibuprofen at 10 mg/kg, or indomethacin at 0.1 mg/kg and gentamicin at 2.5 mg/kg for the first 5 days of their postnatal life and were euthanized at 14 days of age at completion of glomerulogenesis. The rat pups were given the drugs for the first 5 days of their postnatal life, which covers a more extensive period of glomerulogenesis (24 to 30 weeks of gestation). There was no difference between treatment groups in total number of glomeruli per kidney. Significantly fewer glomeruli per gram of kidney were noted in those rat pups that had received indomethacin or ibuprofen. The body and kidney weights of those rat pups given the higher doses of indomethacin and ibuprofen were significantly greater (13 to 25%) than those of control pups and those given the lower dose of indomethacin. The results should be interpreted with caution, due to the timing of the drug regimens and the time the rats were euthanized at 14 days of life. The reduction in the number of glomeruli per gram of kidney may indicate augmented growth of nephron tubules and/or collecting ducts and/or be a consequence of edema secondary to drug exposure (Kent et al., 2009). Another study investigated the type of renal changes found on light (LM) and electron microscopy (EM) following administration of indomethacin, ibuprofen, and gentamicin in a neonatal rat model. Rat pups were exposed to indomethacin or ibuprofen and/or gentamicin antenatally for 5 days before birth or postnatally for 5 days from day 1 of life, and pups were euthanized at 14 days of age. LM examination in all indomethacin- and ibuprofen-treated pups both antenatally and postnatally showed vacuolization of the proximal tubular epithelium, interstitial edema, and intratubular protein deposition but no significant glomerular changes. EM examination showed pleomorphic mitochondria and loss of microvilli in the renal tubules. The glomeruli showed extensive foot process effacement and irregularities of the glomerular basement membrane. EM changes were most marked in pups treated antenatally with ibuprofen, and indomethacin with gentamicin postnatally. Indomethacin, ibuprofen, and gentamicin cause significant change in glomerular and tubular structure in this neonatal rat model (Kent et al., 2007).

EFFECTS OF COX-2 s-NSAIDs ON THE KIDNEY

Complete COX-2 gene disruption results in severe nephropathy (Morham et al., 1995). Gross pathological examination of kidneys from 8-week-old male COX-2 homozygous mutant mice revealed that the kidneys were small and pale and had a granular appearance of the capsular surface. Kidneys from wild-type (WT) animals were normal in appearance. Light microscopic examination of kidneys from WT male and female mice revealed no histologic abnormalities. In contrast, the kidneys of all adult homozygous mice had mild to marked renal lesions. Mild renal lesions of the nephropathy were characterized by multifocal areas of abnormal subcapsular parenchyma comprised of small immature glomeruli and tubules, consistent with nephron hypoplasia. In some cases, the cortex appeared thin, and the numbers of glomeruli were reduced in comparison with WT kidneys. Glomeruli that are not within the hypoplastic region of the kidney were frequently found to be enlarged. Other histopathological findings included cortical areas of tubular atrophy and regeneration, protein and cellular casts within tubular lumens, tubular dilation, interstitial inflammation and fibrosis, and papillary mineralization. In general, these histopathological changes were more severe in male than in female homozygous mutants, and lesion severity increased with advancing age. The 8-week-old homozygous mutant mice kidneys had few small scattered foci of tubular atrophy and interstitial fibrosis that were not present in the 6-week-old homozygous mutant mice. The single homozygous mutant male that survived to necropsy at 16 weeks of age had severe focal interstitial fibrosis and tubular atrophy, associated with focal segmental and global glomerular sclerosis. Renal arteries and arterioles were unremarkable. To ascertain the general time frame in which the renal pathology developed, a litter of 3-day-old pups was euthanized and their kidneys were examined. The renal histopathology of the COX-2-deficient mice did not differ from that of the wild type, and the 3-day-old homozygous mice had a normal subcapsular zone of immature nephrogenic tissue and normal numbers of glomeruli in the kidney cortex. This is in contrast with the 6-week and older homozygous mice, which had abnormal nephron hypoplasia in the subcapsular region and abnormally low numbers of glomeruli in the kidney cortex. Thus, COX-2 null mice show postnatal developmental abnormalities in their kidneys that then deteriorate progressively with increasing age (Morham et al., 1995). The renal effects of COX-2 knockdown were investigated in mice. In these mice, COX-2 expression was suppressed to an extent similar to that achieved with COX-2 s-NSAIDs, but was not eliminated. In LPS-challenged macrophages and cytokine-stimulated endothelial cells obtained from these COX-2 knockdown mice, COX-2 expression was reduced 70 to 90%.

These mice developed a mild renal phenotype compared with COX-2 mice possessing an active site mutation with minimal signs of renal dysfunction as measured by inulin clearance and blood urea nitrogen (Seta et al., 2009). These mice have no notable alterations in systemic blood pressure compared to control mice (Seta et al., 2009).

Postnatal (PN) development and progression of renal dysplasia was investigated in COX-2 null mice (Norwood et al., 2000). No differences in renal morphology were detected between COX-$2^{-/-}$ mice and their controls on the day of birth, postnatal day 3 (PN3), and PN7. By PN10, early cystic changes were identifiable in some kidneys from COX-$2^{-/-}$ mice, and the lesion was affecting several different tubule sections and glomeruli, and crowded, small, subcapsular glomeruli were noted. A distinctive lack of the normal proximal tubular mantle was also noted in COX-$2^{-/-}$ animals. Not all PN10 COX-$2^{-/-}$ kidneys exhibited identifiable abnormalities, suggesting that the rate of pathological progression may vary slightly among animals. However, by PN14, kidneys from all COX-$2^{-/-}$ mice could be differentiated from their WT littermates. Kidneys from PN14 and PN28 COX-$2^{-/-}$ mice had progressive outer cortical dysplasia with cystic subcapsular glomeruli, loss of proximal tubular mass and brush-border definition, and tubular atrophy and cyst formation. Hypertrophy of juxtamedullary glomeruli and tubules was observed by PN28. The severity of pathological changes continued to increase with age. Adult (PN42) homozygous null kidneys had profound diffuse tubular dilation and cyst formation, outer cortical glomerular hypoplasia and periglomerular fibrosis, focally variable glomerular sclerosis, inner cortical nephron hypertrophy, and diffuse interstitial fibrosis. No significant inflammatory infiltrate or vascular pathology was seen. Some degree of variability in the severity of adult pathology was noted among littermates, with some animals showing more severe cystic degeneration than others (Norwood et al., 2000).

PGs also play a role in renal pathophysiology by modulating the release of renin. The macula densa in the kidney is involved in regulating afferent arteriolar tone and renin release by sensing alterations in luminal chloride (Persson et al., 1991). Low plasma renin levels have been demonstrated in COX-2-deficient mice (Kim et al., 2007). It has been demonstrated that COX-2 s-NSAIDs can significantly decrease plasma renin levels and blood pressure in rats (Wang et al., 1999; Cheng et al., 2001). COX-derived prostanoids act in an autocrine or paracrine fashion, exert complex and diverse functions, and serve as physiological buffers within the kidney (Hao and Breyer, 2008). COX-derived prostanoids can modulate RBF and GFR. Modulation of renal physiology is driven by COX-2 catalysis from AA of PGE_2 and PGI_2; both are important PGs for the maintenance of RBF. PGI_2 is the main factor involved in endothelial dysfunction in hypertensive rats, whereas other prostanoids are more critically involved in endothelial dysfunction under normotensive conditions (Blanco-Rivero et al., 2005). Some studies have suggested that nitric oxide, in addition to PGI_2, plays a significant role in the restoration of endothelial function in hypertensive patients treated with ACE-Is (Yamanari et al., 2004). Thus, COX-2 metabolites have been implicated to play a role in regulation of renal sodium excretion, renin release, and maintenance of RBF. A high-salt diet markedly increases renal medullary COX-2 expression

(Yang et al., 1998). The importance of renal medullary COX-2 in promoting natriuresis and maintaining blood pressure during high dietary sodium intake has been demonstrated in rats after intramedullary infusion with NS-398, a COX-2 s-NSAID (Zewde and Mattson, 2004). In nonclinical safety studies with celecoxib, transient sodium retention was reported in rats up through 6 weeks of treatment (Khan et al., 2002). Rats given celecoxib for 3 weeks caused a 30-mm Hg increase in systolic blood pressure, whereas the same dose given hypertensive rats increased systolic blood pressure by 33 mmHg (Muscara et al., 2000a). In another study, however, celecoxib did not alter blood pressure in hypertensive rats, and it has been suggested that the natriuretic response to increased blood pressure may be preserved during inhibition of COX-2 (Richter et al., 2004). In rats infused with NS-398 or meloxicam, responses to increased renal interstitial hydrostatic pressure, induced by direct renal interstitial volume expansion and fractional excretion of sodium, were preserved (Gross et al., 1999). Variations between rat strains in the effects of COX-2 inhibitors on blood pressure have been reported. A study comparing the effects of diclofenac, celecoxib, and rofecoxib in Dahl salt-sensitive rats found that after 8 weeks of treatment, celecoxib slightly (4%), but significantly, reduced systemic blood pressure compared to controls and diclofenac and rofecoxib minimally, but significantly increased systemic blood pressure (1%) (Hermann et al., 2005). In another study, COX-2 inhibition in Sprague–Dawley rats had no effect on systemic blood pressure (Vanecková et al., 2005). Beneficial effects of COX-2 inhibition on the renal system have been reported. In male Wistar and stroke-prone spontaneously hypertensive (SHRSP) rats, the postischemic increase in fractional sodium excretion was blunted after celecoxib treatment (Knight and Johns, 2008). A study in male Sprague–Dawley rats found that celecoxib might reduce proteinuria in nephritic syndrome without impairing renal function (Lee et al., 2009). Streptozotocin-diabetic rats treated with a COX-2 s-NSAID, SC-58236, showed a marked reduction in albuminuria, a reduction in kidney weight-to-body weight ratio, and TGFβ excretion and a marked decrease in the urinary excretion of TNFα (Quilley et al., 2011).

Nimesulide, a COX-2 s-NSAID, infusion (infused as a continuous bolus dose) in anesthetized mongrel dogs with normal sodium intake had no effects on mean arterial pressure (MAP) or renal hemodynamics, but did reduce urinary sodium excretion and urine flow rate. However, during low sodium intake, nimesulide infusion increased arterial pressure and decreased RBF and urine flow rate. Urinary PGE_2 excretion was decreased, but plasma renin activity did not change (Rodriguez et al., 2000). In dogs given nimesulide, a moderate elevation in blood pressure was observed after intravenous infusion of glucose (Brands et al., 2001). Nimesulide given as a single intravenous bolus dose of 3 mg/kg nimesulide elicited a decrease in MAP in both induced and noninduced Cyp1a1–Ren2 transgenic hypertensive rats. However, the decrease in MAP was greater in hypertensive rats than in normotensive rats. Nimesulide also decreased GFR and RPF in hypertensive rats but did not alter GFR or RPF in normotensive rats (Opay et al. 2006). NS-398 given intraperitoneally in a 5-mg/kg per day dose significantly ameliorated an increase in systolic blood pressure and proteinuria and decreased portal hypertension in Sprague–Dawley rats (Tsugawa et al., 1999; Tomida et al., 2003).

In a rat model of hypertension, it has been suggested that COX-2 plays a significant role in the development of two-kidney/one-clip and one-kidney/one-clip renovascular hypertension (Okumura et al., 2002). PGs seem to contribute to portal hypertension, and both COX-1 and COX-2 blockades reduced the portal pressure and visceral blood flow volume of the portal vein and gastric mucosa in rats (Tsugawa et al., 1999). Studies in COX-1 null mice under conditions of reduced sodium intake have revealed a mild but significant reduction in blood pressure over control mice (Athirakul et al., 2001). The COX-2 s-NSAID SC-58236 had no effect on blood pressure in a 5/6 nephroectomized rat hypertension model, but reduced blood pressure and plasma renin activity in a renovascular rat hypertension model (Wang et al., 1999,2000). Another COX-2 s-NSAID, MF-tricyclic, did not decrease RBF and had no effect on blood pressure in normal and volume-depleted dogs (Black et al., 1998).

In salt-depleted subjects, celecoxib was given for 7 days in 200- and 400-mg twice-daily doses. Celecoxib caused sodium and potassium retention and had no effect on systemic blood pressure (Rossat et al., 1999). Short-term and transient decreases in RBF and GFR were found at the highest dose of 400 mg on day 1 (Rossat et al., 1999). In another study, no individual age category of normotensive subjects had blood pressure or renal function effects significantly altered by short-term treatment with standard doses of celecoxib (200 mg twice a day) (Dilger et al., 2002). In a study in healthy elderly volunteers placed on a fixed sodium intake, rofecoxib at 50 mg once daily for 14 days did not affect GRR and caused a transient retention of sodium, reduced sodium excretion by 20% during the first 3 days of drug administration, and then returned to normal (Catella-Lawson et al., 1999). The effects of rofecoxib on renal function were investigated in the elderly, who received a low-sodium diet. In the first study, single doses of rofecoxib of 250 mg were administered to 15 patients. In the second study, multiple doses of rofecoxib at 12.5 or 25 mg/day were administered to 60 patients. The single and multiple doses of rofecoxib decreased GFR and serum and urinary sodium and potassium values (Swan et al., 2000). The safety and tolerability profile of orally administered meloxicam 15 mg given once daily over a 28-day treatment period was investigated in renally impaired patients with rheumatic disease. No adverse events related to the urinary system or increases in serum urea or potassium were recorded (Bevis et al., 1996). It is implied that COX-2 inhibition would disrupt the antihypertensive effects of ACE-I. However, high doses of celecoxib (400 mg twice a day) showed no significant effect on the activity of the ACE-I lisinopril (White et al., 2002). In another study, celecoxib had no significant change in blood pressure during a 24-h ambulatory blood pressure outcome, but increased systolic blood pressure at peak in the presence of ACE-I, a trandolapril (Izhar et al., 2004).

POTENTIAL MECHANISMS OF NSAID EFFECTS ON THE RENAL SYSTEM

Both COX-2 s-NSAIDs and ns-NSAIDs could contribute to changes in RBF and fluid retention imbalance. Several potential mechanisms of the effects of ns-NSAIDs on renal pathophysiology have been proposed (Radi, 2009). These mechanisms include (1) inhibition of COX, which reduces systemic and renal PG synthesis (Meade et al., 1993; Wen, 1997); (2) inhibition of PGs by NSAIDs counteracting the hypotensive effects of ACE-I, which blocks ACE and successively increases bradykinin levels, allowing increased renal release of vasodilatory and natriuretic PGs (Blumberg et al., 1977; Linz et al., 1995); (3) alterations in systemic vascular resistance and cardiac output (de Leeuw, 1996); and (4) alterations in the renin–angiotensin–aldosterone system (Lopez-Ovejero et al., 1978; Sahloul et al., 1990; Houston, 1991; Wen, 1997). It is interesting that aldosterone can promote endothelial dysfunction independent of blood pressure through COX-2-derived PGI_2 vasoconstriction in both normotensive and hypertensive conditions in rats (Blanco-Rivero et al., 2005).

Structurally different COX-2 inhibitors vary in their potency as inhibitors of the COX-2 enzyme (Clark et al., 2004). Selectivity and potency of COX-2 s-NSAIDs may differ between tissues and types of cells from the same tissue. Rofecoxib appears more selective and potent than celecoxib (Clark et al., 2004). In addition, differences in metabolism could contribute to effects on blood pressure regulation. Rofecoxib is a sulfone that is not well distributed in tissues and is metabolized principally by cytosolic reduction. Therefore, the NSAID chemotype (i.e., chemically distinct) is an important consideration. Cystosol reductase metabolizes a number of vasoactive hormones in addition to COX. It has been suggested that COX-2 may competitively inhibit the metabolism of these hormones, particularly aldosterone, significant in fluid and sodium retention, both important in blood pressure regulation (Liew and Krum, 2002). Thus, inhibition of carbonic anhydrase leads to diuretic action that partially offsets blood pressure elevations in the kidney. In addition, rofecoxib can readily decompose to reactive metabolites that may account for peroxyl radical formation (Reddy and Corey, 2005). Celecoxib is a sulfonamide that is well distributed in tissues and is metabolized by the cytochrome P450 system (Tacconelli et al., 2004). Additional studies designed exclusively to evaluate the influence of structural differences, selectivity, and potency are needed to delineate the potential contribution of NSAIDs to these effects.

CONCLUSIONS

It is important to consider the renal anatomical, developmental, and physiological interspecies differences when considering the pathophysiological effects of COX inhibition on the kidney. In the kidney, COX-1 is expressed constitutively and in the most abundant isoform in various microanatomical locations, while COX-2 is usually expressed constitutively in the macula densa across various species and up-regulated during various pathological renal conditions. Studies show that COX-2 s-NSAIDs can, like ns-NSAIDs, affect renal function adversely in some experimental models. However, in some models, COX-2 s-NSAIDs have had no effect or have had beneficial effects such as decreased plasma renin activity or decreased renal injury. Several mechanisms are proposed for the effects on NSAIDs on the renal system. Further studies are needed to better understand the potential for chemotype-mediated renal effects.

REFERENCES

Abe K, Fujino Y, Sakakibara T. The effect of prostaglandin E1 during cardiopulmonary bypass on renal function after cardiac surgery. Eur J Clin Pharmacol 1993;45:217–220.

Abe T, Takeuchi K, Takahashi N, Tsutsumi E, et al. Rat kidney thromboxane receptor: molecular cloning, signal transduction and intrarenal expression localization. J Clin Invest 1995;96:657–664.

Abrahamson DR. Glomerulogenesis in the developing kidney. Semin Nephrol 1991;11:375–389.

Ahmadizadeh M, Echt R, Kuo CH, Hook JB. Sex and strain differences in mouse kidney: Bowman's capsule morphology and susceptibility to chloroform. Toxicol Lett 1984;20:161–171.

Ailabouni W, Eknoyan G. Nonsteroidal anti-inflammatory drugs and acute renal failure in the elderly. A risk-benefit assessment. Drugs Aging 1996;9:341–351.

Alique M, Calleros L, Luengo A, Griera M, et al. Changes in extracellular matrix composition regulate cyclooxygenase-2 expression in human mesangial cells. Am J Physiol Cell Physiol 2011;300: C907–C918.

Anderson RJ, Berl T, McDonald KM, Schrier RW. Evidence for an in vivo antagonism between vasopressin and prostaglandin in the mammalian kidney. J Clin Invest 1975;56:420–426.

Apiwattanakul N, Sekine T, Chairoungdua A, Kanai Y, et al. Transport properties of nonsteroidal anti-inflammatory drugs by organic anion transporter 1 expressed in *Xenopus laevis* oocytes. Mol Pharmacol 1999;55:847–854.

Araujo M, Welch WJ. Cyclooxygenase 2 inhibition suppresses tubuloglomerular feedback: roles of thromboxane receptors and nitric oxide. Am J Physiol Renal Physiol 2009;296:F790–F794.

Asano K, Taniguchi S, Nakao A, Maruyama T, et al. Distribution of thromboxane A2 receptor mRNA along the mouse nephron segments. Biochem Biophys Res Commun 1996;226:613–617.

Athirakul K, Kim HS, Audoly LP, Smithies O, et al. Deficiency of COX-1 causes natriuresis and enhanced sensitivity to ACE inhibition. Kidney Int 2001;60:2324–2329.

Audoly LP, Ruan X, Wagner VA, Goulet JL, et al. Role of EP(2) and EP(3) PGE(2) receptors in control of murine renal hemodynamics. Am J Physiol Heart Circ Physiol 2001;280:H327–H333.

Banks RO, Jacobson ED. Renal vasodilation with ureteral occlusion and prostaglandins: attenuation by histamine H1 antagonists. Am J Physiol 1985;249:F851–F857.

Bassan H, Trejo L, Kariv N, et al. Experimental intrauterine growth retardation alters renal development. Pediatr Nephrol 2000;15:192–195.

Bek M, Nusing RM, Kowark P, Henger A, et al. Characterization of prostanoid receptors in podocytes. J Am Soc Nephrol 1999;10:2084–2093.

Bennett CM, Brenner BM, Berliner RW. Micropuncture study of nephron function in the rhesus monkey. J Clin Invest 1968;47:203–216.

Bennett WM, Henrich WL, Stoff JS. The renal effects of nonsteroidal anti-inflammatory drugs: summary and recommendations. Am J Kidney Dis 1996;28:S56–S62.

Bergelin I, Karlsson B. Colostrum, serum and kidney proteins in the urine of the developing neonatal pig. Int J Biochem 1974;5:885–894.

Bevis PJ, Bird HA, Lapham G. An open study to assess the safety and tolerability of meloxicam 15mg in subjects with rheumatic disease and mild renal impairment. Br J Rheumatol 1996;35:56–60.

Black SC, Brideau C, Cirino M, Belley M, et al. Differential effect of a selective cyclooxygenase-2 inhibitor versus indomethacin on renal blood flow in conscious volume depleted dogs. J Cardiovasc Pharmacol 1998;32:686–894.

Blanco-Rivero J, Cachofeiro V, Lahera V, Aras-Lopez R, et al. Participation of prostacyclin in endothelial dysfunction induced by aldosterone in normotensive and hypertensive rats. Hypertension 2005;46:107–112.

Blasingham MC, Nasjletti A. Differential renal effects of cyclooxygenase inhibition in sodium-replete and sodium-deprived dog. Am J Physiol 1980;239:F360–F365.

Blumberg AL, Denny SE, Marshall GR, Needleman P. Blood vessel–hormone interactions: angiotensin, bradykinin, and prostaglandins. Am J Physiol 1977;232:H305–H310.

Bonilla-Felix M, Jiang W. Expression and localization of prostaglandin EP3 receptor mRNA in the immature rabbit kidney. Am J Physiol 1996;271(1Pt 2):F30–F36.

Bonvalet JP, Pradelles P, Farman N. Segmental synthesis and actions of prostaglandins along the nephron. Am J Physiol 1987;253:F377–F387.

Brands MW, Hailman AE, Fitzgerald SM. Long-term glucose infusion increases arterial pressure in dogs with cyclooxygenase-2 inhibition. Hypertension 2001;37:733–738.

Brater DC, Anderson S, Baird B, Campbell WB. Effects of ibuprofen, naproxen, and sulindac on prostaglandins in men. Kidney Int 1985;27:66–73.

Braun U. Ultrasonographic examination of the right kidney in cows. Am J Vet Res 1991;52:1933–1929.

Bresnahan BA, Le Breton GC, Lianos EA. Localization of authentic thromboxane A2/prostaglandin H2 receptor in the rat kidney. Kidney Int 1996;49:1207–1213.

Breyer MD, Breyer RM. G protein-coupled prostanoid receptors and the kidney. Annu Rev Physiol 2001;63:579–605.

Breyer MD, Jacobson HR, Davis LS, Breyer RM. In situ hybridization and localization of mRNA for the rabbit prostaglandin EP3 receptor. Kidney Int 1993;44:1372–1378.

Breyer MD, Davis L, Jacobson HR, Breyer RM. Differential localization of prostaglandin E receptor subtypes in human kidney. Am J Physiol 1996;270:F912–F918.

Brezis M, Rosen S. Hypoxia of the renal medulla: its implications for disease. N Engl J Med 1995;332:647–655.

Campean V, Theilig F, Paliege A, Breyer M, et al. Key enzymes for renal prostaglandin synthesis: sitespecific expression in rodent kidney (rat, mouse). Am J Physiol Renal Physiol 2003;285:F19–F32.

Casotti G, Braun EJ. Structure of the glomerular capillaries of the domestic chicken and desert quail. J Morphol 1995;224:57–63.

Catella-Lawson F, McAdam B, Morrison BW, Kapoor S, et al. Effects of specific inhibition of cyclooxygenase-2 on sodium balance, hemodynamics, and vasoactive eicosanoids. J Pharmacol Exp Ther 1999;289:735–741.

Cattell V, Smith J, Cook HT. Prostaglandin E1 suppresses macrophage infiltration and ameliorates injury in an experimental model of macrophage-dependent glomerulonephritis. Clin Exp Immunol 1990;79:260–265.

Chamaa NS, Mosig D, Drukker A, Guignard JP. The renal hemodynamic effects of ibuprofen in the newborn rabbit. Pediatr Res 2000;48:600–605.

Cheeke PR, Amberg JW. Comparative calcium excretion by rats and rabbits. J Anim Sci 1973;37: 450–454.

Chen J, Zhao M, He W, Milne GL, et al. Increased dietary NaCl induces renal medullary PGE2 production and natriuresis via the EP2 receptor. Am J Physiol Renal Physiol 2008;295:F818–F825.

Cheng H, Wang S, Jo YI, Hao CM, et al. Overexpression of cyclooxygenase-2 predisposes to podocyte injury. J Am Soc Nephrol 2007;18:551–559.

Cheng H, Fan X, Moeckel GW, Harris RC. Podocyte COX-2 exacerbates diabetic nephropathy by increasing podocyte (pro)renin receptor expression. J Am Soc Nephrol 2011;22:1240–1251.

Cheng HF, Wang JL, Zhang MZ, Miyazaki Y, et al. Angiotensin II attenuates renal cortical cyclooxygenase-2 expression. J Clin Invest 1999;103:953–961.

Cheng HF, Wang JL, Zhang MZ, Wang SW, et al. Genetic deletion of COX-2 prevents increased renin expression in response to ACE inhibition. Am J Physiol Renal Physiol 2001;280:F449–F456.

Cheng X, Zhang H, Lee HL, Park JM. Cyclooxygenase-2 inhibitor preserves medullary aquaporin-2 expression and prevents polyuria after ureteral obstruction. J Urol 2004;172:2387–2390.

Chonko AM. Urate secretion in isolated rabbit renal tubules. Am J Physiol 1980;239:F545–F551.

Chou SY, Cai H, Pai D, Mansour M, et al. Regional expression of cyclooxygenase isoforms in the rat kidney in complete unilateral ureteral obstruction. J Urol 2003;170:1403–1408.

Christensen EI, Verroust PJ. Megalin and cubilin, role in proximal tubule function and during development. Pediatr Nephrol 2002;17:993–999.

Cihlar T, Lin DC, Pritchard JB, Fuller MD, et al. The antiviral nucleotide analogs cidofovir and adefovir are novel substrates for human and rat renal organic anion transporter 1. Mol Pharmacol 1999;56: 570–580.

Cirino M, Morton H, MacDonald C, Hadden J, et al. Thromboxane A2 and prostaglandin endoperoxide analogue effects on porcine renal blood flow. Am J Physiol 1990;258:F109–F114.
Clark DW, Layton D, Shakir SA. Do some inhibitors of COX-2 increase the risk of thromboembolic events? Linking pharmacology with pharmacoepidemiology. Drug Saf 2004;27:427–456.
Clive DM, Stoff JS. Renal syndromes associated with nonsteroidal antiinflammatory drugs. N Engl J Med 1984;310:563–572.
Colletti AE, Vogl HW, Rahe T, Zambraski EJ. Effects of acetaminophen and ibuprofen on renal function in anesthetized normal and sodium-depleted dogs. J Appl Physiol 1999;86:592–597.
Corvilain J, Abramo M, Bergans A. Some effects of human growth hormone on renal hemodynamics and on tubular phosphate transport in man. J Clin Invest 1962;41:1230–1235.
Cox PG, Moons WM, Russel FG, van Ginneken CA. Renal disposition and effects of naproxen and its *l*-enantiomer in the isolated perfused rat kidney. J Pharmacol Exp Ther 1990;255:491–496.
Cremades A, Vicente-Ortega V, Peñafiel R. Influence of murine renal sexual dimorphism on amiloride-induced hyperkalemia. Nephron Physiol 2003;95:57–66.
Currie MG, Needleman P. Renal arachidonic acid metabolism. Annu Rev Physiol 1984;46:327–341.
Deeley RG, Westlake C, Cole SP. Transmembrane transport of endo- and xenobiotics by mammalian ATP-binding cassette multidrug resistance proteins. Physiol Rev 2006;86:849–899.
de Leeuw PW. Nonsteroidal anti-inflammatory drugs and hypertension. The risks in perspective. Drugs 1996;51:179–187.
DeMaria AN, Weir MR. Coxibs—beyond the GI tract: renal and cardiovascular issues. J Pain Symptom Manag 2003;25:S41–S49.
DeRubertis FR, Craven PA. Eicosanoids in the pathogenesis of the functional and structural alterations of the kidney in diabetes. Am J Kidney Dis 1993;22:727–735.
Dilger K, Herrlinger C, Peters J, Seyberth HW, et al. Effects of celecoxib and diclofenac on blood pressure, renal function, and vasoactive prostanoids in young and elderly subjects. J Clin Pharmacol 2002;42:985–994.
Dinchuk JE, Car BD, Focht RJ, Johnston JJ, et al. Renal abnormalities and an altered inflammatory response in mice lacking cyclooxygenase II. Nature 1995;378:406–409.
Dixey JJ, Noormohamed FH, Lant AF, Brewerton DA. The effects of naproxen and sulindac on renal function and their interaction with hydrochlorothiazide and piretanide in man. Br J Clin Pharmacol 1987;23:55–63.
Dresser MJ, Leabman MK, Giacomini KM. Transporters involved in the elimination of drugs in the kidney: organic anion transporters and organic cation transporters. J Pharm Sci 2001;90:397–421.
Dressler GR. The cellular basis of kidney development. Annu Rev Cell Dev Biol 2006;22:509–529.
Duggin GG. Mechanisms in the development of analgesic nephropathy. Kidney Int 1980;18:553–561.
Dunn MJ, Simonson M, Davidson EW, Scharschmidt LA, et al. Nonsteroidal anti-inflammatory drugs and renal function. J Clin Pharmacol 1988;28:524–529.
Edwards RM. Effects of prostaglandins on vasoconstrictor action in isolated renal arterioles. Am J Physiol Renal Physiol 1985;248:F779–F784.
El-Achkar TM, Plotkin Z, Marcic B, Dagher PC. Sepsis induces an increase in thick ascending limb COX-2 that is TLR4 dependent. Am J Physiol Renal Physiol 2007;293:F1187–F1196.
el-Galley RE, Keane TE. Embryology, anatomy, and surgical applications of the kidney and ureter. Surg Clin North Am 2000;80:381–401.
Eriksson LO, Beermann B, Kallner M. Renal function and tubular transport effects of sulindac and naproxen in chronic heart failure. Clin Pharmacol Ther 1987;42:646–654.
Eriksson LO, Sturfelt G, Thysell H, Wollheim FA. Effects of sulindac and naproxen on prostaglandin excretion in patients with impaired renal function and rheumatoid arthritis. Am J Med 1990;89:313–321.
Evan AP Jr, Stoeckel JA, Loemker V, et al. Development of the intrarenal vascular system of the puppy kidney. Anat Rec 1979;194:187–199.
Faerch M, Christensen JH, Rittig S, Johansson JO, et al. Diverse vasopressin V2 receptor functionality underlying partial congenital nephrogenic diabetes insipidus. Am J Physiol Renal Physiol 2009;297:F1518–F1525.
Farquhar WB, Morgan AL, Zambraski EJ, Kenney WL. Effects of acetaminophen and ibuprofen on renal function in the stressed kidney. J Appl Physiol 1999;86:598–604.

Fernandez-Tome M, Kraemer L, Federman SC, Favale N, et al. COX-2-mediated PGD2 synthesis regulates phosphatidylcholine biosynthesis in rat renal papillary tissue. Biochem Pharmacol 2004;15:245–254.

Ferreri NR, An SJ, McGiff JC. Cyclooxygenase-2 expression and function in the medullary thick ascending limb. Am J Physiol 1999;277:F360–F368.

Fisher JW, Koury S, Ducey T, Mendel S. Erythropoietin production by interstitial cells of hypoxic monkey kidneys. Br J Haematol 1996;95:27–32.

Fleming EF, Athirakul K, Oliverio MI, Key M, et al. 1998. Urinary concentrating function in mice lacking EP3 receptors for prostaglandin E2. Am J Physiol Renal Physiol 1998;275: F955–F961.

Fouser L, Avner ED. Normal and abnormal nephrogenesis. Am J Kidney Dis 1993;21:64–70.

Fox J, Anderson LC, Loew FM, Quimby FW. *Laboratory Animal Medicine*, 2nd ed. San Diego, CA: Academic Press, 2002.

Friis C. Postnatal development of the pig kidney: ultrastructure of the glomerulus and the proximal tubule. J Anat 1980;130:513–526.

Frolich JC, Hollifield JW, Dormois JC, Frolich BL, et al. Suppression of plasma renin activity by indomethacin in man. Cir Res 1976;39:447–452.

Fukasawa H, Bornheimer S, Kudlicka K, Farquhar MG. Slit diaphragms contain tight junction proteins. J Am Soc Nephrol 2009;20:1491–1503.

Garella S, Matarese RA. Renal effects of prostaglandins and clinical adverse effects of nonsteroidal anti-inflammatory agents. Medicine 1984;63:165–181.

Gasser B, Mauss Y, Ghnassia JP, et al. A quantitative study of normal nephrogenesis in the human fetus: its implication in the natural history of kidney changes due to low obstructive uropathies. Fetal Diagn Ther 1993;8:371–384.

Gomez RA, Sequeira Lopez ML, Fernandez L, et al. The maturing kidney: development and susceptibility. Renal Fail 1999;21:283–291.

Gong R, Morris DJ, Brem AS. Human renal 11beta-hydroxysteroid dehydrogenase 1 functions and co-localizes with COX-2. Life Sci 2008;82:631–637.

González AA, Céspedes C, Villanueva S, Michea L, et al. E Prostanoid-1 receptor regulates renal medullary alphaENaC in rats infused with angiotensin II. Biochem Biophys Res Commun 2009;389:372–377.

Gooch K, Culleton BF, Manns BJ, Zhang J, et al. NSAID use and progression of chronic kidney disease. Am J Med 2007;120:280, e1–e7.

Gottschalk CW, Mylle M. Evidence that the mammalian nephron functions as a countercurrent multiplier system. Science 1958;128:594.

Gross DM, Brookins J, Fink GD, Fisher JW. Effects of prostaglandins A2, E2 and F2 alpha on erythropoietin production. J Pharmacol Exp Ther 1976;198:489–496.

Gross JM, Dwyer JE, Knox FG. Natriuretic response to increased pressure is preserved with COX-2 inhibitors. Hypertension 1999;34:1163–1167.

Guan Y, Zhang Y, Breyer RM, Fowler B, et al. Prostaglandin E2 inhibits renal collecting duct Na^+ absorption by activating the EP1 receptor. J Clin Invest 1998;102:194–201.

Guan Y, Zhang Y, Wu J, Qi Z, et al. Antihypertensive effects of selective prostaglandin E2 receptor subtype 1 targeting. J Clin Invest 2007;117:2496–2505.

Gwak HS, Oh JH, Han HK. Effects of non-steroidal anti-inflammatory drugs on the pharmacokinetics and elimination of acyclovir in rats. J Pharm Pharmacol 2005;57:393–398.

Hallesy DW, Shott LD, Hill R. Comparative toxicology of naproxen. Scand J Rheumatol Suppl 1973;2:20–28.

Hao CM, Breyer MD. Physiological regulation of prostaglandins in the kidney. Annu Rev Physiol 2008;70:357–377.

Harding P, Sigmon DH, Alfie ME, Huang PL, et al. Cyclooxygenase-2 mediates increased renal renin content induced by low-sodium diet. Hypertension 1997;29:297–302.

Harirforoosh S, Jamali F. Effect of nonsteroidal anti-inflammatory drugs with varying extent of COX-2–COX-1 selectivity on urinary sodium and potassium excretion in the rat. Can J Physiol Pharmacol 2005;83:85–90.

Harris RC, Breyer MD. Physiological regulation of cyclooxygenase-2 in the kidney. Am J Physiol Renal Physiol 2001;281:F1–F11.

Harris RC, McKanna JA, Akai Y, Jacobson HR, Dubois RN, Breyer MD. Cyclooxygenase-2 is associated with the macula densa of rat kidney and increases with salt restriction. J Clin Invest 1994;94:2504–2510.

Hartleroad JY, Beharry KD, Hausman N, et al. Effect of maternal administration of selective cyclooxygenase (COX)-2 inhibitors on renal size, growth factors, proteinases, and COX-2 secretion in the fetal rabbit. Biol Neonate 2005;87:246–253.

Hartner A, Goppelt-Struebe M, Hilgers KF. Coordinate expression of cyclooxygenase-2 and renin in the rat kidney in renovascular hypertension. Hypertension 1998;31:201–205.

Hasan J, Beharry KD, Gharraee Z, Stavitsky Y, et al. Early postnatal ibuprofen and indomethacin effects in suckling and weanling rat kidneys. Prostaglandins Other Lipid Mediat 2008;85:81–88.

Hasler U, Vinciguerra M, Vandewalle A, Martin PY, et al. Dual effects of hypertonicity on aquaporin-2 expression in cultured renal collecting duct principal cells. J Am Soc Nephrol 2005;16:1571–1582.

Hassid A, Konieczkowski M, Dunn MJ. Prostaglandin synthesis in isolated rat kidney glomeruli. Proc Natl Acad Sci USA 1979;76:1155–1159.

Hébert RL, Jacobson HR, Breyer MD. Prostaglandin E2 inhibits sodium transport in rabbit cortical collecting duct by increasing intracellular calcium. J Clin Invest 1991;87:1992–1998.

Hébert RL, Jacobson HR, Fredin D, Breyer MD. Evidence that separate PGE2 receptors modulate water and sodium transport in rabbit cortical collecting duct. Am J Physiol 1993;265:F643–F650.

Heise G, Grabensee B, Schrör K, Heering P. Different actions of the cyclooxygenase 2 selective inhibitor flosulide in rats with passive Heymann nephritis. Nephron 1998;80:220–226.

Herman MB, Rajkhowa T, Cutuli F, Springate JE, et al. Regulation of renal proximal tubule Na,K-ATPase by prostaglandins. Am J Physiol Renal Physiol 2010 Feb 3.

Hermann M, Shaw S, Kiss E, Camici G, et al. Selective COX-2 inhibitors and renal injury in salt-sensitive hypertension. Hypertension 2005;45:193–197.

Hinchliffe SA, Lynch MR, Sargent PH, et al. The effect of intrauterine growth retardation on the development of renal nephrons. Br J Obstet Gynaecol 1992;99:296–301.

Hoffmann KL, Wood AK, McCarthy PH. Ultrasonography of the equine neonatal kidney. Equine Vet J 2000;32:109–113.

Holt S, Marley R, Fernando B, Harry D, et al. Acute cholestasis-induced renal failure: effects of antioxidants and ligands for the thromboxane A2 receptor. Kidney Int 1999;55:271–277.

Hommel E, Mathiesen E, Larsen SA, Edsberg B, et al. Effects of indomethacin on kidney function in type 1 (insulin-dependent) diabetic patients with nephropathy. Diabetologia 1987;30:78–81.

Horster M. Nephron function and perinatal homeostasis. Ann Rech Vet 1977;8:468–482.

Horster MF, Braun GS, Huber SM. Embryonic renal epithelia: induction, nephrogenesis, and cell differentiation. Physiol Rev 1999;79:1157–1191.

Hosoyamada M, Sekine T, Kanai Y, Endou H. Molecular cloning and functional expression of a multispecific organic anion transporter from human kidney. Am J Physiol 1999;276:F122–F128.

Houston MC. Nonsteroidal anti-inflammatory drugs and antihypertensives. Am J Med 1991;90: 42S–47S.

Huerta C, Castellsague J, Varas-Lorenzo C, García Rodríguez LA. Nonsteroidal anti- inflammatory drugs and risk of ARF in the general population. Am J Kidney Dis 2005;45:531–539.

Imig JD. Eicosanoids and renal vascular function in diseases. Clin Sci 2006;111:21–34.

Ito S, Carretero OA, Abe K, Beierwaltes WH, et al. Effect of prostanoids on renin release from rabbit isolated afferent arterioles with and without attached macula densa. Kidney Int 1989;35:1138–1144.

Izhar M, Alausa T, Folker A, Hung E, et al. Effects of COX inhibition on blood pressure and kidney function in ACE inhibitor-treated blacks and hispanics. Hypertension 2004;43:573–577.

Jackson EK, Oates JA, Branch RA. Indomethacin decreases arterial blood pressure and plasma renin activity in rats with aortic ligation. Circ Res 1981;49:180–185.

Jelkmann W. Erythropoietin: structure, control of production, and function. Physiol Rev 1992;72: 449–489.

Jensen BL, Mann B, Skott O, Kurtz A. Differential regulation of renal prostaglandin receptor mRNAs by dietary salt intake in the rat. Kidney Int 1999;56:528–537.

Johnson AG, Nguyen TV, Day RO. Do nonsteroidal anti-inflammatory drugs affect blood pressure? A meta-analysis. Ann Intern Med 1994;121:289–300.

Johnston HH, Herzog JP, Lauler DP. Effect of prostaglandin E1 on renal hemodynamics, sodium and water excretion. Am J Physiol 1967;213:939–946.

Kaplan BS, Restaino I, Raval DS, Gottlieb RP, Bernstein J. Renal failure in the neonate associated with in utero exposure to non-steroidal anti-inflammatory agents. Pediatr Nephrol 1994;8:700–704.

Kawada N, Dennehy K, Solis G, et al. TP receptors regulate renal hemodynamics during angiotensin II slow pressor response. Am J Physiol Renal Physiol 2004;287:F753–F759.

Kawada N, Solis G, Ivey N, et al. Cyclooxygenase-1-deficient mice have high sleep-to-wake blood pressure ratios and renal vasoconstriction. Hypertension 2005;45:1131–1138.

Kelley VE, Winkelstein A, Izui S, Dixon FJ. Prostaglandin E1 inhibits T-cell proliferation and renal disease in MRL/1 mice. Clin Immunol Immunopathol 1981;21:190–203.

Kennedy CR, Xiong H, Rahal S, Vanderluit J, et al. Urine concentrating defect in prostaglandin EP1-deficient mice. Am J Physiol Renal Physiol 2007;292:F868–F875.

Kent AL, Maxwell LE, Koina ME, Falk MC, et al. Renal glomeruli and tubular injury following indomethacin, ibuprofen, and gentamicin exposure in a neonatal rat model. Pediatr Res 2007;62:307–312.

Kent AL, Douglas-Denton R, Shadbolt B, Dahlstrom JE, et al. Indomethacin, ibuprofen and gentamicin administered during late stages of glomerulogenesis do not reduce glomerular number at 14 days of age in the neonatal rat. Pediatr Nephrol 2009;24:1143–1149.

Khan KN, Venturini CM, Bunch RT, et al. Interspecies differences in renal localization of cyclooxygenase isoforms: implications in nonsteroidal antiinflammatory drug-related nephrotoxicity. Toxicol Pathol 1998;26:612–620.

Khan KN, Stanfield KM, Trajkovic D, Knapp DW. Expression of cyclooxygenase-2 in canine renal cell carcinoma. Vet Pathol 2001a;38:116–119.

Khan KN, Stanfield KM, Dannenberg A, Seshan SV, et al. Cyclooxygenase-2 expression in the developing human kidney. Pediatr Dev Pathol 2001b;4:461–466.

Khan N, Hard G, Radi Z. Chapter 11: New horizons in toxicity prediction: a comprehensive overview of the current status and application of predictive toxicology. In: In Vivo Approaches: Renal Toxicity. San Diego, CA: Academic Press, 2011.

Khan NK, Alden CL, Chapter 29: Kidney. In: Haschek WM, Rousseaux CG (eds), *Handbook of Toxicologic Pathology*, 2nd ed. San Diego, CA: Academic Press, 2002.

Kim SM, Chen L, Mizel D, Huang YG, et al. Low plasma renin and reduced renin secretory responses to acute stimuli in conscious COX-2-deficient mice. Am J Physiol Renal Physiol 2007;292:F415–F422.

Kleinman LI, Reuter JH. Maturation of glomerular blood flow distribution in the new-born dog. J Physiol 1973;228:91–103.

Knight S, Johns EJ. Renal functional responses to ischaemia-reperfusion injury in normotensive and hypertensive rats following non-selective and selective cyclooxygenase inhibition with nitric oxide donation. Clin Exp Pharmacol Physiol 2008;35:11–16.

Komers R, Lindsley JN, Oyama TT, et al. Immunohistochemical and functional correlations of renal cyclooxygen- ase-2 in experimental diabetes. J Clin Invest 2001;107:889–898.

Komhoff M, Grone HJ, Klein T, Seyberth HW, et al. Localization of cyclooxygenase-1 and -2 in adult and fetal human kidney: implication for renal function. Am J Physiol 1997;272:460–468.

Komhoff M, Jeck ND, Seyberth HW, et al. Cyclooxygenase-2 expression is associated with the renal macula densa of patients with Bartter-like syndrome. Kidney Int 2000;58:2420–2424.

Kommareddy M, McAllister RM, Ganjam VK, Turk JR, et al. Upregulation of cyclooxygenase-2 expression in porcine macula densa with chronic nitric oxide synthase inhibition. Vet Pathol 2011; 1125–1133.

Kong WX, Ma J, Gu Y, Yang HC, Zuo YQ, Lin SY. Renal protective effects of specific cyclooxygen ase-2 inhibitor in rats with subtotal renal ablation. Zhonghua Nei Ke Za Zhi 2003;42:186–190.

Kotnik P, Nielsen J, Kwon TH, Krzisnik C, et al. Altered expression of COX-1, COX-2, and mPGES in rats with nephrogenic and central diabetes insipidus. Am J Physiol Renal Physiol 2005;288:F1053–F1068.

Kreisberg JI, Karnovsky MJ, Levine L. Prostaglandin production by homogeneous cultures of rat glomerular epithelial and mesangial cells. Kidney Int 1982;22:355–359.

Lasaitiene D, Chen Y, Guron G, Marcussen N, et al. Perturbed medullary tubulogenesis in neonatal rat exposed to renin–angiotensin system inhibition. Nephrol Dial Transplant 2003;18:2534–2541.

Lauridsen TG, Vase H, Starklint J, Graffe CC, et al. Increased renal sodium absorption by inhibition of prostaglandin synthesis during fasting in healthy man: a possible role of the epithelial sodium channels. BMC Nephrol 2010;11:28.

Lebel M, Grose JH. Renal eicosanoids and renal hemodynamics in early borderline hypertension. Clin Invest Med 1991;14:525–534.

Lee DW, Kwak IS, Lee SB, Song SH, et al. Effects of celecoxib and nordihydroguaiaretic acid on puromycin aminonucleoside-induced nephrosis in the rat. J Korean Med Sci 2009;24 Suppl:S183–S188.

Lee JB, Crowshaw K, Takman BH, Attrep KA. The identification of prostaglandins E(2), F(2alpha) and A(2) from rabbit kidney medulla. Biochem J 1967;105:1251–1260.

Lierz M. Avian renal disease: pathogenesis, diagnosis, and therapy. Vet Clin North Am Exot Anim Pract 2003;6:29–55.

Liew D, Krum H. The role of aldosterone receptor blockade in the management of cardiovascular disease. Curr Opin Invest Drugs 2002;3:1468–1473.

Linz W, Wiemer G, Gohlke P, Unger T, et al. Contribution of kinins to the cardiovascular actions of angiotensin-converting enzyme inhibitors. Pharmacol Rev 1995;47:25–24.

Lopez-Ovejero JA, Weber MA, Drayer JI, Sealey JE, et al. Effects of indomethacin alone and during diuretic or beta-adrenergic blockade therapy on blood pressure and the renin system in essential hypertension. Clin Sci Mol Med 1978;55:203S–205S.

Lumbers ER. Functions of the renin–angiotensin system during development. Clin Exp Pharmacol Physiol 1995;22:499–505.

MacDonald MS, Emery JL. The late intrauterine and postnatal development of human renal glomeruli. J Anat 1959;93:331–340.

Makino H, Tanaka I, Mukoyama M, et al. Prevention of diabetic ephropathy in rats by prostaglandin E receptor EP1-selective antagonist. J Am Soc Nephrol 2002;13:1757–1765.

Manohar M, Goetz TE, Griffin R, Sullivan E. Pulmonary vascular pressures of strenuously exercising thoroughbreds after administration of phenylbutazone. Am J Vet Res 1996;57:1354–1358.

Matlhagela K, Taub M. Involvement of EP1 and EP2 receptors in the regulation of the Na,K-ATPase by prostaglandins in MDCK cells. Prostaglandins Other Lipid Mediat 2006;79:101–113.

Meade EA, Smith WL, DeWitt DL. Differential inhibition of prostaglandin endoperoxide synthase (cyclooxygenase) isozymes by aspirin and other non-steroidal anti-inflammatory drugs. J Biol Chem 1993;268:6610–6614.

Mené P. Physiology and pathophysiology of the mesangial cell. Nefrologia 1996;XVISupl 3.

Mené P, Dunn MJ. Eicosanoids and control of mesangial cell contraction. Circ Res 1988;62:916–925.

Meskell M, Ettarh R. Immunohistochemical localisation of renal cyclooxygenase-1 expression in non-steroidal anti-inflammatory drug-treated mice. Exp Toxicol Pathol 2011;63:39–42.

Meteyer CU, Rideout BA, Gilbert M, Shivaprasad HL, et al. Pathology and proposed pathophysiology of diclofenac poisoning in free-living and experimentally exposed oriental white-backed vultures (*Gyps bengalensis*). J Wildl Dis 2005;41:707–716.

Moore L, Williams R, Staples A. Glomerular dimensions in children under 16 years of age. J Pathol 1993;171:145–150.

Morath R, Klein T, Seyberth HW, Nüsing RM. Immunolocalization of the four prostaglandin E2 receptor proteins EP1, EP2, EP3, and EP4 in human kidney. J Am Soc Nephrol 1999;10:1851–1860.

Morello JP, Bichet DG. Nephrogenic diabetes insipidus. Annu Rev Physiol 2001;63:607–630.

Morham SG, Langenbach R, Loftin CD, et al. Prostaglandin synthase 2 gene disruption causes severe renal pathology in the mouse. Cell 1995;83:473–482.

Mulato AS, Ho ES, Cihlar T. Nonsteroidal anti-inflammatory drugs efficiently reduce the transport and cytotoxicity of adefovir mediated by the human renal organic anion transporter 1. J Pharmacol Exp Ther 2000;295:10–15.

Murakami M, Nakatani Y, Tanioka T, Kudo I. Prostaglandin E synthase. Prostaglandins Other Lipid Mediat 2002;68–69:383–399.

Murata T, Ushikubi F, Matsuoka T, et al. Altered pain perception and inflammatory response in mice lacking prostacyclin receptor. Nature 1997;388:678–682.

Murray MD, Brater DC, Tierney WM, Hui SL, et al. Ibuprofen-associated renal impairment in a large general internal medicine practice. Am J Med Sci 1990;299:222–229.

Muscara MN, Vergnolle N, Lovren F, Triggle CR, et al. Selective cyclo-oxygenase-2 inhibition with celecoxib elevates blood pressure and promotes leukocyte adherence. Br J Pharmacol 2000a;129:1423–1430.

Muscara MN, McKnight W, Lovren F, Triggle CR, et al. Antihypertensive properties of a nitric oxide–releasing naproxen derivative in two-kidney, one-clip rats. Am J Physiol 2000b;279: H528–H535.

Mutoh K, Watanabe S, Wakuri H. Histochemical study on the gland of pelvis renalis in the horse. Kitasato Arch Exp Med 1992;65:117–122.

Nagai J, Taogoshi T, Tokunaga A, Nishikawa H, et al. Characterization of prostaglandin E1 transport in rat renal brush-border membrane. Drug Metab Pharmacokinet 2006;21:186–193.

Nagao T, Nagamatsu T, Suzuki Y. Effect of DP-1904, a thromboxane A2 synthase inhibitor, on passive Heymann nephritis in rats. Eur J Pharmacol 1996;316:73–80.

Nagao T, Koseki J, Suzuki Y, Nagamatsu T. Thromboxane A_2 causes retarded clearance of aggregated protein in glomeruli of nephritic mice. Eur J Pharmacol 2001;413:271–279.

Nantel F, Meadows E, Denis D, Connolly B, et al. Immunolocalization of cyclooxygenase-2 in the macula densa of human elderly. FEBS Lett 1999;457:475–477.

Naruse K, Fujieda M, Miyazaki E, Hayashi Y, et al. An immunohistochemical study of developing glomeruli in human fetal kidneys. Kidney Int 2000;57:1836–1846.

Neuhofer W, Holzapfel K, Fraek ML, Ouyang N, et al. Chronic COX-2 inhibition reduces medullary HSP70 expression and induces papillary apoptosis in dehydrated rats. Kidney Int 2004;65:431–441.

Nielsen S, Chou CL, Marples D, Christensen EI, et al. Vasopressin increases water permeability of kidney collecting duct by inducing translocation of aquaporin-CD water channels to plasma membrane. Proc Natl Acad Sci USA 1995;90:11663–11667.

Nørregaard R, Jensen BL, Topcu SO, Wang G, et al. Urinary tract obstruction induces transient accumulation of COX-2-derived prostanoids in kidney tissue. Am J Physiol Regul Integr Comp Physiol 2010;298:R1017–R1025.

Norwood VF, Morham SG, Smithies O. Postnatal development and progression of renal dysplasia in cyclooxygenase-2 null mice. Kidney Int 2000;58:2291–2300.

Novy MJ. Effects of indomethacin on labor, fetal oxygenation, and fetal development in rhesus monkeys. Adv Prostaglandin Thromboxane Res 1978;4:285–300.

Ogawa M, Hirawa N, Tsuchida T, et al. Urinary excretions of lipocalin-type prostaglandin D2 synthase predict the development of proteinuria and renal injury in OLETF rats. Nephrol Dial Transplant 2006;21:924–934.

Ogawa T, Tomomasa T, Hikima A, Kobayashi Y, et al. Developmental changes in cyclooxygenase mRNA expression in the kidney of rats. Pediatr Nephrol 2001;16:618–622.

Ogawa T, Tomomasa T, Morikawa A. Developmental changes in cyclo-oxygenase mRNA induction by hypoxia in rat kidney. Pediatr Int 2002;44:675–679.

Ohta KY, Imamura Y, Okudaira N, Atsumi R, Inoue K, Yuasa H. Functional characterization of multidrug and toxin extrusion protein 1 as a facilitative transporter for fluoroquinolones. J Pharmacol Exp Ther 2009;328:628–634.

Okumura T, Hayashi I, Ikezawa T, et al. Cyclooxygenase-2 inhibitors attenuate increased blood pressure in renovascular hypertensive models, but not in deoxycorticosterone-salt hypertension. Hypertension Res 2002;25:927–938.

Okumura M, Imanishi M, Okamura M, et al. Role for thromboxane A2 from glomerular thrombi in nephropathy with type 2 diabetic rats. Life Sci 2003;72:2695–2705.

Omote H, Hiasa M, Matsumoto T, Otsuka M, Moriyama Y. The MATE proteins as fundamental transporters of metabolic and xenobiotic organic cations. Trends Pharmacol Sci 2006;27:587–593.

Opay AL, Mouton CR, Mullins JJ, Mitchell KD. Cyclooxygenase-2 inhibition normalizes arterial blood pressure in CYP1A1-REN2 transgenic rats with inducible ANG II- dependent malignant hypertension. Am J Physiol Renal Physiol 2006;29:F612–F618.

Ortiz PA. cAMP increases surface expression of NKCC2 in rat thick scending limbs: role of VAMP. Am J Physiol 2006;290:F608–F616.

Ozer MK, Asci H, Oncu M, Calapoglu M, et al. Effects of misoprostol on cisplatin-induced renal damage in rats. Food Chem Toxicol 2011;49:1556–1559.

Pai HC. Dissection of nephrons from the human kidney. J Anat 1935;69(Pt 3): 344–349.

Parfentjev IA, Perlzweig WA. The composition of the urine of white mice. J Biol Chem 1933;100: 551–555.

Passmore AP, Copeland S, Johnston GD. The effects of ibuprofen and indomethacin on renal function in the presence and absence of frusemide in healthy volunteers on a restricted sodium diet. Br J Clin Pharmacol 1990;29:311–319.

Pattaragarn A, Alon US. Treatment of congenital nephrogenic diabetes insipidus by hydrochlorothiazide and cyclooxygenase-2 inhibitor. Pediatr Nephrol 2003;18:1073–1076.

Paulose-Ram R, Hirsch R, Dillon C, Losonczy K, et al. Prescription and non-prescription analgesic use among the US adult population: results from the third National Health and Nutrition Examination Survey (NHANES III). Pharmacoepidemiol Drug Saf 2003;12:315–326.

Pavenstädt H, Kriz W, Kretzler M. Cell biology of the glomerular podocyte. Physiol Rev 2003;83:253–307.

Perazella MA. Drug-induced hyperkalemia: old culprits and new offenders. Am J Med 2000;109: 307–314.

Perez G, McGuckin J. Cellular localiation of prostaglandin A2 in the rat kidney. Prostaglandins 1972;2:393–398.

Persson AE, Salomonsson M, Westerlund P, Greger R, et al. Macula densa cell function. Kidney Int Suppl 1991;32:S39–S44.

Peti-Peterdi J, Komlosi P, Fuson AL, et al. 2003. Luminal NaCl delivery regulates basolateral PGE2 release from macula densa cells. J Clin Invest 2003;112:76–82.

Pfeiffer EW. Comparative anatomical observations of the mammalian renal pelvis and medulla. J Anat 1968;102:321–331.

Plantinga L, Grubbs V, Sarkar U, Hsu CY, et al. Nonsteroidal anti-inflammatory drug use among persons with chronic kidney disease in the United States. Ann Fam Med 2011;9:423–430.

Quilley J, Santos M, Pedraza P. Renal protective effect of chronic inhibition of COX-2 with SC-58236 in streptozotocin-diabetic rats. Am J Physiol Heart Circ Physiol 2011;300:H2316–H2322.

Radi ZA. Pathophysiology of cyclooxygenase inhibition in animal models. Toxicol Pathol 2009;37:34–46.

Radi ZA, Ostroski R. Pulmonary and cardiorenal cyclooxygenase-1 (COX-1), -2 (COX-2), and microsomal prostaglandin E synthase-1 (mPGES-1) and -2 (mPGES-2) expression in a hypertension model. Mediators Inflamm 2007;2007:85091.

Ragolia L, Palaia T, Hall CE, et al. Accelerated glucose intolerance, nephropathy, and atherosclerosis in prostaglandin D2 synthase knock-out mice. J Biol Chem 2005;280:29946–29955.

Rao PS, Cavanagh D, Dietz JR, Marsden KA, et al. Dose-dependent effects of prostaglandin D2 on hemodynamics, renal function and blood gas analyses. Am J Obstet Gynecol 1987;156:843–851.

Reddy LR, Corey EJ. Facile air oxidation of the conjugate base of rofecoxib (Vioxx), a possible contributor to human toxicity. Tetrahedron Lett 2005;46:927–929.

Richter CM, Godes M, Wagner C, Maser-Gluth C, et al. Chronic cyclooxygenase-2 inhibition does not alter blood pressure and kidney function in renovascular hypertensive rats. J Hypertens 2004;22:191–198.

Rink JC, Vu HT, Sánchez Alvarado A. The maintenance and regeneration of the planarian excretory system are regulated by EGFR signaling. Development 2011;138:3769–3780.

Robertson JL. Chemically induced glomerular injury: a review of basic mechanisms and specific xenobiotics. Toxicol Pathol 1998;26:64–72.

Roch-Ramel F. Renal excretion of uric acid in mammals. Clin Nephrol 1979;12:1–6.

Roch-Ramel F, White F, Vowles L, Simmonds HA, et al. Micropuncture study of tubular transport of urate and PAH in the pig kidney. Am J Physiol 1980;239:F107–F112.

Rodriguez F, Llinas MT, Gonzalez JD, Rivera J, et al. Renal changes induced by a cyclooxygenase-2 inhibitor during normal and low sodium intake. Hypertension 2000;36:276–281.

Rodriguez JA, Vio CP, Pedraza PL, McGiff JC, et al. Bradykinin regulates cyclooxygenase-2 in rat renal thick ascending limb cells. Hypertension 2004;44:230–235.

Romagnani P. Parietal epithelial cells: their role in health and disease. Contrib Nephrol 2011;169:23–36.

Rossat J, Maillard M, Nussberger J, Brunner HR, et al. Renal effects of selective cyclooxygenase-2 inhibition in normotensive salt-depleted subjects. Clin Pharmacol Ther 1999;66:76–84.

Rossert J. Drug-induced acute interstitial nephritis. Kidney Int 2001;60:804–187.

Rutkai I, Feher A, Erdei N, Henrion D, et al. Activation of prostaglandin E2 EP1 receptor increases arteriolar tone and blood pressure in mice with type 2 diabetes. Cardiovasc Res 2009;83:148–154.

Sáez F, Reverte V, Salazar F, et al. Hypertension and sex differences in the age-related renal changes when cyclooxygenase-2 activity is reduced during nephrogenesis. Hypertension 2009;53:331–337.

Sahloul MZ, al-Kiek R, Ivanovich P, Mujais SK. Nonsteroidal anti-inflammatory drugs and antihypertensives. Cooperative malfeasance. Nephron 1990;56:345–352.

Schlondorff D. Renal complications of nonsteroidal anti-inflammatory drugs. Kidney Int 1993; 44:643–653.

Schneider A, Harendza S, Zahner G, Jocks T, et al. Cyclooxygenase metabolites mediate glomerular monocyte chemoattractant protein-1 formation and monocyte recruitment in experimental glomerulonephritis. Kidney Int 1999;55:430–441.

Schneider A, Zhang Y, Zhang M, et al. Membrane-associated PGE synthase-1 (mPGES-1) is coexpressed with both COX-1 and COX-2 in the kidney. Kidney Int 2004;65:1205–1213.

Schnermann J. Cyclooxygenase-2 and macula densa control of renin secretion. Nephrol Dial Transplant 2001; 1735–1738.

Schölkens BA. Antihypertensive effect of prostacyclin (PGI2) in experimental hypertension and its influence on plasma renin activity in rats. Prostaglandins Med 1978;1:359–372.

Sebeková K, Ramuscak A, Boor P, Heidland A, et al. The selective TP receptor antagonist, S18886 (terutroban), attenuates renal damage in the double transgenic rat model of hypertension. Am J Nephrol 2008;28:47–53.

Sellers RS, Senese PB, Khan KN. Interspecies differences in the nephrotoxic response to cyclooxygenase inhibition. Drug Chem Toxicol 2004;27:111–122.

Seta F, Chung AD, Turner PV, Mewburn JD, et al. Renal and cardiovascular characterization of COX-2 knockdown mice. Am J Physiol Regul Integr Comp Physiol 2009;296:R1751–R1760.

Singer I, Rotenberg D, Puschett JB. Lithium-induced nephrogenic diabetes insipidus: in vivo and in vitro studies. J Clin Invest 1972;51:1081–1091.

Skov K, Nyengaard JR, Korsgaard N, et al. Number and size of renal glomeruli in spontaneously hypertensive rats. J Hypertens 1994;12:1373–1376.

Smeets B, Angelotti ML, Rizzo P, et al. Renal progenitor cells contribute to hyperplastic lesions of podocytopathies and crescentic glomerulonephritis. J Am Soc Nephrol 2009;20:2593–2603.

Smith WL, Bell TG. Immunohistochemical localization of the prostaglandin-forming cyclooxygenase in renal cortex. Am J Physiol Renal Physiol 1978;235:F451–F457.

Solomon S. Developmental changes in nephron number, proximal tubular length and superficial nephron glomerular filtration rate of rats. J Physiol 1977;272:573–589.

Speziale MV, Allen RG, Henderson CR, Barrington KJ, et al. Effects of ibuprofen and indomethacin on the regional circulation in newborn piglets. Biol Neonate 1999;76:242–252.

Sraer JD, Adida C, Peraldi MN, Rondeau E, et al. Species-specific properties of the glomerular mesangium. J Am Soc Nephrol 1993;3:1342–1350.

Stanfield KM, Bell RR, Lisowski AR, English ML, et al. Expression of cyclooxygenase-2 in embryonic and fetal tissues during organogenesis and late pregnancy. Birth Defects Res A Clin Mol Teratol 2003;67:54–58.

Stock JL, Shinjo K, Burkhardt J, Roach M, et al. The prostaglandin E2 EP1 receptor mediates pain perception and regulates blood pressure. J Clin Invest 2001;107:325–331.

Stokes JB, Kokko JP. Inhibition of sodium transport by prostaglandin E2 across the isolated, perfused rabbit collecting tubule. J Clin Invest 1977;59:1099–1104.

Suganami T, Mori K, Tanaka I, et al. Role of prostaglandin E receptor EP1 subtype in the development of renal injury in genetically hypertensive rats. Hypertension 2003;42:1183–1190.

Sugimoto Y, Namba T, Shigemoto R, Negishi M, et al. Distinct cellular localization of mRNAs for three subtypes of prostaglandin E receptor in kidney. Am J Physiol 1994;266(Renal Fluid Electrolyte Physiol 35):F823–F828.

Sun D, Liu CX, Ma YY, Zhang L. Protective effect of prostaglandin E1 on renal microvascular injury in rats of acute aristolochic acid nephropathy. Renal Fail 2011;33:225–232.

Svendsen KB, Bech JN, Sørensen TB, Pedersen EB. A comparison of the effects of etodolac and ibuprofen on renal haemodynamics, tubular function, renin, vasopressin and urinary excretion of albumin and alpha-glutathione-*S*-transferase in healthy subjects: a placebo-controlled cross-over study. Eur J Clin Pharmacol 2000;56:383–388.

Swan SK, Rudy DW, Lasseter KC, Ryan CF, et al. Effect of cyclooxygenase-2 inhibition on renal function in elderly persons receiving a low-salt diet: a randomized, controlled trial. Ann Intern Med 2000;133:1–9.

Syal A, Schiavi S, Chakravarty S, Dwarakanath V, et al. Fibroblast growth factor-23 increases mouse PGE2 production in vivo and in vitro. Am J Physiol Renal Physiol 2006;290:F450–F455.

Szabó AJ, Muller V, Chen GF, et al. Nephron number determines susceptibility to renal mass reduction-induced CKD in Lewis and Fisher 344 rats: implications for development of experimentally induced chronic allograft nephropathy. Nephrol Dial Transplant 2008;23:2492–2495.

Tacconelli S, Capone ML, Patrignani P. Clinical pharmacology of novel selective COX-2 inhibitors. Curr Pharm Des 2004;10:589–601.

Takahashi N, Takeuchi K, Abe T, et al. Immunolocalization of rat thromboxane receptor in the kidney. Endocrinology 1996;137:5170–5173.

Takami H, Naramoto A, Shigematsu H, Ohno S. Ultrastructure of glomerular basement membrane by quick-freeze and deep-etch methods. Kidney Int 1991;3, 9: 659–664.

Takeda T. Podocyte cytoskeleton is connected to the integral membrane protein podocalyxin through Na^+/H^+-exchanger regulatory factor 2 and ezrin. Clin Exp Nephrol 2003;7:260–269.

Tamma G, Wiesner B, Furkert J, Hahm D, et al. The prostaglandin E2 analog sulprostone antagonizes vasopressin-induced antidiuresis through activation of Rho. J Cell Sci 2003;116:3285–3294.

Tan SY, Shapiro R, Franco R, Stockard H, et al. Indomethacin-induced prostaglandin inhibition with hyperkalemia. Ann Intern Med 1979;90:783–785.

Theilig F, Debiec H, Nafz B, Ronco P, et al. Renal cortical regulation of COX-1 and functionally related products in early renovascular hypertension (rat). Am J Physiol Renal Physiol 2006;291:F987–F994.

Timpl, R. Recent advances in the biochemistry of glomerular basement membrane. Kidney Int 1986;30:293–298.

Tomasoni S, Noris M, Zappella S, Gotti E, et al. Upregulation of renal and systemic cyclooxygenase-2 in patients with active lupus nephritis. J Am Soc Nephrol 1998;9:1202–1212.

Tomida T, Numaguchi Y, Matsui H, et al. Altered expression of prostacyclin synthase in a subset of the thick ascending limb cells and mesangial cells in 5/6- nephrectomized rats. Hypertens Res Clin Exp 2001;24:411–419.

Tomida T, Numaguchi Y, Nishimoto Y, Tsuzuki M, et al. Inhibition of COX-2 prevents hypertension and proteinuria associated with a decrease of 8-iso-PGF2alpha formation in L-NAME-treated rats. J Hypertens 2003;21:601–609.

Torsello G, Schrör K, Szabó Z, Kutkuhn B, et al. Effects of prostaglandin E1 (PGE1) on experimental renal ischaemia. Eur J Vasc Surg 1989;3:5–13.

Tsugawa K, Hashizume M, Migou S, Kishihara F, et al. A selective cyclo-oxygenase-2 inhibitor, NS-398, may improve portal hypertension without inducing gastric mucosal injury. J Gastroenterol Hepatol 1999;14:642–651.

Uehara Y, Makino H, Seiki K, Urade Y L-PGDS Clinical Research Group of Kidney. Urinary excretions of lipocalin-type prostaglandin D synthase predict renal injury in type-2 diabetes: a cross-sectional and prospective multicentre study. Nephrol Dial Transplant 2009;24:475–482.

Ullrich KJ. Renal transporters for organic anions and organic cations. Structural requirements for substrates. J Membr Biol 1997;158:95–107.

Upton RA, Buskin JN, Williams RL, Holford NH, et al. Negligible excretion of unchanged ketoprofen, naproxen, and probenecid in urine. J Pharm Sci 1980;69:1254–1257.

Usberti M, Dechaux M, Guillot M, Seligmann R, et al. Renal prostaglandin E_2 in nephrogenic diabetes insipidus: effects of inhibition of prostaglandin synthesis by indomethacin. J Pediatr 1980;97:476–478.

van der Heijden AJ, Provoost AP, Nauta J, Grose W, et al. Renal functional impairment in preterm neonates related to intrauterine indomethacin exposure. Pediatr Res 1988;24:644–648.

van der Heijden BJ, Carlus C, Narcy F, et al. Persistent anuria, neonatal death, and renal microcystic lesions after prenatal exposure to indomethacin. Am J Obstet Gynecol 1994;171:617–623.

Vanecková I, Skaroupková P, Dvorák P, Certíková Chábová V, et al. Effects of sodium restriction and cyclooxygenase-2 inhibition on the course of hypertension, proteinuria and cardiac hypertrophy in Ren-2 transgenic rats. Physiol Res 2005;54:17–24.

Vermillion ST, Newman RB. Recent indomethacin tocolysis is not associated with neonatal complications in preterm infants. Am J Obstet Gynecol 1999;181:1083–1086.

Veron D, Reidy KJ, Bertuccio C, Teichman J, et al. Overexpression of VEGF-A in podocytes of adult mice causes glomerular disease. Kidney Int 2010;77:989–999.

Vieux R, Desandes R, Boubred F, Semama D, et al. Ibuprofen in very preterm infants impairs renal function for the first month of life. Pediatr Nephrol 2010;25:267–274.

Villa E, Garcia-Robles R, Haas J, Romero JC. Comparative effect of PGE2 and PGI2 on renal function. Hypertension 1997;30:664–666.

Villar D, Buck WB, Gonzalez JM. Ibuprofen, aspirin and acetaminophen toxicosis and treatment in dogs and cats. Vet Hum Toxicol 1998;40:156–162.

Vitzthum H, Abt I, Einhellig S, Kurtz A. Gene expression of prostanoid forming enzymes along the rat nephron. Kidney Int 2002;62:1570–1581.

Vukicevic S, Simic P, Borovecki F, et al. Role of EP2 and EP4 receptor-selective agonists of prostaglandin E_2 in acute and chronic kidney failure. Kidney Int 2006;70:1099–1106.

Walker AM, Oliver J. Methods for the collection of fluid from single glomeruli and tubules of the mammalian kidney. Am J Physiol 1941;134:562–579.

Walker LA, Whorton AR, Smigel M, France R, et al. Antidiuretic hormone renal increases prostaglandin synthesis in vivo. Am J Physiol 1978;235:F180–F185.

Wang JL, Cheng HF, Zhang MZ, McKanna JA, Harris RC. Selective increase of cyclooxygenase-2 expression in a model of renal ablation. Am J Physiol 1998;275:F613–F622.

Wang JL, Cheng HF, Harris RC. Cyclooxygenase-2 inhibition decreases renin content and lowers blood pressure in a model of renovascular hypertension. Hypertension 1999;34:96–101.

Wang JL, Cheng HF, Shappell S, Harris RC. A selective cyclooxygenase-2 inhibitor decreases proteinuria and retards progressive renal injury in rats. Kidney Int 2000;57:2334–2342.

Wang W, Zolty E, Falk S, Summer S, et al. Prostacyclin in endotoxemia-induced acute kidney injury: cyclooxygenase inhibition and renal prostacyclin synthase transgenic mice. Am J Physiol Renal Physiol 2007;293:F1131–F1136.

Warden DH, Stokes JB. EGF and PGE2 inhibit rabbit CCD Na^+ transport by different mechanisms: PGE2 inhibits $Na^+ - K^+$ pump. Am J Physiol 1993;264:F670–F677.

Watson JF. Potassium reabsorption in the proximal tubule of the dog nephron. J Clin Invest 1966;45:1341–1348.

Wen SF. Nephrotoxicities of nonsteroidal anti-inflammatory drugs. J Formos Med Assoc 1997;96: 157–171.

Whelton A. Nephrotoxicity of nonsteroidal anti-inflammatory drugs: physiologic foundations and clinical implications. Am J Med 1999;106:13S–24S.

Whelton A, Hamilton CW. Nonsteroidal anti-inflammatory drugs: effects on kidney function. J Clin Pharmacol 1991;31:588–598.

White WB, Faich G, Whelton A, Maurath C, et al. Comparison of thromboembolic events in patients treated with celecoxib, a cyclooxygenase-2 specific inhibitor, versus ibuprofen or diclofenac. Am J Cardiol 2002;89:425–430.

Whorton AR, Smigel M, Oates JA, Frölich JC. Regional differences in prostacyclin formation by the kidney: Prostacyclin is a major prostaglandin of renal cortex. Biochim Biophys Acta 1978;529: 176–180.

Wicks LE. Sex and proteinuria I mice. Proc Soc Exp Biol Med 1941;48:395–400.

Wiseman EH, Reinert H. Anti-inflammatory drugs and renal papillary necrosis. Agents Actions 1975;5:322–325.

Xu S, Jiang B, Maitland KA, et al. The thromboxane receptor antagonist S18886 attenuates renal oxidant stress and proteinuria in diabetic apolipoprotein E-deficient mice. Diabetes 2006;55:110–119.

Yamanari H, Nakamura K, Kakishita M, Ohe T. Effects of cyclooxygenase inhibition on endothelial function in hypertensive patients treated with angiotensin-converting enzyme inhibitors. Clin Cardiol 2004;27:523–527.

Yang C, Sorokin A. Upregulation of fibronectin expression by COX-2 is mediated by interaction with ELMO1: upregulation of fibronectin expression by COX-2 is mediated by interaction with ELMO1. Cell Signal 2011;23:99–104.

Yang G, Chen L, Zhang Y, Zhang X, et al. Expression of mouse membrane-associated prostaglandin E_2 synthase-2 (mPGES-2) along the urogenital tract. Biochim Biophys Acta 2006;1761:1459–1468.

Yang T, Singh I, Pham H, Sun D, et al. Regulation of cyclooxygenase expression in the kidney by dietary salt intake. Am J Physiol Renal Physiol 1998;274:F481–F489.

Yang T, Sun D, Huang YG, Smart A, et al. Differential regulation of COX-2 expression in the kidney by lipopolysaccharide: role of CD14. Am J Physiol 1999;277:F10–F16.

Yang T, Endo Y, Huang YG, et al. Renin expression in COX-2-knockout mice on normal or low-salt diets. Am J Physiol Renal Physiol 2000;279:F819–F825.

Yokoyama C, Yabuki T, Shimonishi M, et al. Prostacyclin-deficient mice develop ischemic renal disorders, including nephrosclerosis and renal infarction. Circulation 2002;106:2397–2403.

Yoshida T, Kameda H, Ichikawa Y, Tojo T, et al. Improvement of renal function with a selective thromboxane A2 synthetase inhibitor, DP-1904, in lupus nephritis. J Rheumatol 1996;23:1719–1724.

Zahner G, Schaper M, Panzer U, Kluger M, et al. Prostaglandin EP2 and EP4 receptors modulate expression of the chemokine CCL2 (MCP-1) in response to LPS-induced renal glomerular inflammation. Biochem J 2009;422:563–570.

Zewde T, Mattson DL. Inhibition of cyclooxygenase-2 in the rat renal medulla leads to sodium-sensitive hypertension. Hypertension 2004;44:424–428.

Zhang A, Dong Z, Yang T. Prostaglandin D2 inhibits TGF-b1-induced epithelial-to- mesenchymal transition in MDCK cells. Am J Physiol Renal Physiol 2006;291:F1332–F1342.

Zhang MZ, Yao B, Fang X, Wang S, et al. Intrarenal dopaminergic system regulates renin expression. Hypertension 2009;53:564–570.

Zhang Y, Schneider A, Rao R, Lu WJ, et al. 2003. Genomic structure and genitourinary expression of mouse cytosolic prostaglandin E_2 synthase gene. Biochim Biophys Acta 2003;1634:15–23.

Zidar N, Odar K, Glavac D, Jerse M, et al. Cyclooxygenase in normal human tissues: Is COX-1 really a constitutive isoform, and COX-2 an inducible isoform? J Cell Mol Med 2009;13:3753–3763.

Zimran A, Kramer M, Plaskin M, Hershko C. Incidence of hyperkalaemia induced by indomethacin in a hospital population. BMJ 1985;29:107–108.

Zook TE, Strandhoy JW. Mechanisms of the natriuretic and diuretic effects of prostaglandin F2 alpha. J Pharmacol Exp Ther 1981;217:674–680.

CHAPTER 4

CARDIOVASCULAR SYSTEM

INTRODUCTION

The conversion of arachidonic acid (AA) into prostaglandin H_2(PGH_2) is catalyzed by cyclooxygenase (COX). PGH_2 is subsequently converted by synthases to various prostanoids, such as PGE_2, PGI_2, $PGF_{2\alpha}$, and thromboxane A_2(TXA_2) (Eling et al., 1990; Dannenberg et al., 2001). These prostanoids play a role (1) in cardiovascular homeostasis and vascular tone (Fitzgerald, 2004) and (2) in altering renal blood flow, glomerular filtration rate (GFR), and electrolytes (Radi, 2009). COX-1 isoenzyme is expressed constitutively in most normal tissues; COX-2 is highly induced by proinflammatory mediators in the setting of inflammation, injury, and pain (Fig. 4-1) (Parente and Perretti, 2003; Radi, 2009; Sellers et al., 2010). Inhibitors of COX activity include nonselective nonsteroidal anti-inflammatory drugs (ns-NSAIDs) such as naproxen, ibuprofen, diclofenac, and phenybutazone and COX-2 selective NSAIDs (COX-2 s-NSAIDs) such as firocoxib (Previcox), robenacoxib (Onsior), and celecoxib (Celebrex) (Radi and Khan, 2006; Joubert, 2009; Radi, 2009; Sellers et al., 2010). In therapeutic doses, ns-NSAIDs inhibit both COX-1 and COX-2 (Fig. 4-1), whereas COX-2 s-NSAIDs inhibit predominantly COX-2 (Krotz et al., 2005; Sellers et al., 2010).

The potential for cardiovascular events associated with COX-2 selective and ns-NSAID inhibitors, particularly in people with underlying cardiovascular disease, has been addressed in the last few years (Krotz et al., 2005). Nonclinical studies failed to identify these risks in a variety of animal species and models. Although a variety of experimental models have assessed the effects of both selective and ns-NSAIDs, the outcome in these models has been variable (Sellers et al., 2010). In this chapter we discuss the expression of COX-1 and COX-2 in the cardiovascular system, the role of prostaglandins and COX in cardiovascular homeostasis and pathophysiology, and the effects on the cardiovascular system of COX-2 s-NSAIDs and ns-NSAIDs.

Comparative Pathophysiology and Toxicology of Cyclooxygenases, First Edition. Zaher A. Radi.

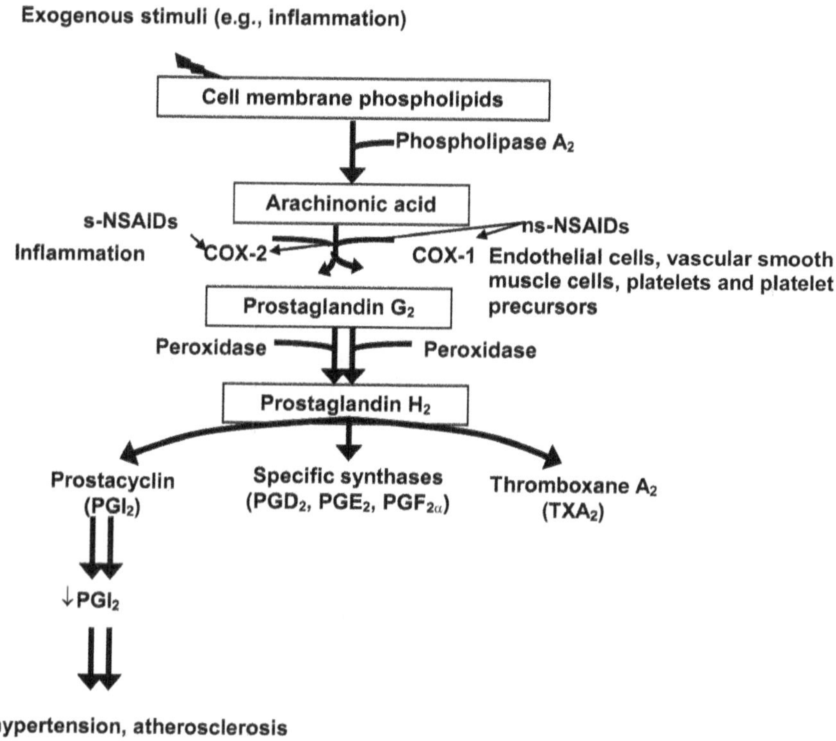

FIGURE 4-1 Prostaglandin synthesis: the role of prostaglandins (PGs) in the cardiovascular system in health and disease and the effects of nonselective and nonsteroidal anti-inflammatory drugs (ns-NSAIDs) and COX-2 s-NSAIDs. Following an exogenous stimulus (e.g., inflammation), cell membrane phospholipid is liberated to arachidonic acid (AA) by phospholipase A_2. Two cyclooxygenases, COX-1 and COX-2, catalyze the conversion of AA into various PGs. COX-1 is expressed in endothelial cells, platelets, platelet precursors, and vascular smooth muscle cell, whereas COX-2 expression is up-regulated during inflammatory or neoplastic conditions. Nonselective NSAIDs (e.g., carprofen, etodolac, flunixin meglumine, ketoprofen, indomethacin, meloxicam, phenylbutazone) inhibit COX-1 and COX-2; COX-2 s-NSAIDs (e.g., celecoxib, deracoxib, lumiracoxib, robenacoxib, rofecoxib) spare COX-1 and inhibit only COX-2. [The final, definitive version of this table was published in R. A. Sellers, Z. A. Radi, and N. K. Khan, *Veterinary Pathology*, 47(4), 2010, pp. 601–613. Copyright © Sage Publications. All rights reserved.]

COMPARATIVE PHYSIOLOGICAL AND ANATOMICAL ASPECTS OF THE CARDIOVASCULAR SYSTEM

The Heart

Understanding the comparative physiological and anatomical aspects of cardiovascular function in diverse species is important for an understanding of the pathological and toxicological consequences of COX inhibition in the cardiovascular system. The heart is a hollow muscular pump that is located within the mediastinum. The main function of the heart is to send oxygenated arterial blood to the body and pump deoxygenated blood into the pulmonary circulation. As arteries enter tissues, they become smaller (arterioles), and from there they become capillaries to provide nutrients and oxygen and remove carbon dioxide. The shape of the normal heart varies from elongated (horse heart) to ovoid (dog heart) to trapezoidal (human) (Crick et al., 1998). The heart of a normal human female is smaller than that of a normal male. Anatomically, the heart is enclosed in a double-walled fibroserous inelastic sac called the pericardium or pericardial sac. The heart wall consists of three layers: epicardium, endocardium, and myocardium (Gavaghan, 1998). The outermost layer of the heart, the epicardium, also known as the visceral pericardium, consists of epithelial cells that form a serous membrane that covers the entire heart (the inner surface of the pericardial sac). The outer surface of the pericardial sac is called the parietal pericardium. Humans have a much thicker pericardium than that of animals (Holt, 1970). The innermost layer of the heart is known as the endocardium. It is a serous membrane that lines the inner surface of the heart, its valves, and the chordae tendineae, which are the cords that connect the free edges of the atrioventricular valves with the papillary muscles. The papillary muscles are muscle eminences on the walls of the ventricles. The endocardium is continuous with the intima (e.g., the inner lining of arteries). The thick middle layer of the heart is the muscular layer known as the myocardium. It is responsible for the major pumping action of the ventricles. The myocardial cells have an intrinsic ability to contract in the absence of stimuli (i.e., automaticity) and in a rhythmic manner (i.e., rhythmicity), and to transmit nerve impulses (i.e., conductivity).

Rhythmicity is the regular generation of an action potential (i.e., transmission of electrical impulses) by the conduction system of the heart. The heart has four chambers: the right and left atriums and the right and left ventricles. The four valves of the heart—the mitral, tricuspid, pulmonary, and aortic valves—are involved in the filling and pumping action of the heart (Zimmerman, 1966; Gavaghan, 1998). The mitral or bicuspid valve is located between the left atrium and the left ventricle. It has two leaves that slightly overlap each other when the valve is closed. The tricuspid (i.e., three cusp) valve lies between the right atrium and the right ventricle. It has three valves, which are thinner than those of the mitral valve. The leaves of both valves are attached to strong fibrous strands called the chordae tendineae. These cords arise from the trabeculae carnae muscle bundles in the inner ventricles. Two groups of papillary muscles arise from the trabeculae carneae in the left ventricle and three arise in the right ventricle. The aortic (separates the left ventricle from the ascending aorta) and pulmonary valves are called semilunar (i.e., half moon) valves because they have three cusps that are cuplike in nature. The atrioventricular (i.e., mitral and tricuspid) valves prevent backflow of blood from the ventricles into the atria during systole. The semilunar (i.e., aortic and pulmonary) valves prevent backflow from the aorta and pulmonary artery during diastole (Gavaghan, 1998). Cardiac valve injury can lead to valve regurgitation, where blood flows in a reverse direction (valve insufficiency), or stenosis, where blood outflow is obstructed (Cannistra, 2005).

The contraction and relaxation of cardiac muscle is mediated by the sliding of interdigitating thick filaments and thin filaments. Thick myofilaments are composed of a protein called myosin. Thin myofilaments are composed of three proteins: actin, troponin, and tropomysin. The control of cardiac muscle function is Ca^{2+}-based and resides in the thin filament. This process involves cardiac troponin (cTn) complex in interaction with tropomyosin (Tm). Cardiac Tn is a complex composed of three subunits, each of which interacts with the other two. Tropomyosin is a long, flexible coil that is in equilibrium among three states of myosin–actin–troponin binding: blocked (Ca^{2+} not bound to troponin C, myosin not bound to actin), closed (Ca^{2+} bound to troponin C, myosin weakly bound to actin), and open (Ca^{2+} bound to troponin C, myosin strongly bound to actin). Subtle changes in the structure and protein contacts within cardiac troponin lead to significant changes throughout the complex which alter Tm dynamics and cardiac troponin–actin interactions (Manning et al., 2011).

Cardiovascular Blood Flow

Blood flow through the heart follows a specific route. The right side of the heart receives venous blood from the body (deoxygenated blood) and sends it to the lung, while the left side receives arterial blood from the lung (oxygenated blood) and sends it to the body. When the blood returns from the systemic circulation to the right atrium, it goes through the tricuspid valve to the right ventricle. From the right ventricle, blood is ejected through the pulmonary valve to the lungs to get oxygenated. The oxygenated blood returns from the lungs to the left atrium, and from there goes through the mitral valve to the left ventricle. Blood is pumped through the aortic valve to the aorta and the systemic circulation (Rudolph, 1985; Reece, 2007). The fetal circulation is significantly different from the adult circulation (Murphy, 2005). In the fetus, gas exchange and oxygenation occurs in the placenta, which receives deoxygenated blood from the fetal systemic organs and returns its oxygen-rich venous drainage to the fetal systemic arterial circulation. Deoxygenated blood gets to the placenta via the umbilical arteries and is returned to the fetus in the umbilical vein. In addition, in the fetal cardiovascular system, most of the highly oxygenated blood is delivered to the myocardium and brain. Because of the preferential streaming of oxygenated blood and the presence of intracardiac and extracardiac shunts, the fetal circulation has been defined as "shunt-dependent" circulation (Murphy, 2005).

Cardiac Conduction System and Blood Pressure

The cardiac conduction system generates and transmits impulses that stimulate myocardial contraction. The components of the cardiac conduction system are (1) the sinoatrial (SA) node, (2) the atrioventricular (AV) node and bundle, and (3) the right and left bundle branches (RBB and LBB) (Lev, 1964). The cardiac cycle is defined as the period from the beginning of one heartbeat to the beginning of the next beat. It includes the systole (i.e., heart chamber contraction during emptying), the diastole (i.e., relaxation of heart chamber during filling), and a short pause called the diastasis cordis (when both atria and ventricles relax) (Gavaghan, 1998). Systolic blood pressure (SBP) is the pressure within the arterial system at the peak of the systole. The SBP is controlled by the stroke volume of the heart, the rate of systolic ejection, and the distensibility of the arterial tree (Koch-Weser, 1973). Diastolic blood pressure (DBP), when the heart is relaxed, is controlled by peripheral resistance. Elevated SBP is a recognized risk factor for cardiovascular complications among older patients, but when this elevation is due to a stiffened arterial tree, DBP is necessarily reduced (Smulyan and Safar, 1997). If arteries lose their elasticity due to pathological conditions (e.g., sclerosis), BP rises (i.e., hypertension) during both systole and diastole. The systolic/diastolic BP values vary across adult species: human (120/70 mmHg), rhesus monkey (130/100 mmHg), dog (120/70 mmHg), horse (130/95 mmHg), cow (140/80 mmHg), cat (140/90 mmHg), rat (110/70 mmHg), and mouse (111/80 mmHg) (Smith and Ansevin, 1957; Detweiler, 1993). In the fetus of humans, BP in the aorta is approximately 30 mmHg at 20 weeks of gestation and increases to approximately 45 mmHg at 40 weeks of gestation (Struijk et al., 2008). Intrauterine recording of the intraventricular pressure in human fetuses has been evaluated. There was an increase in ventricular systolic and end diastolic pressures with advancing gestation. Systemic systolic pressure increased from 15 to 20 mmHg at 16 weeks to 30 to 40 mmHg at 28 weeks. No differences between left and right ventricular pressures were noted, and atrial pressures were equal and remained constant (Johnson et al., 2000). There are species differences in the fetal circulation. The human fetus appears to circulate less blood through the placenta, to shunt less through the ductus venosus and foramen ovale, but to direct more blood through the lungs than does the fetal sheep (Kiserud and Acharya, 2004). In addition, the ductus arteriosus, a short temporary vessel in the fetus that connects the pulmonary trunk and the aorta and conducts most of the blood directly from the right ventricle to the aorta, bypassing the lungs, is important in maintaining adequate aortic circulation in the fetus. After birth, constriction of the ductus arteriosus takes place and PGE_2 plays a major role in ductal patency (Clyman et al., 1978). Furthermore, PGE_1 can play a significant role in the regulation of ductus arteriosus tone in the elevated oxygen environment of the newborn as well as in the low-oxygen environment of the fetus (Clyman et al., 1977).

Hypertension is defined as a sustained increase in systemic arterial BP. Cardiac output (CO), the volume of blood ejected per minute, is defined as being the product of stroke volume and heart rate. Stroke volume is determined by (1) preload (volume of blood that fills the ventricles during diastole and degree of stretch of myocardial fibers), (2) afterload (resistance or pressure the ventricles must overcome to eject blood out), and (3) contractility (force generated by the myocardium when it contracts) (Gavaghan, 1998). Cardiac hypertrophy is an adaptive response of the heart to an imposed load. Two distinct patterns of hypertrophy are discussed, pathological and physiological, which differ in their mechanical and biochemical features (Scheuer and Buttrick, 1987). Several signaling molecules, such as endothelin, angiotensin II and cardiotrophin-1 have been proposed as the mediators promoting pressure overload–induced cardiac hypertrophy (Ito et al., 1994; Senbonmatsu et al., 2000; Takimoto et al., 2002; van de Schans et al., 2007). In the adult circulation, the stroke volume of the right ventricle should equal that of the left ventricle. In the fetus, because of the intracardiac and extracardiac shunting, the stroke volume of the fetal left ventricle is not equal to the stroke volume of the right ventricle. The right ventricle receives about 65% of the venous return and the left ventricle about 35% (Murphy, 2005). Mean arterial BP is determined by the product of CO and total peripheral resistance.

Several physiological factors contribute to arterial BP reflex control, including adrenergic receptors, force of contraction, antidiuretic hormone (ADH) activity, venous capacitance, filling pressure, intravascular volume, stroke volume, and heart rate (Bakris and Mensah, 2003). Thus, factors that may contribute to the pathophysiology of hypertension are increased sympathetic nervous system activity, overproduction of sodium-retaining hormones and vasoconstrictors, increased or inappropriate renin secretion with resulting increased production of Ang II and aldosterone, genetics, environmental factors, deficiencies of vasodilators, vascular reactivity, vascular remodeling, increased peripheral resistance, renal parenchymal and nervous stimulation, production of inflammatory mediators, and endothelial dysfunction (Bakris and Mensah, 2003; Vargas Alarcón, 2006). In cats, chronic kidney disease, adrenal diseases, and hyperthyroidism are the most frequent causes of systemic hypertension (Brown et al., 2007). A diagnosis of hypertension should be considered strongly in a cat with Doppler forelimb SBP $\geq$160 mmHg or oscillometric tail cuff SBP $\geq$140 to 160 mmHg (Stepien, 2011). In dogs, hyperadrenocorticism (i.e., Cushing's disease; an endocrine disorder characterized by chronic elevations in the circulating concentrations of adrenal cortisol hormone) and renal failure are the primary causes of hypertension (Lien et al., 2010). The chromaffin cells of the adrenal gland synthesize and secrete various catecholamines, including epinephrine and norepinephrine, which lead to a variety of physiological effects, including increased cardiac contraction and vasoconstriction. Pheochromocytoma, a tumor of the adrenal gland, can lead in several species to hypertension, which is related to the rapid and massive release of catecholamines from the tumor (Prejbisz et al., 2011). In addition, reversible dilated or hypertrophic cardiomyopathy is a well-established cardiac manifestation of pheochromocytoma in humans (Prejbisz et al., 2011). There are species differences in sensitivity to hypertension. For example, the rhesus monkey is more susceptible than the dog to the morbid effects (cardiac hypertrophy) of hypertension (Frank, 1963). Dogs exhibited marked differences in susceptibility or resistance to the development of chronic hypertension (Schroeder and Goldman, 1952). Approximately two-thirds of dogs return to normotension within a year after constriction of one renal artery (Schroeder and Goldman, 1952). Furthermore, in certain breeds of dogs (i.e., the terrier breeds), hypertension was easily induced experimentally (Schroeder and Goldman, 1952). In nonhuman primates, experimental renal hypertension developed consistently after unilateral renal artery constriction and was maintained for 19 to 35 months and became more marked after contralateral renal constriction or nephrectomy (Frank, 1963).

Heart Rate

Heart rate (HR) is the frequency of cardiac cycle and is measured by the number of beats per minute. Generally, small animals have a higher heart rate than that of larger animals (Reece, 2007). Heart rate values vary significantly among species: human (60 to 90 beats/min), rhesus monkey (180 to 210 beats/min), horse (32 to 44 beats/min), cow (60 to 70 beats/min), dog (70 to 120 beats/min), and cat (111 to 130 beats/min) (Reece, 2007). In the mouse, the heart rate ranges from 310 to 840 beats/min, and there is a wide variation in HR and BP among mouse strains (Fox et al., 2002). When the HR is too slow, it is termed bradycardia, and when HR is too fast, it is termed tachycardia. Heart rate is regulated by the autonomic nervous system. Sympathetic stimulation enhances myocardial performance and increases HR activities (i.e., rate and force of contraction, rate of impulse conduction, and amount of coronary blood flow), and parasympathetic (vagal) stimulation decreases all HR activities (McCorry, 2007; Reece, 2007). HR regulation can vary between neonates and adults. Pharmacological blockade of sympathetic (propranolol) and parasympathetic (atropine) efferent nerves can be used to assess effects on HR. The autonomic control of resting HR was assessed using atropine and propranolol in neonatal (2 to 3 week old) male rhesus monkeys. It was found that the autonomic control of resting HR is mediated by the autonomic nervous system and the sympathetic system, the latter having a dominant influence. This is in contrast to the adult rhesus, where the parasympathetic nervous system controls resting HR (Goldberg and Moberg, 1985). The sympathetic nervous system predominates during emergency "fight-or-flight," with the primary neurotransmitter being norepinephrine (McCorry, 2007). Main receptors for norepinephrine are adrenergic ($\beta 1, \alpha 1$, and $\alpha 2$). Beta 1 receptors are the primary adrenergic receptor on the heart (a small percentage of the adrenergic receptors on the myocardium are $\beta 2$). Both subtypes of beta receptors on the heart are excitatory, and stimulation leads to an increase in cardiac output, lipolysis, or renin (Bakris and Mensah, 2003; McCorry, 2007). In sympathetic nervous stimulation, norepinephrine or catecholamine bind to β-adrenergic receptors on the cell membrane of cardiac cells, leading to increased cellular calcium influx and subsequently increased contractile strength of the heart (McCorry, 2007). The parasympathetic nervous system predominates during resting conditions, with the primary neurotransmitter being acetylcholine (McCorry, 2007). Main receptors for acetylcholine are cholinergic (nicotinic and muscarinic) (McCorry, 2007). In vagal stimulation, acetylcholine is released and binds to nicotinic receptors, causing rapid increase in the cellular permeability to sodium and calcium ions, which causes depolarization (McCorry, 2007). Parasympathetic control of the heart is attenuated in heart failure (HF). In experimentally induced HF in dogs, vagal transmission is reduced, acetylcholinesterase activity is decreased, and muscarinic receptor density and composition are altered (Dunlap et al., 2003).

The Frank–Starling law of the heart describes the length–tension relationship and relates resting muscle length, expressed as the volume of blood in the heart at the end of diastole, to tension generation (Shiels and White, 2008). Heart failure is characterized by a lower length–tension curve. The Frank–Starling law of the heart applies to all classes of vertebrates and is found in both adult and embryonic myocardium (Asnes et al., 2006; Shiels and White, 2008). The cellular mechanisms underlying the Frank–Starling response include three cooperative mechanisms: (1) an increase in myofilament sensitivity for calcium, (2) decreased myofilament lattice spacing, and (3) increased thin filament (Shiels and White, 2008). In the Laplace law or surface tension law, the tension in the wall (T) is directly proportional to intraventricular pressure (P) and internal radius (R) ($T = P \times R$) and inversely to the wall thickness (Strain and Olson, 1975). Dilated cardiomyopathy is an example of Laplace's law, in which the heart chambers become dilated and the radius of ventricle increases. As a consequence of this and to create the same pressure during blood ejection, cardiac muscle develops more wall tension. In hypertension, all parts of the Laplace law equation apply (i.e., intraventricular pressure changes, and with left ventricular hypertrophy, both internal radius and wall thickness are altered) (James et al., 2000). Laplace's law has been used widely to explain the constancy of arterial wall thickness to the lumen radius needed to maintain constant stress in the wall of the pressurized cylinder in a wide range of circumstances. The relationship among carotid BP, arterial geometry, and wall stress was investigated to determine the impact of hypertension, smoking, and their interaction on these relationships (Liang et al., 2001). The investigators found that carotid artery wall remodeling follows Laplace's law. However, this was insufficient to prevent an increase in circumferential stress in hypertensive patients. Smoking did not influence the lumen-to-wall ratio but has a significant effect on wall stiffness (Liang et al., 2001).

Renin–Angiotensin–Aldosterone in the Cardiovascular System

The renin–angiotensin–aldosterone system (RAAS) plays an important role in the control of the cardiovascular homeostatic physiological processes by regulating vascular tone, BP, and fluid volume. Angiotensin II (Ang II) is a physiologically active component of the RAAS, produced via an enzymatic cascade that begins with angiotensinogen (AGT) (synthesized mainly in the liver) cleaving renin, aspartyl protease enzyme, to form Ang I, which is then cleaved by the angiotensin-converting enzyme (ACE) to form Ang II. Ang II has two subtypes of receptors, AT_1 and AT_2, but the AT_1 receptor subtype drives most of the biological functions of Ang II. Ang II causes vasoconstriction, most pronounced in arteries and arterioles and less in veins, directly by activating Ang II type 1 (AT_1) receptors on vascular smooth muscle. Ang II also affects fluid volume via AT_1 receptor activation in the renal proximal tubule, resulting in renal sodium and water reabsorption. Ang II regulates fluid balance by stimulating aldosterone secretion from the adrenal glands (Meune et al., 2003; Mehta and Griendling, 2007; Radi and Ostroski, 2007). Ang II is also a growth promoter in cardiovascular tissues and has been implicated in cardiovascular system inflammation, endothelial dysfunction, atherosclerosis, hypertension, and congestive heart failure (Mehta and Griendling, 2007). In vascular smooth muscle cell contraction, AT_1 receptors can, via G-protein signaling, mobilize calcium from the sarcoplasmic reticulum and promote the interaction of actin and myosin filaments to allow for contraction and migration of cells. Ang II has been shown to phosphorylate and activate phospholipase A_2, which leads to the production of arachidonic acid (AA) and its metabolites. In addition, Ang II can maintain a balance between vasoconstriction and vasodilation in various vascular beds (Mehta and Griendling, 2007). One of the major pathways activated by G-protein interaction with AT_1 receptors led to the activation of protein kinase C (PKC) and the ERK pathway, which are implicated in contraction as well as cellular growth (Mehta and Griendling, 2007). In addition, Ang II, via the AT_1, activates intracellular signaling pathways that promote atherothrombosis through inflammation, endothelial dysfunction, growth, altered fibrinolysis, and potentiation of low-density lipoprotein (LDL) oxidation (Jacoby and Rader, 2003). Thus, RAAS contributes directly to coronary ischemic events via atherosclerosis, altered postinfarct remodeling, and reduced fibrinolysis.

Inhibition of the RAAS has antiatherosclerotic activity, independent of BP, which has been demonstrated in animal models of atherosclerosis (Chobanian, 1990; Miyazaki et al., 1999; Jacoby and Rader, 2003). Chymase is stored intracellularly in the secretary granules of mast cells and is released upon mast cell activation. In the cardiovascular system in dogs, monkeys, hamsters, and humans, chymase is a local Ang II-generating enzyme (Kunori et al., 2005). Interestingly, chymase inhibition using TY-51469, a novel chymase inhibitor, at a dose 0.1 mg/kg per day, reduced the progression to heart failure in an autoimmune myocarditis Lewis rat model (Palaniyandi et al., 2007). PGs are modulators of physiological functions and contribute to renin release, regulation of hemodynamics, salt balance, and BP control via mechanisms involving the regulation of vascular tone (Meune et al., 2003; Mehta and Griendling, 2007; Radi and Ostroski, 2007).

Natriuretic Peptides in the Cardiovascular System

Natriuretic peptides (NPs) are hormones involved in the regulation of sodium excretion (natriuresis), diuresis, vasodilation, reflex tachycardia inhibition, water balance, blood volume, and decreased systemic vascular resistance and arterial blood pressure (Woodard and Rosado, 2008). NPs include atrial (ANP), brain NP (BNP), C-type NP (CNP), dendroaspis-type NP (DNP), and urodilatin. All the natriuretic peptides have been shown to induce vasorelaxant effects in the coronary circulation (Woodard and Rosado, 2008). The actions of these NPs are mediated by the specific binding to three cell surface receptors: type A natriuretic peptide receptor (NPR-A), type B natriuretic peptide receptor (NPR-B), and type C natriuretic peptide receptor (NPR-C) (Woodard and Rosado, 2008). In the heart of rhesus monkey, APRA mRNA, using in situ hybridization, is expressed most abundantly in endocardial endothelial cells of the right and left atrium and right ventricle, while APRC mRNA is expressed most prominently in endocardial endothelial cells of all four heart chambers but was also found throughout the myocardium only in the right atrium (Wilcox et al., 1991). In the rat heart, distribution of the APRA and APRB receptor transcripts was relatively homogeneous; in contrast, the C-type mRNA was concentrated principally in the atria, with no difference between the left and right sides of the heart. In human heart, APRA and APRB receptors were found after amplification of the left (but not the right) ventricle, and APRC receptor expression was similar in both ventricles (Nunez et al., 1992). ANP is synthesized and secreted by normal adult mammalian atrial cardiocytes in response to wall stretch resulting from BP and volume loading and causes natriuresis, diuresis, and inhibition of smooth muscle contraction, and aldosterone and renin release (Ruskoaho et al., 1987). The expression of ANP is low in the ventricles in the physiological state (Woodard and Rosado, 2008). Increased ANP levels have been demonstrated in heart failure in humans (Burnett et al., 1986), in Wistar–Kyoto rats that spontaneously develop biventricular hypertrophy (Lee et al., 1988), in cats with congestive heart failure (Connolly et al., 2009), in dogs with mitral valve disease and congestive heart failure (Tarnow et al., 2009), and upon volume and pressure overload in mice (Mori et al., 2004). BNP is a cardiac hormone secreted predominantly from the ventricle, and its synthesis, secretion, and clearance differs from those of ANP (Mukoyama et al., 1991). BNP has emerged as a useful biomarker for the diagnosis and prognosis of adult patients and children with heart failure (Ekure et al., 2011). The cardiovascular effects of CNP are more likely to be mediated by local effects on blood vessel or by central actions on vasopressin and adrenocorticotropin release (Woodard and Rosado, 2008). DNP is an endogenous human natriuretic peptide that relaxes human arteries more than veins (Best et al., 2002). Urodilatin is synthesized in kidney tubular cells and is secreted into kidney tubules. It has been claimed that urodilatin has beneficial effects in patients with congestive heart failure and lowered preload, increased diuresis and natriuresis, decreased systemic vascular resistance, increased cardiac index, and lowered cardiac filling (Kentsch et al., 1992; Elsner et al., 1995; Mitrovic et al., 2006). The correlation between PGs and NPs in the cardiovascular system has been investigated. Using perfused beating rat atria, the effects of various PGs on the regulation of ANP and signaling molecules involved in PG-mediated ANP secretion were investigated. The study revealed that $PGF_{2\alpha}$ and PGD_2, but not PGE_2, PGJ_2, PGI_2, and TBA_2, increase ANP secretion and positive inotropy through the FP receptor. $PGF_{2\alpha}$ was found to be the most potent stimulator and secretagog of ANP secretion through a signaling pathway involving phospholipase C (PLC), inositol 3-phosphate (IP_3), PKC, and myosin light-chain kinase (MLCK). In two-kidney/one-clip hypertensive rat atria, $PGF_{2\alpha}$-induced ANP secretion was attenuated, whereas COX-2 inhibitor nimesulide-induced ANP secretion was augmented. The expression of COX-2 was increased and FP receptor mRNA was decreased in hypertrophied atria. The authors conclude that the regulation of ANP secretion by COX-2 and $PGF_{2\alpha}$ is due to increased expression of COX-2 and decreased expression FP receptor, which may partly relate to the development of renal hypertension (Bai et al., 2009).

Cardiovascular Research and Animal Models

In cardiovascular function and disease research studies, each experimental animal model, each animal species, and each endpoint has its own advantages and disadvantages, and appreciating these will help the investigator select the most appropriate animal model for the particular question under investigation (Hearse and Sutherland, 2000). In addition, nonclinical-to-clinical translation and relevance to the human condition is an important factor to consider in cardiovascular research (Hearse and Sutherland, 2000). A cautious approach is mandatory when experimental findings in animal models are extrapolated to human hypertension (Lerman et al., 2005). In hypertension cardiovascular research, the first animal model of hypertension was in dogs and was introduced in 1934 and was induced by unilateral constriction of the renal artery (2K1C model) (Goldblatt et al., 1934). Hypertension models fall into a variety of categories and include renovascular (e.g., two kidneys, one clip), renal injury (e.g., partial or total nephrectomy), vasoactive intervention (e.g., pressor prostaglandins), endocrine and dietary (e.g., Dahl salt-sensitive), neurogenic (e.g., baroreceptor denervation), and phenotype-driven (e.g., RAAS, Dahl salt-sensitive rat, SHR-stroke prone) (Lerman et al., 2005). In animal models, the most commonly used indirect method for monitoring BP is the cuff technique, in which BP is measured in a tail or limb by determining the cuff pressure at which changes in blood flow occur during occlusion or release of the cuff. It is important that investigators also use optimal methods for measuring BP. The investigator needs to keep in mind that many environmental factors can affect BP, including, but not limited to, ambient room temperature, light cycle, noise levels, duration of human contact, number of animals per caging unit, proximity to other animals undergoing experimental procedures, cage unit size and design, and access to supplemental items such as toys, treadmills, and hiding spaces within the cage unit (Kurtz et al., 2005). In addition, systemic anesthetics can also exert major effects on cardiovascular function and should be avoided whenever possible (Kurtz et al., 2005). In atherosclerosis and dyslipidemia cardiovascular research, animal models should share the pathophysiology of the disease with humans (Moghadasian et al., 2001). In addition, interspecies differences in cholesterol are important considerations. For example, many characteristics typical of human serum LDL are found in those of the pig, rhesus monkey, and baboon (Chapman and Goldstein, 1976). Unlike humans and several other animals, wild-type (WT) mice and rats do not possess plasma cholesteryl ester transfer protein (CETP), and approximately 70% of their plasma total cholesterol is found in high-density lipoprotein (HDL) particles and therefore are resistant to atherogenesis (Moghadasian et al., 2001). Hyperlipidemia alone is not sufficient to result in atherogenesis in rats. The poor response to dietary cholesterol supplementation is a limitation for using mice in experimental atherogenesis (Moghadasian et al., 2001). However, unlike their WT counterparts, transgenic rodent models (e.g., Apo E knockout mouse) provide investigators with an excellent opportunity to study the interactions of gene and environment in atherogenesis (Moghadasian et al., 2001). Rabbits share with humans several aspects of lipoprotein metabolism, such as similarities in composition of apolipoprotein B–containing lipoproteins (Chapman, 1980). Unlike humans, rabbits need a very high plasma cholesterol to induce atherosclerosis, are hepatic lipase-deficient, and do not have an analog of human Apo A-II (Chapman, 1980; Warren et al, 1991; Moghadasian et al., 2001). Pigeons (e.g., White Carneau strain) have relatively high levels of plasma cholesterol, most of it in HDL particles (St. Clair, 1983; Jerome and Lewis, 1985). In dogs, spontaneous atherosclerosis is rare (Moghadasian et al., 2001). Hypothyroidism has been associated with atherosclerosis in dogs (Liu et al., 1986).

Generally, the heart of pig is anatomically similar to that of humans, a notable exception being the presence of a left azygous (hemiazygous) vein, which drains the intercostal system into the coronary sinus (Swindle et al., 1986). The pig cardiovascular system has many features that make pigs an attractive model of human cardiovascular disease. Features of the pig cardiovascular system include (1) very similar cardiac output to that of humans (Tranquilli et al., 1982), (2) a high heart rate (112.0 ± 3.5 beats/min) (Poletto et al., 2011), (3) blood pressure (mean arterial pressure of 112.8 ± 2.8 mmHg, SBP of 134.0 ± 3.3 mmHg, and DBP of 102.2 ± 2.6 mmHg) (Poletto et al., 2011), (4) a prominent left azygous vein entered on the left side of the heart drained via the coronary sinus instead of the precava (Crick et al., 1998), (5) a vaso vasorum in the aorta, (6) coronary artery anatomy and distribution of blood supply more similar to those of humans than those of other mammals (Weaver et al., 1986), (7) a coronary system similar to that of 90% of the human population in anatomy and function and no preexisting collateral vessels in the myocardium, (8) heart weight/body weight ratio in young pigs similar to that in normal adult human (5 g/kg) (Hughes, 1986), (9) similar chemical, immunological, physical, and physiological properties of the major apolipoprotein in human and pig HDL (Jackson et al., 1973), and (10) the physicochemical characteristics of LDL most closely resembling those of humans (Chapman and Goldstein, 1976).

Cardiovascular Atherosclerosis

In humans and nonhuman primates, atherosclerosis is one of the most important contributors to cardiovascular disease (Sellers et al., 2010). Atherosclerosis is a complex inflammatory disease of medium-sized and large vessels and is characterized by thickening and hardening of the arteries. Vascular thickening and hardening results from the accumulation of lipids, lipid-laden macrophages, matrix, mineral, and other inflammatory cells, as well as smooth muscle cell proliferation (Moubayed et al., 2007). Pathologically, atherosclerosis begins with endothelial cell dysfunction and inflammation, and thus it is considered to be predominantly an inflammatory disease. During the early initiation of atherosclerotic plaques, lipid accumulation, leukocyte recruitment, and expression of proinflammatory cytokines in the vascular wall with subsequent cytokine, cell signaling, and enzyme production promote plaque growth (Ross, 1993; Libby, 2002; Bonetti et al., 2003; Cipollone and Fazia, 2006). Animal models of atherosclerosis have shed light on the pathogenesis of this disease and the roles of COX-1 and COX-2 on both the development and progression of atherosclerotic plaques (Sellers et al., 2010). Mammalian species other than nonhuman primates do not develop atherosclerosis spontaneously under normal conditions, and atherosclerosis in nonhuman primates is generally associated with aging (Sellers et al., 2010). This species difference certainly presents limitations in the evaluation of cardiovascular risks associated with atherosclerosis in the typical nonclinical toxicology species and in the relatively young primates used in such nonclinical toxicology studies (Sellers et al., 2010). Animal models of atherosclerosis include the PGI_2 receptor–deficient mouse (*Ptgir*$^{-/-}$), apolipoprotein E–deficient mouse (*Apoe*$^{-/-}$), the LDL receptor–deficient mouse (*Ldlr*$^{-/-}$), the ApoE*3-Leiden transgenic mouse [*Tg(APOE*3,APOC1)2Lmh*], as well as high-fat-diet-induced atherosclerosis in hamsters and Watanabe rabbits (Kowala et al., 1993; Kobayashi et al., 2004; Egan et al., 2005; Zadelaar et al., 2007; van der Hoorn et al., 2008).

COX-1 AND COX-2 EXPRESSION IN THE CARDIOVASCULAR SYSTEM

COX-1 is the predominant COX isoform that is expressed constitutively in the normal vascular endothelium and smooth muscles (Smith et al., 2000; Koki et al., 2002; Parente et al., 2003; Zidar et al., 2007, 2009; Streicher and Wang, 2008; Sellers et al., 2010; Radi and Ackermann, 2011). In endothelial cells prepared from rabbit or pig aortas, COX-1 activity supports prevention of platelet adhesion via the production of the antithrombotic PG prostacyclin (Moncada et al., 1976). Endothelial cells line the luminal surface of blood vessels and are exposed continuously to hemodynamic shear stress. The level of hemodynamic shear stress endothelial cells experience varies from region to region within the vasculature, and physiologic fluid shear stress has been proposed to have an atheroprotective effects. Atherosclerotic lesions develop invariably in regions with lowered shear stress or vortices, whereas the regions of increased shear stress are protected (Cheng et al., 2006). In areas of high laminar shear stress, endothelial cells are elongated, aligned, and protected from inflammation (Potter et al., 2011). COX isoform expression under static or shear stress was investigated using porcine aortic endothelial cells (PAECs) (Potter et al., 2011). PAECs cultured under static or shear stress conditions readily expressed detectable levels of COX-1, with low levels of COX-2 in all regions of the well. Shear stress did not affect the relative expression of COX-1 and COX-2 immunoreactivity in cells at either 24 h or 7 days. Endothelial cell medium contained low levels of the prostacyclin metabolite 6-keto-$PGF_{1\alpha}$ when cells were grown under static or sheared conditions for up to 7 days. Interestingly, levels of COX-1 or COX-2 did not vary significantly in aligned versus nonaligned endothelial cells (Potter et al., 2011). In another study, human umbilical vein endothelial cells (HUVECs) were exposed to uniform laminar shear stress for 6 h, and COX-1 and COX-2 expression was investigated. COX-2 protein was induced significantly, whereas COX-1 and the downstream synthases were not modulated significantly (Di Francesco et al., 2009). Thus, COX-1 is the predominant COX isoform in the healthy peripheral vasculature (Sellers et al., 2010).

COX-1 expression in the heart of adult humans and animals is in both endothelial cells and vascular smooth muscle (Stanfield et al., 2001; Koki et al., 2002; Radi and Ostroski, 2007; Zadelaar et al., 2007; Zidar et al., 2007; Sellers et al., 2010). In normal human heart, COX-1 is expressed in endothelial and smooth muscle cells of blood vessels and in endothelial cells of the endocardium. In myocardial infarction in humans, COX-1 is expressed in inflammatory cells as well as in myofibroblasts and capillaries of granulation and fibrous tissue (Zidar et al., 2007). In Finnish domestic pig normal heart, COX-1 mRNA was detected in the left ventricular wall (Uotila et al., 2001).

In a renin–angiotensinogen transgenic mouse model of human hypertension, COX-1 expression was noted in vascular endothelial cells and vascular smooth muscle cells of the heart, while cardiac myocytes lacked COX-1 expression (Radi and Ostroski, 2007). In the aorta of spontaneously hypertensive rats, endothelium-dependent contractions were associated with multiple dysfunctions in both the endothelial and the smooth muscle cells. The endothelial dysfunction was due to the release of endothelium-derived contracting factors that counteracted the vasodilator effect of nitric oxide, with no or minor alteration of its production (Félétou et al., 2009). Endothelial cell dysfunction included abnormal calcium handling, increased COX-1 expression, enhanced production of reactive oxygen species, and increased prostacyclin synthase expression (Félétou et al., 2009). COX isoform gene expression was quantified by real-time PCR using isolated endothelial cells and smooth muscle cells (SMCs) from the aorta of Wistar–Kyoto and spontaneously hypertensive rats (Tang and Vanhoutte, 2008). Aging caused overexpression of COX-1, COX-2, endothelial nitric oxide synthase, thromboxane A (TXA) synthase, PGD synthase, mPGE synthase-2, and PGF synthase in endothelial cells and COX-1 and PGE_2 receptors in SMC. In hypertension, the expression of COX-1, PGI synthase, TXA synthase, and PGD synthase were augmented in endothelial cells and PGD2, EP_3, and EP_4 receptors in SMC. The authors conclude that in aging and hypertension, at least in Wistar–Kyoto and the spontaneously hypertensive rat model, the endothelium has greater propensity to release COX-derived vasoconstrictive prostanoids (Tang and Vanhoutte, 2008). In dogs, COX-1 expression in the heart has been localized to the fibrous connective tissue of the tricuspid valve and chordae tendinae and in the microvasculature of the heart, aorta, and renal artery (Stanfield et al., 2001). In the normal aortic endothelium of male beagle dogs and humans, high levels of COX-1 and PGI synthase (PGIS) expression was observed in the endothelium. In addition, intense COX-1/PGIS double labeling, indicating PGIS and COX-1 colocalization was detected throughout the endothelium of both species. Quantification of the immunolabeling in the dog indicated that 55% of the aortic endothelium exhibited COX-1 expression compared to 8% for COX-2 (Kawka et al., 2007). In Sprague–Dawley rats, COX-1 has been identified in the endothelial cells of the heart but not in cardiomyocytes (Sellers et al., 2010). There are strain differences in cardiac COX expression. No COX-1 mRNA expression was noted in Fisher 344 and brown Norway strains of rats (Okamoto and Hino, 2000). In the fetal pig, the ductus arteriosus contains COX-1 almost exclusively (Guerguerian et al., 1998).

Cyclooxygenase activity is induced in human atherosclerosis (Belton et al., 2003). Prostaglandins (PGs) promote atherosclerosis by altering the inflammatory response and expression of matrix metalloproteinases (MMPs) as well as by acting as a mitogen for various cell types, including vascular smooth muscle cells (VSMCs) (Sellers et al., 2010). Up-regulation of MMP-2 and MMP-9 in plaques has been demonstrated to be mediated in part by COX-2- and COX-1-derived PGE_2 (Cipollone and Fazia, 2006), and probably contribute to plaque rupture (Sellers et al., 2010). COX-1 mRNA expression was investigated in atherosclerotic aortic plaques and adjacent normal aorta samples obtained from postmortem human cases. COX-1 mRNA levels were only 1.1-fold higher in plaque tissue. There was no relationship between gender, age, or cause of death and COX-2 or COX-1 levels (McGeer et al., 2002). COX-1 expression was analyzed in biopsies from human atherosclerotic carotid arteries and from healthy mammary arteries and saphenous veins. COX-1 was found in the endothelium in healthy and in atherosclerotic vessels (Stemme et al., 2000). TXA_2 increases in atherosclerosis, and platelet COX-1 plays an important role in the early stages of lesion development in the $ApoE^{-/-}$ knockout mouse model of atherosclerosis (Belton et al., 2003). Suppression of TXA_2 biosynthesis retards atherogenesis (Cyrus et al., 2007).

In contrast to COX-1, COX-2 is generally not expressed in the majority of normal endothelial or vascular smooth muscle cells; however, it is induced rapidly with vessel trauma or inflammation (Smith et al., 2000; Dannenberg et al., 2001; Koki et al., 2002; Zidar et al., 2007, 2009; Sellers et al., 2010). The 5′ flanking region of the COX-2 gene contains transcription factor binding sites for nuclear factor κB (NFκB), nuclear factor for interleukin-6/CCAAT enhancer-binding protein (NF-IL6/C/EBP), and cyclic AMP response element (CRE) (Kirtikara et al., 2000; Smith et al., 2000; Streicher and Wang, 2008). This is consistent with a COX-2 role in inflammation. Cyclic stretch induces COX-2 gene expression in vascular endothelial cells via activation of NFκB (Zhao et al., 2009). Blood vessels are constantly subjected to mechanical forces in the form of stretch, encompassing cyclic mechanical strain due to the pulsatile nature of blood flow and shear stress. Significant variations in mechanical forces, of a physiological or physiopathological nature, occur in vivo (Lehoux et al., 2006). Exposure of human umbilical vein endothelial cells to laminar shear stress in the physiological range up-regulated the expression of COX-2 (Inoue et al., 2002). Cardiac hypertrophy is an increase in the size of individual myofiber and, consequently, in the size of the heart. In response to pathophysiological stress, the adult heart undergoes hypertrophic enlargement. Although cardiac hypertrophy is initially a compensatory response, sustained hypertrophy is a leading predictor for the development of heart failure. At the cellular and molecular level, disease-related stimuli invoke endocrine, paracrine, and autocrine regulatory circuits, which directly influence cardiomyocyte hypertrophy, in part, through membrane-bound G-protein-coupled receptors and receptor tyrosine kinases. These membrane receptors activate intermediate signal transduction pathways within the cytoplasm, such as mitogen-activated protein kinases (MAPKs), protein kinase C (PKC), and calcineurin, which directly modify transcriptional regulatory factors, promoting alterations in cardiac gene expression (Bueno and Molkentin, 2002). Interestingly, stimulation of either PKC or Ras signaling enhances MAPK activity, which, in turn, activates transcription of COX-2 (Chun and Surh, 2004). In fact, PKC-ϵ was found to play a significant role in COX-2 expression in mice after ischemic preconditioning (Xuan et al., 2005). COX-2 expression in the normal heart is limited. In histologically normal coronary arteries of humans (Baker et al., 1999) and in nonatheromatous aorta of humans (Schönbeck et al., 1999), undetectable or little COX-2 protein immunostaining was observed.

COX-2 expression, using immunohistochemistry, was found to be restricted in aortic vascular endothelial cells in the normal beagle dog and Sprague–Dawley rat, and no expression has been identified in the human aorta (Kawka et al., 2007). In the cardiomyocytes of normal adult humans, COX-2 expression was either not present or it was present in occasional myocyes (Zidar et al., 2007). Extensive COX-1 and COX-2 immunohistochemical staining was observed in heart cardiomyocytes of humans, Wistar rats, and CD-1 mice (Testa et al., 2007). Such extensive staining may be explained by the use of a rabbit polyclonal COX-1 and COX-2 antibody in the immunohistochemicl staining of heart sections. Rabbit polyclonal antibodies stain the myocardium extensively, regardless of antigen, and are thus felt to be unreliable for the immunohistochemical evaluation of rodent hearts (Sellers et al., 2010).

COX-2 is readily observed in pathological cardiac conditions and injured cardiac tissues both in humans and in animal models of myocardial disease. Myocardial infarction (MI) is caused by interruption of coronary blood flow to the myocardium, which if persistent would lead to myocyte necrosis and chronic heart failure. COX-2 expression was induced in cardiomyocytes in MI in humans (Zidar et al., 2007). COX-2 mRNA and protein expression was increased in the left ventricular wall after cardioplegic arrest in the heart of pigs (Uotila et al., 2001). In the heart of rabbits, brief episodes of myocardial ischemia with reperfusion resulted in a rapid increase in COX-2 mRNA levels followed by a robust induction of COX-2 protein 24 h later. The enhanced expression of COX-2 was associated with a marked increase in the myocardial content of PGE_2 and 6-keto-$PGF_{1\alpha}$ (Shinmura et al., 2000). Immunohistochemical evaluation of COX-1 and COX-2 expression in human hearts with postmyocardial infarction was investigated. COX-2 staining was observed through the entire peri-infarct region, and uniformly from the epicardial to the endocardial layer, being present in the vast majority of cardiomyocytes. In addition, COX-2 expression in cardiomyocytes occurs 10 days to two months after acute myocardial infarction and is associated with an increased rate of postinfarction myocardial apoptosis (Abbate et al., 2004). In human patients who have suffered from an MI, COX-2 protein expression was induced in cardiomyocytes as well as in interstitial inflammatory cells and in capillaries and myofibroblasts in granulation tissue (Zidar et al., 2007). In a Lewis rat model of chronic MI and heart failure, strong immunohistochemical COX-2 protein expression was observed mainly in the cardiomyocytes, macrophages, endocardium, and vascular endothelial cells in the infarcted myocardium and in the cardiomyocytes and endothelial cells of the noninfarcted myocardium (Saito et al., 2004). The strongest expression of COX-2 in all cells was seen 2 weeks after coronary artery ligation, and decreased slightly thereafter except in macrophages, where it was prominent throughout the study (Saito et al., 2004). Because COX-2 is an inducible enzyme in settings of inflammation, it is not surprising that COX-2 expression has been identified in the hearts of patients with septicemia and inflammation of heart myocardium, myocarditis (Cuenca et al., 2006; Streicher et al., 2008). In the region of atherosclerotic plaques, COX-2 expression is present in endothelial cells, VSMCs, and macrophages (Belton et al., 2003). COX-2 mRNA expression was investigated in atherosclerotic aortic plaques and adjacent normal aorta samples obtained from postmortem human cases.

COX-2 levels were 4.8-fold higher in plaque tissue than in normal artery (McGeer et al., 2002). COX-2 imunoreactivity was seen in macrophages and foam cells, intimal and medial smooth muscle cells, and endothelial cells of vasa vasorum in transplant atherosclerosis (Baker et al., 1999). As in humans, mouse and rabbit models of atherosclerosis have COX-2 expression in fatty streaks and atherosclerotic vessels (Belton et al., 2000; Hong et al., 2000; Stemme et al., 2000; Wong et al., 2001; Burleigh et al., 2005b). These data indicate that in animal models of atherosclerosis, the expression of COX isoforms is similar to that of humans (Sellers et al., 2010).

PATHOPHYSIOLOGICAL ROLE OF PROSTAGLANDINS IN THE CARDIOVASCULAR SYSTEM

The role of prostanoids in homeostasis and vascular tone has been well established. The effects of prostanoids are often opposing, which is important for maintaining a balance between vasoconstriction and vasodilation and anti- and prothrombotic effects (Sellers et al., 2010). Platelets are essential cells for hemostasis, in which they aggregate and cooperate with the coagulation system. The vasoconstriction and vasodilation and anti- and prothrombotic balance are mediated through PGI_2 and platelet-derived TXA_2, and to a lesser extent, PGE_2. The balance between PGI_2 and TXA_2 is a critical factor determining a thrombotic tendency. TXA_2 is a largely platelet-derived (to a lesser extent, monocytes) potent vasoconstrictor (Rossi et al., 1996). Because TXA_2 is platelet derived, it is primarily COX-1-dependent. Platelets express only COX-1 in both humans and nonclinical species; platelets have neither baseline nor induced expression of COX-2 (Kay-Mugford et al., 2000; Zidar et al., 2007). TXA_2 causes local vasoconstriction, platelet aggregation, and smooth muscle cell proliferation. TXA_2 actions are mediated by G-protein-coupled thromboxane–prostanoid (TP) receptors. Infusion of the TP receptor agonist U-46619 causes transient increases in blood pressure followed by cardiovascular collapse in wild-type (WT) mice, but U-46619 caused no hemodynamic effect in $TP^{-/-}$ mice. These knockout (KO) $TP^{-/-}$ mice have a mild bleeding disorder and altered vascular responses to TXA_2 and arachidonic acid (Thomas et al., 1998). During myocardial ischemia, TXA_2 is released in large quantities by activated platelets in the coronary circulation of patients with coronary artery disease (Hirsh et al., 1981). Injections of the TXA_2 mimetic U-46619 at 10 and 20 μg into the left atrium of anesthetized rabbits evoked decreases in HR and arterial BP by stimulation of cardiac vagal afferent nerves (Wacker et al., 2002). Receptors for prostanoids on platelets include the EP_3 receptor for which the natural agonist is PGE_2 produced in atherosclerotic plaques, and EP_3 has been implicated in atherothrombosis (Heptinstall et al., 2008). DG-041, a direct-acting EP_3 antagonist, potentiated the protective effects of PGE_2 on platelet aggregation by inhibiting the proaggregatory effect via EP_3 stimulation (Heptinstall et al., 2008). PGI_2, which is primarily COX-2 derived, is a vasodilator with antithrombotic effects and is expressed primarily by endothelial and vascular smooth muscle cells (Dannenberg et al., 2001; Koki et al., 2002; Zadelaar et al., 2007; Streicher et al., 2008).

The interplay between TXA_2 and PGI_2 in platelet–vascular wall interaction regulation has been studied in rodent models. PGI_2 is important for balancing the prothrombotic and vasoconstrictive effects of TXA_2 (Fitzgerald, 2004; Kearney et al., 2004). The exact source of baseline PGI_2 is uncertain, as there is little to no COX-2 expression in normal human endothelium (Koki et al., 2002). However, COX-2 induction in damaged or inflamed endothelial cells promotes production of prostacyclin, which tempers prothrombotic events at those sites. While PGE_2 can cause vasodilation and can modulate the effects of TXA_2 on platelet aggregation, it does not participate directly in homeostasis (Sellers et al., 2010). Mice lacking the receptor for PGI_2 (IP KO) develop salt-sensitive hypertension, cardiac hypertrophy, and severe cardiac fibrosis. Coincidental deletion of the TP receptor does not prevent the development of hypertension, but cardiac hypertrophy is ameliorated and fibrosis is prevented in IPTP double knockouts (Francois et al., 2005). Decreased susceptibility to renovascular hypertension has been reported in mice lacking the receptor for PGI_2 (Fujino et al., 2004). While a balance exists between TXA_2 and PGI_2 to control thrombosis, studies in PGI_2 receptor–deficient (*Ptgir*$^{-/-}$) and TXA_2-overexpressing mice (B6;129Sv-*Tbxa2*^{Y385F}) have demonstrated that shifting the balance between these factors does not result in increased platelet aggregation or thrombosis under normal physiological conditions (Murata et al., 1997; Cheng et al., 2002). It is likely, therefore, that other homeostatic mechanisms play a role on clotting under normal conditions (Sellers et al., 2010). Such molecules would include nitric oxide, carbon monoxide, and CD39, to name only a few, which play an important role not only in endothelial protection but also in pro- and antithrombotic events and vascular tone (Marcus et al., 2002).

With injury, however, the importance of a balance between PGI_2 and TXA_2 to modulate platelet aggregation becomes apparent. Vascular injury up-regulates COX-2 activity and therefore local PGI_2 production. If PGI_2 receptor–deficient mice are subjected to vascular injury, they are significantly more prone to thrombotic events than are WT controls (Murata et al., 1997; Cheng et al., 2002). Similar to PGI_2 receptor–deficient mice, mice overexpressing the TXA_2 receptor displayed increased neointimal hyperplasia, intimal cell proliferation, and platelet activation after mechanical artery injury (Cheng et al., 2002). These data indicate that the importance of a balance between PGI_2 and TXA_2 is essential, particularly with vascular trauma, although redundant controls via other molecules may predominate under normal physiological conditions (Sellers et al., 2010).

Genetically engineered and knockout mice have allowed a better understanding of the role of prostanoids and COX in the cardiovascular system (Sellers et al., 2010). Using mice null for COX-2 to evaluate cardiac function and histology has had limited value because COX-2 (*Ptgs2*) null mice often die at birth and rarely survive to 6 months of age (Dinchuk et al., 1995). These null mice develop myocardial fibrosis. However, because the mice have significantly underdeveloped kidneys and poor renal function, discerning whether this cardiac change is primary or secondary is challenging (Sellers et al., 2010). The use of site-specific and inducible gene modulating technology has promoted a better understanding of organ and time-specific changes in gene expression, particularly for genes whose mutation results in early death (Sellers et al., 2010). The role of COX-2 in the adult mouse heart was investigated using a cardiomyocyte-specific inducible deletion of *Ptgs2*. Two weeks after the cardiomyocyte-specific deletion of COX-2 in postweanling animals, the mice had reduced systolic ventricular function, decreased heart rate, increased susceptibility to induced ventricular arrhythmias, and decreased exercise tolerance (Wang et al., 2009). When afterload was increased through aortic banding, the null mice had greater cardiomyocyte hypertrophy and fibrosis than those of their WT littermates (Wang et al., 2009). Interestingly, Connexin 43, an important gap junction protein, was reduced in the COX-2-deficient cardiomyocytes compared to littermates, and fibroblasts obtained from these hearts had greater expression of atrial natriuretic peptide (ANP), B-type natriuretic peptide (BNP), and MMP-9. A study by Seta et al. (2009) utilized a complete disruption of the murine gene encoding COX-2 (*Ptgs2)* knockdown approach in an effort to mimic the effect of COX-2 s-NSAID inhibition. The study revealed that mice with marked (but not complete) COX-2 inhibition had greater platelet aggregation along vessel walls after $FeCl_3$-induced vascular injury than that of their (WT) littermates. Ex vivo platelet activation studies demonstrated no significant difference in platelet activation by thrombin or TXA_2 production between the COX-2 knockdown and the WT littermates. Thus, the increased platelet aggregation was probably attributable to COX-2-related changes in the vascular wall physiology. In addition, these mice had no alterations in the mean arterial pressure, suggesting that under normal conditions, reductions in COX-2 do not notably alter blood pressure in mice (Seta et al., 2009).

PGE_2, derived from both COX-1 and COX-2, promotes vasodilatation and although not by itself thrombogenic promotes the thrombogenic potential of TXA_2 (Gross et al., 2007; Sellers et al., 2010). PGE_2 was released from both ischemic and nonischemic cardiac regions during myocardial ischemia in anesthetized mongrel dogs (Berger et al., 1976). The receptors for PGE_2, EP_{1-4}, are G-protein-coupled receptors. Each of these receptors has differing signaling cascades (Degousee et al., 2008). The highest expression level of EP_4 mRNA was found in cardiac ventricle in mice (Xiao et al., 2004). The histopathological features of myocardial reperfusion injury induced by temporary occlusion of the coronary artery were first described in 1960 in the dog. Cardiac pathological features included cell swelling, contracture of myofibrils, disruption of the sarcolemma, and the appearance of intramitochondrial calcium phosphate particles (Jennings et al., 1960). The cardiac ventricle in mice has the highest expression level of EP_4 mRNA (Xiao et al., 2004). The role of PGE_2 via its receptor subtype EP_4 in cardiac ischemia with reperfusion (I/R) injury has been examined in $EP_4^{-/-}$-KO mice. $EP_4^{-/-}$ mice had larger infarct size than WT mice in an I/R model (Xiao et al., 2004). In addition, isolated and perfused $EP_4^{-/-}$ hearts had greater functional and biochemical derangements than WT hearts in response to I/R (Xiao et al., 2004). However, a different cardiac effect was noted if EP_4 function was deleted only in cardiomyocytes. Cardiac remodeling and function were investigated in cardiac-specific EP_4 receptor–KO mice. No difference was noted in infarct size between KO mice and controls. The KO mice showed less myocyte cross-sectional area and interstitial collagen fraction than those of controls and reduced ejection fraction. Cardiomyocyte–EP_4

deletion decreased cardiac hypertrophy, fibrosis, and activation of STAT-3. However, cardiac function was unexpectedly worsened in these KO mice. Thus, cardiac myocyte EP_4 may have a cardioprotective role in cardiac hypertrophy via activation of STAT-3 (Qian et al., 2008). EP_4 knockdown in cardiac myocytes in aged male KO mice is, in part, shown to be associated with increased fibrosis, reduced ejection fraction, and dilated cardiomyopathy (Harding et al., 2010). EP_4 has been found to mediate regulation of inflammatory cytokines in macrophages and neutrophils in vitro by PGE_2 (Sakamoto et al., 2004). An EP_4 agonist, ONO-AE1-329, was administered to rats at one of three concentrations (1, 3, or 10 μg/kg via hourly bolus intravenous injection). Infusion of the EP_4 agonist at 3 and 10 μg/kg per hour attenuated lipopolysaccharide (LPS)-induced hypotension and hyporeactivity to norepinephrine. In addition, the EP_4 agonist significantly attenuated the LPS-induced increases in serum concentrations of TNFα and IL-6. Furthermore, EP_4 agonist attenuated increases in left ventricular and aortic expression of mRNAs encoding TNFα and inducible nitric oxide (Sakamoto et al., 2004). In another study, an EP_4-selective agonist, EP4RAG, was administered to rats at doses of 1 or 3 mg/kg, and the effects on cardiac function after myocardial I/R injury were investigated. EP4RAG significantly reduced myocardial infarction area and ischemia and improved left ventricular contraction and dilatation compared with that of the vehicle control. In addition, EP4RAG attenuated the recruitment of inflammatory cells, especially macrophages, and cardiac interstitial fibrosis (Hishikari et al., 2009).

Interestingly, high concentrations of PGE_2 will inhibit platelet aggregation by activating the PGI_2 receptor (IP) nonspecifically. However, low concentrations of PGE_2 potentiate the activation of partially activated platelets, increasing platelet aggregation. This potentiation is primarily through activation of the PGE_2 receptor EP_3 (Gross et al., 2007). Interestingly, prostaglandins may also play a role in myocardial hypertrophy. Overexpression of the PGE_2 receptor gene *Ptger3* in mice using the α-myosin heavy-chain promoter resulted in marked myocardial hypertrophy with increased expression of TGFβ, ANP, cardiac ankyrin repeat protein, and connective tissue growth factor with increased interstitial fibrosis in TG mice compared to their WT littermates (Meyer-Kirchrath et al., 2009). These mice had marked increases in end diastolic and end systolic volumes at 5 to 6 weeks of age compared to their WT littermates. In addition, the left ventricular ejection fraction was markedly decreased in the TG mice, suggesting reductions in contractility. Hypertrophic effects of *Ptger3* overexpression in these mice was thought to be mediated by increased activation of the calcineurin signaling pathway. Thus, PGE_2 may be a factor in the development of myocardial hypertrophy. However, to identify whether the reduction of PGE_2 signaling could cause myocardial atrophy and dysfunction, deletion of *Ptger3* in the heart would be required. In another study, the effect of cardiospecific overexpression of EP_3 was studied. In the hearts of these mice, ischemic contracture was markedly delayed compared with WT hearts, and the ischemia-induced increase in left ventricular end-diastolic pressure was reduced by 55%. Creatine kinase and lactate dehydrogenase release were decreased significantly, by 85% and 73%, respectively, compared with WT hearts (Martin et al., 2005).

The importance of PGs in the development of atherosclerosis has been demonstrated using several animal models of atherosclerosis. Increases in PGI_2 are the result of both COX isoforms in the endothelium, although much of the PGI_2 increase in atherosclerosis has been attributed to COX-2 up-regulation at the site of the plaque (Belton et al., 2000). Studies by Kobayashi et al. (2004) using mice mutant for both *Apoe* and either the PGI receptor (*Ptgir*) or the TXA_2 receptor (*Tbxa2r*) has helped to elucidate the importance of prostacyclin and thromboxane in the development of atherosclerosis (Sellers et al., 2010). In *Apoe*$^{-/-}$ *Ptgir*$^{-/-}$ mice, atherogenesis was accelerated compared to *Apoe*$^{-/-}$ mice. In contrast, in *Apoe*$^{-/-}$ *Tbxa2r*$^{-/-}$ mice atherogenesis was delayed compared to *Apoe*$^{-/-}$ mice. The PGI_2 system is essential for endothelial progenitor cells (EPCs) to accomplish their function, and it plays a critical role in the regulation of vascular remodeling in mice (Kawabe et al., 2010). Injury-induced vascular proliferation and platelet activation were enhanced in mice that are genetically deficient in the PGI_2 receptor (IP) but were depressed in mice genetically deficient in the TXA_2 receptor (TP) or treated with a TP antagonist (Cheng et al., 2002). These data collectively indicate that PGI_2 is protective against the atherosclerosis, whereas TXA_2 is important in the initiation and progression of atherosclerosis (Kobayashi et al., 2004; Egan et al., 2005). This is due, in part, to modulation of platelet aggregation and leukocyte–endothelial cell interactions through alterations in the adhesion molecules ICAM-1 and PECAM-1. Studies in *Apoe*$^{-/-}$ *Ptgs1*$^{-/-}$ (COX-1) mice fed a 1% cholesterol diet for 8 weeks showed decreased atherosclerotic lesion development compared to control mice (McClelland et al., 2009). These data implicate COX-1 in the early pathogenesis of atherosclerosis. Interestingly, platelet adhesion to vessel walls was not notably different between double mutant and controls under normal conditions; however, reductions in platelet adhesion were evident after vascular ligation injury in the *Apoe*$^{-/-}$ *Ptgs1*$^{-/-}$ mice compared to controls. This suggests that COX-1 is also important in promoting platelet adhesion to damaged vessel walls and is probably mediated through TXA_2 production (Sellers et al., 2010). Studies in hamsters have also demonstrated that the fact that PGI_2 imparts atheroprotective properties and that PGI_2 analogs retarded atherogenesis in hamsters is consistent with results in genetically engineered mouse studies (Kowala et al., 1993).

Systemic inflammation induces various adaptive responses, including inflammatory tachycardia. The role of various prostanoids in inflammatory tachycardia was examined using mice lacking each prostanoid receptor (Takayama et al., 2005). The tachycardia induced in WT mice by injection of LPS was greatly attenuated in TP- or FP-deficient mice and was completely absent in mice lacking both TP and FP. A TP agonist and $PGF_{2\alpha}$ increased the beating rate of the isolated atrium of WT mice via TP and FP, respectively. In addition, the inflammatory cytokine-induced increase in the beating rate in vitro was significantly diminished in atria from mice lacking either TP or FP. The beta-blocker propranolol did not block the LPS-induced increase in heart rate in WT mice. Inflammatory tachycardia was caused by direct action on the heart of TXA_2 and $PGF_{2\alpha}$ formed under systemic inflammatory conditions (Takayama et al., 2005). A pattern of LPS-induced tachycardia was noted in mice with impaired febrile response and lacking EP_3 (Ushikubi et al., 1998). This suggests that febrile response and inflammatory tachycardia are independent events. Interestingly, the effects of NSAIDs on inflammatory tachycardia were examined in humans in a randomized double-blind placebo-controlled trial of intravenous ibuprofen [10 mg/kg body weight (maximal dose, 800 mg) given every 6 h in eight doses] in 455 patients who had sepsis, defined as fever, tachycardia, tachypnea, and acute failure of at least one organ system. Ibuprofen treatment reduced levels of PGI_2 and TXA_2 and decreased fever, tachycardia, and oxygen consumption (Bernard et al., 1997).

EFFECTS OF NSAIDs ON THE CARDIOVASCULAR SYSTEM

The COX-1-inhibitory effects of NSAIDs have been associated with decreased platelet aggregation, and these effects have been studied extensively (Campbell et al., 2007). The effects of COX-2 inhibitors, however, have been mixed and often contradictory. In nonclinical studies, this may be due at least in part to variations in study design, genetic backgrounds, drug exposure, and the environment (Sellers et al., 2010). In addition, under normal physiological conditions, alterations in platelet aggregation would probably not be identified with COX-2 inhibition, as was demonstrated in studies of PGI_2 receptor–KO and TXA_2-overexpressing mice (Sellers et al., 2010). The ex vivo antiplatelet effects of COX-2 s- and ns-NSAIDs were investigated in healthy volunteers to determine whether NSAIDs antagonize the effect of aspirin (Gladding et al., 2008). All ns-NSAIDs tested (buprofen, indomethacin, naproxen, and tiaprofenic acid) blocked the antiplatelet effect of aspirin. The COX-2 s-NSAIDs sulindac and celecoxib did not demonstrate any significant antiplatelet effect or reduce the antiplatelet effect of aspirin. In another study, the effects of 200 mg twice daily celecoxib on the inhibition of platelet COX-1 activity by aspirin (100 mg daily) were investigated in human patients with osteoarthritis and stable ischemic heart disease. Celecoxib did not interfere with the inhibition of platelet COX-1 activity and function by aspirin despite a comparable suppression of COX-2 ex vivo in these patients (Renda et al., 2006). Thus, in studies in healthy humans and normal animal models it is not anticipated that COX-2 inhibition would cause significant reductions in PGI_2 or TXA_2 levels in vascular tissues (Sellers et al., 2010). It is also not anticipated that significant COX-2 inhibition would alter platelet aggregation ex vivo.

EFFECTS OF COX-1 AND COX-2 INHIBITION ON MYOCARDIAL ISCHEMIA, INFARCTION, AND THROMBOSIS

Animal models of myocardial infarction generally fall into four main categories: myocardial infarction (MI), ischemia with reperfusion (I/R), I/R with ischemic preconditioning, and I/R with postconditioning (Ferdinandy et al., 2007). Myocardial infarction is caused by interruption of coronary blood flow to the myocardium. Reperfusion induces an exaggerated hydrolysis of cell membrane phospholipids, causing an unphysiological rise in unesterified arachidonic acid associated with enhanced phospholipase A_2 (PLA_2) activity. Phospholipases A_2 are ubiquitous enzymes that hydrolyze the *sn*-2-acyl bond of phospholipids present in cell membranes and lipoproteins and yield free fatty acids and lysophospholipids, precursors of various bioactive lipid mediators (Balsinde et al., 2002). Myocardial phospholipids serve as primary reservoirs of arachidonic acid (AA), which is liberated through the rate-determining hydrolytic action of cardiac phospholipases A_2. A predominant PLA_2 in myocardium is calcium-independent phospholipase $A_{2\beta}$ ($iPLA_{2\beta}$), which through its calmodulin and ATP-binding domains is regulated by alterations in local cellular calcium concentrations and cardiac bioenergetic status, respectively (Jenkins et al., 2009). PLA_2 have been implicated in playing a critical role in the pathogenesis of myocardial ischemia (Jenkins et al., 2009). Group V secretory phospholipase A_2 ($sPLA_2$-V) is expressed strongly in the heart and has been found to play an important role in the pathogenesis of myocardial I/R injury in mice, at least partly in concert with the activation of cytosolic PLA_2 (Yano et al., 2011). Myocardial injury and echocardiographic LV dysfunction after I/R were reduced in $sPLA_2$-$V^{-/-}$ mice, including a reduction in myocardial apoptosis (Yano et al., 2011). The accumulation of AA and the extent of reperfusion injury have been shown to be closely related. The release of the cytosolic enzyme lactate dehydrogenase (LDH) during reperfusion is a measure of loss of cellular integrity. Prolonged duration of ischemia leads to a substantial reduction in ATP (<25% of preischemic value) and accumulation of adenosine and inosine in the myocardium (De Windt et al., 2001).

Contradictory results have been noted in animal models that investigate the effects of COX inhibition in myocardial ischemia (Tables 4-1 to 4-3) (Sellers et al., 2010). The majority of animal models evaluating the cardiac effects of COX inhibition have utilized either MI (complete coronary artery ligation) or I/R (coronary artery ligation for several hours followed by reperfusion) with or without ischemic preconditioning. While reperfusion after ischemia is necessary for tissue survival, reperfusion also results in the release of reactive oxygen species, which may incite further cell death (Sellers et al., 2010). The duration of the ischemic event prior to reperfusion affects the degree of tissue injury (Ferdinandy et al., 2007). Summaries of the drugs, types of studies, and outcomes are given in Tables 4-1 to 4-3 (Sellers et al., 2010). In a study by Inserte et al. (2009), transgenic (TG) mice carrying human COX-2 specifically overexpressed in cardiomyocytes were subjected to transient coronary artery occlusion (without preconditioning) followed by reperfusion. Histopathological and echocardiographic analysis revealed no differences in the phenotype of COX-2 TG and WT mice. Improved contractility during the reperfusion phase and decreased hypercontracture, infarct area, and LDH release were noted compared to that in the WT mice. The magnitude of this effect was similar to that seen with ischemic preconditioning. In both WT and COX-2-overexpressing mice, levels of PGE_2, PGI_2, and $PGF_{2\alpha}$ were increased but not the level of TXA_2. However, only PGE_2 was significantly higher in TG mice than in WT mice, suggesting that this molecule is important in the prevention of cell death during I/R

TABLE 4-1 Examples of Myocardial Infarction (MI) and Ischemia with Reperfusion (I/R) in Various Animal Models with the COX-1 NSAID Acetylsalicylic Acid

Animal model	Model	Study duration	Findings vs. controls	Reference
Rat: Sprague–Dawley	MI	2 weeks	No difference in MI size Increased MI wall thickness No loss of ischemic preconditioning protection	Alhaddad et al., 1995
Dog	I/R	2 h	No difference in MI size	Alcindor et al., 2004
	MI	6 weeks	No difference in MI size	Brown et al., 1983
	MI	Acute	Decreased infarct size	Such et al., 1983
Rat	MI	3 months	No difference in MI size	Saito et al., 2000
Guinea pig	I/R	Isolated perfused heart	Greater dysfunction	Heindl and Becker, 2001
Rabbit	I/R	Isolated heart	Greater dysfunction	Rossoni et al., 2002
Pig Landrace	MI	3 h	No difference in MI size Increased premature ventricular contractions	Wainwright et al., 2002
Pig	I/R	3 h	No difference in MI size	Jalowy et al., 1998

Source: The final, definitive version of this table was published in R. A. Sellers, Z. A. Radi, and N. K. Khan, *Veterinary Pathology*, 47(4), 2010, pp. 601–613.

events. These data provided direct evidence that constitutive COX-2 expression in cardiomyocytes has a cardioprotective effect that is independent of its influence on hemodynamic factors or on blood constituents (Inserte et al., 2009). PGE_2 receptor (*Ptger3*)–overexpressing mice subjected to ischemia with reperfusion also demonstrated significant protection against ischemia-associated creatine phosphokinase (CPK) release and left ventricular contracture (Meyer-Kirchrath et al., 2009). The authors hypothesized that this is related to the blunting of ischemia-induced endogenous catecholamines by attenuation of the cAMP-mediated calcium influx into cardiomyocytes. However, induced MI in microsomal prostaglandin E_2 synthase 1 (*Ptegs1*) null mice revealed similar effects on post-MI survival, infarct size, or CPK and cardiac troponin I levels compared to control animals, indicating that inflammation-induced PGE_2 may not contribute significantly to the outcome in MI

TABLE 4-2 Findings of Myocardial Infarction (MI) and Ischemia with Reperfusion (I/R) in Various Animal Models with Use of Nonselective Nonsteroidal Anti-inflammatory Drugs

ns-NSAID	Animal model	Model	Study duration	Findings controls	Reference
Azapropazone	Rat	MI	48 h	Decreased infarct size	Estrela et al., 1987
	Rabbit	I/R	60 min	Decreased infarct size	Mousa et al., 1990
	Dog	I/R	5 h	Limitation of infarct size	Mousa et al., 1989
Ibuprofen	Rat	Isoprenaline hydrochloride induced	1,5, 12, 21 days	Increased infarct size	Lal and Sharma, 1992
	Cat	MI	24 h	Decreased infarct size	Flynn et al., 1984
	Dog	MI	Acute	No change in infarct size	Such et al., 1983
		MI	6 weeks	Decreased infarct size; increased scar thinning	Brown et al., 1983
Meclofenamate	Dog	MI	Acute	Decreased infarct size	Such et al., 1983
	Rabbit	I/R	Isolated heart	Greater dysfunction than control	Rossoni et al., 2002
Meloxicam	Rabbit	I/R	Isolated heart	No change in infarct size	Liu et al., 1992
	Guinea pig	I/R	Isolated perfused heart	No difference	Heindl and Becker, 2001

Naproxen	Dog	MI	3 days	No difference in infarct size	Bolli et al., 1981
	Rabbit	I/R	Isolated heart	No difference	Aitchison and Coker, 1999
Indomethacin	Rabbit	I/R	3 days	No difference in infarct size	Shizukuda et al., 1991
		I/R	Isolated rabbit heart	No difference in infarct size	Aitchison and Coker 1999
	Pig	I/R	2 h	No difference in infarct size Abolished weak ischemic preconditioning	Gres et al., 2002
		I/R	2 h	No difference in infarct size	Jalowy et al., 1998
		I/R	3 h	Increased infarct size	Hohlfeld et al., 1993
	Dog	MI	6 h	No difference in infarct size	Fiedler and Mardin, 1985
		MI	2 days	Increased infarct size	Jugdutt et al., 1979
		MI	7 days	Increased infarct size, decreased scar thinning	Hammerman et al., 1983
		MI	6 weeks	Increased scar thinning	Such et al., 1983
Sulfinpyrazon	Dog	MI	3 days	No difference in infarct size	Bolli et al., 1981
	Rat	MI	3 weeks	Decreased infarct size	Innes and Weisman, 1981

Source: The final, definitive version of this table was published in R. A. Sellers, Z. A. Radi, and N. K. Khan, *Veterinary Pathology*, 47(4), 2010, pp. 601–613.

TABLE 4-3 Findings of Myocardial Infarction (MI) and Ischemia with Reperfusion (I/R) in Various Animal Models with COX-2 Selective NSAID Use

COX-2 s-NSAID	Animal model	Model	Study duration	Findings vs. controls	Reference
Celecoxib	Pig	MI	6 weeks	LV rupture	Timmers et al., 2007
	Dog	I/R	3 h	No difference in infarct size Abolished ischemic preconditioning	Alcindor et al., 2004
	Rabbit	I/R	Isolated heart	Greater dysfunction than control	Rossoni et al., 2002
Parecoxib	Mice: B6	MI	7 days	↓ apoptosis in peri-infarct region	Abbate et al., 2007
	Rat: Wistar	MI	7 days	Attenuation of LV dysfunction vs. control ↓ apoptosis in peri-infarct region	Straino et al., 2007
DFU	Rat: Lewis	MI	3 months	Improved LV function and ↓ infarct size	Saito et al., 2004
	Rat: Lewis	MI	3 days	↓ infarct size	Saito et al., 2003
DuP-697	Rabbit	I/R	Isolated heart	Greater dysfunction than control	Rossoni et al., 2002
NS-398	Mice: B6/J	MI	2 weeks	No difference in infarct size	LaPointe et al., 2004
	Mice: B6129F2/J	I/R	24 h	No difference in infarct size Abolished ischemic preconditioning	Guo et al., 2000
Merck tricyclic	Mice: CD-1	Adriamycin-induced heart failure	35 days	Attenuated progression of heart failure	Delgado et al., 2004
Rofecoxib	Mice	MI	2 weeks	No difference in infarct size Abolished ischemic preconditioning	LaPointe et al., 2004
	Dog	MI	30 min	No different from control	Carnieto et al., 2009
SC-236	Rat: Sprague–Dawley	Doxirubicin	Acute	Aggravated acute cardiac injury	Dowd et al., 2001
SC-58125	Guinea pig	I/R	Isolated perfused heart	No effect on cardiac function	Heindl and Becker, 2001

Source: The final, definitive version of this table was published in R. A. Sellers, Z. A. Radi, N. K. Khan, *Veterinary Pathology*, 47(4), 2010, pp. 601–613.

(Wu et al., 2009). Adiponectin (APN) is an adipocyte-specific cytokine that suppresses inflammatory responses and myocardial production of TNFα in vitro and in vivo (Shibata et al., 2005). The role of COX-2 signaling in the cardioprotective action of adiponectin in vivo was investigated. The COX-2 s-NSAID NS-398 was given daily via intraperitoneal injections to mice in a 5-mg/kg dose from 3 days before surgery until study termination. NS-398 did not affect infarct size in APN-KO mice, but abrogated the infarct-sparing actions of exogenous adiponectin by 53% in WT and 48% in APN-KO mice. These data indicated that COX-2-dependent signaling contributes to the protective action of adiponectin in myocardial I/R injury through the suppression of inflammatory cytokines and improved cell survival (Shibata et al., 2005).

The findings from COX inhibition in acute MI studies can be divided into short- and long-term findings, with most studies having a duration of less than 7 days (Sellers et al., 2010). Myocardial infarction and congestive heart failure (CHF) were induced in Lewis rats by ligating of the left coronary artery. Treatment of rats with the COX-2 s-NSAID 5,5-dimethyl-3-(3-fluorophenyl)-4-(4-methylsulfonyl-2(5H)-fluranone (DFU), at 5 mg/kg per day for 3 days and three months revealed that DFU-treated rats had a smaller infarct size, with significant improvement in left ventricular (LV) end diastolic pressure and LV systolic pressure over that of vehicle-treated animals (Saito et al., 2003, 2004). Improved cardiac function and cardioprotective effects were noted in Wistar rats given the COX-2 s-NSAID parecoxib 0.75 mg/kg intraperitoneally for 5 days beginning 2 days postcoronary artery occlusion. These rats had smaller end-diastolic and end-systolic diameters and greater fractional shortening. Systolic thickness in the anterior infarct wall was also significantly greater in the parecoxib-treated animals, whereas the posterior wall was not significantly affected. Aneurysmal dilatation of the left ventricle was more frequent in saline-treated versus parecoxib-treated animals. In addition, parecoxib treatment was associated with lower apoptotic rates and preservation of arteriolar density in the peri-infarct area (Straino et al., 2007). The effects of COX-2 inhibition after prolonged heart failure were investigated in Wistar rats. These rats underwent coronary artery ligation to induce ischemic congestive heart failure. Rats were treated one year after the ligation with daily intraperitoneal parecoxib injections at a dose of 0.75 mg/kg per day for 7 days. An echocardiogram to measure end-diastolic diameter showed no changes, and there was an increase in myocardial function in the peri-infarct area (Abbate et al., 2007). In another study, parecoxib was given to mice at an intraperitoneal dose of 0.75 mg/kg per day for 7 days. Parecoxib reduced mortality, improved ventricular function, reduced LV cavity size, and reduced peri-infarct apoptosis in this mouse model of acute MI (Salloum et al., 2009).

Dogs, rats, and mice that undergo several cycles of ischemic preconditioning (4 to 5 min of ischemia followed by 4 to 5 min of reperfusion repeated four or five times) have reduced myocardial injury when this preconditioning is followed by prolonged ischemia with reperfusion (Fiedler et al., 1985; Guo et al., 2000; Alcindor et al., 2004; Sellers et al., 2010). The ischemic preconditioning effect is lost if celecoxib or NS-398 is administered prior to surgery, while the protective effect of preconditioning on the myocardium is not lost with pretreatment with

ASA (Guo et al., 2000; Alcindor et al., 2004). This demonstrates the importance of COX-2, not COX-1, in ischemic preconditioning associated myocardial protection. Studies of the effect of COX inhibition in I/R without preconditioning are also limited. Many of these studies focus on the early or immediate effects of COX inhibition in reperfusion rather than on the long-term consequences in remodeling (Sellers et al., 2010). A study in dogs treated with the COX-2 s-NSAID rofecoxib prior to 3 h of cardiac ischemia followed by 30 min of reperfusion showed no detectable differences in CPK-MB or troponin I, biomarkers of cardiac injury, compared to controls (Carnieto et al., 2009), suggesting that acute injury is not altered in I/R injury with COX-2 inhibition. A longer-term study in rats exposed to 8 h of cardiac ischemia followed by 2 weeks of reperfusion and ASA therapy revealed no differences in infarct size or septal or left ventricular wall thickness compared to vehicle-treated animals (Alhaddad et al., 1995).

Landrace pigs given 10 mg/kg aspirin (ASA) orally for 7 days before MI showed no difference in infarct size or heart function compared to vehicle-treated animals. However, those treated with a 60-mg/kg oral dose of a nitro derivative of ASA, NCX 4016, did have reduced infarct size following combined ischemia and reperfusion compared to ASA- or vehicle-treated pigs (Wainwright et al., 2002). In dogs, the effects of ASA on cardiac infarct were investigated after intravenous (IV) administration of aspirin at a dose of 3 mg/kg 10 min before occlusion of the left anterior descending coronary artery. Aspirin-treated dogs did not differ from control dogs in percent ventricle at risk, percent infarct weight/left ventricle, or percent infarct weight/weight of ventricle at risk (Bonow et al., 1981). In an acute model of transmural infarction in anesthetized open-chest dogs, the effects of various COX inhibitors on infarct size were evaluated. ASA, 30 mg/kg IV, and meclofenamic acid, 1 mg/kg IV, decreased infarct size without altering other parameters except for an increase in retrograde flow produced by ASA (Such et al., 1983). Indomethacin, 10 mg/kg IV, increased infarct size and ectopic rate and decreased retrograde flow while developing greater tachycardia and less hypotension than in controls (Such et al., 1983). In another study, mongrel dogs underwent 180 min of left anterior descending coronary artery occlusion followed by 30 min of reperfusion. Dogs were treated with rofecoxib at a single 50-mg/day dose that was given orally with their chow during the 48 h preceding the surgical procedure. Rofecoxib was not associated with early detrimental effects on the hemodynamic profile or the gross extent of infarction (Carnieto et al., 2009). Experimental studies to assess the effects of ibuprofen on MI and scar thinning were conducted. Four groups of dogs were studied: a control (untreated) group; ibuprofen 12.5 mg/kg IV 15 min and 6, 12, 18, and 24 h after occlusion (high dose); ibuprofen 12.5 mg/kg IV 15 min and 3 h after occlusion (low dose); and aspirin 30 mg/kg IV 15 min and 3 h after proximal left anterior descending coronary artery occlusion for 6 weeks. Assessment of infarct size was determined at 6 weeks post-MI and revealed no difference in infarct size from control animals in ASA-treated dogs and reduced infarct size in ibuprofen-treated dogs (Brown et al., 1983). Dogs treated with ibuprofen for up to 24 h or ASA for up to 3 h after MI had increased scar thinning 6 weeks after MI compared to vehicle-treated animals (Brown et al., 1983).

The effects of thrombus formation can be devastating and lead to myocardial infarction and stroke. The effects of NSAIDs on platelet aggregation and thrombosis have been investigated in human and animal models. The effects of celecoxib on COX-1-dependent platelet TXA_2 and on systemic biosynthesis of PGI_2 production were investigated in healthy volunteers. Celecoxib was given at doses of 100, 400, or 800 mg and ibuprofen in an 800-mg dose. Ibuprofen (but not celecoxib) significantly inhibited TXA_2-dependent aggregation. Neither NSAID altered platelet aggregation induced by a thromboxane mimetic, U46619. Ibuprofen reduced serum TXB_2 and urinary excretion of the major TX metabolite (McAdam et al., 1999). Thus, celecoxib up to 800 mg did not inhibit COX-1-dependent platelet aggregation and function but suppressed excretion of both the hydration product, 6-keto $PGF_{1\alpha}$, and the major urinary metabolite of prostacyclin, 2,3 dinor-6-keto $PGF_{1\alpha}$ (PGI-M). Prostacyclin receptor–deficient mice show increased thrombosis (Murata et al., 1997). In an African green monkey thrombosis model, animals were exposed for 5 days to placebo, rofecoxib, celecoxib, naproxen, or aspirin and were evaluated following electrolytic injury to the carotid artery and jugular vein. Neither rofecoxib nor celecoxib had any effect on either arterial or venous thrombosis compared to placebo. Animals treated with naproxen, however, demonstrated significantly prolonged times to both arterial and venous thrombosis compared to placebo. The time to thrombosis in the aspirin group had a significantly prolonged time to arterial thrombosis and a numerically increased time to venous thrombosis. These data suggest that COX-2 s-NSAIDs have no effect on thrombosis, and ns-NSAIDs exhibit antithrombotic effects similar to those of aspirin (FDA, 2005).

Rofecoxib (Vioxx) and celecoxib (Celebrex) were the first COX-2 s-NSAID inhibitors to reach the market. Celecoxib and rofecoxib were approved in 1998 and 1999, respectively. Celecoxib exhibits anti-inflammatory, analgesic, and antipyretic activities in animal models and was approved in the United States in December 1998 and in Europe (Sweden) in December 2000 for the indications of symptomatic relief in the treatment of osteoarthritis (OA) and rheumatoid arthritis (RA) in humans. In subsequent years, the U.S. Food and Drug Administration (FDA) placed a boxed warning on all prescription NSAIDs. In humans, the cardiovascular (CV) events (MI, stroke, CV deaths, and peripheral events) of rofecoxib were analyzed in a randomized long-term Vioxx gastrointestinal outcomes research (VIGOR) clinical trial. In the VIGOR study, rofecoxib was used at a daily dose of 50 mg ($n = 4047$) and naproxen was used at a daily dose of 100 mg ($n = 4029$). Compared with naproxen, rofecoxib showed increased CV risk events (MI, unstable angina, ischemic stroke, and transient ischemic attacks) (FDA, 2001). In the Adenomatous Polyp Prevention on Vioxx (APPROVe) multicenter randomized placebo-controlled double-blind trial, a total of 2586 patients with a history of colorectal adenomas underwent randomizations in which 1287 received 25 mg of rofecoxib daily, and 1299 received placebo. Thrombotic, MI, and ischemic cerebrovascular events were increased in the rofecoxib groups (Bresalier et al., 2005). In a retrospective population-based cohort study of elderly adults (<65 years old), the influence of various COX-2 s- and ns-NSAIDs on the risk for a first myocardial infarction was assessed. An increased risk for acute MI in users of rofecoxib with no history of MI was noted (Lévesque et al., 2005).

The effects of long-term treatment (up to one year) with celecoxib and the ns-NSAIDs ibuprofen and diclofenac were investigated in the Celecoxib Long-Term Arthritis Safety Study (CLASS) (White et al., 2002). Celecoxib was given 400 mg twice daily ($n = 2320$), ibuprofen 800 mg three times daily, or diclofenac 75 mg twice daily ($n = 3987$). The incidences of serious CV thromboembolic events were similar, and not significantly different, between celecoxib and ns-NSAID (combined or individually). Therefore, rates of serious adverse cardiac events, regardless of causality, were not different between the treatment groups (White et al., 2002). In another study, CV events (MI, stroke, or heart failure) among 2035 patients with a history of colorectal neoplasia were analyzed. This Adenoma Prevention with Celecoxib (APC) study compared the efficacy and safety of 200 mg of celecoxib twice daily, 400 mg of celecoxib twice daily, and placebo in reducing the occurrence of adenomatous polyps in the colon and rectum one and three years after endoscopic polypectomy. CV events were noted in 7 of 679 patients in the placebo group compared with 16 of 685 patients who received 200 mg of celecoxib twice daily and with 23 of 671 patients who received 400 mg of celecoxib twice daily (Solomon et al., 2005). Thus, a CV risk was shown in patients with colorectal neoplasia receiving celecoxib after approximately one year of continuous treatment. However, differing results of CV event were noted in the Prevention of Colorectal Sporadic Adenomatous Polyps (PreSAP) clinical trial, in which celecoxib was given daily in a single 400-mg dose (Arber et al., 2006). Adjudicated serious CV events occurred in 2.5% of subjects in the celecoxib group and 1.9% of those in the placebo group (Arber et al., 2006). A metaanalysis of randomized clinical trials of the CV events risk in patients receiving celecoxib was conducted. Such analyses failed to demonstrate an increased CV risk with celecoxib relative to placebo and demonstrated a comparable rate of CV events with celecoxib treatment compared with ns-NSAIDs (White et al., 2007). There is an ongoing study to assess the long-term CV safety of the COX-2 s-NSAID celecoxib versus ns-NSAIDs. The Prospective Randomized Evaluation of Celecoxib Integrated Safety Versus Ibuprofen or Naproxen (PRECISION) trial is designed to evaluate the CV safety of celecoxib, ibuprofen, and naproxen. PRECISION is a randomized double-blind triple-dummy multinational multicenter study and will include approximately 20,000 patients with symptomatic osteoarthritis or rheumatoid arthritis at high risk for, or with, established cardiovascular disease. The primary endpoint of the study is the composite of cardiovascular death, nonfatal myocardial infarction, or nonfatal stroke (Becker et al., 2009).

Collectively, nonclinical studies suggest that COX-2 inhibition may reduce the inflammatory response and infarct size at the site of acute MI. COX-1 inhibition may have little impact on short- and long-term effects of MI. The reduction in local inflammation may be acutely valuable but may have long-term consequences in tissue breakdown and scar formation (Sellers et al., 2010). While the findings after long-term inhibition of COX-2 after infarction are variable, there is a trend in these studies indicating decreased collagen density and normal wall remodeling at the infarct site (Sellers et al., 2010). COX-2 is an important mediator in the protective effect of preconditioning in I/R studies. However, neither COX-1 nor COX-2 inhibition appears to affect the outcome in I/R studies significantly in several animal species (Sellers et al., 2010). No evidence of COX-2 s-NSAIDs prothrombotic potential was observed in acute animal models of thrombosis.

EFFECTS OF COX-1 AND COX-2 INHIBITION ON CORONARY BLOOD FLOW AND BLOOD PRESSURE

The effects of COX inhibition on local blood flow in the coronary artery have been evaluated in dog models. These models have evaluated the ability of COX inhibitors to modulate AA- and acetylcholine (ACH)-induced coronary vasodilatation and blood flow. The effects of celecoxib and aspirin administration on left circumflex coronary artery blood flow and thrombosis were evaluated in a dog model (Hennan et al., 2001). Before the administration of celecoxib, the intracoronary administration of 30 and 100 μg of AA or 100 and 300 ng of ACH, an endothelium-dependent vasodilator, increased coronary blood flow. In the same animal, 60 min after intravenous administration of 3 mg/kg of celecoxib, the vasodilator responses to 30 and 100 μg of AA were decreased, whereas the responses to ACH remained unchanged. Time controls were used to evaluate coronary responses over the course of the experiment, and no significant changes in the vasodilator responses to AA or ACH were observed (Hennan et al., 2001). Intracoronary administration of AA (3 to 300 μg) caused a dose-related increase in arterial volume blood flow, characterized as a dose-dependent increase in peak flow, followed by a dose-related prolongation in the vasodilator response. Both the peak response to AA and the duration of the response were suppressed equally by SC-560, a COX-1 inhibitor, at 0.3 mg/kg IV, nimesulide (5 mg/kg IV), and naproxen (3 mg/kg IV). The changes in volume blood flow to graded 10- to 300-ng doses of ACH were not altered after the administration of SC-560, nimesulide, or naproxen (Hong et al., 2008). Thus, AA-induced coronary blood flow is attenuated by pretreatment with ASA, nimesulide, naproxen, and celecoxib (Hennan et al., 2001; Hong et al., 2008). In contrast, coronary vasodilation was induced by direct intracoronary injection of ACH and AA in control animals. This ACH-induced coronary vasodilatation (nitric oxide–dependent pathway) in dogs was unaffected by pretreatment with 10 mg/kg naproxen and 15 mg/kg SC-58236, an experimental COX-2 s-NSAID, and SC-58236 failed to attenuate significantly either AA- or ACH-induced vasodilatation (Gross and Moore, 2004). These data suggest that the effects of AA, at least in the dog, are primarily COX-1 dependent.

As described previously, COX-1 and COX-2 have effects on localized vascular bed pressure. Systemic BP is maintained primarily by the RAAS, and hypertension is often due to abnormal renal vascular control homeostasis (Radi, 2009). Genetically engineered mice with reductions ("knockdown") in COX-2 expression to levels interpreted to be similar to that which would be induced by COX-2 s-NSAID administration had no alterations in mean arterial pressure, heart rate, or diastolic or systolic pressure in knockdown mice compared to WT littermates (Seta et al., 2009; Sellers et al., 2010). These two studies indicate that COX-2 is essential for normal cardiac function, but reductions in COX-2 expression (compared to deletion of COX-2 expression) do not significantly affect myocardial function (Sellers et al., 2010). Mice that overexpress COX-2 in the heart demonstrate no differences in cardiac function, weight, or histopathology under normal conditions (Inserte et al., 2009). *Drosophila* serves as a model for human cardiac pharmacology (Wolf et al., 2006). Celecoxib effects on cardiac rhythm in *Drosophila* hearts and rat cardiomyocytes (the cardiomyocytes were obtained from rat fetuses at gestation stage 20 just 1 day before birth) were investigated. Heart rate in *Drosophila* was reduced significantly by exposure to 3 μM celecoxib. Celecoxib has been shown to predispose to arrhythmia in *Drosophila* hearts and in rat neonatal cardiomyocytes in vitro by modulating delayed rectifier K^+ channel function. However, this has been attributed to an off-target effect independent of COX pathways because *Drosophila* lacks cyclooxygenases (Frolov et al., 2008). Studies in COX-1 null mice revealed that BP in COX-$1^{-/-}$ mice was near normal and that under conditions of reduced sodium intake these mice have a mild, but significant reduction in BP over control mice (Athirakul et al., 2001). The data in this study indicate that COX-1 is important in maintaining BP indirectly by maintaining sodium retention (Sellers et al., 2010). In addition, blood pressure has a circadian pattern, with reductions in BP during sleep (Kawada et al., 2005). COX-1 null mice have impaired reductions in BP during sleep (Kawada et al., 2005), indicating that COX-1 mediates a suppressed sympathetic nervous activity and enhanced nitric oxide and may be important in maintaining normal circadian rhythms in blood pressure. When COX-2 expression is suppressed to an extent similar to that achieved with COX-2 s-NSAIDs, COX-2-knockdown mice have no notable alterations in systemic BP compared to those in control mice (Seta et al., 2009). These data indicate that under normal physiological conditions, COX effects on systemic blood pressure are probably mediated primarily through its actions on renal sodium retention and excretion (Sellers et al., 2010).

The effect of COX inhibition on blood pressure in nonclinical species has been variable. Studies in normal animals indicate that ns-NSAIDs and/or COX-2 s-NSAIDs have little effect on systemic arterial BP (Manohar et al., 1996; Black et al., 1998; Birck et al., 2000; Brands et al., 2001). While blood pressure studies in dogs appear to be consistent, there is reported variation between rat strains in the effects of COX-2 inhibition on blood pressure (Sellers et al., 2010). A study compared the effects of diclofenac, celecoxib, and rofecoxib in Dahl salt-sensitive rats. Rat were fed 4% NaCl-enriched chow for 8 weeks, and from week 6 to week 8, NSAIDs were given in food: diclofenac (6 mg/kg per day), rofecoxib (2 mg/kg per day), celecoxib (25 mg/kg per day). By 5 and 8 weeks, BP increase in DS rats was more pronounced with rofecoxib and diclofenac treatment, but was attenuated in the celecoxib-treated rats. After 8 weeks of treatment, celecoxib slightly (approximately 4%), but significantly, reduced systemic BP compared to controls, and diclofenac and rofecoxib minimally but significantly increased systemic BP (approximately a 1% increase) (Hermann et al., 2005). In addition, celecoxib, but not rofecoxib or diclofenac, normalized proteinuria, decreased renal injury, and reduced cellular inflammation in this rat model of salt-sensitive hypertension (Hermann et al., 2005). In another study, COX-2 inhibition for 26 days using NS-398 at 1 mg/kg per day in Hannover Sprague–Dawley rats had no effect on systemic BP and MAP, and the rats remained normotensive throughout the study (Vaneckova et al., 2005). In a 2K1C model of renovascular hypertension using Wistar–Kyoto (WKY) or Sprague–Dawley rats, COX-2 inhibition using 1 and 15 mg/kg celecoxib per day and 1 mg/kg NS-398 per day, has been shown to be effective in the kidney, and up to 6 weeks of treatment resulted in no significant effect on BP (Hartner et al., 2003; Richter et al., 2004). In contrast, COX-2 inhibition has been found to increase BP in normotensive and hypertensive Wistar rats (Muscara et al., 2000; Höcherl et al., 2002b). Daily oral treatment of rofecoxib at a dose of 10 or 30 mg/kg resulted in a significant increase in systolic BP in normotensive WKY rats after 1 week compared to vehicle-treated rats. Rofecoxib at a dose of 3 mg/kg increased BP after 2 weeks, whereas 1 mg/kg of rofecoxib increased systolic BP after 3 weeks of the beginning of the treatment compared to vehicle-treated rats. At the end of the treatment the mean systolic BP of rofecoxib-treated rats was 15, 17, 21, and 23 mmHg higher than that of vehicle-treated rats, respectively (Höcherl et al., 2002b). Hypertensive rats were treated orally with celecoxib at 10 mg/kg, a dose selected based on inflammatory PGE_2 synthesis and inhibition of COX-2 activity, and BP measurements were performed using the tail cuff method prior to initiating treatment and at 1-week intervals for a total of 3 weeks. Hypertension was induced through the addition of *N*-nitro-L-arginine methyl ester (L-NAME) to the drinking water. Two weeks after beginning treatment with L-NAME, the rats began to receive celecoxib for 3 weeks. Daily treatment with celecoxib resulted in a significant increase in BP over that in the vehicle-treated group at all three time points, the greatest increase being seen at the end of the fifth week of the study (mean increase of 33 ± 2 mmHg) (Muscara et al., 2000).

It may be that some COX-1 inhibition along with COX-2 may be important to prevent thrombotic events in at-risk populations (Krotz et al., 2005). Salt-sensitive (DS) rats were fed a high-sodium diet (4% NaCl) for 56 days. From days 35 to 56, diclofenac (6 mg/kg per day), rofecoxib (2 mg/kg per day), or celecoxib (25 mg/kg per day) was added to the chow. Blood pressure increased with the sodium diet in the DS groups, which was more pronounced after diclofenac and rofecoxib treatment but was slightly decreased by celecoxib. In addition, celecoxib improved endothelial dysfunction and reduced oxidative stress (Hermann et al., 2003). This may point to differential effects of COX-2 s-NSAIDs in salt-induced hypertension (Hermann et al., 2003). Cardiac dysfunction and BP were evaluated in male Sprague–Dawley rats undergoing IV infusion of lipopolysaccharides. Rats were divided into two groups: the control group and the treated group, which received rofecoxib at 20 mg/kg. A significant decrease in BP and myocardial dysfunction was observed in the control group, whereas no changes were observed in the rofecoxib-treated group (Höcherl et al., 2002a).

In normotensive human patients, blood pressure and renal function were not altered significantly by short-term treatment with standard doses of celecoxib 200 mg twice a day (Dilger et al., 2002). The contribution of COX-2 activity to endothelial dysfunction was investigated in human patients with hypertension. Brachial artery vasodilator function was evaluated using ultrasound in 29 hypertensive patients before and after treatment with celecoxib at 200 mg in a randomized double-blind study. Brachial artery flow–mediated dilation improved from a baseline 3 h after the first dose and after 1 week of treatment with celecoxib (Widlansky et al., 2003). Thus, COX-2 may contribute to endothelial dysfunction in hypertension, and treatment with a COX-2 s-NSAID could have a beneficial effect (Widlansky et al., 2003). The effects of celecoxib 400 mg daily for 2 weeks on endothelial function, measured by high-resolution ultrasound, in patients with severe coronary artery disease were studied in a double-blind placebo-controlled crossover design. In addition, plasma levels of oxidized LDL (ox-LDL) and high-sensitivity C-reactive protein (hs-CRP), both of which have been implicated in atherogenesis and represent sensitive markers of oxidative stress and low-grade chronic inflammation, respectively, were studied. Celecoxib significantly improved endothelial function, and both ox-LDL and hs-CRP were lower after celecoxib therapy (Chenevard et al., 2003). The effects on 24-h mean systolic BP of celecoxib (200 mg once daily), rofecoxib (25 mg once daily), and naproxen (500 mg twice daily) treatment for 12 weeks was investigated in patients with type 2 diabetes, hypertension, and osteoarthritis in a double-blind randomized trial. The mean 24-h systolic BP following 6 weeks of therapy was increased significantly by rofecoxib, but not by celecoxib or naproxen (Sowers et al., 2005).

The effects of COX inhibition on aged humans and animals may be of greater significance clinically (Sellers et al., 2010). Studies of the aortic endothelium from aged rats demonstrated increased *Ptsg1* (COX-1) and *Ptsg2* (COX-2) gene expression (Kim et al., 2001; Tang et al., 2008). In the rat, COX-2 mRNA and protein levels in the heart increased with aging, whereas those of COX-1 showed no change (Kim et al., 2001). In the aorta of Wistar–Kyoto rats, aging caused overexpression of COX-1 and COX-2 in endothelial cells and of COX-1 and PGE_2 and EP_4 receptors in smooth muscle cells. Hypertensive rats had increased expression of COX-1 in endothelial cells and PGD_2, EP_3, and EP_4 receptors in smooth muscle cells (Tang et al., 2008). Thus, aortic endothelial cells from hypertensive rats had increased COX-1 gene expression, and this increased COX-1 expression may explain why, in aging, endothelial cells tend to secrete more COX-derived vasoconstrictive prostanoids than in younger animals (Kim et al., 2001; Tang et al., 2008). Studies in aged animals are uncommon; thus, studies of the effects of COX inhibition in aged animals are limited (Sellers et al., 2010). Interestingly, one study found that in aged humans (mean age 66 ± 3 years) with coronary artery disease, COX-2 inhibition was associated with improved coronary artery endothelium-dependent vasodilation (Chenevard et al., 2003). Some studies have indicated that COX inhibition may attenuate the effects of some antihypertensive therapeutics. In a 6-week randomized parallel-group double-blind trial of COX-2 inhibitors in human patients taking antihypertensives, 200 mg/day celecoxib or 25 mg/day rofecoxib increased both systolic and diastolic BP, but to different severities depending on the drug (e.g., the effects on blood BP were greater in patients treated with rofecoxib than in those treated with celecoxib) (Whelton et al., 2001, 2002). Patients on ACE inhibitors treated with high doses of celecoxib had no statistically significant effects on BP (White et al., 2002).

EFFECTS OF COX-1 AND COX-2 INHIBITION ON ATHEROGENESIS

Studies in atherosclerotic mouse models have been divided into evaluation of initiation and evaluation of progression (Sellers et al., 2010). Studies in *Apoe*-null mice have demonstrated that COX-1 inhibition appears consistently to modify the initiation and/or progression of atherosclerosis (antiatherogenic) (Belton et al., 2003; Cyrus et al., 2006). Furthermore, the importance of TXA_2 and COX-1 in atherogenesis is evident in low-density lipoprotein receptor–deficient mice on a high-fat diet (*Ldlr* deletion mice) treated with TXA_2 receptor and COX-1 inhibitors. When treated with COX-1 inhibitor, these mice using SC-560 at 15 mg/kg per day and the TP antagonist BM-573 at 10 mg/L for up to 12 weeks have reductions in atherogenesis compared to untreated animals (Cyrus et al., 2007). In cholesterol-fed rabbits, 25 mg/day indomethacin, an ns-NSAID, given for 12 weeks significantly reduced platelet aggregation and urinary TXB_2, 6-keto-$PGF_{1\alpha}$, and $PGF_{2\alpha}$ excretion, reduced the progression of atherosclerotic lesions, and improved endothelium-mediated vascular responses (Srisawat et al., 2003). COX-2 inhibition with 5 mg/kg rofecoxib once daily for 3 days promoted early atherosclerotic lesion formation in *Ldlr*-deficient mice, and COX-2 inhibition for 3 weeks with Merck Frosst (MF) tricyclic at 12 or 24 mg/kg per day increased both the early atherosclerotic lesion area and atherosclerotic plaque destabilization in *Apoe*-deficient mice (Rott et al., 2003; Burleigh et al., 2005b). In *Apoe*-deficient mice, COX-2 inhibition using SC-236 at 15 mg/kg per day demonstrated that despite selective reduction in PGI_2 generation, there was no effect on lesion development or platelet interactions with the vessel wall (Belton et al., 2003). Other studies demonstrated that chronic (up to four months) administration of COX-2 s-NSAIDs (celecoxib and MF-tricyclic) did not influence or have an effect on the composition of advanced atherosclerotic lesions in the same *Apoe* mouse strain (Pratico et al., 2001; Olesen et al., 2002; Bea et al., 2003). Treatment of *Apoe*-deficient mice with 900 ppm of celecoxib for 16 weeks led to a significant, 81% reduction in atherosclerotic lesions in the proximal aortas (Jacob et al., 2008). In another *Apoe*-deficient mouse atherosclerosis model, celecoxib was provided at a dose of 75 mg/kg (according to the authors, this dose is equivalent to a daily dose of 400 mg for a 60-kg person) and/or atorvastatin at a dose of 20 mg/kg body weight (according to the authors, this dose is equivalent to a daily dose of 80 mg for a 60-kg person) for 12 weeks (Raval et al., 2010). Diet supplemented with a combination of atorvastatin and celecoxib led to a significant reduction in established plaque in the aortas of these mice. Thus, combining statins with COX-2 inhibitors provides a greater benefit in reducing the size of established plaque than either atorvastatin or celecoxib alone, which appears to be due, in part, to a reduction in inflammation and adhesion molecules (Raval et al., 2010). In a carotid artery balloon injury Sprague–Dawley rat model, 50 mg/kg celecoxib was administered daily by oral gavage for 3 days before the balloon injury and were continued for 2 weeks after injury. At 2 weeks after injury, the vehicle-treated group had abundant neointimal hyperplasia. The celecoxib-treated group showed significant suppression of neointimal hyperplasia. In contrast, the aspirin-treated group (50 mg/kg) showed no significant reduction in neointimal hyperplasia. The levels of phosphorylated Akt and GSK−3β, both molecules that regulate cellular processes such as apoptosis, proliferation, migration, and survival, were significantly suppressed in the celecoxib-treated group compared with the vehicle- and aspirin-treated groups (Yang et al., 2004). Ultimately, the data from animal models suggest that while COX-2 is highly expressed in atherosclerotic plaques, it probably plays a lesser role in lesion development and progression than does COX-1. In healthy human volunteers, sulindac and celecoxib had no effect on the antithrombotic effect of ASA (Wilner et al., 2002; Gladding et al., 2008).

CONCLUSIONS

COX-1 and COX-2 isoforms are essential in normal cardiovascular system physiology. Postmarketing withdrawal of some COX-2 s-NSAIDs (e.g., rofecoxib and lumiracoxib) due to concerns of CV complications has raised a question as to why no adverse CV findings were identified in nonclinical studies prior to clinical use. There are drug-specific differences that may have played a role in the CV mechanism and the magnitude of CV risk. Chemically, rofecoxib and celecoxib belonged to a diaryl-substituted class of compounds. However, celecoxib is a sulfonamide-substituted 1,5-diaryl pyrazole compound, whereas rofoecoxib is a methylsulfonylphenyl compound. Extensive review of nonclinical toxicology studies, which included hemodynamic and detailed morphological assessments, revealed no treatment-related adverse findings in cardiovascular function or histopathology at a broad range of doses of celecoxib, valdecoxib, and rofecoxib after repeated dosing in multiple rodent and nonrodent species (Sellers et al., 2010). The two-year chronic toxicity studies indicated that standard nonclinical safety pharmacology and toxicology studies in multiple species at broad dose ranges and up to two years in duration did not predict the adverse CV safety events identified in long-term clinical studies (Sellers et al., 2010). NSAID effects on infarct size and MI in animal models have been variable. Animal models of cardiovascular diseases and genetically engineered mice have provided some indirect information in understanding potential functions of COX-mediated PGs in homeostasis and disease (Sellers et al., 2010). However, whereas disease models may be considered in assessing the overall safety profile of a new therapeutic candidate, species and strain variations in response may not be predictive of human risk (Sellers et al., 2010).

REFERENCES

Abbate A, Santini D, Biondi-Zoccai GG, Scarpa S, et al. Cyclo-oxygenase-2 (COX-2) expression at the site of recent myocardial infarction: Friend or foe? Heart 2004;90:440–443.

Abbate A, Salloum FN, Ockaili RA, Fowler AA 3rd, et al. Improvement of cardiac function with parecoxib, a cyclo-oxygenase-2 inhibitor, in a rat model of ischemic heart failure. J Cardiovasc Pharmacol 2007; 49:416–418.

Aitchison KA, Coker SJ. Cyclooxygenase inhibition converts the effect of nitric oxide synthase inhibition from infarct size reduction to expansion in isolated rabbit hearts. J Mol Cell Cardiol 1999;31:1315–1324.

Alcindor D, Krolikowski JG, Pagel PS, Warltier DC, et al. Cyclooxygenase-2 mediates ischemic, anesthetic, and pharmacologic preconditioning in vivo. Anesthesiology 2004;100:547–554.

Alhaddad IA, Tkaczevski L, Siddiqui F, Mir R, Brown EJ Jr. Aspirin enhances the benefits of late reperfusion on infarct shape: a possible mechanism of the beneficial effects of aspirin on survival after acute myocardial infarction. Circulation 1995;91:2819–2823.

Arber N, Eagle CJ, Spicak J, Rácz I, et al. Celecoxib for the prevention of colorectal adenomatous polyps. N Engl J Med 2006;355:885–895.

Asnes CF, Marquez JP, Elson EL, Wakatsuki T. Reconstitution of the Frank–Starling mechanism in engineered heart tissues. Biophys J 2006;91:1800–1810.

Athirakul K, Kim HS, Audoly LP, Smithies O, et al. Deficiency of COX-1 causes natriuresis and enhanced sensitivity to ACE inhibition. Kidney Int 2001;60:2324–2329.

Bai G, Gao S, Shah A, Yuan K, et al. Regulation of ANP secretion from isolated atria by prostaglandins and cyclooxygenase-2. Peptides 2009;30:1720–1728.

Baker CS, Hall RJ, Evans TJ, Pomerance A, et al. Cyclooxygenase-2 is widely expressed in atherosclerotic lesions affecting native and transplanted human coronary arteries and colocalizes with inducible nitric oxide synthase and nitrotyrosine particularly in macrophages. Arterioscler Thromb Vasc Biol 1999;19:646–655.

Bakris GL, Mensah GA. Pathogenesis and clinical physiology of hypertension. Curr Probl Cardiol 2003;28:137–155.

Balsinde J, Winstead MV, Dennis EA. Phospholipase A(2) regulation of arachidonic acid mobilization. FEBS Lett 2002;531:2–6.

Bea F, Blessing E, Bennett BJ, Kuo CC, et al. Chronic inhibition of cyclooxygenase-2 does not alter plaque composition in a mouse model of advanced unstable atherosclerosis. Cardiovasc Res 2003;60:198–204.

Becker MC, Wang TH, Wisniewski L, Wolski K, et al. Rationale, design, and governance of prospective randomized evaluation of celecoxib integrated safety versus ibuprofen or naproxen (PRECISION), a cardiovascular end point trial of nonsteroidal anti-inflammatory agents in patients with arthritis. Am Heart J 2009;157:606–612.

Belton O, Byrne D, Kearney D, Leahy A, et al. Cyclooxygenase-1 and -2-dependent prostacyclin formation in patients with atherosclerosis. Circulation 2000;102:840–845.

Belton OA, Duffy A, Toomey S, Fitzgerald DJ. Cyclooxygenase isoforms and platelet vessel wall interactions in the apolipoprotein E knockout mouse model of atherosclerosis. Circulation 2003;108: 3017–3023.

Berger HJ, Zaret BL, Speroff L, Cohen LS, et al. Regional cardiac prostaglandin release during myocardial ischemia in anesthetized dogs. Circ Res 1976;38:566–571.

Bernard GR, Wheeler AP, Russell JA, Schein R, et al. The effects of ibuprofen on the physiology and survival of patients with sepsis. The Ibuprofen in Sepsis Study Group. N Engl J Med 1997;336: 912–918.

Best PJ, Burnett JC, Wilson SH, Holmes DR Jr, et al. Dendroaspis natriuretic peptide relaxes isolated human arteries and veins. Cardiovasc Res 2002;55:375–384.

Birck R, Krzossok S, Knoll T, Braun C, et al. Preferential COX-2 inhibitor, meloxicam, compromises renal perfusion in euvolemic and hypovolemic rats. Exp Nephrol 2000;8:173–180.

Black SC, Brideau C, Cirino M, Belley M, et al. Differential effect of a selective cyclooxygenase-2 inhibitor versus indomethacin on renal blood flow in conscious volume-depleted dogs. J Cardiovasc Pharmacol 1998;32:686–694.

Bolli R, Goldstein RE, Davenport N, Epstein SE. Influence of sulfinpyrazone and naproxen on infarct size in the dog. Am J Cardiol 1981;47:841–847.

Bonetti PO, Lerman LO, Lerman A. Endothelial dysfunction: a marker of atherosclerotic risk. Arterioscler Thromb Vasc Biol 2003;23:168–175.

Bonow RO, Lipson LC, Sheehan FH, Capurro NL, et al. Lack of effect of aspirin on myocardial infarct size in the dog. Am J Cardiol 1981;47:258–264.

Brands MW, Hailman AE, Fitzgerald SM. Long-term glucose infusion increases arterial pressure in dogs with cyclooxygenase-2 inhibition. Hypertension 2001;37:733–738.

Bresalier RS, Sandler RS, Quan H, Bolognese JA, et al. Cardiovascular events associated with rofecoxib in a colorectal adenoma chemoprevention trial. N Engl J Med 2005;352:1092–1102.

Brown EJ Jr, Kloner RA, Schoen FJ, Hammerman H, et al. Scar thinning due to ibuprofen administration after experimental myocardial infarction. Am J Cardiol 1983;51:877–883.

Brown S, Atkins C, Bagley R, Carr A, et al. Guidelines for the identification, evaluation, and management of systemic hypertension in dogs and cats. J Vet Intern Med 2007;21:542–558.

Bueno OF, Molkentin JD. Involvement of extracellular signal-regulated kinases 1/2 in cardiac hypertrophy and cell death. Circ Res 2002;91:776–781.

Burleigh ME, Babaev VR, Patel MB, Crews BC, et al. Inhibition of cyclooxygenase with indomethacin phenethylamide reduces atherosclerosis in apoE-null mice. Biochem Pharmacol 2005a;70:334–342.

Burleigh ME, Babaev VR, Yancey PG, Major AS, et al. Cyclooxygenase-2 promotes early atherosclerotic lesion formation in ApoE-deficient and C57BL/6 mice. J Mol Cell Cardiol 2005b;39:443–452.

Burnett JC Jr, Kao PC, Hu DC, Heser DW, et al. Atrial natriuretic peptide elevation in congestive heart failure in the human. Science 1986;231:1145–1147.

Campbell CL, Smyth S, Montalescot G, Steinhubl Sr. Aspirin dose for the prevention of cardiovascular disease: a systematic review. J Am Med Assoc 2007;297:2018–2024.

Cannistra AJ. Adult valvular heart disease: a practical approach. Prim Care 2005;32:1109–1114.

Carnieto A Jr, Dourado PM, da Luz PL, Chagas AC. Selective cyclooxygenase-2 inhibition protects against myocardial damage in experimental acute ischemia. Clinics (Sao Paulo) 2009;64:245–252.

Chapman MJ. Animal lipoproteins: chemistry, structure, and comparative aspects. J Lipid Res 1980;21:789–853.

Chapman MJ, Goldstein S. Comparison of the serum low density lipoprotein and of its apoprotein in the pig, rhesus monkey and baboon with that in man. Atherosclerosis 1976;25:267–291.

Chenevard R, Hurlimann D, Bechir M, Enseleit F, et al. Selective COX-2 inhibition improves endothelial function in coronary artery disease. Circulation 2003;107:405–409.

Cheng C, Tempel D, van Haperen R, van der Baan A, et al. Atherosclerotic lesion size and vulnerability are determined by patterns of fluid shear stress. Circulation 2006;113:2744–2753.

Cheng Y, Austin SC, Rocca B, Koller BH, et al. Role of prostacyclin in the cardiovascular response to thromboxane A2. Science 2002;296:539–541.

Chobanian AV. The effects of ACE inhibitors and other antihypertensive drugs on cardiovascular risk factors and atherogenesis. Clin Cardiol 1990;13:VII43–VII48.

Chun KS, Surh YJ. Signal transduction pathways regulating cyclooxygenase-2 expression: potential molecular targets for chemoprevention. Biochem Pharmacol 2004;68:1089–10100.

Cipollone F, Fazia ML. Cyclooxygenase-2 inhibition: vascular inflammation and cardiovascular risk. Curr Atheroscler Rep 2006;8:245–251.

Clyman RI, Heymann MA, Rudolph AM. Ductus arteriosus responses to prostaglandin E1 at high and low oxygen concentrations. Prostaglandins 1977;13:219–223.

Clyman RI, Mauray F, Roman C, Rudolph AM. PGE_2 is a more potent dilator of the lamb ductus arteriosus than is either PGI_2 or 6-keto $PGF_{1\alpha}$. Prostaglandins 1978;16:259–264.

Connolly DJ, Soares Magalhaes RJ, Fuentes VL, Boswood A, et al. Assessment of the diagnostic accuracy of circulating natriuretic peptide concentrations to distinguish between cats with cardiac and non-cardiac causes of respiratory distress. J Vet Cardiol 2009;11:S41–S50.

Crick SJ, Sheppard MN, Ho SY, Gebstein L, et al. Anatomy of the pig heart: comparisons with normal human cardiac structure. J Anat 1998;193:105–119.

Cuenca J, Martin-Sanz P, Alvarez-Barrientos AM, Bosca L, et al. Infiltration of inflammatory cells plays an important role in matrix metalloproteinase expression and activation in the heart during sepsis. Am J Pathol 2006;169:1567–1576.

Cyrus T, Yao Y, Tung LX, Pratico D. Stabilization of advanced atherosclerosis in low-density lipoprotein receptor-deficient mice by aspirin. Atherosclerosis 2006;184:8–14.

Cyrus T, Yao Y, Ding T, Dogne JM, et al. Thromboxane receptor blockade improves the antiatherogenic effect of thromboxane A2 suppression in LDLR KO mice. Blood 2007;109:3291–3296.

Dannenberg AJ, Altorki NK, Boyle JO, Dang C, et al. Cyclo-oxygenase 2: a pharmacological target for the prevention of cancer. Lancet Oncol 2001;2:544–551.

Degousee N, Fazel S, Angoulvant D, Stefanski E, et al. Microsomal prostaglandin E2 synthase-1 deletion leads to adverse left ventricular remodeling after myocardial infarction. Circulation 2008;117: 1701–1710.

Delgado RM 3rd, Nawar MA, Zewail AM, Kar B, et al. Cyclooxygenase-2 inhibitor treatment improves left ventricular function and mortality in a murine model of doxorubicin-induced heart failure. Circulation 2004;109:1428–1433.

Detweiler DK. Control mechanisms of the circulatory system. In: Swenson MJ, Reece WO (eds), *Duke's Physiology of Domestic Animals*, 11th ed. Ithaca, NY: Cornell University Press, 1993, p. 185.

De Windt LJ, Willems J, Roemen TH, Coumans WA, et al. Ischemic-reperfused isolated working mouse hearts: membrane damage and type IIA phospholipase A2. Am J Physiol Heart Circ Physiol 2001;280:H2572–H2580.

Di Francesco L, Totani L, Dovizio M, Piccoli A, et al. Induction of prostacyclin by steady laminar shear stress suppresses tumor necrosis factor-alpha biosynthesis via heme oxygenase-1 in human endothelial cells. Circ Res 2009;104:506–513.

Dilger K, Herrlinger C, Peters J, Seyberth HW, et al. Effects of celecoxib and diclofenac on blood pressure, renal function, and vasoactive prostanoids in young and elderly subjects. J Clin Pharmacol 2002;42:985–994.

Dinchuk JE, Car BD, Focht RJ, Johnston JJ, et al. Renal abnormalities and an altered inflammatory response in mice lacking cyclooxygenase II. Nature 1995;378:406–409.

Dowd NP, Scully M, Adderley SR, Cunningham AJ, et al. Inhibition of cyclooxygenase-2 aggravates doxorubicin-mediated cardiac injury in vivo. J Clin Invest 2001;108:585–590.

Dunlap ME, Bibevski S, Rosenberry TL, Ernsberger P. Mechanisms of altered vagal control in heart failure: influence of muscarinic receptors and acetylcholinesterase activity. Am J Physiol Heart Circ Physiol 2003;285:H1632–H1640.

Egan KM, Wang M, Fries S, Lucitt MB, et al. Cyclooxygenases, thromboxane, and atherosclerosis: plaque destabilization by cyclooxygenase-2 inhibition combined with thromboxane receptor antagonism. Circulation 2005;111:334–342.

Ekure EN, Okoromah CA, Ajuluchukwu JN, Mbakwem A, et al. Diagnostic usefulness of N-terminal pro-brain natriuretic peptide among children with heart failure in a tertiary hospital in Lagos, Nigeria. West Afr J Med 2011;30:29–34.

Eling TE, Thompson DC, Foureman GL, Curtis JF, et al. Prostaglandin H synthase and xenobiotic oxidation. Annu Rev Pharmacol Toxicol 1990;30:1–45.

Elsner D, Muders F, Müntze A, Kromer EP, et al. Efficacy of prolonged infusion of urodilatin [ANP-(95–126)] in patients with congestive heart failure. Am Heart J 1995;129:766–773.

Estrela JM, Montoro JB, Vina JR, Vina J. Glutathione metabolism under the influence of hydroperoxides in the lactating mammary gland of the rat: effect of glucose and extracellular ATP. Biosci Rep 1987;7:23–31.

FDA. Cardiovascular safety review of rofecoxib. http://www.fda.gov/ohrms/dockets/ac/01/briefing/3677b2_06_cardio.pdf. FDA Advisory Committee, Rockville, MD, 2001.

FDA Advisory Committee Meeting. Merck Briefing Document. Feb 2005.

Félétou M, Verbeuren TJ, Vanhoutte PM. Endothelium-dependent contractions in SHR: a tale of prostanoid TP and IP receptors. Br J Pharmacol 2009;156:563–574.

Ferdinandy P, Schulz R, Baxter GF. Interaction of cardiovascular risk factors with myocardial ischemia/reperfusion injury, preconditioning, and postconditioning. Pharmacol Rev 2007;59: 418–458.

Fiedler VB, Mardin M. Effects of nafazatrom and indomethacin on experimental myocardial ischemia in the anesthetized dog. J Cardiovasc Pharmacol 1985;7:983–989.

Fitzgerald GA. Prostaglandins: modulators of inflammation and cardiovascular risk. J Clin Rheumatol 2004;10:S12–S17.

Flynn PJ, Becker WK, Vercellotti GM, Weisdorf DJ, et al. Ibuprofen inhibits granulocyte responses to inflammatory mediators. A proposed mechanism for reduction of experimental myocardial infarct size. Inflammation 1984;8:33–44.

Fox J, Anderson LC, Loew FM, Quimby FW. *Laboratory Animal Medicine*, 2nd ed. San Diego, CA: Academic Press, 2002.

Francois H, Athirakul K, Howell D, Dash R, et al. Prostacyclin protects against elevated blood pressure and cardiac fibrosis. Cell Metab 2005;2:201–207.

Frank MH. Renin in experimental renal hypertension in monkeys. Circ Res 1963;12:241–255.

Frolov RV, Berim IG, Singh S. Inhibition of delayed rectifier potassium channels and induction of arrhythmia: a novel effect of celecoxib and the mechanism underlying it. J Biol Chem 2008;283: 1518–1524.

Fujino T, Nakagawa N, Yuhki K, Hara A, et al. Decreased susceptibility to renovascular hypertension in mice lacking the prostaglandin I2 receptor IP. J Clin Invest 2004;114:805–812.

Gavaghan M. Cardiac anatomy and physiology: a review. AORN J 1998;67:802–822.

Gladding PA, Webster MW, Farrell HB, Zeng IS, et al. The antiplatelet effect of six non-steroidal anti-inflammatory drugs and their pharmacodynamic interaction with aspirin in healthy volunteers. Am J Cardiol 2008;101:1060–1063.

Goldberg JM, Moberg GP. Autonomic control of heart rate in the neonatal rhesus monkey. J Med Primatol 1985;14:19–27.

Goldblatt H, Lynch J, Hanzal RF, Summerville WW. Studies on experimental hypertension: I. The production of persistent elevation of systolic blood pressure by means of renal ischemia. J Exp Med 1934;59:347–379.

Gres P, Schulz R, Jansen J, Umschlag C, et al. Involvement of endogenous prostaglandins in ischemic preconditioning in pigs. Cardiovasc Res 2002;55:626–632.

Gross GJ, Moore J. Effect of COX-1/COX-2 inhibition versus selective COX-2 inhibition on coronary vasodilator responses to arachidonic acid and acetylcholine. Pharmacology 2004;71:135–142.

Gross S, Tilly P, Hentsch D, Vonesch JL, et al. Vascular wall–produced prostaglandin E2 exacerbates arterial thrombosis and atherothrombosis through platelet EP3 receptors. J Exp Med 2007;204:311–320.

Guerguerian AM, Hardy P, Bhattacharya M, Olley P, et al. Expression of cyclooxygenases in ductus arteriosus of fetal and newborn pigs. Am J Obstet Gynecol 1998;179:1618–1626.

Guo Y, Bao W, Wu WJ, Shinmura K, et al. Evidence for an essential role of cyclooxygenase-2 as a mediator of the late phase of ischemic preconditioning in mice. Basic Res Cardiol 2000;95:479–484.

Hammerman H, Kloner RA, Schoen FJ, Brown EJ Jr, et al. Indomethacin-induced scar thinning after experimental myocardial infarction. Circulation 1983;67:1290–1295.

Harding P, Yang XP, Yang J, Shesely E, et al. Gene expression profiling of dilated cardiomyopathy in older male EP4 knockout mice. Am J Physiol Heart Circ Physiol 2010;298:H623–H632.

Hartner A, Cordasic N, Goppelt-Struebe M, Veelken R, et al. Role of macula densa cyclooxygenase-2 in renovascular hypertension. Am J Physiol Renal Physiol 2003;284:F498–F502.

Hearse DJ, Sutherland FJ. Experimental models for the study of cardiovascular function and disease. Pharmacol Res 2000;41:597–603.

Heindl B, Becker BF. Aspirin, but not the more selective cyclooxygenase (COX)-2 inhibitors meloxicam and SC 58125, aggravates postischaemic cardiac dysfunction, independent of COX function. Naunyn Schmiedebergs Arch Pharmacol 2001;363:233–240.

Hennan JK, Huang J, Barrett TD, Driscoll EM, et al. Effects of selective cyclooxygenase-2 inhibition on vascular responses and thrombosis in canine coronary arteries. Circulation 2001;104:820–825.

Heptinstall S, Espinosa DI, Manolopoulos P, Glenn JR, et al. DG-041 inhibits the EP3 prostanoid receptor: a new target for inhibition of platelet function in atherothrombotic disease. Platelets 2008;19:605–613.

Hermann M, Camici G, Fratton A, Hurlimann D, et al. Differential effects of selective cyclooxygenase-2 inhibitors on endothelial function in salt-induced hypertension. Circulation 2003;108:2308–2311.

Hermann M, Shaw S, Kiss E, Camici G, et al. Selective COX-2 inhibitors and renal injury in salt-sensitive hypertension. Hypertension 2005;45:193–197.

Hirsh PD, Hillis LD, Campbell WB, Firth BG, et al. Release of prostaglandins and thromboxane into the coronary circulation in patients with ischemic heart disease. N Engl J Med 1981;304:685–691.

Hishikari K, Suzuki J, Ogawa M, Isobe K, et al. Pharmacological activation of the prostaglandin E2 receptor EP4 improves cardiac function after myocardial ischaemia/reperfusion injury. Cardiovasc Res 2009;81:123–132.

Höcherl K, Dreher F, Kurtz A, Bucher M. Cyclooxygenase-2 inhibition attenuates lipopolysaccharide-induced cardiovascular failure. Hypertension 2002a;40:947–953.

Höcherl K, Endemann D, Kammerl MC, Grobecker HF, et al. Cyclo-oxygenase-2 inhibition increases blood pressure in rats. Br J Pharmacol 2002b;136:1117–1126.

Hohlfeld T, Strobach H, Schrör K. Stimulation of endogenous prostacyclin protects the reperfused pig myocardium from ischemic injury. J Pharmacol Exp Ther 1993;264:397–405.

Holt JP. The normal pericardium. Am J Cardiol 1970;26:455–465.

Hong BK, Kwon HM, Lee BK, Kim D, et al. Coexpression of cyclooxygenase-2 and matrix metalloproteinases in human aortic atherosclerotic lesions. Yonsei Med J 2000;41:82–88.

Hong TT, Huang J, Barrett TD, Lucchesi BR. Effects of cyclooxygenase inhibition on canine coronary artery blood flow and thrombosis. Am J Physiol Heart Circ Physiol 2008;294:H145–H155.

Hughes HC. Swine in cardiovascular research. Lab Anim Sci 1986;36:348–350.

Innes IR, Weisman H. Reduction in the severity of myocardial infarction by sulfinpyrazone. Am Heart J 1981;102:153–157.

Inoue H, Taba Y, Miwa Y, Yokota C, et al. Transcriptional and posttranscriptional regulation of cyclooxygenase-2 expression by fluid shear stress in vascular endothelial cells. Arterioscler Thromb Vasc Biol 2002;22:1415–1420.

Inserte J, Molla B, Aguilar R, Traves PG, et al. Constitutive COX-2 activity in cardiomyocytes confers permanent cardioprotection constitutive COX-2 expression and cardioprotection. J Mol Cell Cardiol 2009;46:160–168.

Ito H, Hiroe M, Hirata Y, Fujisaki H, et al. Endothelin ET_A receptor antagonist blocks cardiac hypertrophy provoked by hemodynamic overload. Circulation 1994;89:2198–2203.

Jackson RL, Baker HN, Taunton OD, Smith LC, et al. A comparison of the major apolipoprotein from pig and human high density lipoproteins. J Biol Chem 1973;248:2639–2644.

Jacob S, Laury-Kleintop L, Lanza-Jacoby S. The select cyclooxygenase-2 inhibitor celecoxib reduced the extent of atherosclerosis in apo $E^{-/-}$ mice. J Surg Res 2008;146:135–142.

Jacoby DS, Rader DJ. Renin-angiotensin system and atherothrombotic disease: from genes to treatment. Arch Intern Med 2003;163:1155–1164.

Jalowy A, Schulz R, Dörge H, Behrends M, et al. Infarct size reduction by AT1-receptor blockade through a signal cascade of AT2-receptor activation, bradykinin and prostaglandins in pigs. J Am Coll Cardiol 1998;32:1787–1796.

James MA, Saadeh AM, Jones JV. Wall stress and hypertension. J Cardiovasc Risk 2000;7:187–190.

Jenkins CM, Cedars A, Gross RW. Eicosanoid signalling pathways in the heart. Cardiovasc Res 2009;82:240–249.

Jennings RB, Sommers HM, Smyth GA, Flack HA, et al. Myocardial necrosis induced by temporary occlusion of a coronary artery in the dog. Arch Pathol 1960;70:68–78.

Jerome WG, Lewis JC. Early atherogenesis in White Carneau pigeons: II. Ultrastructural and cytochemical observation. Am J Pathol 1985;119:210–222.

Johnson P, Maxwell DJ, Tynan MJ, Allan LD. Intracardiac pressures in the human fetus. Heart 2000;84:59–63.

Joubert KE. The effects of firocoxib (Previcox) in geriatric dogs over a period of 90 days. J S Afr Vet Assoc 2009;80:179–184.

Jugdutt BI, Hutchins GM, Bulkley BH, Pitt B, et al. Effect of indomethacin on collateral blood flow and infarct size in the conscious dog. Circulation 1979;59:734–743.

Kawabe J, Yuhki K, Okada M, Kanno T, et al. Prostaglandin I2 promotes recruitment of endothelial progenitor cells and limits vascular remodeling. Arterioscler Thromb Vasc Biol 2010;30:464–470.

Kawada N, Solis G, Ivey N, Connors S, et al. Cyclooxygenase-1-deficient mice have high sleep-to-wake blood pressure ratios and renal vasoconstriction. Hypertension 2005;45:1131–1138.

Kawka DW, Ouellet M, Hetu PO, Singer II, et al. Double-label expression studies of prostacyclin synthase, thromboxane synthase and COX isoforms in normal aortic endothelium. Biochim Biophys Acta 2007;1771:45–54.

Kay-Mugford PA, Benn SJ, LaMarre J, Conlon PD. Cyclooxygenase expression in canine platelets and Madin–Darby canine kidney cells. Am J Vet Res 2000;61:1512–1516.

Kearney D, Byrne A, Crean P, Cox D, et al. Optimal suppression of thromboxane A(2) formation by aspirin during percutaneous transluminal coronary angioplasty: no additional effect of a selective cyclooxygenase-2 inhibitor. J Am Coll Cardiol 2004;43:526–531.

Kentsch M, Ludwig D, Drummer C, Gerzer R, et al. Haemodynamic and renal effects of urodilatin bolus injections in patients with congestive heart failure. Eur J Clin Invest 1992;22:662–669.

Kim JW, Baek BS, Kim YK, Herlihy JT, et al. Gene expression of cyclooxygenase in the aging heart. J Gerontol A Biol Sci Med Sci 2001;56:B350–B355.

Kirtikara K, Raghow R, Laulederkind SJ, Goorha S, et al. Transcriptional regulation of cyclooxygenase-2 in the human microvascular endothelial cell line, HMEC-1: control by the combinatorial actions of AP2, NF-IL-6 and CRE elements. Mol Cell Biochem 2000;203:41–51.

Kiserud T, Acharya G. The fetal circulation. Prenat Diagn 2004;24:1049–1059.

Kobayashi T, Tahara Y, Matsumoto M, Iguchi M, et al. Roles of thromboxane A(2) and prostacyclin in the development of atherosclerosis in apoE-deficient mice. J Clin Invest 2004;114:784–794.

Koch-Weser J. The therapeutic challenge of systolic hypertension. N Engl J Med 1973;289:481–483.

Koki A, Khan NK, Woerner BM, Dannenberg AJ, et al. Cyclooxygenase-2 in human pathological disease. Adv Exp Med Biol 2002;507:177–184.

Kowala MC, Mazzucco CE, Hartl KS, Seiler SM, et al. Prostacyclin agonists reduce early atherosclerosis in hyperlipidemic hamsters. Octimibate and BMY 42393 suppress monocyte chemotaxis, macrophage cholesteryl ester accumulation, scavenger receptor activity, and tumor necrosis factor production. Arterioscler Thromb 1993;13:435–444.

Krotz F, Schiele TM, Klauss V, Sohn HY. Selective COX-2 inhibitors and risk of myocardial infarction. J Vasc Res 2005;42:312–324.

Kunori Y, Muroga Y, Iidaka M, Mitsuhashi H, et al. Species differences in angiotensin II generation and degradation by mast cell chymases. J Recept Signal Transduct Res 2005;25:35–44.

Kurtz TW, Griffin KA, Bidani AK, Davisson RL, et al. Recommendations for blood pressure measurement in humans and experimental animals: Part 2. Blood pressure measurement in experimental animals: a statement for professionals from the Subcommittee of Professional and Public Education of the American Heart Association Council on High Blood Pressure Research. Arterioscler Thromb Vasc Biol 2005;25:e22–e33.

Lal A, Sharma ML. Failure to reduce experimental myocardial infarct size with ibuprofen pre-treatment in rats. Indian J Physiol Pharmacol 1992;36:133–134.

LaPointe MC, Mendez M, Leung A, Tao Z, et al. Inhibition of cyclooxygenase-2 improves cardiac function after myocardial infarction in the mouse. Am J Physiol Heart Circ Physiol 2004;286: H1416–H1424.

Lee RT, Bloch KD, Pfeffer JM, Pfeffer MA, et al. Atrial natriuretic factor gene expression in ventricles of rats with spontaneous biventricular hypertrophy. J Clin Invest 1988;81:431–434.

Lehoux S, Castier Y, Tedgui A. Molecular mechanisms of the vascular responses to haemodynamic forces. J Intern Med 2006;259:381–392.

Lerman LO, Chade AR, Sica V, Napoli C. Animal models of hypertension: an overview. J Lab Clin Med 2005;146:160–173.

Lev M. The normal anatomy of the conduction system in man and its pathology in atrioventricular block. Ann NY Acad Sci 1964;111:817–829.

Lévesque LE, Brophy JM, Zhang B. The risk for myocardial infarction with cyclooxygenase-2 inhibitors: a population study of elderly adults. Ann Intern Med 2005;142:481–489.

Liang YL, Shiel LM, Teede H, Kotsopoulos D, et al. Effects of blood pressure, smoking, and their interaction on carotid artery structure and function. Hypertension 2001;37:6–11.

Libby P. Inflammation in atherosclerosis. Nature 2002;420:868–784.

Lien YH, Hsiang TY, Huang HP. Associations among systemic blood pressure, microalbuminuria and albuminuria in dogs affected with pituitary- and adrenal-dependent hyperadrenocorticism. Acta Vet Scand 2010;52:61.

Liu GS, Stanley AW, Downey J. Cyclooxygenase products are not involved in the protection against myocardial infarction afforded by preconditioning in rabbit: cyclooxygenase pathway's involvement in preconditioning. Am J Cardiovasc Pathol 1992;4:157–164.

Liu SK, Tilley LP, Tappe JP, Fox PR. Clinical and pathologic findings in dogs with atherosclerosis: 21 cases (1970–1983). J Am Vet Med Assoc 1986;189:227–232.

Manning EP, Tardiff JC, Schwartz SD. A model of calcium activation of the cardiac thin filament. Biochemistry 2011;50:7405–7413.

Manohar M, Goetz TE, Griffin R, Sullivan E. Pulmonary vascular pressures of strenuously exercising thoroughbreds after administration of phenylbutazone. Am J Vet Res 1996;57:1354–1358.

Marcus AJ, Broekman MJ, Pinsky DJ. COX inhibitors and thromboregulation. N Engl J Med 2002;347:1025–1026.

Martin M, Meyer-Kirchrath J, Kaber G, Jacoby C, et al. Cardiospecific overexpression of the prostaglandin EP3 receptor attenuates ischemia-induced myocardial injury. Circulation 2005;112: 400–406.

McAdam BF, Catella-Lawson F, Mardini IA, Kapoor S, et al. Systemic biosynthesis of prostacyclin by cyclooxygenase (COX)-2: the human pharmacology of a selective inhibitor of COX-2. Proc Natl Acad Sci USA 1999;96:272–277.

McClelland S, Gawaz M, Kennerknecht E, Konrad CS, et al. Contribution of cyclooxygenase-1 to thromboxane formation, platelet–vessel wall interactions and atherosclerosis in the ApoE null mouse. Atherosclerosis 2009;202:84–91.

McCorry LK. Physiology of the autonomic nervous system. Am J Pharm Educ 2007;71:78.

McGeer PL, McGeer EG, Yasojima K. Expression of COX-1 and COX-2 mRNAs in atherosclerotic plaques. Exp Gerontol 2002;37:925–929.

Meune C, Mourad JJ, Bergmann JF, Spaulding C. Interaction between cyclooxygenase and the renin–angiotensin–aldosterone system: rationale and clinical relevance. J Renin Angiotensin Aldosterone Syst 2003;4:149–154.

Mehta PK, Griendling KK. Angiotensin II cell signaling: physiological and pathological effects in the cardiovascular system. Am J Physiol Cell Physiol 2007;292:C82–C97.

Meyer-Kirchrath J, Martin M, Schooss C, Jacoby C, et al. Overexpression of prostaglandin EP3 receptors activates calcineurin and promotes hypertrophy in the murine heart. Cardiovasc Res 2009;81:310–318.

Mitrovic V, Seferovic PM, Simeunovic D, Ristic AD, et al. Haemodynamic and clinical effects of ularitide in decompensated heart failure. Eur Heart J 2006;27:2823–3282.

Miyazaki M, Sakonjo H, Takai S. Anti-atherosclerotic effects of an angiotensin converting enzyme inhibitor and an angiotensin II antagonist in cynomolgus monkeys fed a high-cholesterol diet. Br J Pharmacol 1999;128:523–529.

Moghadasian MH, Frohlich JJ, McManus BM. Advances in experimental dyslipidemia and atherosclerosis. Lab Invest 2001;81:1173–1183.

Moncada S, Gryglewski R, Bunting S, Vane Jr. An enzyme isolated from arteries transforms prostaglandin endoperoxides to an unstable substance that inhibits platelet aggregation. Nature 1976;263:663–665.

Mori T, Chen YF, Feng JA, Hayashi T, et al. Volume overload results in exaggerated cardiac hypertrophy in the atrial natriuretic peptide knockout mouse. Cardiovasc Res 2004;61:771–779.

Moubayed SP, Heinonen TM, Tardif JC. Anti-inflammatory drugs and atherosclerosis. Curr Opin Lipidol 2007;18:638–644.

Mousa SA, Cooney JM, Thoolen MJ, Timmermans PB. Myocardial cytoprotective efficacy of azapropazone in a canine heart model of regional ischemia and reperfusion. J Cardiovasc Pharmacol 1989;14:542–548.

Mousa SA, Brown R, Thoolen MJ, Smith RD. Evaluation of the effect of azapropazone on neutrophil migration in regional myocardial ischaemia/reperfusion injury in rabbits. Br J Pharmacol 1990;100:379–382.

Mukoyama M, Nakao K, Hosoda K, Suga S, et al. Brain natriuretic peptide as a novel cardiac hormone in humans. Evidence for an exquisite dual natriuretic peptide system, atrial natriuretic peptide and brain natriuretic peptide. J Clin Invest 1991;87:1402–1142.

Murata T, Ushikubi F, Matsuoka T, Hirata M, et al. Altered pain perception and inflammatory response in mice lacking prostacyclin receptor. Nature 1997;388:678–682.

Murphy PJ. The fetal circulation. Contin Educ Anaesth Crit Care Pain 2005;5:107–112.

Muscara MN, Vergnolle N, Lovren F, Triggle CR, et al. Selective cyclo-oxygenase-2 inhibition with celecoxib elevates blood pressure and promotes leukocyte adherence. Br J Pharmacol 2000;129:1423–1430.

Nunez DJ, Dickson MC, Brown MJ. Natriuretic peptide receptor mRNAs in the rat and human heart. J Clin Invest 1992;90:1966–1971.

Okamoto T, Hino O. Expression of cyclooxygenase-1 and -2 mRNA in rat tissues: tissue-specific difference in the expression of the basal level of mRNA. Int J Mol Med 2000;6:455–457.

Olesen M, Kwong E, Meztli A, Kontny F, et al. No effect of cyclooxygenase inhibition on plaque size in atherosclerosis-prone mice. Scand Cardiovasc J 2002;36:362–367.

Palaniyandi SS, Nagai Y, Watanabe K, Ma M, et al. Chymase inhibition reduces the progression to heart failure after autoimmune myocarditis in rats. Exp Biol Med (Maywood) 2007;232:1213–1221.

Parente L, Perretti M. Advances in the pathophysiology of constitutive and inducible cyclooxygenases: two enzymes in the spotlight. Biochem Pharmacol 2003;65:153–159.

Poletto R, Janczak AM, Marchant-Forde RM, Marchant-Forde JN, et al. Identification of low and high frequency ranges for heart rate variability and blood pressure variability analyses using pharmacological autonomic blockade with atropine and propranolol in swine. Physiol Behav 2011;103:188–196.

Potter CM, Lundberg MH, Harrington LS, Warboys CM, et al. Role of shear stress in endothelial cell morphology and expression of cyclooxygenase isoforms. Arterioscler Thromb Vasc Biol 2011;31:384–391.

Pratico D, Tillmann C, Zhang ZB, Li H, et al. Acceleration of atherogenesis by COX-1-dependent prostanoid formation in low density lipoprotein receptor knockout mice. Proc Natl Acad Sci USA 2001;98:3358–3363.

Prejbisz A, Lenders JW, Eisenhofer G, Januszewicz A. Cardiovascular manifestations of phaeochromocytoma. J Hypertens 2011;29(11): 2049–2060.

Qian JY, Harding P, Liu Y, Shesely E, et al. Reduced cardiac remodeling and function in cardiac-specific EP4 receptor knockout mice with myocardial infarction. Hypertension 2008;51:560–566.

Radi ZA. Pathophysiology of cyclooxygenase inhibition in animal models. Toxicol Pathol 2009;37:34–46.

Radi ZA, Ackermann MR. Ontogeny of pulmonary cyclooxygenase-1 (COX-1) and -2 (COX-2). Pathophysiology 2011;18:215–219.

Radi ZA, Khan NK. Effects of cyclooxygenase inhibition on the gastrointestinal tract. Exp Toxicol Pathol 2006;58:163–173.

Radi ZA, Ostroski R. Pulmonary and cardiorenal cyclooxygenase-1 (COX-1), -2 (COX-2), and microsomal prostaglandin E synthase-1 (mPGES-1) and -2 (mPGES-2) expression in a hypertension model. Mediators Inflamm 2007;2007:85091.

Raval M, Frank PG, Laury-Kleintop L, Yan G, et al. Celecoxib combined with atorvastatin prevents progression of atherosclerosis. J Surg Res 2010;163:e113–e122.

Reece WO. *Functional Anatomy and Physiology of Domestic Animals*, 4th ed. Ames, IA: Iowa State University Academic Press, 2007.

Renda G, Tacconelli S, Capone ML, Sacchetta D, et al. Celecoxib, ibuprofen, and the antiplatelet effect of aspirin in patients with osteoarthritis and ischemic heart disease. Clin Pharmacol Ther 2006;80:264–274.

Richter CM, Godes M, Wagner C, Maser-Gluth C, et al. Chronic cyclooxygenase-2 inhibition does not alter blood pressure and kidney function in renovascular hypertensive rats. J Hypertens 2004;22:191–198.

Ross R. Rous-Whipple Award Lecture. Atherosclerosis: a defense mechanism gone awry. Am J Pathol 1993;143:987–1002

Rossi P, Kuukasjarvi P, Riutta A, Salenius JP, et al. Prostacyclin and thromboxane A2 synthesis are increased in acute lower limb ischaemia. Prostaglandins Leukot Essent Fatty Acids 1996;55:433–436.

Rossoni G, Muscara MN, Cirino G, Wallace JL. Inhibition of cyclo-oxygenase-2 exacerbates ischaemia-induced acute myocardial dysfunction in the rabbit. Br J Pharmacol 2002;135:1540–1546.

Rott D, Zhu J, Burnett MS, Zhou YF, et al. Effects of MF-tricyclic, a selective cyclooxygenase-2 inhibitor, on atherosclerosis progression and susceptibility to cytomegalovirus replication in apolipoprotein-E knockout mice. J Am Coll Cardiol 2003;41:1812–1819.

Rudolph AM. Distribution and regulation of blood flow in the fetal and neonatal lamb. Circ Res 1985;57:811–821.

Ruskoaho H, Lang RE, Toth M, Ganten D, et al. Release and regulation of atrial natriuretic peptide (ANP). Eur Heart J 1987;8:99–109.

Saito T, Rodger IW, Hu F, Shennib H, et al. Inhibition of cyclooxygenase-2 improves cardiac function in myocardial infarction. Biochem Biophys Res Commun 2000;273:772–775.

Saito T, Rodger IW, Shennib H, Hu F, et al. Cyclooxygenase-2 (COX-2) in acute myocardial infarction: cellular expression and use of selective COX-2 inhibitor. Can J Physiol Pharmacol 2003;81:114–119.

Saito T, Rodger IW, Hu F, Robinson R, et al. Inhibition of COX pathway in experimental myocardial infarction. J Mol Cell Cardiol 2004;37:71–77.

Sakamoto A, Matsumura J, Mii S, Gotoh Y, et al. A prostaglandin E2 receptor subtype EP4 agonist attenuates cardiovascular depression in endotoxin shock by inhibiting inflammatory cytokines and nitric oxide production. Shock 2004;22:76–81.

Salloum FN, Hoke NN, Seropian IM, Varma A, et al. Parecoxib inhibits apoptosis in acute myocardial infarction due to permanent coronary ligation but not due to ischemia-reperfusion. J Cardiovasc Pharmacol 2009;53:495–498.

Scheuer J, Buttrick P. The cardiac hypertrophic responses to pathologic and physiologic loads. Circulation 1987;75:I63–168.

Schönbeck U, Sukhova GK, Graber P, Coulter S, et al. Augmented expression of cyclooxygenase-2 in human atherosclerotic lesions. Am J Pathol 1999;155:1281–1291.

Schroeder HA, Goldman ML. Arterial hypertension in dogs: I. Methods. Circulation 1952;5:730–741.

Sellers RS, Radi ZA, Khan NK. Pathophysiology of cyclooxygenases in cardiovascular homeostasis. Vet Pathol 2010;47:601–613.

Senbonmatsu T, Ichihara S, Price E, Gaffney FA, et al. Evidence for angiotensin II type 2 receptor-mediated cardiac myocyte enlargement during in vivo pressure overload. J Clin Invest 2000;106:R25–R29.

Seta F, Chung AD, Turner PV, Mewburn JD, et al. Renal and cardiovascular characterization of COX-2 knockdown mice. Am J Physiol Regul Integr Comp Physiol 2009;296:R1751–R1760.

Shibata R, Sato K, Pimentel DR, Takemura Y, et al. Adiponectin protects against myocardial ischemia–reperfusion injury through AMPK- and COX-2-dependent mechanisms. Nat Med 2005;11: 1096–1103.

Shiels HA, White E. The Frank–Starling mechanism in vertebrate cardiac myocytes. J Exp Biol 2008;211:2005–2013.

Shinmura K, Tang XL, Wang Y, Xuan YT, et al. Cyclooxygenase-2 mediates the cardioprotective effects of the late phase of ischemic preconditioning in conscious rabbits. Proc Natl Acad Sci USA 2000;97:10197–10202.

Shizukuda Y, Miura T, Ishimoto R, Itoya M, et al. Effect of prednisolone on myocardial infarct healing: characteristics and comparison with indomethacin. Can J Cardiol 1991;7:447–454.

Smith CC, Ansevin A. Blood pressure of the normal rhesus monkey. Proc Soc Exp Biol Med 1957;96:428–432.

Smith WL, DeWitt DL, Garavito RM. Cyclooxygenases: structural, cellular, and molecular biology. Annu Rev Biochem 2000;69:145–182.

Smulyan H, Safar ME. Systolic blood pressure revisited. J Am Coll Cardiol 1997;29:1407–1413.

Solomon SD, McMurray JJ, Pfeffer MA, Wittes J, et al. Cardiovascular risk associated with celecoxib in a clinical trial for colorectal adenoma prevention. N Engl J Med 2005;352:1071–1080.

Sowers JR, White WB, Pitt B, Whelton A, et al. The Effects of cyclooxygenase-2 inhibitors and nonsteroidal anti-inflammatory therapy on 24-hour blood pressure in patients with hypertension, osteoarthritis, and type 2 diabetes mellitus. Arch Intern Med 2005;165:161–168.

Srisawat S, Phivthong-Ngam L, Unchern S, Chantharaksri U, et al. Improvement of vascular function by chronic administration of a cyclo-oxygenase inhibitor in cholesterol-fed rabbits. Clin Exp Pharmacol Physiol 2003;30:405–412.

Stanfield KM, Khan KN, Gralinski MR. Localization of cyclooxygenase isozymes in cardiovascular tissues of dogs treated with naproxen. Vet Immunol Immunopathol 2001;80:309–314.

St. Clair RW. Metabolic changes in the arterial wall associated with atherosclerosis in the pigeon. Fed Proc 1983;42:2480–2485.

Stemme V, Swedenborg J, Claesson H, Hansson GK. Expression of cyclooxygenase-2 in human atherosclerotic carotid arteries. Eur J Vasc Endovasc Surg 2000;2:146–152.

Stepien RL. Feline systemic hypertension: diagnosis and management. J Feline Med Surg 2011;13: 35–43.

Strain RE, Olson WH. Selective damage of large diameter peripheral nerve fibers by compression: an application of Laplace's law. Exp Neurol 1975;47:68–80.

Straino S, Salloum FN, Baldi A, Ockaili RA, et al. Protective effects of parecoxib, a cyclo-oxygenase-2 inhibitor, in postinfarction remodeling in the rat. J Cardiovasc Pharmacol 2007;50:571–577.

Streicher JM, Wang Y. The role of COX-2 in heart pathology. Cardiovasc Hematol Agents Med Chem 2008;6:69–79.

Struijk PC, Mathews VJ, Loupas T, Stewart PA, et al. Blood pressure estimation in the human fetal descending aorta. Ultrasound Obstet Gynecol 2008;32:673–681.

Such L, Morcillo EJ, Fortana A, Alberola A, et al. Effects of anti-inflammatory drugs in a model of acute transmural infarction in the dog. J Pharmacol 1983;14:283–293.

Swindle MM, Horneffer PJ, Gardner TJ, Gott VL, et al. Anatomic and anesthetic considerations in experimental cardiopulmonary surgery in swine. Lab Anim Sci 1986;36:357–361.

Takayama K, Yuhki K, Ono K, Fujino T, et al. Thromboxane A2 and prostaglandin F2alpha mediate inflammatory tachycardia. Nat Med 2005;11:562–566.

Takimoto Y, Aoyama T, Iwanaga Y, Izumi T, et al. Increased expression of cardiotrophin-1 during ventricular remodeling in hypertensive rats. Am J Physiol 2002;282:H896–H901.

Tang EH, Vanhoutte PM. Gene expression changes of prostanoid synthases in endothelial cells and prostanoid receptors in vascular smooth muscle cells caused by aging and hypertension. Physiol Genomics 2008;32:409–418.

Tarnow I, Olsen LH, Kvart C, Hoglund K, et al. Predictive value of natriuretic peptides in dogs with mitral valve disease. Vet J 2009;180:195–201.

Testa M, Rocca B, Spath L, Ranelletti FO, et al. Expression and activity of cyclooxygenase isoforms in skeletal muscles and myocardium of humans and rodents. J Appl Physiol 2007;103:1412–1418.

Thomas DW, Mannon RB, Mannon PJ, Latour A, et al. Coagulation defects and altered hemodynamic responses in mice lacking receptors for thromboxane A2. J Clin Invest 1998;102:1994–2001.

Timmers L, Sluijter JP, Verlaan CW, Steendijk P, et al. Cyclooxygenase-2 inhibition increases mortality, enhances left ventricular remodeling, and impairs systolic function after myocardial infarction in the pig. Circulation 2007;115:326–332.

Tranquilli WJ, Parks CM, Thurmon JC, Benson GJ, et al. Organ blood flow and distribution of cardiac output in nonanesthetized swine. Am J Vet Res 1982;43:895–897.

Uotila P, Saraste A, Vähäsilta T, Kentala E, et al. Stimulated expression of cyclooxygenase-2 in porcine heart after bypass circulation and cardioplegic arrest. Eur J Cardiothorac Surg 2001;20:992–995.

Ushikubi F, Segi E, Sugimoto Y, Murata T, et al. Impaired febrile response in mice lacking the prostaglandin E receptor subtype EP3. Nature 1998;395:281–284.

van der Hoorn JW, Jukema JW, Bekkers ME, Princen HM, et al. Negative effects of rofecoxib treatment on cardiac function after ischemia-reperfusion injury in APOE3Leiden mice are prevented by combined treatment with thromboxane prostanoid-receptor antagonist S18886 (terutroban). Crit Care Med 2008;36:2576–2582.

van de Schans VA, van den Borne SW, Strzelecka AE, Janssen BJ, et al. Interruption of Wnt signaling attenuates the onset of pressure overload-induced cardiac hypertrophy. Hypertension 2007;49:473–480.

Vaneckova I, Skaroupkova P, Dvorak P, Certiková Chábová V, et al. Effects of sodium restriction and cyclooxygenase-2 inhibition on the course of hypertension, proteinuria and cardiac hypertrophy in Ren-2 transgenic rats. Physiol Res 2005;54:17–24.

Vargas Alarcón G. Physiopathogenesis of hypertension. Arch Cardiol Mex 2006;76 Suppl 2:S157–S160.

Wacker MJ, Tehrani RN, Smoot RL, Orr JA. Thromboxane A(2) mimetic evokes a bradycardia mediated by stimulation of cardiac vagal afferent nerves. Am J Physiol Heart Circ Physiol 2002;282:H482–H490.

Wainwright CL, Miller AM, Work LM, Del Soldato P. NCX4016 (NO-aspirin) reduces infarct size and suppresses arrhythmias following myocardial ischaemia/reperfusion in pigs. Br J Pharmacol 2002;135:1882–1888.

Wang D, Patel VV, Ricciotti E, Zhou R, et al. Cardiomyocyte cyclooxygenase-2 influences cardiac rhythm and function. Proc Natl Acad Sci USA 2009;106:7548–7552.

Warren RJ, Ebert DL, Mitchell A, Barter PJ. Rabbit hepatic lipase cDNA sequence: Low activity is associated with low messenger RNA levels. J Lipid Res 1991;32:1333–1339.

Weaver ME, Pantely GA, Bristow JD, Ladley HD. A quantitative study of the anatomy and distribution of coronary arteries in swine in comparison with other animals and man. Cardiovasc Res 1986;20:907–917.

Whelton A, Fort JG, Puma JA, Normandin D, et al. Cyclooxygenase-2–specific inhibitors and cardiorenal function: a randomized, controlled trial of celecoxib and rofecoxib in older hypertensive osteoarthritis patients. Am J Ther 2001;8:85–95.

Whelton A, White WB, Bello AE, Puma JA, et al. Effects of celecoxib and rofecoxib on blood pressure and edema in patients > or = 65 years of age with systemic hypertension and osteoarthritis. Am J Cardiol 2002;90:959–963.

White WB, Faich G, Whelton A, Maurath C, et al. Comparison of thromboembolic events in patients treated with celecoxib, a cyclooxygenase-2 specific inhibitor, versus ibuprofen or diclofenac. Am J Cardiol 2002;89:425–230.

White WB, West CR, Borer JS, Gorelick PB, et al. Risk of cardiovascular events in patients receiving celecoxib: a meta-analysis of randomized clinical trials. Am J Cardiol 2007;99:91–98.

Widlansky ME, Price DT, Gokce N, Eberhardt RT, et al. Short- and long-term COX-2 inhibition reverses endothelial dysfunction in patients with hypertension. Hypertension 2003;42:310–315.

Wilcox JN, Augustine A, Goeddel DV, Lowe DG. Differential regional expression of three natriuretic peptide receptor genes within primate tissues. Mol Cell Biol 1991;11:3454–3462.

Wilner KD, Rushing M, Walden C, Adler R, et al. Celecoxib does not affect the antiplatelet activity of aspirin in healthy volunteers. J Clin Pharmacol 2002;42:1027–1030.

Wolf MJ, Amrein H, Izatt JA, Choma MA, et al. *Drosophila* as a model for the identification of genes causing adult human heart disease. Proc Natl Acad Sci USA 2006;103:1394–1399.

Wong E, Huang JQ, Tagari P, Riendeau D. Effects of COX-2 inhibitors on aortic prostacyclin production in cholesterol-fed rabbits. Atherosclerosis 2001;157:393–402.

Woodard GE, Rosado JA. Natriuretic peptides in vascular physiology and pathology. Int Rev Cell Mol Biol 2008;268:59–93.

Wu D, Mennerich D, Arndt K, Sugiyama K, et al. The effects of microsomal prostaglandin E synthase-1 deletion in acute cardiac ischemia in mice. Prostaglandins Leukot Essent Fatty Acids 2009;81:31–33.

Xiao CY, Yuhki K, Hara A, Fujino T, et al. Prostaglandin E2 protects the heart from ischemia–reperfusion injury via its receptor subtype EP4. Circulation 2004;109:2462–2468.

Xuan YT, Guo Y, Zhu Y, Wang OL, et al. Role of the protein kinase C-epsilon-Raf-1-MEK-1/2-p44/42 MAPK signaling cascade in the activation of signal transducers and activators of transcription 1 and 3 and induction of cyclooxygenase-2 after ischemic preconditioning. Circulation 2005;112:1971–1978.

Yang HM, Kim HS, Park KW, You HJ, et al. Celecoxib, a cyclooxygenase-2 inhibitor, reduces neointimal hyperplasia through inhibition of Akt signaling. Circulation 2004;110:301–308.

Yano T, Fujioka D, Saito Y, Kobayashi T, et al. Group V secretory phospholipase A2 plays a pathogenic role in myocardial ischaemia–reperfusion injury. Cardiovasc Res 2011;90:335–343.

Zadelaar S, Kleemann R, Verschuren L, de Vries–Van der Weij J, et al. Mouse models for atherosclerosis and pharmaceutical modifiers. Arterioscler Thromb Vasc Biol 2007;27:1706–1721.

Zhao H, Hiroi T, Hansen BS, Rade JJ. Cyclic stretch induces cyclooxygenase-2 gene expression in vascular endothelial cells via activation of nuclear factor kappa-beta. Biochem Biophys Res Commun 2009;389:599–601.

Zidar N, Dolenc-Strazar Z, Jeruc J, Jerse M, et al. Expression of cyclooxygenase-1 and cyclooxygenase-2 in the normal human heart and in myocardial infarction. Cardiovasc Pathol 2007;16:300–304.

Zidar N, Odar K, Glavac D, Jerse M, et al. Cyclooxygenase in normal human tissues: Is COX-1 really a constitutive isoform, and COX-2 an inducible isoform? J Cell Mol Med 2009;13:3753–3763.

Zimmerman J. The functional and surgical anatomy of the heart. Ann R Coll Surg Engl 1966;39:348–366.

CHAPTER 5

OCULAR SYSTEM

INTRODUCTION

Ocular tissues respond to physiological and pathophysiological stimuli via activation of phospholipases and subsequent release of biologically active metabolites such as prostaglandins (PGs) from membrane phospholipids (Bazan et al., 1997; Radi and Render, 2008). PGs have receptors localized in a variety of ocular tissues, and affect the physiology and pathophysiology of the eye (Hardy et al., 1994; Radi and Render, 2008). Cyclooxygenase isoenzymes (COX-1 and -2) catalyze the conversion of arachidonic acids (AAs) to PGs (Fig. 5-1) (Radi and Render, 2008). The COX-1 isoenzyme is expressed in a variety of ocular tissues, and this expression is variable among different animal species (Radi and Render, 2008). In inflamed tissues, eicosanoids are thought to play a significant role in angiogenesis (Yamada et al., 1999). COX-2 and vascular endothelial growth factor (VEGF) affect vascular permeability and angiogenesis and may be induced by hyperglycemia and tissue hypoxia (Bonazzi et al., 2000; Wilkinson-Berka, 2004). COX-2 inhibition has been associated with attenuating angiogenesis and down-regulating VEGF and bFGF (Masferrer et al., 1999). Among their many uses, NSAIDs are employed for various ophthalmic indications. In this chapter we discuss the expression of COX-1 and COX-2 in the ocular system, the role of prostaglandins and COX in ocular physiology and pathophysiology, and the effects of COX-2 s-NSAIDs and ns-NSAIDs in the treatment of ophthalmic conditions.

COMPARATIVE PHYSIOLOGICAL AND ANATOMICAL ASPECTS OF THE OCULAR SYSTEM

The ocular (visual) system is one of the most complex organs in the body. Understanding the comparative physiological and anatomical aspects of the ocular system is important for an understanding of the pathophysiological consequences of COX inhibition on this system. The eyes allow people and animals to see and interpret their surroundings by collecting light and focusing it onto the retina. The major layers of the eye are the (1) cornea, (2) anterior chamber, (3) lens, (4) vitreous, (5) retina, and (6) optic nerve. The cornea is a transparent, highly specialized tissue

Comparative Pathophysiology and Toxicology of Cyclooxygenases, First Edition. Zaher A. Radi.

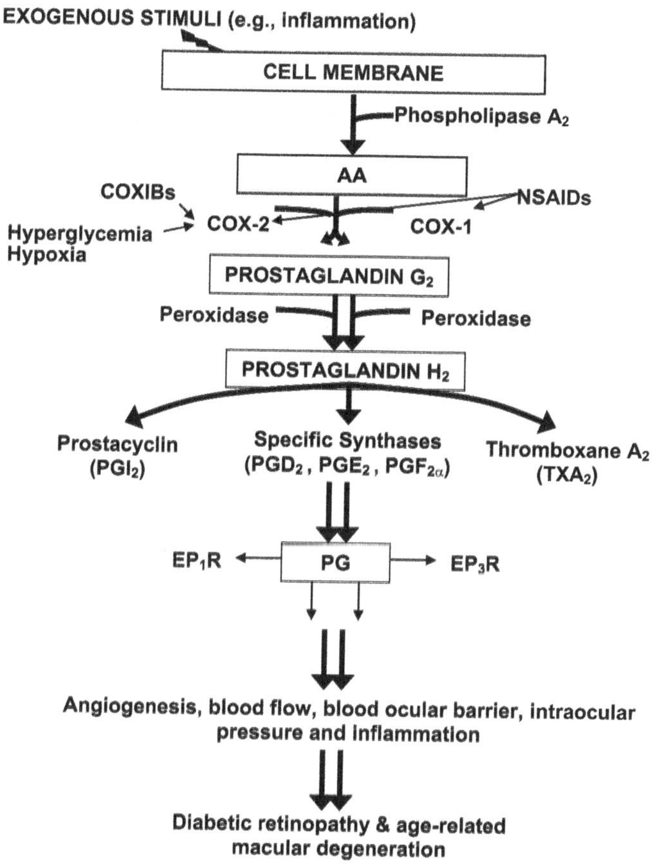

FIGURE 5-1 Role of prostaglandins (PGs) in the eye in health, disease, and the effects of COX-2 s-NSAIDs and ns-NSAIDs. Following an exogenous stimulus (e.g., inflammation), cell membrane phospholipid is liberated to arachidonic acid (AA) by phospholipase A_2. Both COX-1 and COX-2 catalyze the conversion of AA into various PGs. COX-1 is the predominant isoform in the normal eye that is expressed in a variety of ocular tissues, while COX-2 expression is up-regulated during pathological conditions (e.g., hyperglycemia and hypoxia). Nonselective NSAIDs (e.g., diclofenac, fluriprofen, ketorolac, indomethacin) inhibit COX-1 and COX-2, while COX-2 s-NSAIDs (e.g., Celecoxib, Rofecoxib) spare COX-1 and inhibit COX-2. Hyperglycemia and hypoxia can induce COX-2 and participate in vascular permeability. PGs can alter the blood–ocular barrier, affect intraocular pressure, produce miosis and conjunctival hyperemia and retinal and corneal angiogenesis, affect the integrity of the retinal blood–aqueous barrier, and mediate ocular allergic reactions and pain responses. [Reprinted with permission from Z. A. Radi and J. A. Render, *Ocular Pharmacology and Therapeutics*, 24(2), pp. 141–151. Copyright © 2008 Mary Ann Liebert, Inc. Publications. All rights reserved.]

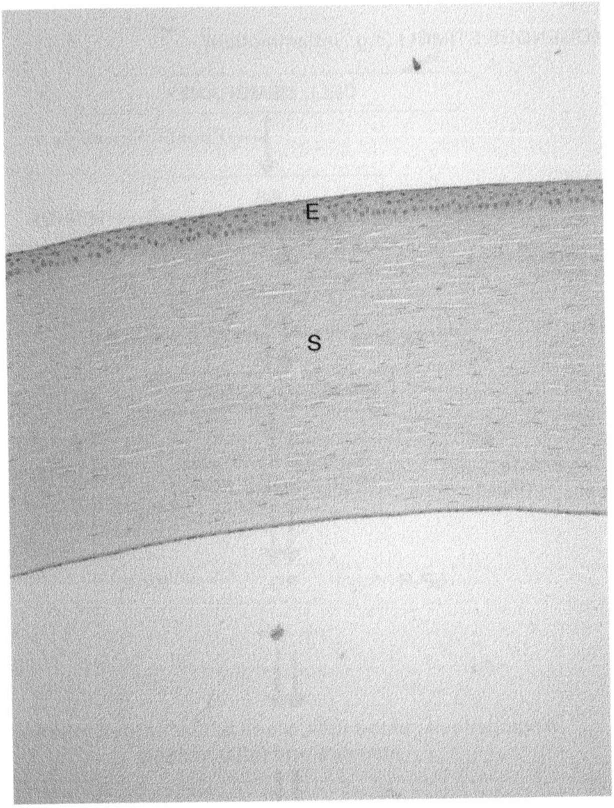

FIGURE 5-2 Normal nonhuman primate cornea anatomic structures. Note epithelium (E) and stroma (S). Hematoxylin and eosin stain, original magnification ×10.

that must maintain transparency to visible light, regularly refract visible light and resist adverse external forces such as dehydration, microbial invasion, and trauma. Structurally, the cornea is an avascular (devoid of blood vessels) convex uniform nonpigmented, nonkeratinized tissue composed histologically of five layers from the outside to the inside: the epithelium, Bowman's layer, the stroma, Descemet's membrane, and the endothelium (Fig. 5-2). The cornea receives nutrition from tears, aqueous humor, and capillaries in the limbus. The corneal epithelium is the barrier of the cornea to the exterior and is a tight epithelium with a relatively low ionic conductance (Ehlers and Hjortdal, 2005). The anterior part of the stroma, known as Bowman's membrane, is acellular, and by light microscopy it appears homogeneous. Bowman's membrane is transparent, continuous, and homogeneous and is prominent in human and nearly all nonhuman primate (except the lemur) corneas (Merindano et al., 2002). Bowman's layer is absent in all carnivores (Merindano et al., 2002). The fibrils in Bowman's membrane are thinner (20 nm) than those of the stroma, and accordingly, show a higher type V/type I collagen ratio (Ehlers and Hjortdal, 2005). Corneal stroma comprises nearly 90% of corneal thickness, is very poorly cellular, and is composed predominantly of highly ordered extracellular type I collagen fibrils with interspersed cells, the keratocytes. Keratocytes are the predominant cell type in the stroma and play a role in maintaining its organization. These stellar-shaped cells contact each other by long cytoplasmic extensions (morphologic and functional syncytium) and also interact with the corneal epithelium (Willoughby et al., 2010). Type I collagen fibrils are uniform in size and spacing, specialization that allows the transmission of visible light. The endothelium and Descemet's membrane regulate the hydration status of the stroma, and the epithelium and Bowman's layer provide protection against the environment (Holland and Mannis, 2002). A large decrease in central endothelial cell density occurs during maturation of the cornea in several species (Bahn et al., 1986). During the first months of life, central endothelial cell counts of developing corneas of cat, dog, and rabbit decrease rapidly. This rapid decrease in endothelial cell density correlates with growth of the cornea to adult size. Despite a wide variation in corneal size, central endothelial cell counts of adult cat, cow, deer, dog, pig, rabbit, and human corneas were similar (2500 cells/mm^2). Hypertrophy of individual cells is primarily responsible for this adult cell density of 2500 cells/mm^2 for these species (Bahn et al., 1986).

Embryologically, the human cornea can be identified at an early stage of development (week 6) (Ehlers and Hjortdal, 2005). At birth the cornea is relatively large compared to the eyeball. The cornea is one of the most densely innervated tissues in the body. It is estimated that there are approximately 7000 nocireceptors/mm^2 in the human cornea (Ehlers and Hjortdal, 2005). A tear film covers the surface of the epithelium. This thin layer of fluid maintains the optical quality of the cornea (Ehlers and Hjortdal, 2005). The cornea may also be divided into a central part and a peripheral or limbal zone (Ehlers and Hjortdal, 2005). The cornea is avascular and the branches of the anterior ciliary arteries stop at the limbus, where they form arcades that supply the peripheral cornea. Thus, the peripheral and central cornea is very distinct in terms of physiology and pathology (Willoughby et al., 2010). The epithelium and endothelium utilize most of the corneal oxygen and receive the most oxygen from the atmosphere. Oxygen diffuses through the tear film, and aerobic metabolism expels CO_2 into the atmosphere. Thus, the cornea is metabolically active and has three exchange routes for metabolites and accumulated breakdown products: the anterior epithelial surface, the posterior endothelial surface, and the peripheral limbus (Ehlers and Hjortdal, 2005). The cornea is the main route for drug transport to the anterior chamber (Willoughby et al., 2010). The abundant presence of hydrated collagen in the stroma may hamper the diffusion of highly lipophilic drugs (Willoughby et al., 2010). The endothelium is more permeable and allows the passage of hydrophilic drugs and macromolecules between the aqueous layer and the stroma due to the presence of "leaky tight junctions" called desmosomes or macula adherens. The passage of topical ocular drugs through the corneal route depends on their lipophilicity, molecular weight, charge, and degree of ionization. Small lipophilic drugs in particular can permeate easily through the cornea (Willoughby et al., 2010). There are species differences in corneal anatomy and physiology. The relative size of the cornea varies among animal species living under different conditions, being generally large in night-living (nocturnal) animals and small in those active in daylight. Corneal thickness is different across species. Corneal thickness in various species is as follows: human (mean central corneal thickness is 0.548 ± 0.037 mm) (Rüfer et al., 2009), rhesus nonhuman primate (mean central corneal thickness is 0.468 ± 0.02 mm) (Ollivier et al., 2003), cynomolgus nonhuman primate (0.4 mm) (Ehlers and Hjortdal, 2005), horse (0.786 mm) (Plummer et al., 2003), dog (0.562 ± 0.06 mm) (Gilger et al., 1993), cat (0.578 ± 0.06 mm) (Gilger et al., 1993), Fisher 344 rat (0.15 ± 0.02 mm) (Tuft et al., 1986), ox (0.8 mm) (Ehlers and Hjortdal, 2005), rabbit (0.4 mm) (Ehlers and Hjortdal, 2005), and C57BL/6J mouse (0.083 mm) (Dimasi et al., 2011). Corneal thickness is of importance for the estimation of intraocular pressure (IOP) and in the diagnosis of corneal and systemic disorders. Pathological conditions can affect corneal thickness. Sustained ocular hypertension in humans causes an increase in corneal thickness (Ventura et al., 2001). Increased corneal thickness can adversely affect IOP measurements (Brusini et al., 2000). Corneal endothelium was evaluated in patients with diabetes mellitus types 1 and 2. There was a statistically significant decrease in endothelial cell density (ECD) in type 1 disease compared with healthy subjects. Corneas from diabetic patients were thicker than normal and the HbA1c level was correlated inversely with the ECD and the mean endothelial cell area. In addition, significant correlation was observed between the endothelial morphology and grade of diabetic retinopathy (Módis et al., 2010).

The cornea is highly innervated and heals rapidly without scarring. Corneal wound healing is a complex physiological process and plays a critical role in the maintenance of corneal structural integrity and clarity. Several growth factors and cytokines are released following an epithelial injury in the cornea and contribute to corneal healing pathophysiology. These factors play essential roles in epithelial–stromal interaction leading to successful corneal wound healing. Growth factors such as keratinocyte growth factor (KGF) and hepatocyte growth factor (HGF) are believed to be produced by keratocytes, while IL-1 and platelet-derived growth factor (PDGF) are secreted by the epithelium. Other growth factors, such as the epidermal growth factor (EGF) family, insulin-like growth factor (IGF), and transforming growth factor beta (TGFβ), regulate both the epithelium and stroma, and the crosstalk among various growth factors determines the outcome of a corneal epithelial wound (Yu et al., 2010). EGFR is generated in response to cell injury to intracellular signaling pathways, particularly extracellular signal-regulated kinase (ERK) and phosphatidylinositol 3-kinase (PI_3K), and regulates corneal epithelial wound healing. In cultured porcine corneas, inhibition of ERK and PI_3K pathways impaired ex vivo epithelial wound healing (Xu et al., 2004). Other growth factors, such as insulin, IGF, and HGF, are known to transactivate EGFR (Yu et al., 2010). Cellular components such as ATP and lysophosphatidic acid (LPA) released from injured cells act as "alarmins" to initiate cell response by transactivating EGFR and to signal potential further damage to the cornea, such as infection. Thus, EGFR represents a pivotal point of cell signaling accessible to a variety of stimuli in response to pathophysiological challenge in the cornea (Yu et al., 2010). There are species differences in reactions to corneal injury and in the healing process. The horse had the longest healing time, whereas cow corneal healing is rapid. Corneal neovascularization occurs following corneal trauma or chemical injury. It has been shown that vascular endothelial growth factor (VEGF) is an important mediator of angiogenesis in a rat model of corneal neovascularization (Edelman et al., 1999). Vascularization can enhance the corneal healing process and increase the likelihood of scarring.

The anterior chamber (AC) of the eye is the front part of the eye between the cornea and the iris. The AC is filled with aqueous humor, an optically clear, colorless, watery, and slightly alkaline fluid. The aqueous humor (AH) provides nutrition to the cornea and lens and removes waste products. Several immunosuppressive substances are present within the normal aqueous humor, such as TGFβ, which inhibits antigen processing and presentation, and suppresses both T-lymphocyte activation and certain aspects of nonspecific inflammation (Streilein and Cousins, 1990). Recent data show that AH suppressed lineage commitment and acquisition of the Th1 and Th17 effector function of naive T cells, manifested as a reduction in lineage-specific transcription factors and cytokines. AH promoted its massive conversion to $Foxp3^{+}$ regulatory T-cells (Tregs), which expressed CD25, glucocorticoid-induced tumor necrosis factor receptor family-related gene (GITR), cytotoxic T-lymphocyte antigen 4 (CTLA-4), and CD103 and were functionally suppressive. TGFβ and retinoic acid were both needed and synergized for Treg conversion by AH, with TGFβ-enhancing T-cell expression of retinoic acid receptor α (Zhou et al., 2011). Other immunosuppressive factors found in the low-molecular-weight fraction (< 5 kDa) of aqueous humor include neuropeptide α-melanocyte-stimulating hormone (MSHα), which suppresses immunogenic inflammation, neuropeptide vasoactive intestinal peptide (VIP), and somatostatin (Taylor et al., 1994; Taylor and Yee, 2003). There is a delicate balance between the production and drainage of the aqueous humor, which maintains the normal physiological range of intraocular pressure at 10 to 20 mmHg (Philipp et al., 1990). In certain pathological conditions, aqueous humor accumulates within the eye and IOP increases, which can lead to glaucoma (i.e., increased pressure inside the eye due to the decreased aqueous humor outflow) (Philipp et al., 1990). Nonhuman primates have been used as a model for ocular hypertension, which leads to optic neuropathy, which closely reflects the optic neurodegeneration associated with human glaucoma (Rasmussen and Kaufman, 2005). Gluacoma is one of the most frequent ophthalmic diseases in dogs and is characterized by high IOP (> 25 to 30 mmHg in dogs and > 31 mmHg in cats).

The lens is a biconvex transparent structure that lies posterior to the iris. It is avascular, lacks a nerve supply, and contains a high concentration of protein, carbonic anhydrase, and glutathione (Friedland and Maren, 1984; Giblin, 2000). The normal human lens contains about 65% water and 35% organic matter, the latter being mainly (greater than 90%) structural proteins (Lerman, 1983). Lens thickness across species is 4 mm (human), 2.98 mm (monkey), 7.85 mm (dog), 8.5 mm (cat), and 3.87 mm (rat) (Vakkur and Bishop, 1963; Massof and Chang, 1972; Lapuerta and Schein, 1995; Mutti et al., 1999). The lens is attached to the ciliary body by zonules. Contraction and relaxation cause changes in the lens configuration, increasing and decreasing its focusing power. This process called accommodation is necessary in focusing on both near and distant objects. The lens of a human eye is only twice as thick as that of the mouse (Zhou et al.,

2007). Glutathione functions as an essential antioxidant vital for maintenance of the lens's transparency (Giblin, 2000). The lens is surrounded by a lens capsule. The lens depends on the aqueous humor and vitreous for nutrition and elimination of waste products. Sodium–potassium–adenosine triphosphatase (Na,K-ATPase) regulates electrolyte concentrations in the lens (Tseng and Tang, 1999). Hexokinase enzyme contributes to glucose metabolism in the lens. In the mouse lens, high levels of hexokinase mRNAs were noted relative to aldose reductase mRNAs. This favors metabolism of glucose via the glycolytic pathway rather than the sorbitol pathway in the mouse; while the opposite is true for the rat lens (Wen and Bekhor, 1993). Cataract is a clouding of the lens. The word *cataract* is derived from the Greek word *cataractos,* which means "waterfall." Cataracts are most commonly caused by chemical changes within the lens of the eye. Diabetic cataracts are caused by an elevation of polyols within the lens of the eye catalyzed by the enzyme aldose reductase (Head, 2001). Pathophysiological mechanisms of cataract formation include deficient glutathione levels, which lead to defective antioxidant defense system within the lens. Nutrients that can increase glutathione levels and activity include lipoic acid, vitamins E and C, and selenium. Cataract patients also tend to be deficient in vitamin A and the carotenes lutein and zeaxanthin (Head, 2001). Lens epithelia from patients with senile cataracts have a decreased amount of Na,K-ATPase α-subunit with increased cataract severity (Tseng and Tang, 1999). It has been demonstrated that the rat lens microsomal preparations possess the ability to convert exogenous arachidonic acid into PGE_2 and $PGF_{2\alpha}$ (Keeting et al., 1985).

The vitreous humor is a hydrogel that fills the space between the lens and the retina. More than 95% of the content of the gelatinous vitreous body by weight is water (Kleinberg et al., 2011). The volume of the vitreous in adult mammals varies: 1 to 2 mL in rabbit, 3.2 mL in pig, 3 to 4 mL in monkey, and 4 to 5 mL in humans (Kleinberg et al., 2011). The vitreous of rabbit has a high concentration of total protein (2.2 mg/mL) (Manzanas et al., 1992). Cattle, lamb, pig, and dog contain much less protein in the vitreous than do nonhuman primates (Chen and Chen, 1981). Physiological levels (approximately 100 pg/mL) of PGs were found in vitreous from human eyes undergoing cataract extraction. Eyes with vitreous hemorrhage, retinal detachment, or cystoid macular edema had similarly low levels. Vitreous PGs were mildly elevated in trauma and endophthalmitis and markedly elevated in aphakic bullous keratopathy (Thomas et al., 1985). Transmission electron microscopy of the human vitreous revealed the presence of parallel collagen fibrils packed in bundles and no membraneous structures (Sebag and Balazs, 1989). Thus, the gelatinous nature of the vitreous body is related to hyaluronic acid–collagen complex, with aggregation of collagen fibrils into bundles (Sebag and Balazs, 1989; Kleinberg et al., 2011). The presence of both hyaluronic acid (HA) and collagen determine the viscoelastic properties of the vitreous (Kleinberg et al., 2011). The amount of vitreal HA varies across species. Rhesus monkeys and human vitreal HA concentrations are similar (approximately 200 μg/mL) (Kleinberg et al., 2011). In other species, concentrations range from 20 to 60 μg/mL in the dog, cat, rabbit, and pig (Weber et al., 1982; Kleinberg et al., 2011) to approximately 470 μg/mL in cattle (Gherezghiher et al., 1987). At birth, the human vitreous is entirely gelatinous. However, within the human life span, a linear increase in the volume of the liquid portion of the central vitreous is observed (Kleinberg et al., 2011). In rabbits, a central structure of the vitreous body is defined by a well-demarcated central channel called Cloquet's canal, a remnant of the hyaloid artery (Los et al., 1999). A typical well-defined Cloquet's canal is not normally observed in humans, although a similar arrangement of vitreous fibers is present (Kleinberg et al., 2011). Although the central vitreous is a completely gelatinous structure at birth in primates, a progressive liquefaction develops with aging (Kleinberg et al., 2011). Age-related liquefaction of the vitreous gel has been associated with lens opacification (Harocopos et al., 2004). Rabbits, cats, dogs, sheep, and cattle show relatively small changes with age in the vitreous body (Kleinberg et al., 2011). By contrast, in rhesus monkeys and humans, vitreous liquefies over time, with the portion of the liquefied tissue increasing with age (Denlinger et al., 1980; Los, 2008; Kleinberg et al., 2011). The rabbit is a frequently used animal model for studies on vitreous function and pathobiology (Los et al., 1999). However, the pig and nonhuman primate could be better animal models than the rabbit (Kleinberg et al., 2011). Drugs can be removed from the vitreous through diffusion into the anterior chamber or by the blood–retinal barrier (Willoughby et al., 2010).

The light continues through the vitreous humor back to a clear focus on the retina, behind the vitreous. The retina is the most complex ocular structure and is composed of 10 layers: retinal pigmented epithelium (RPE), inner and outer segments of rods and cones, outer limiting membrane, outer nuclear layer, inner nuclear layer, inner plexiform layer, ganglion cell layer, nerve fiber layer, and inner limiting membrane. The RPE is a monolayer of cuboidal epithelial cells intercalated between the photoreceptors and choriocapillaries (Willoughby et al., 2010). RPE cells rest on the Bruch's membrane, a trilaminar structure consisting of the basement membrane of RPE cells, an intervening elastocollagen layer, and an outer layer derived from the basal layer of the choriocapillaris. This layering is more evident in nonhuman primates than in mice (Zeiss, 2010). Numerous pigment (melanin and lipofuscin) granules are present in the apical cytoplasm of RPE cells. Functions of RPE include the maintenance of photoreceptor function (phagocytosis of photoreceptor waste), retinal adhesion, storage and metabolism of vitamin A, the production of growth factors, and blood–retinal barrier (BRB) function (Willoughby et al., 2010). The retina is exposed to high levels of light throughout the day, and RPE prevent potential noxious effects of such light exposure. Light filtration by RPE pigments functions as a preventive mechanism against photooxidation, and RPE cells can repair considerable damage to DNA, proteins, and lipids (Kevany and Palczewski, 2010). The RPE and the retinal capillary endothelium are the main barriers for systemically administered drugs (Willoughby et al., 2010). The BRB selectively controls the traverse of substances and pharmaceuticals after systemic and periocular administration to the retina (Willoughby et al., 2010). The BRB has a functional outer barrier that is formed by the RPE and an inner barrier formed by the endothelial cells of retinal vessels (Willoughby et al., 2010). Transcellular passive permeation is the main route for the inward/outward traverse of small molecules across the BRB, whereas the paracellular permeability of RPE is quite low. An inverse correlation exists between the molecular weight and permeability (Willoughby et al., 2010). Choroid/PRE-specific PGD_2 synthase gene was found to be expressed in a much higher level in mouse than in human and guinea pig (Zhou et al., 2007).

The process of vision begins with collection of light by photoreceptor cells in the retina (Kevany and Palczewski, 2010). Photoreceptor cells are comprised of two types: rods, which govern vision in low-light settings, and cones, which collect photons in ambient light and discern color differences (Kevany and Palczewski, 2010). In most mammals, rods are the dominant population throughout the retina (Szél et al., 1996). In humans, rods are approximately 20 times more abundant than cones. The average human retina contains 4.6 million cones (4.08 to 5.29 million). Peak foveal cone density averages 199,000 cones/mm^2 and is highly variable between individuals (100,000 to 324,000 cones/mm^2). The average human retina contains 92 million rods (77.9 to 107.3 million) (Curcio et al., 1990). In nonhuman primates (i.e., pigtail macaques), the retina contains an average of 3.1 million cones (2.8 to 3.3 million), with an average peak foveal cone density of 210,000 cones/mm^2 (190,000 to 260,000 cones/mm^2) (Packer et al., 1989). There are three types of photopigments: red, green, and blue. Damage to the fovea and/or macular area causes reduced color and central vision. The photoreceptors are responsible for phototransduction, the conversion of light into an electrical signal (Willoughby et al., 2010). Rod cells are very sensitive to light but have a low photon saturation threshold, whereas cones have an extremely high saturation threshold but a relatively low sensitivity. Each rod and cone cell consists of four distinct components: a synaptic terminal, an outer segment (OS), an inner segment, and a cilium connecting the outer and inner segments (Kevany and Palczewski, 2010). Photoreceptor cells maintain a roughly constant length by continuously generating new outer segments from their base while simultaneously releasing mature outer segments engulfed by the RPE (Kevany and Palczewski, 2010). The density of rods and cones varies between different regions of the retina. In humans, about 50% of the cones are located in the central 30% of the visual field, corresponding roughly with the macula (Willoughby et al., 2010). There are species differences in retinal maturity. In humans, the retina is mature at birth. In dogs, the rod and cone inner and outer segments are first observed microscopically in the third week after birth. Further differentiation of the rod and cone segments and other retinal layers occurs from week 5 to week 8 (Gum et al., 1984). In cats, retinal maturation process is presumably finished between the fourth and fifth months of life (Vogel, 1978).

The macula lutea is an area in the retina between the temporal vascular arcades containing xanthophyll pigment. Histologically, the macula has several layers of ganglion cells, whereas in the surrounding peripheral retina the ganglion cell layer is only one cell thick. The excavation near the center of the macula is called the fovea (Willoughby et al., 2010). The macula mediates high-acuity central vision and is present in primates as well as in some birds and reptiles (Zeiss, 2010). The mouse lacks a macula and has a much lower proportion of cones in its retina (Zeiss, 2010). Age-related macular degeneration (AMD) is a degenerative progressive condition of the retinal pigment epithelium, its supporting basement membrane, and the overlying photoreceptors. AMD is regarded as the leading cause of blindness in humans older than 65 years. AMD results from variable contributions of age, environment, and genetic predisposition (Zeiss, 2010). Normal aging is accompanied by thickening of the Bruch's membrane and minor accumulation within the macula of small discrete sub-RPE deposits known as hard drusen. The biochemical composition of drusen includes complement components such as complement C3 and C5b-9 terminal complexes, immunoglobulins, apolipoproteins B and E, fibrinogen, vitronectin, and amyloid β (Zeiss, 2010). AMD is characterized by fragmentation and calcification of Bruch's membrane, as well as accumulation of soft drusen (Zeiss, 2010). Rodents and nonhuman primates are the primary models in which choroidal neovascularization is induced and antiangiogenic therapies of AMD are tested (Zeiss, 2010). Although the mouse does not have a macula, it is the most commonly utilized animal model of AMD because (1) the histological anatomy of the primary site of AMD (RPE, Bruch's membrane, and choriocapillary layer) is well preserved across species, and (2) the mouse is genetically manipulable and is thus the most practical model in which to test the numerous genetic alterations associated with AMD (Zeiss, 2010). Macular degeneration occurs in rhesus and cynomolgus macaques. In both, retinal lesions are characterized by drusen accumulation in the central retina. However, the onset of fundus changes and inheritance patterns differ in these two primate species. Macular degeneration in the cynomolgus macaque appears to be a model for early-onset maculopathies rather than AMD. In contrast, adult-onset macular degeneration in rhesus macaques resembles human AMD (Zeiss, 2010). COX-2 immunohistochemical expression was evaluated in sections of choroidal neovascular membranes excised from 16 patients with AMD. Increased expression of COX-2 in human choroidal neovascular membranes was noted and may suggest a possible role for cyclooxygenases in age-related macular degeneration pathogenesis (Maloney et al., 2009).

The optic nerve carries electrical signals from the retina to the brain visual cortex region for processing. In the normal adult human optic nerve, there are an estimated 1.2 million axons per nerve (Balazsi et al., 1984). In normal cynomolgus monkeys, the mean total number of fibers in the optic nerve is 1.2 million. The mean diameter of axons is 0.8 μm (Sanchez et al., 1986). The optic nerve is visible on ophthalmic examination and corresponds to the physiologic blind spot. Other anatomical components of the eye are the conjunctiva, meibomian glands, the uveal tract, and the choroid. The conjunctiva is a vascular mucous membrane that lines the upper and lower eyelids. The portion covering the sclera is termed the bulbar conjunctiva, and that covering the inner lid is termed the palpebral conjunctiva. The conjunctiva provides mucous to lubricate the surface, which is secreted by the goblet cells. Membrane-associated mucins such as MUC1 and MUC4 decorate the membranes of human conjunctival epithelial cells (Berry et al., 2000). The conjunctiva contains lymphoid tissue, consisting of intraepithelial lymphocytes, subepithelial lymphoid follicles, and adjacent lymphatics and blood vessels. This conjunctiva-associated lymphoid tissue (CALT) plays a key role in protection of the ocular surface by initiating and regulating immune responses (Steven and Gebert, 2009). Meibomian glands are oil-secreting sebaceous glands composed of secretory acini that are connected via smaller ductules to the larger long straight central duct. The numerous secretory acini, which have an elongated or spherical shape approximately 150 to 200 μm in diameter, are completely filled with secretory cells called meibocytes. The meibomian glands of the human have a dense meshwork of unmyelinated nerve fibers (nerve plexus) around the acini (Knop et al., 2011).

The uveal tract consists of the iris, ciliary body, and the choroid. The uveal tract forms the central vascular bed for the eye. The iris and ciliary body are referred to as the anterior uvea, and the choroid is referred to as the posterior uvea. The uveal tract contains rich networks of both resident macrophages and MHC class II dendritic cells, which act as sentinels for capturing and sampling bloodborne and intraocular antigens. Large numbers of mast cells are present in the choroid of most species but are virtually absent from the anterior uvea in many laboratory animals. However, the human iris does contain mast cells. Small numbers of what are presumed to be trafficking lymphocytes are present in the uveal tract of normal eyes (McMenamin, 1997). The center aperture is the pupil. Through dilation and constriction, the pupil regulates the amount of light sent to the retina. The ciliary body is continuous with the iris and directly adjacent to the sclera. The ciliary processes, which produce aqueous humor, arise from the ciliary body. Fine suspensory ligaments of lens called the zonules of Zinn arise from ciliary processes. These ligaments connect the ciliary body to the lens (Davanger and Ringvold, 1977). The ciliary muscle is composed of radial, longitudinal, and circular fibers. The circular fibers help in contracting and relaxing, which in turn increases and decreases tension on the zonules, leading to lens accommodation (Hiraoka et al., 2010).

ROLE OF PROSTAGLANDINS IN THE OCULAR SYSTEM

Prostaglandins (PGs) are arachidonic acid (AA) metabolites that have receptors localized in a variety of ocular tissues and affect the physiology and pathophysiology of the ocular system (Fig. 5-1 and Table 5-1) (Hardy et al., 1994; Radi and Render, 2008). PGs can lower ocular pain threshold; disrupt the integrity of the retinal blood–aqueous barrier; increase the permeability of the blood–ocular barrier, increase or decrease intraocular pressure (IOP), produce miosis (constriction of the pupil of the eye), cause conjunctival hyperemia, increase retinal and corneal angiogenesis, exacerbate photophobia, be associated with a stinging or burning sensation, cause periocular hyperpigmentation, cause eyelash growth, and/or mediate ocular allergic reactions and pain responses (Kosaka et al., 1995; Damm et al., 2001; Schalmus, 2003; Wilkinson-Berke, 2004; Radi and Render, 2008; Hommer, 2010; Pirie et al., 2011). The IOP-lowering action of several established and potential antiglaucomatous drugs, including latanoprost (Kashiwagi and Tsukahara, 2003), brimonidine (Sponsel et al., 2002), epinephrine (Kaplan-Messas et al., 2003), and cannabinoids (Green et al., 2001), is mediated, at least in part, through the release of endogenous PGs. Decreased outflow of aqueous humor through the trabecular meshwork and Schlemm's canal characterizes primary open-angle glaucoma (OAG), which results in ocular hypertension. Latanoprost, a PG analog, has been used as an IOP agent for glaucoma and for the treatment of OAG in humans because of its ability to cause increased uveoscleral outflow (Camras, 1996; Camras et al.,

TABLE 5-1 Comparative Expression of Various Prostaglandins and their Receptors in the Eye

Ocular compartment	Species	PGs receptor type(s) and subtype(s)	References
Cornea, Epithelium	Pig	EP_1, EP_2, EP_3	Zhao et al., 1995
Endothelium	Rabbit	PGE_2	Joyce et al., 1995
	Pig	EP_2, EP_3	Zhao et al., 1995; Zhao and Shichi, 1995
Sclera	Human	EP_1, EP_2	Anthony et al., 2001
Lens, epithelium	Pig	EP_3	Zhao et al., 1995
Iris Muscle	Human	PGE_2, $PGF_{2\alpha}$	Matsuo and Cynader, 1992
	Pig	EP_1	Zhao et al., 1995
Epithelium	Human	PGE_2, $PGF_{2\alpha}$	Matsuo and Cynader, 1992
Ciliary body Epithelium	Pig	EP_3	Asakura et al., 1992; Zhao et al., 1995
	Human	EP_4	Crider and Sharif, 2001
Muscle	Mouse	PGE_2, EP_1, EP_2, EP_3, EP_4	Takamatsu et al., 2000
	Monkey	PGE_1, PGE_2	Yamaji et al., 2005
	Pig	EP_2	Zhao et al., 1995
	Human	PGE_2, $PGF_{2\alpha}$	Matsuo and Cynader, 1992; 1993
Trabecular meshwork	Human	EP_2, EP_4	Crider and Sharif, 2001; Kamphuis et al., 2004
	Pig	EP_2	Zhao et al., 1995
Retina	Human	PGE_2, $PGF_{2\alpha}$	Matsuo and Cynader, 1992
	Pig	EP_3	Zhao et al., 1995
Blood vessels	Pig	$PGF_{2\alpha}$	Hardy et al., 1998
	Bovine	PGI_2, PGE_2, $PGF_{2\alpha}$	Kulkarni and Payne, 1997
	Rabbit	$PGF_{2\alpha}$	Poyer et al., 1992

Source: Reprinted with permission from Z. A. Radi and J.A. Render, *Ocular Pharmacology and Therapeutics*, 24(2), pp. 141–151.

1996; Mishima et al., 1997; Noecker et al., 2003; Maihofner et al., 2001). Effects of PG analogs on IOP lowering in vivo were typically observed 4 to 24 h after treatment in beagle dogs and cynomolgus monkeys (Reitmair et al., 2010). Conjunctival hyperemia is a common side effect in patients with ocular hypertension given FP receptor agonists and PGs analog (Honrubia et al., 2009).

Different portions of the eye have different capacities to synthesize PGs (Table 5-1) (Kulkarni and Srinivasan, 1989; Radi and Render, 2008). For example, the cornea, lens, and retina have less capacity to synthesize COX products from AA than that of the conjunctiva and anterior uvea (Kulkarni and Srinivasan, 1989). Human retina has considerably less ability to metabolize AA into COX activity

(Kulkarni and Srinivasan, 1989). The anterior uvea of human, monkey, and rabbit can synthesize PGE_3 and PGD_3 (Kulkarni and Srinivasan, 1989). Conjunctiva and eyelid tissues of cynomolgus and rhesus monkeys can synthesize detectable amounts of COX products. In comparison, cynomolgus anterior uvea can synthesize a considerably smaller amount of COX products (Kulkarni et al., 1987). In the ciliary processes of pigs, PGs synthesis occurs in both nonpigmented and pigmented epithelial cells (Asakura et al., 1992). In humans, COX-2 is expressed constitutively in the nonpigmented secretory epithelium (NPE) of the ciliary body, but is lost completely in the NPE of patients with end-stage OAG (Maihöfner et al., 2001). In addition, significantly fewer PGE_2 levels were present in aqueous humor of patients with OAG than in cataract patients (Maihöfner et al., 2001). Certain PGs, such as $PGF_{2\alpha}$, are specifically localized on the retinal blood vessels in the neonatal pig and along with PGI_2 and PGE_2 are synthesized by bovine retinal microvascular endothelial cells (Poyer et al., 1992; Kulkarni and Payne, 1997; Hardy et al., 1998).

In different mammalian species, there are significant differences in the responsiveness of the iris sphincters to the miotic effects, constriction of the pupil of the eye, of PGs (Miranda and Bito, 1989). In cats and dogs, the iris sphincter contracts in response to $PGF_{2\alpha}$, but not in response to other PGs, at concentrations that can be expected to occur under in vivo physiological conditions. In both cats and dogs, $PGF_{2\alpha}$, but none of the other PGs tested, yielded a dose-dependent full miosis when applied topically to intact eyes (Miranda and Bito, 1989). In contrast to the iris of cats, the bovine iris showed low sensitivity and low specificity and exhibited in vitro responses to high concentrations of several PGs (Miranda and Bito, 1989). In several other species (rabbits, rhesus and cynomolgus monkeys, baboons, humans) in vivo and in vitro studies showed that the iris sphincter does not exhibit similar miotic responses to any of the PGs that had been studied (Miranda and Bito, 1989). Therefore, these differences must be taken into consideration when comparing the effects of PGs on the iris sphincter across species.

Once synthesized, various PGs affect function by acting through specific receptors that are localized in various ocular tissues. The ciliary muscle is a well-documented center of PG receptor localization (Radi and Render, 2008). PGE_2 is produced by many cells of the body, including fibroblasts, macrophages, and dendritic cells. PGE_2 biological effects are mediated via four G-protein-coupled receptor subtypes (EP), which mediate stimulation of phosphoinositol turnover with elevation in intracellular calcium, activation of adenylyl cyclase activity resulting in elevation of intracellular cyclic AMP, or inhibition of adenylyl cyclase (Hinz et al., 2000). The capacity of ocular tissues to synthesize PGE_2 was initially determined on the basis of [^{14}C]AA conversion in vitro and in vivo (Kulkarni et al., 1984; Bazan et al., 1985a; Preud'homme et al., 1985). Both the iris and the retina of the rabbit released PGE_2, $PGF_{2\alpha}$, and thromboxane (TX) B_2 when incubated in vitro, and PGE_2 was the major COX product formed by each tissue (Preud'homme et al., 1985). The metabolism of radiolabeled AA in epithelium, stroma, and endothelium was studied in normal and cryogenically lesioned rabbit corneas (Bazan et al., 1985b). Two hours after injury, COX products in the epithelium increased, particularly $PGF_{2\alpha}$. Prostaglandin levels in the stroma rose rapidly after injury and remained elevated for 15 days. In the endothelium, increases in $PGF_{2\alpha}$ and PGE_2 were the most prominent effects of injury (Bazan et al., 1985b). Chronic or acute self-resolving inflammation was induced in mice by corneal suture or epithelial abrasion, respectively. Corneal levels of endogenous PGE_2 formation increased 4.3 and 6.6-fold after 2 and 4 days of suture-induced chronic inflammation, respectively, and remained elevated through day 7. The increase in endogenous PGE_2 formation correlated with pronounced neovascularization. In contrast, basal corneal levels of PGE_2 demonstrated no significant increase throughout the course of injury and wound healing during an acute, self-resolving epithelial abrasion (Liclican et al., 2010). PGE_2 has four receptor subtypes (EP_1, EP_2, EP_3, and EP_4). In the ciliary muscle of mice, consistent EP_1 and EP_4 receptors of mRNA expression were present in the ciliary muscle, whereas EP_2 and EP_3 signals were not detected (Takamatsu et al., 2000). The gene regulatory networks engaged in mediating the physiological effects on EP_2 and EP_4 receptor stimulation were investigated using cultured human ciliary smooth muscle cells. The effects of EP_2 and EP_4 receptor stimulation on gene regulation were found to be generally very similar and mediated predominantly through the cAMP signaling pathway (Reitmair et al., 2010). Topical application of PGE_1 and PGE_2 attenuated ciliary muscle contraction in monkeys (Yamaji et al., 2005). In the human eye, receptors for both $PGF_{2\alpha}$ and PGE_2 are colocalized in the ciliary and iridal sphincter muscles (Matsuo and Cynader, 1992). EP_2 receptor is expressed in the iris sphincter muscle and ciliary muscle in pigs, while the EP_1 receptor is expressed in both the sphincter and dilator iris muscle (Zhao et al., 1995). In the human eye, the highest expression of EP_1 receptor mRNA and protein is found in the epithelia of the cornea, conjunctiva, lens, ciliary body, trabecular cells, iris vessels, and retinal ganglion cells (Schlötzer-Schrehardt et al., 2002). In addition, the expression of EP_1 mRNA receptors was found in human ciliary muscles, iris vasculature, and in the ganglion and inner and outer nuclear layers of human retina (Mukhopadhyay et al., 2001). Remodeling of the extracellular matrix (ECM) of the ciliary body involves matrix metalloproteinases (MMPs) which are released by ciliary smooth muscle cells. Some studies revealed that human ciliary smooth muscle cells can secrete MMP-1, MMP-2, MMP-3, and MMP-9 and that exposure to $PGF_{2\alpha}$ increased the production of all four MMPs (Weinreb et al., 1997). In pigs, EP_1 is found in the corneal epithelium and ciliary NPE (Abran et al., 1995; Zhao et al., 1995). In C57BL/6 [wild-type (WT)] mice or FP, EP_1, EP_2, and EP_3 receptor-deficient mice, the ocular hypotensive effects and changes of outflow after selective agonists of PGE_2 receptor subtype (EP_1, EP_2, EP_3, and EP_4) treatment were investigated. The EP_2 agonist ONO-AE1-259-01 and EP_4 agonist ONO-AE1-329 significantly reduced IOP in a dose-dependent manner, whereas the EP_1 agonist ONO-DI-004 and EP_3 agonist ONO-AE-248 showed no effect. The reduction in IOP with ONO-AE1-259-01 was eliminated completely in EP_2-knockout mice, whereas the reduction in IOP in other knockout mice was similar to that observed in wild-type mice. Thus, EP_2 and EP_4 receptors mediated IOP reduction in mice, whereas the contribution of EP_1 and EP_3 receptors was insignificant. In addition, the EP_2 and EP_4 receptor-mediated mechanisms of IOP reduction were different from those mediated by the FP receptor (Saeki et al., 2009).

EP_2 is associated with the corneal epithelium and endothelium, lens epithelium, and ciliary NPE (Matsuo and Cynader, 1993; Abran et al., 1995; Zhao and Shichi, 1995; Anthony et al., 2001; Crider and Sharif, 2001). EP_2 receptor agonists cause vasodilatation and disrupt the blood–aqueous barrier (Kitagawa et al., 2001; Bhattacherjee et al., 2002). In the human eye, EP_2 receptor mRNA and protein labeling was most prominent in the corneal epithelium and choriocapillaries (Schlötzer-Schrehardt et al., 2002). There are some regional differences in the intensity of immunofluorescence between EP receptor subtypes between human and mouse eye. In the human eye, immunoreactivity for EP_2 receptors in the conjunctiva and trabecular meshwork is stronger than that in the mouse eye (Biswas et al., 2004). EP_1 and EP_2 receptor localization in the anterior uvea of mice and humans is very similar to that seen in the pig (Zhao et al., 1995; Biswas et al., 2004). EP_2 is associated with the trabecular meshwork in humans and pigs (Zhao et al., 1995; Kamphuis et al., 2004). EP_1 and EP_2 receptors are also found in the scleral fibroblasts of humans (Anthony et al., 2001). The influence of PGE_2 on the expression of COX-2 in human NPE cells was studied. Data have shown that PGE_2 at physiologically relevant concentrations induced COX-2 expression via a mechanism involving EP_2 and EP_4 receptor activation and phosphorylation of the mitogen-activated protein kinases (MAPKs) p38 and p42/44 (Rösch et al., 2005). The long-term one-year effects of topical treatment with 0.1% EP_2 agonist AH13205 on the anterior eye segment was investigated in cynomolgus monkeys. In the ciliary muscle, there was a significant increase in optically empty spaces between muscle bundles in the anterior portion of the longitudinal and the reticular ciliary muscle compared with untreated and vehicle-treated control animals. Within these spaces, significantly more myelinated nerve fiber bundles were found in drug-treated than in normal control animals. The ciliary epithelium had a normal appearance (Richter et al., 2003). In another study in cynomolgus monkeys, after a one-year treatment with 0.1% butaprost, an EP_2 receptor agonist, the morphology of the ciliary muscle showed increased spaces between ciliary muscle bundles and the formation of new outflow channels (Nilsson et al., 2006).

In pigs, EP_3 is localized to the corneal epithelium and endothelium, the lens epithelium, and the pigmented ciliary epithelium, bipolar cell, horizontal cell, amacrine cell, ganglion cell, and Müller cells (Zhao et al., 1995). EP_3 is localized to all uveal tissues in pigs (Zhao and Shichi, 1995). In the human eye, EP_3 and EP_4 receptors mRNA and protein labeling were observed primarily in the corneal endothelium and keratocytes, trabecular cells, ciliary epithelium, and conjunctival and iridal stroma cells. In addition, EP_3 was found in retinal Müller cells (Schlötzer-Schrehardt et al., 2002). Localized receptors of EP_4 have been identified in the trabecular meshwork and ciliary nonpigmented epithelial cells in the human eye (Matsuo and Cynader 1993; Anthony et al., 2001; Crider and Sharif, 2001). EP_4 receptors mediate the disruption of the blood–aqueous barrier induced by EP receptor agonists and paracentesis (Bhattacherjee et al., 2002). In the human eye, $PGF_{2\alpha}$ and PGE_2 receptors are expressed in the iris epithelium and retina (Matsuo and Cynader, 1992). In pigs, EP_3 is localized in the retinal photoreceptor (Zhao and Shichi, 1995). In newborn pigs, inhibition of COX-1 with valeryl salicylate caused increased density of PG receptors, EP_1, EP_3, and $F_{2\alpha}$ in retinal vessels (Hardy et al.,

1998). In mice, EP_3 was constitutively expressed in conjunctival epithelium on the ocular surface (Ueta et al., 2009).

Besides the different areas of PGs receptor localization, there are also species differences in ocular anatomy and reactions of the eye to given exogenous PGs (Radi and Render, 2008). For example, the rabbit eye is most sensitive to exogenous PGs (e.g., $PGF_{2\alpha}$) and the nonhuman primate eye is least sensitive (Bito, 1984; DeSantis and Sallee, 1989; Poyer et al., 1992). Exogenous $PGF_{2\alpha}$ and its analogs, PhXA85 and latanoprost, stimulated the formation of PGE_2, PGD_2, and $PGF_{2\alpha}$, in iris and ciliary muscles isolated from cat, bovine, rabbit, dog, rhesus monkey, and human (Yousufzai et al., 1996).

In addition to their part in normal ocular physiology, PGs play a role in topical and intraocular injury and inflammation (Radi and Render, 2008). PGs are released into the aqueous humor in response to various intraocular conditions, such as paracentesis, mechanical or laser injury to the iris, and experimental immunogenic and nonimmunogenic ocular inflammation (Bhattacherjee et al., 1980). Following corneal surface injury, the production of PGs (PGI_2, $PGF_{2\alpha}$, PGE_2, and PGD_2) by the corneal epithelium increases, and topical application of PGE_1 causes conjunctival changes that include vasodilatation (erythema), increased capillary permeability (chemosis), ocular discharge, decreased palpebral aperture, and occasional miosis (Eakins et al., 1972; Miller et al., 1973; Hall and Jaitly, 1977; Radi and Render, 2008). PGE_2 is synthesized by corneal endothelial cells, and its synthesis is increased in response to injury to the corneal endothelium of rabbits, and the amount in aqueous humor is increased following corneal wounding or inflammation in the anterior chamber (Joyce et al., 1995; Jumblatt, 1994; Jumblatt and Willer, 1996). Corneal regulation of EP expression during inflammatory or reparative response was investigated in a mouse model (Liclican et al., 2010). Chronic or acute self-resolving inflammation was induced in mice by corneal suture or epithelial abrasion, respectively. Interestingly, acute abrasion injury did not alter PGE_2 or EP levels. However, in chronic inflammation, the expression of COX-2, EP_2, and EP_4 was up-regulated and correlated with increased corneal PGE_2 levels and marked neovascularization. PGE_2 topical treatment, 100 ng applied three times a day for up to 7 days, exacerbated neutrophilic infiltration and the angiogenic response to chronic inflammation but did not affect wound healing or neutrophilic infiltration after epithelial abrasion. Exacerbated inflammatory neovascularization with PGE_2 treatment was independent of the vascular endothelial growth factor (VEGF) circuit but was associated with a significant induction of the eotaxin–CCR3 axis (Liclican et al., 2010). The expression of EP_3 protein was investigated in the conjunctiva of patients with various ocular surface diseases such as Stevens–Johnson syndrome (SJS), toxic epidermal necrolysis (TEN), chemical eye burns, Mooren's ulcers, and ocular cicatricial pemphigoid (OCP) (Ueta et al., 2011). EP_3 was expressed in the conjunctival epithelium of patients with chemical eye burns and Mooren's ulcer and in normal human conjunctival epithelium. However, it was markedly down-regulated in the conjunctival epithelium of SJS/TEN and OCP patients (Ueta et al., 2011). In an experimental allergic conjunctivitis (EAC) mouse model, PGE_2 acts on EP_3 in conjunctival epithelium and down-regulates the progression of EAC (Ueta et al., 2009). A selective EP_4 receptor agonist, 3,7-dithia PGE_1, and an isopropyl ester prodrug reduced IOP in monkeys. A single dose of 3,7-dithia PGE_1 isopropyl ester at 0.01% or 0.1% decreased IOP in the glaucomatous monkey in the range 40 to 50% (Woodward et al., 2009). The nonclinical safety profile of a topically administered selective EP_4 receptor agonist, PF-04475270, was investigated in beagle dogs. Treatment with PF-04475270 in dogs significantly decreased the IOP by 5 to 6 mmHg when given 0.25 μg/eye or by 7 mmHg when given at $\geq$1 μg/eye. In addition, PF-04475270 induced corneal neovascularization at doses of $\geq$1.0 μg/eye and conjunctival hyperemia or red eye at doses of $\geq$0.25 μg/eye, and these findings were not reversible during the 1- or 4-week recovery period (Aguirre et al., 2009).

$PGF_{2\alpha}$ is a powerful ocular hypotensive agent in rabbits, cats, dogs, monkeys, and humans (Kerstetter et al., 1988; Hardy et al., 1998; Poyer et al., 1992; Sarchahi et al., 2012). The most frequent side effect observed during daily topical treatment with latanoprost, a $PGF_{2\alpha}$ analog, in dogs was miosis (Sarchahi et al., 2012). Central corneal thickness was decreased after topical treatment with latanoprost in patients with OAG (Sawada et al., 2012). High mRNA and protein expression, using immunohistochemistry and in situ hybridization, of $PGF_{2\alpha}$ receptor (FP) in cynomolgus monkey eye was found in the cornea, conjunctiva, and iridial epithelium, ciliary muscle, and ciliary processes. In the human eye, the highest expression of FP receptor protein was found in the corneal epithelium, in ciliary epithelium, in the circular portion of ciliary muscle, and in iris stromal and smooth muscle cells (Schlötzer-Schrehardt et al., 2002). Iridial and choroidal melanocytes, the retina, and the optic nerve expressed lower levels of both FP receptor message and protein (Ocklind et al., 1996). Human ciliary smooth muscle cells express functional FP receptors whose activation up-regulates the mRNA and protein expression levels of early growth response factor-1 and connective tissue growth factor, important transcriptional activators of downstream genes involved in tissue remodeling and angiogenesis (Hutchinson et al., 2010). Topical treatment with 2 μg of $PGF_{2\alpha}$ isopropyl ester twice daily for 5 days reduced IOP in cynomolgus monkeys by increasing uveoscleral outflow (Gabelt and Kaufman, 1989; Sagara et al., 1999). This increased uveoscleral outflow is thought to result from the dilatation of the intramuscular spaces of ciliary muscle bundles (Lütjen-Drecoll and Tamm, 1988; Yamaji et al., 2005). PGE_2 also reduced IOP when applied topically by the relaxation of the precontracted ciliary muscle (Takamatsu et al., 2000).

In male Wistar rats, intravenous infusions of prostacyclin (PGI_2) and PGE_2 dilated retinal blood vessels, increased fundus blood flow, and decreased systemic blood pressure in a dose-dependent manner. These vasodilatory effects of PGs on retinal arterioles were greater than those on retinal venules (Mori et al., 2007). Similarly, an EP_2 receptor agonist, ONO-AE1-259-01, dilated retinal blood vessels and increased fundus blood flow and decreased systemic blood pressure (Mori et al., 2007). In streptozotocin-induced diabetic rats, insulin-deficient diabetic rats had significant increases in retinal PGE_2 at two months of diabetes (Kern et al., 2007). Prostacyclin-stimulating factor (PSF) acts on vascular endothelial cells to stimulate the synthesis of PGI_2. PSF is associated primarily with retinal vessels (Hata et al., 2000). PGI_2 increased retinal blood flow (RBF) in control and diabetic rats. The early drop in RBF during the initial weeks after diabetes induction in rats, as well as the later increase in RBF, both correlated with levels of retinal PSF. Thus, PSF can induce PGI_2, retinal vascular dilation, and increased RBF, and those alterations in retinal PSF expression may explain the biphasic changes in RBF observed in diabetes, at least in rats (Hata et al., 2000). The reduced blood flow may be associated with platelet thrombi in retinal capillaries of diabetic patients, a finding that may be prevented by antithrombotic drugs such as aspirin (The DAMAD Study Group, 1989; Boeri et al., 2001). The development of retinal hemorrhages and acellular capillaries in diabetic dogs was inhibited by aspirin (Kern and Engerman, 2001). In addition, aspirin decreased retinal vascular leakage in diabetic Long–Evans rats (Joussen et al., 2002). In rats with diabetic retinopathy, salicylates inhibited diabetes-induced loss of neuronal cells from the ganglion cell layer (Zheng et al., 2007). It has been suggested that nitric oxide synthase (NOS) and COX-2 work together to contribute to retinal cell death in diabetes and to the development of diabetic retinopathy and that inhibition of retinopathy by aminoguanidine or aspirin is due at least in part to inhibition of this NO/COX-2 axis (Du et al., 2004).

COX-1 EXPRESSION IN THE EYE UNDER NORMAL AND PATHOLOGICAL CONDITIONS

Normally, COX-1 is expressed in various anatomical locations in the eye (Figs. 5-3 to 5-5) (Radi and Render, 2008). This COX-1 expression is variable among different animal species (Table 5-2) (Radi and Render, 2008). The comparative cyclooxygenase product synthesis was investigated in albino rabbit, dog (puppy and adult), cat, cow, and human ocular tissues (Kulkarni et al., 1984). Albino

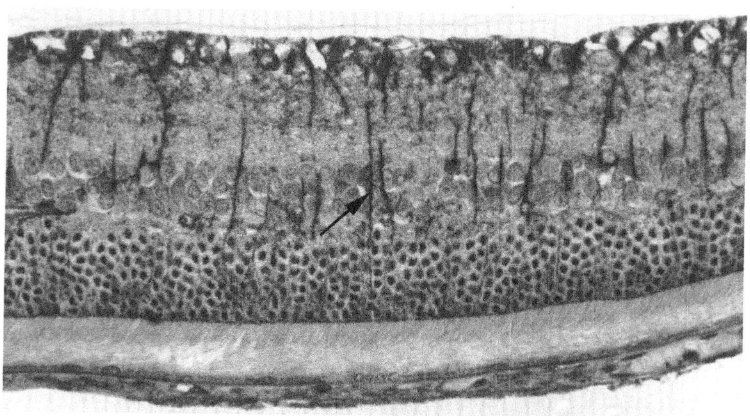

FIGURE 5-3 COX-1 expression in the normal retina of a rat. Note prominent expression of Müller cell processes (arrow). Immunohistochemical stain, original magnification ×40. [Reprinted with permission from Z. A. Radi and J. A. Render, *Ocular Pharmacology and Therapeutics*, 24(2), pp. 141–151. Copyright © 2008 Mary Ann Liebert, Inc. Publications. All rights reserved.]

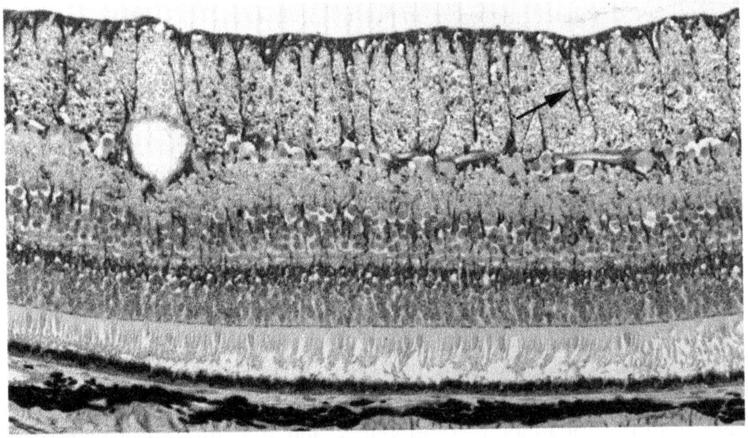

FIGURE 5-4 COX-1 expression in the normal retina of a monkey. Note prominent expression of Müller cell processes (arrow). Immunohistochemical stain, original magnification ×40. [Reprinted with permission from Z. A. Radi and J. A. Render, *Ocular Pharmacology and Therapeutics*, 24(2), pp. 141–151. Copyright © 2008 Mary Ann Liebert, Inc. Publications. All rights reserved.]

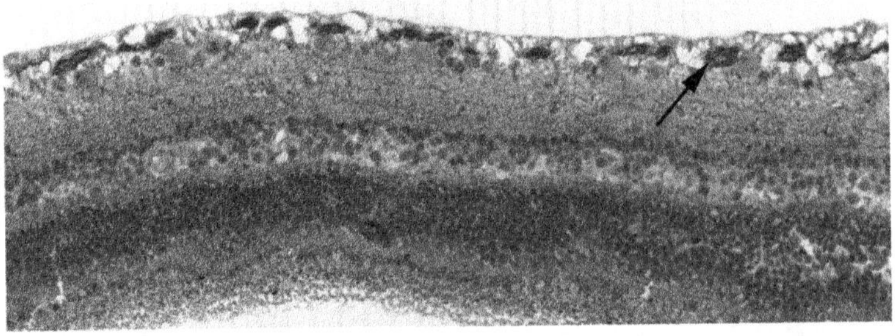

FIGURE 5-5 COX-1 expression in the normal retina of a pig. Note prominent ganglion cell expression (arrow). Immunohistochemical stain, original magnification ×40. [Reprinted with permission from Z. A. Radi and J. A. Render, *Ocular Pharmacology and Therapeutics*, 24(2), pp. 141–151. Copyright © 2008 Mary Ann Liebert, Inc. Publications. All rights reserved.]

rabbit conjunctiva synthesized a substantial amount (47%) of cyclooxygenase products, while cat and dog (puppy as well as adult) conjunctival tissues synthesized smaller but significant amounts of cyclooxygenase products (Kulkarni et al., 1984). In the anterior uvea, variable COX-1 expression exists among species (Radi and Render, 2008). Cyclooxygenase activity in the anterior uvea was greatest in the cat, followed by the albino rabbit, which was greater than that in the dog, followed by humans, and the least activity was noted in the cow. Interestingly, cow anterior uvea synthesized $PGF_{2\alpha}$ and 6-keto-$PGF_{1\alpha}$ (a stable metabolite of PGI_2), while human anterior uvea synthesized all cyclooxygenase products from 14-C-PGH_2 (Kulkarni et al., 1984). In New Zealand albino rabbits,

TABLE 5-2 Comparative COX-1 Expression in the Eye

Ocular compartment	Species	References
Conjunctiva	Cat, dog, rabbit	Kulkarni et al., 1984
Cornea	Rabbit	Amico et al., 2004; Miyamoto et al., 2004
Anterior uvea	Cat, dog, ox, human, rabbits	Kulkarni et al., 1984
Iris	Rabbit	Damm et al., 2001
Ciliary body, epithelium	Human	Maihöfner et al., 2001
Retina		
Ganglion cells	Human, mouse, pig, rat	Ju et al., 2002, personal communication
Astrocytes	Human	Ju et al., 2002
Microglia	Human, mouse, rat	Ju et al., 2002
Horizontal cells	Mouse	Ju et al., 2002
Photoreceptors	Mouse	Ju et al., 2002
Blood vessels	Human	Ju et al., 2002
Müller cell processes	Rat, monkey	Personal communication
Optic nerve, glial cells	Human	Neufeld et al., 1997

Source: Reprinted with permission from Z. A. Radi and J. A. Render *Ocular Pharmacology and Therapeutics*, 24(2), pp. 141–151.

COX-1 is normally expressed, in mRNA and protein, throughout all layers of the cornea (and particularly in the endothelium) (Amico et al., 2004). This COX-1 expression was unaffected in all corneal tissues following corneal injury (Amico et al., 2004). In another study in albino rabbits, the COX-1 expression pattern in corneal cells after photorefractive keratectomy (PRK) and laser in situ keratomileusis (LASIK) procedures was studied. COX-1 protein was not detected immunohistochemically in corneal tissue during the healing intervals after both procedures (Miyamoto et al., 2004). Using real-time RT-PCR and Western blot analysis, the presence of COX-1 in freshly excised iris and ciliary body tissue from adult New Zealand white albino rabbits was demonstrated and NPE of the ciliary body did not show COX-1 immunoreactivity (Damm et al., 2001). In normal eyes of humans, ocular COX-1 expression was confined largely to the NPE of the ciliary body (Maihöfner et al., 2001). The expression of COX-1 in different retinal layers in various species was as follows: outer segments of photoreceptor cells (mouse), microglia (mouse, rat, human), astrocytes (human), horizontal cells (mouse), retinal ganglion cells (mouse, rat, human, and pig) (Fig. 5-5) (Radi and Render, 2008), amacrine cells (mouse, rat, human), and vascular cells. No COX-1 expression was present in the nerve fiber layer (human) and in Müller cell processes in rats and monkeys (Figs. 5-3 and 5-4) (Neufeld et al., 1997; Ju and Neufeld, 2002; Radi and Render, 2008). COX-1 was localized exclusively to the prelaminar and lamina cribrosa regions of the optic nerve head. No COX-1 staining was observed in the nerve fiber layer or the myelinated optic nerve. COX-1 was associated with the astrocytes of the glial columns and the cribriform plates, but not with the endothelia lining the capillaries (Neufeld et al., 1997).

Ocular COX-1 expression may increase in some pathological conditions. In human eyes with congenital juvenile and angle-closure glaucoma, COX-1 expression remained unchanged in both the ciliary body, NPE, and in other sites of the eye (Maihöfner et al., 2001). In glaucoma in humans, more astrocytes stained with COX-1 than in normals, and staining was intensely perinuclear (Neufeld et al., 1997). COX-1 mRNA and protein expression were investigated in an induced conjunctivitis rat model. λ-Carrageenan (500 mg/eye) or bacterial lipopolysaccharide (LPS; 3 mg/eye) was injected into rat conjunctiva to induce conjunctivitis. COX-1 mRNA did not change by 24 hours after injection (Oka et al., 2004). COX-1 expression was studied in the cornea of mice with herpes simplex virus-1 (HSV-1) keratitis (Biswas et al., 2005). COX-1 mRNA was found to be expressed constitutively in naive corneas, and after HSV-1 infection no significant increase in the expression level was observed (Biswas et al., 2005). In dogs, COX-1 was expressed by ocular malignant melanomas (Pires et al., 2010). Using immunohistochemistry, COX-1 expression was evaluated in the cornea, eyelid, and third eyelid of healthy horses and those affected with squamous cell carcinoma (SCC). COX-1 was significantly greater in equine corneas with SCC than in control corneas. No significant differences in COX-1 immunoreactivity were detected in eyelid and third-eyelid SCC compared with site-matched control tissues (McInnis et al., 2007).

COX-2 EXPRESSION IN THE EYE UNDER NORMAL AND PATHOLOGICAL CONDITIONS

COX-2 expression is generally up-regulated during pathological conditions such as in ocular inflammation or injury (Table 5-3) (Ottino and Bazan, 2000a; Radi and Render, 2008). In an induced conjunctivitis rat model, COX-2 mRNA and protein expression were investigated. λ-Carrageenan (500 mg/eye) or bacterial lipopolysaccharide (LPS; 3 mg/eye) was injected into rat conjunctiva to induce conjunctivitis. In the carrageenan-injected model, COX-2 mRNA expression in the inflamed conjunctiva was significantly increased by 2 h and gradually increased until 24 h and COX-2 protein expression was increased markedly 4 h after injection of carrageenan. In the LPS-injected model, COX-2 mRNA expression was increased 1 h after injection, peaked at 2 h, and decreased at 4 h, and COX- 2 protein expression increased 2 h after injection of LPS (Oka et al., 2004). In mice with herpes simplex virus-1 (HSV-1) keratitis, corneal COX-2 expression was evaluated (Biswas et al., 2005). Infected mice had higher COX-2 mRNA and PGE_2 protein levels at different days compared to naive cornea. The levels of COX-2 mRNA and PGE_2 protein were below the limit of detection in naive corneas (Biswas et al., 2005). It was found that IL-1β induces COX-2 expression in mice cornea after ocular HSV-1 infection (Biswas et al., 2005). Thus, the likely source of COX-2, at least initially after HSV-1 infection, was corneal stromal fibroblasts stimulated by proinflammatory cytokines, such as IL-1β, induced by HSV-1 replication (Biswas

TABLE 5-3 Comparative COX-2 Expression in the Eye Under Normal and Pathological Conditions

Ocular compartment	Species (pathological condition)	References
Conjunctiva	Mouse (keratitis), rat (inflammation)	Oka et al., 2004; Biswas et al., 2005
Cornea	Dog (glaucoma), rabbit (normal and after keratectomy)	Marshall et al., 2004; Sellers et al., 2004
Sclera	Dog (glaucoma)	Marshall et al., 2004
Anterior uvea	Rabbit (glaucoma)	Chang et al., 1997
Iris	Rabbit (normal)	Damm et al., 2001
Ciliary body	Dog (normal and glaucoma), human (normal), rabbit (normal)	Damm et al., 2001; Maihöfner et al., 2001; Marshall et al., 2004
Trabecular meshwork	Dog (glaucoma)	Marshall et al., 2004,
Retina	Newborn pig (normal), rat and pig (after ischemia)	Degi et al., 2001; Ju et al., 2002
Ganglion cells	Mouse (normal), pig (normal), rat (normal)	Ju et al., 2002, personal communication
Astrocytes	Mouse and rat (ischemic proliferative retinopathy) human (diabetic retinopathy)	Sennlaub et al., 2003
Amacrine cells	Mouse (normal), rat (normal)	Ju et al., 2002
Outer plexiform layer	Mouse (normal), human (normal), rat (normal)	Ju et al., 2002
Retinal pigmented epithelium	Rat (phagocytosis)	Ershov and Bazan, 1999
Optic nerve head	Human (glaucoma)	Neufeld et al., 1997

Source: Reprinted with permission from Z. A. Radi and J. A. Render, *Ocular Pharmacology and Therapeutics*, 24(2), pp. 141–151.

et al., 2005). Chronic or acute self-resolving inflammation was induced in mice by corneal suture or epithelial abrasion, respectively. COX-2, EP_2, and EP_4 expression were up-regulated with chronic inflammation and correlated with increased corneal PGE_2 formation and marked neovascularization (Liclican et al., 2010). There is variable COX-2 expression across different species (Radi and Render, 2008). In dogs with keratitis, the most common inflammatory eye disease in dogs, and in glaucomatous eyes, COX-2 expression was observed in all corneal layers (epithelium, stromal cells, and endothelium) (Marshall et al., 2004; Sellers et al., 2004). COX-2 immunoreactivity was also noted in the stromal and epithelial cells of the iris and the stromal cells of the trabecular meshwork (Sellers et al., 2004). No COX-2 expresssion was present in the cornea of normal dogs (Sellers et al., 2004; Marshall et al., 2004). COX-2 is expressed strongly in stromal keratocytes of normal New Zealand rabbit corneas (Amico et al., 2004). In rabbits, COX-2 levels increase significantly after injury and, at least in the epithelium, COX-2 becomes the predominant COX isoform at an early stage of healing (Amico et al., 2004). The

COX-2 expression pattern in rabbit corneal cells after photorefractive keratectomy (PRK) was compared with laser in situ keratomileusis (LASIK) (Miyamoto et al., 2004). After PRK, the central and peripheral corneal epithelia COX-2 expression was up-regulated at three days; the central epithelium was positive at 4 weeks. Central and peripheral epithelia COX-2 levels were negative three months later. After LASIK, the central epithelium on the corneal flap had COX-2 up-regulated after 1 and 2 weeks and was negative at 4 weeks (Miyamoto et al., 2004). Keratocytes around the stromal incision between the flap and the stromal bed up-regulated COX-2 and returned to negative at 3 months. Increased expression of COX-2 occurred in the central uninjured epithelium and stromal keratocytes 3 to 14 days after a LASIK procedure in rabbits (Miyamoto et al., 2004). Ocular COX-2 expression appears to be regulated by glucocorticoids (Maihöfner et al., 2001).

COX-2 plays an important role in corneal healing by stimulating corneal neovascularization and angiogenesis that occurs after injury, trauma, infection, or surgery (Masferrer et al., 1999; Wilson et al., 2003; Wilkinson-Berka, 2004; Radi and Render, 2008). Several mediators, such as IL-1, nitric oxide (NO), VEGF, and TNF, contribute to the processes of corneal healing by inducing COX-2 (Bazan et al., 1985a; Wilson et al., 2003; Du et al., 2004; Wilkinson-Berka, 2004). The initial stages of corneal repair are mediated by PGs generated by COX-2 which influence the recruitment of inflammatory cells and the presence of edema and pain (Davies et al., 1984; Amico et al., 2004; Radi and Render, 2008).

In normal human eyes, COX-2 expression was confined primarily to the NPE of the ciliary body (Maihöfner et al., 2001). In normal dogs, only minimal COX-2 expression was noted in the ciliary epithelium (Marshall et al., 2004). COX-2 expression was lost completely in the nonpigmented ciliary epithelium of humans with end-stage primary open-angle glaucoma (Maihöfner et al., 2001). In dogs, COX-2 expression was observed in the ciliary body, trabecular meshwork of the filtration angle, angular aqueous plexus, and sclera of dogs with glaucoma (Marshall et al., 2004). COX-2 is present in the iris and ciliary body of normal rabbits, whereas no expression is present in the cornea or anterior chamber in normal dogs (Damm et al., 2001; Sellers et al., 2004). In rabbits, COX-2 was induced in the anterior uvea after surgery for glaucoma (Chang et al., 1997).

Normal expression of COX-2 in the retina is variable among species and is detectable in the retina of newborn pigs but not in the adult rat and monkey (Figs. 5-6 to 5-8) (Degi et al., 2001; Radi and Render, 2008). COX-2 is expressed in nerve processes of the outer plexiform layer (mouse and rat), certain amacrine cells (mouse and rat), retinal ganglion cells (mouse, rat, pig) (Fig. 5-8) (Radi and Render, 2008), and blood vessels (mouse), whereas in the human retina, COX-2 is reported only in nerve processes of the outer plexiform layer (Ju et al., 2002; Wilkinson-Berka et al., 2003). A recent study reported COX-2 expression in various normal human ocular locations (Wang et al., 2011). Moderate COX-2 expression was observed in the epithelial and endothelium cells of the cornea. Strong expression of COX-2 was found in the nonpigmented epithelial cells of the ciliary body. Moderate to strong COX-2 expression was observed in various layers of the retina, with the most abundant COX-2 expression being detected in the photoreceptor inner segment, the external limiting membrane, and the outer plexiform layer (Wang et al.,

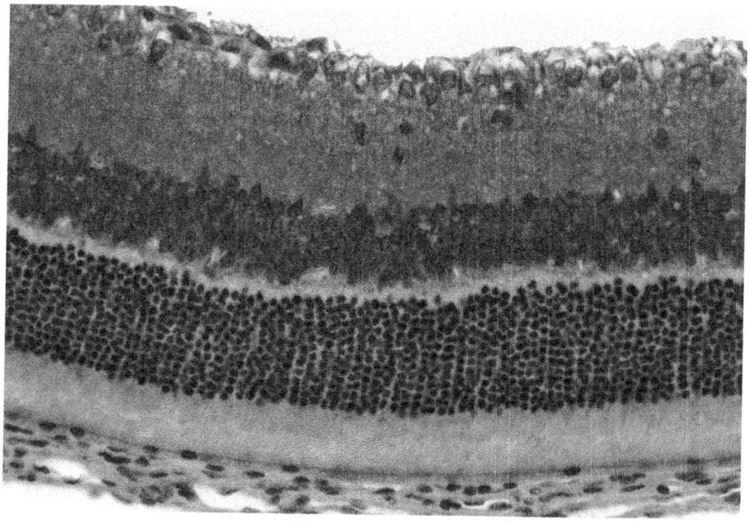

FIGURE 5-6 COX-2 expression in the normal retina of a rat. Immunohistochemical stain, original magnification ×40. [Reprinted with permission from Z. A. Radi and J. A. Render, *Ocular Pharmacology and Therapeutics*, 24(2), pp. 141–151. Copyright © 2008 by Mary Ann Liebert, Inc. Publications. All rights reserved.]

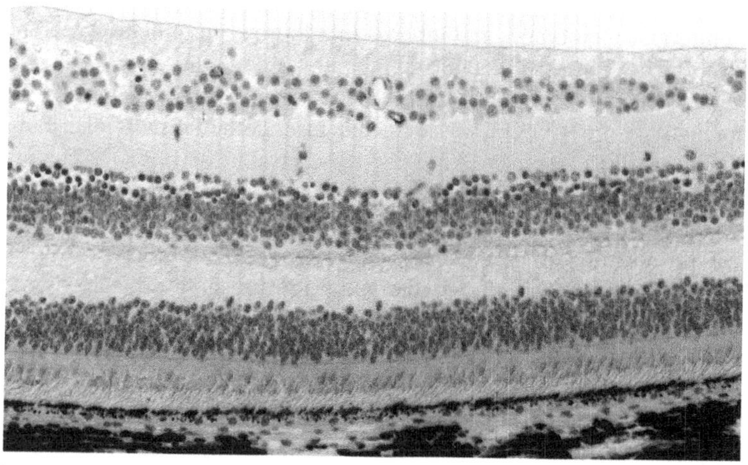

FIGURE 5-7 COX-2 expression in the normal retina of a monkey. Immunohistochemical stain, original magnification ×40. [Reprinted with permission from Z. A. Radi and J. A. Render, *Ocular Pharmacology and Therapeutics*, 24(2), pp. 141–151. Copyright © 2008 Mary Ann Liebert, Inc. Publications. All rights reserved.]

2011). Following transient ischemia, increased COX-2 retinal expression was noted in the rat and pig (Degi et al., 2001; Ju et al., 2002). COX-2 is not expressed in the normal human optic nerve head, but a few COX-2-positive cells were found in the prelaminar, lamina cribrosa, and postlaminar regions of the glaucomatous optic nerves. Positive staining for COX-2 was not associated with microglia (Neufeld

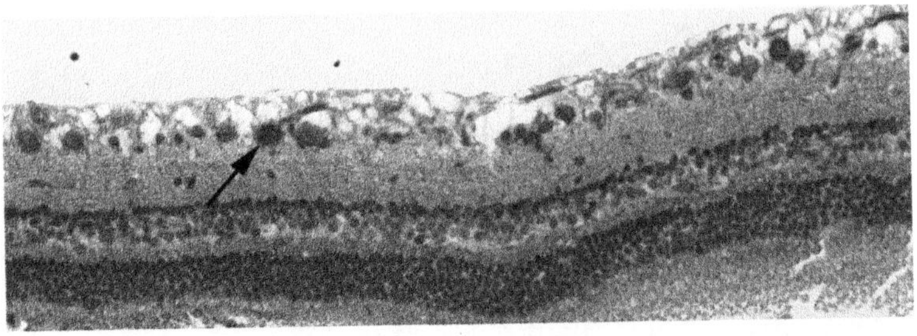

FIGURE 5-8 COX-2 expression in the normal retina of a pig. Note prominent ganglion cell expression (arrow). Immunohistochemical stain, original magnification ×40. [Reprinted with permission from Z. A. Radi and J. A. Render, *Ocular Pharmacology and Therapeutics*, 24(2), pp. 141–151. Copyright © 2008 Mary Ann Liebert, Inc. Publications. All rights reserved.]

et al., 1997). An ocular condition called idiopathic epiretinal membrane (IERM) can be observed in the eye with no history of trauma, ocular inflammatory diseases, or any other type of ocular disease. Histologically, the IERM is composed of nonvascularized tissue growth along the inner limiting membrane (ILM) on the retinal surface (Kase et al., 2010). The number of COX-2-immunopositive cells was significantly higher in the IERM than that in the ILM (Kase et al., 2010). This suggests that COX-2 may play a potential role in the formation of avascular and vascularized epiretinal membranes in humans. In addition, epiretinal membranes which are formed in proliferative diabetic retinopathy (PDR) are composed primarily of neovascular stromal tissue (El-Asrar et al., 2008; Kase et al., 2010). Cytoplasmic immunoreactivity for COX-2 was detected in endothelial cells distributed in the PDR membranes (El-Asrar et al., 2008; Kase et al., 2010). COX-2 expression in retinal astrocytes was increased in human diabetic retinopathy and in mouse and rat models of ischemic proliferative retinopathy (Sennlaub et al., 2003). In a murine model of retinopathy of prematurity, equivalent COX-2 mRNA expression was observed in room air and after hyperoxia exposure at postnatal days 12 (P12), P14, and P17. Immunolocalization of COX-2 in the mouse retina was found to be similar to that in humans. In normoxic mice at P14, COX-2 protein was expressed significantly in retinal pigment epithelial cells in both the outer photoreceptor segment and the inner plexiform layer. In the posthyperoxia-exposed retinas, COX-2 expression was detected at P14 predominantly in astrocytes (Sennlaub et al., 2003). Phagocytosis of rod photoreceptor outer segments in rats was increased selectively with increased COX-2 expression in retinal pigmented epithelial cells (Ershov et al., 1999). Retinal COX-1 expression, on the other hand, did not change in diabetic rats (Fang et al., 1997).

COX-2 expression was investigated in various species and in various ocular neoplasms. In humans, COX-2 expression was lower in adenoma of nonpigmented ciliary body epithelium compared to the expression of the normal epithelium and negative in all medulloepithelioma cases of the nonpigmented ciliary body (Wang et al., 2011). No expression of COX-2 was noted in one case of retinoblastoma and decreased expression in other retinoblastoma cases compared to COX-2 expression in the normal outer layers of retina (Wang et al., 2011). In another study, COX-2-positive immunoreactions were observed in 96% of retinoblastoma specimens from human patients (Karim et al., 2000). COX-2 expression was found in 58% of the human uveal melanoma cases (Figueiredo et al., 2003). Up-regulation of COX-2 expression appears to be associated with poor prognosis in uveal melanoma in humans (Cryan et al., 2008). In dogs, COX-2 expression was noted in ocular malignant melanoma (Paglia et al., 2009; Pires et al., 2010). In horses, ocular SCC and carcinoma in situ have significantly higher levels of COX-2 expression than that of normal control ocular tissue. However, the percentage of cells expressing positive immunohistochemical staining was consistently low (Smith et al., 2008).

EFFECTS OF COX-2 s-NSAIDs IN OPHTHALMOLOGY

During inflammatory tissue response, eicosanoids are thought to play a significant role in angiogenesis (Yamada et al., 1999). The ophthalmic uses of celecoxib mentioned in this chapter are investigational (for research purposes only) and are not within the labeled indications. The role of COX-2 in corneal angiogenic response was studied in a rat model. Angiogenesis was induced in the rat cornea by chemical cautery, and NS-398, a COX-2 s-NSAID, was applied topically three times daily for 4 days and neovascularization was quantitated (Yamada et al., 1999). COX-2 induction in cauterized rat corneas increased the level of eicosanoids, which resulted in corneal angiogenesis, and NS-398 significantly inhibited corneal neovascularization (Yamada et al., 1999). COX-2 and VEGF affect vascular permeability, and angiogenesis and can be induced by ocular tissue hypoxia and hyperglycemia (Bonazzi et al., 2000; Wilkinson-Berka, 2004). In addition, NO has been implicated in VEGF-mediated vascular permeability and angiogenesis (Wilkinson-Berka, 2004). In rabbit corneal epithelial cells, COX-2 mRNAs were increased by hypoxia (Bonazzi et al., 2000). In a rat model, COX-2 inhibition using SC-236 has been associated with attenuating angiogenesis and down-regulating VEGF and bFGF (Masferrer et al., 1999). Selective NSAIDs appear to have a great potential in inhibiting corneal angiogenesis and treating diabetic retinopathy and in intravitreal neovascularization associated with ischemic proliferative retinopathy in humans (Sennlaub et al., 2003). Diabetic persons are usually susceptible to diabetic retinopathy, which could lead to blindness due to retinal neovascularization, and ischemia is a common precursor to neovascularization in retinal diseases (Ayalasomayajula et al., 2004). The ex vivo effects of celecoxib on retinas from diabetic Sprague–Dawley rats were studied. Higher synthesis of PGE_2 was noted in diabetic rat retinas compared to normal rat retinas, and expression was elevated within a week of inducing diabetes in rats (Ayalasomayajula et al., 2004). In this ex vivo system, the rate of PGE_2 synthesis in diabetic rat retinas was inhibited significantly by celecoxib (1 μM), indicating the potential value of using celecoxib to treat diabetic retinopathy (Ayalasomayajula et al., 2004). Increased COX-2 expression leads to an increased production of PGs, which induce VEGF expression. This signaling pathway can effectively be blocked by administering celecoxib. In isolated rat retinas, diabetes-induced PGE_2 expression was inhibited significantly by treatment with 1 μM of celecoxib (Ayalasomayajula et al., 2004). Macular edema is the most common cause of visual impairment in diabetic patients leading to diabetic macular edema (DME). The important pathophysiology of DME is the loss of retinal capillary pericytes, resulting in increased vascular permeability (Ciulla et al., 2003). VEGF, a potent inducer of vascular permeability, is increased markedly in diabetic retina, DME, age-related macular degeneration (ARMD), and corneal neovascularization (Adamis et al., 1994; Aiello et al., 1994; Lopez et al., 1996; Cursiefen et al., 2000; Adamis, 2002). Bevacizumab, a humanized monoclonal antibody against VEGF, also binds and inhibits all the biologically active forms of VEGF (Presta et al., 1997). Bevacizumab has been used for intravitreal injection in patients with ARMD, DME, and iris neovascularization and was shown to provide beneficial effects in such patients (Haritoglou et al., 2006; Oshima et al., 2006; Yoganathan et al., 2006; Ozkiriş, 2009). Increased PGE_2 can lead to up-regulation of VEGF, which is responsible for the initiation of diabetic retinopathy (Cheng et al., 1998; Frank, 2004). Vascular hyperpermeability and neovascularization lead to vision loss in various ocular pathologies, such as diabetic retinopathy and ARMD (Castro et al., 2004). Histologically, there are vascular lesions in the early stages of diabetic retinopathy in both humans and animal models. These vascular lesions are characterized by the presence of saccular capillary microaneurysms, pericyte deficient capillaries, and obliterated and degenerate capillaries (Kern, 2007). These degenerate capillaries are not perfused, so increases in their frequency represent reductions in retinal perfusion. Thus, many of the defects that develop in retinas as a result of diabetes are consistent with a diabetes-induced inflammatory response. These inflammatory changes apparently are important in the pathogenesis of diabetic retinopathy, since inhibition of this inflammatory cascade at any of multiple steps can inhibit the early stages of diabetic retinopathy (notably, degeneration of retinal capillaries) in animal models. Findings of diabetes-induced inflammatory changes, generally, in the human eye also, are consistent with the postulate that inflammatory processes contribute to the development of diabetic retinopathy (Kern, 2007). Elevated VEGF expression is correlated with diabetic blood–retinal barrier breakdown and ischemia-related neovascularization in animal models and humans (Adamis, 2002). In a cynomolgus monkey model of experimental iris neovascularization, VEGF mRNA was increased in the aqueous fluid, and the aqueous VEGF mRNA levels changed synchronously and proportionally with the severity of neovascularization after retinal ischemia induced by laser occlusion of all branch retinal veins (Miller et al., 1994).

COX-2 s-NSAIDs (e.g., celecoxib, rofecoxib, deracoxib, valdecoxib, lumaricoxib, etoricoxib) spare COX-1 and inhibit COX-2. Retinopathy of prematurity (ROP) is a vascular biphasic disease of the eye unique to preterm infants (Heckmann, 2008). In the first phase of the disease, relative hyperoxia results in vaso-obliteration and vessel loss. The second phase is characterized by hypoxia-induced neovascularization, resulting in retinal detachment and blindness. Oxygen-dependent VEGF and oxygen-independent insulin-like growth factor (IGF-1) have been identified as important factors in the pathogenesis of ROP (Heckmann, 2008). In a mouse model of ROP, rofecoxib was given intraperitoneally at 15 mg/kg body weight from postnatal day 12 (P12) to P17, and the effects on angiogenesis as measured by count blood vessel profiles (BVPs) in the inner retina (inner limiting membrane, ganglion cell layer, and inner plexiform layer) were evaluated (Wilkinson-Berka et al., 2003). Rofecoxib attenuated retinal angiogenesis that accompanied retinopathy of prematurity (Wilkinson-Berka et al., 2003). In a rabbit model of argon laser–induced iritis, the effects of COX-2 s-NSAIDs on hyaluronan responses were investigated. SC-236 and rofecoxib were administered orally at 6 and 1.5 mg/kg per day, respectively. Both COX-2 s-NSAIDs prolonged trauma-induced elevations of hyaluronan in the iris of rabbits (Koralewska-Makar et al., 2003). Celecoxib administered twice daily via oral gavage at 50 mg/kg for 7 days inhibited retinal VEGF expression and vascular leakage in streptozotocin-induced diabetic rat model (Ayalasomayajula and Kompella, 2003).

The ocular disease state can affect the ocular pharmacokinetics of drugs following topical and systemic delivery. For example, results from several studies in animal models showed that ocular inflammation can cause increased aqueous and vitreous levels of antibiotics following topical and systemic delivery (Ozturk et al., 2000; Yağci et al., 2007). These increased aqueous and vitreous levels are related to the alteration in blood–retinal and blood–aqueous barriers in the diseased ocular state (Cheruvu et al., 2009). The effect of diabetes and the resulting breakdown of the blood–retinal barrier on drug transscleral delivery are poorly understood. In addition, differences in pigmentation and the relationship to blood–retinal barrier breakdown between rat strains are also poorly understood. Thus, the effects of diabetes on the transscleral delivery of celecoxib and the relationship of difference in ocular pharmacokinetics were investigated using diabetic albino [Sprague–Dawley (SD)] and pigmented [brown Norway (BN)] rats. The transscleral retinal delivery of celecoxib was studied in albino (SD) and pigmented (BN) rats. Diabetes (two-month duration) resulted in higher blood–retinal barrier leakage in diabetic SD and BN rats than that in controls. Transscleral retinal and vitreal delivery of celecoxib, 3 mg/rat, was significantly higher in diabetic rats of both strains. Such an increase in retinal delivery of celecoxib due to diabetes (i.e., breakdown of the blood–retinal barrier) was higher in pigmented rats than in albino rats. Thus, higher delivery of celecoxib in diabetic animals compared to control animals can be attributed to the disruption of the blood–retinal barrier because of the disease state (i.e., diabetes) (Cheruvu et al., 2009). In a rat model of angiogenesis, corneal blood vessel formation was suppressed by celecoxib but not by the COX-1 inhibitor SC-560 (Masferrer et al., 2000). Celecoxib caused a substantial reduction in the number and length of sprouting capillaries and dose-dependently inhibited the angiogenic response with an ED_{50} of 0.3 mg/kg per day, and with a maximal inhibitory activity of 80% at a dose of 30 mg/kg per day in rats (Masferrer et al., 2000). The COX-2 s-NSAID SC-236 given orally at a dose of 10 mg/kg per day (the treatment was started a day before corneal infection and continued until day 10 after infection) resulted in diminished angiogenic responses and significantly reduced levels of VEGF at 3 and 7 days after infection and milder herpetic stromal keratitis in mice (Biswas et al., 2005). These data strongly suggested that COX-2 s-NSAIDs have an antiangiogenic response. The numbers of corneas and ganglia containing infectious herpes simplex virus type 1 were significantly lower in the celecoxib-treated mice than in the placebo-treated mice (Gebhardt et al., 2005). Therefore, it has been suggested that COX-2 s-NSAIDs can suppress hyperthermic stress-induced herpes viral reactivation in the nervous system (Gebhardt et al., 2005). The COX-2 s-NSAID SC-58236 prevented apoptotic cell death and was neuroprotective against loss of retinal ganglion cells after ischemia (Ju et al., 2003). The COX-2 s-NSAID NS-398 attenuated spontaneous and TNFα-induced increases in nonrapid eye movement sleep in rabbits (Yoshida et al., 2003).

In humans, cases of acute, reversible disturbances of vision associated with the use of celecoxib or rofecoxib were reported in the Intensive Medicines Monitoring Program in New Zealand (Coulter and Clark, 2004). These were identified from three different databases using strict selection criteria. Temporary blindness, visual field defect, scotoma, teichopsia, blurred vision, decreased vision, and abnormal vision were reported effects seen with the use of celecoxib and rofecoxib in New Zealand (Coulter and Clark, 2004). In the study celecoxib was used at doses that ranged from 100 to 400 mg/day for different indications, such as rheumatoid arthritis, osteoarthritis, knee surgery, back pain, gout, psoriatic arthropathy, or elbow pain. Rofecoxib in this study was used at doses that ranged from 12.5 to 50 mg/day for different indications, such as arthritis, shoulder pain, lower back pain, osteoarthritis, or pain in limb. The reactions had a mean onset time of 9.5 days and recovery occurred within 3 days following withdrawal of the drug. The reactions did not appear to be related to age, gender, dose, or indication for use. These reversible effects may be related to interference with retinal blood supply (Coulter and Clark, 2004). Vitreous floaters and conjunctival hemorrhage ocular adverse events have been reported, according to the celecoxib package insert. The route of administration could affect the drug availability into ocular tissues. For example, retinal availability of celecoxib was severalfold higher following periocular or subconjunctival administration compared to intraperitoneal or systemic administration (Ayalasomayajula and Kompella, 2004). The periocular effects of celecoxib–poly(lactide-*co*-glycolide) (PLGA) microparticles were tested using an ARPE-19 cell line (a spontaneously arising human retinal pigment epithelial cell line) and in a streptozotocin diabetic rat model (Amrite et al., 2006). The in vitro release of celecoxib from the PLGA particles was performed with dialysis membrane bag suspension of either plain celecoxib (20 μg) or celecoxib–PLGA particles containing 20 μg of celecoxib. Nanomolar concentrations of celecoxib reduced VEGF mRNA and VEGF protein secretion. Periocular celecoxib microparticles were shown to be useful sustained drug delivery systems for inhibiting diabetes-induced elevations in PGE_2, VEGF, and blood–retinal barrier leakage. The periocular celecoxib–PLGA microparticles were demonstrated to be safe and did not cause any damage to the retina (Amrite et al., 2006).

EFFECTS ns-NSAIDs IN OPHTHALMOLOGY

Topically applied ns-NSAIDs (e.g., diclofenac, flurbiprofen, suprofen, indomethacin) inhibit COX-1 and COX-2 and have been used in the management of various ophthalmic conditions, such as seasonal allergic conjunctivitis, photophobia to manage pain after corneal refractive surgery, postoperative inflammation after cataract surgery, postoperative topical analgesia, prevention and treatment of cystoid macular edema (CME), maintenance of intraoperative mydriasis, and prophylaxis of surgical miosis (Frucht and Zauberman, 1984; Sakamoto et al., 1995; Gaynes and Fiseella, 2002; Schalnus, 2003; Takahashi et al., 2003; Radi and Render, 2008). Bromfenac, an ns-NSAID 0.09% ophthalmic solution, is used for the treatment of postoperative inflammation and reduction of ocular pain in subjects who have undergone cataract extraction (Stewart et al., 2007). Other ophthalmic conditions, such as diabetic retinopathy and AMD, may also be treated with ns-NSAIDs (Hariprasad et al., 2007). In veterinary medicine, systemic ns-NSAIDs (e.g., carprofen, flunixin meglumine) are used to treat uveal conditions (Giuliano, 2004). Acetylsalicylic acid, phenylbutazone, and flunixin meglumine have been the most commonly prescribed ns-NSAIDs for dogs, cats, and horses by veterinary ophthalmologists (Giuliano, 2004). Topically applied suspensions of diclofenac, flurbiprofen, and suprofen are effective in preventing blood–aqueous barrier disruption after paracentesis in dogs (Ward, 1996). Intravenous administration of flunixin meglumine significantly limited miosis and reduced the production of PGE_2 in the aqueous humor after acute inflammation induction in dogs (Millichamp et al., 1991). Tepoxalin was more effective than other ns-NSAIDs, carprofen or meloxicam, for controlling the production of PGE_2 in dogs with experimentally induced uveitis (Gilmour and Lehenbauer, 2009). In rabbits, intravenous administration of a COX-1 selective inhibitor, FR122047, inhibited the aqueous flare induced by PGE_2 (Abe et al., 2004). In dogs, topically applied flurbiprofen, an ns-NSAID, caused a decrease in aqueous outflow that was more marked in the inflammed eyes (Millichamp et al., 1991). In normal beagle dogs, the effects of topically applied flurbiprofen 0.03% and latanoprost 0.005%, alone or in combination, were studied (Pirie et al., 2011). Flurbiprofen resulted in a mean IOP elevation of 1.1 mmHg in the treated eye compared with the control eye. No effect on pupil size, conjunctival hyperemia, or aqueous flare was noted. Latanoprost resulted in a mean IOP reduction of 3.4 mmHg. Combined latanoprost and flurbiprofen resulted in a mean IOP reduction of 2.7 mmHg. Miosis was noted in the treated eyes during both latanoprost periods. Thus, concurrent administration of latanoprost and flurbiprofen resulted in a 20.41% reduction in the ocular hypotensive effect relative to latanoprost therapy alone (Pirie et al., 2011).

In diabetic beagle dogs, sulindac oral treatment with 10 mg/kg per day (this dose was based on previous studies using sulindac on diabetic animals) decreased retinal capillary basement membrane thickening and diabetic retinopathy (Gardiner et al., 2003). In nonhuman primates, continuous infusion of 25 μg/day of indomethacin via cannula that was implanted into the vitreous cavity for 14 days inhibited laser-induced subretinal neovascularization and reduced vascular leakage (Sakamoto et al., 1995). Topically applied indomethacin reduced uveitis in rabbits (Yamauchi et al., 1979) and markedly inhibited the breakdown of the blood–aqueous barrier after paracentesis or argon laser photocoagulation of the iris in rabbits (van Delft et al., 1987). When rat corneas with central epithelial abrasions were treated with indomethacin and flurbiprofen in organ culture, corneal reepithelialization was not altered (Gupta et al., 1993). In newborn pigs, ibuprofen treatment at 30 mg/kg given intravenously reduced the basal concentrations of all prostanoids to nearly undetectable levels, prevented their changes during hypotension and hypertension, and enhanced retinal and choroidal blood flow autoregulation (Chemtob et al., 1991). Human infants treated with indomethacin have a lower incidence of retinopathy of prematurity (Gersony et al., 1983).

The systemic absorption of topically applied ophthalmic NSAIDs is generally minimal. However, systemic absorption may occur from mucosal surfaces of the nasolacrimal outflow system. Intraocular diclofenac and flurbiprofen concentrations in human aqueous humor were measured following topical application. Flurbiprofen at 0.03% and diclofenac at 0.1% after a single topical application reached peak aqueous concentrations of 60 and 82 ng/mL at 2.0 and 2.4 h, respectively (Ellis et al., 1994). Diclofenac concentration remained above 20 ng/mL for over 4 h and was detectable after 24 h. Flurbiprofen was undetectable after 7.25 h (Ellis et al., 1994). Topical ophthalmic ns-NSAIDs can lead to local irritations that result in transient burning, stinging, conjunctival hyperemia, and corneal anesthesia (Schalnus, 2003). Diclofenac has been associated with the development of corneal ulcers in patients after cataract surgery or photorefractive keratectomy (Price, 2000). Shimazaki et al. (1995) have reported the development of a postkeratoplasty persistent epithelial defect of the corneal epithelium with the use of diclofenac sodium. In addition, indolent corneal ulceration, perforation, and full-thickness corneal melt can be serious complications associated with the use of topical ns-NSAIDs (Gaynes and Fiscella, 2002; Mian et al., 2006; Isawi and Dhaliwal, 2007). Administration of 0.1% dicofenac sodium eye drops three times daily did not cause significant abnormalities in the corneal epithelial structure or function of patients who had undergone small-incision cataract surgery (Shimazaki et al., 1996).

Indomethacin when applied topically three times daily for 4 days inhibited corneal neovascularization in a rat model (Yamada et al., 1999). Etodolac given in a 5-mg/kg dose once daily for 7 or 14 days suppressed choroidal neovascularization in a mice model (Takahashi et al., 2004). Continuous infusion of indomethacin into the vitreous cavity reduces subretinal neovascularization in monkeys by reducing new blood vessel growth (Frucht and Zauberman, 1984; Sakamoto et al., 1995; Yamada et al., 1999). The topical ns-NSAID nepafenac inhibits choroidal neovascularization and ischemia-induced retinal neovascularization by decreasing production of VEGF in mice (Takahashi et al., 2003). Ketorolac 0.4% inhibited choroidal neovascularization by suppression of retinal VEGF in a rat model (Kim et al., 2010). Corneal neovascularization was induced by alkaline cautery in Sprague–Dawley rats and measured after 4 days. This cautery-induced corneal angiogenesis was partially inhibited by indomethacin, 3.5 or 7 mg/kg per day, and NS-398, 10 or 20 mg/kg per day. The COX-1 selective inhibitor SC-560 given at 10 or 20 mg/kg per day did not inhibit angiogenesis (Castro et al., 2004). In addition, both indomethacin and NS-398 suppressed cautery-induced VEGF protein increase in the cornea (Castro et al., 2004). The effectiveness of a topical 0.1% diclofenac and a topical steroidal drug (betamethasone 0.1%) in preventing CME and blood–aqueous barrier disruption after small-incision cataract surgery was examined in a multicenter prospective randomized comparison study.

Five weeks after surgery, the incidence of fluorescein angiographic CME was lower in the diclofenac group than in the betamethasone group (Asano et al., 2008).

CONCLUSIONS

The ocular anatomical and physiological interspecies differences are important to look at closely when considering the pathophysiological effects of COX inhibition on the ocular system. COX-1 is constitutively expressed in the ocular system, while COX-2 is usually up-regulated during various ophthalmic pathological conditions. In several experimental nonclinical models of different ocular diseases, s-NSAIDs have shown beneficial ophthalmic effects. Topically applied ns-NSAIDs have been used in the management of various ophthalmic conditions. The impact of ocular pathology on topically applied NSAIDs pharmacokinetic properties needs to be considered.

REFERENCES

Abe T, Hayasaka Y, Zhang XY, Hayasaka S. Effects of intravenous administration of FR122047 (a selective cyclooxygenase 1 inhibitor) and FR188582 (a selective cyclooxygenase 2 inhibitor) on prostaglandin-E2-induced aqueous flare elevation in pigmented rabbits. Ophthalmic Res 2004;36:321–326.

Abran D, Li DY, Varma DR, Chemtob S. Characterization and ontogeny of PGE2 and PGF2 alpha receptors on the retinal vasculature of the pig. Prostaglandins 1995;50:253–267.

Adamis AP. Is diabetic retinopathy an inflammatory disease? Br J Ophthalmol 2002;86:363–365.

Adamis AP, Miller JW, Bernal MT, D'Amico DJ, et al. Increased vascular endothelial growth factor levels in the vitreous of eyes with proliferative diabetic retinopathy. Am J Ophthalmol 1994;118:445–450.

Aguirre SA, Huang W, Prasanna G, Jessen B. Corneal neovascularization and ocular irritancy responses in dogs following topical ocular administration of an EP4-prostaglandin E2 agonist. Toxicol Pathol 2009;37:911–920.

Aiello LP, Avery RL, Arrigg PG, Keyt BA, et al. Vascular endothelial growth factor in ocular fluid of patients with diabetic retinopathy and other retinal disorders. N Engl J Med 1994;331:1480–1487.

Amico C, Yakimov M, Catania MV, Giuffrida R, et al. Differential expression of cyclooxygenase-1 and cyclooxygenase-2 in the cornea during wound healing. Tissue Cell 2004;36:1–12.

Amrite AC, Ayalasomayajula SP, Cheruvu NP, Kompella UB. Single periocular injection of celecoxib–PLGA microparticles inhibits diabetes-induced elevations in retinal PGE2, VEGF, and vascular leakage. Invest Ophthalmol Vis Sci 2006;47:1149–1160.

Anthony TL, Lindsey JD, Aihara M, Weinreb RN. Detection of prostaglandin EP(1), EP(2), and FP receptor subtypes in human sclera. Invest Ophthalmol Vis Sci 2001;42:3182–3186.

Asakura T, Sano N, Shichi H. Prostaglandin synthesis and accumulation by porcine ciliary epithelium. J Ocul Pharmacol 1992;8:333–341.

Asano S, Miyake K, Ota I, Sugita G, et al. Reducing angiographic cystoid macular edema and blood–aqueous barrier disruption after small-incision phacoemulsification and foldable intraocular lens implantation: multicenter prospective randomized comparison of topical diclofenac 0.1% and betamethasone 0.1%. J Cataract Refract Surg 2008;34:57–63.

Ayalasomayajula SP, Kompella UB. Celecoxib, a selective cyclooxygenase-2 inhibitor, inhibits retinal vascular endothelial growth factor expression and vascular leakage in a streptozotocin-induced diabetic rat model. Eur J Pharmacol 2003;458:283–289.

Ayalasomayajula SP, Kompella UB. Retinal delivery of celecoxib is several-fold higher following subconjunctival administration compared to systemic administration. Pharm Res 2004;21:1797–1804.

Ayalasomayajula SP, Amrite AC, Kompella UB. Inhibition of cyclooxygenase-2, but not cyclooxygenase-1, reduces prostaglandin E2 secretion from diabetic rat retinas. Eur J Pharmacol 2004;498:275–278.

Bahn CF, Glassman RM, MacCallum DK, Lillie JH, et al. Postnatal development of corneal endothelium. Invest Ophthalmol Vis Sci 1986;27:44–51.

Balazsi AG, Rootman J, Drance SM, Schulzer M, et al. The effect of age on the nerve fiber population of the human optic nerve. Am J Ophthalmol 1984;97:760–766.

Bazan HE, Birkle DL, Beuerman RW, Bazan NG. Inflammation-induced stimulation of the synthesis of prostaglandins and lipoxygenase- reaction products in rabbit cornea. Curr Eye Res 1985a;4:175–179.

Bazan HE, Birkle DL, Beuerman R, Bazan NG. Cryogenic lesion alters the metabolism of arachidonic acid in rabbit cornea layers. Invest Ophthalmol Vis Sci 1985b;26:474–480.

Bazan NG., Allan G. Signal transduction and gene expression in the eye: a contemporary view of the pro-inflammatory, anti-inflammatory and modulatory roles of prostaglandins and other bioactive lipids. Surv Ophthalmol 1997;41 Suppl 2:S23–S34.

Berry M, Ellingham RB, Corfield AP. Membrane-associated mucins in normal human conjunctiva. Invest Ophthalmol Vis Sci 2000;41:398–403.

Bhattacherjee P. Prostaglandins and inflammatory reactions in the eye. Methods Find Exp Clin Pharmacol 1980;2:17–31.

Bhattacherjee P, Mukhopadhyay P, Tilley SL, Koller BH, et al. Blood–aqueous barrier in prostaglandin EP_2 receptor knockout mice. Ocular Immunol Inflammat 2002;10:187–196.

Biswas PS, Banerjee K, Kim B, Kinchington PR, et al. Role of inflammatory cytokine-induced cyclooxygenase 2 in the ocular immunopathologic disease. J Virol 2005;79:10589–10600.

Biswas S, Bhattacherjee P, Paterson CA. Prostaglandin E2 receptor subtypes, EP1, EP2, EP3 and EP4 in human and mouse ocular tissues: a comparative immunohistochemical study. Prostaglandins Leukot Essent Fatty Acids 2004;71:277–288.

Bito LZ. Species differences in the responses of the eye to irritation and trauma: a hypothesis of divergence in ocular defense mechanisms, and the choice of experimental animals for eye research. Exp Eye Res 1984;39:807–829.

Boeri D, Maiello M, Lorenzi M. Increased prevalence of microthromboses in retinal capillaries of diabetic individuals. Diabetes 2001;50:1432–1439.

Bonazzi A, Mastyugin V, Mieyal PA, Dunn MW, et al. Regulation of cyclooxygenase-2 by hypoxia and peroxisome proliferators in the corneal epithelium. J Biol Chem 2000;275:2837–2844.

Brusini P, Miani F, Tosoni C. Corneal thickness in glaucoma: An important parameter? Acta Ophthalmol Scand Suppl 2000;232:41–42.

Camras CB. Comparison of latanoprost and timolol in patients with ocular hypertension and glaucoma: a six-month masked, multicenter trial in the United States. The United States Latanoprost Study Group. Ophthalmology 1996;103:138–147.

Camras CB, Alm A, Watson P, Stjernschantz J. Latanoprost, a prostaglandin analog, for glaucoma therapy: efficacy and safety after 1 year of treatment in 198 patients. Latanoprost Study Groups. Ophthalmology 1996;103:1916–1924.

Castro, MR, Lutz D, Edelman JL. Effect of COX inhibitors on VEGF-induced retinal vascular leakage and experimental corneal and choroidal neovascularization. Exp Eye Res 2004;79:275–285.

Chemtob S, Beharry K, Rex J, Chatterjee T, et al. Ibuprofen enhances retinal and choroidal blood flow autoregulation in newborn piglets. Invest Ophthalmol Vis Sci 1991;32:1799–1807.

Chang MS, Tsai JC, Yang R, DuBois RN, et al. Induction of rabbit cyclooxygenase 2 in the anterior uvea following glaucoma filtration surgery. Curr Eye Res 1997;16:1147–1151.

Chen CH, Chen SC. Studies on soluble proteins of vitreous in experimental animals. Exp Eye Res 1981;32:381–388.

Cheng T, Cao W, Wen R, Steinberg RH, et al. Prostaglandin E2 induces vascular endothelial growth factor and basic fibroblast growth factor mRNA expression in cultured rat Müller cells. Invest Ophthalmol Vis Sci 1998;39:581–591.

Cheruvu NP, Amrite AC, Kompella UB. Effect of diabetes on transscleral delivery of celecoxib. Pharm Res 2009;26:404–414.

Ciulla TA, Amador AG, Zinman B. Diabetic retinopathy and diabetic macular edema: pathophysiology, screening, and novel therapies. Diabetes Care 2003;26:2653–2664.

Coulter DM, Clark DW. Disturbance of vision by COX-2 inhibitors. Expert Opin Drug Saf 2004;3:607–614.

Crider JY, Sharif NA. Functional pharmacological evidence for EP2 and EP4 prostanoid receptors in immortalized human trabecular meshwork and non-pigmented ciliary epithelial cells. J Ocul Pharmacol Ther 2001;17:35–46.

Cryan LM, Paraoan L, Hiscott P, Damato BE, et al. Expression of COX-2 and prognostic outcome in uveal melanoma. Curr Eye Res 2008;33:177–184.

Curcio CA, Sloan KR, Kalina RE, Hendrickson AE. Human photoreceptor topography. J Comp Neurol 1990;292:497–523.

Cursiefen C, Rummelt C, Kuchle M. Immunohistochemical localization of vascular endothelial growth factor, transforming growth factor alpha, and transforming growth factor beta1 in human corneas with neovascularization. Cornea 2000;19:526–533.

Damm J, Rau T, Maihöfner C, Pahl A, et al. Constitutive expression and localization of COX-1 and COX-2 in rabbit iris and ciliary body. Exp Eye Res 2001;72:611–621.

Davanger M, Ringvold A. The ciliary body and the suspension of the lens in a monkey (*Cercopithecus aethiops*): a scanning electron microscopic study. Acta Ophthalmol (Copenh) 1977;55:965–975.

Davies P, Bailey PJ, Goldenberg MM, Ford-Hutchinson AW. The role of arachidonic acid oxygenation products in pain and inflammation. Annu Rev Immunol 1984;2:335–357.

Degi R, Thore C, Bari F, Thrikawala N, et al. Ischemia increases prostaglandin H synthase 2 levels in retina and visual cortex in piglets. Graefes Arch Clin Exp Ophthalmol 2001;239:59–65.

Denlinger JL, Eisner G, Balazs EA. Age-related changes in the vitreus and lens of rhesus monkeys (*Macaca mulatta*). Exp Eye Res 1980;31:67–79.

DeSantis LL Jr, Sallee VL. Comparison of the effects of prostaglandins and their esters on blood–aqueous barrier integrity and intraocular pressure in rabbits. Prog Clin Biol Res 1989;312: 379–386.

Dimasi DP, Hewitt AW, Kagame K, Ruvama S, et al. Ethnic and mouse strain differences in central corneal thickness and association with pigmentation phenotype. PLoS One 2011;6:e22103.

Du Y, Sarthy VP, Kern TS. Interaction between NO and COX pathways in retinal cells exposed to elevated glucose and retina of diabetic rats. Am J Physiol Regul Integr Comp Physiol 2004;287: R735–R741.

Eakins KE, Whitelocke RA, Perkins ES, Bennett A, et al. Release of prostaglandins in ocular inflammation in the rabbit. Nat New Biol 1972;239:248–249.

Edelman JL, Castro MR, Wen Y. Correlation of VEGF expression by leukocytes with the growth and regression of blood vessels in the rat cornea. Invest Ophthalmol Vis Sci 1999;40:1112–1123.

Ehlers N, Hjortdal J. The cornea: epithelium and stroma. Adv Organ Biol 2005;10:83–111.

El-Asrar AM, Missotten L, Geboes K. Expression of cyclo-oxygenase-2 and downstream enzymes in diabetic fibrovascular epiretinal membranes. Br J Ophthalmol 2008;92:1534–1539.

Ellis PP, Pfoff, DS, Bloedow DC, Riegel M. Intraocular diclofenac and flurbiprofen concentrations in human aqueous humor following topical application. J Ocul Pharmacol 1994;10:677–682.

Ershov AV, Bazan NG. Induction of cyclooxygenase-2 gene expression in retinal pigment epithelium cells by photoreceptor rod outer segment phagocytosis and growth factors. J Neurosci Res 1999;58:254–261.

Fang C, Jiang Z, Tomlinson DR. Expression of constitutive cyclo-oxygenase (COX-1) in rats with streptozotocin-induced diabetes; effects of treatment with evening primrose oil or an aldose reductase inhibitor on COX-1 mRNA levels. Prostaglandins Leukot Essent Fatty Acids 1997;56:157–163.

Figueiredo A, Caissie AL, Callejo SA, McLean IW, et al. Cyclooxygenase-2 expression in uveal melanoma: novel classification of mixed-cell-type tumours. Can J Ophthalmol 2003;38:352–356.

Frank RN. Diabetic retinopathy. N Engl J Med 2004;350:48–58.

Friedland BR, Maren TH. The role of carbonic anhydrase in lens ion transport and metabolism. Ann NY Acad Sci 1984;429:582–586.

Frucht J, Zauberman H. Topical indomethacin effect on neovascularisation of the cornea and on prostaglandin E2 levels. Br J Ophthalmol 1984;68:656–659.

Gabelt BT, Kaufman PL. Prostaglandin F2 alpha increases uveoscleral outflow in the cynomolgus monkey. Exp Eye Res 1989;49:389–402.

Gardiner TA, Anderson HR, Degenhardt T, Thorpe SR, et al. Prevention of retinal capillary basement membrane thickening in diabetic dogs by a nonsteroidal anti-inflammatory drug. Diabetologia 2003;46:1269–1275.

Gaynes BI, Fiscella R. Topical nonsteroidal anti-inflammatory drugs for ophthalmic use: a safety review. Drug Saf 2002;25:233–250.

Gebhardt BM, Varnell ED, Kaufman HE. Inhibition of cyclooxygenase-2 synthesis suppresses herpes simplex virus type 1 reactivation. J Ocul Pharmacol Ther 2005;21:114–120.

Gersony WM, Peckham GJ, Ellison RC, Miettinen OS, et al. Effects of indomethacin in premature infants with patent ductus arteriosus: results of a national collaborative study. J Pediatr 1983;102:895–906.

Gherezghiher T, Koss MC, Nordquist RE, Wilkinson CP. Analysis of vitreous and aqueous levels of hyaluronic acid: application of high-performance liquid chromatography. Exp Eye Res 1987;45: 347–349.

Giblin FJ. Glutathione: a vital lens antioxidant. J Ocul Pharmacol Ther 2000;16:121–135.

Gilger BC, Wright JC, Whitley RD, McLaughlin SA. Corneal thickness measured by ultrasonic pachymetry in cats. Am J Vet Res 1993;54:228–230.

Gilmour MA, Lehenbauer TW. Comparison of tepoxalin, carprofen, and meloxicam for reducing intraocular inflammation in dogs. Am J Vet Res 2009;70:902–907.

Giuliano EA. Nonsteroidal anti-inflammatory drugs in veterinary ophthalmology. Vet Clin North Am Small Anim Pract 2004;34:707–723.

Green K, Kearse EC, McIntyre OL. Interaction between Δ^9-tetrahydrocannabinol and indomethacin. Ophthalmic Res 2001: 33:217–220.

Gum GG, Gelatt KN, Samuelson DA. Maturation of the retina of the canine neonate as determined by electroretinography and histology. Am J Vet Res 1984;45:1166–1171.

Gupta AG, Hirakata A, Proia AD. Effect of inhibitors of arachidonic acid metabolism on corneal reepithelialization in the rat. Exp Eye Res 1993;56:701–708.

Hall DW, Jaitly KD. Inflammatory responses of the rabbit eye to prostaglandins. Agents Actions Suppl 1977;2:123–133.

Hardy P, Abran D, Li DY, Fernandez H, et al. Free radicals in retinal and choroidal blood flow autoregulation in the piglet: interaction with prostaglandins. Invest Ophthalmol Vis Sci 1994;35:580–591.

Hardy P, Bhattacharya M, Abran D, Peri KG, et al. Increases in retinovascular prostaglandin receptor functions by cyclooxygenase-1 and -2 inhibition. Invest Ophthalmol Vis Sci 1998;39:1888–1898.

Hariprasad SM, Callanan D, Gainey S, He YG, et al. Cystoid and diabetic macular edema treated with nepafenac 0.1%. J Ocul Pharmacol Ther 2007;23:585–590.

Haritoglou C, Kook D, Neubauer A, Wolf A, et al. Intravitreal bevacizumab (Avastin) therapy for persistent diffuse diabetic macular edema. Retina 2006;26:999–1005.

Harocopos GJ, Shui YB, McKinnon M, Holekamp NM, et al. Importance of vitreous liquefaction in age-related cataract. Invest Ophthalmol Vis Sci 2004;45:77–85.

Hata Y, Clermont A, Yamauchi T, Pierce EA, et al. Retinal expression, regulation, and functional bioactivity of prostacyclin-stimulating factor. J Clin Invest 2000;106:541–550.

Head KA. Natural therapies for ocular disorders, part two: cataracts and glaucoma. Altern Med Rev 2001;6:141–166.

Heckmann M. Pathogenesis of retinopathy of prematurity. Ophthalmologe 2008;105: 1101–1107.

Hinz B, Brune K, Pahl A. Cyclooxygenase-2 expression in lipopolysaccharide-stimulated human monocytes is modulated by cyclic AMP, prostaglandin E(2), and nonsteroidal anti-inflammatory drugs. Biochem Biophys Res Commun 2000;278:790–796.

Hiraoka M, Inoue K, Ohtaka-Maruyama C, Ohsako S, et al. Intracapsular organization of ciliary zonules in monkey eyes. Anat Rec (Hoboken) 2010;293:1797–1804.

Holland EJ, Mannis MJ. *Ocular Surface Disease: Medical and Surgical Management*. New York: Springer-Verlag, 2002.

Hommer A. A review of preserved and preservative-free prostaglandin analogues for the treatment of open-angle glaucoma and ocular hypertension. Drugs Today (Barc) 2010;46:409–416.

Honrubia F, García-Sánchez J, Polo V, de la Casa JM, et al. Conjunctival hyperaemia with the use of latanoprost versus other prostaglandin analogues in patients with ocular hypertension or glaucoma: a meta-analysis of randomised clinical trials. Br J Ophthalmol 2009;93:316–321.

Hutchinson AJ, Coons SC, Chou CL, Xu W, et al. Induction of angiogenic immediate early genes by activation of FP prostanoid receptors in cultured human ciliary smooth muscle cells. Curr Eye Res 2010;35:408–418.

Isawi H, Dhaliwal DK. Corneal melting and perforation in Stevens–Johnson syndrome following topical bromfenac use. J Cataract Refract Surg 2007;33:1644–1646.

Joussen AM, Poulaki V, Mitsiades N, Kirchhof B, et al. Nonsteroidal anti-inflammatory drugs prevent early diabetic retinopathy via TNF-alpha suppression. FASEB J 2002;16:438–440.

Joyce NC, Joyce SJ, Powell SM, Pierce EA, et al. EGF and PGE2: effects on corneal endothelial cell migration and monolayer spreading during wound repair in vitro. Curr Eye Res 1995;14:601–609.

Ju WK, Neufeld AH. Cellular localization of cyclooxygenase-1 and cyclooxygenase-2 in the normal mouse, rat, and human retina. J Comp Neurol 2002;452:392–399.

Ju WK, Kim KY, Neufeld AH. Increased activity of cyclooxygenase-2 signals early neurodegenerative events in the rat retina following transient ischemia. Exp Eye Res 2003;77:137–145.

Jumblatt MM. Autocrine regulation of corneal endothelium by prostaglandin E2. Invest Ophthalmol Vis Sci 1994;35:2783–2790.

Jumblatt MM, Willer SS. Corneal endothelial repair: regulation of prostaglandin E2 synthesis. Invest Ophthalmol Vis Sci 1996;37:1294–1301.

Kamphuis W, Schneemann A, Shichi H, Broersma L, et al. Immunolocalization of prostanoid EP receptor isotypes in human trabecular meshwork. Curr Eye Res 2004;29:17–26.

Kaplan-Messas A, Naveh N, Avni I, Marshall J. Ocular hypotensive effects of cholinergic and adrenergic drugs may be influenced by prostaglandins E_2 in the human and rabbit eye. Eur J Ophthalmol 2003;13:18–23.

Karim MM, Hayashi Y, Inoue M, Imai Y, et al. COX-2 expression in retinoblastoma. Am J Ophthalmol 2000;129:398–401.

Kase S, Saito W, Ohno S, Ishida S. Cyclo-oxygenase-2 expression in human idiopathic epiretinal membrane. Retina 2010;30:719–723.

Kashiwagi K, Tsukahara S. Effect of non-steroidal anti-inflammatory ophthalmic solution on intraocular pressure reduction by latanoprost. Br J Ophthalmol 2003;87:297–301.

Keeting PE, Lysz TW, Centra M, Fu SC. Prostaglandin biosynthesis in the rat lens. Invest Ophthalmol Vis Sci 1985;26:1083–1086.

Kern TS. Contributions of inflammatory processes to the development of the early stages of diabetic retinopathy. Exp Diabetes Res 2007;2007:95103.

Kern TS, Engerman RL. Pharmacological inhibition of diabetic retinopathy: aminoguanidine and aspirin. Diabetes 2001;50:1636–1642.

Kern TS, Miller CM, Du Y, Zheng L, et al. Topical administration of nepafenac inhibits diabetes-induced retinal microvascular disease and underlying abnormalities of retinal metabolism and physiology. Diabetes 2007;56:373–379.

Kerstetter JR, Brubaker RF, Wilson SE, Kullerstrand LJ. Prostaglandin F2 alpha- 1-isopropylester lowers intraocular pressure without decreasing aqueous humor flow. Am J Ophthalmol 1988;105:30–34.

Kevany BM, Palczewski K. Phagocytosis of retinal rod and cone photoreceptors. Physiology (Bethesda) 2010;25:8–15.

Kim SJ, Toma HS, Barnett JM, Penn JS. Ketorolac inhibits choroidal neovascularization by suppression of retinal VEGF. Exp Eye Res 2010;91:537–543.

Kitagawa K, Hayasaka S, Nagaki Y, Watanabe K. Effects of tetramethylpyrazine on prostaglandin E(2) and prostaglandin E(2) receptor agonist induced disruption of blood aqueous barrier in pigmented rabbits. Jpn J Ophthalmol 2001;45:227–232.

Kleinberg TT, Tzekov RT, Stein L, Ravi N, Kaushal S. Vitreous substitutes: a comprehensive review. Surv Ophthalmol 2011;56:300–323.

Knop E, Knop N, Millar T, Obata H, et al. The International Workshop on Meibomian Gland Dysfunction: Report of the Subcommittee on Anatomy, Physiology, and Pathophysiology of the Meibomian Gland. Invest Ophthalmol Vis Sci 2011;52:1938–1978.

Koralewska-Makar A, Johnsson C, Bruun A, Stenevi U, et al. COX-2 inhibitors prolong trauma-induced elevations of iris hyaluronan. J Ocul Pharmacol Ther 2003;19:385–395.

Kosaka T, Mishima HK, Kiuchi Y, Kataoka K. The effects of prostaglandins on the blood–ocular barrier. Jpn J Ophthalmol 1995;39:368–376.

Kulkarni P, Payne S. Eicosanoids in bovine retinal microcirculation. J Ocul Pharmacol Ther 1997;13:139–149.

Kulkarni PS, Srinivasan BD. Cyclooxygenase and lipoxygenase pathways in anterior uvea and conjunctiva. Prog Clin Biol Res 1989;312:39–52.

Kulkarni PS, Fleisher L, Srinivasan BD. The synthesis of cyclooxygenase products in ocular tissues of various species. Curr Eye Res 1984;3:447–452.

Kulkarni PS, Kaufman PL, Srinivasan BD. Cyclo-oxygenase and lipoxygenase pathways in cynomolgus and rhesus monkey conjunctiva, anterior uvea and eyelids. Curr Eye Res 1987;6:801–808.

Lapuerta P, Schein SJ. A four-surface schematic eye of macaque monkey obtained by an optical method. Vision Res 1995;35:2245–2254.

Lerman S. An experimental and clinical evaluation of lens transparency and aging. J Gerontol 1983;38:293–301.

Liclican EL, Nguyen V, Sullivan AB, Gronert K. Selective activation of the prostaglandin E2 circuit in chronic injury-induced pathologic angiogenesis. Invest Ophthalmol Vis Sci 2010;51:6311–6320.

Lopez PF, Sippy BD, Lambert HM, Thach AB, et al. Transdifferentiated retinal pigment epithelial cells are immunoreactive for vascular endothelial growth factor in surgically excised age-related macular degeneration-related choroidal neovascular membranes. Invest Ophthalmol Vis Sci 1996;37:855–868.

Los LI. The rabbit as an animal model for post-natal vitreous matrix differentiation and degeneration. Eye (Lond) 2008;22:1223–1232.

Los LI, van Luyn MJ, Nieuwenhuis P. Organization of the rabbit vitreous body: lamellae, Cloquet's channel and a novel structure, the "alae canalis Cloqueti." Exp Eye Res 1999;69:343–350.

Lütjen-Drecoll E, Tamm E. Morphological study of the anterior segment of cynomolgus monkey eyes following treatment with prostaglandin F2 alpha. Exp Eye Res 1988;47:761–769.

Maihöfner C, Schlotzer-Schrehardt U, Guhring H, Zeilhofer HU, et al. Expression of cyclooxygenase-1 and -2 in normal and glaucomatous human eyes. Invest Ophthalmol Vis Sci 2001;42:2616–2624.

Maloney SC, Fernandes BF, Castiglione E, Antecka E, et al. Expression of cyclooxygenase-2 in choroidal neovascular membranes from age-related macular degeneration patients. Retina 2009;29: 176–180.

Manzanas LL, Pastor JC, Munoz R, Girbes T. Intraocular irrigating solutions and vitrectomy-related changes (in protein, lactic and ascorbic acid) in rabbit vitreous. Ophthalmic Res 1992;24:61–67.

Marshall JL, Stanfield KM, Silverman L, Khan KN. Enhanced expression of cyclooxygenase-2 in glaucomatous dog eyes. Vet Ophthalmol 2004;7:59–62.

Masferrer JL, Koki A, Seibert K. COX-2 inhibitors. A new class of antiangiogenic agents. Ann NY Acad Sci 1999;889:84–86.

Masferrer JL, Leahy KM, Koki AT, Zweifel BS, et al. Antiangiogenic and antitumor activities of cyclooxygenase-2 inhibitors. Cancer Res 2000;60:1306–1311.

Massof RW, Chang FW. A revision of the rat schematic eye. Vision Res 1972;12:793–796.

Matsuo T, Cynader MS. Localisation of prostaglandin F2 alpha and E2 binding sites in the human eye. Br J Ophthalmol 1992;76:210–213.

Matsuo T, Cynader MS. The EP2 receptor is the predominant prostanoid receptor in the human ciliary muscle. Br J Ophthalmol 1993;77:110–114.

McInnis CL, Giuliano EA, Johnson PJ, Turk Jr. Immunohistochemical evaluation of cyclooxygenase expression in corneal squamous cell carcinoma in horses. Am J Vet Res 2007;68:165–170.

McMenamin PG. The distribution of immune cells in the uveal tract of the normal eye. Eye (Lond) 1997;11:183–193.

Merindano MD, Costa J, Canals M, Potau JM, et al. A comparative study of Bowman's layer in some mammals: relationships with other constituent corneal structures. Eur J Anat 2002;6:133–139.

Mian SI, Gupta A, Pineda R 2nd. Corneal ulceration and perforation with ketorolac tromethamine (Acular) use after PRK. Cornea 2006;25:232–234.

Miller JD, Eakins KE, Atwal M. The release of PGE2-like activity into aqueous humor after paracentesis and its prevention by aspirin. Invest Ophthalmol 1973;12:939–942.

Miller JW, Adamis AP, Shima DT, D'Amore PA, et al. Vascular endothelial growth factor/vascular permeability factor is temporally and spatially correlated with ocular angiogenesis in a primate model. Am J Pathol 1994;145:574–584.

Millichamp NJ, Dziezyc J, Rohde BH, Chiou GC, et al. Acute effects of anti-inflammatory drugs on neodymium:yttrium aluminum garnet laser-induced uveitis in dogs. Am J Vet Res 1991;52: 1279–1284.

Miranda OC, Bito LZ. The putative and demonstrated miotic effects of prostaglandins in mammals. Prog Clin Biol Res 1989;312:171–195.

Mishima HK, Kiuchi Y, Takamatsu M, et al. Circadian intraocular pressure management with latanoprost: diurnal and nocturnal intraocular pressure reduction and increased uveoscleral outflow. Surv Ophthalmol 1997;41 Suppl 2:S139–S144.

Miyamoto T, Saika S, Okada Y, Kawashima Y, et al. Expression of cyclooxygenase-2 in corneal cells after photorefractive keratectomy and laser in situ keratomileusis in rabbits. J Cataract Refract Surg 2004;30:2612–2617.

Módis L Jr, Szalai E, Kertész K, Kemény-Beke A, et al. Evaluation of the corneal endothelium in patients with diabetes mellitus type I and II. Histol Histopathol 2010;25:1531–1537.

Mori A, Saito M, Sakamoto K, Narita M, et al. Stimulation of prostanoid IP and EP(2) receptors dilates retinal arterioles and increases retinal and choroidal blood flow in rats. Eur J Pharmacol 2007;570: 135–141.

Mukhopadhyay M, Blan L, Bhattacherjee P, Paterson CA. Localization of EP1 and FP receptors in human ocular tissues by in situ hybridization. Invest Ophthalmol Vis Sci 2001;42:424–428.

Mutti DO, Zadnik K, Murphy CJ. Naturally occurring vitreous chamber-based myopia in the Labrador retriever. Invest Ophthalmol Vis Sci 1999;40:1577–1584.

Neufeld AH, Hernandez MR, Gonzalez M, Geller A. Cyclooxygenase-1 and cyclooxygenase-2 in the human optic nerve head. Exp Eye Res 1997;65:739–745.

Nilsson SF, Drecoll E, Lütjen-Drecoll E, Toris CB, et al. The prostanoid EP2 receptor agonist butaprost increases uveoscleral outflow in the cynomolgus monkey. Invest Ophthalmol Vis Sci 2006;47:4042–4049.

Noecker RS, Dirks MS, Choplin NT, Bernstein P, et al. A six-month randomized clinical trial comparing the intraocular pressure-lowering efficacy of bimatoprost and latanoprost in patients with ocular hypertension or glaucoma. Am J Ophthalmol 2003;135:55–63.

Ocklind A, Lake S, Wentzel P, Nistér M, Stjernschantz J. Localization of the prostaglandin F2 alpha receptor messenger RNA and protein in the cynomolgus monkey eye. Invest Ophthalmol Vis Sci 1996;37:716–726.

Oka T, Shearer T, Azuma M. Involvement of cyclooxygenase-2 in rat models of conjunctivitis. Curr Eye Res 2004;29:27–34.

Ollivier FJ, Brooks DE, Komaromy AM, Kallberg ME, et al. Corneal thickness and endothelial cell density measured by non-contact specular microscopy and pachymetry in rhesus macaques (*Macaca mulatta*) with laser-induced ocular hypertension. Exp Eye Res 2003;76:671–677.

Oshima Y, Sakaguchi H, Gomi F, Tano Y. Regression of iris neovascularization after intravitreal injection of bevacizumab in patients with proliferative diabetic retinopathy. Am J Ophthalmol 2006;142:155–158.

Ottino P, Bazan HEP. Cyclooxygenase-2 (COX-2) metabolites produced by platelet-activating factor (PAF) stimulation in corneal epithelial cells activate urokinase type plasminogen activator (uPA) and matrix metalloproteinases (MMP)-1 and 9. Invest Ophthalmol Vis Sci 2000;41:S906.

Ozkiriş A. Intravitreal bevacizumab (Avastin) for primary treatment of diabetic macular oedema. Eye (Lond) 2009;23:616–620.

Ozturk F, Kurt E, Inan U, Kortunay S, et al. The effects of prolonged acute use and inflammation on the ocular penetration of topical ciprofloxacin. Int J Pharm 2000;204:97–100.

Packer O, Hendrickson AE, Curcio CA. Photoreceptor topography of the retina in the adult pigtail macaque (*Macaca nemestrina*). J Comp Neurol 1989;288:165–183.

Paglia D, Dubielzig RR, Kado-Fong HK, Maggs DJ. Expression of cyclooxygenase-2 in canine uveal melanocytic neoplasms. Am J Vet Res 2009;70:1284–1290.

Philipp W, Klima G, Miller K. Clinicopathological findings 11 months after implantation of a functioning aqueous-drainage silicone implant. Graefes Arch Clin Exp Ophthalmol 1990;228:481–486.

Pires I, Garcia A, Prada J, Queiroga FL. COX-1 and COX-2 expression in canine cutaneous, oral and ocular melanocytic tumours. J Comp Pathol 2010;143:142–149.

Pirie CG, Maranda LS, Pizzirani S. Effect of topical 0.03% flurbiprofen and 0.005% latanoprost, alone and in combination, on normal canine eyes. Vet Ophthalmol 2011;14:71–79.

Plummer CE, Ramsey DT, Hauptman JG. Assessment of corneal thickness, intraocular pressure, optical corneal diameter, and axial globe dimensions in miniature horses. Am J Vet Res 2003;64:661–665.

Poyer JF, Gabelt B, Kaufman PL. The effect of topical PGF2 alpha on uveoscleral outflow and outflow facility in the rabbit eye. Exp Eye Res 1992;54:277–283.

Presta LG, Chen H, O'Connor SJ, Chisholm V, et al. Humanization of an antivascular endothelial growth factor monoclonal antibody for the therapy of solid tumors and other disorders. Cancer Res 1997;57:4593–4599.

Preud'homme Y, Demolle D, Boeynaems JM. Metabolism of arachidonic acid in rabbit iris and retina. Invest Ophthalmol Vis Sci 1985;26:1336–1342.

Price FW. New pieces for the puzzle: nonsteroidal anti-inflammatory drugs and corneal ulcers. J Cataract Refract Surg 2000;26:1263–1265.

Radi ZA, Render JA. The pathophysiologic role of cyclo-oxygenases in the eye. J Ocul Pharmacol Ther 2008;24:141–151.

Rasmussen CA, Kaufman PL. Primate glaucoma models. Glaucoma 2005;14:311–314.

Reitmair A, Lambrecht NW, Yakubov I, Nieves A, et al. Prostaglandin E2 receptor subtype EP2- and EP4-regulated gene expression profiling in human ciliary smooth muscle cells. Physiol Genomics 2010;42:348–360.

Richter M, Krauss AH, Woodward DF, Lütjen-Drecoll E. Morphological changes in the anterior eye segment after long-term treatment with different receptor selective prostaglandin agonists and a prostamide. Invest Ophthalmol Vis Sci 2003;44:4419–4426.

Rösch S, Ramer R, Brune K, Hinz B. Prostaglandin E2 induces cyclooxygenase-2 expression in human non-pigmented ciliary epithelial cells through activation of p38 and p42/44 mitogen-activated protein kinases. Biochem Biophys Res Commun 2005;338:1171–1178.

Rüfer F, Sander S, Klettner A, Frimpong-Boateng A, et al. Characterization of the thinnest point of the cornea compared with the central corneal thickness in normal subjects. Cornea 2009;28:177–180.

Saeki T, Ota T, Aihara M, Araie M. Effects of prostanoid EP agonists on mouse intraocular pressure. Invest Ophthalmol Vis Sci 2009;50:2201–2208.

Sagara T, Gaton DD, Lindsey JD, Gabelt BT, et al. Topical prostaglandin F2 alpha treatment reduces collagen types I, III, and IV in the monkey uveoscleral outflow pathway. Arch Ophthalmol 1999;117:794–801.

Sakamoto T, Soriano D, Nassaralla J, Murphy TL, et al. Effect of intravitreal administration of indomethacin on experimental subretinal neovascularization in the subhuman primate. Arch Ophthalmol 1995;113:222–226.

Sanchez RM, Dunkelberger GR, Quigley HA. The number and diameter distribution of axons in the monkey optic nerve. Invest Ophthalmol Vis Sci 1986;27:1342–1350.

Sarchahi AA, Abbasi N, Gholipour MA. Effects of an unfixed combination of latanoprost and pilocarpine on the intraocular pressure and pupil size of normal dogs. Vet Ophthalmol 2012;15 Suppl 1:64–70.

Sawada A, Yamamoto T, Takatsuka N. Randomized crossover study of latanoprost and travoprost in eyes with open-angle glaucoma. Graefes Arch Clin Exp Ophthalmol 2012;250:123–129.

Schalnus R. Topical nonsteroidal anti-inflammatory therapy in ophthalmology. Ophthalmologica 2003;217:89–98.

Schlötzer-Schrehardt U, Zenkel M, Nüsing RM. Expression and localization of FP and EP prostanoid receptor subtypes in human ocular tissues. Invest Ophthalmol Vis Sci 2002;43:1475–1487.

Sebag J, Balazs EA. Morphology and ultrastructure of human vitreous fibers. Invest Ophthalmol Vis Sci 1989;30:1867–1871.

Sellers RS, Silverman L, Khan KN. Cyclooxygenase-2 expression in the cornea of dogs with keratitis. Vet Pathol 2004;41:116–121.

Sennlaub F, Valamanesh F, Vazquez-Tello A, El-Asrar AM, et al. Cyclooxygenase-2 in human and experimental ischemic proliferative retinopathy. Circulation 2003;108:198–204.

Shimazaki J, Saito H, Yang HY, Toda I, et al. Persistent epithelial defect following penetrating keratoplasty: an adverse effect of diclofenac eyedrops. Cornea 1995;14:623–627.

Shimazaki J, Fujishima H, Yagi Y, Tsubota K, et al. Effects of diclofenac eye drops on corneal epithelial structure and function after small-incision cataract surgery. Ophthalmology 1996;103:50–57.

Smith KM, Scase TJ, Miller JL, Donaldson D, et al. Expression of cyclooxygenase-2 by equine ocular and adnexal squamous cell carcinomas. Vet Ophthalmol 2008;11:8–14.

Sponsel WE, Paris G, Trigo Y, Pena M, et al. Latanoprost and brimonidine: therapeutic and physiologic assessment before and after oral nonsteroidal anti-inflammatory therapy. Am J Ophthalmol 2002;133:11–18.

Steven P, Gebert A. Conjunctiva-associated lymphoid tissue: current knowledge, animal models and experimental prospects. Ophthalmic Res 2009;42:2–8.

Stewart RH, Grillone LR, Shiffman ML, Donnenfeld ED, et al. The systemic safety of bromfenac ophthalmic solution 0.09%. J Ocul Pharmacol Ther 2007;23:601–612.

Streilein JW, Cousins SW. Aqueous humor factors and their effect on the immune response in the anterior chamber. Curr Eye Res 1990;9 Suppl: 175–182.

Szél A, Röhlich P, Caffé AR, van Veen T. Distribution of cone photoreceptors in the mammalian retina. Microsc Res Tech 1996;35:445–462.

Takahashi K, Saishin Y, Saishin Y, Mori K, et al. Topical nepafenac inhibits ocular neovascularization. Invest Ophthalmol Vis Sci 2003;44:409–415.

Takahashi H, Yanagi Y, Tamaki Y, Uchida S, et al. COX-2-selective inhibitor, etodolac, suppresses choroidal neovascularization in a mice model. Biochem Biophys Res Commun 2004;325: 461–466.

Takamatsu M, Hotehama Y, Goh Y, Mishima HK. Localization of prostaglandin E receptor subtypes in the ciliary body of mouse eye. Exp Eye Res 2000;70:623–628.

Taylor AW, Yee DG. Somatostatin is an immunosuppressive factor in aqueous humor. Invest Ophthalmol Vis Sci 2003;44:2644–2649.

Taylor AW, Streilein JW, Cousins SW. Immunoreactive vasoactive intestinal peptide contributes to the immunosuppressive activity of normal aqueous humor. J Immunol 1994;153:1080–1086.

The DAMAD Study Group. Effect of aspirin alone and aspirin plus dipyridamole in early diabetic retinopathy: a multicenter randomized controlled clinical trial. Diabetes 1989;38:491–498.

Thomas MA, O'Grady GE, Swartz SL. Prostaglandin levels in human vitreous. Br J Ophthalmol 1985;69:275–279.

Tseng SH, Tang MJ. Na,K-ATPase in lens epithelia from patients with senile cataracts. J Formos Med Assoc 1999;98:627–632.

Tuft SJ, Williams KA, Coster DJ. Endothelial repair in the rat cornea. Invest Ophthalmol Vis Sci 1986;27:1199–1204.

Ueta M, Matsuoka T, Narumiya S, Kinoshita S. Prostaglandin E receptor subtype EP3 in conjunctival epithelium regulates late-phase reaction of experimental allergic conjunctivitis. J Allergy Clin Immunol 2009;123:466–471.

Ueta M, Sotozono C, Yokoi N, Inatomi T, et al. Prostaglandin E receptor subtype EP3 expression in human conjunctival epithelium and its changes in various ocular surface disorders. PLoS One 2011;6:e25209.

Vakkur GJ, Bishop PO. The schematic eye in the cat. Vision Res 1963;61:357–381.

van Delft JL, van Haeringen NJ, Glasius E, Barthen ER, et al. Comparison of the effects of corticosteroids and indomethacin on the response of the blood–aqueous barrier to injury. Curr Eye Res 1987;6:419–425.

Ventura AC, Böhnke M, Mojon DS. Central corneal thickness measurements in patients with normal tension glaucoma, primary open angle glaucoma, pseudoexfoliation glaucoma, or ocular hypertension. Br J Ophthalmol 2001;85:792–795.

Vogel M. Postnatal development of the cat's retina: a concept of maturation obtained by qualitative and quantitative examinations. Albrecht Von Graefes Arch Klin Exp Ophthalmol 1978;208:93–107.

Wang J, Wu Y, Heegaard S, Kolko M. Cyclooxygenase-2 expression in the normal human eye and its expression pattern in selected eye tumours. Acta Ophthalmol 2011;89:681–685.

Ward DA. Comparative efficacy of topically applied flurbiprofen, diclofenac, tolmetin, and suprofen for the treatment of experimentally induced blood–aqueous barrier disruption in dogs. Am J Vet Res 1996;57:875–878.

Weber H, Landwehr G, Kilp H, Neubauer H. The mechanical properties of the vitreous of pig and human donor eyes. Ophthalmic Res 1982;14:335–343.

Weinreb RN, Kashiwagi K, Kashiwagi F, Tsukahara S, et al. Prostaglandins increase matrix metalloproteinase release from human ciliary smooth muscle cells. Invest Ophthalmol Vis Sci 1997;38:2772–2780.

Wen Y, Bekhor I. Levels of expression of hexokinase, aldose reductase and sorbitol dehydrogenase genes in lens of mouse and rat. Curr Eye Res 1993;12:323–332.

Wilkinson-Berka JL. Vasoactive factors and diabetic retinopathy: vascular endothelial growth factor, cycoloxygenase-2 and nitric oxide. Curr Pharm Des 2004;10:3331–3348.

Wilkinson-Berka JL, Alousis NS, Kelly DJ, Gilbert RE, et al. COX-2 inhibition and retinal angiogenesis in a mouse model of retinopathy of prematurity. Invest Ophthalmol Vis Sci 2003;44:974–979.

Willoughby CE, Ponzin D, Ferrari S, Lobo A, et al. Anatomy and physiology of the human eye: effects of mucopolysaccharidoses disease on structure and function: a review. Clin Exp Ophthalmol 2010;38:2–11.

Wilson SE, Netto M, Ambrosio R Jr. Corneal cells: chatty in development, homeostasis, wound healing, and disease. Am J Ophthalmol 2003;136:530–536.

Woodward DF, Nilsson SFE, Toris CB, Kharlamb AB, et al. Prostanoid EP4 receptor stimulation produces ocular hypotension by a mechanism that does not appear to involve uveoscleral outflow. Invest Ophthalmol Vis Sci 2009;50:3320–3328.

Xu KP, Riggs A, Ding Y, Yu FS. Role of ErbB2 in corneal epithelial wound healing. Invest Ophthalmol Vis Sci 2004;45:4277–4283.

Yagci R, Oflu Y, Dincel A, Kaya E, et al. Penetration of second-, third-, and fourth-generation topical fluoroquinolone into aqueous and vitreous humour in a rabbit endophthalmitis model. Eye 2007;21:990–994.

Yamada M, Kawai M, Kawai Y, Mashima Y. The effect of selective cyclooxygenase-2 inhibitor on corneal angiogenesis in the rat. Curr Eye Res 1999;19:300–304.

Yamaji K, Yoshitomi T, Ishikawa H, Usui S. Prostaglandins E1 and E2, but not F2 alpha or latanoprost, inhibit monkey ciliary muscle contraction. Curr Eye Res 2005;30:661–665.

Yamauchi H, Iso T, Iwao J, Iwata H. The role of prostaglandins in experimental ocular inflammations. Agents Actions 1979;9:280–283.

Yoganathan P, Deramo VA, Lai JC, Tibrewala RK, et al. Visual improvement following intravitreal bevacizumab (Avastin) in exudative age-related macular degeneration. Retina 2006;26:994–998.

Yoshida H, Kubota T, Krueger JM. A cyclooxygenase-2 inhibitor attenuates spontaneous and TNF-alpha-induced non-rapid eye movement sleep in rabbits. Am J Physiol Regul Integr Comp Physiol 2003;285:R99–R109.

Yousufzai SY, Ye Z, Abdel-Latif AA. Prostaglandin F2 alpha and its analogs induce release of endogenous prostaglandins in iris and ciliary muscles isolated from cat and other mammalian species. Exp Eye Res 1996;63:305–310.

Yu FS, Yin J, Xu K, Huang J. Growth factors and corneal epithelial wound healing. Brain Res Bull 2010;81:229–235.

Zeiss CJ. Animals as models of age-related macular degeneration: an imperfect measure of the truth. Vet Pathol 2010;47:396–413.

Zhao C, Shichi H. Immunocytochemical localization of prostaglandin E2 receptor subtypes in porcine ocular tissues: II. Nonuveal tissues. J Ocul Pharmacol Ther 1995;11:437–445.

Zhao C, Fujimoto N, Shichi H. Immunocytochemical localization of prostaglandin E2 receptor subtypes in porcine ocular tissues: I. Uveal tissues. J Ocul Pharmacol Ther 1995;11:421–435.

Zheng L, Howell SJ, Hatala DA, Huang K, et al. Salicylate-based anti-inflammatory drugs inhibit the early lesion of diabetic retinopathy. Diabetes 2007;56:337–345.

Zhou R, Horai R, Mattapallil MJ, Caspi RR. A new look at immune privilege of the eye: dual role for the vision-related molecule retinoic acid. J Immunol 2011;187:4170–4177.

Zhou X, Wang W, Lu F, Hu S, et al. A comparative gene expression profile of the whole eye from human, mouse, and guinea pig. Mol Vis 2007;13:2214–2221.

CHAPTER 6

RESPIRATORY SYSTEM

INTRODUCTION

Respiratory tissues respond to physiological and pathophysiological stimuli via activation of phospholipases and subsequent release of biologically active metabolites such as prostaglandins (PGs) from membrane phospholipids (Radi and Ackermann, 2011). PGs have receptors localized in a variety of respiratory tissues and affect its physiology and pathophysiology (Rolin et al., 2006). Cyclooxygenase isoenzymes (COX-1 and COX-2) catalyze the conversion of arachidonic acids (AA) to PGs (Radi and Ackermann, 2011). The COX-1 isoenzyme is expressed in a variety of respiratory tissues, and this expression is variable among different animal species (Radi and Ostroski, 2007; Radi and Ackermann, 2011). Prostanoids, generated by COX, are present in high concentrations during various pulmonary disease conditions (Jackson et al., 1993; Fulkerson et al., 1996; Vigano et al., 1998). COX-2 selective nonsteroidal anti-inflammatory drugs (COX-2 s-NSAIDs) and nonselective nonsteroidal anti-inflammatory drugs (ns-NSAIDs) inhibit COX isoenzymes. In this chapter we discuss the expression of COX-1 and COX-2 in the respiratory system, the role of prostaglandins and COX in respiratory physiology and pathophysiology, and the effects of COX-2 s-NSAIDs and ns-NSAIDs in pulmonary conditions.

COMPARATIVE PHYSIOLOGICAL AND ANATOMICAL ASPECTS OF THE RESPIRATORY SYSTEM

Understanding the comparative physiological and anatomical aspects of the respiratory system is important for an understanding of the pathophysiological consequences of COX inhibition on this system. The lungs are the principal structures of the respiratory system. The respiratory system has conductive and transitional systems (Mead, 1961). The conductive system includes nasal cavity, pharynx, larynx, trachea, and bronchi. Conducting airways are lined by pseudostratified ciliated epithelium that transports the air–surface liquid (ASL) toward the pharynx or nares (Ackermann et al., 2010). ASL is produced by secretion products from goblet cells and submucosal glands. Within the ASL are mucins as well as antimicrobial proteins (e.g., lysozyme, lactoferrin, surfactant proteins A and D, SLPI, PLUNC, and

Comparative Pathophysiology and Toxicology of Cyclooxygenases, First Edition. Zaher A. Radi.

others); and antimicrobial peptides (defensins and RNAase 7), which are produced by ciliated epithelial cells. The oxidative defense system (ODS) is composed of lactoperoxidase (LPO) produced by submucosal glands, dual function oxidases 1 and 2 (Duox1 and Duox2) produced by epithelia, and thiocyanate (SCN^-) transport proteins produced by epithelia. Duox1 and 2 produce hydrogen peroxide (H_2O_2), which, in the presence of SCN^-, is converted to $OSCN^-$, which has potent antimicrobial activity (Ackermann et al., 2010). The transitional system is composed of bronchioles and an alveolar gas exchange system (Mead, 1961; Massaro and Massaro, 1996). The lung alveolar system is the largest surface of the human body that is exposed to the environment, covering up to approximately 120 m^2 (Johnson et al., 2006). Thus, these systems are susceptible to injury after exposure to microbes and toxicants. The primary functions of the respiratory system are gas exchange, maintaining acid–base balance, olfaction, and protection against noxisous agents (Haschek et al., 2009). Gas exchange is the main function of the pulmonary system, and this involves three processes: ventilation, perfusion, and diffusion of gases across the air–blood barrier (Haschek et al., 2009). In the alveolar zone, O_2 and CO_2 move across the air–blood barrier via diffusion. Ventilation brings O_2 to the alveoli and removes CO_2. A respiratory cycle consists of an inspiratory phase followed by an expiratory phase.

Many interspecies differences exist in the respiratory system. Extensive bronchioles are present in dogs, cats, and rhesus monkeys, while minimal bronchioles are present in humans, rats, mice, horses, and pigs (Haschek et al., 2009). The pleura, the cellular layer that envelopes the lung surface, of rats, mice, dogs, cats, and rhesus monkeys are thin, and are thick in humans, horses and pigs (Haschek et al., 2009). Cattle have a relatively long tracheobronchial tree which increases the amount of dead space volume in comparison to dogs, pigs, and horses (Ackermann et al., 2010). Increased dead space affects the amount of fresh oxygen that can be delivered to the lung and increases the risk of alveolar hypoventilation with partial obstruction (Ackermann et al., 2010). Anatomical and physiological features of the respiratory system that predispose cattle to respiratory infections include the fact that: (1) cattle have a relatively increased area of dead space and also have a right tracheal bronchus; (2) their lungs have interlobular septae with limited interdependence, increased resistance, and decreased compliance; and (3) collateral ventilation is reduced due to a lack of (a) bronchoalveolar communication (channels of Lambert), (b) alveolar pores (pores of Kohn), and (c) interbronchiolar connections (channels of Martin). Due to the lack of collateral ventilation, atelectasis occurs readily and these areas of lung remain consolidated and lack functional gaseous exchange. Regions of atelectasis are also under hypoxic vasoconstriction whereby arterioles shunt blood flow away from these areas to better perfuse regions with adequate gaseous exchange (Ackermann et al., 2010). Thus, these anatomical and physiological conditions increase the demands for protective immune responses. The nasal turbinates or scrolls are relatively simple in humans and nonhuman primates, a double scroll is present in rodents, and a complex membraneous branched scroll is present in dogs (Haschek et al., 2009).

There are approximately 40 cell types in the respiratory system (Haschek et al., 2009). Highly differentiated type I alveolar epithelial cells that do not divide line approximately 90 to 95% of the alveolar surfaces. These cells prevent the leakage of fluid across the alveolar wall (Haschek et al., 2009). Type II alveolar cuboidal epithelial cells, which cover approximately 2 to 4% of the internal surface of the lung, synthesize and secrete pulmonary surfactants and are involved in the reepithalization process after lung injury (Johnson et al., 2006; Haschek et al., 2009). Surfactants protect alveolar walls and enhance phagocytosis. Surfactant reduces alveolar surface tension at low lung volume, thereby reducing the work of breathing and preventing lung collapse. The surface tension-reducing function of surfactant stems from the interaction of phospholipid, neutral lipid, and two hydrophobic proteins, surfactant protein (SP)-B and SP-C. In addition, surfactant plays a key role in innate immunity (Mulugeta and Beers, 2006). Surfactant protein C (SP-C), a highly hydrophobic protein found in pulmonary surfactant, is synthesized exclusively in alveolar type II pneumocytes as a 21-kDa integral membrane precursor protein (Mulugeta and Beers, 2006). SP-C enhances the adsorption and spreading of phospholipids at the air–liquid interface thereby promoting the surface tension–lowering properties of surfactant and probably playing a fundamental role in pulmonary immunology both during acute lung infection and during chronic lung disease (Mulugeta and Beers, 2006). SP-C participates in the pulmonary host defense system through direct interaction with infectious agents and pattern recognition molecules (e.g., CD14) (Mulugeta and Beers, 2006). Surfactant protein-A (SP-A) production by lung epithelia increases markedly in fetal lung near term and thus roughly delineates the level of lung maturation at birth. SP-A production can also be increased with viral pneumonia and enter the pulmonary lymphatic vessels, where it is carried by the thoracic duct to the systemic blood circulation. Thus, serum SP-A measurement has the potential to be a marker of respiratory tract maturation and pneumonia severity (Ackermann et al., 2010).

Shortly before birth the fetal lung converts from fluid secretion to fluid reabsorption. Efficient gas exchange after birth depends on regulation of the amount of fluid in the thin (average, 0.2 μm) liquid layer lining the alveolar epithelium (Bastacky et al., 1995). The alveolar epithelium is the major site of Na^+ transport and fluid absorption in the adult lung (Johnson et al., 2006). Type I pneumocytes contain several types of cation channels (e.g., K^+ channels) and at least one type of anion channel [e.g., cystic fibrosis transmembrane regulator (CFTR)]. There is no evidence that type II pneumocytes contain K^+ or cyclic nucleotide-gated (CNG) channels (Johnson et al., 2006). Sodium ion transport occurs across the entire alveolar epithelium (type I and II pneumocytes) rather than only at specialized anatomical locations (i.e., across type II pneumocytes). Sodium ion is absorbed from the apical surface of both type I and II pneumocytes via epithelial Na^+ channels (ENaC) and CNG channels and transported into the interstitial space by Na^+,K^+-ATPase. K^+ may be transported via K^+ channels located on the apical surface of type II pneumocytes cells or through basolateral K^+ channels in type II pneumocytes (Johnson et al., 2006). If net ion transport is from the apical surface to the interstitium, an osmotic gradient would be created that would direct water transport either through aquaporins or by diffusion. Electroneutrality is conserved with Cl^- movement via CFTR in both type I and II pneumocyte cells (Johnson et al., 2006).

Located within the alveolar interstitium are alveolar macrophages. Pulmonary intravascular macrophages (PIMs), fixed macrophages that reside within pulmonary capillaries and have specialized junctional complexes, are present in cattle, pigs, horses, and cats and absent in nonhuman primates, dogs, and rodents (Longworth, 1997; Brain et al., 1999; Haschek et al., 2009). While one study demonstrated the presence of PIMs is human lung (Dehring and Wismar, 1989), a morphometric study of human lung did not show macrophages or macrophage-like cells in the pulmonary capillaries (Zeltner et al., 1987). There are marked species differences in PIMs' contribution to the mononuclear phagocytic system. Ruminants, pigs, and cats have extensive resident PIMs that avidly remove circulating particles from the blood, whereas rodents, monkeys, and chickens do not (Brain et al., 1999). In rodents, monkeys, and chickens, the removal of pulmonary particles is done primarily by monocytes and neutrophils (Brain et al., 1999). PIMs are large (20 to 80 μm in diameter) mature macrophages that are attached to the pulmonary capillary endothelium and have characteristic morphological features of differentiated macrophages (Brain et al., 1999). PIMs are involved in phagocytosis and removal of endotoxin and particles from the circulation. PIMs may play a role in normal removal of effete erythrocytes or fibrin and cell debris during inflammation (Brain et al., 1999). Rats and rabbits are quite resistant to endotoxin-induced acute lung injury, but sheep, goats, and pigs have commonly been used as animal models of acute lung injury. The increased sensitivity of sheep, goats, and pigs to endotoxin is probably due to the presence of abundant PIMs (Brain et al., 1999). In a study of freshly prepared PIMs (obtained from pigs) that were stimulated in vitro with LPS (10 μg/mL), COX-2 expression was investigated. Results showed that LPS resulted in the up-regulation of the expression of COX-2 in the PIMs (Chen et al., 2003). In many species, activation of macrophages occurs in a classical, proinflammatory/cell-mediated Th1-type response induced by interferon gamma and TNF, an alternative Th2-type response induced by IL-4, and an anti-inflammatory regulatory response induced by IL-10 (Ackermann et al., 2010). In cattle; however, the Th1/Th2 paradiagm is not as clearly distinct functionally as in other species. Despite this, alveolar and PIMs of the bovine do function in the generation of cell-mediated, humoral, and regulatory (inhibitory) responses. In allergic (Th2-type) conditions, eosinophilic and mast cell infiltration occurs with release of granule content. Once established, inhaled antigen can cross-link IgE receptors on mast cells and basophils, resulting in cellular activation (Ackermann et al., 2010).

Bronchiolar nonciliated Clara cells are involved in the detoxification of xenobiotics. Clara cells also produce CC10, an immunomodulatory protein, and express cytochrome P450 enzymes that biometabolize toxins. After pulmonary injury, type II and Clara cells proliferate and replace damaged cells (Ackermann et al., 2010). Clara cells' morphological characteristics are abundance of agranular endoplasmic reticulum (AER) and numerous membrane-bound ovoid granules. There is considerable interspecies (adult male rabbits, guinea pigs, rats, hamsters, and mice) variation in abundance, size, and morphology of the granules, in abundance and distribution of AER, and in mitochondrial morphology. Granules were most abundant in the rat and least in the hamster, largest in the hamster, and of uniform electron density except in the guinea pig (Plopper et al., 1980). Another pulmonary cell type located within the interstitium is the fibroblast, which produces matrix components, enzymes, and PGs and is involved in pulmonary fibrosis (Haschek et al., 2009). Both the sympathetic and parasympathetic nervous systems provide innervations to the pulmonary system, including bronchial smooth muscle, blood vessel, submucosal glands, and lymphatics (Haschek et al., 2009). Of the immunological compartments of the respiratory system that participate in the humoral immunity of the pulmonary system is mucosa-associated lymphoid tissue (MALT). MALT has IgA as the predominant class of opsinizing antibodies that phagocytose and eliminate inhaled particles (Haschek et al., 2009).

Pulmonary epithelial cells play an important role in airway homeostasis, represent the first line of defense in infection, and perform many important biological functions (Takizawa, 1998; Holgate et al., 2009; Holgate, 2011). Epithelial cells are central to respiratory system sensing of the external environment (Ackermann et al., 2010). Most particulate matter and microbial agents are removed from the inhaled air in the nares, nasal conchae, and trachea, leaving the deeper lung sterile and relatively free of particulate material. Once through the ASL, these substances, along with mists, vapors, and gases, can bind the lung epithelia and trigger activation, cell injury, metaplasia, or cell death. Microbial agents produce a number of conserved molecular patterns, termed pathogen-associated molecular patterns (PAMPs), which include substances such as teichoic acid from gram-positive bacteria, lipopolysaccharide (LPS) from gram-negative bacteria, cytokine–phosphate–guanine (CpG) DNA, single- and double-stranded RNA (dsRNA), flaggelin, fungal zymosan, and lipopeptides from mycobacteria (Ackermann et al., 2010). Bronchial epithelial cells are known to have an integral role in the airway defense mechanism via the mucociliary system as well as mechanical barriers (Radi and Ackermann, 2011). In addition, bronchial epithelial cells produce and release biologically active compounds, including lipid mediators, growth factors, and a variety of cytokines important in the pathogenesis of airway disorders (Brannon et al., 1998; Takizawa, 2005). Neutrophils, alveolar and PIMs, α/β T (CD4 and CD8) cells and γ/δ T cells, along with B, NK, and dendritic cells, are effector cells present along the respiratory tree for induction of adaptive immune responses (Ackermann et al., 2010). With inflammation, neutrophils and macrophages are present in the alveolar lumen, bronchioles, and airways. Other pulmonary immune responses include pattern recognition receptors as well as cytokine, interferon, and chemokine responses (Ackermann et al., 2010). Of special interest in cattle are γ/δ T-cells because newborn calves have an unusually high number of circulating γ/δ T-cells and also because in ruminants these cells express WC1 antigen (Ackermann et al., 2010). γ/δ T-cells account for 60% of the peripheral blood mononuclear cells in young calves. $WC1^{+}\gamma/\delta$ T-cells are considered an inflammatory population, whereas $CD1^{-}\gamma/\delta$ T-cells are regulatory with myeloid cell features (Ackermann et al., 2010). In genetically susceptible human patients with asthma, an impaired epithelial barrier function renders the airways vulnerable to early viral infection, and this in turn provides the stimulus to prime immature dendritic cells toward directing a Th2 response and local allergen sensitization. Continued epithelial susceptibility to environmental insults such as viral, allergen, and pollutant exposure and impaired repair responses leads to asthma persistence and provides a mediator and growth factor microenvironment that leads to the persistence of inflammation and airway wall remodeling. Increased deposition of matrix in the epithelial lamina reticularis provides evidence for ongoing epithelial barrier dysfunction, while physical distortion of the epithelium consequent upon repeated bronchoconstriction provides additional stimuli for remodeling (Holgate, 2011).

Immune-mediated hypersensitivity diseases or allergies are common respiratory diseases, with type I (immediate hypersensitivity) and type III (Arthus type) being most common in the pulmonary system (Haschek et al., 2009). Pathologic manifestations of rhinitis and asthma can be the result of type I hypersensitivity, while extrinsic allergic alveolitis results from type III hypersensitivity and is related to antigen antibody deposition in the lung with complement activation (Haschek et al., 2009). Several factors, including gender, age, genetics, nutrition, smoking status, and preexisting pulmonary disease, can affect respiratory system response to injury (Haschek et al., 2009). Pathologically, asthma is a chronic airway inflammatory disease dominated by a Th2-type pattern and can be divided into intrinsic (idiosyncratic) or extrinsic disease (Haschek et al., 2009). Antigen-presenting cells such as airway dendritic cells (DCs) play a critical role in initiating and regulating early inflammatory events and mounting the adaptive immune response (Holgate et al., 2009; Holgate, 2011). Tracheal epithelial cells secrete soluble factors that down-regulate TNFα and IL-12p40 secretion by bone marrow–derived DCs but up-regulate IL-10 and arginase-1. Airway tracheal epithelial cells constitutively release PGE_2. Blocking the synthesis of PGs within airway epithelial cells relieved DCs from inhibition (Schmidt et al., 2011). The airway epithelium is crucial to both the origins and progression of asthma as well as playing a key role in asthma exacerbations (Holgate et al., 2009). The physical and functional barriers of the airway epithelium are both defective in asthma, with disrupted tight junctions, reduced antioxidant activity, and impaired innate immunity. Damage to the epithelium results in augmented expression of epidermal growth factor (EGF) receptors, but because of the reduced ability of signaling through these to repair the epithelium through "primary intention" in response to the levels of EGFR ligands available, the epithelium enters into "frustrated repair (or repair by secondary intention)," leading to enhanced profibrotic growth factor release and a tendency toward epithelial–mesenchymal transition (Holgate et al., 2009). Thus, the defect in epithelial tight junction assembly in asthma may be the result of EGFR signaling and may represent another feature of the chronic wound process (Holgate et al., 2009).

ROLE OF PROSTAGLANDINS IN THE RESPIRATORY SYSTEM

Prostaglandins have been implicated in lung pathophysiology (Radi and Ackermann, 2011). PGs are synthesized from arachidonic acid (AA) by enzymatic reactions such as cyclooxygenases (COXs), specifically COX-1 and COX-2 (Radi and Ackermann, 2011). COX-1 is generally found in most normal body tissues and is involved in pulmonary physiological processes. COX-2 is expressed in normal tissues at very low levels and is highly induced by proinflammatory mediators in the setting of inflammation (Radi, 2009). PGs mediate various pulmonary functions (i.e., vascular vasodilatation during the perinatal, late fetal, and early postnatal periods) (Brannon et al., 1998; Lassus et al., 2000). The increased PG synthesis that occurs during gestation and in the newborn may modulate surfactant synthesis during the late fetal and early newborn periods (Pace-Asciak, 1977; Powell and Solomon, 1978; Acarregui et al., 1990). Microsomal fraction of sheep lung possesses active prostaglandin synthase, prostacyclin synthase, and thromboxane synthase activities (Tai et al., 1978). Fetal sheep lung homogenates from all ages (40 days to term) formed both PGE_2 and $PGF_{2\alpha}$ (Pace-Asciak, 1977). $PGF_{2\alpha}$ was formed in the lung to a greater extent than PGE_2 by day 40. PGE_2 increased with increasing age until at term, at which time the ratio of both prostaglandins approached unity (Pace-Asciak, 1977). Total prostaglandin biosynthesis (E_2 and $F_{2\alpha}$) rose gradually with age (approximately a threefold increase between day 40 and term). Prostaglandin catabolic activity rose sharply at day 90 (approximately threefold), with a maximum around day 110 (approximately four fold), decreasing to day 40 levels by term (day 143) (Pace-Asciak, 1977). Fetal nonhuman primate, baboon, and lung EP receptor expression was investigated. The mRNA levels, protein localization, and abundance of all four PGE_2 receptors by real-time polymerase chain reaction (PCR), immunohistochemistry, and Western blot were studied. EP receptors were widely distributed in bronchiolar epithelium, bronchiolar smooth muscle, and endothelium and media of blood vessels, but not in alveoli (Schmitz et al., 2007). The role of PGs on SP-A gene expression in human fetal lung was studied in vitro (Acarregui et al., 1990). Dexamethasone resulted in a marked decrease in the levels of the PGE_2 and $PGF_{2\alpha}$, the prostacyclin metabolite, 6-keto-$PGF_{1\alpha}$, and the thromboxane A_2 metabolite, thromboxane B_2 (TXB_2). Indomethacin caused a pronounced decrease in the levels of these PGs and had a marked ability to reduce SP-A mRNA levels in human fetal lung in vitro (Acarregui et al., 1990). Indomethacin effects were associated with a 73% reduction in cAMP formation and were prevented by simultaneous incubation with dibutyryl cAMP or with PGE_2. PGE_2 markedly increased cAMP formation by the human fetal lung tissue incubated in the absence or presence of indomethacin. The effects of dexamethasone and indomethacin were also observed on two morphological indices of lung differentiation: alveolar lumenal volume density and lamellar body volume density. PGE_2 significantly increased the lumenal volume density of human fetal lung explants (Acarregui et al., 1990). Levels of various PGs were studied in fetal and adult bovine and rabbit lung homogenates. Adult bovine lungs were very active in converting AA (100 μg/g tissue) to both PGE_2 (10.7 μg/g tissue) and TXB_2 (6.2 μg/g tissue) (Powell and Solomon, 1978). Smaller amounts of $PGF_{2\alpha}$ (0.9 μg/g) and 6-oxo-$PGF_{1\alpha}$ were formed. Fetal bovine lungs during the third trimester of pregnancy were active in converting AA to PGE_2, but formed very little TXB_2, $PGF_{2\alpha}$, or 6-oxo-$PGF_{1\alpha}$. Rabbit lung homogenates converted AA (100 μg/g) mainly to PGE_2, both before and after birth. The amount of PGE_2 formed increased during gestation to a maximum of about 6 μg/g of tissue at 28 days of gestation. It then decreased to a minimum (1.5 μg/g), which was observed

8 days after birth, followed by an increase to about 4 μg/g in older rabbits (Powell and Solomon, 1978).

PGs regulate vascular tone (Holtzman, 1991). High arterial blood circulating levels of prostacyclin (PGI_2) are present in humans and cats (Gryglewski, 1980). In the lungs, generation of PGI_2 can be increased by angiotensin II, bradykinin, and AA, provided that low concentrations of these substances are infused into the pulmonary artery. Air pulmonary emboli and mechanical hyperventilation stimulate the lungs to generate more prostacyclin. Respiratory stimulants such as lobeline or almitrine are also effective PGI_2 releasers from the lung. It is proposed that this para-endocrine function of the lung protects coronary and cerebral arteries against thrombosis and atherosclerosis (Gryglewski, 1980). To test whether vasodilator PGI_2 was synthesized by the lungs in response to vasoconstriction, vasoconstriction induced by angiotensin II in isolated rat lungs before and after inhibition of PG synthetase was studied (Voelkel et al., 1981). It was found that 6-keto-$PGF_{1\alpha}$ was the major AA metabolite released during pulmonary vasoconstriction. Thus, PGI_2 is produced by the lungs in response to vasoconstriction. Hypoxia by itself seems not to be the adequate stimulus for enhanced lung PGI_2 formation (Voelkel et al., 1981). Up-regulation of COX-2 in the lung has been demonstrated in a hypertensive mouse model that expresses the human renin and angiotensinogen genes (Radi and Ostoroski, 2007). Inhaled PGI_2 analogs have been proposed for the treatment of pulmonary hypertension (Gessler et al., 2011). Inhaled treprostinil sodium, a prostacyclin analog, is the most recent agent to receive U.S. Food and Drug Administration (FDA) approval for the treatment of pulmonary arterial hypertension (Ferrantino and White, 2011).

The effects of acetylcholine (Ach), a pulmonary vasodilator, infusion on pulmonary perfusion pressure, vascular responsiveness, and lung PGI_2 production were investigated (Feddersen et al., 1986a). The prolonged effect of depression of subsequent hypoxic and angiotensin II–induced vasoconstrictions, but not the immediate vasoconstriction response of Ach, was associated with an increase in perfusate 6-keto-$PGF_{1\alpha}$ concentration. The Ach-stimulated increase in 6-keto-$PGF_{1\alpha}$ production was inhibited by meclofenamate, an ns-NSAID. Thus, in isolated rat lungs Ach caused immediate vasodilation and prolonged time-dependent depression of vascular responsiveness (Feddersen et al., 1986a). Another study tested whether diminished Ach vasodilation would result from damage of lung vascular endothelium and whether this response could be used as an indication of endothelial injury in a rat model (Feddersen et al., 1986b). No difference between the prolonged depression of vascular responsiveness to hypoxia or angiotensin II and pulmonary vascular responses to ACh were found (Feddersen et al., 1986b). In isolated rat lung perfused with a cell- and protein-free physiological salt solution, the effects of AA on pulmonary vasodilation were studied (Feddersen et al., 1990). When pulmonary vascular tone was elevated by hypoxia, bolus injection of a large dose of AA (75 μg) caused transient vasoconstriction followed by vasodilation. Dosages of less than 7.5 μg of AA caused vasodilation only when injected during hypoxic vasoconstriction and with subsequent blunting of either angiotensin II–or hypoxia-induced pulmonary vasoconstriction. Only the higher doses of AA (7.5 and 75 μg), but not the lower doses (7.5 to 750 ng), caused increases in effluent 6-keto-$PGF_{1\alpha}$, TXB_2, and PGE_2 and $PGF_{1\alpha}$, with 6-keto-$PGF_{1\alpha}$ being the major PG product (Feddersen et al., 1990). The effects of PGI_2 on pulmonary function were investigated in a swine model of septic shock. PGI_2 did not mediate blood gas changes, alter pulmonary hemodynamics, or lead to platelet abnormalities in this porcine septic shock model. Antiprostacyclin antibody infusion did not change the pulmonary hypertension and hypoxemia, and infusion of PGI_2 to pathophysiological blood concentrations did not reproduce such changes (Tran et al., 2001). Inhalation of PGI_2 (28 ng/kg per minute) over a period of 8 h in healthy lambs does not produce major side effects or acute pulmonary toxicity (Habler et al., 1996).

PGs are released during various pulmonary pathological conditions (Fulkerson et al., 1996; Vigano et al., 1998; Profita et al., 2003; Rolin et al., 2006; Radi and Ackermann, 2011). Tumor necrosis factor alpha (TNFα) has been proposed as a mediator of endotoxin-induced lung injury. TNFα induces rapid release of TXA_2, PGE_2, and PGI_2 (Wheeler et al., 1992). TNFα contributes to the pathophysiology of adult respiratory distress syndrome (ARDS) (Arias-Diaz et al., 1994). Aerosolized prostacyclin improved oxygenation in children with acute lung injury (Dahlem et al., 2004). The pulmonary effects of PGE_1 in a 30-ng/kg dose were studied in severely ill surgical patients with ARDS. PGE_1 markedly decreased pulmonary artery pressure, pulmonary and systemic vascular resistance indices, and venous pressures while increasing cardiac output, arterial PO_2, oxygen delivery, and oxygen consumption (Shoemaker and Appel, 1986). In another study, the physiological and side effects of PGE_1 were investigated in patients with ARDS who did not have significant renal or hepatic dysfunction. These patients were given PGE_1 by continuous central venous infusion (30 ng/kg per minute). PGE_1 caused significant systemic vasodilation and possibly decreased intrapulmonary polymorphonuclear leukocyte sequestration. Howerver, PGE_1 did not influence multiple-system organ failure or mortality of patients with ARDS without renal or hepatic dysfunction (Russell et al., 1990). The safety and efficacy of an intravenous liposomal dispersion of PGE_1 for the treatment of patients with ARDS was studied in a randomized prospective multicenter double-blind placebo-controlled phase III clinical trial. PGE_1 accelerated improvement in indexes of oxygenation but did not decrease the duration of mechanical ventilation and did not improve 28-day survival (Abraham et al., 1999).

Mucus production is a primary feature of bronchial asthma (Izuhara et al., 2009). Goblet cells are major mucus-producing cells, and goblet cell hyperplasia (GCH) is one feature of airway remodeling, defined as airway structural changes. Th2-type cells play critical roles in this remodeling process, and among Th2-type cytokines, interleukin-13 (IL-13) in particularl is a central mediator for GCH (Izuhara et al., 2009). Mouse calcium-activated chloride channel-3 (mCLCA-3; gob-5)/human CLCA-1 acts as a downstream molecule of Th2 cytokines, IL-4/IL-9/IL-13 signals, playing an important role in mucus production (Izuhara et al., 2009). Pendrin, an anion transporter, is induced by IL-13 and causes mucus production in airway epithelial cells (Izuhara et al., 2009). PGs have been shown to play a role in bronchial mucus secretion (Tamaoki et al., 1992), Emerging data strongly suggest that PGs collaborate with cytokines and critically regulate T-cell proliferation, differentiation, and functions (Sakata et al., 2010). PGE_2 facilitated Th1 cell differentiation and Th17 cell expansion in collaboration with IL-12 and IL-23, respectively, and these PGE_2 actions contributed to the development of immune diseases mediated by these Th subsets (Sakata et al., 2010). Furthermore, studies using receptor-deficient mice have revealed that PGD_2 and PGI_2 contributed to regulation of immune diseases of the Th2 type, such as allergic asthma (Sakata et al., 2010). In a sensitized BALB/c mouse model of asthma, treatment with a PGD_2 receptor (DP1) agonist or DP1 agonist-treated DCs resulted in an increase in $Foxp3^+$ $CD4^+$ regulatory T cells that suppressed inflammation in an IL-10-dependent way. These effects of DP1 agonist on DCs were mediated by cAMP-dependent protein kinase A. In addition, activation of DP1 by an endogenous ligand inhibited airway inflammation. Triggering DP1 on DCs is an important mechanism that induced regulatory T cells and controlled the extent of airway inflammation (Hammad et al., 2007). Selective DP1 agonists that exploit the full anti-inflammatory effect on airway DCs could be a potential asthma therapeutic strategy. Thus, there are emerging roles for PGs in T cell-mediated immunity and immune diseases of the respiratory system.

PGE_2 is largely produced by epithelial and airway smooth muscle cells, but all cells in the airways have the capacity to release prostanoids (Montuschi and Barnes, 2002). PGE_2 has bronchoprotective and inhibitory effects on inflammatory cells at concentrations known to occur in the airways (Pavord and Tattersfield, 1995). Marked increases in PGE_2 synthesis and up-regulation of COX-2 have been reported in acid-induced epithelial injury (Bonnans et al., 2006). Cystic fibrosis (CF) is one of the most common autosomal recessive diseases among Caucasians caused by a mutation in the CF transmembrane conductance regulator (CFTR) gene. CFTR has been reported to be a negative regulator of PGE_2-mediated inflammatory response, a defect of which may result in excessive activation of NFκB, leading to overproduction of PGE_2 as seen in inflammatory CF tissues (Chen et al., 2012). The effects of inhaled PGE_2 on allergen-induced airway responses and inflammation were investigated in patients with mild asthma. Inhaled PGE_2 attenuated allergen-induced airway responses, hyperresponsiveness, and inflammation when given immediately before inhaled allergen (Gauvreau et al., 1999). The effect of inhaled PGE_2 on allergen-induced asthma was examined in a double-blind crossover study of eight subjects with asthma. Subjects in study 1 inhaled 100 μg of PGE_2 5 min before a normal saline challenge, and FEV1 and bronchial reactivity to methacholine was measured at intervals for 7 h. In study 2, which had a similar design, subjects inhaled the same dose of PGE_2 or placebo followed by a dose of allergen previously shown to cause a fall in FEV1 of 20% or more. PGE_2 inhibited early and late response to allergen (Pavord et al., 1993). PGE_2 was a bronchodilator in healthy subjects but caused either dilation or constriction (the direction of the change in specific airway conductance being consistent for each subject) in asthamatics (Mathé and Hedqvist, 1975). Impaired production of PGE_2 has been proposed to contribute to the pathogenesis of asthma (Pavord and Tattersfield, 1995). PGE_2-like immunoreactivity was measurable in exhaled breath condensate (EBC) in healthy subjects and patients with asthma (Montuschi and Barnes, 2002).

PGE_2 inhibits fibroblast proliferation and collagen production. PGE_2 synthesis by fibroblasts is induced by profibrotic mediators, including transforming growth factor beta (TGFβ) (Keerthisingam et al., 2001). Cyclooxygenase-2 deficiency resulted in loss of the antiproliferative response to TGFβ in human fibrotic lung fibroblasts and promoted bleomycin-induced pulmonary fibrosis in mice (Keerthisingam et al., 2001). Dysregulation of the PGE_2 synthesis–response axis has been identified in human pulmonary fibrotic diseases and implicated in the pathogenesis of animal models of lung parenchymal fibrosis (Wilborn et al., 1995a; Stumm et al., 2011). Reduced expression of COX-2 and PGE_2 in fibroblasts from patients with idiopathic pulmonary fibrosis (IPF) is one factor that promotes fibroblast survival in the fibrotic lung (Maher et al., 2010). IPF is a lung disease that is characterized by excessive proliferation of fibroblasts. Down-regulation of PGE receptor 2 gene (PTGER2) and consequent PGE_2 resistance are both mediated by DNA hypermethylation. Increased Akt signal transduction is a novel mechanism that promotes DNA hypermethylation during pulmonary fibrogenesis in humans and mice (Huang et al., 2010).

There is a homeostatic balance within the lung that requires crosstalk between alveolar epithelial cells, fibroblasts, and inflammatory cells (Bozyk and Moore, 2011). Alveolar epithelial cells produce PGE_2, which limits fibroproliferation and promotes appropriate alveolar epithelial repair. In the normal lung, PGE_2 signals via EP_2 receptor-mediated increases in cAMP that induce fibroblast apoptosis and limit myofibroblast transformation, proliferation, and collagen secretion (Bozyk and Moore, 2011). In pulmonary fibrosis, the homeostatic balance is altered and PGE_2 production becomes limited via inflammatory mediators and epigenetic silencing of the COX-2 promoter (Bozyk and Moore, 2011). Furthermore, fibrotic fibroblasts lose EP_2 receptor expression and may lose expression of the downstream effectors protein kinase A (PKA) and phospho-cAMP response element binding (CREB) (Bozyk and Moore, 2011). Fibrosis is a male-predominant disease, and at least one study has suggested that male gender is associated with reduced EP_2 and EP_4 levels and reduced PGE_2 production in splenic macrophages after trauma (Stapleton et al., 2004). PGE inhalation therapy may be the best option for patients with lung fibrosis, but nebulization of PGE_2 or EP_2 agonists is not practical (Bozyk and Moore, 2011).

Aerosolized PGE_2 prevented experimentally induced airway response to allergen challenge in asthmatics (Herrerias et al., 2009). The effects of PGE_2 on airway inflammation were investigated in a house dust mite–sensitive mouse (HDM) model, a model that reproduces the spontaneous exposure of allergic asthma patients to aeroallergens (Herrerias et al., 2009). Exogenous PGE_2 reduced by half eosinophilic infiltration in HDM-sensitized mice and led to a strong reduction in airway Th2 cytokine expression (Herrerias et al., 2009). The EP_3 receptor has been identified to mediate PGE_2-induced depolarization of sensory nerves in humans, mice, and guinea pigs. Thus, PGE_2 mediates coughs via the EP_3 receptor (Maher et al., 2009). The receptor involved in PGE_2-induced relaxation in guinea pig, murine, monkey, rat, and human airways was studied in vitro. It was found that EP_2 receptor mediates PGE_2-induced relaxation of guinea pig, murine and monkey trachea and that the EP_4 receptor mediates PGE_2-induced relaxation of the rat trachea. Thus, EP_4 receptor was proposed as a new target for bronchodilator therapy (Buckley et al., 2011).

Chronic obstructive pulmonary disease (COPD) is a slowly progressive, chronic, and complex airway disease involving several types of inflammatory cells (e.g., alveolar macrophages, dendritic cells, neutrophils, T-lymphocytes) and mediators. A major cell type in COPD pathophysiology is alveolar macrophages. Some of the major stimuli that activate macrophages are bacterial endotoxin and proinflammatory cytokines such as IL-1β and TNFα (Murugan and Peck, 2009). Thus, alveolar macrophage plays a role in pulmonary defense (Wilborn et al., 1995b; Radi et al., 2010; Radi and Ackermann, 2011). Alveolar macrophages have the capacity to generate PGs. Alveolar macrophages secrete PGE_2 and may contribute to the increased PGE_2 in the exhaled breath of COPD patients (Montuschi et al., 2003). There is increased COX-2 expression in alveolar macrophages from patients with COPD compared to normal control subjects (Taha et al., 2000). TXA_2 biosynthesis is enhanced in patients with COPD and may be influenced by arterial oxygen tension changes (Davi et al., 1997).

PGs are involved in surfactant homeostasis (Oyarzun and Clements, 1978; Arias-Diaz et al., 1994; Wilborn et al., 1995b). In vitro PGE_2 induces surfactant production via EP_1 and cyclic adenosine monophosphate (cAMP)-coupled EP_2 and EP_4 receptors (Schmitz et al., 2007). Human fetal lung in vitro secretes into the culture medium relatively large amounts of PGE_2 and $PGF_{2\alpha}$ and of the PGI_2 and TXA_2 metabolites (Mendelson et al., 1991). In addition, indomethacin, an ns-NSAID, markedly inhibited SP-A gene expression and cAMP formation by human fetal lung in culture. Furthermore, the inhibitory effect of indomethacin on SP-A gene expression can be prevented by simultaneous incubation with either PGE_2 (Mendelson et al., 1991). It has been shown that PGE_1, PGE_2, and $PGF_{2\alpha}$ are involved in controlling alveolar surfactant during increased minute ventilation in rabbit lung (Oyarzun and Clements, 1978).

The kinetics of PGE_2 production and the expression of COX enzymes during the course of experimental pulmonary tuberculosis was studied in a BALB/c mouse model infected via the trachea with *Mycobacterium tuberculosis* (Rangel Moreno et al., 2002). The PGE_2 quantification in lung homogenates showed stable low concentrations during the first month of infection, followed by a significant twofold increase at day 60 and a fourfold increase at day 90 postinfection compared to day 28 of infection. These results correlated with the kinetics of the gene expression of COX-1, COX-2, and PGE synthase, which showed a progressive increase that reached a peak at day 90 of infection. Moreover, the expression of IL-17 was detected only during the advanced phase of infection. Macrophages were the cells that most commonly expressed COX-2, PG synthase, and PGE_2. At the beginning of the infection, days 1 and 3, these macrophages showed the morphological features of activated macrophages (large cells with abundant and compact cytoplasm) and were located in the alveolar–capillary interstitium, where they constituted $20 \pm 7\%$ of the total inflammatory cells. Granulomas started to form after 2 weeks of infection. These granulomas had activated macrophages with immunoreactivity to COX-2 and PGE, but during the entire course of the infection they were maintained at a stable percentage from 13 to $16 \pm 7\%$. In contrast, macrophages embedded in the pneumonic areas which showed positive immunostaining to PGE_2 and COX-2 showed vacuolated cytoplasm, and corresponded to the $17 \pm 5\%$ and $25 \pm 6\%$ of the total inflammatory cells at days 60 and 90, respectively. Considering that these pneumonic areas increased progressively during infection, it was evident that these vacuolated cells were the most important source of PGE_2 during the advanced phase of the infection. In addition, these pneumonic areas also showed COX-1 immunostained cells, but they were monocytes located in the alveolar–capillary interstitium. Thus, PGE_2 contributed significantly to the pathogenesis of pulmonary tuberculosis, and high PGE_2 concentrations during the late phase of the disease contribute to down-regulate cell-mediated immunity, permitting disease progression (Rangel Moreno et al., 2002).

Pulmonary epithelial cells play an integral role in airway homeostasis, perform numerous biological functions, and represent the first line of defense against infection (Takizawa, 1998). The bovine lung anatomical and physiological features predispose cattle to various respiratory infections (Ackermann et al., 2010). Parainfluenza virus (PI) and bovine respiratory syncytial virus (RSV) are major pulmonary pathogens of cattle, which all produce some type of PAMP, recognized by epithelia, alveolar macrophages, and intravascular macrophages (Ackermann et al., 2010). In addition, both RSV and PI are pathogens of human infants and perinatal lambs (Radi et al., 2010; Radi and Ackermann, 2011). RSV and PI are leading viral causes of respiratory disease in infants in their first year of life and cause morbidity and mortality in transplant patients and older adults (Hall, 2001; Thompson et al., 2003; Radi and Ackermann, 2011). Bovine and ovine strains of both RSV and paramyxoviruses have high homology to human strains and induce identical pulmonary lesions (Cutlip and Lehmkuhl, 1979; Lehmkuhl and Cutlip, 1983; Hall, 2001; Grubor et al., 2004; Meyerholz et al., 2004; Johnson et al., 2007; Radi, 2009; Radi et al., 2010). Several proinflammatory cytokines and chemokines can be detected in patients with RSV and PI infection (Bonville et al., 1999; Hornsleth et al., 2001; Graham et al., 2002; Bennett et al., 2007). Increased levels of PGE_2 in the plasma or endotracheal aspirates has been demonstrated in animals and infants infected with RSV (Gershwin et al., 1989; Sznajer et al., 2004). Infection of a human alveolar epithelial transformed cell line (A549 cells) with live RSV substantially increased production of PGE_2 (Bryan et al., 2005). In addition, RSV induced PGE_2 production in human alveolar type II-like epithelial cells (Liu et al., 2005).

Mast cell–derived PGD_2 is one of the essential modulators of eosinophilic airway inflammation in asthma and allergic rhinitis (Shiraishi et al., 2005; Uller et al., 2007). Chemoattractant receptor homologous molecule expressed on Th2 cells (CRTh2), a high-affinity receptor for PGD_2, mediates trafficking of Th2-cells, mast cells, and eosinophils to inflammatory sites, and has been proposed as a target for treatment of allergic airway diseases (Uller et al., 2007). The antiallergic efficacy of a highly selective small-molecule CRTh2 antagonist has been demonstrated in a mouse asthma model (Uller et al., 2007). Matsuoka et al. (2000) demonstrated that PGD_2 plays a key role in allergic asthma by using mice deficient in the prostanoid DP receptor. Sensitization and aerosol challenge with ovalbumin of the homozygous mutant ($DP^{-}/^{-}$) mice induced increases in the serum concentration of immunoglobulin E (IgE) similar to those in wild-type mice. However, the concentrations of Th2 cytokines and the extent of lymphocyte accumulation in the lung of OVA-challenged $DP^{-}/^{-}$ mice were greatly reduced compared with those in wild-type animals. In addition, $DP^{-}/^{-}$ mice had only marginal infiltration of eosinophils and failed to develop airway hyperreactivity. Thus, PGD_2 functions as a mast cell–derived mediator to trigger asthmatic responses. PGD_2 overproduction has been demonstrated in patients with ARDS and acute extrinsic allergic alveolitis (Higashi et al., 2005).

COX-1 EXPRESSION IN THE RESPIRATORY SYSTEM UNDER NORMAL AND PATHOLOGICAL CONDITIONS

Normally, COX-1 is expressed in many tissues throughout the body, including the lung (Figs. 6-1 and 6-2A) (Radi et al., 2010; Radi and Ackermann, 2011). COX-1 isoenzyme is expressed in a variety of respiratory tissues, and this expression is variable among different animal species. In ovine lungs, COX-1 is the main isoform that is expressed in endothelial cells and airway epithelium and is involved in vasodilatation, bronchodilation, and surfactant synthesis (Brannon et al., 1998). The ontogeny of COX-1 expression was investigated in the ovine lungs at time points of ontologic development that include preterm, term, and after birth (of full-gestation lambs) (Table 6-1) (Radi and Ackermann, 2011). Moderate to strong COX-1 expression was present in all pulmonary microanatomic locations from preterm, full-term, and postnatal lambs, with the exception of bronchial and bronchiolar epithelial cells, which had minimal expression (Radi and Ackermann, 2011). Other studies in the ovine lung have showed the lack of significant changes in COX-1 expression between the fetal preterm to term (125 to 135 days of gestation) and postnatal (7 days old) (Brannon et al., 1998). A significant maturational increase in COX-1 expression was demonstrated in the airway epithelium only in the postnatal lung of lambs that were 1 to 4 months old (Brannon et al., 1998). In neonatal lambs, COX-1 expression in alveolar macrophages was noted (Wilborn et al., 1995b; Radi and Ackermann, 2011). Thus, COX-1 is expressed constitutively in the lungs from preterm, full-term, and after-birth lambs in various microanatomical pulmonary locations, and there is differential expression of COX-1 in the developing lung (Radi and Ackermann, 2011).

COX-1 expression in vascular endothelium and alveolar epithelial cells has been shown in nonhuman primates and ovine lung (Khan et al., 2000; Radi and Ackermann, 2011). In lungs from adult rats and normal adult humans, COX-1 was expressed in bronchiolar epithelium, bronchiolar smooth muscle, alveolar macrophages, endothelial cells, and vascular smooth muscle cells (Brannon et al., 1994, 1998; Ermert et al., 1998; Hasturk et al., 2002; Radi and Ackermann, 2011). COX-1 was present in vascular endothelium and alveolar epithelial cells in the nonhuman primate (Khan et al., 2000). In lungs of normal mice, positive immunohistochemical COX-1 staining was observed in the bronchiolar epithelium (Clara cells), in the smooth muscle underlying the bronchiolar epithelium, in alveolar macrophages, and in the alveolar epithelium (Bauer et al., 2000). Another study

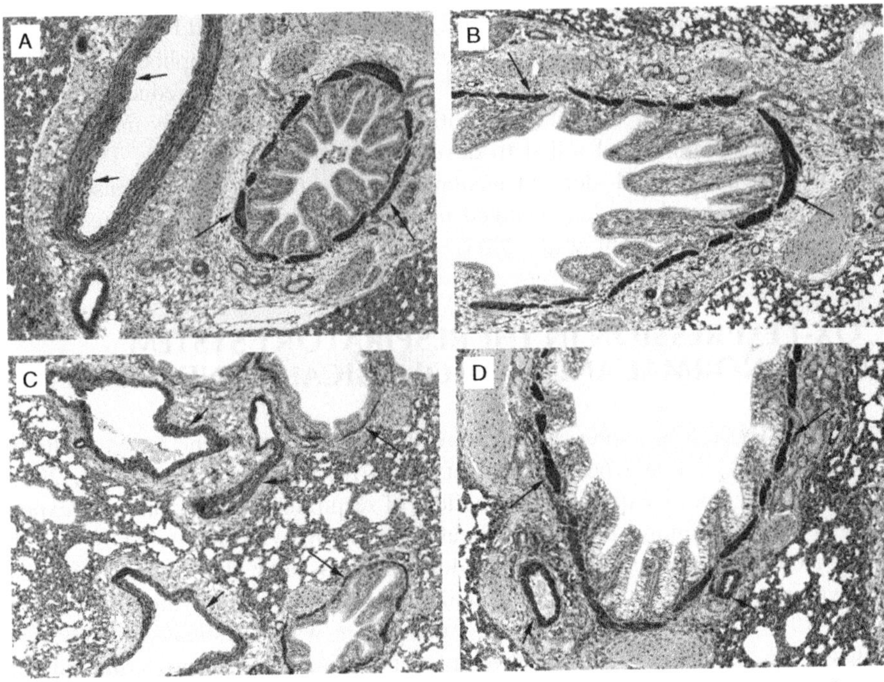

FIGURE 6-1 (A) COX-1 expression in the lung from a preterm lamb at 115 days of gestation. Strong expression in bronchial smooth muscle cells (long arrows) and vascular smooth muscle and endothelial cells (short arrows). (B) COX-1 expression in the lung from a preterm lamb at 130 days of gestation. Strong expression in bronchial smooth muscle cells (long arrows) and moderate expression in alveolar septa (short arrows). (C) COX-1 expression in the lung from a term lamb at 145 days of gestation. Strong expression in bronchial smooth muscle cells (long arrows) and vascular smooth muscle cells (short arrows). (D) COX-1 expression in the lung from a postnatal lamb. Strong expression in bronchial smooth muscle cells (long arrows) and vascular smooth muscle cells (short arrows). Immunohistochemical stain, original magnification 10×. (From Z. A. Radi and M. R. Ackermann, *Pathophysiology*, 18, pp. 215–219. Copyright © 2011 Elsevier Ireland Ltd. All rights reserved.)

investigated the role of COX-1 in the lung using tissue from genetically modified mice lacking functional COX-1 (COX-$1^{-/-}$). Bronchi from wild-type mice contained predominantly COX-1 immunoreactivity, while bronchi from COX-$1^{-/-}$ mice were hyperresponsive to bronchoconstrictors (Harrington et al., 2008). Thus, COX-1 was shown to be the functional isoform present in the airway with regard to bronchoconstriction.

In normal human lung, strong COX-1 immunostaining was noted in airway (bronchial) epithelial cells and in endothelial cells of pulmonary vessels (Ermert et al., 2003). Other studies have demonstrated COX-1 expression in various pulmonary microanatomical locations in normal human lungs (Hasturk et al., 2002). The allele frequencies of 20 polymorphisms in the promoter and coding region of the COX-1 gene was studied in random Australian Caucasian subjects, and four common polymorphisms were studied further in relation to asthma and atopy. This

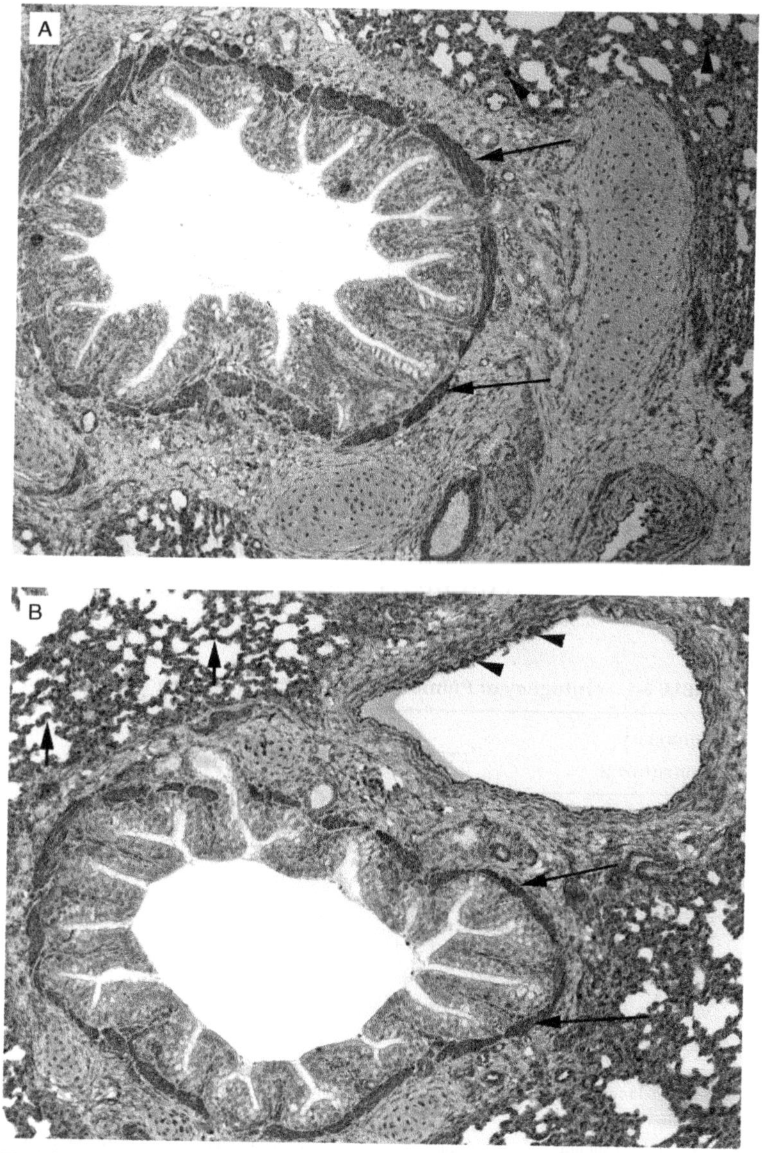

FIGURE 6-2 (A) COX-1 expression in the lung from a control lamb. Strong expression in bronchial smooth muscle cells (long arrows) and macrophages (arrowheads) can be seen (immunohistochemical stain, original magnification 10×). (B) COX-1 expression in the lung from a RSV-infected lamb. Strong expression in bronchial smooth muscle cells (long arrows), alveolar septa (short arrows), and vascular endothelial cells (arrowheads) can be seen (immunohistochemical stain, original magnification 10×). (C) COX-1 expression in the lung from a PI3-infected lamb. Strong expression in bronchial and bronchiolar smooth muscle cells (long arrows) and vascular endothelial cells (arrowheads) can be seen (immunohistochemical stain, original magnification 10×). (From Z. A. Radi, D. K. Meyerholz and M. R. Ackermann, *Viral Immunology*, 23, pp. 43–48. Copyright © 2010 Mary Ann Liebert, Inc. Publications. All rights reserved.)

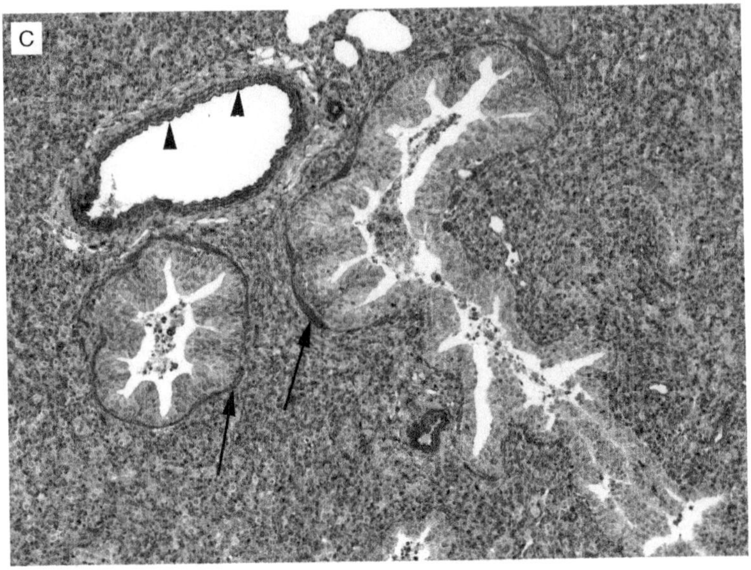

FIGURE 6-2 (*Continued*)

TABLE 6-1 Ontogeny of Pulmonary COX-1 Expression and Distribution[a]

Pulmonary compartment	Preterm 115 days	Preterm 130 days	Term 145 days	Postnatal 15 days
Alveolar septa	+++	+++	+++	+++
Bronchial epithelium	+	+	+	+
Bronchial SMC	++++	++++	++++	++++
Bronchiolar epithelium	+	+	+	+
Bronchiolar SMC	++++	++++	++++	++++
Alveolar macrophages	+++	+++	+++	+++
Vascular EC	++++	++++	++++	++++
Vascular SMC	++++	++++	++++	++++

Source: From Z. A. Radi and M. R. Ackermann, *Pathophysiology*, 18, pp. 215–219.

[a] +, minimal staining; ++, mild staining; +++, moderate staining; ++++, strong staining; EC, endothelial cells; SMC, smooth muscle cells.

study did not find any associations of any of the four individual polymorphisms or their haplotypes with asthma or asthma severity, suggesting that despite their potential functional importance, these promoter and coding region polymorphisms in the COX-1 gene are unlikely to play a role in asthma. However, the results from the study did indicate that the c.22TT genotype may confer a small protective effect against atopy (Shi et al., 2005). COX-1 expression in cultured airway tissue from "aspirin-sensitive" and control patients was investigated. It was shown that despite detectable levels of COX-2, COX-1 was functionally predominant in the airways (Harrington et al., 2008). The expression of COX-1 and COX-2 has been studied in the bronchial mucosa of 10 normal and 18 asthmatic subjects, 11 of whom had aspirin-sensitive asthma (ASA) and seven had non-aspirin-sensitive asthma (NASA). significant increase in the epithelial and submucosal cellular expression of COX-2, but not of COX-1, was noted in asthmatic patients. There was no significant difference in the total number of cells staining for either COX-1 or COX-2 between subjects with ASA and NASA, but the number and percentage of mast cells that expressed COX-2 was increased significantly in individuals with ASA (Sousa et al., 1997). Thus, both isoforms of COX are present in the airways of patients with aspirin-sensitive asthma.

Some studies have demonstrated increased COX-1 in pulmonary neoplastic and nonneoplastic conditions. COX-1 was highly expressed in lung tumor epithelial cell hyperplasias and adenomas in mice (Bauer et al., 2000). COX-1 expression was investigated in human non-small-cell lung cancer (NSLC) (Ermert et al., 2003). Only 4 of 15 adenocarcinomas and the chondrohamartomas showed a moderate COX-1 immunostaining intensity. No COX-1 expression was noted in squamous cell carcinomas and in vascular smooth muscle cells or endothelial cells of pulmonary vessels located within the tumor or adjacent to the tumor (Ermert et al., 2003). COX-1 mRNA expression in NSLC via in situ hybridization was absent or below the level of detection. COX-1 immunohistochemistry demonstrated uniform faint cytoplasmic staining in tumor cells and stromal inflammatory cells (Watkins et al., 1999). The COX biosynthetic pathway may act synergistically and enhance tumor angiogenesis, the expression of angiogenic factors, and metastases in patients with small cell lung cancer (Yoshimoto et al., 2005). COX-1 mRNA levels were significantly higher in the lung parenchyma of COPD patients than in control subjects (Roca-Ferrer et al., 2011). COX-1 expression was examined in nasal biopsies from patients with allergic rhinitis (AR), seasonal AR (SAR), and perennial AR (PAR). No significant differences were noted in the lamina propria in immunostaining for COX-1. Within the epithelium, increased expression of COX-1 was observed in PAR and SAR biopsies and was associated with more intraepithelial mast cells in groups (Westergren et al., 2009). COX-1 mRNA expression was investigated in an ovalbumin-sensitized and ovalbumin-challenged guineapig lung model. The levels of COX-1 mRNA remained unchanged (Oguma et al., 2002). COX-1 expression in bronchiolar epithelial cells was significantly lower in IPF and sarcoidosis than in controls (Petkova et al., 2003). Increased COX expression has been demonstrated in various viral infections (Fig. 6-2B and C) (Steer and Corbett, 2003; Radi et al., 2010). Therefore, prostanoids are important mediators in both normal and disease pulmonary functions (Radi and Ackermann, 2011). In a study in a neonatal lamb model of viral infection, all microanatomical locations in control, RSV, and PI3-infected lungs had some degree of COX-1 expression (Table 6-2) (Radi et al., 2010). A slightly higher COX-1 expression was present in bronchial and bronchiolar epithelial cells from RSV-infected lungs compared to controls of PI3-infected lungs (Radi et al., 2010). Thus, COX-1 expression is not altered after RSV or PI3 pulmonary infection (Radi et al., 2010).

TABLE 6-2 Pulmonary Cellular Expression of COX-1 Following RSV and PI3 Infection[a]

Pulmonary compartment	Control	RSV	PI3
Alveolar septa	+++	++++	+++
Bronchial epithelium	+	++	+
Bronchial SMC	++++	++++	++++
Bronchiolar epithelium	+	++	+
Bronchiolar SMC	++++	++++	++++
Macrophages	++++	++++	++++
Vascular EC	++++	++++	++++
Vascular SMC	++++	++++	++++

Source: From Z. A. Radi, D. K. Meyerholz, and M. R. Ackermann, *Viral Immunology*, 23, pp. 43–48.

[a] +, minimal expression; ++, mild expression; +++, moderate expression; ++++, strong expression; EC, endothelial cells; SMC, smooth muscle cells.

COX-2 EXPRESSION IN THE RESPIRATORY SYSTEM UNDER NORMAL AND PATHOLOGICAL CONDITIONS

The COX-2 isoenzyme is not generally expressed in normal respiratory tissues, but is up-regulated during pathological conditions (Fig. 6-3) (Radi and Ackermann, 2011). Airway epithelium forms a continuous barrier that limits the access of luminal substances to the systemic circulation (Wallace, 2006; Radi and Ackermann, 2011). In a human airway epithelial cell culture system, COX-2 induction was observed in the setting of inflammatory cytokine stimulation (Watkins et al., 1997). No COX-2 expression was present in lungs of preterm lambs (Radi and Ackermann, 2011). The lack of inducible COX-2 expression in the lung of preterm lambs may be attributed to the sterility of the fetal environment (Radi and Ackermann, 2011). However, minimal COX-2 expression was present in bronchial and bronchiolar epithelial cells and the alveolar septa and macrophages 15 days postnatally in the ovine lung (Table 6-3) (Radi and Ackermann, 2011). Other studies noted low levels of COX-2 in untreated human airway epithelial cells in culture (Mitchell et al., 1994). In vitro studies using cultured human bronchial epithelial cells have demonstrated that these cells express COX-2 constitutively (Asano et al., 1996). The biological significance of COX-2 expression in airway epithelia in term and 15-day-old lambs may be related to pulmonary epithelial cells direct regulatory function via an action of cytokines released from alveolar macrophages on specific epithelial cell receptors (Radi and Ackermann, 2011). Airway epithelial cells express a number of cytokines (e.g., IL-1β) that are known to induce COX-2 (Devalia and Davies, 1993). Thus, epithelial cells may generate such cytokines, which then act in an autocrine fashion, leading to COX-2 expression up-regulation (Devalia and Davies, 1993). In human fetal lung, COX-2 is present in the perinatal

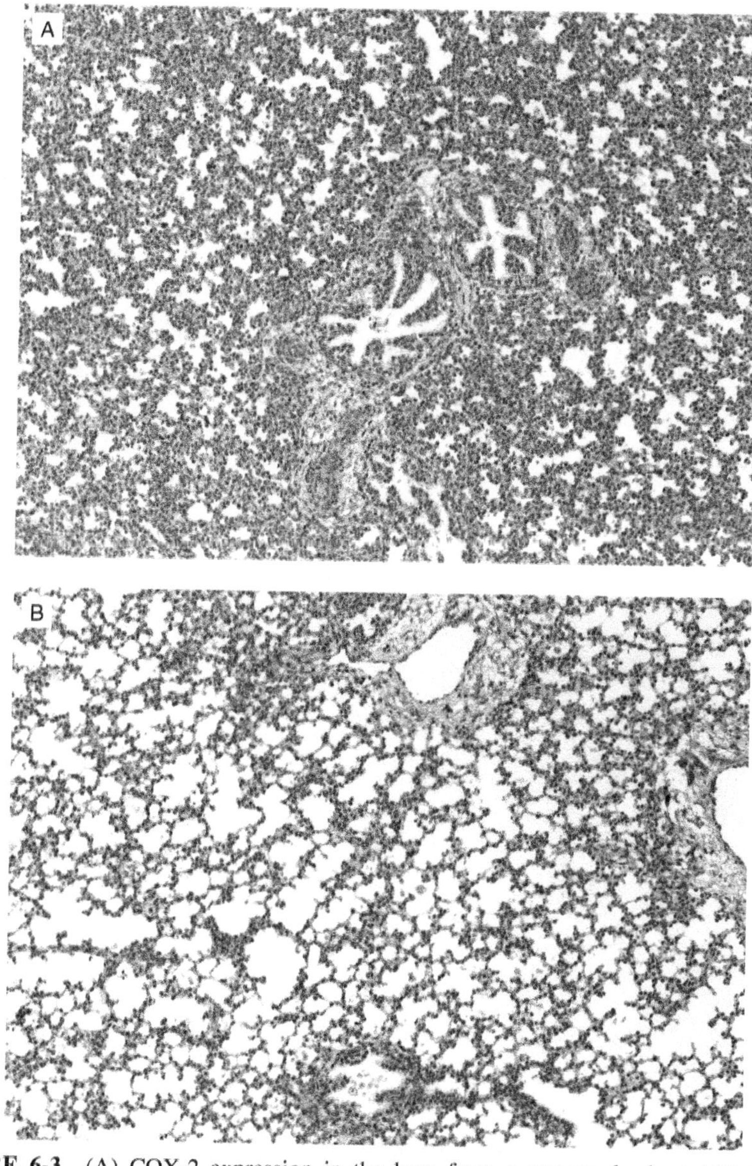

FIGURE 6-3 (A) COX-2 expression in the lung from a preterm lamb at 115 days of gestation. No expression in any pulmonary microanatomic location. (B) COX-2 expression in the lung from a preterm lamb at 130 days of gestation. No expression in any pulmonary microanatomic location. (C) COX-2 expression in the lung from a term lamb at 145 days of gestation. Mild expression in bronchial (long arrows) and bronchiolar (short arrows) epithelial cells. (D) COX-2 expression in the lung from a postnatal lamb. Mild expression in bronchial and bronchiolar epithelial cells (long arrows). Immunohistochemical stain, original magnification 10×. (From Z. A. Radi and M. R. Ackermann, *Pathophysiology*, 18, pp. 215–219. Copyright © 2011 Elsevier Ireland Ltd. All rights reserved.)

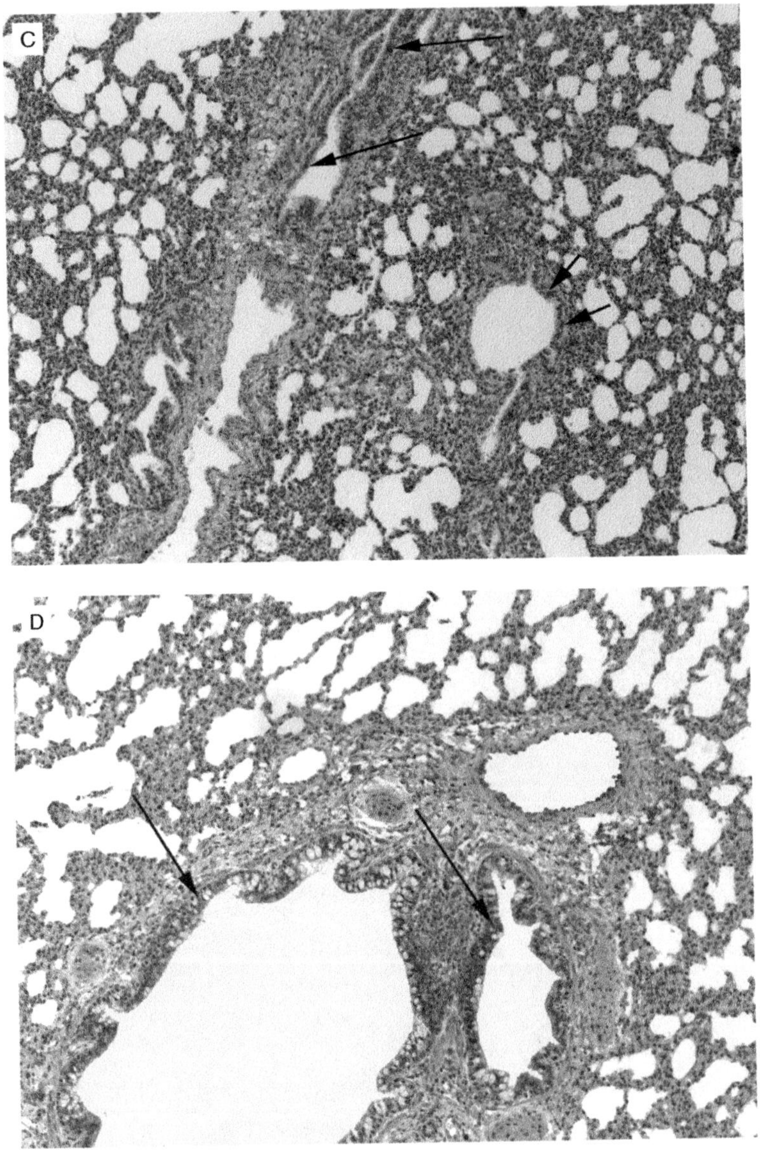

FIGURE 6-3 (*Continued*)

lung of neonates from the gestational age of 16 weeks to term (Lassus et al., 2000). COX-2 staining was found exclusively in the alveolar and bronchial epithelial cells, with none in the interstitium, inflammatory cells, or endothelium. Bronchial staining was localized in ciliated epithelial cells. Alveolar staining was seen in cells resembling type II pneumocytes and in most of the cells lining the alveolar epithelium. In preterm infants, alveolar staining was most intense in areas with hyaline membranes. In term infants, the pattern of alveolar staining was similar to that

TABLE 6-3 Ontogeny of Pulmonary COX-2 Expression and Distribution[a]

Pulmonary compartment	Preterm		Term	Postnatal
	115 days	130 days	145 days	15 days
Alveolar septa	–	–	–	–
Bronchial epithelium	–	–	+/++	+/++
Bronchial SMC	–	–	–	–
Bronchiolar epithelium	–	–	+/++	+/++
Bronchiolar SMC	–	–	–	–
Alveolar macrophages	–	–	–	++
Vascular EC	–	–	–	–
Vascular SMC	–	–	–	–

Source: From Z. A. Radi and M. R. Ackermann, *Pathophysiology*, 18, pp. 215–219.

[a] –, no staining; +, minimal staining; ++, mild staining; EC, endothelial cells; SMC, smooth muscle cells.

seen in preterm infants, but with a lower proportion of positive cells (Lassus et al., 2000). In bronchopulmonary dysplasia (BPD), no staining appeared in alveolar epithelial cells. In three cases with BPD and with clinically and microscopically verified pneumonic infection at the time of death, alveolar staining showed a pattern similar to that in preterm infants. In some samples, an infiltration of granulocytes was apparent, and in these cells an occasional positivity was observable (Lassus et al., 2000).

In lungs of adult normal subjects, COX-2 expression was found in bronchial epithelium (Demoly et al., 1997). Bronchial epithelial cells play an integral role in the airway defense mechanism via the mucociliary system, as well as mechanical barriers (Radi et al., 2010). In addition, bronchial epithelial cells produce and release biologically active compounds, including lipid mediators, growth factors, and a variety of cytokines important in the pathogenesis of airway disorders (Takizawa, 2005; Radi et al., 2010). Some in vitro studies have demonstrated that cultured human bronchial epithelial cells express COX-2 constitutively (Asano et al., 1996). Pulmonary vascular endothelial cells lacked COX-2 expression under basal conditions in humans and lambs (Jackson et al., 1993; Radi and Ackermann, 2011).

RSV, PI1, and PI3 cause respiratory infection and hospitialization in young children and animals (Hall, 2001; Radi et al., 2010). Marked increase in COX-2 expression was noted in pulmonary bronchial and bronchiolar epithelium and macrophages following RSV or PI3 infection in a study in neonatal lambs (Fig. 6-4 and Table 6-4) (Radi et al., 2010). Minimal COX-2 expression was present in macrophages and bronchial and bronchiolar epithelial cells from control animals, but was increased in RSV and in a PI3-infected neonatal lamb model (Radi et al., 2010). In an experimental mouse model of acute lung, significant induction of COX-2 expression was observed after injury (Fukunaga et al., 2005). It has been shown using in vitro human alveolar type II-like epithelial cells that RSV infection

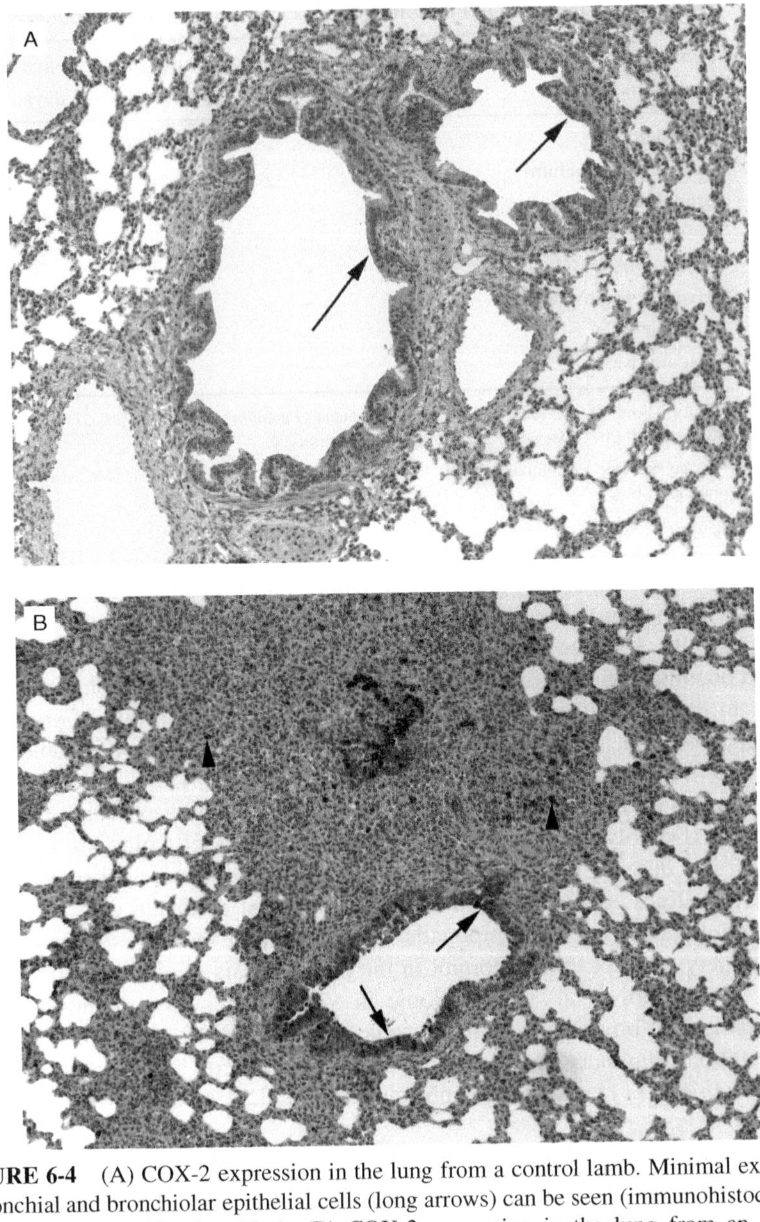

FIGURE 6-4 (A) COX-2 expression in the lung from a control lamb. Minimal expression in bronchial and bronchiolar epithelial cells (long arrows) can be seen (immunohistochemical stain, original magnification 10×). (B) COX-2 expression in the lung from an RSV-infected lamb. Strong expression in bronchiolar epithelial cells (long arrows) and macrophages (arrowheads) can be seen (immunohistochemical stain, original magnification 10×). (C) COX-2 expression in the lung from a PI3-infected lamb. Strong expression in bronchial epithelial cells (long arrows) and macrophages (arrowheads) can be seen (immunohistochemical stain, original magnification 10×). (From Z. A. Radi, D. K. Meyerholz, and M. R. Ackermann, *Viral Immunology*, 23, pp. 43–48. Copyright © 2010 Mary Ann Liebert, Inc. Publications. All rights reserved.)

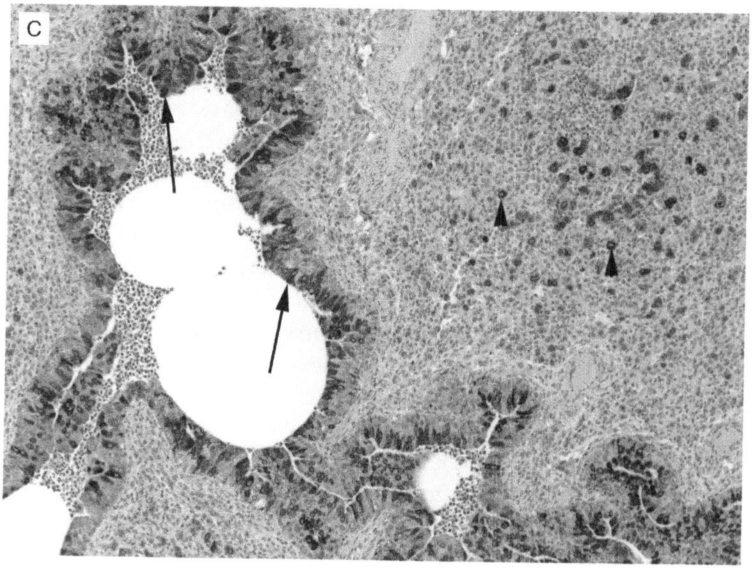

FIGURE 6-4 *(Continued)*

TABLE 6-4 Pulmonary Cellular Expression of COX-2 Following RSV and PI3 Infection[a]

Pulmonary compartment	Control	RSV	PI3
Alveolar septa	–	–	–
Bronchial epithelium	+	+++/++++	++++
Bronchial SMC	–	–	–
Bronchiolar epithelium	+	+++/++++	+++/++++
Bronchiolar SMC	–	–	–
Macrophages	+	++++	++++
Vascular EC	–	–	–
Vascular SMC	–	–	–

Source: From Z. A. Radi, D. K. Meyerholz, and M. R. Ackermann, *Viral Immunology*, 23, pp. 43–48. Copyright © 2010 Mary Ann Liebert, Inc. Publications. All rights reserved.

[a] –, no staining; +, minimal expression; ++, mild expression; +++, moderate expression; ++++, strong expression; EC, endothelial cells; SMC, smooth muscle cells.

induces a time-dependent increase in COX-2 expression, is a potent inducer of PGE_2, and that viral replication is required for PGE_2 secretion (Liu et al., 2005). The cotton rat is a powerful model for studying RSV infection in vivo (Richardson et al., 2005). In a cotton rat model, increased COX-2 protein expression was noted 1 day post-RSV virus infection and persisted on day 2. Thus, RSV induced

COX-2 mRNA expression in lungs of infected cotton rats. Lung epithelial cells and macrophages (peritoneal and bronchioalveolar) are potential sources of COX-2 and PG during infection with RSV (Richardson et al., 2005). COX-2 was immunoprecipitated in all infected animals during the first 7 days postinfection with RSV. There were two peaks of immunoreactive COX-2. The first was detected as early as 6 h postinfection, which decreased slightly after 12 h, but it remained detectable throughout day 3 postinfection (Richardson et al., 2005). A second peak was seen on day 5 postinfection and correlated with the peak of lung pathology during primary RSV infection in cotton rats. Thus, COX-2 is strongly induced in vivo during primary RSV infection in the lung (Richardson et al., 2005; Radi et al., 2010).

Alveolar macrophages are a major source of PGE_2 (Hempel et al., 1994). Marked increase in COX-2 expression in macrophages occurred after RSV infection (Richardson et al., 2005; Radi et al., 2010). Human cord blood–derived macrophages and dendritic cells have been shown to secrete PGE_2 following exposure to RSV (Bartz et al., 2002). Increased COX-2 levels in macrophages have been demonstrated in humans after rhinovirus infection (Seymour et al., 2002). Macrophages are postulated to be the primary cell type expressing COX-2 at sites of inflammation (Seibert et al., 1994). In a mouse model of oxygen-induced ARDS, COX-2 was expressed in alveolar macrophages (Adawi et al., 1998). Thus, COX-2 expression is up-regulated in bronchiolar and bronchial epithelial cells and macrophages after RSV and PI3 infection, and pulmonary epithelia along with macrophages are important microanatomical compartments regulating host inflammatory response during viral infection (Radi et al., 2010).

COX-2 and vascular endothelial growth factor (VEGF) affect vascular permeability and angiogenesis and may be induced by hyperglycemia and tissue hypoxia (Wilkinson-Berka, 2004). COX-2 inhibition has been associated with attenuating angiogenesis and down-regulating VEGF and bFGF (Masferrer et al., 1999). In inflamed tissues, eicosanoids are thought to play a significant role in angiogenesis (Yamada et al., 1999). Total COX-2 mRNA expression in cancer cells and surrounding stromal cells correlated strongly and positively with VEGF mRNA expression, intratumoral microvessel counts, and adverse prognosis in NSCLC patients (Yuan et al., 2005).

COX-2 expression in pulmonary neoplasia is variable depending on the type and grade of the tumor. The expression of COX-2 mRNA and protein was studied in human lung adenocarcinoma, squamous cell carcinoma, and small cell lung cancer (Wolff et al., 1998). Ninety percent of lung tumors (e.g., squamous cell carcinomas and adenocarcinomas) expressed COX-2 at a moderate to strong level, which was significantly different from the negligible expression in distant nonneoplastic epithelium (Soslow et al., 2000). COX-2 mRNA steady-state levels were high in well-differentiated adenocarcinoma samples but low in poorly differentiated adenocarcinoma, squamous cell carcinoma, and small cell lung cancer. Immunohistochemistry showed COX-2 staining in 19 of 21 adenocarcinomas. However, well-differentiated adenocarcinomas had more COX-2 staining than did poorly differentiated adenocarcinomas. Expression of the COX-2 protein was also seen in all 11 squamous cell carcinomas studied, although the level of staining seemed to be less than that in the adenocarcinomas. Small cell lung cancer specimens stained with a relatively weak intensity (Wolff et al., 1998). There were no differences in the COX-2 mRNA levels between squamous carcinoma and adenocarcinoma. COX-2 protein levels were significantly higher in tumors than in lung parenchyma and airways. COX-2 protein levels were higher in adenocarcinoma than in squamous carcinoma (Roca-Ferrer et al., 2011). HuR is an mRNA-binding protein that binds to labile transcripts containing AU-rich elements, such as mRNAs for protooncogenes, cytokines, and cytokine-response genes. Cytoplasmic HuR expression was associated with COX-2 expression in pulmonary squamous cell carcinomas. Thus, COX-2 could be a potential target of the mRNA-stabilizing activity of cytoplasmic HuR, resulting in COX-2 overexpression. Increased COX-2 expression is associated with active conversion of survivin from nuclear to cytoplasmic, and low expression of nuclear survivin, which contributes to the antiapoptotic effect and overproliferation of pulmonary squamous cell carcinomas (Kim et al., 2011). In another study, increased COX-2 expression occurred frequently in human lung cancers and specifically in adenocarcinomas (Hida et al., 1998). Scattered weak COX-2 reactivity was seen in normal airway epithelial cells, and markedly up-regulated COX-2 expression was detected in about one-third of atypical adenomatous hyperplasias and carcinoma in situ specimens, and a significant increase in COX-2 expression was observed in 70% of invasive adenocarcinoma cases (Hida et al., 1998).

COX-2 expression in bronchiolar epithelial cells was significantly lower in IPF and sarcoidosis than in controls. No significant difference was found in COX-2 expression between macrophages in IPF and control sections, but COX-2 was reduced in macrophages in sarcoidosis compared with controls (Petkova et al., 2003). Atypical alveolar epithelium, which associates with asbestosis and idiopathic fibrosing alveolitis and is considered to be a precursor lesion for lung cancer, expressed the COX-2 protein (Wolff et al., 1998). In the lung parenchyma of COPD patients, COX-2 mRNA levels were increased significantly compared with those of control subjects (Roca-Ferrer et al., 2011). COX-2 expression was examined in nasal biopsies from patients with allergic rhinitis (AR), seasonal AR (SAR), and perennial AR (PAR). No significant differences in the lamina propria in immunostaining for COX-2 were noted among groups (Westergren et al., 2009). A significant increase in the epithelial and submucosal cellular expression of COX-2 has been reported in asthmatic patients (Sousa et al., 1997). The 72-h kinetics of COX-2 mRNA expression was investigated in an ovalbumin-sensitized and ovalbumin-challenged guinea pig lung model. The sensitized animals showed a robust and transient induction of COX-2 mRNA expression within 1 h after ovalbumin challenge (Oguma et al., 2002). In a human CF bronchial epithelia cell line, increased COX-2 expression and overproduction of PGE_2 were noted (Chen et al., 2012).

EFFECTS OF COX-2 s-NSAIDs ON THE RESPIRATORY SYSTEM

Because of the role COX-2 plays in the pathophysiology of several pulmonary tumors, COX-2 s-NSAIDs have been proposed as potential therapeutic agents.

Lung tumor prevention with NSAIDs involves both the induction of apoptosis and the inhibition of COX-2 expression (Yao et al., 2000). COX-2 was characterized in several human epithelial cancers. COX-2 was detected in 85% of the hyperproliferating, dysplastic, and neoplastic epithelial cells and in the existing and angiogenic vasculature within and adjacent to hyperplastic/neoplastic lesions (Koki et al., 2002). This implied COX-2 may play an important role during premalignant hyperproliferation as well as the later stages of invasive carcinoma and metastasis in various human epithelial cancers (Koki et al., 2002). PGE_2 production by A549 non-small-cell lung cancer (NSCLC) cells was elevated up to 50-fold in response to IL-1β. Reversal of IL-1β-induced PGE_2 production in A549 cells was achieved by specific pharmacological or antisense oligonucleotide inhibition of COX-2 activity or expression. In contrast, specific COX-1 inhibition was not effective (Huang et al., 1998). The role of host-derived COX-1 and COX-2 in tumor growth was investigated using grafted Lewis lung carcinoma (LLC) cells into either wild-type (WT), COX-$1^{-/-}$, or COX-$2^{-/-}$ C57BL/6 mice (Williams et al., 2000). LLC cells grown rapidly as solid tumors when implanted in C57BL/6 mice. Tumor growth was markedly attenuated in COX-$2^{-/-}$ but not in COX-$1^{-/-}$ or WT mice. Treatment of WT C57BL/6 mice bearing LLC tumors with celecoxib also reduced tumor growth. A decrease in vascular density was observed in tumors grown in COX-$2^{-/-}$ mice compared with those in WT mice. In addition, COX-$2^{-/-}$ mouse fibroblasts had a 94% reduction in their ability to produce the proangiogenic factor VEGF. Furthermore, treatment of WT mouse fibroblasts with the s-NSAID SC-58635 reduced VEGF production by 92% (Williams et al., 2000). Thus, inhibition of the COX-2/VEGF-dependent pathway is effective in tumor-associated angiogenesis, tumor growth, and metastasis (Yoshida et al., 2003).

Nimesulide inhibited proliferation of NSCLC cell lines in vitro in a dose-dependent manner, in part by inducing apoptosis even at clinically achievable low concentrations (Hida et al., 2000). In another experiment using NSCLC cell lines, celecoxib treatment decreased cell survival, activated caspase cascades, and increased DNA fragmentation. In addition, celecoxib treatment induced the expression of death receptors, particularly that of DR5. Thus, induction of apoptosis using celecoxib in human NSCLC is through the extrinsic death receptor pathway (Liu et al., 2004). The molecular effects of a derivative of celecoxib, 2,5-dimethylcelecoxib (DMC), on NSCLC cells were investigated. DMC sensitizes human NSCLC cells to tumor necrosis factor–related apoptosis-inducing ligand (TRAIL)–induced apoptosis via induction of DR5 and down-regulation of cellular FLICE-inhibitory protein (c-FLIP) (Chen et al., 2007). Rofecoxib, a COX-2 s-NSAID, significantly attenuated growth of murine Lewis lung carcinoma cells (3LL) in vitro. It caused considerable dose- and time-dependent inhibition of proliferation in cancerous cells and, similarly, induced dose-dependent apoptosis (Qadri et al., 2002). There is compelling experimental evidence that s-NSAIDs have potential for the chemoprevention of lung cancers. Molecular studies revealed that COX-2 overexpression is a prominent feature of premalignant and malignant neoplasms (Harris, 2009). Evidence is accumulating that carcinogenesis often evolves as a progressive series of highly specific cellular and molecular changes in response to induction of constitutive overexpression of COX-2 and the PG cascade in the "inflammogenesis of cancer" (Harris, 2009). Lung adenocarcinoma cell line growth was inhibited in vitro by JTE-522, a COX-2 s-NSAID, as a single or adjunct agent of conventional anticancer drugs (e.g., docetaxel, vinorelbine). In addition, the combined use of JTE-522 with conventional anticancer drugs was shown to exhibit significant enhancement in treatment efficacy using an in vivo athymic nude mouse model with xenografted lung adenocarcinoma cells (Hida et al., 2002). NS-398, a COX-2 s-NSAID, inhibited tumor multiplicity and and returned plasma PGE_2 to basal levels observed in untreated mice in a A/J female mouse model of lung tumorigenesis (Rioux and Castonguay, 1998).

The effects of COX-2 s-NSAID celecoxib or rofecoxib use on lung cancer were studied in a case–control study. Use of either drug for more than one year produced a significant reduction (60%) in the risk of lung cancer (Harris et al., 2007). The effects of celecoxib on lung cancer prevention were investigated in a randomized double-blind placebo-controlled trial (Mao et al., 2011). Former-smoker subjects (age $\geq$45, $\geq$30 pack-years of smoking, $\geq$1 year of sustained abstinence from smoking) received celecoxib (400 mg by mouth twice daily). The impact of celecoxib on cellular and molecular events associated with lung cancer pathogenesis was assessed. The primary endpoint was the bronchial Ki-67 labeling index (Ki-67 LI) after six months of treatment. Decreased Ki-67 LI correlated with a reduction and/or resolution of lung nodules on computed tomography. Celecoxib significantly reduced plasma C–reactive protein and IL-6 mRNA and protein and increased 15(*S*)-hydroxyeicosatetraenoic acid levels in bronchoalveolar lavage samples (Mao et al., 2011). In another study, celecoxib decreased Ki-67 labeling by 3.85% in former smokers and by 1.10% in current smokers. This was a significantly greater reduction than that seen with placebo after adjusting for metaplasia and smoking status. Celecoxib (400 mg twice daily) was biologically active in the bronchial epithelium of current and former smokers (Kim et al., 2010). In a study of populations of patients with advanced NSCLC, the addition of celecoxib (400 mg twice daily) to docetaxel did not seem to improve the outcome compared with the single agent docetaxel (Gadgeel et al., 2008). However, in another study in patients with stage IIIb/IV NSCLC who had pathological confirmation and no prior chemotherapy, the effect of celecoxib was evaluated. Treatment consisted of docetaxel and carboplatin every 3 weeks for five cycles. Patients were randomly assigned to receive celecoxib 400 mg twice daily. It was found that in advanced NSCLC, celecoxib does not improve survival (Groen et al., 2011).

The role of COX-2 in pulmonary inflammation was investigated. Allergen-induced pulmonary inflammation and airway hyperresponsiveness were studied in WT mice and in COX-$1^-/^-$ and COX-$2^-/^-$ mice. After allergen exposure, immunized WT mice had mild pulmonary inflammation, mostly in the perivascular/peribronchial regions, characterized by the presence of lymphoid aggregates, focal areas of consolidation with mononuclear and polymorphonuclear leukocytes, occasional large eosinophilic alveolar macrophages, and rare multinucleated giant cells (Gavett et al., 1999). In contrast, lungs from allergic COX-$1^-/^-$ mice had severe inflammation that not only involved perivascular and peribronchial regions, but also extended into the airspace. Lymphoid aggregates were identifiable around most blood vessels and bronchi and often involved the smallest vessels. Airway epithelium was thickened, alveoli contained large numbers of large eosinophilic macrophages and eosinophils, and multinucleated giant cells were numerous (Gavett et al., 1999). Lungs from allergic COX-2^{-}/$^{-}$ mice had pulmonary inflammation that was less intense than that in allergic COX-1^{-}/$^{-}$ mice but more severe than that in allergic WT mice. Compared with allergic COX-1^{-}/$^{-}$ mice, the inflammation observed in allergic COX-2^{-}/$^{-}$ mice was more focal and confined primarily to interstitial regions surrounding vessels and bronchi. Allergic COX-2^{-}/$^{-}$ mice showed loss of fine alveolar septae in comparison with allergic WT mice, and remaining septae were thickened. These alterations in parenchymal alveolar structure were comparable to those observed in allergic COX-1^{-}/$^{-}$ mice. Airway mucus production was more pronounced in allergic COX-1^{-}/$^{-}$ and COX-2^{-}/$^{-}$ mice than in allergic WT or nonimmunized mice of any genotype. Both allergic COX-1^{-}/$^{-}$ and COX-2^{-}/$^{-}$ mice exhibited decreased baseline respiratory system compliance, whereas only allergic COX-1^{-}/$^{-}$ mice showed increased baseline resistance and responsiveness to methacholine (Gavett et al., 1999). COX inhibition during allergic sensitization with ovalbumin in a murine model leads to an increase in type 2 cytokines (i.e., IL-5 and IL-13) (Peeble et al., 2002). In a BALB/c mouse model, COX inhibition with either a COX-1 (SC-58560) or a COX-2 (SC-58236) inhibitor during allergen sensitization augmented production of IL-13, increased lung eosinophilia, and increased airway hyperresponsiveness (Peeble et al., 2002). In this experiment, the mice were exposed to aerosols of 1% ovalbumin on days 14 through 21 for 40 min each day. Methacholine challenges were performed on day 22 (Peeble et al., 2002). In an experiment in ovalbumin-sensitized and ovalbumin-challenged male Hartley guinea pig model, the effects of NS-398 and JTE-522 administration on pulmonary allergic inflammation during airway challenge were investigated (Oguma et al., 2002). COX-2 s-NSAID treatment did not modify the enhanced airway responsiveness to histamine observed after allergen sensitization and challenge, but it did significantly reduce the number of eosinophils and neutrophils in bronchoalveolar lavage fluid (Oguma et al., 2002). In an experimental dog model of acute lung injury (ALI), the effects of SC-236, a COX-2 s-NSAID, on pulmonary perfusion pattern were evaluated. ALI was induced using oleic acid (OA), and SC-236 was given 50 min before OA. SC-236 prevented the loss of perfusion redistribution associated with endotoxin (Gust et al., 1999).

Treatment for 1 week with oral celecoxib (200 mg twice daily) did not affect bronchial responsiveness in asthmatic patients (Dicpinigaitis, 2001). Aspirin-exacerbated respiratory disease (AERD) is a distinct clinical entity that is characterized by aspirin-induced respiratory reactions, asthma, nasal polyposis, and chronic hyperplastic eosinophilic sinusitis (CHES) (Stevenson and Szczeklik, 2006). Etoricoxib was well tolerated, without any signs of immediate or delayed hypersensitivity in aspirin and NSAID-induced asthmatic patients (El Miedany et al., 2006). Rofecoxib (25 mg) was well tolerated and basal FEV1 remained unchanged during the period of observation in patients with aspirin-induced asthma (Micheletto et al., 2006).

EFFECTS OF ns-NSAIDs ON THE RESPIRATORY SYSTEM

The effects of indomethacin on endothelial dysfunction during lipopolysaccharide (LPS)-induced sepsis were investigated in an isolated rat lung model. Rats were administered intraperitoneally indomethacin at 5 or 10 mg/kg. Three hours later the rats were anesthetized, the lungs were isolated, and pulmonary vasoreactivity was assessed with bradykinin or receptor-independent hypoxic pulmonary vasoconstriction. Indomethacin significantly attenuated bradykinin-induced vasoconstriction in septic isolated rat lungs (Fischer et al., 2000). In an adult New Zealand white rabbit smoke inhalation model, the effect of ibuprofen on extravascular lung water accumulation was studied. Rabbits were anaesthetized and intubated, and each ibuprofen-treated animal received a dose of 50 mg/kg either intraperitoneally or intravenously. Peak carboxyhemoglobin levels as well as CO half-lives were not significantly different between ibuprofen-treated groups and the controls. However, ibuprofen treatment resulted in significantly decreased lung water. Thus, ibuprofen promoted an early reduction in lung water accumulation (Stewart et al., 1990). The effects of ibuprofen (35 mg/kg orally twice daily) on chronic *Pseudomonas* endobronchial infection and inflammation were investigated in a rat model. The area of lung inflammation 14 days after animal inoculation with *Pseudomonas* was significantly less in animals treated with ibuprofen than in animals given placebo. Ibuprofen did not increase the pulmonary burden of *Pseudomonas*, and the ibuprofen-treated infected animals gained weight better than did placebo-treated controls (Konstan et al., 1990). The postnatal pulmonary effects of diclofenac sodium were studied in the rat offspring treated with diclofenac sodium during pregnancy. Pregnant female rats were given diclofenac sodium at 1 mg/kg, which was injected intraperitoneally during gestational days 5 to 19. Histopathological examination of the lung tissues of 4- and 20-week-old rats revealed no significant differences between males and females in both the control and diclofenac sodium-treated rats (Ragbetli et al., 2011).

Ingestion of aspirin can result in a spectrum of upper and/or lower respiratory reactions, to include rhinitis, conjunctivitis, laryngospasm, and bronchospasm (Lee and Stevenson, 2011). Aspirin and other ns-NSAIDs that inhibit COX-1 induce unique nonallergic reactions, consisting of attacks of rhinitis and asthma. These hypersensitivity reactions occur in a subset of asthmatic subjects (Stevenson and Szczeklik, 2006). Samter and Beers first described a clinical syndrome known as Samter's triad, which includes asthma, aspirin sensitivity, and nasal polyps (Zeitz, 1988). Symptoms can occur within minutes to hours after aspirin ingestion and include angioedema, facial flushing, and conjunctivitis. Aspirin- and ns-NSAID-induced bronchospasm has been reported (Lee and Stevenson, 2011; Sun and Pei, 2011). Aspirin-exacerbated respiratory disease (AERD) is a clinical tetrad of nasal polyps, chronic hypertrophic eosinophilic sinusitis, asthma, and sensitivity to aspirin and other NSAIDs (Lee and Stevenson, 2011). AERD has been described as Samter's triad. AERD affects 0.3 to 0.9% of the general population, but the prevalence rises to 10 to 20% of asthmatics and up to 30 to 40% in those asthmatics with nasal polyposis (Lee and Stevenson, 2011). AERD is more commonly reported in females. Nasal ketorolac administration has been shown to be a reasonably accurate and safe method for diagnosing AERD (White et al., 2006). Alteration in the function of PGE_2 receptors (particularly EP_2) might contribute to the inability of PGE_2 to sustain the blockade. From mast cells, PGD_2 is oversynthesized, and histamine and tryptase are released during reactions (Stevenson and Szczeklik, 2006). Short term treatment with ibuprofen reduced PGE_2 concentrations in EBC in patients with COPD (Montuschi et al., 2005). Ibuprofen decreased exhaled PGE_2 concentrations in patients with stable COPD (Montuschi et al., 2001).

Although this is not an approved indication, ibuprofen has been shown in some clinical trials to slow progression of lung disease in patients with cystic fibrosis (CF), particularly in children and adolescents with mild to moderate disease. The use of ibuprofen in over 1000 children followed by the CF Foundation Patient Registry supported the finding that ibuprofen therapy slowed disease progression as measured by forced expiratory volume (FEV_1) (Konstan et al., 2007). In a two-year multicenter double-blind placebo-controlled trial conducted in 12 Canadian centers, a total of 142 patients age 6 to 18 years with mild CF disease were randomized to receive high-dose ibuprofen (20 to 30 mg/kg twice daily, adjusted to a peak serum concentration of 50 to 100 μg/mL). Those patients in the high-dose ibuprofen group exhibited a significant reduction in the rate of decline of forced vital capacity percent predicted, but not forced expiratory volume in 1 min % (Lands et al., 2007). In a retrospective study, pulmonary function decline and hospitalization rates were compared among 51 children and adolescents treated with ibuprofen versus 39 untreated patients. The medical records of all patients, ages 5 to 18 years, were followed at the CF center from 1995 to 2002. Pulmonary function decline and hospitalization rates for each group were compared, examining both intent to treat and patients who continued therapy for at least four years. No differences in either outcome measure were observed in the intent-to-treat group or those who completed at least four years of therapy. Neither use of high-dose ibuprofen nor its cessation resulted in a significant change in the rate of decline in pulmonary function or influenced hospitalization rates. Matching based on age, gender, pancreatic sufficiency status, pretreatment FEV_1, and *Pseudomonas* status failed to show differences in FEV_1 decline between the treated and untreated groups. A confounding factor in this retrospective study was the high and disproportionate use of systemic corticosteroids among these patients (Fennell et al., 2007).

CONCLUSIONS

There are important anatomical and physiological interspecies differences in the respiratory system that can affect the pathophysiological effects noted after COX inhibition. The normal ontogeny of COX-1 and COX-2 expression varies across species during pulmonary development and in pathological conditions that affect the lung. In experimental nonclinical studies, NSAIDs have shown some potential for the chemoprevention of some lung cancers.

REFERENCES

Abraham E, Baughman R, Fletcher E, Heard S, et al. Liposomal prostaglandin E1 (TLC C-53) in acute respiratory distress syndrome: a controlled, randomized, double-blind, multicenter clinical trial. TLC C-53 ARDS Study Group. Crit Care Med 1999;27:1478–1485.

Acarregui MJ, Snyder JM, Mitchell MD, Mendelson CR. Prostaglandins regulate surfactant protein A (SP-A) gene expression in human fetal lung in vitro. Endocrinology 1990;127:1105–1113.

Ackermann MR, Derscheid R, Roth JA. Innate immunology of bovine respiratory disease. Vet Clin North Am Food Anim Pract 2010;26:215–228.

Adawi A, Zhang Y, Baggs R, Finkelstein J, et al. Disruption of the CD40–CD40 ligand system prevents an oxygen-induced respiratory distress syndrome. Am J Pathol 1998;152:651–657.

Arias-Diaz J, Vara E, Garcia C, Balibrea JL. Tumor necrosis factor-alpha-induced inhibition of phosphatidylcholine synthesis by human type II pneumocytes is partially mediated by prostaglandins. J Clin Invest 1994;94:244–250.

Asano K, Lilly CM, Drazen JM. Prostaglandin G/H synthase-2 is the constitutive and dominant isoform in cultured human lung epithelial cells. Am J Physiol 1996;271:L126–L131.

Bartz H, Buning-Pfaue F, Turkel O, Schauer U. Respiratory syncytial virus induces prostaglandin E2, IL-10 and IL-11 generation in antigen presenting cells. Clin Exp Immunol 2002;129:438–445.

Bastacky J, Lee CY, Goerke J, Koushafar H, et al. Alveolar lining layer is thin and continuous: low-temperature scanning electron microscopy of rat lung. J Appl Physiol 1995;79:1615–1628.

Bauer AK, Dwyer-Nield LD, Malkinson AM. High cyclooxygenase 1 (COX-1) and cyclooxygenase 2 (COX-2) contents in mouse lung tumors. Carcinogenesis 2000;21:543–550.

Bennett BL, Garofalo RP, Cron SG, Hosakote YM, et al. Immunopathogenesis of respiratory syncytial virus bronchiolitis. J Infect Dis 2007;195:1532–1540.

Bonnans C, Fukunaga K, Levy MA, Levy BD. Lipoxin A4 regulates bronchial epithelial cell responses to acid injury. Am J Pathol 2006;168:1064–1072.

Bonville CA, Rosenberg HF, Domachowske JB. Macrophage inflammatory protein-1alpha and RANTES are present in nasal secretions during ongoing upper respiratory tract infection. Pediatr Allergy Immunol 1999;10:39–44.

Bozyk PD, Moore BB. Prostaglandin E2 and the pathogenesis of pulmonary fibrosis. Am J Respir Cell Mol Biol 2011;45:445–452.

Brain JD, Molina RM, DeCamp MM, Warner AE. Pulmonary intravascular macrophages: their contribution to the mononuclear phagocyte system in 13 species. Am J Physiol 1999;276:L146–L154.

Brannon TS, North AJ, Wells LB, Shaul PW. Prostacyclin synthesis in ovine pulmonary artery is developmentally regulated by changes in cyclooxygenase-1 gene expression. J Clin Invest 1994;93: 2230–2235.

Brannon TS, MacRitchie AN, Jaramillo MA, Sherman TS, et al. Ontogeny of cyclooxygenase-1 and cyclooxygenase-2 gene expression in ovine lung Am J Physiol 1998;274:L66–L71.

Bryan DL, Hart P, Forsyth K, Gibson R. Modulation of respiratory syncytial virus–induced prostaglandin E2 production by n-3 long-chain polyunsaturated fatty acids in human respiratory epithelium. Lipids 2005;40:1007–1011.

Buckley J, Birrell MA, Maher SA, Nials AT, et al. EP4 receptor as a new target for bronchodilator therapy. Thorax 2011;66:1029–1035.

Chen J, Jiang XH, Chen H, Guo JH, et al. CFTR negatively regulates cyclooxygenase-2- PGE(2) positive feedback loop in inflammation. J Cell Physiol 2012;227:2759–2766.

Chen S, Liu X, Yue P, Schönthal AH, et al. CCAAT/enhancer binding protein homologous protein-dependent death receptor 5 induction and ubiquitin/proteasome-mediated cellular FLICE-inhibitory protein down-regulation contribute to enhancement of tumor necrosis factor-related apoptosis-inducing ligand-induced apoptosis by dimethyl-celecoxib in human non small-cell lung cancer cells. Mol Pharmacol 2007;72:1269–1279.

Chen ZT, Li SL, Cai EQ, Wu WL, et al. LPS induces pulmonary intravascular macrophages producing inflammatory mediators via activating NF-kappaB. J Cell Biochem 2003;89:1206–1214.

Cutlip RC, Lehmkuhl HD. Lesions in lambs experimentally infected with bovine respiratory syncytial virus. Am J Vet Res 1979;40:1479–1482.

Dahlem P, van Aalderen WM, de Neef M, Dijkgraaf MG, et al. Randomized controlled trial of aerosolized prostacyclin therapy in children with acute lung injury. Crit Care Med 2004;32: 1055–1060.

Davi G, Basili S, Vieri M, Cipollone F, et al. Enhanced thromboxane biosynthesis in patients with chronic obstructive pulmonary disease. The Chronic Obstructive Bronchitis and Haemostasis Study Group. Am J Respir Crit Care Med 1997;156:1794–1799.

Dehring DJ, Wismar BL. Intravascular macrophages in pulmonary capillaries of humans. Am Rev Respir Dis 1989;139:1027–1029.

Demoly P, Jaffuel D, Lequeux N, Weksler B, et al. Prostaglandin H synthase 1 and 2 immunoreactivities in the bronchial mucosa of asthmatics. Am J Respir Crit Care Med 1997;155:670–675.

Devalia JL, Davies RJ. Airway epithelial cells and mediators of inflammation. Respir Med 1993;87: 405–408.

Dicpinigaitis PV. Effect of the cyclooxygenase-2 inhibitor celecoxib on bronchial responsiveness and cough reflex sensitivity in asthmatics. Pulm Pharmacol Ther 2001;14:93–97.

El Miedany Y, Youssef S, Ahmed I, El Gaafary M. Safety of etoricoxib, a specific cyclooxygenase-2 inhibitor, in asthmatic patients with aspirin-exacerbated respiratory disease. Ann Allergy Asthma Immunol 2006;97:105–109.

Ermert L, Ermert M, Goppelt-Struebe M, Walmrath D, et al. Cyclooxygenase isoenzyme localization and mRNA expression in rat lungs. Am J Respir Cell Mol Biol 1998;18:479–488.

Ermert L, Dierkes C, Ermert M. Immunohistochemical expression of cyclooxygenase isoenzymes and downstream enzymes in human lung tumors. Clin Cancer Res 2003;9:1604–1610.

Feddersen CO, Mathias MM, McMurtry IF, Voelkel NF. Acetylcholine induces vasodilation and prostacyclin synthesis in rat lungs. Prostaglandins 1986a;31:973–987.

Feddersen CO, McMurtry IF, Henson P, Voelkel NF. Acetylcholine-induced pulmonary vasodilation in lung vascular injury. Am Rev Respir Dis 1986b;133:197–204.

Feddersen CO, Chang S, Czartalomna J, Voelkel NF. Arachidonic acid causes cyclooxygenase-dependent and -independent pulmonary vasodilation. J Appl Physiol 1990;68:1799–1808.

Fennell PB, Quante J, Wilson K, Boyle M, et al. Use of high-dose ibuprofen in a pediatric cystic fibrosis center. J Cyst Fibros 2007;6:153–158.

Ferrantino M, White RJ. Inhaled treprostinil sodium for the treatment of pulmonary arterial hypertension. Expert Opin Pharmacother 2011;12:2583–2593.

Fischer LG, Hollmann MW, Horstman DJ, Rich GF. Cyclooxygenase inhibitors attenuate bradykinin-induced vasoconstriction in septic isolated rat lungs. Anesth Analg 2000;90:625–631.

Fukunaga K, Kohli P, Bonnans C, Fredenburgh LE, et al. Cyclooxygenase 2 plays a pivotal role in the resolution of acute lung injury. J Immunol 2005;174:5033–5039.

Fulkerson WJ, MacIntyre N, Stamler J, Crapo JD. Pathogenesis and treatment of the adult respiratory distress syndrome. Arch Intern Med 1996;156:29–38.

Gadgeel SM, Wozniak A, Ruckdeschel JC, Heilbrun LK, et al. Phase II study of docetaxel and celecoxib, a cyclooxygenase-2 inhibitor, in elderly or poor performance status (PS2) patients with advanced non-small cell lung cancer. J Thorac Oncol 2008;3:1293–1300.

Gauvreau GM, Watson RM, O'Byrne PM. Protective effects of inhaled PGE2 on allergen-induced airway responses and airway inflammation. Am J Respir Crit Care Med 1999;159:31–36.

Gavett SH, Madison SL, Chulada PC, Scarborough PE, et al. Allergic lung responses are increased in prostaglandin H synthase-deficient mice. J Clin Invest 1999;104:721–732.

Gershwin LJ, Giri SN, Stewart RS, Chen J. Prostaglandin and thromboxane concentrations in plasma and lung lavage fluids during sequential infection of vaccinated and nonvaccinated calves with bovine respiratory syncytial virus. Am J Vet Res 1989;50:1254–1262.

Gessler T, Seeger W, Schmehl T. The potential for inhaled treprostinil in the treatment of pulmonary arterial hypertension. Ther Adv Respir Dis 2011;5:195–206.

Graham BS, Rutigliano JA, Johnson TR. Respiratory syncytial virus immunobiology and pathogenesis. Virology 2002;297:1–7.

Groen HJ, Sietsma H, Vincent A, Hochstenbag MM, et al. Randomized, placebo-controlled phase III study of docetaxel plus carboplatin with celecoxib and cyclooxygenase-2 expression as a biomarker for patients with advanced non-small-cell lung cancer: the NVALT-4 study. J Clin Oncol 2011;29:4320–4326.

Grubor B, Gallup JM, Meyerholz DK, Crouch E, et al. Enhanced surfactant protein and defensin mRNA levels and reduced viral replication during paramyxoviral pneumonia in neonatal lambs. Clin Diag Lab Immunol 2004;11:599–607.

Gryglewski RJ. The lung as a generator of prostacyclin. Ciba Found Symp 1980;78:147–164.

Gust R, Kozlowski JK, Stephenson AH, Schuster DP. Role of cyclooxygenase-2 in oleic acid–induced acute lung injury. Am J Respir Crit Care Med 1999;160:1165–1170.

Habler O, Kleen M, Zwissler B, Pusch R, et al. Inhalation of prostacyclin (PGI2) for 8 hours does not produce signs of acute pulmonary toxicity in healthy lambs. Intensive Care Med 1996;22:426–433.

Hall CB. Respiratory syncytial virus and parainfluenza virus. N Engl J Med 2001;344:1917–1928.

Hammad H, Kool M, Soullié T, Narumiya S, et al. Activation of the D prostanoid 1 receptor suppresses asthma by modulation of lung dendritic cell function and induction of regulatory T cells. J Exp Med 2007;204:357–367.

Harrington LS, Lucas R, McMaster SK, Moreno L, et al. COX-1, and not COX-2 activity, regulates airway function: relevance to aspirin-sensitive asthma. FASEB J 2008;22:4005–4010.

Harris RE. Cyclooxygenase-2 (COX-2) blockade in the chemoprevention of cancers of the colon, breast, prostate, and lung. Inflammopharmacology 2009;17:55–67.

Harris RE, Beebe-Donk J, Alshafie GA. Reduced risk of human lung cancer by selective cyclooxygenase 2 (COX-2) blockade: results of a case–control study. Int J Biol Sci 2007;3:328–334.

Haschek WM, Rousseaux CG, Wallig MA (editors). Chapter 6: Respiratory system. In: *Fundamentals of Toxicologic Pathology*. Elsevier Science Publishing Inc. 2009, pp. 93–133.

Hasturk S, Kemp B, Kalapurakal SK, Kurie JM, et al. Expression of cyclooxygenase-1 and cyclooxygenase-2 in bronchial epithelium and nonsmall cell lung carcinoma. Cancer 2002;94:1023–1031.

Hempel SL, Monick MM, Hunninghake GW. Lipopolysaccharide induces prostaglandin H synthase-2 protein and mRNA in human alveolar macrophages and blood monocytes. J Clin Invest 1994;93: 391–396.

Herrerias A, Torres R, Serra M, Marco A, et al. Subcutaneous prostaglandin E(2) restrains airway mast cell activity in vivo and reduces lung eosinophilia and Th(2) cytokine overproduction in house dust mite–sensitive mice. Int Arch Allergy Immunol 2009;149:323–332.

Hida T, Yatabe Y, Achiwa H, Muramatsu H, et al. Increased expression of cyclooxygenase 2 occurs frequently in human lung cancers, specifically in adenocarcinomas. Cancer Res 1998;58:3761–3764.

Hida T, Kozaki K, Muramatsu H, Masuda A, et al. Cyclooxygenase-2 inhibitor induces apoptosis and enhances cytotoxicity of various anticancer agents in non-small cell lung cancer cell lines. Clin Cancer Res 2000;6:2006–2011.

Hida T, Kozaki K, Ito H, Miyaishi O, et al. Significant growth inhibition of human lung cancer cells both in vitro and in vivo by the combined use of a selective cyclooxygenase 2 inhibitor, JTE-522, and conventional anticancer agents. Clin Cancer Res 2002;8:2443–2447.

Higashi A, Higashi N, Tsuburai T, Takeuchi Y, et al. Involvement of eicosanoids and surfactant protein D in extrinsic allergic alveolitis. Eur Respir J 2005;26:1069–1073.

Holgate ST. The sentinel role of the airway epithelium in asthma pathogenesis. Immunol Rev 2011;242:205–219.

Holgate ST, Roberts G, Arshad HS, Howarth PH, et al. The role of the airway epithelium and its interaction with environmental factors in asthma pathogenesis. Proc Am Thorac Soc 2009;6:655–659.

Holtzman MJ. Arachidonic acid metabolism: implications of biological chemistry for lung function and disease. Am Rev Respir Dis 1991;143:188–203.

Hornsleth A, Loland L, Larsen LB. Cytokines and chemokines in respiratory secretion and severity of disease in infants with respiratory syncytial virus (RSV) infection: cytokines and chemokines in respiratory secretion and severity of disease in infants with respiratory syncytial virus (RSV) infection. J Clin Virol 2001;21:163–170.

Huang M, Stolina M, Sharma S, Mao JT, et al. Non-small cell lung cancer cyclooxygenase-2-dependent regulation of cytokine balance in lymphocytes and macrophages: up-regulation of interleukin 10 and down-regulation of interleukin 12 production. Cancer Res 1998;58:1208–1226.

Huang SK, Fisher AS, Scruggs AM, White ES, et al. Hypermethylation of PTGER2 confers prostaglandin E2 resistance in fibrotic fibroblasts from humans and mice. Am J Pathol 2010;177: 2245–2255.

Izuhara K, Ohta S, Shiraishi H, Suzuki S, et al. The mechanism of mucus production in bronchial asthma. Curr Med Chem 2009;16:2867–2875.

Jackson BA, Goldstein RH, Roy R, Cozzani M, et al. Effects of transforming growth factor beta and interleukin-1 beta on expression of cyclooxygenase 1 and 2 and phospholipase A2 mRNA in lung fibroblasts and endothelial cells in culture. Biochem Biophys Res Commun 1993;197:1465–1474.

Johnson JE, Gonzales RA, Olson SJ, Wright PF, et al. The histopathology of fatal untreated human respiratory syncytial virus infection. Mod Pathol 2007;20:108–119.

Johnson MD, Bao HF, Helms MN, Chen XJ, et al. Functional ion channels in pulmonary alveolar type I cells support a role for type I cells in lung ion transport. Proc Natl Acad Sci USA 2006;103:4964–4969.

Keerthisingam CB, Jenkins RG, Harrison NK, Hernandez-Rodriguez NA, et al. Cyclooxygenase-2 deficiency results in a loss of the anti-proliferative response to transforming growth factor-beta in human fibrotic lung fibroblasts and promotes bleomycin-induced pulmonary fibrosis in mice. Am J Pathol 2001;158:1411–1422.

Khan KN, Stanfield K, Trajkovic D, Harris RK. Cyclooxygenase-2 expression in inflammatory lung lesions of nonhuman primates. Vet Pathol 2000;37:512–516.

Kim ES, Hong WK, Lee JJ, Mao L, et al. Biological activity of celecoxib in the bronchial epithelium of current and former smokers. Cancer Prev Res (Phila) 2010;3:148–159.

Kim GY, Lim SJ, Kim YW. Expression of HuR, COX-2, and survivin in lung cancers; cytoplasmic HuR stabilizes cyclooxygenase-2 in squamous cell carcinomas. Mod Pathol 2011;24:1336–1347.

Koki AT, Khan NK, Woerner BM, Seibert K, et al. Characterization of cyclooxygenase-2 (COX-2) during tumorigenesis in human epithelial cancers: evidence for potential clinical utility of COX-2 inhibitors in epithelial cancers. Prostaglandins Leukot Essent Fatty Acids 2002;66:13–18.

Konstan MW, Vargo KM, Davis PB. Ibuprofen attenuates the inflammatory response to *Pseudomonas aeruginosa* in a rat model of chronic pulmonary infection: implications for antiinflammatory therapy in cystic fibrosis. Am Rev Respir Dis 1990;141:186–192.

Konstan MW, Schluchter MD, Xue W, Davis PB. Clinical use of ibuprofen is associated with slower FEV_1 decline in children with cystic fibrosis. Am J Respir Crit Care Med 2007;176:1084–1089.

Lands LC, Milner R, Cantin AM, Manson D, et al. High-dose ibuprofen in cystic fibrosis: Canadian safety and effectiveness trial. J Pediatr 2007;151:249–254.

Lassus P, Wolff H, Andersson S. Cyclooxygenase-2 in human perinatal lung. Pediatr Res 2000;47: 602–605.

Lee RU, Stevenson DD. Aspirin-exacerbated respiratory disease: evaluation and management. Allergy Asthma Immunol Res 2011;3:3–10.

Lehmkuhl HD, Cutlip RC. Experimental parainfluenza type 3 infection in young lambs: clinical, microbiological, and serological response. Vet Microbiol 1983;8:437–442.

Liu T, Zaman W, Kaphalia BS, Ansari GA, et al. RSV-induced prostaglandin E2 production occurs via cPLA2 activation: role in viral replication. Virology 2005;343:12–24.

Liu X, Yue P, Zhou Z, Khuri FR, et al. Death receptor regulation and celecoxib-induced apoptosis in human lung cancer cells. J Natl Cancer Inst 2004;96:1769–1780.

Longworth KE. The comparative biology of pulmonary intravascular macrophages. Front Biosci 1997;2: d232–d241.

Maher SA, Birrell MA, Belvisi MG. Prostaglandin E2 mediates cough via the EP3 receptor: implications for future disease therapy. Am J Respir Crit Care Med 2009;180:923–928.

Maher TM, Evans IC, Bottoms SE, Mercer PF, et al. Diminished prostaglandin E2 contributes to the apoptosis paradox in idiopathic pulmonary fibrosis. Am J Respir Crit Care Med 2010;182:73–82.

Mao JT, Roth MD, Fishbein MC, Aberle DR, et al. Lung cancer chemoprevention with celecoxib in former smokers. Cancer Prev Res (Phila) 2011;4:984–993.

Masferrer JL, Koki A, Seibert K. COX-2 inhibitors: a new class of antiangiogenic agents. Ann NY Acad Sci 1999;889:84–86.

Massaro GD, Massaro D. Formation of pulmonary alveoli and gas-exchange surface area: quantitation and regulation. Annu Rev Physiol 1996;58:73–92.

Mathé AA, Hedqvist P. Effect of prostaglandins F2 alpha and E2 on airway conductance in healthy subjects and asthmatic patients. Am Rev Respir Dis 1975;111:313–320.

Matsuoka T, Hirata H, Tanaka Y, Takahashi Y, et al. Prostaglandin D2 as a mediator of allergic asthma. Science 2000;287:2013–2017.

Mead J. Mechanical properties of lungs. Physiol Rev 1961;41:281–330.

Mendelson CR, Acarregui MJ, Odom MJ, Boggaram V. Developmental and hormonal regulation of surfactant protein A (SP-A) gene expression in fetal lung. J Dev Physiol 1991;15:61–69.

Meyerholz DK, Grubor B, Fach SJ, Sacco RE, et al. Reduced clearance of respiratory syncytial virus in a preterm lamb model. Microbes Infect 2004;6:1312–1319.

Micheletto C, Tognella S, Guerriero M, Dal Negro R. Nasal and bronchial tolerability of Rofecoxib in patients with aspirin induced asthma. Eur Ann Allergy Clin Immunol 2006;38:10–14.

Mitchell JA, Belvisi MG, Akarasereenont P, Robbins RA, et al. Induction of cyclo-oxygenase-2 by cytokines in human pulmonary epithelial cells: regulation by dexamethasone. Br J Pharmacol 1994;113:1008–1014.

Montuschi P, Barnes PJ. Exhaled leukotrienes and prostaglandins in asthma. J Allergy Clin Immunol 2002;109:615–620.

Montuschi P, Ciabattoni G, Kharitonov SA, et al. Non selective cyclo-oxygenase inhibition decreases exhaled prostaglandin E_2 in patients with chronic obstructive pulmonary disease. Am J Respir Crit Care Med 2001;163:A908.

Montuschi P, Kharitonov SA, Ciabattoni G, Barnes PJ. Exhaled leukotrienes and prostaglandins in COPD. Thorax 2003;58:585–588.

Montuschi P, Macagno F, Parente P, Valente S, et al. Effects of cyclo-oxygenase inhibition on exhaled eicosanoids in patients with COPD. Thorax 2005;60:827–833.

Mulugeta S, Beers MF. Surfactant protein C: its unique properties and emerging immunomodulatory role in the lung. Microbes Infect 2006;8:2317–2323.

Murugan V, Peck MJ. Signal transduction pathways linking the activation of alveolar macrophages with the recruitment of neutrophils to lungs in chronic obstructive pulmonary disease. Exp Lung Res 2009;35:439–485.

Oguma T, Asano K, Shiomi T, Fukunaga K, et al. Cyclooxygenase-2 expression during allergic inflammation in guinea-pig lungs. Am J Respir Crit Care Med 2002;165:382–386.

Oyarzun MJ, Clements JA. Control of lung surfactant by ventilation, adrenergic mediators, and prostaglandins in the rabbit. Am Rev Respir Dis 1978;117:879–891.

Pace-Asciak CR. Prostaglandin biosynthesis and catabolism in the developing fetal sheep lung. Prostaglandins 1977;13:649–660.

Pavord ID, Tattersfield AE. Bronchoprotective role for endogenous prostaglandin E2. Lancet 1995; 345:436–438.

Pavord ID, Wong CS, Williams J, Tattersfield AE. Effect of inhaled prostaglandin E2 on allergen-induced asthma. Am Rev Respir Dis 1993;148:87–90.

Peebles RS Jr, Hashimoto K, Morrow JD, Dworski R, et al. Selective cyclooxygenase-1 and -2 inhibitors each increase allergic inflammation and airway hyperresponsiveness in mice. Am J Respir Crit Care Med 2002;165:1154–1160.

Petkova DK, Clelland CA, Ronan JE, Lewis S, Knox AJ. Reduced expression of cyclooxygenase (COX) in idiopathic pulmonary fibrosis and sarcoidosis. Histopathology 2003;43:381–386.

Plopper CG, Mariassy AT, Hill LH. Ultrastructure of the nonciliated bronchiolar epithelial (Clara) cell of mammalian lung: I. A comparison of rabbit, guinea pig, rat, hamster, and mouse. Exp Lung Res 1980;1:139–154.

Powell WS, Solomon S. Biosynthesis of prostaglandins and thromboxane B2 by fetal lung homogenates, Prostaglandins 1978;15:351–364.

Profita M, Sala A, Bonanno A, Riccobono I, et al. Increased prostaglandin E2 concentrations and cyclooxygenase-2 expression in asthmatic subjects with sputum eosinophilia. J Allergy Clin Immunol 2003;112:709–716.

Qadri SS, Wang JH, Redmond KC, O'Donnell AF, et al. The role of COX-2 inhibitors in lung cancer. Ann Thorac Surg 2002;74:1648–1652.

Radi ZA. Pathophysiology of cyclooxygenease inhibition in animal models. Toxicol Pathol 2009;37:34–46.

Radi ZA, Ackermann MR. Ontogeny of pulmonary cyclooxygenase-1 (COX-1) and -2 (COX-2). Pathophysiology 2011;18:215–219.

Radi ZA, Ostroski O, Pulmonary and cardiorenal cyclooxygenase-1 (COX- 1), -2 (COX-2), and microsomal prostaglandin E synthase-1 (mPGES-1) and -2 (mPGES-2) expression in a hypertension model. Mediators Inflamm 2007;2007:85091.

Radi ZA, Meyerholz DK, Ackermann MR. Pulmonary cyclooxygenase-1 (COX-1) and -2 (COX-2) cellular expression and distribution following respiratory syncytial virus and parainfluenza virus infection. Viral Immunol 2010;23:43–48.

Ragbetli C, Ilhan F, Aydinlioğlu A, Kara M, et al. A histological investigation concerning the effects of diclofenac sodium to the lung in 4- and 20-week-old rats treated prenatally. J Matern Fetal Neonatal Med 2011;24:208–212.

Rangel Moreno J, Estrada García I, De La Luz García Hernández M, et al. The role of prostaglandin E2 in the immunopathogenesis of experimental pulmonary tuberculosis. Immunology 2002;106:257–266.

Richardson JY, Ottolini MG, Pletneva L, et al. Respiratory syncytial virus (RSV) infection induces cyclooxygenase 2: a potential target for RSV therapy. J Immunol 2005;174:4356–4364.

Rioux N, Castonguay A. Prevention of NNK-induced lung tumorigenesis in A/J mice by acetylsalicylic acid and NS-398. Cancer Res 1998;58:5354–5360.

Roca-Ferrer J, Pujols L, Agusti C, Xaubet A, et al. Cyclooxigenase-2 levels are increased in the lung tissue and bronchial tumors of patients with chronic obstructive pulmonary disease. Arch Bronconeumol 2011;47:584–589.

Rolin S, Masereel B, Dogne JM. Prostanoids as pharmacological targets in COPD and asthma. Eur J Pharmacol 2006;533:89–100.

Russell JA, Ronco JJ, Dodek PM. Physiologic effects and side effects of prostaglandin E1 in the adult respiratory distress syndrome. Chest 1990;97:684–692.

Sakata D, Yao C, Narumiya S. Emerging roles of prostanoids in T cell–mediated immunity. IUBMB Life 2010;62:591–596.

Schmidt LM, Belvisi MG, Bode KA, Bauer J, et al. Bronchial epithelial cell–derived prostaglandin E2 dampens the reactivity of dendritic cells. J Immunol 2011;186:2095–2105.

Schmitz T, Cox LA, Li C, Levine BA, et al. Prostaglandin E2 receptor expression in fetal baboon lung at 0.7 gestation after betamethasone exposure. Pediatr Res 2007;61:421–426.

Seibert K, Zhang Y, Leahy K, Hauser S, et al. Pharmacological and biochemical demonstration of the role of cyclooxygenase 2 in inflammation and pain. Proc Natl Acad Sci USA 1994;91:12013–12017.

Seymour ML, Gilby N, Bardin PG, et al. Rhinovirus infection increases 5-lipoxygenase and cyclooxygenase-2 in bronchial biopsy specimens from nonatopic subjects. J Infect Dis 2002;185:540–544.

Shi J, Misso NL, Duffy DL, Bradley B, et al. Cyclooxygenase-1 gene polymorphisms in patients with different asthma phenotypes and atopy. Eur Respir J 2005;26:249–256.

Shiraishi Y, Shiraishi Y, Asano K, Nakajima T, et al. Prostaglandin D2–induced eosinophilic airway inflammation is mediated by CRTH2 receptor. J Pharmacol Exp Ther 2005;312:954–960.

Shoemaker WC, Appel PL. Effects of prostaglandin E1 in adult respiratory distress syndrome. Surgery 1986;99:275–283.

Soslow RA, Dannenberg AJ, Rush D, Woerner BM, et al. COX-2 is expressed in human pulmonary, colonic, and mammary tumors. Cancer 2000;89:2637–2645.

Sousa A, Pfister R, Christie PE, Lane SJ, et al. Enhanced expression of cyclooxygenase isoenzyme 2 (COX-2) in asthmatic airways and its cellular distribution in aspirin-sensitive asthma. Thorax 1997;52:940–945.

Stapleton PP, Strong VE, Freeman TA, Winter J, et al. Gender affects macrophage cytokine and prostaglandin E2 production and PGE2 receptor expression after trauma. J Surg Res 2004;122: 1–127.

Steer SA, Corbett JA. The role and regulation of COX-2 during viral infection. Viral Immunol 2003;16:447–460.

Stevenson DD, Szczeklik A. Clinical and pathologic perspectives on aspirin sensitivity and asthma. J Allergy Clin Immunol 2006;118:773–786.

Stewart RJ, Yamaguchi KT, Knost PM, Mason SW, et al. Effects of ibuprofen on pulmonary oedema in an animal smoke inhalation model. Burns 1990;16:409–413.

Stumm CL, Wettlaufer SH, Jancar S, Peters-Golden M. Airway remodeling in murine asthma correlates with a defect in PGE2 synthesis by lung fibroblasts. Am J Physiol Lung Cell Mol Physiol 2011;301:L636–L644.

Sun W, Pei L. Flurbiprofen axetil, nasal polyps, and status asthmaticus: an unusual case report. J Asthma 2011;48:647–649.

Sznajer Y, Westcott JY, Wenzel SE, Mazer B, et al. Airway eicosanoids in acute severe respiratory syncytial virus bronchiolitis. J Pediatr 2004;145:115–118.

Taha R, Olivenstein R, Utsumi T, Ernst P, et al. Prostaglandin H synthase 2 expression in airway cells from patients with asthma and chronic obstructive pulmonary disease. Am J Respir Crit Care Med 2000;161:636–640.

Tai HH, Yuan B, Wu AT. Transformation of arachidonate into 6-oxoprostaglandin F1 alpha, thromboxane B2 and prostaglandin E2 by sheep lung microsomal fraction. Biochem J 1978;170:441–444.

Takizawa H. Airway epithelial cells as regulators of airway inflammation. Int J Mol Med 1998;1:367–378.

Takizawa H. Bronchial epithelial cells in allergic reactions. Curr Drug Targets Inflamm Allergy 2005;4:305–311.

Tamaoki J, Chiyotani A, Kobayashi K, Sakai N, et al. Effect of indomethacin on bronchorrhea in patients with chronic bronchitis, diffuse panbronchiolitis, or bronchiectasis. Am Rev Respir Dis 1992;145:548–552.

Thompson WW, Shay DK, Weintraub E, Brammer L, et al. Mortality associated with influenza and respiratory syncytial virus in the United States. J Am Med Assoc 2003;289:179–186.

Tran HS, Quinn JV, Puc MM, Woolley DS, et al. Prostacyclin is neither sufficient alone nor necessary to cause pulmonary dysfunction: results from infusions of prostacyclin and antiprostacyclin antibody in porcine septic shock. Crit Care Med 2001;29:1445–1451.

Uller L, Mathiesen JM, Alenmyr L, Korsgren M, et al. Antagonism of the prostaglandin D2 receptor CRTH2 attenuates asthma pathology in mouse eosinophilic airway inflammation. Respir Res 2007;8:16.

Vigano T, Toia A, Crivellari MT, Galli G, et al. Prostaglandin synthetase inhibition and formation of lipoxygenase products in immunologically challenged normal human lung parenchyma. Eicosanoids 1998;1:73–77.

Voelkel NF, Gerber JG, McMurtry IF, Nies AS, et al. Release of vasodilator prostaglandin, PGI2, from isolated rat lung during vasoconstriction. Circ Res 1981;48:207–213.

Wallace JL. Commonality of defensive roles of COX-2 in the lung and gut. Am J Pathol 2006;168:1060–1063.

Watkins DN, Garlepp MJ, Thompson PJ. Regulation of the inducible cyclooxygenase pathway in human cultured airway epithelial (A549) cells by nitric oxide. Br J Pharmacol 1997;121:1482–1488.

Watkins DN, Lenzo JC, Segal A, Garlepp MJ, et al. Expression and localization of cyclo-oxygenase isoforms in non-small cell lung cancer. Eur Respir J 1999;14:412–418.

Westergren VS, Wilson SJ, Penrose JF, Howarth PH, et al. Nasal mucosal expression of the leukotriene and prostanoid pathways in seasonal and perennial allergic rhinitis. Clin Exp Allergy 2009;39:820–828.

Wheeler AP, Hardie WD, Bernard GR. The role of cyclooxygenase products in lung injury induced by tumor necrosis factor in sheep. Am Rev Respir Dis 1992;145:632–639.

White A, Bigby T, Stevenson D. Intranasal ketorolac challenge for the diagnosis of aspirin-exacerbated respiratory disease. Ann Allergy Asthma Immunol 2006;97:190–195.

Wilborn J, Crofford LJ, Burdick MD, Kunkel SL, et al. Cultured lung fibroblasts isolated from patients with idiopathic pulmonary fibrosis have a diminished capacity to synthesize prostaglandin E2 and to express cyclooxygenase-2. J Clin Invest 1995a;95:1861–1868.

Wilborn J, DeWitt DL, Peters-Golden M. Expression and role of cyclooxygenase isoforms in alveolar and peritoneal macrophages. Am J Physiol 1995b;268:L294–L301.

Wilkinson-Berka JL. Vasoactive factors and diabetic retinopathy: vascular endothelial growth factor, cycoloxygenase-2 and nitric oxide. Curr Pharm Des 2004;10:3331–3348.

Williams CS, Tsujii M, Reese J, Dey SK, et al. Host cyclooxygenase-2 modulates carcinoma growth. J Clin Invest 2000;105:1589–1594.

Wolff H, Saukkonen K, Anttila S, Karjalainen A, et al. Expression of cyclooxygenase-2 in human lung carcinoma. Cancer Res 1998;58:4997–5001.

Yamada M, Kawai M, Kawai Y, Mashima Y. The effect of selective cyclooxygenase-2 inhibitor on corneal angiogenesis in the rat. Curr Eye Res 1999;19:300–304.

Yao R, Rioux N, Castonguay A, You M. Inhibition of COX-2 and induction of apoptosis: two determinants of nonsteroidal anti-inflammatory drugs' chemopreventive efficacies in mouse lung tumorigenesis. Exp Lung Res 2000;26:731–742.

Yoshida S, Amano H, Hayashi I, Kitasato H, et al. COX-2/VEGF-dependent facilitation of tumor-associated angiogenesis and tumor growth in vivo. Lab Invest 2003;83:1385–1394.

Yoshimoto A, Kasahara K, Kawashima A, Fujimura M, et al. Characterization of the prostaglandin biosynthetic pathway in non-small cell lung cancer: a comparison with small cell lung cancer and correlation with angiogenesis, angiogenic factors and metastases. Oncol Rep 2005;13:1049–1057.

Yuan A, Yu CJ, Shun CT, Luh KT, et al. Total cyclooxygenase-2 mRNA levels correlate with vascular endothelial growth factor mRNA levels, tumor angiogenesis and prognosis in non-small cell lung cancer patients. Int J Cancer 2005;115:545–555.

Zeitz HJ. Bronchial asthma, nasal polyps, and aspirin sensitivity: Samter's syndrome. Clin Chest Med 1988;9:567–576.

Zeltner TB, Caduff JH, Gehr P, Pfenninger J, et al. The postnatal development and growth of the human lung: I. Morphometry. Respir Physiol 1987;67:247–267.

CHAPTER 7

CURRENT RESEARCH STRATEGIES FOR DESIGNING SAFER NSAIDs

NITRIC OXIDE–RELEASING NSAIDs

The nitric oxide (NO) synthase system has roles in gastrointestinal (GI) inflammation and mucosal defense (Nishio et al., 2006). NO released from vascular epithelium, epithelial cells of GI tract, and sensory nerves can influence many of the same components of mucosal defense as do prostaglandins (PGs) (Pajdo et al., 2011). A novel class of nitric oxide–releasing NSAID (NO-NSAID) derivatives has been described which exert similar anti-inflammatory activities in experimental animals but produce significantly less GI injury than the parent NSAID from which they are derived (Wallace et al., 1994; Elliott et al., 1995; Davies et al., 1997; Coruzzi et al., 2002; Stefano and Distrutti, 2007). A Wistar rat model of GI injury used to determine if a nitroxybutyl ester derivative of naproxen was less ulcerogenic to the GI tract than was its parent ns-NSAID, naproxen (Davies et al., 1997). In the acute gastric damage model, rats were given oral doses of naproxen at 40 or 80 mg/kg or equimolar doses of NO-naproxen at 58 or 116 mg/kg and were euthanized 3 h after dosing (Davies et al., 1997). Extensive gastric hemorrhage and ulceration were noted in the naproxen-dosed rats, whereas the stomachs from the NO-naproxen-dosed rats were similar to those of controls. In the small intestinal damage model, rats were given oral twice-daily doses of naproxen at 20 mg/kg or equimolar doses of NO-naproxen at 29 mg/kg for 1 week. At the end of this dosing period, the doses of naproxen and NO-naproxen were increased to 30 and 44 mg/kg and dosing continued for 7 days. At the end of the second week, the doses of naproxen and NO-naproxen were increased to 45 and 66 mg/kg; however, the rats dosed with the 45 mg/kg naproxen became moribund by day 3 of further dosing, and animals were euthanized on day 18 (Davies et al., 1997). Rats dosed with naproxen had penetrating intestinal ulcers, whereas rats dosed with NO-naproxen did not have intestinal ulcers (Davies et al., 1997). AZD3582 [4-(nitro-oxy)butyl-(2*S*)-2-(6-methoxy-2-naphthyl)propanoate] is the first cyclooxygenase (COX) inhibiting nitric oxide donator (CINOD) to be studied in large clinical trials. AZD3582 donates

Comparative Pathophysiology and Toxicology of Cyclooxygenases, First Edition. Zaher A. Radi.

NO and inhibits both COX-1 and COX-2 (Lohmander et al., 2005). The GI safety and efficacy of AZD3582 were studied in patients with hip or knee osteoarthritis (OA). AZD3582 was given at 750 mg twice daily and naproxen at 500 mg twice daily in a double-blind study. The primary endpoint was the 6-week incidence of endoscopic gastroduodenal ulcers (diameter $\geq$3 mm). Compared with baseline, significantly fewer ulcers and erosions developed in stomach and stomach–duodenum combined, and fewer erosions developed in stomach, duodenum, and both combined on AZD3582 than on naproxen. The GI symptom rating scale (GSRS) reflux and abdominal pain subscale scores were lower for AZD3582 than for naproxen and AZD3582 was as effective as naproxen (Lohmander et al., 2005). In another clinical study, the efficacy, safety, and tolerability of AZD3582 were compared with that of rofecoxib, naproxen, and placebo in patients with knee OA. Patients who experienced increased pain on withdrawal of analgesia were randomized to receive AZD3582 125, 375, or 750 mg twice a day; rofecoxib 25 mg once a day; naproxen 500 mg twice a day; or placebo for 6 weeks. AZD3582 375 and 750 mg twice a day were superior to placebo and as effective as rofecoxib 25 mg/day in treating the signs and symptoms of knee OA (Schnitzer et al., 2005). AZD3582 has been the first, and so far the only, CINOD investigated extensively in clinical trials. Despite its promising profile, approval of this drug was recently rejected by the U.S. Food and Drug Administration because of the lack of long-term controlled studies (Fiorucci and Santucci, 2011).

The pathogenesis of nitric oxide GI tract protective effects is related to (1) decreased leukocyte activation and infiltration, (2) decreased reactive oxygen species formation, (3) maintenance of GI blood flow, (4) increased GI mucus gel thickness, and (5) modulation of intestinal permeability (Brown et al., 1992; Davies et al., 1997; Fiorucci and Antonelli, 2006). In experimental models, NO-releasing aspirin derivative effects included inhibition of leukocyte adherence, preservation of blood flow during ischemic events, more potent inhibition of platelet aggregation, reduction of blood pressure in hypertension, and inhibition of vascular smooth muscle proliferation (Wallace et al., 2002). In general, NO-releasing aspirin derivatives are more potent anti-inflammatory and antithrombotic agents than is the parent drug (Wallace et al., 2002). NO-aspirin can be metabolized by esterases (primarily in the liver and in plasma) to yield the parent drug plus the spacer attached to the NO-releasing moiety. Subsequent metabolism of the latter leads to the liberation of NO, which can occur many hours after oral administration of the drug (Wallace et al., 2002). Lipoxins were recently considered as another group of lipid mediators that can protect the stomach similarly to NO donors (Pajdo et al., 2011). Aspirin-triggered lipoxin synthesis via COX-2 acts to reduce the severity of GI damage induced by NSAIDs. Lipoxin analogs may prove to be useful for preventing mucosal injury and for modulating mucosal inflammation (Pajdo et al., 2011). NO-releasing derivatives of NSAIDs offer great potential as GI-sparing and antithrombotic drugs with reduced toxicities (Wallace et al., 1999). NCX-4016 is a NO-releasing derivative of aspirin with antiplatelet activity. The effect of NCX-4016 on GI mucosa and platelet functions was evaluated in healthy human volunteers (Fiorucci et al., 2003). Forty healthy subjects were randomly allocated to receive 7 days of treatment with NCX-4016 (400 and 800 mg twice daily) or equimolar doses of aspirin (200 and 420 mg twice daily). Upper endoscopies were performed before and at the end of the treatment period, and GI lesions were graded. Basal and posttreatment platelet aggregation in response to arachidonic acid (AA) and serum thromboxane (TX) B_2 (TXB) and AA-stimulated platelet TXB_2 production were investigated. NCX-4016 was virtually devoid of gastric and duodenal toxicity and inhibited AA-induced platelet aggregation as well as serum TXB_2 and platelet TXB_2 generation induced by AA to the same extent as aspirin (Fiorucci et al., 2003). Therefore, several NSAIDs, such as aspirin, naproxan, diclofenac, ketoprofen, and ibuprofen, have been coupled to a nitroxybutyl or a nitrosothiol NO moiety to generate new chemical entities (Fiorucci and Antonelli, 2006; Stefano and Distrutti, 2007; Fiorucci and Santucci, 2011; Pajdo et al., 2011). The design and synthesis of hybrid aspirin, ibuprofen, and indomethacin derivatives coupled with novel NO-donating prodrugs have been described (Abdellatif et al., 2009).

HYDROGEN SULFIDE–RELEASING NSAIDs

H_2S plays a significant role in regulating gastric mucosal blood flow and is involved in the maintenance of gastric mucosal integrity in experimental rodent models exposed to NSAIDs (Fiorucci et al., 2006). Aspirin given at 30 mg/kg orally induced extensive gastric damage in the rat stomach, and this was accompanied by infiltration of granulocytes into gastric tissue and increased expression of TNFα mRNA (Fiorucci et al., 2006). Pretreatment with an H_2S donor (NaHS; 100 μmol/kg) reduced gastric damage and inflammation (Fiorucci et al., 2006). Deficiencies of H_2S synthesis may contribute to the pathogenesis of several gastrointestinal disorders (Fiorucci et al., 2006). In addition, H_2S have an overlapping actions with NO and PGs in terms of modulating mucosal defense and resolution of inflammation (Wallace, 2010). The GI safety and anti-inflammatory effects of a novel "HS-NSAID" (ATB-337) that consists of diclofenac linked to a H_2S-releasing moiety was evaluated in Wistar rats (Wallace et al., 2007). Rats were dosed orally with diclofenac (10, 27, and 50 μmol/kg) or ATB-337 (equimolar doses), and 3 h later the rats were killed for blind assessment of gastric damage (Wallace et al., 2007). Oral administration of diclofenac resulted in the development of dose-related hemorrhagic erosions in the rat stomach. In contrast, oral administration of ATB-337 at doses equimolar to those of diclofenac did not produce any hemorrhagic erosions (Wallace et al., 2007). Thus, H_2S-NSAIDs are a potential new class of anti-inflammatory and analgesic agents with the potential to spare the GI and cardiovascular system (Fiorucci and Santucci, 2011).

NSAIDs ASSOCIATED WITH ZWITTERIONIC PHOSPHOLIPIDS

Zwitterionic are neutral compounds with a positive and a negative electrical charge at different locations. The lipid bilayer bending modulus, characterized by thermal undulations, is often affected by the presence of membrane active molecules (Boggara et al., 2010). The intercalation of phospholipid bilayers with small molecules is expected to lead to significant changes in the mechanical properties of the lipid bilayer and is typically quantified by the bilayer bending modulus (Boggara et al., 2010). In addition, properties such as headgroup charges, hydration, and bilayer thickness play a role in bilayer mechanical properties (Boggara et al., 2010). NSAIDs are known to have strong interactions with lipid membranes (Boggara and Krishnamoorti, 2010). The effect of ibuprofen, an ns-NSAID, on the radius of small unilamellar vesicles of 1,2-dimyristoyl-*sn*-glycero-3-phosphatidylcholine (DMPC) and their bilayer structure was studied as a function of pH (ranging from 2 to 8) and drug-to-lipid mole ratio (Boggara and Krishnamoorti, 2010). Ibuprofen was found to affect the bilayer structure significantly at all pH values, whatever the charge state of the drug. At low pH values, the ibuprofen reduced the bilayer thickness, induced fluidlike behavior, and changed headgroup hydration (Boggara and Krishnamoorti, 2010). In another study, the effect of ibuprofen on the bending modulus of phospholipid membranes was studied as a function of pH and temperature. Ibuprofen was found to lower the bending modulus at all pH values (Boggara et al., 2010). Salicylic acid and related NSAIDs may alter the stability of membranes, inducing the formation of unstable pores that may lead to back-diffusion of luminal acid and membrane rupture, which explains the associated GI mucosal damage (Lichtenberger et al., 2006). Thus, interactions between NSAIDs and the lipid membrane and the protonation and deprotonation of NSAIDs with pH can affect the incorporation of NSAIDs into the lipid membrane and alter the mechanical properties of the membrane (Boggara et al., 2010). It has been proposed that direct interactions of ns-NSAIDs with zwitterionic phospholipids (i.e., GI tract lining) are responsible for GI toxicity (Lichtenberger et al., 2001). Therefore, combining NSAIDs with zwitterionic phospholipids could be a strategy to prevent the interaction of hydrophobic portion of cells with NSAID drugs, which may help in reducing toxicity. For example, the potential of omeprazole, a proton pump inhibitor, to interfere with the bioavailability of aspirin administered to rats either alone or complexed with the zwitterionic phospholipid dipalmitoylphosphatidylcholine (DPPC) was investigated (Giruad et al., 1997). DPPC increased the lipid solubility and gastric permeability of NSAIDs. Combination of acetylsalicylic acid and DPPC retained analgesic and anti-inflammatory effects while causing significantly less GI damage (Giruad et al., 1997). Preassociating a number of NSAIDs with exogenous zwitterionic phospholipids prevented an increase in the surface wettability of the mucus gel layer and protected rats against the GI side effects of NSAIDs while enhancing their lipid permeability and antipyretic and anti-inflammatory activity (Lichtenberger et al., 1995). Thus, the GI toxicity of NSAIDs can be reduced markedly when the drug is chemically preassociated with a zwitterionic phospholipid such as DPPC before intragastric administration (Lichtenberger et al., 1996). Aspirin's gastric toxicity is dependent, in part, on its ability to attenuate the stomach's surface hydrophobic barrier. This adverse drug effect can be circumvented by the administration of phosphatidylcholine (PC)-associated aspirin to maintain the stomach's hydrophobic properties (Lichtenberger et al.,

2007). Lipid-based PC-NSAID formulations that are showing encouraging preclinical and clinical observations are being validated for their GI safety and therapeutic efficacy and constitute a novel strategy to reduce GI toxicity (Lichtenberger et al., 2009).

CHIRAL NSAIDs

Left and right hands are similar but are mirror images of each other. This phenomenon of chemical "handedness" is called chirality, from *cheiro*, Greek for "hand" (Sweetman, 2003). A chiral molecule is one that lacks an internal plane of symmetry and thus has a nonsuperimposable mirror image. The opposing pairs of isomers are referred to collectively as enantiomers from the Greek *enantios*, meaning "opposite" (Sweetman, 2003). The conversion of the *R* to the *S* enantiomer is extensive in some rodents, such as rats and mice, and in monkeys, but is low in guinea pigs, rabbits, gerbils, and humans. In mice, the *R* enantiomer of ketoprofen undergoes about 60% partial unidirectional metabolic inversion to the corresponding *S* enantiomer. This is carried out by a stereoselective microsomal or mitochondrial acyl-CoA synthetase (Sweetman, 2003). The *R*-to-*S* conversion also occurs with flurbiprofen, although only to the extent of about 25%. The *R* isomer of ketoprofen has no NSAID activity, highlighting the importance of the three-dimensional structure in cyclooxygenase inhibition (Sweetman, 2003). It is thought that the inhibitory effect of chiral NSAIDs on the synthesis of PGs and hence their efficacy and toxicity are due primarily to the *S* enantiomer, although the other isomer is safer (Davies et al., 1996). The effects of three NSAIDs (flurbiprofen, ibuprofen, and ketoprofen) on small intestinal permeability were investigated in rats. Single doses of each NSAID were administered orally as either the racemate or the *R* or *S* enantiomer, the enantiomer dose being half that of the racemate. Each treatment caused a significant increase in intestinal permeability above that seen in untreated rats. The *R* enantiomers of all three NSAIDs increased small intestinal permeability significantly above baseline. *S*-Flurbiprofen, used at one-half the dose of the racemate, increased permeability to a magnitude similar to that of the racemate (Davies et al., 1996). Some experiments in animals show that the *S* isoform leads to GI mucosal damage, whereas *R* has substantially less propensity to do so. The intestinal ulcerogenic effects of single oral doses of *S*-ketoprofen were compared with racemic ketoprofen in the small intestine and cecum of rats (Cabré et al., 1998). *S*-Ketoprofen had no significant ulcerogenic effect at 10 or 20 mg/kg. However, racemic ketoprofen was clearly ulcerogenic in the small intestine and cecum at 40 mg. *R*-Ketoprofen at 20 mg/kg does not show any effect in the cecum and only limited ulcerogenesis in the small intestine: The ulcerogenic action of racemic ketoprofen can be interpreted as a synergism between *S*- and *R*-ketoprofen (Cabré et al., 1998). The stereoselective inhibition of COX-2 by chiral NSAIDs, ketoprofen, flurbiprofen, and ketorolac has been investigated (Carbaza et al., 1996). It was found that chiral NSAIDs inhibited, with comparable stereoselectivity, both COX-2 and COX-1 isoenzymes, and that the inhibition of COX-2 observed previously for racemic NSAIDs should be attributed almost exclusively to their *S*-enantiomers (Carbaza et al., 1996). The comparative pathogenesis of intestinal injury induced by racemic ketoprofen and its enantiomers was evaluated in re-fed rats. Racemic ketoprofen and *R*-ketoprofen dose-dependently caused similar and multiple lesions in the mid-jejunum significantly higher than those observed with *S*-ketoprofen (Nieto et al., 2002).

TREFOIL PEPTIDES

Trefoil peptides are a family of cysteine-containing protective peptides normally secreted in the GI tract. Trefoil peptides are expressed aberrantly by a wide range of human GI inflammatory conditions. They impart protection from injury to the GI mucosa by possible interaction with mucin glycoproteins. Trefoil peptides influence epithelial cell migration and mucosal restitution following injury (Longman et al., 1999). Trefoil peptides have been implicated in the formation of new blood vessels during normal and pathophysiological processes linked to wound healing, inflammation, and cancer progression in the digestive mucosa and other human solid tumors associated with aberrant expression of trefoil peptides (Rodrigues et al., 2003). Mice lacking the intestinal trefoil factor had impaired mucosal healing and died from extensive colitis after oral administration of dextran sulfate sodium (Mashimo et al., 1995). In a rat model, gastric injury was induced by either intragastric absolute ethanol or subcutaneous indomethacin at 20 mg/kg. Rat intestinal trefoil factor (ITF) was administered at different doses and time points before or after injury. ITF markedly protected against both ethanol- and indomethacin-induced gastric injury when given up to 2 h before injury. ITF did not alter gastric pH (Babyatsky et al., 1996). Trefoil peptide (TFF) is expressed in normal epithelium of the stomach, and expression is lost in gastric adenocarcinomas. The antral and pyloric gastric mucosa of TFF-null mice was dysfunctional and exhibited severe hyperplasia and dysplasia. All homozygous mutant mice developed antropyloric adenoma, and 30% developed multifocal intraepithelial or intramucosal carcinomas (Lefebvre et al., 1996). The GI tract COX-2 expression and the effects of celecoxib on gastric adenoma were studied in TFF1-deficient mice (Saukkonen et al., 2003). Strong COX-2 mRNA and protein expression were noted in the pyloric adenomas of the $TFF1^{-/-}$ mice. Nonneoplastic GI tissues of wild-type or $TFF1^{-/-}$ mice expressed low or undetectable levels of COX-2. Chronic oral administration of celecoxib (1600 ppm for three months) caused ulceration and inflammation of the adenoma, but not of the nonneoplastic GI tissue, in all $TFF1^{-/-}$ mice treated. All untreated $TFF1^{-/-}$ mice had an adenoma, but none demonstrated the combination of ulceration and inflammation. Thus, COX-2 is expressed in gastric adenomas of $TFF1^{-/-}$ mice, and s-NSAIDs disturb the integrity of the adenoma by promoting ulceration and inflammation (Saukkonen et al., 2003).

NOVEL MECHANISM OF ACTION NSIADs

An alternative and novel pathway of AA metabolism is the leukotriene pathway. Inhibition of 5-lipoxygenase (5-LOX) inhibits the synthesis of leukotrienes

(Radmark, 2000). A balanced inhibition of both the COX and LOX pathways may provide superior anti-inflammatory effects and reduce the adverse side effects associated with NSAID use (Fiorucci et al., 2001). Thus, developing NSAID compounds with a novel mechanism of action and an improved safety profile compared to standard NSAIDs is one strategy that has been considered (Laufer, 2001; Tries et al., 2002b). Compounds designed to block both COX and 5-lipooxygenase (5-LOX) were evaluated in animal models and have shown to have anti-inflammatory and antipyretic activity with a better GI safety profile (Hidaka et al., 1984; Wong et al., 1992; Argentieri et al., 1994). The use of these compounds was stopped due to their liver toxicity. However, the liver toxicity was not related to the pharmacological mode of action of inhibiting COX and 5-LOX, but to common structural features of the molecules (Laufer, 2001). These compounds are redox-active and all have a di-*tert*-butyl moiety or are hydroxamic acids. Both COX and 5-LOX are redox enzymes containing an iron ion at the active site. Therefore, redox active compounds and iron chelators, such as hydroxamic acids, interfere with both enzymes, and they also interfere with other enzyme systems in the liver, leading to hepatotoxicity (Laufer, 2001). A competitive COX-1, COX-2, and 5-LOX inhibitor with a nonredox substrate named ML3000 ([2,2-dimethyl-6-(4 chlorophenyl)-7-phenyl-2,3-dihydro-1*H*-pyrrolizine-5-yl]acetic acid) was discovered and designed (Laufer, 2001). ML3000 (also called licofelone) has analgesic, anti-inflammatory, antipyretic, antiplatelet, and antibronchoconstrictive activity, and minimal GI side effects (Laufer, 2001; Tries et al., 2002a). The tolerability of licofelone, assessed by gastro- and duodenoscopy, was tested in a randomized trial in healthy volunteers who received licofelone 200 mg (twice daily) licofelone 400 mg (twice daily), or naproxen 500 mg (twice daily) for 4 weeks (Bias et al., 2004). Licofelone was associated with significantly superior gastric tolerability and a lower incidence of ulcers then was naproxen (Bias et al., 2004). Licofelone (200 mg twice a day) and naproxen (500 mg twice a day) were equally effective in reducing osteoarthritis (OA) symptoms; however, licofelone significantly reduced cartilage volume loss over time in patients with knee OA (Raynauld et al., 2009). Licofelone phase III clinical trials have been completed in OA patients. Tepoxalin [5-(4-chlorophenyl)-*N*-hydroxy-1-(4-methoxyphenyl)-*N*-methyl-1*H*-pyrazole-3-propanamide] is an orally active anti-inflammatory agent which inhibits both COX and 5-LOX activities. The oral toxicity profile of tepoxalin (Zubrin) was evaluated in 1- and 6-month rat (up to 50 mg/kg per day) and dog (up to 150 mg/kg, twice daily) studies (Knight et al., 1996). Analgesic nephropathy syndrome (i.e., papillary edema or necrosis, cortical tubular dilatation) was seen at $\geq$15 mg/kg. Gastrointestinal erosions and ulcers were seen in female rats given 40 mg/kg per day for six months. Small pyloric ulcerations were seen in dogs in doses of 100 and 300 mg/kg per day up to six months. The no-effect doses in rats were 5 mg/kg per day and in dogs were 20 mg/kg per day (Knight et al., 1996). Zubrin is approved for the treatment of OA in dogs. Structure-based virtual screening has been used to identify thiazolidinones ring scaffolds as dual COX/LOX inhibitors (Geronikaki et al., 2008).

mPGES-1 INHIBITORS

Microsomal PGE synthase-1 (mPGES-1) is a member of the membrane-associated proteins involved in the eicosanoid and glutathione metabolism (MAPEG) superfamily with the ability to catalyze the conversion of PGH_2 to PGE_2 (Jakobsson et al., 1999). mPGES-1-deficient mice (mPGES-$1^{-/-}$) are viable and fertile, develop normally compared with wild-type controls, and could not be distinguished from wild-type controls in their general behavior, appearance, body weight, or various tissue histological or hematological parameters (Trebino et al., 2003). In addition, mPGES-$1^{-/-}$ mice displayed a marked reduction in inflammatory responses compared with mPGES-$1^{+/+}$ mice. Edema formation in antigen- or saline-injected mPGES-$1^{-/-}$ mice was similar to that of saline-injected mPGES-$1^{+/+}$ mice (Trebino et al., 2003). mPGES-$1^{-/-}$ mice displayed significant reduction in severity and incidence of collagen-induced arthritis disease compared with wild-type controls (Trebino et al., 2003). Disruption of mPGES-1 retarded atherogenesis and did not affect blood pressure in mPGES-$1^{-/-}$ mice, which were crossed into a low-density lipoprotein receptor knockout ($LDLR^{-/-}$) mouse model (Wang et al., 2006). The effects of genetic mPGES-1 loss on myocardial damage after coronary occlusion were studied in mPGES-$1^{-/-}$ mice (Wu et al., 2009). Reduced myocardial damage was noted in mice lacking mPGES-1 (Wu et al., 2009). Genetic deletion of mPGES-1 suppressed intestinal tumorigenesis in an $Apc^{\Delta 14/+}$ mouse model (Nakanishi et al., 2008). Reduced size and numbers of preneoplastic aberrant crypt foci (ACF) were observed in this $Apc^{\Delta 14/+}$ mouse model after treatment with the colon carcinogen azoxymethane (Nakanishi et al., 2008). Thus, mPGES-1 inhibitors are attractive therapeutic targets for the treatment of inflammatory pain and as chemotherapeutic agents. The effects of a novel and selective mPGES-1 inhibitor, MF63 [2-(6-chloro-1*H*-phenanthro-[9,10-d]imidazol-2-yl)isophthalonitrile], were tested in animal models of inflammation (Xu et al., 2008). MF63 did not cause GI side effects that can be seen with NSAIDs, such as mucosal erosions in knockout mice or non-human primates, although it markedly inhibited PGE_2 synthesis (Xu et al., 2008). Ring templates, including fatty acid derivatives, have been developed as mPGES-1 and dual mPGES-1/5-LOX inhibitors (Friesen and Mancini, 2008). Pirinixic acid derivatives is a novel class of dual mPGES-1/5-LO inhibitors with a promising pharmacological profile and a potential for therapeutic use (Koeberle et al., 2008).

REFERENCES

Abdellatif KR, Chowdhury MA, Dong Y, Das D, et al. Dinitroglyceryl and diazen-1-ium- 1,2-diolated nitric oxide donor ester prodrugs of aspirin, indomethacin and ibuprofen: synthesis, biological evaluation and nitric oxide release studies. Bioorg Med Chem Lett 2009;19:3014–3018.

Argentieri DC, Ritchie DM, Ferro MP, Kirchner T, et al. Tepoxalin: a dual cyclooxygenase/5-lipoxygenase inhibitor of arachidonic acid metabolism with potent anti-inflammatory activity and a favorable gastrointestinal profile. J Pharmacol Exp Ther 1994;271:1399–1408.

Babyatsky MW, deBeaumont M, Thim L, Podolsky DK. Oral trefoil peptides protect against ethanol- and indomethacin-induced gastric injury in rats. Gastroenterology 1996;110:489–497.

Bias P, Buchner A, Klesser B, Laufer S. The gastrointestinal tolerability of the LOX/COX inhibitor, licofelone, is similar to placebo and superior to naproxen therapy in healthy volunteers: results from a randomized, controlled trial. Am J Gastroenterol 2004;99:611–618.

Boggara MB, Faraone A, Krishnamoorti R. Effect of pH and ibuprofen on the phospholipid bilayer bending modulus. J Phys Chem B 2010;114:8061–8066.

Boggara MB, Krishnamoorti R. Small-angle neutron scattering studies of phospholipid–NSAID adducts. Langmuir 2010;26:5734–5745.

Brown JF, Hanson PJ, Whittle BJ. Nitric oxide donors increase mucus gel thickness in rat stomach. Eur J Pharmacol 1992;223:103–104.

Cabré F, Fernández F, Zapatero MI, Araño A, et al. Intestinal ulcerogenic effect of S(+)- ketoprofen in the rat. J Clin Pharmacol 1998;38:27S–32S.

Carbaza A, Cabré F, Rotten E, Gómez M, et al. Stereoselective inhibition of cyclooxygenase by chiral nonsteroidal antiinflammatory drugs. J Clin Pharmacol 1996; 505–512.

Coruzzi G, Coppelli G, Spaggiari S, Cavestro GM, et al. Gastroprotective effects of amtolmetin guacyl: a new non-steroidal anti-inflammatory drug that activates inducible gastric nitric oxide synthase. Dig Liver Dis 2002;34:403–410.

Davies NM, Wright MR, Russell AS, Jamali F. Effect of the enantiomers of flurbiprofen, ibuprofen, and ketoprofen on intestinal permeability. J Pharm Sci 1996;85:1170–1173.

Davies NM, Seth AG, Appley, McKnight W, et al. No-naproxan vs. naproxan: ulcerogenic, analgesic and anti-inflammatory effects. Alim Pharmacol Ther 1997;11:69–79.

Elliott SN, McKnight W, Cirino G, Wallace JL. A nitric oxide-releasing nonsteroidal anti- inflammatory drug accelerates gastric ulcer healing in rats. Gastroenterology 1995;109:524–530.

Fiorucci F, Distrutti E, Cirino G, Wallace JL. The emerging roles of hydrogen sulphide in the gastrointestinal tract and liver. Gastroenterology 2006;131:259–271.

Fiorucci S, Antonelli E. NO-NSAIDs: from inflammatory mediators to clinical readouts. Inflamm Allergy Drug Targets 2006;5:121–131.

Fiorucci S, Santucci L. Hydrogen sulfide-based therapies: focus on H2S releasing NSAIDs. Inflamm Allergy Drug Targets 2011;10:133–140.

Fiorucci S, Meli R, Bucci M, Cirino G. Dual inhibitors of cyclooxygenase and 5- lipoxygenase: a new avenue in anti-inflammatory therapy? Biochem Pharmacol 2001;62:1433–1438.

Fiorucci S, Santucci L, Gresele P, Faccino RM, et al. Gastrointestinal safety of NO- aspirin (NCX-4016) in healthy human volunteers: a proof of concept endoscopic study. Gastroenterology 2003;124: 600–607.

Friesen RW, Mancini JA. Microsomal prostaglandin E2 synthase-1 (mPGES-1): a novel anti-inflammatory therapeutic target. J Med Chem 2008;51:4059–4067.

Geronikaki AA, Lagunin AA, Hadjipavlou-Litina DI, Eleftheriou PT, et al. Computer-aided discovery of anti-inflammatory thiazolidinones with dual cyclooxygenase/lipoxygenase inhibition. J Med Chem 2008;51:1601–1609.

Giruad MN, Sandujask, Felder TB, Illich PA, et al. The effect of omeprazole on the bioavailability of unmodified and phospholid complexed aspirin in rats. Alimert Pharmacol Ther 1997;11:899–906.

Hidaka T, Hosoe K, Ariki Y, Takeo K, et al. Pharmacological properties of a new anti- inflammatory compound, alpha-(3,5-di-*tert*-butyl-4-hydroxybenzylidene)-gamma- butyrolactone (KME-4), and its inhibitory effects on prostaglandin synthetase and 5-lipoxygenase. Jpn J Pharmacol 1984;36:77–85.

Jakobsson PJ, Thoren S, Morgenstern R, Samuelsson B. Identification of human prostaglandin E synthase: a microsomal, glutathione-dependent, inducible enzyme, constituting a potential novel drug target. Proc Natl Acad Sci USA 1999;96:7220–7225.

Knight EV, Kimball JP, Keenan CM, Smith IL, et al. Preclinical toxicity evaluation of tepoxalin, a dual inhibitor of cyclooxygenase and 5-lipoxygenase, in Sprague–Dawley rats and beagle dogs. Fundam Appl Toxicol 1996;33:38–48.

Koeberle A, Zettl H, Greiner C, Wurglics M, et al. Pirinixic acid derivatives as novel dual inhibitors of microsomal prostaglandin E2 synthase-1 and 5-lipoxygenase. J Med Chem 2008;51:8068–8076.

Laufer S. Discovery and development of ML3000. Inflammopharmacology 2001;1:101–112.

Lefebvre O, Chenard MP, Masson R, Linares J, et al. Gastric mucosa abnormalities and tumorigenesis in mice lacking the pS2 trefoil protein. Science 1996;274:259–262.

Lichtenberger LM, Wang ZM, Romero JJ, Ulloa C, et al. Non-steroidal anti-inflammatory drugs (NSAIDs) associate with zwitterionic phospholipids: insight into the mechanism and reversal of NSAID-induced gastrointestinal injury. Nat Med 1995;1:154–158.

Lichtenberger LM, Ulloa C, Vanous AL, Romero JJ, et al. Zwitterionic phospholipids enhance aspirin's therapeutic activity, as demonstrated in rodent model systems. J Pharmacol Exp Ther 1996;277:1221–1227.

Lichtenberger LM, Romero JJ, de Ruijter WM, Behbod F, et al. Phosphatidylcholine association increases the anti-inflammatory and analgesic activity of ibuprofen in acute and chronic rodent models of joint inflammation: relationship to alterations in bioavailability and cyclooxygenase-inhibitory potency. J Pharmacol Exp Ther 2001;298:279–287.

Lichtenberger LM, Zhou Y, Dial EJ, Raphael RM. NSAID injury to the gastrointestinal tract: evidence that NSAIDs interact with phospholipids to weaken the hydrophobic surface barrier and induce the formation of unstable pores in membranes. J Pharm Pharmacol 2006;58:1421–1428.

Lichtenberger LM, Romero JJ, Dial EJ. Surface phospholipids in gastric injury and protection when a selective cyclooxygenase-2 inhibitor (Coxib) is used in combination with aspirin. Br J Pharmacol 2007;150:913–919.

Lichtenberger LM, Barron M, Marathi U. Association of phosphatidylcholine and NSAIDs as a novel strategy to reduce gastrointestinal toxicity. Drugs Today (Barc) 2009;45:877–890.

Lohmander LS, McKeith D, Svensson O, Malmenäs M, et al. A randomised, placebo controlled, comparative trial of the gastrointestinal safety and efficacy of AZD3582 versus naproxen in osteoarthritis. Ann Rheum Dis 2005;64:449–456.

Longman RJ, Thomas MG, Poulsom R. Trefoil peptides and surgical disease. Br J Surg 1999; 86:740–748.

Mashimo HM, Wu DC, Podolasky DK, Fishman MC. Impaired defenses of intestinal mucosa in mice lacking in intestinal foil peptides. Science 1995;274:262–265.

Nakanishi M, Montrose DC, Clark P, Nambiar PR, et al. Genetic deletion of mPGES-1 suppresses intestinal tumorigenesis. Cancer Res 2008;68:3251–3259.

Nieto AI, Cabré F, Moreno FJ, Alarcón de la Lastra C. Mechanisms involved in the attenuation of intestinal toxicity induced by (S)-(+)-ketoprofen in re-fed rats. Dig Dis Sci 2002;47:905–913.

Nishio H, Hayashi Y, Terashima S, Takeuchi K. Role of endogenous nitric oxide in mucosal defense of inflamed rat stomach following iodoacetamide treatment. Life Sci 2006;79:1523–1530.

Pajdo R, Brzozowski T, Szlachcic A, Konturek PC, et al. Lipoxins, the novel mediators of gastroprotection and gastric adaptation to ulcerogenic action of aspirin. Curr Pharm Des 2011;17:1541–1551.

Radmark OP. The molecular biology and regulation of 5-lipoxygenase. Am J Respir Crit Care Med 2000;161:S11–S15.

Raynauld JP, Martel-Pelletier J, Bias P, Laufer S, et al. Protective effects of licofelone, a 5-lipoxygenase and cyclo-oxygenase inhibitor, versus naproxen on cartilage loss in knee osteoarthritis: a first multicentre clinical trial using quantitative MRI. Ann Rheum Dis 2009;68:938–947.

Rodrigues S, Van Aken E, Van Bocxlaer S, Attoub S, et al. Trefoil peptides as proangiogenic factors in vivo and in vitro: implication of cyclooxygenase-2 and EGF receptor signaling. FASEB J 2003;17:7–16.

Saukkonen K, Tomasetto C, Narko K, Rio MC, et al. Cyclooxygenase-2 expression and effect of celecoxib in gastric adenomas of trefoil factor 1-deficient mice. Cancer Res 2003;63:3032–3036.

Schnitzer TJ, Kivitz AJ, Lipetz RS, Sanders N, et al. Comparison of the COX-inhibiting nitric oxide donator AZD3582 and rofecoxib in treating the signs and symptoms of osteoarthritis of the knee. Arthritis Rheum 2005;53:827–837.

Stefano F, Distrutti E. Cyclo-oxygenase (COX) inhibiting nitric oxide donating (CINODs) drugs: a review of their current status. Curr Top Med Chem 2007;7:277–282.

Sweetman BJ. Development and use of the quick acting chiral NSAID dexketoprofen trometamol (keral). Acute Pain 2003;4:109–115.

Trebino CE, Stock JL, Gibbons CP, Naiman BM, et al. Impaired inflammatory and pain responses in mice lacking an inducible prostaglandin E synthase. Proc Natl Acad Sci USA 2003;100:9044–9049.

Tries S, Laufer S, Radziwon P, Breddin HK. Antithrombotic and platelet function inhibiting effects of ML3000, a new antiinflammatory drug with Cox/5-LOX inhibitory activity. Inflamm Res 2002a;51:129–134.

Tries S, Neuper TW, Laufer S. The mechanism of action of new compound ML3000 (locofelone): inhibition of 5-LO COX-1, and COX-2. Inflamm Res 2002b;51:135–43.

Wallace JL. Physiological and pathophysiological roles of hydrogen sulfide in the gastrointestinal tract. Antioxid Redox Signal 2010;12:1125–1133.

Wallace JL, Reuter B, Cicala C, McKnight W, et al. A diclofenac derivative without ulcerogenic properties. Eur J Pharmacol 1994;257:249–255.

Wallace JL, Del Soldato P, Cirino G, Muscará MN. Nitric oxide-releasing NSAIDs: GI- safe antithrombotics. IDrugs 1999;2:321–326.

Wallace JL, Ignarro LJ, Fiorucci S. Potential cardioprotective actions of no-releasing aspirin. Nat Rev Drug Discov 2002;1:375–382.

Wallace JL, Caliendo G, Santagada V, Cirino G, et al. Gastrointestinal safety and anti- inflammatory effects of a hydrogen sulphide-releasing diclofenac derivative in the rat. Gastroenterology 2007;132:261–271.

Wang M, Zukas AM, Hui Y, Ricciotti E, et al. Deletion of microsomal prostaglandin E synthase-1 augments prostacyclin and retards atherogenesis. Proc Natl Acad Sci USA 2006;103:14507–14512.

Wong S, Lee SJ, Frierson MR 3rd, Proch J, et al. Antiarthritic profile of BF-389: a novel anti-inflammatory agent with low ulcerogenic liability. Agents Actions 1992;37:90–98.

Wu D, Mennerich D, Arndt K, Sugiyama K, et al. Comparison of microsomal prostaglandin E synthase-1 deletion and COX-2 inhibition in acute cardiac ischemia in mice. Prostaglandins Other Lipid Mediat 2009;90:21–25.

Xu D, Rowland SE, Clark P, Giroux A, et al. MF63 [2-(6-chloro-1*H*-phenanthro[9,10- d]imidazol-2-yl)-isophthalonitrile], a selective microsomal prostaglandin E synthase-1 inhibitor, relieves pyresis and pain in preclinical models of inflammation. J Pharmacol Exp Ther 2008;326:754–763.

INDEX

Comparative Pathophysiology and Toxicology of Cyclooxygenases, First Edition. Zaher A. Radi.

Printed and bound by CPI Group (UK) Ltd, Croydon, CR0 4YY
16/09/2026
14963456-0003

MODELING AND HIGH-PERFORMANCE CONTROL OF ELECTRIC MACHINES

Books in the IEEE Press Series on Power Engineering

Rating of Electric Power Cables in Unfavorable Thermal Environments
George J. Anders

Power System Protection
P. M. Anderson

Understanding Power Quality Problems: Voltage Sags and Interruptions
Math H. J. Bollen

Modeling and High-Performance Control of Electric Machines
John Chiasson

Electric Power Applications of Fuzzy Systems
Edited by M. E. El-Hawary

Principles of Electric Machines with Power Electronic Applications,
Second Edition
M. E. El-Hawary

Pulse Width Modulation for Power Converters: Principles and Practice
D. Grahame Holmes and Thomas Lipo

Analysis of Electric Machinery and Drive Systems, Second Edition
Paul C. Krause, Oleg Wasynczuk, and Scott D. Sudhoff

Risk Assessment for Power Systems: Models, Methods, and Applications
Wenyan Li

Optimization Principles: Practical Applications to the Operations and Markets of the Electric Power Industry
Narayan S. Rau

Electric Economics: Regulation and Deregulation
Geoffrey Rothwell and Tomas Gomez

Electric Power Systems: Analysis and Control
Fabio Saccomanno

Electrical Insulation for Rotating Machines: Design, Evaluation, Aging, Testing, and Repair
Greg Stone, Edward A. Boulter, Ian Culbert, and Hussein Dhirani

MODELING AND HIGH-PERFORMANCE CONTROL OF ELECTRIC MACHINES

JOHN CHIASSON

IEEE PRESS SERIES ON POWER ENGINEERING

IEEE Press Series on Power Engineering
Mohamed E. El-Hawary, *Series Editor*

IEEE

The Institute of Electrical and Electronics Engineers, Inc., New York

WILEY-INTERSCIENCE

A JOHN WILEY & SONS, INC., PUBLICATION

Published by John Wiley & Sons, Inc., Hoboken, New Jersey.
Published simultaneously in Canada.

Library of Congress Cataloging-in-Publication Data:

Chiasson, John Nelson.
Modeling and high performance control of electric machines / John Chiasson.
p. cm. — (IEEE Press series on power engineering)
Includes bibliographical references and index.
ISBN 0-471-68449-X (cloth)
1. Electric machinery—Automatic control—Mathematical models. I. Title. II. Series.
TK2181.C43 2005
621.31'042—dc22 2004021739

10 9 8 7 6 5 4 3 2

To My Parents

and to

Marc Bodson

Pour son soutien et son aide jour après jour durant une période très difficile de ma carrière

Amro El-Jaroudi
Mahmoud El Nokali

إِلَى اصدقَائِي الذينَ وَقَفُوا بِجَانِبي في تِلكَ الأَوقَاتِ العَصيبَةِ مِن حَيَاتي

James B. Lieber

על שתמיד האמנת בי ועל עזרתך הרבה בשעתי הקשה בקריירה האקדמית שלי

Contents

[1] Sections marked with an asterisk (*) may be skipped without loss of continuity.

II AC Machine Theory 235

Preface

This book is intended to be an exposition of the modeling and control of electric machines, specifically, the direct current (DC) machine and the alternating current (AC) machines consisting of the induction motor, the permanent magnet (PM) synchronous motor, and the brushless DC motor. The particular emphasis here is on techniques used for high-performance applications, that is, applications that require both rapid and precise control of position, speed, and/or torque. Traditionally, DC motors were reserved for high-performance applications (positioning systems, rolling mills, traction drives, etc.) because of their relative ease of control compared to AC machines. However, with the advances in control methods, computing capability, and power electronics, AC motors continue to replace DC motors in high-performance applications. The intent here is to carefully derive the mathematical models of the AC machines and show how these mathematical models are used to design control algorithms that achieve high performance.

Electric machines are a particularly fascinating application of basic electricity and magnetism. The presentation here relies heavily on these basic concepts from Physics to develop the models of the motors. Specifically, Faraday's law ($\xi = -d\phi/dt$, where $\phi = \int_S \vec{\mathbf{B}} \cdot d\vec{\mathbf{S}}$), the magnetic force law ($\vec{\mathbf{F}} = i\vec{\ell} \times \vec{\mathbf{B}}$ or, $\vec{\mathbf{F}} = q\vec{\mathbf{v}} \times \vec{\mathbf{B}}$), Gauss's law ($\oint \vec{\mathbf{B}} \cdot d\vec{\mathbf{S}} = 0$), Ampère's law ($\oint \vec{\mathbf{H}} \cdot d\vec{\ell} = i_{\text{free}}$), the relationship between $\vec{\mathbf{B}}$ and $\vec{\mathbf{H}}$, properties of magnetic materials, and so on are reviewed in detail and used extensively to derive the currently accepted nonlinear differential equation models of the various AC motors. The author made his best attempt to make the modeling assumptions as clear as possible and to consistently show that the magnetic and electric fields satisfy Maxwell's equations (as, of course, they must). The classical approach to teaching electric machinery is to present their equivalent circuit models and to analyze these circuit models ad nauseam. Further, the use of the basic Physics of electricity and magnetism to explain their operation is minimized if not omitted. However, the equivalent circuit is a result of assuming constant-speed operation of the machine and computing the sinusoidal steady-state solution of the nonlinear differential equation model of the machine. Here, the emphasis is on explaining how the machines work using fundamental concepts from electricity and magnetism, and on the derivation of their nonlinear differential equation models. The derivation of the corresponding equivalent circuit assuming steady-state conditions is then straightforward.

Electric machines also provide fascinating examples to illustrate concepts from electromagnetic field theory (in contrast to electricity and magnetism). In particular, the way the electric and magnetic fields change as one goes between reference frames that are in relative motion are vividly illustrated using AC machines. For this reason, optional sections are included to show how the electric and magnetic fields change as one goes between a coordinate system attached to the stator to a coordinate system that rotates with the rotating magnetic field produced by the stator currents or a frame attached to the rotor. Also given in an optional section is the derivation of the axial electric and azimuthal magnetic fields in the air gap.

This is also a book on the control of electric machines based on their differential equation models. With the notable exception of the sinusoidal steady-state analysis of the induction motor in Chapter 7, very little attention is given to the classical equivalent circuits as these models are valid only in steady state. Rather, the differential equation models are used as the basis to develop the notions of field-oriented control, input-output linearization, flux observers, least-squares identification methods, state feedback trajectory tracking, and so on. This is a natural result of the emphasis here on high-performance control methods (e.g., field-oriented control) as opposed to classical methods (e.g., V/f, slip control, etc.).

There are of course many good books in the area of electric machines and their control. The author owes a debt of gratitude to Professor W. Leonhard for his book [1] (see the most recent edition [2]), from which he was educated in the modeling and control of electric drives. The present book is narrower in focus with an emphasis on the modeling and operation of electric machines based on elementary classical physics and an emphasis on high-performance control methods using a state-space formulation. The books by P. C. Krause [3] and P. C. Krause et al. [4] are complete in their derivation of the mathematical models of electric machines while C. B. Gray [5] presents electromagnetic theory in the context of electric machines. A comprehensive treatment using SIMULINK to simulate electric machinery is given in C-M. Ong's book [6]. The graduate level books by D. W. Novotny and T. A. Lipo [7], P. Vas [8], J. M. D. Murphy and F. G. Turnbull [9], I. Boldea and S. A. Nasar [10], B. Adkins and R. G. Harley [11], A. M. Trzynadlowski [12], M. P. Kazmierkowski and H. Tunia [13], B. K. Bose [14], and R. Krishnan [15] all cover the modeling and control of electric machines while the books by R. Ortega et al. [16], D. M. Dawson et al. [17], and F. Khorrami et al. [18] emphasize advanced control methods.

The introductory-level books by S. J. Chapman [19], H. Woodson and J. Melcher [20], L. W. Matsch and J. D. Morgan [21], G. McPherson and R. D. Laramore [22], D. V. Richardson [23], P. C. Krause and O. Wasynczuk [24], N. Mohan [25], G. R. Slemon and A. Straughn [26], J. Sokira and W. Jaffe [27], G. J. Thaler and M. L. Wilcox [28], V. Deltoro [29], M. El-Hawary [30], P. C. Sen [31], and G. R. Slemon [32] are among the many books on electric machines from which this author has benefited.

The beautifully written textbooks *PSSC Physics* by the Physical Science Curriculum Study [33], *Physics* by D. Halliday and R. Resnick [34], *Principles of Electrodynamics* by M. Schwartz [35], and *Electromagnetic Fields* by R. K. Wangness [36] are used as references for the theory of electricity and magnetism.

This book borrows from these above works and hopefully makes its own contribution to the literature on electric machines.

Part I of the book consists of the first three chapters. Chapters 1 and 2 present a detailed review of the basic concepts of electricity and magnetism in the context of DC machines and an introduction to control methods, respectively, which will be used extensively in the remaining chapters. The third chapter on magnetic fields and magnetic materials is intended to be a detailed introduction to the subject. For example, most textbooks assume that the reader understands Ampère's law in the form $\oint \vec{\mathbf{H}} \cdot d\vec{\ell} = i_{\text{free}}$ and that $\vec{\mathbf{B}} = \mu\vec{\mathbf{H}}$ in (soft) magnetic materials, yet it is the experience of the author that students do not have a fundamental understanding of these concepts.

These first three chapters are elementary in nature and were written to be accessible to undergraduates. The reason for this is that often control engineers do not have any background in electric machinery while power/electric-machine engineers often do not have any background in basic state-space concepts of control theory. Consequently, it is hoped that these chapters can bring the reader "up to speed" in these areas.

Chapter 1 reviews the basic ideas of electricity and magnetism that are needed to model electric machines. In particular, the notions of magnetic fields, magnetic force and Faraday's law are reviewed by using them to derive the standard model of a DC motor.

Chapter 2 provides an elementary introduction to the control techniques required for the high-performance control of electric machines. This includes an elementary presentation of state feedback control, observers, and identification theory as applied to DC machines to prepare the reader for the subsequent chapters.

Chapter 3 goes into the modeling of magnetic materials in terms of their use in electric machines. The fundamental result of this chapter is the modification of Ampère's law $\oint_C \vec{\mathbf{B}} \cdot d\vec{\ell} = \mu_0 i$ so that it is valid in the presence of magnetic material. This introduces the magnetic intensity field $\vec{\mathbf{H}}$ and its relationship to magnetic induction field $\vec{\mathbf{B}}$ via the magnetization vector $\vec{\mathbf{M}}$ to obtain the more general version of Ampère's law $\oint \vec{\mathbf{H}} \cdot d\vec{\ell} = i_{\text{free}}$. All of this requires a significant discussion of the modeling of magnetic materials. The approximation $\vec{\mathbf{H}} = 0$ in magnetic materials is discussed, and then it is shown how this approximation along with Ampère's law can be used to find the radial component of $\vec{\mathbf{B}}$ in the air gap of electric machines. Also presented is Gauss's law for $\vec{\mathbf{B}}$; this leads to the notion of conservation of flux, as well as the fact that $\vec{\mathbf{B}}$ is normal to the surface of soft magnetic materials. This chapter should be read, but the reader should not get "bogged down" in the chapter. Rather, the main results should be remembered.

Part II consists of Chapters 4 through 10 and presents the modeling and control of AC machines.

Chapter 4 uses the results of Chapters 1 and 3 to explain how a radially directed rotating magnetic field can be established in the air gap of AC machines. In particular, the notions of distributed windings and of sinusoidally wound turns (phase windings) are explained. Ampère's law is then used to show that a sinusoidal (spatially) distributed radial magnetic field is established in the air gap by the currents in the phase windings. The concept of flux linkage in distributed windings is explained, and the chapter ends with an optional section on the azimuthal magnetic field in the air gap.

Chapter 5 explains the fundamental Physics behind the working of induction and synchronous machines. Specifically, this chapter uses a simplified model of the induction motor and shows how voltages and currents are induced in the rotor loops by the rotating magnetic field established by the stator currents. Then it is shown how torque is produced on these induced currents by the same stator rotating magnetic field that induced them introducing the idea of slip. Similarly, the synchronous machine is analyzed to show how the rotating radial magnetic field established by the stator currents produces torque on a rotor carrying constant current.

An optional section on the microscopic point of view of the Physics of the induction motor is also presented. This includes a discussion of how the electric and magnetic fields change as one goes between coordinate systems that are rotating with respect to each other and how one reinterprets the Physics of the machine's operation. The chapter ends with another optional section of the steady-state behavior of an induction machine with a squirrel cage rotor.

Chapter 6 derives the systems of differential equations that mathematically model the two-phase induction and synchronous machines. The concept of leakage is presented and accounted for in the derived models. These models are the accepted models used throughout the literature and form the basis for high-performance control of these machines. In an optional section it is shown that the stator and rotor *magnetic fields* of an induction motor rotate synchronously together as they do in a synchronous machine. The chapter ends with another optional section on the concepts of field energy and co-energy, and how the expression for the torque of an electric machine can be derived using these notions.

Chapter 7 presents the derivation of the models of three-phase AC machines and their two-phase equivalent models. These derivations readily follow from the results of Chapter 6. The classical steady-state analysis of the induction motor is also presented including its equivalent circuit. The chapter ends with a discussion of why the standard power system is an AC sinusoidal three-phase 60-Hz (or 50-Hz) system.

Chapter 8 covers the control of induction motors presenting both field-oriented control and input-output linearization control. Flux observers, field weakening, and speed observers are also presented along with experimental results. The chapter ends with an optional section on how to identify the induction motor parameters using a nonlinear least-squares technique.

Chapter 9 covers the control of synchronous motors describing field-oriented control, field weakening, speed observers and identification methods. The operation and modeling of permanent magnet stepping motors is also covered.

Chapter 10 covers the modeling and control of PM synchronous motors with trapezoidal back emf, which are also known as brushless DC (BLDC) motors.

The logical dependence of the chapters is shown in the block diagram below assuming that the optional sections are *not* covered.

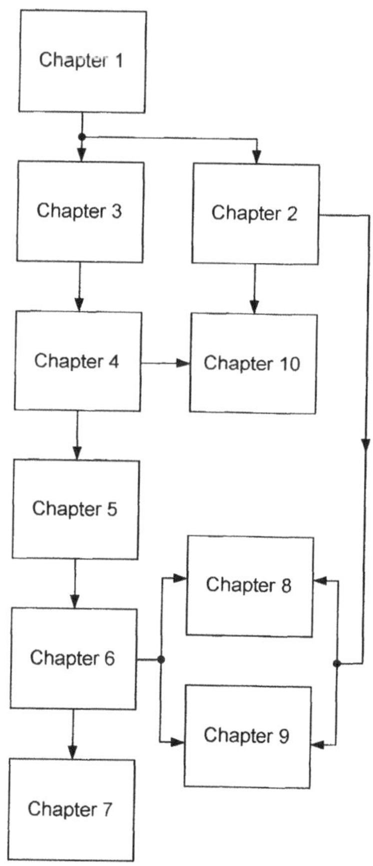

Logical dependence of the chapters.

Finally, the author's intent for this book was for the reader to understand how electric machines are modeled and to understand the basic techniques in their control. The references at the end of the book are only those directly referenced in the book and are *not* representative of (nor give proper recognition to) the many important contributions made by researchers throughout the world. The reader is referred to Professor Leonhard's book [2] for a much more extensive reference list.

Comments on the Use of the Book

In using this book in a one-semester graduate-level course, the following material was usually covered:

Chapter 1, Sections 1.1–1.7
Chapter 2, Sections 2.1–2.4
Chapter 3, Sections 3.1–3.4
Chapter 4, Sections 4.1–4.5
Chapter 5, Sections 5.1–5.3
Chapter 6, Sections 6.1–6.10
Chapter 7, Sections 7.1–7.3
Chapter 8, Sections 8.1–8.3
Chapter 9, Section 9.1

Sections marked with an asterisk (*) may be omitted without loss of continuity. Some of these optional sections assume familiarity with Maxwell's equations in *differential* form.

Acknowledgments

There are many people that I would to acknowledge for their help in my work in the control of electric machines.

I did my Ph.D. thesis in Algebraic Systems Theory which is quite far from the area of Electric Machines. Nevertheless, I am most grateful to my advisor, Professor E. Bruce Lee, who has always shown me his kindness over the years.

Professor Edward W. Kamen (formerly at the University of Pittsburgh) along with Mr. Stephen Botos (President of Aerotech, Inc.) were instrumental through their enthusiasm and financial support at the University of Pittsburgh, through which many of the results presented here were funded. I am very grateful to Mike Aiello (chief design engineer at Aerotech) for both designing and building our hardware platform, resulting in a successful set of experiments.

I would like to thank The Oak Ridge National Laboratory in particular, Don Adams and Laura Marlino, for funding the recent results on the identification of the induction motor parameters presented in Chapter 8.

Shortly after I arrived at the University of Pittsburgh, I started to work with Marc Bodson (formerly at Carnegie Mellon University) and I want to express my deep gratitude to him for this collaboration, which led to many of the new theoretical and experimental results presented in Chapters 8 and 9.

Also, shortly after I arrived at the University of Tennessee, I began to work with Leon M. Tolbert, and I am very grateful for this collaboration as well.

Mohamed Zribi was the first student I worked with in this area, leading to a paper on feedback linearization control of stepper motors. Ron Rekowski suffered through our first attempts to do some experiments for which I am grateful. My Ph.D. student Bob Novotnak continued this work and did the experiments on the control of the stepper and induction motors presented in Chapters 8 and 9 of this book. Through his skill we were able to obtain experimental results demonstrating very high performance. Jennifer Stephans (Marc Bodson's student) did the early work on identification for the induction motor while my student Kaiyu Wang did the identification experiments for the induction motor given in Chapter 8. Andy Blauch (Marc Bodson's student) did the identification experiments for the PM stepper motor presented in Chapter 9.

I first taught (and learned!) induction motor control using the book by Professor Werner Leonhard, and I am very grateful for his writing that book. I later had the opportunity to visit his institute in Braunschweig Germany from which I left with even more enthusiasm for the field.

I would like to thank the many students who suffered under the early versions of this book, or suffered with me as their advisor, or both. These include Mohamed Zribi, Bob Novotnak, Eric Shook, Ron Rekowski, Walt Barie, Joe Matesa, Chellury Sastry, Gary Campbell, David Schuerer, Atul Chaudhari, Jason Mueller, Samir Mehta, Pete Hammond, Dick Osman, Jim Short, Vincent Allarouse, Marc Aiello, Sean West, Baskar Vairamohan, Yinghui Lu, Zhong Tang, Mengwei Li, Kaiyu Wang, Yan Xu, Madhu Chinthavali, Jianqing Chen, Nivedita Alluri, Zhong Du, Faisal Khan, Pankaj Pandit, Hui Zhang, Wenjuan Zhang, Ben Sooter, Keith McKenzie, Rebin Zhou, Ann Chee Tan, Jesse Richmond, SeongTaek Lee, and all my other students.

I am very grateful to Thomas Keller for teaching me about real time simulators.

I would like to thank my colleagues Leon Tolbert, Saul Gelfand, Joachim Böcker, Miguel Vélez-Reyes, Michel Fliess, Thomas Keller, George Verghese, Jeff Lang, David Taylor, Ray DeCarlo, Mark Spong, Steve Yurkovich, Jessy Grizzle, Henk Nijmeijer, Chaouki Abdallah, Doug Birdwell, Gerardo Espinosa-Pérez, Romeo Ortega, Yih-Choung Yu, Gerardo Escobar-Valderrama, Daniel Campos-Delgado, Ricardo Fermat-Flores, Jesús Alvarez, Jesús Leyva-Ramos, Jeffrey Mayer, Kai Mueller, Henrik Mosskull, Stanislaw Żak, Samer Saab, Burak Özpineci, and Alex Stanković for their words of encouragement.

This book was written in LaTeX using SCIENTIFIC WORD from Mackichan Software (see *http://www.mackichan.com*). I would especially like to thank John McKendrick and Alan Green of Mackichan Software for their help with my many questions.

I would like thank my editor Valerie Moliere as well as my production editor Lisa Vanhorn of John Wiley & Sons for stepping me through the process of getting this book published. Bob Golden, who copy edited the manuscript of this book, is gratefully acknowledged for fixing many errors and inconsistencies.

I am very grateful to Sharon Katz for her drawings of Figures 1.8(a)–(d), 1.25, 1.41–1.47, 9.31, 9.34, 9.35(a)–(e) and for her help, suggestions and encouragement of the artwork in this book and for actually setting me on the path to getting the figures drawn. I would also like to thank Bret Wilfong, who drew Figures 4.1, 5.4(b), 5.7(b), and 5.8(b).

John Wiley & Sons maintains an ftp site at

ftp://ftp.wiley.com/public/sci_tech_med/high_performance_control

for downloading an errata sheet for the book. Instructors, upon obtaining password privileges, will also be able to download the simulation files that go with this textbook. A solutions manual is available to instructors by contacting their local Wiley representative.

Any comments, criticisms, and corrections are most welcome and may be sent to the author at *chiasson@ieee.org*.

John Chiasson

MODELING AND HIGH-PERFORMANCE CONTROL OF ELECTRIC MACHINES

Part I

DC Machines, Controls, and Magnetics

1

The Physics of the DC Motor

The principles of operation of a direct current (DC) motor are presented based on fundamental concepts from electricity and magnetism contained in any basic physics course. The DC motor is used as a concrete example for reviewing the concepts of magnetic fields, magnetic force, Faraday's law, and induced electromotive forces (emf) that will be used throughout the remainder of the book for the modeling of electric machines. All of the Physics concepts referred to in this chapter are contained in the book *Physics* by Halliday and Resnick [34].

1.1 Magnetic Force

Motors work on the basic principle that magnetic fields produce forces on wires carrying a current. In fact, this experimental phenomenon is what is used to define the magnetic field. If one places a current carrying wire between the poles of a magnet as in Figure 1.1, a force is exerted on the wire. Experimentally, the magnitude of this force is found to be proportional to both the amount of current in the wire and to the length of the wire that is between the poles of the magnet. That is, F_{magnetic} is proportional to ℓi. The direction of the magnetic field $\vec{\mathbf{B}}$ at any point is defined to be the direction that a small compass needle would point at that location. This direction is indicated by arrows in between the north and south poles in Figure 1.1.

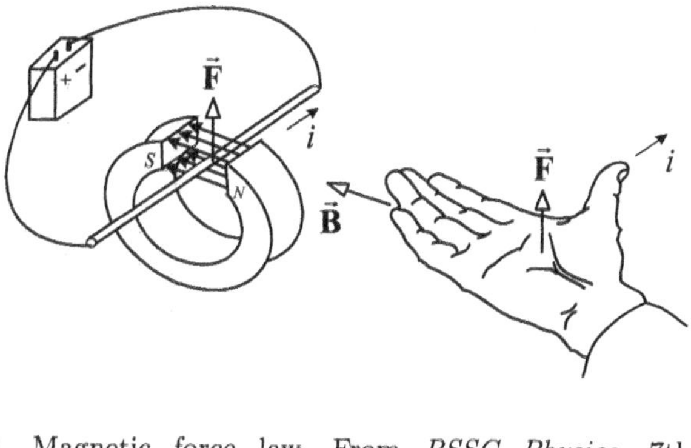

FIGURE 1.1. Magnetic force law. From *PSSC Physics*, 7th edition, by Haber-Schaim, Dodge, Gardner, and Shore, published by Kendall/Hunt, 1991.

With the direction of $\vec{\mathbf{B}}$ perpendicular to the wire, the strength (magnitude) of the *magnetic induction field* $\vec{\mathbf{B}}$ is defined to be

$$B = |\vec{\mathbf{B}}| \triangleq \frac{F_{\text{magnetic}}}{\ell i}$$

where F_{magnetic} is the magnetic force, i is the current, and ℓ is the length of wire perpendicular to the magnetic field carrying the current. That is, B is the proportionality constant so that $F_{\text{magnetic}} = i\ell B$. As illustrated in Figure 1.1, the direction of the force can be determined using the right-hand rule. Specifically, using your right hand, point your fingers in the direction of the magnetic field and point your thumb in the direction of the current. Then the direction of the force is out of your palm.

Further experiments show that if the wire is parallel to the $\vec{\mathbf{B}}$ field rather than perpendicular as in Figure 1.1, then no force is exerted on the wire. If the wire is at some angle θ with respect to $\vec{\mathbf{B}}$ as in Figure 1.2, then the force is proportional to the *component* of $\vec{\mathbf{B}}$ perpendicular to the wire; that is, it is proportional to $B_\perp = B\sin(\theta)$. This is summarized in the *magnetic force law*: Let $\vec{\boldsymbol{\ell}}$ denote a vector whose magnitude is the length ℓ of the wire in the magnetic field and whose direction is defined as the positive direction of current in the bar; then the magnetic force on the bar of length ℓ carrying the current i is given by

$$\vec{\mathbf{F}}_{\text{magnetic}} = i\vec{\boldsymbol{\ell}} \times \vec{\mathbf{B}}$$

or, in scalar terms, $F_{\text{magnetic}} = i\ell B\sin(\theta) = i\ell B_\perp$. Again, $B_\perp \triangleq B\sin(\theta)$ is the component of $\vec{\mathbf{B}}$ perpendicular to the wire.[1]

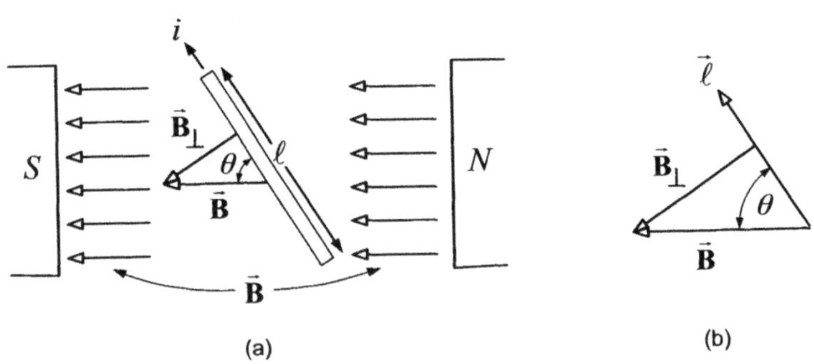

FIGURE 1.2. Only the component $B_\perp$ of the magnetic field which is perpendicular to the wire produces a force on the current.

[1] Motors are designed so that the conductors are perpendicular to the external magnetic field.

Example *A Linear DC Machine* [19]

Consider the simple linear DC machine in Figure 1.3 where a sliding bar rests on a simple circuit consisting of two rails. An external magnetic field is going through the loop of the circuit up out of the page indicated by the $\otimes$ in the plane of the loop. Closing the switch results in a current flowing around the circuit and the external magnetic field produces a force on the bar which is free to move. The force on the bar is now computed.

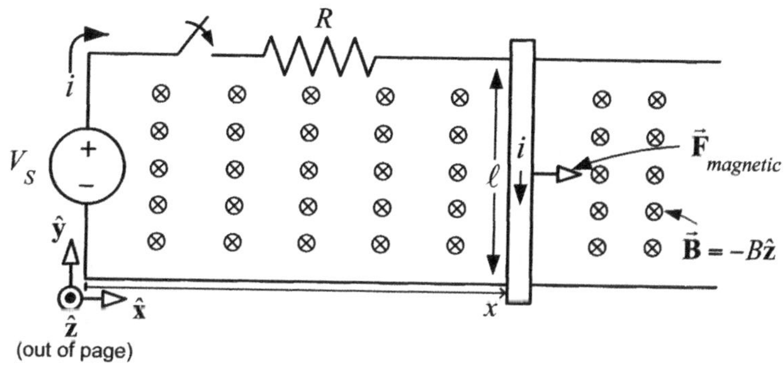

FIGURE 1.3. A linear DC motor.

The magnetic field is constant and points into the page (indicated by $\otimes$) so that written in vector notation, $\vec{\mathbf{B}} = -B\hat{\mathbf{z}}$ with $B > 0$. By the right hand rule, the magnetic force on the sliding bar points to the right. Explicitly, with $\vec{\boldsymbol{\ell}} = -\ell\hat{\mathbf{y}}$, the force is given by

$$\begin{aligned}\vec{\mathbf{F}}_{\text{magnetic}} &= i\vec{\boldsymbol{\ell}} \times \vec{\mathbf{B}} = i(-\ell\hat{\mathbf{y}}) \times (-B\hat{\mathbf{z}}) \\ &= i\ell B\hat{\mathbf{x}}.\end{aligned}$$

To find the equations of motion for the bar, let f be the coefficient of viscous (sliding) friction of the bar so that the friction force is given by $F_f = -f dx/dt$. Then, with m_ℓ denoting the mass of the bar, Newton's law gives

$$i\ell B - f dx/dt = m_\ell d^2x/dt^2.$$

Just after closing the switch at $t = 0$, but before the bar starts to move, the current is $i(0^+) = V_S(0^+)/R$. However, it turns out that as the bar moves the current does *not* stay at this value, but instead decreases due to electromagnetic induction. This will be explained later.

1.2 Single-Loop Motor

As a first step to modeling a DC motor, a simplistic single-loop motor is considered. It is first shown how torque is produced and then how the current in the single loop can be reversed (commutated) every half turn to keep the torque constant.

1.2.1 *Torque Production*

Consider the magnetic system in Figure 1.4, where a cylindrical core is cut out of a block of a permanent magnet and replaced with a soft iron core. The term "soft" iron refers to the fact that material is easily magnetized (a permanent magnet is referred to as "hard" iron).

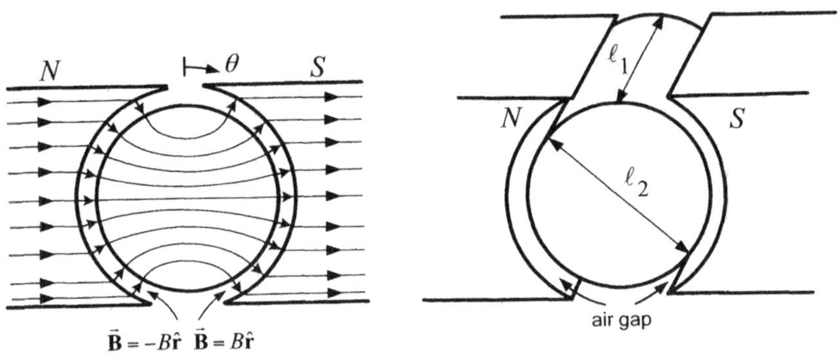

FIGURE 1.4. Soft iron cylindrical core placed inside a hollowed out permanent magnet to produce a radial magnetic field in the air gap.

An important property of soft magnetic materials is that the magnetic field at the surface of such materials tends to be normal (perpendicular) to the surface. Consequently, the cylindrical shape of the surfaces of the soft iron core and the stator permanent magnet has the effect of making the field in the air gap *radially* directed; furthermore, it is reasonably constant (uniform) in magnitude. A mathematical description of the magnetic field in the air gap due to the permanent magnet is simply

$$\vec{\mathbf{B}} = \begin{cases} +B\hat{\mathbf{r}} & \text{for } 0 < \theta < \pi \\ -B\hat{\mathbf{r}} & \text{for } \pi < \theta < 2\pi \end{cases}$$

where $B > 0$ is the magnitude or strength of the magnetic field and θ is an arbitrary location in the air gap.[2]

Figure 1.5 shows a rotor loop wound around the iron core of Figure 1.4. The length of the rotor is ℓ_1 and its diameter is ℓ_2. The torque on this rotor loop is now calculated by considering the magnetic forces on sides a and a' of the loop. On the other two sides of the loop, that is, the front and

[2] Actually it will be shown in a later chapter that the magnetic field must be of the form $\vec{\mathbf{B}} = \pm B(r_0/r)\hat{\mathbf{r}}$ in the air gap, that is, it varies as $1/r$ in the air gap. However, as the air gap is small, the $\vec{\mathbf{B}}$ field is essentially constant across the air gap.

back sides, the magnetic field has negligible strength so that no significant force is produced on these sides. As illustrated in Figure 1.5(b), the rotor angular position is taken to be the angle θ_R from the vertical to side a of the rotor loop.

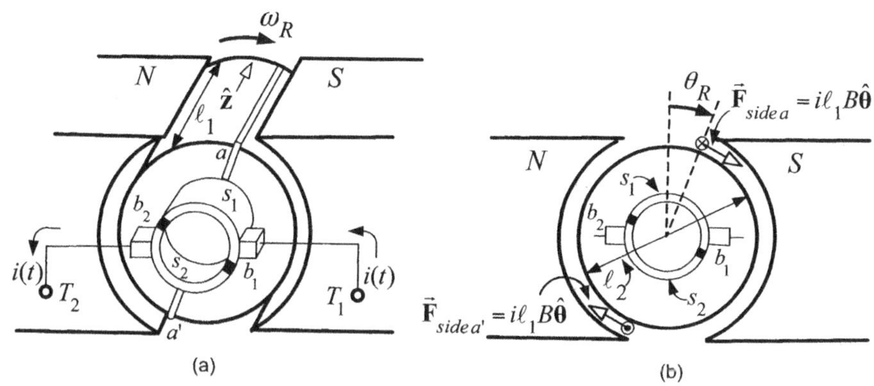

FIGURE 1.5. A single-loop motor. From *Electromagnetic and Electromechanical Machines*, 3rd edition, L. W. Matsch and J. Derald Morgan, 1986. Reprinted by permisson of John Wiley & Sons.

Figure 1.6 shows the cylindrical coordinate system used in Figure 1.5. Here $\hat{\mathbf{r}}, \hat{\boldsymbol{\theta}}, \hat{\mathbf{z}}$ denote unit cylindrical coordinate vectors. The unit vector $\hat{\mathbf{z}}$ points along the rotor axis into the paper in Figure 1.5(b), $\hat{\boldsymbol{\theta}}$ is in the direction of increasing θ, and $\hat{\mathbf{r}}$ is in the direction of increasing r.

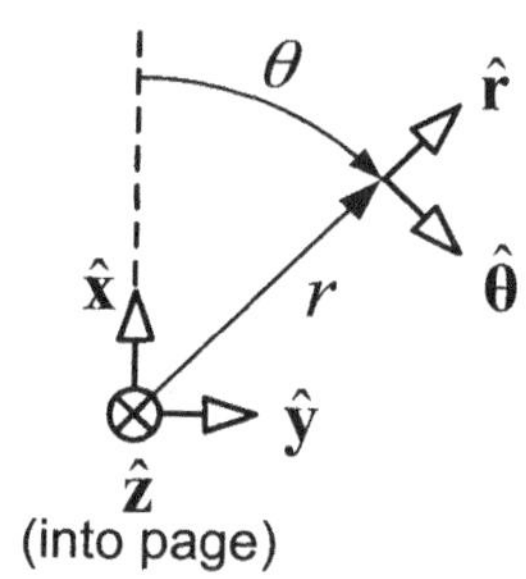

FIGURE 1.6. Cylindrical coordinate system used in Figure 1.5.

Referring back to Figure 1.5, for $i > 0$, the current in side a of the loop is going into the page (denoted by $\otimes$) and then comes out of the page (denoted by $\odot$) on side a'. Thus, on side a, $\vec{\boldsymbol{\ell}} = \ell_1 \hat{\mathbf{z}}$ (as $\vec{\boldsymbol{\ell}}$ points in the direction of positive current flow) and the magnetic force $\vec{\mathbf{F}}_{\text{side } a}$ on side a

is then

$$\begin{aligned}\vec{\mathbf{F}}_{\text{side } a} &= i\vec{\boldsymbol{\ell}}\times\vec{\mathbf{B}}\\ &= i(\ell_1\hat{\mathbf{z}})\times(B\hat{\mathbf{r}})\\ &= i\ell_1 B\hat{\boldsymbol{\theta}}\end{aligned}$$

which is tangential to the motion as shown in Figure 1.5(b). The resulting torque is

$$\begin{aligned}\vec{\boldsymbol{\tau}}_{\text{side } a} &= (\ell_2/2)\hat{\mathbf{r}}\times\vec{\mathbf{F}}_{\text{side } a}\\ &= (\ell_2/2)i\ell_1 B\hat{\mathbf{r}}\times\hat{\boldsymbol{\theta}}\\ &= (\ell_2/2)i\ell_1 B\hat{\mathbf{z}}.\end{aligned}$$

Similarly, the magnetic force on side a' of the rotor loop is

$$\begin{aligned}\vec{\mathbf{F}}_{\text{side } a'} &= i\vec{\boldsymbol{\ell}}\times\vec{\mathbf{B}}\\ &= i(-\ell_1\hat{\mathbf{z}})\times(-B\hat{\mathbf{r}})\\ &= i\ell_1 B\hat{\boldsymbol{\theta}}\end{aligned}$$

so that the corresponding torque is then

$$\begin{aligned}\vec{\boldsymbol{\tau}}_{\text{side } a'} &= (\ell_2/2)\hat{\mathbf{r}}\times\vec{\mathbf{F}}_{\text{side } a'}\\ &= (\ell_2/2)i\ell_1 B\hat{\mathbf{r}}\times\hat{\boldsymbol{\theta}}\\ &= (\ell_2/2)i\ell_1 B\hat{\mathbf{z}}.\end{aligned}$$

The total torque on the rotor loop is then

$$\begin{aligned}\vec{\boldsymbol{\tau}}_m &= \vec{\boldsymbol{\tau}}_{\text{side } a}+\vec{\boldsymbol{\tau}}_{\text{side } a'}\\ &= 2(\ell_2/2)i\ell_1 B\hat{\mathbf{z}}\\ &= \ell_1\ell_2 Bi\hat{\mathbf{z}}.\end{aligned}$$

The torque points along the z axis, which is the axis of rotation. In scalar form,

$$\tau_m = K_T i$$

where $K_T \triangleq \ell_1\ell_2 B$. The force is proportional to the strength B of magnetic field $\vec{\mathbf{B}}$ in the air gap due to the permanent magnet.

In order to increase the strength of the magnetic field in the air gap, the permanent magnet can be replaced with a soft iron material with wire wound around the periphery of the magnetic material as shown in Figure 1.7(a). This winding is referred to as the *field winding*, and the current it carries is called the *field current*. In normal operation, the field current is held constant. The strength of the magnetic field in the air gap is then proportional to the field current i_f at lower current levels (i.e., $B = K_f i_f$) and then saturates as the current increases. This may be written as $B =$

$f(i_f)$ where $f(\cdot)$ is a saturation curve satisfying $f(0) = 0, f'(0) = K_f$ as shown in Figure 1.7(b).

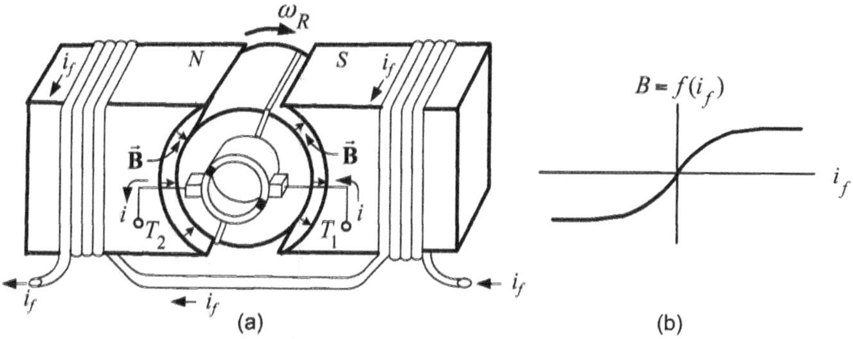

FIGURE 1.7. (a) DC motor with a field winding. (b) Radial magnetic field strength in the air gap produced by the field current.

1.2.2 Commutation of the Single-Loop Motor

The above derivation for the torque $\tau_m = K_T i$ assumes that the current in the side of the rotor loop[3] under the south pole face is into the page and the current in the side of the loop under the north pole face is out of the page as in Figure 1.8(a). In order to make this assumption valid, the direction of the current in the loop must be changed each time the rotor loop passes through the vertical.

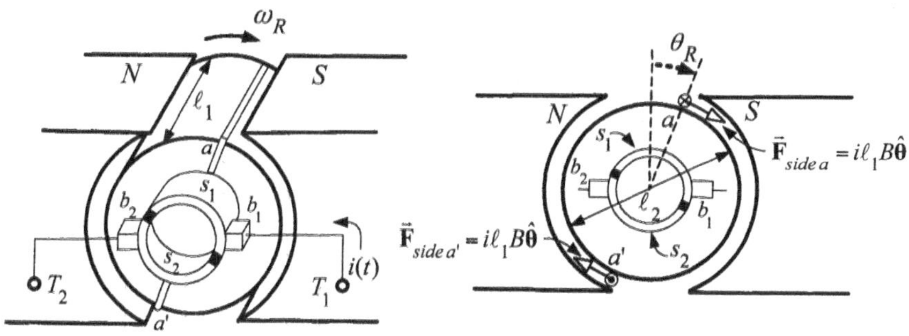

FIGURE 1.8. (a) $0 < \theta_R < \pi$. From *Electromagnetic and Electromechanical Machines*, 3rd edition, L. W. Matsch and J. Derald Morgan, 1986. Reprinted by permisson of John Wiley & Sons.

[3]The rotor loop is also referred to as the *armature* winding and the current in it as the *armature* current.

The process of changing the direction of the current is referred to as *commutation* and is done at $\theta_R = 0$ and $\theta_R = \pi$ through the use of the slip rings s_1, s_2 and brushes b_1, b_2 drawn in Figure 1.8. The slip rings are rigidly attached to the loop and thus rotate with it. The brushes are fixed in space with the slip rings making a sliding electrical contact with the brushes as the loop rotates.

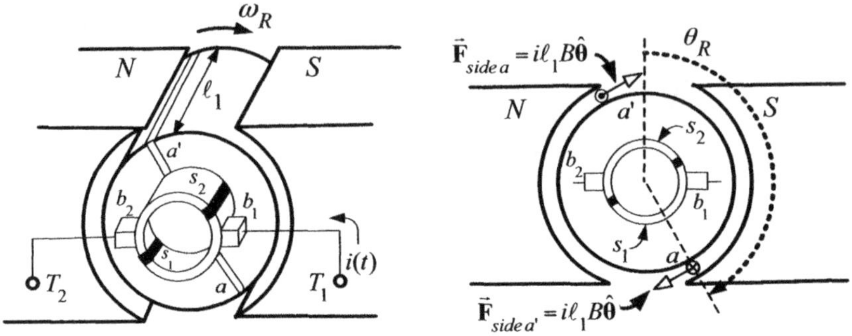

Figure 1.8(b) Rotor loop just prior to commutation where $0 < \theta_R < \pi$.

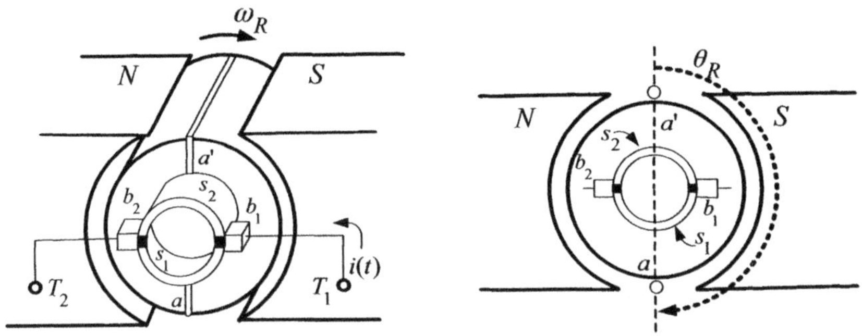

Figure 1.8(c) The ends of the rotor loop are shorted when $\theta_R = \pi$.

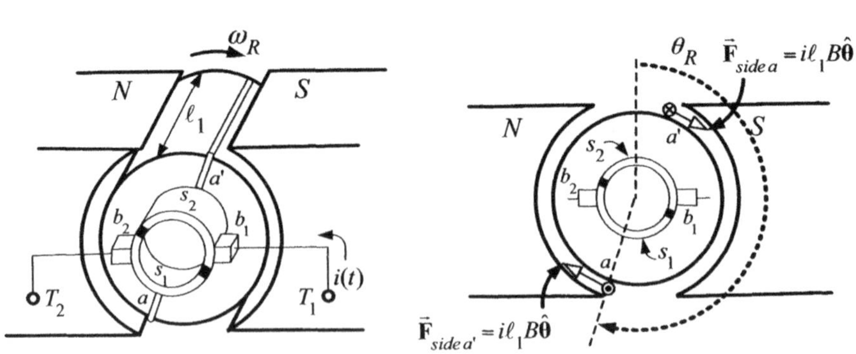

Figure 1.8(d) Rotor loop just after commutation where $\pi < \theta_R < 2\pi$.

To see how the commutation of the current is accomplished using the brushes and slip rings, consider the sequence of Figures 1.8(a)–(d). As shown in Figure 1.8(a), the current goes through brush b_1 into the slip ring s_1. From there, it travels down (into the page $\otimes$) side a of the loop, comes back up side a' (out of the page $\odot$) into the slip ring s_2, and, finally, comes out the brush b_2. Note that side a of the loop is under the south pole face while side a' is under the north pole face. Figure 1.8(b) shows the rotor loop just before commutation where the same comments as in Figure 1.8(a) apply.

Figure 1.8(c) shows that when $\theta_R = \pi$, the slip rings at the ends of the loop are shorted together by the brushes forcing the current in the loop to drop to zero. Subsequently, as shown in Figure 1.8(d), with $\pi < \theta_R < 2\pi$, the current is now going through brush b_1 into slip ring s_2. From there, the current travels down (into the page $\otimes$) side a' of the loop and comes back up (out of the page $\odot$) side a. In other words, the current has *reversed* its direction in the loop from that in Figures 1.8(a) and 1.8(b). This is precisely what is desired, as side a is now under the north pole face and side a' is under the south pole face. As a result of the brushes and slip rings, the current direction in the loop is reversed every half-turn.

1.3 Faraday's Law

Figure 1.9 shows a magnet moving upwards into a wire loop producing a changing magnetic flux in the loop.

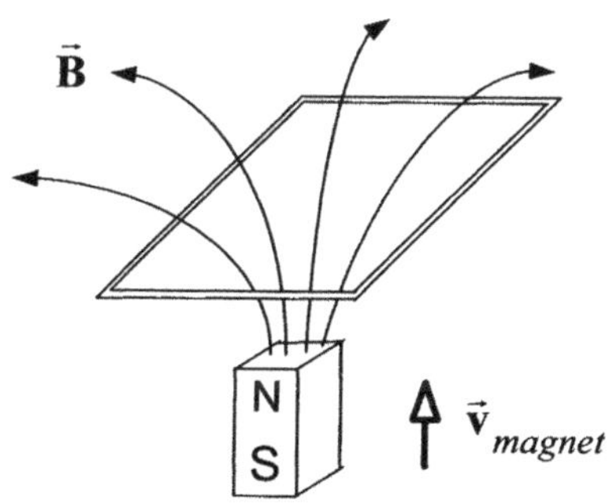

FIGURE 1.9. A magnet moving upwards produces a changing flux in the loop which in turn results in an induced emf and current in the loop.

Recall that a changing flux within a loop produces an induced *electro-*

motive force (emf) ξ in the loop according to Faraday's law.[4] That is,

$$\xi = -d\phi/dt$$

where

$$\phi = \int_S \vec{\mathbf{B}} \cdot d\vec{\mathbf{S}}$$

is the flux in the loop and S is any surface with the loop as its boundary. Faraday's law is now reviewed in some detail.

1.3.1 *The Surface Element Vector* $d\vec{\mathbf{S}}$

The surface element $d\vec{\mathbf{S}}$ is a vector whose magnitude is a differential (small) element of area dS and whose direction is normal (perpendicular) to the surface element. As there are two possibilities for the normal to the surface, one must choose the normal in a consistent manner. In particular, depending on the particular normal chosen, a convention is used to characterize the positive and negative directions of travel around the surface boundary. To describe this, consider Figure 1.10(a) which shows a small surface element with the normal direction taken to be up in the positive z direction. In this case, with $\hat{\mathbf{n}} = \hat{\mathbf{z}}$, $dS = dxdy$, the surface element vector is defined by

$$d\vec{\mathbf{S}} \triangleq dxdy\hat{\mathbf{z}}.$$

The corresponding direction of travel around the surface boundary is indicated by the curved arrow in the figure.

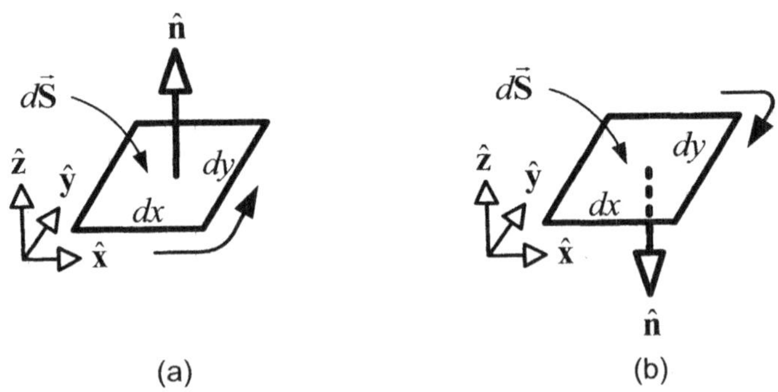

FIGURE 1.10. (a) Positive direction of travel around a surface element with the normal up. (b) Positive direction of travel around a surface element with the normal down.

[4] ξ is the Greek letter "xi" and is pronounced "ksi".

In Figure 1.10(b) a surface element with the normal direction taken to be down in the negative z direction is shown. In this case $\hat{\mathbf{n}} = -\hat{\mathbf{z}}, dS = dxdy$ so that the surface element vector is defined as

$$d\vec{\mathbf{S}} = -dxdy\hat{\mathbf{z}}.$$

The direction of positive travel around the surface element is indicated by the curved arrow in Figure 1.10(b) and is opposite to that of Figure 1.10(a).

As illustrated in Figure 1.10, the vector differential surface element $d\vec{\mathbf{S}}$ is defined to be a vector whose magnitude is the area of the differential surface element and whose direction is normal to the surface. One may choose either normal, and the corresponding direction of positive travel around the surface is then determined.

Two surface elements may be connected together as in Figure 1.11 and travel around the total surface is defined as shown. Note that along the common boundary of the two joined surface elements, the directions of travel "cancel" out each other, resulting in a net travel path around both surface elements. The normals for the surface elements must both be up or both be down; that is, the normal must be continuous as one goes from one surface element to the next.

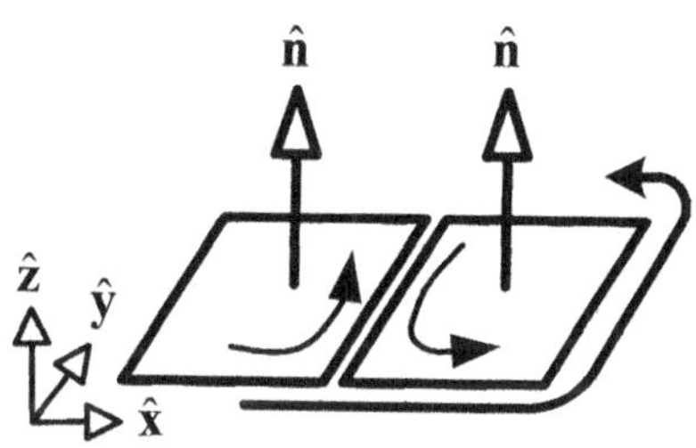

FIGURE 1.11. Positive direction of travel around two joined surface elements.

1.3.2 *Interpreting the Sign of* ξ

The interpretation of positive and negative values of the induced electromotive force ξ is now explained. Faraday's law says that the induced emf (voltage) in a loop is given by

$$\xi = -d\phi/dt$$

where

$$\phi = \int_S \vec{\mathbf{B}} \cdot d\vec{\mathbf{S}}.$$

If $\xi > 0$, the induced emf will force current in the positive direction of travel around the surface while if $\xi < 0$, the induced emf will force current in the opposite direction. As illustrated in problems 1 and 2, this sign convention for Faraday's law is just a precise mathematical way of describing Lenz's law: *"In all cases of electromagnetic induction, an induced voltage will cause a current to flow in a closed circuit in such a direction that the magnetic field which is caused by that current will oppose the change that produced the current"* (pages 873–877 of Ref. [34]).

Faraday's law is now illustrated by some examples. Specifically, it is used to compute the induced emf in the linear DC machine, the induced emf in the single-loop machine and the self-induced voltage in the single-loop machine.

1.3.3 Back Emf in a Linear DC Machine

Figure 1.12 shows the linear DC machine where the back emf it generates is now computed. The magnetic field is constant and points into the page, that is, $\vec{\mathbf{B}} = -B\hat{\mathbf{z}}$, where $B > 0$. The magnetic force on the bar is $\vec{\mathbf{F}}_{\text{magnetic}} = i\ell B\hat{\mathbf{x}}$. To compute the induced voltage in the loop of the circuit, let $\hat{\mathbf{n}} = \hat{\mathbf{z}}$ be the normal to the surface so that $d\vec{\mathbf{S}} = dxdy\hat{\mathbf{z}}$, where $dS = dxdy$.

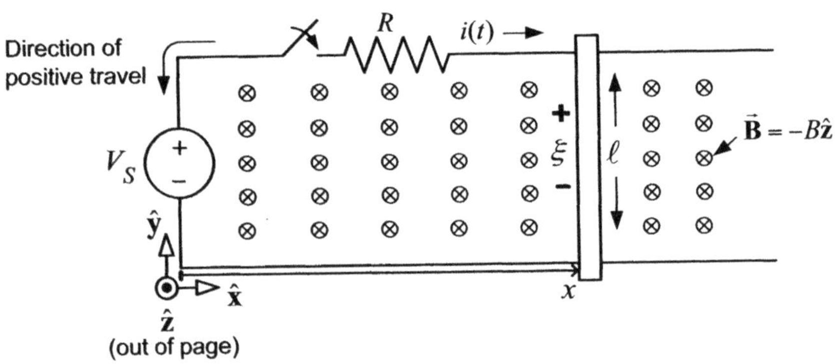

FIGURE 1.12. With $d\vec{\mathbf{S}} = dxdy\hat{\mathbf{z}}$, the direction of positive travel around the flux surface is in the counterclockwise direction.

Then

$$\phi = \int_S \vec{\mathbf{B}} \cdot d\vec{\mathbf{S}} = \int_0^\ell \int_0^x (-B\hat{\mathbf{z}}) \cdot (dxdy\hat{\mathbf{z}}) = \int_0^\ell \int_0^x -Bdxdy = -B\ell x.$$

The induced (back) emf is therefore given by

$$\xi = -d\phi/dt = -d(-B\ell x)/dt = B\ell v.$$

In the flux computation, the normal for the surface was taken to be in $+\hat{\mathbf{z}}$ direction. By putting together the differential flux surfaces $d\vec{\mathbf{S}}$ in a fashion similar to Figure 1.11, the positive direction of travel around the surface is counterclockwise around the loop as indicated in Figure 1.12. Here the sign conventions for source voltage V_S and the back emf ξ are opposite so that, as the back emf $\xi = B\ell v > 0$, it is *opposing* the applied source voltage V_S.

Remark $\phi = -B\ell x$ is the flux in the circuit due to the *external* magnetic field $\vec{\mathbf{B}} = -B\hat{\mathbf{z}}$. There is also a flux $\psi = Li$ due to the current i in the circuit. For this example, the inductance is small and one just sets $L = 0$.

Electromechanical Energy Conversion

As the back emf $\xi = Blv$ opposes the current i, electrical power is being absorbed by this back emf. Specifically, the electrical power absorbed by the back emf is $i\xi = iB\ell v$ while the mechanical power produced is $F_{\text{magnetic}}v = i\ell Bv$. That is, the electrical power absorbed by the back emf reappears as mechanical power, as it must by conservation of energy. Another way to view this is to note that $V_S i$ is the electrical power delivered by the source and, as $V_S - B\ell v = Ri$, one may write

$$V_S i = Ri^2 + i(B\ell v) = Ri^2 + F_{\text{magnetic}}v.$$

In words, the power from the source $V_S i$ is dissipated as heat in the resistance R while the rest is converted into mechanical power.

Equations of Motion for the Linear DC Machine

The equations of motion for the bar in the linear DC machine are now derived. With the inductance L of the circuit loop taken to be zero, m_ℓ the mass of the bar, f the coefficient of viscous friction, it follows that

$$\begin{aligned} V_S - B\ell v &= Ri \\ m_\ell \frac{dv}{dt} &= i\ell B - fv. \end{aligned}$$

Eliminating the current i, one obtains

$$m_\ell \frac{d^2x}{dt^2} = \ell B(V_S - B\ell v)/R - fv = -\left(\frac{B^2\ell^2}{R} + f\right)\frac{dx}{dt} + \frac{\ell B}{R}V_S$$

or

$$m_\ell \frac{d^2x}{dt^2} + \left(\frac{B^2\ell^2}{R} + f\right)\frac{dx}{dt} = \frac{\ell B}{R}V_S.$$

This is the equation of motion for the bar with V_S as the control input and the position x at the measured output.

1.3.4 Back Emf in the Single-Loop Motor

The back emf induced in the single loop motor by the external magnetic field of the permanent magnet is now computed. To do so, consider the flux surface for the rotor loop shown in Figure 1.13. The surface is a half-cylinder of radius $\ell_2/2$ and length ℓ_1 with the rotor loop as its boundary. The cylindrical surface is in the air gap, where the magnetic field is known to be radially directed and constant in magnitude, that is,

$$\vec{\mathbf{B}} = \begin{cases} +B\hat{\mathbf{r}} & \text{for } 0 < \theta < \pi \\ -B\hat{\mathbf{r}} & \text{for } \pi < \theta < 2\pi. \end{cases} \tag{1.1}$$

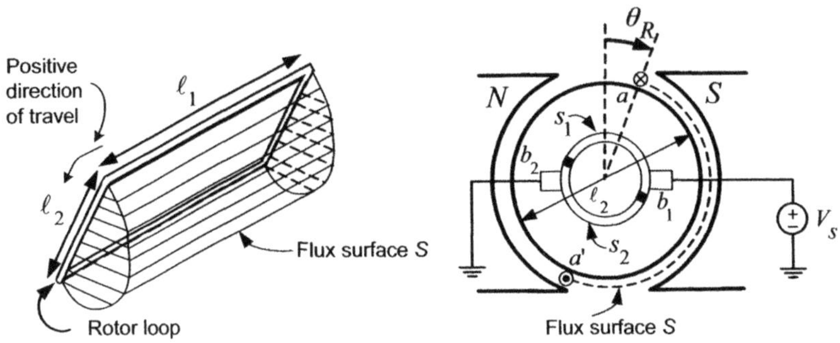

FIGURE 1.13. Flux surface for the single loop motor.

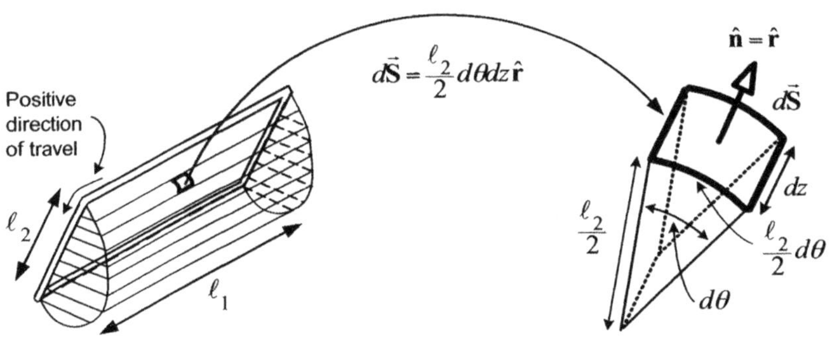

FIGURE 1.14. Surface element vector for the flux surface of Figure 1.13. The positive direction of travel around this surface is indicated by the curved arrow.

On the cylindrical part of the surface, the surface element is chosen as

$$d\vec{\mathbf{S}} = (\ell_2/2) d\theta dz \hat{\mathbf{r}}$$

which is directed outward from the axis of the cylinder as illustrated in Figure 1.14. The corresponding direction of positive travel is also indicated in Figure 1.14. On the two ends (half-disks) of the cylindrical surface, the $\vec{\mathbf{B}}$ field is quite weak making the flux through these two half-disks negligible.

Then, neglecting the flux through the two ends of the surface, the flux $\phi(\theta_R)$ for $0 < \theta_R < \pi$ is given by

$$\begin{aligned}\phi(\theta_R) &= \int_S \vec{\mathbf{B}} \cdot d\vec{\mathbf{S}} \\ &= \int_0^{\ell_1}\int_{\theta=\theta_R}^{\theta=\pi} (B\hat{\mathbf{r}}) \cdot (\frac{\ell_2}{2} d\theta dz \hat{\mathbf{r}}) + \int_0^{\ell_1}\int_{\theta=\pi}^{\theta=\pi+\theta_R} (-B\hat{\mathbf{r}}) \cdot (\frac{\ell_2}{2} d\theta dz \hat{\mathbf{r}}) \\ &= \int_0^{\ell_1}\int_{\theta=\theta_R}^{\theta=\pi} B\frac{\ell_2}{2} d\theta dz + \int_0^{\ell_1}\int_{\theta=\pi}^{\theta=\pi+\theta_R} -B\frac{\ell_2}{2} d\theta dz \\ &= \frac{\ell_1\ell_2 B}{2}(\pi - \theta_R) - \frac{\ell_1\ell_2 B}{2}\theta_R \\ &= -\ell_1\ell_2 B\left(\theta_R - \frac{\pi}{2}\right). \end{aligned} \tag{1.2}$$

This derivation is based on the fact that the $\vec{\mathbf{B}}$ field is directed radially outward over the length $(\ell_2/2)(\pi - \theta_R)$ and radially inward over the length $(\ell_2/2)\theta_R$ (see Figure 1.13). In problem 7, the reader is asked to show that

$$\phi(\theta_R) = -\ell_1\ell_2 B\left(\theta_R - \pi/2 - \pi\right) \text{ for } \pi < \theta_R < 2\pi. \tag{1.3}$$

A plot of the flux versus the rotor angle θ_R is given in Figure 1.15.

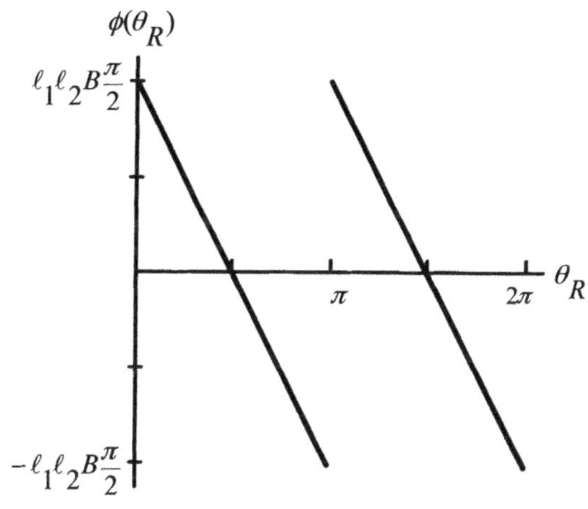

FIGURE 1.15. The rotor flux $\phi(\theta_R)$ due to the external magnetic field vs. θ_R.

Equations (1.2) and (1.3) may be combined into one expression as[5]

$$\phi(\theta_R) = -\ell_1\ell_2 B\left(\theta_R \bmod \pi - \frac{\pi}{2}\right) \tag{1.4}$$

[5] $\theta_R \bmod \pi$ is the remainder after θ_R is divided by π. For example, $\theta_R = 5\pi/2 = 4 \times \pi + \pi/2$ so that $5\pi/2 \bmod \pi = \pi/2$.

which is a correct expression for any angle θ_R. By (1.2) and (1.3), the induced emf in the rotor loop is calculated as

$$\xi = -\frac{d\phi}{dt} = (\ell_1\ell_2 B)\frac{d\theta_R}{dt} = K_b\omega_R$$

where $K_b \triangleq \ell_1\ell_2 B$ is called the back emf constant.

The total emf in the rotor loop due to the voltage source V_S and external magnetic field is $V_S - K_b\omega_R$. How does one know to subtract ξ from the applied voltage V_S? As shown in Figure 1.14, the positive direction of travel around the loop is in *opposition* to V_S, so that if $\xi > 0$, it is opposing the applied voltage V_S. The standard terminology is to call $\xi \triangleq K_b\omega_R$ the *back emf* of the motor.

1.3.5 Self-Induced Emf in the Single-Loop Motor

The computation of the flux in the rotor loop produced by its own (armature) *current* is now done. To do so, consider the flux surface shown in Figure 1.16.

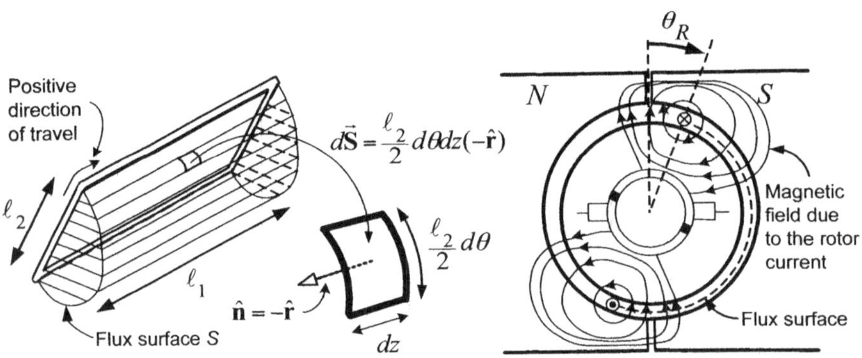

FIGURE 1.16. Computation of the inductance of the rotor loop. The surface element vector is $d\vec{\mathbf{S}} = -r_R d\theta dz \hat{\mathbf{r}}$ with a resulting positive direction of travel as indicated by the curved arrow. This direction coincides with the direction of positive current, that is, $i > 0$.

With reference to Figure 1.16, note that the magnetic field on the flux surface *due to the armature current* has the form

$$\vec{\mathbf{B}}(r_R, \theta - \theta_R, i) = iK(r_R, \theta - \theta_R)(-\hat{\mathbf{r}})$$

where

$$\begin{aligned} K(r_R, \theta - \theta_R) &> 0 \quad \text{for } 0 \le \theta - \theta_R \le \pi \\ K(r_R, \theta - \theta_R) &< 0 \quad \text{for } \pi \le \theta - \theta_R \le 2\pi. \end{aligned}$$

The exact expression for $K(r_R, \theta - \theta_R)$ is not easy to compute, but it is not needed for the analysis here. Rather, the point is that with $i > 0$, the magnetic field $\vec{\mathbf{B}}(r_R, \theta - \theta_R, i)$ due to the current in the rotor loop is radially in on the flux surface shown in Figure 1.16, that is, for $\theta_R \le \theta \le \theta_R + \pi$. For convenience, the surface element is chosen to be $d\vec{\mathbf{S}} = r_R d\theta dz(-\hat{\mathbf{r}})$ so that positive direction of travel around the surface coincides with the positive direction of the current i in the loop. The flux ψ in the rotor loop is then computed as[6]

$$\begin{aligned}
\psi(i) &= \int_S \vec{\mathbf{B}} \cdot d\vec{\mathbf{S}} = \int_{\theta_R}^{\theta_R+\pi} \int_0^{\ell_1} iK(r_R, \theta - \theta_R)\,(-\hat{\mathbf{r}}) \cdot (-r_R d\theta dz \hat{\mathbf{r}}) \\
&= i \int_{\theta_R}^{\theta_R+\pi} \int_0^{\ell_1} K(r_R, \theta - \theta_R) r_R d\theta dz \\
&= Li
\end{aligned}$$

where

$$L \triangleq \int_{\theta_R}^{\theta_R+\pi} \int_0^{\ell_1} K(r_R, \theta - \theta_R) r_R d\theta dz > 0. \tag{1.5}$$

This last equation just says the flux in the loop (due to the current in the loop) is proportional to the current i in the loop. The proportionality constant L is the called the *inductance* of the loop.[7] If $-d\psi/dt = -Ldi/dt > 0$, then the induced emf will force current into the page $\otimes$ on side a and out of the page $\odot$ of side a' in Figure 1.16. That is, this induced emf has the same sign convention as the armature current i and the source voltage V_S.

With the rotor locked at some angle θ_R so that the external magnetic field cannot induce an emf in the rotor loop, the equation describing the current i in the rotor loop is given by Kirchhoff's voltage law

$$V_S - Ri - L\frac{di}{dt} = 0$$

or

$$V_S = Ri + L\frac{di}{dt}.$$

Here R is the resistance of the loop and V_S is the source voltage applied to the loop. The loop and its equivalent circuit are shown in Figure 1.17.

---

[6]The notation ψ is used to distinguish this flux from the flux ϕ in the loop due to the *external* permanent magnet. However, the total flux using an inward normal would be $\psi - \phi$ as the *outward* normal was used to compute ϕ in Section 1.3.4.

[7]It appears from equation (1.5) and Figure 1.16 that L can vary with θ_R. However, in an actual motor, there are loops spread evenly around the complete periphery of the rotor and, due to symmetry, the total self-inductance does not depend on θ_R.

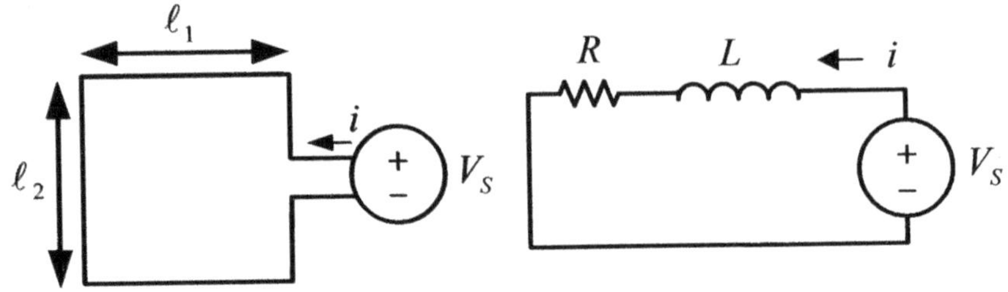

FIGURE 1.17. Left: Rotor loop. Right: Equivalent circuit.

The reader should convince himself/herself that Lenz's law holds as it must. For example, suppose a voltage $V_S > 0$ is applied to the loop resulting in both $i > 0$ and $di/dt > 0$, that is, the flux $\psi = Li$ is positive and increasing. The induced voltage is $-Ldi/dt < 0$ and opposes the current i producing the increasing flux $\psi = Li$. In this circumstance, the voltage source V_S is forcing the current i against this induced voltage $-Ldi/dt$ and the power absorbed by the induced voltage is $-iLdi/dt = -d\left(\frac{1}{2}Li^2\right)/dt$. This power is stored in the energy $\frac{1}{2}Li^2$ of the magnetic field surrounding the loop.

Arcing Between the Commutator and Brushes

Suppose the single-loop motor is rotating at constant speed ω_0 with a constant current i_0 in the rotating loop. Let L be the inductance of the loop. Now, every half-turn, the current in the loop reverses direction as shown in Figures 1.8(b)–(d). During this commutation, the current in the loop goes from i_0 to 0 to $-i_0$ (or vice versa) with a corresponding change in the loop's flux given by $\Delta\psi = L(-i_0) - Li_0 = -2Li_0$. By Faraday's law, the self-induced emf is then $-\Delta\psi/\Delta t = 2Li_0/\Delta t$ where Δt is the time for the current to change direction. Note that this time Δt decreases as the motor speed increases, so that, even if L is small, the induced emf in the loop (due to the reversal of current in the loop) can be quite large at high motor speeds. Large electric fields are produced by the induced voltage Ldi/dt when the loop is shorted which in turn ionizes the surrounding air. As the free electrons collide and recombine with the ionized air, light is given off and seen as arcing or sparking. These large voltages which cause the arcing between the slip rings and the brushes can damage the brushes as well as produce unwanted transient currents in the armature circuit.

1.4 Dynamic Equations of the DC Motor

Based on the simple single-loop DC motor analyzed above, the complete set of equations for a DC motor can be found. The total emf (voltage) in the loop due to the voltage source V_S, the external permanent magnet and the changing current i in the rotor loop is

$$V_S - K_b\omega_R - L\frac{di}{dt}.$$

This voltage goes into building up the current in the loop against the loop's resistance, that is,

$$V_S - K_b\omega_R - L\frac{di}{dt} = Ri$$

or

$$L\frac{di}{dt} = -Ri - K_b\omega_R + V_S.$$

This relationship is often illustrated by the equivalent circuit given in Figure 1.18. Recall that the torque τ_m on the loop due to the external magnetic field acting on the current in the loop is

$$\tau_m = K_T i$$

where $K_T \triangleq \ell_1\ell_2 B$ is called the torque constant. By connecting a shaft and gears to one end of the loop, this motor torque can be used to do work (lift weight, etc.). Let $-f\omega_R$ model the friction torque (due to the brushes, bearings, etc.) where f is the coefficient of viscous friction and let τ_L be the load torque (e.g., due to a weight being lifted).

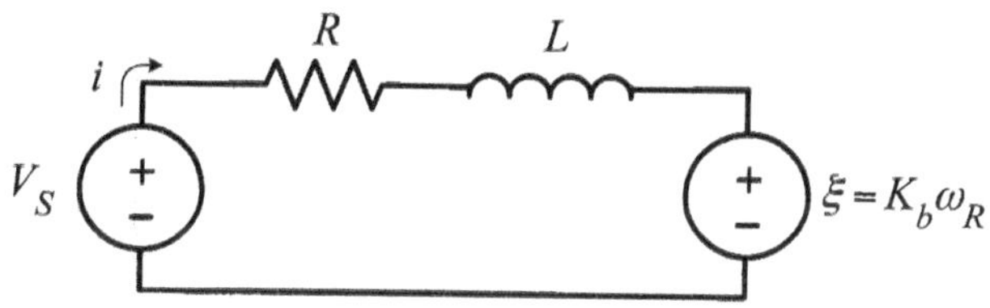

FIGURE 1.18. Equivalent circuit of the armature electrical dynamics.

Then, by Newton's law,

$$\tau_m - \tau_L - f\omega_R = J\frac{d\omega_R}{dt}$$

where J is the moment of inertia of the rotor (See the appendix of this chapter). The system of equations characterizing the DC motor is then

$$\begin{aligned} L\frac{di}{dt} &= -Ri - K_b\omega_R + V_S \\ J\frac{d\omega_R}{dt} &= K_T i - f\omega_R - \tau_L \\ \frac{d\theta_R}{dt} &= \omega_R. \end{aligned} \tag{1.6}$$

A picture of a DC motor servo system and its associated schematic is shown in Figure 1.19. In the schematic, R is the resistance of the rotor loop, L is the inductance of the rotor loop, $\xi = K_b\omega_R$ is the back emf, $\tau_m = K_T i$ is the motor torque, J is the rotor moment of inertia, and f is the coefficient of viscous friction. The positive directions for τ_m, θ_R, and τ_L are indicated by the curved arrows. The fact that the curved arrow for τ_L is opposite to that of τ_m just means that if the load torque is positive then it opposes a positive motor torque τ_m.

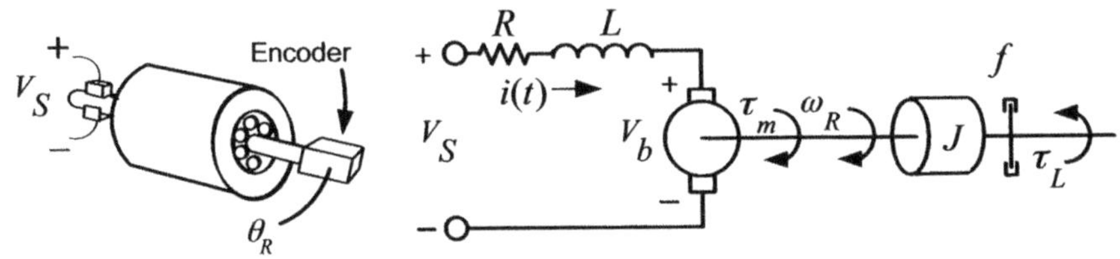

FIGURE 1.19. DC motor drawing and schematic.

Electromechanical Energy Conversion

The mechanical power produced by the DC motor is $\tau_m\omega_R = K_T i\omega_R = i\ell_1\ell_2 B\omega_R$ while the electrical power absorbed by the back emf is $i\xi = iK_b\omega_R = i\ell_1\ell_2 B\omega_R$. The fact that $K_T = K_b = \ell_1\ell_2 B$ must be for conservation of energy to hold. That is, the electrical power absorbed by the back emf equals (is converted to) the mechanical power produced. Another way to view this energy conversion is to write the electrical equation as

$$V_S = Ri + L\frac{di}{dt} + \xi.$$

The power out of the voltage source $V_S(t)$ is given by

$$\begin{aligned} V_S(t)i(t) &= Ri^2(t) + Li\frac{di}{dt} + iK_b\omega_R \\ &= Ri^2 + \frac{d}{dt}\left(\frac{1}{2}Li^2\right) + K_T i\omega_R \\ &= Ri^2 + \frac{d}{dt}\left(\frac{1}{2}Li^2\right) + \tau_m\omega_R. \end{aligned}$$

Thus the power $V_S(t)i(t)$ delivered by the source goes into heat loss in the resistance R, into stored magnetic energy in the inductance L of the loop and the amount $i\xi$ goes into the mechanical energy $\tau_m\omega_R$.

Remark *Voltage and Current Limits*

The amount of voltage V_S that may be applied to the input terminals T_1, T_2 of the motor is limited by capabilities of the amplifier supplying the voltage, that is, $|V_S| \le V_{\max}$. Let $V_c(t)$ be the voltage commanded to the amplifier, then the actual voltage V_S out of the amplifier to the motor is limited by $V_{\max}$ as illustrated in Figure 1.20.

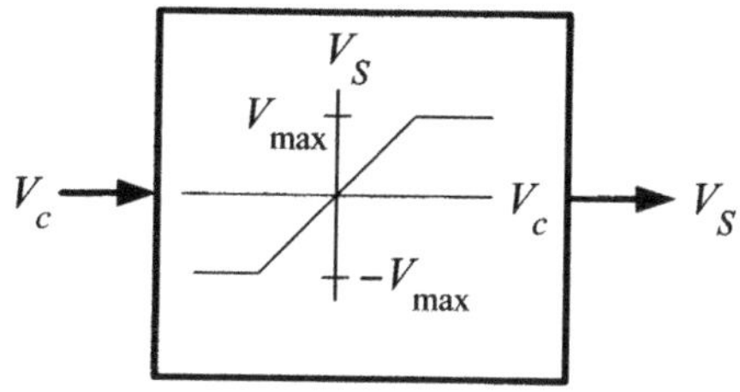

FIGURE 1.20. Saturation model of an amplifier.

In addition, there is a limit to the amount of current the rotating loop can handle before overheating or causing problems with commutation as previously mentioned. Typically there are two current limits (ratings), the *continuous* current limit $I_{\max\_cont}$ and the *peak* current limit $I_{\max\_peak}$. The continuous current limit $I_{\max\_cont}$ is the amount of current the motor can handle if left in use indefinitely. That is, the amount of heat dissipated in the rotor windings due to ohmic losses is equal to the amount of heat taken away by thermal conduction through the brushes and thermal convection with the air so as to be in a thermal equilibrium. The peak current limit $I_{\max\_peak}$ is the amount of current the motor can handle for short periods of time (typically only a few seconds).

1.5 Microscopic Viewpoint

Additional insight into the back emf ξ is found by calculating it from a microscopic point of view using the ideas given in Ref. [34] (page 887). To illustrate this approach, the back emf in the linear DC machine is recomputed from the microscopic point of view. To proceed, recall that the magnetic force on a charged particle q is $\vec{\mathbf{F}}_{\text{magnetic}} = q\vec{\mathbf{v}} \times \vec{\mathbf{B}}$, where $\vec{\mathbf{v}}$ is the velocity of the charge (see Ref. [34], page 816).

Example *A Linear DC Machine*

In this example, the linear DC machine is reanalyzed from the microscopic point of view. As before, $\vec{\mathbf{B}} = -B\hat{\mathbf{z}}$ where $B > 0$. Suppose the motor (bar) is moving to the right with a constant speed v_m. Each charge q in the sliding bar has total velocity $\vec{\mathbf{v}} = v_m\hat{\mathbf{x}} - v_d\hat{\mathbf{y}}$, where v_d is the drift speed of the charges down the wire. The magnetic force on the charge q is

$$\vec{\mathbf{F}}_{\text{magnetic}} = q\vec{\mathbf{v}}\times\vec{\mathbf{B}} = q(v_m\hat{\mathbf{x}} - v_d\hat{\mathbf{y}}) \times (-B\hat{\mathbf{z}}) = qv_mB\hat{\mathbf{y}} + qv_dB\hat{\mathbf{x}}.$$

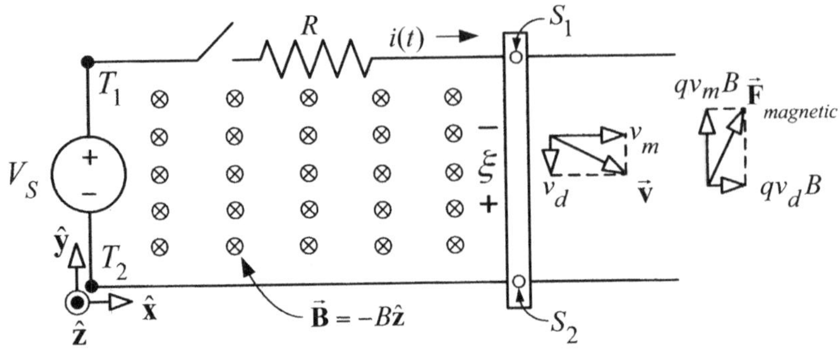

FIGURE 1.21. Linear DC machine.

Now, the component of force $qv_dB\hat{\mathbf{x}}$ perpendicular to the bar causes the bar to move to the right and the component $qv_mB\hat{\mathbf{y}}$ along the bar opposes the current flow. The source voltage V_S sets up an electric field $\vec{\mathbf{E}}_S$ in the bar to overcome the magnetic force $qv_mB\hat{\mathbf{y}}$ so as to make the current flow (setup the drift velocity v_d of the charge carriers against the resistance of the bar). In more detail, with T_1 and T_2 the upper and lower terminals of the source voltage, respectively, and S_1 and S_2 the upper and lower sliding contact points, respectively, the source voltage is given as

$$V_S = \int_{\mathrm{T}_1-\mathrm{S}_1-\mathrm{S}_2-\mathrm{T}_2} \vec{\mathbf{E}}_S \cdot d\vec{\ell}.$$

The quantity $q\vec{\mathbf{E}}_S$ is the force on each charge carrier and qV_S is the energy given to the charge carrier by the source voltage as the charge goes around the loop. There is also a component of the magnetic force on the charge carrier that opposes the electric field $\vec{\mathbf{E}}_S$. The energy per unit charge ξ that the magnetic force takes from the charge carrier as it goes down the bar from S_1 to S_2 is given by

$$\begin{aligned}\xi &= \frac{1}{q}\int_{S_1}^{S_2} \vec{\mathbf{F}}_{\text{magnetic}} \cdot d\vec{\ell} = \frac{1}{q}\int_{S_1}^{S_2} q(\vec{\mathbf{v}}\times\vec{\mathbf{B}}) \cdot d\vec{\ell} = \int_0^{\ell} (v_dB\hat{\mathbf{x}}+v_mB\hat{\mathbf{y}}) \cdot (-d\ell\hat{\mathbf{y}}) \\ &= -v_mB\ell.\end{aligned}$$

The fact that ξ is *negative* just indicates that the magnetic force is taking energy out of the charge carrier as it goes down the bar from S_1 to S_2. (Note: The sign convention for ξ shown in Figure 1.21 is reversed from that of Figure 1.12.) This energy per unit charge ξ taken from each charge carrier by the magnetic force as it goes around the loop is called the induced emf. The voltage V_S was computed by integrating $\vec{\mathbf{E}}_S$ in the *clockwise* direction $\mathrm{T}_1-\mathrm{S}_1-\mathrm{S}_2-\mathrm{T}_2$ around the loop, and ξ by integrating $\vec{\mathbf{v}}\times\vec{\mathbf{B}}$ also in the *clockwise* direction down the bar from S_1 to S_2; that is, they both have the same sign convention. This is in contrast to the macroscopic case where V_S and ξ had opposite sign conventions resulting in $\xi = v_m B\ell$ being positive. However, the same (physical) result occurs as in the macroscopic case.

In general, the emf in a loop is defined as the integral of the force per unit charge around the loop. The total emf is the sum of the source voltage and the induced emf. This total emf goes into producing the current, that is,

$$V_S + \xi = V_S - v_m \ell B = Ri$$

where an identical equation was found in the macroscopic case using Faraday's law. The *total* magnetic force on all the charge carriers in the bar in the $\hat{\mathbf{x}}$ direction is given by

$$q(NS\ell)v_d B\hat{\mathbf{x}}$$

where N is the number of charge carriers/volume and S is the cross sectional area of the sliding bar. That is, $NS\ell$ is the total number of charge carriers in the sliding bar each experiencing the force $qv_d B\hat{\mathbf{x}}$. As illustrated in Figure 1.22, in a time Δt, the charges in the volume $NS(v_d\Delta t)$ have moved along the conductor past the point P in Figure 1.22.

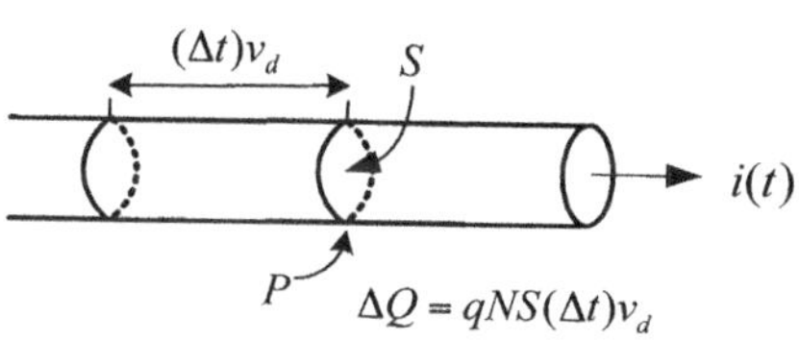

FIGURE 1.22. In the time Δt, the amount of charge $\Delta Q = qNS(v_d\Delta t)$ has moved past the point P resulting in the current $i = \Delta Q/\Delta t = qNSv_d$ in the bar.

That is, the amount of charge $\Delta Q = qNS(v_d\Delta t)$ has moved past the point P in the time Δt resulting in the current $i = \Delta Q/\Delta t = qNSv_d$ in the conductor. Consequently, the total magnetic force on the bar may be rewritten as

$$q(NS\ell)v_d B\hat{\mathbf{x}} = (qNSv_d)\ell B\hat{\mathbf{x}} = i\ell B\hat{\mathbf{x}}$$

which is identical to the expression derived from the macroscopic point of view.

1.5.1 *Microscopic Viewpoint of the Single-Loop DC Motor*

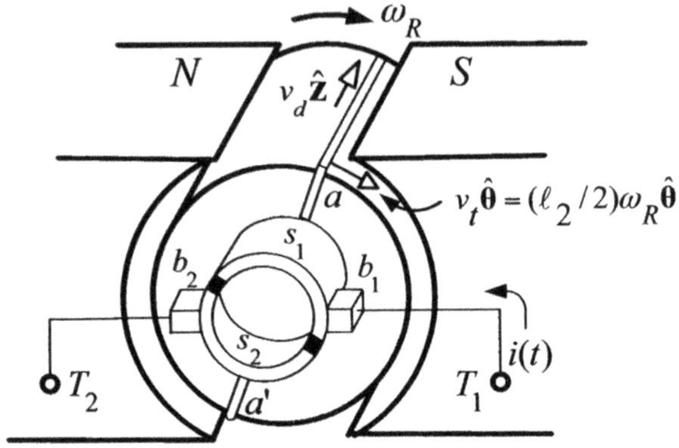

FIGURE 1.23. Single loop DC motor. Adapted from Ref. [21].

The back emf in the single-loop DC motor of Figure 1.23 is now computed from the microscopic point of view. With the loop rotating at the angular speed ω_R, the velocity of the charge carriers that make up the current is given by

$$\vec{\mathbf{v}} = \begin{cases} v_t\hat{\boldsymbol{\theta}} + v_d\hat{\mathbf{z}} & \text{for side } a \\ v_t\hat{\boldsymbol{\theta}} - v_d\hat{\mathbf{z}} & \text{for side } a' \end{cases}$$

where v_d is the drift speed of the charge carriers along the wire and $v_t = (\ell_2/2)\omega_R$ is the tangential velocity due to the rotating loop. Recall that the drift speed has the same sign as the current, that is, $v_d > 0$ for $i > 0$. Also recall that the angular velocity is written as $\vec{\boldsymbol{\omega}}_R = \omega_R\hat{\mathbf{z}}$ where $\hat{\mathbf{z}}$ is the axis the motor is turning about. The magnetic force per unit charge $\vec{\mathbf{F}}_{\text{magnetic}}/q$ on the charge carriers on the axial sides of the loop is

$$\vec{\mathbf{F}}_{\text{magnetic}}/q = \vec{\mathbf{v}}\times\vec{\mathbf{B}}$$

where

$$\vec{\mathbf{v}}\times\vec{\mathbf{B}} = \begin{cases} (v_t\hat{\boldsymbol{\theta}}+v_d\hat{\mathbf{z}}) \times (+B)\hat{\mathbf{r}} = -v_tB\hat{\mathbf{z}}+v_dB\hat{\boldsymbol{\theta}} & \text{for side } a \\ (v_t\hat{\boldsymbol{\theta}}-v_d\hat{\mathbf{z}}) \times (-B)\hat{\mathbf{r}} = +v_tB\hat{\mathbf{z}}+v_dB\hat{\boldsymbol{\theta}} & \text{for side } a' \end{cases}$$

or

$$\vec{\mathbf{v}}\times\vec{\mathbf{B}} = \begin{cases} v_dB\hat{\boldsymbol{\theta}} - \omega_R(\ell_2/2)B\hat{\mathbf{z}} & \text{for side } a \\ v_dB\hat{\boldsymbol{\theta}} + \omega_R(\ell_2/2)B\hat{\mathbf{z}} & \text{for side } a'. \end{cases}$$

The component $v_dB\hat{\boldsymbol{\theta}}$ is what produces the torque. In more detail, with N the number of charge carriers/unit volume, S the cross-sectional area of the wire loop, and ℓ_1 the axial length of the loop, the quantity $NS\ell_1$ is the total number of charge carriers of each side of the loop and the current due to the movement of these charges is $i = qNSv_d$ (see Figure 1.22). The total tangential forces on the axial sides of the rotor loop are given by

$$\begin{aligned}\vec{\mathbf{F}}_{\text{side } a} &= (qNS\ell_1)v_dB\hat{\boldsymbol{\theta}} = i(t)\ell_1B\hat{\boldsymbol{\theta}}\\ \vec{\mathbf{F}}_{\text{side } a'} &= (qNS\ell_1)v_dB\hat{\boldsymbol{\theta}} = i(t)\ell_1B\hat{\boldsymbol{\theta}}.\end{aligned}$$

The torque is then

$$\vec{\tau} = \frac{\ell_2}{2}\hat{\mathbf{r}}\times\vec{\mathbf{F}}_{\text{side } a} + \frac{\ell_2}{2}\hat{\mathbf{r}}\times\vec{\mathbf{F}}_{\text{side } a'} = 2(\frac{\ell_2}{2}\hat{\mathbf{r}}) \times (i\ell_1B\hat{\boldsymbol{\theta}}) = i\ell_1\ell_2B\hat{\mathbf{z}}$$

which is the same result as in the macroscopic case.

It is now shown that the $\hat{\mathbf{z}}$ component of the magnetic force produces the back emf. The $\hat{\mathbf{z}}$ component of $\vec{\mathbf{F}}_{\text{magnetic}}$, that is, along the axial sides of the loop, is given by

$$(\vec{\mathbf{F}}_{\text{magnetic}}/q)_z\hat{\mathbf{z}} = \begin{cases} -\omega_R(\ell_2/2)B\hat{\mathbf{z}} & \text{for side } a\\ +\omega_R(\ell_2/2)B\hat{\mathbf{z}} & \text{for side } a'.\end{cases}$$

As shown in Figure 1.23, this component of the magnetic force per unit charge $(\vec{\mathbf{F}}_{\text{magnetic}}/q)_z$ opposes the electric field $\vec{\mathbf{E}}_S$ set up in the loop by the applied armature voltage V_S. The relationship between V_S and $\vec{\mathbf{E}}_S$ is

$$V_S = \int_{T_1}^{T_2} \vec{\mathbf{E}}_S \cdot d\vec{\ell}$$

where

$$d\vec{\ell} = \begin{cases} +d\ell\hat{\mathbf{z}} & \text{for side } a\\ -d\ell\hat{\mathbf{z}} & \text{for side } a'.\end{cases}$$

Then, with ξ denoting the integral of the magnetic force per unit charge from T_1 to T_2, one has

$$\begin{aligned}\xi &\triangleq \int_{T_1}^{T_2} (\vec{\mathbf{F}}_{\text{magnetic}}/q)\cdot d\vec{\ell}\\ &= \int_{\text{side } a} (-\omega_R(\ell_2/2)B\hat{\mathbf{z}})\cdot(d\ell\hat{\mathbf{z}}) + \int_{\text{side } a'} (\omega_R(\ell_2/2)B\hat{\mathbf{z}})\cdot(-d\ell\hat{\mathbf{z}})\\ &= \int_{\ell=0}^{\ell=\ell_1} -\omega_R(\ell_2/2)Bd\ell + \int_{\ell=0}^{\ell=\ell_1} -\omega_R(\ell_2/2)Bd\ell\\ &= -\omega_R(\ell_2/2)B\ell_1 - \omega_R(\ell_2/2)B\ell_1\\ &= -\ell_1\ell_2B\omega_R.\end{aligned}$$

In this example, the (back) emf ξ is due to the magnetic force while V_S is due to the electric field set up by the voltage source. As the induced emf ξ and the source voltage V_S have the same sign convention, the minus sign in the expression for ξ shows that it opposes the applied armature voltage V_S. The total emf (voltage) in the loop is

$$V_S + \xi = V_S - \ell_1 \ell_2 B \omega_R.$$

Finally, the equation governing the current in the rotor loop is

$$L\frac{di}{dt} + Ri = V_S - \ell_1 \ell_2 B \omega_R.$$

This is the same physical result as shown in the macroscopic case using Faraday's law. However, here the induced emf $\xi = -\ell_1 \ell_2 B \omega_R$ is negative because it was chosen to have the same sign convention as V_S (i.e., both V_S and ξ are positive going from T_1 to T_2). This is in contrast to the macroscopic case in which they had opposite sign conventions so that $\xi = \ell_1 \ell_2 B \omega_R$ was positive, but still opposed V_S.

Remark *Voltage and Emf*

The *electromotive force* or *emf* between two points in a circuit is the integral of the total force per unit charge along the circuit between those two points.[8] The force per unit charge can be due to an electric field, a magnetic field or both. The term voltage (drop) was originally reserved for the integral of the *electric field* between the two points. However, this distinction is usually not made and the two terms (voltage and emf) are used interchangeably.

1.5.2 Drift Speed

Above, it was shown that the drift speed of the charge carriers making up the current is given by $v_d(t) = i(t)/(qNS)$. As explained in Ref. [34] (page 781), this motion (drift speed) of the charges in the conductor is caused by the electric field setup in the conductor by the voltage source and/or induced emfs in the conducting circuit. This electric field is pushing the charges along the conductor against the internal resistance of the conductor. In metals, the outer valence electrons are free to move about the lattice of the metal and are called *conduction electrons*. For example, in copper there is one valence electron per atom and the other 28 electrons remain bound to the copper nucleus. Consequently, as there are 8.4×10^{22} atoms/cm^3 in copper, there are $N = 8.4 \times 10^{22}$ electrons/cm^3 that can move freely within the copper lattice to make up the current in the wire. To consider a simple numerical example, let $i = 10$ A, $S = 0.1$ cm^2 so that with $q = 1.6 \times 10^{-19}$ coulomb/electron, the corresponding drift speed is

$$\begin{aligned} v_d &= \frac{i(t)}{qNS} = \frac{10 \text{ A}}{(1.6 \times 10^{-19} \frac{\text{coulomb}}{\text{electron}})(8.4 \times 10^{22} \frac{\text{electrons}}{\text{cm}^3})(0.1 \text{ cm}^2)} \\ &= 0.74 \text{ cm/sec.} \end{aligned}$$

---

[8] Note that the electromotive force is *not* a force, but rather an energy per unit charge.

That is, it takes $1/(0.74 \text{ cm/sec}) = 1.35$ sec for the charge carrier to travel one centimeter. However, it should be noted that when a voltage/emf is applied to a circuit, the corresponding electric field is setup around the circuit at a speed close to the speed of light. This is analogous to applying pressure to a long tube of water. The pressure wave is transmitted down the tube rapidly (at the speed of sound in water) while the water itself moves much slower [34].

1.6 Tachometer for a DC Machine*[9]

A tachometer is a device for measuring the speed of a DC motor by putting out a voltage proportional to the motor's speed. A tachometer for the simple linear DC machine is considered first.

1.6.1 Tachometer for the Linear DC Machine

Figure 1.24 shows a tachometer added to the linear DC machine. The magnetic field in the DC motor is $\vec{\mathbf{B}}_1 = -B_1\hat{\mathbf{z}}$ with $B_1 > 0$ while in the tachometer it is $\vec{\mathbf{B}}_2 = -B_2\hat{\mathbf{z}}$ with $B_2 > 0$.

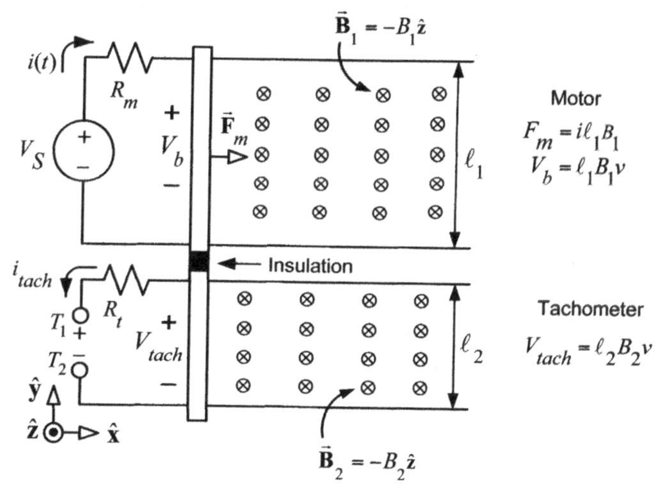

FIGURE 1.24. DC tachometer (generator).

The two bars are rigidly connected together by the insulating material. The motor force (the magnetic force on the upper bar) is $F_m = i\ell_1 B_1$, and

[9] Sections marked with an asterisk (*) may be skipped without loss of continuity.

the induced (back) emf in the motor is $\xi = V_b = B_1 \ell_1 v$, where v is the speed of the motor (bar).

The induced (back) emf in the tachometer is given by $\xi = V_{tach} = B_2 \ell_2 v$ so that by measuring the voltage between the terminals T_1 and T_2, the speed v of the motor can be computed. Note that the tachometer and motor have the same physical structure. In fact, the tachometer is nothing more than a generator putting out a voltage proportional to the speed.

1.6.2 *Tachometer for the Single-Loop DC Motor*

A tachometer for the single loop DC motor is constructed by attaching another loop to the shaft and rotating it an external magnetic field to act as a DC generator. That is, the changing flux in the tachometer loop produces (generates) an induced emf according to Faraday's law and this emf is proportional to the shaft's speed. To see this, consider Figure 1.25, where a motor loop is driven by a voltage V_S and, attached to the same shaft, is a second loop called a tachometer. Both loops rotate in an external radial magnetic field which is not shown in Figure 1.25, but is shown for the tachometer loop in Figure 1.26. It is important to point out that no voltage is applied to the terminals T_1 and T_2 of the tachometer as was the case for the motor. Instead, the voltage V_{tach} between the terminals T_1 and T_2 of the tachometer is measured (this voltage is proportional to the motor speed ω_R).

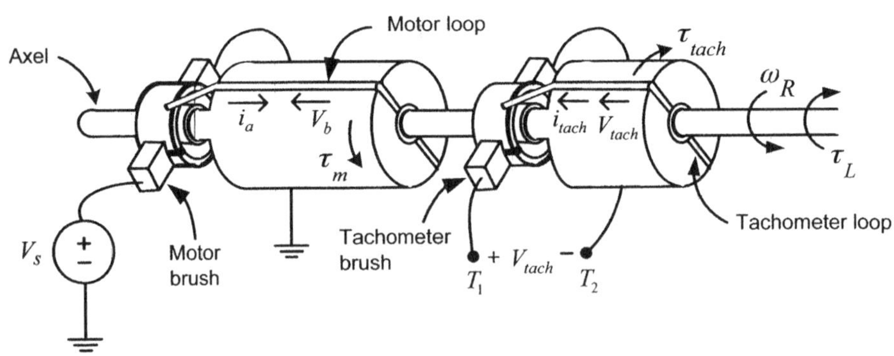

FIGURE 1.25. Single loop motor and tachometer. Drawn by Sharon Katz.

Specifically, in the same way the back emf was computed for the DC motor, one can calculate the flux in the loop of the tachometer due to the external magnetic field. This computation is (see Figure 1.26)

$$\begin{aligned}
\phi &= \int_S \vec{\mathbf{B}} \cdot d\vec{\mathbf{S}} \\
&= \int_0^{\ell_1} \int_{\theta_R}^{\pi} (B\hat{\mathbf{r}}) \cdot (\frac{\ell_2}{2} d\theta dz \hat{\mathbf{r}}) + \int_0^{\ell_1} \int_{\pi}^{\pi+\theta_R} (-B\hat{\mathbf{r}}) \cdot (\frac{\ell_2}{2} d\theta dz \hat{\mathbf{r}}) \\
&= \int_0^{\ell_1} \int_{\theta_R}^{\pi} B\frac{\ell_2}{2} d\theta dz + \int_0^{\ell_1} \int_{\pi}^{\pi+\theta_R} -B\frac{\ell_2}{2} d\theta dz \\
&= (\ell_1 \ell_2 B/2)(\pi - \theta_R) - (\ell_1 \ell_2 B/2)\theta_R \\
&= -\ell_1 \ell_2 B \theta_R + (\ell_1 \ell_2 B/2)\pi.
\end{aligned}$$

The induced emf is then

$$V_{\text{tach}} = -d\phi/dt = (\ell_1 \ell_2 B) d\theta_R/dt = K_{b\_\text{tach}} \omega_R$$

where $K_{b\_\text{tach}} = \ell_1 \ell_2 B$ is a constant depending on the dimensions of the tachometer rotor and the strength of the external magnetic field of the tachometer. This shows that the voltage between the terminals T_1 and T_2 is proportional to the angular speed and therefore can be used to measure the speed.

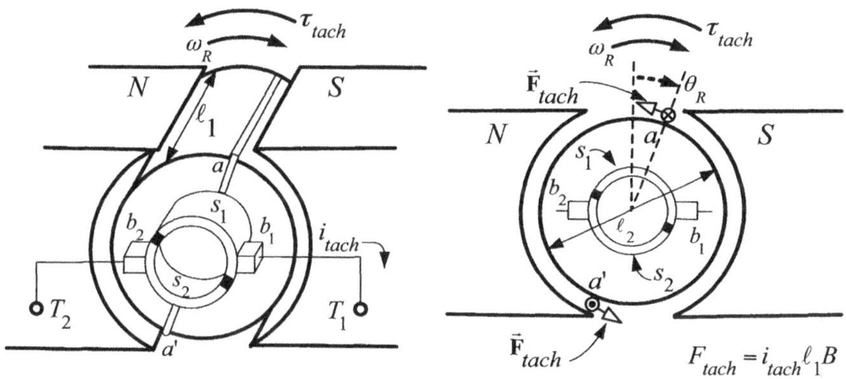

FIGURE 1.26. Cutaway view of the DC tachometer. From *Electromagnetic and Electromechanical Machines,* 3rd edition, L. W. Matsch and J. Derald Morgan, 1986. Reprinted by permisson of John Wiley & Sons.

1.7 The Multiloop DC Motor*

The above single loop motor of Figure 1.5 was used to illustrate the basic Physics of the DC motor. However, it is not a practical motor. The first thing that must be done is to add more loops to extract more torque from the machine. Further, in the single-loop motor, the magnetic field produced by the current in the *rotor* is an external magnetic field acting on the field windings. As the loop rotates, this magnetic field results in a changing flux in the field windings, which in turn induces an emf in the field windings. This emf is referred to as the *armature reaction* and makes it difficult to maintain a constant field current. (The term armature refers to the rotating current loop, and reaction refers to the induced emf in the field windings produced by the rotor current.) The problem of armature reaction can be alleviated by adding more loops to the motor.

1.7.1 Increased Torque Production

Figure 1.27 below shows the addition of several loops to the motor with each loop similar in form to the loop in Figure 1.5. As shown in the figure, there are now eight slots in the rotor with two loops placed in each pair of slots that are 180° apart for a total of eight loops.

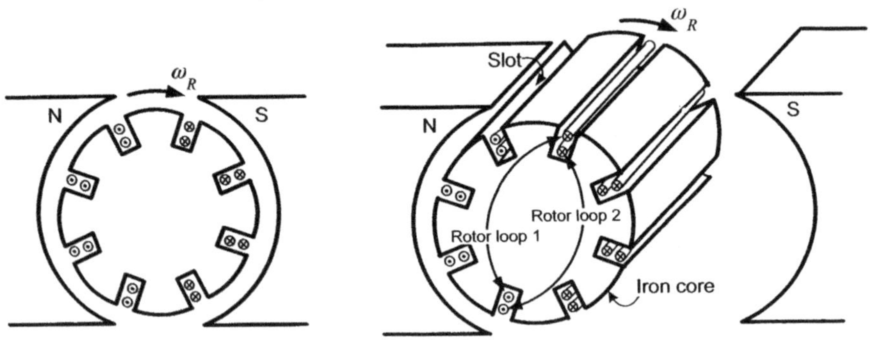

FIGURE 1.27. A multiloop armature for a DC motor.

The torque on the rotor is now $\tau_m = n\ell_1\ell_2 Bi$, where $n = 8$ is the number of rotor loops and B is the strength of the radial magnetic field in the air gap produced by the external magnetic field. Of course, some method must be found to ensure the current in each loop is reversed every half-turn so that (for positive torque) all the loop sides under the south pole face will have their current going into the page $\otimes$ and all the loop sides under the north pole face will have their current coming out of the page $\odot$. This process is referred to as *commutation* and is considered next.

1.7.2 Commutation of the Armature Current

As seen in Figure 1.27, as a rotor loop rotates clockwise past the vertical position, the current in the top side of the loop must change direction from coming out of the page to going into the page. That is, each rotor loop must have the current in it reversed every half-turn. This is done using a commutator which is illustrated in Figure 1.28 for the rotor shown in Figure 1.27. The commutator for this rotor consists of 8 copper segments (labeled a–h in 1.30(a)) which are separated by insulating material. By connecting each of the ends of the rotor loops of Figure 1.27 to the appropriate copper segments of the commutator, the current will be reversed every half-turn as it rotates past the vertical. To explain all of this, consider Figure 1.30(a), which shows explicitly how the ends of the rotor loops are connected to the segments of the commutator. The eight rotor loops of Figure 1.27 are labeled as $1-1'$,..., $8-8'$ in Figure 1.30(a) with the ends of each such loop electrically connected (soldered) to a particular pair of commutator segments. For example, the ends of loop $1-1'$ are connected to commutator segments a and b, respectively. The commutator and rotor loops all rotate together rigidly while the two brushes (labeled b_1 and b_2) remain stationary. The brushes are typically made of a carbon material and are mechanically pressed against the commutator surface, making electrical contact.[10] That is, as the commutator rotates, the particular segment that is rubbing against the brush makes electrical contact.

FIGURE 1.28. Commutator for the rotor in Figure 1.27.

Figure 1.29 is a photograph of the rotor of an actual DC motor with a tachometer.

[10]The figure shows a gap between the brushes and the commuator, but this was done for illustration and there is no gap in reality. Also, for illustrative purposes, the brushes are shown inside the commutator when in fact they are normally pressed against the commutator from the outside.

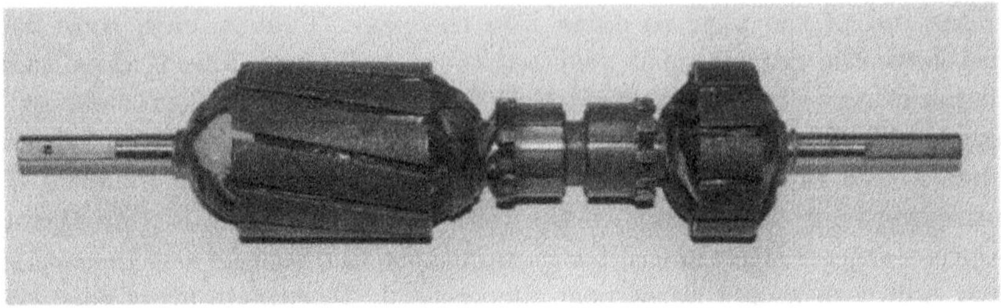

FIGURE 1.29. Photo of the rotor of a DC motor (left) and its tachometer (right). Note that the slots for the windings of the DC motor are skewed (see problem 9). Photo courtesy of Professor J. D. Birdwell of the University of Tennessee.

As previously explained, to obtain positive torque, it must be that whenever a side of a loop is under a south pole face, the current must be into the page ($\otimes$) and the other side of the loop, which is under the north pole face, must have its current out of the page ($\odot$). When the loop side rotates from being under one pole face to the other pole face, the current in that loop must be reversed (commutated). The mechanism of how this connection between the armature loops, the commutator and the brushes can reverse the current in each rotor loop every half turn is now explained.

With reference to Figure 1.30(a), the armature current i enters brush b_1 and into commutator segment c. By symmetry, half of this armature current (i.e., $i/2$) goes through loop $3-3'$ into commutator segment d, then through loop $4-4'$ into commutator segment e, then through loop $5-5'$ into commutator segment f, then through loop $6-6'$ into commutator segment g, and, finally, out through brush b_2. This path (circuit) of the current is denoted in **bold**. Similarly, there is a parallel path for the other half of the current armature current. Specifically, the other half of the armature current $i/2$ goes through loop $2'-2$ into commutator segment b, then through loop $1'-1$ into commutator segment a, then through loop $8'-8$ into commutator segment h, then into loop $7'-7$ into commutator segment g, and finally, out through brush b_2. This path (circuit) is denoted without bold. So, for the rotor in the position shown in Figure 1.30(a), there are two parallel circuits from b_1 to b_2 each made up of four loops connected in series and each circuit carries half of the armature current. The sides of the loops under the south pole face have their current into the page while the other side of these loops (which are under the north pole face) have their current out of the page so that positive torque is produced.

The sides of the loops in Figure 1.30(a) are 45° apart. Figure 1.30(b) shows the rotor turned 45°/2 with respect to Figure 1.30(a). In this case, brush b_1 shorts the two commutator segment b and c together while the brush b_2 shorts together the two commutator segments f and g. The ends of loop $2-2'$ are connected to commutator segments b and c (which are now shorted together) so that the current in this loop is now zero. Similarly, the ends of loop 6–6′ are connected to commutator segments f and g and the current in this loop is also zero. For the remaining loops, $i/2$ goes through loop 3–3′ into commutator segment d, then through loop 4–4′ into commutator segment e, then into loop 5–5′ into commutator segment f, and finally, out brush b_2. These loops are denoted in **bold** in the figure. Similarly, $i/2$ goes through loop 1–1′ into commutator segment a, then through loop 8′–8 into commutator segment h, then into loop 7′–7 into commutator segment g, and finally out brush b_2.

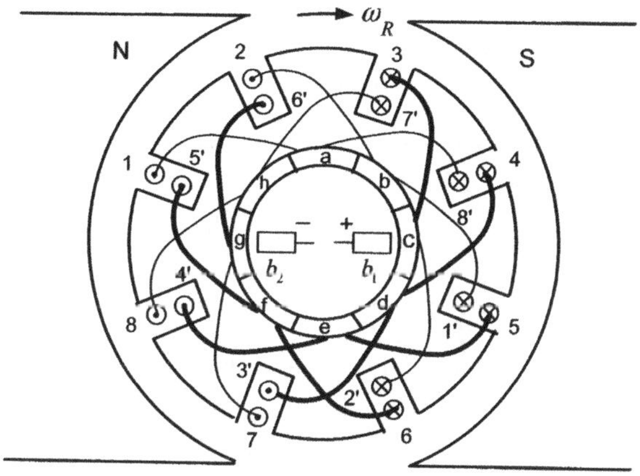

FIGURE 1.30. (a). Rotor loops and commutator for 4 sets of rotor loops. Brushes remained fixed in space, that is, they do not rotate. From *Electric Machinery Fundamentals*, 2nd edition by S. J. Chapman, McGraw-Hill 1991. Reproduced with permission of McGraw-Hill Companies.

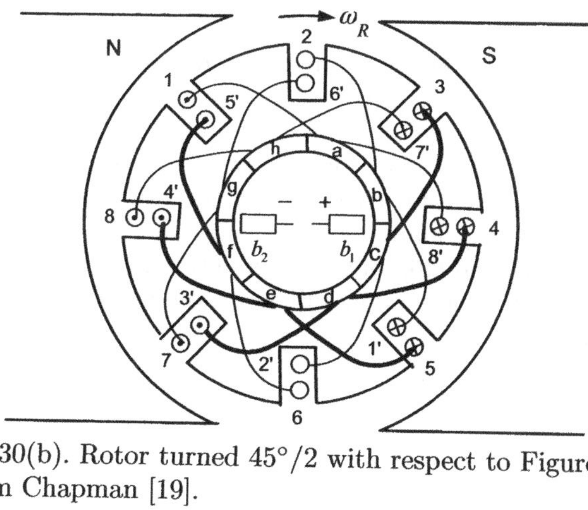

Figure 1.30(b). Rotor turned 45°/2 with respect to Figure 1.30(a). Adapted from Chapman [19].

The motor continues to rotate and consider it now after it has moved additional $45°/2$ so that it has the position shown in Figure 1.30(c). In this case, the current enters brush b_1 and into commutator segment b. By symmetry, half the current $i/2$ goes through loop $2{-}2'$ into commutator segment c, then through loop $3{-}3'$ into commutator segment d, then through loop $4 - 4'$ into commutator segment e, then through loop $5{-}5'$ into commutator segment f, and finally out through brush b_2. This path (circuit) of the current is denoted in **bold**. Similarly, the other half of the current goes through loop $1'{-}1$ into commutator segment a then through loop $8'{-}8$ into commutator segment h then through loop $7'{-}7$ into commutator segment g then into loop $6'{-}6$ into commutator segment f and finally, out through brush b_2. This path (circuit) is denoted without using bold.

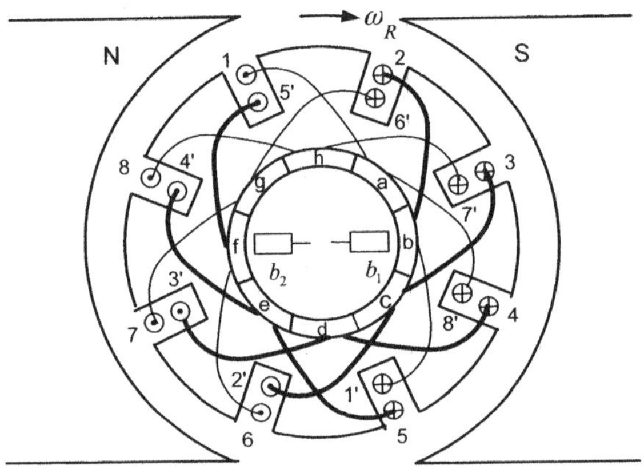

Figure 1.30(c). Rotor turned 45° with respect to Figure 1.30(a). Adapted from Chapman [19].

As the sequence of figures 1.30(a)–(c) show, the current in loops $2{-}2'$ and $6{-}6'$ were reversed as these two loops rotated past the vertical position. In summary, there are two parallel paths, each consisting of four loops, and when any loop goes to the vertical position, the current in that loop is reversed. In this way, all sides of the loop under the south pole have their current going into the page and all sides under the north pole have their current coming out of the page for positive torque production.

Remark The scheme for current commutation presented here is from [19]. However, there are many other schemes and the reader is referred to Refs. [19], [21], [24], and [26] for an introduction to other schemes. See Ref. [24] for a discussion of how commutation is often carried out in small PM DC motors.

Stator Iron Construction

A more realistic depiction of the stator for a (single pole pair) permanent magnet DC motor is shown in Figure 1.31. The radial magnetic field in the air gap is produced by the two semicircular-shaped permanent magnets.

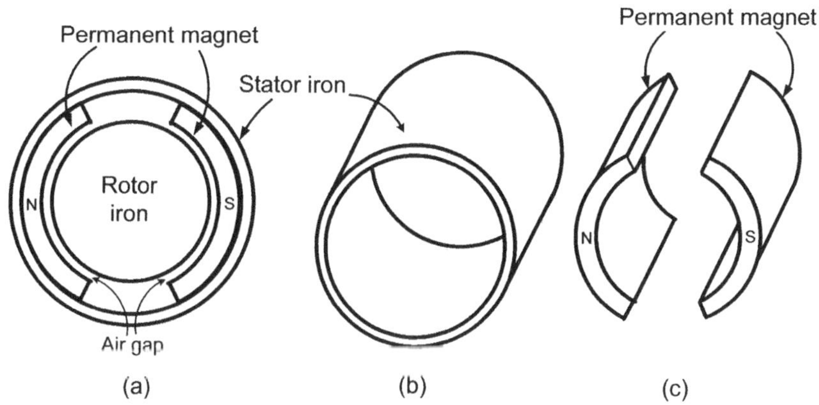

FIGURE 1.31. (a) Stator of a PM DC motor. (b) Stator iron. (c) Stator permanent magnet.

In the case where the DC motor has a field winding, a more realistic depiction of the stator iron (single pole pair) is shown in Figure 1.32 (compare with Figure 1.7).

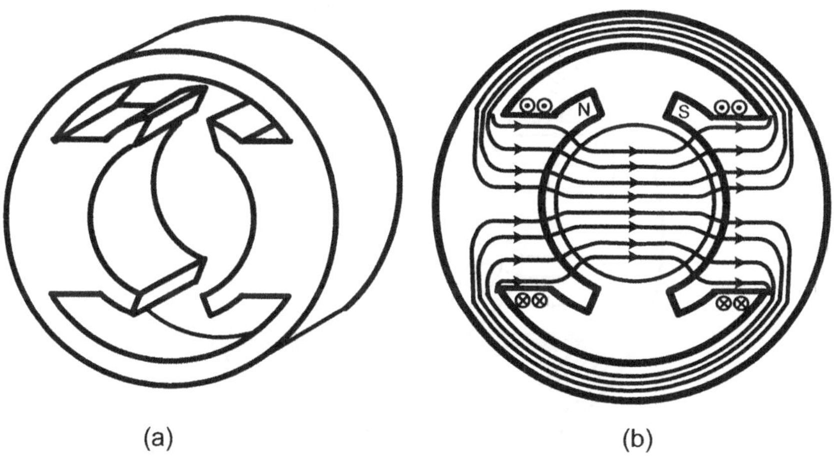

FIGURE 1.32. (a) Stator iron for a wound field DC motor. (b) Cross-sectional view of the stator iron with the field windings.

1.7.3 Armature Reaction

Figure 1.33 shows current in a field winding which is used to magnetize the iron in a fashion similar to that shown in Figure 1.7. Closed curves are drawn in Figure 1.33 to show the magnetic field distribution in the iron and air gap due to just the field current. Note that the magnetic field tends to be in the horizontal direction as it goes inside the field windings. Figure 1.34 shows the magnetic field distribution in the iron and air gap due to just the armature current. By having many equally spaced loops on the rotor, the magnetic field distribution due to the armature current will be as shown in Figure 1.34 for *any* rotor position. Note that this magnetic field tends to be vertical in the iron core inside the field windings (the field windings are not shown in Figure 1.34). In other words, in the stator iron core within the field windings, the magnetic field due to the armature current is perpendicular to the flux surfaces of the field windings. As a result, any changing magnetic field due to a changing armature current will not induce voltages inside the field windings. *Armature reaction* refers to the voltage induced in the field winding by the magnetic field of the armature current. (This is undesirable as such a voltage would cause the current in the field winding to change, and therefore not be constant as the analysis up to this point has assumed.) The symmetric placement of rotor loops around the periphery of the rotor essentially eliminates the armature reaction.

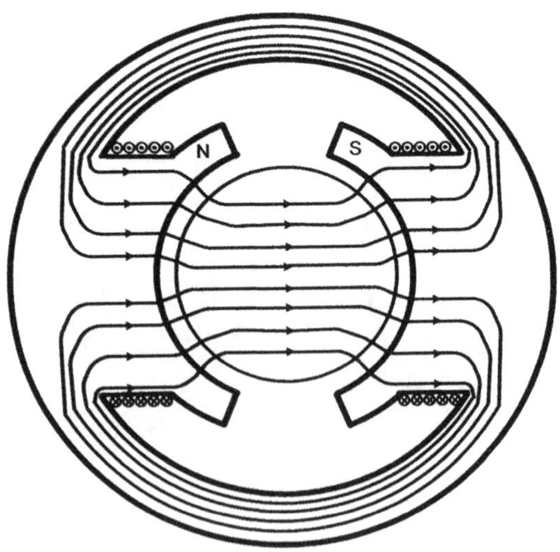

FIGURE 1.33. Magnetic field due to the field current. Adapted from Ref. [23].

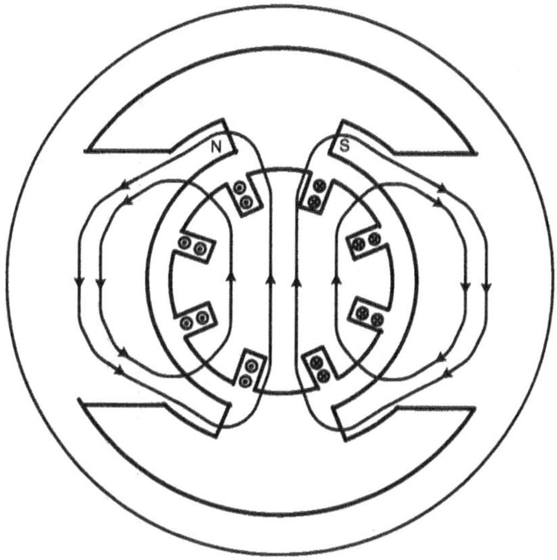

FIGURE 1.34. Magnetic field distribution due to the armature current. Adapted from Ref. [23].

Finally, Figure 1.35 shows the sum of the two magnetic fields.

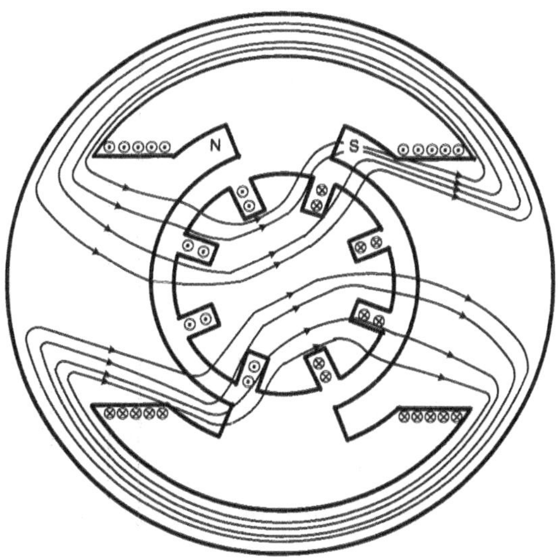

FIGURE 1.35. Magnetic field due to field and armature currents. Adapted from Ref. [23].

1.7.4 Field Flux Linkage and the Air Gap Magnetic Field

In the separately excited DC motor of Figure 1.36, it is convenient to have an expression relating the radial magnetic field B in the air gap to the total flux in the field windings.

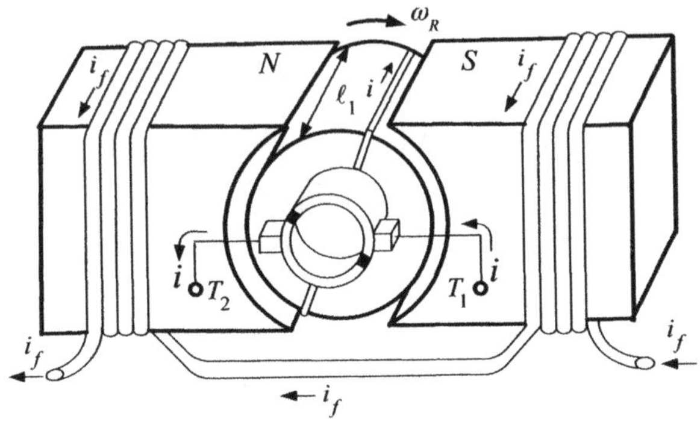

FIGURE 1.36. Separately excited DC motor. Adapted from Ref. [21].

The magnetic field lines are shown in Figure 1.4. Figure 1.37 shows a closed flux surface which is now used to derive a relationship between the radial magnetic field B in the air gap and the field flux linkage λ_f.

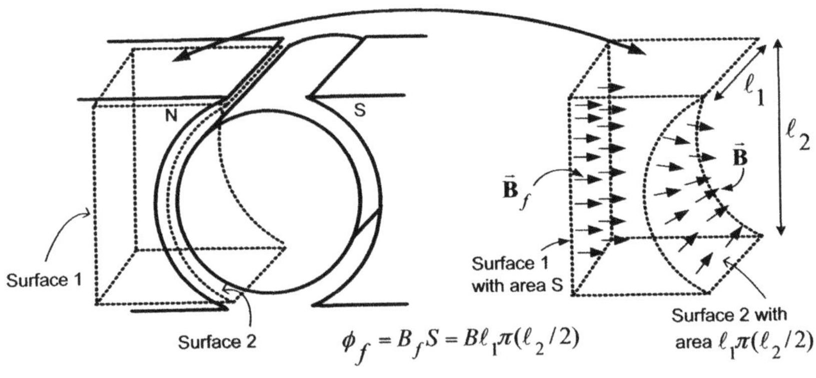

FIGURE 1.37. The flux through surface 1 is $B_f S$ and the flux through surface 2 is $\ell_1\pi(\ell_2/2)B$. To a good approximation, these two fluxes are equal.

The field flux *linkage* λ_f is defined by

$$\lambda_f(i_f) = N_f\phi_f = N_f S B_f(i_f)$$

where N_f is the total number of field windings, S is the cross-sectional area of the iron core of the field winding and $B_f(i_f)$ is the magnetic field produced *inside the magnetic material* of the field circuit due to the field current i_f. With this definition of flux linkage, the total induced emf in the field windings produced by a changing B_f is given by

$$\xi_{\text{fieldwinding}} = -\frac{d\lambda_f(i_f)}{dt} = -N_f\frac{d\phi_f}{dt}.$$

With the two flux surfaces shown in Figure 1.37, it turns out that *conservation of flux*[11] implies the flux through the surface 1 in the iron core equals the flux through the half cylindrical surface 2 in the air gap. That is,

$$\phi_f = SB_f(i_f) = \ell_1\pi(\ell_2/2)B(i_f)$$

where $\ell_2/2$ is the radius of the rotor, ℓ_1 is the axial length of the rotor and $B(i_f)$ is the radial magnetic field *in the air gap* due to the field current i_f. Consequently,

$$B(i_f) = \frac{SB_f(i_f)}{\pi\ell_1\ell_2/2} = \frac{\lambda_f(i_f)}{N_f\pi\ell_1\ell_2/2}$$

is an expression for the radial magnetic field $B(i_f)$ *in the air gap* in terms of the flux linkage $\lambda_f(i_f) = N_f\ell_1\ell_2B_f(i_f)$ in the field windings.

1.7.5 Armature Flux Due to the External Magnetic Field

In what follows, the multiloop motor of Figure 1.30 is considered in which the armature circuit consists of two parallel circuits each having n loops. Let θ_R be referenced relative to loop $1{-}1'$ so that $\theta_R = 0$ corresponds to loop $1{-}1'$ being vertical. (Recall in Figure 1.5 that $\theta_R = 0$ corresponded to loop $a{-}a'$ being vertical.) For the single loop motor, the flux $\phi_{sl}(i_f, \theta_R)$ in the loop $1{-}1'$ of the rotor due to the external magnetic field $B(i_f)$ is

$$\phi_{sl}(i_f, \theta_R) = -\ell_1\ell_2B(i_f)(\theta_R \bmod \pi - \pi/2).$$

In general, let $\theta_k = 0$ correspond to loop $k{-}k'$ being vertical so that the position of the kth rotor loop may be written as

$$\theta_k = \theta_R + (k-1)\pi/n \qquad \text{for } k = 1, ..., n$$

where n is the number of loops connected in series in each parallel circuit ($n = 4$ in Figure 1.30). The flux in the kth rotor loop is [see equation (1.4)]

$$\begin{aligned}\phi_k(i_f, \theta_R) &= -\ell_1\ell_2B(i_f)\left(\theta_k \bmod \pi - \frac{\pi}{2}\right)\\ &= -\ell_1\ell_2B(i_f)\left(\left\{\theta_R + (k-1)\frac{\pi}{n}\right\}\bmod \pi - \frac{\pi}{2}\right).\end{aligned}$$

[11]See problem 6 and Chapter 3.

The total flux linkage $\lambda(i_f, \theta_R)$ in the n rotor loops is then

$$\begin{aligned}\lambda(i_f,\theta_R) &= \sum_{k=1}^{n} \phi_k(i_f,\theta_R)\\ &= -\sum_{k=1}^{n} \ell_1\ell_2 B(i_f)\left(\left\{\theta_R + (k-1)\frac{\pi}{n}\right\} \bmod \pi - \frac{\pi}{2}\right). \quad (1.7)\end{aligned}$$

In Figures 1.30(a)–(c), *each* of the two parallel sets of n rotor loops has the flux linkage $\lambda(i_f, \theta_R)$ in it.

Recall that the sign convention for the fluxes $\phi_k(i_f, \theta_R)$ is such that if $-d\phi_k(i_f, \theta_R)/dt > 0$, then it is acting in the direction opposite to positive current flow, that is, its sign convention is opposite to that of applied voltage V_S. As a consequence, the sum

$$V_S - \left(-\frac{d\lambda(i_f,\theta_R)}{dt}\right) = V_S + \frac{d\lambda(i_f,\theta_R)}{dt}$$

is the total emf in the loop due to the applied voltage and the external magnetic field.

For $n = 4$, the flux linkage $\lambda(i_f, \theta_R)$ is plotted as a function of the rotor position θ_R in Figure 1.38. Note that $\partial\lambda(i_f, \theta_R)/\partial\theta_R = -n\ell_1\ell_2 B(i_f)$ which is proportional to the number of rotor loops n and to the strength of the external magnetic field strength $B(i_f)$ in the air gap.

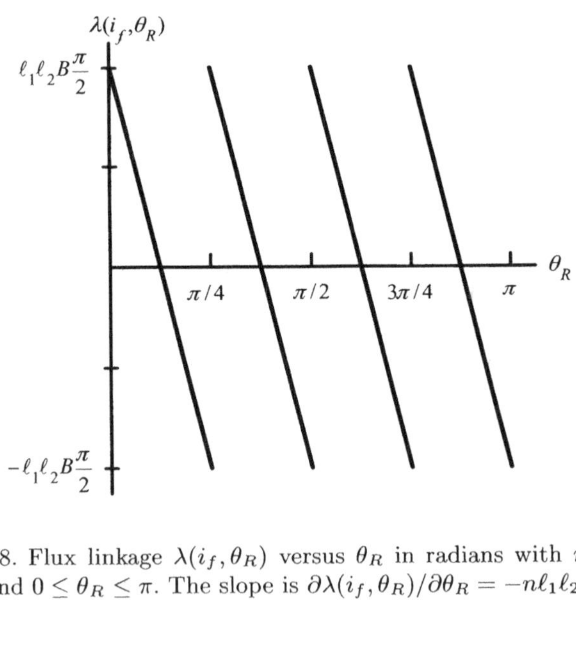

FIGURE 1.38. Flux linkage $\lambda(i_f, \theta_R)$ versus θ_R in radians with $n = 4$ sets of rotor loops and $0 \le \theta_R \le \pi$. The slope is $\partial\lambda(i_f, \theta_R)/\partial\theta_R = -n\ell_1\ell_2 B(i_f)$.

In the case of a permanent magnet stator, the armature flux linkage is simply given by

$$\begin{aligned}\lambda(\theta_R) &= \sum_{k=1}^{n}\phi_k(\theta_R)\\ &= -\sum_{k=1}^{n}\ell_1\ell_2 B\Big(\Big\{\theta_R + (k-1)\frac{\pi}{n}\Big\}\bmod \pi - \frac{\pi}{2}\Big) \qquad (1.8)\end{aligned}$$

where $B > 0$ is the strength of the radial magnetic field in the air gap produced by the permanent magnet of the stator.

1.7.6 Equations of the PM DC Motor

In the multiloop motor considered here, the armature circuit consists of two parallel circuits each having n loops. That is, there are a total of $2n$ loops on the rotor as each of the two parallel circuits has a loop at the same location on the rotor. Let L denote the total inductance of the n rotor loops making up either of the two parallel circuits. Then, if a current $i/2$ is in each parallel circuit, each circuit will have a flux linkage of $Li/2$ due to its current and an *additional* flux linkage of $Li/2$ due to the current in the other circuit. This is simply because the two sets of parallel loops (windings) are wound around the rotor core together so that they can be considered to be perfectly (magnetically) coupled.

The total flux linkage in the n loops making up either of the parallel circuits is now computed. To proceed, let i be the current into the armature so that $i/2$ is the current in the n loops of each parallel circuit. The quantity $Li/2$ is the flux linkage due to the current $i/2$ in the loops and an additional flux linkage of $Li/2$ is produced in these same loops by the current $i/2$ in the *other* parallel circuit for a total flux of Li. By equation (1.8), $\lambda(\theta_R)$ is the flux linkage in the n loops due to the *external* magnetic field produced by the permanent magnet of the stator. Recall that the sign convention for the induced emf $-d\lambda(\theta_R)/dt$ is opposite to that of $-d(Li)/dt$. Specifically (see Section 1.3.5), the normal to the flux surface was taken to be *radially in* to compute the flux Li while in Section 1.3.4 the surface normal was taken to be *radially out* to compute the flux $\lambda(\theta_R)$. Simply writing $-\lambda(\theta_R)$ then gives the flux due to the external magnetic field with the surface normal radially in and the induced voltages by both of these changing fluxes will then have the same sign convention as the applied armature voltage. Using the same sign convention as the applied armature voltage V_S, the total flux linkage in each parallel circuit of the armature may be written as

$$Li - \lambda(\theta_R).$$

Let R_1 denote the resistance of the n loops connected in series making up either of the two parallel circuits. The equation describing the electrical dynamics of the current in each of the parallel circuits is found by applying Kirchhoff's voltage law to obtain

$$-\frac{d}{dt}\Big(Li - \lambda(\theta_R)\Big) - R_1 i/2 + V_S = 0$$

where V_S is the applied voltage to the armature. Finally, defining $R \triangleq R_1/2$, the equation describing the electrical dynamics of the armature circuit is

$$L\frac{di}{dt} = -Ri + \frac{d\lambda(\theta_R)}{dt} + V_S. \tag{1.9}$$

The quantity $d\lambda(\theta_R)/dt$ can be expanded to obtain

$$\frac{d\lambda(\theta_R)}{dt} = \frac{\partial\lambda(\theta_R)}{\partial\theta_R}\omega_R = -n\ell_1\ell_2 B\omega_R = -K_b\omega_R \tag{1.10}$$

where $K_b \triangleq n\ell_1\ell_2 B$.

Each loop carries the current $i/2$ so that the torque produced by the two sides of each loop is $2\,(\ell_2/2)\,(i/2)\ell_1 B$. As there are $2n$ loops, the total torque is

$$\tau_m = 2n\ell_1\ell_2 B(i/2) = n\ell_1\ell_2 Bi = K_T i \tag{1.11}$$

where $K_T \triangleq n\ell_1\ell_2 B = K_b$. Finally, using (1.9) and (1.11), the complete set of equations for the PM DC motor is then

$$L\frac{di}{dt} = -Ri - K_b\omega_R + V_S \tag{1.12}$$

$$J\frac{d\omega_R}{dt} = K_T i - f\omega_R - \tau_L \tag{1.13}$$

$$\frac{d\theta_R}{dt} = \omega_R \tag{1.14}$$

where V_S is the applied armature voltage, τ_L is the load torque on the motor, f is the coefficient of viscous friction and J is the rotor's moment of inertia.

1.7.7 *Equations of the Separately Excited DC Motor*

The separately excited DC motor has an additional equation compared to the PM DC motor which describes the flux/current in the field winding. Similar to the case of the PM DC motor, the total flux linkage in either parallel circuit of the armature windings is given by

$$Li - \lambda(i_f, \theta_R)$$

where $\lambda(i_f, \theta_R)$ now depends on i_f as given by (1.7), and L is the inductance the n loops connected in series making up either parallel circuit. Again, R_1 denotes the resistance of either circuit and Kirchhoff's voltage law gives

$$-\frac{d}{dt}\Big(Li - \lambda(i_f, \theta_R)\Big) - R_1 i/2 + V_S = 0$$

where V_S is the applied voltage to the armature. Finally, defining $R \triangleq R_1/2$, the equation describing the electrical dynamics of the armature circuit is

$$L\frac{di}{dt} = -Ri + \frac{d\lambda(i_f, \theta_R)}{dt} + V_S. \tag{1.15}$$

The quantity $d\lambda(i_f, \theta_R)/dt$ can be expanded to obtain

$$\frac{d\lambda(i_f, \theta_R)}{dt} = \frac{\partial\lambda}{\partial i_f}\frac{di_f}{dt} + \frac{\partial\lambda}{\partial\theta_R}\omega_R = \frac{\partial\lambda}{\partial i_f}\frac{di_f}{dt} - n\ell_1\ell_2 B(i_f)\omega_R. \tag{1.16}$$

Here $B(i_f)$ is the strength of the radial magnetic field *in the air gap* produced by the current in the field windings. The flux linkage in the field windings is $\lambda_f(i_f) = N_f S B_f(i_f)$ where $B_f(i_f)$ is the magnetic field strength *in the iron core* of the field, N_f is the number of field windings, and S is the cross sectional area of this iron core. The flux $SB_f(i_f)$ in each of the field windings goes through the air gap and, by conservation of flux (see Section 1.7.4), $SB_f(i_f) = \ell_1(\pi\ell_2/2)B(i_f)$ so that the radial magnetic field in the air gap produced by the field current is

$$B(i_f) = \frac{N_f}{N_f}\frac{SB_f(i_f)}{\pi\ell_1\ell_2/2} = \frac{\lambda_f(i_f)}{N_f\pi\ell_1\ell_2/2}.$$

Consequently,

$$n\ell_1\ell_2 B(i_f) = n\ell_1\ell_2\frac{\lambda_f(i_f)}{N_f\pi\ell_1\ell_2/2} = K_m\lambda_f(i_f)$$

where

$$K_m \triangleq \frac{2}{\pi}\frac{n}{N_f}.$$

As a result, equation (1.16) may be rewritten as

$$\frac{d\lambda(i_f, \theta_R)}{dt} = \frac{\partial\lambda}{\partial i_f}\frac{di_f}{dt} - K_m\lambda_f(i_f)\omega_R. \tag{1.17}$$

Each loop carries the current $i/2$ so that the torque produced on each of the two sides of the loop is $2\,(\ell_2/2)\,(i/2)\ell_1 B(i_f)$. As there are $2n$ loops, the total torque is

$$\tau_m = 2n\ell_1\ell_2 B(i_f)(i/2) = n\ell_1\ell_2 B(i_f)i = K_m\lambda_f(i_f)i \tag{1.18}$$

using the above expressions for K_m and $\lambda_f(i_f)$. Finally, using (1.15), (1.16), and (1.18), the mathematical model of the separately excited *DC* motor is given by

$$L\frac{di}{dt} = -Ri + \frac{\partial\lambda}{\partial i_f}\frac{di_f}{dt} - K_m\lambda_f(i_f)\omega_R + V_S \tag{1.19}$$

$$J\frac{d\omega_R}{dt} = K_m\lambda_f(i_f)i - \tau_L \tag{1.20}$$

$$\frac{d\lambda_f(i_f)}{dt} = -R_f i_f + V_f \tag{1.21}$$

where V_f is the applied voltage to the field, R_f is the resistance of the field winding, τ_L is the load torque on the motor and J is the rotor's moment of inertia.

If the iron in the field is not in saturation, then one may write

$$\lambda_f(i_f) = L_f i_f$$

and the dynamic model simplifies to

$$L\frac{di}{dt} = -Ri + \frac{\partial\lambda}{\partial i_f}\frac{di_f}{dt} - K_m L_f i_f \omega_R + V_S \quad (1.22)$$

$$J\frac{d\omega_R}{dt} = K_m L_f i_f i - \tau_L \quad (1.23)$$

$$L_f\frac{di_f}{dt} = -R_f i_f + V_f. \quad (1.24)$$

Under normal operating conditions, the term $\dfrac{\partial\lambda}{\partial i_f}\dfrac{di_f}{dt}$ is typically negligible and the model then reduces to[12]

$$L\frac{di}{dt} = -Ri - K_m L_f i_f \omega_R + V_S \quad (1.25)$$

$$J\frac{d\omega_R}{dt} = K_m L_f i_f i - \tau_L \quad (1.26)$$

$$L_f\frac{di_f}{dt} = -R_f i_f + V_f. \quad (1.27)$$

A typical mode of operation is to use the field voltage V_f to hold the field current i_f constant at some value. Then the field flux is of course constant and $L_f di_f/dt = 0$. However, in the first equation (1.19) (or 1.25), it is seen that the back emf is $-K_m\lambda_f\omega_R$ (or $-K_m L_f i_f \omega_R$) which increases in proportion to the speed. The input voltage V_S must be at least large enough to overcome the back emf in order to maintain the armature current. To have the motor achieve higher speeds within the voltage limit $|V_S| \le V_{\max}$, *field weakening* is employed. This is accomplished by using the input voltage V_f to decrease the field flux $\lambda_f = L_f i_f$ at higher speeds usually according to a flux reference of the form

$$\lambda_{ref} = \begin{cases} \lambda_{f0} & |\omega_R| \le \omega_{base} \\ \lambda_{f0}\dfrac{\omega_{base}}{|\omega_R|} & |\omega_R| \ge \omega_{base}. \end{cases} \quad (1.28)$$

This is illustrated in Figure 1.39. For $\omega > \omega_{base}$, field weakening results in the back emf $-K_m\lambda_f\omega_R = -K_m\lambda_{f0}\frac{\omega_{base}}{|\omega_R|}\omega_R = -K_m\lambda_{f0}\omega_{base}\text{sign}(\omega_R)$ being constant. The trade-off is that the torque $K_m\lambda_f i = K_m\lambda_{f0}\frac{\omega_{base}}{|\omega_R|} i$ is now

[12] It will be seen later that field-oriented control of an induction motor results in a mathematical model of the induction motor that looks similar to these equations!

less for the same armature current i due to the decrease in the field flux linkage λ_f. If the armature resistance is negligible and i_f is constant, then the *base speed* ω_{base} is defined to be the speed satisfying $K_m\lambda_f\omega_{base} = V_{\max}$. Otherwise, the base speed is chosen to be somewhat smaller to account for the Ri and $\frac{\partial\lambda}{\partial i_f}\frac{di_f}{dt}$ voltage drops [see equation (1.19)].

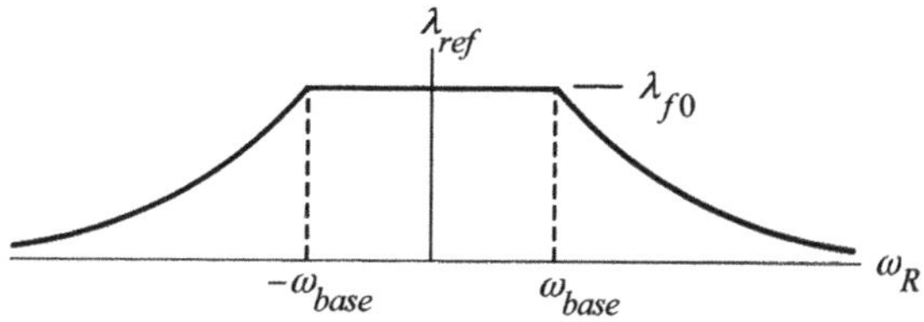

FIGURE 1.39. Flux reference for field weakening.

Appendices

Rotational Dynamics

The equations of motion of a rigid body that is constrained to rotate about a fixed axis are reviewed here briefly. Consider a cylinder which is constrained to rotate about a fixed axis as shown below.

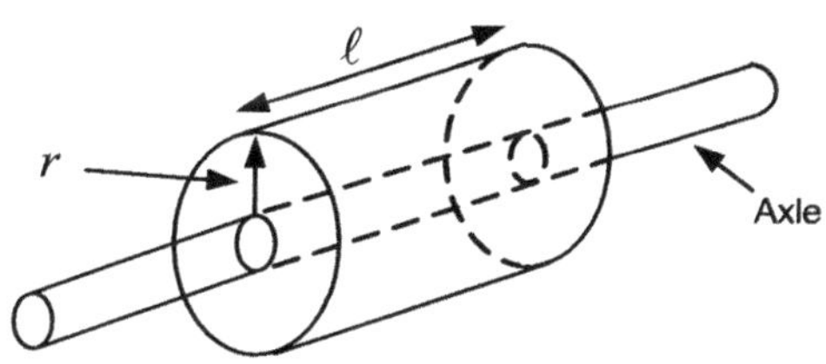

FIGURE 1.40. Cylinder constrained to rotate about a fixed axis.

The approach here is to obtain the equations of motion of the cylinder by first computing obtaining an expression for its kinetic energy. To do so, denote the angular speed of the cylinder by ω and the mass density of the material making up the cylinder by ρ. Then consider the cylinder to be made up of n small pieces of material Δm_i where the ith piece has mass

$$\Delta m_i = \rho r \Delta\theta \Delta\ell \Delta r.$$

This is illustrated in Figure 1.41.

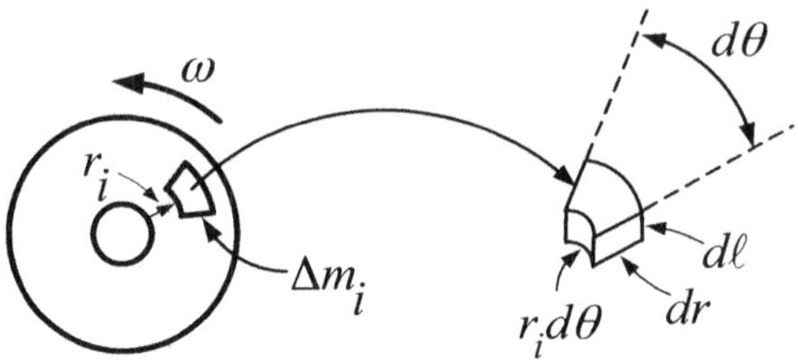

FIGURE 1.41. Cylinder is considered to be made up of small masses Δm_i. Drawn by Sharon Katz.

Each piece of mass Δm_i is rotating at the same angular speed ω so that the *linear* speed of Δm_i is $v_i = r_i\omega$ where r_i is the distance of Δm_i from the axis of rotation. The kinetic energy KE_i of Δm_i is given by

$$KE_i = \frac{1}{2}\Delta m_i v_i^2 = \frac{1}{2}\Delta m_i (r_i\omega)^2 .$$

The total kinetic energy is then

$$KE = \sum_{i=1}^{n} (KE)_i = \sum_{i=1}^{n} \frac{1}{2}\Delta m_i v_i^2 = \sum_{i=1}^{n} \frac{1}{2}\Delta m_i (r_i\omega)^2 = \frac{1}{2}\omega^2 \sum_{i=1}^{n} \Delta m_i r_i^2.$$

Dividing the cylinder into finer and finer pieces so that $n \to \infty$ and $\Delta m_i \to 0$, the sum

$$\sum_{i=1}^{n} \Delta m_i r_i^2$$

becomes the integral

$$J = \iiint_{cylinder} r^2 dm.$$

The quantity J is called the *moment of inertia* and the kinetic energy of the cylinder may now be written as

$$KE = \frac{1}{2}J\omega^2.$$

Assuming the axle radius is zero, the moment of inertia of the cylinder is computed to be

$$J = \int_0^R \int_0^\ell \int_0^{2\pi} r^2 \rho r d\theta d\ell dr = \frac{1}{2}(\pi R^2 \ell \rho) R^2 = \frac{1}{2} M R^2$$

where M is the total mass of the cylinder.

The above expression for the kinetic energy is now used to derive a relationship between torque and angular acceleration. Recall from elementary mechanics that the work done on a mass by an external force equals the change in its kinetic energy. In particular, consider an external force $\vec{\mathbf{F}}$ acting on the cylinder as shown in Figure 1.42.

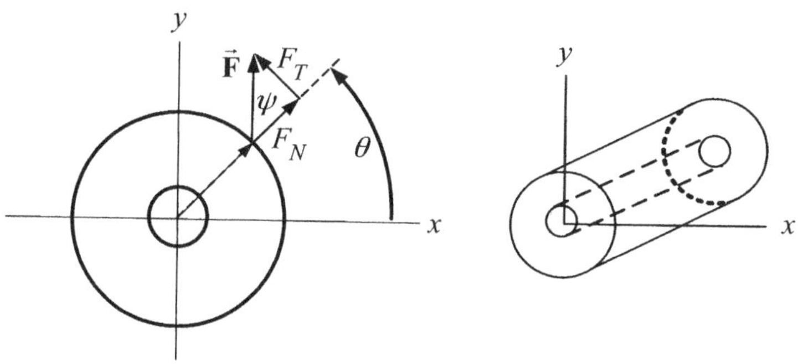

FIGURE 1.42. Force $\vec{\mathbf{F}}$ applied to the cylinder is resolved into a normal and tangential component. Drawn by Sharon Katz.

The cylinder is on an axle and therefore constrained to rotate about the z axis. Figure 1.42 shows the force $\vec{\mathbf{F}}$ applied to the cylinder at the position (r, θ) (in polar coordinates) resolved into a tangential component F_T and a normal component F_N so that $\vec{\mathbf{F}} = F_N\hat{\mathbf{r}} + F_T\hat{\boldsymbol{\theta}}$. The torque is defined as

$$\vec{\boldsymbol{\tau}} \triangleq \vec{\mathbf{r}} \times \vec{\mathbf{F}} = r\hat{\mathbf{r}} \times \left(F_N\hat{\mathbf{r}} + F_T\hat{\boldsymbol{\theta}}\right) = rF_T\hat{\mathbf{z}} = rF\sin(\psi)\hat{\mathbf{z}} \qquad (1.29)$$

where ψ is the angle between $\vec{\mathbf{r}}$ and $\vec{\mathbf{F}}$. (By definition, the magnitude of the cross product $\vec{\mathbf{r}} \times \vec{\mathbf{F}}$ is defined as $rF\sin(\psi)$ and the direction of $\vec{\mathbf{r}} \times \vec{\mathbf{F}}$ is perpendicular to both $\vec{\mathbf{r}}$ and $\vec{\mathbf{F}}$ along the axis of rotation determined by the right hand rule[13]). As $F_T = F\sin(\psi)$ is the tangential component, the torque may be rewritten as

$$\vec{\boldsymbol{\tau}} = \tau\hat{\mathbf{z}} = rF_T\hat{\mathbf{z}}$$

or, in scalar form, as

$$\tau = rF_T.$$

The motivation for the definition of torque as given by (1.29) is that it is the cause of *rotational* motion. Specifically, the rotational motion about an axis is caused by the applied tangential force F_T and the further away from the axis of rotation that the tangential force F_T is applied, the easier it is to get rotational motion. That is, the torque (cause of rotational motion)

[13] Using your right hand, curl your fingers in the direction from the first vector $\vec{\mathbf{r}}$ to the second vecotor $\vec{\mathbf{F}}$. Then your thumb points in the direction of $\vec{\mathbf{r}} \times \vec{\mathbf{F}}$.

increases if either r or F_T increases which corresponds to one's experience (e.g., opening doors).

To summarize, $\vec{\tau}$ is a vector pointing along the axis of rotation and the magnitude is given by

$$|\vec{\tau}| = |\tau| = |rF_T|.$$

(Recall that the angular velocity vector $\vec{\omega} = \omega\hat{\mathbf{z}}$ also points along the axis of rotation where ω is the angular speed.)

With $d\vec{s} \triangleq ds\hat{\boldsymbol{\theta}} = rd\theta\hat{\boldsymbol{\theta}}$, the change in work done on the cylinder by the external force $\vec{\mathbf{F}}$ is

$$dW = \vec{\mathbf{F}} \cdot d\vec{s} = F_T r d\theta = \tau d\theta$$

where $\tau \triangleq rF_T$. Dividing by dt, the power (rate of work) delivered to the cylinder is given by

$$\frac{dW}{dt} = \tau\frac{d\theta}{dt} = \tau\omega.$$

As the rate of work done equals the rate of change of kinetic energy, it follows that

$$\frac{dW}{dt} = \frac{d}{dt}\left(\frac{1}{2}J\omega^2\right) = \tau\frac{d\theta}{dt}$$

or

$$J\omega\frac{d\omega}{dt} = \tau\omega.$$

This gives the fundamental relationship between torque and angular acceleration:

$$\tau = J\frac{d\omega}{dt}.$$

That is, the applied torque equal the moment of inertia times the angular acceleration. This is the basic equation for rigid body rotational dynamics.

Viscous Rotational Friction

Almost always there are frictional forces, and therefore, frictional torques acting between the axle and the bearings.[14] This is illustrated in Figure 1.43.

[14] An interesting exception are magnetic bearings where the axel is levitated by magnetic fields so that there is no mechanical contact.

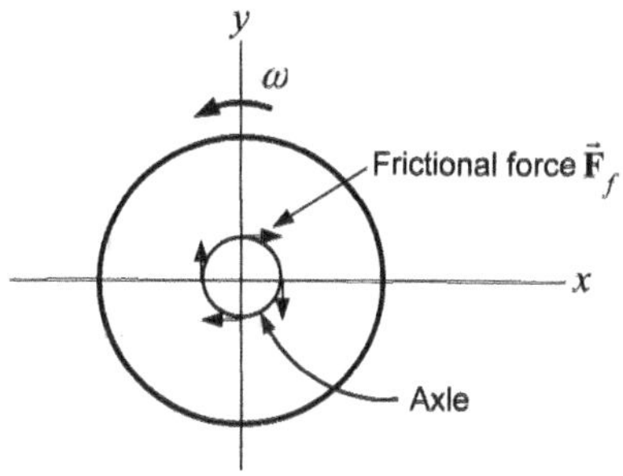

FIGURE 1.43. Viscous friction torque.

Often the frictional force is proportional to the angular speed and this model of friction is called *viscous friction* which is expressed mathematically as

$$\vec{\tau} = -f\vec{\omega} = -f\omega\hat{\mathbf{z}}$$

or, in scalar form,

$$\tau = -f\omega$$

where $f > 0$ is the *coefficient of viscous friction.*

Sign Convention for Torque

Suppose the axis of rotation is along the z axis. Recall that the definition of torque is

$$\vec{\tau} \triangleq \vec{\mathbf{r}} \times \vec{\mathbf{F}} = rF\sin(\psi)\hat{\mathbf{z}} = rF_T\hat{\mathbf{z}}$$

where ψ is the angle between $\vec{\mathbf{r}}$ and $\vec{\mathbf{F}}$. The magnitude of the cross product $\vec{\mathbf{r}} \times \vec{\mathbf{F}}$ is $rF\sin(\psi)$ and the direction of $\vec{\mathbf{r}} \times \vec{\mathbf{F}}$ is perpendicular to both $\vec{\mathbf{r}}$ and $\vec{\mathbf{F}}$ along the z axis. In engineering applications, the systems are designed so that the applied force is tangential to the rotational motion, i.e., $\psi = \pi/2$, $F = F_T$, and

$$\vec{\tau} \triangleq \tau\hat{\mathbf{z}} = rF\hat{\mathbf{z}}.$$

If $\tau = rF > 0$ then the torque will cause the cylinder to rotate around the z axis in the direction indicated by the curved arrow. On the other hand, if $\tau = rF < 0$ then the torque will cause the cylinder to rotate around the z axis in the direction *opposite* to that indicated by the curved arrow. Typically in engineering texts, the sign convention for torque is indicated by a curved arrow as shown in Figure 1.44. (Physics texts prefer to write $\vec{\tau} \triangleq \tau\hat{\mathbf{z}}$.)

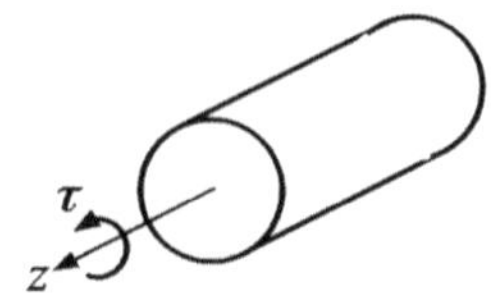

FIGURE 1.44. Sign convention for torque.

Gears

Using the elementary rigid body dynamics developed in the previous appendix, the model of a two gear system illustrated in Figure 1.45 below is now developed.

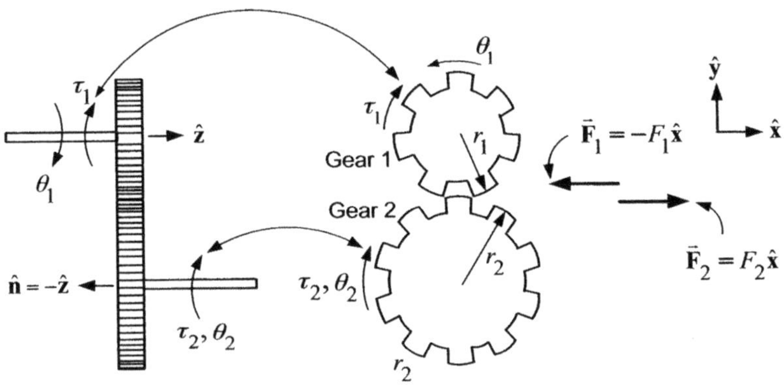

FIGURE 1.45. Two gear system. Drawn by Sharon Katz.

This presentation is from that given in Professor Ogata's book [37]. In Figure 1.45,

τ_1 is the torque exerted on gear 1 by gear 2.
F_1 is the force exerted on gear 1 by gear 2.
τ_2 is the torque exerted on gear 2 by gear 1.
F_2 is the force exerted on gear 2 by gear 1.
θ_1 is the angle rotated by gear 1.
θ_2 is the angle rotated by gear 2.
n_1 is the number of teeth on gear 1.
n_2 is the number of teeth on gear 2.
r_1 is the radius of gear 1.
r_2 is the radius of gear 2.

Let $\vec{\mathbf{F}}_1 \triangleq F_1(-\hat{\mathbf{x}})$ so that if $F_1 > 0$, the force is in the $-\hat{\mathbf{x}}$ direction as shown in Figure 1.45. Also, let $\vec{\mathbf{r}}_1 \triangleq r_1(-\hat{\mathbf{y}})$ so that $\vec{\boldsymbol{\tau}}_1 = \vec{\mathbf{r}}_1 \times \vec{\mathbf{F}}_1 =$

$r_1F_1(-\hat{\mathbf{y}})\times(-\hat{\mathbf{x}}) = r_1F_1(-\hat{\mathbf{z}}) = r_1F_1\hat{\mathbf{n}}$. That is, $\vec{\boldsymbol{\tau}}_1 = \tau_1\hat{\mathbf{n}}$ where $\tau_1 = r_1F_1$ and $\hat{\mathbf{n}} \triangleq -\hat{\mathbf{z}}$ is a unit vector. Similarly, let $\vec{\mathbf{F}}_2 \triangleq F_2\hat{\mathbf{x}}$ so that if $F_2 > 0$, the force is in the $\hat{\mathbf{x}}$ direction. Writing $\vec{\mathbf{r}}_2 \triangleq r_2\hat{\mathbf{y}}$ it follows that $\vec{\boldsymbol{\tau}}_2 = \vec{\mathbf{r}}_2 \times \vec{\mathbf{F}}_2 = r_2F_2(-\hat{\mathbf{z}}) = -\tau_2\hat{\mathbf{z}} = \tau_2\hat{\mathbf{n}}$ with $\tau_2 = r_2F_2$. The reason that $\vec{\mathbf{r}}_1, \vec{\boldsymbol{\tau}}_1, \vec{\mathbf{F}}_1$ are referred to the basis vectors $-\hat{\mathbf{x}}, -\hat{\mathbf{y}}, -\hat{\mathbf{z}}$ while $\vec{\mathbf{r}}_2, \vec{\boldsymbol{\tau}}_2, \vec{\mathbf{F}}_2$ are referred to the basis vectors $\hat{\mathbf{x}}, \hat{\mathbf{y}}, \hat{\mathbf{z}}$ is so that there will no minus signs in the gear relationships to be derived below.

Algebraic Relationships Between Two Gears

There are three important algebraic relationships between the gears.

1. The gears have different radii, but the teeth on each gear are the same size so that they will mesh together properly. Consequently, the number of teeth on the surface of gears is proportional to the radius of the gears so that, for example, if $r_2 = 2r_1$, then $n_2 = 2n_1$. In general,
$$\frac{r_2}{r_1} = \frac{n_2}{n_1}.$$
2. By Newton's third law, the forces $\vec{\mathbf{F}}_1 = F_1(-\hat{\mathbf{x}}), \vec{\mathbf{F}}_2 \triangleq F_2\hat{\mathbf{x}}$ are equal in magnitude, but opposite in direction so that $F_2 = -(-F_1) = F_1$. Thus, as $\tau_1 = r_1F_1$ and $\tau_2 = r_2F_2$ it follows that
$$\frac{\tau_2}{\tau_1} = \frac{r_2}{r_1}.$$
3. As the teeth on each gear are meshed together at the point of contact, the distance traveled along the surface of the gears is the same. In other words, $\theta_1 r_1 = \theta_2 r_2$ or
$$\frac{\theta_2}{\theta_1} = \frac{r_1}{r_2}$$

The first two algebraic relationships can be summarized as

$$\frac{\tau_2}{\tau_1} = \frac{r_2}{r_1} = \frac{n_2}{n_1}$$

and these ratios are easily remembered by thinking of gear 2 as larger in radius than gear 1. Then the number of teeth on gear 2 must also be larger (because its circumference is larger) and the torque on gear 2 is also larger (because its radius is larger).

The last algebraic relationship is summarized as

$$\frac{r_1}{r_2} = \frac{\theta_2}{\theta_1} = \frac{\omega_2}{\omega_1}$$

but it is more easily remembered by writing $\theta_1 r_1 = \theta_2 r_2$ which just states the distance traveled along the surface of each gear is the same as they are meshed together.

Dynamic Relationship Between Two Gears

Consider the two gear system shown in Figure 1.46 below. The motor torque τ_m acts on gear 1 and the torque τ_L is a load torque acting on gear 2.

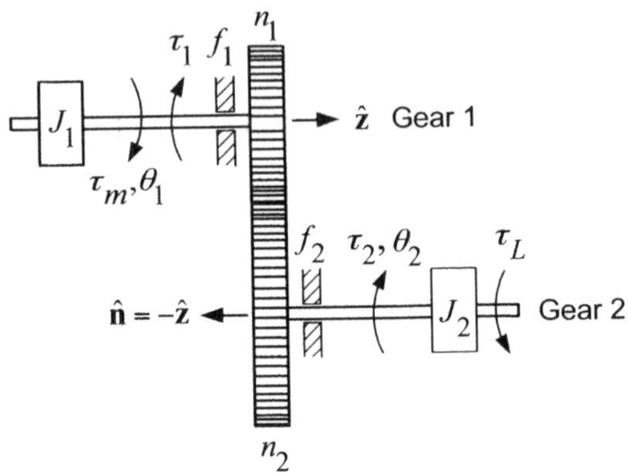

FIGURE 1.46. Dynamic equations for a two gear system. Drawn by Sharon Katz.

In Figure 1.46,

J_1 is the moment of inertia of the motor shaft.
J_2 is the moment of inertia of the output shaft.
f_1 is the viscous friction coefficient of the motor shaft.
f_2 is the viscous friction coefficient of the output shaft.
θ_1 is the angle rotated by gear 1.
θ_2 is the angle rotated by gear 2.
ω_1 is the angular speed of gear 1.
ω_2 is the angular speed of gear 2.
τ_1 is the torque exerted on gear 1 by gear 2.
τ_2 is the torque exerted on gear 2 by gear 1.

The sign conventions for the torques $\tau_m, \tau_1, \tau_2, \tau_L$ are indicated in Figure 1.46. In particular, if $\tau_m > 0, \tau_1 > 0$ then they oppose each other and similarly, if $\tau_2 > 0, \tau_L > 0$ then these two torques oppose each other. A load torque is illustrated in Figure 1.47 in which the load torque on gear 2 is $\tau_L = r_2 mg$ with r_2 the radius of the pick up reel (gear 2).

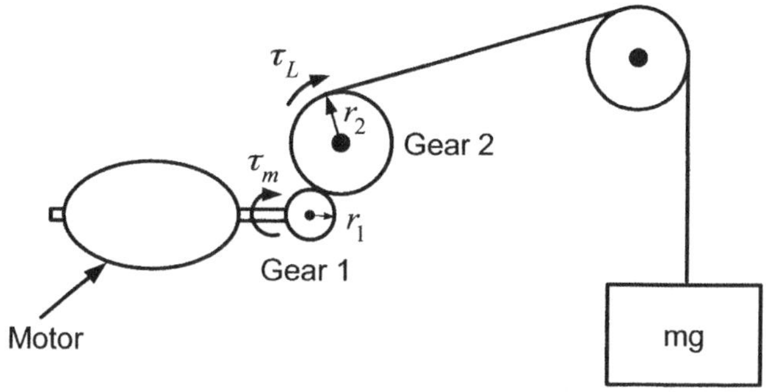

FIGURE 1.47. Illustration of load torque. Drawn by Sharon Katz.

The above development is now put together to write down differential equations that characterize the dynamic behavior of the gears. Recall that the fundamental equation of rigid body dynamics is given by

$$\tau = J\frac{d\omega}{dt}$$

where τ is the *total* torque on the rigid body, J is the moment of inertia of the rigid body and $d\omega/dt$ is its angular acceleration about the fixed axis of rotation. Applying this relationship, the equations of motion for the two gears are then

$$\begin{aligned} \tau_m - \tau_1 - f_1\omega_1 &= J_1\frac{d\omega_1}{dt} \\ \tau_2 - \tau_L - f_2\omega_2 &= J_2\frac{d\omega_2}{dt}. \end{aligned} \tag{1.30}$$

Typically, the input (motor) torque τ_m is known and, the output position θ_2 and speed ω_2 are measured. Consequently, the variables τ_1, τ_2, ω_1 need to be eliminated which is done as follows:

$$\begin{aligned} \tau_2 &= \frac{n_2}{n_1}\tau_1 \\ &= \frac{n_2}{n_1}\left(\tau_m - f_1\omega_1 - J_1\frac{d\omega_1}{dt}\right) \\ &= \frac{n_2}{n_1}\left(\tau_m - f_1\left(\frac{n_2}{n_1}\omega_2\right) - J_1\frac{d}{dt}\left(\frac{n_2}{n_1}\omega_2\right)\right) \\ &= \frac{n_2}{n_1}\tau_m - \left(\frac{n_2}{n_1}\right)^2 f_1\omega_2 - \left(\frac{n_2}{n_1}\right)^2 J_1\frac{d\omega_2}{dt}. \end{aligned} \tag{1.31}$$

Substituting this expression for τ_2 into the second equation of (1.30) results in

$$\frac{n_2}{n_1}\tau_m - \left(\frac{n_2}{n_1}\right)^2 f_1\omega_2 - \left(\frac{n_2}{n_1}\right)^2 J_1\frac{d\omega_2}{dt} - \tau_L - f_2\omega_2 = J_2\frac{d\omega_2}{dt}$$

Rearranging, the desired result is

$$\frac{n_2}{n_1}\tau_m = \left(J_2 + (n_2/n_1)^2 J_1\right)\frac{d\omega_2}{dt} + \left(f_2 + (n_2/n_1)^2 f_1\right)\omega_2 + \tau_L. \quad (1.32)$$

Let $n = n_2/n_1$ denote the *gear ratio*, $J \triangleq J_2 + n^2 J_1$ denote the total inertia *reflected to output shaft* and $f \triangleq f_2 + n^2 f_1$ denote the total viscous friction coefficient *reflected to the output shaft*, equation (1.32) can be written succinctly as

$$n\tau_m = J\frac{d\omega_2}{dt} + f\omega_2 + \tau_L. \quad (1.33)$$

The net effect of the gears is to increase the motor torque from τ_m on the motor shaft to $n\tau_m$ on the output shaft, to add the quantity $n^2 J_1$ to the inertia of the output shaft and to add $n^2 f_1$ to the viscous friction coefficient of the output shaft.

Remark

Everything could have been referred to the motor shaft instead of the output (load) shaft. To do so, simply substitute $\omega_2 = (n_1/n_2)\omega_1$ into (1.32) to obtain

$$\frac{n_2}{n_1}\tau_m = \left(J_2 + \left(\frac{n_2}{n_1}\right)^2 J_1\right)\frac{d\left(\frac{n_1}{n_2}\omega_1\right)}{dt} + \left(f_2 + \left(\frac{n_2}{n_1}\right)^2 f_1\right)\left(\frac{n_1}{n_2}\omega_1\right) + \tau_L$$

Multiply both sides by n_1/n_2 results in

$$\tau_m = \left(J_2 + \left(\frac{n_2}{n_1}\right)^2 J_1\right)\left(\frac{n_1}{n_2}\right)^2\frac{d\omega_1}{dt} + \left(f_2 + \left(\frac{n_2}{n_1}\right)^2 f_1\right)\left(\frac{n_2}{n_1}\right)^2\omega_1 + \frac{n_1}{n_2}\tau_L$$

or, finally, the desired form is

$$\tau_m = \left(\left(\frac{n_1}{n_2}\right)^2 J_2 + J_1\right)\frac{d\omega_1}{dt} + \left(\left(\frac{n_1}{n_2}\right)^2 f_2 + f_1\right)\omega_1 + \frac{n_1}{n_2}\tau_L. \quad (1.34)$$

In this formulation, the load torque on the input shaft is *reduced* by n_1/n_2 from that on the output shaft, and $(n_1/n_2)^2 J_2$ has been added to the inertia of the motor shaft and $(n_1/n_2)^2 f_2$ has been added to the viscous friction coefficient of the motor shaft.

Problems

Faraday's Law and Induced Electromotive Force (emf)

Problem 1 Faraday's Law

Consider Figure 1.48 where a magnet is moving up *into a square planar loop of copper wire.*

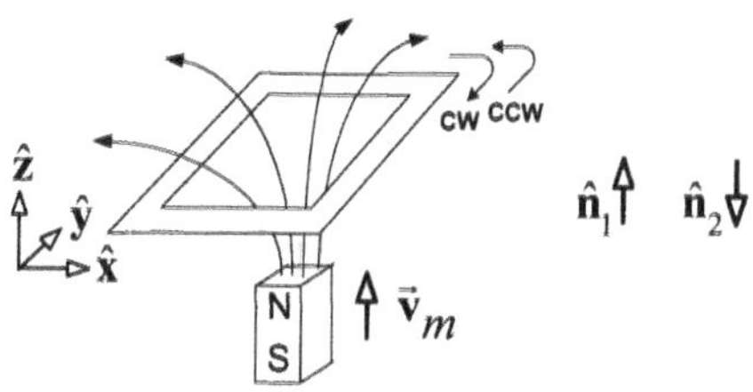

FIGURE 1.48. Induced emf in a loop due to a moving magnet.

(a) Using the normal $\hat{\mathbf{n}}_1$, is the flux in the loop produced by the magnet increasing or decreasing?

(b) Using the normal $\hat{\mathbf{n}}_1$, what is the direction of positive travel around the surface whose boundary is the loop (clockwise or counterclockwise)?

(c) What is the direction of the induced current in Figure 1.48 (clockwise or counterclockwise)? Does the induced current produce a change in the flux in the loop that opposes the change in flux produced by the magnet?

(d) Using the normal $\hat{\mathbf{n}}_2$, is the flux increasing or decreasing?

(e) Using the normal $\hat{\mathbf{n}}_2$, what is the direction of positive travel around the surface whose boundary is the loop (clockwise or counterclockwise)?

(f) What is the direction of the induced current in Figure 1.48 (clockwise or counterclockwise)? Does the induced current produce a change in the flux in the loop that opposes the change in flux produced by the magnet?

Problem 2 Faraday's Law

Consider Figure 1.49 where a magnet is moving down *away from a square planar loop of copper wire.*

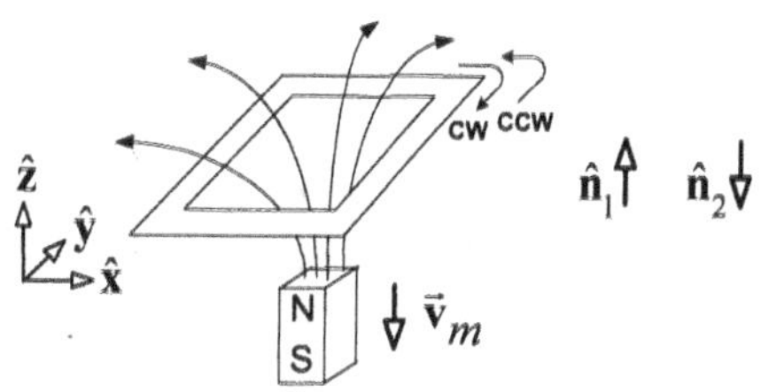

FIGURE 1.49. Induced emf in a loop due to a moving magnet.

(a) Using the normal $\hat{\mathbf{n}}_1$*, is the flux in the loop produced by the magnet increasing or decreasing?*

(b) Using the normal $\hat{\mathbf{n}}_1$*, what is the direction of positive travel around the surface whose boundary is the loop (clockwise or counterclockwise)?*

(c) What is the direction of the induced current in Figure 1.49 (clockwise or counterclockwise)? Does the induced current produce a change in the flux in the loop that opposes the change of flux produced by the magnet?

(d) Using the normal $\hat{\mathbf{n}}_2$*, is the flux increasing or decreasing?*

(e) Using the normal $\hat{\mathbf{n}}_2$*, what is the direction of positive travel around the surface whose boundary is the loop (clockwise or counterclockwise)?*

(f) What is the direction of the induced current in Figure 1.49 (clockwise or counterclockwise)? Does the induced current produce a change in the flux in the loop that opposes the change of flux produced by the magnet?

Problem 3 The Linear DC Motor

Consider the simple linear DC motor of Figure 1.3. Take the normal to the surface enclosed by the loop taken to be $\hat{\mathbf{n}} = -\hat{\mathbf{z}}$.

(a) What is the flux through the surface?

(b) What is the direction of positive travel around this flux surface?

(c) What is the induced emf ξ *in the loop in terms of* B, ℓ *and the speed* v *of the bar?*

(d) Do V_S *and* ξ *have the same sign convention? Explain why* ξ *is now negative. With* R *the resistance of the circuit and the inductance* $L = 0$*, write down the differential equations for the current* i *in the machine and the speed* v *of the bar.*

Problem 4 The Linear DC Motor

Consider the simple linear motor in Figure 1.50 where the magnetic field $\vec{\mathbf{B}} = B\hat{\mathbf{z}}$ $(B > 0)$ *is up out of the page.*

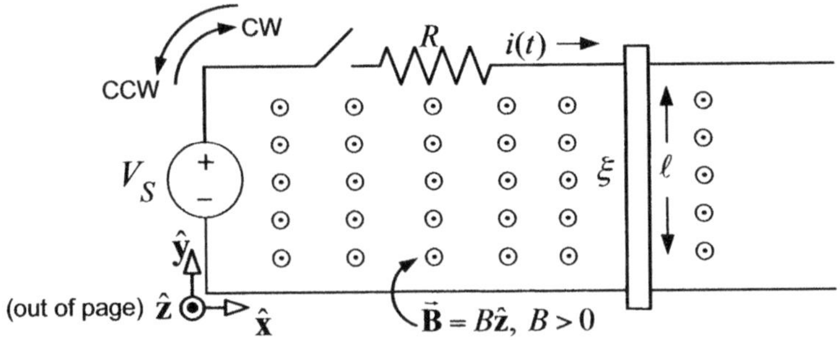

FIGURE 1.50. Linear DC machine with $\vec{\mathbf{B}} = B\hat{\mathbf{z}}$, $B > 0$.

Closing the switch causes a current to flow in the wire loop.

(a) What is the magnetic force $\vec{\mathbf{F}}_{magnetic}$ *on the sliding bar in terms of* B, i, *and* ℓ*? Give both the magnitude and direction of* $\vec{\mathbf{F}}_{magnetic}$.

(b) Take the normal to the surface enclosed by the loop to be $\hat{\mathbf{n}} = \hat{\mathbf{z}}$. *What is the flux through the surface?*

(c) What is the induced emf ξ *in the loop in terms of* B, ℓ, *and the speed* v *of the bar?*

(d) What is the sign convention for the induced emf ξ *drop around the loop? (That is, if* $\xi > 0$, *would it act to push current in the clockwise or counterclockwise direction?)*

(e) Do V_S *and* ξ *have the same sign convention? Draw + and − signs above and below* ξ *to indicate the sign convention for* ξ.

Problem 5 Back Emf in the Single-Loop Motor

Consider the single loop motor with the flux surface as indicated in Figure 1.51. A voltage source connected to the brushes is forcing current down side a $(\otimes)$ *and up side* a' $(\odot)$.

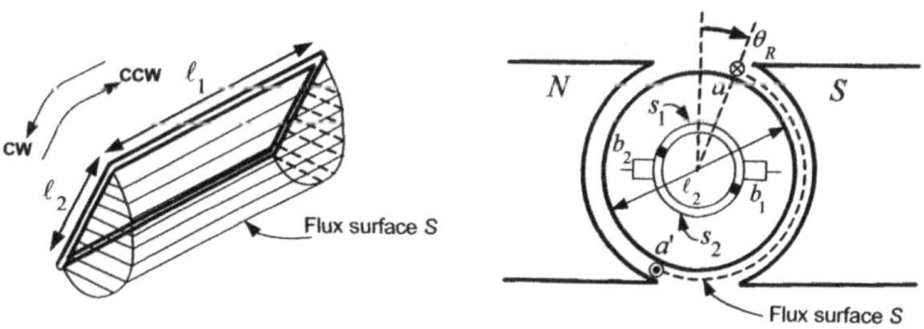

FIGURE 1.51. Computing the flux with $\hat{\mathbf{n}} = -\hat{\mathbf{r}}$.

(a) With the motor at the angular position θ_R *shown, that is, with* $0 < \theta_R < \pi$, *and using the* inward *normal* $(\hat{\mathbf{n}} = -\hat{\mathbf{r}})$, *compute the flux through the surface in terms of the magnitude* B *of the radial magnetic field in the air gap, the axial length* ℓ_1 *of the motor, the diameter* ℓ_2 *of the motor and the angle* θ_R *of the rotor.*

(b) What is the positive direction of travel around the flux surface S *(CW or CCW)?*

(c) What is the emf ξ *induced in the rotor loop? What is the sign convention for the induced emf* ξ *drop around the loop? (That is, if* $\xi > 0$, *would it act to push current in CW direction or the CCW direction?) Do* V_S *and* ξ *have the same sign convention? Explain why* ξ *is now negative. Draw an equivalent circuit of the form in Figure 1.18 for the rotor loop current.*

Problem 6 Gauss's Law and Conservation of Flux

The flux surface in Figure 1.13 was chosen as the half-cylindrical surface with two half disks at either end because the B *field is known on the cylindrical surface being given by equation (1.1) and can be taken to be zero on the two half disks. If the flux surface had been taken to be a flat planar surface with the rectangular loop as its boundary, then it would not be clear how to compute the flux on this surface as the B field there is unknown. Show, using Gauss's law*

$$\phi = \oint_S \vec{\mathbf{B}} \cdot d\vec{\mathbf{S}} \equiv 0,$$

that both surfaces give the same flux. In general, by Gauss's law, one can compute the flux using any surface as long as its boundary is the loop.

Problem 7 Flux in the Single-Loop DC Motor

Figure 1.52 shows the rotor loop with $\pi < \theta_R < 2\pi$.

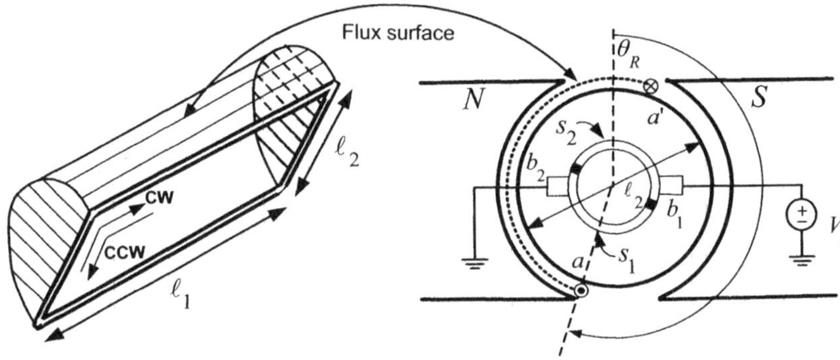

FIGURE 1.52. Rotor loop where $\pi < \theta_R < 2\pi$.

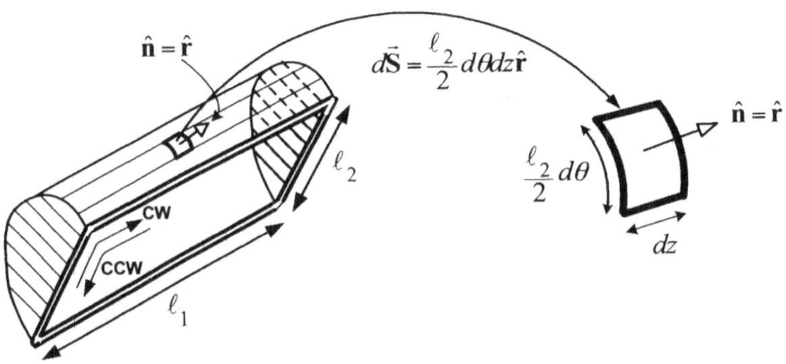

FIGURE 1.53. Flux surface with the normal radially out.

(a) Using the flux surface shown in Figure 1.53 with $d\vec{\mathbf{S}} = (\ell_2/2)d\theta dz\hat{\mathbf{r}}$, *show that* $\phi(\theta_R) = (\theta_R - \pi/2 - \pi)\ell_1\ell_2 B$ *for* $\pi < \theta_R < 2\pi$. *Plot* $\phi(\theta_R)$ *for* $0 < \theta_R < 2\pi$ *(note that* $\phi(\theta_R)$ *for* $0 < \theta_R < \pi$ *is computed in the text). Compute the back emf* ξ *and give its sign convention, that is, if* $\xi = -d\phi(\theta_R)/dt > 0$, *will it force current in the CW or the CCW direction?*

Do ξ and V_S have the same sign convention? (Yes! Explain!) Draw an equivalent circuit diagram of the form of Figure 1.18 to illustrate the sign convention.

(b) Using the flux surface shown in Figure 1.54 with $d\vec{\mathbf{S}} = (\ell_2/2)d\theta dz\,(-\hat{\mathbf{r}})$, *show that* $\phi(\theta_R) = -(\theta_R - \pi/2 - \pi)\,\ell_1\ell_2 B$ *for* $\pi < \theta_R < 2\pi$.

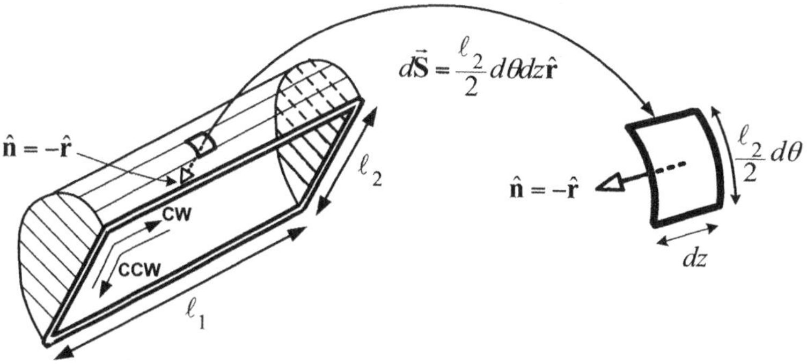

FIGURE 1.54. Flux surface with the normal radially in.

Compute the back emf and give its sign convention, that is, if $-d\phi(\theta_R)/dt > 0$ *will it force current in the CW or the CCW direction. Do ξ and V_S have the same sign convention? (No, they are opposite! Explain!) Draw an equivalent circuit diagram of the form of Figure 1.18 to illustrate the sign convention.*

(c) Note that in part (b) the normal to the flux surface is taken to be radially in while with $0 < \theta_R < \pi$ *it was taken to be radially out (see Figure 1.14). Explain why reversing the normal of the flux surface each half turn in this way results in an equivalent circuit of the form shown in Figure 1.18 which is valid for all rotor angular positions*

Hint: Note that the + side of V_S is now electrically connected to side a' of the loop through brush b_1, while when $0 < \theta_R < \pi$ in Figure 1.14, the + side of V_S was electrically connected to side a of the loop through brush b_1. That is, the sign convention for V_S in the loop changes every half-turn. Thus it is also necessary to change the sign convention for the flux and therefore for ξ every half-turn in order that the sign conventions for V_S and ξ have the same relationship to each other for all θ_R.

(d) Show by taking $d\vec{\mathbf{S}} = (\ell_2/2)d\theta dz\hat{\mathbf{r}}$ *for* $0 < \theta_R < \pi$ *and* $d\vec{\mathbf{S}} = (\ell_2/2)d\theta dz\,(-\hat{\mathbf{r}})$ *for* $\pi < \theta_R < 2\pi$, *that* $\phi(\theta_R) = -(\theta_R \bmod \pi - \pi/2)\,\ell_1\ell_2 B$ *for all θ_R and that* $\xi = -d\phi(\theta_R)/dt$ *has the sign convention given in Figure 1.18 for all θ_R. Verify that the plot of $\phi(\theta_R)$ in Figure 1.15 is correct.*

Problem 8 Simulation of the DC Motor

Let $V_{max} = 40$ *V*, $I_{max} = 5$ *A*, $K_b = K_T = 0.07$ *V/rad/sec (= N-m/A)*, $J = 6\times10^{-5}$ *kg-m*^{2}, $R = 2$ *ohms*, $L = 2$ *mH, and* $f = 0.0004$ *N-m/rad/sec. Develop a simulation of the DC motor that includes the motor model given by (1.6) and a voltage saturation model of the amplifier as illustrated in Figure 1.20. Put a step input of* $V_S(t) = 10$ *V into the motor and plot out (a)* $\theta(t)$, *(b)* $\omega(t)$, *(c)* $i(t)$, *and (d)* $V_S(t)$.

Multiloop Motor

Problem 9 Skewing of the Rotor Sides

Figure 1.29 shows the sides of the rotor loops for the motor (but not the tachometer!) are not straight in the axial direction, but instead are skewed. Can you think of a reason why this is done?

Problem 10 Neutral Plane and Brush Shifting

In the commutation scheme for the multiloop motor, it was shown that when a rotor loop was perpendicular to the brushes, the current in the loop was shorted out (brought to zero). Consequently, it is highly desirable that the total induced voltage in that loop be as close to zero as possible to prevent arcing. Figure 1.35 shows the total B *field distribution in the DC machine. If the armature current were zero, then the field would be horizontal as in Figure 1.33. At very high armature currents (e.g., in large machines used in heavy industry), the field is skewed as shown in Figure 1.35. The neutral plane is the plane cutting through the axis of the rotor for which the* total B *field is perpendicular to the plane. Let* $\psi_k(i, i_f, \theta_R)$ *be the total flux in the* kth *rotor loop due to* both *the field current and the armature current.*

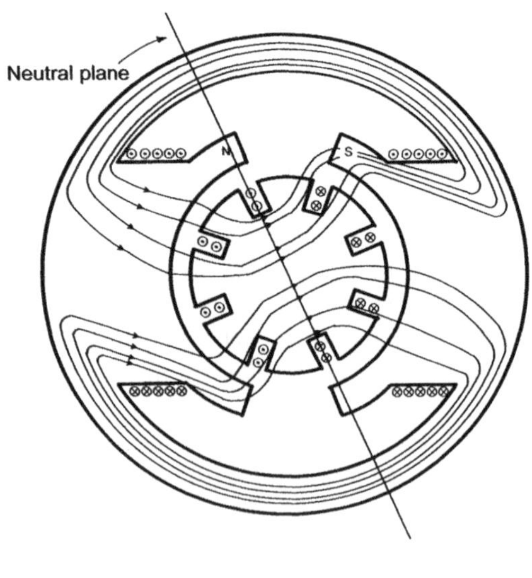

FIGURE 1.55. Magnetic field due to both the field and armature currents.

(a) Explain why $\psi_k(i, i_f, \theta_R)$ is a maximum (or minimum) as a function of the rotor position θ_R when the kth *loop coincides with the neutral plane, or equivalently, why $-\partial\psi_k(i, i_f, \theta_R)/\partial\theta_R = 0$ when the* kth *loop coincides with the neutral plane. (This is seen more easily if one takes the flux surface to be the plane of the loop rather than the cylindrical surface.)*

(b) As explained in Section 1.7.2, the current in a rotor loop is commutated when the plane of the loop is perpendicular to the line containing the two brushes. Consider the situation in which the brushes of Figure 1.30(a) are rotated counterclockwise (shifted) so that the plane *of the loop undergoing commutation is coincident with the neutral plane as shown in Figure 1.56. (After the brushes are rotated, they are held fixed at that position.)*

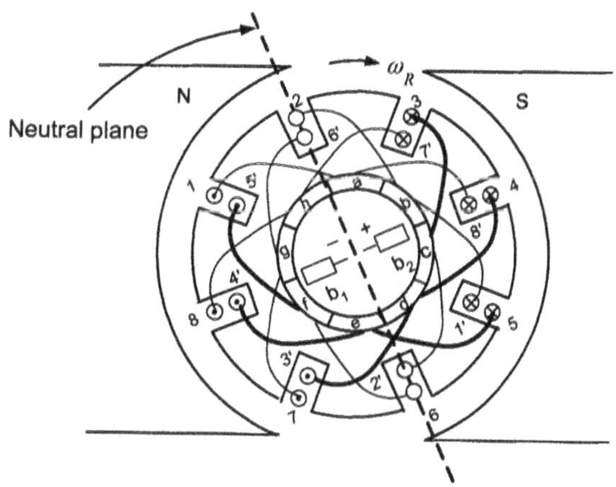

FIGURE 1.56. Rotating the brushes so that commutation in the loop occurs when the loop is aligned with the neutral plane.

Show then that the induced voltage in the kth *loop when it undergoes commutation is*

$$-\frac{d\psi_k(i, i_f, \theta_R)}{dt} = -\frac{\partial\psi_k(i, i_f, \theta_R)}{\partial i}\frac{di}{dt} - \frac{\partial\psi_k(i, i_f, \theta_R)}{\partial i_f}\frac{di_f}{dt}$$

(c) Explain why shifting the brushes alleviates arcing.

(d) Does the amount that the brushes are to be rotated depend on the amount of armature current?

Problem 11 Flux in the Multiloop Motor

With $\lambda(\theta_R)$ given by (1.8) and $n = 4$, write a program to plot $\lambda(\theta_R)/(\ell_1\ell_2 B)$ versus θ_R to obtain a figure similar to that of Figure 1.38.

Separately Excited, Series and Shunt DC Machines

Problem 12 Conservation of Energy in the Separately Excited DC Motor
Using the equations (1.19), (1.20) and (1.21) of the separately excited DC motor show that energy conservations holds. Give a physical interpretation to the various expressions.

Problem 13 Series DC Motor [1]
Consider a separately excited DC motor in which the terminal T_1 of the armature is connected to the terminal T_2' of the field circuit and a single voltage source is applied between the two remaining terminals T_1' and T_2. This configuration is referred to as a series DC motor *and is illustrated in Figure 1.57. This type of connection is often used when the motor is employed as a traction drive, that is, for subways, trolley cars, and so on. An equivalent circuit for the series DC motor is given in Figure 1.58 where $\lambda_f(i_f) = L_f i_f$ is used.*

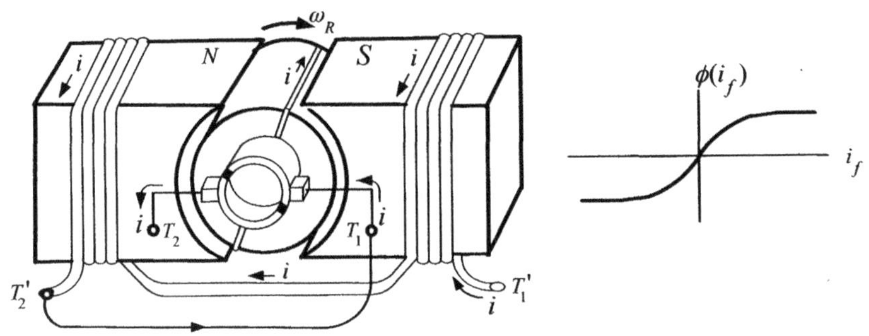

FIGURE 1.57. Series-connected DC motor. Adapted from Ref. [21].

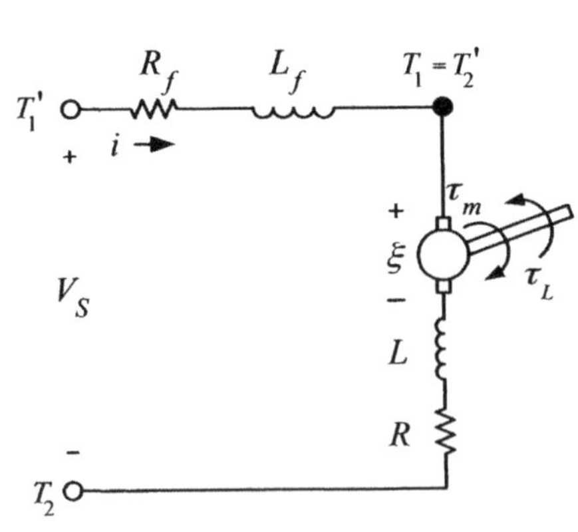

FIGURE 1.58. Equivalent circuit for a series DC motor.

(a) Starting from the model (1.22), (1.23), (1.24) of the separately excited DC motor, derive the set of differential equations that characterize the series connected DC motor.

(b) Show that the torque cannot change sign, i.e., it is always positive or always negative.

(c) What must be done to change the sign of the torque?

Problem 14 Shunt DC Motor [3]

Consider a separately excited DC motor in which the terminal T_1 of the armature is connected to the terminal T_1' of the field circuit and similarly, the other two terminals T_2, T_2' are connected together as indicated in Figure 1.59. This configuration is referred to as a shunt DC motor *for which an equivalent circuit is shown in Figure 1.60. The resistance R_{adj} is an adjustable resistance added in series with the field winding and aids in the control of the motor.*

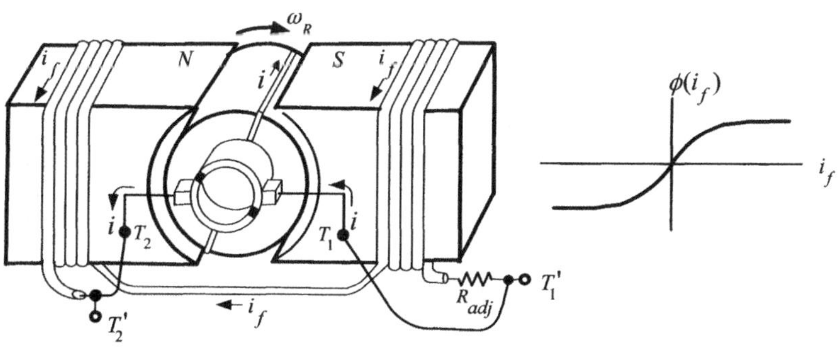

FIGURE 1.59. Shunt connected DC motor. Adapted from Ref. [21].

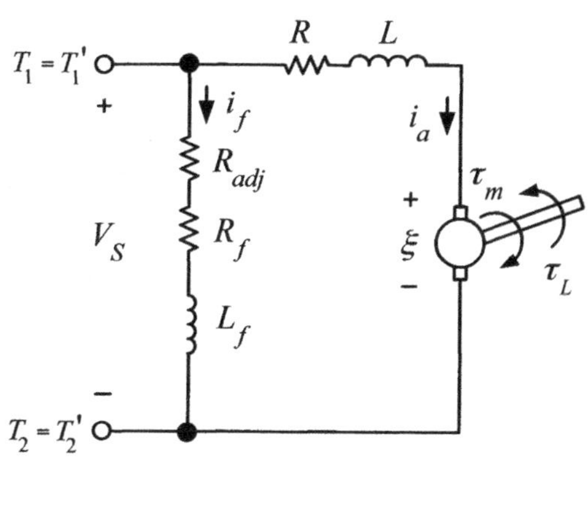

FIGURE 1.60. Equivalent circuit for a shunt DC motor.

Let $\lambda(i_f) = L_f i_f$ and using the model (1.22), (1.23), (1.24) of the separately excited DC motor, derive the complete set of equations that characterize the shunt DC motor.

Simple AC Generators

Problem 15 A Three-Phase Generator [38]
Consider a simplified model of a three-phase generator shown in Figures 1.61 and 1.62.

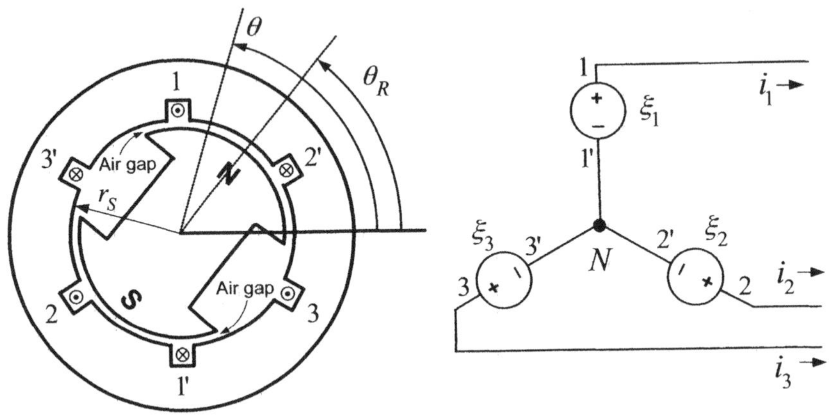

FIGURE 1.61. Simplified three-phase permanent magnet generator.

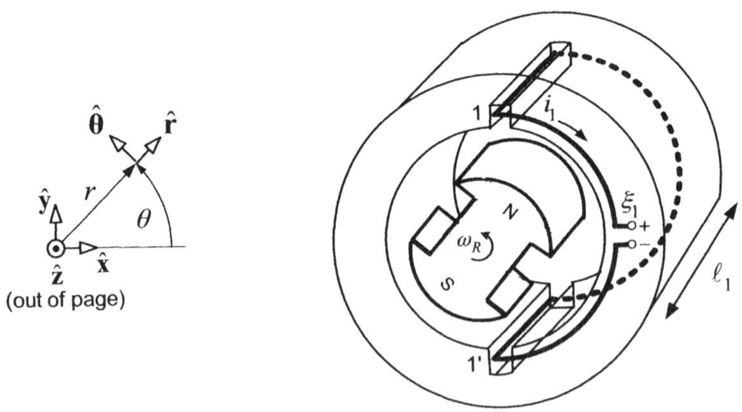

FIGURE 1.62. Half-cylindrical-shaped winding of phase 1.

The rotor consists of a two-pole (one north and one south) permanent magnet with the pole faces shaped so that the radial magnetic field in the air gap due to the rotor's permanent magnet is given by

$$\vec{\mathbf{B}}(r,\theta) = B_{R\max}\frac{r_S}{r}\cos(\theta - \theta_R)\hat{\mathbf{r}}$$

where $B_{R\max} > 0, r_S = \ell_2/2 + g$ *is the radius of the inside surface of the stator iron and* θ, θ_R *are defined as in Figure 1.61. One of the stator phases (consisting of a single loop) is shown in Figure 1.62 where the rotor and stator sizes are distorted for expository reasons.*

(a) Using a half-cylinder flux surface whose boundary is the stator loop and whose surface normal is $\hat{\mathbf{n}} = \hat{\mathbf{r}}$, *compute the flux* ϕ_1 *in stator loop 1 and the voltage* $\xi_1 = -d\phi_1/dt$ *induced in this loop.*

(b) Also, compute the fluxes ϕ_2, ϕ_3 *and the induced voltages* ξ_2, ξ_3 *in phases 2 and 3, respectively.*

(c) With the ends $1', 2', 3'$ *tied together (called the neutral point), show that if the machine is rotating at constant angular speed* ω_R, *this results in a three-phase generator producing three sinusoidal voltages which are identical except being* $120°$ *out of phase with each other.*

(d) It is shown in Chapters 4 and 5 that the magnetic field due to the rotating permanent magnet rotor produces an axial electrical field *in the air gap given by*

$$\vec{\mathbf{E}}_R(\theta - \theta_R) = \omega_R B_{R\max} r_S \cos(\theta - \theta_R)\,\hat{\mathbf{z}}.$$

Show that

$$\xi_1 = \int_{1'}^{1} \vec{\mathbf{E}}_R(\theta - \theta_R) \cdot d\vec{\ell}.$$

Problem 16 A Two-Phase Four-Pole Generator

To introduce a two-phase four-pole *generator, consider first Figure 1.63(a), which shows a two-phase* two-pole *generator. Its rotor is a two-pole (one north and one south) permanent magnet whose pole faces are shaped so that the radial air gap magnetic field is given by*

$$\vec{\mathbf{B}}(r,\theta) = B_{R\max}\frac{r_S}{r}\cos(\theta - \theta_R)\hat{\mathbf{r}}$$

where $B_{R\max} > 0, r_S = \ell_2/2 + g$ *is the radius of the inside surface of the stator iron, and* θ, θ_R *are defined as in Figure 1.63(a). The stator phases are wound in a similar fashion to those of problem 15, and the voltages induced in the stator phases are computed similarly.*

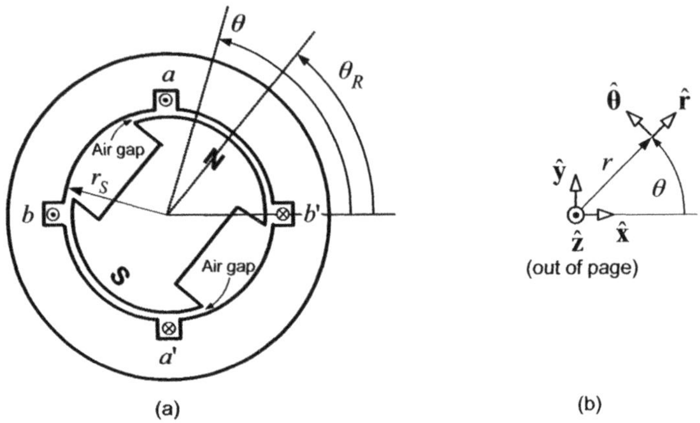

FIGURE 1.63. (a) A two-phase two-pole machine. (b) Polar coordinate system.

Now consider a two-phase four-pole *generator illustrated in Figure 1.64(a). The rotor is a* four-pole *(two north poles and two south poles) permanent magnet whose pole faces are shaped so that the radial air gap magnetic field is given by*

$$\vec{\mathbf{B}}(r,\theta) = B_{R\max}\frac{r_S}{r}\cos\left(n_p(\theta-\theta_R)\right)\hat{\mathbf{r}}.$$

Here $B_{R\max} > 0$, $r_S = \ell_2/2 + g$ *is the radius of the inside surface of the stator iron,* θ_R *is defined as in Figure 1.64(a), and* n_p $(= 2$ *in Figure 1.64)* *is the number of* pole pairs. *Figure 1.64(c) is a perspective view showing how phase a is wound while Figure 1.64(d) is a cross-sectional view of the same phase.*

(a) Using an outward normal for the flux surface ($\hat{\mathbf{n}} = \hat{\mathbf{r}}$), *compute the flux linkage* λ_a *in phase a due to the permanent magnet rotor. (The flux linkage is the sum of the fluxes in the two loops* $a_1 - a_1'$ *and* $a_2 - a_2'$ *making up phase a.)*

(b) Does the positive direction of travel around each of the two flux surfaces of phase a coincide with the positive direction of current?

(c) Compute the voltage ξ_a *induced in phase a by the permanent magnet rotor.*

(d) Using the expression for ξ_a *in phase a from part (c), and the fact that phase b is rotated* $\pi/(2n_p)$ *radians from phase a, give the expression for* ξ_b.

(e) It is shown in Chapters 4 and 5 that the magnetic field due to the rotating permanent magnet rotor produces an axial electrical field *in the air gap given by*

$$\vec{\mathbf{E}}_R\left(\theta-\theta_R\right) = \omega_R B_{R\max} r_S \cos\left(n_p(\theta-\theta_R)\right)\hat{\mathbf{z}}.$$

Show that

$$\xi_a = \int_{a'}^{a} \vec{\mathbf{E}}_R(\theta - \theta_R) \cdot d\vec{\ell}.$$

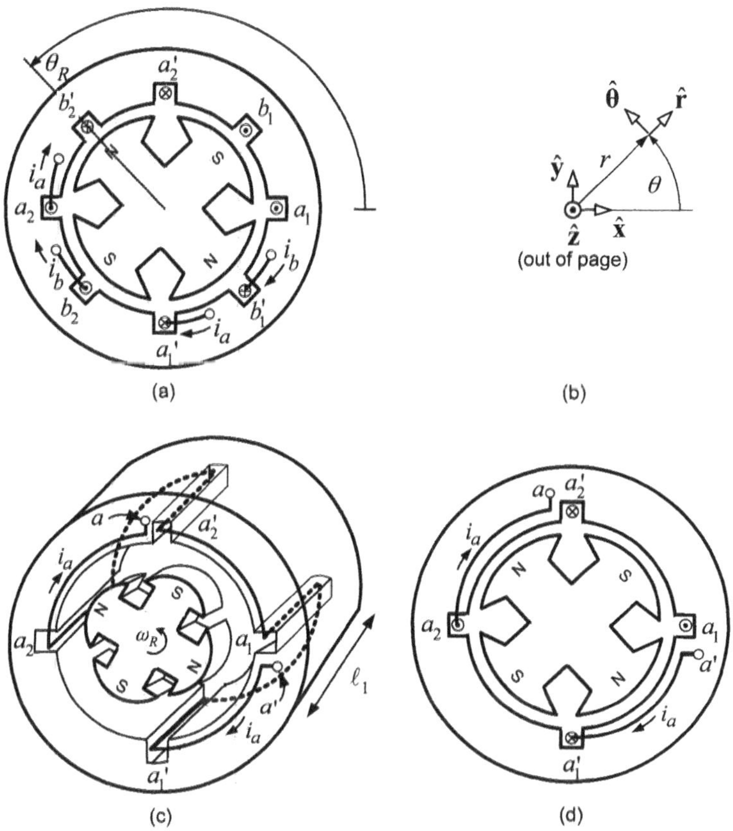

FIGURE 1.64. (a) A two-phase four-pole machine. (b) Polar coordinate system. (c) A perspective view of the phase a winding. (d) Cross-sectional view of the phase a winding.

Problem 17 Generator/Motor

Do problem 6 in Chapter 5. (The background in this chapter is sufficient to work this problem!)

2

Feedback Control

The DC motor model developed in the previous chapter is used to illustrate basic control techniques. The techniques developed here will be straightforwardly generalized for the AC motor case.

2.1 Model of a DC Motor Servo System

A DC motor servo (positioning) system typically consists of a DC motor, amplifier and sensors for position and current measurements. The interest here is to understand how to model this system for control purposes. Recall the dynamic equations of the DC motor given as

$$
\begin{aligned}
L\frac{di}{dt} &= -Ri(t) - K_b\omega(t) + v(t) \\
J\frac{d\omega}{dt} &= -f\omega(t) + K_T i(t) - \tau_L(t) \\
\frac{d\theta}{dt} &= \omega(t)
\end{aligned}
\tag{2.1}
$$

where, in this chapter, θ and ω replace θ_R and ω_R, respectively. Also, $-f\omega$ models the viscous friction torque on the motor due to both the bearings and to the brushes rubbing against the commutator. The voltage v is commanded to the motor through a power amplifier. The amplifier is limited in how much voltage it can actually put out. This limit is denoted by $V_{\max}$ in Figure 2.1.

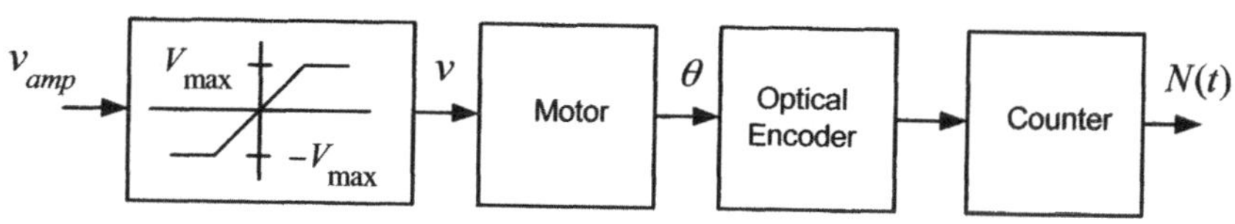

FIGURE 2.1. Open-loop DC motor.

A common position sensor used in industry is an optical encoder, which is illustrated in Figure 2.2(a) [39]. As shown, the optical encoder consists of a set of windows spaced equally around a circular disk with a light source shining through the window when it is aligned with the source. A detector puts out a high voltage when there is light and a low voltage otherwise.

For the setup of Figure 2.2(a), there are 12 windows (lines or slots) so that for every complete revolution of the circular disk (i.e., of the motor), there will be 12 pulses. Using digital electronic circuitry, one can detect a pulse going high or low so that in one revolution with 12 pulses, there will be a total of 24 times that a pulse went either high or low. Note that each time a pulse goes high or low, the motor has rotated $2\pi/24$ or $360°/24 = 15°$. By simply counting the number N of rising and falling edges of the pulse, one can obtain the position of the rotor to within 15°.

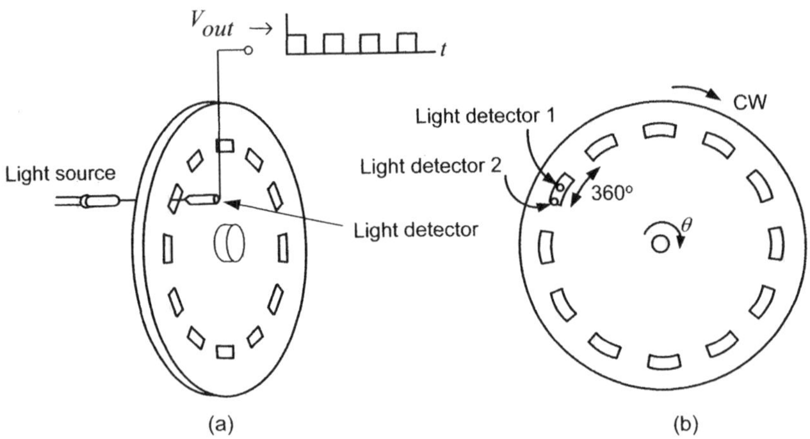

FIGURE 2.2. Schematic diagram of an optical encoder. (a) Encoder disk with 12 windows. Reprinted with permission from *Mechatronics: An Integrated Approach* by Clarence W. deSilva, Copyright CRC Press, Boca Raton, Florida [40]. (b) A second light detector added 90° from the first one in order to detect the direction of rotation.

In order to detect the direction of rotation, two light detectors are used as shown in Figure 2.2(b). In detail, the length of the windows is the same as the length of the distance between windows. The two light detectors are placed a distance of 1/2 of a window length apart. One period of the voltage waveform coming out of the light detector corresponds to the distance from the beginning of one window to the next and this cycle of the voltage waveform is considered to be 360° [see Figure 2.2(b)]. Consequently, the two light detectors are considered to be 90° apart and are said to be in *quadrature*.

Figure 2.3(a) shows the voltage waveforms out of the two light detectors when the rotor is turning *clockwise*. Note that the voltage of light detector 1 is 90° (1/4 cycle) behind that of light detector 2; that is, the voltage from light detector 2 goes high first and then a quarter of a cycle later, the voltage from light detector 1 goes high. However, if the rotor is turning *counterclockwise*, then the voltage of light detector 2 is 90° behind that of light detector 1 as shown in Figure 2.3(b). Consequently, the encoder electronics detects the relative phase of the two light detector voltage signals and uses it to determine whether the rising or falling edges of a pulse should increase the count (clockwise motion) or decrease the count (counterclockwise motion). For example, one could monitor light detector 2. If light detector1 goes high before light detector 2 goes low as in Figure 2.3(a), then the motion is clockwise. Otherwise, if light detector 1 goes low before light detector 2 goes low as in Figure 2.3(b), the motion is counterclockwise.

If an optical encoder has N_w windows (lines/slots), then there are $2N_w$ rising and falling edges per revolution giving a resolution of $2\pi/(2N_w)$ radians. Problem 4 shows how using the output of both light detectors, a resolution of $2\pi/(4N_w)$ radians is achieved. So, for example, with $N_w = 500$, the resolution of the encoder is $2\pi/2000$ radians or $360°/2000 = 0.18°$.

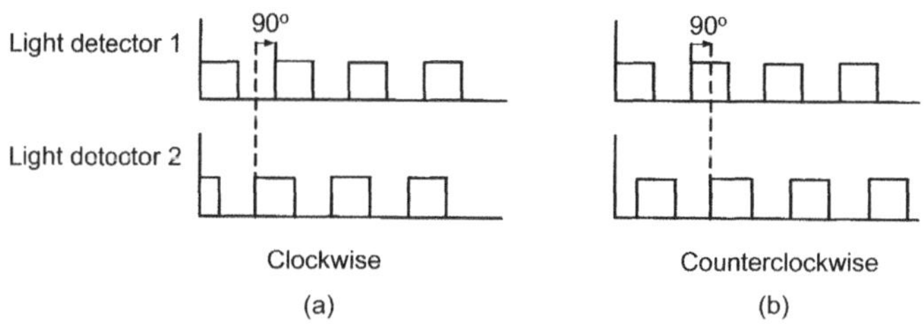

FIGURE 2.3. (a) Detector outputs for clockwise rotation. (b) Detector outputs for counterclockwise rotation.

Figure 2.4 shows a plot of the position $\theta(t)$ and the corresponding encoder

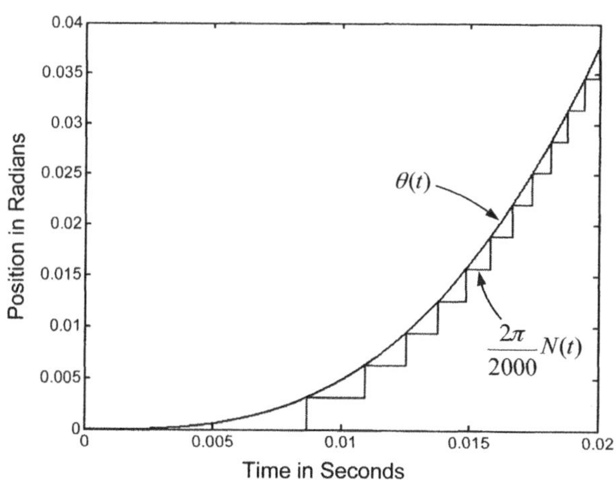

FIGURE 2.4. Plot of $\theta(t)$ and the encoder output $(2\pi/2000)N(t)$.

output $(2\pi/2000)N(t)$. Due to the way the encoder works, if $\theta(t)$ is increasing, the position output of the encoder is always less than or equal to the actual position.

Current Command

The input to the motor is the voltage v. However, from the above equations, the torque equation is $Jd\omega/dt = -f\omega + K_T i - \tau_L$, where the motor torque $K_T i$ is proportional to the current. Thus, it would be convenient if the current was the input since one could easily specify the motor's torque by specifying the current. Typically, to get around the fact that the voltage is the input, one designs an inner current control loop that allows direct current command. That is, the voltage is forced by a controller to go to whatever value necessary to obtain the desired current. To understand how this is done, consider the equations (2.1) in the Laplace domain given by

$$\begin{aligned} i(s) &= \frac{-K_b\omega(s) + V(s)}{sL + R} \\ \omega(s) &= \frac{K_T i(s) - \tau_L(s)}{sJ + f} \\ \theta(s) &= \frac{1}{s}\omega(s). \end{aligned}$$

These algebraic relationships are illustrated in the block diagram shown in Figure 2.5.

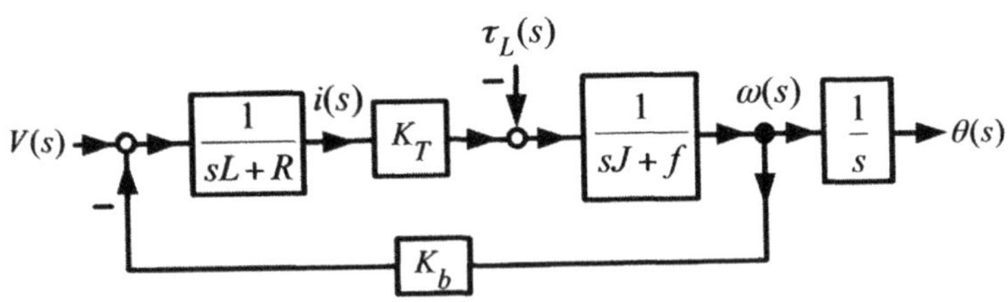

FIGURE 2.5. Block diagram of a DC motor.

Often the approximation that $L = 0$ is made to simplify the analysis of the system. However, a standard approach in industry is to put the amplifier into a *current command* mode which results in making the effect of L negligible. To do so, the current is fed back (typically using analog electronics) through a proportional controller as shown in Figure 2.6.

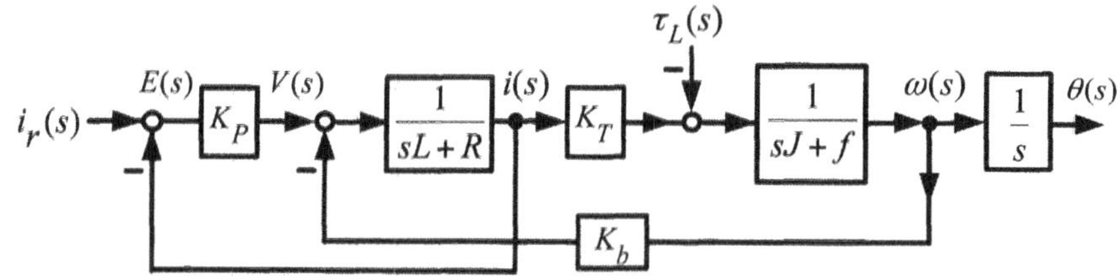

FIGURE 2.6. DC motor with an inner current control loop.

Here, $i_r(s)$ is the Laplace transform of the reference current, $i(s)$ is the Laplace transform of the actual current in the motor, and $K_P > 0$ is a proportional gain. The transfer function $G(s) = \omega(s)/i_r(s)$ is easily found using the block diagram reduction method. In order to do so, note that this block diagram is equivalent to that of Figure 2.7.

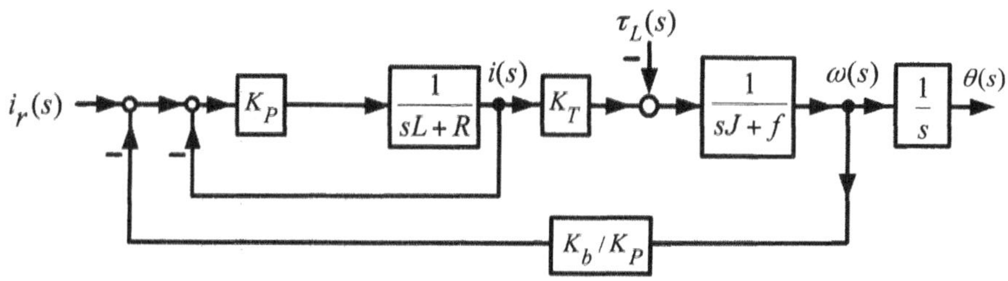

FIGURE 2.7. Equivalent block diagram.

The block diagram of Figure 2.7 is then easily seen to simplify to that of Figure 2.8.

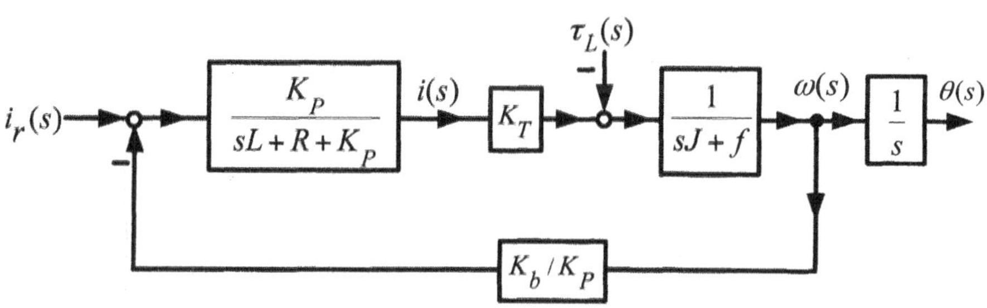

FIGURE 2.8. Simplified block diagram.

As a result, with $\tau_L(s) \equiv 0$,

$$\begin{aligned} G(s) &\triangleq \omega(s)/i_r(s) \\ &= \frac{K_P K_T}{(sL+R+K_P)(sJ+f)+K_T K_b} \\ &= \frac{K_T}{(\frac{sL+R}{K_P}+1)(sJ+f)+K_T K_b/K_P}. \end{aligned}$$

Using high-gain feedback, that is, letting $K_P \to \infty$, $G(s)$ reduces to

$$G(s) = \omega(s)/i_r(s) = \frac{K_T}{(sJ+f)}. \tag{2.2}$$

In other words, if the gain K_P can be made large enough, the actual current $i(t)$ can be forced to track $i_r(t)$ quite fast. However, note that one cannot make the gain K_P arbitrarily large. This is easily seen by noting that the voltage commanded into the amplifier is

$$v(t) = K_P(i_r(t) - i(t))$$

which, for large gains K_P, could be greater than $V_{\max}$, causing the amplifier to saturate.

In summary, with a good current controller, the voltage $v(t)$ is automatically adjusted to force $i(t) \to i_r(t)$ fast enough that the reference (commanded) current $i_r(t)$ can be considered to be equal to the actual motor current $i(t)$. That is, for all practical purposes, the dynamics from $i_r(t)$ to $i(t)$ can be neglected allowing one to consider the input $i_r(t)$ as equal to the motor current $i(t)$. The following reduced-order system can then be used to design the controller:

$$\begin{aligned} \frac{d\omega}{dt} &= (K_T/J)i_r(t) - (f/J)\omega(t) - \tau_L/J \\ \frac{d\theta}{dt} &= \omega. \end{aligned}$$

The corresponding block diagram of the reduced-order system is shown in Figure 2.9.

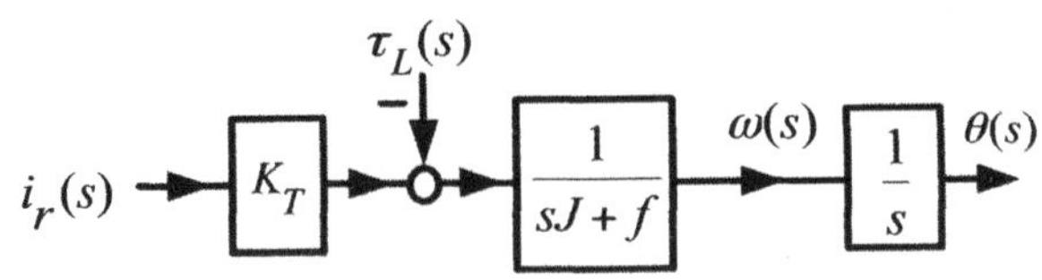

FIGURE 2.9. Reduced-order model of a DC motor.

Speed Controller

Using the reduced-order model, it is straightforward to design a simple proportional speed controller as illustrated in Figure 2.10.

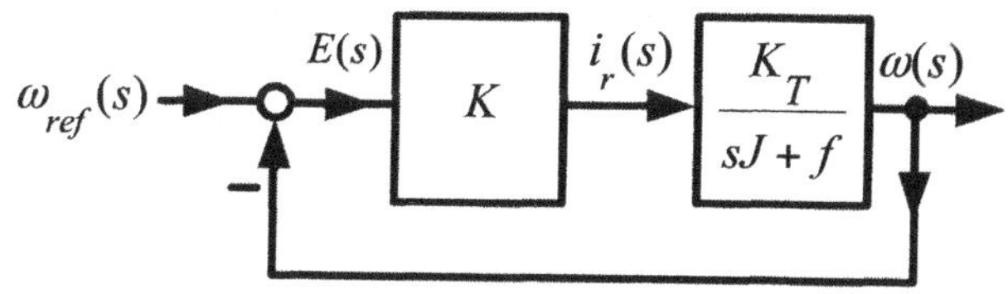

FIGURE 2.10. Simple speed controller for a DC motor.

From the block diagram of Figure 2.10, it follows that

$$\begin{aligned}\omega(s)/\omega_{ref}(s) &= \frac{KK_T/(sJ+f)}{1+KK_T/(sJ+f)} = \frac{KK_T/J}{s+(f+K_TK)/J}\\ &\approx \frac{KK_T/J}{s+K_TK/J}\\ &= \frac{1}{\tau_m s+1}\end{aligned}$$

where $\tau_m \triangleq \dfrac{1}{K_TK/J}$ and $f << K_TK$ were used. This is an example of the classical approach to control and is described in more detail in the Appendix of this chapter. However, in this chapter, the state-space approach to feedback control is pursued.

2.2 Speed Estimation

Typically, in a DC motor drive system, the current, voltage, and position are available by direct measurement. However, the speed is not usually measured directly. In this section, two approaches to estimating the speed are given.

2.2.1 Backward Difference Estimation of Speed

The optical encoder gives the position measurement, but not the speed of the motor. However, one can use this measurement to deduce the speed. The most straightforward way is to compute the *backward difference* of the position and divide by the sample period, that is,

$$\hat{\omega}_{bd}(kT) \triangleq \frac{2\pi}{2000}\left(\frac{N(kT)-N(kT-T)}{T}\right)$$

where T is the time between samples and $N(kT)$ is the optical encoder count at time kT.

The error in estimating the speed by differentiation of the position measurements can be found as follows: At any discrete time $kT, N(kT)$ is in error by at most one encoder count. In particular, $N(kT)$ can be only too small by at most one encoder count ($N(kT)$ is never too large because of the way the encoder works). Thus, with $\theta(kT)$ the true position of the motor in radians,

$$\theta(kT) = \frac{2\pi}{2000}N(kT) + \frac{2\pi}{2000}e(kT)$$

where $e(kT)$ represents the *positive fractional* count that the encoder cannot sense; that is, $0 \leq e(kT) < 1$ for all k. Then, the speed may be written as

$$\begin{aligned}\omega(kT) &= \left(\frac{\theta(kT) - \theta(kT-T)}{T}\right) \\ &= \frac{2\pi}{2000}\left(\frac{N(kT) - N(kT-T)}{T}\right) + \frac{2\pi}{2000}\left(\frac{e(kT) - e(kT-T)}{T}\right)\end{aligned}$$

where, as $0 \leq e(kT) \leq 1$ and $0 \leq e((k-1)T) \leq 1$, it follows that $|e(kT) - e(kT-T)| \leq 1$.

It is now straightforward to compute a bound on the error in estimating the speed. As the speed estimate is given by

$$\hat{\omega}_{bd}(kT) = \frac{2\pi}{2000}\left(\frac{N(kT) - N(kT-T)}{T}\right)$$

and the difference $e(kT) - e(kT-T)$ is bounded by ± 1,

$$|\text{Error in } \hat{\omega}_{bd}(kT)| = \frac{2\pi}{2000}\left|\frac{e(kT) - e(kT-T)}{T}\right| \leq \frac{2\pi}{2000}\frac{1}{T}.$$

As the sampling rate increases (T gets smaller), the error gets larger. Of course, as the sampling rate decreases, the approximation

$$\omega(kT) \approx \left(\frac{\theta(kT) - \theta(kT-T)}{T}\right)$$

becomes less and less valid. The choice of T is a trade-off between the error and the accuracy of the finite difference in approximating the derivative. One way to decrease this error would be to use an encoder with higher resolution. Such encoders are typically more expensive and cannot operate at higher speeds (as the speed increases, a large number of pulses are coming in so fast that the pulse detection circuitry cannot keep up).

Figure 2.11 is a plot of the speed estimated by the backward difference method. In this example, the sample period is $T = 0.5$ msec and the encoder has a resolution of $2\pi/2000$ so that error bound is $(2\pi/2000)/T = 6.28$ radians/sec.

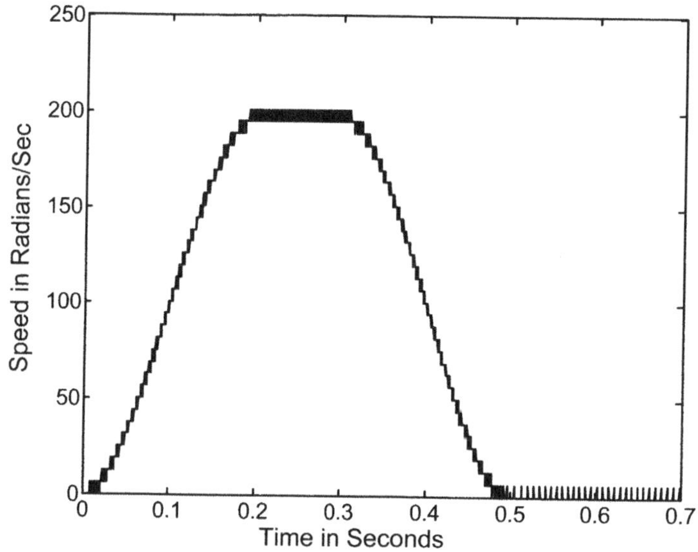

FIGURE 2.11. Plot of the speed computed using the backward difference. The error bound in this example is $(2\pi/2000)/(0.0005) = 6.28$ radians/sec.

2.2.2 Estimation of Speed Using an Observer

The differentiation of the position output from the optical encoder can result in a low-resolution, noisy estimate of the speed. Here a different approach is considered. With $\tau_L = 0$ (the case $\tau_L \neq 0$ is explored in problem 13), the system equations for the DC motor are

$$\begin{aligned} \frac{d\theta}{dt} &= \omega \\ \frac{d\omega}{dt} &= -(f/J)\omega + (K_T/J)i(t). \end{aligned} \tag{2.3}$$

Now, consider an *observer* defined by

$$\begin{aligned} \frac{d\hat{\theta}}{dt} &= \hat{\omega} + \ell_1(\theta - \hat{\theta}) \\ \frac{d\hat{\omega}}{dt} &= (K_T/J)i(t) - (f/J)\hat{\omega} + \ell_2(\theta - \hat{\theta}) \end{aligned} \tag{2.4}$$

where $\theta(t) = (2\pi/2000)N(t)$ is the (discretized) position measurement from the encoder and $\hat{\theta}, \hat{\omega}$ are the *estimates* of the position and speed, respectively. If ℓ_1,ℓ_2 are chosen correctly, the solution $\hat{\omega}$ to (2.4) gives a good estimate of the speed. To motivate this approach, let $\ell_1 = 0$ and $\ell_2 = 0$ so that (2.4) becomes

$$\begin{aligned} \frac{d\hat{\theta}}{dt} &= \hat{\omega} \\ \frac{d\hat{\omega}}{dt} &= (K_T/J)i(t) - (f/J)\hat{\omega} \end{aligned} \tag{2.5}$$

and thus (2.5) is just a real-time *simulation* of the motor. That is, the actual current in the motor is sampled and brought into a computer processor, the equations (2.5) are integrated in *realtime* by the microprocessor (initialized with the same values of speed and position as the actual motor) and the solution $\hat{\omega}$ is then used as the estimate of the speed. However, there are problems with using (2.5) to estimate the speed. If there is any disturbance that causes the actual motor speed to change, then $\hat{\omega}$ will no longer be correct as such a disturbance is *not* modeled in (2.5) and thus $\hat{\omega}$ will not change to respond to it. After a disturbance has acted on the motor (e.g., suppose the motor shaft is bumped at some time $t = t_d$), its effect can be mathematically modeled as a change of initial conditions for equations (2.3). That is, the motor is described by the system of equations (2.3), but the initial condition $\omega(t_d+)$ is unknown ($\theta(t_d+)$ is measured and thus known at t_d). This is where the error term $\theta - \hat{\theta}$ in (2.4) becomes so important. For example, suppose $\hat{\omega} < \omega$ for a period of time resulting in $\hat{\theta} < \theta$ (as $\hat{\theta}$ is computed by integrating $\hat{\omega}$). Then, with $\ell_2 > 0$, the error term $\ell_2(\theta - \hat{\theta}) > 0$ will cause $\hat{\omega}$ to increase in (2.4) reducing the error $\omega - \hat{\omega}$.

A mathematical argument is now presented to show that, by choosing the gains ℓ_1, ℓ_2 appropriately, the estimate $\hat{\omega}(t)$ from (2.4) can be made to converge to $\omega(t)$ as fast as desired *without* knowledge of the initial condition for ω. To this end, let

$$\begin{aligned} e_1(t) &= \theta(t) - \hat{\theta}(t) \\ e_2(t) &= \omega(t) - \hat{\omega}(t) \end{aligned}$$

so that the dynamic equations for $e_1(t)$ and $e_2(t)$ are found by simply subtracting (2.4) from (2.3) to get

$$\begin{aligned} \frac{de_1}{dt} &= e_2 - \ell_1 e_1 \\ \frac{de_2}{dt} &= -(f/J)e_2 - \ell_2 e_1 \end{aligned} \tag{2.6}$$

Notice that the motor torque terms $(K_T/J)i(t)$ in (2.3) and (2.4) canceled each other out.

This system of differential equations can be solved by computing the Laplace transform of (2.6) to obtain

$$\begin{aligned} sE_1(s) - e_1(0) &= E_2(s) - \ell_1 E_1(s) \\ sE_2(s) - e_2(0) &= -(f/J)E_2(s) - \ell_2 E_1(s). \end{aligned}$$

In matrix notation, this is written as

$$\begin{bmatrix} s+\ell_1 & -1 \\ \ell_2 & s+f/J \end{bmatrix} \begin{bmatrix} E_1(s) \\ E_2(s) \end{bmatrix} = \begin{bmatrix} e_1(0) \\ e_2(0) \end{bmatrix}.$$

Multiplying both sides of this matrix equation on the left by the inverse of the 2×2 matrix gives

$$\begin{bmatrix} E_1(s) \\ E_2(s) \end{bmatrix} = \frac{1}{s^2 + (\ell_1 + f/J)s + \ell_2 + \ell_1(f/J)} \begin{bmatrix} s+f/J & 1 \\ -\ell_2 & s+\ell_1 \end{bmatrix} \begin{bmatrix} e_1(0) \\ e_2(0) \end{bmatrix}.$$

In particular,

$$E_2(s) = \frac{-\ell_2 e_1(0) + (s+\ell_1)e_2(0)}{s^2 + (\ell_1 + f/J)s + \ell_2 + \ell_1(f/J)}.$$

It is required that $\omega(t) - \hat{\omega}(t) = e_2(t) = \mathcal{L}^{-1}(E_2(s)) \to 0$ as $t \to \infty$ for any values of $e_1(0)$ and $e_2(0)$. To do this, choose ℓ_1 and ℓ_2 so that

$$\begin{aligned} s^2 + (\ell_1 + f/J)s + \ell_2 + \ell_1(f/J) &= (s+p_1)(s+p_2) \\ &= s^2 + (p_1+p_2)s + p_1 p_2 \end{aligned}$$

with $p_1 > 0, p_2 > 0$. That is, choose

$$\begin{aligned} \ell_1 &= p_1 + p_2 - f/J \\ \ell_2 &= p_1 p_2 - \ell_1(f/J). \end{aligned}$$

Then

$$\begin{aligned} E_2(s) &= \frac{-\ell_2 e_1(0) + (s+\ell_1)e_2(0)}{(s+p_1)(s+p_2)} \\ &= \frac{A}{(s+p_1)} + \frac{B}{(s+p_2)} \end{aligned}$$

so that

$$e_2(t) = Ae^{-p_1 t} + Be^{-p_2 t} \to 0$$

as $t \to \infty$ for any $e_1(0) = \theta(0) - \hat{\theta}(0)$ and $e_2(0) = \omega(0) - \hat{\omega}(0)$.

Remark The importance of the error term in the observer equations should be understood. Note that one would typically know that the initial speed is zero and the initial position is also known (it is measured) so that $e_1(0) = \theta(0) - \hat{\theta}(0) = 0$ and $e_2(0) = \omega(0) - \hat{\omega}(0) = 0$. Consequently, even with $\ell_1 = \ell_2 = 0$, the solution to (2.6) would be $e_1(t) = e_2(t) \equiv 0$. However, in practice if the observer was implemented with the gains equal to zero, it would not work. For example, there could be a torque disturbance acting on the motor for a short period of time. Specifically, let

$$\tau_L(t) = \begin{cases} \tau_{L0} & \text{for } t_1 \leq t \leq t_2 \\ 0 & \text{otherwise.} \end{cases}$$

Then, the second equation of (2.3) is

$$\frac{d\omega}{dt} = -(f/J)\omega + (K_T/J)i(t) - \tau_L(t)/J.$$

During the time this disturbance torque acts on the motor, the observer's speed estimate given by (2.4) will diverge from the true value since the observer equations (2.4) do not have this disturbance modeled. However, after the disturbance has quit acting on the motor at time t_2, the observers equation are again valid and the error term $e_1(t) = \theta(t) - \hat{\theta}(t)$ will force the estimate $\hat{\omega}(t) \to \omega(t)$. That is, at time t_2, the motor and observer are correctly given by (2.3) and (2.4), respectively. Thus, the error system is correctly given by (2.5) with the "initial condition" $\hat{\omega}(t_2)$ at time t_2 *unknown*. As shown above, in spite of $\hat{\omega}(t_2)$ being unknown, the error $e_2(t) = \omega(t) - \hat{\omega}(t) \to 0$ if the gains ℓ_1, ℓ_2 are chosen properly.

Discretization of the Observer Equations

Using a simple Euler integration routine, the observer equations (2.4) may be implemented in discrete time as

$$\begin{aligned} \hat{\theta}\left((k+1)T\right) &= \hat{\theta}\left(kT\right) + T\hat{\omega}(kT) + T\ell_1 e_1(kT) \\ \hat{\omega}\left((k+1)T\right) &= \hat{\omega}\left(kT\right) - T(f/J)\hat{\omega}(kT) + T(K_T/J)i(kT) + T\ell_2 e_1(kT). \end{aligned} \tag{2.7}$$

However, the trapezoidal method described in Section 2.7.9 provides a more accurate method of numerical integration for a given sample period T.

2.3 Trajectory Generation

In this section, a simple way to generate a position and speed reference for a point-to-point move is developed. Assuming the load torque is zero, the equations for the position and speed of the motor are

$$\begin{aligned}\frac{d\theta}{dt} &= \omega \\ \frac{d\omega}{dt} &= -(f/J)\omega + (K_T/J)i_r.\end{aligned} \tag{2.8}$$

For a point-to-point move, it is required that the position reference trajectory $\theta_{ref}(t)$ satisfy $\theta_{ref}(0) = 0$ and $\theta_{ref}(t_f) = \theta_f$ where t_f is the final time and θ_f is the final desired position, that is, going from the "point" 0 to the "point" θ_f. There are many ways to do this, but consider the simple symmetric trajectory shown in Figure 2.12.

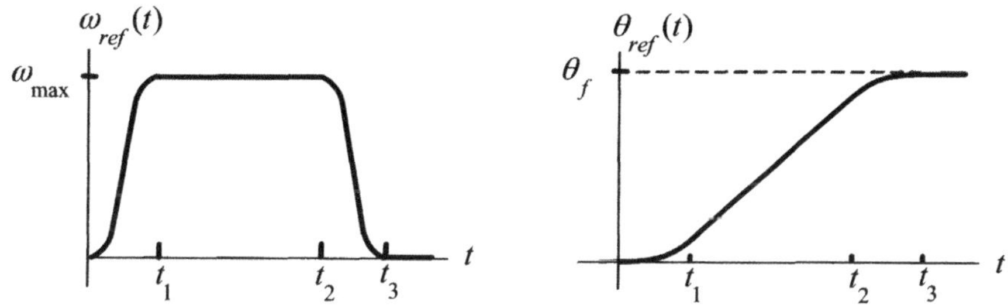

FIGURE 2.12. Reference speed and position profiles.

In order to have a smooth trajectory, that is, the acceleration reference is a continuous function of time, it is required that

$$\begin{array}{cc} \omega_{ref}(0) = 0 & \dot\omega_{ref}(0) = 0 \\ \omega_{ref}(t_1) = \omega_{\max} & \dot\omega_{ref}(t_1) = 0 \\ \omega_{ref}(t) = \omega_{\max} & t_1 \le t \le t_2 \\ \omega_{ref}(t_2) = \omega_{\max} & \dot\omega_{ref}(t_2) = 0 \\ \omega_{ref}(t_3) = 0 & \dot\omega_{ref}(t_3) = 0. \end{array}$$

It is also required that $t_3 - t_2 = t_1$ (i.e., $t_3 = t_1 + t_2$) so that the time required to decelerate the motor is the same as the amount of time to accelerate it. The speed reference is made symmetric by setting

$$\omega_{ref}(t) = \omega_{ref}(t_3 - t)$$

for $t_2 \le t \le t_3$. As the final position is to be θ_{ref}, the following must also hold:

$$\int_0^{t_f} \omega_{ref}(\tau)d\tau = \theta_f.$$

There are still many ways to define a reference trajectory and still satisfy all these conditions. To be specific, consider a polynomial reference trajectory given by

$$\omega_{ref}(t) = c_1 t^2 + c_2 t^3 \quad \text{for } 0 \le t \le t_1$$

which clearly satisfies $\omega_{ref}(0) = 0, \dot{\omega}_{ref}(0) = 0$. The conditions at t_1 become

$$\begin{aligned}\omega_{ref}(t_1) &= c_1 t_1^2 + c_2 t_1^3 = \omega_{\max} \\ \dot{\omega}_{ref}(t_1) &= 2c_1 t_1 + 3c_2 t_1^2 = 0\end{aligned}$$

or

$$\begin{bmatrix} t_1^2 & t_1^3 \\ 2t_1 & 3t_1^2 \end{bmatrix} \begin{bmatrix} c_1 \\ c_2 \end{bmatrix} = \begin{bmatrix} \omega_{\max} \\ 0 \end{bmatrix}.$$

This has the unique solution

$$\begin{bmatrix} c_1 \\ c_2 \end{bmatrix} = \frac{1}{t_1^4} \begin{bmatrix} 3t_1^2 & -t_1^3 \\ -2t_1 & t_1^2 \end{bmatrix} \begin{bmatrix} \omega_{\max} \\ 0 \end{bmatrix} = \begin{bmatrix} +3\omega_{\max}/t_1^2 \\ -2\omega_{\max}/t_1^3 \end{bmatrix}.$$

The distance traveled at time t_1 is then

$$\begin{aligned}\theta_{ref}(t_1) &= \int_0^{t_1} \omega_{ref}(\tau)d\tau = c_1 t_1^3/3 + c_2 t_1^4/4 = \frac{3\omega_{\max}}{t_1^2}\frac{t_1^3}{3} - \frac{2\omega_{\max}}{t_1^3}\frac{t_1^4}{4} \\ &= \frac{\omega_{\max} t_1}{2}.\end{aligned}$$

The speed reference trajectory is

$$\omega_{ref}(t) = \begin{cases} c_1 t^2 + c_2 t^3 & 0 \le t \le t_1 \\ \omega_{\max} & t_1 \le t \le t_2 \\ c_1(t_3 - t)^2 + c_2(t_3 - t)^3 & t_2 \le t \le t_3. \end{cases}$$

As the total distance traveled at time t_3 must be θ_f, it follows that

$$\begin{aligned}\theta_f &= \int_0^{t_3} \omega_{ref}(\tau)d\tau = 2\theta_{ref}(t_1) + \omega_{\max}(t_2 - t_1) \\ &= 2\frac{\omega_{\max} t_1}{2} + \omega_{\max}(t_2 - t_1) \\ &= \omega_{\max} t_2.\end{aligned}$$

This then puts a constraint on the choices of $\omega_{\max}$ and t_2. For example, if θ_f is specified, then $\omega_{\max} = \theta_f/t_2$.

The position reference is just the integral of the speed reference. Consequently,

$$\theta_{ref}(t) = \begin{cases} c_1 t^3/3 + c_2 t^4/4 & 0 \le t \le t_1 \\ \omega_{\max} t_1/2 + \omega_{\max}(t - t_1) & t_1 \le t \le t_2 \\ \omega_{\max} t_2 - c_1(t_3 - t)^3/3 - c_2(t_3 - t)^4/4 & t_2 \le t \le t_3. \end{cases}$$

Finally, the reference acceleration is just the derivative of the speed reference given by

$$\alpha_{ref}(t) = \begin{cases} 2c_1 t + 3c_2 t^2 & 0 \le t \le t_1 \\ 0 & t_1 \le t \le t_2 \\ -2c_1(t_3 - t) - 3c_2(t_3 - t)^2 & t_2 \le t \le t_3. \end{cases}$$

The corresponding current reference i_{ref} is then found setting ω equal ω_{ref} in (2.8) and solving for the current to obtain

$$i_{ref}(t) \triangleq \frac{J\alpha_{ref}(t) + f\omega_{ref}(t)}{K_T}.$$

By design, the acceleration, speed, and position references are continuous functions of time. Figure 2.13 is a plot of the acceleration reference $\alpha_{ref}(t)$ and the jerk reference $j_{ref} \triangleq d\alpha_{ref}/dt$, where it is seen that j_{ref} is not continuous at $t = 0, t_1, t_2$, and t_3. In the figure

$$\begin{aligned} \alpha_{\max} &= (3/2)\omega_{\max}/t_1 \\ j_{\max} &= -6\omega_{\max}/t_1^2. \end{aligned}$$

Problem 17 considers a trajectory design for which the jerk is continuous.

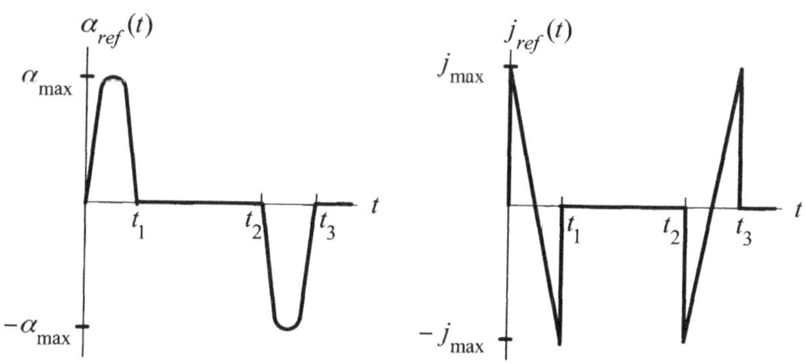

FIGURE 2.13. Left: α_{ref} versus time. Right: j_{ref} versus time.

Specifying a Reference Trajectory

A typical scenario is that θ_f is given and that one chooses t_1 and t_2 with $t_1 < t_2$. Then t_3 and $\omega_{\max}$ are specified by $t_3 = t_2 + t_1$ and $\omega_{\max} = \theta_f/t_1$, respectively. This results in a mechanical reference trajectory

$$\Big(\theta_{ref}(t), \omega_{ref}(t), \alpha_{ref}(t), j_{ref}(t)\Big).$$

The electrical reference trajectory is then determined by

$$\begin{aligned} i_{ref}(t) &\triangleq \frac{J\alpha_{ref}(t) + f\omega_{ref}(t)}{K_T} \\ v_{ref}(t) &\triangleq L\frac{di_{ref}(t)}{dt} + Ri_{ref}(t) + K_b\omega_{ref}(t). \end{aligned}$$

How does one choose t_1, t_2? One typically wants to rotate the motor as fast as possible and therefore chooses t_1 and t_2 to be small. However, the smaller the values of t_1 and t_2, the larger the peak values of $\omega_{ref}(t), \alpha_{ref}(t)$ and therefore the peak values of $i_{ref}(t)$ and $v_{ref}(t)$. It is then a little trial and error to choose values for t_1 and t_2 and check that $|i_{ref}| \leq I_{\max}$, $|v_{ref}| \leq V_{\max}$.

2.4 Design of a State Feedback Tracking Controller

Assuming current command, a *state-space* model of the DC motor is

$$\begin{aligned} \frac{d\theta}{dt} &= \omega \\ \frac{d\omega}{dt} &= (K_T/J)i_r - (f/J)\omega - \tau_L/J \end{aligned} \tag{2.9}$$

where the *state variables* are θ, ω, the *input* is i_r and the *disturbance* is τ_L. The terminology "state-space" model refers to having a system of first-order differential equations, that is, a first-order derivative on one side of the equation and a (possibly) nonlinear function of the state variables, inputs, and disturbances on the other side of the equation.

The reference trajectory and reference input are chosen to satisfy

$$\begin{aligned} \frac{d\theta_{ref}}{dt} &= \omega_{ref} \\ \frac{d\omega_{ref}}{dt} &= (K_T/J)i_{ref} - (f/J)\omega_{ref}. \end{aligned} \tag{2.10}$$

Define

$$\begin{aligned} e_1(t) &= \theta_{ref}(t) - \theta(t) \\ e_2(t) &= \omega_{ref}(t) - \omega(t) \end{aligned}$$

so that subtracting (2.9) from (2.10) results in the error system

$$\begin{aligned} \frac{de_1}{dt} &= e_2 \\ \frac{de_2}{dt} &= -(f/J)e_2 + \tau_L/J + w \end{aligned} \tag{2.11}$$

with

$$w \triangleq \frac{K_T}{J}(i_{ref} - i_r). \tag{2.12}$$

Here, i_r is the input the control engineer is free to specify. However, the approach here is to specify w such that $e_1(t) \to 0$ and $e_2(t) \to 0$ as $t \to \infty$. Then i_r is chosen so that (2.12) holds. Specifically, w is specified by

$$w = -\left(K_0 \int_0^t e_1(\tau)d\tau + K_1 e_1(t) + K_2 e_2(t)\right) \tag{2.13}$$

which results in accurate tracking of the trajectory (2.10) provided that the gains are chosen appropriately. By (2.12), this then requires choosing the input i_r as

$$i_r = i_{ref} + \frac{J}{K_T}\left(K_0 \int_0^t e_1(\tau)d\tau + K_1 e_1(t) + K_2 e_2(t)\right). \tag{2.14}$$

To find the gains, define

$$e_0(t) \triangleq \int_0^t e_1(\tau)d\tau$$

so that with the state feedback (2.13) and the system (2.11), one may write

$$\begin{aligned} \frac{de_0}{dt} &= e_1 \\ \frac{de_1}{dt} &= e_2 \\ \frac{de_2}{dt} &= -(f/J)e_2 - K_0 e_0 - K_1 e_1 - K_2 e_2 + \tau_L/J. \end{aligned} \tag{2.15}$$

To solve (2.15), its Laplace transform is computed as

$$\begin{aligned} sE_0(s) - e_0(0) &= E_1(s) \\ sE_1(s) - e_1(0) &= E_2(s) \\ sE_2(s) - e_2(0) &= -K_0 E_0(s) - K_1 E_1(s) - (K_2 + f/J)\, E_2(s) + \tau_L(s)/J. \end{aligned}$$

In matrix notation, one has

$$\begin{bmatrix} s & -1 & 0 \\ 0 & s & -1 \\ K_0 & K_1 & s + f/J + K_2 \end{bmatrix} \begin{bmatrix} E_0(s) \\ E_1(s) \\ E_2(s) \end{bmatrix} = \begin{bmatrix} e_0(0) \\ e_1(0) \\ e_2(0) \end{bmatrix} + \begin{bmatrix} 0 \\ 0 \\ \tau_L(s)/J \end{bmatrix} \tag{2.16}$$

where the inverse of the 3×3 matrix on the left-hand side of (2.16) is given by

$$\frac{1}{s^3 + (K_2 + f/J)\, s^2 + K_1 s + K_0} \times$$
$$\begin{bmatrix} s^2 + (K_2 + f/J)\, s + K_1 & s + K_2 + f/J & 1 \\ -K_0 & s^2 + (K_2 + f/J)\, s & s \\ -K_0 s & -(K_1 s + K_0) & s^2 \end{bmatrix}.$$

Let $\tau_L = 0$ (the case $\tau_L \neq 0$ is explored in problems 21 and 22). Multiplying both sides of (2.16) on the left by this inverse matrix, the Laplace transform of the state variables are found to be

$$
\begin{aligned}
E_0(s) &= \frac{\left(s^2 + (K_2 + f/J)\, s + K_1\right) e_0(0) + (s + K_2 + f/J)\, e_1(0) + e_2(0)}{s^3 + (K_2 + f/J)\, s^2 + K_1 s + K_0} \\
E_1(s) &= \frac{-K_0 e_0(0) + \left(s^2 + (K_2 + f/J)\, s\right) e_1(0) + s e_2(0)}{s^3 + (K_2 + f/J)\, s^2 + K_1 s + K_0} \qquad (2.17) \\
E_2(s) &= \frac{-K_0 s e_0(0) - (K_1 s + K_0)\, e_1(0) + s^2 e_2(0)}{s^3 + (K_2 + f/J)\, s^2 + K_1 s + K_0}.
\end{aligned}
$$

Note that all three Laplace transforms have the same denominator given by $s^3 + (K_2 + f/J)\, s^2 + K_1 s + K_0$. This is the *characteristic equation* of the system. Using the gains K_0, K_1, and K_2, the roots of the characteristic equation may be assigned to any desired values. For example, let

$$
\begin{aligned}
a(s) &= (s + r_1)(s + r_2)(s + r_3) \\
&= s^3 + (r_1 + r_2 + r_3)s^2 + (r_1 r_2 + r_1 r_3 + r_2 r_3)s + r_1 r_2 r_3
\end{aligned}
$$

be the desired characteristic polynomial and then choose the gains as

$$
\begin{aligned}
K_2 &= r_1 + r_2 + r_3 - f/J \\
K_1 &= r_1 r_2 + r_1 r_3 + r_2 r_3 \qquad (2.18) \\
K_0 &= r_1 r_2 r_3.
\end{aligned}
$$

With this choice of gains, the closed-loop poles are $p_1 = -r_1$, $p_2 = -r_2$, and $p_3 = -r_3$. Assuming these poles are distinct, $E_1(s)$ in (2.17) becomes

$$
\begin{aligned}
E_1(s) &= \frac{-K_0 e_0(0) + \left(s^2 + (K_2 + f/J)\, s\right) e_1(0) + s e_2(0)}{(s + r_1)(s + r_2)(s + r_3)} \\
&= \frac{A}{s + r_1} + \frac{B}{s + r_2} + \frac{C}{s + r_3}
\end{aligned}
$$

where A, B, and C are constants. The inverse Laplace transform of $E_1(s)$ is then

$$
e_1(t) = Ae^{-r_1 t} + Be^{-r_2 t} + Ce^{-r_3 t} \to 0
$$

as $t \to \infty$. The further the closed-loop poles are in the left-half plane, the faster $e_1(t) \to 0$ and thus, the faster $\theta(t) \to \theta_{ref}(t)$. However, note that the further in the left-half plane the closed-loop poles are chosen, the larger r_1, r_2, and r_3 and, thus, the larger the gains K_0, K_1 and K_2. The difficulty with choosing large values for the feedback gains is that the resulting feedback

$$
i_r = i_{ref} + \frac{J}{K_T}\left(K_0 \int_0^t e_1(\tau)d\tau + K_1 e_1(t) + K_2 e_2(t)\right)
$$

can be quite large and saturate the amplifier,[1] even if the errors e_0, e_1, and e_2 are small. The complete system is illustrated in the block diagram of 2.14.

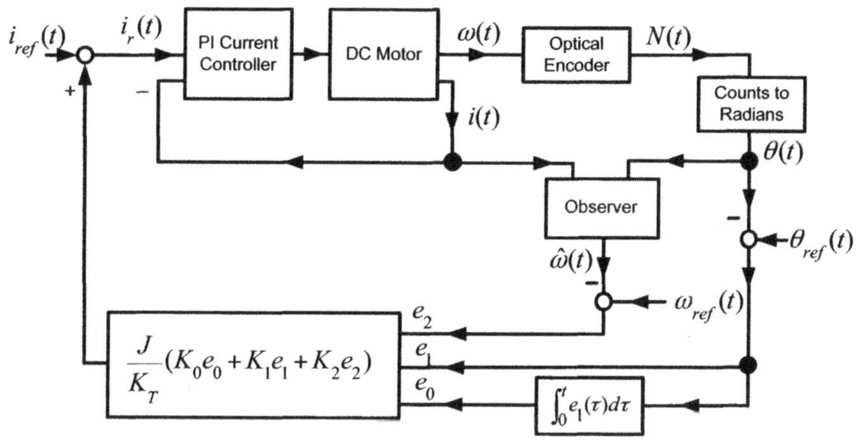

FIGURE 2.14. Block diagram for the state-space controller.

Note that if the desired characteristic equation had been of the form

$$a(s) = (s + r_1)(s^2 + 2\zeta\omega_n s + \omega_n^2)$$

where $0 < \zeta < 1, \omega_n > 0$, then it is again easy to find the corresponding gains. It is up to the control engineer to choose the closed-loop pole locations, or equivalently, the gains. Typically, if the poles are not far enough in the left-half plane, the response is too slow. On the other hand, if the poles are chosen too far in the left-half plane, the amplifier might saturate due to the large value of the gains. This procedure of varying the location of the closed-loop poles by varying the gains is referred to as "tuning the system".

The control designer also must choose the observer poles $-p_1, -p_2$. One typically wants $\hat{\omega}(t) \to \omega(t)$ much faster than the rate at which $\omega_{ref}(t) - \hat{\omega}(t) \to 0$. This is simply because one wants the estimate $\hat{\omega}(t)$ to close to the value $\omega(t)$ so that feeding back $K_1(\omega_{ref}(t) - \hat{\omega}(t))$ will be essentially the same as feeding back $K_1(\omega_{ref}(t) - \omega(t))$. In practice, the observer poles $-p_i$ are taken to be 5–10 times further in the left-half plane than the controller poles $-r_i$.

---

[1]That is, a current which is larger than the amplifier can produce would be commanded to the amplifier. The voltage output of the amplifier will increase (in order to increase the current) until it reaches the maximum possible value $V_{\max}$ and then it will saturate at this maximum value.

2.5 Nested Loop Control Structure*

A nested loop control structure is shown in Figure 2.15. This controller has two nested[2] PID (*proportional, integral plus derivative*) control loops and is common in industrial applications. Part of its popularity is due to the simplicity in tuning (adjusting) the gains of the controller. Specifically, with the position loop gains set to zero, the gains $K_{\omega P}$ and $K_{\omega I}$ of the speed loop can be tuned first online until a satisfactory speed response is obtained.[3] With these values set, the gains $K_{\theta P}$ and $K_{\theta I}$ of the position loops are then tuned until a satisfactory position response is obtained. The signal denoted as $\bar{\omega}_{ref}$ (the output of the position loop) is referred to as the "speed reference" since this would be the speed reference if the position loop was removed.[4]

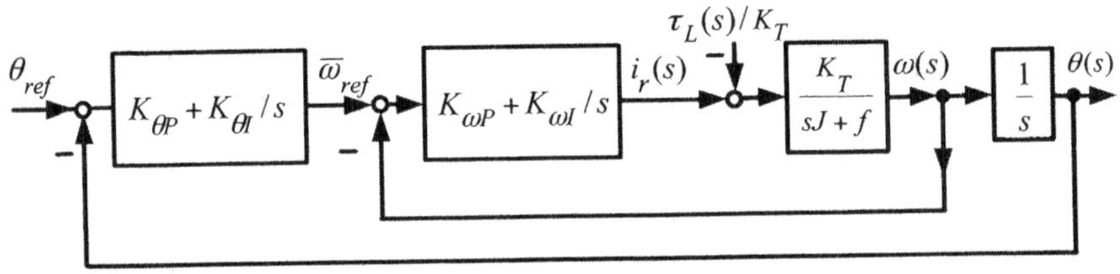

FIGURE 2.15. Nested loop control structure. The gains of the speed loop are tuned first and then the gains of the position loop are tuned.

Taking the transfer function of the system equations

$$\begin{aligned}\frac{d\omega}{dt} &= \frac{K_T}{J}i_r - \frac{f}{J}\omega - \frac{\tau_L}{J}\\ \frac{d\theta}{dt} &= \omega\end{aligned}$$

gives

$$\begin{aligned}\omega(s) &= G_m(s)\left(i_r(s) - \frac{\tau_L(s)}{K_T}\right)\\ \theta(s) &= \frac{1}{s}\omega(s)\end{aligned}$$

where

$$G_m(s) \triangleq \frac{K_T/J}{s+f/J}. \tag{2.19}$$

---

[2] The terminology "nested" just refers to one of the loops being inside the other.

[3] That is, the response from $\bar{\omega}_{ref}(t) \to \omega(t)$.

[4] However, if the position loop is intact, then there is no guarantee that $\bar{\omega}_{ref}$ (defined as the output of the position loop) equals $\omega_{ref} \triangleq d\theta_{ref}/dt$.

To simplify the discussion, let $K_{\omega I} = 0$ (see problem 24) and substitute

$$i_r(s) = K_{\omega P}\left(\bar{\omega}_{ref}(s) - \omega(s)\right)$$

into the previous expression for $\omega(s)$ to obtain

$$\omega(s) = G_m(s)\Big(K_{\omega P}\left(\bar{\omega}_{ref}(s) - \omega(s)\right)\Big) - G_m(s)\frac{\tau_L(s)}{K_T}. \qquad (2.20)$$

The output $\omega(s)$ in terms of the transfer functions from $\bar{\omega}_{ref}(s)$ to $\theta(s)$ and $\tau_L(s)$ to $\theta(s)$ is then

$$\omega(s) = \frac{K_{\omega P}G_m(s)}{1+K_{\omega P}G_m(s)}\bar{\omega}_{ref}(s) - \frac{G_m(s)}{1+K_{\omega P}G_m(s)}\frac{\tau_L(s)}{K_T}. \qquad (2.21)$$

The position output is then simply

$$\theta(s) = \frac{1}{s}\frac{K_{\omega P}G_m(s)}{1+K_{\omega P}G_m(s)}\bar{\omega}_{ref}(s) - \frac{1}{s}\frac{G_m(s)}{1+K_{\omega P}G_m(s)}\frac{\tau_L(s)}{K_T}.$$

With $E(s) \triangleq \theta_{ref}(s) - \theta(s)$ and $\bar{\omega}_{ref}(s) = \left(\frac{K_{\theta P}s+K_{\theta I}}{s}\right)E(s)$, it follows that

$$\begin{aligned} E(s) &\triangleq \theta_{ref}(s) - \theta(s) \\ &= \theta_{ref}(s) - \left(\frac{1}{s}\frac{K_{\omega P}G_m(s)}{1+K_{\omega P}G_m(s)}\frac{K_{\theta P}s+K_{\theta I}}{s}E(s)\right. \\ &\qquad \left. - \frac{1}{s}\frac{G_m(s)}{1+K_{\omega P}G_m(s)}\frac{\tau_L(s)}{K_T}\right) \end{aligned}$$

or

$$\begin{aligned} E(s) \triangleq\ & \frac{1}{1+\frac{1}{s}\frac{K_{\omega P}G_m(s)}{1+K_{\omega P}G_m(s)}\left(\frac{K_{\theta P}s+K_{\theta I}}{s}\right)}\theta_{ref}(s) \\ & + \frac{1}{1+\frac{1}{s}\frac{K_{\omega P}G_m(s)}{1+K_{\omega P}G_m(s)}\left(\frac{K_{\theta P}s+K_{\theta I}}{s}\right)}\frac{1}{s}\frac{G_m(s)}{1+K_{\omega P}G_m(s)}\frac{\tau_L(s)}{K_T}. \end{aligned}$$

Then as

$$\frac{K_{\omega P}G_m(s)}{1+K_{\omega P}G_m(s)} = \frac{K_{\omega P}K_T/J}{s+f/J+K_{\omega P}K_T/J}$$

one may write $E(s) = G_1(s)\theta_{ref}(s) + G_2(s)\tau_L(s)$, where

$$\begin{aligned} G_1(s) &\triangleq \frac{1}{1+\frac{1}{s}\frac{K_{\omega P}K_T/J}{s+f/J+K_{\omega P}K_T/J}\left(\frac{K_{\theta P}s+K_{\theta I}}{s}\right)} \\ &= \frac{s^2\left(s+f/J+K_{\omega P}K_T/J\right)}{s^2\left(s+f/J+K_{\omega P}K_T/J\right)+K_{\omega P}K_T/J\left(K_{\theta P}s+K_{\theta I}\right)} \\ &= \frac{s^2\left(s+f/J+K_{\omega P}K_T/J\right)}{s^3+s^2\left(f/J+K_{\omega P}K_T/J\right)+\left(K_{\theta P}K_{\omega P}K_T/J\right)s+K_{\theta I}K_{\omega P}K_T/J} \end{aligned}$$

and

$$G_2(s) \triangleq \frac{1}{1+\dfrac{1}{s}\dfrac{K_{\omega P}G_m(s)}{1+K_{\omega P}G_m(s)}\left(\dfrac{K_{\theta P}s+K_{\theta I}}{s}\right)}\frac{1}{s}\frac{G_m(s)}{1+K_{\omega P}G_m(s)}\frac{1}{K_T}$$

$$= \frac{1}{1+\frac{1}{s}\frac{K_{\omega P}K_T/J}{(s+f/J+K_{\omega P}K_T/J)}\left(\frac{K_{\theta P}s+K_{\theta I}}{s}\right)}\frac{1}{s}\frac{K_T/J}{s+f/J+K_{\omega P}K_T/J}\frac{1}{K_T}$$

$$= \frac{s^2}{s^2\left(s+f/J+K_{\omega P}K_T/J\right)+\left(K_{\omega P}K_T/J\right)\left(K_{\theta P}s+K_{\theta I}\right)}\frac{1}{s}\frac{1}{J}$$

$$= \frac{s/J}{s^3+s^2\left(f/J+K_{\omega P}K_T/J\right)+\left(K_{\theta P}K_{\omega P}K_T/J\right)s+K_{\theta I}K_{\omega P}K_T/J}.$$

Both $G_1(s)$ and $G_2(s)$ have the same denominator (characteristic polynomial) given by

$$s^3+s^2\left(f/J+K_{\omega P}K_T/J\right)+\left(K_{\theta P}K_{\omega P}K_T/J\right)s+K_{\theta I}K_{\omega P}K_T/J.$$

Suppose it is desired to have the closed-loop poles at $-r_1, -r_2$, and $-r_3$ with $r_1 > 0, r_2 > 0, r_3 > 0$. The closed-loop characteristic polynomial is then of the form

$$(s+r_1)(s+r_2)(s+r_3) = s^3+(r_1+r_2+r_3)s^2+(r_1r_2+r_1r_3+r_2r_3)s+r_1r_2r_3.$$

This requires setting

$$\begin{aligned} f/J+K_{\omega P}K_T/J &= r_1+r_2+r_3 \\ K_{\theta P}K_{\omega P}K_T/J &= r_1r_2+r_1r_3+r_2r_3 \\ K_{\theta I}K_{\omega P}K_T/J &= r_1r_2r_3. \end{aligned} \tag{2.22}$$

Solving (2.22) for the gains gives

$$\begin{aligned} K_{\omega P} &= \frac{r_1+r_2+r_3-f/J}{K_T/J} \\ K_{\theta P} &= \frac{r_1r_2+r_1r_3+r_2r_3}{K_{\omega P}K_T/J} \\ K_{\theta I} &= \frac{r_1r_2r_3}{K_{\omega P}K_T/J}. \end{aligned} \tag{2.23}$$

With this choice of gains,[5] the error becomes

$$\begin{aligned} E(s) &= E_1(s)+E_2(s) \qquad (2.24) \\ &= \frac{s^2\left(s+f/J+K_{\omega P}K_T/J\right)}{(s+r_1)(s+r_2)(s+r_3)}\theta_{ref}(s)+\frac{s/J}{(s+r_1)(s+r_2)(s+r_3)}\tau_L(s). \end{aligned}$$

---

[5]However, as noted in the beginning of this section, the gains for this controller are not usually chosen according to (2.23). Rather, $K_{\omega P}$ is varied until a satisfactory speed response is obtained. With $K_{\omega P}$ fixed at this value, the gains $K_{\theta P}$ and $K_{\theta I}$ are varied until the position response is satisfactory.

The objective here is to have $e(t) = \mathcal{L}^{-1}\{E(s)\} \to 0$ as $t \to \infty$. To study this, expressions for $\theta_{ref}(s)$ and $\tau_L(s)$ are needed. To proceed, consider a trajectory where it is desired to have the motor rotate from $\theta(0) = 0$ to $\theta(t_f) = \theta_f$. A reference trajectory to do this is given in Figure 2.16. The position reference $\theta_{ref}(t)$ is given by

$$\theta_{ref}(t) = \begin{cases} \dfrac{\omega_{\max}}{2t_1}t^2 & 0 \le t \le t_1 \\ \omega_{\max}t_1/2 + \omega_{\max}(t - t_1) & t_1 \le t \le t_2 \\ \omega_{\max}t_1/2 + \omega_{\max}(t_2 - t_1) + \dfrac{\omega_{\max}t_1^2}{2t_1} - \dfrac{\omega_{\max}}{2t_1}(t_f - t)^2 & t_2 \le t \le t_f \\ \omega_{\max}t_2 & t \ge t_f \end{cases} \tag{2.25}$$

where $\omega_{\max} = \theta_f/t_2$ and $t_f \triangleq t_2 + t_1$. The corresponding speed reference $\omega_{ref}(t)$ is given by

$$\omega_{ref}(t) = \begin{cases} \dfrac{\omega_{\max}}{t_1}t & 0 \le t \le t_1 \\ \omega_{\max} & t_1 \le t \le t_2 \\ \dfrac{\omega_{\max}}{t_1}(t_f - t) & t_2 \le t \le t_f \\ 0 & t \ge t_f \end{cases} \tag{2.26}$$

and is shown in Figure 2.16(b).

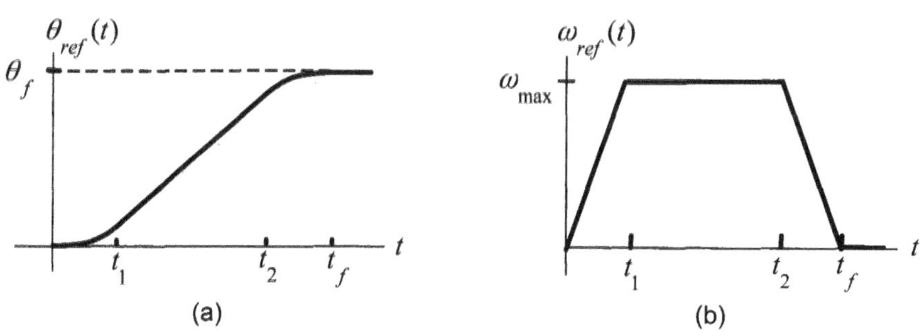

FIGURE 2.16. (a) Position reference. (b) Speed reference.

The expression for $\theta_{ref}(t)$ can also be obtained by simply integrating the speed reference $\omega_{ref}(t) \triangleq d\theta_{ref}(t)/dt$. Note that although $\omega_{ref}(t)$ is continuous, $\alpha_{ref}(t) = d\omega_{ref}(t)/dt$ is not and $j_{ref}(t) = d\alpha_{ref}(t)/dt$ has impulses at $t = 0, t_1, t_2, t_f$. The various components of $\theta_{ref}(t)$ have Laplace transforms given by [$u_s(t)$ is the unit step function]

$$\begin{aligned}
\mathcal{L}\left\{\omega_{\max}\frac{t^2}{2t_1}u_s(t)\right\} &= \frac{\omega_{\max}}{t_1}\frac{1}{s^3} \\
\mathcal{L}\left\{\left(\omega_{\max}\frac{t_1}{2}+\omega_{\max}(t-t_1)\right)u_s(t)\right\} &= -\omega_{\max}\frac{t_1}{2}\frac{1}{s}+\omega_{\max}\frac{1}{s^2} \\
\mathcal{L}\left\{\left(\omega_{\max}t_2-\omega_{\max}\frac{(t_f-t)^2}{2t_1}\right)u_s(t)\right\} &= \omega_{\max}t_2\frac{1}{s} \\
&\quad -\omega_{\max}\left(\frac{t_f^2}{2t_1}\frac{1}{s}-\frac{t_f}{t_1}\frac{1}{s^2}+\frac{1}{t_1}\frac{1}{s^3}\right) \\
\mathcal{L}\left\{\omega_{\max}t_2u_s(t)\right\} &= \omega_{\max}t_2\frac{1}{s}.
\end{aligned}$$

Finally, the Laplace transform of a *constant* load torque $\tau_L(t) = \tau_{L0}u_s(t)$ acting on the motor is

$$\mathcal{L}\{\tau_L(t)\} = \frac{\tau_{L0}}{s}.$$

The interest here is to analyze the system for tracking and disturbance rejection, that is, to see if it can track $\theta_{ref}(t)$ while a constant load disturbance τ_{L0} acts on the motor.

From the above Laplace transform expressions, $\theta_{ref}(s)$ and $\tau_L(s)$ are of the form (more precisely, see problem 26)

$$\begin{aligned}
\theta_{ref}(s) &= \frac{A}{s}+\frac{B}{s^2}+\frac{C}{s^3} \\
\tau_L(s) &= \tau_{L0}\frac{1}{s}.
\end{aligned}$$

for some constants A, B, and C. Substituting into the above expression for

$E(s)$ gives

$$\begin{aligned} E(s) &= E_1(s) + E_2(s) \\ &= \frac{s^2\left(s + f/J + K_{\omega P}K_T/J\right)}{(s+r_1)(s+r_2)(s+r_3)}\left(\frac{A}{s} + \frac{B}{s^2} + \frac{C}{s^3}\right) \\ &\quad + \frac{s}{(s+r_1)(s+r_2)(s+r_3)}\frac{1}{J}\frac{\tau_{L0}}{s} \\ &= \frac{s + f/J + K_{\omega P}K_T/J}{(s+r_1)(s+r_2)(s+r_3)}\left(As + B + \frac{C}{s}\right) \\ &\quad + \frac{1}{(s+r_1)(s+r_2)(s+r_3)}\frac{\tau_{L0}}{J} \\ &= \frac{c_1}{s+r_1} + \frac{c_2}{s+r_2} + \frac{c_3}{s+r_3} + \frac{c_4}{s} + \frac{d_1}{s+r_1} + \frac{d_2}{s+r_2} + \frac{d_3}{s+r_3} \end{aligned}$$

where

$$\begin{aligned} c_4 &= \lim_{s\to 0} sE_1(s) = \frac{f/J + K_{\omega P}K_T/J}{r_1 r_2 r_3}C \\ &= \frac{r_1 + r_2 + r_3}{r_1 r_2 r_3}C \neq 0 \text{ if } C \neq 0. \end{aligned}$$

Note that $C \neq 0$ if there is a parabolic term in the input (i.e., $t^2/2$). However, if r_1, r_2 and r_3 are large so that the closed-loop poles are far in the left-half plane, then c_4 will be small. In the time domain,

$$\begin{aligned} e(t) &= c_1 e^{-r_1 t} + c_2 e^{-r_2 t} + c_3 e^{-r_3 t} + c_4 u_s(t) + d_1 e^{-r_1 t} + d_2 e^{-r_2 t} + d_3 e^{-r_3 t} \\ &\to c_4 \text{ as } t \to \infty. \end{aligned}$$

This control structure cannot track parabolic type inputs (i.e., $t^2/2$) with zero steady-state error, but if the poles are far enough in the left-half plane, then it will track such an input for all *practical* purposes. Thus good tracking of the reference input with a constant load torque on the motor can be obtained if the poles can be put far enough in the left-half plane (i.e., if the gains can be chosen large enough without causing the amplifier to saturate).

Using the state space approach, it was shown that *any* reference trajectory of the form $\theta_{ref}(t), \omega_{ref}(t) = d\theta_{ref}(t)/dt, \alpha_{ref}(t) = d\omega_{ref}(t)/dt$ with reference input $i_{ref}(t) = \left(J\alpha_{ref}(t) + f\omega_{ref}(t)\right)/K_T$ can be tracked with zero steady-state error even with a constant load torque acting on the motor. Recall that the reference trajectory used for the state feedback controller had a *continuous* acceleration and the corresponding position reference had the polynomial terms $t^3/3$ and $t^4/4$ with Laplace transforms $2/s^4, 6/s^5$, respectively. The PID nested loop controller above cannot track this reference profile with zero steady-state error.

Another important comparison with the state feedback controller is that of setting the gains. The relationship between the feedback gains and the coefficients of the desired closed-loop characteristic polynomial is always linear in the case of the state feedback controller as illustrated in (2.18). However, in the case of the feedback structure of Figure 2.15, the relationship between the gains and the closed-loop poles is nonlinear [see equation (2.22) and problem 23]. Though (2.22) was readily solvable for the gains, higher-order systems will not be (see problems 24 and 27).

2.6 Identification of the DC Motor Parameters*

Recall the mathematical model of the DC motor given by

$$\begin{aligned} L\frac{di}{dt} &= -Ri(t) - K_b\omega(t) + v(t) \\ J\frac{d\omega}{dt} &= -f\omega(t) + K_T i(t) \\ \frac{d\theta}{dt} &= \omega(t). \end{aligned} \tag{2.27}$$

In order to design a controller based on these equations, the values of the motor parameters $L, R, K_b = K_T, J$, and f need to be found. This can be done by an experiment in which a voltage is commanded into the amplifier of the motor and the measured data $v(t), i(t)$, and $\omega(t)$ is then used to determine the parameters. To understand how this is accomplished, the first two equations above are rewritten as

$$\begin{bmatrix} di/dt & i(t) & \omega(t) & 0 & 0 \\ 0 & 0 & -i(t) & d\omega/dt & \omega(t) \end{bmatrix} \begin{bmatrix} L \\ R \\ K_T \\ J \\ f \end{bmatrix} = \begin{bmatrix} v(t) \\ 0 \end{bmatrix}. \tag{2.28}$$

This is a system of two linear algebraic equations in the unknowns L, R, K_T, J, and f. The coefficients of this system of linear equations are found from the measured/calculated data $\theta(t), \omega(t), d\omega/dt, i(t), di/dt, v$.

If the model of the motor is precisely given by (2.27) and the quantities $\theta(t), i(t)$, and $v(t)$ can be measured exactly along with the derivatives $d\theta/dt, d\omega/dt$, and di/dt computed exactly, then equation (2.28) must hold for all time t. Let $\omega(nT)$ denote the speed at time nT, $i(nT)$ denote the current at time nT, $di(nT)/dt$ the derivative of the current at time nT, and so on. Define

$$
\begin{aligned}
W(nT) &\triangleq \begin{bmatrix} \frac{di}{dt}(nT) & i(nT) & \omega(nT) & 0 & 0 \\ 0 & 0 & -i(nT) & \frac{d\omega}{dt}(nT) & \omega(nT) \end{bmatrix} \in \mathbb{R}^{2\times 5} \\
y(nT) &\triangleq \begin{bmatrix} v(nT) \\ 0 \end{bmatrix} \in \mathbb{R}^2 \\
K &\triangleq \begin{bmatrix} L \\ R \\ K_T \\ J \\ f \end{bmatrix} \in \mathbb{R}^5
\end{aligned}
$$

so that equation (2.28) above may be written in the form

$$W(nT)K = y(nT). \tag{2.29}$$

W is referred to as the *regressor* matrix. The desire here is to find the constant vector K that satisfies this for all n! To do so, multiply both sides of (2.29) by $W^T(nT)$ to obtain

$$W^T(nT)W(nT)K = W^T(nT)y(nT) \tag{2.30}$$

where

$$
\begin{aligned}
&W^T(nT)W(nT) \\
&= \begin{bmatrix} \frac{di(nT)}{dt} & 0 \\ i(nT) & 0 \\ \omega(nT) & -i(nT) \\ 0 & \frac{d\omega(nT)}{dt} \\ 0 & \omega(nT) \end{bmatrix} \begin{bmatrix} \frac{di(nT)}{dt} & i(nT) & \omega(nT) & 0 & 0 \\ 0 & 0 & -i(nT) & \frac{d\omega(nT)}{dt} & \omega(nT) \end{bmatrix} \\
&= \begin{bmatrix} (di/dt)^2 & idi/dt & \omega di/dt & 0 & 0 \\ idi/dt & i^2 & \omega i & 0 & 0 \\ \omega di/dt & \omega i & \omega^2 + i^2 & -id\omega/dt & -\omega i \\ 0 & 0 & -id\omega/dt & (d\omega/dt)^2 & \omega d\omega/dt \\ 0 & 0 & -\omega i & \omega d\omega/dt & \omega^2 \end{bmatrix}_{|\, t = nT}
\end{aligned} \tag{2.31}
$$

and

$$W^T(nT)y(nT) = \begin{bmatrix} di(nT)/dt & 0 \\ i(nT) & 0 \\ \omega(nT) & -i(nT) \\ 0 & d\omega(nT)/dt \\ 0 & \omega(nT) \end{bmatrix} \begin{bmatrix} v(nT) \\ 0 \end{bmatrix} = \begin{bmatrix} v(nT)di(nT)/dt \\ v(nT)i(nT) \\ v(nT)\omega(nT) \\ 0 \\ 0 \end{bmatrix}.$$

Note that $W^T(nT)W(nT) \in \mathbb{R}^{5\times 5}$; that is, it is a square matrix. It would then be convenient if it were possible to multiply both sides of (2.30) by $\left(W^T(nT)W(nT)\right)^{-1}$ to solve for K. However, $W^T(nT)W(nT)$ is *never* invertible! To see this, note that if one multiplies the first row of $W^T(nT)W(nT)$ by $-i(n)$, the second row by $di(nT)/dt$ and adds them, the result is zero. That is,

$$\begin{bmatrix} -i(nT) & di(nT)/dt & 0 & 0 & 0 \end{bmatrix} W^T(nT)W(nT) \equiv \begin{bmatrix} 0 & 0 & 0 & 0 & 0 \end{bmatrix}. \tag{2.32}$$

Now, if $W^T(n)W(nT)$ were invertible, one could multiply both sides of (2.32) by this inverse to obtain

$$\begin{bmatrix} -i(nT) & di(nT)/dt & 0 & 0 & 0 \end{bmatrix} W^T(n)W(nT)\left(W^T(nT)W(nT)\right)^{-1} = \begin{bmatrix} 0 & 0 & 0 & 0 & 0 \end{bmatrix}\left(W^T(n)W(nT)\right)^{-1}$$

or

$$\begin{bmatrix} -i(nT) & di(nT)/dt & 0 & 0 & 0 \end{bmatrix} = \begin{bmatrix} 0 & 0 & 0 & 0 & 0 \end{bmatrix}$$

which is a contradiction, as the current does not have to be zero.

Something else has to be done. As equation (2.30) must hold for all n with K a fixed parameter vector, sum up equation (2.30) for N time instants to obtain

$$\left(\sum_{n=1}^{N} W^T(nT)W(nT)\right)K = \sum_{n=1}^{N} W^T(nT)y(nT) \tag{2.33}$$

Define

$$\begin{aligned} R_W &\triangleq \sum_{n=1}^{N} W^T(nT)W(nT) \in \mathbb{R}^{5\times 5} \\ R_{Wy} &\triangleq \sum_{n=1}^{N} W^T(nT)y(nT) \in \mathbb{R}^{5\times 1} \end{aligned} \tag{2.34}$$

Then rewrite (2.33) as

$$R_W K = R_{Wy}. \tag{2.35}$$

Suppose the matrix sum $R_W \triangleq \sum_{n=1}^{N} W^T(nT)W(nT)$ is invertible, then multiply both sides of (2.35) by $R_W^{-1} \triangleq \left(\sum_{n=1}^{N} W^T(nT)W(nT)\right)^{-1}$ to obtain K as

$$K = R_W^{-1} R_{Wy}. \tag{2.36}$$

The key to this method is making sure R_W is invertible. This requires choosing a suitable input to the motor, that is, any arbitrary input will not work. For example, suppose $v(t) \equiv 0$ so that $i(t) \equiv 0$ and thus $\omega(t) \equiv 0$. In this case, $W(nT) \equiv 0$ for all n so that $R_W \triangleq \sum_{n=1}^{N} W^T(nT)W(nT) \equiv 0$ and is therefore *not* invertible. As a less trivial example, suppose $i(t) = i_0$ is constant. Then, equation (2.31) shows that the first row of $W^T(nT)W(nT)$ is identically zero for all n and therefore, the first row of R_W is identically zero for all n so that R_W is *not* invertible. Finally, if an experiment was performed where $\omega(t) \equiv \omega_0$ is constant, then the fourth row of R_W is identically zero and thus again, R_W is *not* invertible. It is up to the control engineer to design an input so that the invertibility condition holds. If such an input results in R_W being invertible, then one says the system has been *sufficiently excited.*

2.6.1 Least-Squares Approximation

The above analysis was based on the equation $W(nT)K = y(nT)$ being true for all n. However, in the "real-world" that engineers work, this is never true. The model (2.27) is not an exact description of the motor (i.e., it was derived assuming ideal magnetic materials, etc.), the voltage $v(t)$, current $i(t)$ and position $\theta(t)$ cannot be measured perfectly, and the derivatives $\omega(t) = d\theta/dt, d\omega/dt$, and di/dt cannot be computed exactly. Thus, there will not be a parameter vector K that satisfies $W(nT)K = y(nT)$ for all n. One can still run an experiment and collect the data $v(t), i(t)$, and $\theta(t)$ to compute R_W and R_{Wy}, and thus determine

$$\hat{K} = R_W^{-1} R_{Wy}.$$

The key question is then "How well does $\hat{K} = R_W^{-1} R_{Wy}$ satisfy $W(nT)K = y(nT)$ for all n?" To answer this question, let

$$\begin{aligned} W(nT) &\triangleq \begin{bmatrix} di(nT)/dt & i(nT) & \omega(nT) & 0 & 0 \\ 0 & 0 & -i(nT) & d\omega(nT)/dt & \omega(nT) \end{bmatrix} \in \mathbb{R}^{2\times 5} \\ y(nT) &\triangleq \begin{bmatrix} v(nT) \\ 0 \end{bmatrix} \in \mathbb{R}^2 \end{aligned}$$

and define the error

$$e(nT) \triangleq y(nT) - W(nT)K \in \mathbb{R}^2.$$

The problem is then one of finding the value of K that makes this difference as small as possible for all n. Specifically, the goal is to find the value of K that minimizes the squared error given by

$$\begin{aligned}
E^2(K) &\triangleq \sum_{n=1}^{N} (y(nT) - W(nT)K)^T (y(nT) - W(nT)K) \\
&= \sum_{n=1}^{N} (y(nT) - \hat{y}(nT))^T (y(nT) - \hat{y}(nT)) \\
&= \sum_{n=1}^{N} (y_1(nT) - \hat{y}_1(nT))^2 + (y_2(nT) - \hat{y}_2(nT))^2 \\
&= \sum_{n=1}^{N} \left(e_1(nT)^2 + e_2(nT)^2\right)
\end{aligned}$$

where $e_1(nT) \triangleq y_1(nT) - \hat{y}_1(nT), e_2(nT) \triangleq y_2(nT) - \hat{y}_2(nT)$.

In the jargon of identification theory, $y(nT)$ is considered the *output* while $\hat{y}(nT) = W(nT)K$ is the *predicted output* based on K as the estimate of the parameters. Consequently,

$$e(nT) \triangleq y(nT) - W(nT)K = y(nT) - \hat{y}(nT) \in \mathbb{R}^2$$

is the *error* and

$$E^2(K) \triangleq \sum_{n=1}^{N} (y(nT) - W(nT)K)^T (y(nT) - W(nT)K) \tag{2.37}$$

is the total *squared error*. If a K can be found that minimizes equation (2.37), it is referred to as the *least-squares* estimate.

It is now shown that there is a unique solution and it equals $R_W^{-1} R_{Wy}$. To do so, expand equation (2.37) above to obtain [recall the matrix fact $(AB)^T = B^T A^T$]

$$\begin{aligned}
E^2(K) = \sum_{n=1}^{N} \Big(& y^T(nT)y(nT) - y^T(nT)W(nT)K - K^T W^T(nT)y(nT) \\
&+ K^T W^T(nT)W(nT)K\Big) \\
= \sum_{n=1}^{N} & y^T(nT)y(nT) - \left(\sum_{n=1}^{N} y^T(nT)W(nT)\right) K \\
&- K^T \left(\sum_{n=1}^{N} W^T(nT)y(nT)\right) + K^T \left(\sum_{n=1}^{N} W^T(nT)W(nT)\right) K.
\end{aligned}$$

Adding to the definitions given in (2.34), define

$$
\begin{aligned}
R_{yW} &\triangleq \sum_{n=1}^{N} y^T(nT)W(nT) \in \mathbb{R}^{1\times 5} \\
R_y &\triangleq \sum_{n=1}^{N} y^T(nT)y(nT) \in \mathbb{R}
\end{aligned}
$$

where it is clear that $R_{yW} = R_{Wy}^T$. Then $E^2(K)$ is written compactly as

$$
\begin{aligned}
E^2(K) &= R_y - R_{yW}K - K^T R_{Wy} + K^T R_W K \\
&= R_y - R_{yW}R_W^{-1}R_{Wy} + \left(K - R_W^{-1}R_{Wy}\right)^T R_W \left(K - R_W^{-1}R_{Wy}\right).
\end{aligned}
\tag{2.38}
$$

For a fixed N, the matrices R_y, R_{yW}, R_{Wy}, and R_W are constant and do not depend K. Also, R_W is a *symmetric positive semidefinite* matrix.

Digression [41] A matrix Q is *symmetric* if $Q^T = Q$. A symmetric matrix $Q \in \mathbb{R}^{m\times m}$ is *positive semidefinite* if for all $x \in \mathbb{R}^m$, $x^TQx \geq 0$. Further, Q is *positive definite* if $x^TQx \geq 0$ for all $x \in \mathbb{R}^m$ *and* $x^TQx = 0$ if and only if x is the zero vector (i.e., $x = 0$).

For example, let

$$
Q_1 = \begin{bmatrix} 1 & 0 \\ 0 & 2 \end{bmatrix}
$$

and note that $Q_1 = Q_1^T$. Also,

$$
x^TQ_1x = \begin{bmatrix} x_1 & x_2 \end{bmatrix}^T \begin{bmatrix} 1 & 0 \\ 0 & 2 \end{bmatrix} \begin{bmatrix} x_1 \\ x_2 \end{bmatrix} = x_1^2 + 2x_2^2 \geq 0
$$

for all $x \in \mathbb{R}^2$ and the only way it can equal zero is if $x_1 = 0, x_2 = 0$. That is, Q_1 is positive definite.

As another example, let

$$
Q_1 = \begin{bmatrix} 0 & 0 \\ 0 & 2 \end{bmatrix}
$$

and note that $Q_2 = Q_2^T$. Also,

$$
x^TQ_2x = \begin{bmatrix} x_1 & x_2 \end{bmatrix}^T \begin{bmatrix} 0 & 0 \\ 0 & 2 \end{bmatrix} \begin{bmatrix} x_1 \\ x_2 \end{bmatrix} = 2x_2^2 \geq 0
$$

for all $x \in \mathbb{R}^2$. Thus Q_2 is positive semidefinite. However, in this case, the nonzero vector $x = \begin{bmatrix} 1 & 0 \end{bmatrix}^T$ makes $x^TQ_2x = 0$; that is, Q_2 is not positive definite.

As a final example, let

$$Q_3 = \begin{bmatrix} -1 & 0 \\ 0 & 2 \end{bmatrix}$$

and note that $Q_3 = Q_3^T$. Also,

$$x^T Q_3 x = \begin{bmatrix} x_1 & x_2 \end{bmatrix}^T \begin{bmatrix} -1 & 0 \\ 0 & 2 \end{bmatrix} \begin{bmatrix} x_1 \\ x_2 \end{bmatrix} = -x_1^2 + 2x_2^2$$

In this case, $x^T Q_3 x$ can be positive (e.g., $x = [0\ 1]^T$) or negative (e.g., $x = [1\ 0]^T$) depending on the chosen value of the vector x. Consequently, Q_3 is neither positive definite nor positive semidefinite.

■

To show that R_W is positive semidefinite, for any $x \in \mathbb{R}^5$, it follows that

$$\begin{aligned} x^T W^T(nT) W(nT) x &= (W(nT)x)^T (W(nT)x) = z^T(nT) z(nT) \\ &= z_1^2(nT) + z_2^2(nT) \geq 0 \end{aligned}$$

where $z(nT) \triangleq W(nT)x \in \mathbb{R}^2$. Thus,

$$\begin{aligned} x^T R_W x &= x^T \left(\sum_{n=1}^{N} W^T(nT) W(nT) \right) x \\ &= \sum_{n=1}^{N} x^T W^T(nT) W(nT) x \geq 0. \end{aligned}$$

The control engineer will design the experiment, that is, the specification of $v(t)$ and $i(t)$, so that R_W is invertible. It turns out that if a symmetric, positive semidefinite matrix is also *invertible*, then it must be positive definite. Consequently, R_W is then positive definite so that second term in (2.38) satisfies

$$\left(K - R_W^{-1} R_{Wy}\right)^T R_W \left(K - R_W^{-1} R_{Wy}\right) \geq 0$$

for all $K \in \mathbb{R}^5$. This equals zero if and only if

$$K - R_W^{-1} R_{Wy} = \begin{bmatrix} 0 & 0 & 0 & 0 & 0 \end{bmatrix}^T.$$

That is, by inspection of equation (2.38), $E^2(K)$ is minimized for $K = \hat{K} = R_W^{-1} R_{Wy}$! Thus, choosing $K = \hat{K} = R_W^{-1} R_{Wy}$ results in the *least*-squared error. The " ^ " in $\hat{K}$ is used to denote an *estimate* of K. However, the least squares estimate is an *optimal* estimate as it *minimizes* the least-squares error. Then the convention is to use an asterisk $*$ to denote an optimal estimate and so $K^* \triangleq R_W^{-1} R_{Wy}$ is now used.

2.6.2 Error Index

How good is the least-squares estimate? Well, the exact value of K is not known so the error between the "exact" value of the parameter vector and its estimate (i.e., $K - K^*$) is unknown. However, an indication of how good the estimate K^* is can be found by comparing it to a given (and thus known) value of K. Specifically, if $K = 0$, then the squared error is $E(0) = R_y$ as seen by putting $K = 0$ in equation (2.38) above. Using the least-squares estimate K^*, that is, setting $K = K^* \triangleq R_W^{-1}R_{Wy}$ in equation (2.38), the error is given by

$$E^2(K)_{|K=K^*\triangleq R_W^{-1}R_{Wy}} = R_y - R_{yW}R_W^{-1}R_{Wy}.$$

This is called the *residual* error; that is, it is the total squared error after using the value of K that minimizes the squared error. As R_W is positive definite, it turns out that its inverse R_W^{-1} must also be positive definite. Further, $R_{yW} \in \mathbb{R}^{1\times 5}, R_{Wy} \in \mathbb{R}^{5\times 1}$ with $R_{yW} = R_{Wy}^T$, it follows that $R_{yW}R_W^{-1}R_{Wy} \geq 0$ so that $E^2(K^*) = R_y - R_{yW}R_W^{-1}R_{Wy} \leq R_y = E^2(0)$. As a result, the quantity

$$\frac{E^2(K^*)}{E^2(0)} = \frac{R_y - R_{yW}R_W^{-1}R_{Wy}}{R_y} \leq 1.$$

The ratio $E^2(K^*)/E^2(0)$ is a measure of the minimum squared error relative to the squared error obtained from taking the parameter vector K to be the zero vector. By taking the square root, a measure of the relative error rather than squared error is obtained. This motivates the definition of the so-called *error index* as

$$\text{Error Index} \triangleq \sqrt{\frac{E^2(K^*)}{E^2(0)}} = \sqrt{\frac{R_y - R_{yW}R_W^{-1}R_{Wy}}{R_y}} \leq 1.$$

Note that if the error index is close to 1, then the estimate is not much better than taking all the parameter values equal to zero! Thus, the error index must be much less than one for the estimate to be of any value. If the error index is close to one, then one would suspect that the original model of the system is incorrect.

2.6.3 Parametric Error Indices

In addition to the overall error index, it is important to know how sensitive the error is to each parameter. That is, does a change δK_i in the ith parameter K_i produce a large or small change in $E^2(K)$? A small change indicates that the parameter estimate could vary greatly without a large change in the residual error. Thus, the accuracy of the parameter estimate of K_i would be in doubt.[6] On the other hand, a large change indicates that the residual error is very sensitive to changes in the parameter K_i so that such a parameter estimate may be considered more accurate.

In addressing the issue of sensitivity of the full parametric vector K^* to errors, the method in Ref. [42] is considered. If follows from (2.38) that K^* satisfies

$$\left.\frac{\partial E^2(K)}{\partial K}\right|_{K=K^*} = 2R_W\left(K - R_W^{-1}R_{Wy}\right)\Big|_{K=K^*} = 0 \tag{2.39}$$

so that, in particular,

$$\left.\frac{\partial E^2(K)}{\partial K_i}\right|_{K=K^*} = 0. \tag{2.40}$$

Therefore, it is not possible to use the derivative $\partial E^2(K)/\partial K_i|_{K=K^*}$ of the residual error as a measure of how sensitive the error is with respect to K_i as it is always zero.

An alternative is to define δK as the variation in K such that the increase of error is equal to the residual error $E^2(K^*)$ itself. To explain, recall that

$$\begin{aligned} E^2(K^*+\delta K) &= R_y - 2R_{Wy}^T(K^*+\delta K) + (K^*+\delta K)^T R_W (K^*+\delta K) \\ &= E^2(K^*) + \delta K^T R_W \delta K. \end{aligned} \tag{2.41}$$

The δK's that lead to a doubling of the residual error satisfy

$$\delta K^T R_W \delta K = E^2(K^*). \tag{2.42}$$

This is illustrated in Figure 2.17.

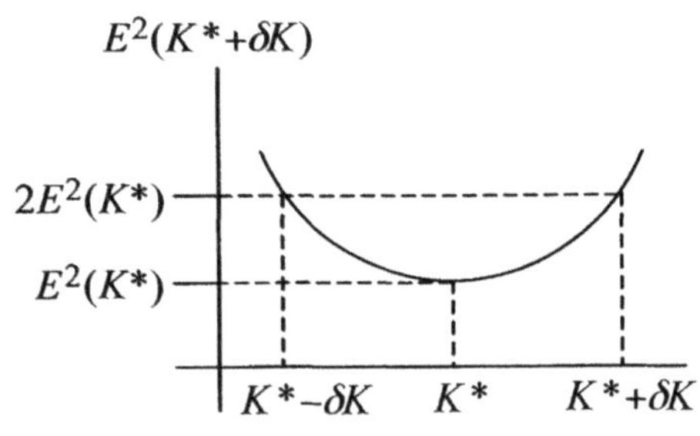

FIGURE 2.17. Illustrating the choice of δK that doubles the minimum mean square error.

The points $\delta K \in \mathbb{R}^5$ that satisfy (8.116) define an *ellipsoid* as illustrated in Figure 2.18.

[6] In other words, if $K_i + \delta K_i$ produces essentially the same value of $E^2(K)$ as K_i, then one is not sure if $K_i + \delta K_i$ or K_i is the better estimate.

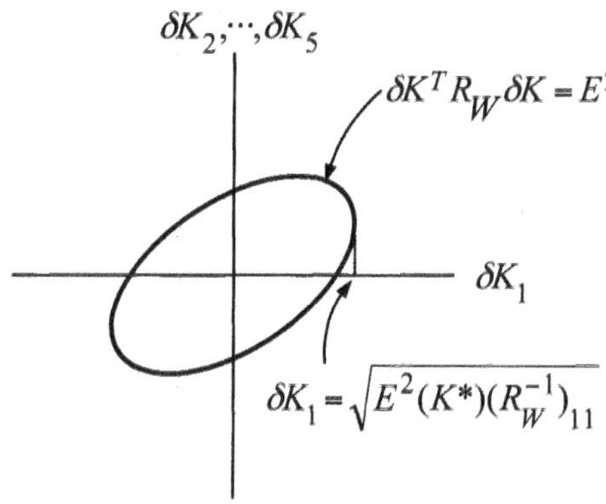

FIGURE 2.18. $\delta K_1 = \sqrt{E^2(K^*)(R_W)_{11}}$ is the largest value of δK_1 that is on the ellipsoid $\delta K^T R_W \delta K = E^2(K^*)$. This is defined as the *parametric error index* for K_1.

Define the *parametric error index* associated to the parameter K_i as the *maximum* value of δK_i such that (8.116) is satisfied. That is, δK_i is the solution to a constrained maximization problem that requires maximizing the quantity

$$\delta K_i$$

subject to the constraint

$$\delta K^T R_W \delta K = E^2(K^*).$$

This is simply the largest possible value for δK_i that would result in $E^2(K^* + \delta K) = 2E^2(K^*)$. This constrained optimization problem can be solved as an unconstrained maximization problem with the help of Lagrange multipliers. To do so, let (see Ref. [42])

$$C(\delta K, \lambda) \triangleq \delta K_i + \lambda\Big(E^2(K^*) - \delta K^T R_W \delta K\Big) \tag{2.43}$$

with respect to δK and λ. To fix ideas, let $i = 1$, and successively differentiate (2.43) with respect to $\delta K_1, \ldots, \delta K_5, \lambda$ so that at the maximum point of (2.43), it follows that

$$\begin{aligned}
\frac{\partial C(\delta K,\lambda)}{\partial \delta K_1} &= 1-2\lambda\begin{bmatrix} 1 & 0 & 0 & 0 & 0 \end{bmatrix} R_W \delta K = 0\\
\frac{\partial C(\delta K,\lambda)}{\partial \delta K_2} &= -2\lambda\begin{bmatrix} 0 & 1 & 0 & 0 & 0 \end{bmatrix} R_W \delta K = 0\\
\frac{\partial C(\delta K,\lambda)}{\partial \delta K_3} &= -2\lambda\begin{bmatrix} 0 & 0 & 1 & 0 & 0 \end{bmatrix} R_W \delta K = 0\\
\frac{\partial C(\delta K,\lambda)}{\partial \delta K_4} &= -2\lambda\begin{bmatrix} 0 & 0 & 0 & 1 & 0 \end{bmatrix} R_W \delta K = 0\\
\frac{\partial C(\delta K,\lambda)}{\partial \delta K_5} &= -2\lambda\begin{bmatrix} 0 & 0 & 0 & 0 & 1 \end{bmatrix} R_{W_w} \delta K = 0\\
\frac{\partial C(\delta K,\lambda)}{\partial \lambda} &= E^2(K^*) - \delta K^T R_W \delta K = 0.
\end{aligned}$$

This can then be rewritten as

$$\begin{bmatrix} 1\\0\\0\\0\\0 \end{bmatrix} - 2\lambda R_W \delta K = \begin{bmatrix} 0\\0\\0\\0\\0 \end{bmatrix}. \tag{2.44}$$

Solving for δK gives

$$\delta K = \frac{1}{2\lambda} R_W^{-1} \begin{bmatrix} 1\\0\\0\\0\\0 \end{bmatrix}. \tag{2.45}$$

To compute λ, multiply both sides of (2.44) by δK^T to obtain

$$\delta K_1 = 2\lambda \delta K^T R_W \delta K = 2\lambda E^2(K^*)$$

and rearranging results in

$$2\lambda = \delta K_1 / E^2(K^*).$$

Substitute this expression for 2λ back into (2.45) to obtain

$$\delta K = \frac{E^2(K^*)}{\delta K_1} R_W^{-1} \begin{bmatrix} 1\\0\\0\\0\\0 \end{bmatrix}. \tag{2.46}$$

With $\left(R_W^{-1}\right)_{11}$ denoting the $(1,1)$ element of the matrix R_W^{-1}, the first row of (2.46) gives

$$\delta K_1 = \frac{E^2(K^*)}{\delta K_1}\left(R_W^{-1}\right)_{11}$$

or

$$\delta K_1 = \sqrt{E^2(K^*)(R_W^{-1})_{11}}$$

where the + sign was chosen to maximize δK_1. In general (see Ref. [42])

$$\delta K_i = \sqrt{E^2(K^*)(R_W^{-1})_{ii}}. \tag{2.47}$$

The parametric error index δK_i indicates the maximum amount by which K_i, the ith component of K, could vary without causing more than a doubling of the residual error. A large parametric error index indicates that the parameter estimate could vary greatly without a large change in the residual error and thus making the accuracy of the parameter estimates suspect. On the other hand, a small parametric error indicates that the residual error is very sensitive to the changes in the parameter estimates and that the parameter estimates may be considered to be more accurate. In any case, the error indices should not be considered as actual errors, but rather as orders of magnitude of the errors to be expected, to guide the identification process and to warn about unreliable results.

The choice of a parametric error index as corresponding to a doubling of the residual error is arbitrary. A different level of residual error would lead to a scaling of all the components of the parametric error index by a common factor.

2.7 Filtering of Noisy Signals*

This section gives a concise discussion of the various ideas in the area of signals and systems that are needed to analyze/process data for implementation of identification and control algorithms. This material is covered in any basic "signals and systems" book such as references [43][44][45].

Recall that the identification algorithm required computing derivatives of the measured data. However, these measured signals typically contain high-frequency low-amplitude noise and the differentiation of such a signal amplifies the noise! That is, a measured signal $m(t)$ is often of the form $m(t) = s(t) + n(t)$, where $s(t)$ is the signal of interest and $n(t)$ is high-frequency low-amplitude noise. For example, Figure 2.19 is a plot of the simulated position measurement of a motor as given by an optical encoder. This signal $m(t) = (2\pi/2000)\,N(t)$ is the measured signal and appears to be right on top of the true signal $s(t) = \theta(t)$.

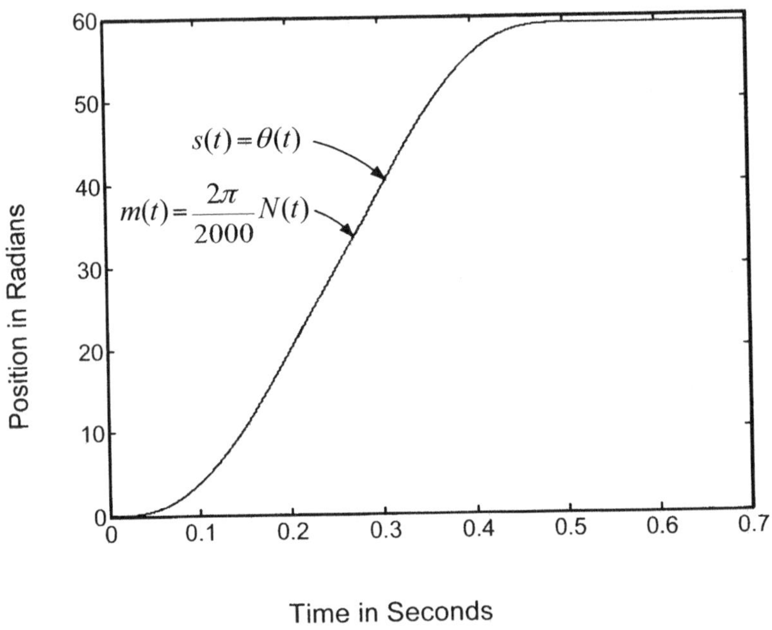

FIGURE 2.19. $\theta(t)$ and the encoder output $(2\pi/2000)N(t)$ in radians versus time in seconds.

Figure 2.20 is a magnification of Figure 2.19 clearly illustrating the discrete nature of the encoder output. The smooth curve $s(t) = \theta(t)$ is the "actual" position of the motor from the simulation and again the measured signal is $m(t) = (2\pi/2000)N(t)$.

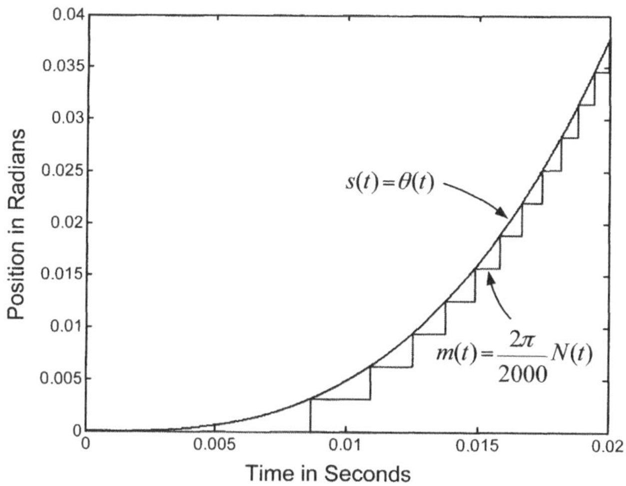

FIGURE 2.20. $\theta(t)$ and the encoder output $(2\pi/2000)N(t)$ in radians vs. time in seconds. Magnification of Figure 2.19.

As this "data set" is from a simulation, the true signal $s(t)$ is known and thus one can calculate the noise $n(t)$ as $n(t) = m(t) - s(t) = (2\pi/2000)N(t) - \theta(t)$, so that $m(t) = s(t) + n(t)$. Figure 2.21 is a plot of the noise $n(t) = (2\pi/2000)N(t) - \theta(t)$, where it is seen that $0 \leq n(t) \leq 3.14 \times 10^{-3}$ radians, that is, it is quite small.

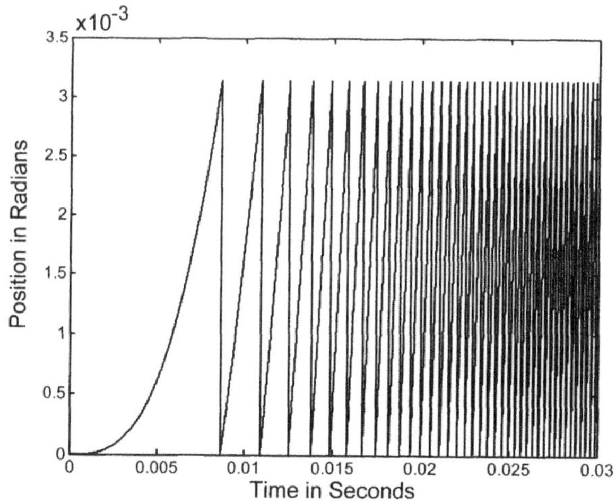

FIGURE 2.21. Position error/noise computed as $n(t) = (2\pi/2000)N(t) - \theta(t)$. The encoder error is bounded by $2\pi/2000 = 3.14 \times 10^{-3}$ radians.

To obtain an estimate of $\dot{s}(t)$, the measured signal $m(t) = s(t) + n(t)$ can be numerically differentiated to obtain

$$\frac{s(t) - s(t-T) + n(t) - n(t-T)}{T} = \dot{s}(t) + \dot{n}(t).$$

However, as the difference $n(t) - n(t-T)$ may be of the same order (or higher) of magnitude as $s(t) - s(t-T)$, it follows that $n(t)$ can be of the same order of magnitude as $s(t)$. Going back to the encoder example of Section 2.2, the encoder resolution (error) is $2\pi/2000$ radians where 2000 is the number of counts per revolution so that the noise (error) in the position measurement is bounded by $2\pi/2000$. Computing the speed by numerically differentiating the encoder output results in a tight error bound given by

$$\frac{2\pi}{2000T}$$

where T is the sample period. The sample period T is usually small (on the order of a millisecond) so that $1/T$ is large showing that differentiation amplifies the noise. This is illustrated in Figure 2.22 which was found by differentiating $m(t) = (2\pi/2000)N(t)$ from Figure 2.19 with $T = 0.5$ msec.

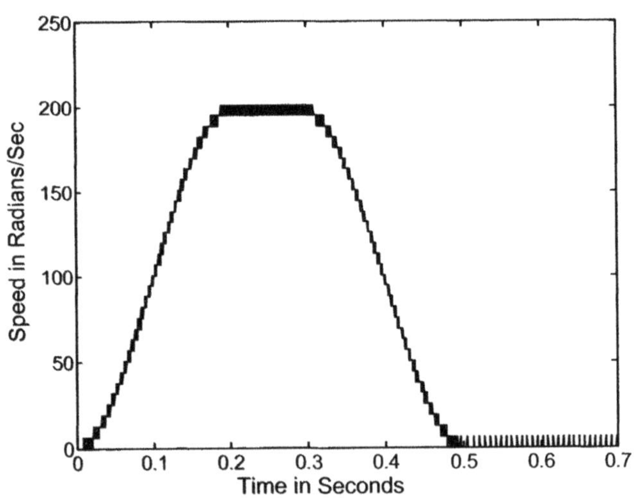

FIGURE 2.22. Estimated angular speed in rads/sec versus time in seconds. Here ω is estimated as $(\theta(nT) - \theta((n-1)T))/T$ with $T = 0.5$ msec resulting in a speed error bound of $\frac{2\pi}{2000T} = 6.28$ radians/sec.

A common way to remove additive high frequency noise is to use a low-pass filter that works as long as the signal $s(t)$ has a much lower frequency content than the noise $n(t)$. For example, let $s(t) = \sin(10t)$ and $n(t) = 0.01\sin(1000t)$ so that $m(t) = s(t) + n(t)$ is the measured signal. Then $\dot{m}(t) = \dot{s}(t) + \dot{n}(t) = 10\cos(10t) + 10\cos(1000t)$. Although $|n(t)| \ll |s(t)|$, the differentiated noise $\dot{n}(t)$ is of the same order of magnitude as the differentiated signal $\dot{s}(t)$. However, by setting the cutoff of the low-pass filter at (say) 100 rad/sec, then $s(t)$ will make it through the filter essentially unchanged and $n(t)$ will be filtered out. Before going into detail of how to design such a filter, some basic facts about filtering are reviewed.

2.7.1 Filter Representations

Let $m(t)$ be the input to the filter and $y(t)$ the output. There are three useful ways to characterize (represent) a causal filter:

(1) Impulse response $g(t)$ so that $y(t) = \int_{-\infty}^{t} g(t-\tau)m(\tau)d\tau$.

(2) Transfer function $G(s)$ so that $Y(s) = G(s)M(s)$
where $Y(s) = \int_0^\infty e^{-st}y(t)dt, G(s) = \int_0^\infty e^{-st}g(t)dt, M(s) = \int_0^\infty e^{-st}m(t)dt$.

(3) Differential equation $\dot{x}(t) = Ax(t) + bm(t), y(t) = cx(t) + dm(t)$
where $x \in \mathbb{R}^n, y \in \mathbb{R}, m \in \mathbb{R}, A \in \mathbb{R}^{n\times n}, b \in \mathbb{R}^n, c \in \mathbb{R}^{1\times n}$.

Example
Consider an impulse response given by

$$g(t) = \begin{cases} e^{-at} & t \geq 0 \\ 0 & t < 0. \end{cases}$$

In terms of $g(t)$, the filter is given by the convolution integral

$$y(t) = \int_{-\infty}^{t} g(t-\tau)m(\tau)d\tau = \int_{-\infty}^{t} e^{-a(t-\tau)}m(\tau)d\tau.$$

The corresponding transfer function is

$$\begin{aligned} G(s) &= \int_0^\infty e^{-st}e^{-at}dt = \frac{1}{s+a} \\ Y(s) &= \frac{1}{s+a}M(s). \end{aligned}$$

The state space (differential equation) representation is

$$\begin{aligned} \dot{x}(t) &= -ax(t) + m(t) \\ y(t) &= x(t). \end{aligned}$$

2.7.2 *Causality*

A more general representation of a linear filter is $y(t) = \int_{-\infty}^{\infty} g(t-\tau)m(\tau)d\tau$. However, recall that for a filter to be causal, the output $y(t)$ of the filter can only depend on the input $m(\tau)$ up to time t, that is, on $m(\tau)$ for $-\infty < \tau < t$. As a consequence of this causality requirement, in the representation $y(t) = \int_{-\infty}^{\infty} g(t-\tau)m(\tau)d\tau$, it follows that $g(t) \equiv 0$ for $t < 0$. If this were not the case, then for some $\tau > t$ (i.e., $t - \tau < 0$), $g(t-\tau)$ would not be zero and $g(t-\tau)m(\tau)d\tau$ would be the contribution of the *input* at time $\tau > t$ to the output response $y(t)$ at time t! Consequently, a causal filter may be written as $y(t) = \int_{-\infty}^{\infty} g(t-\tau)m(\tau)d\tau = \int_{-\infty}^{t} g(t-\tau)m(\tau)d\tau$ as $g(t-\tau) = 0$ for $\tau > t$.

As just explained, the impulse response model provides a simple way of characterizing the causality of a filter. The differential equation representation provides the means of actually *implementing* a filter as will be shown below. On the other hand, the transfer function representation is particularly easy to work with as it allows one to just do algebraic manipulations as opposed to working with a differential equation or a convolution integral. It is also quite useful for filter design and specification as is shown in the next section.

2.7.3 *Frequency Response*

Consider a filter represented as $y(t) = \int_{-\infty}^{\infty} g(t-\tau)m(\tau)d\tau$ with an input given by $m(t) = e^{st}$ for $-\infty < t < \infty$ and $s = \sigma + j\omega$ is *constant*. Then, the response is given by

$$\begin{aligned} y(t) &= \int_{-\infty}^{\infty} g(t-\tau)e^{s\tau}d\tau = \int_{-\infty}^{\infty} g(t-\tau)e^{-s(t-\tau)}e^{st}d\tau \\ &= e^{st}\int_{-\infty}^{\infty} g(t-\tau)e^{-s(t-\tau)}d\tau = e^{st}\int_{-\infty}^{\infty} g(u)e^{-su}du \\ &= e^{st}\int_{0}^{\infty} g(u)e^{-su}du \\ &= G(s)e^{st}. \end{aligned}$$

In particular, if $s = j\omega$ so that $m(t) = e^{j\omega t} = \cos(\omega t) + j\sin(\omega t)$,

$$\begin{aligned} y(t) &= G(j\omega)e^{j\omega t} = |G(j\omega)|\, e^{j\angle G(j\omega)}e^{j\omega t} \\ &= |G(j\omega)|\left(\cos(\omega t + \angle G(j\omega)) + j\sin(\omega t + \angle G(j\omega))\right). \end{aligned}$$

Considering just the real part, the input $m(t) = \cos(\omega t)$ for $-\infty < t < \infty$ results in the output $y(t) = |G(j\omega)|\cos(\omega t + \angle G(j\omega))$ for $-\infty < t < \infty$.

2.7.4 Low-Pass Filters with Linear Phase

Suppose for a given *bandwidth* specification ω_b, one is able to design a filter $G(j\omega)$ such that its magnitude response satisfies

$$\begin{aligned} |G(j\omega)| &\leq 1 && \text{for all } \omega \\ |G(j\omega)| &\approx 1 && \text{for } 0 \leq \omega \leq \omega_b < \omega_c \\ &\approx 0 && \text{for } \omega \geq 2\omega_c \end{aligned}$$

where ω_c is referred to as the *cutoff frequency* (see Figure 2.23). Further, suppose the phase response of the filter has the property that

$$\angle G(j\omega) \approx -\gamma\omega \text{ for } 0 \leq \omega \leq \omega_b \qquad \text{where } \gamma > 0.$$

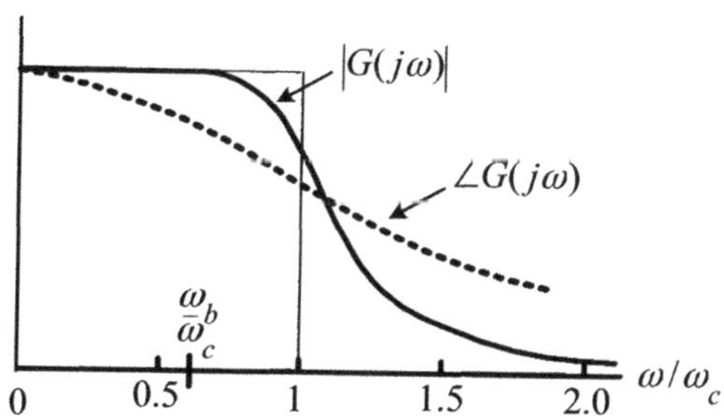

FIGURE 2.23. Magnitude and phase of a low-pass filter.

Then, if the input to this filter is $m(t) = \cos(\omega t)$ with $\omega \leq \omega_b$, the output is given by

$$\begin{aligned} y(t) &= |G(j\omega)| \cos(\omega t + \angle G(j\omega)) \\ &\approx \cos(\omega t - \gamma\omega) \\ &= \cos(\omega(t - \gamma)) \end{aligned}$$

that is, the output is the same as the input except delayed by the time γ. Carrying this idea further, consider the input $m(t) = A_1 \cos(\omega_1 t) + A_2 \cos(\omega_2 t)$ where $\omega_1 < \omega_2 \leq \omega_b < \omega_c$. Then, the corresponding output is

$$\begin{aligned} y(t) &= |G(j\omega_1)| A_1 \cos(\omega_1 t + \angle G(j\omega_1)) + |G(j\omega_2)| A_2 \cos(\omega_2 t + \angle G(j\omega_2)) \\ &\approx A_1 \cos(\omega_1(t - \gamma)) + A_2 \cos(\omega_2(t - \gamma)) \\ &= m(t - \gamma). \end{aligned}$$

Again, the output of the filter is the same as the input except delayed by the time γ.

The usual application of a low-pass filter is to the following situation: There is a signal $s(t)$ with high-frequency additive noise $n(t)$ corrupting it.

That is, the signal measured is $m(t) = s(t)+n(t)$ and the desire is to extract the signal $s(t)$ from $m(t)$. For example, let $s(t) = A_1\cos(\omega_1 t)+A_2\cos(\omega_2 t)$, $n(t) = N_0\cos(\omega_N t)$ with $\omega_1 < \omega_2 << \omega_N$ and $m(t) = s(t) + n(t)$. Suppose a low-pass linear-phase filter can be found (as described above) with $\omega_1 < \omega_2 \le \omega_b < \omega_c << \omega_N$. Then the output of the filter is given by

$$\begin{aligned} y(t) &= |G(j\omega_1)|\, A_1\cos(\omega_1 t + \angle G(j\omega_1)) + |G(j\omega_2)|\, A_2\cos(\omega_2 t + \angle G(j\omega_2)) \\ &\quad + |G(j\omega_N)|\, N_0\cos(\omega_N t + \angle G(j\omega_N)) \\ &\approx A_1\cos\left(\omega_1(t-\gamma)\right) + A_2\cos\left(\omega_2(t-\gamma)\right) \\ &= s(t-\gamma) \end{aligned} \tag{2.48}$$

as $|G(j\omega_N)| \approx 0$. The desired signal has been extracted from the noise, but it is delayed by γ seconds.

2.7.5 *Distortion*

The magnitude specification of the filter as $|G(j\omega)| \approx 1$ for $0 \le \omega \le \omega_b < \omega_c$ and $|G(j\omega)| \approx 0$ for $\omega \ge 2\omega_c$ is necessary to remove the high-frequency noise. The angle specification of $\angle G(j\omega) \approx -\gamma\omega$ for $0 \le \omega \le \omega_b < \omega_c$, $\gamma > 0$ (linear phase) is required to prevent distortion of the signal. That is, suppose $G(j\omega)$ did not have linear phase so that $\angle G(j\omega_1) \approx -\gamma_1\omega_1, \angle G(j\omega_2) \approx -\gamma_2\omega_2$ with $\gamma_1 \ne \gamma_2$, then with $m(t) = A_1\cos(\omega_1 t) + A_2\cos(\omega_2 t) + N_0\cos(\omega_N t)$ applied to the filter, the output would be

$$\begin{aligned} y(t) &= |G(j\omega_1)|\, A_1\cos(\omega_1 t + \angle G(j\omega_1)) + |G(j\omega_2)|\, A_2\cos(\omega_2 t + \angle G(j\omega_2)) \\ &\quad + |G(j\omega_N)|\, N_0\cos(\omega_N t + \angle G(j\omega_N)) \\ &\approx A_1\cos\left(\omega_1(t-\gamma_1)\right) + A_2\cos\left(\omega_2(t-\gamma_2)\right) \\ &\ne s(t-\gamma) \end{aligned}$$

for some $\gamma > 0$. That is, *distortion* in the signal results since the output of the filter is not an exact (delayed) reproduction of the desired signal.

2.7.6 *Low-Pass Filtering of High-Frequency Noise*

Now consider an arbitrary signal $s(t)$ where the measured signal is given by $m(t) = s(t)+n(t)$ and the noise $n(t)$ is assumed to have a much higher frequency content than $s(t)$. To understand the filter's effect on these signals, Fourier transform theory is used. The Fourier transforms of these signals are

$$\begin{aligned}
M(j\omega) &= \int_{-\infty}^{+\infty} e^{-j\omega t} m(t) dt \\
S(j\omega) &= \int_{-\infty}^{+\infty} e^{-j\omega t} s(t) dt \\
N(j\omega) &= \int_{-\infty}^{+\infty} e^{-j\omega t} n(t) dt.
\end{aligned}$$

where $M(j\omega) = S(j\omega) + N(j\omega)$. Their inverse Fourier transforms are given by

$$\begin{aligned}
m(t) &= \frac{1}{2\pi} \int_{-\infty}^{+\infty} e^{j\omega t} M(j\omega) d\omega \\
s(t) &= \frac{1}{2\pi} \int_{-\infty}^{+\infty} e^{j\omega t} S(j\omega) d\omega \\
n(t) &= \frac{1}{2\pi} \int_{-\infty}^{+\infty} e^{j\omega t} N(j\omega) d\omega.
\end{aligned}$$

Recall that $\frac{1}{2\pi} M(j\omega) d\omega$ is interpreted as the frequency content of $m(t)$ between ω and $\omega + d\omega$. As shown in (2.48), when $\frac{1}{2\pi} e^{j\omega t} M(j\omega) d\omega$ is passed through the filter, the output is $\frac{1}{2\pi} e^{j\omega t} G(j\omega) M(j\omega) d\omega$. Consequently, when $m(t)$ is input to the filter, the output $y(t)$ of the filter is given by

$$\begin{aligned}
y(t) &= \frac{1}{2\pi} \int_{-\infty}^{+\infty} e^{j\omega t} G(j\omega) M(j\omega) d\omega \\
&= \frac{1}{2\pi} \int_{-\infty}^{+\infty} e^{j\omega t} G(j\omega) S(j\omega) d\omega + \frac{1}{2\pi} \int_{-\infty}^{+\infty} e^{j\omega t} G(j\omega) N(j\omega) d\omega.
\end{aligned}$$

As explained above, to have a low-pass filter remove the noise, the frequency content of $n(t)$ must be much higher than that of $s(t)$. Thus, assuming that

$S(j\omega) \approx 0$ for $|\omega| > \omega_b$ and $|N(j\omega)| \approx 0$ for $|\omega| < 2\omega_c$,

$$\begin{aligned}
y(t) &= \frac{1}{2\pi}\int_{-\infty}^{\infty} e^{j\omega t}G(j\omega)S(j\omega)d\omega + \frac{1}{2\pi}\int_{-\infty}^{\infty} e^{j\omega t}G(j\omega)N(j\omega)d\omega \\
&= \frac{1}{2\pi}\int_{-\omega_b}^{+\omega_b} e^{j\omega t}G(j\omega)S(j\omega)d\omega + \frac{1}{2\pi}\int_{-\infty}^{-2\omega_c} e^{j\omega t}G(j\omega)N(j\omega)d\omega \\
&\quad + \frac{1}{2\pi}\int_{2\omega_c}^{\infty} e^{j\omega t}G(j\omega)N(j\omega)d\omega \text{ (as } |N(j\omega)| \approx 0 \text{ for } |\omega| < 2\omega_c) \\
&= \frac{1}{2\pi}\int_{-\omega_b}^{+\omega_b} e^{j\omega t}G(j\omega)S(j\omega)d\omega \text{ as } |G(j\omega)| \approx 0 \text{ for } |\omega| > 2\omega_c \\
&= \frac{1}{2\pi}\int_{-\omega_b}^{+\omega_b} e^{j\omega(t-\gamma)}S(j\omega)d\omega \left(\begin{array}{c} \text{as } G(j\omega) = |G(j\omega)|\, e^{j\angle G(j\omega)} \\ \approx 1e^{-j\gamma\omega} \text{ for } |\omega| \le \omega_b \end{array} \right) \\
&= \frac{1}{2\pi}\int_{-\infty}^{+\infty} e^{j\omega(t-\gamma)}S(j\omega)d\omega \text{ as } S(j\omega) \approx 0 \text{ for } |\omega| > \omega_b \\
&= s(t-\gamma).
\end{aligned}$$

The effect of passing $m(t)$ through the filter is to reproduce $s(t)$, that is, remove the noise from $m(t)$ delayed by γ seconds.

Remark

The "ideal" low-pass filter is specified by

$$G(j\omega) = \begin{cases} 1 & \text{for } |\omega| < \omega_c \\ 0 & \text{for } |\omega| > \omega_c. \end{cases}$$

That is, the frequency content of a signal below ω_c passes through unaffected while the frequency content of the signal above ω_c is removed. However, the ideal filter is neither causal nor stable (see problem 34).

2.7.7 *Butterworth Filters*

The class of filters known as *Butterworth Filters* have, to a good approximation, the two necessary properties; that is, the magnitude response has the low-pass characteristic along with a linear-phase. Below are the magnitude (Figure 2.24) and phase plots (Figure 2.25) for Butterworth filters of orders $n = 1, 2, 4, 6, 8, 10$, where ω_c denotes the filter cutoff frequency.

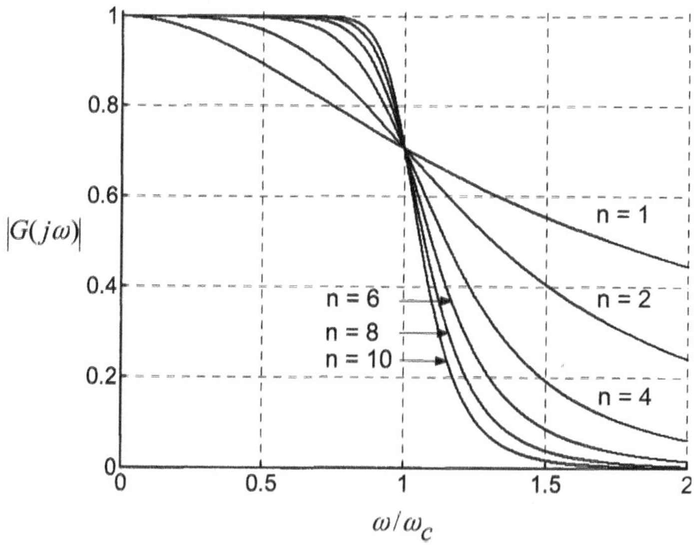

FIGURE 2.24. Plots of the magnitude versus ω/ω_c for Butterworth filters of order $n = 1, 2, 4, 6, 8, 10$.

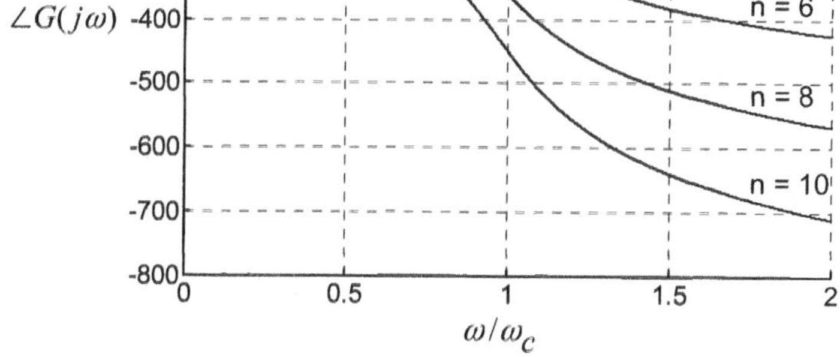

FIGURE 2.25. Plots of the phase in degrees versus ω/ω_c for Butterworth filters of order $n = 1, 2, 4, 6, 8, 10$.

Note that for Butterworth filters of order $n = 4$ or greater,

$$\begin{aligned} |G_n(j\omega)| &\approx 1 \quad \text{for } 0 \le \omega \le \omega_b = 0.6\omega_c \\ &\approx 0 \quad \text{for } \omega \ge 2\omega_c \end{aligned}$$

and

$$\angle G_n(j\omega) \approx -\gamma_n \omega \text{ for } 0 \le \omega \le \omega_b = 0.6\omega_c$$

where the delays $\gamma_n = m_n/\omega_c$ ($m_n \triangleq$ the slope of the $\angle G_n(j\omega)$ vs. ω/ω_c) are monotonically increasing. In other words, $\gamma_1 < \gamma_2 < \gamma_4 < \gamma_6 < \gamma_8 < \gamma_{10}$ so that the time delay increases as the order of the filter increases.

Let's consider the specific example of a third-order Butterworth filter given by

$$\begin{aligned} G_3(s) &= \frac{1}{(s/\omega_c + 1)\left((s/\omega_c)^2 + s/\omega_c + 1\right)} \\ &= \frac{\omega_c^3}{(s+\omega_c)(s^2 + \omega_c s + \omega_c^2)}. \end{aligned}$$

Note that $|G_3(j\omega)|_{\omega=0} = 1$ and $|G_3(j\omega)|_{\omega=\omega_c} = 1/\sqrt{2}$. By direct computation (tedious!), one can verify that $d^k \left|G_3(j\omega)\right| / d\omega^k \mid_{\omega=0} = 0$ for $k = 1, 2, 3, 4, 5$ and $d^6 \left|G_3(j\omega)\right| / d\omega^6 \mid_{\omega=0} = -(1/2)(6!)$. The third-order Butterworth filter is the only third-order, causal, rational (i.e., ratio of two polynomials) filter that has its first five derivatives zero at $\omega = 0$. For this reason, the Butterworth filter is known as being *maximally flat*. In general, if $G_n(j\omega)$ denotes an nth-order Butterworth filter, then $d^k \left|G_n(j\omega)\right| / d\omega^k_{|\omega=0} = 0$ for $k = 1, 2, 3, 4, 5, ..., 2n - 1$ and
$d^{2n} \left|G_n(j\omega)\right| / d\omega^{2n} \mid_{\omega=0} = -(1/2)(2n!)$.

2.7.8 *Implementation of the Filter*

To actually implement the filter, a differential equation representation of the filter is used. For a transfer function of the form

$$G(s) = \frac{b_1 s^2 + b_2 s + b_3}{s^3 + a_1 s^2 + a_2 s + a_3},$$

a state space representation (called the control canonical form) is given by

$$\begin{aligned} \dot{x}(t) &= \begin{bmatrix} 0 & 1 & 0 \\ 0 & 0 & 1 \\ -a_3 & -a_2 & -a_1 \end{bmatrix} x(t) + \begin{bmatrix} 0 \\ 0 \\ 1 \end{bmatrix} m(t) \\ y(t) &= \begin{bmatrix} b_3 & b_2 & b_1 \end{bmatrix} x(t) \end{aligned}$$

which may be written in a more compact form as

$$\begin{aligned} \dot{x}(t) &= Ax(t) + bm(t) \\ y(t) &= cx(t) \end{aligned}$$

where

$$A = \begin{bmatrix} 0 & 1 & 0 \\ 0 & 0 & 1 \\ -a_3 & -a_2 & -a_1 \end{bmatrix}, \; b = \begin{bmatrix} 0 \\ 0 \\ 1 \end{bmatrix}, \; c = \begin{bmatrix} b_3 & b_2 & b_1 \end{bmatrix}.$$

The characteristic polynomial of the system is given by

$$\begin{aligned} \det(sI - A) &= \det \begin{bmatrix} s & -1 & 0 \\ 0 & s & -1 \\ a_3 & a_2 & s + a_1 \end{bmatrix} \\ &= s \begin{vmatrix} s & -1 \\ a_2 & s + a_1 \end{vmatrix} - (-1) \begin{vmatrix} 0 & -1 \\ a_3 & s + a_1 \end{vmatrix} + 0 \begin{vmatrix} 0 & s \\ a_3 & a_2 \end{vmatrix} \\ &= s^3 + a_1 s^2 + a_2 s + a_3 \end{aligned}$$

and the transfer function is computed as

$$\begin{aligned} Y(s) &= G(s) = c(sI - A)^{-1} b \\ &= \begin{bmatrix} b_3 & b_2 & b_1 \end{bmatrix} \begin{bmatrix} s & -1 & 0 \\ 0 & s & -1 \\ a_3 & a_2 & s + a_1 \end{bmatrix}^{-1} \begin{bmatrix} 0 \\ 0 \\ 1 \end{bmatrix} \\ &= \begin{bmatrix} b_3 & b_2 & b_1 \end{bmatrix} \frac{\begin{bmatrix} \times & \times & 1 \\ \times & \times & s \\ \times & \times & s^2 \end{bmatrix}}{s^3 + a_1 s^2 + a_2 s + a_3} \begin{bmatrix} 0 \\ 0 \\ 1 \end{bmatrix} \\ &= \frac{b_1 s^2 + b_2 s + b_3}{s^3 + a_1 s^2 + a_2 s + a_3} \end{aligned}$$

where the "×" means the specific value of each of these elements is not needed. For the third-order Butterworth filter given by

$$\begin{aligned} G_3(s) &= \frac{\omega_c^3}{(s + \omega_c)(s^2 + \omega_c s + \omega_c^2)} \\ &= \frac{\omega_c^3}{s^3 + 2\omega_c s^2 + 2\omega_c^2 s + \omega_c^3} \end{aligned}$$

direct computation shows that a state-space representation is given by

$$\begin{aligned} \dot{x}(t) &= \begin{bmatrix} 0 & 1 & 0 \\ 0 & 0 & 1 \\ -\omega_c^3 & -2\omega_c^2 & -2\omega_c \end{bmatrix} x(t) + \begin{bmatrix} 0 \\ 0 \\ 1 \end{bmatrix} m(t) \\ y(t) &= \begin{bmatrix} \omega_c^3 & 0 & 0 \end{bmatrix} x(t). \end{aligned} \tag{2.49}$$

2.7.9 *Discretization of Differential Equations*

In order to implement a filter on a computer, the continuous time differential equation representing the filter must be discretized, that is, put into a discrete-time equivalent. There are many ways to do this, but only the trapezoidal method (also known as the bilinear transformation) is discussed here.

Consider $\dot{y}(t) = m(t)$ or $y(t) = \int_0^t m(\tau)d\tau$ and let the sample period be T. The trapezoidal method of integration uses the average value of the integrand over the sample period to compute the area under the curve. This is illustrated in Figure 2.26.

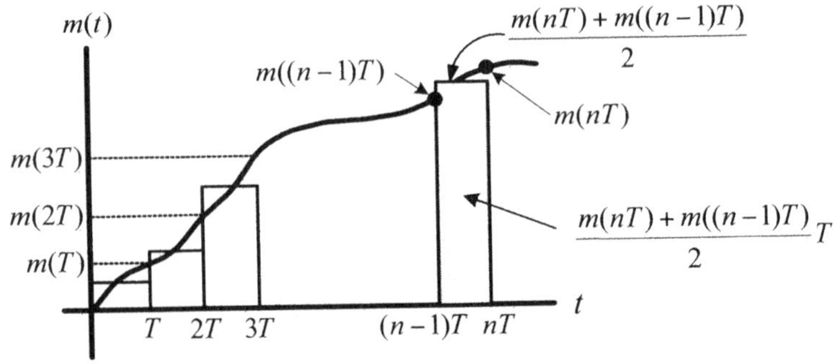

FIGURE 2.26. Trapezoidal method of integration.

Denote the area under the curve $m(t)$ at time nT by $y(nT)$. The average value of $m(t)$ between the times $(n-1)T$ and nT is $\frac{m(nT)+m((n-1)T)}{2}$ so that the approximate area under the curve in this same time period is $\frac{m(nT)+m((n-1)T)}{2}T$. Thus, one can (approximately) find the area under $m(t)$ at any time nT as

$$y(nT) = y((n-1)T) + \frac{m(nT) + m((n-1)T)}{2}T. \tag{2.50}$$

Let

$$\begin{aligned} Y(z) &= \sum_{n=0}^{\infty} y(nT)z^{-n} \\ M(z) &= \sum_{n=0}^{\infty} m(nT)z^{-n} \end{aligned}$$

be the z transform of $\{Y(nT)\}$ and $\{M(nT)\}$, respectively. Take the z

transform of (2.50) to obtain

$$\begin{aligned}\sum_{n=0}^{\infty} y(nT)z^{-n} &= z^{-1}\sum_{n=0}^{\infty} y((n-1)T)z^{-(n-1)} + \\ &\quad \frac{T}{2}\left(\sum_{n=0}^{\infty} m(nT)z^{-n} + z^{-1}\sum_{n=0}^{\infty} m((n-1)T)z^{-(n-1)}\right).\end{aligned}$$

Note that with $y(-1) = 0$,

$$\begin{aligned}\sum_{n=0}^{\infty} y((n-1)T)z^{-(n-1)} &= \sum_{n=1}^{\infty} y((n-1)T)z^{-(n-1)} \\ &= \sum_{k=0}^{\infty} y(kT)z^{-k} \\ &= Y(z).\end{aligned}$$

Similarly, with $m(-1) = 0$, $\sum_{n=0}^{\infty} m((n-1)T)z^{-(n-1)} = M(z)$ so that

$$Y(z) = z^{-1}Y(z) + \frac{T}{2}\left(M(z) + z^{-1}M(z)\right)$$

or

$$\frac{Y(z)}{M(z)} = \frac{T}{2}\frac{1+z^{-1}}{1-z^{-1}}$$

where the right-hand side is known as the *bilinear transformation*. The general case is now considered where the state-space representation

$$\dot{x}(t) = Ax(t) + bm(t)$$

of the filter is used. To numerically integrate this differential equation, the trapezoidal integration algorithm is used as follows:

$$\begin{aligned}x(nT) &= x((n-1)T) + T\frac{\dot{x}(nT) + \dot{x}((n-1)T)}{2} \\ &= x((n-1)T) \\ &\quad + \frac{T}{2}\left(Ax(nT) + bm(nT) + Ax((n-1)T) + bm((n-1)T)\right).\end{aligned}$$

Take the z transform of both sides to obtain

$$X(z) = z^{-1}X(z) + \frac{T}{2}\left(AX(z) + Az^{-1}X(z)\right) + \frac{T}{2}\left(bM(z) + bz^{-1}M(z)\right)$$

or

$$\begin{aligned}(1-z^{-1})X(z) - \frac{T}{2}(1+z^{-1})AX(z) &= \frac{T}{2}(1+z^{-1})bM(z) \\ \frac{2}{T}\frac{1-z^{-1}}{1+z^{-1}}X(z) - AX(z) &= bM(z) \\ \left(\left(\frac{2}{T}\frac{1-z^{-1}}{1+z^{-1}}\right)I - A\right)X(z) &= bM(z).\end{aligned}$$

Then, as $Y(z) = cX(z)$,

$$H(z) \triangleq \frac{Y(z)}{M(z)} = c\left(\left(\frac{2}{T}\frac{1-z^{-1}}{1+z^{-1}}\right)I - A\right)^{-1} b.$$

The remarkable result is that the discrete-time transfer function is found by simply evaluating the *continuous-time* transfer function $G(s) = c(sI - A)^{-1}b$ at $s = \frac{2}{T}\frac{1-z^{-1}}{1+z^{-1}}$!

Going back to the third-order Butterworth filter

$$G_3(s) = \frac{\omega_c^3}{s^3 + 2\omega_c s^2 + 2\omega_c^2 s + \omega_c^3},$$

the discrete-time transfer function for this filter (corresponding to integrating the differential equation (2.49) using the trapezoidal approximation) is given by

$$\begin{aligned} H(z) &= G(s)\big|_{s=\frac{2}{T}\frac{1-z^{-1}}{1+z^{-1}}} \\ &= \frac{\omega_c^3}{s^3 + 2\omega_c s^2 + 2\omega_c^2 s + \omega_c^3}\Big|_{s=\frac{2}{T}\frac{1-z^{-1}}{1+z^{-1}}} \\ &= \frac{b_1 + b_2 z^{-1} + b_3 z^{-2} + b_4 z^{-3}}{a_1 + a_2 z^{-1} + a_3 z^{-2} + a_4 z^{-3}} \end{aligned}$$

where

$$\begin{aligned} b_1 &= \omega_c^3 T^3, b_2 = 3\omega_c^3 T^3, b_3 = 3\omega_c^3 T^3, b_4 = \omega_c^3 T^3 \\ a_1 &= \omega_c^3 T^3 + 4\omega_c^2 T^2 + 8\omega_c T + 8 \\ a_2 &= 3\omega_c^3 T^3 + 4\omega_c^2 T^2 - 8\omega_c T - 24 \\ a_3 &= 3\omega_c^3 T^3 - 4\omega_c^2 T^2 - 8\omega_c T + 24 \\ a_4 &= \omega_c^3 T^3 - 4\omega_c^2 T^2 + 8\omega_c T - 8. \end{aligned} \tag{2.51}$$

Usually, the filter coefficients are normalized so that leading coefficient in the denominator is 1. The coefficients are then given by $a_1' = 1$, $a_2' = a_2/a_1$, $a_3' = a_3/a_1$, $a_4' = a_4/a_1$, $b_1' = b_1/a_1$, $b_2' = b_2/a_1$, $b_3' = b_3/a_1$, $b_4' = b_4/a_1$.

In order to implement this discrete-time filter, it needs to be reformulated as a discrete-time difference equation. This can be done either as a input-output difference equation or as a state-space difference equation. Both of these representations are considered in the next section.

2.7.10 *Digital Filtering*

A general second-order digital filter may be represented in input-output form as ($a_1 = 1$)

$$\begin{aligned} y(nT) =& b_1 m(nT) + b_2 m((n-1)T) + b_3 m((n-2)T) \\ &- a_2 y((n-1)T) - a_3 y((n-2)T) \end{aligned} \tag{2.52}$$

That is, to get the output of the filter at time nT, one needs to know the current input $m(nT)$, have stored the previous two inputs $m((n-1)T)$, $m((n-2)T)$ and outputs $y((n-1)T), y((n-2)T)$, and then carry out the calculation. Rewrite (2.52) as

$$y(nT) + a_2 y((n-1)T) + a_3 y((n-2)T) = b_1 m(nT)) + b_2 m((n-1)T + b_3 m((n-2)T). \qquad (2.53)$$

To compute the discrete transfer function, multiply both sides by z and sum from $n = 0$ to $n = \infty$ to obtain

$$\sum_{n=0}^{\infty} y(nT)z^{-n} + a_2 z^{-1} \sum_{n=0}^{\infty} y((n-1)T)z^{-(n-1)} + a_3 z^{-2} \sum_{n=0}^{\infty} y((n-2)T)z^{-(n-2)}$$
$$= b_1 \sum_{n=0}^{\infty} m(nT)z^{-n} + b_2 z^{-1} \sum_{n=0}^{\infty} m((n-1)T)z^{-(n-1)} + b_3 z^{-2} \sum_{n=0}^{\infty} m((n-2)T)z^{-(n-2)}.$$

Let $Y(z) = \sum_{n=0}^{\infty} y(nT)z^{-n}$ so that with $y(-1) = 0$ it follows that

$$\sum_{n=0}^{\infty} y((n-1)T)z^{-(n-1)} = \sum_{n=1}^{\infty} y((n-1)T)z^{-(n-1)} = \sum_{k=0}^{\infty} y((k)T)z^{-k} = Y(z).$$

In addition, if $y(-2) = 0$, it then follows that $Y(z) = \sum_{n=0}^{\infty} y((n-2)T)z^{-(n-2)}$. Similarly, with $m(-1) = m(-2) = 0$, it follows that

$$\begin{aligned} M(z) &= \sum_{n=0}^{\infty} m(nT)z^{-n} = z^{-1} \sum_{n=0}^{\infty} m((n-1)T)z^{-(n-1)} \\ &= z^{-2} \sum_{n=0}^{\infty} m((n-2)T)z^{-(n-2)}, \end{aligned}$$

Using these expressions, (2.53) becomes

$$Y(z) = a_2 z^{-1} Y(z) + a_3 z^{-2} Y(z) = b_1 M(z) + b_2 z^{-1} M(z) + b_3 z^{-2} M(z)$$

or

$$H(z) = \frac{b_1 + b_2 z^{-1} + b_3 z^{-2}}{1 + a_2 z^{-1} + a_3 z^{-2}}. \qquad (2.54)$$

Thus, the filter represented by this transfer function may be implemented on a computer using the difference equation (2.52).

Example Consider the computation of speed by differentiation of position:

$$\omega_{bd}(nT) = \frac{\theta(nT) - \theta((n-1)T)}{T}$$

Let $m(nT) = \theta(nT)$ and $y(nT) = \omega(nT)$ so that this becomes

$$y(nT) = \frac{1}{T}m(nT) - \frac{1}{T}m((n-1)T)$$

that is, $b_1 = 1/T, b_2 = -1/T, a_1 = 1, a_2 = 0$. The transfer function is given by

$$\frac{Y(z)}{M(z)} = \frac{b_1 + b_2 z^{-1}}{1}.$$

2.7.11 *State-Space Representation*

Consider a third-order discrete-time transfer function of the form

$$\begin{aligned}
H(z) &= \frac{Y(z)}{M(z)} = \frac{b_0 + b_1 z^{-1} + b_2 z^{-2} + b_3 z^{-3}}{1 + a_1 z^{-1} + a_2 z^{-2} + a_3 z^{-3}} \\
&= \frac{b_0 z^3 + b_1 z^2 + b_2 z + b_3}{z^3 + a_1 z^2 + a_2 z + a_3} = \frac{b_0\left(z^3 + a_1 z^2 + a_2 z + a_3\right)}{z^3 + a_1 z^2 + a_2 z + a_3} \\
&\quad + \frac{(b_1 - b_0 a_1)z^2 + (b_2 - b_0 a_2)z + (b_3 - b_0 a_3)}{z^3 + a_1 z^2 + a_2 z + a_3} \\
&= b_0 + \frac{(b_1 - b_0 a_1)z^2 + (b_2 - b_0 a_2)z + (b_3 - b_0 a_3)}{z^3 + a_1 z^2 + a_2 z + a_3}. \qquad (2.55)
\end{aligned}$$

With

$$\begin{aligned}
A &= \begin{bmatrix} 0 & 1 & 0 \\ 0 & 0 & 1 \\ -a_3 & -a_2 & -a_1 \end{bmatrix}, \quad b = \begin{bmatrix} 0 \\ 0 \\ 1 \end{bmatrix} \\
c &= \begin{bmatrix} b_3 - b_0 a_3 & b_2 - b_0 a_2 & b_1 - b_0 a_1 \end{bmatrix} \\
d &= b_0
\end{aligned}$$

a state-space realization of this transfer function is given by

$$\begin{aligned}
x((n+1)T)) &= Ax(nT) + bm(nT), \; x(nT) \in \mathbb{R}^3 \\
y(nT) &= cx(nT) + dm(nT), \; y(nT) \in \mathbb{R}.
\end{aligned} \qquad (2.56)$$

To see this, the z transform of (2.56) is now computed to obtain the transfer function. Proceeding, let

$$X(z) = \sum_{n=0}^{\infty} x(nT) z^{-n} \in \mathbb{R}^3$$

and note that

$$\begin{aligned}\sum_{n=0}^{\infty} x\left((n+1)T\right) z^{-n} &= z\sum_{n=0}^{\infty} x\left((n+1)T\right) z^{-(n+1)} \\ &= z\left(\sum_{n=-1}^{\infty} x\left((n+1)T\right) z^{-(n+1)} - x(0)\right) \\ &= z\left(\sum_{k=0}^{\infty} x(kT) z^{-k} - x(0)\right) \\ &= zX(z) - zx(0).\end{aligned}$$

Thus, taking the z transform of (2.56) results in

$$zX(z) - zx(0) = AX(z) + bM(z).$$

Solve for $X(z)$ to obtain

$$X(z) = (zI - A)^{-1}bM(z) + (zI - A)^{-1}zx(0)$$

or

$$Y(z) = cX(z) + dM(z) = \left(c(zI - A)^{-1}b + d\right) M(z) + z(zI - A)^{-1}x(0).$$

With $x(0) = 0$, the transfer function is

$$\frac{Y(z)}{M(z)} = c(zI - A)^{-1}b + d$$

where

$$\begin{aligned}&c(zI - A)^{-1}b + d \\ &= \begin{bmatrix} b_3 - b_0a_3 & b_2 - b_0a_2 & b_1 - b_0a_1 \end{bmatrix} \begin{bmatrix} z & -1 & 0 \\ 0 & z & -1 \\ a_3 & a_2 & z + a_1 \end{bmatrix}^{-1} \begin{bmatrix} 0 \\ 0 \\ 1 \end{bmatrix} + b_0 \\ &= \begin{bmatrix} b_3 - b_0a_3 & b_2 - b_0a_2 & b_1 - b_0a_1 \end{bmatrix} \frac{\begin{bmatrix} \times & \times & 1 \\ \times & \times & z \\ \times & \times & z^2 \end{bmatrix}\begin{bmatrix} 0 \\ 0 \\ 1 \end{bmatrix}}{z^3 + a_1z^2 + a_2z + a_3} + b_0\end{aligned}$$

where the "×" denotes values that do not affect the final result. Simplifying, this becomes

$$\begin{aligned}
c(zI-A)^{-1}b+d &= \frac{(b_1-b_0a_1)z^2+(b_2-b_0a_2)z+(b_3-b_0a_3)}{z^3+a_1z^2+a_2z+a_3}+b_0 \\
&= \frac{b_0\left(z^3+a_1z^2+a_2z+a_3\right)}{z^3+a_1z^2+a_2z+a_3} \\
&\quad + \frac{(b_1-b_0a_1)z^2+(b_2-b_0a_2)z+(b_3-b_0a_3)}{z^3+a_1z^2+a_2z+a_3} \\
&= \frac{b_0z^3+b_1z^2+b_2z+b_3}{z^3+a_1z^2+a_2z+a_3} \\
&= \frac{b_0+b_1z^{-1}+b_2z^{-2}+b_3z^{-3}}{1+a_1z^{-1}+a_2z^{-2}+a_3z^{-3}}.
\end{aligned}$$

In summary, the state-space representation (2.56) can used to implement the digital filter (2.55) in real time on a computer.

2.7.12 *Noncausal Filtering*

It has been shown how to use a Butterworth filter to realize a filter with low-pass magnitude characteristics and a linear phase. One can get around the problem of requiring a linear phase by filtering the signal twice in an appropriate manner as is now described.

Let $m(t) = s(t) + n(t)$ be the input to a filter whose transfer function is $G(s)$ and let the output be $y(t)$ so that $y(t) = \int_{-\infty}^{t} g(t-\tau)m(\tau)d\tau$. Also, let T_f be a large final time such that $[0, T_f]$ contains the period of time over which the signal $s(t)$ is nonzero. The plots of $m(t), s(t)$ and $y(t)$ are given below in Figure 2.27. Note that the filtered signal $y(t)$ is shifted to the right from $s(t)$ corresponding to a time delay. Next, let $y_b(t) \triangleq y(T_f - t)$ (the subscript b stands for backwards) and then pass this signal through $G(s)$ once again and call the output $y_{ffb}(t)$ (the ff stands for *filter-filter*, i.e., double filtering). Finally, let $y_{ff}(t) \triangleq y_{ffb}(T_f - t)$. Figure 2.28 illustrates these different signals.

It is now shown that $y_{ff}(t)$ is a good estimate of $s(t)$. To explain how this double filtering works, the transfer function from m to y is now computed. The filter may be represented as a convolution or in the s domain as

$$\begin{aligned}
y(t) &= \int_0^t g(t-\tau)m(\tau)d\tau \\
Y(s) &= G(s)M(s)
\end{aligned}$$

where $G(j\omega) = |G(j\omega)|\, e^{j\angle G(j\omega)}$.

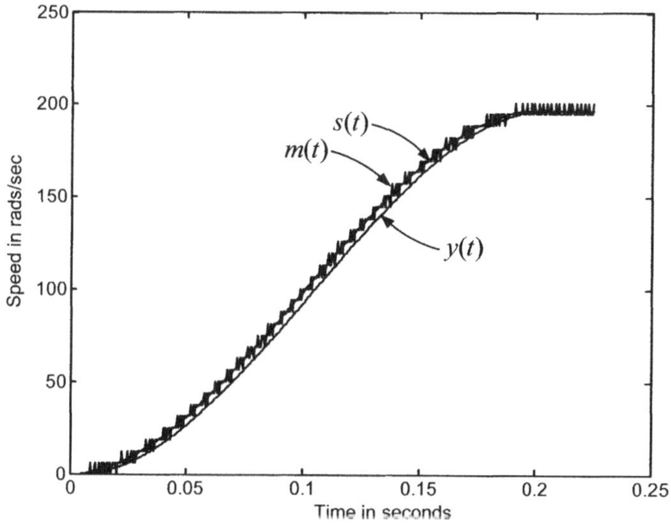

FIGURE 2.27. $m(t) = (\theta(nT) - \theta((n-1)T))/T$ with $T = 0.5$ msec is the measured speed. $s(t)$ is the "true" speed from a simulation and $y(t)$ is the filtered version of $m(t)$. Note that $y(t)$ is delayed with respect to $s(t)$.

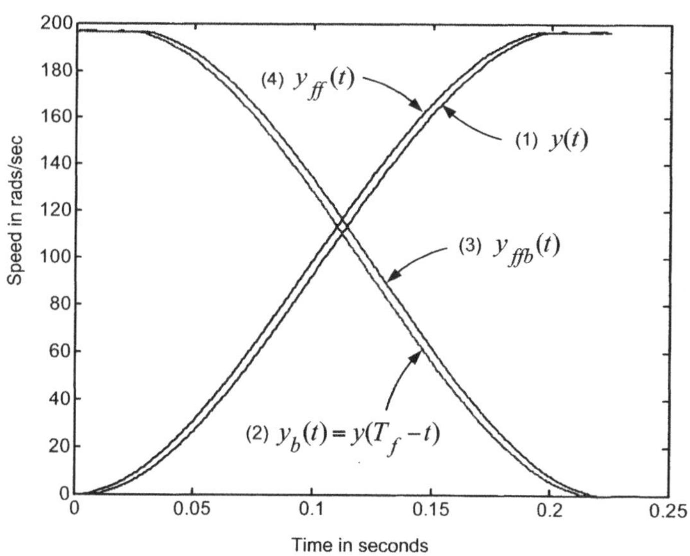

FIGURE 2.28. $y(t), y_b(t) = y(T_f - t), y_{ffb}(t), y_{ff}(t)$ versus t.

Then from the convolution representation

$$y_{ffb}(t) = \int_0^t g(t-\tau)y_b(\tau)d\tau = \int_0^t g(t-\tau)y(T_f-\tau)d\tau$$

it follows that

$$Y_{ffb}(s) = G(s)Y_b(s)$$

where[7]

$$\begin{aligned} Y_b(s) &= \int_0^{T_f} e^{-st}y(T_f-t)dt = -\int_{T_f}^0 e^{-s(T_f-u)}y(u)du \\ &= e^{-sT_f}\int_0^{T_f} e^{-(-s)u}y(u)du \\ &= e^{-sT_f}Y(-s) \\ &= e^{-sT_f}G(-s)M(-s). \end{aligned}$$

Thus,

$$Y_{ffb}(s) = G(s)Y_b(s) = e^{-sT_f}G(s)G(-s)M(-s).$$

Then, with

$$y_{ff}(t) = y_{ffb}(T_f - t)$$

it follows that[8]

$$\begin{aligned} Y_{ff}(s) &= e^{-sT_f}Y_{ffb}(-s) = e^{-sT_f}\left(e^{sT_f}G(-s)G(s)M(s)\right) \\ &= G(-s)G(s)M(s) \end{aligned}$$

or

$$Y_{ff}(j\omega) = G(-j\omega)G(j\omega)M(j\omega).$$

As

$$\begin{aligned} G(j\omega) &= |G(j\omega)|\, e^{j\angle G(j\omega)} \\ G(-j\omega) &= |G(-j\omega)|\, e^{j\angle G(-j\omega)} = |G(j\omega)|\, e^{-j\angle G(j\omega)} \end{aligned}$$

it follows that

$$\begin{aligned} Y_{ff}(j\omega) &= G(-j\omega)G(j\omega)M(j\omega) \\ &= |G(j\omega)|\, e^{-j\angle G(j\omega)}\, |G(j\omega)|\, e^{j\angle G(j\omega)}M(j\omega) \\ &= |G(j\omega)|^2\, M(j\omega). \end{aligned}$$

That is, there is no phase shift from $m(t)$ to $y_{ff}(t)$ and thus no distortion. Of course, this filter is noncausal as all of the data $m(t)$ must be collected before the computations can be done.

---

[7] Recall that by assumption $y(t) = 0$ for $t < 0$ and $t > T_f$, which implies $y(T_f - t) = 0$ for $t < 0$ and $t > T_f$.

[8] If $\mathcal{L}\{x(t)\} = X(s)$, then $\mathcal{L}\{x(-t)\} = X(-s)$ and $\mathcal{L}\{x(t-t_0)\} = e^{-t_0 s}X(s)$ so that $\mathcal{L}\{x(-(t-t_0))\} = e^{-t_0 s}\mathcal{L}\{x(-t)\} = e^{-t_0 s}X(-s)$.

Appendix - Classical Feedback Control

Tracking and Disturbance Rejection

In order to explain the ideas of tracking and disturbance rejection, consider the block diagram of Figure 2.9 which is redrawn in a standard form in Figure 2.29.

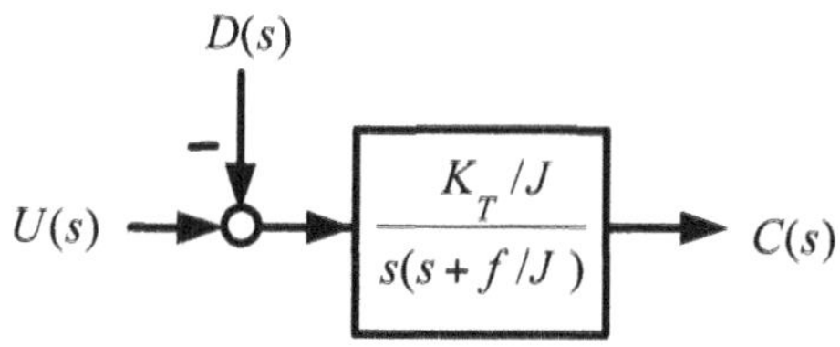

FIGURE 2.29. DC motor block diagram.

In Figure 2.29 $U(s) \triangleq i_r(s), C(s) \triangleq \theta(s)$, and $D(s) \triangleq \tau_L(s)/K_T$. Note that, though this system is mathematically equivalent to the system of Figure 2.9, there is not a one-to-one physical equivalence. Specifically, the disturbance $D(s) = \tau_L(s)/K_T$ has the units of amperes, which is consistent with the above drawing, as it is being subtracted from the input *current* $U(s) = i_r(s)$. This is not to say that the load is a current, but rather, if a current given by $-\tau_L/K_T$ is input to the motor, then the effect on the output position $c(t)$ is the same as the actual load torque.

A block diagram for a simple proportional control system for the motor is given in Figure 2.30, where $R(s) \triangleq \theta_{ref}(s)$ is the reference input and $G_m(s) \triangleq \dfrac{K_T/J}{s\,(s+f/J)}$.

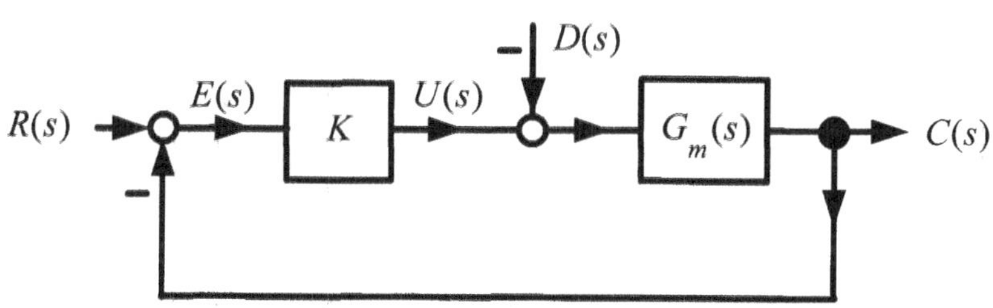

FIGURE 2.30. Simple proportional control system.

A sensor is used to feed back the motor position to obtain the error signal

$$E(s) \triangleq R(s) - C(s).$$

A simple proportional controller is specified by

$$U(s) = KE(s)$$

that is, the current commanded to the motor $[U(s) = i_r(s)]$ is proportional to the difference between the reference (desired) position and the actual position.

The simple proportional controller is often replaced by a more general controller (compensator) specified by a transfer function $G_c(s) = b_c(s)/a_c(s)$ where the subscript "c" stands for controller (or compensator). In fact, the whole point of automatic control theory is to provide a methodology on how to specify $G_c(s)$. A general block diagram form for a control system is given in Figure 2.31.

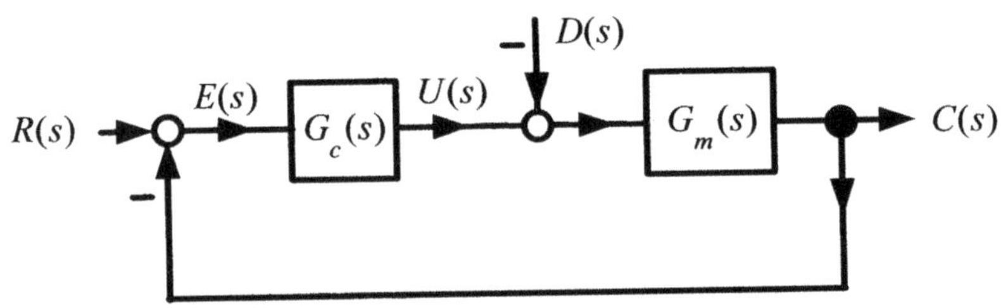

FIGURE 2.31. Standard block diagram of a control system.

To obtain a convenient expression for the transfer function from $R(s)$ and $D(s)$ to the output $C(s)$, compute

$$C(s) = (G_c(s)E(s) - D(s))\,G_m(s) = (G_c(s)\,(R(s) - C(s)) - D(s))\,G_m(s).$$

This is rearranged to obtain

$$C(s)\,(1 + G_c(s)G_m(s)) = G_c(s)G_m(s)R(s) - D(s)G_m(s)$$

or

$$C(s) = \frac{G_c(s)G_m(s)}{1 + G_c(s)G_m(s)}R(s) - \frac{G_m(s)}{1 + G_c(s)G_m(s)}D(s). \tag{2.57}$$

The error $E(s)$ is given by

$$E(s) = R(s) - C(s) = \frac{1}{1 + G_c(s)G_m(s)}R(s) + \frac{G_m(s)}{1 + G_c(s)G_m(s)}D(s). \tag{2.58}$$

Typically, the compensator $G_c(s)$ is chosen to achieve two objectives:

1. *Tracking* If the reference position is set as $r(t) = 30°$ $(R(s) = 30°/s)$, then it is required that $c(t) \to 30°$ as $t \to \infty$, i.e., the output must track the input

2. *Disturbance Rejection* If the reference position is set as $r(t) = 30°$, then it is required that $c(t) \to 30°$ as $t \to \infty$ despite any load on the motor. For example, if a robot arm is to move 30°, it must do so no matter how much weight it is carrying.

PID Control of a Simple DC Servo Motor

The ideas of tracking and disturbance rejection are now illustrated through a series of simple examples. This is done by considering a fixed physical system given by $G_m(s) = \dfrac{1}{s(s+1)}$, which could be the model of a DC servo motor system. Various controllers are considered for both tracking and disturbance rejection.

Tracking Control

A simple proportional controller given by $G_c(s) = K$ is considered first.

Example *Tracking a step input*

Consider the unity feedback system control system of Figure 2.32 where $G_m(s) = \dfrac{1}{s(s+1)}$, $D(s) = 0$ and a constant gain controller.

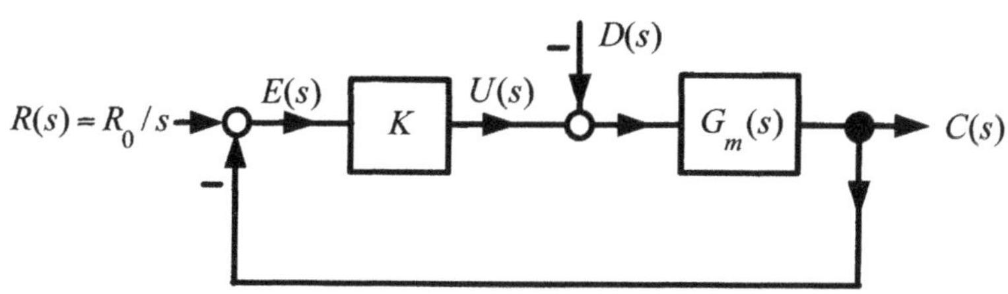

FIGURE 2.32. Tracking a step input.

The output and error transfer functions are, respectively,

$$\begin{aligned} C(s) &= \frac{KG_m(s)}{1+KG_m(s)}R(s) \\ E(s) &= \frac{1}{1+KG_m(s)}R(s). \end{aligned}$$

Suppose the objective is to track a step input. Let $u_s(t)$ denote the unit step function and let $r(t) = R_0 u_s(t)$ so that $R(s) = R_0/s$. The objective is to force the error $e(t) \to 0$ as $t \to \infty$ as this means that $c(t) \to R_0$. Evaluating $E(s)$, one obtains

$$E(s) = \frac{1}{1+K\dfrac{1}{s(s+1)}}\frac{R_0}{s} = \frac{s(s+1)}{s(s+1)+K}\frac{R_0}{s} = \frac{s+1}{s^2+s+K}R_0.$$

Note that the poles of the forward open-loop transfer function $KG_m(s) = K\frac{1}{s(s+1)}$ reappear in the numerator of $E(s)$ after clearing the fractions in the denominator. In particular, the "s" in $K\frac{1}{s(s+1)}$ cancelled the "s" in the denominator of $R(s)$. For $K > 0$, $s^2 + s + K$ is a stable polynomial so that $sE(s) = \frac{s(s+1)}{s^2+s+K}R_0$ is stable. By the final value theorem,

$$e(\infty) = \lim_{s\to 0} sE(s) = \lim_{s\to 0} s\frac{s(s+1)}{s(s+1)+K}\frac{R_0}{s} = 0.$$

Where does stability come in? Let $K = 1$ so that

$$\begin{aligned} s^2+s+1 &= \left(s - \left(-1/2 + j\sqrt{3}/2\right)\right)\left(s - \left(-1/2 - j\sqrt{3}/2\right)\right) \\ &= (s-p_1)(s-p_2) \end{aligned}$$

with $p_1 = -1/2 + j\sqrt{3}/2$ and $p_2 = p_1^* = -1/2 - j\sqrt{3}/2$. Then a partial fraction expansion of $E(s)$ gives

$$E(s) = \frac{s(s+1)}{s(s+1)+K}\frac{R_0}{s} = \frac{s+1}{(s-p_1)(s-p_2)}R_0 = \frac{\beta}{(s-p_1)} + \frac{\beta^*}{(s-p_2)}$$

where $\beta = \beta_1 + j\beta_2 = |\beta|\, e^{j\angle\beta}$ and $|\beta|^2 = \beta_1^2 + \beta_2^2$, $\angle\beta = \tan^{-1}(\beta_2/\beta_1)$.

In the time domain,

$$\begin{aligned} e(t) &= \beta e^{p_1 t} + \beta^* e^{p_2 t} \\ &= |\beta|\, e^{j\angle\beta} e^{-(1/2)t + j(\sqrt{3}/2)t} + |\beta|\, e^{-j\angle\beta} e^{-(1/2)t - j(\sqrt{3}/2)t} \\ &= 2|\beta|\, e^{-(1/2)t}\cos\left((\sqrt{3}/2)t + \angle\beta\right) \to 0 \text{ as } t\to\infty. \end{aligned}$$

Recall that the poles of the closed-loop transfer function determine the form of the transient response. As the closed-loop poles p_1 and p_2 are in the open left-half plane (LHP), the transients die out. The real part of the closed-loop poles ($\text{Re}\{p_1\} = \text{Re}\{p_2\} = -1/2$) determine how fast the transient $e(t)$ dies out and typically one chooses the amplifier gain K such that the poles of the closed-loop system are as far as possible in the LHP.

Example *Tracking a ramp input*

Consider the same system as in the previous example, but let the input be a ramp function. Specifically, the input is given by $r(t) = \omega_0 t$ where ω_0 is a constant ($R(s) = \omega_0/s^2$) and $D(s) = 0$. The error $E(s)$ is then

$$E(s) = \frac{1}{1 + K\frac{1}{s(s+1)}}\frac{\omega_0}{s^2} = \frac{s(s+1)}{s(s+1)+K}\frac{\omega_0}{s^2} = \frac{s+1}{s^2+s+K}\frac{\omega_0}{s}.$$

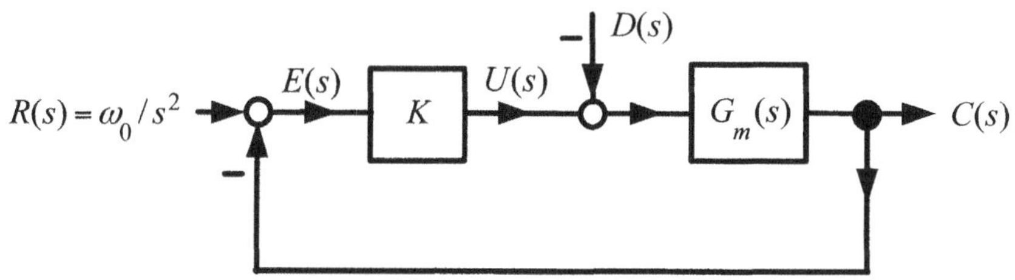

FIGURE 2.33. Tracking a ramp input.

Note again that the pole at $s = 0$ of $KG_m(s)$ cancelled one of the poles at $s = 0$ of $R(s)$. As $s^2 + s + K$ is stable for $K > 0$, $sE(s) = \dfrac{s+1}{s^2+s+K}\omega_0$ is stable. By the final value theorem,

$$e(\infty) = \lim_{s\to 0} sE(s) = \lim_{s\to 0} \frac{s+1}{s(s+1)+K}\omega_0 = \frac{\omega_0}{K}.$$

Consequently, asymptotic tracking is not achieved as $e(t)$ does not go to 0 as $t \to \infty$. However, the motor does follow the input with a finite error which can be made small if K can be made large.

Where does stability come in? Let $K = 1$ so that $p_i = \left(-1 \pm j\sqrt{3}\right)/2$ and

$$\begin{aligned} E(s) &= \frac{s(s+1)}{s(s+1)+K}\frac{\omega_0}{s^2} = \frac{s+1}{s(s-p_1)(s-p_2)}\omega_0 \\ &= \frac{A}{s} + \frac{\beta}{(s-p_1)} + \frac{\beta^*}{(s-p_2)} \\ &= \frac{\omega_0}{s} + \frac{\beta}{(s-p_1)} + \frac{\beta^*}{(s-p_2)} \end{aligned}$$

for some complex constant β and $\omega_0 = \lim_{s\to 0} sE(s)$. Then

$$e(t) = \omega_0 u_s(t) + \beta e^{p_1 t} + \beta^* e^{p_2 t} \to \omega_0$$

as $t \to \infty$, as the poles are in the open left-half plane ($\mathrm{Re}(p_1) = \mathrm{Re}(p_2) < 0$). Again, as the closed-loop poles are in the left-half plane, the transients die out.

The simple proportional controller is not adequate to track a ramp input $r(t) = \omega_0 t$ with zero steady-state error. The answer to achieving zero steady-state error for this system lies in considering a different controller!

Example *Integral Controller*

Consider the controller $G_c(s) = K/s$; that is, an integrator.

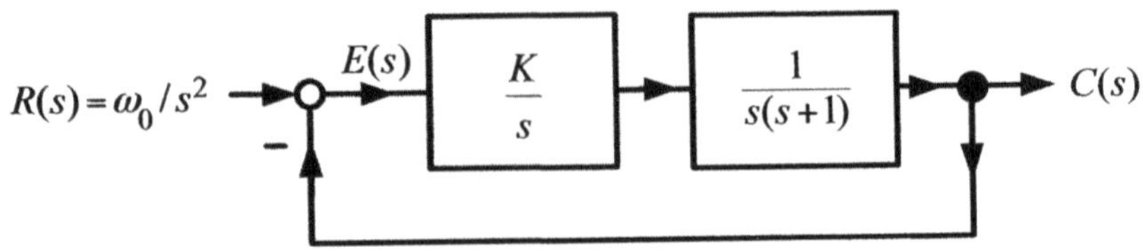

FIGURE 2.34. An integral controller.

Then

$$E(s) = \frac{1}{1+\dfrac{K}{s}\dfrac{1}{s(s+1)}}\frac{\omega_0}{s^2} = \frac{s^2(s+1)}{s^2(s+1)+K}\frac{\omega_0}{s^2} = \frac{s+1}{s^3+s^2+K}\omega_0.$$

where note that the denominator (poles) of $\dfrac{K}{s}\dfrac{1}{s(s+1)}$ reappeared in the numerator of $E(s)$ after clearing fractions. As a consequence, the s^2 in the denominator of the forward open-loop transfer function $\dfrac{K}{s}\dfrac{1}{s(s+1)}$ is in the numerator of $E(s)$ and cancels the s^2 in the denominator of $R(s)$. If s^3+s^2+K were stable, then by the final value theorem,

$$e(\infty) = \lim_{s\to 0} sE(s) = \lim_{s\to 0} s\frac{s+1}{s^3+s^2+K}\omega_0 = 0.$$

The difficulty here is that $s^3+s^2+K = s^3+s^2+0s+K$ is never stable for any value of K as a necessary condition for stability is that all the coefficients of the closed-loop characteristic polynomial be positive. To emphasize the stability aspect, let $K=1$ for which s^3+s^2+1 has roots $-1.47, 0.23\pm j0.79$. The error response is found by doing a partial fraction expansion of $E(s)$ as

$$\begin{aligned} E(s) &= \frac{s+1}{(s+1.47)(s-0.23-j0.79)(s-0.23+j0.79)}\omega_0 \\ &= \frac{A}{s+1.47} + \frac{\beta}{s-0.23-j0.79} + \frac{\beta^*}{s-0.23+j0.79} \end{aligned}$$

for some complex constant β. The time response then has the form

$$\begin{aligned} e(t) &= Ae^{-1.47t} + \beta e^{0.23t}e^{j0.79t} + \beta^* e^{0.23t}e^{-j0.79t} \\ &= Ae^{-1.47t} + 2\left|\beta\right| e^{0.23t}\cos(0.79t + \angle\beta). \end{aligned}$$

The error $e(t)$ does not go to zero due to the unstable closed-loop poles at $-0.23 \pm j0.79$. Remember, the fact that $lim_{s\to 0}\ sE(s) = 0$ says nothing about the steady-state error unless $sE(s)$ is stable!

As this example demonstrates, the difficult problem is not in getting $\lim_{s\to 0} sE(s) = 0$, but instead, it is in making the closed-loop system stable. As shown above, a constant gain (proportional) controller will give closed-loop stability, but not $lim_{s\to 0}\ sE(s) = 0$. On the other hand, an integral controller gives $lim_{s\to 0} sE(s) = 0$, but not closed-loop stability. Let's combine the two and see if that will work. Specifically, let

$$G_c(s) = K\frac{s+\alpha}{s} = K + \frac{\alpha K}{s}.$$

This is called a proportional (K) plus integral ($\alpha K/s$) compensator or PI compensator.

Example *PI controller*

Again, consider the same system as in the previous examples, except use a PI compensator.

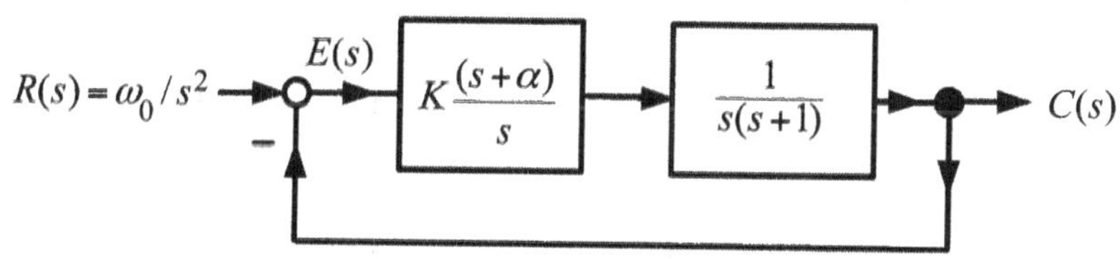

FIGURE 2.35. A proportional plus integral (PI) controller.

From the block diagram, the error $E(s)$ is seen to be given by

$$\begin{aligned} E(s) &= \frac{1}{1 + K\frac{s+\alpha}{s}\frac{1}{s(s+1)}}R(s) = \frac{s^2(s+1)}{s^2(s+1) + K(s+\alpha)}\frac{\omega_0}{s^2} \\ &= \frac{s+1}{s^3 + s^2 + Ks + \alpha K}\omega_0. \end{aligned}$$

Again note that denominator of the open-loop system $K\frac{s+\alpha}{s}\frac{1}{s(s+1)}$ reappears in the numerator of $E(s)$. Consequently, the factor of s^2 in the denominator of $G_c(s)G_m(s)$ cancels the "s" in the denominator of $R(s)$. Then, *if* $s^3 + s^2 + Ks + \alpha K$ is stable, the final value theorem gives

$$e(\infty) = \lim_{s\to 0} sE(s) = \lim_{s\to 0} s\frac{s+1}{s^3 + s^2 + Ks + \alpha K}\omega_0 = 0.$$

To check stability of $s^3 + s^2 + Ks + \alpha K$, the Routh test is used [37][46][47]. The Routh table is

s^3	1	K
s^2	1	αK
s	$K - \alpha K$	0
s^0	αK	

The first column is positive if and only if $\alpha K > 0$ and $K - \alpha K = K(1-\alpha) > 0$ or

$$K > 0 \text{ and } 0 < \alpha < 1.$$

For example, with $K = 1$ and $\alpha = 1/2$, the closed-loop characteristic equation is

$$s^3 + s^2 + s + 1/2 = (s + 0.65)(s + 0.176 - j0.861)(s + 0.176 + j0.861).$$

Then, letting $r = -0.65, p_i = -0.176 \pm j0.861$, the partial fraction expansion of $E(s)$ gives

$$E(s) = \frac{s+1}{(s-r)(s-p_1)(s-p_2)}\omega_0 = \frac{A}{s-r} + \frac{\beta}{s-p_1} + \frac{\beta^*}{s-p_2}.$$

The corresponding time response is then

$$\begin{aligned} e(t) &= Ae^{-0.65t} + \beta e^{-0.176t} e^{j0.861t} + \beta^* e^{-0.176t} e^{-j0.861t} \\ &= Ae^{-0.65t} + 2\,|\beta|\, e^{-0.176t} \cos(0.861t + \angle\beta) \to 0 \text{ as } t \to \infty. \end{aligned}$$

In general, adding integrators to the compensator tends to destabilize the closed-loop system while adding a proportional (constant gain) compensator tends to stabilize the closed-loop system. In the above example, an integrator is added to obtain zero steady-state error and the proportional compensator is added to make the closed-loop system stable.

Disturbance Rejection

Consider now the problem of having the motor to rotate R_0 radians with an external load acting on it (Figure 2.37 illustrates such cases).

Example *Constant Load Torque with a Proportional Controller*
Consider the control system in the block diagram below. Let

$$G_m(s) = \frac{1}{s(s+1)}$$

and let the disturbance be the constant load torque

$$\tau_L(t) = \tau_0 u_s(t)$$

so that $D(s) = \dfrac{\tau_L(s)}{K_T} = \dfrac{\tau_0}{K_T}\dfrac{1}{s}$.

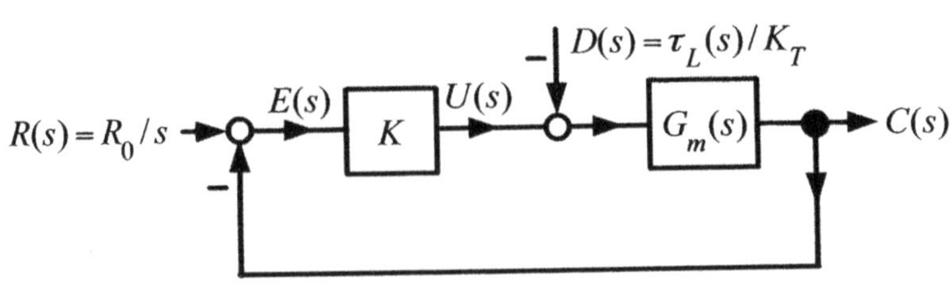

FIGURE 2.36. A proportional controller with a load acting on the system.

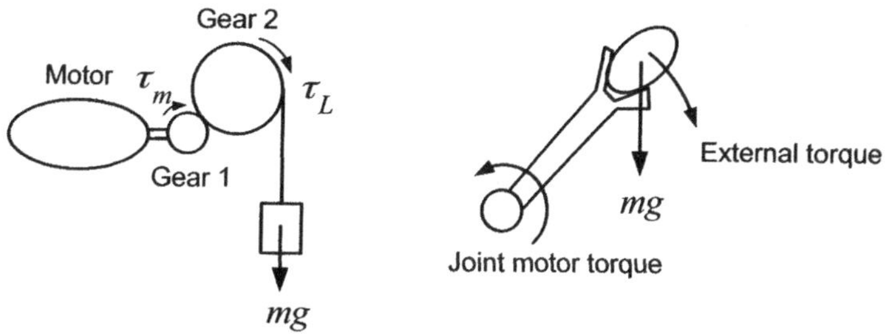

FIGURE 2.37. Examples of torque loads on DC motors.

It was shown in (2.58) that $E(s) = E_R(s) + E_D(s)$ where

$$E_R(s) \triangleq \frac{1}{1 + KG_m(s)} R(s)$$

and

$$E_D(s) \triangleq \frac{G_m(s)}{1 + KG_m(s)} D(s).$$

It was also shown that $\mathcal{L}^{-1}\{E_R(s)\} = e_R(t) \to 0$ as long as $K > 0$. The interest here is disturbance rejection, that is, whether or not $e_D(t) = \mathcal{L}^{-1}\{E_D(s)\} \to 0$ as $t \to \infty$. The Laplace transform of $e_D(t)$ is

$$E_D(s) = \frac{G_m(s)}{1 + KG_m(s)} D(s) = \frac{\dfrac{1}{s(s+1)}}{1 + K\dfrac{1}{s(s+1)}} \frac{\tau_L(s)}{K_T} = \frac{1}{s^2 + s + K} \frac{\tau_L(s)}{K_T}.$$

In words, $E_D(s)$ is the error in the position response due to the load torque. With a constant load torque $\tau_L(s) = \tau_0/s$ and, as $s^2 + s + K$ is stable for $K > 0$, the final position error due to the load torque is

$$e_D(\infty) = \lim_{s \to 0} s \frac{1}{s^2 + s + K} \frac{\tau_0}{K_T} \frac{1}{s} = \frac{\tau_0}{KK_T}.$$

The error

$$r(t) - c(t) = e(t) = e_R(t) + e_D(t) \to e_R(\infty) + e_D(\infty) = 0 + \frac{\tau_0}{KK_T}$$

as $t \to \infty$. Rearranging, the steady-state output is

$$c(\infty) = r(\infty) - e(\infty) = R_0 - \frac{\tau_0}{KK_T}.$$

The conclusion here is that the final output position depends on the value of the load torque τ_0. This is usually not acceptable as typically the load torque is unknown and it is important to precisely position the motor regardless of the load.[9] As is now shown, a PI controller can achieve zero steady-state error for this system even with unknown constant load torques acting on the system.

Example *Constant Load Torque with a PI Controller*

Let $G_c(s) = K_p + K_I/s = K\dfrac{s+\alpha}{s}$ where $K_p = K$ and $K_I = \alpha K$.

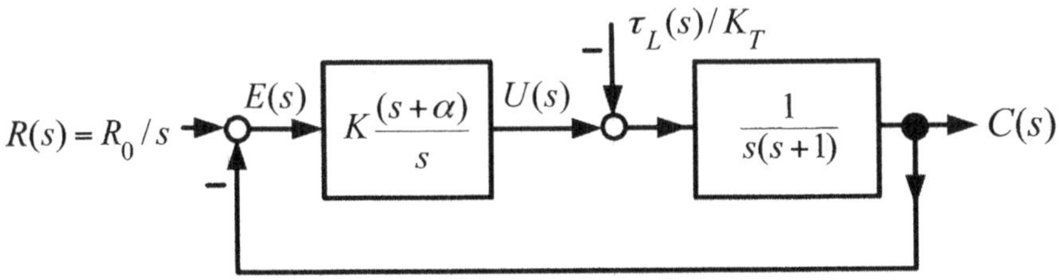

FIGURE 2.38. PI controller for disturbance rejection.

The error due to the disturbance is calculated as

$$\begin{aligned} E_D(s) &= \frac{G_m(s)}{1+G_c(s)G_m(s)}D(s) = \frac{\dfrac{1}{s(s+1)}}{1+K\dfrac{s+\alpha}{s}\dfrac{1}{s(s+1)}}\frac{\tau_0}{K_T}\frac{1}{s} \\ &= \frac{s}{s^2(s+1)+K(s+\alpha)}\frac{\tau_0}{K_T}\frac{1}{s} = \frac{1}{s^3+s^2+Ks+\alpha K}\frac{\tau_0}{K_T}. \end{aligned}$$

Note that, in contrast to the tracking case, only the denominator of $G_c(s)$ (i.e., s) reappears in the numerator after clearing the fractions in $E_D(s)$. This "s" cancels the $1/s$ in $\tau_L(s)$. Now, by the Routh test, $s^3+s^2+Ks+\alpha K$ is stable for $K > 0$ and $0 < \alpha < 1$, and so $sE_D(s)$ is also stable. Again, by the final value theorem,

$$e_D(\infty) = \lim_{s\to 0} sE_D(s) = \lim_{s\to 0} s\frac{1}{s^3+s^2+Ks+\alpha K}\frac{\tau_0}{K_T} = 0.$$

Consequently, the load torque τ_0 has no effect on the final position.

Interpretation of the PI Controller

For the system of Figure 2.38, it was shown that K and α can be chosen so that the closed-loop system is stable and $e(\infty) = \lim_{s\to 0} sE(s) = 0$. $U(s) = i_r(s)$ is the input current to the motor and is given by

$$U(s) = K\frac{s+\alpha}{s}E(s) = KE(s) + \frac{\alpha K}{s}E(s).$$

[9] Note that if the gain K can be made large, then the disturbance term $\tau_0/(K_T K)$ perhaps could be made negligible.

In the time domain,

$$u(t) = Ke(t) + \alpha K \int_0^t e(t)dt.$$

The limiting value of $u(t)$ is

$$u(\infty) = \lim_{s\to 0} sU(s) = \alpha K E(0)$$

which is due only to the output of the integrator. With $R(s) = R_0/s$ and $\tau_L(s) = \tau_0/s$, the error is given by

$$E(s) = \frac{s^2(s+1)}{s^3+s^2+Ks+\alpha K}\frac{R_0}{s} + \frac{s}{s^2(s+1)+K(s+\alpha)}\frac{\tau_0}{K_T}\frac{1}{s}$$

and, therefore, $E(0) = \tau_0/(\alpha K K_T)$. Consequently, $u(\infty) = \tau_0/K_T$ which shows that the output of the integrator goes to exactly the value needed to cancel the load disturbance.

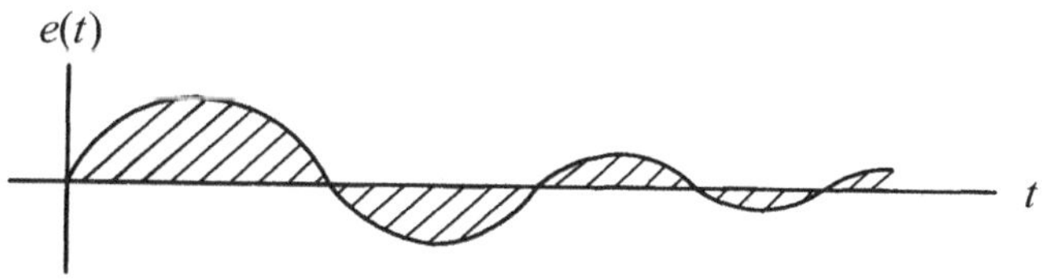

FIGURE 2.39. The error $e(t) \to 0$. The shaded area is $\int_0^\infty e(t)dt$ and $u(\infty) = \alpha K \int_0^\infty e(t)dt = \tau_0/K_T$.

Summary of the PI Controller for a DC Servo

It has been shown that the controller $G_c(s) = K(s+\alpha)/s$ with $K > 0$ and $0 < \alpha < 1$, will force the motor with transfer function $G_m(s) = \dfrac{1}{s(s+1)}$ to track inputs of the form $r(t) = R_0 + \omega_0 t$ with zero steady-state error in spite of any constant load disturbances. The control system designer then wants to choose K and α so that the closed-loop poles are far in the left-half plane in order have the transients die out quickly. Remember, the closed-loop poles determine the transient response! Recall from (2.57) and (2.58) that

$$\begin{aligned} C(s) &= \frac{G_c(s)G_m(s)}{1+G_c(s)G_m(s)}R(s) - \frac{G_m(s)}{1+G_c(s)G_m(s)}\frac{\tau_L(s)}{K_T} \\ E(s) &= \frac{1}{1+G_c(s)G_m(s)}R(s) + \frac{G_m(s)}{1+G_c(s)G_m(s)}\frac{\tau_L(s)}{K_T}. \end{aligned}$$

Evaluating the error $E(s)$ gives

$$E(s) = \frac{s^2(s+1)}{s^3+s^2+Ks+\alpha K}\left(\frac{R_0}{s} + \frac{\omega_0}{s^2}\right) + \frac{K_L s}{s^3+s^2+Ks+\alpha K}\frac{\tau_0}{K_T}\frac{1}{s}.$$

The gains K and α are chosen so that the poles of the closed-loop characteristic polynomial

$$s^3 + s^2 + Ks + \alpha K = (s - r)(s - p_1)(s - p_2)$$

satisfy $r < 0, \mathrm{Re}(p_1) = \mathrm{Re}(p_2) < 0$. The error $E(s)$ is then

$$E(s) = \frac{(s+1)(R_0 s + \omega_0)}{(s-r)(s-p_1)(s-p_2)} + \frac{1}{(s-r)(s-p_1)(s-p_2)}\frac{\tau_0}{K_T}$$

or, in the time domain,

$$e(t) = Ae^{rt} + Be^{p_1 t} + B^* e^{p_2 t} + (Ce^{rt} + De^{p_1 t} + D^* e^{p_2 t})K_L \tau_0$$

so that

$$e(t) \to 0 \text{ as } t \to \infty$$

as $r < 0, \mathrm{Re}(p_1) = \mathrm{Re}(p_2) < 0$.

Proportional Plus Integral Plus Derivative (PID) Control

A proportional plus integral plus derivative (PID) controller is defined by

$$G_c(s) = (K_P + K_I/s + K_D s) = \frac{K_D s^2 + K_P s + K_I}{s}$$

or, in the time domain,

$$v_a(t) = Ke(t) + \alpha K \int_0^t e(t)dt + K_D \frac{de}{dt}.$$

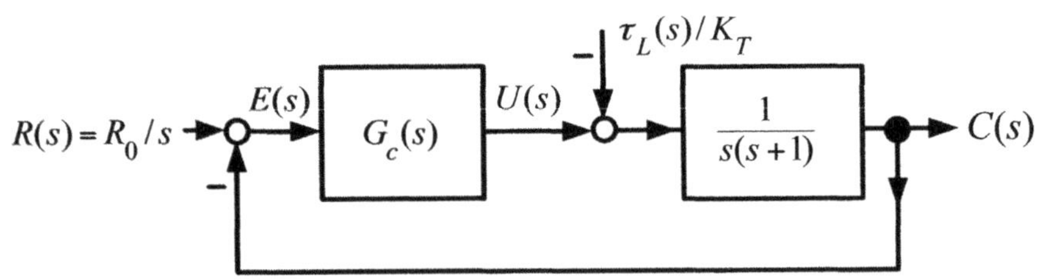

FIGURE 2.40. A proportional plus integral plus derivative controller.

The advantages of Proportional plus Integral (PI) controllers of the form $K_P + K_I/s$ has been discussed. The derivative control term $K_D s$ is used to force the transients to die out faster. That is, as will be shown below, the term $K_D s$ allows the control designer to put the closed-loop poles further in the left-half plane. However, before this is done, some practical issues concerning the derivative controller $K_D s$ are considered.

Practical Problem with Derivative Controllers

Differentiation of a signal with noise amplifies the noise!! Most signals contain a high-frequency low-amplitude noise. For example, a DC power supply takes a 60-Hz AC signal and rectifies it into a DC voltage. Due to this process (Full-Wave Rectification) there is a small amplitude 120 Hz signal (i.e., noise) added onto the DC output.

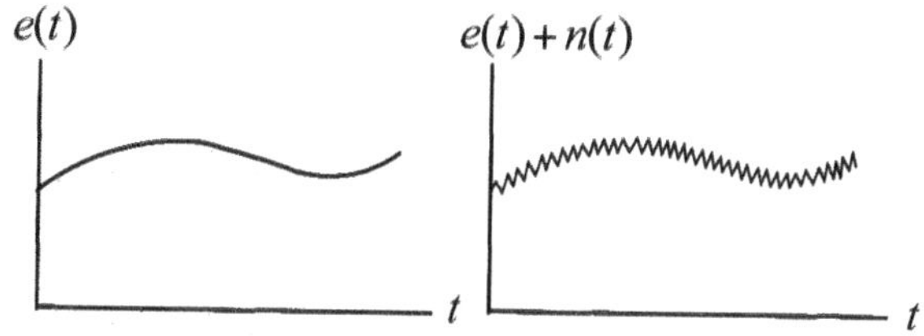

FIGURE 2.41. Left: Error signal. Right: Error signal containing high-frequency low-amplitude noise.

If one measures the error signal $e(t)$, the signal $e(t) + n(t)$ is actually obtained where $n(t)$ is a noise term. Numerically differentiating this signal gives the approximation

$$\dot{e}(t) + \dot{n}(t) \cong \frac{e(t) - e(t - \Delta t) + n(t) - n(t - \Delta t)}{\Delta t}.$$

As the difference $n(t) - n(t - \Delta t)$ may be of the same order (or higher) of magnitude as $e(t) - e(t - \Delta t)$, it follows that $\dot{n}(t)$ can be of the same order of magnitude as $\dot{e}(t)$. For example, let this noise[10] be given by $n(t) = 0.01\sin(2\pi(120t)) = 0.01\sin(754t)$. Thus, even though $n(t)$ is quite small, $\dot{n}(t) = 7.54\sin(754t)$ is large and can swamp out the value of $e(t)$.

Practical Implementation of Derivative Feedback for DC Motors

Consider the control structure shown in Figure 2.42. Using a tachometer, a direct measurement of the speed $\omega(t)$ is obtained which is then used as a feedback signal. K_t is an adjustable gain set by the control system designer. Note that, in contrast to the general PID controller structure originally introduced above, the tachometer is feeding back only the derivative of the output $\dot{c}(t)$ not the derivative of the error $\dot{e}(t)$.

[10]This could be a model for the noise on the DC bus of an amplifier when it uses a full-wave rectifier to convert the 120-Hz AC outlet to DC.

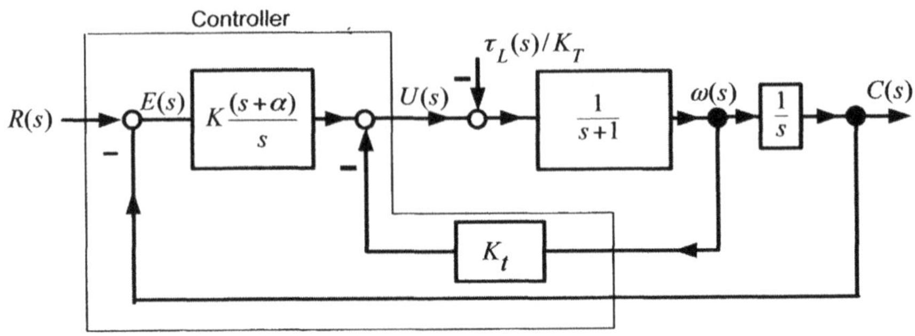

FIGURE 2.42. Implementation of a derivative controller using a tachometer.

To compute the transfer functions from $R(s)$ and $D(s)$ to the error $E(s)$, the block diagram of Figure 2.42 is reduces to

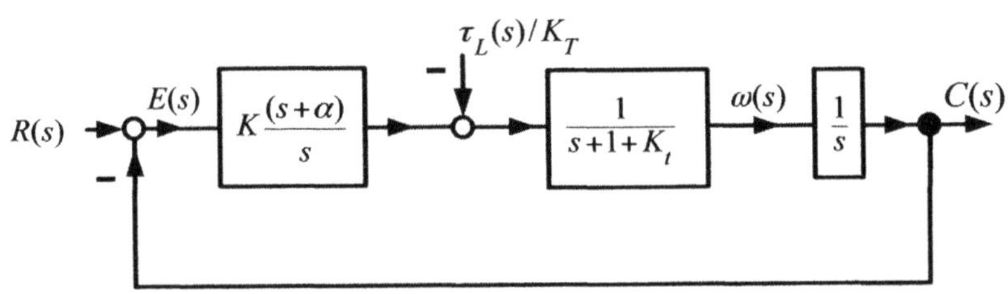

FIGURE 2.43. Block diagram reduction of Figure 2.42.

Thus $E(s) = E_R(s) + E_D(s)$ where

$$\begin{aligned} E_R(s) &= \frac{1}{1 + K\frac{s+\alpha}{s}\frac{1}{s+1+K_t}\frac{1}{s}}R(s) \\ &= \frac{s^2\,(s+1+K_t)}{s^3 + (1+K_t)\,s^2 + Ks + \alpha K}R(s) \end{aligned}$$

and

$$\begin{aligned} E_D(s) &= \frac{\frac{1}{s+1+K_t}\frac{1}{s}}{1 + K\frac{s+\alpha}{s}\frac{1}{s+1+K_t}\frac{1}{s}}D(s) \\ &= \frac{s}{s^3 + (1+K_t)\,s^2 + Ks + \alpha K}D(s). \end{aligned}$$

As the parameters K_t, K, and α are chosen by the control engineer, the coefficients of the closed-loop characteristic polynomial $s^3 + (1+K_t)\,s^2 + Ks + \alpha K$ may be chosen arbitrarily. In other words, the location of the closed-loop poles can be placed anywhere in the left-half plane by the control system designer.

In summary, a PID controller for a DC servomotor allows one to achieve tracking of step and ramp inputs, and rejection of constant load torque disturbances, with the closed-loop poles placed in any desired location. Typically, one desires the closed-loop poles to be as far in the left-half plane as possible so that the transients die out quickly. However, this typically means the control gains must be quite large. To see this, consider the above DC servomotor with tachometer feedback. As derived above, the error $E_R(s)$ is given by

$$E_R(s) = \frac{s^2\left(s+1+K_t\right)}{s^3+\left(1+K_t\right)s^2+Ks+\alpha K}R(s).$$

Suppose it is desired to put the closed-loop poles at $-10, -10+j10$ and $-10-j10$. Set

$$\begin{aligned} s^3+\left(1+K_t\right)s^2+Ks+\alpha K &= (s+10)\left(s-(-10+j10)\right)\left(s-(-10-j10)\right) \\ &= (s+10)(s^2+20s+200) \\ &= s^3+30s^2+400s+2000. \end{aligned}$$

That is, $1+K_t=30$ or $K_t=29, K=400$, and $\alpha K=2000$ or $\alpha=5$.

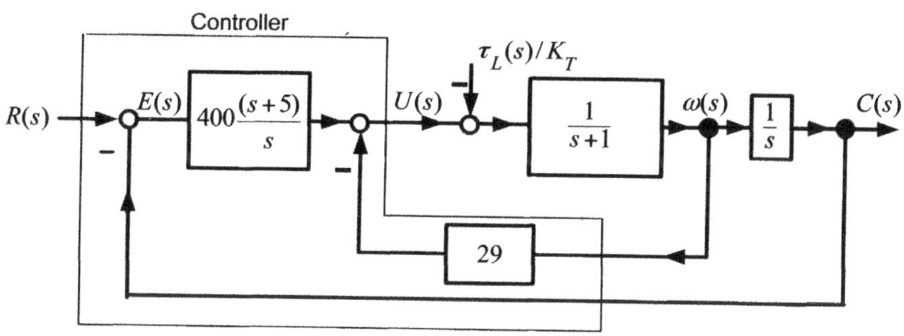

FIGURE 2.44. Example of a PID controller design.

At $t=0^+$, $c(0^+)=0, \omega(0^+)=0$.Then, $e(0^+)=r(0^+)-c(0^+)=R_0$, where $r(t)=R_0u_s(t)$. The input at $t=0^+$ is then given by

$$u(0^+)=400e(0^+)+2000\int_0^{0^+}e(t)dt-29\omega(0^+)=400R_0.$$

This shows that right after the reference input $r(t)=R_0u_s(t)$ is applied, the output of the controller (the current commanded to the motor amplifier) is $400R_0$ amperes. In particular, if $R_0=1$ radian, then the amplifier is required to put out 400 A which would most certainly be beyond the limit of the amplifier. The point here is that, in practice, one cannot place the closed-loop poles arbitrarily, but is limited by physical constraints such as the amplifier limits. These physical constraints were not included in the model on which the design was based.

General Theory of Tracking and Disturbance Rejection

With the above examples as background, a general approach to tracking and disturbance rejection in now presented. First, some definitions are required.

Definition *Type Number of a System*

Let

$$G(s) = \frac{b(s)}{s^j \bar{a}(s)}$$

where $\bar{a}(0) \neq 0$. Then $G(s)$ is said to be a type j systems which simply means it has j poles at the origin.

Examples

$$G(s) = \frac{1}{s^2} \qquad \text{Type 2} \quad \bar{a}(s) = 1$$

$$G(s) = \frac{1}{s(s+1)^2} \qquad \text{Type 1} \quad \bar{a}(s) = (s+1)^2$$

$$G(s) = \frac{1}{(s+2)} \qquad \text{Type 0} \quad \bar{a}(s) = (s+2)$$

$$G(s) = \frac{s+1}{s(s^2+2s+2)} \qquad \text{Type 1} \quad \bar{a}(s) = s^2+2s+2$$

Definition *Type Number of an Input*

If the input to a system $R(s)$ has j poles at the origin (i.e., at $s = 0$) and the rest of the poles of $R(s)$ are stable, then $R(s)$ is said to be a type j input.

Examples

$$R(s) = \frac{1}{s} \qquad \text{Type 1}$$

$$R(s) = \frac{1}{s^2} \qquad \text{Type 2}$$

$$R(s) = \frac{1}{s^j} \qquad \text{Type } j$$

$$R(s) = \frac{1}{s(s+2)} \qquad \text{Type 1}$$

$$R(s) = \frac{1}{s(s-2)} \qquad \text{Type number not assigned (unstable pole at 2)}$$

Remark Consider $R(s) = \dfrac{1}{s(s+2)} = \dfrac{1}{2}\dfrac{1}{s} - \dfrac{1}{2}\dfrac{1}{s+2}$, so that $r(t) = 1/2 - (1/2)e^{-2t}$. To asymptotically track this $r(t)$, the output signal $c(t)$ must just go to $1/2$ as $t \to \infty$ as $-(1/2)e^{-2t}$ dies out. However, if $R(s) = \dfrac{1}{s(s-2)} = -\dfrac{1}{2}\dfrac{1}{s} + \dfrac{1}{2}\dfrac{1}{s-2}$, then $r(t) = -1/2 + (1/2)e^{2t}$ so that $c(t)$ must asymptotically got to $-1/2+(1/2)e^{2t}$ as $t \to \infty$ in order to track the input.

Consider now the general tracking and disturbance rejection problem as setup in the block diagram of Figure 2.45.

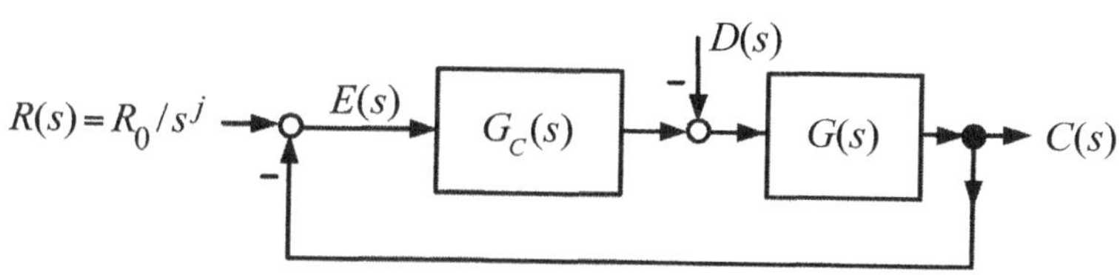

FIGURE 2.45. Block diagram for tracking a type j input.

Let $G_c(s) = Kb_c(s)/a_c(s)$ and $G(s) = b(s)/a(s)$ be the transfer functions of the controller and plant (physical system), respectively. First, the disturbance is taken to be zero and the asymptotic tracking of type j inputs is considered. Then the asymptotic rejection of type j disturbances is addressed.

Theorem *Tracking with Zero Steady-State Error*

Let $R(s) = R_0/s^j$ be a type j input so $r(t) = \dfrac{R_0 t^{j-1}}{(j-1)!}$ and with $D(s) \equiv 0$, the error $e(t)$ satisfies

$$\lim_{t\to\infty} e(t) = \lim_{t\to\infty} (r(t) - c(t)) = 0$$

if and only if the closed-loop system is stable; that is, the roots of

$$a_c(s)a(s) + Kb_c(s)b(s) = 0$$

are in the open left-half plane *and* the type number of the forward open-loop transfer function $G_c(s)G(s)$ is j or greater.

Remark The input is $R(s) = R/s^j$, and the second condition says that the forward open-loop transfer function $G_c(s)G(s)$ must also contain at least a factor of $1/s^j$

Proof

(if) Assume that the conditions hold. To show that $e(t) \to 0$ as $t \to \infty$, consider the error $E(s)$ given by

$$\begin{aligned} E(s) &= \frac{1}{1+G_c(s)G(s)}R(s) = \frac{1}{1+K\dfrac{b_c(s)}{a_c(s)}\dfrac{b(s)}{a(s)}}\frac{R_0}{s^j} \\ &= \frac{a_c(s)a(s)}{a_c(s)a(s)+Kb_c(s)b(s)}\frac{R_0}{s^j}. \end{aligned}$$

By hypothesis, $G_c(s)G(s)$ is at least type j, so that $a_c(s)a(s) = s^j\bar{a}(s)$. Using this expression, the s^j terms cancel resulting in

$$E(s) = \frac{\bar{a}(s)}{a_c(s)a(s)+Kb_c(s)b(s)}R_0.$$

Also, by hypothesis, the roots of the closed-loop characteristic polynomial

$$a_c(s)a(s) + Kb_c(s)b(s) = (s-p_1)\cdots(s-p_n)$$

satisfy $\mathrm{Re}(p_i) < 0$ for $i = 1, \ldots n$. A partial fraction expansion[11] of $E(s)$ then gives

$$E(s) = \frac{\bar{a}(s)}{(s-p_1)\cdots(s-p_n)}R_0 = \frac{A_1}{s-p_1} + \frac{A_2}{s-p_2} + \cdots + \frac{A_n}{s-p_n}$$

or, in the time domain,

$$e(t) = A_1e^{p_1t} + \cdots A_ne^{p_nt}.$$

As $\mathrm{Re}(p_i) < 0$ for $i = 1, \ldots, n$, $e(t) \to 0$ as $t \to \infty$.

(only if) Here it must be shown that if $e(t) \to 0$ as $t \to \infty$, then

$$a_c(s)a(s) = s^j\bar{a}(s)$$

[11]This is assuming the poles are distinct. If they are not, the proof is easily modified with the same final result.

where $\bar{a}(s)$ is a polynomial and the closed-loop characteristic polynomial is stable, that is,

$$a_c(s)a(s) + Kb_c(s)b(s) = (s - p_1)\cdots(s - p_n)$$

satisfies $\mathrm{Re}(p_i) < 0$ for $i = 1, ..., n$. Proceeding, the error $e(t)$ is the inverse Laplace transform of

$$E(s) = \frac{a_c(s)a(s)}{a_c(s)a(s) + Kb_c(s)b(s)} \frac{R_0}{s^j}. \tag{2.59}$$

Computing $e(t)$ using a partial fraction expansion of $E(s)$, it is clear that $e(t) \to 0$ as $t \to \infty$ only if $E(s)$ is asymptotically stable; that is, $E(s)$ must have all of its poles in the open left-half plane. Now, by equation (2.59), $E(s)$ will have j poles at the origin (unstable) due to the input $R(s) = R_0/s^j$ unless $a_c(s)a(s) = s^j\bar{a}(s)$. It is given that $e(t) \to 0$ as $t \to \infty$, so it follows that $a_c(s)a(s)$ *must* contain a factor s^j. Thus, $E(s)$ is of the form

$$E(s) = \frac{\bar{a}(s)}{a_c(s)a(s) + Kb_c(s)b(s)} R_0.$$

Further, by doing a partial-fraction expansion of this last expression for $E(s)$, it is seen that $\mathrm{Re}(p_i) < 0$ for $i = 1, ..., n$ in order that $e(t) \to 0$ as $t \to \infty$.

■

Consider now the disturbance rejection problem. The block diagram below illustrates the setup.

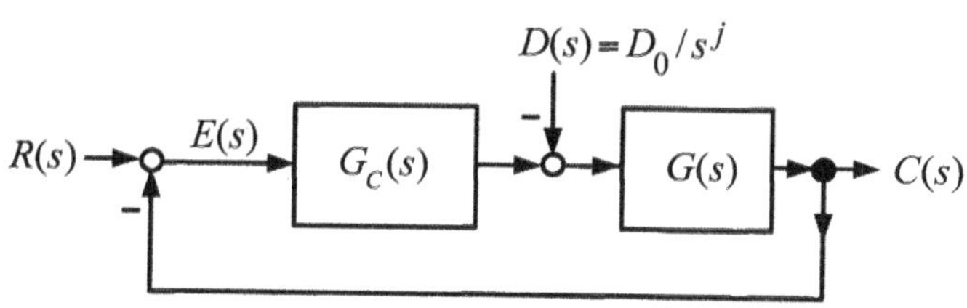

FIGURE 2.46. Block diagram with a type j disturbance.

As before, $G(s) = b(s)/a(s)$ and $G_c(s) = Kb_c(s)/a_c(s)$. The error $E(s)$ is then

$$E(s) = \frac{1}{1 + G_c(s)G(s)} R(s) + \frac{G(s)}{1 + G_c(s)G(s)} D(s).$$

The error $E_D(s)$ due to the disturbance is

$$E_D(s) = \frac{G(s)}{1+G_c(s)G(s)}D(s) = \frac{K\dfrac{b(s)}{a(s)}}{1+K\dfrac{b_c(s)}{a_c(s)}\dfrac{b(s)}{a(s)}}D(s)$$

$$= \frac{a_c(s)b(s)}{a_c(s)a(s)+Kb_c(s)b(s)}D(s).$$

Consider a type j disturbance $D(s) = D_0/s^j$ and denote $e(t) = \mathcal{L}^{-1}\{E_D(s)\}$. The following theorem shows how such a disturbance signal can be rejected asymptotically.

Theorem *Disturbance Rejection with Zero Steady-State Error*
The error $e_D(t)$ satisfies

$$\lim_{t\to\infty} e_D(t) = 0$$

if and only if the closed-loop system is stable; that is, the roots of

$$a_c(s)a(s) + Kb_c(s)b(s) = 0$$

are all in the open left-half plane *and* the type number of the compensator transfer function $G_c(s)$ is j or greater.

Remark In contrast to the tracking problem where $G_c(s)G(s)$ must be type j, the compensator $G_c(s)$ *by itself* must contain at least the factor $1/s^j$.

Proof
By the above,

$$E_D(s) = \frac{a_c(s)b(s)}{a_c(s)a(s)+Kb_c(s)b(s)}\frac{D_0}{s^j}.$$

It is assumed here that $b(0) \neq 0$ which is true of most physical systems.[12] Thus $a_c(s)$ must have a factor of s^j to cancel the denominator of $D(s)$ in order to achieve zero steady-state error. With $a_c(s) = s^j\bar{a}(s)$, it follows that

$$E_D(s) = \frac{\bar{a}(s)b(s)}{(s-p_1)\cdots(s-p_n)}D_0 = \frac{A_1}{s-p_1} + \frac{A_2}{s-p_2} + \cdots + \frac{A_n}{s-p_n}.$$

The time response

$$e_D(t) = A_1e^{p_1t} + \cdots + A_ne^{p_nt} \to 0$$

as $t \to \infty$ if and only if $\mathrm{Re}(p_i) < 0$ for $i = 1, ..., n$.

■

---

[12] If $b(s) = s^k\bar{b}(s), \bar{b}(0) \neq 0$, then $G_c(s)$ need only be type $j-k$ so that $a_c(s)b(s) = s^{j-k}\bar{a}_c(s)s^k\bar{b}(s) = \bar{a}_c(s)\bar{b}(s)$.

Internal Model Principle

The proofs of the tracking and disturbance rejection theorems suggest how one can achieve the same results with more general types of reference and disturbance signals. This is illustrated in the following example and is known as the *internal model principle.*

Example *Tracking a Sinusoid*

Consider a sinusoidal input given by $r(t) = R_0 \sin(t)$ or $R(s) = \dfrac{R_0}{s^2+1}$. Let $G(s) = \dfrac{1}{s(s+1)}$ and the problem is to design a compensator so that the output (asymptotically) tracks the input signal. The block diagram of the system is shown in Figure 2.47.

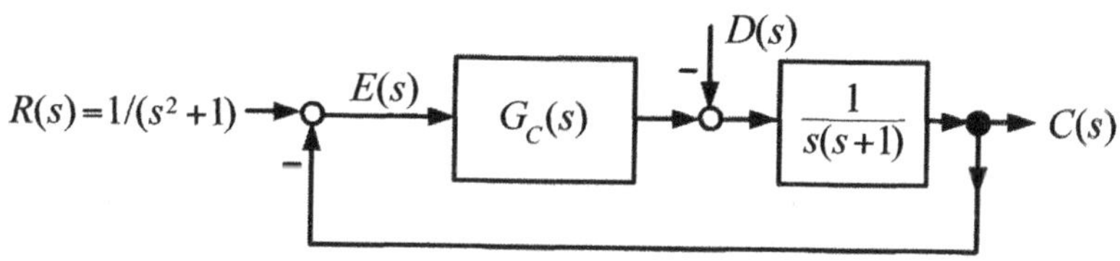

FIGURE 2.47. Ilustrating the internal model principle by tracking a sinusoidal input.

Consider the compensator

$$G_c(s) = K\frac{s+\alpha}{s^2+1}$$

which was chosen to have a denominator that was the *same* as that of the input. The reason for this is clear when the error $E(s)$ is computed as

$$\begin{aligned}
E(s) &= \frac{1}{1+G_c(s)G(s)}R(s)\\
&= \frac{1}{1+K\dfrac{s+\alpha}{s^2+1}\dfrac{1}{s(s+1)}}\frac{R_0}{s^2+1}\\
&= \frac{s(s+1)(s^2+1)}{s^4+s^3+s^2+(K+1)s+\alpha K}\frac{R_0}{s^2+1}\\
&= \frac{s(s+1)}{s^4+s^3+s^2+(K+1)s+\alpha K}R_0.
\end{aligned}$$

That is, the factor s^2+1 in the denominator of $G_c(s)$ reappears in the numerator of $E(s)$ to cancel the factor s^2+1 in the denominator of $R(s)$. Then $e(t) \to 0$ if the closed-loop characteristic polynomial is stable. The

Routh table for $s^4 + s^3 + s^2 + (K+1)s + \alpha K$ is

$$\begin{array}{llll}
s^4 & 1 & 1 & \alpha K \\
s^3 & 1 & K+1 & 0 \\
s^2 & \dfrac{1-(K+1)}{1} = -K & \alpha K & 0 \\
s & \dfrac{-K(K+1)-\alpha K}{-K} = \dfrac{K(K+1+\alpha K)}{K} & 0 & \\
1 & \alpha K & &
\end{array}$$

Stability requires $-K > 0, K + 1 + \alpha > 0$ and $\alpha K > 0$ or

$$K < 0, \quad -(K+\alpha) < 1, \ \alpha < 0.$$

For example, choose $\alpha = -0.25$ and $K = -0.6$ so that the closed-loop characteristic polynomial is $s^4 + s^3 + s^2 + 0.4s + 0.15$ which has roots $-0.17 \pm j0.63, -0.33 \pm j0.49$. The error response is

$$e(t) = 4.7e^{-0.17t}\sin(0.63t + 0.37) - 4.1e^{-0.33t}\sin(0.49t + 0.4) \to 0$$

as $t \to \infty$. Note, however, the transients die out slowly.

Remark General procedures exist that allow one to design a compensator that asymptotically tracks any given input signal and, at the same time, arbitrarily place the closed-loop poles in the open left-half plane. The reader is referred to Refs. [47][48][49][50] for this theory.

This example and the previous results can be summarized as the *Internal Model Principle.* This just says that in order to (asymptotically) track a given reference signal, the forward open-loop transfer function $G_c(s)G(s)$ must contain the same *unstable* poles as the reference signal and the closed-loop system must be stable. (The terminology "internal model" refers to the poles of the transfer function as they determine the form of the time response of a transfer function model.)

Similarly, in order to achieve (asymptotic) steady-state rejection of a given disturbance signal, the forward open-loop controller (compensator) $G_c(s)$ must (by itself) contain the same unstable poles as the disturbance signal along with the closed-loop system being stable.

Problems

PI Current Controller for a DC Motor

Problem 1 Simulation of the DC Motor

Let $V_{max} = 40$ *V,* $I_{max} = 5$ *A,* $K_b = K_T = 0.07$ *V/rad/sec (= N-m/A),* $J = 6 \times 10^{-5}$ *kg-m*^{2}*,* $R = 2$ *ohms,* $L = 2$ *mH, and* $f = 0.0004$ *N-m/rad/sec. Develop a simulation of the DC motor that includes the motor and amplifier (with saturation) as well as an encoder with 2000 pulses/rev. Put a step input of* $v(t) = 10$ *V into the motor and plot out (a)* $\theta(t) = \frac{2\pi}{2000}N(t)$ *(*$N(t)$ *is the encoder output), (b)* $\omega(t)$*, (c)* $i(t)$*, and (d)* $v(t)$*. Also, zoom in on* $\theta(t)$ *to show the staircase waveform similar to that shown in Figure 2.4. (Hint: If it does not show a staircase waveform, then the step size in the simulator's integration algorithm may be too large.)*

Problem 2 Load-Torque Transfer Function

(a) Compute the transfer function from $\tau_L(s)$ *to* $\theta(s)$ *in Figure 2.5.*

(b) Compute the transfer function from $\tau_L(s)$ *to* $\theta(s)$ *in Figure 2.6. What does it reduce to as* $K_P \to \infty$*?*

Problem 3 Transfer Function $i(s)/i_r(s)$

With $\tau_L(s) = 0$ *in Figure 2.6, find the transfer function* $i(s)/i_r(s)$*. What does it reduce to as* $K_p \to \infty$*?*

Problem 4 Encoder Resolution

It was shown that if there are N_w *windows (lines/slots) equally spaced around the encoder disk (wheel), then there are* N_w *pulses per revolution so that there are* $2N_w$ *rising and falling edges of the output signal. With the length of the windows the same as the length between the window, the resolution of the encoder is then* $2\pi/(2N_w)$ *radians. A second light detector is added to determine the direction of rotation. Show that by using the outputs of both light detectors, a resolution of* $2\pi/(4N_w)$ *radians can be achieved. Hint: Add the two waveforms of Figure 2.3(a) [or Figure 2.3(b)] and show that there is a rising or falling edge every* $2\pi/(4N_w)$ *radians.*

Problem 5 Load-Torque Transfer Function

(a) Find the transfer function from $\tau_L(s)$ *to* $\theta(s)$ *in Figure 2.6.*

(b) What does it reduce to as $K_P \to \infty$*?*

Problem 6 High-Gain Current Control

Consider a PI current controller given by $K_P + K_I/s = K_P(s+\alpha)/s$ *where* $K_I = \alpha K_P$*. This is illustrated in the block diagram given in Figure 2.48.*

(a) Using the block diagram reduction method, find $G(s) = \omega(s)/i_r(s)$*. Let* $K_P \to \infty$ *and show that* $G(s) = \omega(s)/i_r(s) \to K_T/(sJ + f)$*.*

(b) With $\tau_L(s) = 0$ *in Figure 2.48, find the transfer function* $i(s)/i_r(s)$*. What does it reduce to as* $K_p \to \infty$*?*

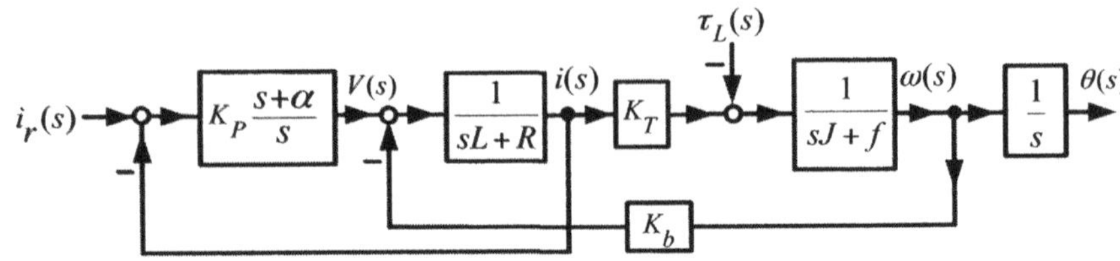

FIGURE 2.48. PI current controller.

Problem 7 Amplifier Saturation

With $i(t) = i_0$ constant and $\tau_L = 0$, find the steady-state values of the speed $\omega(\infty)$ and voltage $v(\infty)$ in terms of i_0 and the motor parameters. With $i_0 = 1$ A, $V_{max} = 40$ V and the values of the motor parameters in problem 1, can this steady-state speed be achieved? Explain. Redo this exercise with $i_0 = 4$ A. Note: You can easily answer this question using the original state space equations:

$$
\begin{aligned}
L\frac{di}{dt} &= -Ri(t) - K_b\omega(t) + v(t) \\
J\frac{d\omega}{dt} &= -f\omega(t) + K_T i(t) \\
\frac{d\theta}{dt} &= \omega(t).
\end{aligned}
$$

Problem 8 PI Current Controller

Add the PI current controller to the DC motor simulation of problem 1. Also, add a saturation block to the input of the PI current controller that limits the command current $i_r(t)$ between $\pm I_{max}$. Let $i_r(t)$ be a square wave current reference where one period is specified by

$$
i_r(t) = \begin{cases} +I_1 & 0 \le t \le 0.25 \\ -I_1 & 0.25 \le t \le 0.50. \end{cases}
$$

Try $I_1 = 0.1$ A and $I_1 = 1.0$ A in a simulation. Plot (a) $i(t)$ and $i_r(t)$ on the same plot and (b) $i_r(t) - i(t)$ on a separate plot. How well does $i(t)$ respond to step changes in $i_r(t)$?

Problem 9 Current Controller Tracking

With the parameter values as given in problem 1, this motor is capable of turning 180° in 0.065 sec (i.e., 65 msec). Consequently, the current controller must respond significantly faster than this if $i(t)$ is to be considered as the input, that is, if the dynamics $i_r(t) \to i(t)$ are to be neglected. To test the current controller under these conditions, consider $i(t) = i_0 \sin(t/T)$ with $i_0 = 1$ A, $T = 2\pi/\left(65 \times 10^{-3}\right)$ sec in the simulation. How well does $i(t)$ track this reference? Use your "best" controller design (i.e., your choice of K_P and α) and plot $i_r(t)$ and $i(t)$ on the same graph. Also plot out $v(t)$ to *show it does not saturate. If you have problems with saturation, reduce your gains until you can achieve "good" tracking without exceeding the voltage limits.*

Problem 10 Proportional Speed Control

Find the error transfer function $E(s)/\omega_{ref}(s)$ for the system in Figure 2.10.

(a) With $e(t) = \mathcal{L}^{-1}\{E(s)\}$ and $\omega_{ref}(s) = \omega_0/s$, show that the steady-state error $e(\infty) \approx (f/KK_T)\omega_0$ for K large.

(b) Using the parameter values given in problem 1 and K chosen such that $KK_T/J = 100$, compute $e(t) = \mathcal{L}^{-1}\{E(s)\}$ with $\omega_{ref}(t) = 100\sin\left(\frac{2\pi}{0.2}t\right)$.

Estimating the Motor Speed

Problem 11 Backward Difference Speed Estimator

(a) Simulate the backward difference method of computing speed with $T = 0.0002$ sec ($f_s = 5000$). Let $i_r = 1$ A and plot $\omega(kT)$ and $\hat{\omega}_{bd}(kT)$. What are some of the problems with this way of measuring speed? Repeat with $i_r = 0.5\sin\left((2\pi/0.2)\,t\right)$.

(b) With $i_r = 1$ A, use the output of a simulation to plot the speed and speed error to show it is bounded by $(2\pi/2000)/T$ when using the backward difference approximation to compute the speed. Do this for $T = 0.001$ sec and $T = 0.0002$ sec.

Problem 12 Recovery from a Load-Torque Disturbance

Add the load torque disturbance

$$\tau_L(t) = \begin{cases} 0 & 0.0 \le t \le 0.5 \\ \tau_{L0} & 0.5 \le t \le 0.6 \\ 0 & 0.6 \le t \end{cases}$$

to the torque equation of simulation of the DC motor. Run the simulation with $i_r = 1$ A and $\tau_{L0} = 0.035$ N-m. Plot $\hat{\omega}(kT), \hat{\omega}_{bd}(kT)$ and $\omega(kT)$ on the same graph. Redo this with $\ell_1 = 0, \ell_2 = 0$. Do you see a difference?

Problem 13 Speed and Load-Torque Estimator

(a) Consider the modification of the speed observer to estimate any constant (or, slowly varying) load-torque. That is, consider the situation where there is an unknown constant *load torque acting on the motor. This may be modeled as*

$$\begin{aligned} \frac{d\theta}{dt} &= \omega \\ \frac{d\omega}{dt} &= (K_T/J)i - (f/J)\omega - \tau_L/J \\ \frac{d(\tau_L/J)}{dt} &= 0 \end{aligned}$$

Let a speed and load-torque estimator be given by

$$\begin{aligned}\frac{d\hat{\theta}}{dt} &= \hat{\omega} + \ell_1(\theta - \hat{\theta})\\ \frac{d\hat{\omega}}{dt} &= (K_T/J)i - \hat{\tau}_L/J - (f/J)\hat{\omega} + \ell_2(\theta - \hat{\theta})\\ \frac{d(\hat{\tau}_L/J)}{dt} &= 0 + \ell_3(\theta - \hat{\theta}).\end{aligned}$$

Define the estimation errors to be $e_1 = \theta - \hat{\theta}, e_2 = \omega - \hat{\omega}$ *and* $e_3 = \tau_L/J - \hat{\tau}_L/J$. *Subtract the above sets of equations to the find the system of equations describing the dynamics of observer error* $e(t) = (e_1(t), e_2(t), e_3(t))$. *Find* $E_1(s), E_2(s)$, *and* $E_3(s)$ *and show that the observer gains* ℓ_1, ℓ_2, *and* ℓ_3 *can be chosen so that* $e_2(t) \to 0$, $e_3(t) \to 0$ *as* $t \to \infty$ *but that* $e_1(t) \nrightarrow 0$.

(b) Add the load-torque disturbance

$$\tau_L(t) = \begin{cases} 0 & 0.0 \le t \le 0.5 \\ \tau_{L0} & 0.5 \le t \le 1.0 \end{cases}$$

to the torque equation of simulation of the DC motor. Run the simulation with $i_r = 1$ *A and* $\tau_{L0} = 0.035$ *N-m. Plot* $\hat{\omega}(kT), \hat{\omega}_{bd}(kT)$, *and* $\omega(kT)$ *on the same graph. Redo this with* $\ell_1 = 0, \ell_2 = 0, \ell_3 = 0$. *Do you see a difference?*

Problem 14 Speed Observer Versus Backward Differentiation

Implement the observer (2.7) in the DC motor simulation and compare it with the calculation $\hat{\omega}_{bd}(kT) \triangleq \frac{2\pi}{2000}\left(\frac{N(kT)-N(kT-T)}{T}\right)$. *Specifically, let* $i_r = 0.1$ *A*, $\omega(0) = \hat{\omega}(0) = 0$ *and plot* $\omega(kT)$ *and its estimates* $\hat{\omega}(kT), \hat{\omega}_{bd}(kT)$. *You need to choose values for* ℓ_1 *and* ℓ_2*! Repeat with* $i_r = 0.5\sin((2\pi/0.2)t)$.

Problem 15 Position Cannot Be Estimated from Speed Measurements

Suppose a tachometer is available to measure speed, but there is no position measurement. Can the position be estimated from the speed measurements? Well, the answer is no. An easy way to understand this is to realize that $\theta(t) = \theta(0) + \int_0^t \omega(\tau)d\tau$, *but that* $\frac{d}{dt}\left(c + \int_0^t \omega(\tau)d\tau\right) = \omega(t)$ *for* any *constant c. That is, there are an infinite number of position signals with the same speed signal. Consequently, the speed signal does not contain enough information to obtain an estimate for the position. From the point of view of observer theory, consider a position estimator given by*

$$\begin{aligned}\frac{d\hat{\theta}}{dt} &= \hat{\omega} + \ell_1(\omega - \hat{\omega})\\ \frac{d\hat{\omega}}{dt} &= (K_T/J)i - (f/J)\hat{\omega} + \ell_2(\omega - \hat{\omega}).\end{aligned}$$

Show that ℓ_1, ℓ_2 *cannot be chosen so that* $\hat{\theta} \to \theta$.

Problem 16 Discretization of the Observer Equations

Discretize the speed observer equations

$$\begin{aligned}\frac{d\hat{\theta}}{dt} &= \hat{\omega} + \ell_1(\theta - \hat{\theta}) \\ \frac{d\hat{\omega}}{dt} &= (K_T/J)i - (f/J)\hat{\omega} + \ell_2(\theta - \hat{\theta})\end{aligned}$$

using the trapezoidal *method of integration.*

Trajectory Generation

Problem 17 Trajectory Design

Design a trajectory that also has a continuous jerk rate, that is, a trajectory that satisfies

$$\begin{array}{ccc} \omega_{ref}(0) = 0 & \dot{\omega}_{ref}(0) = 0 & \ddot{\omega}_{ref}(0) = 0 \\ \omega_{ref}(t_1) = \omega_{max} & \dot{\omega}_{ref}(t_1) = 0 & \ddot{\omega}_{ref}(t_1) = 0 \\ \omega_{ref}(t) = \omega_{max} & \text{for} & t_1 \le t \le t_2 \\ \omega_{ref}(t_2) = \omega_{max} & \dot{\omega}_{ref}(t_2) = 0 & \ddot{\omega}_{ref}(t_2) = 0 \\ \omega_{ref}(t_3) = 0 & \dot{\omega}_{ref}(t_3) = 0 & \ddot{\omega}_{ref}(t_3) = 0 \end{array}$$

Design of a State Feedback Tracking Controller

Problem 18 Simulation of a Fast Point-to-Point Move

Implement the complete state feedback controller of Figure 2.14 in continuous time using the parameter values of problem 1. Specifically, use the speed observer given in equation (2.4), the feedback controller given in equation (2.14) and the trajectory generator of Section 2.3. Be sure to have a saturation block for the current commanded into the PI current controller and for the voltage out of the amplifier. Design a trajectory such that the motor turns $\theta_f = \pi$ radians in under 70 *msec (by choosing the time t_1 to reach maximum speed and the time t_2 to begin decelerating) and does not violate the voltage or current constraints. Find the control gains K_0, K_1 and K_2 (by choosing r_1, r_2, r_3) and the observer gains ℓ_1, ℓ_2 (by choosing p_1, p_2) to track the trajectory. Plot (a) $\theta(t), \theta_{ref}(t)$, (b) $\hat{\omega}(t), \omega(t), \omega_{ref}(t)$, (c) $i_r(t), i_{ref}(t)$, (d) $\theta_{ref}(t) - \theta(t)$, (e) $\omega_{ref}(t) - \hat{\omega}(t)$, (g) $i_r(t) - i(t)$.*

Problem 19 Simulation Study of Feedback

In order to understand the importance of the controller and reference inputs, carry out the following simulation runs using the simulation setup of problem 18 except in the trajectory generator, use $\theta_f = 30$ radians, $\omega_{max} = 100$ rads/sec, and $t_1 = 0.2$ and run the simulation for 0.6 *sec.*

(a) With $K_0 = 0, K_1 = 0, K_2 = 0$, $\omega(0) = 0, \theta(0) = 0, \tau_L = 0$ run the simulation with $i_r = i_{ref}$ and plot $\theta(t), \theta_{ref}(t), \hat{\omega}(t), \omega(t), \omega_{ref}(t), i_r(t)$, $i_{ref}(t)$ as well as $\theta_{ref}(t) - \theta(t), \omega(t) - \hat{\omega}(t), \omega_{ref}(t) - \omega(t), i_r(t) - i(t)$. This is referred to as running the system open-loop.

(b) Redo (a) with $\omega(0) = 0, \theta(0) = 0.1, \tau_L = 0$.
(c) Redo (a) with $\omega(0) = 5, \theta(0) = 0, \tau_L = 0$.
(d) Redo (a) with $\omega(0) = 0, \theta(0) = 0,\ \tau_L = 0.035$ *N-m.*
(e) With $r_1 = 100, r_2 = 100, r_3 = 100$ *find the corresponding gains* K_0, K_1, K_2 *using given by equation (2.18). Run the simulation with these gains for the following cases:*
(i) Set $\omega(0) = 0, \theta(0) = 0$ *and* $\tau_L = 0$.
(ii) Set $\omega(0) = 0, \theta(0) = 0.1$ *and* $\tau_L = 0$.
(iii) Set $\omega(0) = 0, \theta(0) = 0$ *and* $\tau_L = 0.035$ *N-m.*
For each case, plot $\theta(t), \theta_{ref}(t), \hat{\omega}(t), \omega(t), \omega_{ref}(t), i_r(t), i_{ref}(t)$ *as well as* $\theta_{ref}(t) - \theta(t), \omega(t) - \hat{\omega}(t), \omega_{ref}(t) - \omega(t), i_r(t) - i(t)$.
(f) Remove the integrator term in the feedback expression (2.13) (or (2.14)) to study its effect. That is, substitute $w = K_1 e_1(t) + K_2 e_2(t)$ *for* w *into (2.11) and choose the gains as* $K_2 = r_1 + r_2 - f/J,\ K_1 = r_1 r_2$ *so that the closed-loop characteristic equation is then* $(s + r_1)(s + r_2)$. *With* $r_1 = 100, r_2 = 100$, *redo cases (i), (ii) and (iii) of part (e).*

Problem 20 Final Value Theorem

Let $E(s) = n(s)/d(s)$ *with* $\deg\{n(s)\} < \deg\{d(s)\}$ *and* $e(t) \triangleq \mathcal{L}^{-1}\{E(s)\}$. *The final value theorem says*

If $sE(s)$ is stable, *then*

$$e(\infty) = \lim_{s \to 0} sE(s).$$

(a) Let

$$E(s) = \frac{1}{s(s+1)(s+2)}$$

and compute the final value two ways: (1) Do a partial fraction expansion to compute the inverse Laplace transform $e(t)$ *and let* $t \to \infty$ *and (2) Use the final value theorem.*

(b) Let

$$E(s) = \frac{1}{s(s-1)(s+2)}$$

and do a partial fraction expansion to compute the inverse Laplace transform $e(t)$ *and let* $t \to \infty$. *How does this compare with computing* $\lim_{s \to 0} sE(s)$*? Explain.*

(c) Let

$$E(s) = \frac{n(s)}{d(s)}$$

where $\deg\{n(s)\} < \deg\{d(s)\}$. *Prove the final value theorem by doing a partial fraction expansion of* $E(s)$. *To simplify the presentation of the partial fraction expansion, assume that the poles of* $E(s)$ *are distinct, that is,* $d(s)$ *has no repeated roots. [Hint: Note then that* $\lim_{s \to 0} sE(s)$ *is the coefficient of the* $1/s$ *term in the partial fraction expansion of* $E(s)$.*]*

Problem 21 State Feedback with a Constant Load Torque

(a) For the case where $\tau_L \neq 0$, give the appropriately modified version of (2.17). Also, with $\tau_L(s) = \tau_{L0}/s$, find the steady-state values $e_0(\infty), e_1(\infty)$, and $e_2(\infty)$ using the final-value theorem. With the constant load-torque disturbance acting on the motor, explain why the steady-state error in speed and position are both zero, that is, $e_2(\infty) = 0$ and $e_1(\infty) = 0$, but that $e_0(\infty) \neq 0$. With

$$i_{fb}(t) \triangleq \frac{J}{K_T}\left(K_0 \int_0^t e_1(\tau)d\tau + K_1 e_1(t) + K_2 e_2(t)\right)$$

Show that $i_{fb}(\infty) = \tau_{L0}/K_T$ and give an interpretation to this value.

(b) Consider state feedback control, but now without *the integrator feedback. That is, consider the system [compare with (2.15)]*

$$\begin{aligned}
\frac{de_1}{dt} &= e_2 \\
\frac{de_2}{dt} &= -(f/J)e_2 - K_1 e_1 - K_2 e_2 + \tau_L/J.
\end{aligned}$$

With $\tau_L(s) = \tau_{L0}/s$, find $E_1(s)$ and $E_2(s)$. Show that $e_2(t) \to 0$, but $e_1(t) \nrightarrow 0$ as $t \to \infty$. Comparing with your solution to part (a), what can you conclude about using integrator feedback?

Problem 22 State Feedback with a Ramp Load Torque

With $\tau_L(s) = \tau_{L0}/s^2$, that is, $\tau_L(t) = \tau_{L0}t$, find $e_0(\infty)$, $e_1(\infty)$, $e_2(\infty)$. Explain your answers.

Problem 23 State Feedback Design with Voltage as Input

Design a state feedback voltage controller for the DC motor based on the model

$$\begin{aligned}
\frac{di}{dt} &= -(R/L)i(t) - (K_b/L)\omega(t) + v(t)/L \\
\frac{d\omega}{dt} &= -(f/J)\omega(t) + (K_T/J)i(t) - \tau_L(t)/J \\
\frac{d\theta}{dt} &= \omega(t).
\end{aligned}$$

The reference trajectory satisfies

$$\begin{aligned}
\frac{di_{ref}}{dt} &= -(R/L)i_{ref}(t) - (K_b/L)\omega_{ref}(t) + v_{ref}(t)/L \\
\frac{d\omega_{ref}}{dt} &= -(f/J)\omega_{ref}(t) + (K_T/J)i_{ref}(t) \\
\frac{d\theta_{ref}}{dt} &= \omega_{ref}(t).
\end{aligned}$$

Let

$$\begin{aligned} e_1(t) &= \theta_{ref}(t) - \theta(t) \\ e_2(t) &= \omega_{ref}(t) - \omega(t) \\ e_3(t) &= i_{ref}(t) - i(t) \\ u(t) &= (v_{ref}(t) - v(t))/L \end{aligned}$$

and consider a controller of the form

$$u = -\left(K_0 \int_0^t e_1(\tau)d\tau + K_1 e_1(t) + K_2 e_2(t) + K_3 e_3(t)\right).$$

With $\tau_L(t) = \tau_{L0} u_s(t)$, *for* $i = 0, 1, 2, 3$ *compute* $E_i(s)$ *and* $e_i(\infty)$. *In particular, does the steady-state position error* $e_1(t)$ *go to zero as* $t \to \infty$*?*

Nested Loop Controllers

Problem 24 Nested PI Controllers

Consider the PID nested loop controller given in Figure 2.49.

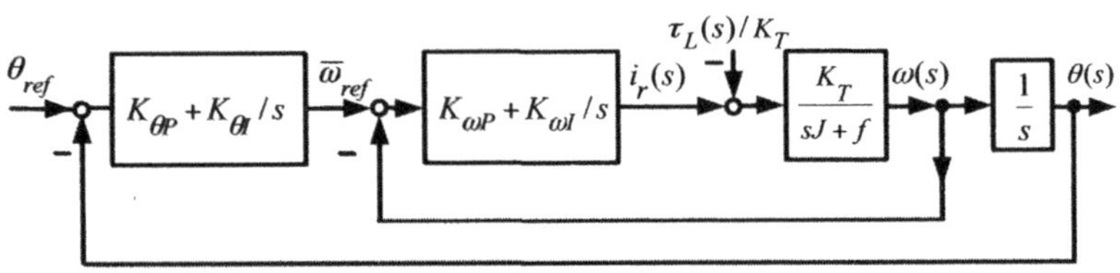

FIGURE 2.49. Control structure where the gains of the speed loop are tuned first and then the gains of the position loops are tuned. The speed reference $\bar{\omega}_{ref}$ is the output of the position loop rather than the stored value of $\omega_{ref} = d\theta_{ref}/dt$.

(a) With $\theta_{ref}(t) = (t^2/2)\, u_s(t)$ $(\theta_{ref}(s) = 1/s^3)$, $\tau_L \equiv 0$ *and* $K_{\omega I} = 0$, *can the other gains be chosen so that steady-state error is zero for this type of input? Explain by computing* $E(s)/\theta_{ref}(s)$ *with* $\tau_L \equiv 0$ *and* $K_{\omega I} = 0$.

(b) With $\theta_{ref}(t) \equiv 0, \tau_L(t) = \tau_0 t$ $(\tau_L(s) = \tau_0/s^2)$ *and* $K_{\omega I} \neq 0$, *can the gains be chosen such that steady-state error is zero for this type of load torque disturbance? Explain by computing* $E(s)/\theta_{ref}(s)$ *with* $\theta_{ref}(t) \equiv 0$. *How does your answer change if* $K_{\omega I} = 0$*?*

(c) With $\tau_L \equiv 0$, *can this controller track* $\theta_{ref}(t) = \cos(\omega_0 t)$ *or, equivalently,* $\theta_{ref}(s) = \omega_0/\left(s^2 + \omega_0^2\right)$ *with zero steady-state error? Explain.*

(d) Repeat (a), (b), and (c) with $K_{\omega I} \neq 0$.

Problem 25 Simulation of the Nested Loop Controller

Do a simulation of the complete system (i.e., motor, controller, and trajectory) given in problem 24. Use the same motor parameters as in problem 1. Take $\theta_f = \pi$ *radians and find the control gains* $K_{\theta P}, K_{\theta I}, K_{\omega P}$

($K_{\omega I} = 0$), ω_{max}, and t_1 for the trajectory (2.25) such that the motor turns π radians as fast as possible without violating the voltage or current constraints. Plot (a) $\theta(t), \theta_{ref}(t)$, (b) $\hat{\omega}(t), \omega(t), \omega_{ref}(t)$, (c) $i_r(t), i_{ref}(t)$, (d) $\theta_{ref}(t) - \theta(t)$, (e) $\omega(t) - \hat{\omega}(t)$, (f) $\omega_{ref}(t) - \omega(t)$, (g) $i_r(t) - i(t)$. Compare with problem 18. In particular, can this move be carried out in under 70 msec?

Problem 26 Laplace Transform of the Reference Trajectory

The reference trajectory given in (2.25) can be rewritten as (u_s is the unit step function)

$$\begin{aligned}
\theta_{ref}(t) &= \omega_{max}\frac{t^2}{2t_1}\left(u_s(t) - u_s(t-t_1)\right) \\
&+ \left(\omega_{max}\frac{t_1}{2} + \omega_{max}(t-t_1)\right)\left(u_s(t-t_1) - u_s(t-t_2)\right) \\
&+ \left(\omega_{max}t_2 - \omega_{max}\frac{(t_f-t)^2}{2t_1}\right)\left(u_s(t-t_2) - u_s(t-t_f)\right) \\
&+ \omega_{max}t_2 u_s(t-t_f)
\end{aligned}$$

where $\omega_{max} = \theta_f/t_2$ to ensure $\theta_{ref}(t_f) = \theta_f$.

Show that the Laplace transform of $\theta_{ref}(t)$ is given by

$$\begin{aligned}
\mathcal{L}\{\theta_{ref}(t)\} &= \frac{\omega_{max}}{t_1}\frac{1}{s^3} - \left(\frac{\omega_{max}t_1}{2}\frac{1}{s} + \omega_{max}\frac{1}{s^2} + \frac{\omega_{max}}{t_1}\frac{1}{s^3}\right)e^{-t_1 s} \\
&+ \left(\frac{\omega_{max}t_1}{2}\frac{1}{s} + \frac{\omega_{max}}{s^2}\right)e^{-t_1 s} - \left(\left(-\frac{\omega_{max}t_1}{2} + \omega_{max}t_2\right)\frac{1}{s} + \frac{\omega_{max}}{s^2}\right)e^{-t_2 s} \\
&+ \left(\left(\omega_{max}t_2 - \frac{\omega_{max}\left(t_2-t_f\right)^2}{2t_1}\right)\frac{1}{s} - \frac{\omega_{max}(t_2-t_f)}{t_1}\frac{1}{s^2} - \frac{\omega_{max}}{t_1}\frac{1}{s^3}\right)e^{-t_2 s} \\
&- \left(\omega_{max}t_2\frac{1}{s} - \frac{\omega_{max}}{t_1}\frac{1}{s^3}\right)e^{-t_f s} + \omega_{max}t_2\frac{1}{s}e^{-t_f s}.
\end{aligned}$$

Explain why the presence of the terms $e^{-t_1 s}, e^{-t_2 s}$, and $e^{-t_f s}$ does not affect the analysis of the steady-state error given in the text.

Problem 27 Nested Loop Control Structure With Voltage Input

In this problem, the PI current loop of the amplifier is included in the analysis of a nested loop controller for a DC motor. The PI current controller in Figure 2.50 can be equivalently considered a torque controller as $\tau = K_T i$. First, both the speed and position loops are opened in order to tune (choose) the torque (current) controller gains. After tuning the gains for this controller, the gains of the speed controller are tuned with the input to the torque controller being the output of the speed controller. Finally, the gains of the position loop are tuned with the input to the speed controller being the output of the position controller. Use the full motor model (2.1) in the analysis below.

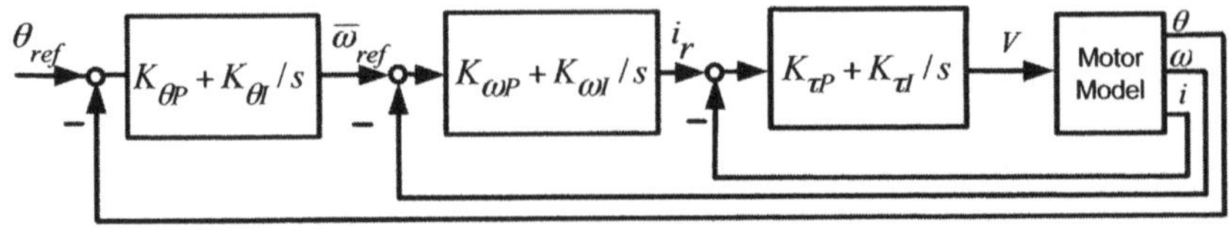

FIGURE 2.50. The gains of the torque (current) loop are tuned first. Next the gains of the speed loop are tuned and finally the gains of the position loop are tuned.

(a) Find the transfer functions $G_1(s), G_2(s)$ *such that the error* $E(s) = G_1(s)\theta_{ref}(s) + G_2(s)\tau_L(s)$.

(b) Can the gains be chosen to arbitrarily place the closed-loop poles?

(c) What kinds of inputs can be tracked? In particular, what is the largest power n such that

$$\theta_{ref}(t) = \frac{t^{n-1}}{(n-1)!}u_s(t)$$

or, equivalently,

$$\theta_{ref}(s) = \mathcal{L}\left\{\frac{t^{n-1}}{(n-1)!}u_s(t)\right\} = 1/s^n$$

can be tracked with zero steady-state error? What is the largest value of n if it is only required to have constant steady-state error?

(d) What kinds of disturbances can be rejected? That is, with

$$\tau_L(t) = \tau_{L0}\left(t^{m-1}/(m-1)!\right)u_s(t)$$

or

$$\mathcal{L}\left\{\frac{t^{m-1}}{(m-1)!}u_s(t)\right\} = 1/s^m,$$

what is the largest value of m such that there is no steady-state error due to this disturbance?

Identification of the Motor Parameters

Problem 28 Identification of the Motor Parameters

With a PI current controller implemented, run the DC motor simulation with the current reference given by

$$i_r(t) = \begin{cases} 2.5\sin\left(\dfrac{2\pi}{0.1}t\right) & \text{for } 0 \le t \le 0.5 \\ 2.5\sin\left(\dfrac{2\pi}{0.05}t\right) & \text{for } 0.5 \le t \le 1. \end{cases}$$

Store $\theta(nT)$ in counts (from the encoder model), $i(nT)$ and $v(nT)$ with a sample period of $T = 0.001$ sec.

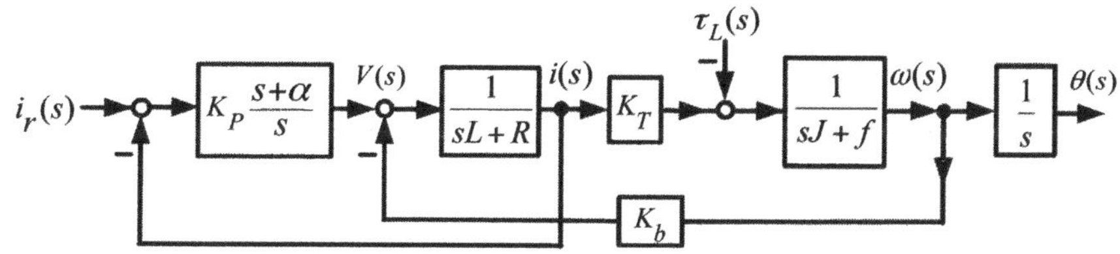

FIGURE 2.51. Block diagram of an open-loop DC motor with a current-command controller.

(a) Save the "data" from the simulation into a file. Using this "data set", write a program to implement the least-squares identification algorithm given in the text. It will be necessary to compute $\omega = d\theta/dt, d\omega/dt$ and di/dt using the "data" from the simulation. Due to the encoder resolution, $d\theta/dt$ and $d\omega/dt$ are noisy and di/dt is also noisy. It will be necessary to filter these computed quantities. Show the double filtering method (e.g., the `filtfilt` *command in* MATLAB*) with a second order Butterworth filter will work with an appropriate choice of the cutoff frequency. Compare your parameter estimates to the actual values in the simulation. Be sure to compute the error index and the parametric error indices of your estimates.*

(b) Do the identification of the parameters using "ideal" data by saving $d\theta/dt, d\omega/dt$ and di/dt from the simulation (rather than computing them offline) and show that this results in almost perfect agreement with the parameter values in the simulation. Compute the error index and the parametric error indices for this case. This is a way to check your identification program for errors.

Problem 29 Insufficient Excitation

Redo problem 28 with $i_r = i_0$ where $i_0 = 2.5$ A. Does the identification procedure work? Explain why or why not.

Problem 30 Completing the Square

Show that equation (2.38) is true.

Problem 31 Matrix Transpose Property

Show by direct computation that

$$\left(y^T(nT)W(nT)\right)^T = W^T(nT)y(nT).$$

Use this to show that $R_{yW} = R_{Wy}^T$.

Problem 32 Parametric Error Indices [42]

Choose

$$C(\delta K, \lambda) \triangleq (\delta K_i)^2 + \lambda\left(E^2(K^*) - \delta K^T R_w \delta K\right)$$

and show that one still obtains

$$\delta K_i = \sqrt{E^2(K^*)(R_w^{-1})_{ii}}.$$

for the maximum.

Filtering of Noisy Signals

Problem 33 Transfer Functions

(a) Starting from the representation $y(t) = \int_{-\infty}^{t} g(t-\tau)m(\tau)d\tau$, *take the Laplace transform of both sides to show that* $Y(s) = G(s)M(s)$.

(b) Starting from the representation $\dot{x}(t) = Ax(t) + bm(t)$, $y(t) = cx(t) + dm(t)$ *show that* $G(s) = c(sI - A)^{-1}b$, *where* $A \in \mathbb{R}^{n \times n}, b \in \mathbb{R}^n, c \in \mathbb{R}^{1 \times n}$.

Problem 34 Ideal Filters

Consider a unit impulse response given by

$$h(t) \triangleq \frac{\sin(\omega_c t)}{\pi t}$$

so with input $m(t)$, *the output* $y(t)$ *is defined to be*

$$y(t) = \int_{-\infty}^{\infty} h(\tau)m(t-\tau)d\tau.$$

This integral is defined as the Cauchy principal value, that is,

$$y(t) = \int_{-\infty}^{\infty} h(\tau)m(t-\tau)d\tau \triangleq \lim_{T\to\infty} \int_{-T}^{T} h(\tau)m(t-\tau)d\tau.$$

(a) With $m(t) = \cos(\omega t)$, *compute* $y(t) = h(t) * \cos(\omega t)$ *directly by evaluating*

$$y(t) = \lim_{T\to\infty} \int_{-T}^{T} \frac{\sin(\omega_c \tau)}{\pi\tau} \cos(\omega(t-\tau))d\tau$$

to show

$$y(t) = \lim_{T\to\infty} \int_{-T}^{T} \frac{\sin(\omega_c \tau)}{\pi\tau} \cos\left(\omega(t-\tau)\right)dt = \begin{cases} \cos(\omega t) & \text{for } |\omega| < \omega_c \\ \frac{1}{2}\cos(\omega t) & \text{for } \omega = \omega_c \\ 0 & \text{for } |\omega| > \omega_c. \end{cases}$$

This is the reason it is called an ideal filter as it passes any frequency below ω_c *unchanged and any frequency above* ω_c *is zeroed out.*

Why is it necessary to use the Cauchy principal value in the definition of the integral?

(b) Compute the Fourier transform of $h(t)$ in the Cauchy principal value sense, that is, show

$$H(j\omega) = \mathfrak{F}\left\{\frac{\sin(\omega_c t)}{\pi t}\right\} \triangleq \lim_{T\to\infty}\int_{-T}^{T} e^{-j\omega t}\frac{\sin(\omega_c \tau)}{\pi\tau}dt = \begin{cases} 1 & \text{for } |\omega| < \omega_c \\ \frac{1}{2} & \text{for } \omega = \omega_c \\ 0 & \text{for } |\omega| > \omega_c. \end{cases}$$

(c) Is this filter causal? Show why or why not.

(d) Is this filter stable? Hint: Consider the input signal

$$m(t) = m_1(t)m_2(t).$$

where $m_1(t)$ is a periodic function with period $2\pi/\omega_c$ specified on one period by

$$m_1(t) = \begin{cases} -1 & \text{for } 0 \le t \le \pi/\omega_c \\ 1 & \text{for } \pi/\omega_c < t \le 2\pi/\omega_c \end{cases}$$

and $m_2(t)$ is defined by

$$m_2(t) = \begin{cases} -1 & \text{for } t \ge 0 \\ 1 & \text{for } t < 0. \end{cases}$$

This input signal is bounded as $|m(t)| = 1$. Compute the response to this input at time $t = 0$; that is, evaluate

$$y(t) = \lim_{T\to\infty}\int_{-T}^{T}\frac{\sin(\omega_c\tau)}{\pi\tau}m(t-\tau)d\tau$$

for $t = 0$ to obtain

$$y(0) = \int_{-\infty}^{\infty}\left|\frac{\sin(\omega_c\tau)}{\pi\tau}\right|d\tau = \infty.$$

(e) Evaluate the inverse Fourier transform of $H(j\omega)$ from part (b) as

$$h(t) = \frac{1}{2\pi}\int_{-\infty}^{\infty}e^{j\omega t}H(j\omega)d\omega$$

to get back $h(t) = \sin(\omega_c t)/(\pi t)$.

Problem 35 Butterworth Filters

For the third-order Butterworth filter, find the value of γ_n that best approximates $\angle G_n(j\omega) \approx -\gamma_n\omega$. For what range of values of ω is this approximation a good one?

Problem 36 Butterworth Filters from MATLAB

(a) With $\omega_c = 2\pi f_c$, $f_c = 100$ Hz and $T = 1/5000$ sec, explicitly compute the coefficients of a third-order Butterworth filter using (2.51).

(b) In MATLAB, *use the* `butter` *command to design a third-order Butterworth filter with a cutoff frequency of* 100 *Hz and a sample rate of* 5000 *Hz. That is, use the command* `[b,a] = butter(3,`f_c`/(`f_s`/2))` *and* MATLAB *will return two vectors of the form* `b = [`b_1 b_2 b_3 b_4 `]` *and* `a = [1` a_2 a_3 a_4 `]` *which contain the coefficients of the filter.*

(c) Compare the coefficients obtained from MATLAB *with those obtained from the above design. Are they the same? Do you need to normalize the coefficients obtained using (2.51)?*

Recall that the cutoff frequency in MATLAB *is specified as* $f_{cmatlab} = f_c/(f_s/2)$ *where* $f_s = 5000$ *Hz. The* MATLAB *command is then* `[a,b] = butter(3,100/2500)`.

Problem 37 The `filter` Command in MATLAB

(a) Use the MATLAB *command* `butter` *to design a third-order Butterworth filter with* $f_s = 5000$ *Hz and* $f_c = 100$ *Hz* ($\omega_c = 2\pi \times 100$ *rads/sec). That is, use the command* `[b,a] = butter(3,`f_c`/(`f_s`/2))` *and* MATLAB *will return two vectors of the form* `b = [`b_1 b_2 b_3 b_4 `]` *and* `a = [1` a_2 a_3 a_4 `]` *which contain the coefficients of the filter.*

(b) Write a MATLAB *program to implement the state space equations (2.56) from* $n = 0$ *to* $n = N$ *using the coefficients given in part (a). With* $T = 1/f_s$*, run your program on some particular input data file* $\{m(nT)\}$ *with* $x(0) = 0$ *and plot out the response* $y(nT)$ *versus time* nT.

(c) Using the filter designed in part (a) and the MATLAB *command* `filter` *to filter the same input file used in part (b), that is,* `y2 = filter(b,a,m)`. *Plot* $y_2(nT)$ *versus time* nT. *Compare* y *from part (b) to* y_2.

Problem 38 Signal Flow Diagrams

Consider the second-order discrete-time transfer function

$$H(z) = \frac{Y(z)}{M(z)} = \frac{b_0 + b_1 z^{-1} + b_2 z^{-2}}{1 + a_1 z^{-1} + a_2 z^{-2}} = \frac{b_0 z^2 + b_1 z + b_2}{z^2 + a_1 z + a_2}$$

(a) A signal flow diagram for this system is shown in Figure 2.52. Find the transfer function $Y(z)/M(z)$ *from the signal flow (block) diagram. Give a discrete-time state space realization of the transfer function based on the signal flow diagram.*

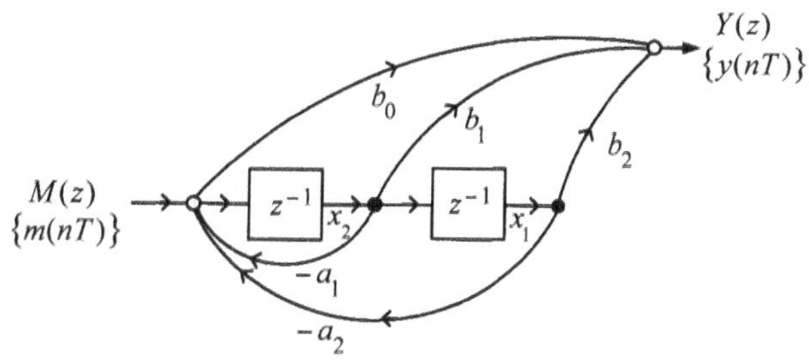

FIGURE 2.52. Signal flow diagram for a second-order filter.

(b) Show that the signal flow (block) diagram in Figure 2.53 is equivalent to the one in part (a) by finding a discrete-time state space realization based on the signal flow diagram and then computing the transfer function $Y(z)/M(z)$ from the state space realization.

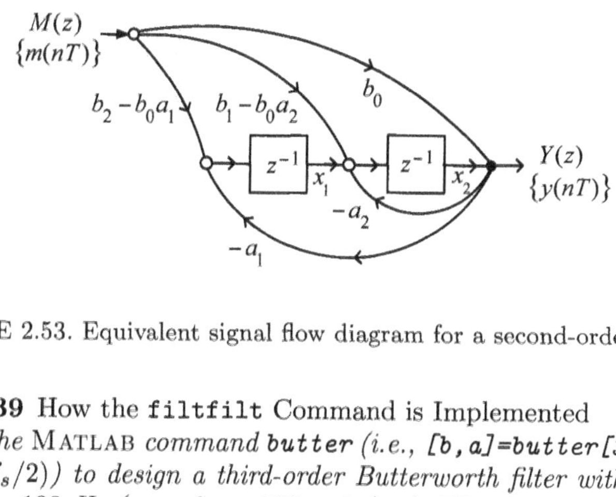

FIGURE 2.53. Equivalent signal flow diagram for a second-order filter.

Problem 39 How the **`filtfilt`** Command is Implemented

(a) Use the MATLAB *command* **`butter`** *(i.e.,* **`[b,a]=butter[3,`**w_n**`]`**, *where* $w_n = f_c/(f_s/2)$*) to design a third-order Butterworth filter with* $f_s = 5000$ *Hz and* $f_c = 100$ *Hz (*$\omega_c = 2\pi \times 100$ *rads/sec). Write a* MATLAB *program to implement this filter using the state space equations (2.56) (from* $n = 0$ *to* $n = N$*) and the coefficients given by the* **`butter`** *command. Run your program on some particular input data file* $\{m(nT)\}$ *with* $x(0) = 0$ *and plot out the response* $y(nT)$ *versus time* nT*. Store the final value of* x*, that is,* $x(NT)$*.*

(b) Take the output $\{y(nT), n = 0, ..., N\}$ *of the filter in part (a), and compute* $\{y_b(nT), n = 0, ..., N\} \triangleq \{y((N-n)T), n = 0, ..., N\}$*. Use this as input to the (state space) filter designed in part (a), except* initialize *the filter with the final conditions* $x(NT)$ *found in part (a). Let* $\{y_{ffb}(nT), n = 0, ..., N\}$ *be the output of this second filtering process and finally compute* $\{y_{ff}(nT), n = 0, ..., N\} \triangleq \{y_{ffb}((N-n)T), n = 0, ..., N\}$*. Plot* $y_{ff}(nT)$ *for* $n = 0, ..., N$ *versus time* nT*.*

(c) Compare the output $y_{ff}(nT)$ from part (b) with the output using the MATLAB *command* **filtfilt** *as* $\mathtt{y}_{ff2} \triangleq \mathtt{filtfilt[b,a,m]}$, *where* **[b,a]=butter[3,** w_n**]** *with* $w_n = f_c/(f_s/2) = 100/(5000/2)$. *(They should be the same!)*

Control of Separately Excited, Series and Shunt DC motors

Problem 40 Armature Control of a Separately Excited DC Motor [2][51]

Recall from the previous chapter, the full mathematical model for the separately excited DC motor is

$$\begin{aligned} L\frac{di}{dt} &= -Ri + \frac{\partial\lambda}{\partial i_f}\frac{di_f}{dt} - K_m\lambda_f(i_f)\omega_r + v \\ J\frac{d\omega}{dt} &= K_m\lambda_f(i_f)i - \tau_L \\ \frac{d\lambda_f(i_f)}{dt} &= -R_f i_f + v_f \end{aligned}$$

where $\lambda(i_f,\theta)$ *is the flux linkage in the armature loops due to the field current* i_f *and* $\lambda_f(i_f)$ *is the flux linkage in the field windings due to the field current* i_f.

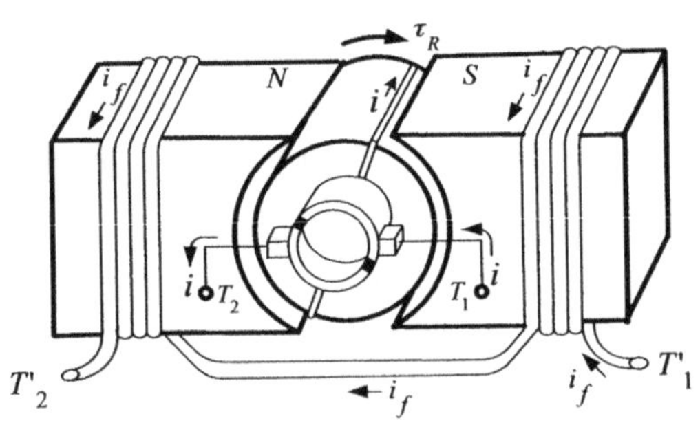

FIGURE 2.54. DC motor with a field winding.

In normal operation of a separately excited DC motor, di_f/dt *varies slowly and the* $\dfrac{\partial\lambda}{\partial i_f}\dfrac{di_f}{dt}$ *term is often neglected. With* $\lambda_f(i_f) = f(i_f)$ *the saturation (magnetization) curve for the field flux linkage so that* $i_f = f^{-1}(\lambda_f)$, *the model written in terms of the state variables* (i, ω, λ_f) *be-*

comes[13]

$$
\begin{aligned}
L\frac{di}{dt} &= -Ri - K_m\lambda_f\omega + v \\
J\frac{d\omega}{dt} &= K_m\lambda_f i - \tau_L \\
\frac{d\lambda_f}{dt} &= -R_f f^{-1}(\lambda_f) + v_f.
\end{aligned}
$$

Interestingly, it will be shown in a later chapter that a field-oriented control of an induction motor results in a model of this form! With $x_1 = i, x_2 = \lambda_f, x_3 = \omega, u_1 = v/L$ *and* $u_2 = v_f$*, the model becomes*

$$
\begin{aligned}
\frac{dx_1}{dt} &= -\frac{R}{L}x_1 - \frac{K_m}{L}x_2x_3 + u_1 \\
\frac{dx_2}{dt} &= -R_f f^{-1}(x_2) + u_2 \\
\frac{dx_3}{dt} &= \frac{K_m}{J}x_2x_1 - \frac{\tau_L}{J}.
\end{aligned}
$$

(a) Show that the nonlinear transformation

$$
\begin{aligned}
z_1 &= x_3 \\
z_2 &= (K_m/J)x_2x_1 \\
z_3 &= x_2
\end{aligned}
$$

results in the system being equivalent to

$$
\begin{aligned}
\frac{dz_1}{dt} &= z_2 - \tau_L/J \\
\frac{dz_2}{dt} &= g(x) + u_1(K_m/J)x_2 + (K_m/J)x_1u_2 \\
\frac{dz_3}{dt} &= -R_f f^{-1}(x_2) + u_2
\end{aligned}
$$

where $g(x) \triangleq -\frac{RK_m}{JL}x_1x_2 - \frac{K_mR_f}{J}f^{-1}(x_2)x_2 - \frac{K_m^2}{JL}x_2^2x_3$.
(b) Show that the nonlinear feedback

$$
\begin{bmatrix} u_1 \\ u_2 \end{bmatrix} = \begin{bmatrix} (K_m/J)x_2 & (K_m/J)x_1 \\ 0 & 1 \end{bmatrix}^{-1} \times \left(-\begin{bmatrix} g(x) \\ -R_f f^{-1}(x_2) \end{bmatrix} + \begin{bmatrix} v_1 \\ v_2 \end{bmatrix} \right)
$$

---

[13] In the case that there is no magnetic saturation, $\lambda_f(i_f) = L_f i_f$ so that $i_f = \lambda_f/L_f$.

results in the linear *system*

$$\begin{aligned} \frac{dz_1}{dt} &= z_2 - \tau_L/J \\ \frac{dz_2}{dt} &= v_1 \\ \frac{dz_3}{dt} &= v_2. \end{aligned}$$

This is referred to as exact linearization *or* feedback linearization.

(c) Show that if the load torque is constant, *then choosing*

$$\begin{aligned} v_1 &= -K_0 \int_0^t (z_1 - \omega_{ref})dt - K_1(z_1 - \omega_{ref}) - K_2(z_2 - \alpha_{ref}) + \frac{d}{dt}\alpha_{ref} \\ v_2 &= -K_f(z_3 - \lambda_{f\_ref}) + d\lambda_{f\_ref}/dt \end{aligned}$$

with the gains chosen appropriately results in $z_1 \to \omega_{ref}$ *and* $z_3 \to \lambda_{f\_ref}$.

(d) Draw a block diagram for the complete (i.e., nonlinear and linear feedback) control system.

Remarks: *If position control is desired, one simply appends* $dz_0/dt = z_1$ *to the system where* $z_0 = \theta$. *The feedback is singular only if* $x_2 = \lambda_f = 0$. *However, one chooses the input* v_2 *to ensure* $\lambda_f > 0$. *The importance of this approach is that the control of the speed through the input* v_1 *is decoupled from the control of the flux* ($z_2 = x_2 = \lambda_f$) *through the input* v_2. *The linear control techniques of this chapter are directly applicable to this linear model.*

Problem 41 Field Control of a Separately Excited DC Motor [51][52]

Another approach to controlling a separately excited DC motor is the so-called method of field control. *In this approach the armature voltage is held constant while the field voltage is varied to control the speed. This is often used in large DC drives (e.g., rolling mills) since the armature circuit can require* 2000-3000 A *and* 250 V *while the field circuit may require only* 20 A. *Consequently, it is of great economic advantage to use an amplifier that only has to handle the* 20 A *of the field circuit together with a large generator at constant voltage to produce the armature current. A mathematical model for a field controlled motor model in terms of the state variables* (i, ω, λ_f) *is then*

$$\begin{aligned} L\frac{di}{dt} &= -Ri - K_m\lambda_f\omega + V_{a0} \\ J\frac{d\omega}{dt} &= K_m\lambda_f i - \tau_L \\ \frac{d\lambda_f}{dt} &= -R_f f^{-1}(\lambda_f) + v_f \end{aligned}$$

where V_{a0} *is* constant *(typically the output of a generator).*

(a) Consider the nonlinear transformation [52]

$$\begin{aligned} z_1 &= \left(Li^2 + J\omega^2\right)/2 \\ z_2 &= i\left(-Ri - K_m\lambda_f\omega + V_{a0}\right) + \omega K_m\lambda_f i = -Ri^2 + iV_{a0} \\ z_3 &= \left(-2Ri + V_{a0}\right)\left(-Ri - K_m\lambda_f\omega + V_{a0}\right)/L \end{aligned}$$

and show that the system equations become

$$\begin{aligned} \frac{dz_1}{dt} &= z_2 \\ \frac{dz_2}{dt} &= z_3 - \tau_L/J \\ \frac{dz_3}{dt} &= f(i,\omega,\lambda_f) + g(i,\omega,\lambda_f)v_f. \end{aligned}$$

Here $f(i,\omega,\lambda_f), g(i,\omega,\lambda_f)$ are nonlinear functions of the state (i,ω,λ_f). Show that application of the feedback

$$v_f = \left(-f(i,\omega,\lambda_f) + u_f\right)/g(i,\omega,\lambda_f)$$

results in a linear system. This is referred to as exact linearization *or a* feedback linearization *controller.*

(b) When is $g(i,\omega,\lambda_f)$ singular (i.e., when does $g(i,\omega,\lambda_f) = 0$)? Can this singularity be avoided in normal operation of the motor?

(c) Show how to pick the gains of the controller

$$\begin{aligned} u_f &= -K_0\int_0^t (z_1 - z_{1ref})dt - K_1(z_1 - z_{1ref}) - K_2\left(z_2 - z_{2ref}\right) \\ &\quad -K_3\left(z_3 - z_{3ref}\right) + dz_{3ref}/dt \end{aligned}$$

such that $z_1 - z_{1ref} \to 0$ with a constant load torque.

(d) What are some practical problems with this approach to controlling the machine? (Hint: How does one choose $z_{1ref}, z_{2ref}, z_{3ref}$ so that $\omega \to \omega_{ref}$?)

Problem 42 Series Connected DC Motor [2][51][53][54]

A DC motor in which the field circuit is connected in series with the armature circuit is referred to as a series DC *motor. This connection is shown in the equivalent circuit of the series DC motor shown in Figure 2.55 and is typically used in applications that require high torque at low speed such as traction drives (e.g., subway trains) [2]. Comparing with the separately excited motor of Figure 2.54, it is seen that the terminal T_2' is connected to the terminal T_1 with the voltage v applied between terminals T_1' and T_2. The resistor R_p put in parallel with the field circuit is switched in at higher speeds to achieve field weakening (see problem 43).*

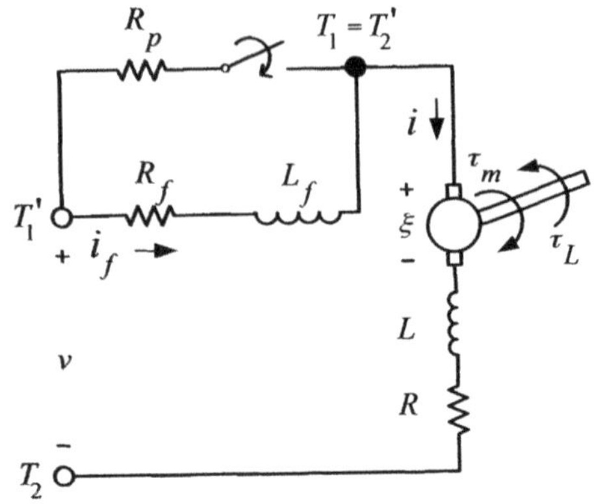

FIGURE 2.55. Equivalent circuit for a series DC motor.

With the switch for R_P open, the equations describing electrical circuit of the field and armature winding is

$$-\frac{d\left(Li - \lambda(i,\theta) + \lambda_f(i)\right)}{dt} - (R_f + R)\, i + v = 0.$$

Using the fact that $\partial\lambda(i_f,\theta)/\partial\theta = -K_m\lambda_f(i_f)$ and that the torque is given by $K_m\lambda_f(i_f)i$, the equations of the series connected DC motor are then

$$\begin{aligned}\frac{d\left(Li + \lambda_f(i)\right)}{dt} &= -(R_f + R)\, i + \frac{\partial\lambda}{\partial i_f}\frac{di_f}{dt} - K_m\lambda_f(i)\omega + v \\ J\frac{d\omega}{dt} &= K_m\lambda_f(i)i - \tau_L.\end{aligned}$$

In the case of the series winding, the armature flux Li can be neglected compared to the field flux $\lambda_f(i)$. Also $\frac{\partial\lambda}{\partial i_f}\frac{di_f}{dt}$ is neglected to obtain

$$\begin{aligned}\frac{d\lambda_f(i)}{dt} &= -(R_f + R)\, i - K_m\lambda_f(i)\omega + v \\ J\frac{d\omega}{dt} &= K_m\lambda_f(i)i - \tau_L.\end{aligned}$$

(In the case of no magnetic saturation in the field winding $\lambda_f(i) = L_f i$.)

(a) Show that the nonlinear transformation

$$\begin{aligned}z_1 &= \omega \\ z_2 &= (K_m/J)\lambda_f(i)i\end{aligned}$$

gives

$$\begin{aligned}\frac{dz_1}{dt} &= z_2 \\ \frac{dz_2}{dt} &= f(i,\omega) + g(i,\omega)v - \tau_L/J.\end{aligned}$$

What are the expressions for $f(i,\omega), g(i,\omega)$*?*

(b) When is $g(i,\omega)$ *singular? Can this be avoided in normal operation? Show that the nonlinear feedback*

$$v_f = \frac{-f(i,\omega)+u}{g(i,\omega)}$$

results in the linear system

$$\begin{aligned} \frac{dz_1}{dt} &= z_2 \\ \frac{dz_2}{dt} &= u - \tau_L/J. \end{aligned}$$

This is called a feedback linearizing *controller or an* exact linearization *controller.*

(c) Show how to choose the gains for the controller

$$u = K_0\int_0^t (\omega_{ref}(\tau) - \omega(\tau)d\tau + K_1(\omega_{ref}-\omega) + K_2(\alpha_{ref}-\alpha) + \frac{d}{dt}\alpha_{ref}$$

so that $\omega_{ref} - \omega \to 0$ *with a constant load torque.*

(d) Draw a complete block diagram of the system including the nonlinear and linear feedback.

Problem 43 Field Weakening Control for a Series DC Motor [2][53]

Consider the series connected DC motor described in problem 42 running at high speed with the switch closed and $\lambda_f > 0$*. The current in the field winding is now less that the armature current* ($i_f < i_a$) *so that the field flux linkage* $\lambda_f(i_f)$ *and therefore the back emf* $K_m\lambda_f(i_f)\omega$ *is now less than it would have been carrying full armature current. Consequently, this provides a way to achieve higher speeds than otherwise without having the back emf exceed the voltage limit. This is referred to as field weakening as the field flux linkage* $\lambda_f(i_f)$ *is now "weakened" compared to the value* $\lambda_f(i)$ *with full armature current.*

(a) Using the state variables (i, λ_f, ω)*, derive a nonlinear state space model of the system.*

(b) Let $L = 0$ *and use the equation for the current* i *(setting* $Ldi/dt = 0$*) to solve for* $i = (v + R_p i_f - K_m\lambda_f\omega)/(R+R_p)$*. Use this to write a second order state space model in terms of the state variables* (λ_f, ω)*.*

(c) Consider an input-output linearization controller of the form $v = -R_p i_f + K_m\lambda_f\omega + u(R+R_p)/(K_m\lambda_f/J)$ *where* u *is a new input. Explain how to control the speed and why the flux must satisfy* $0 < c_1 < \lambda_f < c_2$*, that is, it must be bounded from below and above.*

Problem 44 Speed and Load-Torque Observer for the Series DC Motor [51][53][54]

An estimator for the speed and load torque of a series DC motor can be constructed using only a measurement of the current. To see this, consider the equations of the series DC motor with $L = 0$

$$\begin{aligned}\frac{d\lambda_f}{dt} &= -(R_f + R)i - K_m\lambda_f\omega + v\\ J\frac{d\omega}{dt} &= K_m\lambda_f i - \tau_L.\end{aligned}$$

Model the load torque as an unknown constant so that the equation

$$\frac{d(\tau_L/J)}{dt} = 0$$

is appended to the system.

(a) Show that the nonlinear change of coordinates

$$\begin{aligned}z_1 &= \ln(\lambda_f)\\ z_2 &= \omega\\ z_3 &= \tau_L/J\end{aligned}$$

results in

$$\begin{aligned}\dot{z} &= Az + h(i_a, i_f, \lambda_f, V)\\ y &= cz\end{aligned}$$

where

$$z = \begin{bmatrix} z_1\\ z_2\\ z_3\end{bmatrix},\ A = \begin{bmatrix} 0 & -K_m & 0\\ 0 & 0 & -1\\ 0 & 0 & 0\end{bmatrix},\ c = \begin{bmatrix} 1 & 0 & 0\end{bmatrix}$$

and

$$h(i, i_f, \lambda_f, v) = \begin{bmatrix} \left(-R_f i_f - Ri + v\right)/\lambda_f\\ (K_m/J)\lambda_f i\\ 0\end{bmatrix}.$$

(b) Show that the gain vector $\ell = \begin{bmatrix} \ell_1 & \ell_2 & \ell_3\end{bmatrix}^T$ *can be chosen so that the observer defined by*

$$\begin{aligned}d\hat{z}/dt &= A\hat{z} + h(i, i_f, \lambda_f, v) + \ell(y - \hat{y})\\ \hat{y} &= c\hat{z}\end{aligned}$$

results in $\hat{z} \to z$ *and, in particular,* $\hat{\omega} \to \omega$ *and* $\hat{\tau}_L/J \to \tau_L/J$.

(c) Is there a practical problem with using such an observer? Explain. (Hint: When is $h(i, i_f, \lambda_f, v)$ *singular?)*

Problem 45 Speed and Load-Torque Observer for the Field-Weakened Series DC Motor [53]

Consider the state space model using the state variables (i, λ_f, ω) *as developed in problem 43. Add the first two equations together with* $L = 0$ *to show that*

$$\begin{aligned} \frac{d\lambda_f}{dt} &= -R_f i_f - Ri - K_m \lambda_f \omega + v \\ J\frac{d\omega}{dt} &= K_m \lambda_f i - \tau_L. \end{aligned}$$

Similar to the construction in problem 44, show that a speed and load-torque observer can be designed.

Problem 46 Shunt-Connected DC Machine [51][55]

A shunt-connected DC motor is one in which the armature and field circuits are connected in parallel. Specifically, using the separately excited DC motor of Figure 2.54, one adds an adjustable resistance R_{adj} *to the field circuit and then ties together terminals* T_1 *and* T_1' *as well as the terminals* T_2 *and* T_2' *resulting in the equivalent circuit shown in Figure 2.56.*

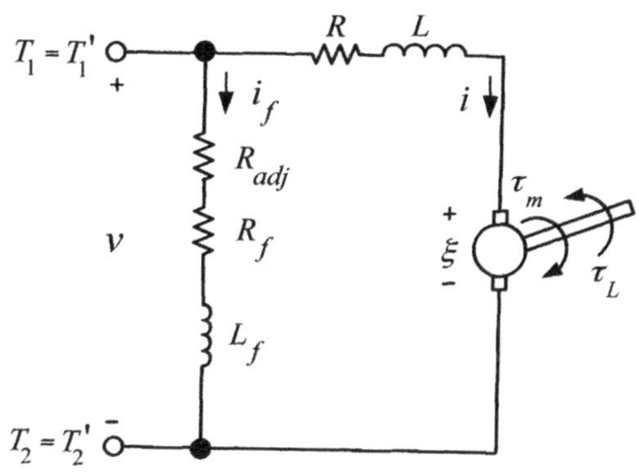

FIGURE 2.56. Equivalent circuit for a shunt DC motor.

The mathematical model is found to be

$$\begin{aligned} L\frac{di}{dt} &= -Ri - K_m \lambda_f \omega + v \\ \frac{d\lambda_f}{dt} &= -R_F f^{-1}(\lambda_f) + v \\ J\frac{d\omega}{dt} &= K_m \lambda_f i - \tau_L \end{aligned}$$

where $R_F = R_f + R_{adj}$. *As in the case of the series DC motor, the armature inductance is negligible so that setting* $L = 0$ *and* $i = (v - K_m \lambda_f \omega)/R$

results in the reduced-order system

$$\begin{aligned}\frac{d\lambda_f}{dt} &= -R_F f^{-1}(\lambda_f) + v\\ \frac{d\omega}{dt} &= -\frac{K_m^2}{JR}\lambda_f^2\omega + \frac{K_m}{JR}\lambda_f v - \frac{\tau_L}{J}.\end{aligned}$$

(a) Show that the nonlinear transformation [where $g(\lambda_f) \triangleq f^{-1}(\lambda_f) = i_f$*]*

$$\begin{aligned}z_1 &= \omega - \frac{K_m}{2JR}\lambda_f^2\\ z_2 &= -\frac{K_m^2}{JR}\lambda_f^2\omega + \frac{R_F K_m}{JR}\lambda_f g(\lambda_f).\end{aligned}$$

results in the system dynamics

$$\begin{aligned}\frac{dz_1}{dt} &= z_2\\ \frac{dz_2}{dt} &= \alpha(\omega,\lambda_f) + \beta(\omega,\lambda_f)v - \frac{\tau_L}{J}.\end{aligned}$$

Show that application of the feedback

$$v = \frac{-\alpha(\omega,\lambda_f) + u}{\beta(\omega,\lambda_f)}$$

results in a linear system. This is called a feedback linearizing controller.

(b) Compute the expressions for $\alpha(\omega,\lambda_f), \beta(\omega,\lambda_f)$ *for the special case of linear magnetics, that is,* $\lambda_f \triangleq L_f i_f$*. In particular, show that*

$$\beta(\omega,\lambda_f) \triangleq \frac{K_m\lambda_f}{JR}\left(\frac{2R_F}{L_f} - 2K_m\omega - \frac{K_m^2}{JR}\lambda_f^2\right).$$

It turns out that the condition $\beta(\omega,\lambda_f) \neq 0$ *is satisfied for a region of the state space for which one would operate such a motor [55].*

(c) Show how to pick the gains of the controller

$$u_f = -K_0\int_0^t (z_1 - z_{1ref})dt - K_1(z_1 - z_{1ref}) - K_2(z_2 - z_{2ref}) + \frac{dz_{2ref}}{dt}$$

such that $z_1 - z_{1ref} \to 0$ *with a constant load torque.*

(d) What are some practical problems with this approach to controlling the machine? (Hint: How does one choose z_{1ref}, z_{2ref} *so that* $\omega \to \omega_{ref}$*?)*

Problem 47 Input-Output Linearization of a Shunt DC Motor [51][55]

An alternative to the feedback linearizing controller for the shunt DC motor (see problem 46) is an input-output linearization *controller.*

(a) Show that applying the feedback

$$v = \frac{(K_m^2/JR)\lambda_f^2\omega + u}{(K_m/JR)\lambda_f}$$

to the mathematical model in problem 46 results in the system

$$\begin{aligned}\frac{d\lambda_f}{dt} &= -R_F f^{-1}(\lambda_f) + K_m\lambda_f\omega + \frac{JR}{K_m\lambda_f}u \\ \frac{d\omega}{dt} &= u - \tau_L/J\end{aligned}$$

which is still a nonlinear system. The controller is nonsingular for $\lambda_f > 0$.

(b) The stability of the first equation, the so-called zero dynamics, *can be ascertained by multiplying it through by* λ_f *to obtain*

$$\begin{aligned}\frac{d\lambda_f^2}{dt} &= -R\lambda_f f^{-1}(\lambda_f) - K_m\lambda_f^2\omega + \frac{JR}{K_m}u \\ &= -\left(\frac{R}{L_f} - K_m\omega\right)\lambda_f^2 + \frac{JR}{K_m}u\end{aligned}$$

where the second line assumes linear magnetics. Show that λ_f *is bounded if* u *is bounded and* $\omega < \omega_{max} \triangleq R/(K_m L_f)$. *(Note that the adjustable resistor allows the designer to accommodate any specified maximum speed.)*

(c) Show how to pick the gains of the controller

$$u = -K_0\int_0^t(\omega - \omega_{ref})dt - K_1(\omega - \omega_{ref}) + \frac{d\omega_{ref}}{dt}$$

such that $\omega - \omega_{ref} \to 0$ *with a constant load torque.*

3

Magnetic Fields and Materials

The goal of this chapter is to present a concise, but relatively complete, introduction to the modeling of magnetic materials. The magnetic *intensity* field $\vec{\mathbf{H}}$ is introduced as way to modify Ampère's law $\oint \vec{\mathbf{B}} \cdot d\vec{\ell} = \mu_0 i_{\text{enclosed}}$ so that it is valid in the presence of magnetic materials. The difference between the magnetic *intensity* field $\vec{\mathbf{H}}$ and the magnetic *induction* field $\vec{\mathbf{B}}$ is explained. It is shown that $\vec{\mathbf{B}} = \mu_0 \vec{\mathbf{H}}$ in air while in ideal magnetic materials $\vec{\mathbf{H}} \equiv \mathbf{0}$ with $\vec{\mathbf{B}} \neq \mathbf{0}$. The distinction between "soft" and "hard" magnetic materials is explained.

3.1 Introduction

In this introductory section, an attempt is made to review the basic concepts of magnetic fields. Magnetic materials include iron (Fe), cobalt (Co), and nickel (Ni). All magnetic fields are produced by the motion of charged particles. The concern here will be with those magnetic fields setup by the flow (motion) of currents in wires and those due to the spin (motion) of electrons in the atoms making up magnetic materials. A common way one is introduced to magnetic fields is in the form of magnetic materials such as iron. That is, pieces of iron can be magnetized so that they attract or repel one another depending on their orientation with respect to each other. In Figure 3.1, the familiar bar magnet is shown.

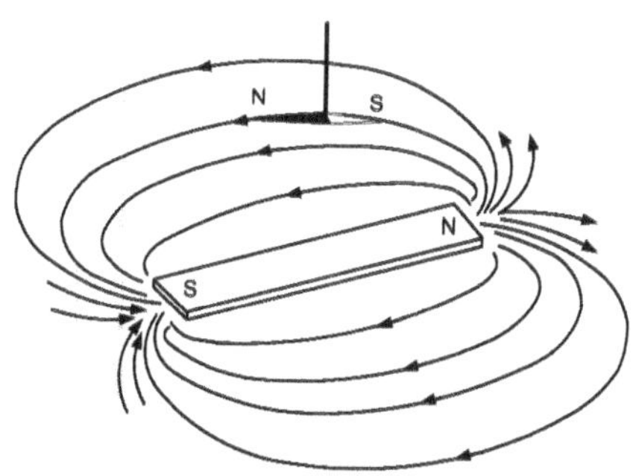

FIGURE 3.1. The direction of the compass needle specifies the direction of the magnetic field. From Figure 11-1 of *PSSC Physics*, 7th edition, by Haber-Schaim, Dodge, Gardner, and Shore, published by Kendall/Hunt, 1991.

The curves shown in the figure are drawn to show the *direction* of the magnetic field. As shown in the figure, the direction of the magnetic field is simply defined to be the direction that a small compass needle (which is just another bar magnet) points. That is, the large bar magnet exerts a force/torque on the compass needle such that in equilibrium it is aligned in a particular direction which is taken by definition to be the direction of the magnetic field at that point.

Figures 3.2 through 3.7 show some simple experiments to demonstrate the direction of the magnetic fields produced by currents and magnetic materials. In these experiments, small iron filings (i.e., compass needles!) are uniformly spread over a piece of smooth paper so that when a magnetic field is applied, the iron filings will align themselves with the direction of the magnetic field. Of course, the field must be strong enough to make the filings turn against the friction force of the paper. In Figure 3.2, the magnetic field of a long straight wire carrying a current is seen to consist of concentric circles around the wire.

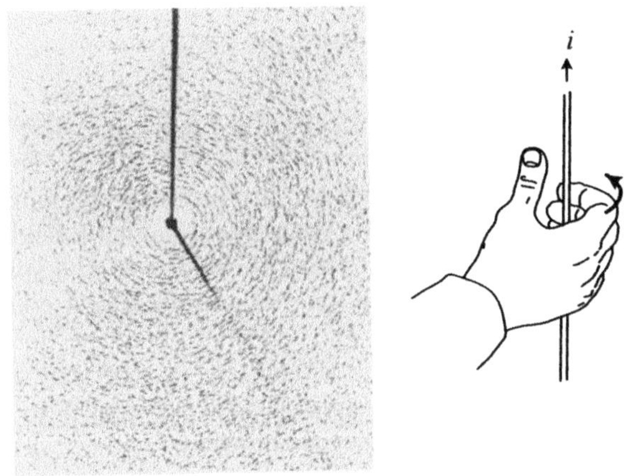

FIGURE 3.2. Magnetic field lines of a long straight wire carrying a current. Near the wire, concentric circles of iron filings are visible. At right is an illustration of the right-hand rule for the direction of the magnetic field lines around a wire carrying a current. From Figures 11-2(b) and 11-8, respectively, of *PSSC Physics*, 7th edition, by Haber-Schaim, Dodge, Gardner, and Shore, published by Kendall/Hunt, 1991.

Close to the wire where the magnetic field is strong, the concentric circles following the magnetic field are visible. However, as the distance from the wire increases, the magnetic field weakens to the point that it cannot align the iron filings against the friction forces of the paper.

Figure 3.2 also illustrates the right-hand rule for the direction of the magnetic field due to current in a wire. Simply, one puts their right thumb along the wire in the direction of the current and the fingers curl around the wire in the direction of the magnetic field.

Figure 3.3 shows another view of Figure 3.2 along with a simple drawing to illustrate the magnetic field lines.

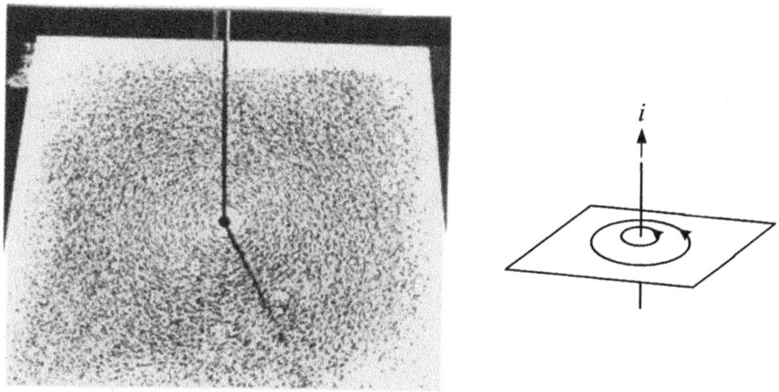

FIGURE 3.3. Iron filing experimental view and drawing of the magnetic field lines produced by a long straight wire carrying a current. From Figures 11-5 and 11-6(a) of *PSSC Physics*, 7th edition, by Haber-Schaim, Dodge, Gardner, and Shore, published by Kendall/Hunt, 1991.

In Figure 3.4, the magnetic field due to a loop carrying a current is shown along with a drawing illustrating the magnetic field lines.

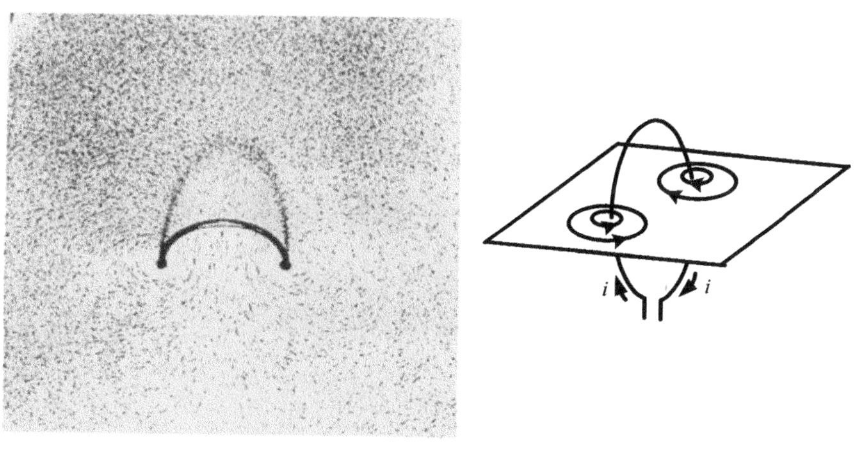

FIGURE 3.4. Iron filing experimental view and drawing of the magnetic field lines produced by a loop of wire carrying a current. From Figure 11-2(c) of *PSSC Physics*, 7th edition, by Haber-Schaim, Dodge, Gardner, and Shore, published by Kendall/Hunt, 1991.

Figure 3.5 depicts the magnetic field produced by a coil of wire carrying a current. Such a coil is referred to as a *solenoid*. Notice that inside the solenoid, the field follows straight lines and then flares out at the ends. Figure 3.6 is another experimental example of a tightly wound short solenoid.

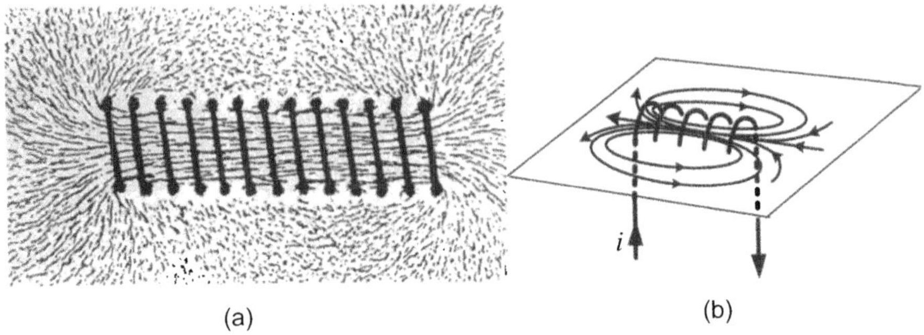

FIGURE 3.5. (a) Experimental view of the magnetic field lines of a solenoid using iron filings. Reprinted from Figure 43 of *Textbook of Physics*. R. Kronig. Editor, Pergamon Press. 1959 with permission from Elsevier. (b) Drawing of the magnetic field lines produced by a coil of wire carrying a current. From Figure 11-6(c) of *PSSC Physics*, 7th edition, by Haber-Schaim, Dodge, Gardner. and Shore, published by Kendall/Hunt, 1991.

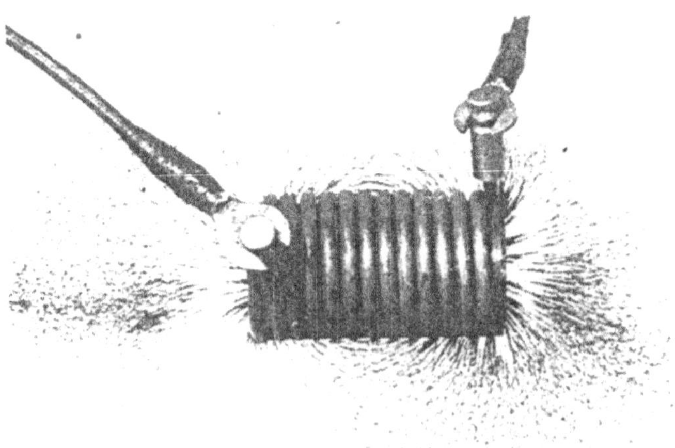

FIGURE 3.6. Experimental view of the magnetic field lines of a tightly wound solenoid using iron filings. From *An Introduction to Electrical Machines and Transformers*, 2nd edition, G. McPherson and R. D. Laramore, 1990. Reprinted by permisson of John Wiley & Sons.

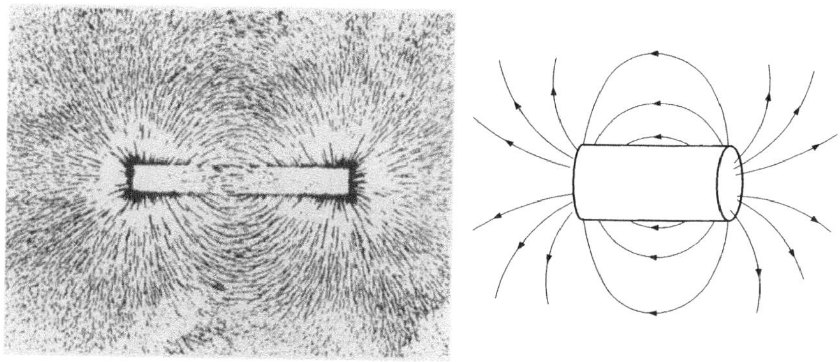

(a) (b)

FIGURE 3.7. Experimental view of the magnetic field lines of a cylindrical bar magnet using iron filings. From Figure 11-2(b) of *PSSC Physics*, 7th edition, by Haber-Schaim, Dodge, Gardner, and Shore, published by Kendall/Hunt, 1991. (b) Drawing of the magnetic field produced by a cylindrical-shaped bar magnet.

Figure 3.7 shows the magnetic field of a cylindrical-shaped bar magnet along with a drawing illustrating the magnetic field lines. Note that the magnetic field lines are similar in shape to the air-filled solenoid coil of Figure 3.5. Figures 3.8(a) and 3.8(b) compare the magnetic field lines of an air filled coil with that of an iron core coil. The magnetic field lines of the coil with the iron core are drawn closer together to illustrate that its magnetic field is much stronger than that of the air filed coil due to the iron. In other words, for the same current in the coil, an iron core coil results in a much stronger magnetic field.

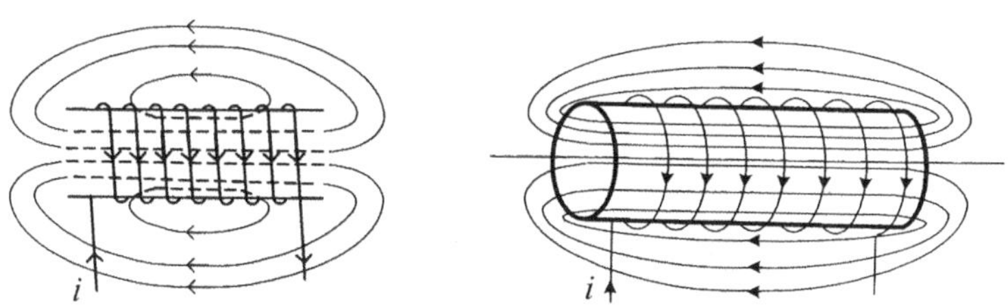

(a) (b)

FIGURE 3.8. (a) Coil with air core. (b) Coil with iron core. From *Electric Machines: Dynamics and Steady State* by G. J. Thaler and M. L. Wilcox, 1966. Reprinted by permission of John Wiley & Sons.

Figure 3.9 illustrates the important property of iron that it can be used to make the magnetic field lines follow a specific path. In this case, the current in the coil magnetizes the iron making a strong magnetic field inside the iron that follows the path of the iron to the air gap. If the air gap is small, the strength of the magnetic field there is approximately the same as in the iron.

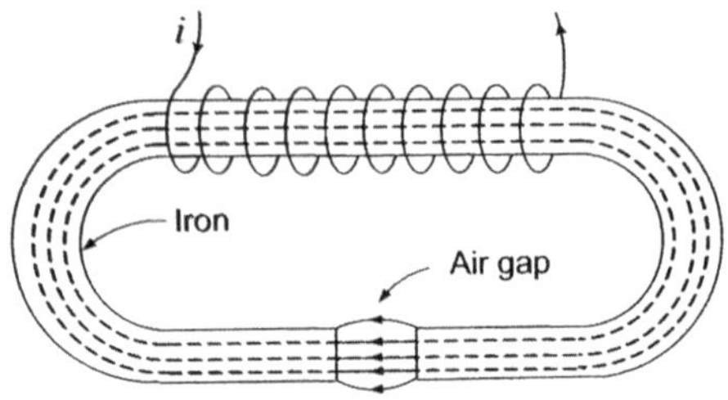

FIGURE 3.9. Iron core with air gap. From *Electric Machines: Dynamics and Steady State* by G. J. Thaler and M. L. Wilcox, 1966. Reprinted by permission of John Wiley & Sons.

An experimental verification of the setup in Figure 3.9 is the photograph in Figure 3.10. Note that the magnetic field does not actually stay confined to the iron core and air gap. However, the large number of iron filings oriented by the magnetic field in and around the air gap compared to those oriented outside the iron core indicates that the magnetic field is much stronger in the core and air gap.

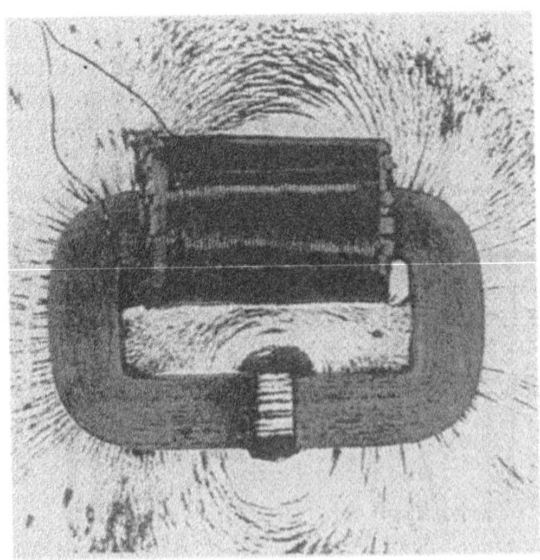

FIGURE 3.10. Experimental verification of the magnetic field produced by the setup of Figure 3.9. From *An Introduction to Electrical Machines and Transformers*, 2nd edition, G. McPherson and R. D. Laramore, 1990. Reprinted by permisson of John Wiley & Sons.

An experimental verification of the magnetic field lines of a horseshoe magnet is given in Figure 3.11(a).

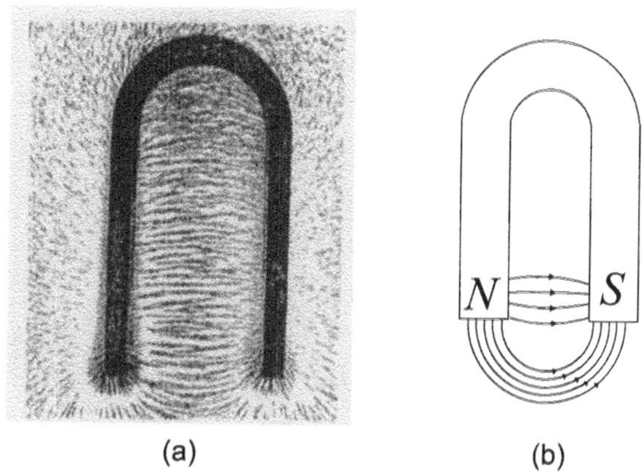

FIGURE 3.11. Magnetic field lines of a horseshoe magnet using iron filings. From *PSSC Physics*, 7th edition, by Haber-Schaim, Dodge, Gardner, and Shore, published by Kendall/Hunt, 1991. (b) Drawing illustrating the magnetic field lines of a horseshoe magnet.

Remark Normally a solenoidal coil is tightly wound so that there is no space between successive turns as shown in Figure 3.10. It was loosely wound in Figure 3.5 so that the orientation of the iron filings inside the coil could be shown. In Figure 3.5, the field outside the coil is too weak to align the iron filings against the friction of the paper, but the direction in this region is indicated in the drawing.

3.2 The Magnetic Field $\vec{\mathbf{B}}$ and Gauss's Law

Before going into the details of modeling magnetic material, the definition of the magnetic field is reviewed along with the concept of conservation of flux.

Recall that the direction of the magnetic field is defined to be the direction that a small compass needle points. The magnitude of the magnetic field is defined in terms of the forces it produces. Magnetism was discovered because naturally occurring magnets (e.g., the earth) exerted forces on each other. In 1819, the Danish physicist Hans Ørsted discovered that magnets can also exert forces on currents in wires and that currents in wires produces magnetic fields. This is illustrated in Figure 3.12, where a wire carrying a current is placed in a magnetic field. The conducting wire is oriented so that it is *perpendicular* to the magnetic field. The experimentally determined rule for determining the direction of the force is shown in the figure as another right-hand rule. Specifically, extend the fingers of the right hand in the direction of the magnetic field and the thumb in the direction of the current. Then the force is in the direction up out of the palm. Another experimental fact is that if the wire was placed *parallel* to the magnetic field, one finds that there is no force. That is, only the component of the magnetic field perpendicular to the current produces a force.

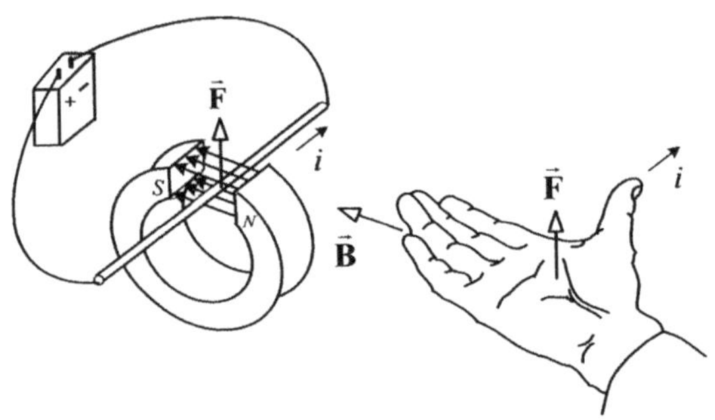

FIGURE 3.12. Direction of the magnetic force on currents. From Figure 11-3 of *PSSC Physics*, 7th edition, by Haber-Schaim, Dodge, Gardner, and Shore, published by Kendall/Hunt, 1991.

It is exactly this force phenomena that is used to define the magnetic field strength. In more detail, consider the experimental setup shown in Figure 3.13 which consists of an air-core solenoid coil carrying a current that produces a nearly constant magnetic field $\vec{\mathbf{B}}$ inside the coil in the direction shown. A wire is strung along a rectangular shaped pivot of width ℓ carrying a current i as shown in the figure. Only the part of the wire of length ℓ at the back end of the rectangular shaped pivot experiences a force since the other two sides are parallel to the field and thus experience no force. One then measures the force $\vec{\mathbf{F}}$. Experimentally, one finds that the force is proportional to the current i in the wire, i.e., if the current is doubled, then the force is doubled. Also, the force is proportional to the length of wire perpendicular to the field. That is, if the width was decreased to $\ell/2$, then one would find that the force is decreased by 1/2 also. Consequently, one needs to divide out the length of the wire and current in the wire to find the intrinsic quantity producing the force. This leads to the following definition of the magnetic field.

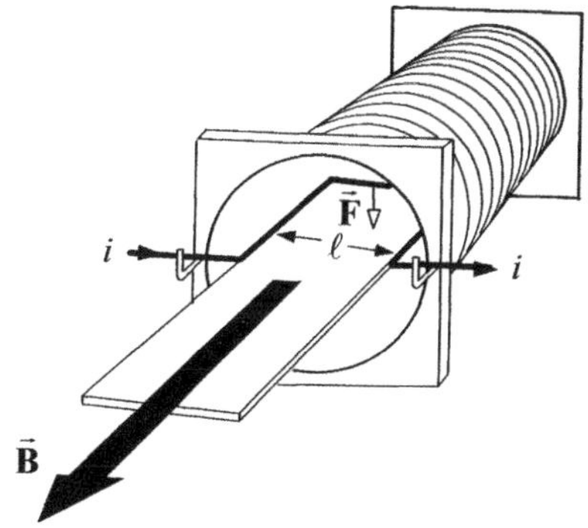

FIGURE 3.13. Apparatus to measure the magnetic field by the force it produces. From Figure 11-12 of *PSSC Physics*, 7th edition, by Haber-Schaim, Dodge, Gardner, and Shore, published by Kendall/Hunt, 1991.

Definition of the Magnetic Induction Field $\vec{\mathbf{B}}$

1. The *direction* of the magnetic field $\vec{\mathbf{B}}$ is defined by the direction a small compass needle points.

2. The *magnitude* $B \triangleq \left|\vec{\mathbf{B}}\right|$ of the magnetic field $\vec{\mathbf{B}}$ is defined by

$$B \triangleq \frac{F}{\ell i}$$

   where F is the force on the conductor of length ℓ carrying the current i. Here, the length of wire ℓ carrying the current i is assumed to be perpendicular to the direction of $\vec{\mathbf{B}}$.

3. The units of the magnetic field are

$$\text{tesla} \triangleq \frac{\text{N}}{\text{m-A}} = \frac{\text{weber}}{\text{m}^2} = 10^4 \text{ gauss.}$$

Magnetic Force Law

As stated above, magnetic fields were discovered because of the forces they produce and this phenomena is used to define the magnetic field. The experiments with magnetic forces and the definition of the magnetic field can be turned around to obtain the magnetic force law. To repeat, the force on the current carrying wire is proportional to the *component* of $\vec{\mathbf{B}}$ perpendicular to the wire. One defines a vector $\vec{\ell}$ whose magnitude is the length ℓ of the wire in the magnetic field and its direction is that of the positive current flow in the wire. Then, the magnetic force on a wire of length ℓ carrying a current i in a magnetic field $\vec{\mathbf{B}}$ is given by

$$\vec{\mathbf{F}} = i\vec{\ell} \times \vec{\mathbf{B}}.$$

Let θ denote the angle between $\vec{\ell}$ and $\vec{\mathbf{B}}$ as shown in Figure 3.14. The magnitude of the force $F = |\vec{\mathbf{F}}|$ may be written as

$$F = i\ell B\sin(\theta) = i\ell B_{\perp}$$

where $B_{\perp} = B\sin(\theta)$ is just the component of $\vec{\mathbf{B}}$ perpendicular to the wire.

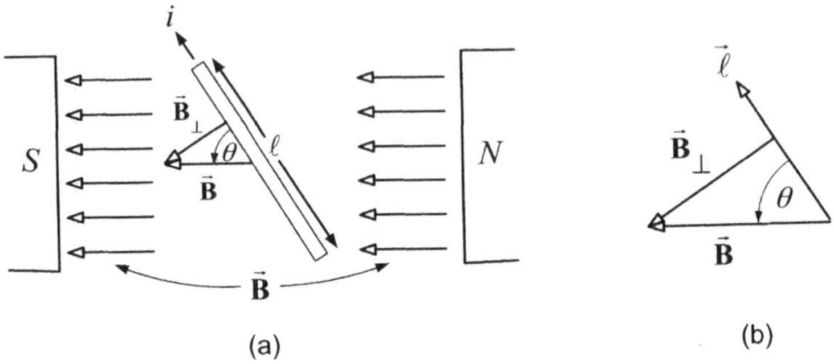

FIGURE 3.14. $F = i\ell B\sin(\theta) = i\ell B_{\perp}$.

3.2.1 *Conservation of Flux*

One of Maxwell's equations is that integral of the flux over a closed surface is zero, that is,

$$\oint_S \vec{\mathbf{B}} \cdot d\vec{\mathbf{S}} \equiv 0.$$

This fundamental property of magnetic fields is referred to as Gauss's law for magnetic fields or as the *conservation of flux*. To explain the notion of conservation of flux, consider the *closed* surface S shown in Figure 3.15 consisting of two disk-shaped ends connected by the tube-shaped surface.

The two disk-shaped surfaces making up the closed surface S are denoted by S_1 and S_2, respectively, while the tube-shaped surface is denoted by S_3. This particular surface S was purposely chosen so that on the surface S_3 the magnetic field $\vec{\mathbf{B}}$ points along the surface S_3 resulting in $\vec{\mathbf{B}} \cdot d\vec{\mathbf{S}} \equiv 0$ there. At all points of the surface S, let $d\vec{\mathbf{S}}$ be the outward normal to the surface. Then, by Gauss's law, it follows that

$$\oint_S \vec{\mathbf{B}} \cdot d\vec{\mathbf{S}} = \int_{S_1} \vec{\mathbf{B}} \cdot d\vec{\mathbf{S}} + \int_{S_2} \vec{\mathbf{B}} \cdot d\vec{\mathbf{S}} + \int_{S_3} \vec{\mathbf{B}} \cdot d\vec{\mathbf{S}} \equiv 0$$

or

$$\int_{S_1} \vec{\mathbf{B}} \cdot d\vec{\mathbf{S}} + \int_{S_2} \vec{\mathbf{B}} \cdot d\vec{\mathbf{S}} \equiv 0$$

as $\vec{\mathbf{B}} \cdot d\vec{\mathbf{S}} \equiv 0$ on S_3.

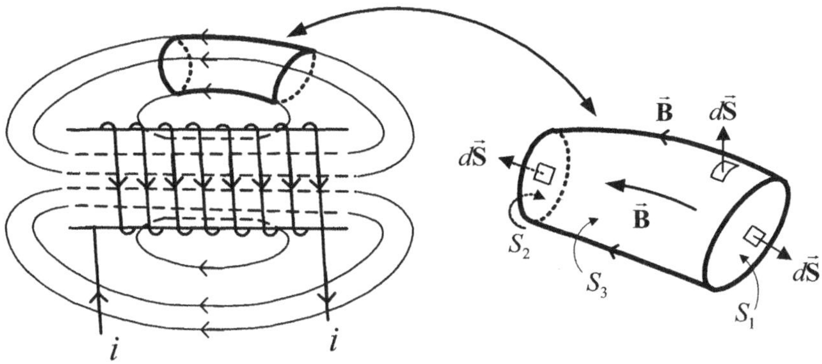

FIGURE 3.15. Conservation of flux in a flux tube.

Then

$$\phi_{S_2} = \int_{S_2} \vec{\mathbf{B}} \cdot d\vec{\mathbf{S}} = -\int_{S_1} \vec{\mathbf{B}} \cdot d\vec{\mathbf{S}} = \int_{S_1} \vec{\mathbf{B}} \cdot d\vec{\mathbf{S}}' = \phi_{S_1} \tag{3.1}$$

where

$$d\vec{\mathbf{S}}' \triangleq -d\vec{\mathbf{S}}$$

is the *inward* normal to the closed surface S on S_1. Equation (3.1) shows that the flux ϕ_{S_1} through surface S_1 *equals* the flux ϕ_{S_2} through the surface S_2 if their surface normals are chosen as above. The volume enclosed by S_1, S_2, S_3 is called a *flux tube* and is characterized by $\vec{\mathbf{B}} \cdot d\vec{\mathbf{S}} \equiv 0$ on the tube shaped part of the surface S_3. This results in the flux through S_1 equaling that through S_2 and is simply a consequence of conservation of flux applied to this closed surface.

In the above derivation, it was crucial that the flux through S_3 was zero. Magnetic materials provide the capability of being able to approximately construct such a situation, that is, of being able to (approximately) direct the magnetic field. This is illustrated in Figure 3.16, where a coil has been wrapped around an iron core containing an air gap. The point here is that the axially directed magnetic field inside the coil produced by the current and magnetized iron follows along the path of the iron to the air gap. That is, to a reasonable approximation, the magnetic material keeps the magnetic field inside the magnetic material following the path of the material. As a result, the tube-shaped surface of the magnetic material in Figure 3.16 forms a surface for which for which $\vec{\mathbf{B}} \cdot d\vec{\mathbf{S}} = \mathbf{0}$[1] so that the magnetic flux through any cross section of the iron tube is approximately

[1]It turns out that there will always be some $\vec{\mathbf{B}}$ field leaving the tube-shaped outer surface of the iron (leakage). However, by magnetizing the material as shown in the figure, the main $\vec{\mathbf{B}}$ field strength goes around to the air gap and the part that leaves the tube-shaped outer surface of the iron is much weaker.

the same throughout the tube and in particular at the air gap. As the figure shows, the magnetic field spreads out across the air gap, but this spreading can be kept small by keeping the air gap small. In summary, magnetic materials give the capability to create a magnetic field/flux at one location and transport it to another location where it can be used, for example, to produce a force on a wire carrying a current. However, it is important to note that all of this is only approximately true in that the magnetic field does not stay completely within the material so that the field/flux is actually weaker at the air gap than the field/flux in the coil.

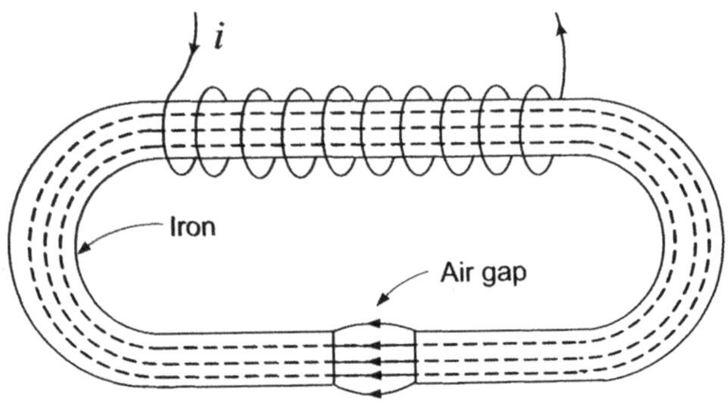

FIGURE 3.16. Creating and directing a magnetic field. From *Electric Machines: Dynamics and Steady State* by G. J. Thaler and M. L. Wilcox, 1966. Reprinted by permission of John Wiley & Sons.

Continuity of the Normal Component of $\vec{\mathbf{B}}$

As another application of Gauss's flux law $\oint_S \vec{\mathbf{B}} \cdot d\vec{\mathbf{S}} \equiv 0$, it is now shown that at the interface (boundary) between air and magnetic material, the normal (perpendicular) component of $\vec{\mathbf{B}}$ to the surface is *continuous* across the boundary. Figure 3.17 shows a piece of magnetic material and a closed cylindrical-shaped surface, of which half is inside the magnetic material and the other half in the air. Let $\vec{\mathbf{B}}_1 = B_{1x}\hat{\mathbf{x}} + B_{1y}\hat{\mathbf{y}} + B_{1z}\hat{\mathbf{z}}$ denote the magnetic field on the surface S_1 in the air and $\vec{\mathbf{B}}_2 = B_{2x}\hat{\mathbf{x}} + B_{2y}\hat{\mathbf{y}} + B_{2z}\hat{\mathbf{z}}$ denote the magnetic field on the surface S_2 just inside the magnetic material. Note that coordinate system has been chosen so that the normal component to the surface is the z component of the field. The conservation of flux $\oint_S \vec{\mathbf{B}} \cdot d\vec{\mathbf{S}} \equiv 0$ is then applied to the closed cylindrical surface S shown in Figure 3.17 consisting of the four pieces S_1 (disk-shaped in air with area πr^2), S_2 (disk-shaped in the magnetic material with area πr^2), S_3 (cylindrical-shaped in air with area $2\pi r(h/2)$), and S_4 (cylindrical-shaped in the magnetic material with area $2\pi r(h/2)$).

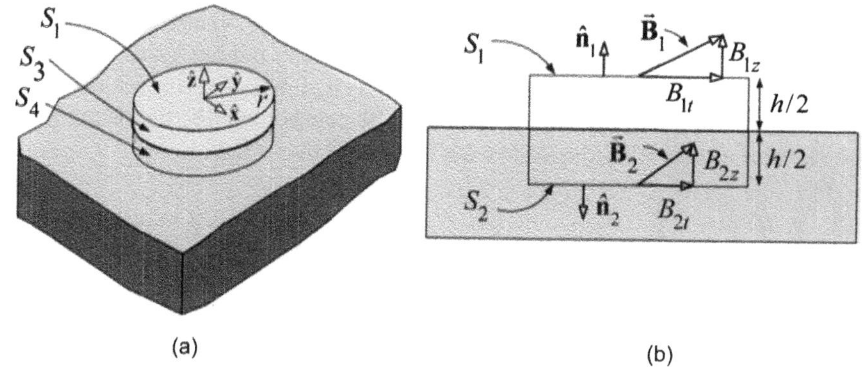

FIGURE 3.17. (a) Closed cylindrical surface at the interface of air and magnetic material (b) Cross-sectional view of the flux surface. As $h \to 0$, $\vec{\mathbf{B}}_1$ is just above the surface and $\vec{\mathbf{B}}_2$ is just below the surface.

It follows that

$$\oint_S \vec{\mathbf{B}} \cdot d\vec{\mathbf{S}} = \int_{S_1} \vec{\mathbf{B}} \cdot d\vec{\mathbf{S}} + \int_{S_2} \vec{\mathbf{B}} \cdot d\vec{\mathbf{S}} + \int_{S_3} \vec{\mathbf{B}} \cdot d\vec{\mathbf{S}} + \int_{S_4} \vec{\mathbf{B}} \cdot d\vec{\mathbf{S}} \equiv 0.$$

With r finite, but small, let the length h of the cylinder go to zero with the consequence that the area of sides S_3 and S_4 go to zero, and the disk-shaped sides S_1 and S_2 are just above and just below the air-magnetic material interface, respectively. Letting $h \to 0$, it follows that

$$\int_{S_1} \vec{\mathbf{B}} \cdot d\vec{\mathbf{S}} + \int_{S_2} \vec{\mathbf{B}} \cdot d\vec{\mathbf{S}} \equiv 0.$$

Evaluating these integrals results in

$$\begin{aligned} \int_{S_1} \vec{\mathbf{B}}_1 \cdot d\vec{\mathbf{S}} &= \int_{S_1} (B_{1z}\hat{\mathbf{z}}) \cdot (dS\hat{\mathbf{z}}) = B_{1z}\pi r^2 \\ \int_{S_2} \vec{\mathbf{B}}_2 \cdot d\vec{\mathbf{S}} &= \int_{S_2} (B_{2z}\hat{\mathbf{z}}) \cdot (-dS\hat{\mathbf{z}}) = -B_{2z}\pi r^2 \end{aligned}$$

so that

$$B_{1z} = B_{2z}.$$

That is, the component of $\vec{\mathbf{B}}$ perpendicular to the surface is *continuous* across the surface.

As an example, consider Figure 3.18 [from Figure 3.19(a)] showing the magnetic fields due to a cylindrical-shaped permanent magnet. For reference, use a cylindrical coordinate system whose origin is at the center of the magnet with the z axis pointing up along the axis of the magnet. Note that on the cylindrical side of the magnet, the tangential component (z

component) of $\vec{\mathbf{B}}$ is *not* continuous. In particular, consider the point P in Figure 3.18 where just inside the surface of the magnet, the tangential component of $\vec{\mathbf{B}}$ is up ($+\hat{\mathbf{z}}$ direction) while just outside the surface, it is down ($-\hat{\mathbf{z}}$ direction). However, the normal component ($\hat{\mathbf{r}}$ direction) must be the same on both sides of the surface by Gauss's law.

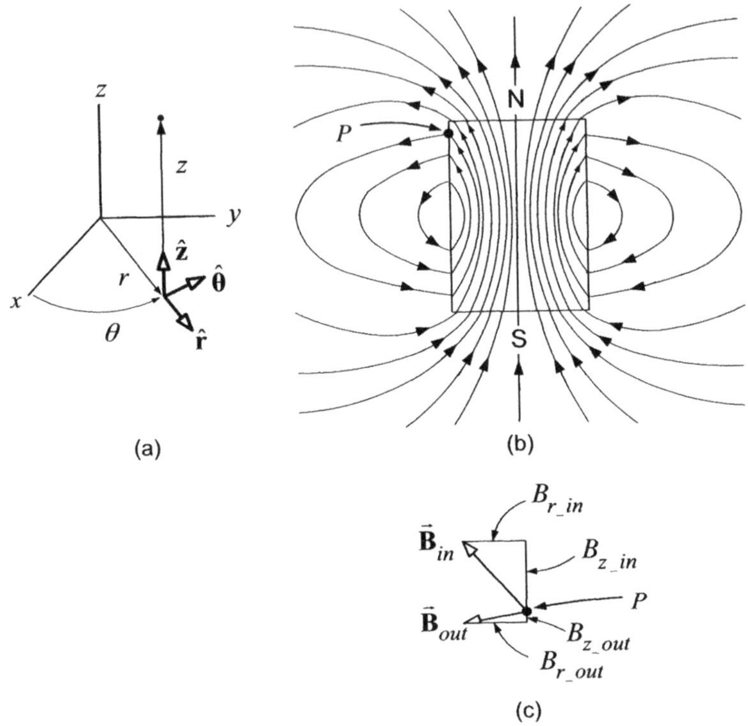

FIGURE 3.18. (a) Cylindrical coordinate system for the magnet. (b) Magnetic field lines for a cylindrical-shaped magnet. At the point P, the normal component ($\hat{\mathbf{r}}$ direction) is continuous while the tangential component ($\hat{\mathbf{z}}$ direction) is not. Adapted from Figure 37-4 of Ref. [34]. (c) $\vec{\mathbf{B}}_{in}$ and $\vec{\mathbf{B}}_{out}$ are, respectively, the magnetic field just inside and just outside the surface at the point P. Note that $B_{r\_in} = B_{r\_out}$, but $B_{z\_in} \neq B_{z\_out}$.

3.3 Modeling Magnetic Materials

Figures 3.19(a) and 3.19(b) compare the magnetic field lines of a cylindrical shaped bar magnet and a cylindrical-shaped air-filled solenoid coil, respectively. The interesting and important observation here is that the magnetic field outside the bar magnet has the *same* shape as that of the coil of wire carrying a current. This leads one to conjecture that the magnetic field inside the bar magnet of Figure 3.19(a) follows straight lines as they do inside the coil of Figure 3.19(b). Perhaps one can picture the bar magnet as having some kind of "bound" current which goes around its periphery producing the bar magnet's field analogous to the current in the coil of the air filled solenoid.

Pursuing this conjecture, consider an iron atom in the bar magnet as a small current loop as in Figure 3.20. That is, the motion of the electrons in the individual iron atoms are considered to be equivalent to a current loop as drawn in Figure 3.20(a). As illustrated in Figures 3.20(a) and 3.20(b), inside the material, the current in the adjacent sides of neighboring loops go in opposite directions so that their effects cancel. However, the part of the loops on the surface of the material have no corresponding neighboring loop to cancel and the net effect is a current around the periphery of the magnetic material as illustrated in Figure 3.20(c). This is a so-called *bound* current, since an individual electron does not go around the surface of the magnet, but stays with the particular iron atom. It is only the net effect of all the current loops being aligned that results in a current around the surface of the bar magnet. This is in contrast to *free* currents associated with current flow in copper conductors where individual electrons actually move down the wire through the crystal lattice of the copper.

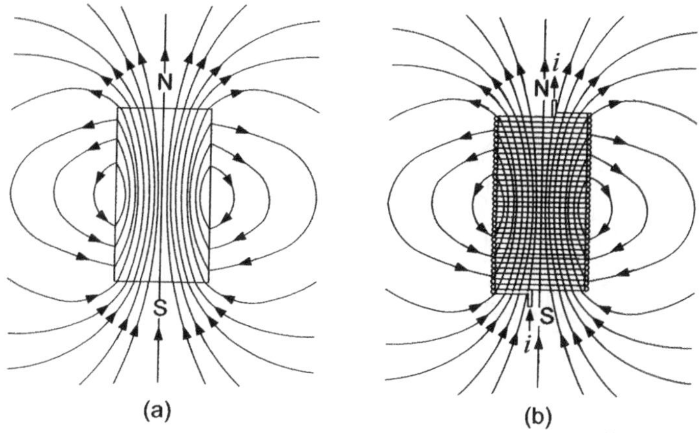

FIGURE 3.19. (a) Magnetic field lines for a bar magnet. (b) Magnetic field lines for a short air-filled solenoid coil. From *Physics*, 2nd edition, by D. Halliday and R. Resnick, 1962. Reprinted by permission of John Wiley & Sons.

To reiterate, the point here is that in *magnetic* materials, the motion (spin) of the electrons in the atoms are able to align themselves as shown in Figure 3.20 so that the current loops are all in the same plane with the currents circulating clockwise. This is in contrast to *nonmagnetic* materials where the current loops are at *random* orientations with respect to each other and are unable to align themselves so their net effect cancels.

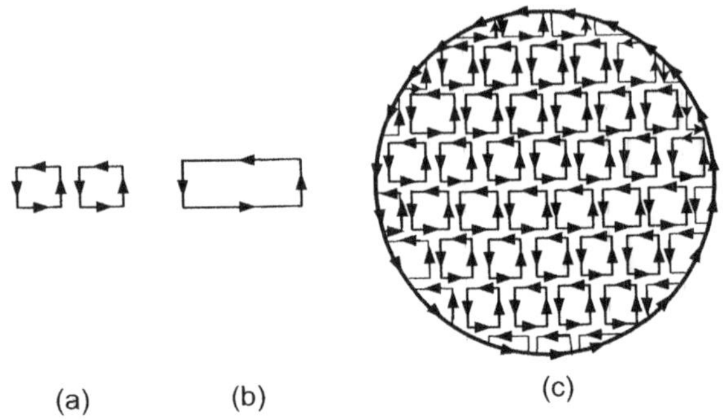

FIGURE 3.20. Viewing the atoms of magnetic materials as current loops which are able to align themselves. From *PSSC Physcis*, 7th edition, by Haber-Schaim, Dodge, Gardner, and Shore, published by Kendall/Hunt, 1991.

Figure 3.21 again illustrates this idea of representing the magnetic material by current loops which results in an equivalent bound current i_b going around the periphery of the cylinder.

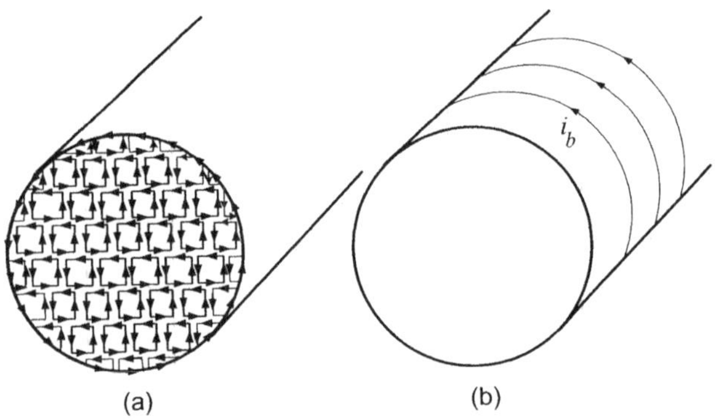

FIGURE 3.21. (a) View of magnetic material consisting of aligned dipole moments. (b) Magnetic material represented by an equivalent bound current i_b going around the periphery of the cylinder that produces the same magnetic field as the magnetic material. This interpretation is motivated by Figure 3.20.

3.3.1 Magnetic Dipole Moments

To develop a mathematical model to go along with the pictures in Figures 3.20 and 3.21, the concept of a magnetic dipole moment is reviewed. Consider a small planar loop of area S with a current i going around the loop such as illustrated in Figure 3.22. The scalar magnetic moment is defined to be

$$m \triangleq iS.$$

The direction of the magnetic moment is taken to be normal to the loop. A *unique* normal $\hat{\mathbf{n}}$ is specified by direction of the current using the right-hand rule. That is, curl the fingers of the right-hand around the loop in the direction of positive current flow and the normal $\hat{\mathbf{n}}$ is defined to be the direction of the extended thumb. Putting this all together, the *magnetic moment* vector is defined as

$$\vec{\mathbf{m}} \triangleq iS\hat{\mathbf{n}}$$

where $\hat{\mathbf{n}}$ is a unit normal to the loop determined by the right-hand rule. Magnetic material (such as iron) is now pictured by modeling the atoms as magnetic moments which are able to *align* themselves in the same direction. As shown in Figure 3.21, the net effect is then a (bound) surface current that produces the external magnetic field analogous to a current-carrying coil with air inside.

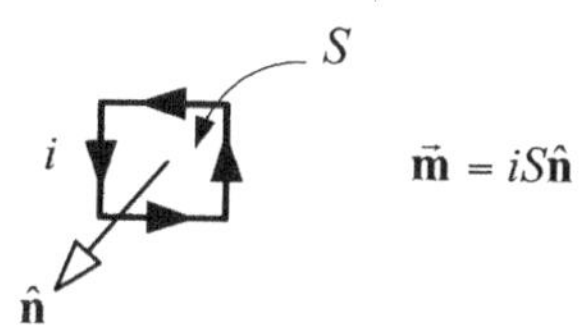

FIGURE 3.22. Small current loop as a magnetic dipole momentt. S is the area of the loop, i is the current in the loop and $\hat{\mathbf{n}}$ is the normal to the loop given by the right-hand rule. The dipole moment is $\vec{\mathbf{m}} \triangleq iS\hat{\mathbf{n}}$.

The model of the iron atom as a magnetic dipole moment will be used to construct a macroscopic picture of the magnetic properties of iron. However, the magnetic property of iron (ferromagnetism) is only explained through Quantum Mechanics. Quoting from Melvin Schwartz [35] "*Iron has two remarkable atomic properties that have far-reaching consequences with respect to its macroscopic magnetic behavior. First, 4 of the 26 electrons in an isolated atom of iron have their intrinsic angular momenta lined up. Second, within solid iron, there are very strong quantum-mechanical forces tending to make the intrinsic angular momenta of neighboring atoms line up. This results in domains of macroscopic size having net magnetizations corresponding to about two aligned electron moments per atom on the average. Applying a magnetic field causes those domains, which are aligned in the same direction as the field, to grow until the iron finally reaches a saturated state of magnetization. Typically, saturation occurs at fields in the neighborhood of 10,000 to 20,000 gauss. Needless to say, removal of the applied field does not lead to complete randomization of the domains. The residual magnetization can be quite large, as in the case of special permanent magnetic alloys, or it can be quite small, as in the case of soft iron. These phenomena are characteristic of ferromagnetism."*

3.3.2 The Magnetization $\vec{\mathbf{M}}$ *and Ampère's Law*

The picture adopted here for iron and other magnetic materials is one in which they consist of small current loops with a magnetic dipole moment associated with each one as illustrated in Figure 3.21. It is emphasized that one could model any material with such current loops, but that in the case of magnetic materials, the loops (dipole moments) align themselves when an external magnetic field is applied. This alignment of the current loops can be viewed as a net bound current which goes around the surface and produces a magnetic field.[2] Nonmagnetic materials have their dipole moments in random orientations with respect to each other and are not able to align to produce any significant magnetic field even under the presence of an external magnetic field. Consequently, there is no net bound surface current or magnetic field. When an external magnetic field is applied to a magnetic dipole moment (loop of current), the moment tends to align itself with the external field. This is illustrated in Figure 3.23, which shows two current loops whose normals $\hat{\mathbf{n}}$ are not aligned with the external magnetic field $\vec{\mathbf{B}}$. In Figure 3.23(a), the forces $\vec{\mathbf{F}}_1$,$\vec{\mathbf{F}}_2$ provide a torque $\vec{\tau}$ to align the dipole moment (i.e., the normal $\hat{\mathbf{n}}$ to the surface of the loop) with the direction of the external field $\vec{\mathbf{B}}$. Similarly, in 3.23(b) the forces $\vec{\mathbf{F}}_3$,$\vec{\mathbf{F}}_4$ provide a torque $\vec{\tau}$ to align the dipole moment with the direction of the external field $\vec{\mathbf{B}}$.

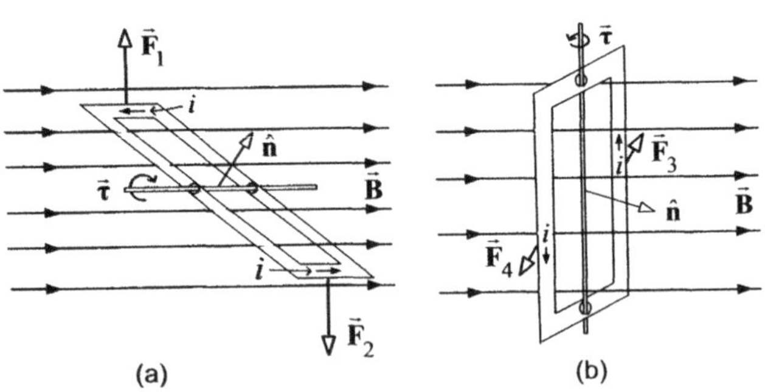

FIGURE 3.23. Magnetic dipole in an external magnetic field. (a) Dipole tilted vertically. (b) Dipole tilted laterally.

[2]This ability of materials such as iron to align their magnetic moments together to form domains which grow and align in the same direction as an externally applied magnetic field is succesfully explained only by quantum mechanics.

As a specific example of this viewpoint, consider the magnetic circuit of Figure 3.16. Here the current in the coil magnetizes the iron, that is, causes the dipole moments of the iron atoms to align themselves with the (external) magnetic field produced by the coil current. This results in a much larger magnetic field in the iron than that which would be produced by the current in the coil alone.

The Magnetization Vector $\vec{\mathbf{M}}$

Define the *magnetization* vector $\vec{\mathbf{M}}$ as follows: At any arbitrary point (x', y', z') inside the magnetic material (see Figure 3.24) define

$$\vec{\mathbf{m}}(x', y', z') \triangleq \text{Average magnetic dipole moment of the atoms in a neighborhood of the point } (x', y', z').$$

$$N_m \triangleq \text{Number of atoms (e.g., iron) per unit volume.}$$

Then, the magnetization $\vec{\mathbf{M}}(x', y', z')$ is defined as

$$\vec{\mathbf{M}}(x', y', z') \triangleq N_m \vec{\mathbf{m}}(x', y', z').$$

The magnitude $M = |\vec{\mathbf{M}}\,|$ of the magnetization vector is simply the magnetic dipole moment per unit volume. The fundamental issue here is to determine the magnetic field $\vec{\mathbf{B}}(x, y, z)$ at the point (x, y, z) outside the material due to the magnetization $\vec{\mathbf{M}}(x', y', z')$ at the point (x', y', z') inside the material. To do so, it is necessary to relate the magnetization vector $\vec{\mathbf{M}}(x', y', z')$ to the magnetic field it produces. First, Ampère's law is reviewed as it is the Maxwell equation that relates currents to magnetic fields.

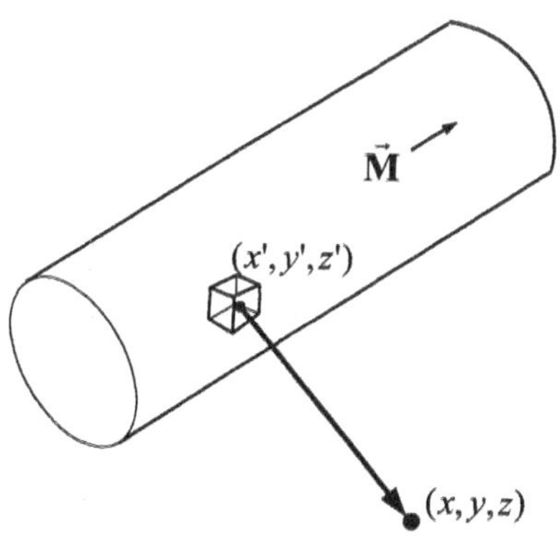

FIGURE 3.24. Magnetic field produced by magnetic material. How does one compute $\vec{\mathbf{B}}$ at (x, y, z) due to $\vec{\mathbf{M}}$ at (x', y', z')?

Ampère's Law

When magnetic materials are *not* present, Ampère's law is

$$\oint_C \vec{\mathbf{B}} \cdot d\vec{\ell} = \mu_0 i_{\text{enclosed}}$$

where i_{enclosed} is the current enclosed by the path C.

Some examples are now given to recall how Ampère's law is used to determine the magnetic field from currents.

Example *Long Straight Wire*

Consider a long straight wire carrying a current as in Figure 3.25(a).

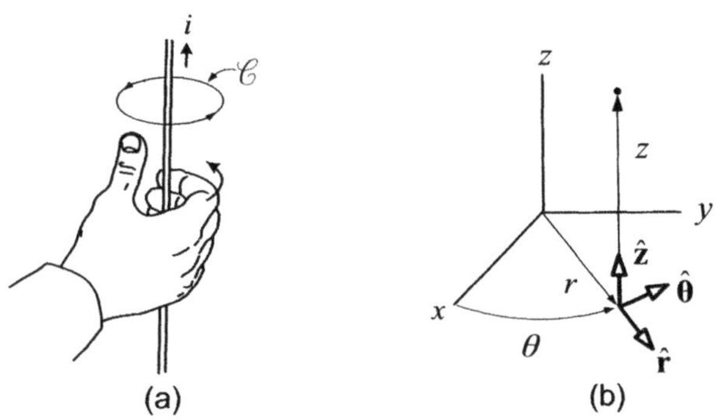

FIGURE 3.25. Sign convention in Ampère's law for positive current through a closed curve C. From *PSSC Physics*, 7th edition, by Haber-Schaim, Dodge, Gardner, and Shore, published by Kendall/Hunt, 1991.

This figure shows a closed curve C with a direction of *positive travel* for going around the curve indicated as well as a direction for *positive current* in the wire. For Ampère's law to apply, these two sign conventions must be related to each other by another right-hand rule. Specifically, as shown in Figure 3.25(a), if one takes their right hand and curls the fingers in the positive direction of travel around the curve, then the thumb points in the positive direction of current through the surface *enclosed* by this curve.

Ampère's law is now used to compute the magnetic field due to a current in a long straight wire. A cylindrical coordinate system whose z axis is along the wire is appropriate for the geometry of this problem [see Figure 3.25(b)]. To proceed, first note that due to symmetry, the $\vec{\mathbf{B}}$ field cannot depend on the axial position z nor the azimuthal position θ so that

$$\vec{\mathbf{B}} = B_r(r)\hat{\mathbf{r}} + B_\theta(r)\hat{\boldsymbol{\theta}} + B_z(r)\hat{\mathbf{z}}.$$

To find the azimuthal component $B_\theta(r)$, apply Ampère's law to the closed curve C in Figure 3.25(b) to obtain

$$\oint_C \vec{\mathbf{B}} \cdot d\vec{\ell} = \int_0^{2\pi} \left(B_\theta \hat{\boldsymbol{\theta}}\right) \cdot \left(r d\theta \hat{\boldsymbol{\theta}}\right) = \int_0^{2\pi} B_\theta r d\theta = 2\pi r B_\theta = \mu_0 i$$

or,

$$B_\theta(r) = \frac{\mu_0 i}{2\pi r}.$$

In problem 3, using Gauss's law for $\vec{\mathbf{B}}$, the reader is asked to show that $B_r(r) \equiv 0$.

It is now shown that $B_z(r) \equiv 0$. To do so, apply Ampère's law to the curve C shown in Figure 3.26, to obtain

$$\begin{aligned} \oint_C \vec{\mathbf{B}} \cdot d\vec{\ell} &= \mu_0 i_{\text{enclosed}} \\ \int_1^2 \vec{\mathbf{B}} \cdot d\vec{\ell} + \int_2^3 \vec{\mathbf{B}} \cdot d\vec{\ell} + \int_3^4 \vec{\mathbf{B}} \cdot d\vec{\ell} + \int_4^1 \vec{\mathbf{B}} \cdot d\vec{\ell} &= 0 \\ \int_1^2 (B_z(r_1)\hat{\mathbf{z}}) \cdot (dh\hat{\mathbf{z}}) + \int_3^4 (B_z(r_2)\hat{\mathbf{z}}) \cdot (-dh\hat{\mathbf{z}}) &= 0 \\ B_z(r_1)h - B_z(r_2)h &= 0 \\ B_z(r_1) &= B_z(r_2) \end{aligned}$$

where the fact that $\int_2^3 \vec{\mathbf{B}} \cdot d\vec{\ell} + \int_4^1 \vec{\mathbf{B}} \cdot d\vec{\ell} = 0$ was used.

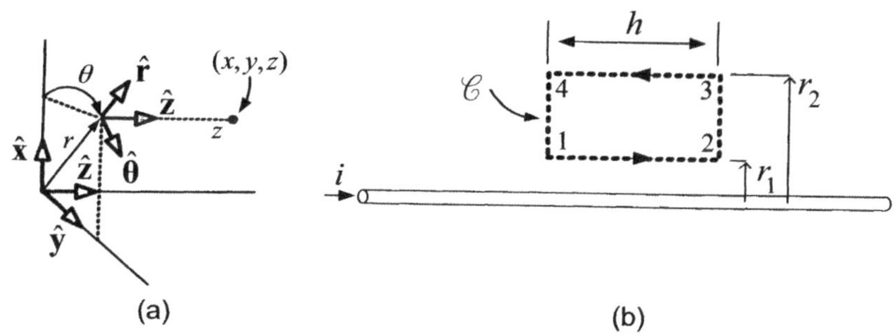

FIGURE 3.26. (a) Cylindrical coordinate system. (b) Long straight wire carrying a current.

That is, the z component of $\vec{\mathbf{B}}$ is constant. However, as one goes to infinity in the radial direction ($r \to \infty$), the magnetic field will be zero as one is infinitely far from the source i of the magnetic field. Consequently, B_z must be zero at infinity and, as it is constant, it must equal zero everywhere. Thus, the form of the magnetic field of an infinitely long straight wire reduces to

$$\vec{\mathbf{B}}(r, \theta, z) = \frac{\mu_0 i}{2\pi r}\hat{\boldsymbol{\theta}}.$$

Example *Ideal Infinitely Long Solenoid* [34]
Recall that a solenoid coil is a long wire which is wrapped in a close-packed helix-shaped coil where the length of the coil is very long compared to its diameter. An ideal solenoid is one in which the helix is infinitely long with no spacing between coils as shown in Figure 3.27.

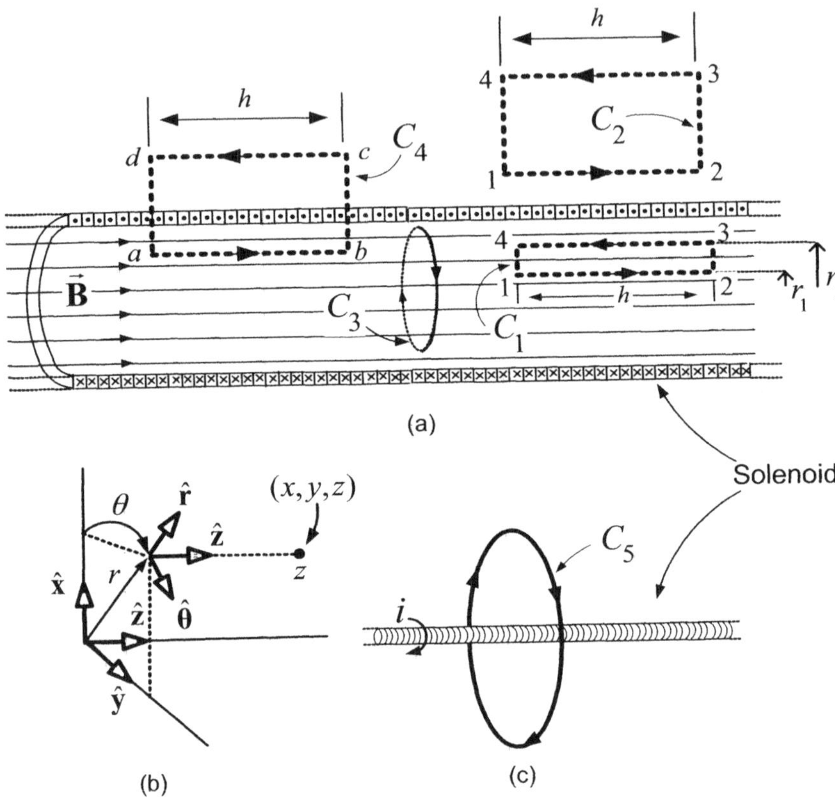

FIGURE 3.27. Infinitely long coil with an air core. (a) Curves C_1, C_2, C_3, and C_4 for applying Ampère's law. (b) Cylindrical coordinate system. (c) Curve C_5 for applying Ampère's law.

In this example, a current is put through the wire and the goal is to compute the magnetic field inside and outside the coil. In this case of an ideal solenoid coil, the magnetic field can be computed using symmetry arguments and Ampère's law. To do so, Figure 3.27 shows five closed curves C_1, C_2, C_3, C_4 and C_5 that are used to apply Ampère's law to compute $\vec{\mathbf{B}}$. Figure 3.27(b) shows a cylindrical coordinate system chosen for the solenoid.

Proceeding, first note that due to symmetry, the $\vec{\mathbf{B}}$ field cannot depend on the axial position z nor on the azimuthal angle θ so that

$$\vec{\mathbf{B}} = B_r(r)\hat{\mathbf{r}} + B_\theta(r)\hat{\boldsymbol{\theta}} + B_z(r)\hat{\mathbf{z}}.$$

The reader is asked in problem 4 to show using Gauss's law for $\vec{\mathbf{B}}$ that $B_r(r) \equiv 0$. To compute B_z inside the solenoid, Ampère's law is applied to the curve C_1 (side *1–2* is a radial distance r_1 from the center of the coil and side *3–4* is a radial distance r_2 from the center of the coil) and one obtains

$$\begin{aligned}
\oint_{C_1} \vec{\mathbf{B}} \cdot d\vec{\ell} &= \mu_0 i_{\text{enclosed}} \\
\int_1^2 \vec{\mathbf{B}} \cdot d\vec{\ell} + \int_2^3 \vec{\mathbf{B}} \cdot d\vec{\ell} + \int_3^4 \vec{\mathbf{B}} \cdot d\vec{\ell} + \int_4^1 \vec{\mathbf{B}} \cdot d\vec{\ell} &= \mu_0 i_{\text{enclosed}} \\
B_z(r_1)h - B_z(r_2)h &= 0 \\
B_z(r_1) &= B_z(r_2)
\end{aligned}$$

where the fact that $\int_2^3 \vec{\mathbf{B}} \cdot d\vec{\ell} + \int_4^1 \vec{\mathbf{B}} \cdot d\vec{\ell} = 0$ was used. Thus, $B_z = B_{z\text{inside}}$ is *constant* inside the solenoid. Similarly, applying Ampère's law to the curve C_2, one finds that $B_z = B_{z\text{outside}}$ is also *constant* outside the solenoid. As one goes to infinity in the radial direction ($r \to \infty$), the magnetic field will be zero as one is infinitely far from the source i of the magnetic field. Consequently, $B_{z\text{outside}}$ must be equal to zero.

Applying Ampère's law to the curves C_3 and C_5, the reader can show that $B_\theta(r) \equiv 0$ inside the solenoid and $B_\theta(r) = \mu_0 i/(2\pi r)$ outside the cylinder (see problem 5). Thus,

$$\vec{\mathbf{B}} = \begin{cases} B_{z\text{inside}}\hat{\mathbf{z}} & \text{inside the solenoid} \\ \dfrac{\mu_0 i}{2\pi r}\hat{\boldsymbol{\theta}} & \text{outside the solenoid.} \end{cases}$$

Finally, to determine $B_{z\text{inside}}$ in terms of the solenoid current i, consider the curve C_4. Let n be the number of coil turns per unit length and i be the current in the coil, so that in the length h of the coil, the current nhi is going through the curve *abcda* in Figure 3.27(a). Applying Ampère's law to the curve C_4, one obtains

$$\begin{aligned}
\oint_{C_4} \vec{\mathbf{B}} \cdot d\vec{\ell} &= \mu_0 i_{\text{enclosed}} \\
\int_a^b \vec{\mathbf{B}} \cdot d\vec{\ell} &= \mu_0 i_{\text{enclosed}} \\
B_{z\text{inside}} h &= \mu_0 n h i \\
B_{z\text{inside}} &= \mu_0 n i.
\end{aligned}$$

The final expression for the magnetic field is given by

$$\vec{\mathbf{B}}(r, \theta, z) = \begin{cases} \mu_0 n i\hat{\mathbf{z}} & \text{inside the solenoid} \\ \dfrac{\mu_0 i}{2\pi r}\hat{\boldsymbol{\theta}} & \text{outside the solenoid.} \end{cases}$$

This expression is consistent with the iron filings experiment of Figure 3.5 which shows the field inside the solenoid is along the axis and outside the solenoid the field is weaker (by a factor of $\frac{n}{2\pi r}$) compared to the field inside.

The quantity $\lambda = ni$ is the surface current (on the surface defined by the coil) per unit length of the coil going around it in the azimuthal ($\hat{\theta}$) direction. It is referred to as the *surface current density* and, for an infinitely long solenoid, $B = \mu_0 \lambda$ inside.

Example *Toroidal Solenoid*

A toroidal solenoidal coil with an air core is shown in Figure 3.28. It has an inside radius of r_1, an outside radius of r_2 and a core diameter of d. Figure 3.28(b) shows an iron filing experiment of the lines of $\vec{\mathbf{B}}$ of an air filled toroidal solenoid. The toroid does not possess as much symmetry as the previous examples making the complete derivation of the magnetic field everywhere more difficult. However, it turns out that outside of the toroid, the magnetic field is quite weak. Inside the toroid, the reader is asked to show that the magnetic field is only nonzero in the azimuthal direction (see problem 6).

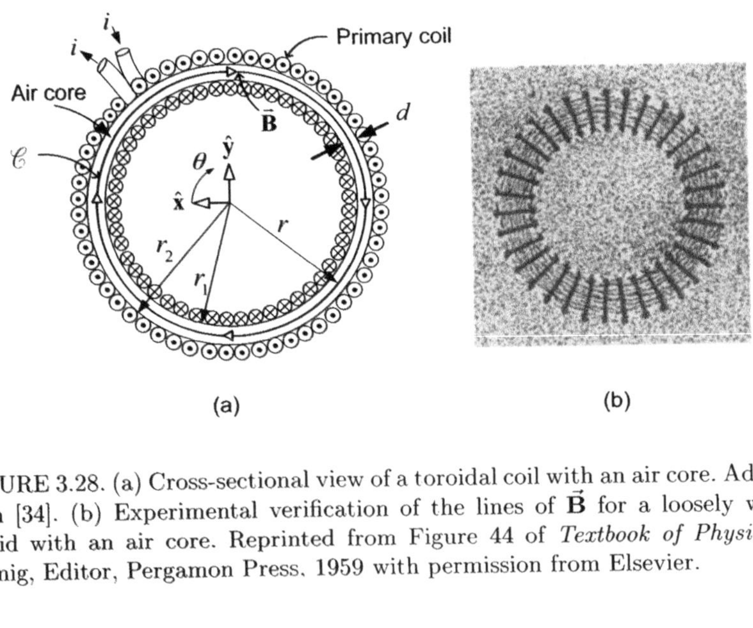

FIGURE 3.28. (a) Cross-sectional view of a toroidal coil with an air core. Adapted from [34]. (b) Experimental verification of the lines of $\vec{\mathbf{B}}$ for a loosely wound toroid with an air core. Reprinted from Figure 44 of *Textbook of Physics*. R. Kronig, Editor, Pergamon Press, 1959 with permission from Elsevier.

Let i denote the current in the toroid wire and n denote the number of loops (turns) per unit length along the radius r_1 of the toroid so that $2\pi r_1 n$ the total number of loops wound around the toroid. Application of

Ampère's law to the circular curve C at radius r gives

$$\begin{aligned}\oint_C \vec{\mathbf{B}} \cdot d\vec{\ell} &= \mu_0 i_{\text{enclosed}} \\ B_\theta 2\pi r &= \mu_0 2\pi r_1 ni \\ B_\theta(r) &= \frac{r_1}{r}\mu_0 ni.\end{aligned}$$

This is valid for $r_1 < r < r_2$ and, if $d = r_2 - r_1 << r_1$, then $r_1/r \approx 1$ and $B \approx \mu_0 ni$ inside the toroid just as in the case of the long straight solenoid.

Remark

In the examples in this chapter using a toroid, it will always be assumed that $d = r_2 - r_1 << r_1$ so that the approximation $r_1/r \approx 1$ is valid.

3.3.3 Relating $\vec{\mathbf{B}}$ *to* $\vec{\mathbf{M}}$

Using Ampère's law, it was shown in the previous example that $B \approx \mu_0 ni$ in an air core toroidal coil. It is now shown how this relationship can be *experimentlally* verified. To do so, a secondary coil is wrapped around the toroid as shown in Figure 3.29 to carry out the experiment.

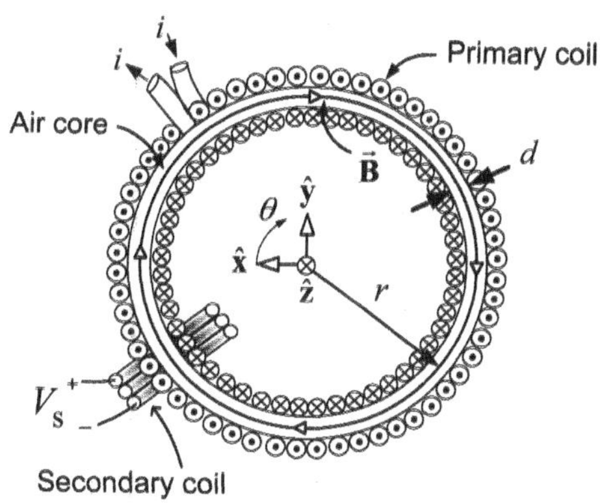

FIGURE 3.29. Experimental setup for determining $\vec{\mathbf{B}}$ versus i in an *air-filled* toroid. Adapted from [34].

The secondary coil is used to measure $\vec{\mathbf{B}} = B\hat{\boldsymbol{\theta}}$ inside the toroid as follows:[3] With N_s the number of turns in the secondary coil, $S = \pi(d/2)^2$ the cross sectional area of the toroidal coil, the flux linkage λ_s through the secondary coil is

$$\lambda_s = N_s B(t) S$$

[3]The subscript θ on B_θ has now been dropped without confusion because the other components of the magnetic field are insignificant (see problem 6).

where, as it is assumed that $d = r_2 - r_1 << r_1$, B is taken to be constant across the cross section S of the toroid as well as in the azimuthal direction. Now, by Faraday's law, the voltage induced in the secondary coil is given by

$$V_s = -\frac{d\lambda_s}{dt} = -\frac{d}{dt}(N_s B(t) S) = -N_s S \frac{dB(t)}{dt}$$

so that

$$B(t) = -\frac{1}{N_s S}\int_0^t V_s(\tau)d\tau. \tag{3.2}$$

By measuring the current $i(t)$ in the primary windings and the voltage $V_s(t)$ between the terminals of the secondary coil, the magnetic field B inside the toroid can be found as a function of the current i. That is, one would ramp up the primary current $i(t)$ from 0 to some $I_{\max}$ as in Figure 3.30(a), while at the same time measure/store the secondary voltage $V_s(t)$ (see Figure 3.30(b). $B(t)$ is then computed using (3.2) as in Figure 3.30(c) and one then plots B versus i as in Figure 3.30(d). In this case in which the toroid is air filled, these measurements simply verify that $B \approx \mu_0 ni$.

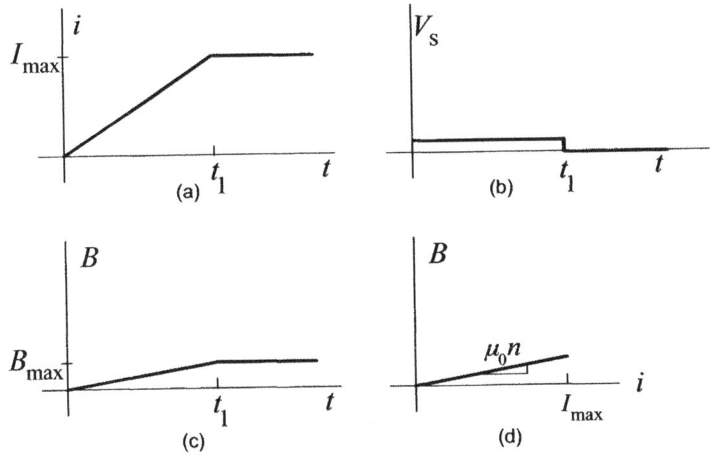

FIGURE 3.30. (a) Current in primary winding. (b) Voltage measured in secondary winding. (c) Magnetic field inside the toroid versus time calculated according to equation (3.2). (d) Magnetic field plotted versus the primary current (slope $\approx \mu_0 n$).

This *same* experimental setup can also be used to measure the magnetic field inside the toroidal coil when it has an *iron* core. The experiment is now repeated, but with the toroid filled with an iron core as shown in Figure 3.31. Of course, there is then a *non zero* magnetization $\vec{\mathbf{M}}$ in the iron core.

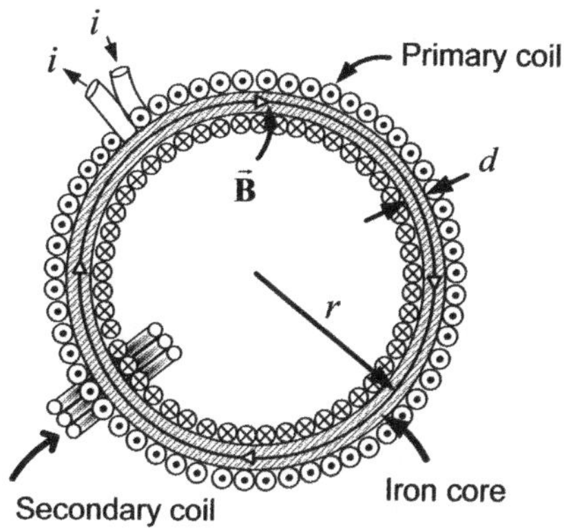

FIGURE 3.31. Iron-filled toroid with secondary coil. Adapted from [34].

The magnetic field strength B inside the toroid is computed as a function of the current i just as before. In this case, the primary current is as shown in Figure 3.32(a) and is the same as in Figure 3.30(a).

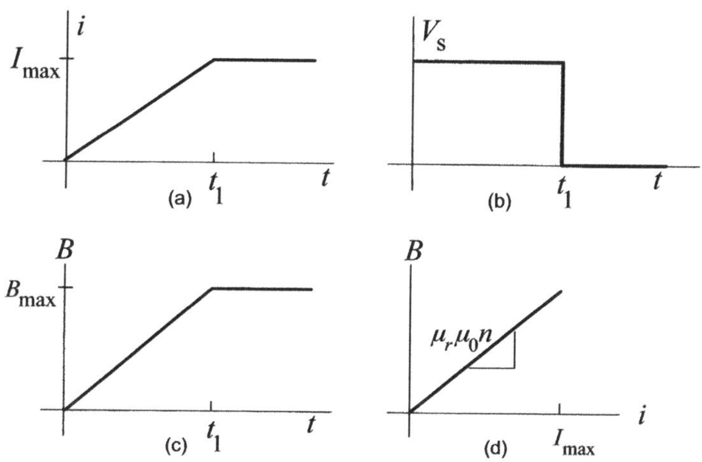

FIGURE 3.32. (a) Current in primary winding. (b) Voltage measured in secondary winding. (c) Magnetic field inside the toroid calculated according to (3.2). (d) Magnetic field plotted versus the primary current (slope $\approx \mu_r \mu_0 ni$).

Again, the voltage induced in the secondary winding is measured [see 3.32(b)] and is now much greater in value than the measured voltage in the case of the air filled toroid. The magnetic field B is again found by integrating the voltage $V_s(t)$ in the secondary coil according to equation (3.2) and is shown in Figure 3.32(c). Finally, the magnetic field B is plotted versus the primary current i as shown in Figure 3.32(d). In this case, with the toroid filled with magnetic material, the slope is greater by a factor μ_r than that of Figure 3.30(d). The factor μ_r is called the *relative permeability* and is typically on the order of 1000. Rewrite B in the iron core

$$B = B_0 + B_M \tag{3.3}$$

where $B_0 = \mu_0 n i$ denotes the magnetic field when no magnetic material is present and B_M represents the additional strength in the magnetic field due to the magnetic material. The strength B of the magnetic field in the toroid is then plotted versus $B_0 = \mu_0 n i$ (rather than i) resulting in a curve of the form shown in Figure 3.33. The slope of the straight line part of the curve is then the relative permeability μ_r so that $B = \mu_r B_0 = \mu_r \mu_0 n i$ for $i \leq i_{\text{sat}}$. If one continues to increase the current above some current level i_{sat}, B no longer increases at the rate $\mu_r \mu_0 n$, but only as $\mu_0 n$.

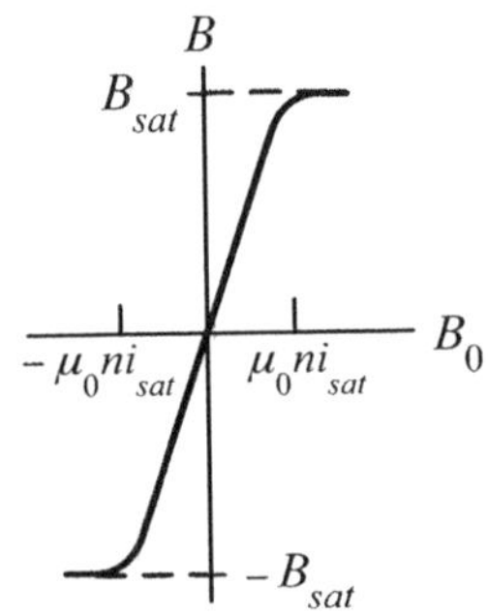

FIGURE 3.33. B versus B_0.

Again, the important point here is that the field B_M is typically on the order of a 1000 times greater than B_0 in good magnetic materials (i.e., $\mu_r > 1000$). This is, of course, the reason magnetic materials are used! This phenomena can be pictured as the magnetic field inside the toroid produced by the current i aligns (magnetizes) the magnetic dipoles moments [see Figures 3.21 and 3.23)] of the iron so that they produce a magnetic field that adds to that produced by the current. As the current in the wire of the toroid is increased, more magnetic dipoles are aligned increasing the B field more. As a result, the magnetic field due to the current and magnetic material is much larger (on the order of a thousand times) than that produced by just the current. As the current increases to a level i_{sat}, all of the dipole moments of the iron atoms are aligned so that increasing the current beyond this level does not result in a large increase in the magnetic field as it did below this level of current.

Reiterating, for the same current in the toroidal coil, the one filled with magnetic material will have a $\vec{\mathbf{B}}$ field that is typically more than a 1000 times that of the coil with only air inside. As will be described shortly, the quantity $H \triangleq B_0/\mu_0 = ni$ is the *magnetic field intensity* and $M \triangleq B_M/\mu_0$ turns out to be the *magnetization* so that

$$B = B_0 + B_M = \mu_0 H + \mu_0 M$$

where $H << M$ as $B_0 << B_M$. Consequently, in soft[4] magnetic materials, one can take $H = 0$ as a reasonable approximation. In other words, in such magnetic materials, the B field is primarily due to the magnetization (alignment of the magnetic dipoles) of the magnetic material rather than the current that magnetizes the magnetic material.

3.4 The Magnetic Intensity Field Vector $\vec{\mathbf{H}}$

Here the definition of the *magnetic intensity* vector $\vec{\mathbf{H}}$ is given in terms of the *magnetization* vector $\vec{\mathbf{M}}$ (defined in Section 3.3.2) and the *magnetic induction* vector $\vec{\mathbf{B}}$. To proceed, consider the toroid in Figure 3.34 which shows a toroid with an iron core 3.34(a) and redrawn in 3.34(b) with the iron core replaced by an equivalent coil carrying the bound current i_b.

Ampère's law must be modified to account for magnetic materials. Recall that when the iron core was not present inside the toroid, Ampère's law showed that $B \approx \mu_0 ni$. However, when the iron core was present, experiments showed that the B field was on the order of a 1000 times greater. The discussion concerning Figure 3.21 showed that the magnetization (dipole moment per unit volume) can be pictured as an equivalent bound current going around the surface of the toroidal core. Following Ref. [34], consider the modification to Ampère's law as

$$\oint_C \vec{\mathbf{B}} \cdot d\vec{\ell} = \mu_0 \left(i_{\text{enclosed}} + i_{M\text{enclosed}}\right) \tag{3.4}$$

where $i_{M\text{enclosed}}$ is the equivalent *bound* current enclosed by the curve C (see Figure 3.34).

As before, it is assumed the core diameter $d = r_2 - r_1 << r_1$ so that $r_1 \approx r \approx r_2$ is valid. With n_b the number of loops/coils per unit length of the *equivalent* coil in Figure 3.34(b), $i_{M\text{enclosed}} = 2\pi r_1 n_b i_b$. The next step is to relate this bound current to the magnetization M. To do so, define the *surface current density*

$$\lambda_M \triangleq \frac{i_{M\text{enclosed}}}{2\pi r_1} = n_b i_b$$

and is simply the (average) amount of bound current per unit length around the toroid coil coming out of (or going into) the page.

---

[4] For now, consider a "soft" magnetic material to be one that has a B versus B_0 curve similar to that of Figure 3.33 and whose slope is orders of magnitude greater than one.

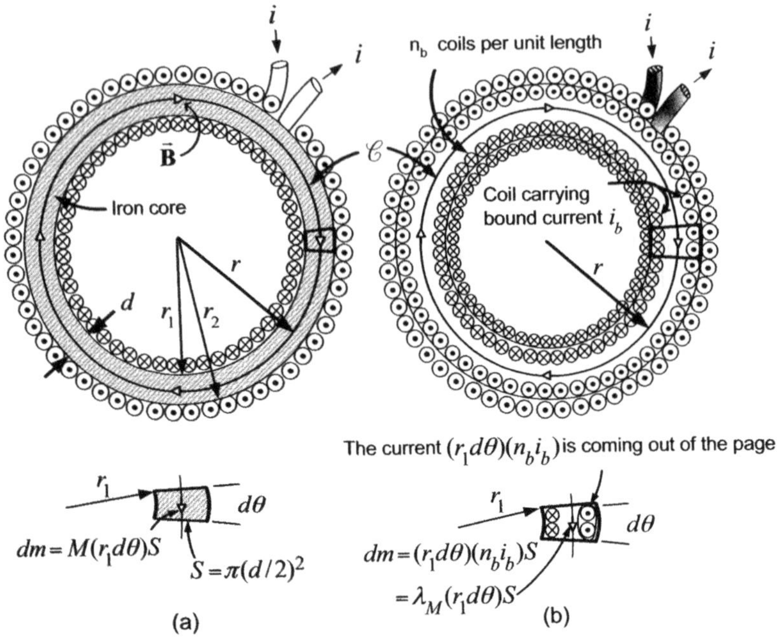

FIGURE 3.34. (a) Toroidal coil with an iron core. Adapted from [34]. (b) Iron core replaced with an equivalent coil carrying the bound current i_b. The bound current produces the same magnetic field as the magnetized iron in Figure 3.34(a).

This is analogous to $\lambda = ni$ being the *free* current in the wire per unit length around the circumference of the toroid. As shown in Section 3.3.1, the magnetic dipole moment is the product of the current around a loop times the area of the loop. The amount of bound current going around the periphery of the toroid between θ and $\theta + d\theta$ is $\lambda_M(r_1 d\theta)$ and, with $S = \pi(d/2)^2$ the cross-sectional area of the toroid, it follows that the magnetic dipole moment dm due to the bound surface current density λ_M in the volume between θ and $\theta + d\theta$ is

$$dm = \lambda_M (r_1 d\theta) S.$$

However, by definition, the magnetization M is the amount of magnetic dipole moment per unit volume, so that the magnetic dipole moment dm in the volume between θ and $\theta + d\theta$ is also given by

$$dm = M(r_1 d\theta)S$$

which results in

$$\lambda_M = M.$$

That is, in this example, the surface current density λ_M is equal to the magnetization M of the iron material. As a result,

$$i_{M\text{enclosed}} = \lambda_M 2\pi r = M 2\pi r = \oint_C \left(M\hat{\boldsymbol{\theta}}\right)\cdot\left(rd\theta\hat{\boldsymbol{\theta}}\right) = \oint_C \vec{\mathbf{M}}\cdot d\vec{\ell}.$$

Equation (3.4) therefore becomes

$$\oint_C \vec{\mathbf{B}}\cdot d\vec{\ell} = \mu_0\left(i_{\text{enclosed}} + \oint_C \vec{\mathbf{M}}\cdot d\vec{\ell}\right).$$

Rearranging this equation gives

$$\oint_C \left(\frac{\vec{\mathbf{B}} - \mu_0\vec{\mathbf{M}}}{\mu_0}\right)\cdot d\vec{\ell} = i_{\text{enclosed}}.$$

This leads to the definition of the *magnetic intensity* field $\vec{\mathbf{H}}$ given by

$$\vec{\mathbf{H}} \triangleq \frac{\vec{\mathbf{B}}}{\mu_0} - \vec{\mathbf{M}}.$$

Ampère's law in its most general form is then

$$\oint_C \vec{\mathbf{H}}\cdot d\vec{\ell} = i_{\text{free}} \tag{3.5}$$

where $i_{\text{free}} = i_{\text{enclosed}}$ represents the enclosed *free* current in the wire in contrast to the *bound* current in the magnetic material. Both $\vec{\mathbf{B}}$ and $\vec{\mathbf{M}}$ were defined in terms of physical phenomena. However, $\vec{\mathbf{H}}$ is defined as the difference between $\vec{\mathbf{B}}/\mu_0$ and $\vec{\mathbf{M}}$ which does not lend itself to a physical interpretation. It is simply defined as a quantity whose line integral around a closed curve depends only on the free current through the curve. The next section shows how $\vec{\mathbf{B}}$ and $\vec{\mathbf{H}}$ can be related experimentally.

3.4.1 The B–H Curve

With the definition $\vec{\mathbf{H}} \triangleq \vec{\mathbf{B}}/\mu_0 - \vec{\mathbf{M}}$, Ampère's law was modified to

$$\oint_C \vec{\mathbf{H}}\cdot d\vec{\ell} = i_{\text{free}}.$$

Ampère's law can be used to calculate $\vec{\mathbf{H}}$, but $\vec{\mathbf{B}}$ is the desired quantity as it produces forces on currents and magnetic materials. As $\vec{\mathbf{M}}$ is typically not known, one cannot compute $\vec{\mathbf{B}}$ from the definition of $\vec{\mathbf{H}}$. As a consequence, empirical/experimental methods are used to relate $\vec{\mathbf{B}}$ and $\vec{\mathbf{H}}$. The empirical method to determine this relationship is typically done using the toroidal coil setup of the previous section. Specifically, consider the setup of Figure 3.31 in which the initial magnetization is zero (this can be done by heating up the iron core so that the magnetic dipole moments become randomly oriented with respect to each other). Application of Ampère's law gives[5]

$$\begin{aligned} \oint_C \vec{\mathbf{H}} \cdot d\vec{\ell} &= i_{\text{free}} \\ H 2\pi r &= 2\pi r_1 n i \\ H &= \frac{r_1}{r} n i \approx n i \end{aligned}$$

so that $H = B_0/\mu_0$ as in equation (3.3). Equation (3.2) given by

$$B(t) = -\frac{1}{N_s S}\int_0^t V_s(\tau)d\tau$$

is again used to find the value of B as a function of $H = ni$ resulting in the so-called $B-H$ curve as shown in Figure 3.35(a). This is now considered in more detail. (Note that as $H = ni$, one can specify the value of H by simply specifying the current i.)

First, the current i is increased from 0 to $I_{\max}$ so that H goes from 0 to $H_{\max} = nI_{\max}$ along the curve $a-b$ in Figure 3.35(a), where $H_{\max} = 100\frac{\text{ampere-turn}}{\text{meter}}$. As the current is increased, the voltage V_s in the secondary coil is measured and B is calculated according to (3.2). Plotting B versus H one obtains the curve $a-b$ shown in Figure 3.35(a). Along the curve $a-b$, $H > 0, B > 0$, and $M > 0$.

Next, H (or equivalently i) is decreased from $H_{\max}$ to 0. Again, B is computed according to (3.2), resulting in the curve $b-c$ shown in Figure 3.35(a). At the point c, $B > 0$ with $H = 0$ and therefore $i = 0$, that is, there is a magnetic field in the core without any current in the coil. In words, the toroidal is core is permanently magnetized. The corresponding value of the magnetic field is denoted B_r and is called the *remanence* or *retentivity*. Along the curve $b-c$, $H > 0, B > 0$, and $M > 0$.

In the next step, H is made negative by putting negative current in the coil ($i < 0$), and B decreases along the curve $c-d$ where at the point d, the magnetic field strength is zero in the core ($B = 0$). The magnetic field produced by the (negative) current in the coil is canceling the magnetic field produced by the aligned dipoles of the magnetic material. At this point, $H = -H_c$ where $H_c > 0$ is called the *coercive force* or *coercivity*.[6] Along the curve $c-d$, $H < 0, B > 0$, and $M > 0$.

Continuing, H is decreased even more (i.e., $i < 0$ is decreased more) resulting in the magnetic dipoles of the material realigning in the opposite

---

[5]By definition $\vec{\mathbf{H}} \triangleq \vec{\mathbf{B}}/\mu_0 - \vec{\mathbf{M}}$ and on physcial grounds $\vec{\mathbf{B}}$ and $\vec{\mathbf{M}}$ point in the same direction which in this case is the $\hat{\boldsymbol{\theta}}$ direction inside the toroid. Consequently, $\vec{\mathbf{H}}$ will also be in the $\hat{\boldsymbol{\theta}}$ direction and may be written in the form $\vec{\mathbf{H}} = H\hat{\boldsymbol{\theta}}$.

[6]H_c is not a force so "coercive force" is not a good name. The value of H_c depends on the particular B-H loop that the system is going through.

direction to add to the magnetic field produced by the current in the coil (see path d–e). At the point e where $H = -H_{\max} = -100\frac{\text{ampere-turn}}{\text{meter}}$, the value of B is the negative of that given at the point b. Along the curve d–e, $H < 0, B < 0$, and $M < 0$.

Finally, in a similar fashion, H is brought from $-H_{\max}$ to 0 to $H_{\max}$ and the plot of B versus H follows the curve e–f–g–b.

In summary, letting H vary from $H_{\max}$ to 0 to $-H_{\max}$ and then from $-H_{\max}$ to 0 back to $H_{\max}$, the plot of B versus H follows the curve b–c–d–e–f–g–b. This is called the *hysteresis loop*. If the hysteresis loop of the iron is quite thin, it is considered to be "soft" iron as in Figure 3.35(b) while if the hysteresis loop is quite thick, then it is considered to be "hard" iron as in Figure 3.35(a).

In the case of soft iron magnetic material such as Figure 3.35(b), the hysteresis loop is thin enough that it can be approximated by a single curve lying only in the first and third quadrants.

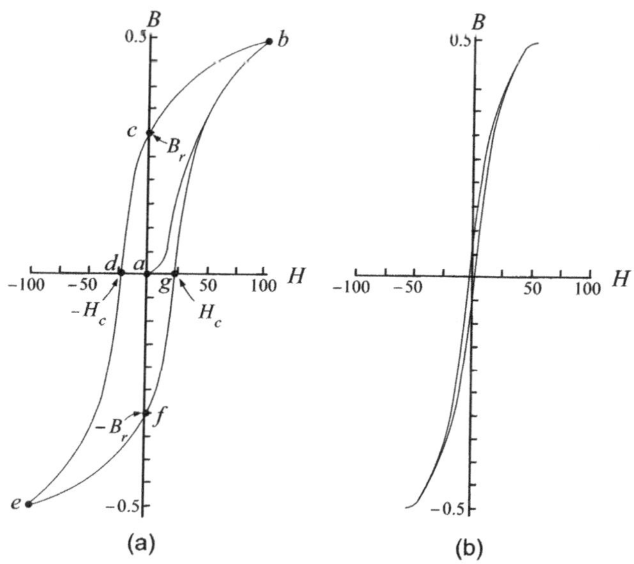

FIGURE 3.35. B versus H curve. (a) Hysteresis curve for a "hard" magnetic material. (b) Hysteresis curve for a "soft" magnetic material.

Figure 3.36 shows experimentally determined B–H curves for the two soft magnetic materials M-19 29-gauge steel and a soft steel iron casting where, as B is an odd function of H, only the first quadrant need be given. Note that at smaller values of H, the B–H curves in Figure 3.36 are approximately linear.

Remark Figure 3.33 would correspond to a soft magnetic material in which the hysteresis can be ignored.

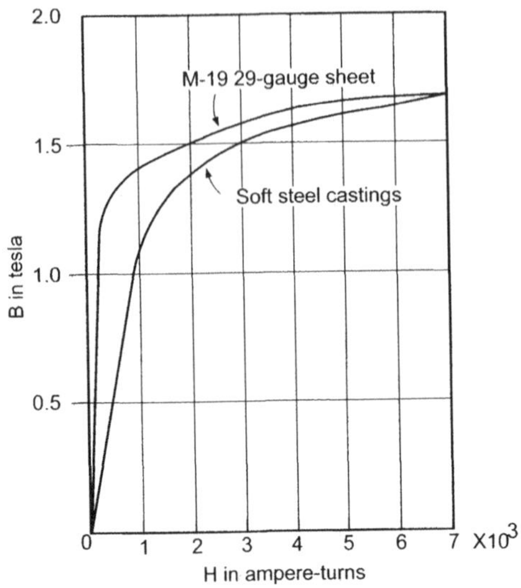

FIGURE 3.36. B (tesla) versus H (ampere-turns) curves for two different soft magnetic materials. From *An Introduction to Electrical Machines and Transformers*, 2nd edition, G. McPherson and R. D. Laramore, 1990. Reprinted by permisson of John Wiley & Sons.

Example

Consider the experimental B versus H curves for two different soft iron materials as given in Figure 3.36. As $\vec{\mathbf{B}}$ and $\vec{\mathbf{M}}$ (and therefore $\vec{\mathbf{H}}$) tend to point in the *same* direction in soft iron, the relationship

$$\vec{\mathbf{B}} = \mu_0 \vec{\mathbf{H}} + \mu_0 \vec{\mathbf{M}}$$

may be written approximately as

$$\vec{\mathbf{B}} = \mu_r \mu_0 \vec{\mathbf{H}}$$

or

$$B = \mu_r \mu_0 H$$

where μ_r is that empirically determined number that makes this equation valid. For example, using the soft steel casting material with i chosen in the primary coil of the toroidal solenoid so that $H = 1000$ ampere-turns, the data curve in Figure 3.36 gives $B = 1.2$ tesla. Thus,

$$\mu_r = \frac{B}{\mu_0 H} = \frac{1.2}{(4\pi \times 10^{-7}) \times 1000} = 955.$$

The important observation from this numerical example is that μ_r is quite large. In a coil carrying a current i with an *air core*,

$$B_{\text{inside}} = \mu_0 H_{\text{inside}} = \mu_0 n i$$

while in an identical coil carrying the same current i with an *iron core*,

$$B_{\text{inside}} = \mu_r \mu_0 H_{\text{inside}} = \mu_r \mu_0 n i.$$

The H field is the *same* in both coils as the current is the same in each coil. However, as μ_r is on the order of 1000, the magnetic field B in the coil with the iron core is much larger. The terminology "ideal" soft magnetic material is one in which $\mu_r = \infty$ while a "good" magnetic material just means μ_r is very large.

3.4.2 Computing $\vec{\mathbf{B}}$ and $\vec{\mathbf{H}}$ in Magnetic Circuits

Based on the previous sections, a simple procedure is now given that gives an accurate enough computation of $\vec{\mathbf{B}}$ and $\vec{\mathbf{H}}$ for modeling the magnetic circuits of electric machines. To start with a simple example, consider the toroidal coil with an iron core shown in Figure 3.37 in which a small air gap of length g has been cut out.

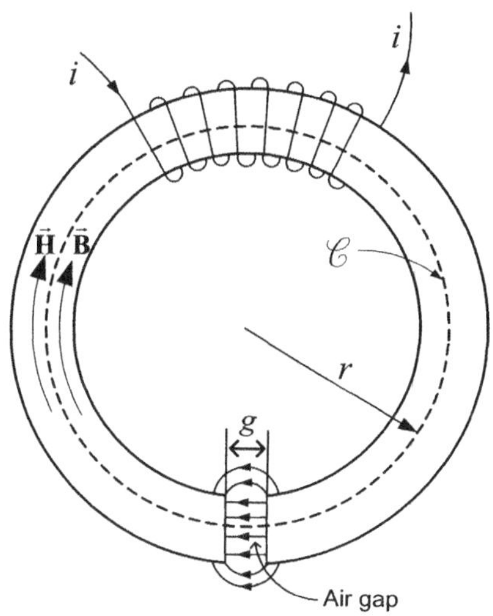

FIGURE 3.37. Coil wrapped around a torus with an air gap. From *Electric Machines: Dynamics and Steady State* by G. J. Thaler and M. L. Wilcox, 1966. Reprinted by permission of John Wiley & Sons.

Ampère's law

$$\oint_C \vec{\mathbf{H}} \cdot d\vec{\ell} = i_{\text{free}}$$

is applied to a circular curve C of radius r in Figure 3.37. In the magnetic material, $\vec{\mathbf{B}} = B_m \hat{\boldsymbol{\theta}}$ while in the air gap $\vec{\mathbf{B}} = B_a \hat{\boldsymbol{\theta}}$. Though azimuthal symmetry does not hold because of the air gap, it turns out that if the air gap is small, it is still approximately true that B_m has a constant value as a function of the azimuthal angle θ in the magnetic material and, similarly, B_a is constant as a function of θ in the air gap. Further, with the standing assumption that the diameter of the core is much smaller than the inner radius of the toroidal coil, B_m and B_a are essentially constant across the cross section of the core. In the same way, $\vec{\mathbf{M}} = M\hat{\boldsymbol{\theta}}$ is constant as a function of θ in the iron core while it is, of course, zero in the air gap. Thus H_m and H_a are also approximately constant as functions of θ and over the cross-sectional area of the core. With N the total number of coil turns, Ampère's law applied to the circular curve C of radius r in Figure 3.37 gives

$$H_m(2\pi r - g) + H_a g = Ni \tag{3.6}$$

with

$$\begin{aligned} B_m &= \mu_0 H_m + \mu_0 M = \mu_r \mu_0 H_m \text{ (from the measured } B\text{–}H \text{ curve)} \\ B_a &= \mu_0 H_a. \end{aligned}$$

It is assumed that the air gap is small enough that B_a is essentially constant across it. Recall that by Gauss's law for $\vec{\mathbf{B}}$, $B_m = B_a$ *at the air-iron interface*. As B_m is approximately constant in the iron and B_a is constant in the air gap, $B_m = B_a$. Thus,

$$\mu_0 H_a = B_a = B_m = \mu_r \mu_0 H_m$$

or

$$H_m = \frac{H_a}{\mu_r}.$$

Again, in good magnetic materials, μ_r is greater than 1000 so that H_m is more than a thousand times smaller than H_a. Consequently, one often takes $H_m = 0$ in good magnetic materials. In this case, taking $H_m = 0$ in (3.6) leads to

$$H_a g = Ni$$

so that

$$B_a = \mu_0 H_a = \frac{\mu_0 Ni}{g} \qquad (H_m = 0). \tag{3.7}$$

This last expression is important because it gives us the magnetic field in air gap in terms of known quantities. The value of B in the air gap is the important quantity since, for example, a current carrying wire can be run through the air gap and the value of B will determine how much force can be exerted on the wire.

It is important to note that to use the approximation $H_m = 0$, one must check that $2\pi r - g$ is not too large. For example, with $\mu_r = 1000$, if $2\pi r - g$ is a thousand times longer than g, then

$$H_m(2\pi r - g) + H_a g = \frac{H_a}{1000} 1000g + H_a g = 2H_a g = Ni$$

or

$$B_a = \mu_0 H_a = \frac{\mu_0 N i}{2g}$$

This large error (a factor of 2) in the calculation of the magnetic field in the air gap due to the fact that $H_m(2\pi r - g)$ is not negligible. However, if the designer of this magnetic circuit made sure that $2\pi r - g = 100g$, then

$$\begin{aligned} H_m(2\pi r - g) + H_a g &= \frac{H_a}{1000} 100g + H_a g \\ &= 1.1 H_a g \end{aligned}$$

resulting in

$$B_a = \mu_0 H_a = \frac{\mu_0 N i}{1.1g}.$$

which is not much different from (3.7) where $H_m = 0$ was assumed.

In summary, B in the air gap is the fundamental quantity of interest as it is the field that can produce forces on currents in wires placed in the air gap, or can induce voltages in wire loops, and so on. Here, Ampère's law

$$\oint_C \vec{\mathbf{H}} \cdot d\vec{\ell} = i_{\text{free}}$$

was used to find a relation between H_m (H in the magnetic material) and H_a (H in the air gap). The free currents (currents in the wires) are known or easily measured. Assuming that the magnetic paths in the iron are not too long, one can use the approximation that $H_m = 0$ in the magnetic material. Then, H_a can be found in the air gap from Ampère's law which in turn is used to obtain the magnetic field *in the air gap* from $B_a = \mu_0 H_a$.

Example (Adapted from Chapman [19])

Following Chapman [19], Figure 3.38 shows a setup to produce the radial magnetic field in the air gap of a separately excited DC motor. The objective here is to determine the strength of the field B in the air gap in terms of the current i_f in the field coil. Ampère's law gives

$$H_m \ell + H_a g + H_m \ell_2 + H_a g = N i_f \tag{3.8}$$

where ℓ is the length of the path in the field iron, ℓ_1 is the length of the rotor, ℓ_2 is the diameter of the rotor iron, and g is the air gap length. Here $N = 200$, $\ell = 0.5$ m, $g = 5 \times 10^{-4}$ m, $\ell_1 = 2.4 \times 10^{-2}$ m, $\ell_2 = 5 \times 10^{-2}$ m, and $S = \ell_1 \ell_2 = 12 \times 10^{-4}$ m^2. Equation (3.8) is an approximation as it is being assumed that H_m is constant in magnitude everywhere inside the magnetic material and that it follows the path of the iron.

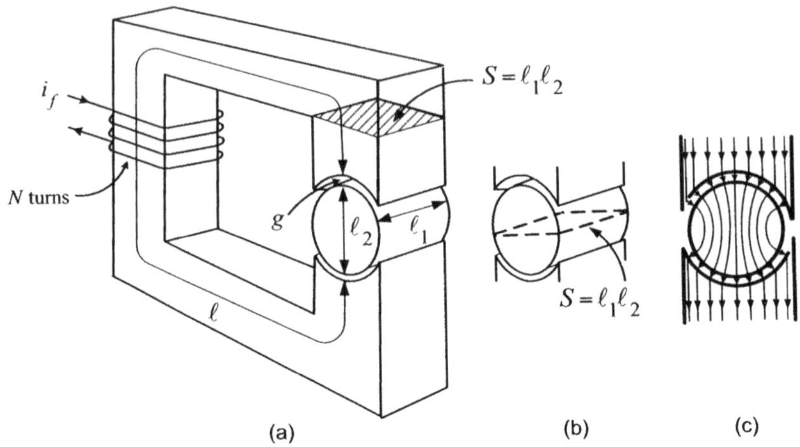

FIGURE 3.38. Magnetic circuit for producing the magnetic field in the air gap of a DC motor. Figure 3.38(a) from *Electric Machinery Fundamentals* by S. J. Chapman, 2nd edition, 1991, McGraw-Hill. Reproduced by permission of McGraw-Hill Companies.

Assuming ideal magnetic material so that H_m is taken to be zero inside the magnetic material,

$$B_a = \mu_0 H_a = \frac{\mu_0 N i_f}{2g} = K_f i_f.$$

Evaluating,

$$B_a = \frac{(4\pi \times 10^{-7}) \times 200}{2 \times 5 \times 10^{-4}\ \text{m}} i_f = 0.25 i_f \text{ tesla}$$

where 1 weber/m^2 = 1 tesla.

Example (Chapman [19])

The previous example is redone without setting $H_m = 0$ in the magnetic material. It is still assumed that H_m is constant in magnitude everywhere inside the magnetic material and that it follows the path of the iron. Suppose that for the iron used in the rotor and the core of the field winding, the relative permeability is $\mu_r = 2000$ so that $B_m = \mu_r \mu_0 H_m$. As before, $N = 200$, $\ell = 0.5$ m, $g = 5 \times 10^{-4}$ m, $\ell_1 = 2.4 \times 10^{-2}$ m, $\ell_2 = 5 \times 10^{-2}$ m, and $S = \ell_1 \ell_2 = 12 \times 10^{-4}$ m^2. The path $\ell + \ell_2 = 55$ cm is 550 times longer than the path $2g = 0.1$ cm in the air gap. From Ampère's law

$$H_m \ell + H_a g + H_m \ell_2 + H_a g = N i_f. \tag{3.9}$$

The problem here is that this is only one equation with two unknowns H_m, H_a. However, another relation is found by making use of the assumption (approximation) that the flux ϕ in the cross-section of the magnetic circuit is constant. That is, it is assumed that the magnetic material essentially keeps the magnetic field either in the core itself or within the air gap in the radial direction. This assumption is usually stated as saying the magnetic circuit has no (flux) *leakage*. In the iron core of the field,

$$\phi = B_m S = \mu_r \mu_0 H_m S$$

where $S = \ell_1 \ell_2$ is the core cross sectional area and B_m is constant in the cross-section. Then

$$H_m \ell = \frac{H_m \ell}{\phi}\phi = \frac{H_m \ell}{\mu_r \mu_0 H_m S}\phi = \frac{\ell}{\mu_r \mu_0 S}\phi = R_c \phi.$$

Here R_c is called the *reluctance* of the core path and it has been used to relate H_m to the flux ϕ. Evaluating this expression for R_c gives

$$R_c = \frac{\ell}{\mu_r \mu_0 S} = \frac{50 \times 10^{-2}\ \text{m}}{(2000)(4\pi \times 10^{-7})(12 \times 10^{-4}\ \text{m}^2)} = 166 \times 10^3\ \frac{\text{A-turns}}{\text{Wb}}.$$

Because of the cylindrical shape of the rotor core and conservation of flux, B_m (and therefore $H_m = \mu_r \mu_0 B_m$) is not constant as one goes through the rotor [see Figure 3.38(c)]. Specifically, $B_{mR} S_R = B_m S$ where $S_R = \pi(\ell_2/2)\ell_1$ is the surface area of the top (or bottom) side of the rotor core, B_{mR} is the magnetic field at the surface of the rotor core and $B_m S$ is the flux computed in the middle of the core [which has the same cross section as the field core as indicated in Figure 3.38(b)]. By conservation of flux, H_m in the rotor core has a minimum value at the (inside) surface of the rotor iron and a maximum value in the middle of the rotor on the surface S in Figure 3.38(b). The maximum value of H_m in the middle of the core is used because it leads to a larger value for the reluctance of the core and therefore a more conservative (higher) estimate of the current needed to set up a specified magnetic field in the air gap. So, with H_m taken as its value in the middle of the core on the surface S in Figure 3.38(b), in the rotor iron

$$H_m \ell_2 = \frac{H_m \ell_2}{\phi}\phi = \frac{H_m \ell_2}{\mu_r \mu_0 H_m S}\phi = \frac{\ell_2}{\mu_r \mu_0 S}\phi = R_r \phi.$$

Evaluating the expression for R_r gives

$$R_r = \frac{\ell_2}{\mu_r \mu_0 S_r} = \frac{5 \times 10^{-2}\ \text{m}}{(2000)(4\pi \times 10^{-7})(12 \times 10^{-4}\ \text{m}^2)} = 16.6 \times 10^3 \frac{\text{A-turns}}{\text{Wb}}.$$

The flux in the air gap is $B_a S_g = \mu_0 H_a S_g$ where $S_g = \pi(\ell_2/2 + g/2)\ell_1 = 19.0\ \text{cm}^2$ is the surface area of the top (bottom) stator side of the air gap. Further this flux is equal to the flux $\phi = B_m S = \mu_r \mu_0 H_m S$ in the iron core. It then follows that

$$H_a g = \frac{H_a g}{\phi}\phi = \frac{H_a g}{\mu_0 H_a S_g}\phi = \frac{g}{\mu_0 S_g}\phi = R_g \phi.$$

Then

$$R_g = \frac{g}{\mu_0 S_g} = \frac{0.05 \times 10^{-2} \text{ m}}{(4\pi \times 10^{-7})(19.0 \times 10^{-4} \text{ m}^2)} = 209.4 \times 10^3 \frac{\text{A-turns}}{\text{Wb}}.$$

With this information, equation (3.9) may be rewritten as

$$(R_c + R_r + 2R_g)\,\phi = Ni_f$$

which evaluates to

$$(166 + 16.6 + 2(209.4)) \times 10^3 \phi = 601.4 \times 10^3 \phi = Ni_f.$$

Solving for ϕ, one obtains

$$\phi = 3.3 \times 10^{-4} i_f \text{ Wb}.$$

The magnetic field in the air gap is then

$$B_a = \frac{\phi}{S_g} = \frac{3.3 \times 10^{-4}}{19.0 \times 10^{-4}} i_f = 0.17 i_f \ \frac{\text{Wb}}{\text{m}^2}$$

which is significantly less than the $0.25 i_f$ Wb/m^2 found assuming $H_m = 0$. However, suppose that the magnetic circuit is redesigned so that $\ell = 10$ cm instead of 50 cm. Then

$$\begin{aligned} R_c &= \frac{\ell}{\mu_r \mu_0 S} = \frac{5 \times 10^{-2} \text{ m}}{(2000)(4\pi \times 10^{-7})(12 \times 10^{-4} \text{ m}^2)} \\ &= 33.2 \times 10^3 \ \frac{\text{A-turns}}{\text{Wb}} \end{aligned}$$

and equation (3.9) reduces to

$$(R_c + R_r + 2R_g)\,\phi = Ni_f$$

and evaluates to

$$\Big(33.2 + 16.6 + 2(209.4)\Big) \times 10^3 \phi = 468.6 \times 10^3 \phi = Ni_f.$$

The flux is then

$$\phi = 4.23 \times 10^{-4} i_f \text{ Wb}$$

resulting in the magnetic field in the air gap being

$$B_a = \frac{\phi}{S_g} = \frac{4.23 \times 10^{-4}}{19.0 \times 10^{-4}} i_f = 0.22 i_f \ \frac{\text{Wb}}{\text{m}^2}$$

which is much closer to $0.25 i_f$ Wb/m^2.

The point of this example is that taking $H_m = 0$ in magnetic materials and then using Ampère's law to find B_a in the air gap depends on the proper design of the magnetic circuit. Machine designers spend a great deal of effort trying to make the reluctance in the magnetic material as small as possible by making the path length small, the cross sectional area large and using highly permeable magnetic material (i.e., μ_r large). Figure 3.38 is not a practical design for the stator iron of a DC motor. Figures 1.31 and 1.32 of Chapter 1 depict actual stator iron cores for a DC motor. It will be assumed in all that follows that good machine designs have resulted in making the approximation $H_m = 0$ valid.

3.4.3 $\vec{\mathbf{B}}$ *is Normal to the Surface of Soft Magnetic Material*

It is now shown that $\vec{\mathbf{B}}$ is normal (perpendicular) to the surface at an air/magnetic-material interface under two assumptions: (1) The magnetic material is assumed to be ideal (soft) magnetic material so that $\mu_r = \infty$, or equivalently, $H = 0$. (2) There are no current carrying wires on the surface.

For example, it is not true for toroids as wire is wrapped around surface of the magnetic material. Also, typically $H \neq 0$ inside a permanent magnet.

Consider the closed rectangular-shaped curve C in Figure 3.39 for which half is in the air and the other half is in the magnetic material.

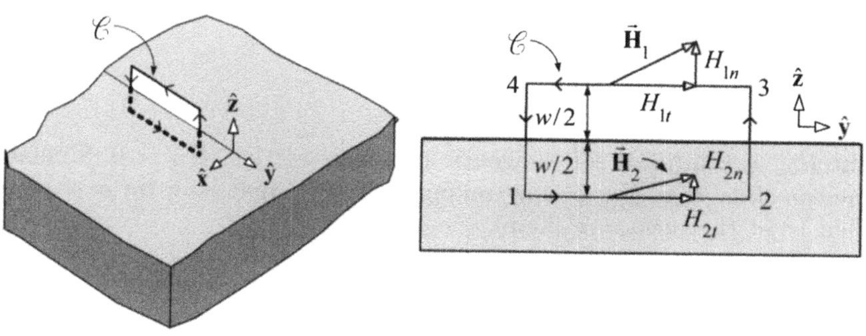

FIGURE 3.39. Closed rectangular-shaped curve enclosing the air and magnetic material. As $w \to 0$, $\vec{\mathbf{H}}_1$ is just above the surface and $\vec{\mathbf{H}}_2$ is just below the surface.

Ampère's law is applied to the rectangular-shaped curve C, which in this instance is given by

$$\oint_C \vec{\mathbf{H}} \cdot d\vec{\ell} = 0 \tag{3.10}$$

as there is no free current, that is, no wires on the surface carrying current. With $\vec{\mathbf{H}}(x, y, z) = H_x\hat{\mathbf{x}} + H_y\hat{\mathbf{y}} + H_z\hat{\mathbf{z}}$ and the coordinate system defined as shown in Figure 3.39, it is seen that $H_z\hat{\mathbf{z}}$ is the part of $\vec{\mathbf{H}}$ normal to the surface while $H_x\hat{\mathbf{x}} + H_y\hat{\mathbf{y}}$ is the tangential component to the surface.

Expanding (3.10) results in

$$\oint_C \vec{\mathbf{H}} \cdot d\vec{\ell} = \int_1^2 \vec{\mathbf{H}} \cdot d\vec{\ell} + \int_2^3 \vec{\mathbf{H}} \cdot d\vec{\ell} + \int_3^4 \vec{\mathbf{H}} \cdot d\vec{\ell} + \int_4^1 \vec{\mathbf{H}} \cdot d\vec{\ell} = 0 \quad (3.11)$$

where side 1–2 is in the magnetic material and side 3–4 is in the air. Sides 2–3 and 4–1 are short segments of length w going between the air and magnetic material. The interest here is in the tangential component of $\vec{\mathbf{H}}$ just above and just below the surface. Making the rectangular curve C very thin ($w \to 0$) so that the lengths of sides 2–3 and 4–1 go to zero, equation (3.11) reduces to

$$\int_1^2 \vec{\mathbf{H}} \cdot d\vec{\ell} + \int_3^4 \vec{\mathbf{H}} \cdot d\vec{\ell} = 0 \quad (3.12)$$

as $w \to 0$. Also, as

$$d\vec{\ell} = \begin{cases} d\ell\hat{\mathbf{y}} & \text{side } 1\text{–}2 \\ -d\ell\hat{\mathbf{y}} & \text{side } 3\text{–}4 \end{cases}$$

it follows that

$$\int_1^2 (H_{1x}\hat{\mathbf{x}} + H_{1y}\hat{\mathbf{y}} + H_{1z}\hat{\mathbf{z}}) \cdot (d\ell\hat{\mathbf{y}}) + \int_3^4 (H_{2x}\hat{\mathbf{x}} + H_{2y}\hat{\mathbf{y}} + H_{2z}\hat{\mathbf{z}}) \cdot (-d\ell\hat{\mathbf{y}})$$
$$= H_{1y}\ell_1 - H_{2y}\ell_1 = 0$$

or

$$H_{2y} = H_{1y}.$$

But $H_{2y} = 0$ in ideal soft magnetic materials so that $H_{1y} = 0$. Similarly, one can show that $H_{1x} = 0$ by taking a curve C that is in the x–z plane. That is, at the surface of the magnetic material

$$\vec{\mathbf{H}}_1(x, y, z) = H_{1z}\hat{\mathbf{z}}$$

and, as this is in the air,

$$\vec{\mathbf{B}}_1 = \mu_0\vec{\mathbf{H}}_1 = \mu_0 H_{1z}\hat{\mathbf{z}}$$

so that $\vec{\mathbf{B}}$ is normal to the surface.

Remark

The magnetic circuit of Figure 3.10 has a soft iron core. Note that the lines of $\vec{\mathbf{B}}$ are perpendicular at the surface of the iron core, but not at the coil surface. Figures 3.7 and 3.11 of the bar and horseshoe magnets, respectively, are permanent (hard) magnets for which the above assumptions do not necessarily hold (in particular, $\vec{\mathbf{H}} = 0$ is not usually true in permanent magnets). For example, note that $\vec{\mathbf{B}}$ is *not* perpendicular to the surface of the cylindrical *permanent* magnet shown in Figure 3.18.

3.5 Permanent Magnets*

Consider the cylindrical-shaped permanent magnet in Figure 3.40 where there is no free current.

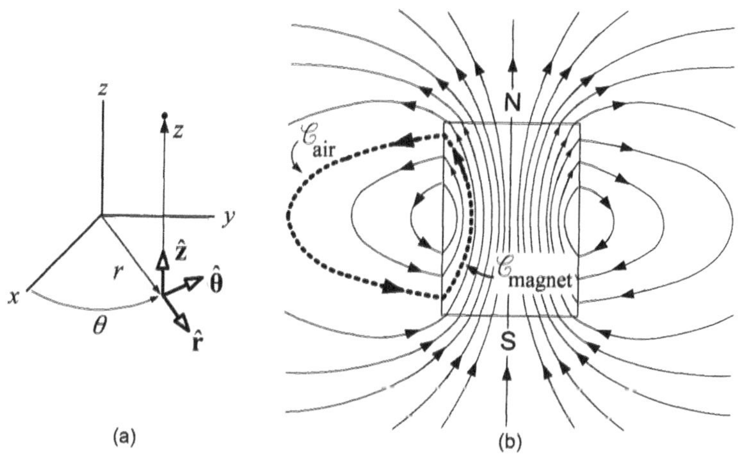

FIGURE 3.40. (a) Cylindrical coordinate system for the bar magnet. (b) Magnetic field lines $\vec{\mathbf{B}}$ around a cylindrical-shaped permanent magnet. Adapted from Figures 37-4(a) and 37-4(b) of Ref. [34].

Ampère's law applied to the path C in the figure gives

$$\oint_C \vec{\mathbf{H}} \cdot d\vec{\ell} = \int_{C_{\text{air}}} \vec{\mathbf{H}} \cdot d\vec{\ell} + \int_{C_{\text{magnet}}} \vec{\mathbf{H}} \cdot d\vec{\ell} = 0$$

where $d\vec{\ell} = dr\hat{\mathbf{r}} + rd\theta\hat{\boldsymbol{\theta}} + dz\hat{\mathbf{z}}$ and the curve has been decomposed into its part in the air and its part in the magnet. As $\vec{\mathbf{H}} = \vec{\mathbf{B}}/\mu_0$ in the air,

$$\int_{C_{\text{air}}} \vec{\mathbf{H}} \cdot d\vec{\ell} > 0.$$

Consequently,

$$\int_{C_{\text{magnet}}} \vec{\mathbf{H}} \cdot d\vec{\ell} < 0$$

in order that $\oint_C \vec{\mathbf{H}} \cdot d\vec{\ell} = 0$. Note that the path C was chosen to follow the closed curve of one of the magnetic lines of $\vec{\mathbf{B}}$. As a result,

$$\oint_C \vec{\mathbf{H}} \cdot d\vec{\ell} = \oint_C \left(H_r\hat{\mathbf{r}} + H_\theta\hat{\boldsymbol{\theta}} + H_z\hat{\mathbf{z}}\right) \cdot d\vec{\ell} = \oint_C (H_r\hat{\mathbf{r}} + H_z\hat{\mathbf{z}}) \cdot d\vec{\ell}$$

as the $\hat{\boldsymbol{\theta}}$ component of $d\vec{\boldsymbol{\ell}}$ is zero. Thus

$$\int_{C_{\text{air}}} (H_r\hat{\mathbf{r}} + H_z\hat{\mathbf{z}})\cdot d\vec{\boldsymbol{\ell}} + \int_{C_{\text{magnet}}} (H_r\hat{\mathbf{r}} + H_z\hat{\mathbf{z}})\cdot d\vec{\boldsymbol{\ell}} = 0. \tag{3.13}$$

Figure 3.40 shows that on C_{air}, $d\vec{\boldsymbol{\ell}}$ and $\vec{\mathbf{H}} = \vec{\mathbf{B}}/\mu_0$ point in the same direction so that the first integral in (3.13) must be positive. As a result, the second integral in (3.13) must be negative meaning that on C_{magnet}, the component of $\vec{\mathbf{H}}$ tangent to the curve C_{magnet} points opposite to $d\vec{\boldsymbol{\ell}}$. In other words, in Figure 3.40, $\vec{\mathbf{H}}$ must be pointing *downwards* inside the magnet while $\vec{\mathbf{B}}$ is pointing *upwards* inside the magnet. Physically, the z components of $\vec{\mathbf{B}}$ and $\vec{\mathbf{M}}$ are in the $+\hat{\mathbf{z}}$ direction inside the magnet which means that $M_z > B_z/\mu_0$ as $H_z = B_z/\mu_0 - M_z < 0$. This is in contrast to the air-filled solenoid coil shown in Figure 3.19(b) where $\vec{\mathbf{B}}$ and $\vec{\mathbf{H}}$ point in the *same* direction inside the coil and $\oint_C \vec{\mathbf{H}}\cdot d\vec{\boldsymbol{\ell}} = i_{\text{free}} \neq 0$ as there is free current on the surface of the solenoid going through the closed curve C. Consequently, in the case of the permanent magnet, the magnetic material is operating in the *second* quadrant of the B–H curve $(B > 0, H < 0)$ at some point *strictly* between points c and d in Figure 3.35(a) while the air-filled solenoid is operating in the *first* quadrant $(B > 0, H > 0)$, at some point strictly between points b and c in Figure 3.35(a). How can this be understood? Consider a process to make a cylindrical magnet as shown in Figure 3.41. To begin, take an iron alloy such as carbon steel (used for permanent magnets) and make it into an oblong shape closed solid as shown in Figure 3.41(a). Note that the straight segments are already "cut out" and will become the cylindrical magnets. Wire is wrapped around this iron core (like a toroid) and the current applied to the winding to magnetize the iron as shown in Figure 3.41(b). Assuming that the core diameter is small compared to the overall length of the core, it is a reasonable approximation to assume that $\vec{\mathbf{B}}$ and $\vec{\mathbf{H}}$ are constant in magnitude over the cross section of the core and that they follow the closed path around the core. As a result,

$$H(2\ell_1 + 2\ell_2) = ni \tag{3.14}$$

where H is the magnitude of $\vec{\mathbf{H}}$ in the core, ℓ_1 is the length of each cylinder and ℓ_2 is the length of each end piece. Using the setup shown in Figure 3.41(b), the current is increased up to the saturation point of the core (point b in Figure 3.42) and then it is brought back down to zero (point c in Figure 3.42). At this point, the magnet is operating at the point c of the B–H curve so that $H = 0$ as $i = 0$ from equation (3.14). The winding is now removed and this does not affect the B–H relationship in the magnetic material of the core as $i = 0$. At each point of the core,

$$\vec{\mathbf{H}} = \frac{\vec{\mathbf{B}}}{\mu_0} - \vec{\mathbf{M}} = 0 \tag{3.15}$$

that is,

$$\vec{\mathbf{B}} = \mu_0 \vec{\mathbf{M}} \quad \text{or} \quad B = \mu_0 M. \tag{3.16}$$

Recall this same result was shown for the toroidal cylindrical core where $B = \mu_0 \lambda_m = \mu_0 M$ with $\lambda_m = M$ an equivalent (bound) surface current density going around core periphery due to the magnetization M (magnetic dipole moment per unit volume).

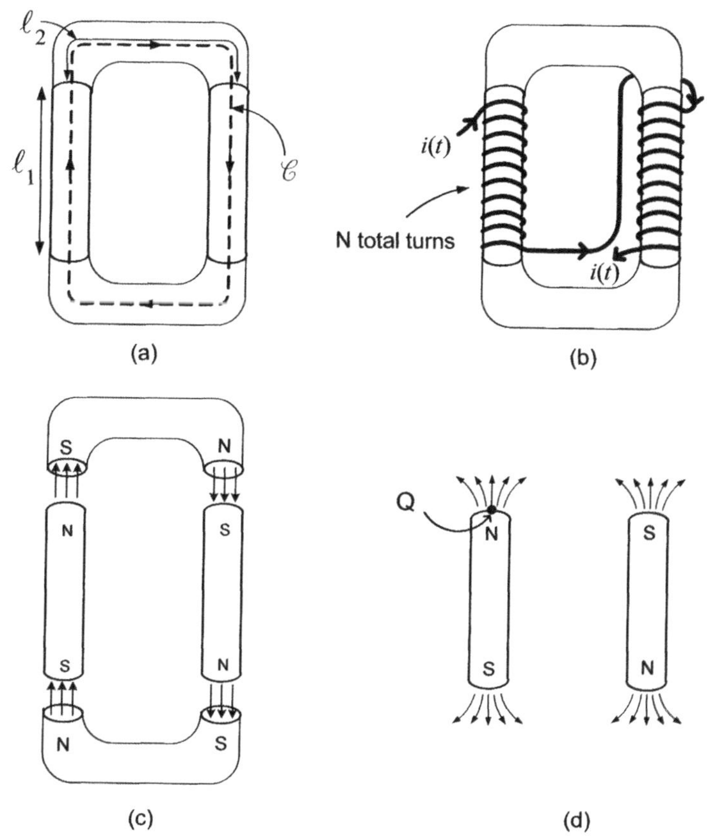

FIGURE 3.41. (a) Oblong shaped iron core with two cylinders cut out. (b) Winding wrapped around the core to magnetize the iron. (c) Current is brought to zero, windings removed and the end pieces removed in the direction shown. (d) Two cylindrical-shaped permanent magnets.

Next, as shown in Figure 3.41(c), the two end pieces of the iron core are pulled away in the direction shown. By pulling apart the end pieces in this manner, the magnetization M in each of the remaining cylindrical cores stays pretty much the same, i.e., the magnetic dipoles of the iron tend to stay aligned. However, with the end pieces removed, one would not expect equation (3.15), or equivalently (3.16), to remain valid for the remaining cylindrical cores. For example, consider the point Q at the end of the cylinder shown in Figure 3.41(d), where now there is only air above the surface instead of magnetic material as before. The B field at any point of the material is due to the magnetic field produced by *all* of the iron atoms in the material, especially those atoms that are close by. Even with the iron uniformly magnetized with the same constant strength M everywhere, at the ends of the cylinder such as the point Q, one would expect that B is reduced (on the order of a factor of 2) as there is only air on one side of the surface rather than magnetized iron as before. Consequently,

$$H = \frac{B}{\mu_0} - M < 0$$

as the value of B at the point Q is reduced after removing the two end pieces even if M remains the same.

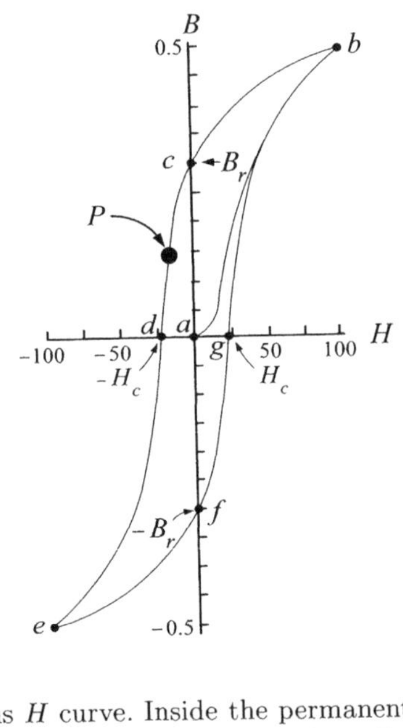

FIGURE 3.42. B versus H curve. Inside the permanent magnet, the material is operating at a point in the second quadrant such as P.

Reiterating, if the magnetization M in the core remains constant and equal to its value before the two end pieces were removed, the H field is now negative in the two cylindrical bar magnets because B is reduced in strength due to the two end pieces being removed. Consequently, as the two end pieces are removed, the operating point for the magnetic material of the two cylindrical magnets moves down the B–H curve from c to a new point (say) P. The closer one is to either end of the cylindrical magnets, the farther down the curve the operating point goes into the second quadrant. On the other hand, at the center of the cylindrical magnets, the operating point is closer to point c and would be exactly at point c if the cylindrical piece was infinitely long.

3.5.1 Hysteresis Loss

Consider a steady-state situation of going around the B–H curve of Figure 3.35(a). That is, after an initial magnetization of the material so that the (B, H) coordinates are at the point b with $i = I_{\max}$, the current is decreased to 0 bringing the system to point c where $H = 0$, then the current is brought down to the point where the B field is zero putting the system at d, and next the current is brought down to $-I_{\max}$ so the system is at e. Going from b to e represents 1/2 of the B–H cycle as indicated in Figure 3.43(a).

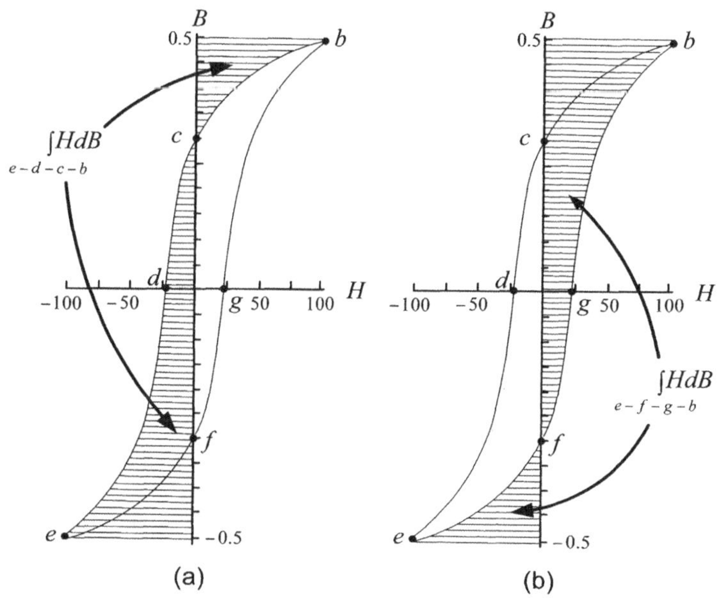

FIGURE 3.43. (a) $\int_{e-d-c-b} HdB$. (b) $\int_{e-f-g-b} HdB$.

The other half is similar. That is, the current is brought back to 0 to put the system at f, next it is increased so that the B field is again zero and the system is at g and finally the current is brought back up to $I_{\max}$ so that the system is back at b (see Figure 3.43(b)).

At any time t, the flux linkage in the primary windings wound around the toroid core is $\lambda(t) = N\phi(t) = NB(t)S$ where N is the total number of primary windings, S is the cross-sectional area of the core and B is the magnetic field in the core. Further, at any time t, $H(t) = Ni(t)/(2\pi r)$ where r is a radius inside the toroid. Suppose that one goes through this complete cycle b–c–d–e–f–g–b in the time T. Then the energy from the source which is providing the current to the coil is given by

$$\begin{aligned} E_{\text{one cycle}} &= \int_{t_1}^{t_1+T} i\dot{\lambda}dt = \int_{t_1}^{t_1+T} \frac{2\pi r H(t)}{N} NS\frac{dB}{dt}dt \\ &= 2\pi r S\int_{t_1}^{t_1+T} H(t)\frac{dB}{dt}dt \end{aligned}$$

By a change of variables, this becomes

$$E_{\text{one cycle}} = V\oint_{\text{b-c-d-e-f-g-b}} HdB$$

where $V = 2\pi r S$ is the volume of the core. In the first half period $T/2$, the energy provided by the source is

$$E_{\text{First half-cycle/unit volume}} = \int_{\text{b-c-d-e}} HdB = -\int_{\text{e-d-c-b}} HdB$$

where $\int_{\text{e-d-c-b}} HdB$ is the area indicated in Figure 3.43(a). Similarly,

$$E_{\text{Second half-cycle/unit volume}} = \int_{\text{e-f-g-b}} HdB$$

which is the area indicated in Figure 3.43(b). The total energy absorbed per cycle is then

$$E_{\text{cycle/unit-volume}} = \int_{\text{e-f-g-b}} HdB - \int_{\text{e-d-c-b}} HdB = \oint_{\text{b-c-d-e-f-g-b}} HdB$$

which is just the area inside the B–H curve. This is the energy required for each cycle to magnetize (align the dipole moments of) the iron along the B–H curve (i.e., rotate the magnetic dipole moments 360°).

As an example, consider the iron core inside the rotor of a DC motor which is constantly being magnetized and demagnetized as it rotates by the external magnetic field. For example, the point P in Figure 3.44 corresponds to the point b in Figure 3.35. After the iron core rotates clockwise 90 degrees, the magnetic material at point P corresponds to the point d in Figure 3.43. After rotating another 90 degrees, the magnetic material at point P corresponds to point e in Figure 3.43, and so on. That is, the magnetic material at each point in the core is going through a complete cycle of the B–H curve each time the core rotates 360 degrees. In this application, it is important that the core be made of soft magnetic material with a very narrow B–H loop so that the hysteresis loss is small.

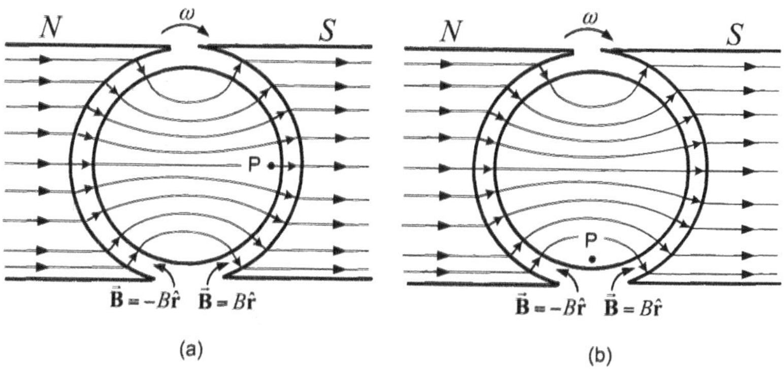

FIGURE 3.44. The magnetic material of the rotor goes through the $B-H$ curve as it rotates.

3.5.2 Common Magnetic Materials

Both soft and hard magnetic materials are used in electric machines. Soft magnetic materials are used as a core structure on which the turns (windings) of the motor are wound (e.g., such as the field circuit of a separately excited DC motor). In such an application, the copper wire of the field winding is wrapped around soft magnetic material, which then results in a much larger radial magnetic field in the air gap than if nonmagnetic material was used. The most common material for the core in an electric machine is an iron-silicon alloy (silicon steel) which contains between 1% and 4% silicon [56]. The relative permeability μ_r of silicon steel ranges from 400 to 800 [56][7] which seems low compared to the gauge steel and iron casting steel of Figure 3.36. However, its low, good mechanical strength and a high Curie Point[8] of 740° C makes it an attractive material [56] and provides a challenge to the machine designer to have short paths for the magnetic field in the iron. It turns out that by special processing, some 4% silicon steels can have a relative permeability 5.5 times greater than the 800 quoted above, that is, up to 4400 [56].

Small permanent magnet synchronous machines often use rare-earth cobalt alloys for the hard magnetic material which has a rather high cost [56]. As a consequence, it is usually bonded to the surface of rotor which is made of soft magnetic material. The rare-earth cobalt alloys have a remanent (permanent) magnetic field between 0.75 and 1.1 tesla, are very hard to demagnetize and have a high Curie temperature [56].

---

[7]Reference [56] gives the permeability range as 5 to 10 kilogauss/oersted $= 0.5 \times 10^4$ to 1.0×10^4 gauss/(10^3ampere/meter) $= 0.5 \times 10^{-3}$ to 1.0×10^{-3} telsa/(ampere/meter). The relative permeability is found by dividing through by $\mu_0 = 4\pi \times 10^{-7}$ henry/meter giving 400 to 800.

[8]This is the temperature at which the material loses it magnetic properties.

Problems

Problem 1 Azimuthal Magnetic Field of a Circular Current Loop

Consider the circular current loop shown in Figure 3.45. Using symmetry arguments and Ampère's law, show that the azimuthal component $B_\theta\hat{\boldsymbol{\theta}}$ *of* $\vec{\mathbf{B}}$ *must be zero everywhere.*

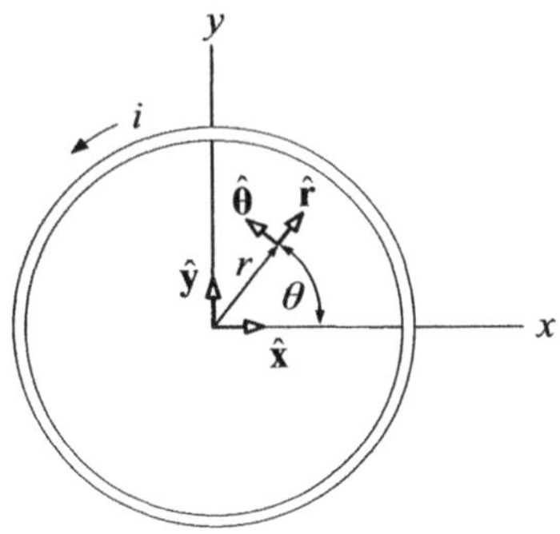

FIGURE 3.45. Circular loop carrying the current i.

Problem 2 Ampère's Law [35]

(a) Figure 3.46 shows a long straight conductor carrying a current I. *The current is uniformly distributed across the cross section so that the current* i *enclosed within a circle of radius* r *is*

$$i = \begin{cases} I\dfrac{r^2}{R^2} & \text{for } 0 \le r \le R \\ \\ I & \text{for } r > R. \end{cases}$$

Use Ampère's law to find the magnetic field B *inside the conductor.*

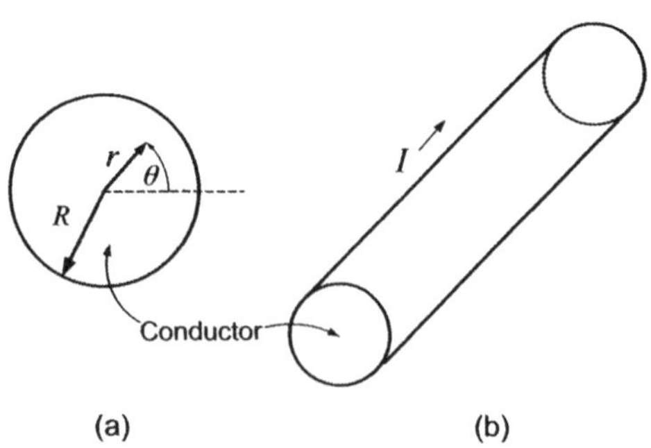

FIGURE 3.46. Long straight wire. (a) Cross-sectional view. (b) Perspective view.

(b) A cylindrical conductor with an off axis cylindrical hole carries a total current I. Find the magnetic field in the cylindrical hole. [Hint: Use part (a) and superposition.]

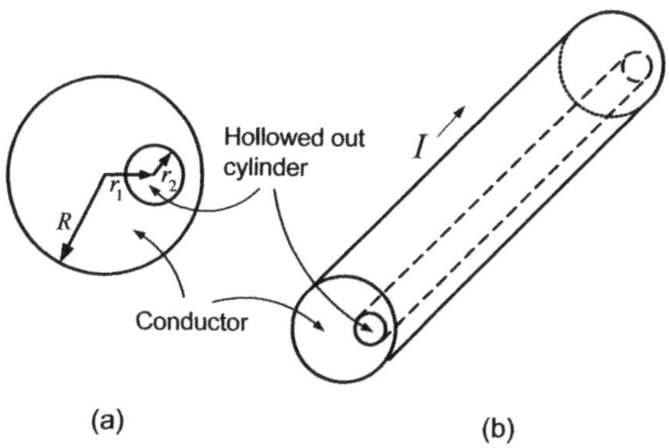

FIGURE 3.47. Conductor with a hollowed cylindrical core. (a) Cross-sectional view. (b) Perspective view.

Problem 3 Radial Magnetic Field of a Long Straight Wire

Apply conservation of flux to the cylindrical-shaped flux surface shown in Figure 3.48 to show that $B_r(r) \equiv 0$ for an infinitely long straight wire. Denote by R the radii of the disk-shaped surfaces S_1, S_2 and let ℓ denote the length of the cylindrical-shaped surface S_3.

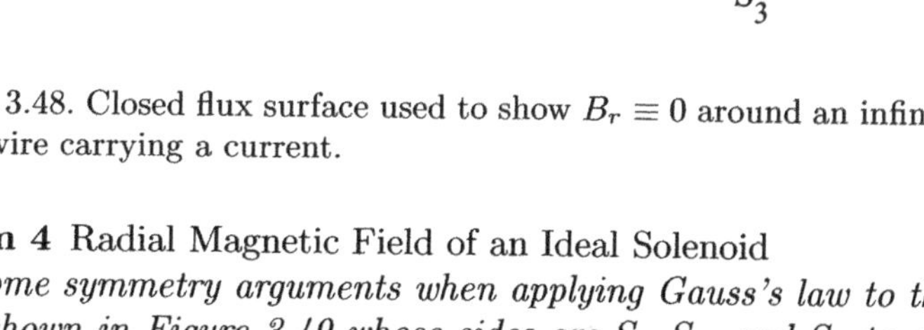

FIGURE 3.48. Closed flux surface used to show $B_r \equiv 0$ around an infinitely long straight wire carrying a current.

Problem 4 Radial Magnetic Field of an Ideal Solenoid

Use some symmetry arguments when applying Gauss's law to the closed surface shown in Figure 3.49 whose sides are S_1, S_2, and S_3 to show that the radial component of $\vec{\mathbf{B}}$ must be zero inside *an infinitely long solenoidal coil. Denote by R the radii of the disk-shaped surfaces S_1 and S_2, and let ℓ denote the length of the cylindrical-shaped surface S_3.*

Using a similar argument, show that the radial component of $\vec{\mathbf{B}}$ outside *of an infinitely long solenoidal coil must also be zero.*

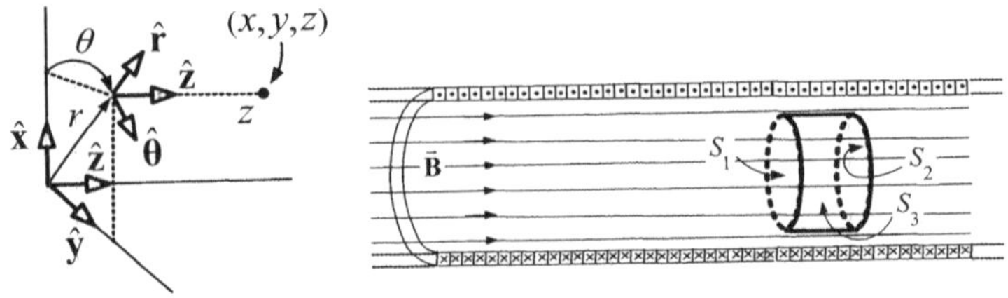

FIGURE 3.49. Using Gauss's law to show $B_r \equiv 0$ inside an ideal solenoid.

Problem 5 Azimuthal Magnetic Field of an Ideal Solenoid

Apply Ampère's law to the curves C_3, C_5 of Figure 3.27 to show that $B_\theta(r) \equiv 0$ inside the solenoid and $B_\theta(r) = \mu_0 i/(2\pi r)$ outside the cylinder.

Problem 6 Magnetic Field in an Ideal Toroidal Coil

(a) In Figure 3.50, apply Ampère's law to the circular curve C and Gauss's law to the closed surface S whose sides are S_1, S_2, and S_3 to show that $\vec{\mathbf{B}}$ only has a nonzero azimuthal component inside *the toroidal coil, that is, the lines of B are concentric circles inside the toroid coil. (Hint: Let $\vec{\mathbf{B}} = B_\rho \hat{\boldsymbol{\rho}} + B_\varphi \hat{\boldsymbol{\varphi}} + B_\theta \hat{\boldsymbol{\theta}}$ using the "toroidal" coordinate system of Figure 3.50 and show that $B_\rho = B_\varphi = 0$.)*

(b) Show that the azimuthal component of $\vec{\mathbf{B}}$ must be zero outside *the toroidal coil.*

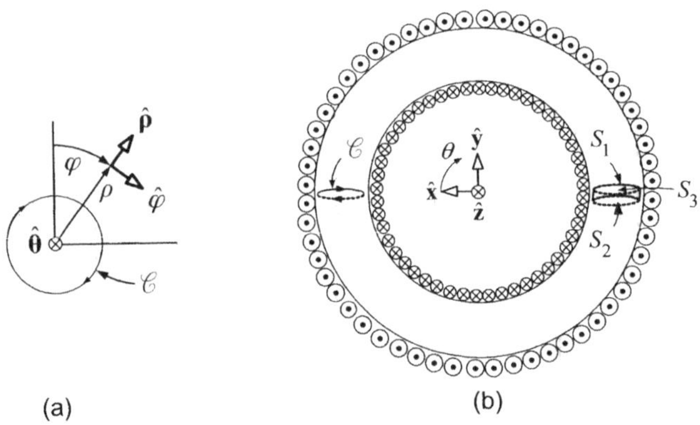

FIGURE 3.50. (a) "Toroidal" coordinate system. $\hat{\boldsymbol{\theta}}$ is into the page. (b) Curve and surface for Ampère's and Gauss's law, respectively, to show that $B_\rho = B_\varphi = 0$.

Problem 7 Equations of the Ideal Transformer from Ampère's Law

Consider an ideal transformer in Figure 3.51 consisting of two sets of turns wound on a single iron core. Assuming ideal soft magnetic material ($H = 0$ in the iron), N_1 turns in phase 1, and N_2 turns in phase 2, apply

Ampère's law to a closed path C in the core of the iron to show that

$$i_1 = -\frac{N_2}{N_1} i_2.$$

Use this result and conservation of power $v_1 i_1 + v_2 i_2 = 0$ *(assuming there are no losses) to show that*

$$v_1 = \frac{N_1}{N_2} v_2.$$

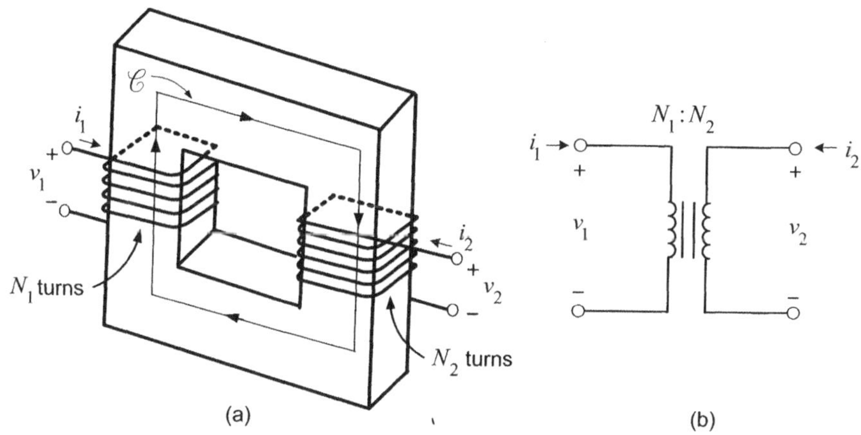

FIGURE 3.51. Derivation of the ideal transformer equation using Ampère's law. Adapted from Figure 20.12 of Kassakian et al. [57].

Problem 8 Equations of the Ideal Transformer from Faraday's Law

Consider an ideal transformer as shown in Figure 3.52 consisting of two sets of turns wound on a single iron core. Assume ideal soft magnetic material ($H = 0$ *in the iron*), N_1 *turns in phase 1 and* N_2 *turns in phase 2. Let* ϕ *be the flux in any cross section of the iron core.*

(a) Compute ϕ *using the normal* $\hat{\mathbf{n}}_1$ *in the turns of phase 1, that is,*

$$\phi = \int_S \vec{\mathbf{B}} \cdot d\vec{\mathbf{S}}$$

with $d\vec{\mathbf{S}} = dS\hat{\mathbf{n}}_1$ *and the corresponding flux linkage* $\lambda_1 = N_1\phi$. *What is the sign convention for the induced voltage* $-d\lambda_1/dt$? *That is, if*

$$-\frac{d\lambda_1}{dt} > 0,$$

will it attempt to force current in the positive direction of i_1 *or the negative direction?*

(b) Use Gauss's law with the iron core defining a flux tube to explain why ϕ*, as computed in part (a), can also be computed using the normal* $\hat{\mathbf{n}}_2$ *to obtain the same value.*

(c) Use the fact that by Faraday's law, $v_1 = -d\lambda_1/dt, v_2 = -d\lambda_2/dt$ *to show that*

$$v_1 = \frac{N_1}{N_2} v_2.$$

(d) Use this result and $v_1 i_1 + v_2 i_2 = 0$ *(assuming there are no losses) to show that*

$$i_1 = -\frac{N_2}{N_1} i_2.$$

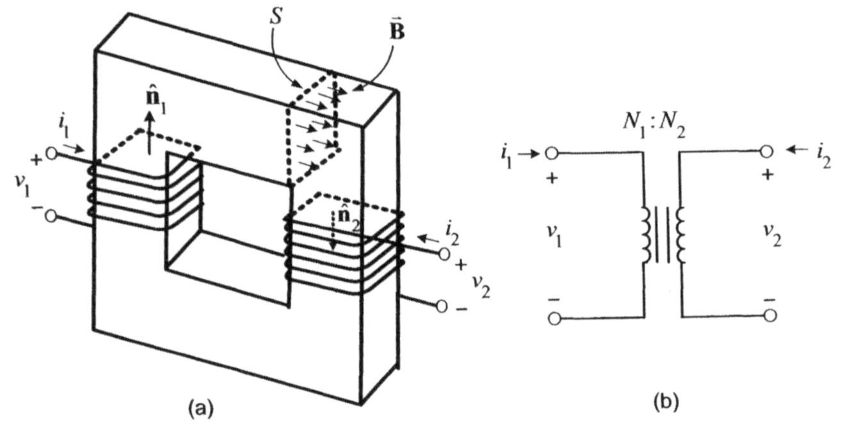

FIGURE 3.52. Derivation of the ideal transformer equations using Faraday's law. Adapted from Figure 20.12 of Kassakian et al. [57].

Problem 9 Magnetizing Inductance in the Transformer Model [25][57]

Consider the transformer in Figure 3.53 to be non ideal in the sense that H *is not zero in the iron; that is, assume that* μ_r *is large, but finite. It is still assumed that the magnetic fields* B *and* H *are confined to the core, follow the path of the core, and are constant in magnitude in the core.*

(a) With a current in phase 1, but phase 2 open-circuited, apply Ampère's law to the curve C *and use this to show that the magnetic field in the core is given by*

$$B = \mu_r \mu_0 H = \frac{\mu_r \mu_0}{\ell_c} N_1 i_m \text{ (phase 2 open-circuited)} \tag{3.17}$$

where i_m *is the current in phase 1 with phase 2 open circuited,* ℓ_c *is the length of the path* C *and* μ_r *is the relative permeability of the iron in the core. Of course the length* ℓ_c *is different in value for different paths. For this* *approximate analysis, one just takes ℓ_c to be some average length around the core.*

(b) Show that the flux ϕ in the core is given by

$$\phi = \frac{\mu_r \mu_0 S}{\ell_c} N_1 i_m. \tag{3.18}$$

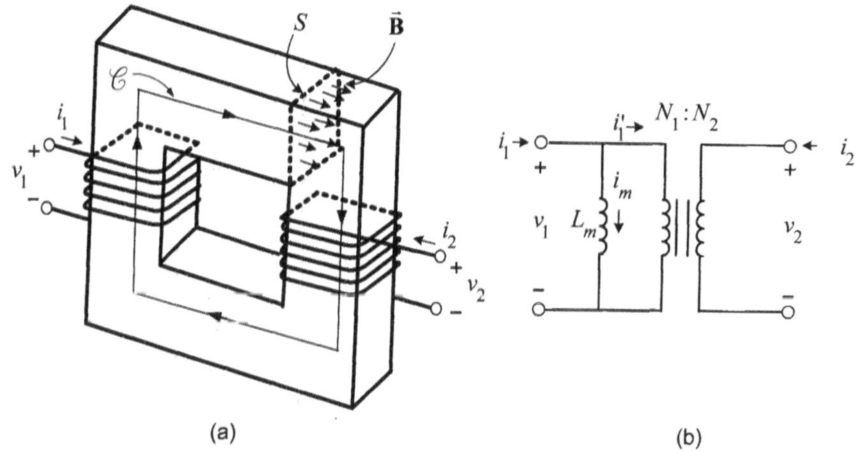

FIGURE 3.53. Magnetizing inductance L_m for a transformer.

By the assumption that the magnetic fields B and H are confined to the core and follow the path of the core, this flux is the same everywhere in the core. The flux linkage in phase 1 is

$$\lambda_1 = N_1 \phi = \frac{\mu_r \mu_0 S}{\ell_c} N_1^2 i_m = L_m i_m.$$

This current i_m (with phase 2 open-circuited) is called the magnetizing current *and is the current required to produce the magnetic field B (by aligning/magnetizing the dipole moments of the iron atoms) given in equation (3.17) and the corresponding flux ϕ in the core as given in equation (3.18). $L_m \triangleq \mu_r \mu_0 N_1^2 S/\ell_c$ is called the* magnetizing inductance *and the equivalent circuit for a transformer with a finite μ_r is shown in Figure 3.53(b).*

(c) By Faraday's law, the flux in the iron core and the applied voltage are related by

$$-\frac{d}{dt}\lambda_1 = -\frac{d}{dt} N_1 \phi = v_1$$

or

$$\phi = \frac{1}{N_1} \int_0^t v_1(\tau) d\tau.$$

That is, the flux ϕ *in the core is completely determined by the applied voltage to the phase. Consider now that phase 2 has a load (e.g., a resistor) attached to it so* i_2 *is nonzero and apply the same voltage as before to phase 1. Explain why the total flux is now*

$$\phi = \frac{\mu_r \mu_0 S}{\ell_c}\left(N_1 i_1 + N_2 i_2\right).$$

(d) Decompose the current i_1 *as* $i_1 = i_m + i_1'$ *where* i_m *is the (magnetizing) current with phase 2 open and* i_1' *is the additional current in phase 1 when phase 2 is not an open circuit. The flux can then be written as*

$$\phi = \frac{\mu_r \mu_0 S}{\ell_c}\left(N_1 i_m + N_1 i_1' + N_2 i_2\right)$$

As the flux depends only on the applied voltage v_1, *explain why*

$$N_1 i_1' + N_2 i_2 \equiv 0$$

is always true.

(e) Explain why it is still true with finite μ_r *that*

$$v_2 = \frac{N_2}{N_1} v_1.$$

(f) Show that in an ideal transformer, the flux $\phi \equiv 0$. *How does this affect the derivation in problem 8?*

Remark *The effect of finite* μ_r *in a transformer is accounted for in the equivalent circuit model by the addition of an inductor with inductance* L_m *in parallel with the ideal transformer. Note that as* $\mu_r \to \infty$, $L_m \to \infty$, *that is, it becomes an infinite impedance and the circuit reduces to the ideal case with* $i_1 = i_1'$. *Note that in the ideal transformer with phase 2 open, applying a voltage* v_1 *to phase 1 would result in* i_1 *also being zero; that is, phase 1 would behave as open circuit due to the magnetic coupling of the two phases. In this nonideal case with phase 2 open, after applying a voltage* v_1, *there is a nonzero current* i_m *in phase 1 which satisfies* $v_1 = L_m di_m/dt$.

Problem 10 B–H Curve

Explain how a transformer can be used to experimentally determine the B–H *curve of its iron core, i.e., the core is never perfectly ideal soft magnetic material and therefore has a hysteresis curve to be measured.*

Problem 11 Magnetic Force [35]

Consider the electromagnetic crane drawn in Figure 3.54. Figure 3.54(a) shows the movable piece a distance x *below the electromagnet. The current in the wire magnetizes the iron core it is wrapped about to produce a magnetic field which crosses the air gaps to magnetize the movable piece attracting it to the electromagnet. Let the iron have permeability* μ_r, *the cross section have area* $S = \ell_1\ell_2$, *the number of turns of the coil be* N, *the path length of the closed curve* C *in Figure 3.54(a) be* $\ell + 2x$ *and* ℓ *in Figure 3.54(b). It is assumed that* B *and* H *are both constant in the iron and (a different) constant in the air gap and that these fields are contained within the iron and air gaps. Compute the force of the electromagnet on the moveable piece by doing the following:*

(a) Use Ampère's law to compute the magnetic field B *in the core (which is equal to the magnetic field strength in the air gap for small* x*) as a function of the current* i *in the winding.*

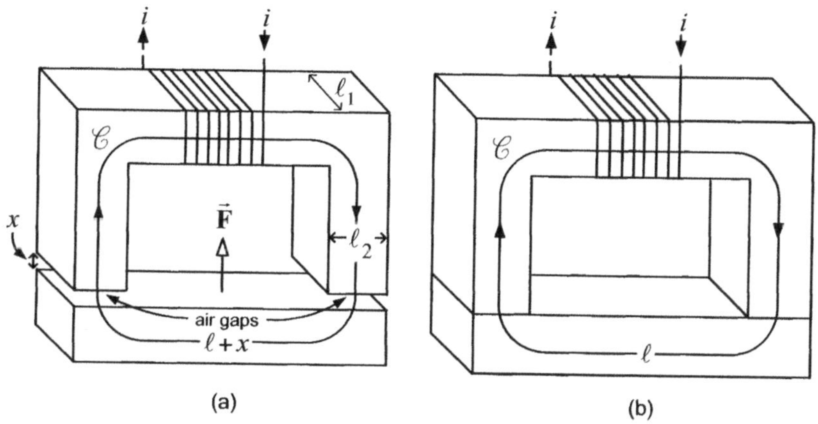

FIGURE 3.54. Electromagnet exerting a force on a rectangular-shaped iron block.

(b) Use the answer of part (a) to show that the flux linkage of the winding is

$$\lambda = N\phi = NBS = \frac{NS\mu_0}{(\ell/\mu_r + 2x)} i = L(x)i$$

where

$$L(x) \triangleq \frac{NS\mu_0}{(\ell/\mu_r + 2x)}.$$

(c) With x *held constant, compute the energy* W *required to bring the current from* 0 *to* I*; that is, compute*

$$W(x, I) = \int_0^t i\frac{d\lambda}{dt} = \int_0^I i d\lambda.$$

This energy from the source current is stored as magnetic energy in the iron core and the air gaps.

(d) With the current held constant at I*, if the moveable piece moves by the force of attraction from* x *to* 0 *(so the change in position is* $-x$*), the energy goes from* $W(x, I)$ *to* $W(0, I)$*. The change in energy equals the work*

done, that is, with x small

$$\left(F\Big|_{x=0} \right) \times (-x) = W(x, I) - W(0, I).$$

Use this to show that the force of attraction at $x = 0$ in Figure 3.54(b) is

$$F\Big|_{x=0} = -\frac{\partial W(x, I)}{\partial x}\Bigg|_{I\ constant}.$$

(e) With the bar at the location x and the current in the winding given by i, explain why the force is given by

$$F(x, i) = -\frac{\partial W(x, i)}{\partial x}$$

even if the current is changing in time. That is, holding the current constant was only required for the derivation of the expression for the force. (Note that the notation "I constant" in $-\frac{\partial W(x,I)}{\partial x}\Big|_{I\ constant}$ is redundant as the partial derivative $\partial/\partial x$ means to take the derivative with the other variable held constant.)

Problem 12 Relative Permeability

Consider a permanent magnet which is placed in another (external) magnetic field. Explain that, as far as this external magnetic field's effect on the permanent magnet, one can take $\mu_r = 1$ in the permanent magnet.

Part II

AC Machine Theory

With the preliminary material presented in the first three chapters, the remaining chapters cover the modeling and control of AC machines. However, before embarking on this odyssey, an elementary notion of how AC machines operate is first presented. This will hopefully help the reader to understand the overall picture as he/she goes through the detailed analysis in the remaining chapters.

Induction Motor

The goal here is to give an elementary idea of how an induction motor works. First, the construction of a simple two-phase machine is described. This is followed by explaining how a radially directed rotating magnetic field is set up in the air gap by the stator currents and how this rotating field induces currents in the rotor loops. Finally, it is described how this same rotating magnetic field produces torque of on the current it induces.

Construction of the Induction Motor

To begin, the construction of a simplistic model of a two-phase induction motor is described.

Figure II.1(a) shows a half-cylindrical-shaped loop, which is wound around a cylindrical-shaped iron core, and denoted as loop a. A second identical loop, denoted as loop b, is then wound 90 degrees from loop a as shown in Figure II.1(b). The currents in loops a and b are denoted as i_{Ra}, i_{Rb}, respectively. The two loops are electrically isolated.

In Figure II.1, the notation $\odot$ means that if $i > 0$, the current is coming *out* of the page while $\otimes$ means that if $i > 0$, then the current is going *into* the page.

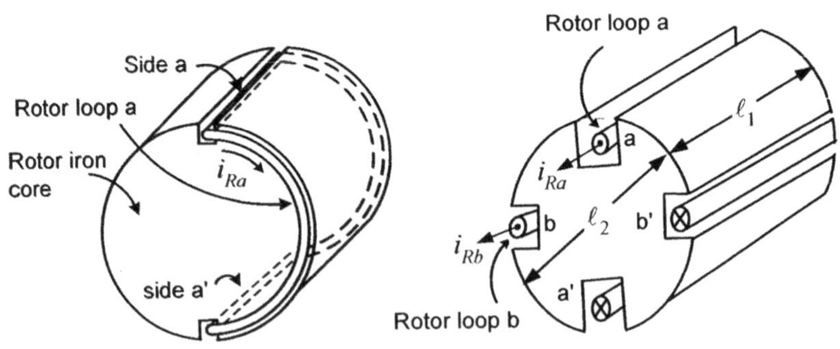

Figure II.1. Rotor of an induction motor.

The stator is constructed similarly. Figure II.2(a) shows stator loop a

that has a half-cylindrical shape and is wound on the inside surface of the stator iron. However, as shown in Figure II.2(a), it also has a voltage source connected. Similarly, Figure II.2(b) shows that stator loop b is identical in form to stator loop a, but is wound 90 degrees from loop a. The applied voltages are denoted as u_{Sa}, u_{Sb} and the corresponding currents are denoted as i_{Sa}, i_{Sb}, respectively. These two loops are electrically isolated.

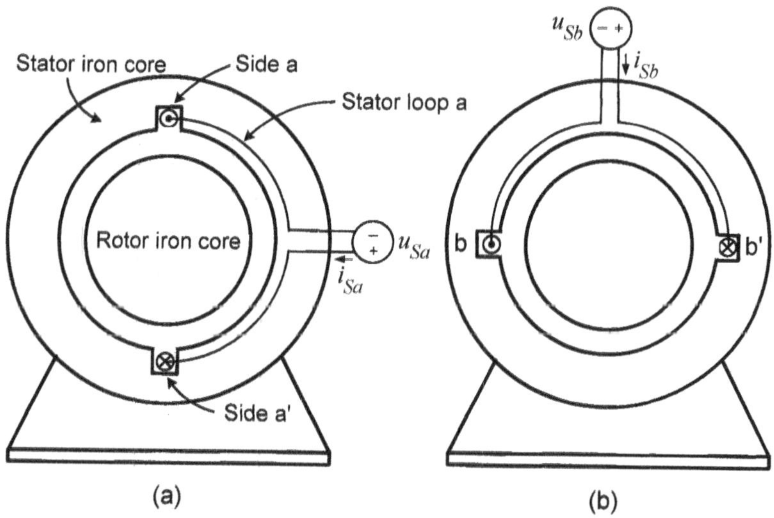

Figure II.2. Induction motor stator. (a) Stator loop a. (b) Stator loop b.

Combining Figures II.1 and II.2, a simple two-phase induction motor is illustrated in Figure II.3. The position of the rotor is located by a line perpendicular to rotor loop a as shown in Figure II.3.

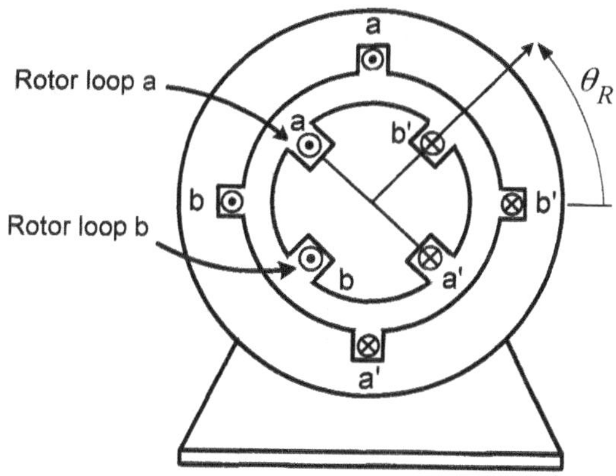

Figure II.3. Cross-sectional view of a simple two-phase induction motor.

Rotating Magnetic Field

Figure II.4(a) shows that the magnetic field in the air gap produced by $i_{Sa} > 0$ (with $i_{Sb} = 0$) is radially out on the right-hand side and radially in on the left-hand side. Figure II.4(b) shows that the directions are simply reversed for $i_{Sa} < 0$.

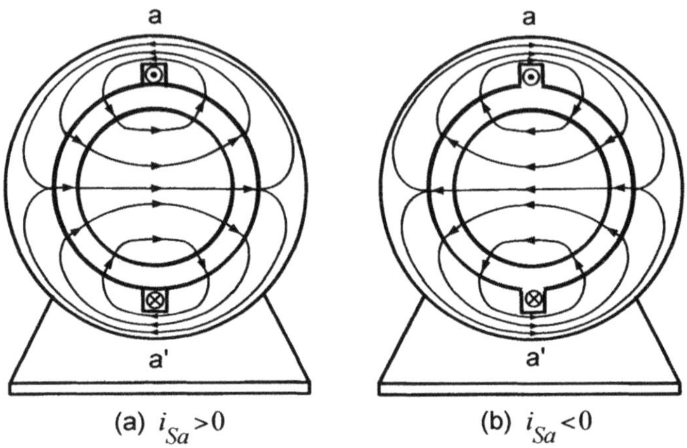

Figure II.4. Magnetic field distribution produced by i_{Sa}.

Similarly, Figure II.5(a) shows that the magnetic field in the air gap produced by $i_{Sb} > 0$ (with $i_{Sa} = 0$) is radially out on the top-half side of the air gap and radially in on the bottom-half side of the air gap. The directions are simply reversed for $i_{Sb} < 0$.

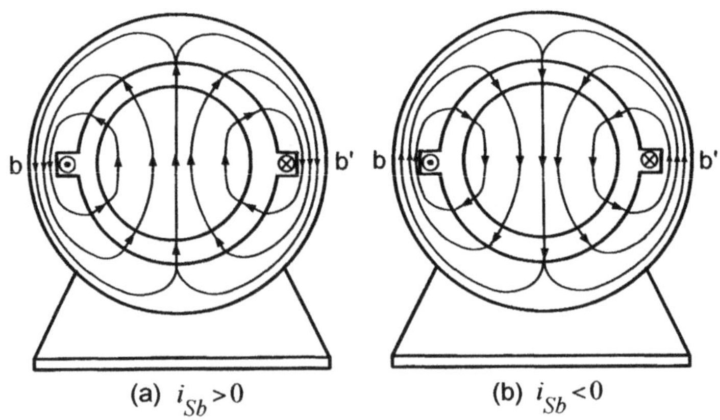

Figure II.5. Magnetic field distribution produced by i_{Sb}.

Combing the magnetic fields of stator loops a and b, a *rotating* magnetic field is created in the air gap by choosing

$$i_{Sa} = I_S \cos(\omega_S t), i_{Sb} = I_S \sin(\omega_S t).$$

This is illustrated in Figure II.6, which shows the total magnetic field $\vec{\mathbf{B}}_S$ due to both i_{Sa} and i_{Sb}.

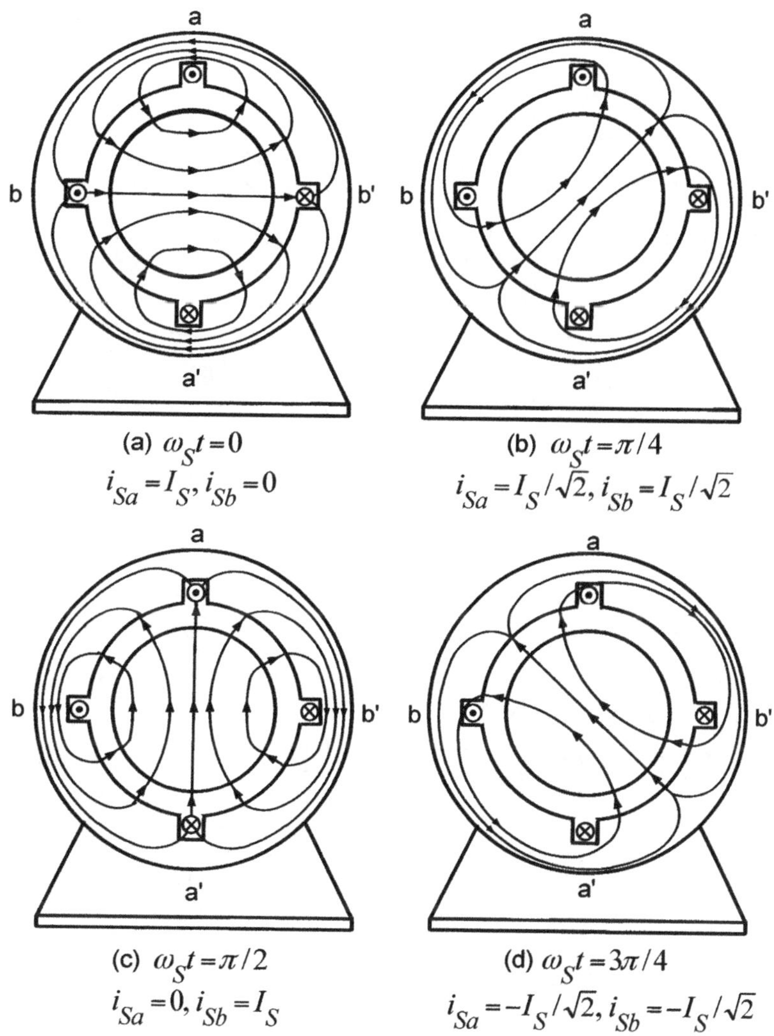

(a) $\omega_S t = 0$, $i_{Sa} = I_S, i_{Sb} = 0$

(b) $\omega_S t = \pi/4$, $i_{Sa} = I_S/\sqrt{2}, i_{Sb} = I_S/\sqrt{2}$

(c) $\omega_S t = \pi/2$, $i_{Sa} = 0, i_{Sb} = I_S$

(d) $\omega_S t = 3\pi/4$, $i_{Sa} = -I_S/\sqrt{2}, i_{Sb} = -I_S/\sqrt{2}$

Figure II.6. Rotating magnetic field.

Induced Rotor Currents

The rotating magnetic field $\vec{\mathbf{B}}_S$ produced by the stator currents produces a changing flux in each of the rotor phases. By Faraday's law, voltages and hence currents, are produced in the rotor phases. This is illustrated in Figure II.7 showing the rotation of the stator magnetic field past rotor loop a (rotor loop b is not shown for clarity). As it rotates past rotor loop a, the changing flux in the loop produces an emf, which in turn produces the current i_{Ra}.

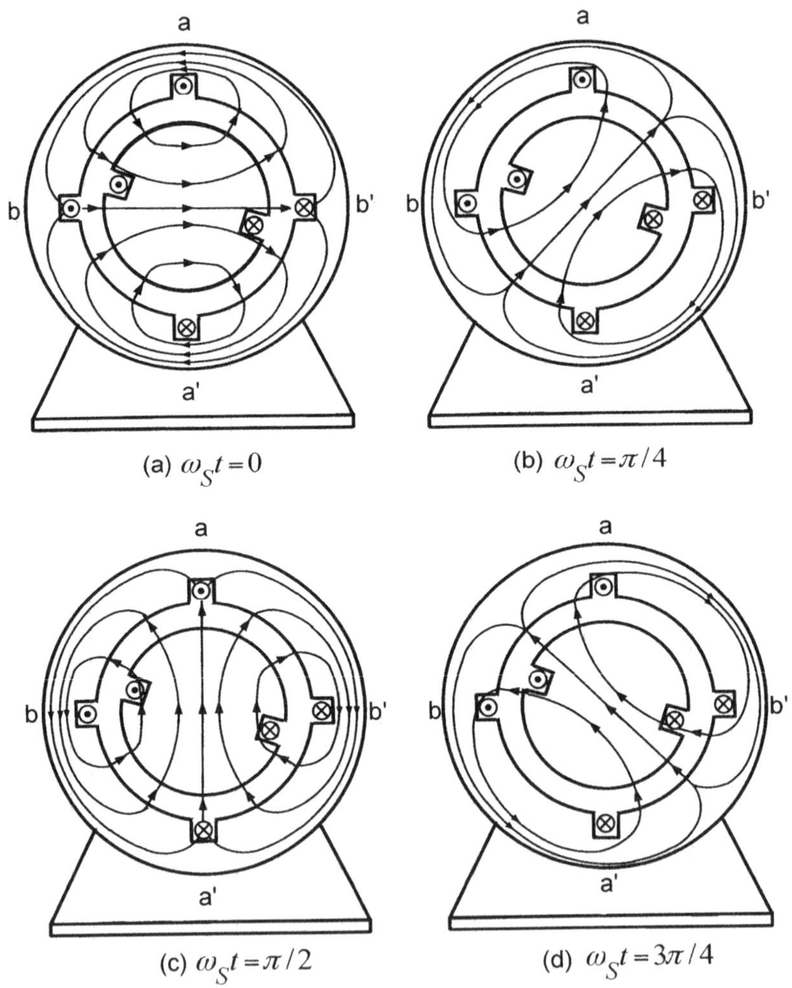

Figure II.7. The changing flux in rotor phase a due to the rotating magnetic field of the stator. Rotor phase b is not shown.

Torque

Recall that from the magnetic force law, the force on side a of rotor loop a is given by

$$F_{\text{side a}} = i_{Ra}\ell_1 B_S$$

where ℓ_1 is the length of the rotor. By the right-hand rule, this force has the direction shown in Figure II.8. The torque on this rotor loop is then

$$\tau_{\text{side a}} = (\ell_2/2) i_{Ra}\ell_1 B_S$$

where $\ell_2/2$ is the radius of the rotor. Similarly, the forces on sides a', b, and b' can be computed and are shown in Figure II.8. With the currents in the rotor phases as shown in Figure II.8, the magnetic force law shows that the sides of each rotor loop have a tangential force pushing the rotor in the counterclockwise direction.[1] That is, there is a torque on the rotor, causing it to turn.

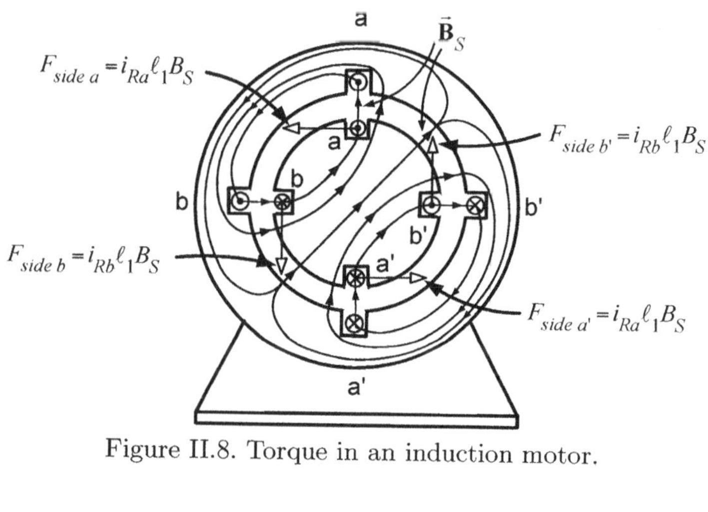

Figure II.8. Torque in an induction motor.

Slip

ω_S is the angular speed of the stator's rotating magnetic field and ω_R is the rotor speed. The slip speed is defined as

$$\omega_{slip} \triangleq \omega_S - \omega_R.$$

It turns out that positive torque production requires $\omega_S > \omega_R$, while to obtain negative torque $\omega_S < \omega_R$ and no torque is produced if $\omega_S = \omega_R$.

[1] That the currents in the rotor loops are in the directions indicated in the figure can only be proven using the analysis developed in later chapters.

That is, $\omega_{slip} \triangleq \omega_S - \omega_R \neq 0$ when producing torque. The induction motor is said to operate *asynchronously* since the rotor angular speed ω_R is not the same as the angular speed ω_S ($\omega_S \neq \omega_R$) of the stator's rotating magnetic field when producing torque. The fact that the slip must be nonzero to get torque is easily explained. Consider the situation shown in Figure II.9 where $\omega_{slip} = 0$ or $\omega_S = \omega_R$. As Figure II.9 shows, the flux in the rotor loop does not change and consequently there is no induced voltage nor current in the rotor loop and, therefore, no torque is produced.

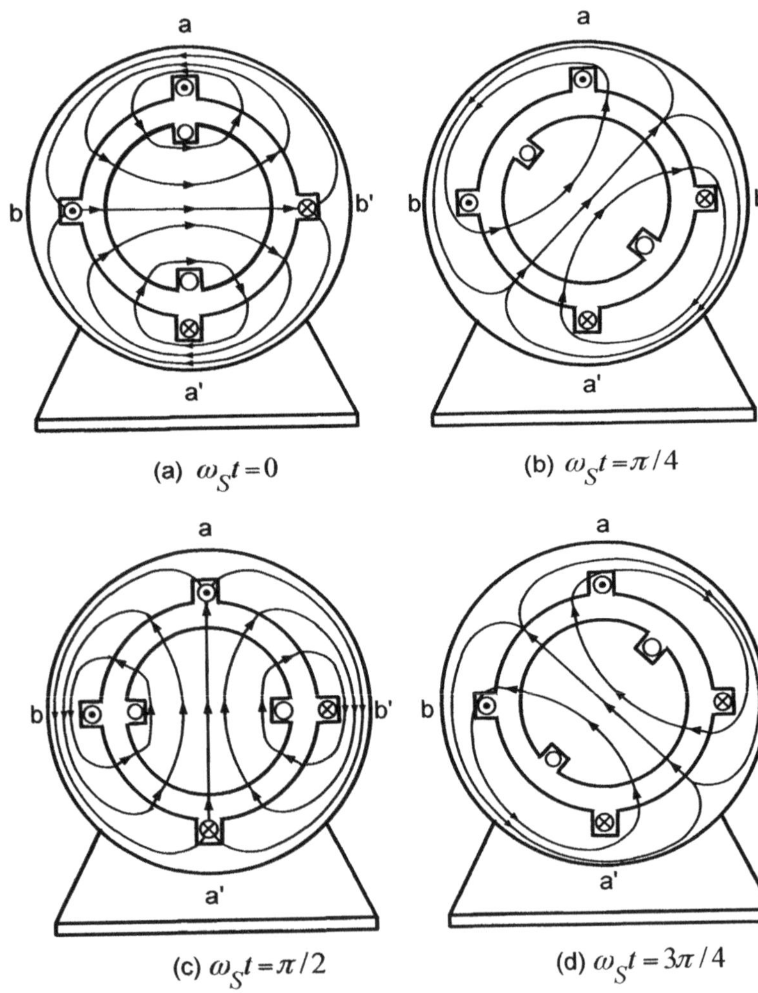

Figure II.9. No changing flux in the rotor loop if $\omega_S = \omega_R$. (Rotor loop b not shown for clarity.)

Summary

Summarizing, the induction motor works as follows:

- A rotating radially directed magnetic field $\vec{\mathbf{B}}_S$ is set up in the air gap between the rotor and stator iron by the stator currents i_{Sa}, i_{Sb}.
- This rotating magnetic field produces a changing flux in the rotor loops (phases).
- By Faraday's law, voltages, and hence currents, are induced in the rotor loops.
- This *same* magnetic field $\vec{\mathbf{B}}_S$ that induced the currents in the rotor loops produces a torque on them.

Permanent Magnet Synchronous Motor

The principle of operation of a synchronous machine is illustrated in Figure II.10, which shows two permanent magnets rotating on different axles. The "stator" permanent magnet is rotated at angular speed ω_S by mechanically turning it on its axle. As it rotates, its north end attracts the south end of the rotor permanent magnet to turn it on its axle. Consequently, they rotate together at the same speed (i.e., $\omega_R = \omega_S$) so that their speeds are *synchronized.*

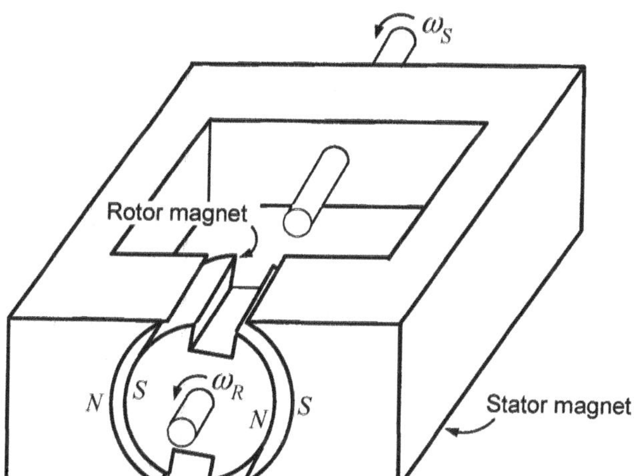

Figure II.10. Two permanent magnets rotating on different axles.

It was shown above that the stator of an induction motor can produce a rotating magnetic field. To make a synchronous motor, one simply removes the rotor from the induction motor and replaces it with a permanent magnet rotor as shown in Figure II.11. As in the case of the induction motor, the stator currents i_{Sa}, i_{Sb} produce a rotating magnetic field. This rotating magnetic field then pulls the rotor magnet along with it at the same angular speed. That is, $\omega_S = \omega_R$ so that the rotor itself and the stator magnetic field rotate together *synchronously*.

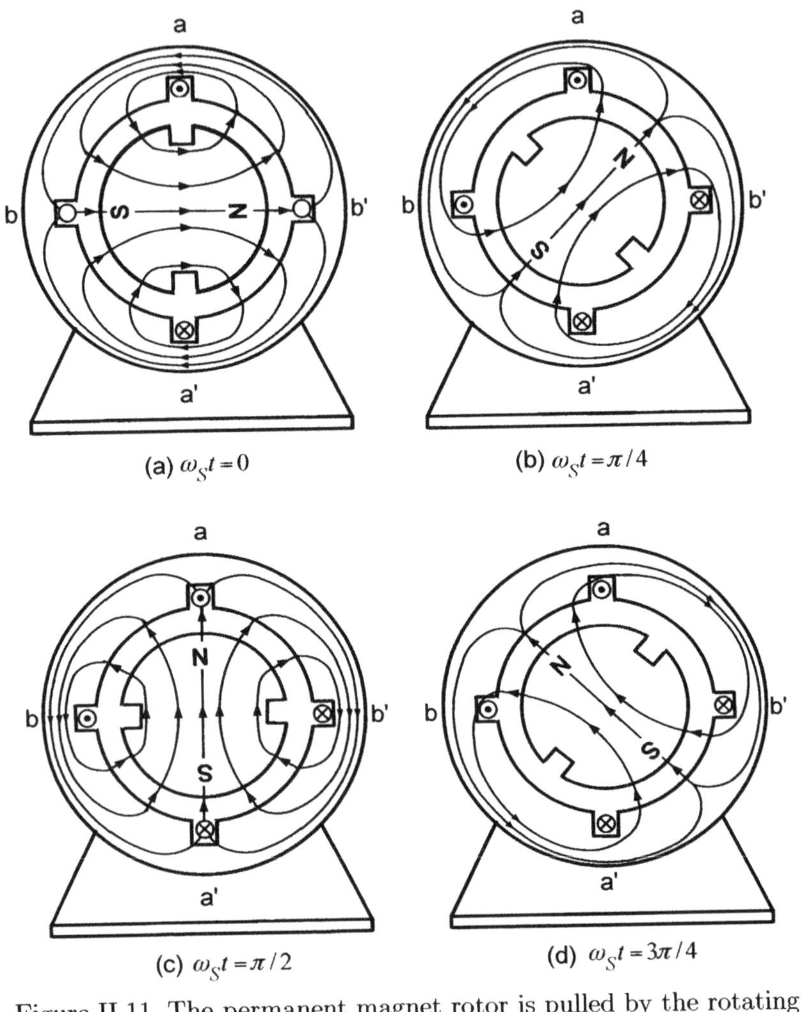

(a) $\omega_S t = 0$ (b) $\omega_S t = \pi/4$ (c) $\omega_S t = \pi/2$ (d) $\omega_S t = 3\pi/4$

Figure II.11. The permanent magnet rotor is pulled by the rotating magnetic field of the stator. (a) $i_{Sa} = I_S$, $i_{Sb} = 0$. (b) $i_{Sa} = I_S/\sqrt{2}$, $i_{Sb} = I_S/\sqrt{2}$. (c) $i_{Sa} = 0$, $i_{Sb} = I_S$. (d) $i_{Sa} = -I_S/\sqrt{2}$, $i_{Sb} = I_S/\sqrt{2}$.

4

Rotating Magnetic Fields

The production of a rotating radially directed magnetic field in the air gap by the stator currents is fundamental to the operation of both induction motors and synchronous motors. Using the assumption of ideal magnetic materials (i.e., materials where $\vec{\mathbf{H}} = \mathbf{0}$), expressions for the radial magnetic field distribution in the air gap of AC machines due to the currents in the stator phases are found. Specifically, a two-phase machine is considered. It is shown that if the stator phases are sinusoidally wound and 90° apart from each other and, with the stator currents given by $i_{Sa} = I_S \cos(\omega_S t)$ and $i_{Sa} = I_S \cos(\omega_S t)$, the result is a rotating radial magnetic field that is sinusoidally distributed in space. Finally, expressions for the azimuthal magnetic field as well as the axial electric field in the air gap are also found.

4.1 Distributed Windings

Before embarking on how a rotating magnetic field is produced, the geometric structure of the windings (loops/turns) on the rotor and stator iron are described. To do so, consider Figure 4.1(a), which shows a cylindrical iron rotor core with six slots in it. A single length of wire is wound around the core forming three rotor loops to make up a phase winding. The top three slots are at $\theta = \pi/3$, $\theta = \pi/2$, and $\theta = 2\pi/3$, and the bottom three slots are at $\theta = 4\pi/3$, $\theta = 3\pi/2$, and $\theta = 5\pi/3$.

In more detail, starting with a single wire, the first loop of the phase is wound starting in the bottom leftmost slot at the point a in Figure 4.1(a) and then follows a semicircular path in the counterclockwise direction to the rightmost top slot. From there, the wire goes down the length of the cylinder to the back side and then follows another semicircular path in the clockwise direction to the back side of the leftmost bottom slot. Next, the wire is brought up the length of the iron to the front side to the point b in the figure. The wire from point a to point b makes up the first *loop* (or *turn* or *winding*) of the phase. Continuing, the wire is extended to point c to start the second loop. As shown in Figure 4.1(b), the second loop goes from point c to point d, making a half-cylindrical-shaped loop in the top and bottom slots. Finally, the wire is extended from point d to point e to start the third loop. The third loop goes from point e to point f as shown in Figure 4.1(c). As stated above, the single piece of wire (consisting of three loops in this example) is referred to as a *phase winding* or simply *phase*.

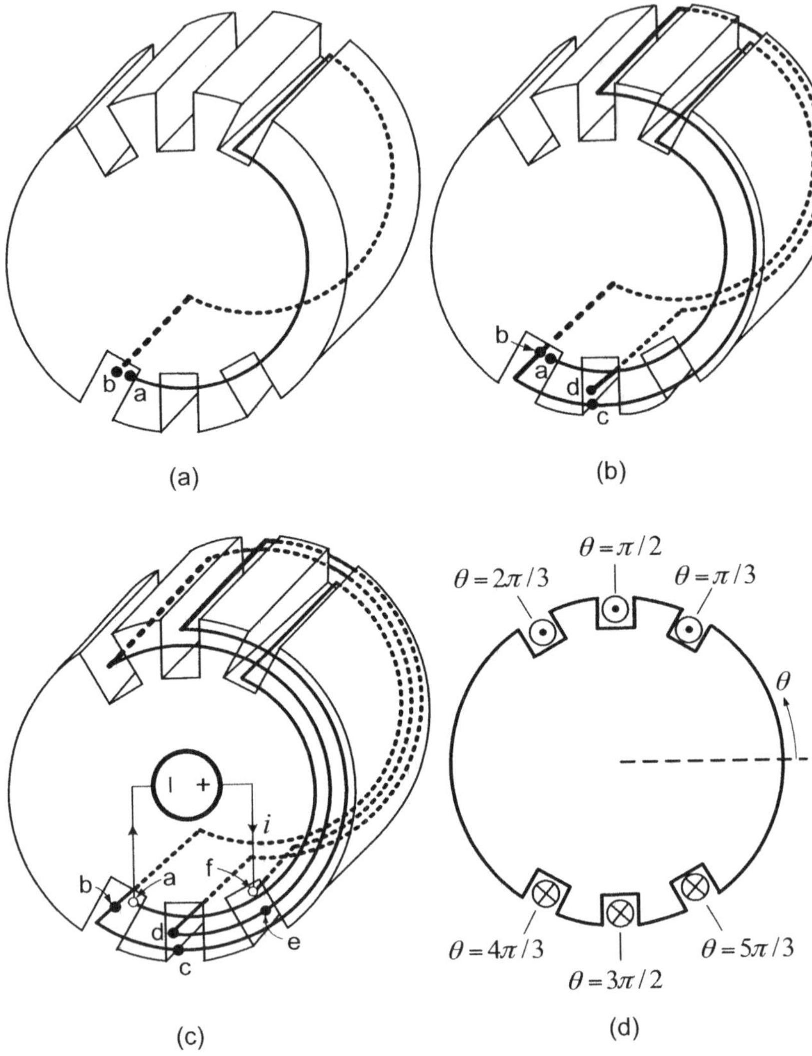

FIGURE 4.1. Distributed phase winding consisting of three loops. (a) First loop of the phase winding from a to b. (b) Second loop of the phase winding from c to d. (c) Third loop of the phase winding from e to f. (d) Cross-sectional view of the rotor phase winding. All three of these half-cylindrical surfaces have the same radii. They are drawn with different radii in the figure so that the reader can tell them apart. Drawn by Bret Wilfong.

This is called a *distributed* phase winding because the loops making up the phase winding are not all in a single pair of slots, but rather, they are distributed over three pairs of slots. The semi-circular sides of each loop are referred to as *end turns*.

Figure 4.1(c) shows that the two ends of the phase (points a and f) are connected to a voltage source, which, if positive, will force current up out of the top sides of the loop and down the bottom side of the loops. This sign convention for the current is also illustrated in Figure 4.1(d), which is a cross section of the rotor. The cross-sectional view of Figure 4.1(d) is the standard way to represent a phase winding. Figure 4.3(a) shows a cross-sectional view of a single phase similar to Figure 4.1(d) except that two loops are now wound in the middle slots rather than just one loop as in Figure 4.1(b).

Figure 4.2 illustrates how electrical power can be brought into the rotor of 4.1. The slip rings 1 and 2 are made of conducting material and are rigidly connected to the rotor. As the rotor turns, they make sliding contact with (slip against) the brushes b_1 and b_2, which are fixed in space. Slip ring 1 is also electrically connected to a wire conductor that goes through slip ring 2 (and is electrically isolated from slip ring 2) and is connected to point f in Figure 4.1(c). Slip ring 1 is electrically connected to a wire conductor which is connected to point a in Figure 4.1(c). In this way, voltage and current can be supplied to the rotor from an outside power source.

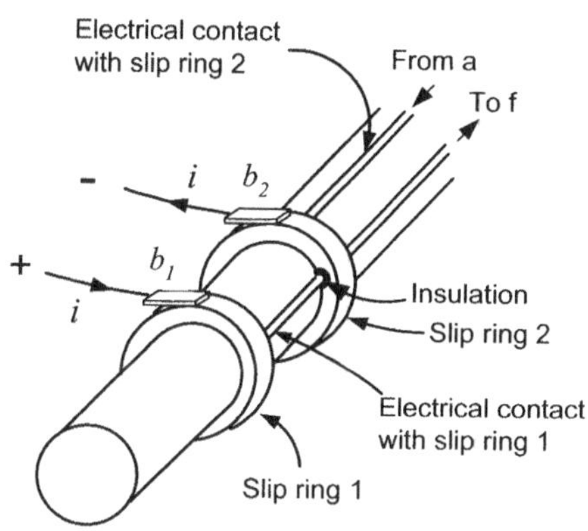

FIGURE 4.2. A set of brushes and slip rings can be used to bring electrical power to the rotor winding.

A distributed *stator* phase winding is wound in similar fashion as the distributed rotor winding of Figure 4.1. In the case of the stator winding, the loops of the phase are wound on the *inside* surface of the stator as shown in Figure 4.3(b). The simple single stator phase of Figure 4.3(b) consists of four turns (loops/windings). Specifically, this phase winding is made by winding a single wire around the inside of the stator iron once at $\theta = \pi/3$ (with the other side of the loop at $4\pi/3$), then twice at $\theta = \pi/2$ (with the other side of the loop at $3\pi/2$), and finally once more at $\theta = 2\pi/3$ (with the other side at $5\pi/3$). The radial air gap distance is denoted as g and an arbitrary point is located using the polar coordinates (r, θ) as shown.

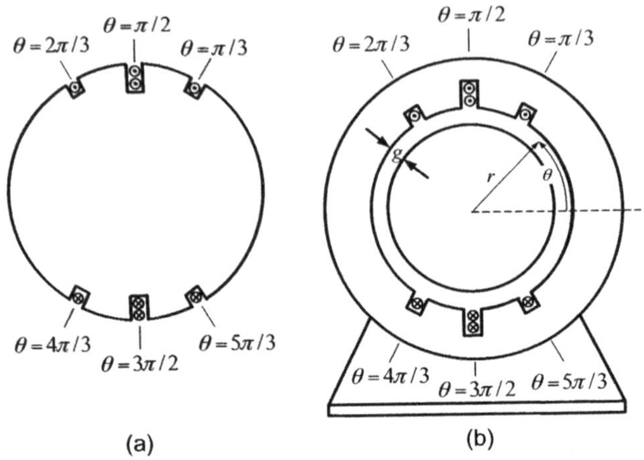

FIGURE 4.3. (a) Distributed rotor phase winding with two loops in the middle slot. (b) Distributed stator phase winding with two loops in the middle slots.

There are other ways to distribute the windings of a phase such as shown in Figure 4.4. Figure 4.4(a) shows a distributed rotor phase winding with one side of a loop at $\theta = \pi/6$ (the other side of the loop at $7\pi/6$), two loop sides at $\theta = \pi/3$ (with the other sides at $4\pi/3$), three loop sides at $\theta = \pi/2$ (with the other sides at $3\pi/2$), two loops at $\theta = 2\pi/3$ (with the other sides at $5\pi/3$) and one loop side at $\theta = 5\pi/6$ (with the other side at $11\pi/6$). Figure 4.4(b) shows a similar winding on the stator.

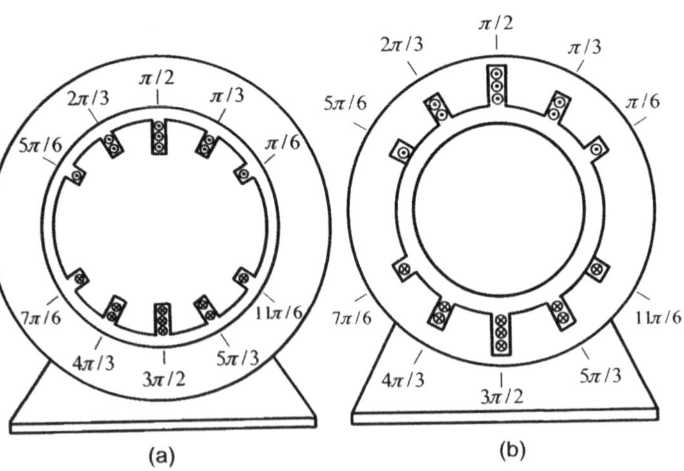

FIGURE 4.4. (a) Distributed rotor winding. (b) Distributed stator winding.

4.2 Approximate Sinusoidally Distributed $\vec{\mathbf{B}}$ Field

The goal here is to compute the radial magnetic field $\vec{\mathbf{B}}$ established *in the air gap* by the current in a distributed winding. To do so, Ampère's law $\oint \vec{\mathbf{H}} \cdot d\vec{\ell} = i_{\text{enclosed}}$ is the key tool that is used. However, to obtain a nice elegant expression for the magnetic field $\vec{\mathbf{B}}$ using Ampère's law, an additional assumption is required. This assumption is that the wires loops in a phase winding have an *infinitesimal* cross-sectional area; that is, the wire has length, but its cross-sectional area is infinitesimal. To emphasize this, Figure 4.3(a) is redrawn in Figure 4.5(a) to indicate that the wire has no width. The two ⊙ are drawn at the top middle slot while only one ⊙ is drawn at either side to indicate that the middle slot has twice as many windings (loops) as either of the other two slots. Similarly, the stator winding of Figure 4.3(b) is redrawn in Figure 4.5(b).

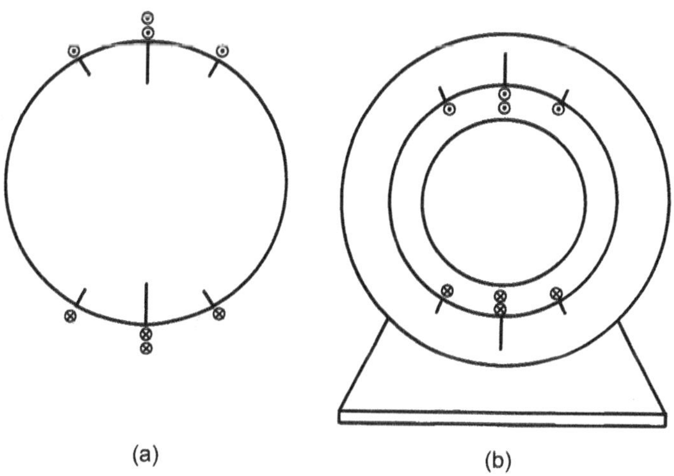

FIGURE 4.5. (a) This is Figure 4.3(a) redrawn to show that the wire has no width. (b) This is Figure 4.3(b) redrawn to show that the wire has no width.

However, even Figure 4.5 does not quite capture what it means for the wire to have a cross-sectional area that is infinitesimal in size because the slots should not have any depth. So, Figure 4.5 is redrawn in Figure 4.6 to illustrate that the depth of the slots are also infinitesimal.

To be consistent with the literature, Figures 4.1(d), 4.3, and 4.4 will be used for illustration, but it is to be understood that the standing assumption is that the wires loops in a phase winding have an *infinitesimally* small cross-sectional area so that the reader should think of Figure 4.6 in the derivations that follow.

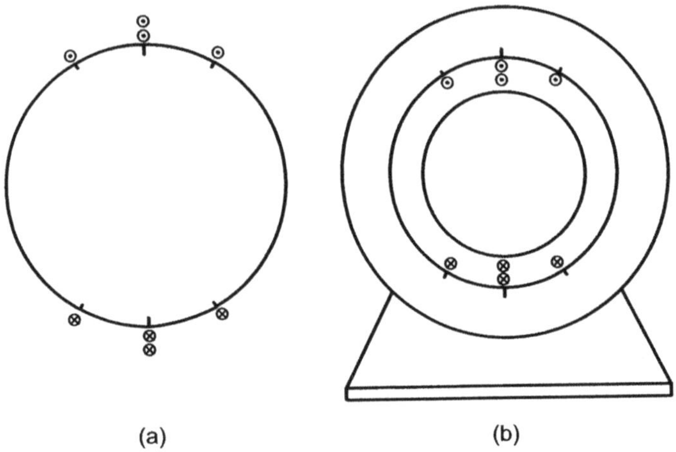

FIGURE 4.6. This is Figures 4.5(a) and 4.5(b) redrawn to show the slots have no depth.

The problem of computing the radial magnetic field established *in the air gap* by the current in a distributed winding is now considered. To fix ideas, the distributed winding shown in Figure 4.3(b) [or equivalently, Figure 4.6(b)] is used. This winding is redrawn in Figure 4.7 where it is now denoted as phase a and the current in it is denoted as i_{Sa}.

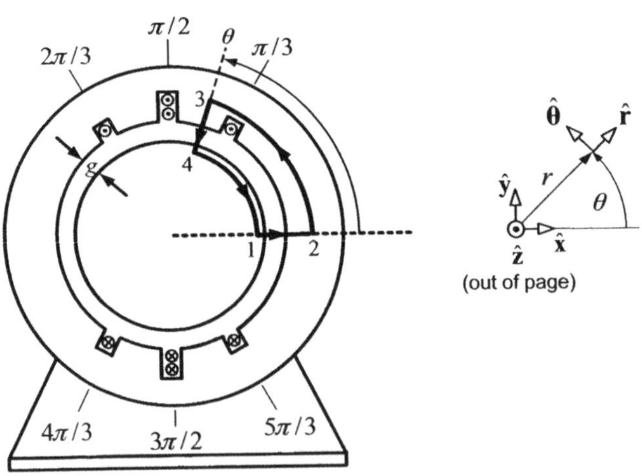

FIGURE 4.7. Use of Ampère's law to determine the radial magnetic field in the airgap due to the current i_{Sa} in stator phase a.

To compute the radial magnetic field $\vec{\mathbf{B}}_{Sa}$ in the air gap produced by i_{Sa}, Ampère's law $\oint \vec{\mathbf{H}} \cdot d\vec{\ell} = i_{\text{enclosed}}$ is applied to the closed path 1–2–3–4–1 shown in Figure 4.7. The assumption that the wire has an infinitesimal cross section means that the sides of the copper loops lie just at the inside surface of the stator iron. In particular, this assumption allows one to take the points $1, 2, 3, 4$ on the path to all be just inside the magnetic material at the air gap/iron interface and still have the closed-path $1-2-3-4-1$ enclose any current in the loop sides between 0 and θ. Further, computing $\int \vec{\mathbf{H}} \cdot d\vec{\ell}$ across any (infinitesimally thin) wire is then equal to zero.[1] With $\vec{\mathbf{H}} \equiv 0$ in the iron and g the air gap length, it follows that

$$
\begin{aligned}
\int_1^2 \vec{\mathbf{H}}_{Sa} \cdot d\vec{\ell} + \int_2^3 \vec{\mathbf{H}}_{Sa} \cdot d\vec{\ell} + \int_3^4 \vec{\mathbf{H}}_{Sa} \cdot d\vec{\ell} + \int_4^1 \vec{\mathbf{H}}_{Sa} \cdot d\vec{\ell} &= i_{\text{enclosed}} \\
\int_1^2 \vec{\mathbf{H}}_{Sa} \cdot d\vec{\ell} + \int_3^4 \vec{\mathbf{H}}_{Sa} \cdot d\vec{\ell} &= i_{\text{enclosed}} \\
\int_1^2 H_{Sa}(0)\hat{\mathbf{r}} \cdot d\ell\hat{\mathbf{r}} + \int_3^4 H_{Sa}(\theta)\hat{\mathbf{r}} \cdot (-d\ell\hat{\mathbf{r}}) &= i_{\text{enclosed}} \\
gH_{Sa}(0) - gH_{Sa}(\theta) &= i_{\text{enclosed}}.
\end{aligned}
\tag{4.1}
$$

In this derivation, it was assumed that H_{Sa} is constant across the air gap, that is, H_{Sa} does not depend on the radial distance r in the air gap. The enclosed current i_{enclosed} between 0 and θ inside the curve $1-2-3-4-1$ of Figure 4.7 is given by

$$
i_{\text{enclosed}} = \begin{cases}
0 & \text{for} \quad 0 \le \theta \le \pi/3 \\
i_{Sa} & \text{for} \quad \pi/3 \le \theta \le \pi/2 \\
3i_{Sa} & \text{for} \quad \pi/2 \le \theta \le 2\pi/3 \\
4i_{Sa} & \text{for} \quad 2\pi/3 \le \theta \le 4\pi/3 \\
3i_{Sa} & \text{for} \quad 4\pi/3 \le \theta \le 3\pi/2 \\
i_{Sa} & \text{for} \quad 3\pi/2 \le \theta \le 5\pi/3 \\
0 & \text{for} \quad 5\pi/3 \le \theta \le 2\pi
\end{cases}
$$

and is plotted in Figure 4.8.

---

[1] This assumption does not appear to be realistic especially in the above figure. Copper is not a magnetic material so $\vec{\mathbf{H}} \neq 0$ inside. The assumption of an "infinitesimal cross section" for the wire allows one to still take $\int \vec{\mathbf{H}} \cdot d\vec{\ell} = 0$ across the wire as its cross-sectional area is infinitesmal. Although on the face of it this assumption may not seem reasonable, it does lead to a model that is not only reasonable, but quite good. Again, the reader is encouraged to think of Figure 4.6(b) in this derivation and all that follow.

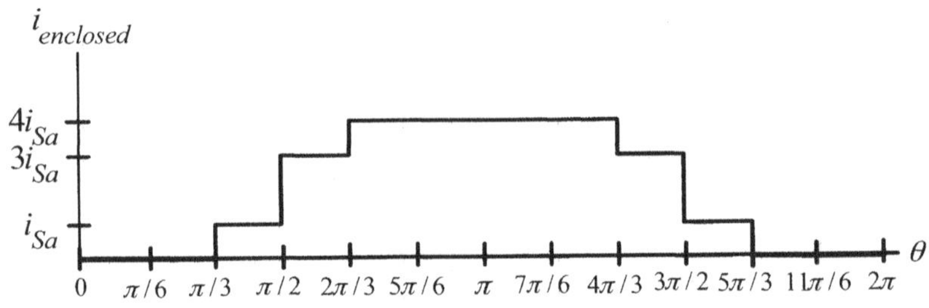

FIGURE 4.8. $i_{enclosed}$ versus θ.

The magnetic intensity is then

$$H_{Sa}(\theta) = H_{Sa}(0) - \frac{i_{\text{enclosed}}(\theta)}{g}$$

where $H_{Sa}(0)$ on the right-hand side must still be determined.

To compute $H_{Sa}(0)$, the conservation of flux law is used. Consider a *closed* cylindrical flux surface whose cylindrical surface is in the air gap and whose disk shaped ends just contain the rotor iron as shown in Figure 4.9. Since this surface is in air, $\vec{\mathbf{B}}_{Sa} = \mu_0 \vec{\mathbf{H}}_{Sa}$ on the surface. On the disk-shaped surfaces at the each end of the cylinder, $\vec{\mathbf{B}}_{Sa} = 0$, while on the cylindrical-shaped part of the surface, $\vec{\mathbf{B}}_{Sa} = B_{Sa}\hat{\mathbf{r}} = \mu_0 \left(H_{Sa}(0) - i_{\text{enclosed}}(\theta)/g \right) \hat{\mathbf{r}}$. Conservation of flux gives

$$\oint \vec{\mathbf{B}} \cdot d\vec{\mathbf{S}} = \int_0^{\ell_1} \int_0^{2\pi} (B_{Sa}(\theta)\hat{\mathbf{r}}) \cdot (r d\theta dz \hat{\mathbf{r}}) = r\ell_1 \int_0^{2\pi} B_{Sa}(\theta) d\theta = 0$$

or, substituting in the above expression for $B_{Sa}(\theta)$, one finds

$$\int_0^{2\pi} \left(\mu_0 H_{Sa}(0) - \mu_0 \frac{i_{\text{enclosed}}(\theta)}{g} \right) d\theta = 0.$$

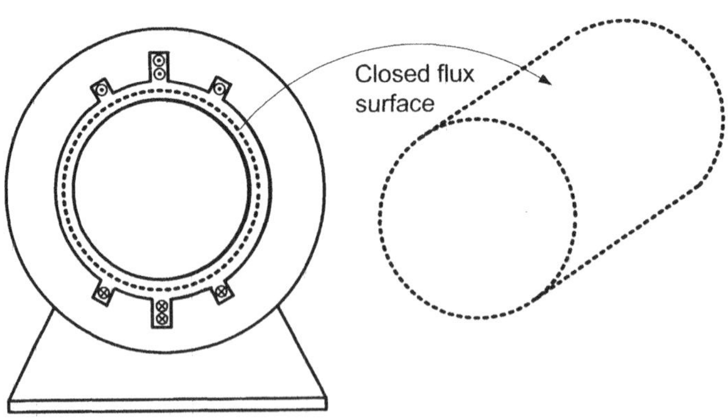

FIGURE 4.9. A closed flux surface in the air gap which contains the rotor.

Rearranging

$$\begin{aligned} H_{Sa}(0) &= \int_0^{2\pi} \frac{i_{\text{enclosed}}(\theta)}{2\pi g} d\theta \\ &= \frac{1}{2\pi g}\left(i_{Sa}\frac{\pi}{6} + 3i_{Sa}\frac{\pi}{6} + 4i_{Sa}\frac{2\pi}{3} + 3i_{Sa}\frac{\pi}{6} + i_{Sa}\frac{\pi}{6}\right) \\ &= \frac{2i_{Sa}}{g}. \end{aligned}$$

So

$$\vec{\mathbf{B}}_{Sa} = B_{Sa}(\theta)\hat{\mathbf{r}} = \mu_0\left(\frac{2i_{Sa}}{g} - \frac{i_{\text{enclosed}}(\theta)}{g}\right)\hat{\mathbf{r}}. \tag{4.2}$$

$B_{Sa}(\theta)$ versus θ is plotted in Figure 4.10.

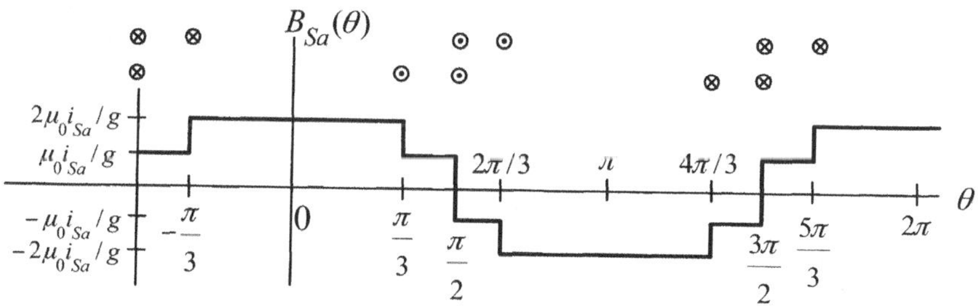

FIGURE 4.10. Magnetic field $B_{Sa}(\theta)$ versus θ due to the current in phase a of a two-phase machine.

Expanding $B_{Sa}(\theta)$ in a Fourier series, one obtains (see problem 3)

$$\begin{aligned} B_{Sa}(\theta) &= \mu_0\frac{i_{Sa}}{g}\frac{4}{\pi} \times \sum_{k=1,3,5,\ldots}^{\infty} \frac{(1+\cos(\frac{k\pi}{6}))}{k}\sin\Big(k(\theta+\frac{\pi}{2})\Big) \\ &= \mu_0\frac{i_{Sa}}{g}\frac{4}{\pi}\left(\frac{2+\sqrt{3}}{2}\sin\Big(\theta+\frac{\pi}{2}\Big) + \frac{1}{3}\sin\Big(3(\theta+\frac{\pi}{2})\Big)\right. \\ &\quad \left. + \frac{1-\sqrt{3}/2}{5}\sin\Big(5(\theta+\frac{\pi}{2})\Big) + \frac{1-\sqrt{3}/2}{7}\sin\Big(7(\theta+\frac{\pi}{2})\Big) + \cdots\right) \\ &= \mu_0\frac{i_{Sa}}{g}\frac{4}{\pi}\Big(1.866\cos(\theta) - 0.333\cos 3(\theta) + 0.0268\cos(5\theta) \\ &\quad - 0.0191\cos(7\theta) + \cdots\Big). \end{aligned} \tag{4.3}$$

To a first approximation, the radial magnetic field distribution $B_{Sa}(\theta)$ may be approximated by its first harmonic as

$$\vec{\mathbf{B}}_{Sa}(\theta) = B_{Sa}(\theta)\hat{\mathbf{r}} \approx 1.866\mu_0\frac{i_{Sa}}{g}\frac{4}{\pi}\cos(\theta)\hat{\mathbf{r}} \tag{4.4}$$

which is a sinusoidal distribution in θ.

A similar procedure can be applied to the stator winding in Figure 4.4(b), resulting in a radial magnetic field distribution in the air gap as given in Figure 4.11 (see problem 4).

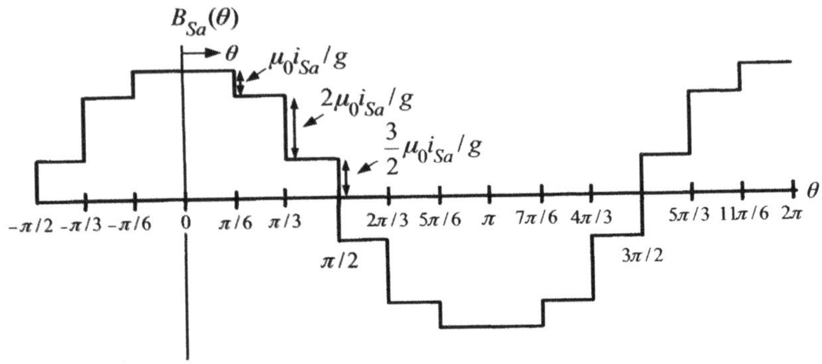

FIGURE 4.11. Radial magnetic field $B_{Sa}(\theta)$ in the air gap for the distributed winding of Figure 4.4(b).

4.2.1 Conservation of Flux and 1/r Dependence

Consider the closed flux surface in the air gap shown in Figure 4.12. This closed surface is composed of two concentric cylindrical-shaped surfaces S_1 and S_2 and four other planar sides S_3, S_4, S_5, and S_6 as illustrated in Figure 4.12. The surface normal on the closed surface is given by

$$d\vec{\mathbf{S}} = \begin{cases} -r_1 d\theta dz\hat{\mathbf{r}} & \text{for } S_1 \\ r_2 d\theta dz\hat{\mathbf{r}} & \text{for } S_2 \\ drdz\hat{\boldsymbol{\theta}} & \text{for } S_3 \\ rd\theta dr\hat{\mathbf{z}} & \text{for } S_4 \\ -drdz\hat{\boldsymbol{\theta}} & \text{for } S_5 \\ -rd\theta dr\hat{\mathbf{z}} & \text{for } S_6 \end{cases}$$

As $\vec{\mathbf{B}}$ is taken to have only a radial component, $\vec{\mathbf{B}}\cdot d\vec{\mathbf{S}} = 0$ on the surfaces S_3, S_4, S_5, S_6 as $\vec{\mathbf{B}}$ is perpendicular (normal) to $d\vec{\mathbf{S}}$ on these surfaces. Consequently, $\int_{S_3}\vec{\mathbf{B}}\cdot d\vec{\mathbf{S}} = \int_{S_4}\vec{\mathbf{B}}\cdot d\vec{\mathbf{S}} = \int_{S_5}\vec{\mathbf{B}}\cdot d\vec{\mathbf{S}} = \int_{S_6}\vec{\mathbf{B}}\cdot d\vec{\mathbf{S}} = 0$. In this case, conservation of flux gives

$$\oint \vec{\mathbf{B}} \cdot d\vec{\mathbf{S}} = \int_{S_1} \vec{\mathbf{B}} \cdot d\vec{\mathbf{S}} + \int_{S_2} \vec{\mathbf{B}} \cdot d\vec{\mathbf{S}} = 0$$

$$\int_0^{\ell_1}\int_{\theta_1}^{\theta_2} (B_{Sa}(\theta)\hat{\mathbf{r}}) \cdot (r_1 d\theta dz(-\hat{\mathbf{r}})) + \int_0^{\ell_1}\int_{\theta_1}^{\theta_2} (B_{Sa}(\theta)\hat{\mathbf{r}}) \cdot (r_2 d\theta dz\hat{\mathbf{r}}) = 0 \tag{4.5}$$

$$-\ell_1 r_1 \int_{\theta_1}^{\theta_2} B_{Sa}(\theta)d\theta + \ell_1 r_2 \int_{\theta_1}^{\theta_2} B_{Sa}(\theta)d\theta = 0$$

$$\ell_1 (r_2 - r_1) \int_{\theta_1}^{\theta_2} B_{Sa}(\theta)d\theta = 0$$

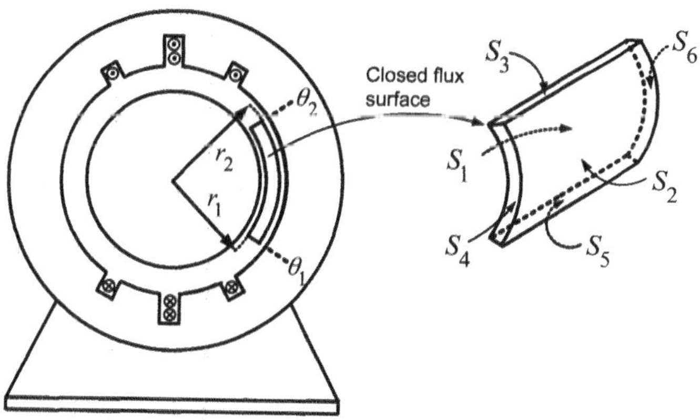

FIGURE 4.12. Closed flux surface to show the $1/r$ dependence of the magnetic field in the air gap.

As $r_1 \neq r_2$, conservation of flux does not hold. The problem here is that the magnetic field B_{Sa} was assumed to be constant as a function of r across the air gap in deriving (4.2) [and therefore, in deriving (4.3) and (4.4)] when in fact it must vary as $1/r$ to satisfy the conservation of flux law. That is, with r_R the radius of the rotor, replace $B_{Sa}(\theta)$ by $B_{Sa}(r,\theta) \triangleq \frac{r_R}{r} B_{Sa}(\theta)$ in (4.5) and $\oint \vec{\mathbf{B}} \cdot d\vec{\mathbf{S}} \equiv 0$ will then hold. The expression for the magnetic field is now

$$\vec{\mathbf{B}}_{Sa}(r,\theta) = B_{Sa}(r,\theta)\hat{\mathbf{r}} = \mu_0 \frac{r_R}{r}\left(\frac{2i_{Sa}}{g} - \frac{i_{\text{enclosed}}(\theta)}{g}\right)\hat{\mathbf{r}}. \tag{4.6}$$

Similarly, the expressions (4.3) and (4.4) are multiplied by the factor r_R/r.

The air gap g is assumed to be small so that

$$\frac{r_R}{r} \approx 1 \text{ for } r_R \leq r \leq r_S = r_R + g.$$

Consequently, the factor r_R/r does not really change the value of the magnetic field in the air gap.[2] However, it is straightforward to see that this expression for $\vec{\mathbf{B}}_{Sa}(r,\theta)$ satisfies conservation of flux for the closed surface given in Figure 4.12 (see problems 5 and 17).

4.2.2 Magnetic Field Distribution Due to the Stator Currents

Consider a second stator phase, denoted as phase b, which is rotated 90° from phase a as drawn in Figure 4.13.

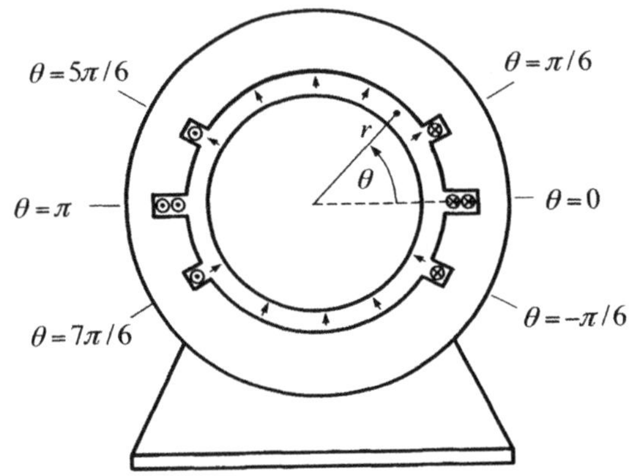

FIGURE 4.13. Phase b windings carrying the current i_{Sb}.

A similar argument to that given for phase a shows that the magnetic field $\vec{\mathbf{B}}_{Sb}$ due to the current i_{Sb} in phase b is found by replacing i_{Sa} by i_{Sb} and θ by $\theta - \pi/2$ to obtain

$$\begin{aligned}\vec{\mathbf{B}}_{Sb}(r,\theta) &= B_{Sb}(r,\theta)\hat{\mathbf{r}} = \mu_0 \frac{i_{Sb}}{g}\frac{r_R}{r}\frac{4}{\pi}\sum_{k=1,3,5,\ldots}^{\infty}\frac{(1+\cos(k\pi/6))}{k}\sin(k\theta)\\ &\approx 1.866\mu_0\frac{i_{Sb}}{g}\frac{r_R}{r}\frac{4}{\pi}\sin(\theta)\hat{\mathbf{r}}.\end{aligned}$$

In summary, the first harmonic approximation for $\vec{\mathbf{B}}_{Sa}(r,\theta)$ and $\vec{\mathbf{B}}_{Sb}(r,\theta)$ are

$$\vec{\mathbf{B}}_{Sa}(r,\theta) = B_{Sa}(r,\theta)\hat{\mathbf{r}} \approx 1.866\mu_0\frac{i_{Sa}}{g}\frac{4}{\pi}\frac{r_R}{r}\cos(\theta)\hat{\mathbf{r}}$$

$$\vec{\mathbf{B}}_{Sb}(r,\theta) = B_{Sb}(r,\theta)\hat{\mathbf{r}} \approx 1.866\mu_0\frac{i_{Sb}}{g}\frac{4}{\pi}\frac{r_R}{r}\sin(\theta)\hat{\mathbf{r}}.$$

---

[2] It would have been just as valid to use r_S/r where r_S is the inside radius of the stator, or a value $r_0, r_R < r_0 < r_S$ satisfying $\int_{r_R}^{r_S}(r_0/r_R)dr = g$ (see problem 11).

4.3 Sinusoidally Wound Phases

In the previous section, the higher-order terms in the expression for the radial magnetic field in the air gap were ignored. In this section, the concept of a sinusoidally wound phase is explained, and it is shown that the higher-order harmonic terms in the Fourier series expansion of $\vec{\mathbf{B}}_{Sa}, \vec{\mathbf{B}}_{Sb}$ are *identically* zero. A sinusoidally wound rotor phase is first described as the corresponding figures are easier to draw in order to get the point across. Next, sinusoidally wound stator phases are described and an expression for the corresponding sinusoidally distributed magnetic fields produced in the air gap by the phase currents are then derived.

4.3.1 Sinusoidally Wound Rotor Phase

To construct a *sinusoidally* wound rotor phase, a single strand of wire is wrapped around the axial length of a cylindrical core of iron, making coil-loops or *turns* according to a *sinusoidal turns density* given by

$$N_{Ra}(\theta - \theta_R) = \frac{N_R}{2}\sin(\theta - \theta_R) \text{ for } \theta_R < \theta < \pi + \theta_R$$

where θ_R is the rotor position. That is, this specifies that the number of the loops (turns) wrapped around the rotor iron core between θ and $\theta + d\theta$ is $N_{Ra}(\theta - \theta_R)d\theta$ [see Figure 4.14(a)].

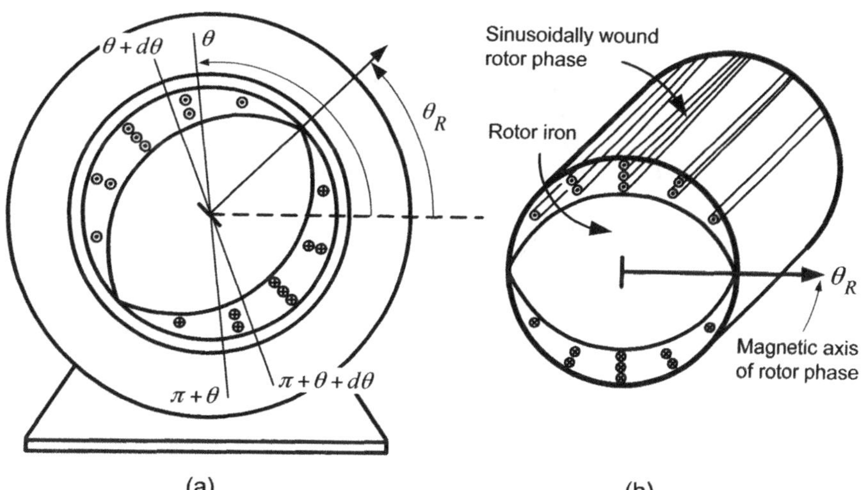

FIGURE 4.14. Sinusoidally wound rotor phase. The number of loops/turns wound between θ and $\theta + d\theta$ is $N_{Ra}(\theta - \theta_R)d\theta$ with the other side of these loops between $\pi + \theta$ and $\pi + \theta + d\theta$. The sine curve envelope drawn around the windings is used to indicate a sinusoidally wound phase. Although not drawn to show this, the cross-sectional area of the sinusoidally wound phase is taken to be infinitesimal.

Consequently, the total number of turns wrapped around the rotor core is then $\int_{\theta_R}^{\theta_R+\pi} N_{Ra}(\theta - \theta_R)d\theta = N_R$. Figure 4.14(a) shows how a sinusoidally wound rotor is indicated schematically. In comparison with Figure 4.4, the slots have been removed in Figure 4.14(a), showing only the loops which are now enclosed in a *sine* curve envelope. This is a standard representation to denote a sinusoidal winding and is also adopted here [3][24].

4.3.2 Sinusoidally Wound Stator Phases

As in the case of the rotor phase, a sinusoidally wound stator phase is constructed by wrapping a single strand of wire around the inside surface of the stator iron to make coil loops or *turns*. For stator phase a, the number of the coil loops at any angle θ is specified by the *sinusoidal turns density*

$$N_{Sa}(\theta) = \frac{N_S}{2}\sin(\theta) \text{ for } 0 < \theta < \pi \tag{4.7}$$

that is, $N_{Sa}(\theta)d\theta$ is the number of turns between θ and $\theta + d\theta$. The total number of turns wrapped around the inside surface of the stator core is $\int_0^\pi N_{Sa}(\theta)d\theta = N_S$. This is illustrated in Figure 4.15.[3] The top side of this winding, where positive current is out of the page, is denoted as side a while the bottom side of the winding, where positive current is into the page, is denoted as side a'.

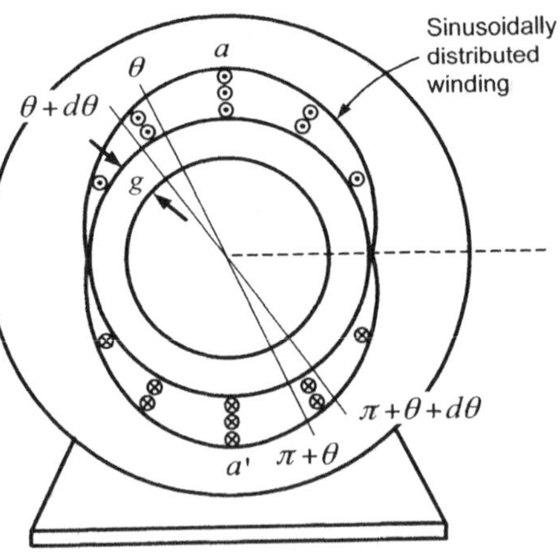

FIGURE 4.15. Sinusoidally wound stator phase a. The number of loops wound between θ and $\theta+d\theta$ is $N_{Sa}(\theta)d\theta$ with the other side of these loops between $\pi+\theta$ and $\pi+\theta+d\theta$.

[3]The figure shows a single phase wound such that there is one loop at $\pi/6$, two loops at $\pi/3$, three loops at $\pi/2$, two loops at $2\pi/3$, and one loop at $5\pi/6$ (see problem 4, where such a winding is analyzed). However, again following Refs. [3] and [24], the sine curve envelope of these loops is used to indicate that it is *sinusoidally* wound.

Similarly, stator phase b consists of a separate sinusoidally wound wire identical in structure to phase a except it is rotated 90° with respect to phase a [see Figure 4.17(b)]. The turns density for phase b is therefore

$$N_{Sb}(\theta) = \frac{N_S}{2}\sin(\theta - \pi/2) \quad \text{for } \pi/2 < \theta < 3\pi/2. \tag{4.8}$$

The number of turns between θ and $\theta + d\theta$ is $N_{Sb}(\theta)d\theta = (N_S/2)\sin(\theta - \pi/2)d\theta$ and the total number of turns making up stator phase b is

$$\int_{\pi/2}^{3\pi/2} N_{Sb}(\theta)d\theta = N_S.$$

Remark Though Figures 4.14 and 4.15 are not drawn to show this, the cross-sectional area of a sinusoidally distributed winding is modeled as being infinitesimal.

4.4 Sinusoidally Distributed Magnetic Fields

Sinusoidally distributed windings (turns density) are used because the current in such a winding produces a sinusoidally distributed (in θ) radial magnetic field in the air gap. This is now explained.

The objective here is to determine the radial magnetic field $\vec{\mathbf{B}}_{Sa}$ established *in the air gap* by the current i_{Sa} in the sinusoidally distributed winding of phase a. To do so, Ampère's law $\oint \vec{\mathbf{H}} \cdot d\vec{\ell} = i_{\text{enclosed}}$ is applied to the closed-path 1–2–3–4–1 indicated in Figure 4.16 to obtain

$$\begin{aligned} \oint \vec{\mathbf{H}}_{Sa} \cdot d\vec{\ell} &= \int_0^{\theta} i_{Sa}(N_S/2)\sin(\theta')d\theta' \\ \int_1^2 \vec{\mathbf{H}}_{Sa} \cdot d\vec{\ell} + \int_3^4 \vec{\mathbf{H}}_{Sa} \cdot d\vec{\ell} &= \int_0^{\theta} i_{Sa}(N_S/2)\sin(\theta')d\theta'. \end{aligned}$$

Evaluating, this becomes

$$\begin{aligned} \int_{\ell=0}^{\ell=g} H_{Sa}(i_{Sa},0)\hat{\mathbf{r}} \cdot (d\ell\hat{\mathbf{r}}) + \int_{\ell=0}^{\ell=g} H_{Sa}(i_{Sa},\theta)\hat{\mathbf{r}} \cdot (-d\ell\hat{\mathbf{r}}) &= -i_{Sa}\frac{N_S}{2}\cos(\theta) \\ &\quad + i_{Sa}\frac{N_S}{2} \end{aligned}$$

or

$$H_{Sa}(i_{Sa},0)g - H_{Sa}(i_{Sa},\theta)g = -i_{Sa}\frac{N_S}{2}\cos(\theta) + i_{Sa}\frac{N_S}{2} \tag{4.9}$$

where it was assumed that $\vec{\mathbf{H}}_{Sa}$ is *constant* across the air gap. Rearranging,

$$H_{Sa}(i_{Sa},\theta) = i_{Sa}\frac{N_S}{2g}\cos(\theta) + H_{Sa}(i_{Sa},0) - i_{Sa}\frac{N_S}{2g}.$$

In this equation, both $H_{Sa}(i_{Sa},\theta)$ and $H_{Sa}(i_{Sa},0)$ are unknown. The conservation of flux law $\oint_S \vec{\mathbf{B}}\cdot d\vec{\mathbf{S}} = 0$ can be used to determine $H_{Sa}(i_{Sa},0)$. To do so, consider a closed surface S that contains the rotor core as shown in Figure 4.16. The cylindrical part of the surface S is in the air gap and has radius r_R so that it just contains the rotor core. The two disk-shaped ends of the surface S are considered to have zero flux through them as $\vec{\mathbf{B}}$ is essentially zero there.

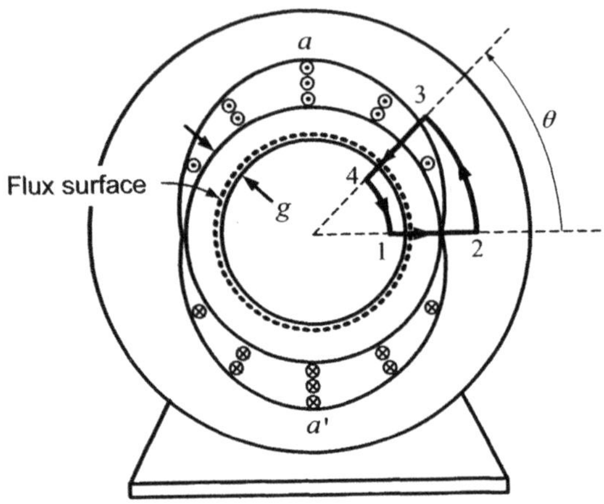

FIGURE 4.16. Use of Ampère's law and Gauss's law to determine the air gap radial magnetic field due to i_{Sa}.

Then

$$\begin{aligned}\oint_S \vec{\mathbf{B}}_{Sa}\cdot d\vec{\mathbf{S}} &= \int_0^{\ell_1}\int_0^{2\pi} B_{Sa}(i_{Sa},\theta)\hat{\mathbf{r}}\cdot(r_R d\theta dz\hat{\mathbf{r}})\\ &= \ell_1 r_R\int_0^{2\pi} B_{Sa}(i_{Sa},\theta)d\theta\\ &= 0.\end{aligned}$$

Now $B_{Sa}(i_{Sa},\theta) = \mu_0 H_{Sa}(i_{Sa},\theta)$ as the flux surface is in the air gap. Thus,

$$\begin{aligned}\int_0^{2\pi} B_{Sa}(i_{Sa},\theta)d\theta &= \int_0^{2\pi}\mu_0\Big(i_{Sa}\frac{N_S}{2g}\cos(\theta) + H_{Sa}(i_{Sa},0) - i_{Sa}\frac{N_S}{2g}\Big)d\theta\\ &= 2\pi\mu_0\Big(H_{Sa}(i_{Sa},0) - i_{Sa}\frac{N_S}{2g}\Big)\\ &= 0\end{aligned}$$

or

$$H_{Sa}(i_{Sa},0) = i_{Sa}\frac{N_S}{2g}.$$

It then follows that

$$\begin{aligned} H_{Sa}(i_{Sa},\theta) &= \frac{N_S}{2g} i_{Sa}\cos(\theta) \\ B_{Sa}(i_{Sa},\theta) &= \frac{\mu_0 N_S}{2g} i_{Sa}\cos(\theta). \end{aligned} \tag{4.10}$$

In applying Ampère's law, it was assumed that $\vec{\mathbf{B}} = \mu_0\vec{\mathbf{H}}$ was constant across the air gap in the radial direction, that is, $\vec{\mathbf{B}}$ did not depend on the cylindrical coordinate r. However, in order to satisfy the conservation of flux $\oint_S \vec{\mathbf{B}}\cdot d\vec{\mathbf{S}} = 0$ in the air gap for a closed surface such as the one in Figure 4.12, it is necessary that $\vec{\mathbf{B}}$ decrease as $1/r$ in the air gap. Therefore, the expressions for H_{Sa}, B_{Sa} are modified by the factor r_R/r so that conservation of flux holds in the air gap.[4] Finally then, the magnetic field $\vec{\mathbf{B}}_{Sa}$ *in the air gap* due to i_{Sa} is given by

$$\vec{\mathbf{B}}_{Sa}(i_{Sa},r,\theta) = \frac{\mu_0 N_S r_R}{2g}\frac{i_{Sa}\cos(\theta)}{r}\hat{\mathbf{r}}.$$

Similarly, for stator phase b which is also sinusoidally wound, but rotated 90° counterclockwise from phase a, the magnetic field $\vec{\mathbf{B}}_{Sb}$ *in the air gap* due to i_{Sb} is given by

$$\begin{aligned} \vec{\mathbf{B}}_{Sb}(i_{Sb},r,\theta) &= \frac{\mu_0 N_S r_R}{2g}\frac{i_{Sb}\cos(\theta-\pi/2)}{r}\hat{\mathbf{r}} \\ &= \frac{\mu_0 N_S r_R}{2g}\frac{i_{Sb}\sin(\theta)}{r}\hat{\mathbf{r}}. \end{aligned}$$

Figure 4.17(a) shows the magnetic field lines due to the current i_{Sa} in phase a (drawn with $i_{Sa} > 0$) while Figure 4.17(b) shows the magnetic field lines due to i_{Sb} in phase b (drawn with $i_{Sb} > 0$). The sinusoidal windings have resulted in sinusoidally distributed radial magnetic fields in the air gap produced by the stator currents.

---

[4]See problem 11 for a discussion how Ampère's law changes when the factor r_R/r is added.

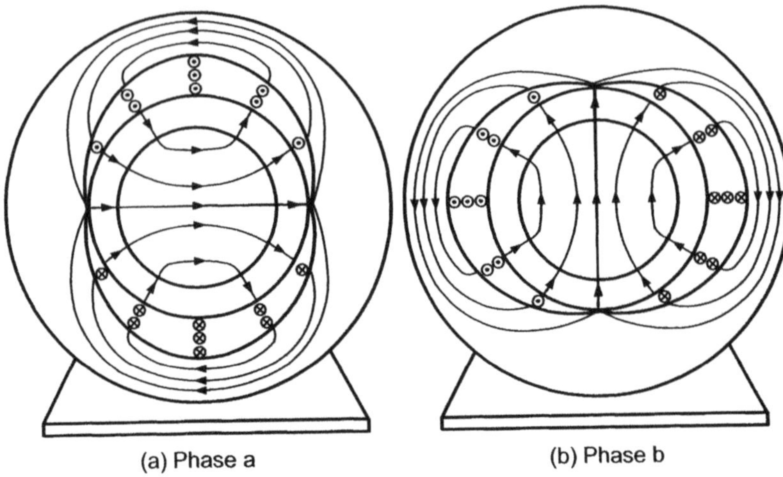

FIGURE 4.17. (a) Magnetic field lines due to the current i_{Sa} in the sinusoidally distributed winding of phase a. (b) Magnetic field lines due to the current i_{Sb} in the sinusoidally distributed winding of phase b.

4.4.1 Sinusoidally Distributed Rotating Magnetic Field

The total magnetic field in the air gap due to the current in both stator phases is then

$$\begin{aligned}\vec{\mathbf{B}}_S(i_{Sa}, i_{Sb}, r, \theta) &= \vec{\mathbf{B}}_{Sa}(i_{Sa}, r, \theta) + \vec{\mathbf{B}}_{Sb}(i_{Sb}, r, \theta) \\ &= \frac{\mu_0 r_R N_S}{2g}\frac{1}{r}\Big(i_{Sa}\cos(\theta) + i_{Sb}\sin(\theta)\Big)\hat{\mathbf{r}}. \quad (4.11)\end{aligned}$$

With $i_{Sa}(t) = I_S\cos(\omega_S t)$ and $i_{Sb}(t) = I_S\sin(\omega_S t)$, the expression (4.11) simplifies to

$$\begin{aligned}\vec{\mathbf{B}}_S(r, \theta, t) &= \frac{\mu_0 r_R N_S I_S}{2g}\frac{1}{r}\Big(\cos(\omega_S t)\cos(\theta) + \sin(\omega_S t)\sin(\theta)\Big)\hat{\mathbf{r}} \\ &= \frac{\mu_0 r_R N_S I_S}{2g}\frac{1}{r}\cos(\theta - \omega_S t)\hat{\mathbf{r}} \\ &= \frac{\mu_0 r_R N_S I_S}{2g}\frac{1}{r}\cos\Big(\theta - \theta_S(t)\Big)\hat{\mathbf{r}} \quad (4.12)\end{aligned}$$

where $\theta_S(t) \triangleq \omega_S t$.

Equation (4.12) is the mathematical representation of a *rotating magnetic field* which rotates at the angular (electrical) frequency ω_S and is illustrated in Figure 4.18. Note that the magnetic field has a fixed distribution *with respect* to θ_S. The angle θ_S is referred to as the *magnetic axis* of the rotating magnetic field.

Remark In Figure 4.18, the sinusoidally distributed stator windings are drawn with four loops in the sine curve envelope rather than with nine loops as in Figure 4.17. This is done for clarity of presentation. The reader should always assume that the sine curve envelope implies that the winding densities for phases a and b are given by (4.7) and (4.8), respectively.

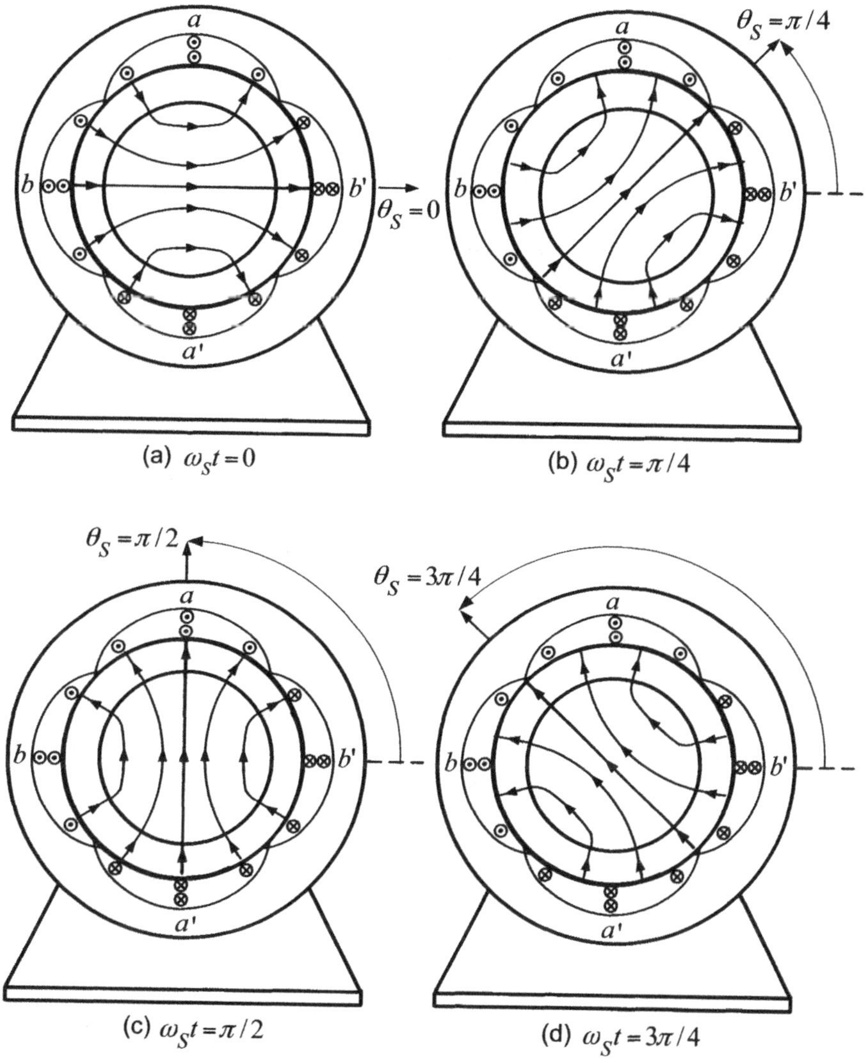

FIGURE 4.18. Sinusoidally distributed rotating radial magnetic field in the air gap. (For convenience, the magnetic field lines inside the stator iron are not drawn.)

4.5 Magnetomotive Force (mmf)

The magnetomotive force (mmf) is defined to be $\Im \triangleq \oint \vec{\mathbf{H}} \cdot d\vec{\ell}$, that is, the integral of $\vec{\mathbf{H}}$ around a *closed* curve. Of course, by Ampère's law,

$$\Im = \oint \vec{\mathbf{H}} \cdot d\vec{\ell} = i_{\text{enclosed}}.$$

Instead of the path used in Figure 4.16, consider applying Ampère's law to the path 1–2–3–4–1 in Figure 4.19 .

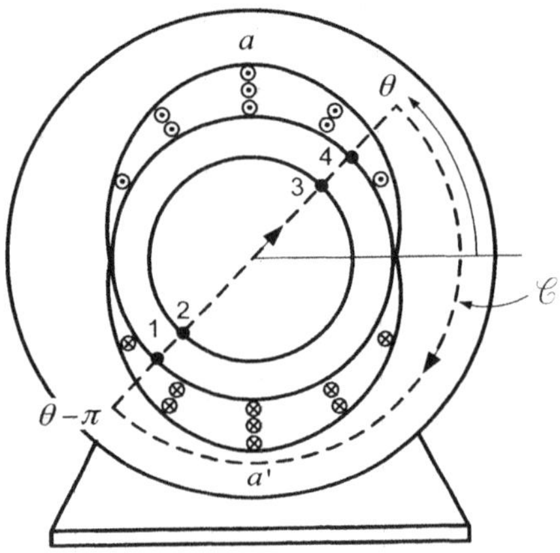

FIGURE 4.19. Sinusoidally wound stator phase.

Now $\vec{\mathbf{H}} = 0$ on the paths 2–3 and 4–1. By symmetry, $\vec{\mathbf{H}}(\theta - \pi) = -\vec{\mathbf{H}}(\theta)$ and, as $d\vec{\ell} = d\ell(-\hat{\mathbf{r}})$ on the path 1–2 while $d\vec{\ell} = d\ell\hat{\mathbf{r}}$ on the path 3–4, it follows that $\int_3^4 \vec{\mathbf{H}} \cdot d\vec{\ell} = \int_1^2 \vec{\mathbf{H}} \cdot d\vec{\ell}$ and so

$$\Im \triangleq \oint \vec{\mathbf{H}} \cdot d\vec{\ell} = 2\int_3^4 \vec{\mathbf{H}} \cdot d\vec{\ell} = 2\int_3^4 (H(\theta)\hat{\mathbf{r}}) \cdot (d\ell\hat{\mathbf{r}}) = 2H(\theta)g.$$

In Figure 4.19 the path is traversed in the clockwise direction in contrast to the path in Figure 4.16 which is traversed in the counterclockwise direction. Consequently, in the application of Ampère's law, the current is now considered positive if it is into the page rather than out of the page. The enclosed current for the path 1–2–3–4–1 is therefore

$$i_{\text{enclosed}} = -\int_{\theta-\pi}^{\theta} i_{Sa}(N_S/2)\sin(\theta')d\theta' = i_{Sa}N_S\cos(\theta).$$

Ampère's law $\oint \vec{\mathbf{H}} \cdot d\vec{\ell} = i_{\text{enclosed}}$ then gives

$$\Im \triangleq \oint \vec{\mathbf{H}} \cdot d\vec{\ell} = i_{Sa} N_S \cos(\theta)$$

or

$$\Im(\theta) = 2H(\theta)g = i_{Sa} N_S \cos(\theta).$$

The interpretation is that the mmf $\Im(\theta) = i_{Sa} N_S \cos(\theta)$ which is produced by the current is "dropped" across the air gap in the amount of $\Im_1(\theta) = H(\theta)g$ across each of the two diametrically opposite sides of the air gap. Then one works with the mmf $\Im$ and refers to an "mmf wave" $\Im_1(\theta) = H(\theta)g$ set up in the air gap, and so on (see, e.g., Refs. [5][19][21][24][26][58]). In contrast, a different emphasis (but with the same results) has been taken here. Specifically, Ampère's law is used to find $\vec{\mathbf{H}}$ in the air gap by computing the mmf and using $\vec{\mathbf{H}} \equiv 0$ in the iron. Then $\vec{\mathbf{B}}$ in the air gap is found simply from $\vec{\mathbf{B}} = \mu_0 \vec{\mathbf{H}}$. In other words, the mmf is only used as a device to compute the magnetic induction field $\vec{\mathbf{B}}$ in the air gap. After this is accomplished, the $\vec{\mathbf{B}}$ field is used exclusively because it is the quantity that produces forces on currents and voltages in coil loops, that is, it is directly related to the physical phenomena. This is perhaps better said in the following quotation from Melvin Schwartz, a 1988 Noble Laureate in Physics:[5]

... we must interject a small bit of philosophy. It is customary to call $\vec{\mathbf{B}}$ *the magnetic induction and* $\vec{\mathbf{H}}$ *the magnetic field strength. We reject this custom inasmuch as* $\vec{\mathbf{B}}$ *is the truly fundamental field and* $\vec{\mathbf{H}}$ *is a subsidiary artifact. We shall call* $\vec{\mathbf{B}}$ *the magnetic field and leave the reader to deal with* $\vec{\mathbf{H}}$ *as he pleases.*

Remark

In this and the previous sections, only the radial component (i.e., in the $\hat{\mathbf{r}}$ direction) of $\vec{\mathbf{B}}$ in the air gap was found. However, Figures 3.2–3.5 indicate that one would expect that an azimuthal component (i.e., in the $\hat{\boldsymbol{\theta}}$ direction) of $\vec{\mathbf{B}}$ would also be present in the air gap. This is indeed the case and it is derived in (the optional) Section 4.6. Only the radial component of $\vec{\mathbf{B}}$ contributes to the machine torque and produces emfs in the windings of the machine, which is the reason that one need only consider this component to derive the mathematical models of AC machines.

---

[5] See M. Schwartz [35], page 156.

4.6 Flux Linkage

The notion of flux $\phi = \int_S \vec{\mathbf{B}} \cdot d\vec{\mathbf{S}}$ is defined for a surface whose boundary is a simple closed curve. After computing the flux, Faraday's law $\xi = -d\phi/dt$ is then used to compute the induced emf (voltage) in the loop. However, phase windings are usually comprised of many windings which are also distributed and one wants to compute the total emf induced in the phase winding. It has been found that the notion of *flux linkage* can be used to do this in a convenient fashion.

To show how flux linkage is used in AC machines, consider Figure 4.20 which shows three loops wound on the inside surface of a stator iron. That is, a single piece of wire is wound around the inside surface where one loop is placed in a slot at $\theta = \pi/3$ (the other side of the loop is in the slot at $\theta = \pi/3 - \pi$) then a second loop is wound at $\theta = \pi/2$ with its other side at $\theta = 3\pi/2$) and finally a third loop is wound at $\theta = 2\pi/3$ (with its other side at $\theta = 5\pi/3$). The flux surface for each loop is taken to coincide with the loop itself as shown in Figure 4.20(b). As this figure shows, the first loop/surface is the path from a to b, the second loop/surface is the path from c to d and the third loop is the path from e to $f = a'$. One end of the wire is labeled a and the other a' and the objective here is to calculate the total emf $\xi_{a-a'}$ induced in the wire by the rotating permanent magnet rotor.

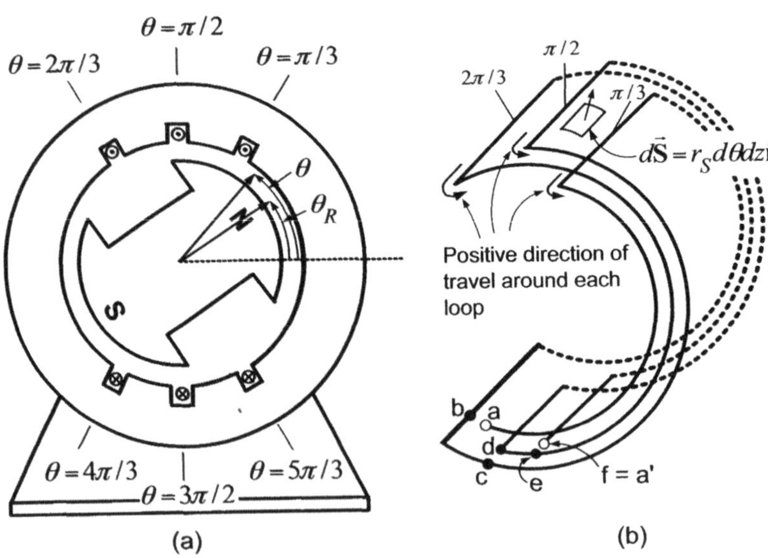

FIGURE 4.20. (a) Distributed winding for stator phase a with a permanent magnet rotor. (b) The flux surfaces for each stator loop.

As before, if the current in the stator winding is positive, the symbol $\odot$ means that it is coming out of the page and the symbol $\otimes$ means that it is going into the page. Here it is assumed that the permanent magnet produces a radial magnetic field in the air gap given by

$$\vec{\mathbf{B}}_R(\theta - \theta_R) = B_{\max}\frac{r_R}{r}\cos(\theta - \theta_R)\hat{\mathbf{r}}$$

where r_R is the radius of the rotor, (r, θ) are the polar coordinates of an arbitrary location in the air gap and θ_R is the rotor angle defined by the center line (magnetic axis) of the rotor's north pole [see Figure 4.20(a)]. This rotor magnet produces a flux in each loop and, for any given rotor position θ_R, this flux is different in the loop at $\theta = \pi/3$, the loop at $\pi/2$, and the loop at $2\pi/3$. Further, if the rotor is moving, then it is producing a changing flux in each of the three loops and therefore, by Faraday's law, this changing flux will produce an emf in each of the three loops. Of course, at any point in time, the emfs will be different in each of the three loops.

To compute these emfs, let

$$d\vec{\mathbf{S}} = r_S d\theta dz\hat{\mathbf{r}} \tag{4.13}$$

where r_S is the radius of the inside surface of the stator iron. Note that with this choice of $d\vec{\mathbf{S}}$, the positive direction of travel around the loop coincides with the positive direction chosen for the current in that loop [see Figure 4.20(b)]. On the inside surface of the rotor, $r = r_S$ so that the flux in the loop whose sides are in the slots at $\theta = \pi/3$ and $\theta = \pi/3 - \pi$ is

$$\begin{aligned}
\phi_{\pi/3} &= \int_{\substack{\text{Loop from} \\ \pi/3-\pi \text{ to } \pi/3}} \vec{\mathbf{B}}_R \cdot d\vec{\mathbf{S}} \\
&= \int_0^{\ell_1}\int_{\theta=\pi/3-\pi}^{\theta=\pi/3} B_{\max}\frac{r_R}{r_S}\cos(\theta - \theta_R)\hat{\mathbf{r}}\cdot(r_S d\theta dz\hat{\mathbf{r}}) \\
&= \ell_1 r_R \int_{\theta=\pi/3-\pi}^{\theta=\pi/3} B_{\max}\cos(\theta - \theta_R)d\theta \\
&= \ell_1 r_R B_{\max}\sin(\theta - \theta_R)d\theta\Big|_{\theta=\pi/3-\pi}^{\theta=\pi/3} \\
&= 2\ell_1 r_R B_{\max}\sin(\pi/3 - \theta_R).
\end{aligned}$$

Then the emf induced in this loop by the magnetic field of the permanent magnet is

$$\xi_{\pi/3} = -\frac{d\phi_{\pi/3}}{dt} = 2\ell_1 r_R B_{\max}\omega_R\cos(\theta_R - \pi/3)$$

where $\omega_R \triangleq d\theta_R/dt$. If $\xi_{\pi/3} > 0$, this emf will force current to go in the positive direction of travel around the loop which coincides with the positive direction of current in that loop.

Again using the outward normal so that $d\vec{\mathbf{S}}$ is still given by equation (4.13) as before, the flux in the loop between $-\pi/2$ to $\pi/2$ is

$$\phi_{\pi/2} = \int_{\substack{\text{Loop from} \\ -\pi/2 \text{ to } \pi/2}} \vec{\mathbf{B}}_R \cdot d\vec{\mathbf{S}} = 2\ell_1 r_R B_{\max} \sin(\pi/2 - \theta_R)$$

and the induced emf in the loop is

$$\xi_{\pi/2} = -\frac{d\phi_{\pi/2}}{dt} = 2\ell_1 r_R B_{\max}\omega_R \cos(\theta_R - \pi/2).$$

If $\xi_{\pi/2} > 0$, it will force current to go in the positive direction of travel around the loop which also coincides with the positive direction of current in that loop.

Finally, with $d\vec{\mathbf{S}}$ given by (4.13) once more,

$$\phi_{2\pi/3} = \int_{\substack{\text{Loop from} \\ 2\pi/3-\pi \text{ to } 2\pi/3}} \vec{\mathbf{B}}_R \cdot d\vec{\mathbf{S}} = 2\ell_1 r_R B_{\max} \sin(2\pi/3 - \theta_R)$$

and

$$\xi_{2\pi/3} = -\frac{d\phi_{2\pi/3}}{dt} = 2\ell_1 r_R B_{\max}\omega_R \cos(\theta_R - 2\pi/3).$$

Again, if $\xi_{2\pi/3} > 0$, it will force current to go in the positive direction of travel around the loop coinciding with the positive direction of current in that loop.

All three loops are connected in series to make up the phase winding and the sign convention for positive travel around each loop was chosen to coincide with the positive direction of current in the phase as illustrated in Figure 4.20(b). That is, the induced emfs in the loops are all in series in the phase winding and, as they all have the *same* sign convention, they can be added up to obtain the total emf in the phase winding. As a result, the total voltage induced in the phase is

$$\begin{aligned}
\xi_{a-a'} &= \xi_{\pi/3} + \xi_{\pi/2} + \xi_{2\pi/3} = 2\ell_1 r_R B_{\max}\omega_R \cos(\theta_R - \pi/3) \\
&\quad + 2\ell_1 r_R B_{\max}\omega_R \cos(\theta_R - \pi/2) + 2\ell_1 r_R B_{\max}\omega_R \cos(\theta_R - 2\pi/3) \\
&= \left(1 + \sqrt{3}\right) 2\ell_1 r_R B_{\max}\omega_R \sin(\theta_R).
\end{aligned}$$

However, consider

$$\begin{aligned}
\xi_{a-a'} &= \xi_{\pi/3} + \xi_{\pi/2} + \xi_{2\pi/3} \\
&= -\left(\frac{d\phi_{\pi/3}}{dt} + \frac{d\phi_{\pi/2}}{dt} + \frac{d\phi_{2\pi/3}}{dt}\right) \\
&= -\frac{d}{dt}\left(\phi_{\pi/3} + \phi_{\pi/2} + \phi_{2\pi/3}\right) \\
&= -\frac{d}{dt}\lambda_{a-a'}
\end{aligned}$$

where

$$\begin{aligned}\lambda_{a-a'} &\triangleq \phi_{\pi/3}+\phi_{\pi/2}+\phi_{2\pi/3}\\ &= 2\ell_1 r_R B_{\max}\sin(\pi/3-\theta_R)+2\ell_1 r_R B_{\max}\sin(\pi/2-\theta_R)\\ &\quad +2\ell_1 r_R B_{\max}\sin(2\pi/3-\theta_R)\\ &= \left(1+\sqrt{3}\right)2\ell_1 r_R B_{\max}\cos(\theta_R)\end{aligned}$$

is defined as the total *flux linkage* in phase $a-a'$. Then the total induced emf $\xi_{a-a'}$ in the phase $a-a'$ is

$$\xi_{a-a'} = -\frac{d\lambda_{a-a'}}{dt} = \left(1+\sqrt{3}\right)2\ell_1 r_R B_{\max}\omega_R\sin(\theta_R).$$

The point here is that one can first sum the fluxes in all the loops of a phase winding (i.e., compute the flux linkage), and then apply Faraday's law to the resulting flux linkage to obtain the total emf in the phase winding. However, as done above, care must be taken to ensure that the flux in each loop is computed in a consistent fashion so that the resulting emfs all have the same sign convention and therefore add up to give the total emf in the phase winding.

4.7 Azimuthal Magnetic Field in the Air Gap*

It was shown above that the radial magnetic field $\vec{\mathbf{B}}_{Sa}$ at a point (r,θ) in the air gap due to the current i_{Sa} in phase a is given by ($r_R=\ell_2/2$)

$$\vec{\mathbf{B}}_{Sa}(i_{Sa},r,\theta) = \frac{\mu_0 r_R N_S}{2gr}i_{Sa}\cos(\theta)\hat{\mathbf{r}} \tag{4.14}$$

where $\hat{\mathbf{r}}$ is a unit vector in the radial direction and g is the radial length of the air gap. However, there *has* to be a component of the magnetic field in the $\hat{\boldsymbol{\theta}}$ direction (azimuthal component) as is now shown by applying Ampère's law to the curve 1–2–3–4–1 shown in Figure 4.21. By Ampère's law,

$$\oint_{1-2-3-4-1}\vec{\mathbf{H}}_{Sa}\cdot d\vec{\ell} = \int_1^2\vec{\mathbf{H}}_{Sa}\cdot d\vec{\ell} = \int_{\theta_1}^{\theta_2}(-i_{Sa})\frac{N_S}{2}\sin(\theta)d\theta$$

as sides 2–3, 3–4, and 4–1 are inside the magnetic material[6] where $\vec{\mathbf{H}}\equiv\mathbf{0}$ and side 1–2 is in the air gap just along the inside surface of the stator. Note

[6] Remember that the cross-sectional area of the windings is modeleded as being infinitesimal.

that for closed curve 1–2–3–4–1 with $0 \le \theta_1 \le \theta_2 \le \pi$, the minus sign is used with the current (i.e., $-i_{Sa}$). This is because the path 1–2–3–4–1 is traversed in the clockwise direction so that Ampère's law assumes that positive current is *into* the page ($\otimes$) while $i_{Sa} > 0$ means the current is *out* of the page ($\odot$).

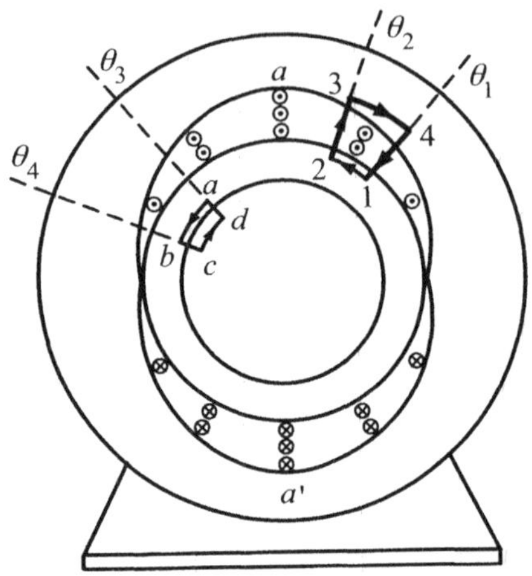

FIGURE 4.21. Closed curves for determining the *azimuthal* component $B_{Sa\theta}$ of the magnetic field produced by i_{Sa} in the air gap using Ampère's law.

With $r_S = r_R + g$, on the path 1–2 the differential line element is $d\vec{\ell} = r_S d\theta\hat{\boldsymbol{\theta}}$ so that the integral $\int_1^2 \vec{\mathbf{H}}_{Sa} \cdot d\vec{\ell}$ evaluates to

$$\int_{\theta_1}^{\theta_2} (H_{Sa\theta}\hat{\boldsymbol{\theta}}) \cdot (r_S d\theta\hat{\boldsymbol{\theta}}) = -\int_{\theta_1}^{\theta_2} i_{Sa}\frac{N_S}{2}\sin(\theta)d\theta$$

for $0 \le \theta_1 \le \theta \le \theta_2 \le \pi$. This must hold for *any* θ_1, θ_2 which implies that

$$H_{Sa\theta}(i_{Sa}, r_S, \theta) = -\frac{N_S}{2r_S}i_{Sa}\sin(\theta)$$

for $0 \le \theta \le \pi$. For $\pi \le \theta_1 \le \theta \le \theta_2 \le 2\pi$, the turns density is $(N_S/2)\,|\sin(\theta)| = -(N_S/2)\sin(\theta)$. However, for this range of θ, $i_{Sa} > 0$ means that the current is *into* the page so that a similar argument shows that

$$H_{Sa\theta}(i_{Sa}, r_S, \theta) = -\frac{N_S}{2r_S}i_{Sa}\sin(\theta)$$

for $\pi \le \theta \le 2\pi$. Combining,

$$B_{Sa\theta}(i_{Sa}, r_S, \theta) = -\frac{\mu_0 N_S}{2r_S}i_{Sa}\sin(\theta)$$

for $0 \leq \theta \leq 2\pi$. This is the *tangential* magnetic field at the inside surface of the stator (i.e., at $r = r_S$). To compute the azimuthal component of $B_{Sa\theta}$ at the surface of the rotor, that is, $B_{Sa\theta}(i_{Sa}, r_R, \theta)$, Ampère's law is applied to the closed path $a-b-c-d-a$ in Figure 4.21. Specifically,

$$\oint_{a\text{-}b\text{-}c\text{-}d\text{-}a} \vec{\mathbf{H}}_{Sa} \cdot d\vec{\ell} = \int_a^b \vec{\mathbf{H}}_{Sa} \cdot d\vec{\ell} = \int_{\theta_3}^{\theta_4} H_{Sa\theta}(i_{Sa}, r_R, \theta) r_R d\theta \equiv 0$$

as the path does not enclose any current. (The sides $b-c$,$c-d$ and $d-a$ are inside the magnetic material of the rotor where $\vec{\mathbf{H}} \equiv 0$ and the path $a-b$ is just on the surface of the rotor in the air gap.) As θ_3, θ_4 are arbitrary, it follows that

$$\begin{aligned} H_{Sa\theta}(i_{Sa}, r_R, \theta) &\equiv 0 \\ B_{Sa\theta}(i_{Sa}, r_R, \theta) &\equiv 0. \end{aligned}$$

The azimuthal component of the magnetic field in the air gap at the *surface* of both the stator and rotor has now been found. To find $\vec{\mathbf{B}}_S$ inside the air gap, let the tangential (azimuthal) component of $\vec{\mathbf{B}}_S$ be of the form

$$B_{Sa\theta}(i_{Sa}, r_S, \theta)\hat{\boldsymbol{\theta}} = -\alpha(r)\frac{\mu_0 N_S}{2r_S} i_{Sa} \sin(\theta)\hat{\boldsymbol{\theta}} \tag{4.15}$$

where

$$\begin{aligned} \alpha(r_S) &= 1 \\ \alpha(r_R) &= 0 \end{aligned} \tag{4.16}$$

as $\alpha(r)$ must satisfy (9.75) in order for (4.15) to hold at the rotor and stator surfaces.

Before the expression for $\alpha(r)$ is derived, it is first shown that the radial component of $\vec{\mathbf{B}}_{Sa}$ given in (4.14) must be modified to account for the azimuthal magnetic field. To see this, recall that Gauss's equation requires that the divergence of $\vec{\mathbf{B}}$ be identically zero. With $\vec{\mathbf{B}} = B_r\hat{\mathbf{r}} + B_\theta\hat{\boldsymbol{\theta}} + B_z\hat{\mathbf{z}}$, the divergence in cylindrical coordinates is given by

$$\nabla \cdot \vec{\mathbf{B}} = \frac{1}{r}\frac{\partial}{\partial r}(rB_r) + \frac{1}{r}\frac{\partial B_\theta}{\partial \theta} + \frac{\partial B_z}{\partial z} \equiv 0.$$

Consider $\vec{\mathbf{B}}_{Sa}$ of the form

$$\begin{aligned} \vec{\mathbf{B}}_{Sa} &= B_{Sar}\hat{\mathbf{r}} + B_{Sa\theta}\hat{\boldsymbol{\theta}} \\ &= B_{Sar}\hat{\mathbf{r}} - \alpha(r)\frac{\mu_0 N_S}{2r_S} i_{Sa} \sin(\theta)\hat{\boldsymbol{\theta}}. \end{aligned}$$

Taking the divergence of this expression results in

$$\frac{1}{r}\frac{\partial}{\partial r}(rB_{Sar}) - \frac{1}{r}\alpha(r)\frac{\mu_0 N_S}{2r_S} i_{Sa}\frac{\partial}{\partial \theta}\sin(\theta) \equiv 0$$

or

$$\frac{\partial}{\partial r}(rB_{Sar}) = \alpha(r)\frac{\mu_0 N_S}{2r_S} i_{Sa}\cos(\theta).$$

Integrating with respect to r, one obtains

$$rB_{Sar} = \frac{\mu_0 N_S}{2r_S} i_{Sa}\cos(\theta)\int_{r_R}^{r}\alpha(r')dr' + f(\theta)$$

where $f(\theta)$ is a "constant of integration". Taking it to be

$$f(\theta) = \frac{\mu_0 r_R N_S}{2g} i_{Sa}\cos(\theta)$$

gives

$$\frac{f(\theta)}{r} = \frac{\mu_0 r_R N_S}{2gr} i_{Sa}\cos(\theta)$$

which is the original expression for the radial component of $\vec{\mathbf{B}}_{Sa}$ given in (4.14). Thus, the radial component of $\vec{\mathbf{B}}_{Sa}$ is now given by

$$B_{Sar} = \frac{\mu_0 r_R N_S}{2g} i_{Sa}\frac{\cos(\theta)}{r}\left(1 + \frac{g}{r_S r_R}\int_{r_R}^{r}\alpha(r')dr'\right).$$

At the rotor surface, this expression for B_{Sar} is the same as that given by (4.14) with $r = r_R$. Consequently, the torque on the rotor will not change due to this change in the radial component of $\vec{\mathbf{B}}$. It is shown below that $0 \le \int_{r_R}^{r}\alpha(r')dr' \le g/2$ for small air gaps so that the change in the radial component is bounded by

$$\frac{g^2}{2r_S r_R} << 1$$

so that it is indeed a small effect.

Summarizing,

$$\begin{aligned}\vec{\mathbf{B}}_{Sa} &= \frac{\mu_0 r_R N_S}{2g} i_{Sa}\frac{\cos(\theta)}{r}\left(1 + \frac{g}{r_S r_R}\int_{r_R}^{r}\alpha(r')dr'\right)\hat{\mathbf{r}} \\ &\quad - \frac{\mu_0 r_R N_S}{2g}\frac{g}{r_S r_R}\alpha(r) i_{Sa}\sin(\theta)\hat{\boldsymbol{\theta}} \end{aligned} \tag{4.17}$$

with

$$\begin{aligned}\alpha(r_S) &= 1 \\ \alpha(r_R) &= 0.\end{aligned}$$

Figure 4.22 indicates the distribution of the azimuthal magnetic field with $i_{Sa} > 0$. Note that the field strength decreases as one moves along the radial direction from the stator side of the air gap to the rotor side. It also decreases as one moves in the θ direction from $\pm\pi/2$ to 0 or π.

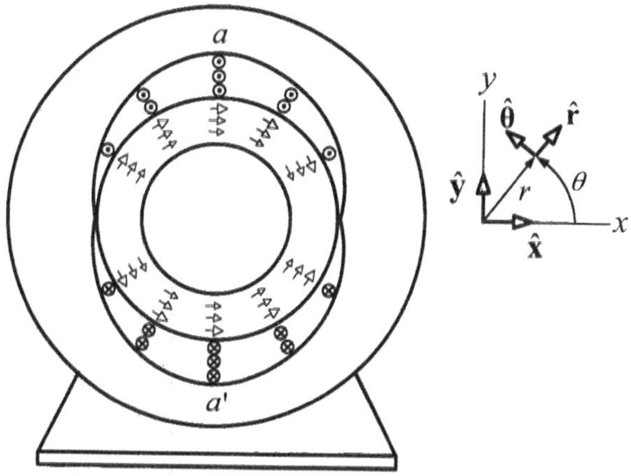

FIGURE 4.22. Distribution of azimuthal magnetic field $B_{Sa\theta}\hat{\boldsymbol{\theta}}$ in the air gap.

Determination of $\alpha(r)$

To determine $\alpha(r)$ for $r_R < r < r_S$, Ampère's law in differential form $\nabla \times \vec{\mathbf{H}} = \vec{\mathbf{J}}_{\text{free}}$ is used. In the air gap, there are no currents so $\nabla \times \vec{\mathbf{H}} = \mathbf{0}$, nor is there any magnetic material so that $\vec{\mathbf{B}} = \mu_0 \vec{\mathbf{H}}$. Consequently, Ampère's law in the air gap is simply

$$\nabla \times \vec{\mathbf{B}} = \mathbf{0}.$$

In cylindrical coordinates, this becomes

$$\begin{aligned}\nabla \times \vec{\mathbf{B}} &= \left(\frac{1}{r}\frac{\partial B_z}{\partial \theta} - \frac{\partial B_\theta}{\partial z}\right)\hat{\mathbf{r}} + \left(\frac{\partial B_r}{\partial z} - \frac{\partial B_z}{\partial r}\right)\hat{\boldsymbol{\theta}} + \frac{1}{r}\left(\frac{\partial}{\partial r}(rB_\theta) - \frac{\partial B_r}{\partial \theta}\right)\hat{\mathbf{z}} \\ &= \mathbf{0}.\end{aligned}$$

By the assumed cylindrical symmetry, $B_{Saz} \equiv 0$ and $B_{Sar}, B_{Sa\theta}$ do not depend on the coordinate z, so that one need only satisfy the z component of $\nabla \times \vec{\mathbf{B}} = \mathbf{0}$, that is,

$$\frac{\partial}{\partial r}(rB_{Sa\theta}) = \frac{\partial B_{Sar}}{\partial \theta}.$$

This becomes

$$-\frac{g}{r_S r_R}\frac{\partial}{\partial r}\Big(r\alpha(r)\Big)\, i_{Sa}\sin(\theta) = -i_{Sa}\frac{\sin(\theta)}{r}\left(1 + \frac{g}{r_S r_R}\int_{r_R}^{r}\alpha(r')dr'\right).$$

Canceling terms and expanding, this reduces to

$$\begin{aligned}
\frac{g}{r_S r_R}\frac{d}{dr}\Big(r\alpha(r)\Big) &= \frac{1}{r}\left(1+\frac{g}{r_S r_R}\int_{r_R}^{r}\alpha(r')dr'\right)\\
\frac{d}{dr}(r\alpha) &= \frac{1}{r}\frac{r_S r_R}{g}+\frac{\int_{r_R}^{r}\alpha(r')dr'}{r}\\
r\alpha+r^2\frac{d\alpha}{dr} &= \frac{r_S r_R}{g}+\int_{r_R}^{r}\alpha(r')dr'.
\end{aligned}$$

Differentiating this last expression with respect to r and rearranging, the differential equation

$$\alpha+r\frac{d\alpha}{dr}+2r\frac{d\alpha}{dr}+r^2\frac{d^2\alpha}{dr^2}=\alpha$$

is obtained, or

$$\frac{d^2\alpha}{dr^2}+\frac{3}{r}\frac{d\alpha}{dr}=0.$$

Solving this differential equation for $d\alpha/dr$ results in

$$\frac{d\alpha}{dr}=c_1 e^{-\int_{r_R}^{r}(3/r')dr'}=c_1 e^{-3\ln(r/r_R)}$$

and therefore

$$\alpha(r)=c_1\int_{r_R}^{r}e^{-3\ln(r'/r_R)}dr'+c_2. \tag{4.18}$$

The condition $\alpha(r_R)=0$ forces $c_2=0$ and $\alpha(r_S)=1$ requires that

$$c_1=\frac{1}{\int_{r_R}^{r_S}e^{-3\ln(r'/r_R)}dr'}\approx\frac{1}{g}.$$

This approximation follows from the assumption of a small air gap as

$$\ln(r/r_R)\approx\ln(1)=0 \quad \text{for } r_R<r<r_S \tag{4.19}$$

so that

$$\int_{r_R}^{r_S}e^{-3\ln(r'/r_R)}dr'\approx\int_{r_R}^{r_S}1dr'=r_S-r_R=g.$$

Substitution of $c_1=1/g, c_2=0$ into (4.18) and invoking the approximation (4.19) gives

$$\begin{aligned}
\alpha(r) &= \frac{1}{g}\int_{r_R}^{r}e^{-3\ln(r'/r_R)}dr'\\
&\approx \frac{1}{g}\int_{r_R}^{r}1dr'\\
&= \frac{r-r_R}{g}.
\end{aligned}$$

In other words, $\alpha(r)$ is approximately a linear function of r in the air gap. Such a linear dependence is pointed out by C. B. Gray [5].

4.7.1 Electric Field $\vec{\mathbf{E}}_{Sa}$

The axial electric field produced by the magnetic field $\vec{\mathbf{B}}_{Sa}$ is now computed. Recall that the magnetic field is given by

$$\begin{aligned}\vec{\mathbf{B}}_{Sa} &= \frac{\mu_0 \ell_2 N_S}{4g} i_{Sa} \frac{\cos(\theta)}{r}\left(1+\frac{g}{r_S r_R}\int_{r_R}^{r}\alpha(r')dr'\right)\hat{\mathbf{r}} \\ &\quad -\frac{\mu_0 \ell_2 N_S}{4g}\frac{g}{r_S r_R}\alpha(r) i_{Sa}\sin(\theta)\hat{\boldsymbol{\theta}}.\end{aligned}$$

The electric field induced by this changing magnetic field is found by solving

$$\nabla\times\vec{\mathbf{E}}_{Sa} = -\frac{\partial\vec{\mathbf{B}}_{Sa}}{\partial t}. \tag{4.20}$$

In cylindrical coordinates, the curl of $\vec{\mathbf{E}}$ is given by

$$\nabla\times\vec{\mathbf{E}} = \left(\frac{1}{r}\frac{\partial E_z}{\partial\theta}-\frac{\partial E_\theta}{\partial z}\right)\hat{\mathbf{r}}+\left(\frac{\partial E_r}{\partial z}-\frac{\partial E_z}{\partial r}\right)\hat{\boldsymbol{\theta}}+\frac{1}{r}\left(\frac{\partial}{\partial r}(rF_\theta)-\frac{\partial E_r}{\partial\theta}\right)\hat{\mathbf{z}}.$$

By the usual symmetry assumptions

$$\frac{\partial E_{Sa\theta}}{\partial z}=0,\quad \frac{\partial E_{Sar}}{\partial z}=0 \tag{4.21}$$

and, as the z component of $\vec{\mathbf{B}}_{Sa}$ is zero (i.e., $B_{Saz}\equiv 0$), it follows that

$$\frac{\partial}{\partial r}(rE_{Sa\theta})-\frac{\partial E_{Sar}}{\partial\theta}\equiv 0. \tag{4.22}$$

Then (4.20) reduces to

$$\nabla\times\vec{\mathbf{E}}_{Sa} = \frac{1}{r}\frac{\partial E_{Saz}}{\partial\theta}\hat{\mathbf{r}}-\frac{\partial E_{Saz}}{\partial r}\hat{\boldsymbol{\theta}} = -\frac{\partial B_{Sar}}{\partial t}\hat{\mathbf{r}}-\frac{\partial B_{Sa\theta}}{\partial t}\hat{\boldsymbol{\theta}}. \tag{4.23}$$

The simplest solution is one of the form

$$\vec{\mathbf{E}}_{Sa} = E_{Saz}\hat{\mathbf{z}},$$

that is, where $E_{Sar}=E_{Sa\theta}\equiv 0$ so that the conditions (4.21) and (4.22) are automatically satisfied.

Solving equation (4.23), one obtains

$$\begin{aligned}\vec{\mathbf{E}}_{Sa} &= E_{Saz}\hat{\mathbf{z}} = -\frac{\mu_0 r_R N_S}{2g}\frac{di_{Sa}}{dt}\sin(\theta)\left(1+\frac{g}{r_S r_R}\int_{r_R}^{r}\alpha(r')dr'\right)\hat{\mathbf{z}} \\ &\approx -\frac{\mu_0 r_R N_S}{2g}\frac{di_{Sa}}{dt}\sin(\theta)\left(1+\frac{(r-r_R)^2}{2r_S r_R}\right)\hat{\mathbf{z}} \\ &\approx -\frac{\mu_0 r_R N_S}{2g}\frac{di_{Sa}}{dt}\sin(\theta)\hat{\mathbf{z}}.\end{aligned}$$

4.7.2 The Magnetic and Electric Fields $\vec{\mathbf{B}}_{Sa}, \vec{\mathbf{E}}_{Sa}, \vec{\mathbf{B}}_{Sb}, \vec{\mathbf{E}}_{Sb}$

The magnetic and electric fields due to the current in phase b are found from their corresponding expressions for phase a by simply replacing θ by $\theta - \pi/2$.

In summary, given two sinusoidally distributed stator windings a and b wound 90 degrees apart, the magnetic and electric fields in the air gap produced by the phase currents i_{Sa}, i_{Sb} in the phase a and b, respectively, are given by

$$\begin{aligned}
\vec{\mathbf{B}}_{Sa} &= \frac{\mu_0 \ell_2 N_S}{4g} i_{Sa} \frac{\cos(\theta)}{r}\left(1 + \frac{g}{r_S r_R}\int_{r_R}^{r} \alpha(r')dr'\right)\hat{\mathbf{r}} \\
&\quad - \frac{\mu_0 \ell_2 N_S}{4g}\frac{g}{r_S r_R}\alpha(r) i_{Sa}\sin(\theta)\hat{\boldsymbol{\theta}} \\
\vec{\mathbf{E}}_{Sa} &= E_{Saz}\hat{\mathbf{z}} = -\frac{\mu_0 r_R N_S}{2g}\frac{di_{Sa}}{dt}\sin(\theta)\left(1 + \frac{g}{r_S r_R}\int_{r_R}^{r} \alpha(r')dr'\right)\hat{\mathbf{z}}
\end{aligned} \tag{4.24}$$

and

$$\begin{aligned}
\vec{\mathbf{B}}_{Sb} &= \frac{\mu_0 r_R N_S}{2g} i_{Sb} \frac{\sin(\theta)}{r}\left(1 + \frac{g}{r_S r_R}\int_{r_R}^{r} \alpha(r')dr'\right)\hat{\mathbf{r}} \\
&\quad + \frac{\mu_0 r_R N_S}{2g}\frac{g}{r_S r_R}\alpha(r) i_{Sb}\cos(\theta)\hat{\boldsymbol{\theta}} \\
\vec{\mathbf{E}}_{Sb} &= E_{Sbz}\hat{\mathbf{z}} = \frac{\mu_0 r_R N_S}{2g}\frac{di_{Sb}}{dt}\cos(\theta)\left(1 + \frac{g}{r_S r_R}\int_{r_R}^{r} \alpha(r')dr'\right)\hat{\mathbf{z}}.
\end{aligned} \tag{4.25}$$

With $i_{Sa}(t) = I_S\cos(\omega_S t)$, $i_{Sb}(t) = I_S\sin(\omega_S t)$ and $\alpha \equiv 0$, the expressions (4.24) and (4.25) simplify to

$$\begin{aligned}
\vec{\mathbf{B}}_S(r,\theta,t) &= \frac{\mu_0 r_R N_S I_S}{2g}\frac{1}{r}\cos(\theta - \omega_S t)\hat{\mathbf{r}} \\
\vec{\mathbf{E}}_S(\theta,t) &= \omega_S\frac{\mu_0 r_R N_S I_S}{2g}\cos(\theta - \omega_S t)\hat{\mathbf{z}}.
\end{aligned} \tag{4.26}$$

Remark

Note that at $r = r_R$, the expressions for the electric and magnetic fields given by (4.24) and (4.25) reduce to their approximate values (i.e., taking $\alpha \equiv 0$). Consequently, the induced emfs in the rotor loops are not affected by neglecting the azimuthal magnetic fields.

Problems

Problem 1 A Two-Phase Stator with Single-Loop Windings

(a) Consider the simple single-loop stator phase shown in Figure 4.23. Show that in the air gap, the magnetic intensity $H(\theta)$ *in the radial direction satisfies*

$$H(\theta) = \begin{cases} H_1 & \textit{for} \quad -\pi/2 < \theta < \pi/2 \\ H_2 & \textit{for} \quad +\pi/2 < \theta < 3\pi/2 \end{cases}$$

where H_1, H_2 *are constants and assuming* H *is constant as a function of* r *across the air gap. (Hint: Apply Ampère's law to the path a–b–c–d–a to show that* $H(\theta) = H_1$ *for* $-\pi/2 < \theta < \pi/2$*, etc.).*

(b) Applying Ampère's law to the path 1–2–3–4–1 in Figure 4.23(a) along with a symmetry argument to show that $H_2 = -H_1$*.*

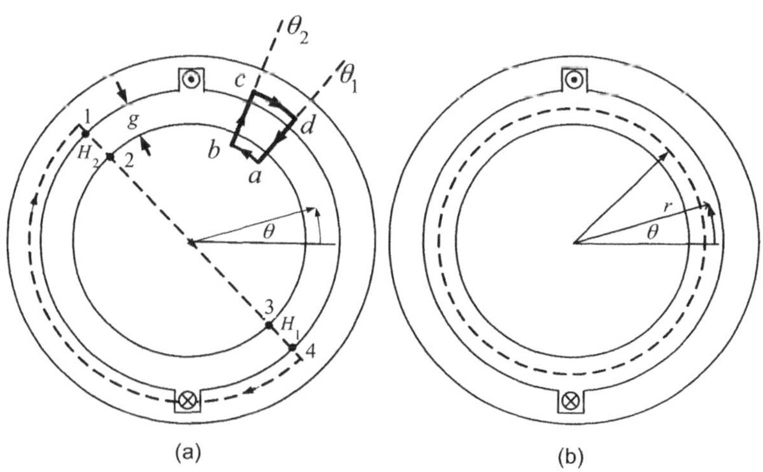

FIGURE 4.23. Single-loop stator phase. (a) Paths for Ampère's law. (b) Flux surface in the air gap. The closed flux surface completely contains the rotor.

(c) Alternatively, rather than using Ampère's law as in part (b), use Gauss's law applied to the surface shown in Figure 4.23(b) to obtain $H_2 = -H_1$*.*

(d) Give an explicit expression for $\vec{\mathbf{B}}$ *in the air gap as a function of* θ *in terms of the current* i *in the winding and the air gap length* g*. Recall that* $\vec{\mathbf{B}} = \mu_0 \vec{\mathbf{H}}$ *in the air gap. This is said to be a* uniformly *distributed magnetic field in* θ*.*

(e) Show that by multiplying the answer for $\vec{\mathbf{B}}$ *in part (d) by* r_R/r *results in the magnetic field satisfying Gauss's law for closed surfaces in the air gap as in Figure 4.12.*

(f) Consider a simple two-phase machine as in Figure 4.24.

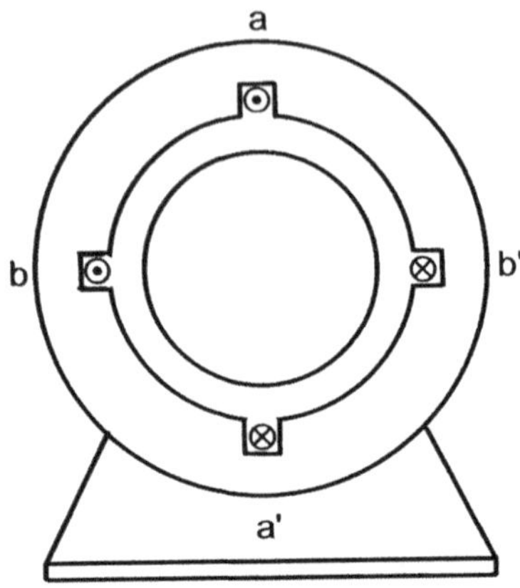

FIGURE 4.24. Simple two-phase machine.

Let the phase currents i_{Sa}, i_{Sb} be periodic with period T as given in Figure 4.25. Show that the magnetic axis of the stator's magnetic field can assume only four angular positions. What are they? In particular, plot the radial air-gap magnetic field for $0 \leq \theta \leq 2\pi$ at $t = 0, T/4, T/2,$ and $3T/4$.

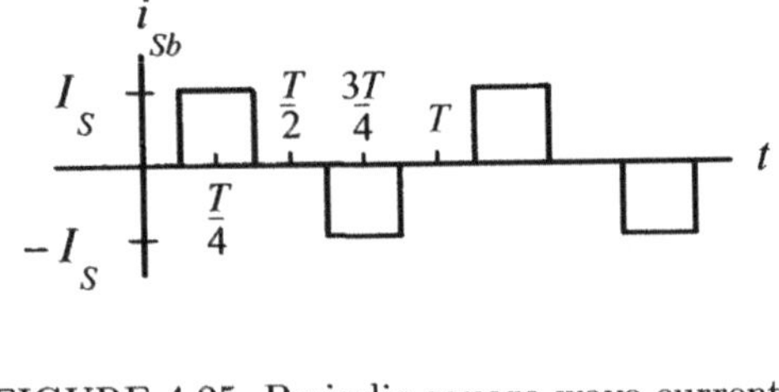

FIGURE 4.25. Periodic square-wave currents.

(g) Repeat part (f) where the phase currents i_{Sa}, i_{Sb} are again periodic with period T with their waveforms as given in Figure 4.26. Show that the magnetic axis of the stator's magnetic field can assume only four angular positions. What are they? In particular, plot the radial air-gap magnetic field for $0 \leq \theta \leq 2\pi$ at $t = T/8, 3T/8, 5T/8,$ and $7T/8$.

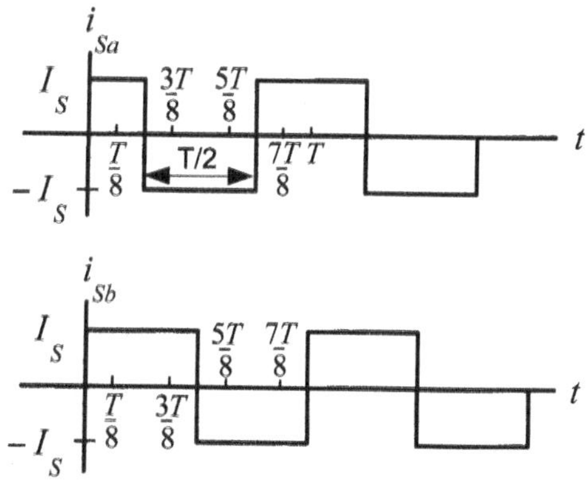

FIGURE 4.26. Periodic square-wave phase currents.

Problem 2 Fourier Series Expansion

Consider the function $F'(\theta')$ shown in Figure 4.27.

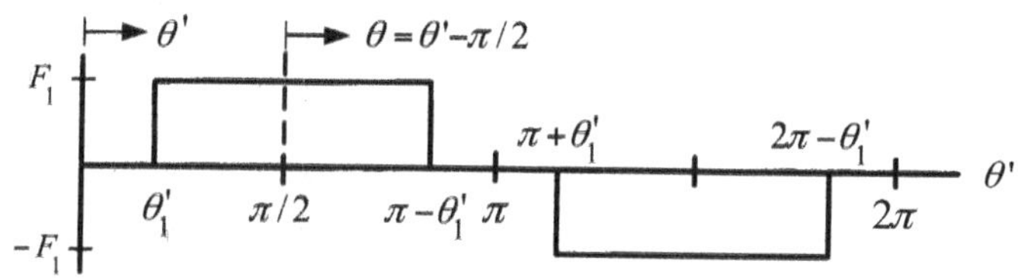

FIGURE 4.27. $F'(\theta')$ versus θ'.

(a) Show that the Fourier series expansion of this function is

$$F'(\theta') = \frac{4}{\pi} F_1 \times \sum_{k=1,3,5,\ldots}^{\infty} \frac{1}{k} \cos(k\theta_1') \sin(k\theta'). \tag{4.27}$$

(b) With respect to the angle $\theta \triangleq \theta' - \pi/2$, show that (4.27) becomes

$$F(\theta) \triangleq F'(\theta + \pi/2) = \frac{4}{\pi} \times \sum_{k=1,3,5,\ldots}^{\infty} F_1 \frac{1}{k} \cos(k\theta_1') \sin(k(\theta + \pi/2)).$$

(c) Use the result of part (b) to show that the Fourier series expansion of the function in Figure 4.28 is

$$F(\theta) = \frac{4}{\pi} \sum_{k=1,3,5,\cdots}^{\infty} \frac{1}{k} \Big(F_1 \cos(k\theta_1') + F_2 \cos(k\theta_2') + F_3 \cos(k\theta_3') \Big) \sin\Big(k\Big(\theta + \frac{\pi}{2}\Big)\Big).$$

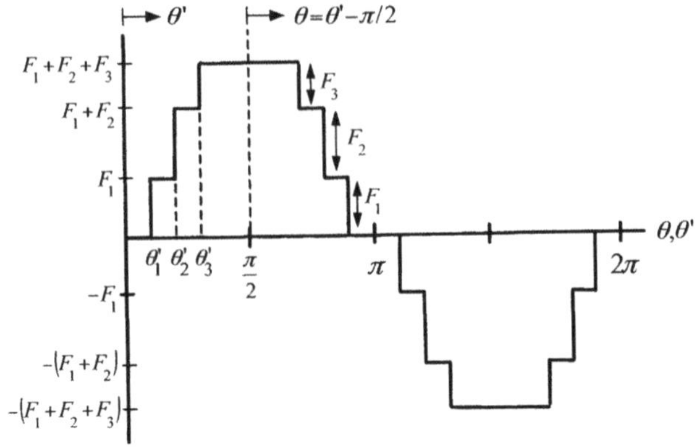

FIGURE 4.28. $F(\theta), F'(\theta')$.

(d) Show that

$$\sin\left(k(\theta+\pi/2)\right) = \begin{cases} \cos(k\theta) & \textit{for} \quad k = 6m-1 \textit{ where } m = 1,2,\ldots \\ -\cos(k\theta) & \textit{for} \quad k = 6m+1 \textit{ where } m = 1,2,\ldots \end{cases}$$

Problem 3 Two-Level Approximate Sinusoidal Winding

Apply the results of problem 2 (with $F_3 = 0$ or $\theta_3 = \pi/2$) to derive the Fourier expansion given in equation (4.3).

Problem 4 Three-Level Approximate Sinusoidal Winding

Consider the approximate sinusoidal winding of Figure 4.29(a) in which a single phase is wound such that there is one loop at $\pi/6$, two loops at $\pi/3$, three loops at $\pi/2$, two loops at $2\pi/3$, and one loop at $5\pi/6$.

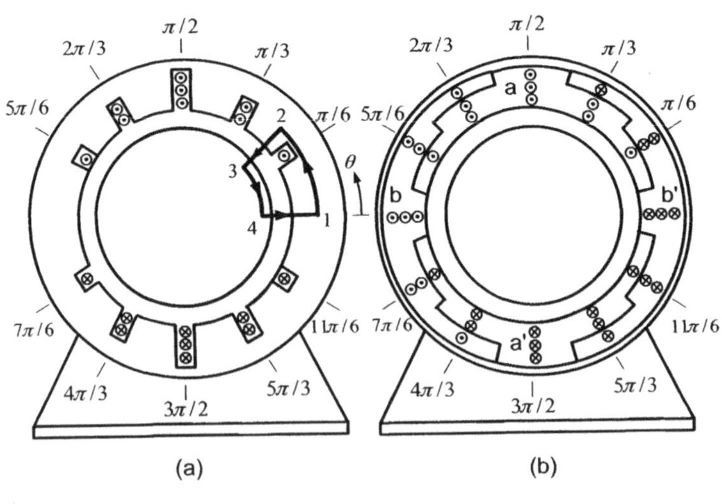

FIGURE 4.29. Approximately sinusoidally wound two-phase stator.

(a) Find the enclosed current $i_{enclosed}$ between 0 and θ for $0 \le \theta \le 2\pi$.

(b) Assuming $\vec{\mathbf{H}}$ is constant across the air gap and the windings/loops have an infinitesimal cross section, show that a straightforward application of Ampère's law gives $H_{Sa}(\theta) = H_{Sa}(0) - i_{enclosed}(\theta)/g$. Use conservation of flux to find $H_{Sa}(0)$. Show that $B_{Sa}(\theta) = \mu_0 H_{Sa}(\theta)$ is given by the plot in Figure 4.11.

(c) With $B_{Sa}(r,\theta) = \mu_0 (r_R/r) H_{Sa}(\theta)$ as given in Figure 4.11, use the results of problem 2 to show that the Fourier series expansion of $B_{Sa}(r,\theta)$ in θ is

$$B_{Sa}(r,\theta) = \mu_0 \frac{i_{Sa}}{g}\frac{4}{\pi}\frac{r_R}{r}$$

$$\times \sum_{k=1,3,5,\ldots}^{\infty} \frac{1}{k}\left(\frac{3}{2}\cos(0) + 2\cos(k\pi/6) + \cos(k\pi/3)\right)\sin(k(\theta+\pi/2) \quad (4.28)$$

(d) Let b_k be the coefficient of $\sin(k(\theta+\pi/2)$ in the Fourier expansion of $B_{Sa}(r,\theta)/\left(\mu_0 \frac{i_{Sa}}{g}\frac{r_R}{r}\right)$ in equation (4.28) and let a_k be the coefficient of $\sin(k(\theta+\pi/2)$ in the Fourier expansion of $B_{Sa}(\theta)/(\mu_0 i_{Sa}/g)$ in equation (4.3). With $a_{\max} = \max_{k=1,\ldots,15}\{a_k\}, b_{\max} = \max_{k=1,\ldots,15}\{b_k\}$, plot the relative harmonic content $a_k/a_{\max}$ and $b_k/b_{\max}$ as a function of k for $k = 1, 2, \ldots, 15$ on separate figures. Note that $a_{\max} = a_1, b_{\max} = b_1$; that is, the maximum is achieved at the fundamental for both expansions. Compute b_1/a_1.

(e) Finally, plot $a_k/a_{\max} - b_k/b_{\max}$ for $k = 1, 2, 3, \ldots, 15$. What is the value of this difference for $k = 5, 7, 11, 13$? What is this difference for $k = 3, 9, 15$?

Comments: Note the reduction in relative harmonic content is only for the triplen harmonics. Also, the fact that $b_1 = 2a_1$ means that, for the same phase current, the number of turns in this (three-level) approximate sinusoidal winding results in doubling the strength of the magnetic field in the air gap compared to the approximate sinusoidal winding with only two levels.

(f) Consider using the winding of Figure 4.29(a) in a two-phase machine as shown in Figure 4.29(b). Use the solution of part (c) to write down the expression for the radial magnetic field in the air gap due to the current i_{Sb} in stator phase b.

Problem 5 Gauss's Law

Show that the expression [see equation (4.6)]

$$\vec{\mathbf{B}}_{Sa}(r,\theta) = B_{Sa}(r,\theta)\hat{\mathbf{r}} = \mu_0 \frac{r_R}{r}\left(\frac{2i_{Sa}}{g} - \frac{i_{enclosed}(\theta)}{g}\right)\hat{\mathbf{r}}$$

for the magnetic field due to the current in an approximate sinusoidal winding satisfies $\oint \vec{\mathbf{B}} \cdot d\vec{\mathbf{S}} = 0$ for the flux surface given in Figure 4.12.

Problem 6 Three-Phase Stator and Spatial Harmonics

Figure 4.30(a) shows a machine with both a three-phase stator and a three-phase rotor. Figure 4.30(b) gives more detail of stator phase 1′−1 where it is shown that one side of a loop is wound at $\pi/3$ (the other side at $4\pi/3$), two loop sides are wound at $\pi/2$ (the other two sides at $3\pi/2$), and, finally, one side of a loop is wound at $2\pi/3$ (the other side at $5\pi/3$).

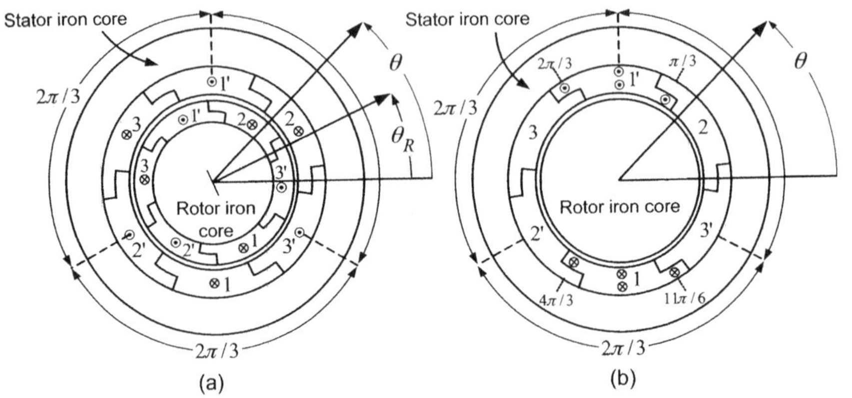

FIGURE 4.30. (a) Three-phase stator and rotor. (b) Three-phase stator with an iron core rotor. Windings shown for stator phase *a* only.

With the assumptions that the conductors have an infinitesimal cross section, the iron has infinite permeability (i.e., $\vec{\mathbf{H}} = 0$ in the iron), and the air gap is small, Ampère's law was used to show that the radial magnetic field in the air gap produced by the current i_{S1} in phase 1′−1 is given by [see equation (4.3)]

$$B_{S_1}(i_{S1},\theta) = \mu_0 \frac{i_{S1}}{g}\frac{4}{\pi} \times \sum_{k=1,3,5,\ldots}^{\infty} \frac{(1+\cos(\frac{k\pi}{6}))}{k} \sin\Big(k(\theta + \frac{\pi}{2})\Big)$$

Let $B_{S_1}(i_{S1},r,\theta) \triangleq (r_R/r)\, B_{S_1}(i_{S1},\theta)$ and similarly for $B_{S_2}(i_{S2},r,\theta)$ and $B_{S_3}(i_{S3},r,\theta)$ so that

$$B_{S_1}(i_{S1},r,\theta) = \mu_0 \frac{i_{S1}}{g}\frac{4}{\pi}\frac{r_R}{r} \times \sum_{k=1,3,5,\ldots}^{\infty} \frac{(1+\cos(\frac{k\pi}{6}))}{k} \sin\Big(k(\theta + \frac{\pi}{2})\Big)$$

$$B_{S_2}(i_{S2},r,\theta) = \mu_0 \frac{i_{S2}}{g}\frac{4}{\pi}\frac{r_R}{r} \times \sum_{k=1,3,5,\ldots}^{\infty} \frac{(1+\cos(\frac{k\pi}{6}))}{k} \sin\Big(k(\theta - \frac{2\pi}{3} + \frac{\pi}{2})\Big)$$

$$B_{S_3}(i_{S3},r,\theta) = \mu_0 \frac{i_{S3}}{g}\frac{4}{\pi}\frac{r_R}{r} \times \sum_{k=1,3,5,\ldots}^{\infty} \frac{(1+\cos(\frac{k\pi}{6}))}{k} \sin\Big(k(\theta - \frac{4\pi}{3} + \frac{\pi}{2})\Big).$$

(a) With the currents balanced, i.e., $i_{S1} + i_{S2} + i_{S3} \equiv 0$*, show that in the Fourier expansion of the total radial magnetic field*

$$\vec{\mathbf{B}}_S = \vec{\mathbf{B}}_{S_1} + \vec{\mathbf{B}}_{S_2} + \vec{\mathbf{B}}_{S_3}$$

the triplen harmonics are zero. (The triplen harmonics are the harmonic terms of order k *with* $k = 3m, m = 1, 2, ...$*)*

(b) Explain why the phase winding of Figure 4.3(b) is really more suited for a three-phase machine rather than a two phase machine. What about the winding of Figure 4.29 (see problem 4)?

(c) Let

$$i_{S1} = I_S\cos(\omega_S t), i_{S2} = I_S\cos(\omega_S t - 2\pi/3), i_{S3} = I_S\cos(\omega_S t - 4\pi/3)$$

in the above expressions for the magnetic fields. Show that the 5*th harmonic component of* $\vec{\mathbf{B}}_S$ *given by*

$$\vec{\mathbf{B}}_{S5}(r,\theta,t) \triangleq \vec{\mathbf{B}}_{S_15}(i_{S1},r,\theta) + \vec{\mathbf{B}}_{S_25}(i_{S2},r,\theta) + \vec{\mathbf{B}}_{S_35}(i_{S3},r,\theta)$$

is a rotating magnetic field in the clockwise (negative) direction at angular speed $-\omega_S/5$*. (Here* $\vec{\mathbf{B}}_{S_15}$ *is the* $k = 5$ *term in the above Fourier series expansion of* $\vec{\mathbf{B}}_{S_1}$*, etc.)*

(d) Repeat part (c) for the 7*th harmonic* $\vec{\mathbf{B}}_{S7}(r,\theta,t)$ *of* $\vec{\mathbf{B}}_S$*. What direction does it rotate? At what angular speed?*

Problem 7 Self-Inductance of a Distributed Winding

Consider the stator phase winding shown in Figure 4.3(b) and redrawn in Figure 4.31. The radial magnetic field due to the current in the phase is shown in Figure 4.10.

(a) Using flux surfaces similar to those of Figure 4.20, compute the total flux linkage in the stator phase winding due to the current in the phase.

(b) What is the self-inductance of the stator phase?

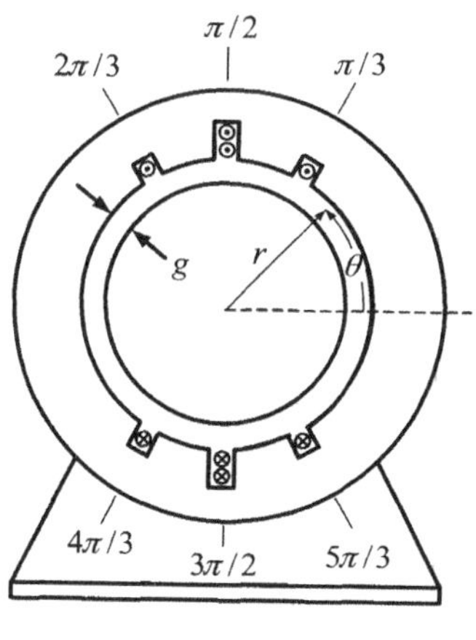

FIGURE 4.31. Computation of the self-inductance of a distributed winding.

Problem 8 Flux Linkage

Figure 4.32 shows a motor with a permanent magnet rotor which has a single stator phase wound as in Figure 4.3(b). Let the magnetic field in the air gap due to the permanent magnet rotor be given by

$$\vec{\mathbf{B}}_R(\theta - \theta_R) = B_{max}\frac{r_R}{r}\cos(\theta - \theta_R)\hat{\mathbf{r}}.$$

(a) Compute the flux in each loop and the total flux linkage in the stator phase.

(b) Find an expression for the emf induced in the stator phase by the PM rotor.

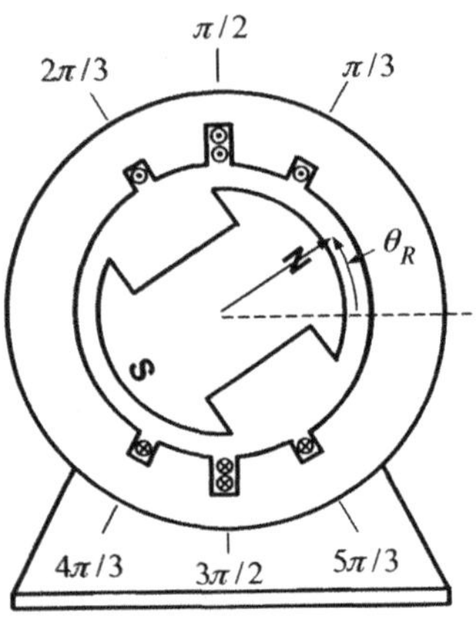

FIGURE 4.32. A motor with a PM rotor and a single rotor phase.

Problem 9 A Two-Phase Machine [4]

Consider a machine with two stator phases wound as shown in Figure 4.33. That is, phase a has one turn at $\pi/4$ (with its other side at $5\pi/4$), two turns at $5\pi/12$ (with the other sides at $17\pi/12$), and so on. Let the current in phase a be denoted as i_{Sa} and the current in phase b be denoted as i_{Sb}.

(a) Compute the radial magnetic field $\vec{\mathbf{B}}_{Sa}$ in the air gap due to the current i_{Sa} in phase a. Express it as a Fourier series.

(b) Compute the flux linkage in phase a due to $\vec{\mathbf{B}}_{Sa}$.

(c) Compute the flux linkage in phase b due to $\vec{\mathbf{B}}_{Sa}$.

(d) Compute the Fourier series expansion of the expression for the magnetic field in part (a).

(e) Compute the ratio of the fundamental of the magnetic field given in part (d) to the fundamental of the magnetic field whose Fourier series is given in equation (4.3) for the winding of Figure 4.7. Also, compute the ratio of the 3rd harmonic to the 1st harmonic for the magnetic field given in part (d) and then compare it to the corresponding ratio for the magnetic

field of equation (4.3). Is the winding of Figure 4.33 or the winding of Figure 4.7 better suited for a two phase machine? Explain.

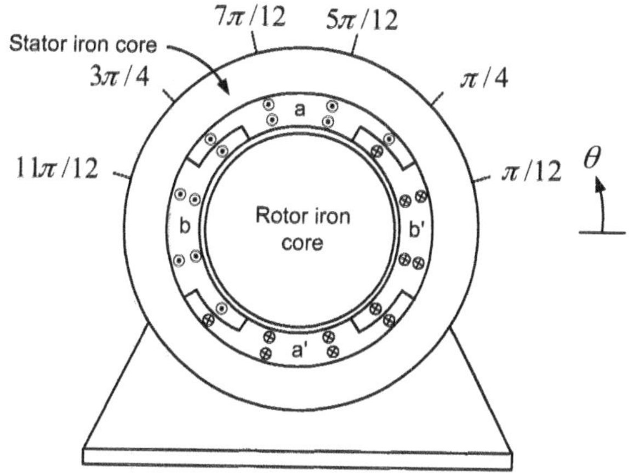

FIGURE 4.33. Stator phases for a single pole pair two-phase machine.

Problem 10 Uniformly Distributed Winding

Consider a machine with a three-phase stator and a rotor of soft iron as shown in Figure 4.34. There are three sets of windings which are uniformly wound on the inside surface of the stator. By uniformly *wound, it is meant that the number of turns between θ and $\theta+d\theta$ is constant independent of the angle θ. Phase 1 with current i_{S1} has its turns uniformly wound between $\pi/3$ and $2\pi/3$ on one side of the stator's inside surface and between $4\pi/3$ and $5\pi/3$ on the other side of the stator. Similarly, phase 2 with current i_{S2} has its turns uniformly wound between π and $4\pi/3$, and between 0 and $\pi/3$. Phase 3 with current i_{S3} has its turns uniformly wound between $2\pi/3$ and π, and between $5\pi/3$ and 2π.*

Let the turns density for stator phase 1 be given by

$$N_{S1}(\theta) = \begin{cases} \dfrac{N_S}{\pi/3} & \textit{for} \quad \dfrac{\pi}{3} \le \theta \le \dfrac{2\pi}{3} \\ \dfrac{N_S}{\pi/3} & \textit{for} \quad \dfrac{4\pi}{3} \le \theta \le \dfrac{5\pi}{3} \\ 0 & \textit{elsewhere.} \end{cases}$$

The total number of windings (turns or loops) making up phase 1 is then

$$\int_{\pi/3}^{2\pi/3} \frac{N_S}{\pi/3} d\theta = N_S.$$

Compute the radial magnetic field $\vec{\mathbf{B}}_{S1}$ in the air gap produced by the current i_{S1} in phase 1.

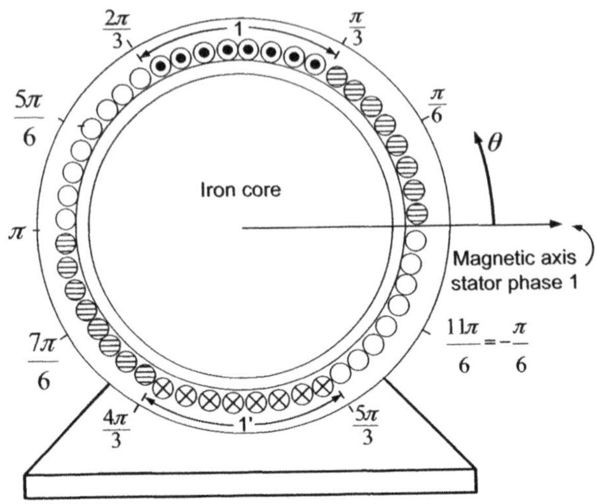

FIGURE 4.34. Uniformly distributed windings.

Problem 11 Modifying Ampère's Law to Account for a r_0/r Dependence

This problem considers how Ampère's law changes when H_{Sa} is assumed to have a factor of the form r_0/r so that conservation of flux holds. Specifically, in Ampère's law as given by equation (4.9), replace $H_{Sa}(i_{Sa}, \theta)$ by

$$H_{Sa}(i_{Sa}, r, \theta) \triangleq (r_0/r) H_{Sa}(i_{Sa}, \theta) \tag{4.29}$$

to obtain

$$\int_{r=r_R}^{r=r_S} H_{Sa}(i_{Sa}, r, 0)\hat{\mathbf{r}} \cdot (dr\hat{\mathbf{r}}) + \int_{r=r_R}^{r=r_S} H_{Sa}(i_{Sa}, r, \theta)\hat{\mathbf{r}} \cdot (-dr\hat{\mathbf{r}})$$
$$= -\frac{N_S i_{Sa}}{2}\cos(\theta) + \frac{N_S i_{Sa}}{2}. \tag{4.30}$$

(a) Show that for the expression

$$\vec{\mathbf{H}}_{Sa}(i_{Sa}, r, \theta) = \frac{N_S}{2g}\frac{r_0}{r} i_{Sa}\cos(\theta)\hat{\mathbf{r}}$$

to still work out the same way as before, that is, for (4.30) to reduce to

$$H_{Sa}(i_{Sa}, r_0, 0)g - H_{Sa}(i_{Sa}, r_0, \theta)g = -i_{Sa}\frac{N_S}{2}\cos(\theta) + i_{Sa}\frac{N_S}{2},$$

then r_0 must be chosen such that $\int_{r=r_R}^{r=r_S} r_0/r dr = g$.

(b) Show that there is an r_0 with $r_R < r_0 < r_S$ such that $\int_{r=r_R}^{r=r_S} r_0/r dr = g$ holds.

(c) Show that (4.30) holds approximately with H_{Sa} given by (4.29) if $r_0 = r_R$ (or $r_0 = r_S$) and $g << r_R$.

Problem 12 Three-Phase, Four-Pole, 36-Slot Stator

Consider a three-phase, four-pole, 36-slot stator as shown in Figure 4.35. The figure illustrates the slots in which the wire making up phase 1 is wound with the (angular) location of the windings as indicated in the figure. Denote the current in this phase as i_{S1}.

(a) Assuming that the cross-sectional area of the slots is infinitesimal, compute the radial magnetic field B_{S1} in the air gap produced by i_{S1} and plot it as a function of θ.

(b) Compute the Fourier series expansion of the radial air-gap magnetic field B_{S1} found in part (a).

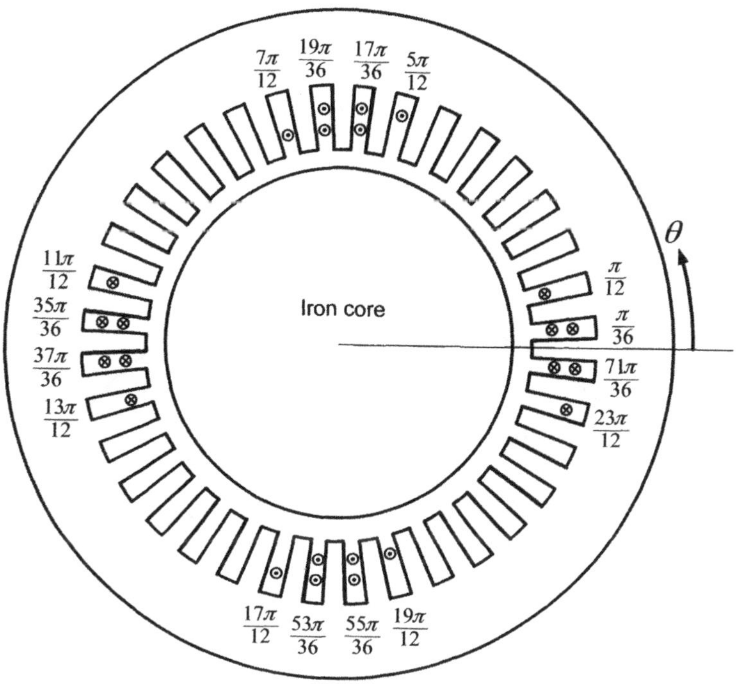

FIGURE 4.35. A three-phase, four-pole, 36-slot stator. Only the winding of stator phase 1 is shown.

Problem 13 Three-Phase Six-Pole Permanent Magnet Machine

Figure 4.36 shows a three-phase, six-pole machine with a permanent magnet rotor. Take $\theta_R = 0$ *for the rotor in the position as drawn. Let the magnetic field in the air gap due to the permanent magnet rotor be given by*

$$\vec{\mathbf{B}}_R(r,\theta) = B_{R\max}\frac{r_R}{r}\cos\Big(n_p(\theta - \theta_R)\Big)$$

where n_p *is the number of pole pairs that in this case is* $n_p = 3$. *Phase 1 of the stator consists of connecting the three single-loop windings* $a_1 - a_1', a_2 - a_2'$, *and* $a_3 - a_3'$ *in series as shown. Each winding encloses a cylindrical-shaped surface spanning* $\pi/3$ *radians (or* $60°$*).*

(a) Compute the flux linkage λ_{S1} *in the phase winding 1 due to the magnetic field of the rotor as a function of the rotor position* θ_R.

(b) Compute the induced voltage in the phase winding 1 due to the magnetic field of the rotor as a function of the rotor position θ_R *and the rotor speed* ω_R.

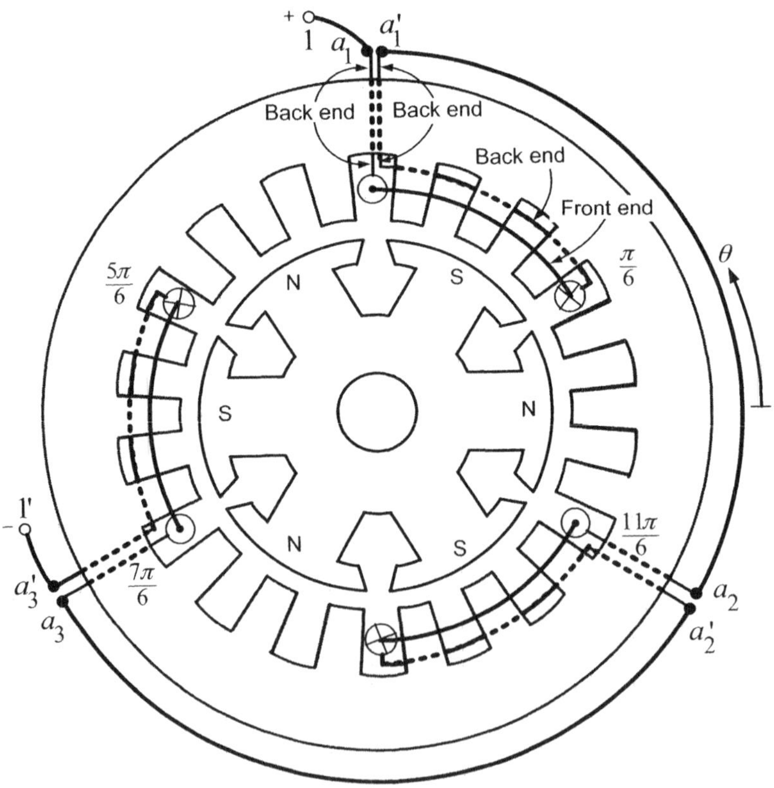

FIGURE 4.36. A three-phase, six-pole permanent magnet machine.

Problem 14 Flux Linkage in a Phase Winding with Short-Pitched Turns

Figure 4.37 shows a motor with a PM rotor and a single stator phase whose turns are short-pitched. By "short-pitched" is meant that each turn does not span 180° *degrees, but rather less than* 180° *such as in this problem where they span only* 150°*. The first loop has axial sides* a_1 *and* a'_1*, the second loop has axial sides* a_2 *and* a'_2*, and so on. Let the magnetic field in the air gap due to the permanent magnet rotor be given by*

$$\vec{\mathbf{B}}_R(\theta - \theta_R) = B_{max}\frac{r_R}{r}\cos(\theta - \theta_R)\hat{\mathbf{r}}.$$

In order to illustrate the windings of a short-pitched phase, Figure 4.38 shows the slots of a cylindrical rotor iron core being wound with four short-pitched turns. The short-pitched windings of the stator phase of Figure 4.37 are wound similarly.

(a) Compute the flux in each loop using a surface normal for whose direction is chosen so that the positive direction of travel around the surface coincides with the direction of positive current flow as given in Figure 4.37.

(b) Compute the flux linkage in the stator phase produced the rotor's magnetic field.

(c) Using Faraday's law, compute the induced emf in the stator phase due to the rotor's magnetic field.

(d) How do the answers to parts (a), (b), and (c) compare to the corresponding values for the full-pitch wound stator of problem 8?

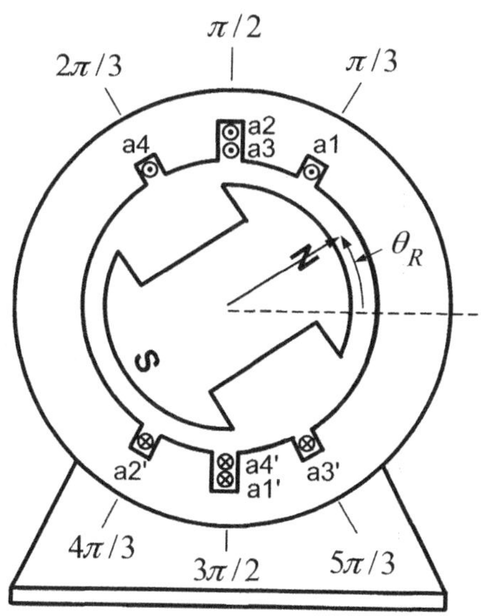

FIGURE 4.37. A short-pitched stator winding.

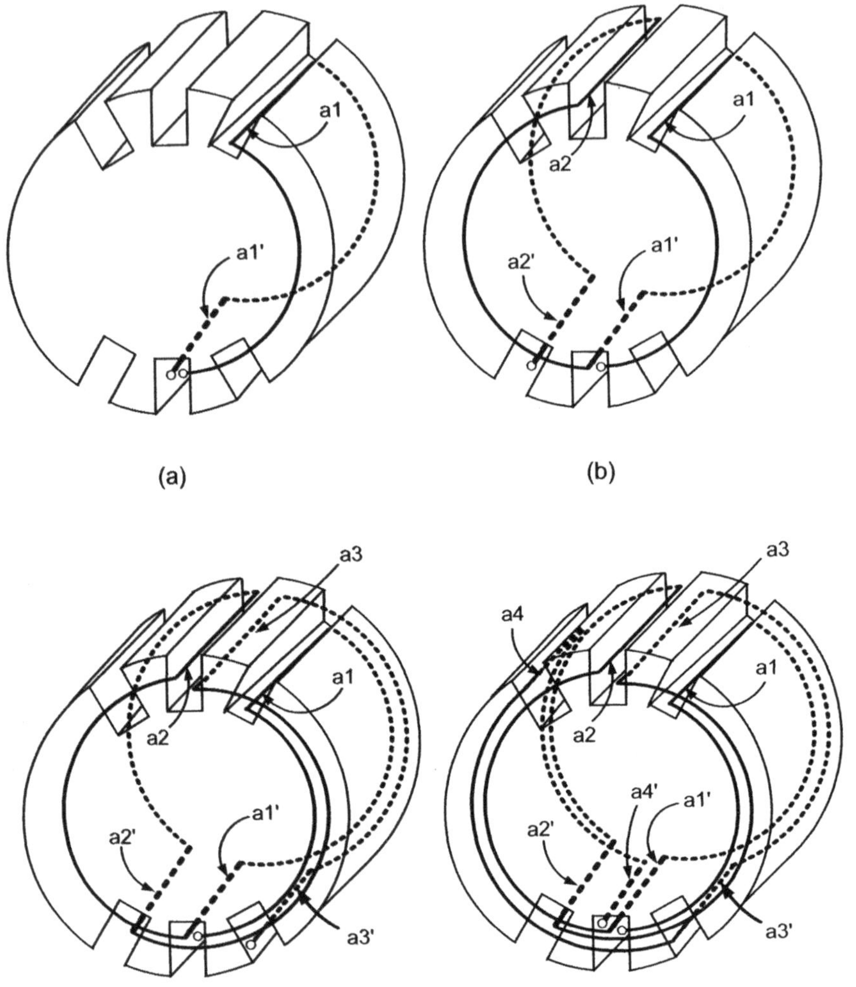

(a) (b) (c) (d)

FIGURE 4.38. Illustration of winding a distributed rotor phase with short-pitched windings. (a) First loop with axial sides a_1 and a_1'. (b) Second loop with axial sides a_2 and a_2'. (c) Third loop with axial sides a_3 and a_3'. (d) Fourth loop with axial sides a_4 and a_4'.

Problem 15 Self-Inductance of a Short-Pitched Phase Winding

Figure 4.39 shows a single stator phase with short-pitched windings (see problem 14 for an explanation of short-pitched windings).

(a) Explain why the radial magnetic field is identical to that drawn in Figure 4.10.

(b) Compute the flux in each loop due to the current in that loop and then the total flux linkage. To ensure that the induced emfs in each loop have the same sign conventions, choose a flux surface normal whose positive direction of travel around the surface coincides with the positive direction of current.

(c) What is the self-inductance of the stator phase? How does it compare with the self-inductance of the stator phase in problem 7?

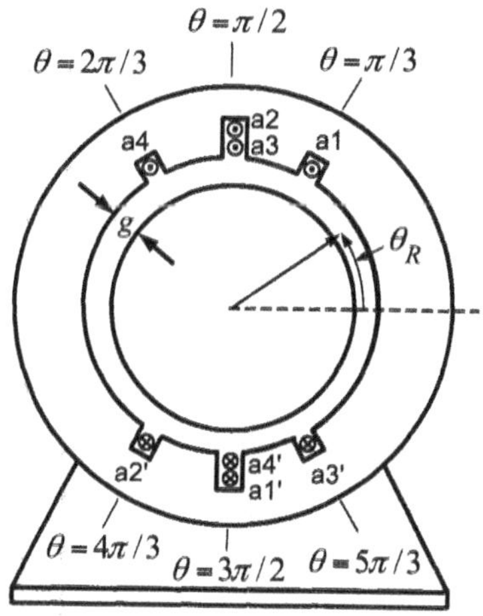

FIGURE 4.39. Self-inductance for a distributed phase winding with fractional pitch turns.

Problem 16 "Long-Pitched" Phase Winding

Consider Figure 4.39 where now each winding loop spans $7\pi/6$ radians (210°) rather than $5\pi/6$ radians (150°) as in Figure 4.38.

(a) Explain why the radial magnetic field is identical to that drawn in Figure 4.10.

(b) Compute the flux in each loop due to the current in that loop and then the total flux linkage. To ensure that the induced emfs in each loop have the same sign conventions, choose a flux surface normal whose positive direction of travel around the surface coincides with the positive direction of current. Without any computation, use Gauss's law for $\vec{\mathbf{B}}$ to show the result must be the same as part (b) of problem 15. Consequently, the inductance of this winding must be the same as that given in part (c) of problem 15.

Problem 17 Divergence of the Magnetic Field

Recall that the divergence of a vector field $\vec{\mathbf{W}}(r,\theta,z)$ *in cylindrical coordinates is given by*

$$\nabla\cdot\vec{\mathbf{W}} = \frac{1}{r}\frac{\partial}{\partial r}(rW_r) + \frac{1}{r}\frac{\partial W_\theta}{\partial\theta} + \frac{\partial W_z}{\partial z}$$

where $\vec{\mathbf{W}} = W_r\hat{\mathbf{r}} + W_\theta\hat{\boldsymbol{\theta}} + W_z\hat{\mathbf{z}}$. *Also, recall that the divergence theorem says that*

$$\oint_S \vec{\mathbf{W}}\cdot d\vec{\mathbf{S}} = \iiint_V \nabla\cdot\vec{\mathbf{W}}\; dV$$

where S *is the closed surface that encloses the volume* V.

With

$$\vec{\mathbf{B}}_{Sa} = B_{Sa}(r,\theta)\hat{\mathbf{r}} = \mu_0\frac{r_R}{r}\left(\frac{2i_{Sa}}{g} - \frac{i_{enclosed}(\theta)}{g}\right)\hat{\mathbf{r}}$$

in the air gap of an approximately sinusoidally wound stator phase, show that $\nabla\cdot\vec{\mathbf{B}}_{Sa}\equiv 0$.

Problem 18 Azimuthal Magnetic Field

Use the expression (4.24) for the radial magnetic field to explicitly show that

$$\oint_S \vec{\mathbf{B}}_{Sa}\cdot d\vec{\mathbf{S}} = 0$$

for the closed surface shown in Figure 4.40 with $r_R \le r_1 < r_2 \le r_S$.

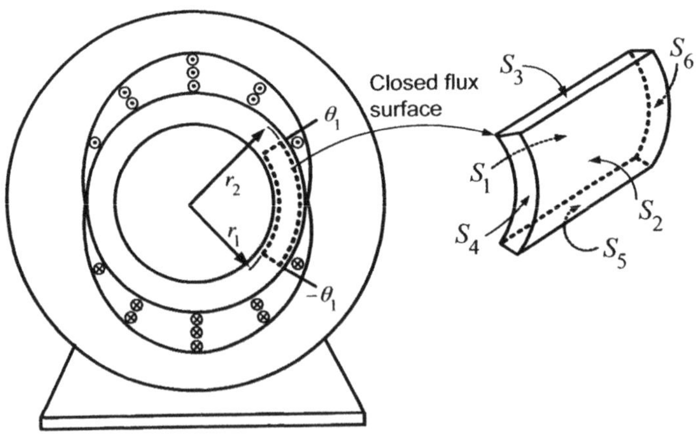

FIGURE 4.40. Gauss's law in the air gap with a nonzero azimuthal magnetic field.

5

The Physics of AC Machines

The goal here is to present the operation of both the induction and synchronous machines using basic concepts in physics. In particular, Faraday's law of induction and the Lorentz force/magnetic force law will be used to analyze the physical basis of operation of these machines [34].

5.1 Rotating Magnetic Field

Assuming ideal magnetic material, that is, $\vec{\mathbf{H}} \equiv 0$ in the iron, and that the stator phases are sinusoidally wound, Ampère's law ($\oint \vec{\mathbf{H}} \cdot d\vec{\ell} = i_{enclosed}$) was used to show that the magnetic field $\vec{\mathbf{B}}_{Sa}$ at any point (r, θ) in the air gap due to the current i_{Sa} in phase a is given by

$$\vec{\mathbf{B}}_{Sa}(i_{Sa}, r, \theta) = \frac{\mu_0 \ell_2 N_S}{4gr} i_{Sa} \cos(\theta)\hat{\mathbf{r}}.$$

Here $\hat{\mathbf{r}}$ is a unit vector in the radial direction, $r_R = \ell_2/2$ is the radius of the rotor and g is the radial length of the air gap. The magnetic field distribution of $\vec{\mathbf{B}}_{Sa}$ is illustrated in Figure 5.1.

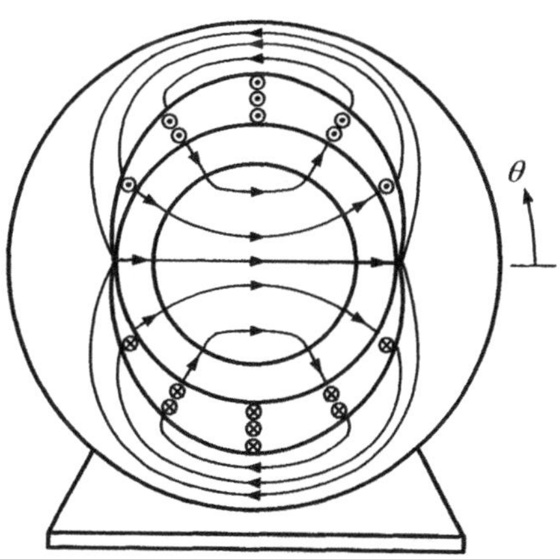

FIGURE 5.1. Radial air-gap magnetic field due to phase current i_{Sa} in a sinusoidally wound stator.

Similarly, the magnetic field $\vec{\mathbf{B}}_{Sb}$ at the point (r, θ) in the air gap due to the current i_{Sb} in phase b is given by

$$\begin{aligned}\vec{\mathbf{B}}_{Sb}(i_{Sb}, r, \theta) &= \frac{\mu_0 \ell_2 N_S}{4gr} i_{Sb} \cos(\theta - \pi/2)\hat{\mathbf{r}} \\ &= \frac{\mu_0 \ell_2 N_S}{4gr} i_{Sb} \sin(\theta)\hat{\mathbf{r}}.\end{aligned}$$

The magnetic field distribution of $\vec{\mathbf{B}}_{Sb}$ is illustrated in Figure 5.2.

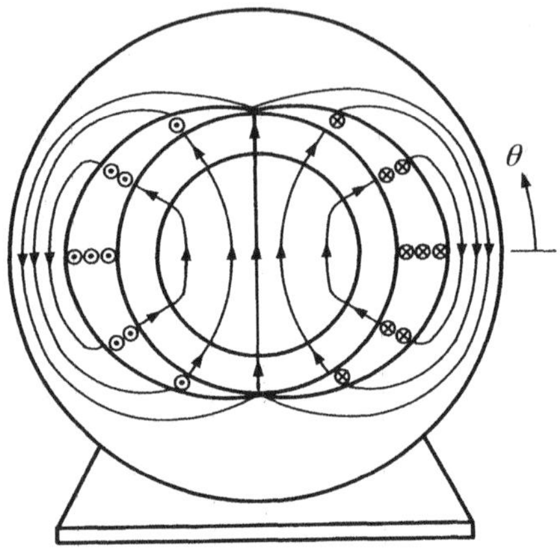

FIGURE 5.2. Radial air-gap magnetic field due to phase current i_{Sb} in a sinusoidally wound stator.

The total magnetic field in the air gap due to the current in both stator phases is then

$$\begin{aligned}\vec{\mathbf{B}}_S(i_{Sa}, i_{Sb}, r, \theta) &= \vec{\mathbf{B}}_{Sa}(i_{Sa}, r, \theta) + \vec{\mathbf{B}}_{Sb}(i_{Sb}, r, \theta) \\ &= \frac{\mu_0 \ell_2 N_S}{4gr}\left(i_{Sa}\cos(\theta) + i_{Sb}\sin(\theta)\right)\hat{\mathbf{r}}. \qquad (5.1)\end{aligned}$$

With $i_{Sa}(t) = I_S\cos(\omega_S t)$, $i_{Sb}(t) = I_S\sin(\omega_S t)$, (5.1) simplifies to

$$\begin{aligned}\vec{\mathbf{B}}_S &= \frac{\mu_0 \ell_2 N_S I_S}{4gr}\left(\cos(\omega_S t)\cos(\theta) + \sin(\omega_S t)\sin(\theta)\right)\hat{\mathbf{r}} \\ &= \frac{\mu_0 \ell_2 N_S I_S}{4gr}\cos(\theta - \omega_S t)\hat{\mathbf{r}} \\ &= \frac{\mu_0 \ell_2 N_S I_S}{4gr}\cos(\theta - \theta_S(t))\hat{\mathbf{r}} \qquad (5.2)\end{aligned}$$

where $\theta_S(t) \triangleq \omega_S t$. This is a *rotating magnetic field* rotating at the electrical frequency ω_S and is illustrated in Figure 5.3. The center line of this rotating magnetic field is referred to as its *magnetic axis* and is located at $\theta_S(t) \triangleq \omega_S t$.

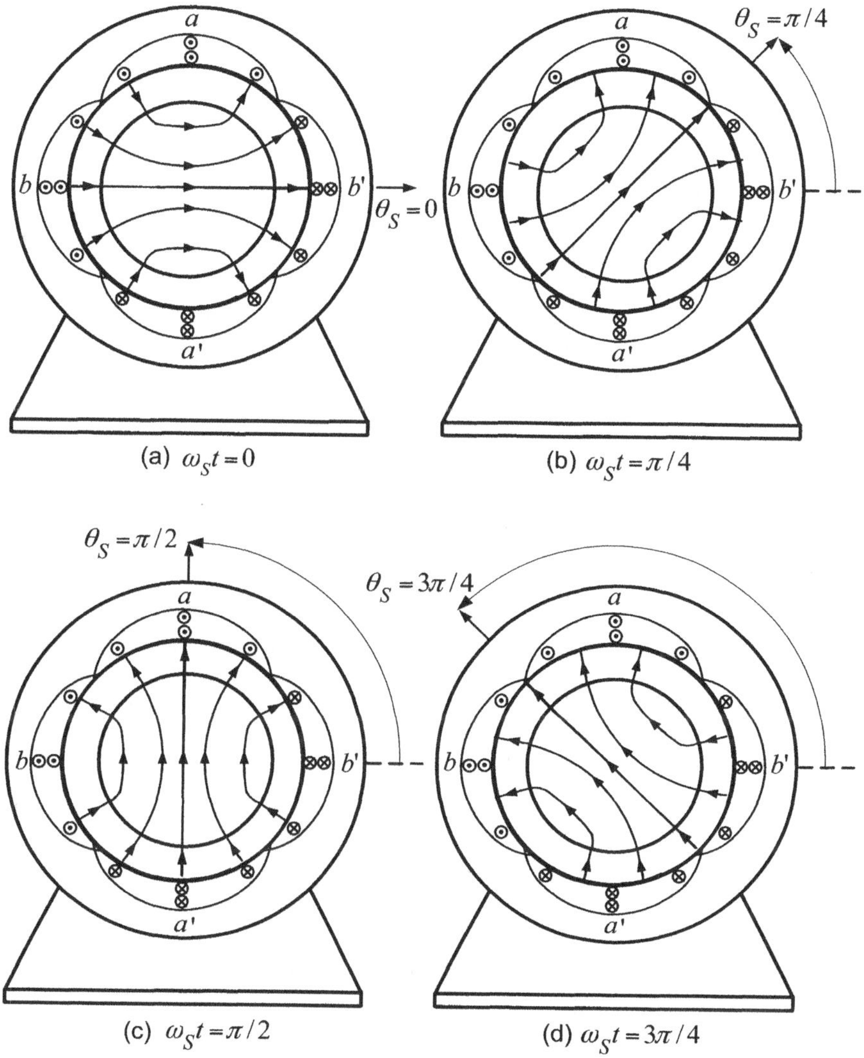

FIGURE 5.3. Sinusoidally distributed rotating radial magnetic field in the air gap. (For convenience, the magnetic field lines are not drawn in the stator iron.)

5.2 The Physics of the Induction Machine

In this section, a simple two-phase induction motor is analyzed in order to understand the elementary Physics which characterizes its operation.

The stator consists of two sinusoidally wound phases 90° apart as described in the previous section. Figure 5.4 shows a loop wound around the surface of the rotor iron denoted as phase a (or simply loop a) and its two axial sides as a and a', respectively.

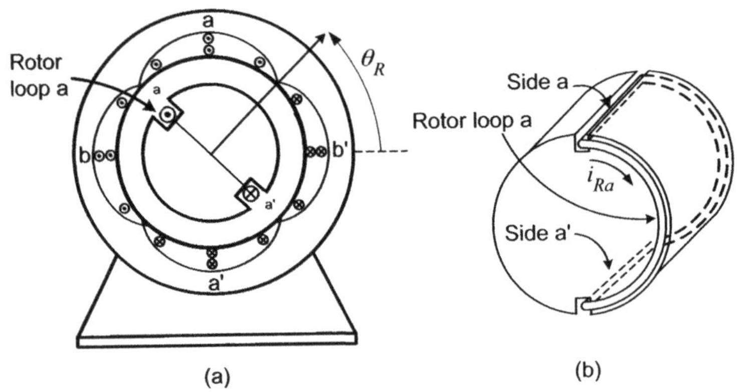

FIGURE 5.4. A single rotor loop and the chosen positive direction of current.

Similarly, a second loop is wound around the rotor iron 90° from (and electrically isolated from) loop a and is denoted as phase b or simply, loop b. This results in the simple two phase machine illustrated in Figure 5.5.

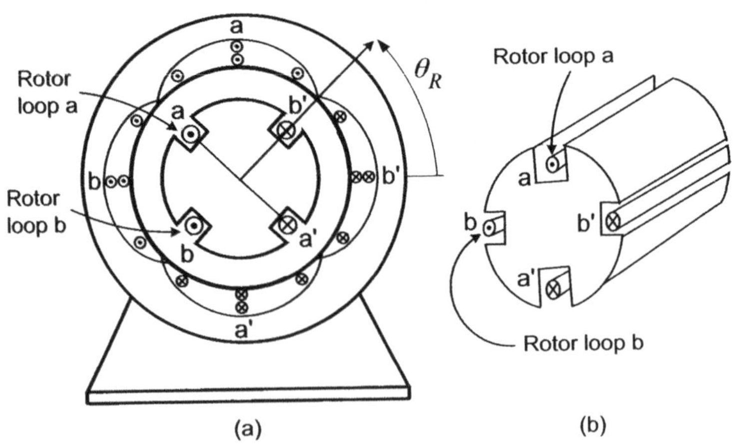

FIGURE 5.5. Basic motor with two single-loop rotor phases a and b.

As above, $\theta_S(t) = \omega_S t$ denotes the angular position of the magnetic axis of the rotating magnetic field produced by the stator currents, $\theta_R(t)$ denotes the angular position of the rotor defined as the normal to rotor loop a [see Figure 5.4(a) or 5.5(a)] and (r, θ) denotes the polar coordinates of an arbitrary point in the air gap. In the remainder of this chapter, it will assumed that the rotor is turning at constant angular speed ω_R with $\theta_R(t) = \omega_R t$.

5.2.1 Induced Emfs in the Rotor Loops

The rotating magnetic field $\vec{\mathbf{B}}_S$ produces a changing flux in the rotor loops so that an *electromotive force (emf)* is induced in each rotor loop according to Faraday's law. To compute the induced emf in rotor phase a, consider a flux surface for rotor loop a consisting of a half-cylinder lying in the air gap whose boundary is the rotor loop itself. This is illustrated in Figure 5.6 where only rotor loop a is shown for clarity.

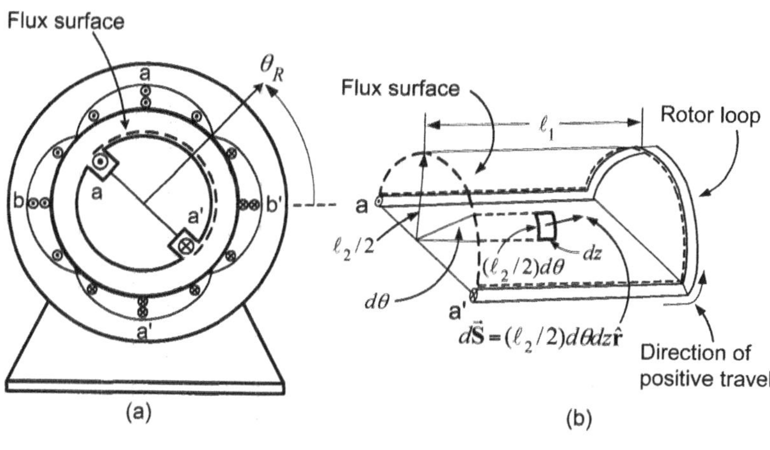

FIGURE 5.6. (a) Flux surface for the rotor winding. (b) Surface element $d\vec{\mathbf{S}}$ and direction of positive travel around the flux surface. (For clarity, only rotor loop a is shown.)

With $d\vec{\mathbf{S}} = (\ell_2/2)d\theta dz\hat{\mathbf{r}}$, the direction of positive travel is indicated in Figure 5.6(b), and it is the same as the direction chosen for positive current flow in the loop. The *flux* λ_{Ra} in rotor loop a due to the stator magnetic field $\vec{\mathbf{B}}_S$ is calculated as follows:

$$\begin{aligned}
\lambda_{Ra} &= \int_S \vec{\mathbf{B}}_S \cdot d\vec{\mathbf{S}} \\
&= \int_{z=0}^{z=\ell_1} \int_{\theta_R(t)-\pi/2}^{\theta_R(t)+\pi/2} \frac{\mu_0 \ell_2 N_S I_S}{4gr} \Big|_{r=\ell_2/2} \cos(\theta - \theta_S(t))\hat{\mathbf{r}} \cdot ((\ell_2/2) d\theta dz \hat{\mathbf{r}}) \\
&= \frac{\mu_0 \ell_1 \ell_2 N_S I_S}{4g} \sin(\theta - \theta_S(t)) \Big|_{\theta_R(t)-\pi/2}^{\theta_R(t)+\pi/2} \\
&= \frac{\mu_0 \ell_1 \ell_2 N_S I_S}{2g} \cos\Big(\theta_S(t) - \theta_R(t)\Big).
\end{aligned}$$

By Faraday's law, the induced *electromotive force* or *emf* in rotor loop a is given by

$$\begin{aligned}
\xi_{Ra} &= -\frac{d\lambda_{Ra}}{dt} \\
&= -\frac{d}{dt}\frac{\mu_0 \ell_1 \ell_2 N_S I_S}{2g} \cos\Big(\theta_S(t) - \theta_R(t)\Big) \\
&= \frac{\mu_0 \ell_1 \ell_2 N_S I_S}{2g} (\omega_S - \omega_R) \sin\Big((\omega_S - \omega_R)t\Big)
\end{aligned} \tag{5.3}$$

where $\omega_R = d\theta_R/dt$. The current i_{Ra} in phase a is then found by solving

$$L_R \frac{di_{Ra}}{dt} = -R_R i_{Ra} + \xi_{Ra}$$

where L_R and R_R are, respectively, the inductance and resistance of the rotor phases. Neglecting the inductance[1] (i.e., set $L_R = 0$), the current in phase a produced by the emf is then

$$i_{Ra}(t) = \xi_{Ra}/R_R. \tag{5.4}$$

[1] See problem 2 for the analysis when $L_R > 0$.

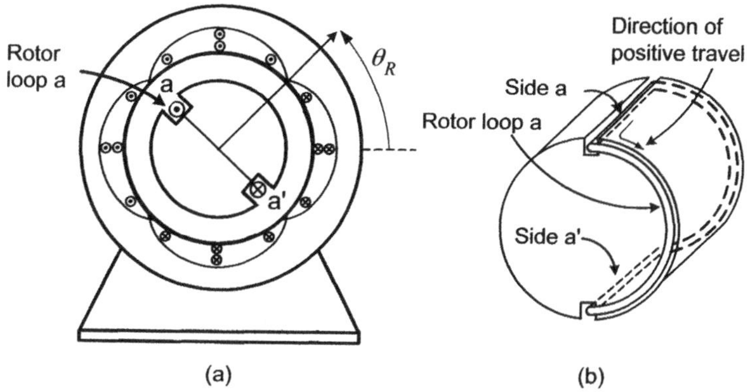

FIGURE 5.7. Induced rotor current i_{Ra} produced by ξ_{Ra}. (Rotor loop b is not shown for clarity.)

Similarly, the flux λ_{Rb}, the induced emf ξ_{Rb}, and the current i_{Rb} in rotor loop b are computed. The flux λ_{Rb} in rotor loop b due to $\vec{\mathbf{B}}_S$ is given by

$$\begin{aligned}
\lambda_{Rb} &= \int_S \vec{\mathbf{B}}_S \cdot d\vec{\mathbf{S}} \\
&= \int_{z=0}^{z=\ell_1} \int_{\theta_R(t)}^{\theta_R(t)+\pi} \frac{\mu_0 \ell_2 N_S I_S}{4gr}\Big|_{r=\ell_2/2} \cos(\theta - \theta_S(t))\, \hat{\mathbf{r}} \cdot ((\ell_2/2) dz d\theta \hat{\mathbf{r}}) \\
&= \frac{\mu_0 \ell_1 \ell_2 N_S I_S}{4g} \sin(\theta - \theta_S(t)) \Big|_{\theta_R(t)}^{\theta_R(t)+\pi} \\
&= \frac{\mu_0 \ell_1 \ell_2 N_S I_S}{2g} \sin(\theta_S(t) - \theta_R(t)).
\end{aligned}$$

The induced *emf* in rotor loop b is then

$$\begin{aligned}
\xi_{Rb}(t) &= -\frac{d\lambda_{Rb}}{dt} \\
&= -\frac{\mu_0 \ell_1 \ell_2 N_S I_S}{2g} (\omega_S - \omega_R) \cos\Big((\omega_S - \omega_R)t\Big)
\end{aligned}$$

where $\omega_R = d\theta_R/dt$. Again, neglecting the inductance of rotor phase b, this emf produces a current i_{Rb} in rotor loop b given by

$$i_{Rb}(t) = \xi_{Rb}(t)/R_R. \tag{5.5}$$

5.2.2 *Magnetic Forces and Torques on the Rotor*

The rotating magnetic field $\vec{\mathbf{B}}_S$ produced by the stator currents produces a magnetic force on the rotor currents it induced. The magnetic force on the rotor and the resulting torque are now computed. Recall that the magnetic force law is simply $\vec{\mathbf{F}}_{magnetic} = i\vec{\boldsymbol{\ell}}\times\vec{\mathbf{B}}$, where $\ell = |\,\vec{\boldsymbol{\ell}}\,|$ is the length of the conductor in the magnetic field, and $\vec{\boldsymbol{\ell}}$ points in the direction of positive current flow. Thus, the magnitude of the force is $F_{magnetic} = i\ell B_{\perp}$ where $B_{\perp}$ is the magnetic field strength perpendicular to the conductor. The magnetic force law is now applied to *side a* of rotor loop a, which is located at $\theta = \theta_R + \pi/2$ [see Figure 5.8(a)]. To do so, recall the symbol ⊙ means that if $i_{Ra} > 0$, the current is coming out of the page so $\vec{\boldsymbol{\ell}} = \ell_1\hat{\mathbf{z}}$ on side a, and compute

$$\begin{aligned}
\vec{\mathbf{F}}_{side-a} &= i_{Ra}\vec{\boldsymbol{\ell}}\times\vec{\mathbf{B}}_S \\
&= i_{Ra}\ell_1\hat{\mathbf{z}}\times\frac{\mu_0\ell_2N_SI_S}{4gr}\Big|_{r=\ell_2/2}\cos(\theta-\theta_S)_{|\theta=\theta_R+\pi/2}\,\hat{\mathbf{r}} \\
&= i_{Ra}\frac{\mu_0\ell_1N_SI_S}{2g}\sin(\theta_S-\theta_R)\hat{\boldsymbol{\theta}}.
\end{aligned}$$

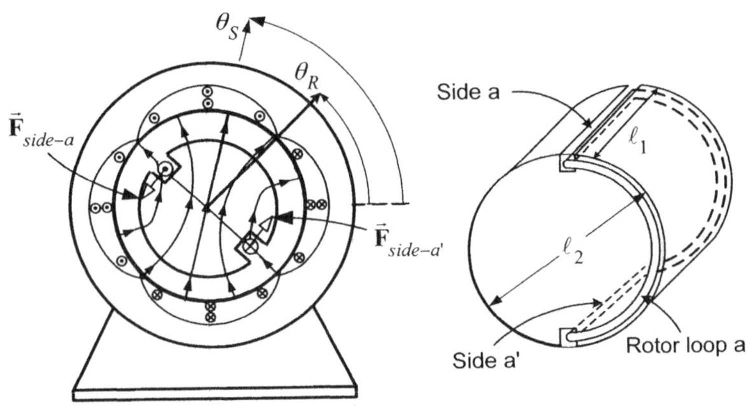

FIGURE 5.8. The torque τ_{Ra} on rotor loop a due to i_{Ra} and $\vec{\mathbf{B}}_S$.

The torque produced by this force is then

$$\begin{aligned}
\vec{\boldsymbol{\tau}}_{side-a} &= \vec{\mathbf{r}}\times\vec{\mathbf{F}}_{side-a} \\
&= (\ell_2/2)\hat{\mathbf{r}}\times i_{Ra}\frac{\mu_0\ell_1N_SI_S}{2g}\sin(\theta_S-\theta_R)\hat{\boldsymbol{\theta}} \\
&= \frac{\mu_0\ell_1\ell_2N_SI_S}{4g}i_{Ra}\sin(\theta_S-\theta_R)\hat{\mathbf{z}}.
\end{aligned}$$

Similarly, the magnetic force law is applied to *side a′* of rotor loop a, which is located at $\theta_R - \pi/2$. Recalling the symbol ⊗ means that if $i_{Ra} > 0$, the current is going into the page so that $\vec{\boldsymbol{\ell}} = -\ell_1\hat{\mathbf{z}}$ for side a'; it follows that

$$\begin{aligned}
\vec{\mathbf{F}}_{side-a'} &= i_{Ra}\vec{\boldsymbol{\ell}}\times\vec{\mathbf{B}}_S \\
&= i_{Ra}(-\ell_1\hat{\mathbf{z}})\times\frac{\mu_0\ell_2 N_S I_S}{4gr}\Big|_{r=\ell_2/2}\cos(\theta-\theta_S)_{|\theta=\theta_R-\pi/2}\hat{\mathbf{r}} \\
&= i_{Ra}\frac{\mu_0\ell_1 N_S I_S}{2g}\sin(\theta_S-\theta_R)\hat{\boldsymbol{\theta}}.
\end{aligned}$$

The *torque* produced by this force on side a' is then

$$\begin{aligned}
\vec{\boldsymbol{\tau}}_{side-a'} &= (\ell_2/2)\hat{\mathbf{r}}\times\vec{\mathbf{F}}_{side-a'} \\
&= \frac{\mu_0\ell_1\ell_2 N_S I_S}{4g}i_{Ra}\sin\Big((\omega_S-\omega_R)t\Big)\hat{\mathbf{z}} \\
&= \vec{\boldsymbol{\tau}}_{side-a}.
\end{aligned}$$

The total torque $\vec{\boldsymbol{\tau}}_{Ra} = \tau_{Ra}\hat{\mathbf{z}}$ on rotor phase a is simply

$$\tau_{Ra} = \tau_{side-a} + \tau_{side-a'} = \frac{\mu_0\ell_1\ell_2 N_S I_S}{2g}i_{Ra}\sin\Big((\omega_S-\omega_R)t\Big). \tag{5.6}$$

Recall that

$$i_{Ra}(t) = \xi_{Ra}(t)/R_R = \frac{\mu_0\ell_1\ell_2 N_S I_S}{2g}\frac{1}{R_R}(\omega_S-\omega_R)\sin\Big((\omega_S-\omega_R)t\Big).$$

Using this expression, the total torque on rotor phase a may be rewritten as

$$\tau_{Ra} = \left(\frac{\mu_0\ell_1\ell_2 N_S I_S}{2g}\right)^2\frac{1}{R_R}(\omega_S-\omega_R)\sin^2\Big((\omega_S-\omega_R)t\Big). \tag{5.7}$$

Similarly, for rotor phase b

$$\begin{aligned}
\vec{\mathbf{F}}_{side-b} &= i_{Rb}\vec{\boldsymbol{\ell}}\times\vec{\mathbf{B}}_S \\
&= i_{Rb}\ell_1\hat{\mathbf{z}}\times\frac{\mu_0\ell_2 N_S I_S}{4gr}\Big|_{r=\ell_2/2}\cos(\theta-\theta_S)_{|\theta=\theta_R+\pi}\hat{\mathbf{r}} \\
&= -i_{Rb}\frac{\mu_0\ell_1 N_S I_S}{2g}\cos(\theta_S-\theta_R)\hat{\boldsymbol{\theta}}
\end{aligned}$$

and, as in the case of loop a, $\vec{\mathbf{F}}_{side-b'} = \vec{\mathbf{F}}_{side-b}$. It then follows that the total torque on rotor phase b is

$$\begin{aligned}
\tau_{Rb} &= \tau_{side-b} + \tau_{side-b'} \\
&= 2\tau_{side-b} \\
&= -\frac{\mu_0\ell_1\ell_2 N_S I_S}{2g}i_{Rb}\cos\Big((\omega_S-\omega_R)t\Big).
\end{aligned}$$

Further, as

$$i_{Rb}(t) = \xi_{Rb}(t)/R_R = -\frac{\mu_0 \ell_1 \ell_2 N_S I_S}{2g}(\omega_S - \omega_R)\cos\Big((\omega_S - \omega_R)t\Big)/R_R$$

the torque on rotor loop b may be rewritten as

$$\tau_{Rb} = \left(\frac{\mu_0 \ell_1 \ell_2 N_S I_S}{2g}\right)^2 \frac{1}{R_R}(\omega_S - \omega_R)\cos^2\Big((\omega_S - \omega_R)t\Big). \tag{5.8}$$

Combining equations (5.7) and (5.8), the *total torque* τ_R is given by

$$\tau_R = \tau_{Ra} + \tau_{Rb} = \left(\frac{\mu_0 \ell_1 \ell_2 N_S I_S}{2g}\right)^2 \frac{1}{R_R}(\omega_S - \omega_R). \tag{5.9}$$

For a constant angular frequency ω_S and a constant angular rotor speed ω_R, the torque is constant.

5.2.3 *Slip Speed*

The stator currents are given by $i_{Sa}(t) = I_S \cos(\omega_S t)$, $i_{Sb}(t) = I_S \sin(\omega_S t)$ and, neglecting the inductance of the rotor phases, equation (5.9) shows that the torque on the rotor is proportional to the *difference* between the angular speed ω_S of the rotating stator magnetic field and the rotor speed ω_R, that is,

$$\tau_R = \left(\frac{\mu_0 \ell_1 \ell_2 N_S I_S}{2g}\right)^2 \frac{1}{R_R}(\omega_S - \omega_R).$$

This difference is called the *slip speed* ω_{slip}, that is,

$$\omega_{slip} \triangleq \omega_S - \omega_R.$$

In particular, if the slip speed is zero, there is *no* torque. The reason is simply that if the slip speed is zero, then the rotating stator magnetic field and the rotor are rotating at the *same* angular rate with the result that the fluxes in the rotor loops are *constant*. As the rate of change of these fluxes is zero, Faraday's law tells us that neither voltages nor currents are induced in the rotor and, without rotor currents, there can be no magnetic force/torque on the rotor. In order to produce torque in an induction machine, the rotor shaft and the stator's magnetic field do not rotate together at the same rate, that is, they are not synchronized. It is for this reason that the induction motor is referred to as an *asynchronous* machine.

5.3 The Physics of the Synchronous Machine

In this section, the physics of the synchronous motor is described where both the salient and nonsalient machines are considered. A *salient* rotor is one in which the rotor is not cylindrically shaped so that air gap does not have a uniform thickness. A *nonsalient* machine is one which is *not* salient meaning it does have a cylindrical rotor with a uniform or constant air-gap thickness.

5.3.1 Two-Phase Synchronous Motor with a Sinusoidally Wound Rotor

A synchronous machine with a sinusoidally wound rotor (also referred to as a nonsalient machine) is considered first. The rotor consists of a single sinusoidally wound phase of wire while the stator is constructed identically to that of the induction motor as illustrated in Figure 5.9. The rotor construction, torque production and energy conversion for this synchronous machine is now described.

Sinusoidally Wound Rotor

Consider a synchronous machine where the stator consists of two sinusoidally wound phases which are 90° apart as shown in Figure 5.9(a). Let the rotor consist of a cylindrical core of iron on which a single phase with a sinusoidal turns density has been wound as illustrated in Figure 5.9(b). That is, the turns density of the single-phase rotor winding is given by

$$N_{RF}(\theta - \theta_R) = \frac{N_F}{2} \left|\sin(\theta - \theta_R)\right|$$

where N_F is the total number of windings in the phase.

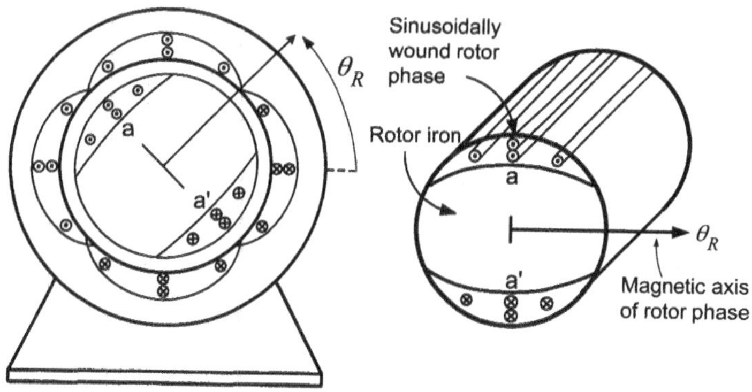

FIGURE 5.9. Single-phase sinusoidally wound rotor.

The single-phase winding on the rotor is usually referred to as the *field* winding and hence the subscript "F". The radial magnetic field in the air-gap due to the rotor current i_F in the sinusoidally wound rotor is (see problem 7)

$$\vec{\mathbf{B}}_R = B_R\hat{\mathbf{r}} = \frac{\mu_0 \ell_2 N_F i_F}{4gr}\cos(\theta - \theta_R)\hat{\mathbf{r}} \tag{5.10}$$

For this machine to operate as a synchronous machine, the rotor must be carrying a *constant current*, that is, $i_F = I_F$ with I_F constant. Again, the subscript "F" is used to denote the rotor current as the *field* current where the term "field" refers to the current being constant. The rotor position is taken to be aligned with the magnetic axis of the rotor as shown in Figure 5.9. Furthermore, as the rotor current is constant, the rotor can be viewed simply as a magnet that rotates at angular speed $\omega_R = d\theta_R/dt$. Figure 5.10 illustrates the radial magnetic field in air gap produced by a constant current I_F in a sinusoidally wound rotor phase.

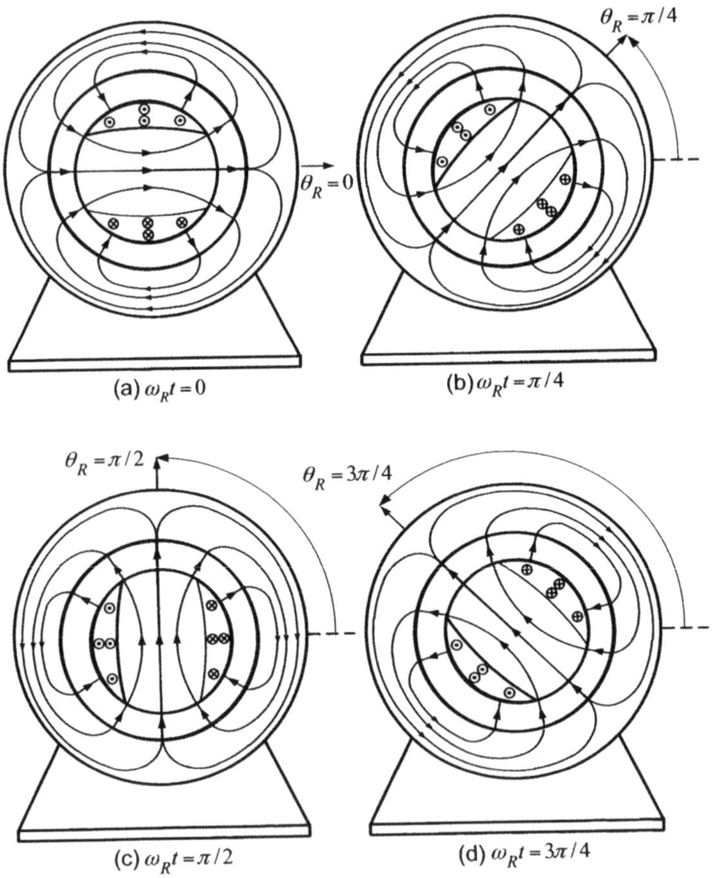

FIGURE 5.10. Magnetic field lines produced by a sinusoidally distributed rotor winding carrying a constant current $i_F = I_F$. (The stator windings are not drawn for clarity.)

To describe the Physics of this machine, consider the motor to be in steady-state operation. By this it is meant that the stator magnetic field $\vec{\mathbf{B}}_S$ and the rotor are rotating at the *same* angular speed (i.e., $\omega_R = \omega_S$), which is the origin of the term *synchronous*. As shown below, with the magnetic axis of $\vec{\mathbf{B}}_S$ given by $\theta_S(t) = \omega_S t$, the rotor position in steady state is written as $\theta_R(t) = \omega_S t - \delta$, where $\delta > 0$ is a constant angle whose value will be shown to depend on how much load (i.e., external torque, friction, etc.) there is on the motor.

Magnetic Forces/Torques

The magnetic field in the air-gap due to the stator currents is given by

$$\vec{\mathbf{B}}_S(i_{Sa}, i_{Sb}, r, \theta) = \frac{\mu_0 \ell_2 N_S}{4gr}\left(i_{Sa}\cos(\theta) + i_{Sb}\sin(\theta)\right)\hat{\mathbf{r}}. \tag{5.11}$$

With the rotor field current $i_F(t) = I_F$ constant, the magnetic field in the air gap due to this constant rotor current is

$$\vec{\mathbf{B}}_R = \frac{\mu_0 \ell_2 N_F I_F}{4gr}\cos(\theta - \theta_R)\hat{\mathbf{r}} \tag{5.12}$$

where θ_R is the rotor position which is aligned with the magnetic axis of the rotor.

The same procedure used to compute the torque produced by the induction motor could also be used. That is, one uses the expression (5.11) for $\vec{\mathbf{B}}_S$ to compute the torque this magnetic field produces on the current in the sinusoidally wound rotor phase. This is explored in problem 9. However, a different approach is taken here that will also work in the case of both a permanent magnet rotor and a salient rotor (see problem 8).

By Newton's third law, the magnetic forces on the rotor currents produced by the rotating stator magnetic field is the *negative* of the forces on the stator currents produced by rotor's magnetic field. The approach here is to compute the torque $\vec{\boldsymbol{\tau}}_S$ exerted on the currents in the sinusoidally wound stator phases by the rotor's magnetic field. Then the rotor torque $\vec{\boldsymbol{\tau}}_R$ is found by $\vec{\boldsymbol{\tau}}_R = -\vec{\boldsymbol{\tau}}_S$.

To proceed, consider Figure 5.11 where at any position θ on the stator, the number of axial sides of stator phase a between θ and $\theta + d\theta$ is $(N_S/2)\,|\sin(\theta)|\,d\theta$. Thus, the total amount of current in the axial sides of stator phase a between θ and $\theta + d\theta$ is $i_{Sa}(N_S/2)\,|\sin(\theta)|\,d\theta$. Using the magnetic force law, the incremental force $d\vec{\mathbf{F}}_{Sa}$ on the currents in the windings of stator phase a between θ and $\theta + d\theta$ is given by

$$d\vec{\mathbf{F}}_{Sa} = \begin{cases} i_{Sa}(t)\dfrac{N_S}{2}\,|\sin(\theta)|\,d\theta(+\ell_1\hat{\mathbf{z}}) \times \vec{\mathbf{B}}_R & 0 \le \theta \le \pi \\[2ex] i_{Sa}(t)\dfrac{N_S}{2}\,|\sin(\theta)|\,d\theta(-\ell_1\hat{\mathbf{z}}) \times \vec{\mathbf{B}}_R & \pi \le \theta \le 2\pi. \end{cases}$$

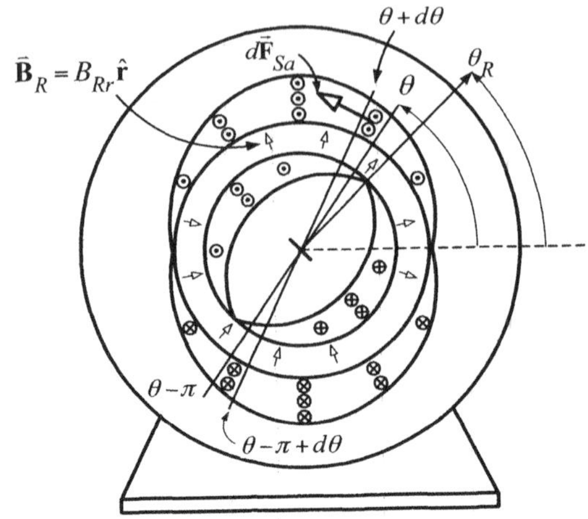

FIGURE 5.11. Computation of force/torque on stator phase a due to the rotor's magnetic field. Stator phase b is not shown for clarity.

As $\sin(\theta)$ is positive in the interval $0 \le \theta \le \pi$ and $|\sin(\theta)| = -\sin(\theta)$ in the interval $\pi \le \theta \le 2\pi$, the incremental force for $0 \le \theta \le 2\pi$ may be rewritten compactly as

$$\begin{aligned}
d\vec{\mathbf{F}}_{Sa} &= \left(i_{Sa}(t)\frac{N_S}{2}\sin(\theta)d\theta\right)(\ell_1\hat{\mathbf{z}}) \times \vec{\mathbf{B}}_R \\
&= \left(i_{Sa}(t)\frac{N_S}{2}\sin(\theta)d\theta\right)(\ell_1\hat{\mathbf{z}}) \times \frac{\mu_0\ell_2 N_F I_F}{4gr}\Big|_{r=r_S}\cos(\theta-\theta_R)\hat{\mathbf{r}} \\
&= \frac{\mu_0\ell_1\ell_2 N_S N_F I_F}{8gr_S} i_{Sa}(t)\sin(\theta)\cos(\theta-\theta_R)d\theta\hat{\mathbf{z}} \times \hat{\mathbf{r}} \\
&= \frac{\mu_0\ell_1\ell_2 N_S N_F I_F}{8gr_S} i_{Sa}(t)\sin(\theta)\cos(\theta-\theta_R)d\theta\hat{\boldsymbol{\theta}}
\end{aligned}$$

where $r_S = \ell_2/2+g$ is the inside radius of the stator. Then, the incremental torque on these windings of phase a is given by

$$\begin{aligned}
d\vec{\tau}_{Sa} &= (r_S\hat{\mathbf{r}}) \times d\vec{\mathbf{F}}_{Sa} \\
&= r_S\hat{\mathbf{r}} \times \left(\frac{\mu_0\ell_1\ell_2 N_S N_F I_F}{8gr_S} i_{Sa}(t)\sin(\theta)\cos(\theta-\theta_R)d\theta\right)\hat{\boldsymbol{\theta}} \\
&= \frac{\mu_0\ell_1\ell_2 N_S N_F I_F}{8g} i_{Sa}(t)\sin(\theta)\cos(\theta-\theta_R)d\theta\hat{\mathbf{z}}.
\end{aligned}$$

The total torque on stator phase a is then

$$\begin{aligned}\vec{\tau}_{Sa}(i_{Sa},\theta_R) &= \int_0^{2\pi}\frac{\mu_0\ell_1\ell_2N_SN_FI_F}{8g}i_{Sa}(t)\sin(\theta)\cos(\theta-\theta_R)d\theta\hat{\mathbf{z}}\\ &= \frac{\mu_0\ell_1\ell_2N_SN_FI_F}{8g}i_{Sa}(t)\int_0^{2\pi}\sin(\theta)\Big(\cos(\theta)\cos(\theta_R)\\ &\quad+\sin(\theta)\sin(\theta_R)\Big)d\theta\hat{\mathbf{z}}\\ &= \frac{\mu_0\ell_1\ell_2N_SN_FI_F}{8g}i_{Sa}(t)\sin(\theta_R)\int_0^{2\pi}\sin^2(\theta)d\theta\hat{\mathbf{z}}\\ &= \frac{\mu_0\pi\ell_1\ell_2N_SN_FI_F}{8g}i_{Sa}(t)\sin(\theta_R)\hat{\mathbf{z}}.\end{aligned}$$

Similarly, with $N_{Sb}(\theta)d\theta = (N_S/2)\left|\sin(\theta-\pi/2)\right|d\theta$ the number of axial sides of the windings of stator phase b between θ and $\theta+d\theta$, the incremental force $d\vec{\mathbf{F}}_{Sb}$ on the current in these windings for $0\le\theta\le 2\pi$ is given by[2]

$$\begin{aligned}d\vec{\mathbf{F}}_{Sb} &= \left(i_{Sb}(t)\frac{N_S}{2}\sin(\theta-\pi/2)d\theta\right)(\ell_1\hat{\mathbf{z}})\times\vec{\mathbf{B}}_R\\ &= \left(i_{Sb}(t)\frac{N_S}{2}\sin(\theta-\pi/2)d\theta\right)(\ell_1\hat{\mathbf{z}})\times\frac{\mu_0\ell_2N_FI_F}{4gr}\Big|_{r=r_S}\cos(\theta-\theta_R)\hat{\mathbf{r}}\\ &= \frac{\mu_0\ell_1\ell_2N_SN_FI_F}{8gr_S}i_{Sb}(t)\sin(\theta-\pi/2)\cos(\theta-\theta_R)d\theta\hat{\mathbf{z}}\times\hat{\mathbf{r}}\\ &= -\frac{\mu_0\ell_1\ell_2N_SN_FI_F}{8gr_S}i_{Sb}(t)\cos(\theta)\cos(\theta-\theta_R)d\theta\hat{\boldsymbol{\theta}}.\end{aligned}$$

Thus, the incremental torque on phase b of the stator is given by

$$\begin{aligned}d\vec{\tau}_{Sb} &= (r_S\hat{\mathbf{r}})\times d\vec{\mathbf{F}}_{Sb}\\ &= (r_S\hat{\mathbf{r}})\times\left(-\frac{\mu_0\ell_1\ell_2N_SN_FI_F}{8gr_S}i_{Sb}(t)\cos(\theta)\cos(\theta-\theta_R)d\theta\right)\hat{\boldsymbol{\theta}}\\ &= -\frac{\mu_0\ell_1\ell_2N_SN_FI_F}{8g}i_{Sb}(t)\cos(\theta)\cos(\theta-\theta_R)d\theta\hat{\mathbf{z}}.\end{aligned}$$

[2] Note that $(N_S/2)\left|\sin(\theta-\pi/2)\right| = (N_S/2)\sin(\theta-\pi/2)$ for $\pi/2\le\theta\le 3\pi/2$ and $(N_S/2)\left|\sin(\theta-\pi/2)\right| = -(N_S/2)\sin(\theta-\pi/2)$ for $-\pi/2\le\theta\le\pi/2$.

The total torque on stator phase b is then

$$\begin{aligned}
\vec{\tau}_{Sb}(i_{Sb},\theta_R) &= \int_0^{2\pi} -\frac{\mu_0\ell_1\ell_2N_SN_FI_F}{8g}i_{Sb}(t)\cos(\theta)\cos(\theta-\theta_R)d\theta\hat{\mathbf{z}} \\
&= -\frac{\mu_0\ell_1\ell_2N_SN_FI_F}{8g}i_{Sb}(t)\int_0^{2\pi}\cos(\theta)\Big(\cos(\theta)\cos(\theta_R) \\
&\quad +\sin(\theta)\sin(\theta_R)\Big)d\theta\hat{\mathbf{z}} \\
&= -\frac{\mu_0\ell_1\ell_2N_SN_FI_F}{8g}i_{Sb}(t)\cos(\theta_R)\int_0^{2\pi}\cos^2(\theta)d\theta\hat{\mathbf{z}} \\
&= -\frac{\mu_0\pi\ell_1\ell_2N_SN_FI_F}{8g}i_{Sb}(t)\cos(\theta_R)\hat{\mathbf{z}}.
\end{aligned}$$

The total torque exerted on the stator by the rotor is then

$$\begin{aligned}
\vec{\tau}_S &= \vec{\tau}_{Sa}+\vec{\tau}_{Sb} \\
&= \frac{\mu_0\pi\ell_1\ell_2N_SN_FI_F}{8g}\left(i_{Sa}(t)\sin(\theta_R)-i_{Sb}(t)\cos(\theta_R)\right)\hat{\mathbf{z}}.
\end{aligned}$$

Consequently, $\vec{\tau}_R=-\vec{\tau}_S$ is the torque exerted on the rotor by the stator and is given by

$$\vec{\tau}_R=\frac{\mu_0\pi\ell_1\ell_2N_SN_FI_F}{8g}\left(-i_{Sa}(t)\sin(\theta_R)+i_{Sb}(t)\cos(\theta_R)\right)\hat{\mathbf{z}}.$$

Define

$$\begin{aligned}
B_{R\max} &\triangleq \frac{\mu_0\ell_2N_FI_F}{4gr}\Big|_{r=\ell_2/2} = \frac{\mu_0N_FI_F}{2g} \\
V_{\text{airgap}} &\triangleq 2\pi(\ell_2/2)\ell_1g=\pi\ell_1\ell_2g
\end{aligned}$$

where $B_{R\max}$ is the maximum values of the rotor's magnetic field at the surface of the rotor[3] and V_{airgap} is the volume of the air gap (assuming g is small). The torque is now rewritten as

$$\tau_R=V_{\text{airgap}}B_{R\max}\frac{N_S}{4g}\left(-i_{Sa}(t)\sin(\theta_R)+i_{Sb}(t)\cos(\theta_R)\right)$$

and the dynamic equation describing the motion of the rotor is then

$$J\frac{d\omega_R}{dt}=V_{\text{airgap}}B_{R\max}\frac{N_S}{4g}\left(-i_{Sa}(t)\sin(\theta_R)+i_{Sb}(t)\cos(\theta_R)\right)-\tau_L.$$

To obtain an expression for this torque in steady state, set

$$i_{Sa}(t)=I_S\cos(\omega_St)=I_S\cos(\theta_S), i_{Sb}(t)=I_S\sin(\omega_St)=I_S\sin(\theta_S)$$

---

[3] See equation (5.12).

where $\theta_S(t) = \omega_S t$ so that

$$\begin{aligned}
\tau_R &= V_{\text{airgap}} B_{R\max} \frac{N_S I_S}{4g} \left(\cos(\theta_S)\sin(\theta_R) - \sin(\theta_S)\cos(\theta_R)\right) \\
&= V_{\text{airgap}} B_{R\max} \frac{N_S I_S}{4g} \sin(\theta_S - \theta_R) \\
&= V_{\text{airgap}} \frac{B_{R\max} B_{S\max}}{2\mu_0} \sin(\theta_S - \theta_R).
\end{aligned}$$

Here

$$B_{S\max} \triangleq \frac{\mu_0 \ell_2 N_S I_S}{4gr}\Big|_{r=\ell_2/2} = \frac{\mu_0 N_S I_S}{2g}$$

is the maximum value of the rotating stator magnetic field at the surface of the rotor [see equation (5.2)]. With

$$\begin{aligned}
\theta_S(t) &= \omega_S t \\
\theta_R(t) &= \omega_R t - \delta,
\end{aligned}$$

it follows that

$$\sin(\theta_S - \theta_R) = \sin(\omega_S t - (\omega_R t - \delta)) = \sin((\omega_S - \omega_R)t + \delta)).$$

Thus, the only way to obtain *constant* torque is with the stator's magnetic field and the rotor rotating at the same constant angular speed (i.e., $\omega_S = \omega_R$). In other words, the stator's rotating magnetic field $\vec{\mathbf{B}}_S$ and the rotor must be rotating *synchronously*. In this case, the expression for the torque simplifies to

$$\tau_R = V_{\text{airgap}} \frac{B_{R\max} B_{S\max}}{2\mu_0} \sin(\delta). \tag{5.13}$$

This last expression leads to a simple picture for the synchronous motor. That is, the stator currents set up a rotating "stator magnet" which pulls along the "rotor magnet" (see Figure II.10). The torque on the "rotor magnet" is proportional to the maximum value of the magnetic field of the rotating "stator magnet", the maximum value of the magnetic field of the "rotor magnet", and the sine of the angle between them.

5.3.2 Emfs and Energy Conversion

The flux linkage in stator phase a due to the rotor's magnetic field $\vec{\mathbf{B}}_R$ is now found in order to compute the induced emf in this phase. Recall that for $0 \le \theta \le \pi$, the number of turns (loops) in stator phase a between θ and $\theta + d\theta$ is $(N_S/2)\sin(\theta)d\theta$. As the cross-sectional area of the wire is assumed negligible, the flux through each of these loops/turns is the same.

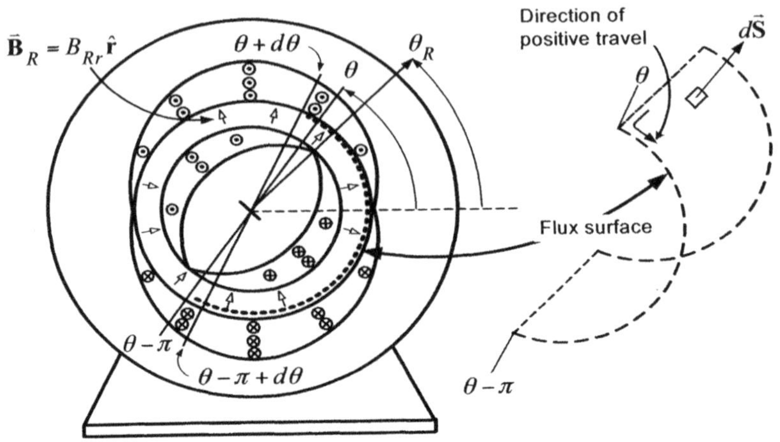

FIGURE 5.12. Computation of the flux in a turn of stator phase a located between θ and $\theta + d\theta$. All the loops between θ and $\theta + d\theta$ have the same flux as the cross-sectional area of the wire is assumed infinitesimal. The normal for the flux surface is taken to be $\hat{\mathbf{n}} = \hat{\mathbf{r}}$. Stator phase b is not shown for clarity.

Denoting the flux in each stator turn/loop between θ and $\theta + d\theta$ by $\phi_{Sa}(\theta)$, this flux is computed as

$$
\begin{aligned}
\phi_{Sa}(\theta) &\triangleq \int_{\substack{\text{A single loop of stator} \\ \text{phase } a \text{ at the angle } \theta}} \vec{\mathbf{B}}_R \cdot d\vec{\mathbf{S}} \\
&= \int_{\theta'=\theta-\pi}^{\theta'=\theta} \int_{z=0}^{z=\ell_1} \left(\frac{\mu_0 N_F \ell_2 I_F}{4 g r_S} \cos(\theta' - \theta_R(t))\hat{\mathbf{r}} \right) \cdot \left(r_S d\theta' dz\, \hat{\mathbf{r}} \right) \\
&= \int_{\theta'=\theta-\pi}^{\theta'=\theta} \frac{\mu_0 N_F \ell_1 \ell_2 I_F}{4g} \cos(\theta' - \theta_R(t)) d\theta' \\
&= \frac{\ell_1 \ell_2}{2} \frac{\mu_0 N_F I_F}{2g} \left(\sin(\theta - \theta_R) - \sin(\theta - \pi - \theta_R) \right) \\
&= \ell_1 \ell_2 B_{R\max} \sin(\theta - \theta_R)
\end{aligned}
\tag{5.14}
$$

where $B_{R\max} = (\mu_0 N_F I_F)/(2g)$.

As there are $(N_S/2)\sin(\theta)d\theta$ turns/loops of stator phase a between θ and $\theta + d\theta$ each containing the flux $\phi_{Sa}(\theta)$, the total flux linkage λ_{Sa} in all the windings making up stator phase a due to the *rotor's* magnetic field is given by

$$\begin{aligned}
\lambda_{Sa}(\theta_R) &\triangleq \int_{\theta=0}^{\theta=\pi} \phi_{Sa}(\theta)\frac{N_S}{2}\sin(\theta)d\theta \\
&= \int_0^{\pi} \ell_1\ell_2 B_{R\max}\sin(\theta-\theta_R)\frac{N_S}{2}\sin(\theta)d\theta \\
&= \ell_1\ell_2 B_{R\max}\frac{N_S}{2}\int_0^{\pi}\sin(\theta)\sin(\theta-\theta_R)d\theta \\
&= \ell_1\ell_2 B_{R\max}\frac{N_S}{2}\frac{\pi}{2}\cos(\theta_R) \\
&= \pi\ell_1\ell_2 g B_{R\max}\frac{N_S}{4g}\cos(\theta_R) \\
&= V_{\text{airgap}} B_{R\max}\frac{N_S}{4g}\cos(\theta_R) \qquad (5.15)
\end{aligned}$$

where $V_{\text{airgap}} \triangleq \pi\ell_1\ell_2 g$. Similarly, denote by $\phi_{Sb}(\theta)$ the flux in windings/turns of stator phase b between θ and $\theta+d\theta$ due to the rotor's magnetic field. It follows that

$$\begin{aligned}
\phi_{Sb}(\theta) &\triangleq \int_{\substack{\text{A single loop of stator}\\ \text{phase } b \text{ at the angle } \theta}} \vec{\mathbf{B}}_R\cdot d\vec{\mathbf{S}} \\
&= \int_{\theta'=\theta-\pi}^{\theta'=\theta}\int_{z=0}^{z=\ell_1}\left(\frac{\mu_0 N_F \ell_2 I_F}{4 g r_S}\cos(\theta'-\theta_R(t))\hat{\mathbf{r}}\right)\cdot\left(r_S d\theta' dz\hat{\mathbf{r}}\right) \\
&= \ell_1\ell_2\frac{\mu_0 N_F I_F}{2g}\sin(\theta-\theta_R) \\
&= \ell_1\ell_2 B_{R\max}\sin(\theta-\theta_R). \qquad (5.16)
\end{aligned}$$

The total flux linkage in stator phase b is then

$$\begin{aligned}
\lambda_{Sb}(\theta_R) &= \int_{\theta=\pi/2}^{\theta=3\pi/2}\phi_{Sb}(\theta)\frac{N_S}{2}\sin(\theta-\pi/2)d\theta \\
&= \int_{\pi/2}^{3\pi/2}\ell_1\ell_2 B_{R\max}\sin(\theta-\theta_R)\frac{N_S}{2}\sin(\theta-\pi/2)d\theta \\
&= -\ell_1\ell_2 B_{R\max}\frac{N_S}{2}\int_{\pi/2}^{3\pi/2}\cos(\theta)\sin(\theta-\theta_R)d\theta \\
&= -\ell_1\ell_2 B_{R\max}\frac{N_S}{2}\left(-\frac{\pi}{2}\sin(\theta_R)\right) \\
&= V_{\text{airgap}} B_{R\max}\frac{N_S}{4g}\sin(\theta_R). \qquad (5.17)
\end{aligned}$$

The quantities λ_{Sa} and λ_{Sb} denote the flux linkages in the stator phases due to the *rotor's* magnetic field.[4] By Faraday's law, the emf's induced in the stator phases due to the rotor's magnetic field (i.e., the *back emfs* in the stator phases) are then

$$\begin{aligned}\xi_{Sa} &= -d\lambda_{Sa}/dt = +V_{\text{airgap}} B_{R\max}\frac{N_S}{4g}\sin(\theta_R)\frac{d\theta_R}{dt} \\ \xi_{Sb} &= -d\lambda_{Sb}/dt = -V_{\text{airgap}} B_{R\max}\frac{N_S}{4g}\cos(\theta_R)\frac{d\theta_R}{dt}.\end{aligned} \tag{5.18}$$

As the normal $\hat{\mathbf{n}} = \hat{\mathbf{r}}$ was chosen for the flux surface to compute (5.14) and therefore for all the loops in (5.15), the induced emf ξ_{Sa} in phase a has the same sign convention as that chosen for the phase current i_{Sa}. Similarly, the induced emf ξ_{Sb} in phase b has the same sign convention as the phase current i_{Sb}.

The electrical power absorbed by the emfs ξ_{Sa}, ξ_{Sb} is

$$\begin{aligned}P_{\text{elec}} &= i_{Sa}(t)\xi_{Sa}(t) + i_{Sb}(t)\xi_{Sb}(t) \\ &= V_{\text{airgap}} B_{R\max}\frac{N_S}{4g}\left(i_{Sa}(t)\sin(\theta_R) - i_{Sb}(t)\cos(\theta_R)\right)\omega_R \\ &= V_{\text{airgap}} B_{R\max}\frac{N_S I_S}{4g}\left(\cos(\omega_S t)\sin(\theta_R) - \sin(\omega_S t)\cos(\theta_R)\right)\omega_R \\ &= V_{\text{airgap}} B_{R\max} B_{S\max}\frac{1}{2\mu_0}\omega_R\sin(\theta_R - \omega_S t) \\ &= \frac{V_{\text{airgap}}}{2\mu_0} B_{S\max} B_{R\max}\omega_R\sin\Big((\omega_R - \omega_S)t - \delta\Big) \\ &= -\frac{B_{S\max} B_{R\max}}{2\mu_0} V_{\text{airgap}}\omega_R\sin(\delta)\end{aligned}$$

where the motor is now assumed to be in steady state so that $\theta_S = \omega_S t$, $\theta_R = \omega_R t - \delta$ with $\omega_S = \omega_R$. By the torque equation (5.13), the mechanical power produced is

$$P_{\text{mech}} = \tau_R\omega_R = \frac{B_{S\max} B_{R\max}}{2\mu_0} V_{\text{airgap}}\sin(\delta)\omega_R$$

so that

$$P_{\text{elec}} + P_{\text{mech}} = 0.$$

That is, the electrical power absorbed in the stator phases reappears as the mechanical power of the motor.

---

[4] Note that as $V_{\text{airgap}} B_{R\max}\frac{N_S}{4g} = \pi\ell_1\ell_2 g\frac{\mu_0 N_F I_F}{2g}\frac{N_S}{4g} = (\mu_0\pi\ell_1\ell_2 N_S N_R/8g)I_F$, these flux linkages may also be written in the form $\lambda_{Sa}(\theta_R) = MI_F\cos(\theta_R), \lambda_{Sb}(\theta_R) = MI_F\sin(\theta_R)$ where, $M \triangleq (\mu_0\pi\ell_1\ell_2 N_S N_R/8g)$ is called the *coefficient of mutual inductance*.

5.3.3 Synchronous Motor with a Salient Rotor

To actually produce a rotor winding for a machine that is sinusoidally wound is more expensive than other types of windings. In the case of the synchronous motor, one can eliminate the need for a sinusoidally wound rotor. Recall that in the above analysis of the synchronous machine, the important fact was that the magnetic field of the rotor was sinusoidally distributed in $\theta - \theta_R$. It is now shown that by putting a *uniformly* wound coil on the rotor and shaping the pole faces of the rotor appropriately, a sinusoidally distributed magnetic field in the air gap can be produced.

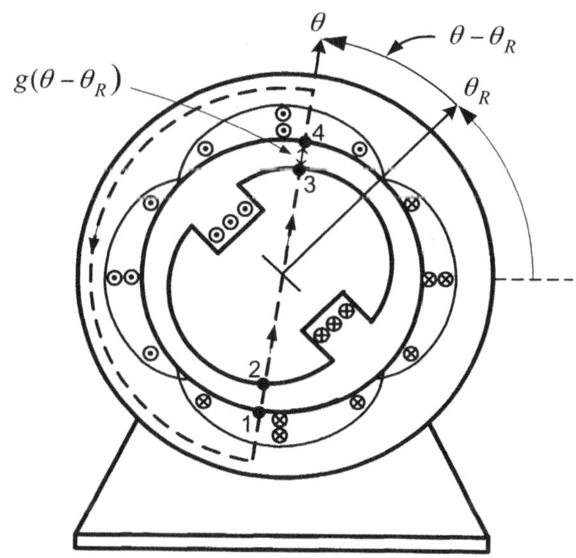

FIGURE 5.13. Synchronous motor with a salient rotor.

To do so, consider Figure 5.13 where the rotor iron is now wrapped with a standard coil with N_F turns and the pole faces have been shaped so that the air gap is no longer uniform, but instead given by

$$g(\theta - \theta_R) \triangleq \frac{g_0}{\cos(\theta - \theta_R)} = g_0 \sec(\theta - \theta_R) \tag{5.19}$$

where g_0 is the minimum air-gap distance. (Obviously, this can only be done approximately as $\cos(\theta - \theta_R) = 0$ for $\theta = \theta_R \pm \pi/2$.) To determine the magnetic field $\vec{\mathbf{B}}_R$ in the air gap due to the rotor, Ampère's law is applied to the path 1–2–3–4–1 above to obtain[5]

$$\int_1^2 \vec{\mathbf{H}}\cdot d\vec{\ell} + \int_3^4 \vec{\mathbf{H}}\cdot d\vec{\ell} = N_F i_F \tag{5.20}$$

where the standard assumption that $\vec{\mathbf{H}} \equiv 0$ in the magnetic material has been used. As

$$d\vec{\ell} = \begin{cases} +dr\hat{\mathbf{r}} & \text{for path } 1\text{–}2 \\ -dr\hat{\mathbf{r}} & \text{for path } 3\text{–}4 \end{cases}$$

equation (5.20) reduces to ($\vec{\mathbf{H}} = H_r\hat{\mathbf{r}}$)

$$H_r(\theta)g(\theta-\theta_R) - H_r(\theta+\pi)g(\theta-\theta_R) = N_F i_F. \tag{5.21}$$

By symmetry, the radial component must satisfy $H_r(\theta) = -H_r(\theta \pm \pi)$ so that (5.21) becomes

$$2H_r(\theta)g(\theta-\theta_R) = N_F i_F.$$

Finally,

$$H_r(\theta) = \frac{N_F i_F}{2g(\theta-\theta_R)} = \frac{N_F i_F}{2g_0}\cos(\theta-\theta_R).$$

Finally, this is multiplied by both μ_0 and r_R/r (to satisfy conservation of flux in the air gap) to obtain

$$\vec{\mathbf{B}}_R(\theta-\theta_R) = \frac{\mu_0 N_F I_F}{2g_0}\frac{r_R}{r}\cos(\theta-\theta_R)\hat{\mathbf{r}}.$$

By shaping the pole faces of the rotor according to equation (5.19), it was found this salient rotor produced the *same* magnetic field in the air gap as a sinusoidally wound rotor. However, as one moves away from the (salient) rotors' magnetic axis, the air gap is no longer small so that the assumptions about the magnetic field due to the rotor current being constant and radially directed across the air gap are less valid. On the other hand, with either a salient or sinusoidally wound rotor, the magnetic field strength of the rotor is quite weak as one moves away from the magnetic axis so that these portions do not contribute significantly to the torque. (The same can be said of the strength of the stator's magnetic field as one moves away from its magnetic axis.) As the angle between the stator's and rotor's magnetic axis is usually small, the region where the assumptions

[5] Only those paths that do not go through the field coil are being considered. At those angles θ corresponding to going through the field coil, it will turn out that the $\vec{\mathbf{B}}$ field due to the current in the coil is weak and can be neglected.

As before, it is assumed that the stator windings have an infinitesimal cross section so that $\int \vec{\mathbf{H}}\cdot d\vec{\ell} = 0$ across the winding because $d\vec{\ell} = 0$.

are violated do not contribute significantly to the torque lending credence to the above analysis.

Remark With $r_R = \ell_2/2$, the expression for $\vec{\mathbf{B}}_S$ in (5.11) can be written as

$$\vec{\mathbf{B}}_S(i_{Sa}, i_{Sb}, r, \theta) = \frac{\mu_0 r_R N_S}{2gr}\Big(i_{Sa}\cos(\theta) + i_{Sb}\sin(\theta)\Big)\hat{\mathbf{r}}.$$

and was derived assuming a uniform air gap. If a salient rotor as in Figure 5.13 is used, then this assumption no longer holds. However, to a reasonable approximation, one can use (5.11) with $g = g_1 > g_0$, that is, some average value g_1 of the air gap length. Further, as the rotor no longer has a circular cross section, one can take $r_R = r_S - g_1$ as an average radius of the rotor.

5.3.4 Armature and Field Windings

Recall that the rotor of a DC motor was referred to as the *armature* winding. This terminology is used to indicate that the current in this winding is alternating current (AC) as every half-turn the commutator and brushes reverse the direction of the current in any given rotor loop. For a synchronous or induction motor, the stator windings carry AC current (in normal operation) so the stator windings are referred to as the armature (see Woodson and Melcher [20], page 121). Further, recall that the stator winding of a separately excited DC motor is called the *field* winding and this terminology is used to indicate that it normally carries a DC current to establish a *constant magnetic field* in the air gap that produces torque on the current in the rotor windings. Similarly, on a wound rotor synchronous machine, the rotor winding is called the *field* winding because it normally operates with a DC current to establish a *constant magnetic field* in the air gap. (This magnetic field is constant for someone rotating with the rotor.)

5.4 Microscopic Viewpoint of AC Machines*

The microscopic viewpoint of the Physics of AC machines is described in this section. This is analogous to the microscopic viewpoint of DC machines given in Chapter 1. The analysis is first done from the point of view of a stationary coordinate system and then from the point of view of a coordinate system rotating with the stator's magnetic field.

5.4.1 *Rotating Axial Electric Field Due to the Stator Currents*

Recall from equation (5.2) that the rotating magnetic field in the air gap produced by the stator currents is

$$\begin{aligned}\vec{\mathbf{B}}_S(r,\theta,t) &= B_{Sr}(r,\theta,t)\hat{\mathbf{r}} \\ &= \frac{\mu_0 \ell_2 N_S I_S}{4gr}\cos(\theta-\omega_S t)\hat{\mathbf{r}}. \end{aligned} \tag{5.22}$$

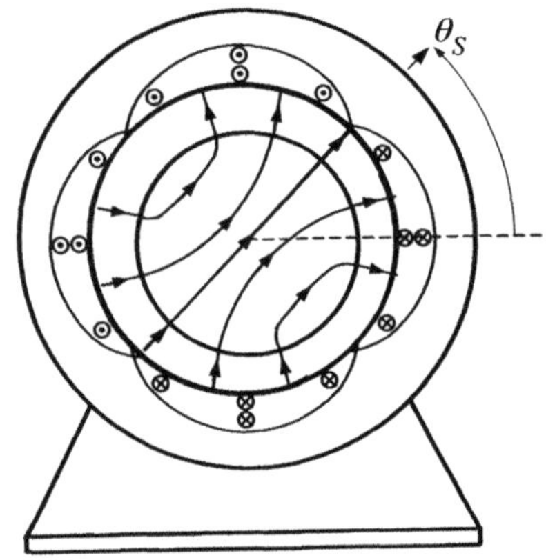

FIGURE 5.14. Rotating magnetic field $\vec{\mathbf{B}}_S$.

In Chapter 4 [see equation (4.26)], it was shown that the axial electric field produced by $\vec{\mathbf{B}}_S$ in equation (5.22) according to

$$\nabla \times \vec{\mathbf{E}}_S = -\frac{\partial \vec{\mathbf{B}}_S}{\partial t}$$

is

$$\vec{\mathbf{E}}_S(\theta,t) = E_{Sz}(\theta,t)\hat{\mathbf{z}} = \frac{\mu_0 \ell_2 N_S I_S}{4g}\omega_S\cos(\theta-\omega_S t)\hat{\mathbf{z}}. \tag{5.23}$$

This axially directed rotating electric field $\vec{\mathbf{E}}_S(\theta,t)$ is illustrated in Figure 5.15. For $\theta_S-\pi/2 \le \theta \le \theta_S+\pi/2$, the electric field is into the page ($\otimes$) while for $\theta_S+\pi/2 \le \theta \le \theta_S+3\pi/2$, the electric field is out of the page ($\odot$).

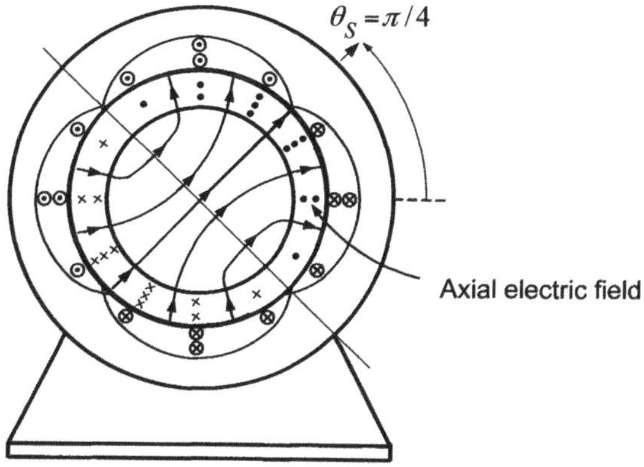

FIGURE 5.15. Rotating axial electric field $\vec{\mathbf{E}}_S = E_{Sz}(\theta, t)\hat{\mathbf{z}}$ induced by the rotating magnetic field $\vec{\mathbf{B}}_S = B_{Sr}(\theta, t)\hat{\mathbf{r}}$.

5.4.2 Induction Machine in the Stationary Coordinate System

Consider the setup of Figure 5.16.

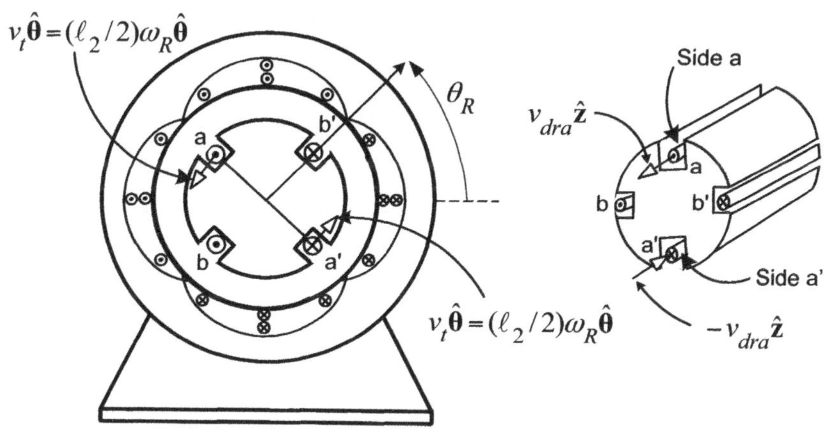

FIGURE 5.16. Simple induction motor with two single-loop rotor phases.

The axial electric field $\vec{\mathbf{E}}_S$ produces a force on charge carriers in the rotor loops. However, the *total* force on a charge carrier in a rotor loop is due to *both* the electric and magnetic fields as given by the Lorentz expression

$$\vec{\mathbf{F}} = q\vec{\mathbf{E}}_S + q\vec{\mathbf{v}}\times\vec{\mathbf{B}}_S$$

where

$$\vec{\mathbf{B}}_S(r,\theta,t) = B_{Sr}(r,\theta,t)\hat{\mathbf{r}} = (\mu_0\ell_2 N_S I_S/4gr)\cos(\theta-\omega_S t)\hat{\mathbf{r}}$$
$$\vec{\mathbf{E}}_S(\theta,t) = E_{Sz}(\theta,t)\hat{\mathbf{z}} = \frac{\mu_0\ell_2 N_S I_S}{4g}\omega_S\cos(\theta-\omega_S t)\hat{\mathbf{z}}. \tag{5.24}$$

To compute $\vec{\mathbf{v}}\times\vec{\mathbf{B}}_S$, let v_{dra} be the (drift) speed of the charge carriers (which make up the current) along rotor loop a. The speed v_{dra} being positive means that the current is positive; that is, coming out of the page $\odot$ on side a and into the page $\otimes$ on side a'. Then, with $v_t = (\ell_2/2)\omega_R$, the speed of the charge carriers in the $\hat{\boldsymbol{\theta}}$ direction, the *total* velocity of the charge carriers along the axial sides of rotor loop a is

$$\vec{\mathbf{v}} = \begin{cases} +v_{dra}\hat{\mathbf{z}} + v_t\hat{\boldsymbol{\theta}} & \text{for side } a \\ -v_{dra}\hat{\mathbf{z}} + v_t\hat{\boldsymbol{\theta}} & \text{for side } a'. \end{cases}$$

The magnetic force per unit charge on the charge carriers on the axial sides of rotor loop a is given by

$$\vec{\mathbf{v}}\times\vec{\mathbf{B}}_S =$$
$$\begin{cases} +v_{dra}B_{Sr}(\ell_2/2,\theta_R+\pi/2,t)\hat{\boldsymbol{\theta}} - v_t B_{Sr}(\ell_2/2,\theta_R+\pi/2,t)\hat{\mathbf{z}} & \text{for side } a \\ -v_{dra}B_{Sr}(\ell_2/2,\theta_R-\pi/2,t)\hat{\boldsymbol{\theta}} - v_t B_{Sr}(\ell_2/2,\theta_R-\pi/2,t)\hat{\mathbf{z}} & \text{for side } a'. \end{cases}$$

Using equation (5.24), it follows that for side a

$$\begin{aligned} B_{Sr}(\ell_2/2,\theta_R+\pi/2,t)\hat{\mathbf{r}} &= \frac{\mu_0 N_S I_S}{2g}\cos(\theta_R+\pi/2-\omega_S t)\hat{\mathbf{r}} \\ &= \frac{\mu_0 N_S I_S}{2g}\sin(\omega_S t-\theta_R)\hat{\mathbf{r}}. \end{aligned}$$

Similarly, on side a'

$$\begin{aligned} B_{Sr}(\ell_2/2,\theta_R-\pi/2,t)\hat{\mathbf{r}} &= \frac{\mu_0 N_S I_S}{2g}\cos(\theta_R-\pi/2-\omega_S t)\hat{\mathbf{r}} \\ &= -\frac{\mu_0 N_S I_S}{2g}\sin(\omega_S t-\theta_R)\hat{\mathbf{r}}. \end{aligned}$$

With $v_t = (\ell_2/2)\omega_R$ and $\theta_R = \omega_R t$, the magnetic force per unit charge in rotor loop a may now be written as

$$\vec{\mathbf{v}}\times\vec{\mathbf{B}}_S =$$
$$\begin{cases} v_{dra}\frac{\mu_0 N_S I_S}{2g}\sin(\omega_S t-\omega_R t)\hat{\boldsymbol{\theta}} - \omega_R\frac{\mu_0\ell_2 N_S I_S}{4g}\sin(\omega_S t-\omega_R t)\hat{\mathbf{z}} & \text{for side } a \\ v_{dra}\frac{\mu_0 N_S I_S}{2g}\sin(\omega_S t-\omega_R t)\hat{\boldsymbol{\theta}} + \omega_R\frac{\mu_0\ell_2 N_S I_S}{4g}\sin(\omega_S t-\omega_R t)\hat{\mathbf{z}} & \text{for side } a'. \end{cases} \tag{5.25}$$

Next, the electric field $\vec{\mathbf{E}}_S$ in rotor loop a is given by

$$\vec{\mathbf{E}}_S = \begin{cases} E_{Sz}(\theta_R + \pi/2, t)\hat{\mathbf{z}} & \text{for side } a \\ E_{Sz}(\theta_R - \pi/2, t)\hat{\mathbf{z}} & \text{for side } a'. \end{cases}$$

Using (5.24) and $\theta_R(t) = \omega_R t$, this becomes for side a

$$\begin{aligned} \vec{\mathbf{E}}_S &= \frac{\mu_0 \ell_2 N_S I_S}{4g}\omega_S \cos(\omega_R t + \pi/2 - \omega_S t)\hat{\mathbf{z}} \\ &= \frac{\mu_0 \ell_2 N_S I_S}{4g}\omega_S \sin(\omega_S t - \omega_R t)\hat{\mathbf{z}} \end{aligned} \tag{5.26}$$

and for side a'

$$\begin{aligned} \vec{\mathbf{E}}_S &= \frac{\mu_0 \ell_2 N_S I_S}{4g}\omega_S \cos(\omega_R t - \pi/2 - \omega_S t)\hat{\mathbf{z}} \\ &= -\frac{\mu_0 \ell_2 N_S I_S}{4g}\omega_S \sin(\omega_S t - \omega_R t)\hat{\mathbf{z}}. \end{aligned} \tag{5.27}$$

The total force per unit charge $\vec{\mathbf{F}}/q = \vec{\mathbf{E}}_S + \vec{\mathbf{v}}\times\vec{\mathbf{B}}_S$ on side a is then

$$\begin{aligned} \vec{\mathbf{F}}_{\text{side } a}/q = {} & \frac{\mu_0 \ell_2 N_S I_S}{4g}\omega_S \sin(\omega_S t - \omega_R t)\hat{\mathbf{z}} + v_{dra}\frac{\mu_0 N_S I_S}{2g} \\ & \times \sin(\omega_S t - \omega_R t)\hat{\boldsymbol{\theta}} - \omega_R \frac{\mu_0 \ell_2 N_S I_S}{4g}\sin(\omega_S t - \omega_R t)\hat{\mathbf{z}} \end{aligned} \tag{5.28}$$

and for side a' it is given by

$$\begin{aligned} \vec{\mathbf{F}}_{\text{side } a'}/q = {} & -\frac{\mu_0 \ell_2 N_S I_S}{4g}\omega_S \sin(\omega_S t - \omega_R t)\hat{\mathbf{z}} + v_{dra}\frac{\mu_0 N_S I_S}{2g} \\ & \times \sin(\omega_S t - \omega_R t)\hat{\boldsymbol{\theta}} + \omega_R \frac{\mu_0 \ell_2 N_S I_S}{4g}\sin(\omega_S t - \omega_R t)\hat{\mathbf{z}}. \end{aligned} \tag{5.29}$$

These last two expressions simplify to

$$\begin{aligned} \vec{\mathbf{F}}_{\text{side } a}/q &= v_{dra}\frac{\mu_0 N_S I_S}{2g}\sin\left((\omega_S - \omega_R)t\right)\hat{\boldsymbol{\theta}} \\ &\quad + \frac{\mu_0 \ell_2 N_S I_S}{4g}(\omega_S - \omega_R)\sin\left((\omega_S - \omega_R)t\right)\hat{\mathbf{z}} \end{aligned} \tag{5.30}$$

and

$$\begin{aligned} \vec{\mathbf{F}}_{\text{side } a'}/q &= v_{dra}\frac{\mu_0 N_S I_S}{2g}\sin\left((\omega_S - \omega_R)t\right)\hat{\boldsymbol{\theta}} \\ &\quad - \frac{\mu_0 \ell_2 N_S I_S}{4g}(\omega_S - \omega_R)\sin\left((\omega_S - \omega_R)t\right)\hat{\mathbf{z}}. \end{aligned} \tag{5.31}$$

The component in $\hat{\boldsymbol{\theta}}$ direction (tangential to the rotor motion) is what produces the torque. With N the number of charge carriers per unit volume in the conductor of rotor phase a, S the cross-sectional area of the loop, it follows that $qNS\ell_1$ is the total number of charge carriers on each axial side of the loop. Therefore, the total tangential magnetic force on these charge carriers is

$$\vec{\mathbf{F}}_{\hat{\boldsymbol{\theta}}} = qNS\ell_1 v_{dra} \frac{\mu_0 N_S I_S}{2g} \sin\left((\omega_S - \omega_R)t\right)\hat{\boldsymbol{\theta}}.$$

Recalling that $i_{Ra} = qNSv_{dra}$ is the rotor current and computing the torque according to $(\ell_2/2)\hat{\mathbf{r}} \times \vec{\mathbf{F}}_{\hat{\boldsymbol{\theta}}}$, the total torque on rotor loop a is

$$\begin{aligned} \vec{\boldsymbol{\tau}}_{Ra} &= \tau_{Ra}\hat{\mathbf{z}} \\ &= \left((\ell_2/2)\hat{\mathbf{r}} \times \vec{\mathbf{F}}_{\hat{\boldsymbol{\theta}}}\right)_{side-a} + \left((\ell_2/2)\hat{\mathbf{r}} \times \vec{\mathbf{F}}_{\hat{\boldsymbol{\theta}}}\right)_{side-a'} \\ &= 2i_{Ra}\ell_1 \frac{\mu_0 N_S I_S}{2g} \sin((\omega_S - \omega_R)t)\,(\ell_2/2)\hat{\mathbf{r}} \times \hat{\boldsymbol{\theta}} \\ &= i_{Ra} \frac{\mu_0 \ell_1 \ell_2 N_S I_S}{2g} \sin((\omega_S - \omega_R)t)\hat{\mathbf{z}}. \end{aligned} \tag{5.32}$$

To evaluate the torque, the current i_{Ra} must be computed. The current in the rotor loop is produced by the total emf in the loop. Note that the electric field $\vec{\mathbf{E}}_S = E_{Sz}\hat{\mathbf{z}}$ induced by the stator's rotating magnetic field is pushing charge out of side a and into side a', and it is being opposed by the axial magnetic force $(\vec{\mathbf{v}}\times\vec{\mathbf{B}}_S)_z\hat{\mathbf{z}}$ as equation (5.28) shows. For example, on side a, the electric field

$$E_{Sz}\hat{\mathbf{z}} = \frac{\mu_0 \ell_2 N_S I_S}{4g}\omega_S \sin((\omega_S - \omega_R)t)\hat{\mathbf{z}}$$

is opposed by the z component of the magnetic force per unit charge

$$(\vec{\mathbf{v}}\times\vec{\mathbf{B}}_S)_z\hat{\mathbf{z}} = -\frac{\mu_0 \ell_2 N_S I_S}{4g}\omega_R \sin((\omega_S - \omega_R)t)\hat{\mathbf{z}}.$$

It follows by (5.30) and (5.31) that the total *axial* or z component of the force per unit charge is

$$(\vec{\mathbf{F}}/q)_z = \begin{cases} \dfrac{\mu_0 \ell_2 N_S I_S}{4g}(\omega_S - \omega_R)\sin((\omega_S - \omega_R)t)\,\hat{\mathbf{z}} & \text{side } a \\ -\dfrac{\mu_0 \ell_2 N_S I_S}{4g}(\omega_S - \omega_R)\sin((\omega_S - \omega_R)t)\,\hat{\mathbf{z}} & \text{side } a'. \end{cases} \tag{5.33}$$

Thus *electromotive force* or *emf* in the rotor loop is simply the integral of the force per unit charge around the loop, that is,

$$\begin{aligned} \xi_{Ra} &\triangleq \oint_{\text{rotor loop } a} (\vec{\mathbf{F}}/q)\cdot d\vec{\boldsymbol{\ell}} \\ &= \oint_{\text{rotor loop } a} (\vec{\mathbf{E}}_S + \vec{\mathbf{v}}\times\vec{\mathbf{B}}_S)\cdot d\vec{\boldsymbol{\ell}} \end{aligned}$$

where

$$d\vec{\ell} = \begin{cases} +dz\hat{\mathbf{z}} & \text{for side } a \\ -dz\hat{\mathbf{z}} & \text{for side } a'. \end{cases}$$

As $d\vec{\ell}$ is aligned with $\pm\hat{\mathbf{z}}$, only the axial component $(\vec{\mathbf{F}}/q)_z$ given in equation (5.33) contributes to the emf.

Evaluating the line integral, the induced emf in rotor phase a is given by

$$\xi_{Ra} = \frac{\mu_0 \ell_1 \ell_2 N_S I_S}{2g}(\omega_S - \omega_R)\sin((\omega_S - \omega_R)t) \tag{5.34}$$

which is the same as in the macroscopic case. Neglecting the rotor inductance so that $i_{Ra} = \xi_{Ra}/R_R$, the torque on rotor loop a is

$$\tau_{Ra} = \frac{1}{R_R}\left(\frac{\mu_0 \ell_1 \ell_2 N_S I_S}{2g}\right)^2 (\omega_S - \omega_R)\sin^2\Big((\omega_S - \omega_R)t\Big). \tag{5.35}$$

A similar analysis shows that

$$\tau_{Rb} = \frac{1}{R_R}\left(\frac{\mu_0 \ell_1 \ell_2 N_S I_S}{2g}\right)^2 (\omega_S - \omega_R)\cos^2\Big((\omega_S - \omega_R)t\Big). \tag{5.36}$$

Rotor Torque Computed from Microscopic Point of View

Adding equations (5.35) and (5.36), the total torque on the rotor is

$$\tau = \tau_{Ra} + \tau_{Rb} = \left(\frac{\mu_0 \ell_1 \ell_2 N_S I_S}{2g}\right)^2 \frac{1}{R_R}(\omega_S - \omega_R)$$

which is the same expression as that computed in the macroscopic case.

Electromechanical Energy Conversion

Define the voltage in rotor loop a produced by the stator electric field $\vec{\mathbf{E}}_S$ as

$$V_{Ra} \triangleq \oint_{\text{rotor loop } a} \vec{\mathbf{E}}_S \cdot d\vec{\ell} = \frac{\mu_0 \ell_1 \ell_2 N_S I_S}{2g}\omega_S \sin\Big((\omega_S - \omega_R)t\Big). \tag{5.37}$$

For example, if the rotor was locked so that $\omega_R = 0$ in (5.37), then this would be the total emf in the rotor loop as can be seen by setting $\omega_R = 0$ in (5.34). However, due to the rotor's motion ($\omega_R \neq 0$), there is an axial (z component) magnetic force per unit charge. Define the "back emf" ζ_{Ra} as the integral of the axial magnetic force per unit charge around the rotor loop, that is,

$$\begin{aligned}
\zeta_{Ra} &\triangleq \oint_{\text{rotor loop } a} (\vec{\mathbf{v}}\times\vec{\mathbf{B}}_S)_z \cdot d\vec{\ell} \\
&= \int_0^{\ell_1} \left(-\omega_R \frac{\mu_0 \ell_2 N_S I_S}{4g} \sin(\omega_S t - \omega_R t)\hat{\mathbf{z}}\right) \cdot (d\ell\hat{\mathbf{z}}) \\
&\quad + \int_0^{\ell_1} \left(\omega_R \frac{\mu_0 \ell_2 N_S I_S}{4g} \sin(\omega_S t - \omega_R t)\hat{\mathbf{z}}\right) \cdot (-d\ell\hat{\mathbf{z}}) \\
&= -\frac{\mu_0 \ell_1 \ell_2 N_S I_S}{2g} \omega_R \sin((\omega_S - \omega_R)t) \,. \qquad (5.38)
\end{aligned}$$

In summary, the total emf ξ_{Ra} in rotor loop a may be written as

$$\begin{aligned}
\xi_{Ra} &= \oint_{\text{rotor loop } a} (\vec{\mathbf{E}}_S + \vec{\mathbf{v}}\times\vec{\mathbf{B}}_S) \cdot d\vec{\ell} = V_{Ra} + \zeta_{Ra} \\
&= \frac{\mu_0 \ell_1 \ell_2 N_S I_S}{2g} (\omega_S - \omega_R) \sin\left((\omega_S - \omega_R)t\right) \qquad (5.39)
\end{aligned}$$

where V_{Ra} is the voltage produced in the rotor loop by the induced electric field $\vec{\mathbf{E}}_S$ and ζ_{Ra} is the back emf produced by the axial magnetic force per unit charge $\vec{\mathbf{v}}\times\vec{\mathbf{B}}_S$. By direct substitution using equations (5.32) and (5.38), it is easy to see that

$$\tau_{Ra}\omega_R + i_{Ra}\zeta_{Ra} = 0.$$

Here $\tau_{Ra}\omega_R$ is the mechanical power produced on rotor loop a while $i_{Ra}\zeta_{Ra}$ is the electrical power absorbed by the back emf. That is, if $\tau_{Ra}\omega_R > 0$ so that mechanical power is being produced, then $i_{Ra}\zeta_{Ra} < 0$ is the electrical power being absorbed by back emf in rotor loop a. Another interesting point is that

$$i_{Ra} V_{Ra}$$

is the total electrical power into rotor loop a from the stator, that is,

$$i_{Ra} V_{Ra} = i_{Ra}\xi_{Ra} - i_{Ra}\zeta_{Ra} = i_{Ra}\xi_{Ra} + \tau_{Ra}\omega_R.$$

In other words, the total power $i_{Ra} V_{Ra}$ into rotor loop a is converted into the electrical power $i_{Ra}\xi_{Ra}$ (ultimately dissipated as heat in the rotor resistance) and into the mechanical power $\tau_{Ra}\omega_R$ $(= -i_{Ra}\zeta_{Ra})$. Similar remarks hold for rotor loop b.

5.4.3 *Faraday's Law and the Integral of the Force per Unit Charge*

In this subsection, it is shown that Faraday's law gives the same emf as integrating the force per unit charge around the loop, that is,

$$\xi = -\frac{d}{dt}\left(\int_S \vec{\mathbf{B}} \cdot d\vec{\mathbf{S}}\right) = \oint_C (\vec{\mathbf{E}} + \vec{\mathbf{v}} \times \vec{\mathbf{B}}) \cdot d\vec{\ell}. \tag{5.40}$$

In equation (5.3), ξ_{Ra} was computed using $\xi_{Ra} = -\frac{d}{dt}\left(\int_S \vec{\mathbf{B}} \cdot d\vec{\mathbf{S}}\right)$ while in equation (5.39), it was computed using $\xi_{Ra} = \oint_C (\vec{\mathbf{E}}_S + \vec{\mathbf{v}} \times \vec{\mathbf{B}}_S) \cdot d\vec{\ell}$ and both methods gave the same result. The objective here is to show that this is true in general.

To show that (5.40) is true in general, let S be a surface whose boundary is the closed loop C in a region of space with an electric field $\vec{\mathbf{E}}$ and a magnetic field $\vec{\mathbf{B}}$. The line integral of the force per unit charge gives

$$\begin{aligned}\oint_C (\vec{\mathbf{E}} + \vec{\mathbf{v}} \times \vec{\mathbf{B}}) \cdot d\vec{\ell} &= \oint_S \left(\nabla \times \vec{\mathbf{E}}\right) \cdot d\vec{\mathbf{S}} + \oint_C \left(\vec{\mathbf{v}} \times \vec{\mathbf{B}}\right) \cdot d\vec{\ell} \\ &= \oint_S \left(-\frac{\partial \vec{\mathbf{B}}}{\partial t}\right) \cdot d\vec{\mathbf{S}} + \oint_C \left(\vec{\mathbf{v}} \times \vec{\mathbf{B}}\right) \cdot d\vec{\ell}\end{aligned}$$

where Stokes theorem

$$\oint_C \vec{\mathbf{E}} \cdot d\vec{\ell} = \oint_S \left(\nabla \times \vec{\mathbf{E}}\right) \cdot d\vec{\mathbf{S}}$$

and Faraday's law in differential form

$$\nabla \times \vec{\mathbf{E}} = -\frac{\partial \vec{\mathbf{B}}}{\partial t}$$

were used [35][36]. Using the fact that

$$\oint_S \left(-\frac{\partial \vec{\mathbf{B}}}{\partial t}\right) \cdot d\vec{\mathbf{S}} = -\frac{\partial}{\partial t}\oint_S \vec{\mathbf{B}} \cdot d\vec{\mathbf{S}}$$

and the vector identity

$$\left(\vec{\mathbf{v}} \times \vec{\mathbf{B}}\right) \cdot d\vec{\ell} = -\vec{\mathbf{B}} \cdot \left(\vec{\mathbf{v}} \times d\vec{\ell}\right)$$

this reduces to

$$\begin{aligned}\oint_C (\vec{\mathbf{E}} + \vec{\mathbf{v}} \times \vec{\mathbf{B}}) \cdot d\vec{\ell} &= -\frac{\partial}{\partial t}\oint_S \vec{\mathbf{B}} \cdot d\vec{\mathbf{S}} + \oint_C \left(\vec{\mathbf{v}} \times \vec{\mathbf{B}}\right) \cdot d\vec{\ell} \\ &= -\frac{\partial}{\partial t}\oint_S \vec{\mathbf{B}} \cdot d\vec{\mathbf{S}} - \oint_C \vec{\mathbf{B}} \cdot \left(\vec{\mathbf{v}} \times d\vec{\ell}\right). \end{aligned} \tag{5.41}$$

Now $\vec{\mathbf{v}}$ is the velocity of the charge carriers in the loop C and note that if the loop itself is not moving, then $\vec{\mathbf{v}}$ is in the same direction as $d\vec{\boldsymbol{\ell}}$ so that $\vec{\mathbf{v}} \times d\vec{\boldsymbol{\ell}} = 0$. In general, with γ the angle between $\vec{\mathbf{v}}$ and $d\vec{\boldsymbol{\ell}}$,

$$| \vec{\mathbf{v}} \times d\vec{\boldsymbol{\ell}} |= v\sin(\gamma)d\ell = v_\perp d\ell$$

where $v_\perp = v\sin(\gamma)$ is the velocity component perpendicular to $d\vec{\boldsymbol{\ell}}$. The quantity $| \vec{\mathbf{v}}\times d\vec{\boldsymbol{\ell}} |$ represents a *change* in the flux surface as is now explained.

To fix ideas, consider a single loop on the rotor of a motor as shown in Figure 5.17.

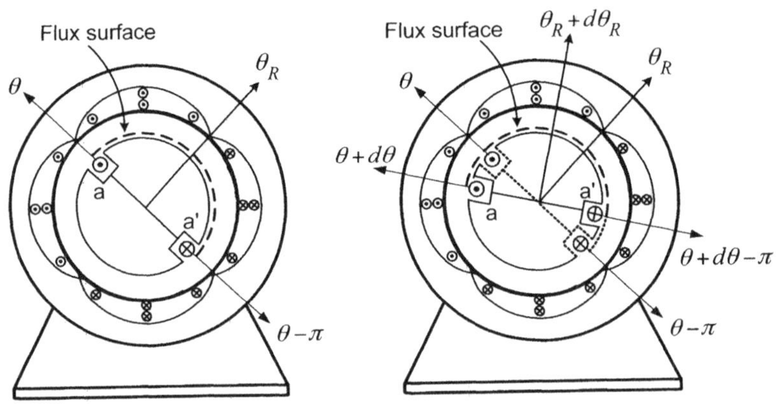

FIGURE 5.17. (a) Flux surface on a moving loop with side a at θ. (b) Loop rotates with side a going from θ to $\theta + d\theta$.

Consider that in the time dt, side a of the loop rotates from $\theta = \theta_R + \pi/2$ to $\theta + d\theta = \theta_R + \pi/2 + d\theta_R$ (with side a' going from $\theta - \pi$ to $\theta + d\theta - \pi$, $d\theta/dt = \omega_R$). With $\vec{\mathbf{B}}$ denoting the magnetic field produced by the stator currents, the quantity

$$\begin{aligned}
dt \oint_C \vec{\mathbf{B}} \cdot \left(\vec{\mathbf{v}} \times d\vec{\boldsymbol{\ell}}\right) &= dt \int_{\text{side } a} B_r(r_R, \theta, t)\hat{\mathbf{r}} \cdot \left(\frac{\ell_2}{2}\omega_R\hat{\boldsymbol{\theta}} \times d\ell\hat{\mathbf{z}}\right) \\
&\quad + dt \int_{\text{side } a'} B_r(r_R, \theta - \pi, t)\hat{\mathbf{r}} \cdot \left(\frac{\ell_2}{2}\omega_R\hat{\boldsymbol{\theta}} \times (-d\ell)\hat{\mathbf{z}}\right) \\
= \int_{\text{side } a} & B_r(r_R, \theta, t)\hat{\mathbf{r}} \cdot \left(\omega_R dt\frac{\ell_2}{2}d\ell\hat{\mathbf{r}}\right) \\
+ \int_{\text{side } a'} & B_r(r_R, \theta - \pi, t)\hat{\mathbf{r}} \cdot \left(-\omega_R dt\frac{\ell_2}{2}d\ell\hat{\mathbf{r}}\right) \\
= \int_{\text{side } a} & \left(B_r(r_R, \theta, t)\frac{\ell_2}{2}\omega_R dt\right) d\ell + \int_{\text{side } a'} \left(-B_r(r_R, \theta - \pi, t)\frac{\ell_2}{2}\omega_R dt\right) d\ell \\
= B_r(r_R, & \theta, t)\ell_1\frac{\ell_2}{2}d\theta - B_r(r_R, \theta - \pi, t)\ell_1\frac{\ell_2}{2}d\theta
\end{aligned}$$

where $d\theta = \omega_R dt$ was used in the last line. Then $B_r(r_R, \theta, t)\ell_1(\ell_2/2)d\theta$ is the additional flux due to the surface area change as side a moved from θ to $\theta + d\theta$ while $-B_r(r_R, \theta - \pi, t)\ell_1(\ell_2/2)d\theta$ is the decrease in flux due to the surface area change as side a' moved from $\theta - \pi$ to $\theta + d\theta - \pi$. That is, the quantity

$$dt \oint_C \left(\vec{\mathbf{v}} \times d\vec{\ell}\right) \cdot \vec{\mathbf{B}} = B_r(r_R, \theta, t)\ell_1 \frac{\ell_2}{2} d\theta - B_r(r_R, \theta - \pi, t)\ell_1 \frac{\ell_2}{2} d\theta \quad (5.42)$$

is the *change in flux* $d\phi$ in the rotor loop due to the rotation of the loop through the angle $d\theta$. In more detail,

$$\phi(\theta, t) \triangleq \oint_S \vec{\mathbf{B}} \cdot d\vec{\mathbf{S}} = \int_{\theta-\pi}^{\theta} \int_0^{\ell_1} B_r(r_R, \theta', t)(\ell_2/2)d\theta'$$

so that

$$\begin{aligned}
\frac{d\phi}{dt} &= \frac{\partial \phi(\theta, t)}{\partial t} + \frac{\partial \phi(\theta, t)}{\partial \theta}\frac{d\theta}{dt} \\
&= \frac{\partial \phi(\theta, t)}{\partial t} + \frac{\partial}{\partial \theta}\left(\int_{\theta-\pi}^{\theta} \int_0^{\ell_1} B_r(r_R, \theta', t)(\ell_2/2)d\theta'\right)\frac{d\theta}{dt} \\
&= \frac{\partial}{\partial t}\oint_S \vec{\mathbf{B}} \cdot d\vec{\mathbf{S}} + \left(B_r(r_R, \theta, t)\ell_1 \frac{\ell_2}{2} - B_r(r_R, \theta - \pi, t)\ell_1 \frac{\ell_2}{2}\right)\frac{d\theta}{dt}
\end{aligned}$$

where the Liebniz rule for differentiation was used. Consequently, by comparison with (5.42), it follows that

$$\oint_C \left(\vec{\mathbf{v}} \times d\vec{\ell}\right) \cdot \vec{\mathbf{B}} = \frac{\partial \phi(\theta, t)}{\partial \theta}\frac{d\theta}{dt}$$

is the rate of change of flux due to the motion of the loop while

$$\frac{\partial}{\partial t}\oint_S \vec{\mathbf{B}} \cdot d\vec{\mathbf{S}} = \frac{\partial \phi(\theta, t)}{\partial t}$$

is the rate of change of flux due to the time-varying magnetic field $\vec{\mathbf{B}}$. By Faraday's law, the induced voltage in the loop is given by

$$\begin{aligned}
\xi &= -\frac{d\phi}{dt} = -\frac{\partial \phi(\theta, t)}{\partial t} - \frac{\partial \phi(\theta, t)}{\partial \theta}\frac{d\theta}{dt} = -\frac{\partial}{\partial t}\oint_S \vec{\mathbf{B}} \cdot d\vec{\mathbf{S}} - \oint_C \left(\vec{\mathbf{v}} \times d\vec{\ell}\right) \cdot \vec{\mathbf{B}} \\
&= \oint_C (\vec{\mathbf{E}} + \vec{\mathbf{v}} \times \vec{\mathbf{B}}) \cdot d\vec{\ell}
\end{aligned}$$

where (5.41) was used.

5.4.4 *Induction Machine in the Synchronous Coordinate System*

Up to this point, the electric and magnetic fields have been viewed from the reference frame of the stator. Often in electric machine theory, concepts are discussed from the point of view of a *rotating* reference frame such as that of the stator's rotating magnetic field which rotates at angular speed ω_S or that of the rotor which rotates at angular speed ω_R. It turns out that the electric and magnetic fields *change* from one moving coordinate system to another. To understand the changes, the microscopic viewpoint is repeated in a coordinate system that rotates with the rotating magnetic field produced by the stator currents. (Problem 17 considers the microscopic point of view in the rotor coordinate system.)

Figure 5.18 shows the rotating stator magnetic field $\vec{\mathbf{B}}_S$. The observer is face down and rotates with the stator magnetic field.

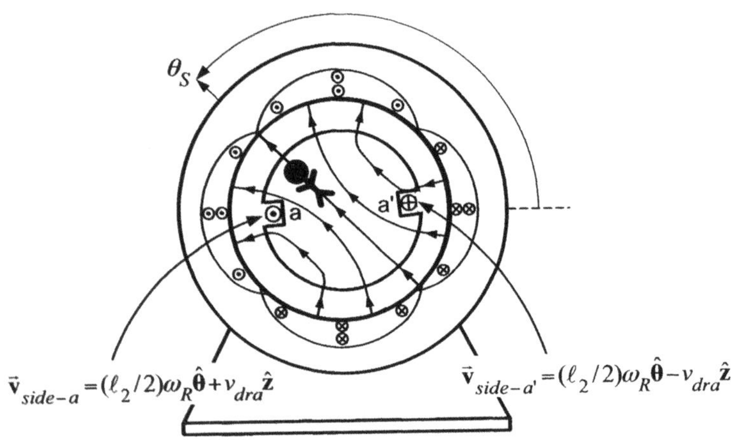

FIGURE 5.18. An observer in the synchronous coordinate system that rotates with $\vec{\mathbf{B}}_S$. The rotating magnetic field produced by the stator currents has angular speed ω_S while the rotor has angular speed ω_R.

In any given reference/coordinate system, the Lorentz force on a charge carrier is given by

$$\vec{\mathbf{F}} = q\left(\vec{\mathbf{E}} + \vec{\mathbf{v}}\times\vec{\mathbf{B}}\right)$$

where $\vec{\mathbf{E}}$ and $\vec{\mathbf{B}}$ are the electric and magnetic fields measured in the particular reference/coordinate system and $\vec{\mathbf{v}}$ is the velocity of the charge carrier as measured in the same coordinate system.

Recall that, in the coordinate system fixed to stator, the electric and magnetic fields in the air gap of the motor produced by the stator currents are

$$\begin{aligned}\vec{\mathbf{E}}_S(\theta,t) &= E_{Sz}(\theta,t)\hat{\mathbf{z}} = \frac{\mu_0\ell_2 N_S I_S}{4g}\omega_S\cos(\theta-\omega_S t)\hat{\mathbf{z}}\\ \vec{\mathbf{B}}_S(r,\theta,t) &= B_{Sr}(r,\theta,t)\hat{\mathbf{r}} = \frac{\mu_0\ell_2 N_S I_S}{4gr}\cos(\theta-\omega_S t)\hat{\mathbf{r}}\end{aligned} \tag{5.43}$$

and the velocity of the charge carriers in the axial sides of rotor loop a is

$$\vec{\mathbf{v}}_{Ra} = \begin{cases} v_{dra}\hat{\mathbf{z}} + v_t\hat{\boldsymbol{\theta}} & \text{side } a\\ -v_{dra}\hat{\mathbf{z}} + v_t\hat{\boldsymbol{\theta}} & \text{side } a'\end{cases}$$

where $v_t = (\ell_2/2)\omega_R$.

Consider a jump into a coordinate system moving (rotating) with the magnetic field $\vec{\mathbf{B}}_S$, that is, a coordinate system that rotates at angular speed ω_S. This referred to as the *synchronous coordinate* coordinate system and is denoted here as the primed (′) coordinate system. How do the above electric and magnetic fields transform (change) when one goes to the rotating (primed) coordinate system? Consider Figure 5.19 depicting two coordinate systems in which the primed (′) coordinate system has velocity $\vec{\mathbf{v}}' = v'\hat{\mathbf{y}}$ with respect to the unprimed coordinate system. According to the Special Theory of Relativity [35][36], the fields transform according to

$$\vec{\mathbf{E}}'_{||} = \vec{\mathbf{E}}_{||} \qquad \vec{\mathbf{B}}'_{||} = \vec{\mathbf{B}}_{||}$$

$$\vec{\mathbf{E}}'_{\perp} = \gamma\left(\vec{\mathbf{E}}_{\perp} + \vec{\mathbf{v}}'\times\vec{\mathbf{B}}_{\perp}\right) \quad \vec{\mathbf{B}}'_{\perp} = \gamma\left(\vec{\mathbf{B}}_{\perp} + (v'/c^2)\times\vec{\mathbf{E}}_{\perp}\right)$$

where $\gamma \triangleq 1/\sqrt{1-(v'/c)^2}$, c is the speed of light, $v' = |\vec{\mathbf{v}}'|$, $\vec{\mathbf{E}}_{||}$ and $\vec{\mathbf{B}}_{||}$ are, respectively, the electric and magnetic fields parallel to the motion (i.e., in the y direction in Figure 5.19) and finally, $\vec{\mathbf{E}}_{\perp}$ and $\vec{\mathbf{B}}_{\perp}$ are, respectively, the electric and magnetic fields perpendicular to the motion (i.e., in an x-z plane of Figure 5.19). Similar comments apply to $\vec{\mathbf{E}}'_{||}$, $\vec{\mathbf{B}}'_{||}$, $\vec{\mathbf{E}}'_{\perp}$, and $\vec{\mathbf{B}}'_{\perp}$.

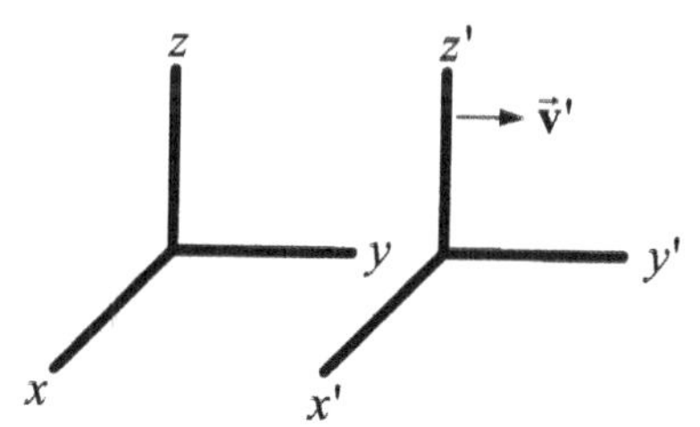

FIGURE 5.19. Two coordinate systems moving with respect to one another.

For the application to electric machines where $(v'/c)^2 \ll 1$, it follows that $\gamma = 1$ can be used as an excellent approximation and the above transformation reduces to

$$\begin{array}{ll} \vec{\mathbf{E}}'_{\|} = \vec{\mathbf{E}}_{\|} & \vec{\mathbf{B}}'_{\|} = \vec{\mathbf{B}}_{\|} \\ \vec{\mathbf{E}}'_{\perp} = \vec{\mathbf{E}}_{\perp} + \vec{\mathbf{v}}' \times \vec{\mathbf{B}}_{\perp} & \vec{\mathbf{B}}'_{\perp} = \vec{\mathbf{B}}_{\perp}. \end{array} \quad (5.44)$$

Returning to the induction motor and the rotating coordinate system of the stator magnetic field, the coordinate transformation from the stationary coordinate system (unprimed) to the rotating coordinate system (primed) of the rotating stator magnetic field is

$$\begin{bmatrix} x' \\ y' \\ z' \end{bmatrix} = \begin{bmatrix} \cos(\omega_S t) & \sin(\omega_S t) & 0 \\ -\sin(\omega_S t) & \cos(\omega_S t) & 0 \\ 0 & 0 & 1 \end{bmatrix} \begin{bmatrix} x \\ y \\ z \end{bmatrix}.$$

In cylindrical coordinates, the transformation is simply

$$\begin{aligned} r' &= r \\ \theta' &= \theta - \omega_S t \\ z' &= z. \end{aligned}$$

Remark Note that a *stationary* point (r', θ') in this rotating coordinate system has velocity $\vec{\mathbf{v}} = r'\omega_S\hat{\boldsymbol{\theta}}$ in the (original unprimed) stator coordinate system. Note also that in this rotating coordinate system, the rotor's angular speed ω'_R is given by $\omega'_R = -(\omega_S - \omega_R) < 0$. This is simply a consequence of the fact that the coordinate system rotating with the stator magnetic field is rotating *faster* than the rotor.

The first step is to find the electric and magnetic fields in this rotating coordinate system. For the induction motor, the fields perpendicular and parallel to the motion are given by

$$\begin{aligned} \vec{\mathbf{E}}_{\perp} &= \vec{\mathbf{E}}_S \\ \vec{\mathbf{E}}_{\|} &= \mathbf{0} \\ \vec{\mathbf{B}}_{\perp} &= \vec{\mathbf{B}}_S \\ \vec{\mathbf{B}}_{\|} &= \mathbf{0} \end{aligned}$$

where $\vec{\mathbf{E}}_S$ and $\vec{\mathbf{B}}_S$ are given in (5.43). At the surface of the rotor, the new coordinate system has velocity $\vec{\mathbf{v}}' = (\ell_2/2)\omega_S\hat{\boldsymbol{\theta}}$ with respect to the original coordinate system so that the electric and magnetic fields in the rotating coordinate system are given by

$$\begin{aligned}\vec{\mathbf{E}}'_S &= \vec{\mathbf{E}}_S + \vec{\mathbf{v}}' \times \vec{\mathbf{B}}_S = E_{Sz}(\theta, t)\hat{\mathbf{z}} + (\ell_2/2)\omega_S \hat{\boldsymbol{\theta}} \times B_{Sr}\hat{\mathbf{r}} \\ &= \frac{\mu_0 \ell_2 N_S I_S}{4g}\omega_S \cos(\theta - \omega_S t)\hat{\mathbf{z}} - (\ell_2/2)\omega_S \frac{\mu_0 \ell_2 N_S I_S}{4g(\ell_2/2)}\cos(\theta - \omega_S t)\hat{\mathbf{z}} \\ &= 0\end{aligned}$$

$$\vec{\mathbf{B}}'_S = B'_{Sr}(r', \theta')\hat{\mathbf{r}} = \vec{\mathbf{B}}_{S}\big|_{r=r', \theta=\theta'+\omega_S t} = \frac{\mu_0 \ell_2 N_S I_S}{4gr'}\cos(\theta')\hat{\mathbf{r}}.$$

The fact that the axial electric field in the rotating coordinate system is zero could have been anticipated. That is, to an observer in the rotating coordinate system, the magnetic field is given by $\vec{\mathbf{B}}'_S = (\mu_0 \ell_2 N_S I_S / 4gr')\cos(\theta')\hat{\mathbf{r}}$ which is *static* in time. Faraday's law $\nabla \times \vec{\mathbf{E}}'_S = -\partial \vec{\mathbf{B}}'_S/\partial t = 0$ then gives $\vec{\mathbf{E}}'_S = \mathbf{0}$ as the solution.

In the rotating coordinate system, the Lorentz force on the charge carriers in rotor loop a is

$$\vec{\mathbf{F}}' = q(\vec{\mathbf{E}}'_S + \vec{\mathbf{v}}'_{Ra} \times \vec{\mathbf{B}}'_S) = q\vec{\mathbf{v}}'_{Ra} \times \vec{\mathbf{B}}'_S$$

where, with $\omega'_R = -(\omega_S - \omega_R)$,

$$\vec{\mathbf{v}}'_{Ra} = \begin{cases} (\ell_2/2)\omega'_R \hat{\boldsymbol{\theta}} + v_{dra}\hat{\mathbf{z}} & \text{side } a \\ (\ell_2/2)\omega'_R \hat{\boldsymbol{\theta}} - v_{dra}\hat{\mathbf{z}} & \text{side } a' \end{cases}$$

is the total velocity of the charge carriers in the axial sides of rotor loop a with respect to the *rotating* coordinate system.

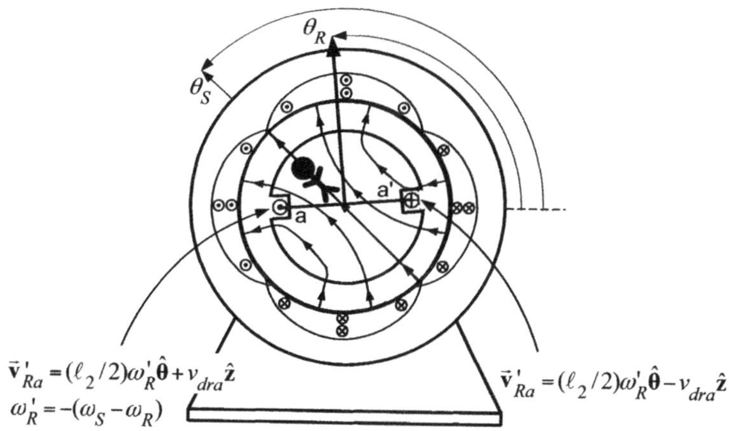

FIGURE 5.20. In the synchronous coordinate system, the rotor has angular speed $\omega'_R \hat{\boldsymbol{\theta}} = -(\omega_S - \omega_R)\hat{\boldsymbol{\theta}}$. Rotor loop b is not drawn for clarity.

Then, with $\vec{\mathbf{B}}'_S = \vec{\mathbf{B}}_S = B_{Sr}\hat{\mathbf{r}}$, $\theta' = \theta_R + \pi/2 - \omega_S t = \omega'_R t + \pi/2$ on side a of the rotor and $\theta' = \theta_R - \pi/2 - \omega_S t = \omega'_R t - \pi/2$ on side a' of the rotor,

$$\vec{\mathbf{v}}'_{Ra} \times \vec{\mathbf{B}}'_S =$$
$$\begin{cases} +v_{dra} B_{Sr}(\ell_2/2, \omega'_R t + \pi/2)\hat{\boldsymbol{\theta}} - (\ell_2/2)\omega'_R B_{Sr}(\ell_2/2, \omega'_R t + \pi/2)\hat{\mathbf{z}} & \text{side } a \\ -v_{dra} B_{Sr}(\ell_2/2, \omega'_R t - \pi/2)\hat{\boldsymbol{\theta}} - (\ell_2/2)\omega'_R B_{Sr}(\ell_2/2, \omega'_R t - \pi/2)\hat{\mathbf{z}} & \text{side } a'. \end{cases}$$

Substituting in the expressions

$$\begin{aligned} B_{Sr}(\ell_2/2, \omega'_R t + \pi/2) &= \frac{\mu_0 \ell_2 N_S I_S}{4g(\ell_2/2)} \cos(\omega'_R t + \pi/2) = -\frac{\mu_0 N_S I_S}{2g} \sin(\omega'_R t) \\ B_{Sr}(\ell_2/2, \omega'_R t - \pi/2) &= \frac{\mu_0 \ell_2 N_S I_S}{4g(\ell_2/2)} \cos(\omega'_R t - \pi/2) = +\frac{\mu_0 N_S I_S}{2g} \sin(\omega'_R t) \end{aligned}$$

results in

$$\begin{aligned} \vec{\mathbf{F}}'_{\text{side } a}/q &= \vec{\mathbf{v}}'_{Ra} \times \vec{\mathbf{B}}'_S \\ &= -v_{dra} \frac{\mu_0 N_S I_S}{2g} \sin(\omega'_R t)\hat{\boldsymbol{\theta}} + (\ell_2/2)\omega'_R \frac{\mu_0 N_S I_S}{2g} \sin(\omega'_R t)\hat{\mathbf{z}} \end{aligned}$$

$$\begin{aligned} \vec{\mathbf{F}}'_{\text{side } a'}/q &= \vec{\mathbf{v}}'_{Ra} \times \vec{\mathbf{B}}'_S \\ &= -v_{dra} \frac{\mu_0 N_S I_S}{2g} \sin(\omega'_R t)\hat{\boldsymbol{\theta}} - (\ell_2/2)\omega'_R \frac{\mu_0 N_S I_S}{2g} \sin(\omega'_R t)\hat{\mathbf{z}}. \end{aligned}$$

With $\omega'_R = -(\omega_S - \omega_R)$, it is seen that the force per unit charge $\vec{\mathbf{F}}'/q = \vec{\mathbf{v}}'_{Ra} \times \vec{\mathbf{B}}'_S$ in this coordinate system equals that obtained in the original coordinates system given in (5.30). This is as it must be since the forces in the two coordinate systems must be equal by the principle of relativity (both Newton's and Einstein's).

The total torque on rotor loop a is then

$$\begin{aligned} \vec{\boldsymbol{\tau}}'_{Ra} &= 2(qNS\ell_1)(-v_{dra}) \frac{\mu_0 N_S I_S}{2g} \sin(\omega'_R t)(\ell_2/2)(\hat{\mathbf{r}} \times \hat{\boldsymbol{\theta}}) \\ &= -i_{Ra} \frac{\mu_0 \ell_1 \ell_2 N_S I_S}{2g} \sin(\omega'_R t)\hat{\mathbf{z}} \end{aligned}$$

where $i_{Ra} = qNSv_{dra}$. Note that, as $\omega'_R = -(\omega_S - \omega_R)$, we may rewrite this as

$$\tau'_{Ra} = i_{Ra} \frac{\mu_0 \ell_1 \ell_2 N_S I_S}{2g} \sin\Big((\omega_S - \omega_R)t\Big)$$

which is the same expression as in (5.6) and (5.32).

Of course, there is a similar expression for the torque on rotor loop b given by

$$\vec{\boldsymbol{\tau}}'_{Rb} = i_{Rb} \frac{\mu_0 \ell_1 \ell_2 N_S I_S}{2g} \cos(\omega'_R t)\hat{\mathbf{z}} = i_{Rb} \frac{\mu_0 \ell_1 \ell_2 N_S I_S}{2g} \cos\Big((\omega_S - \omega_R)t\Big)\hat{\mathbf{z}}.$$

It has been shown that the axial magnetic force per unit charge is given by

$$\left(\vec{\mathbf{v}}'_{Ra} \times \vec{\mathbf{B}}'_S\right)_z = \begin{cases} +\omega'_R \dfrac{\mu_0 \ell_2 N_S I_S}{4g} \sin(\omega'_R t)\hat{\mathbf{z}} & \text{for side } a \\ \\ -\omega'_R \dfrac{\mu_0 \ell_2 N_S I_S}{4g} \sin(\omega'_R t)\hat{\mathbf{z}} & \text{for side } a' \end{cases}$$

and, as the electric field $\vec{\mathbf{E}}'_S = \mathbf{0}$, this is also the total axial force per unit charge $(\vec{\mathbf{F}}/q)_z$. As a result, with

$$d\vec{\ell} = \begin{cases} dz\hat{\mathbf{z}} & \text{for side } a \\ -dz\hat{\mathbf{z}} & \text{for side } a' \end{cases}$$

the emf induced in rotor loop a is simply

$$\begin{aligned} \xi'_{Ra} &= \oint_{\text{rotor loop } a} (\vec{\mathbf{F}}/q)_z \cdot d\vec{\ell} = \ell_1 \omega'_R \frac{\mu_0 \ell_2 N_S I_S}{2g} \sin(\omega'_R t) \\ &= (\omega_S - \omega_R) \frac{\mu_0 \ell_1 \ell_2 N_S I_S}{2g} \sin((\omega_S - \omega_R)t). \end{aligned} \tag{5.45}$$

Similarly,

$$\begin{aligned} \xi'_{Rb} &= \omega'_R \frac{\mu_0 \ell_1 \ell_2 N_S I_S}{2g} \cos(\omega'_R t) \\ &= -(\omega_S - \omega_R) \frac{\mu_0 \ell_1 \ell_2 N_S I_S}{2g} \cos((\omega_S - \omega_R)t). \end{aligned} \tag{5.46}$$

Then, with

$$i_{Ra} = \xi'_{Ra}/R_R \tag{5.47}$$
$$i_{Rb} = \xi'_{Rb}/R_R \tag{5.48}$$

the total torque on the rotor in the primed coordinate system is given by

$$\begin{aligned} \tau' &= \tau'_{Ra} + \tau'_{Rb} = -\omega'_R \left(\frac{\mu_0 \ell_1 \ell_2 N_S I_S}{2g}\right)^2 \frac{1}{R_R} \\ &= \frac{1}{R_R} \left(\frac{\mu_0 \ell_1 \ell_2 N_S I_S}{2g}\right)^2 (\omega_S - \omega_R). \end{aligned}$$

This is the same as the expression (5.9) derived in the stationary coordinate system.

Electromechanical Energy Conversion

It has been shown that, to the observer in Figure 5.20 in the rotating coordinate system, the torque is given by

$$\tau' = -\omega'_R \left(\frac{\mu_0 \ell_1 \ell_2 N_S I_S}{2g}\right)^2 \frac{1}{R_R} > 0$$

However, to this same observer, the angular speed of the rotor is

$$\omega'_R = -(\omega_S - \omega_R) < 0$$

so that power delivered to the rotor is

$$\tau'\omega'_R = -(\omega_S - \omega_R)^2 \left(\frac{\mu_0 \ell_1 \ell_2 N_S I_S}{2g}\right)^2 \frac{1}{R_R} < 0.$$

That is, in the rotating coordinate system of the stator magnetic field (Figure 5.20), the tangential velocity of the rotor loop is in the $-\hat{\boldsymbol{\theta}}$ direction while the above torque is pushing the rotor in the $+\hat{\boldsymbol{\theta}}$ direction. The torque is *slowing* the rotor down from the point of view of the observer *in* the rotating coordinate system of Figure 5.20 and thus taking out mechanical power! Where is this mechanical power going? From the point of view of someone in the rotating coordinate system, this mechanical energy is being converted into electrical power in the rotor. That is, the electric power produced is

$$i_{Ra}\xi'_{Ra} + i_{Rb}\xi'_{Rb} = (\omega_S - \omega_R)^2 \left(\frac{\mu_0 \ell_1 \ell_2 N_S I_S}{2g}\right)^2 \frac{1}{R_R}$$

which is easily verified by substitution of the expressions (5.45), (5.46), (5.47), and (5.48) for the emfs and currents. In summary, to someone in the rotating coordinate system, the induction machine looks like a *generator* rather than a motor.

Magnetic Force and Work

Recall, in the rotating coordinate system of Figure 5.20, the electric field $\vec{\mathbf{E}}'_S = \mathbf{0}$ so that the magnetic force per unit charge on the axial sides of the rotor is then

$$\vec{\mathbf{F}}'/q = \vec{\mathbf{E}}'_S + \vec{\mathbf{v}}'_{Ra} \times \vec{\mathbf{B}}'_S = \vec{\mathbf{v}}'_{Ra} \times \vec{\mathbf{B}}'_S$$

where

$$\vec{\mathbf{v}}'_{Ra} = \begin{cases} (\ell_2/2)\omega'_R\hat{\boldsymbol{\theta}} + v_{dra}\hat{\mathbf{z}} & \text{side } a \\ (\ell_2/2)\omega'_R\hat{\boldsymbol{\theta}} - v_{dra}\hat{\mathbf{z}} & \text{side } a'. \end{cases}$$

Explicitly,

$$\vec{\mathbf{v}}'_{Ra} \times \vec{\mathbf{B}}'_S = \begin{cases} -v_{dra}\dfrac{\mu_0 N_S I_S}{2g}\sin(\omega'_R t)\hat{\boldsymbol{\theta}} + \omega'_R \dfrac{\mu_0 \ell_2 N_S I_S}{4g}\sin(\omega'_R t)\hat{\mathbf{z}} & \text{side } a \\[2ex] -v_{dra}\dfrac{\mu_0 N_S I_S}{2g}\sin(\omega'_R t)\hat{\boldsymbol{\theta}} - \omega'_R \dfrac{\mu_0 \ell_2 N_S I_S}{4g}\sin(\omega'_R t)\hat{\mathbf{z}} & \text{side } a'. \end{cases}$$

The power per unit charge done by the (purely) magnetic force $\vec{\mathbf{F}}' = q(\vec{\mathbf{v}}'_{Ra} \times \vec{\mathbf{B}}'_S)$ is then

$$(\vec{\mathbf{F}}'/q) \cdot \vec{\mathbf{v}}'_{Ra} = (\vec{\mathbf{v}}'_{Ra} \times \vec{\mathbf{B}}'_S) \cdot \vec{\mathbf{v}}'_{Ra} \equiv \mathbf{0}$$

as it must, since the magnetic force on a moving charge is always orthogonal to the velocity of the charge and thus cannot change the energy of the charge carriers. How is the energy transfer in the rotating coordinate system accomplished from mechanical to electrical? Consider another observer in the rotating coordinate system as shown in Figure 5.21(a).

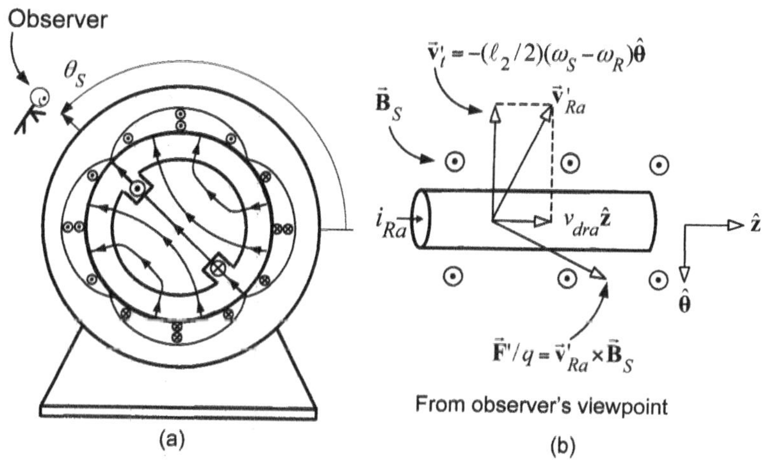

FIGURE 5.21. (a) View by an observer in the synchronous coordinate system. (b) The magnetic force on the rotor loop from the observer's point of view.

The observer is still rotating with the stator magnetic field at angular speed ω_S. The magnetic force opposes the rotor angular velocity and produces the current in the rotor as seen in Figure 5.21(b). From the point of view of the rotating coordinate system, the magnetic force is the mechanism which converts the mechanical (kinetic) energy into electrical energy while not adding any energy (i.e., doing any work). In more detail, the power delivered by the magnetic force on the two sides of rotor loop a is

$$\begin{aligned}
&2(qNS\ell_1)\vec{\mathbf{F}}' \cdot \vec{\mathbf{v}}'_{Ra} = 2(qNS\ell_1)(\vec{\mathbf{v}}'_{Ra} \times \vec{\mathbf{B}}'_S) \cdot \vec{\mathbf{v}}'_{Ra}\\
&= 2(qNS\ell_1)\left(-v_{dra}\frac{\mu_0 N_S I_S}{2g}\sin(\omega'_R t)\hat{\boldsymbol{\theta}} + \omega'_R\frac{\mu_0 \ell_2 N_S I_S}{4g}\sin(\omega'_R t)\hat{\mathbf{z}}\right)\\
&\quad\cdot\left((\ell_2/2)\omega'_R\hat{\boldsymbol{\theta}} + v_{dra}\hat{\mathbf{z}}\right)\\
&= 2\left(-qNS\ell_1 v_{dra}\frac{\mu_0 N_S I_S}{2g}\sin(\omega'_R t)\hat{\boldsymbol{\theta}}\right)\cdot\left((\ell_2/2)\omega'_R\hat{\boldsymbol{\theta}}\right)\\
&\quad + 2\left(\omega'_R\frac{\mu_0 \ell_2 N_S I_S}{4g}\sin(\omega'_R t)\hat{\mathbf{z}}\right)\cdot\left(qNS\ell_1 v_{dra}\hat{\mathbf{z}}\right)
\end{aligned}$$

Rearranging the right hand side of this last equations gives

$$2(qNS\ell_1)\vec{\mathbf{F}}'\cdot\vec{\mathbf{v}}'_{Ra} = -\left(i_{Ra}\frac{\mu_0\ell_1\ell_2N_SI_S}{2g}\sin(\omega'_Rt)\right)\omega'_R\hat{\boldsymbol{\theta}}\cdot\hat{\boldsymbol{\theta}} + \left(\omega'_R\frac{\mu_0\ell_1\ell_2N_SI_S}{2g}\sin(\omega'_Rt)\right)i_{Ra}\hat{\mathbf{z}}\cdot\hat{\mathbf{z}}. \qquad (5.49)$$

However

$$\vec{\mathbf{F}}'\cdot\vec{\mathbf{v}}'_{Ra} = q(\vec{\mathbf{v}}'_{Ra}\times\vec{\mathbf{B}}'_S)\cdot\vec{\mathbf{v}}'_{Ra}\equiv 0$$

as the magnetic force on a charged particle is always perpendicular to its velocity. The coefficient of $\hat{\boldsymbol{\theta}}\cdot\hat{\boldsymbol{\theta}}$ in (5.49) is $\tau'_{Ra}\omega'_R$ - the mechanical power, while the coefficient of $\hat{\mathbf{z}}\cdot\hat{\mathbf{z}}$ is $\xi'_{Ra}i_{Ra}$ - the electrical power in rotor loop a. As these terms are equal in magnitude, but opposite in sign, it is seen that the magnetic force just converts the mechanical energy to electrical energy, but does not add any energy.

5.4.5 Synchronous Machine

The microscopic viewpoint of the synchronous machine is explored in problems 18 and 19 at the end of the chapter.

5.5 Steady-State Analysis of a Squirrel Cage Induction Motor*

Consider the two-loop induction motor of Figure 5.22 where, as before, $\theta_S(t) = \omega_St$ denotes the angular position of the magnetic axis of the rotating magnetic field produced by the stator currents, $\theta_R(t) = \omega_Rt$ denotes the angular position of the rotor defined as the magnetic axis of rotor loop a (the magnetic axis of rotor loop b is at $\theta_R(t)+\pi/2$), and θ denotes an arbitrary angular position in the air gap.

In order to obtain more torque, additional pairs of loops (each pair $\pi/2$ apart) are placed around the complete periphery of the rotor. For example, Figure 5.23 shows a cross section of a rotor with six rotor loops $a_1-a'_1$, $a_2-a'_2$, $a_3-a'_3$, $b_1-b'_1$, $b_2-b'_2$, and $b_3-b'_3$, which are considered as the three *pairs* of rotor loops $\{a_1-a'_1, b_1-b'_1\}$, $\{a_2-a'_2, b_2-b'_2\}$, and $\{a_3-a'_3, b_3-b'_3\}$. In general, the ith rotor loop pair is located by the center line (magnetic axis) of its rotor loop a which is $\theta_{R_i} = \theta_R + \frac{\pi}{2}\frac{i-1}{N_R}$ for $i = 1, ..., N_R$, where N_R is the number of *pairs* of rotor loops.[6] Phase a of

[6] Each loop has four sides. The term *rotor bar* refers to the sides of a loop which go along the length of the rotor. The other two sides of the loop are referred to as the *end turns*. Consequently, there are two rotor bars per loop and so each pair of loops has four rotor bars. With N_R the number of pairs of *loops*, there are then $4N_R$ rotor bars around the periphery of the rotor.

the first rotor loop pair ($i = 1$) is arbitrarily chosen to be the one whose magnetic axis is used to reference the rotor position θ_R. The induced emfs, the currents, and the resulting torque in each pair of loops are now computed, and then the total torque is found by summing the torques on each rotor loop pair.

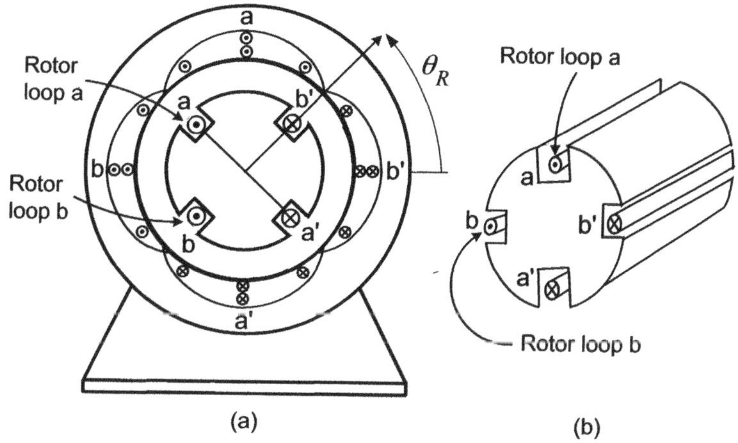

FIGURE 5.22. Induction motor with two single-loop rotor phases a and b.

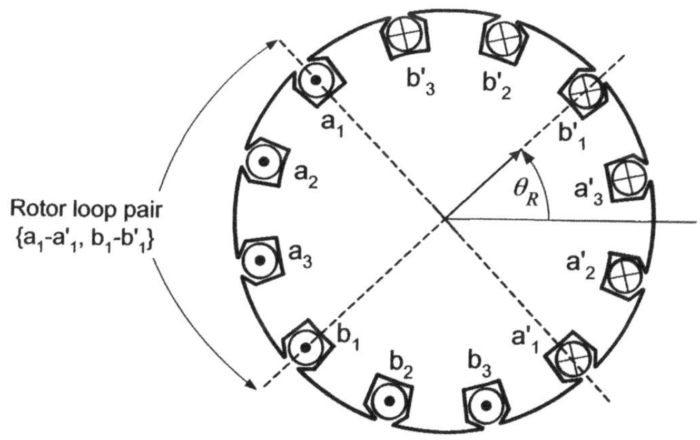

FIGURE 5.23. Rotor with six rotor loops a_1-a_1', a_2-a_2', a_3-a_3', b_1-b_1', b_2-b_2', and b_3-b_3' considered as three pairs of loops $\{a_1-a_1', b_1-b_1'\}$, $\{a_2-a_2', b_2-b_2'\}$, and $\{a_3-a_3', b_3-b_3'\}$. In this figure, there are $N_R = 3$ pairs of rotor loops and θ_R is chosen to coincide with the magnetic axis of rotor loop a_1-a_1'.

5.5.1 Rotor Fluxes, Emfs, and Currents

The *flux* λ_{Ra_i} in phase a of the ith rotor loop pair due to the stator magnetic field $\vec{\mathbf{B}}_S$ is calculated as $\left(\theta_{R_i} = \theta_R + \frac{\pi}{2}\frac{i-1}{N_R}\right)$

$$\begin{aligned}
\lambda_{Ra_i} &= \int_{S_{a_i}} \vec{\mathbf{B}}_S \cdot d\vec{\mathbf{S}} \\
&= \int_{\theta_{R_i}-\pi/2}^{\theta_{R_i}+\pi/2} \frac{\mu_0 \ell_2 N_S I_S}{4gr}\Big|_{r=\ell_2/2} \cos(\theta - \theta_S)\hat{\mathbf{r}}\cdot (\ell_1(\ell_2/2)d\theta\hat{\mathbf{r}}) \\
&= \frac{\mu_0 \ell_1 \ell_2 N_S I_S}{4g} \sin(\theta - \theta_S)\Big|_{\theta_{R_i}-\pi/2}^{\theta_{R_i}+\pi/2} \\
&= \frac{\mu_0 \ell_1 \ell_2 N_S I_S}{2g} \cos\left(\theta_S - \left(\theta_R + \frac{\pi}{2}\frac{i-1}{n}\right)\right). \qquad (5.50)
\end{aligned}$$

By Faraday's law, the induced *electromotive force* (*emf*) in phase a of ith rotor loop pair is given by

$$\begin{aligned}
\xi_{Ra_i} &= -\frac{d\lambda_{Ra_i}}{dt} = -\frac{d}{dt}\frac{\mu_0 \ell_1 \ell_2 N_S I_S}{2g}\cos(\theta_S - \theta_{R_i}) \\
&= \frac{\mu_0 \ell_1 \ell_2 N_S I_S}{2g}(\omega_S - \omega_R)\sin\left(\theta_S - \left(\theta_R + \frac{\pi}{2}\frac{i-1}{n}\right)\right) \qquad (5.51)
\end{aligned}$$

where $\omega_R = d\theta_R/dt$.

Similarly, the flux λ_{Rb_i} in phase b of the ith rotor loop pair is

$$\lambda_{Rb_i} = \int_{S_{b_i}} \vec{\mathbf{B}}_S \cdot d\vec{\mathbf{S}} = \frac{\mu_0 \ell_1 \ell_2 N_S I_S}{2g}\sin\left(\theta_S - \left(\theta_R + \frac{\pi}{2}\frac{i-1}{N_R}\right)\right)$$

and the corresponding induced *emf* ξ_{Rb_i} is

$$\xi_{Rb_i} = -\frac{d\lambda_{Rb_i}}{dt} = -\frac{\mu_0 \ell_1 \ell_2 N_S I_S}{2g}(\omega_S - \omega_R)\cos\left(\theta_S - \left(\theta_R + \frac{\pi}{2}\frac{i-1}{N_R}\right)\right) \qquad (5.52)$$

where $\omega_R = d\theta_R/dt$. Assuming each rotor loop has an inductance L_R and resistance R_R, the equations describing the current dynamics are

$$\begin{aligned}
&L_R\frac{di_{Ra_i}}{dt} + R_R i_{Ra_i} = \xi_{Ra_i}, \\
&\qquad \xi_{Ra_i} = \frac{\mu_0 \ell_1 \ell_2 N_S I_S}{2g}(\omega_S - \omega_R)\sin\left((\omega_S - \omega_R)t - \frac{\pi}{2}\frac{i-1}{N_R}\right)
\end{aligned}$$

and

$$\begin{aligned}
&L_R\frac{di_{Rb_i}}{dt} + R_R i_{Rb_i} = \xi_{Rb_i}, \\
&\qquad \xi_{Rb_i} = -\frac{\mu_0 \ell_1 \ell_2 N_S I_S}{2g}(\omega_S - \omega_R)\cos\left((\omega_S - \omega_R)t - \frac{\pi}{2}\frac{i-1}{N_R}\right).
\end{aligned}$$

A system of the form

$$L\frac{di}{dt} + Ri = A\cos(\omega t + \varphi)$$

with $R/L > 0$ is a stable linear time-invariant system so that its steady-state solution is given by

$$i_{ss}(t) = |G(j\omega)|\, A\cos\left(\omega t + \varphi + \angle G(j\omega)\right) \quad \text{where } G(j\omega) = \frac{1}{R + j\omega L}.$$

So, for $i = 1, ..., N_R$, the *steady-state* solutions for the currents in the ith rotor loop pair are given by

$$\begin{aligned} i_{RaSS_i} &= \frac{(\omega_S - \omega_R)}{\sqrt{(\omega_S - \omega_R)^2 L_R^2 + R_R^2}} \frac{\mu_0 \ell_1 \ell_2 N_S I_S}{2g} \\ &\quad \times \sin\left((\omega_S - \omega_R)t - \frac{\pi}{2}\frac{i-1}{N_R} + \tan^{-1}\left(\frac{(\omega_S - \omega_R)L_R}{R_R}\right)\right) \\ i_{RbSS_i} &= -\frac{(\omega_S - \omega_R)}{\sqrt{(\omega_S - \omega_R)^2 L_R^2 + R_R^2}} \frac{\mu_0 \ell_1 \ell_2 N_S I_S}{2g} \\ &\quad \times \cos\left((\omega_S - \omega_R)t - \frac{\pi}{2}\frac{i-1}{N_R} + \tan^{-1}\left(\frac{(\omega_S - \omega_R)L_R}{R_R}\right)\right). \end{aligned} \tag{5.53}$$

5.5.2 *Rotor Torque*

As previously explained, the rotating magnetic field produced by the stator currents produces a magnetic force on the rotor currents it induced. The magnetic force law is simply $\vec{\mathbf{F}}_{magnetic} = i\vec{\boldsymbol{\ell}}\times\vec{\mathbf{B}}$, where $\ell = |\vec{\boldsymbol{\ell}}|$ is the length of the conductor in the magnetic field, and $\vec{\boldsymbol{\ell}}$ points in the direction of positive current flow. Recalling that $\odot$ means that if $i_{Ra_i} > 0$, the current is coming out of the page, it follows that $\vec{\boldsymbol{\ell}} = \ell_1\hat{\mathbf{z}}$ on side a_i and $\vec{\boldsymbol{\ell}} = -\ell_1\hat{\mathbf{z}}$ on side a_i'. Along with the fact that $\vec{\mathbf{B}}_S(r, \theta+\pi) = -\vec{\mathbf{B}}_S(r,\theta)$, it was shown that the force on each side of the loop is equal (i.e., $\vec{\mathbf{F}}_{a_i} = \vec{\mathbf{F}}_{a_i'}$). Applying the magnetic force law to loop a_i of the ith rotor loop pair, the torque is computed as

$$\vec{\boldsymbol{\tau}}_{a_i} = 2\vec{\mathbf{r}}\times\vec{\mathbf{F}}_{a_i} = 2(\ell_2/2)\hat{\mathbf{r}}\times\vec{\mathbf{F}}_{a_i}$$

where

$$\begin{aligned} \vec{\mathbf{F}}_{a_i} &= i_{Ra_i}\vec{\boldsymbol{\ell}}\times\vec{\mathbf{B}}_S \\ &= i_{Ra_i}\ell_1\hat{\mathbf{z}}\times\frac{\mu_0\ell_2 N_S I_S}{4gr}\Big|_{r=\ell_2/2}\cos(\theta-\theta_S)\big|_{\theta=\theta_{R_i}+\pi/2}\,\hat{\mathbf{r}} \\ &= i_{Ra_i}\frac{\mu_0\ell_1 N_S I_S}{2g}\sin(\theta_S-\theta_{R_i})\hat{\boldsymbol{\theta}}. \end{aligned}$$

The torque is then

$$\vec{\boldsymbol{\tau}}_{a_i} = \frac{\mu_0 \ell_1 \ell_2 N_S I_S}{2g} i_{Ra_i} \sin(\theta_S - \theta_{R_i})\hat{\mathbf{z}}. \tag{5.54}$$

Similarly,

$$\vec{\boldsymbol{\tau}}_{b_i} = 2\vec{\mathbf{r}} \times \vec{\mathbf{F}}_{b_i} = 2(\ell_2/2)\hat{\mathbf{r}} \times \vec{\mathbf{F}}_{b_i}$$

where

$$\begin{aligned}
\vec{\mathbf{F}}_{b_i} &= i_{Rb_i} \vec{\boldsymbol{\ell}} \times \vec{\mathbf{B}}_S \\
&= i_{Rb_i} \ell_1 \hat{\mathbf{z}} \times \frac{\mu_0 \ell_2 N_S I_S}{4gr}\Big|_{r=\ell_2/2} \cos(\theta - \theta_S)\Big|_{\theta=\theta_{R_i}+\pi} \hat{\mathbf{r}} \\
&= -i_{Rb_i} \frac{\mu_0 \ell_1 N_S I_S}{2g} \cos(\theta_S - \theta_{R_i})\hat{\boldsymbol{\theta}}.
\end{aligned}$$

The torque $\vec{\boldsymbol{\tau}}_{b_i}$ becomes

$$\vec{\boldsymbol{\tau}}_{b_i} = -\frac{\mu_0 \ell_1 \ell_2 N_S I_S}{2g} i_{Rb_i} \cos(\theta_S - \theta_{R_i})\hat{\mathbf{z}}.$$

Letting

$$I_R \triangleq \frac{(\omega_S - \omega_R)}{\sqrt{(\omega_S - \omega_R)^2 L_R^2 + R_R^2}} \frac{\mu_0 \ell_1 \ell_2 N_S I_S}{2g} \tag{5.55}$$

in (5.53) and substituting (5.53) into equation (5.54), the total torque on loop $a_i - a_i'$ of the ith rotor loop pair becomes

$$\begin{aligned}
\tau_{Ra_i} &= \frac{\mu_0 \ell_1 \ell_2 N_S I_S}{2g} i_{RaSS_i} \sin\left((\omega_S - \omega_R)t - \frac{\pi}{2}\frac{i-1}{N_R}\right) \\
&= \frac{\mu_0 \ell_1 \ell_2 N_S I_S}{2g} I_R \sin\left((\omega_S - \omega_R)t - \frac{\pi}{2}\frac{i-1}{N_R} + \tan^{-1}\left(\frac{(\omega_S - \omega_R)L_R}{R_R}\right)\right) \\
&\quad \times \sin\left((\omega_S - \omega_R)t - \frac{\pi}{2}\frac{i-1}{N_R}\right)
\end{aligned}$$

Similarly, the total torque on loop $b_i - b_i'$ of the ith rotor loop pair is

$$\begin{aligned}
\tau_{Rb_i} &= -\frac{\mu_0 \ell_1 \ell_2 N_S I_S}{2g} i_{RbSS_i} \cos\left((\omega_S - \omega_R)t - \frac{\pi}{2}\frac{i-1}{N_R}\right) \\
&= \frac{\mu_0 \ell_1 \ell_2 N_S I_S}{2g} I_R \cos\left((\omega_S - \omega_R)t - \frac{\pi}{2}\frac{i-1}{N_R} + \tan^{-1}\left(\frac{(\omega_S - \omega_R)L_R}{R_R}\right)\right) \\
&\quad \times \cos\left((\omega_S - \omega_R)t - \frac{\pi}{2}\frac{i-1}{N_R}\right).
\end{aligned}$$

Using the trigonometric identity $\sin(x)\sin(y) + \cos(x)\cos(y) = \cos(x - y)$ to combine the above results, the *total torque* τ_{R_i} on the ith rotor loop *pair* is given by

$$\tau_{R_i} = \tau_{Ra_i} + \tau_{Rb_i} = \frac{\mu_0 \ell_1 \ell_2 N_S I_S}{2g} I_R \cos\left(\tan^{-1}\left(\frac{(\omega_S - \omega_R)L_R}{R_R}\right)\right).$$

This can be put in a more illuminating form by noting that

$$\cos\left(\tan^{-1}\left(\frac{(\omega_S - \omega_R)L_R}{R_R}\right)\right) = \frac{1}{\sqrt{\left(\frac{(\omega_S - \omega_R)L_R}{R_R}\right)^2 + 1}}$$

and substituting back the expression (5.55) for I_R, the torque can be written as

$$\begin{aligned}
\tau_{R_i} &= \frac{\mu_0 \ell_1 \ell_2 N_S I_S}{2g} I_R \cos\left(\tan^{-1}\left(\frac{(\omega_S - \omega_R)L_R}{R_R}\right)\right) \\
&= \left(\frac{\mu_0 \ell_1 \ell_2 N_S I_S}{2g}\right)^2 \frac{(\omega_S - \omega_R)}{\sqrt{(\omega_S - \omega_R)^2 L_R^2 + R_R^2}} \frac{1}{\sqrt{\left(\frac{(\omega_S - \omega_R)L_R}{R_R}\right)^2 + 1}} \\
&= \left(\frac{\mu_0 \ell_1 \ell_2 N_S I_S}{2g}\right)^2 \frac{1}{L_R} \frac{\frac{(\omega_S - \omega_R)L_R}{R_R}}{\left(\frac{(\omega_S - \omega_R)L_R}{R_R}\right)^2 + 1} \\
&= \left(\frac{\mu_0 \ell_1 \ell_2 N_S I_S}{2g}\right)^2 \frac{1}{L_R} \frac{1}{2} \frac{2}{S/s_p + s_p/S}
\end{aligned}$$

where

$$S \triangleq \frac{\omega_S - \omega_R}{\omega_S} \text{ and } s_p \triangleq \frac{R_R}{\omega_S L_R}.$$

The quantity $S \triangleq \frac{\omega_S - \omega_R}{\omega_S}$ is the *normalized slip*.[7] It is straightforward to show that $\frac{2}{S/s_p + s_p/S}$ has a maximum value of 1 for $S = s_p$ and a minimum value of -1 for $S = -s_p$ so that the torque achieves its maximum (minimum) at $S = s_p (S = -s_p)$.

This derivation shows that, in steady state, each pair of rotor loops has the same torque on it. Consequently, the total torque on the N_R pairs of

[7]Later the quantity $S_p \triangleq \frac{R_R}{\sigma \omega_S L_R}$ will be defined as the *pull out slip* so that $s_p \triangleq \frac{R_R}{\omega_S L_R} = \sigma S_p$ where σ is the so-called *leakage parameter* (also defined later). For the reader familiar with the torque-slip curve of an induction motor, the quantity s_p rather than the pull out slip S_p occurs because impressed currents are used here rather than impressed voltages (see problem 25 of Chapter 7).

loops is

$$\begin{aligned}\tau_R &= \sum_{i=1}^{N_R} \tau_{Ri} = \frac{N_R}{2L_R}\left(\frac{\mu_0 \ell_1 \ell_2 N_S I_S}{2g}\right)^2 \frac{2}{S/s_p + s_p/S} \\ &= \tau_p \frac{2}{S/s_p + s_p/S}\end{aligned}$$

where

$$\tau_p \triangleq \frac{N_R}{2L_R}\left(\frac{\mu_0 \ell_1 \ell_2 N_S I_S}{2g}\right)^2 .$$

For $|S| \ll s_p$, this reduces to

$$\begin{aligned}\tau_R &\approx k(\omega_S - \omega_R) \\ k &\triangleq \frac{N_R}{R_R}\left(\frac{\mu_0 \ell_1 \ell_2 N_S I_S}{2g}\right)^2 .\end{aligned}$$

The analysis was done assuming that each rotor loop was electrically isolated from any other rotor loop. The most common rotor for an induction motor is the *squirrel cage rotor* shown in Figure 5.24. It is simply a set of conducting bars around the periphery of the rotor iron that are connected to conducting end rings at each end. This is in contrast to the rotor of Figure 5.5 where loop $a-a'$ is electrically isolated from loop $b-b'$.

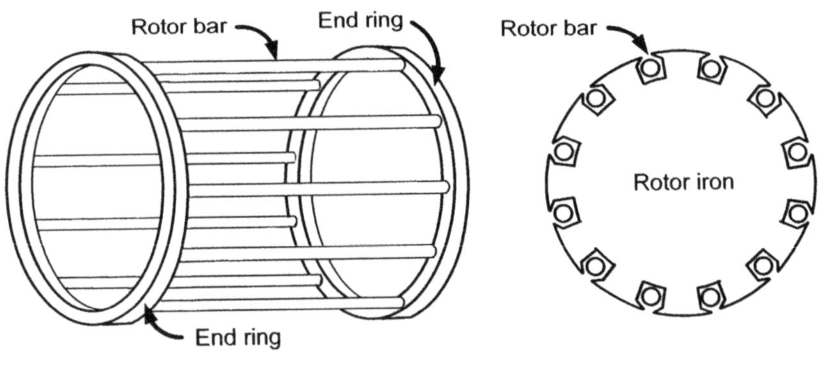

FIGURE 5.24. Squirrel cage rotor.

However, assuming the emf drop across each rotor bar remains the same as in the case when the rotor loops are electrically isolated, the above analysis still applies.[8] That is, consider the squirrel cage to be made up of loops consisting of two rotor bars that are on diametrically opposite sides

[8] This argument is perhaps more convincing from the microscopic point of view. It is shown in an optional section of this chapter that the rotating magnetic field $\vec{\mathbf{B}}_S(r, \theta, t) =$

of the rotor. The flux in each loop depends on the rotor position so that the induced emf in each rotor loop is different. In particular, as one goes from one rotor bar to the next, the emf drop across their lengths are different even though they are connected in parallel. This can be visualized as shown in Figure 5.25 which shows the induced emfs $\xi_{b'_1}$, $\xi_{b'_2}$, $\xi_{b'_3}$, ξ_{a_1}, ξ_{a_2}, ξ_{a_3}, ξ_{b_1}, ξ_{b_2}, ξ_{b_3}, $\xi_{a'_1}$, $\xi_{a'_2}$, $\xi_{a'_3}$ in each rotor bar being in series with the resistance $R_R/2$ of each rotor bar.

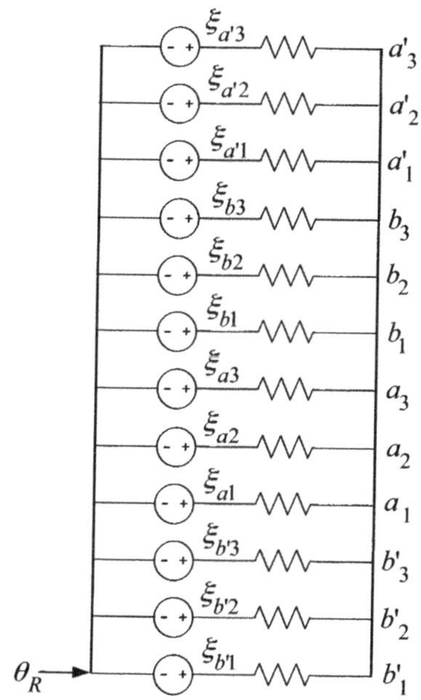

FIGURE 5.25. Equivalent circuit of a squirrel cage rotor. The rotor bars are annotated as in Figure 5.23. Each bar has resistance $R_R/2$ so that the total resistance in a loop consisting of diametrically opposite bars is R_R.

For example, in the two rotor bars a_1, a'_1 (which are on diametrically opposite sides of the rotor), the voltage in the rotor loop made up from

---

$B_r(r,\theta,t)\hat{\mathbf{r}} = \frac{\mu_0 \ell_2 N_S I_S}{4gr}\cos(\theta - \omega_S t)\hat{\mathbf{r}}$ in the air gap produced by the stator currents induces an axial electric field in the air gap (according to $\nabla \times \vec{\mathbf{E}}_S = -\partial \vec{\mathbf{B}}_S/\partial t$) given by $\vec{\mathbf{E}}_S(\theta,t) = E_z(\theta,t)\hat{\mathbf{z}} = \frac{\mu_0 \ell_2 N_S I_S}{4g}\omega_S \cos(\theta - \omega_S t)\hat{\mathbf{z}}$ (see Figure 5.15). The voltage drop across the ith rotor bar produced by this electric field is then $\int_0^{\ell_1}\left(\vec{\mathbf{E}}_S(\theta,t)\Big|_{\theta=\theta_R+\varphi_i}\right)\cdot d\vec{\ell}$ where $\varphi_i = \frac{\pi}{2}\frac{i-1}{N_R}$. One then argues that, as this depends only on the stator quantities, this voltage drop is the same whether or not the rotor bars are electrically connected (shorted) at their ends. The total emf in each bar is this voltage drop plus the back emf $\int_0^{\ell_1}\left(\vec{\mathbf{v}} \times \vec{\mathbf{B}}_S(\theta,t)\Big|_{\theta=\theta_R+\varphi_i}\right)\cdot d\vec{\ell}$ in each rotor bar ($|\vec{\mathbf{v}}| = v = r_R\omega$).

these two sides is[9] $\xi_{a_1} - \xi_{a_1'} = 2\xi_{a_1}$ and this loop has resistance R_R. This emf $\xi_{Ra_1} \triangleq \xi_{a_1} - \xi_{a_1'} = 2\xi_{a_1}$ equals the emf given in equation (5.51) with $i = 1$. Note from Figure 5.23 that the rotor angle θ_R locates the rotor bar b_1' as is also indicated in Figure 5.25.

5.5.3 *Rotor Magnetic Field*

As equation (5.53) shows, the current in the rotor bars of the squirrel cage rotor differs from bar to bar (loop to loop) while the current in each phase of a *sinusoidally wound* rotor loop is the same. Of course, in the squirrel cage rotor, it is arbitrary what is called a phase a loop and what is called a phase b loop. It is more convenient to speak of a current distribution on the surface of the rotor (i.e., in the rotor bars). With $\varphi_i = \frac{\pi}{2}\frac{i-1}{N_R}$, the ith rotor bar is located at $\theta_R(t)+\varphi_i = \omega_R t + \frac{\pi}{2}\frac{i-1}{N_R}$. Combining the expressions (5.53) above for the rotor currents, the steady-state current distribution in the $4N_R$ rotor bars distributed around the periphery of the rotor can be found. Specifically, with I_R given by (5.55), the two equations (5.53) are combined to show that the rotor current in the ith bar at $\theta_R(t) + \varphi_i$ is[10]

$$\begin{aligned}
i_{RSS}(t,\varphi_i) &= I_R \sin\left(\theta_S(t) - (\theta_R(t) + \varphi_i) + \tan^{-1}\left(\frac{(\omega_S - \omega_R)L_R}{R_R}\right)\right) \\
\theta_R(t) + \varphi_i &= \omega_R t + \frac{\pi}{2}\frac{i-1}{N_R} \quad \text{for } i = 1, ..., 4N_R \qquad (5.56) \\
\theta_S(t) &= \omega_S t.
\end{aligned}$$

Applying Ampère's law $\oint \vec{\mathbf{H}}_R \cdot d\vec{\ell} = i_{\text{enclosed}}$ to the closed curve 1–2–3–4–1 in Figure 5.26, the radial magnetic field $B_R(\theta)$ in the air gap due to this rotor current distribution can be found. To do so, first note that for N_R large, the discrete angles $\varphi_i = \frac{\pi}{2}\frac{i-1}{N_R}$ locating the individual rotor bars can be approximated by a continuous angle φ with a continuous current distribution $i_{RSS}(t,\varphi)$. Specifically, the incremental angle from one rotor bar to the next is $d\varphi = \frac{\pi}{2}\frac{1}{N_R}$ so that the enclosed current may be written as

$$\begin{aligned}
i_{\text{enclosed}}(\theta) &= \sum_{\varphi_i=\theta-\theta_R}^{\varphi_i=\theta-\theta_R+\pi} i_{RSS}(t,\varphi_i) = \frac{2N_R}{\pi}\left(\sum_{\varphi_i=\theta-\theta_R}^{\varphi_i=\theta-\theta_R+\pi} i_{RSS}(t,\varphi_i)d\varphi\right) \\
&\approx \frac{2N_R}{\pi}\int_{\varphi=\theta-\theta_R}^{\varphi=\theta-\theta_R+\pi} i_{RSS}(t,\varphi)d\varphi
\end{aligned}$$

---

[9] By symmetry, diametrically opposite sides of the rotor have emfs which are opposite in sign, but equal in magnitude.

[10] As $i = 1, ..., 4N_R$, the angles φ_i go from $\varphi_1 = 0$ to $\varphi_{4N_R} = \frac{\pi}{2}\frac{4N_R-1}{N_R} = 2\pi - \frac{\pi}{2}\frac{1}{N_R}$.

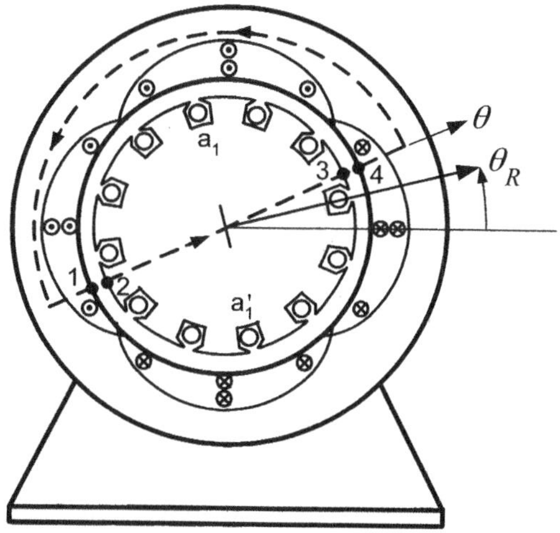

FIGURE 5.26. Path for Ampère's law to compute $\vec{\mathbf{B}}_R$ in the air gap.

where the summation has been approximated by the integral. Ampère's law gives

$$2H_R(\theta)g = i_{\text{enclosed}}(\theta)$$

or

$$H_R(\theta) = \frac{i_{\text{enclosed}}(\theta)}{2g}.$$

The expression (5.56) is substituted for $i_{RSS}(t, \varphi)$ to evaluate $i_{\text{enclosed}}(\theta)$ as

$$\begin{aligned}
i_{\text{enclosed}}(\theta) &= \frac{2N_R}{\pi}\int_{\varphi=\theta-\theta_R}^{\varphi=\theta-\theta_R+\pi} i_{RSS}(\varphi)d\varphi \\
&= -\frac{2N_R}{\pi}I_R\cos\Big(\theta_S(t) - (\theta_R(t)+\varphi) + \tan^{-1}\Big(\frac{(\omega_S-\omega_R)L_R}{R_R}\Big)\Big)\Big|_{\varphi=\theta-\theta_R}^{\varphi=\theta-\theta_R+\pi} \\
&= \frac{4}{\pi}N_R I_R\cos\Big(\theta_S(t) - \theta + \tan^{-1}\Big(\frac{(\omega_S-\omega_R)L_R}{R_R}\Big)\Big).
\end{aligned}$$

The radial magnetic field due to the rotor currents is then

$$\begin{aligned}
\vec{\mathbf{B}}_R(\theta,t) &= B_R(\theta,t)\hat{\mathbf{r}} \\
&= \mu_0 H_R(\theta)\hat{\mathbf{r}} \\
&= \frac{4}{\pi}\frac{\mu_0 N_R I_R}{2g}\cos\Big(\theta - \Big(\theta_S(t) + \varphi_R\Big)\Big)\hat{\mathbf{r}}
\end{aligned}$$

where

$$I_R \triangleq \frac{(\omega_S-\omega_R)}{\sqrt{(\omega_S-\omega_R)^2L_R^2+R_R^2}}\frac{\mu_0\ell_1\ell_2N_SI_S}{2g}$$
$$\varphi_R \triangleq \tan^{-1}\left(\frac{(\omega_S-\omega_R)L_R}{R_R}\right).$$

In chapter 8, it will be shown that the steady-state rotor currents in a *sinusoidally wound* rotor induction motor will produce a rotating magnetic field given by

$$\vec{\mathbf{B}}_R(\theta,t)=\frac{\mu_0N_RI_{R_{\text{wound}}}}{2g}\cos(\theta-(\omega_St+\varphi_{R_{\text{wound}}}))\hat{\mathbf{r}}$$

The only difference in the *form* of these expressions is the factor of $4/\pi$. However, the values of the rotor current $I_{R_{\text{wound}}}$ and the phase angle $\varphi_{R_{\text{wound}}}$ are in general different as they depend on the $L_{R_{\text{wound}}}, R_{R_{\text{wound}}}$ which have different values in a squirrel cage rotor.

5.5.4 Comparison with a Sinusoidally Wound Rotor

It is quite interesting to note that in the squirrel cage rotor, the effective rotor time constant is still L_R/R_R, that is, the time constant due to just a *single* loop. (By "effective" is meant that this is the ratio that appears in the expression for the quantity $s_p = R_R/(\omega_SL_R)$ used in the above expression for the steady-state torque.) A small value for the time constant $T_R = L_R/R_R$ is usually desired since the value of s_p is then large, making $\tau \approx k(\omega_S-\omega_R)$ for a large range of slip speeds. Assuming ideal magnetic material and a small air gap, it is straightforward to show that $L_R = \pi\mu_0\ell_1\ell_2/(4g)$ for each loop of the rotor. Consequently, the time constant is $T_R = L_R/R_R = \pi\mu_0\ell_1\ell_2/(4gR_R)$ where R_R is the resistance of the single loop.

This is in contrast to a sinusoidally wound rotor induction motor (discussed in a later chapter) where the rotor loops of each phase are connected in series resulting in $L_{R_{\text{wound}}} = \pi\mu_0\ell_1\ell_2N_R^2/(8g)$ (N_R is the total number of loops in a sinusoidally wound rotor phase) and $R_{R_{\text{wound}}}$ is the sum of the resistances of each loop in the rotor ($R_{R_{\text{wound}}} = N_RR_1$ with R_1 the resistance in each loop). The time constant is then $T_{R_{wound}} = L_{R_{\text{wound}}}/R_{R_{\text{wound}}} = \pi\mu_0\ell_1\ell_2N_R/(8gR_1)$). The ratio of these two time constants is $T_{R_{wound}}/T_{R_{squirrel}} = N_R/2$ showing that for a squirrel cage rotor with the same number of loops as the wound rotor, the effective time constant can be significantly smaller.

To compare the torque of the two motors, consider the steady-state torque produced by a sinusoidally wound two-phase induction motor. It is shown in Chapter 7 (or see Chapter 10 of [1]) that the expression for this torque is given by ($|\underline{I}_S| = I_S/\sqrt{2}$, $|\underline{I}_R| = I_{R_{\text{wound}}}/\sqrt{2}$)[11]

$$\begin{aligned}\tau_{\text{wound rotor}} &= M\,\text{Im}\{\underline{I}_S \underline{I}_R^*\},\\ &= M\,\text{Im}\left\{\underline{I}_S\left(-\frac{j\omega_S M}{R_R/S + j\omega_S L_R}\underline{I}_S\right)^*\right\}\end{aligned}$$

With $S_p \triangleq \dfrac{R_{R_{\text{wound}}}}{\sigma\omega_S L_{R_{\text{wound}}}}$ and $s_p \triangleq \sigma S_p = \dfrac{R_{R_{\text{wound}}}}{\omega_S L_{R_{\text{wound}}}}$ this becomes

$$\tau_{\text{wound rotor}} = \frac{I_S^2}{2} M\,\text{Im}\left\{\frac{\dfrac{j\omega_S MS}{R_R}\,\dfrac{1+\dfrac{jS}{\sigma S_p}}{}}{\left(1-\dfrac{jS}{\sigma S_p}\right)\left(1+\dfrac{jS}{\sigma S_p}\right)}\right\}.$$

Rearranging, this becomes

$$\begin{aligned}\tau_{\text{wound rotor}} &= \frac{I_S^2}{2} M\,\text{Im}\left\{\frac{\dfrac{j\omega_S M R_R}{S}\left(1+\dfrac{jS}{s_p}\right)}{1+(S/s_p)^2}\right\}\\ &= \frac{I_S^2}{2} M \frac{\omega_S MS/R_{R_{\text{wound}}}}{1+(S/s_p)^2}\\ &= \frac{1}{4L_R} M^2 I_S^2 \frac{2}{s_p/S + S/s_p}.\end{aligned}$$

Finally, using

$$M = \frac{\mu_0 \pi \ell_1 \ell_2 N_S N_R}{8g} = \left(\frac{\pi N_R}{4}\right)\left(\frac{\mu_0 \ell_1 \ell_2 N_S}{g}\right)$$

it follows that

$$\begin{aligned}\tau_{\text{wound rotor}} &= \frac{1}{4L_R}\left(\frac{\pi N_R}{4}\right)^2\left(\frac{\mu_0 \ell_1 \ell_2 N_S I_S}{g}\right)^2 \frac{2}{s_p/S + S/s_p}\\ &= \frac{N_R}{2}\left(\frac{\pi}{4}\right)^2 \frac{N_R}{2L_R}\left(\frac{\mu_0 \ell_1 \ell_2 N_S I_S}{g}\right)^2 \frac{2}{s_p/S + S/s_p}.\end{aligned}$$

In *form*, this differs from the squirrel cage case by the factor $\dfrac{N_R}{2}\left(\dfrac{\pi}{4}\right)^2$. However, also note that $s_p \triangleq \dfrac{R_{R_{\text{wound}}}}{\omega_S L_{R_{\text{wound}}}}$ has a different value than that of the squirrel cage rotor.

---

[11]The magnitude of the currents are written in the (nonstandard) form $|\text{I}_S| = I_S/\sqrt{2}$, $|\text{I}_R| = I_{R_{\text{wound}}}/\sqrt{2}$ so that $I_S, I_{R_{\text{wound}}}$ are peak values (rather than rms values) to facilitate a comparison with the squirrel cage rotor torque.

Problems

Problem 1 Solutions to Linear Time-Invariant Systems
Consider the linear time-invariant system specified by

$$\frac{d^3y}{dt^3} + a_2\frac{d^2y}{dt^2} + a_1\frac{dy}{dt} + a_0y = b_2\frac{d^2u}{dt^2} + b_1\frac{du}{dt} + b_0u. \tag{5.57}$$

The transfer function $G(s)$ *of this system is*

$$G(s) = \frac{b(s)}{a(s)} = \frac{b_2s^2 + b_1s + b_0}{s^3 + a_2s^2 + a_1s + a_0}.$$

(a) Show that if

$$u(t) \triangleq U_0e^{j\omega t} = U_0(\cos(\omega t) + j\sin(\omega t))$$

then

$$y_p(t) = U_0\,|G(j\omega)|\,e^{j(\omega t + \angle G(j\omega))}$$

is a (particular) solution to the differential equation (5.57).
(b) Show that if

$$u(t) \triangleq \mathrm{Re}\left\{U_0e^{j\omega t}\right\} = U_0\cos(\omega t),$$

then

$$y_p(t) = U_0\,|G(j\omega)|\cos(\omega t + \angle G(j\omega))$$

is a particular solution to the differential equation (5.57).
(c) Let

$$a(s) = s^3 + a_2s^2 + a_1s + a_0 = (s - p_1)(s - p_2)(s - p_3)$$

so that p_1, p_2, p_3 *are the poles of* $G(s)$*. To simplify the discussion, assume the* p_i *are distinct.*

With $u(t) \equiv 0$ *and arbitrary initial conditions* $y(0), y'(0), y''(0)$*, show that the solution to (5.57) has the form* $y(t) = A_1e^{p_1t} + A_2e^{p_2t} + A_3e^{p_3t}$.

Next, show that the complete solution to the differential equation (5.57) with arbitrary initial conditions $y(0), y'(0), y''(0)$ *and input* $u(t) = U_0\cos(\omega t)$ *is*

$$y(t) = U_0\,|G(j\omega)|\cos(\omega t + \angle G(j\omega)) + A_1e^{p_1t} + A_2e^{p_2t} + A_3e^{p_3t}$$

where the A_i *are the unique solution to*

$$\begin{bmatrix} y(0) \\ y'(0) \\ y''(0) \end{bmatrix} = \begin{bmatrix} y_p(0) \\ y_p'(0) \\ y_p''(0) \end{bmatrix} + \begin{bmatrix} 1 & 1 & 1 \\ p_1 & p_2 & p_3 \\ p_1^2 & p_2^2 & p_3^2 \end{bmatrix}\begin{bmatrix} A_1 \\ A_2 \\ A_3 \end{bmatrix}.$$

(d) Show that the steady-state solution of the differential equation (5.57) equals $y_p(t)$ *if and only if* $\mathrm{Re}\{p_i\} < 0$ *for* $i = 1, 2, 3$.

(e) Show that the steady-state solution to $dy/dt + ay = u(t)$ *with* $a > 0$ *and* $u(t) = U_0\cos(\omega_0t)$ *is* $y_{SS}(t) = U_0\,|G(j\omega)|\cos(\omega t + \angle G(j\omega))$ *with* $G(s) = 1/(s + a)$.

The Physics of the Induction Machine

Problem 2 Torque Versus Slip for the Induction Motor

The expressions for the rotor currents in equations (5.4) and (5.5) were found neglecting the rotor loop inductance. Assuming the rotor speed is constant and the currents are in sinusoidal steady-state, find an expression for these currents when the rotor loops have resistance R_R and *an inductance* L_R. *Modify the corresponding expression for the torque given in (5.9) to show that it now has the form*

$$\tau = \left(\frac{\mu_0 \ell_1 \ell_2 N_S I_S}{2g}\right)^2 \frac{1}{2L_R} \frac{2}{S/s_p + s_p/S}$$

where $S \triangleq \dfrac{\omega_S - \omega_R}{\omega_S}$ *is the* normalized slip *and* $s_p \triangleq \dfrac{R_R}{\omega_S L_R}$. *The* pull-out slip *is defined by* $S_p \triangleq \dfrac{R_R}{\sigma \omega_S L_R} = s_p/\sigma$ *where* σ *is the so-called leakage factor (see Chapter* 6*).*

Finally, sketch the torque as a function of the normalized slip $S \triangleq (\omega_S - \omega_R)/\omega_S$.

Problem 3 Induction Motor under Load

In problem 2 it is shown that the torque produced by a two-phase induction motor with two rotor loops $\pi/2$ *radians apart and nonzero rotor inductance* L_R *is given by*

$$\tau = \left(\frac{\mu_0 \ell_1 \ell_2 N_S I_S}{2g}\right)^2 \frac{1}{2L_R} \frac{2}{s_p/S + S/s_p} = \tau_p \frac{2}{s_p/S + S/s_p} \tag{5.58}$$

where

$$\begin{aligned} S &\triangleq \frac{\omega_S - \omega_R}{\omega_S} \\ \tau_p &\triangleq \left(\frac{\mu_0 \ell_1 \ell_2 N_S I_S}{2g}\right)^2 \frac{1}{2L_R} \\ s_p &\triangleq \frac{R_R}{\omega_S L_R}. \end{aligned}$$

A plot of τ/τ_p *versus* S/s_p *is given in Figure 5.27. A typical rotor for an induction motor has more than two loops on it. The squirrel cage rotor for an induction motor is shown in Figure 5.28 with a cross-sectional view of the rotor given in Figure 5.28(b). This shows a rotor consisting of* 6 *rotor loops* (12 *sides) that can be viewed as three sets of* 2 *rotor loops. Each set consists of two rotor loops* $\pi/2$ *radians apart.*

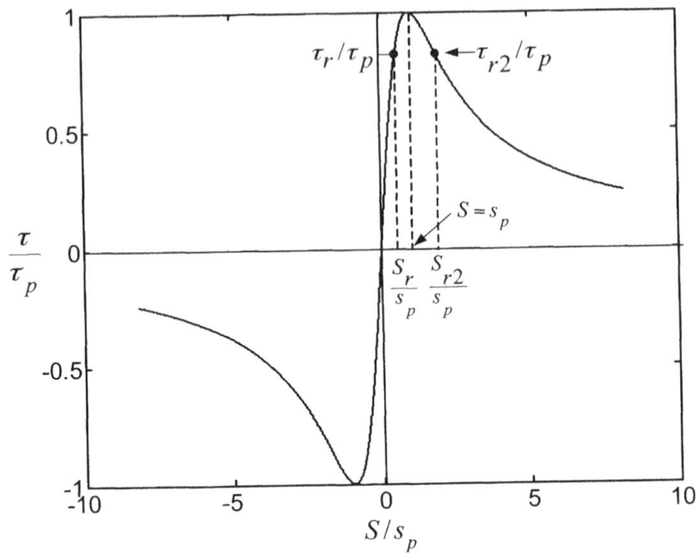

FIGURE 5.27. Normalized torque versus normalized slip.

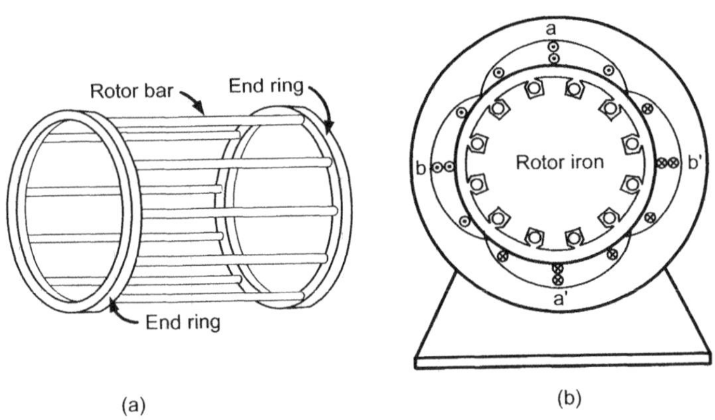

FIGURE 5.28. (a) Squirrel cage rotor for an induction motor. (b) Cross-sectional view.

(a) With the rotor of Figure 5.28(b), explain why the expression (5.58) for the steady-state torque needs to be multiplied by three to obtain the torque of the motor in Figure 5.28(b).

(b) Suppose the induction motor has a load on it and is producing the torque τ_r *that is, it is operating at the point* $(S_r/s_p, \tau_r/\tau_p)$ *shown in Figure 5.27. Next, suppose an additional load is put on the motor so that the total*

load torque τ_L *now satisfies* $\tau_r < \tau_L < \tau_p$. *After the additional load is put on the motor:*

Will the speed increase or decrease?

Will the normalized slip increase or decrease?

Will the motor torque increase to handle the increased load?

(c) Repeat part (b), but with the motor initially operating at $(S_{r2}/s_p, \tau_{r2}/\tau_p)$.

(d) Suppose the induction motor is turned off (no currents applied to the stator phases) so that $\omega_R = 0$, *but it has a load torque* $\tau_L = \tau_r$ *on it (the same* τ_r *as in Figure 5.27). Suppose further that* $s_p = 0.2$ *and that currents of frequency* ω_S *are now applied to the stator phases. Just after applying the stator currents, answer the following:*

What is the value of S/s_p *?*

Mark on Figure 5.27 the operating point of the motor.

Can the motor start with the load torque $\tau_L = \tau_r$ *on it? Explain.*

Problem 4 Torque

In developing the expression (5.6) for the torque on a rotor loop of an induction motor, why wasn't the magnetic force on the two semicircular ends of the rotor loop considered (see Figure 5.8)?

Problem 5 Simple Induction Machine with Three-Phases

Figure 5.29 shows a three phase stator with a rotor iron core.

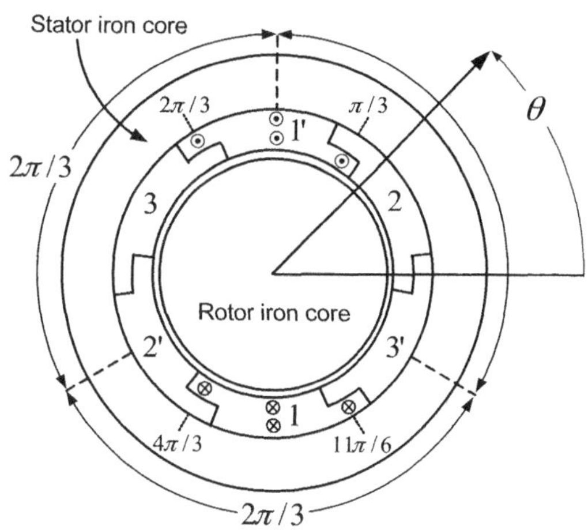

FIGURE 5.29. Three-phase stator. Currents are only shown for phase 1.

Phase 1 *is identical to phase a in a two phase stator. The magnetic field in the air gap due to the current* i_{S1} *in phase* 1 *is then*

$$\vec{\mathbf{B}}_{S1}(i_{S1}, r, \theta) = \frac{\mu_0 \ell_2 N_S}{4gr} i_{S1} \cos(\theta)\hat{\mathbf{r}}.$$

Phase 2 is wound the same way as phase 1 except that is rotated $2\pi/3$ radians with respect to phase 1. The third phase is also wound the same way as phases 1 and 2, but rotated another $2\pi/3$ radians with respect to phase 2. The magnetic field in the air gap due to the current i_{S2} in phase 2 is then

$$\vec{\mathbf{B}}_{S2}(i_{S2}, r, \theta) = \frac{\mu_0 \ell_2 N_S}{4gr} i_{S2} \cos(\theta - 2\pi/3)\hat{\mathbf{r}}$$

and the magnetic field in the air gap due to the current i_{S3} in phase 3 is then

$$\vec{\mathbf{B}}_{S3}(i_{S3}, r, \theta) = \frac{\mu_0 \ell_2 N_S}{4gr} i_{S3} \cos(\theta - 4\pi/3)\hat{\mathbf{r}}.$$

Let the currents be given by

$$\begin{aligned} i_{S1} &= I_S \cos(\omega_S t) \\ i_{S2} &= I_S \cos(\omega_S t - 2\pi/3) \\ i_{S3} &= I_S \cos(\omega_S t - 4\pi/3). \end{aligned}$$

(a) Show that the total magnetic in the air gap may be written as

$$\vec{\mathbf{B}}_S = \vec{\mathbf{B}}_{S1} + \vec{\mathbf{B}}_{S2} + \vec{\mathbf{B}}_{S3} = \frac{\mu_0 \ell_2 N_S I_S}{4gr} \frac{3}{2} \cos(\theta - \omega_S t)\hat{\mathbf{r}}.$$

(b) Is this a sinusoidally distributed magnetic field?
(c) With two rotor loops $\pi/2$ radians apart, what changes (if any) are required to analyze this three-phase induction machine compared to the two-phase induction machine in this chapter?

Problem 6 A PM-Generator/Induction-Motor Machine (motivated by [27])
In this problem, a rather strange type of machine is analyzed. Figure 5.30 shows a permanent magnet rotating inside a hollowed out core of soft iron. (This setup is the reverse of that for the DC motor in Chapter 1, where the rotating part is made of soft iron and the external stationary part is soft iron.) By properly designing the permanent magnet, it can be assumed that the magnetic field in the air gap is given by

$$\vec{\mathbf{B}}_S = B_S \hat{\mathbf{r}} = B_{Smax} \frac{r_R}{r} \cos(\theta - \theta_S)\hat{\mathbf{r}} \tag{5.59}$$

where θ is an arbitrary angular location in the air gap and θ_S locates the north end (magnetic axis) of the permanent magnet as shown in Figure 5.30 (The subscript "S" is used because this magnetic field will take the place of the stator magnetic field discussed in the text.). Figures 5.30(b)–(d) illustrate the radial magnetic field due to the permanent magnet as it rotates from 0 to $\pi/2$ to π.

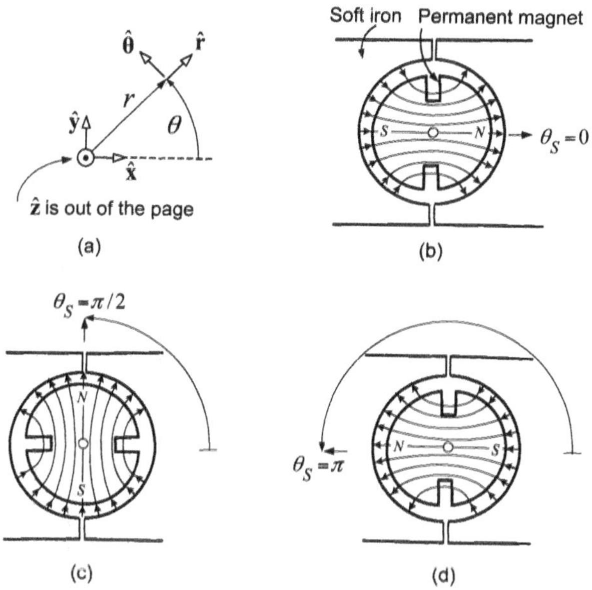

FIGURE 5.30. A permanent magnet rotating inside a hollowed out core of soft iron.

The next part of the construction is to put a conducting loop in the air gap. This is illustrated in Figure 5.31(a), which shows a half-cylindrical-shaped loop placed in the air gap.

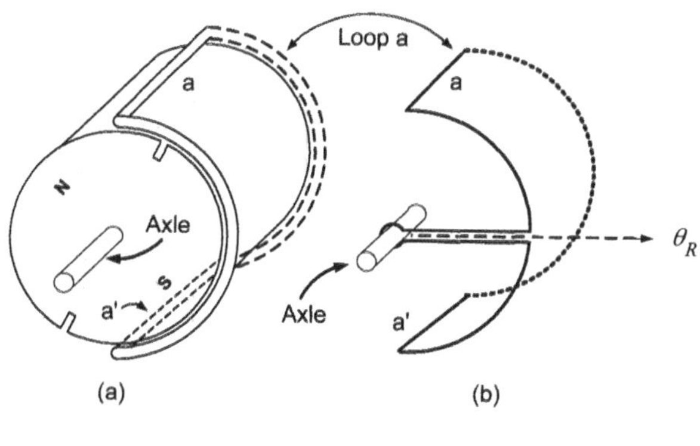

FIGURE 5.31. (a) Loop in the air gap which rotates about the axle independent of the rotating permanent magnet. (b) Illustrating how the loop is mechanically supported so that it can rotate independent of the permanent magnet.

However, this loop is not *attached to the permanent magnet; rather, it rotates independently of the permanent magnet on the same axle. Figure*

5.31(b) illustrates how the loop is mechanically supported so that it can rotate on the axle independent of the permanent magnet. This loop is referred to as rotor loop a–a′ (or rotor phase a) and its position is located by the angle θ_R which is fixed to the loop half way between sides a and a′ as shown in Figure 5.31(b). To complete the description of the construction of this motor, a second rotor loop is added to the system as shown in Figure 5.32. This loop is denoted rotor loop b–b′ (or rotor phase b) and is rotated 90° with respect to loop a–a′. As shown in Figure 5.32(b), a brace is put between the two loops so that they are rigidly held 90° apart. The brace and axle are non conducting so that the two loops are electrically isolated. The pair of loops rotate together on the axle independent of the permanent magnet. This completes the construction of the machine.

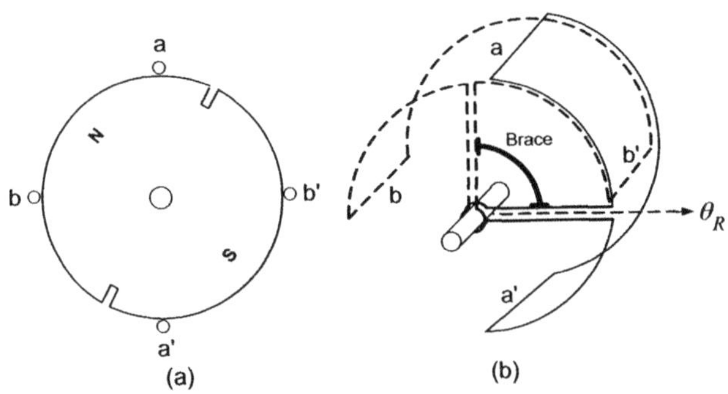

FIGURE 5.32. Two cylindrical rotor loops mechanically held together 90° apart, which rotate together on the axle independently of the rotating permanent magnet.

(a) Let $\theta_S(t) = \omega_S t$ so that the permanent magnet is rotating with constant angular speed ω_S by some external mechanical force. Consequently, a rotating radially directed magnetic field is setup in the air gap of the machine. Compute the fluxes in the two rotor loops due to the rotating magnetic field $\vec{\mathbf{B}}_S$.

(b) With R_R the resistance of a rotor loop and L_R the inductance of the rotor loop, compute the sinusoidal steady-state currents i_{RaSS} and i_{RbSS} in the two rotor loops. Assume the rotor has constant angular velocity ω_R where $0 < \omega_R < \omega_S$.

(c) Compute the torques τ_{Ra} and τ_{Rb} on the two rotor loops due to the magnetic field $\vec{\mathbf{B}}_S$ acting on the current in the rotor sides a, a', b, b' in terms of $i_{Ra}, i_{Rb}, \ell_1, \ell_2, B_{Smax}, \omega_S$ and ω_R.

(d) Substitute the expressions for the currents from part (b) into the expressions for the torques in part (c) and show that the resulting expression has the same form as that derived in problem 2.

Synchronous Machines

Problem 7 Magnetic Field of a Sinusoidally Wound Rotor

Using Ampère's law, derive the expression given in equation (5.10) for the radial magnetic field in the air gap due to the current in the rotor windings. Assume ideal magnetic material and, that the rotor and stator windings have cross-sectional areas that are infinitesimal.

Problem 8 Mathematical Model of a PM Synchronous Motor

Consider a two phase permanent magnet synchronous motor. This is constructed like the wound rotor in Figure 5.33, but the rotor is replaced by a permanent magnet *as shown. Let the magnetic field due to the permanent magnet rotor be given by*

$$\vec{\mathbf{B}}_R = B_R(r,\theta)\hat{\mathbf{r}} = B_{R\,\max}\frac{r_R}{r}\cos(\theta - \theta_R)\hat{\mathbf{r}}$$

and derive the mathematical model of the PM synchronous motor.

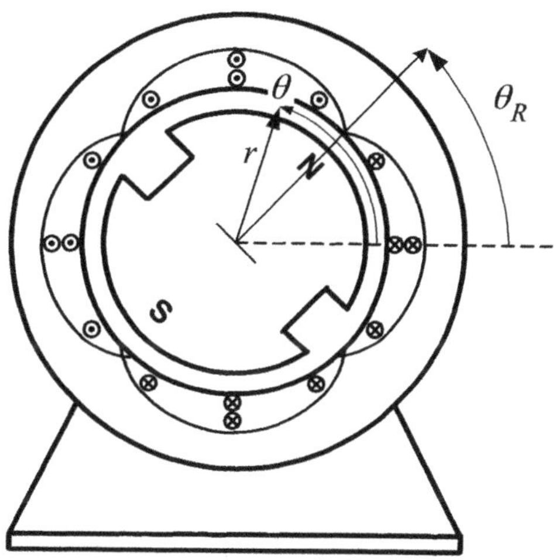

FIGURE 5.33. Two-phase permanent magnet synchronous motor.

Problem 9 Synchronous Motor with a Sinusoidally Wound Rotor

Consider a synchronous motor with a sinusoidally wound rotor as in Figure 5.9. An analytical expression for the torque produced by this motor was derived in the text. However, this expression can also be derived by considering the magnetic force due to the stator magnetic field $\vec{\mathbf{B}}_S$ *acting on the currents in the rotor windings. This problem outlines such an approach.*

(a) Using the magnetic force law, show that the incremental magnetic force $d\vec{\mathbf{F}}_R$ *due to* $\vec{\mathbf{B}}_S$ *acting on the rotor current in the windings of the*

rotor between θ and $\theta + d\theta$ is given by

$$d\vec{\mathbf{F}}_R = \begin{cases} I_F \dfrac{N_F}{2} |\sin(\theta - \theta_R)| d\theta(+\ell_1 \hat{\mathbf{z}}) \times \vec{\mathbf{B}}_S & \theta_R \le \theta \le \theta_R + \pi \\ I_F \dfrac{N_F}{2} |\sin(\theta - \theta_R)| d\theta(-\ell_1 \hat{\mathbf{z}}) \times \vec{\mathbf{B}}_S & \theta_R + \pi \le \theta \le \theta_R + 2\pi \end{cases}$$

or, more simply, for $\theta_R \le \theta \le \theta_R + 2\pi$

$$\begin{aligned} d\vec{\mathbf{F}}_R &= I_F \frac{N_F}{2} \sin(\theta - \theta_R) d\theta(+\ell_1 \hat{\mathbf{z}}) \times \vec{\mathbf{B}}_S \\ &= I_F \frac{N_F}{2} \sin(\theta - \theta_R) d\theta(+\ell_1 \hat{\mathbf{z}}) \times \frac{\mu_0 N_S I_S}{2g} \cos(\theta - \theta_S) \hat{\mathbf{r}} \\ &= \frac{\mu_0 \ell_1 N_S N_F I_S I_F}{4g} \sin(\theta - \theta_R) \cos(\theta - \theta_S) d\theta \hat{\boldsymbol{\theta}}. \end{aligned}$$

(b) Integrate the incremental torque $d\vec{\tau}_R$

$$\begin{aligned} d\vec{\tau}_R &= (\ell_2/2)\hat{\mathbf{r}} \times d\vec{\mathbf{F}}_R \\ &= \frac{\mu_0 \ell_1 \ell_2 N_S N_F I_S I_F}{8g} \sin(\theta - \theta_R) \cos(\theta - \theta_S) d\theta \hat{\mathbf{z}} \end{aligned}$$

over the rotor windings to obtain the total torque $\vec{\tau}_R$ on the rotor.

(c) Assume the motor is in steady-state with $\theta_S = \omega_S t, \theta_R = \omega_R t - \delta$ and $\omega_S = \omega_R$. With

$$\begin{aligned} B_{R\max} &\triangleq \frac{\mu_0 \ell_2 N_F I_F}{4gr} \Big|_{r=\ell_2/2} = \frac{\mu_0 N_F I_F}{2g} \\ V_{airgap} &\triangleq 2\pi(\ell_2/2)\ell_1 g = \pi \ell_1 \ell_2 g \\ B_{S\max} &\triangleq \frac{\mu_0 \ell_2 N_S I_S}{4gr} \Big|_{r=\ell_2/2} = \frac{\mu_0 N_S I_S}{2g} \end{aligned}$$

show that the answer to part (b) reduces to the expression given in (5.13).

Problem 10 Synchronous Motor with a Salient Rotor

Consider a synchronous machine with two sinusoidally wound stator phases and a salient *rotor as in Figure 5.13 of Section 5.3.3. Let $i_{Sa} = I_S \cos(\omega_S t)$, $i_{Sb} = I_S \sin(\omega_S t)$ be the stator currents and $i_F(t) = I_F =$ constant so that the magnetic fields in the air gap are*

$$\begin{aligned} \vec{\mathbf{B}}_S &= \frac{\mu_0 N_S I_S}{2g_1} \frac{r_R}{r} \cos(\theta - \theta_S) \hat{\mathbf{r}} \\ \vec{\mathbf{B}}_R &= \frac{\mu_0 N_F I_F}{2g_0} \frac{r_R}{r} \cos(\theta - \theta_R) \hat{\mathbf{r}} \end{aligned}$$

where $\theta_S(t) = \omega_S t$ denotes angular position of the magnetic axis of the stator magnetic field and $\theta_R(t) = \omega_R t - \delta$ denotes the angular position of *the rotor which is taken to coincide with the rotor's magnetic axis. Here g_1 is the average air gap length ($g_1 > g_0$) and, as the rotor's cross section is not circular, let $r_R \triangleq r_S - g_1$ (see the remark in Section 5.3.3).*

(a) Compute the torque τ_S on the stator phases due to the rotor's magnetic field $\vec{\mathbf{B}}_R$ acting on the stator currents i_{Sa}, i_{Sb}. Use this to find τ_R. Is there anything that has to be done different here than in the sinusoidally wound rotor case? Explain why or why not.

(b) Compute the induced emf's ξ_{Sa} and ξ_{Sb} in the stator phases due to the rotor's magnetic field $\vec{\mathbf{B}}_R$. Is there anything that has to be done different here than in the sinusoidally-wound rotor case? Explain why or why not.

(c) Compute the power absorbed in the stator by these induced emf's and show it equals the negative of the mechanical power produced by the motor. Is there anything that has to be done different here than in the sinusoidally wound rotor case? Explain why or why not.

(d) Is it possible (in a nice analytic fashion) to find the motor's torque directly by computing the torque on the rotor due to the magnetic field $\vec{\mathbf{B}}_S$ acting on the rotor current? Explain why or why not.

Problem 11 Emf in the Rotor Winding of a Synchronous Machine

Consider a synchronous motor with a sinusoidally wound rotor as in Figure 5.9. This problem outlines how to derive an expression for the flux and induced emf in the rotor winding produced by the stator magnetic field.

(a) Compute the flux linkage λ_R in the rotor loops due to the stator magnetic field $\vec{\mathbf{B}}_S$; that is, with

$$\vec{\mathbf{B}}_S(i_{Sa}, i_{Sb}, r, \theta) = \frac{\mu_0 \ell_2 N_S}{4gr}\left(i_{Sa}\cos(\theta) + i_{Sb}\sin(\theta)\right)\hat{\mathbf{r}}$$

compute

$$\lambda_R = \int_{\substack{\text{All loops of}\\ \text{the rotor}}} \vec{\mathbf{B}}_S \cdot d\vec{\mathbf{S}}.$$

(b) Compute the emf induced in the windings, that is,

$$\xi_R = -\frac{d\lambda_R}{dt}.$$

(c) Let $i_{Sa}(t) = I_S\cos(\omega_S t)$, $i_{Sb}(t) = I_S\sin(\omega_S t)$, $\theta_S = \omega_S t$, $\theta_R = \omega_R t - \delta$ with $\omega_S = \omega_R$ and show that in this case

$$\xi_R = -\frac{d\lambda_R}{dt} \equiv 0.$$

Explain why.

Microscopic Viewpoint of Electric Machines

Problem 12 Linear Motor in a Coordinate System Attached to the Bar

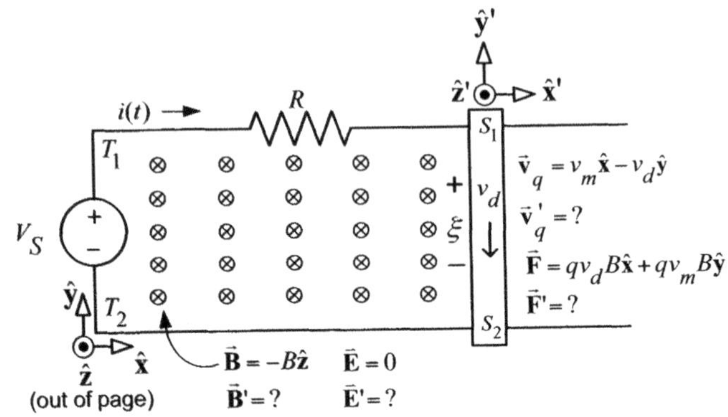

FIGURE 5.34. Linear DC machine. The primed coordinate system is attached to the sliding bar and moves at velocity $v_m\hat{\mathbf{x}}$ with respect to the unprimed coordinate system attached to the rails.

In chapter 1, the linear DC machine was analyzed from the microscopic point of view in a stationary coordinate system of the rails. In the stationary coordinate system, $\vec{\mathbf{B}} = -B\hat{\mathbf{z}}$, $\vec{\mathbf{E}} = 0$ *(*$B > 0$ *and* $\vec{\mathbf{E}}$ *denotes the electric field in the air surrounding the linear motor; there is another electric field* $\vec{\mathbf{E}}_S$ *in the wire setup by the source voltage* V_S*). In this coordinate system, the bar moves in the* $+\hat{\mathbf{x}}$ *direction with a constant speed* v_m *and each charge* q *in the sliding bar has total velocity* $\vec{\mathbf{v}}_q = v_m\hat{\mathbf{x}} - v_d\hat{\mathbf{y}}$ *where* v_d *is the drift speed. Consider the primed (*$'$*) coordinate system shown in Figure 5.34 which is attached to the sliding bar. Redo the analysis from the point of view of this coordinate system. Specifically,*

(a) Compute the electric $\vec{\mathbf{E}}'$ *and magnetic* $\vec{\mathbf{B}}'$ *fields in the reference frame of the moving bar using the transformation (5.44).*

(b) Find the velocity $\vec{\mathbf{v}}_q'$ *of the charge carriers in the sliding bar.*

(c) Find the magnetic force per unit charge $\vec{\mathbf{v}}_q' \times \vec{\mathbf{B}}'$ *on the charge carriers in sliding bar.*

(d) Compute the Lorentz force per unit charge $\vec{\mathbf{F}}'/q \triangleq \vec{\mathbf{E}}' + \vec{\mathbf{v}}_q' \times \vec{\mathbf{B}}'$ *on charge carriers in the sliding bar due to the external electric and magnetic fields. What part of this force is canceled out by* $\vec{\mathbf{E}}_S$*? What is the resulting total force on the sliding bar?*

(e) In this coordinate system, what is the "back electric field" and "back emf"? Show that conservation of energy holds; that is, the energy absorbed by the back emf reappears as mechanical power out.

(f) In the coordinate system of the sliding bar, what is the flux in the loop of the linear motor. Use Faraday's law to compute the emf ξ_{bar} *induced in*

loop of the linear motor. Neglecting the inductance, compute the current as $i_b = \xi_{bar}/R$ where R is the resistance of the bar. What is the net force on the bar in this coordinate system?

Problem 13 Linear Motor in a Coordinate System with Velocity $v_c < v_m$

Repeat problem 12 for the case where the coordinate system is moving at speed v_c, where $v_c < v_m$.

Problem 14 Linear Motor in a Coordinate System with Velocity $v_c > v_m$

Repeat problem 12 for the case where the coordinate system is moving at speed v_c, where $v_c > v_m$.

Problem 15 Stator Electric Field

In the derivation of the stator electric field $\vec{\mathbf{E}}_S(\theta, t)$ in the air gap due to the rotating stator magnetic field, the "constant of integration" was set to zero so that $\vec{\mathbf{E}}_S(\theta, t)$ is given by equation (5.23)

$$\vec{\mathbf{E}}_S(\theta, t) = E_{Sz}(\theta, t)\hat{\mathbf{z}} = \frac{\mu_0 \ell_2 N_S I_S}{4g} \omega_S \cos(\theta - \omega_S t)\hat{\mathbf{z}}.$$

Note that $\vec{\mathbf{E}}_S(\theta, t)$ is a maximum at $\theta = \omega_S t$, a minimum at $\theta = \omega_S t + \pi$ and zero at $\theta = \omega_S t \pm \pi/2$. Similarly, the stator magnetic field

$$\vec{\mathbf{B}}_S(r, \theta, t) = \frac{\mu_0 \ell_2 N_S I_S}{4gr} \cos(\theta - \omega_S t)\hat{\mathbf{r}}$$

is a maximum at $\theta = \omega_S t$, a minimum at $\theta = \omega_S t + \pi$ and zero at $\theta = \omega_S t \pm \pi/2$.

(a) Consider a locked rotor, that is, the rotor is constrained so that it cannot move. Show that the induced emf in a rotor loop is zero at a time t when one side of the loop is located at $\theta = \omega_S t + \pi/2$ (so the other side of the loop is located at $\theta = \omega_S t - \pi/2$), that is, when the flux through the loop is maximum or minimum (see Figure 5.15).

(b) With the rotor locked (i.e., $\omega_R = 0$), the induced emf is equal to

$$\xi_{Ra} = V_{Ra} \triangleq \oint_{rotor\ loop\ a} \vec{\mathbf{E}}_S \cdot d\vec{\ell}.$$

Use this to explain why the stator electric field $\vec{\mathbf{E}}_S(\theta, t)$ must be zero for $\theta = \omega_S t \pm \pi/2$, or equivalently, why must the "constant of integration" be zero (see Figure 5.15).

Problem 16 Induction Motor in the Synchronous Coordinate System

Rederive the emf $\xi'_{Ra} = -d\lambda'_{Ra}/dt$ given in equation (5.45) by computing the flux λ'_{Ra} *in rotor loop a due to the stator magnetic field $\vec{\mathbf{B}}'_S$. That is,*

show that

$$\lambda'_{Ra} = \int_S \vec{\mathbf{B}}'_S \cdot d\vec{\mathbf{S}} = \int_{\theta'=-\omega_R t-\pi/2}^{\theta'=-\omega_R t+\pi/2} \frac{\mu_0 \ell_2 N_S I_S}{4gr}\Big|_{r=\ell_2/2} \cos(\theta')\hat{\mathbf{r}}\cdot\ell_1(\ell_2/2)d\theta'\hat{\mathbf{r}}$$

$$= \frac{\mu_0 \ell_1 \ell_2 N_S I_S}{4g}\cos(-\omega'_R t).$$

Problem 17 Induction Motor in the Rotor Coordinate System

Consider the simple induction motor analyzed in the text. That is, a machine whose stator consists of two sinusoidally wound phases 90° apart and whose rotor consists of two electrically isolated loops also 90° apart. This problem outlines the Physics of the motor from a coordinate system attached to the rotor of the motor; that is, it rotates at an angular speed ω_R *with respect to the stator. Denote this* rotor *coordinate system as the primed (′) coordinate system and,* in this coordinate system, *do the following:*

(a) Compute the electric $\vec{\mathbf{E}}'_S$ *and magnetic* $\vec{\mathbf{B}}'_S$ *fields in the air gap from the electric and magnetic fields* $\vec{\mathbf{E}}_S$ *and* $\vec{\mathbf{B}}_S$ *in the stator coordinate system.*

(b) Find the velocity $\vec{\mathbf{v}}'_{Ra}$ *of the charge carriers on the axial sides of rotor phase/loop a (note that in this coordinate system* $\omega'_R = 0$*).*

(c) Find the magnetic force per unit charge $\vec{\mathbf{v}}'_{Ra} \times \vec{\mathbf{B}}'_S$ *on the charge carriers in the axial sides of rotor phase a.*

(d) Compute the torque τ'_{Ra} *on the rotor loop.*

(e) Compute the emf ξ'_{Ra} *induced in rotor loop a by computing the rate of change of the flux in the loop. Neglecting the rotor inductance, compute the current as* $i_{Ra} = \xi'_{Ra}/R_R$.

(f) Compute the voltage $V'_{Ra} = \oint_{rotor\ loop\ a} \vec{\mathbf{E}}'_S \cdot d\vec{\boldsymbol{\ell}}$ *in rotor loop a produced by the axial electric field* $\vec{\mathbf{E}}'_S$.

(g) Compute the back emf ζ'_{Ra} *by integrating the axial magnetic force per unit charge around rotor loop a. Show that* $\xi'_{Ra} = V'_{Ra} + \zeta'_{Ra}$.

(h) Find the corresponding quantities $\vec{\mathbf{v}}'_{Rb}$, $\vec{\mathbf{v}}'_{Rb} \times \vec{\mathbf{B}}'_S$, τ'_{Rb}, ξ'_{Rb}, i_{Rb}, V'_{Rb}, *and* ζ'_{Rb} *for rotor phase b.*

(i) Show that conservation of energy holds in this coordinate system; that is,

$$i_{Ra}V'_{Ra} + i_{Rb}V'_{Rb} = (i_{Ra}\xi'_{Ra} + i_{Rb}\xi'_{Rb}) - (i_{Ra}\zeta'_{Ra} + i_{Rb}\zeta'_{Rb})$$

where $-(i_{Ra}\zeta'_{Ra} + i_{Rb}\zeta'_{Rb}) = (\tau'_{Ra} + \tau'_{Rb})\omega'_R$ *so that the power absorbed by the back emf is converted into mechanical power.*

Problem 18 Synchronous Machine in the Stationary Coordinate System

Consider a synchronous motor with a sinusoidally wound rotor to be in steady state so $i_F = I_F =$ *constant,* $i_{Sa} = I_S\cos(\omega_S t), i_{Sb} = I_S\sin(\omega_S t)$, $\omega_R = \omega_S$, $\theta_S = \omega_S t$ *and* $\theta_R = \omega_R t - \delta$. *The stator of this machine is identical to that of an induction motor so that the electric and magnetic fields in the air gap due to the stator currents are given by*

$$\begin{aligned}\vec{\mathbf{E}}_S(\theta,t) &= E_{Sz}(\theta,t)\hat{\mathbf{z}} = \frac{\mu_0\ell_2 N_S I_S}{4g}\omega_S\cos(\theta-\omega_S t)\hat{\mathbf{z}}\\ \vec{\mathbf{B}}_S(r,\theta,t) &= B_{Sr}(\theta,t)\hat{\mathbf{r}} = \frac{\mu_0\ell_2 N_S I_S}{4gr}\cos(\theta-\omega_S t)\hat{\mathbf{r}}.\end{aligned}$$

The sinusoidally wound rotor phase has a winding density $(N_F/2)\,|\sin(\theta-\theta_R)|$ *with the constant current* I_F *in it so that the rotor magnetic field in the air gap is*

$$\vec{\mathbf{B}}_R = \frac{\mu_0\ell_2 N_F I_F}{4gr}\cos(\theta-\theta_R)\hat{\mathbf{r}}.$$

Redo the Physics of the synchronous motor from the microscopic point of view by doing the following:

(a) Find the velocity $\vec{\mathbf{v}}_F$ *of the charge carriers on the axial sides of the rotor phase (field winding).*

(b) Find the magnetic force per unit charge $\vec{\mathbf{v}}_F\times\vec{\mathbf{B}}_S$ *on the charge carriers in the axial sides of the rotor phase.*

(c) Find the Lorentz force per unit charge $\vec{\mathbf{F}}/q \triangleq \vec{\mathbf{E}}_S + \vec{\mathbf{v}}_F\times\vec{\mathbf{B}}_S$ *on the charge carriers in the axial sides of the rotor phase.*

(d) Compute the emf $V_F = \oint_{\text{All rotor loops}} \left(\vec{\mathbf{F}}/q\right)\cdot d\vec{\boldsymbol{\ell}}$ *in the rotor windings/loops produced by the Lorentz force* $\vec{\mathbf{F}}/q$*. Can you explain your answer?*

(e) Compute the torque τ_R *on the rotor loop using the magnetic force per unit charge in the* $\hat{\boldsymbol{\theta}}$ *direction.*

(f) Use Faraday's law in differential form $\nabla\times\vec{\mathbf{E}}_R = -\partial\vec{\mathbf{B}}_R/\partial t$ *to find* $\vec{\mathbf{E}}_R$ *in the air gap.*

(g) In the macroscopic approach, it was shown in (5.18) that

$$\begin{aligned}\xi_{Sa} &= -d\lambda_{Sa}/dt\\ \xi_{Sb} &= -d\lambda_{Sb}/dt.\end{aligned}$$

Show that the expressions for the emf's induced in the stator windings by the rotor's magnetic field can also be found by computing

$$\begin{aligned}\xi_{Sa} &= \oint_{\substack{\text{All loops of}\\ \text{stator phase } a}} \vec{\mathbf{E}}_R\cdot d\vec{\boldsymbol{\ell}}\\ \xi_{Sb} &= \oint_{\substack{\text{All loops of}\\ \text{stator phase } b}} \vec{\mathbf{E}}_R\cdot d\vec{\boldsymbol{\ell}}.\end{aligned}$$

Why isn't it necessary to include the magnetic force per unit charge $\vec{\mathbf{v}}_{Sa}\times\vec{\mathbf{B}}_R$ *term in the computation of* ξ_{Sa} *and* ξ_{Sb}*? (*$\vec{\mathbf{v}}_{Sa}$ *is the velocity of charge carriers in the windings of stator phase a.)*

(h) Show that

$$-\frac{d(L_S i_{Sa})}{dt} = \oint_{\substack{\text{All loops of}\\ \text{stator phase } a}} \vec{\mathbf{E}}_S \cdot d\vec{\ell}$$

$$-\frac{d(L_S i_{Sb})}{dt} = \oint_{\substack{\text{All loops of}\\ \text{stator phase } b}} \vec{\mathbf{E}}_S \cdot d\vec{\ell}.$$

That is, compute the line integrals on the right-hand side and show that they are of the form given on the left-hand side with $L_S \triangleq \frac{\mu_0 \ell_1 \ell_2 \pi}{8g} N_S^2$. *(It can be shown that* $L_S i_{Sa}$ *is the flux in stator phase a due to the stator magnetic field* $\vec{\mathbf{B}}_S$*). Why isn't it necessary to include the magnetic force per unit charge* $\vec{\mathbf{v}}_{Sa} \times \vec{\mathbf{B}}_S$ *term in the computation of the line integrals?*

(i) Given that $L_S i_{Sa}$ *is the flux in stator phase a due to the stator magnetic field* $\vec{\mathbf{B}}_S$ *and the answers to the previous parts of this problem, show that the* total *voltage/emf induced in the stator phases may be written as*

$$-\frac{d(\lambda_{Sa} + L_S i_{Sa})}{dt} = \oint_{\substack{\text{All loops of}\\ \text{stator phase } a}} \left(\vec{\mathbf{E}}_R + \vec{\mathbf{E}}_S\right) \cdot d\vec{\ell}$$

$$-\frac{d(\lambda_{Sb} + L_S i_{Sb})}{dt} = \oint_{\substack{\text{All loops of}\\ \text{stator phase } b}} \left(\vec{\mathbf{E}}_R + \vec{\mathbf{E}}_S\right) \cdot d\vec{\ell}.$$

(j) Show that power is conserved; that is,

$$i_{Sa}(t)\xi_{Sa}(t) + i_{Sb}(t)\xi_{Sb}(t) = -\tau_R \omega_R.$$

Problem 19 Synchronous Machine in the Rotor Coordinate System

Consider a synchronous motor with a sinusoidally wound rotor. With the motor in steady state ($\omega_R = \omega_S$)*, consider a coordinate system attached to the rotor so that it is rotating at synchronous speed; that is, it rotates at the angular speed* $\omega_R = \omega_S$ *with respect to the stator. Further, let the coordinate system be aligned with the rotor so that in cylindrical coordinates,*

$$\begin{aligned} r' &= r \\ \theta' &= \theta - \omega_R t \\ z' &= z \end{aligned}$$

where (r, θ, z) *are the cylindrical coordinates in the stationary coordinate system of the stator and* (r', θ', z') *are the cylindrical coordinates in the rotating coordinate system of the* rotor *coordinate system is referred to as the primed (′) coordinate system;* in this coordinate system, *do the following:*

(a) Compute the electric $\vec{\mathbf{E}}'_S$ *and magnetic* $\vec{\mathbf{B}}'_S$ *fields in the air gap from the electric and magnetic fields* $\vec{\mathbf{E}}_S, \vec{\mathbf{B}}_S$ *in the stator coordinate system.*

(b) Compute the electric $\vec{\mathbf{E}}'_R$ *and magnetic* $\vec{\mathbf{B}}'_R$ *fields in the air gap from the electric and magnetic fields* $\vec{\mathbf{E}}_R, \vec{\mathbf{B}}_R$ *in the stator coordinate system.*

(c) Find the velocity $\vec{\mathbf{v}}'_F$ *of the charge carriers on the axial sides of the (field) rotor windings (note that in this coordinate system* $\omega'_R = 0$*).*

(d) Find the magnetic force per unit charge $\vec{\mathbf{v}}'_F \times \vec{\mathbf{B}}'_S$ *on the charge carriers in the axial sides of the rotor.*

(e) Compute the torque τ_R *on the rotor loop.*

(f) Compute the back emf ξ'_{Sa} *induced in stator phase a by computing the rate of change of the flux due to* $\vec{\mathbf{B}}'_R$ *in that phase. Similarly, compute* ξ'_{Sb} *in stator phase b.*

(g) Find the velocity $\vec{\mathbf{v}}'_{Sa}$ *of the charge carriers on the axial sides of the stator phase a. Note that the stator windings are rotating with angular speed* $-\omega_R$ *with respect to the rotor. Similarly, find* $\vec{\mathbf{v}}'_{Sb}$.

(h) Show that the back emf in stator phase a can also be computed as

$$\xi'_{Sa} = \oint_{rotor\ loops} \left(\vec{\mathbf{E}}'_R + \vec{\mathbf{v}}'_{Sa} \times \vec{\mathbf{B}}'_R\right) \cdot d\vec{\boldsymbol{\ell}}$$

Similarly, find ξ'_{Sb}.

(i) Compute the rotor emf $V_F = \oint_{rotor\ loops} \left(\vec{\mathbf{E}}'_S + \vec{\mathbf{v}}'_F \times \vec{\mathbf{B}}'_S\right) \cdot d\vec{\boldsymbol{\ell}}$ *in the rotor loops produced by the stator fields* $\vec{\mathbf{E}}'_S, \vec{\mathbf{B}}'_S$.

(j) Using the $\hat{\boldsymbol{\theta}}$ *component of* $\vec{\mathbf{v}}'_{Sa} \times \vec{\mathbf{B}}'_R$*, compute the torque* τ_{Sa} *exerted on the stator by the rotor. Similarly, compute* τ_{Sb}*. Is* $\tau_R = -(\tau_{Sa} + \tau_{Sb})$*?*

(k) In this coordinate system, $\omega'_R = 0$ *so that no mechanical power is being given to the rotor. Show that conservation of energy holds in this coordinate system, i.e.,* $i_{Sa}\xi'_{Sa} + i_{Sb}\xi'_{Sb} = -(\tau_{Sa} + \tau_{Sb})\omega'_S$ *where* $\omega'_S = -\omega_S$*. From the point of view of this coordinate system, is the machine acting as a motor or generator?*

Problem 20 Emfs in Full-Pitched and Short-Pitched Windings

Figure 5.35 shows a motor with a permanent magnet rotor which has a single stator phase with full-pitched turns, that is, wound with each loop spanning π *radians [see Figures 4.1 and 4.3(b) of Chapter 4]. Let the magnetic field in the air gap due to the permanent magnet rotor be given by*

$$\vec{\mathbf{B}}_R(\theta - \theta_R) = B_{max}\frac{r_R}{r}\cos(\theta - \theta_R)\hat{\mathbf{r}}.$$

(a) Let $\theta_R(t) = \omega_R t$ *and compute the axial electric field* $\vec{\mathbf{E}}_R = E_{Rz}\hat{\mathbf{z}}$ *induced in the air gap by the time varying magnetic field.*

(b) Using the expression derived for the axial electric field in part (a) to compute the emf induced in the stator phase by the PM rotor.

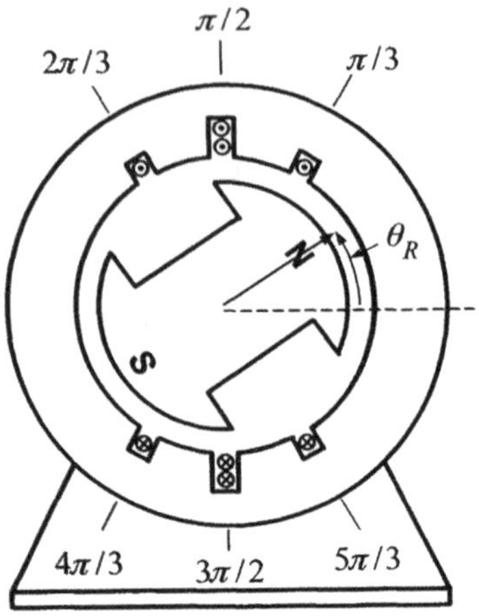

FIGURE 5.35. A machine with a PM rotor and a single stator phase.

(c) Suppose that the stator in part (b) now consists of short-pitched turns as indicated in Figure 5.36. That is, each turn spans only $150°$ *rather than* $180°$ *as in part (b). (See problem 14 of Chapter 4 for a more detailed explanation of short-pitched turns.)*

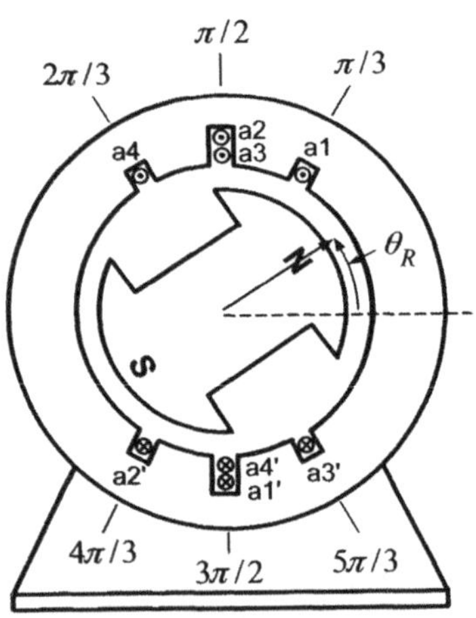

FIGURE 5.36. A motor with a PM rotor and a single stator phase whose turns are short-pitched.

Without any computation, explain why the induced emf in this short-pitched winding must be the same as in the full-pitched winding of part (b).

6

Mathematical Models of AC Machines

In this chapter, the mathematical models of the induction and synchronous machines are derived. The derivations in this chapter will be based on a two-phase machine. In Chapter 7 three-phase machines are considered, and it is shown there that any balanced three-phase induction or synchronous machine has a two-phase equivalent model identical in form to its two-phase counterpart. The models derived in this chapter are standard throughout the literature and form the basis for deriving control algorithms.

The symmetric, two-pole (single pole-pair), two-phase machine shown in Figure 6.1 is used for the derivations that follow. The stator of this machine is constructed with a hollow cylindrical shell of iron wrapped with two sets of sinusoidally wound phases, each with a total of N_S turns and 90° apart. These stator windings are denoted as phases a and b, respectively, carrying the currents i_{Sa} and i_{Sb}, respectively. Stator phase a is connected to a voltage source u_{Sa}, while stator phase b is connected to a voltage source u_{Sb}. The resistance of the windings of the two phases are assumed to be identical and equal to R_S.

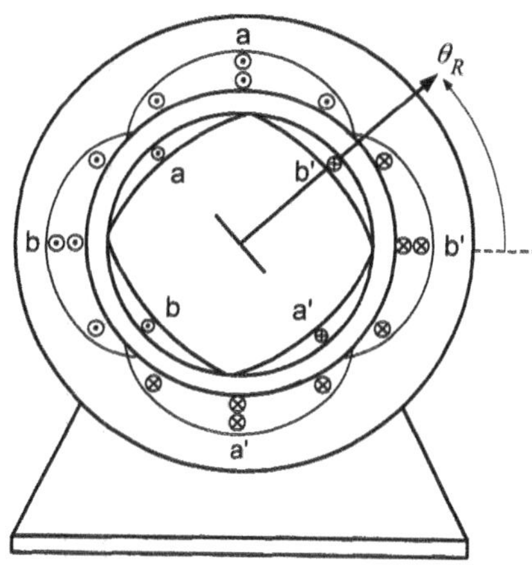

FIGURE 6.1. Representation of a two-phase sinusoidally wound machine.

The rotor consists of a cylindrical core of iron with two sinusoidally wound phases each with a total of N_R turns and 90° apart. These phases are denoted a and b as indicated in Figure 6.1. The currents in the rotor phases are denoted as i_{Ra} and i_{Rb}, respectively and the resistance of each of the rotor phase windings is denoted as R_R. At this point, no assumption is made on whether or not the ends of the rotor phases are shorted or connected to voltage sources. The rotor is located with respect to the stator by the magnetic axis of rotor phase a as shown in Figure 6.1.

The currents in the phases of the machine establish magnetic fields and these magnetic fields in turn produce fluxes in the windings of all four of the machine's phases. In order to obtain a mathematical model of an AC machine, an expression for the total flux linkage in each of the stator and rotor phases due to the currents i_{Sa}, i_{Sb}, i_{Ra}, and i_{Rb} must be found. However, before these flux linkages can be computed, an expression for the magnetic field $\vec{\mathbf{B}}_R(i_{Ra}, i_{Rb}, r, \theta)$ in the air gap produced by the rotor currents is required. An expression for the magnetic field $\vec{\mathbf{B}}_S(i_{Sa}, i_{Sb}, r, \theta)$ in the air gap produced by the stator currents was derived in Chapter 4. Also, the concept of leakage is explained and the expressions for the magnetic fields $\vec{\mathbf{B}}_S, \vec{\mathbf{B}}_R$ are modified to account for it. The modified expressions for the magnetic fields $\vec{\mathbf{B}}_S$ and $\vec{\mathbf{B}}_R$ are then used to compute the flux linkage in each phase of the machine.

6.1 The Magnetic Field $\vec{\mathbf{B}}_R(i_{Ra}, i_{Rb}, r, \theta - \theta_R)$

The radial magnetic field $\vec{\mathbf{B}}_{Ra}$ established *in the air gap* by the current i_{Ra} in the sinusoidally distributed winding of rotor phase a is now computed. The winding/turns density of rotor phase a is $(N_R/2)\,|\sin(\theta - \theta_R)|$ so that the number of turns between θ and $\theta + d\theta$ is $(N_R/2)\,|\sin(\theta - \theta_R)|\,d\theta$. To proceed, Ampère's law $\oint \vec{\mathbf{H}} \cdot d\vec{\ell} = i_{enclosed}$ (with $\vec{\mathbf{H}} \equiv 0$ in the magnetic material) is applied to the closed path 1–2–3–4–1 drawn in Figure 6.2 to obtain

$$\int_4^1 \vec{\mathbf{H}}_{Ra} \cdot d\vec{\ell} + \int_2^3 \vec{\mathbf{H}}_{Ra} \cdot d\vec{\ell} = \int_{\theta'=\theta_R}^{\theta'=\theta} i_{Ra}\frac{N_R}{2}\sin(\theta' - \theta_R)d\theta'.$$

Evaluating gives

$$\int_{\ell=0}^{\ell=g} H_{Ra}(i_{Ra}, \theta_R)\hat{\mathbf{r}} \cdot (d\ell\hat{\mathbf{r}}) + \int_{\ell=0}^{\ell=g} H_{Ra}(i_{Ra}, \theta)\hat{\mathbf{r}} \cdot (-d\ell\hat{\mathbf{r}})$$
$$= -i_{Ra}\frac{N_R}{2}\cos(\theta - \theta_R) + i_{Ra}\frac{N_R}{2}$$

or

$$H_{Ra}(i_{Ra}, \theta_R)g - H_{Ra}(i_{Ra}, \theta)g = -i_{Ra}\frac{N_R}{2}\cos(\theta - \theta_R) + i_{Ra}\frac{N_R}{2}.$$

It was assumed that $\vec{\mathbf{H}}_{Ra}$ is *constant* across the air gap. Rearranging gives

$$H_{Ra}(i_{Ra},\theta) = i_{Ra}\frac{N_R}{2g}\cos(\theta-\theta_R) + H_{Ra}(i_{Ra},\theta_R) - i_{Ra}\frac{N_R}{2g} \qquad (6.1)$$

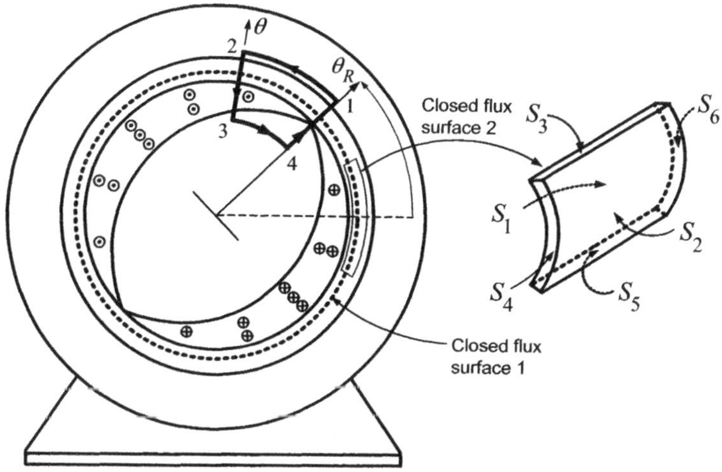

FIGURE 6.2. Use of Ampère's and Gauss's laws to determine B_{Ra} in the air gap due to i_{Ra}. (Rotor phase b is not shown for clarity.)

In this equation, both $H_{Ra}(i_{Ra},\theta)$, $H_{Ra}(i_{Ra},\theta_R)$ are unknown. As in the case of the magnetic field produced by the stator currents, conservation of flux $\oint_S \vec{\mathbf{B}}\cdot d\vec{\mathbf{S}} = 0$ is applied to the closed surface 1 which encloses the rotor (see the dashed circle in Figure 6.2 around the rotor iron) to show that $H_{Ra}(i_{Ra},\theta_R) = i_{Ra}(N_R/2g)$. Then,

$$\begin{aligned} H_{Ra}(i_{Ra},\theta) &= \frac{N_R}{2g} i_{Ra}\cos(\theta-\theta_R) \\ B_{Ra}(i_{Ra},\theta) &= \frac{\mu_0 N_R}{2g} i_{Ra}\cos(\theta-\theta_R). \end{aligned} \qquad (6.2)$$

In applying Ampère's law, it was assumed that $\vec{\mathbf{B}}_{Ra} = \mu_0\vec{\mathbf{H}}_{Ra}$ was constant across the air gap in the radial direction, that is, $\vec{\mathbf{B}}_{Ra}$ does not depend on the cylindrical coordinate r. However, as shown in Chapter 4 for the stator magnetic field, in order that $\vec{\mathbf{B}}_{Ra}$ satisfy the conservation of flux for the closed flux surface 2 shown in Figure 6.2 whose sides are $S_1, S_2, S_3, S_3, S_4, S_5$, and S_6, equation (6.2) is modified by setting

$$\vec{\mathbf{B}}_{Ra}(i_{Ra},r,\theta-\theta_R) = B_{Ra}(i_{Ra},r,\theta-\theta_R)\hat{\mathbf{r}} \triangleq \frac{\mu_0 N_R}{2g}\frac{r_R}{r} i_{Ra}\cos(\theta-\theta_R)\hat{\mathbf{r}} \qquad (6.3)$$

where r_R is the rotor radius.[1]

Similarly, the magnetic field $\vec{\mathbf{B}}_{Rb}$ established in the air gap by the sinusoidally wound rotor phase b (rotated 90° with respect to rotor phase a) carrying the current i_{Rb} is given by

$$\begin{aligned}\vec{\mathbf{B}}_{Rb}(i_{Rb}, r, \theta - \theta_R) &= B_{Rb}(i_{Rb}, r, \theta - \theta_R)\hat{\mathbf{r}} \\ &= \frac{\mu_0 N_R}{2g}\frac{r_R}{r} i_{Rb}\cos(\theta - \pi/2 - \theta_R)\hat{\mathbf{r}} \\ &= \frac{\mu_0 N_R}{2g}\frac{r_R}{r} i_{Rb}\sin(\theta - \theta_R)\hat{\mathbf{r}}. \end{aligned} \tag{6.4}$$

In summary, the magnetic fields due to the stator and rotor currents are

$$\begin{aligned}\vec{\mathbf{B}}_S(i_{Sa}, i_{Sb}, r, \theta) &= \vec{\mathbf{B}}_{Sa}(i_{Sa}, r, \theta) + \vec{\mathbf{B}}_{Sb}(i_{Sb}, r, \theta) \\ &= \frac{\mu_0 r_R N_S}{2g}\frac{1}{r}\left(i_{Sa}\cos(\theta) + i_{Sb}\sin(\theta)\right)\hat{\mathbf{r}} \end{aligned} \tag{6.5}$$

and

$$\begin{aligned}\vec{\mathbf{B}}_R(i_{Ra}, i_{Rb}, r, \theta - \theta_R) &= \vec{\mathbf{B}}_{Ra}(i_{Ra}, r, \theta - \theta_R) + \vec{\mathbf{B}}_{Rb}(i_{Rb}, r, \theta - \theta_R) \\ &= \frac{\mu_0 N_R r_R}{2g}\frac{1}{r}\left(i_{Ra}\cos(\theta - \theta_R) + i_{Rb}\sin(\theta - \theta_R)\right)\hat{\mathbf{r}}, \end{aligned} \tag{6.6}$$

respectively.

6.2 Leakage

Consider Figure 6.3, which shows a single-phase distributed winding on the rotor. It has been assumed up to now that the magnetic field produced by the rotor current is *radially* directed in the air gap so that only the surfaces S_1 and S_2 in Figure 6.3 have nonzero flux. With this assumption and along with the (outward) surface element vectors to the closed surface of Figure 6.3 given by

$$d\vec{\mathbf{S}} = \begin{cases} -r_R d\theta dz\hat{\mathbf{r}} & \text{on } S_1 \\ r_S d\theta dz\hat{\mathbf{r}} & \text{on } S_2 \\ dr dz\hat{\boldsymbol{\theta}} & \text{on } S_3 \\ r d\theta dr\hat{\mathbf{z}} & \text{on } S_4 \\ -dr dz\hat{\boldsymbol{\theta}} & \text{on } S_5 \\ -r d\theta dr\hat{\mathbf{z}} & \text{on } S_6, \end{cases}$$

[1] It is now easy to see that $\oint_S \vec{\mathbf{B}}_{Ra} \cdot d\vec{\mathbf{S}} = 0$ for *any* closed surface in the air gap. Ampere's law will also continue to hold for $g << r_R$ using the approximation $\int_{\ell=0}^{\ell=g} r_R/(r_R + \ell)\, d\ell \approx g$.

it was shown that conservation of flux $\oint_S \vec{\mathbf{B}}_R \cdot d\vec{\mathbf{S}} = 0$ required a $1/r$ dependence by the magnetic field in the air gap (see Chapter 4). That is, up to now, it has been assumed that $\vec{\mathbf{B}}_R \cdot d\vec{\mathbf{S}} \equiv 0$ on the surfaces S_3, S_4, S_5, and S_6 so that the radial magnetic field $\vec{\mathbf{B}}_R$ must decrease as $1/r$ going from S_1 to S_2 in order to satisfy conservation of flux. However, in a real machine, the slots have finite dimensions, the air gap is of finite size and there are end effects at either end of the rotor. As a consequence, the magnetic field does have components in the azimuthal $\hat{\boldsymbol{\theta}}$ direction and axial $\hat{\mathbf{z}}$ direction in addition to the radial direction $\hat{\mathbf{r}}$ so that the surfaces S_3, S_4, S_5, and S_6 also have nonzero flux.

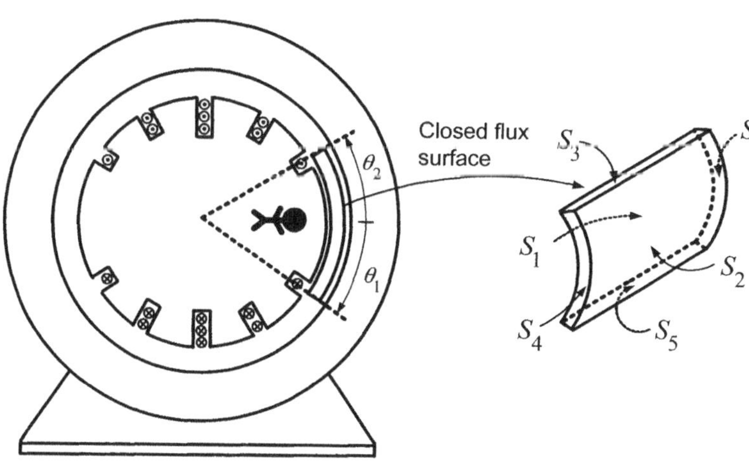

FIGURE 6.3. Closed flux surface to illustrate flux leakage. With the assumption of no leakage, only the surfaces S_1, S_2 have nonzero flux. With leakage, all six surfaces have nonzero flux through them.

To see how this affects the radial magnetic field, Figure 6.4 shows an enlargement of the cross section of Figure 6.3 from the point of view of the observer in Figure 6.3 looking *into* the page. In Figure 6.4, the lines of the magnetic field $\vec{\mathbf{B}}_R$ produced by the rotor currents are shown. At the slot openings, there is no iron so that in this vicinity the magnetic field tends to circle around the rotor winding rather than go across the air gap. Even away from the slots, the magnetic field spreads out from the radial direction due to the finite length of the air gap and the finite (axial) length of the machine. What this all means is that the average radial $\vec{\mathbf{B}}_R$ field on the surface S_2 actually decreases a little more than $1/r$. This now considered in more detail.

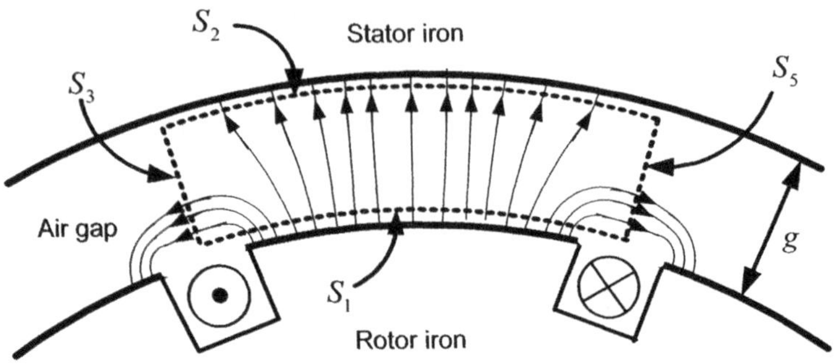

FIGURE 6.4. Cross-sectional view of the flux surface in Figure 6.3. The rotor magnetic field $\vec{\mathbf{B}}_R$ is perpendicular at the surface of the rotor and then spreads out in both the azimuthal and axial directions so the flux it produces at the inside surface of the stator is less that at the surface of the rotor.

To simplify the discussion, it is now assumed that there is no spreading in axial (z) direction so that $\int_{S_4} \vec{\mathbf{B}}_R \cdot d\vec{\mathbf{S}} = \int_{S_6} \vec{\mathbf{B}}_R \cdot d\vec{\mathbf{S}} = 0$. Using an outward normal, $\oint \vec{\mathbf{B}}_R \cdot d\vec{\mathbf{S}} = 0$ is computed on the closed surface shown in Figures 6.3 and 6.4 to obtain

$$\oint \vec{\mathbf{B}}_R \cdot d\vec{\mathbf{S}} = \int_{S_1} \vec{\mathbf{B}}_R \cdot d\vec{\mathbf{S}} + \int_{S_2} \vec{\mathbf{B}}_R \cdot d\vec{\mathbf{S}} + \int_{S_3} \vec{\mathbf{B}}_R \cdot d\vec{\mathbf{S}} + \int_{S_5} \vec{\mathbf{B}}_R \cdot d\vec{\mathbf{S}} = 0.$$

Rearranging, this becomes

$$\int_{S_1} \vec{\mathbf{B}}_R \cdot \left(-d\vec{\mathbf{S}}\right) = \int_{S_2} \vec{\mathbf{B}}_R \cdot d\vec{\mathbf{S}} + \int_{S_3} \vec{\mathbf{B}}_R \cdot d\vec{\mathbf{S}} + \int_{S_5} \vec{\mathbf{B}}_R \cdot d\vec{\mathbf{S}}$$

or

$$\begin{aligned}\int_{S_1} B_R(r_R,\theta,t)\hat{\mathbf{r}} \cdot (r_R d\theta dz\hat{\mathbf{r}}) &= \int_{S_2} B_R(r_S,\theta,t)\hat{\mathbf{r}} \cdot (r_S d\theta dz\hat{\mathbf{r}}) \\ &\quad + \int_{S_3} \vec{\mathbf{B}}_R \cdot d\vec{\mathbf{S}} + \int_{S_5} \vec{\mathbf{B}}_R \cdot d\vec{\mathbf{S}}. \end{aligned} \tag{6.7}$$

The surface S_1 is just at the surface of the rotor and the surface element vector $-d\vec{\mathbf{S}} = r_R d\theta dz\hat{\mathbf{r}}$ points in the positive radial direction on the surface S_1. If there was no spreading, then $\int_{S_1} B_R(r_R,\theta,t)\hat{\mathbf{r}} \cdot (r_R d\theta dz\hat{\mathbf{r}}) = \int_{S_2} B_R(r_S,\theta,t)\hat{\mathbf{r}} \cdot (r_S d\theta dz\hat{\mathbf{r}})$; that is, if both surfaces use the normal $\hat{\mathbf{r}}$, then the flux through S_1 equals the flux through S_2. However, the magnetic field does spread out in azimuthal direction and the fluxes through the surfaces S_2, S_3, and S_5 are then all *positive*. In this case, the radial $\vec{\mathbf{B}}_R$ field on the surface S_2 must be less on average than it was when it was assumed there was no flux through the surfaces S_3 and S_5. This is simply because the flux $\int_{S_1} B_R(r_R,\theta,t)\hat{\mathbf{r}} \cdot (r_R d\theta dz\hat{\mathbf{r}})$ on the left-hand side of (6.7) stays the same whether or not there is spreading[2] and is always equal to the sum of the fluxes on the right-hand side of (6.7).

To model this effect, a parameter κ, with $0 < \kappa < 1$, is introduced so that on the *stator side* of the air gap, the *rotor* magnetic field is given by

$$\vec{\mathbf{B}}_R(i_{Ra}, i_{Rb}, r, \theta - \theta_R)\Big|_{r=r_S} = \kappa\frac{\mu_0 N_R}{2g}\frac{r_R}{r_S}\left(i_{Ra}\cos(\theta - \theta_R) + i_{Rb}\sin(\theta - \theta_R)\right)\hat{\mathbf{r}}.$$

That is, the rotor radial $\vec{\mathbf{B}}_R$ field decreases to $\kappa r_R/r_S$ (rather than just r_R/r_S) in going from the surface of the rotor to the inside surface of the stator. The parameter κ is referred to as a *global coupling factor* [1].

Similarly, on the *rotor* side of the air gap, the *stator* magnetic field is given by

$$\vec{\mathbf{B}}_S(i_{Sa}, i_{Sb}, r, \theta)\Big|_{r=r_R} = \kappa\frac{\mu_0 N_S}{2g}\frac{r_R}{r_R}\left(i_{Sa}\cos(\theta) + i_{Sb}\sin(\theta)\right)\hat{\mathbf{r}}$$

so that the radial dependence of the stator $\vec{\mathbf{B}}_S$ field increases from r_R/r_S to only $\kappa r_R/r_R$ (rather than $r_R/r_R = 1$) in going from the inside surface of the stator to the surface of the rotor. The above example is not the only source of leakage,[3] but all such effects can be modeled for the purposes here by a single parameter κ.

In summary, the coupling factor κ is included in the expression for $\vec{\mathbf{B}}_R$ on the *stator side* of the air gap to account for $\vec{\mathbf{B}}_R$ decreasing at a rate slightly less than $1/r$ (due to leakage) going across the air gap to the stator side. Similarly, κ is included in the expression for $\vec{\mathbf{B}}_S$ on the *rotor side* of the air gap to account for leakage effect going across the air gap.

It may seem rather inconsistent with the development of the mathematical model of the machine to now add in this leakage effect. That is, up to this point, ideal magnetic materials have been assumed, the air gap length has been assumed to be small, and the cross sections of the windings/slots have been assumed to be negligible. However, it will be seen shortly that this leakage factor will be essential to obtain a realistic model.

---

[2] As iron is a good magnetic material, $\vec{\mathbf{B}}_R$ is perpendicular at both the surface of the rotor iron and the stator iron.

[3] For example, a similar analysis can be done to account for the leakage through S_4 and S_6.

6.3 Flux Linkages in AC Machines

Expressions for the flux linkages in both the rotor and stator windings are now computed.

6.3.1 *Flux Linkages in the Stator Phases*

The computation of the total flux linkage in the sinusoidally wound stator phase a is now considered. The flux in stator phase a is due to the magnetic fields produced by both the stator currents i_{Sa} and i_{Sb}, and the rotor currents i_{Ra} and i_{Rb}. That is,

$$\lambda_{Sa}(i_{Ra}, i_{Rb}, i_{Sa}, i_{Sb}, \theta_R) \triangleq \int_{\substack{\text{All loops of} \\ \text{stator phase } a}} (\vec{\mathbf{B}}_S + \vec{\mathbf{B}}_R) \cdot d\vec{\mathbf{S}}.$$

This computation is carried out in two parts by separately considering the flux linkage due to the stator currents and then the flux linkage due to the rotor currents. That is,

$$\lambda_{Sa}(0, 0, i_{Sa}, i_{Sb}, \theta_R) \triangleq \int_{\substack{\text{All loops of} \\ \text{stator phase } a}} \vec{\mathbf{B}}_S \cdot d\vec{\mathbf{S}}$$

and

$$\lambda_{Sa}(i_{Ra}, i_{Rb}, 0, 0, \theta_R) \triangleq \int_{\substack{\text{All loops of} \\ \text{stator phase } a}} \vec{\mathbf{B}}_R \cdot d\vec{\mathbf{S}}$$

so that

$$\lambda_{Sa}(i_{Ra}, i_{Rb}, i_{Sa}, i_{Sb}, \theta_R) = \lambda_{Sa}(0, 0, i_{Sa}, i_{Sb}, \theta_R) + \lambda_{Sa}(i_{Ra}, i_{Rb}, 0, 0, \theta_R).$$

Stator Flux Linkage Produced by the Stator Currents

Consider a single turn (loop) of stator phase a at the angular position θ with $0 \le \theta \le \pi$, that is, one side of the loop is at θ and the other side is at $\theta - \pi$. Choose the flux surface S for this single turn to lie just inside the air gap next to the stator surface as shown in Figure 6.5.

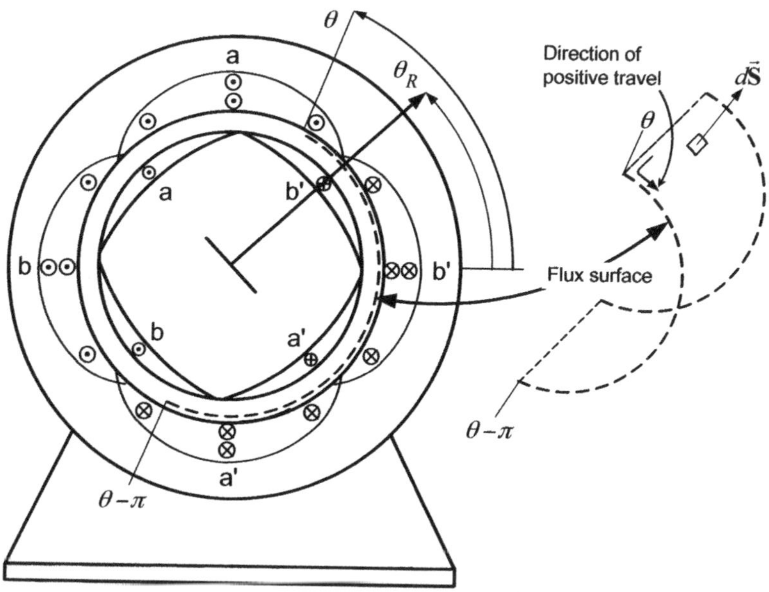

FIGURE 6.5. Surface to compute the flux in a single turn of stator phase a produced by $\vec{\mathbf{B}}_S$.

To compute the flux through the surface, the surface element vector $d\vec{\mathbf{S}} = r_S d\theta' dz\hat{\mathbf{r}}$ is chosen. The flux $\phi_{Sa}(i_{Sa}, i_{Sb}, \theta)$ in each turn of stator phase a at the angular position θ of the stator is then given by

$$\begin{aligned}
&\phi_{Sa}(i_{Sa}, i_{Sb}, \theta) \\
&\quad \triangleq \int_{\theta'=\theta-\pi}^{\theta'=\theta} \int_{z=0}^{z=\ell_1} \frac{\mu_0 r_R N_S}{2g} \frac{1}{r_S} \Big(i_{Sa}\cos(\theta') + i_{Sb}\sin(\theta') \Big) \hat{\mathbf{r}} \cdot (r_S d\theta' dz \hat{\mathbf{r}}) \\
&\quad = \frac{\mu_0 r_R \ell_1 N_S}{2g} \int_{\theta'=\theta-\pi}^{\theta'=\theta} \Big(i_{Sa}\cos(\theta') + i_{Sb}\sin(\theta') \Big) d\theta' \\
&\quad = \frac{\mu_0 r_R \ell_1 N_S}{2g} \Big(i_{Sa}\sin(\theta') - i_{Sb}\cos(\theta') \Big) \Big|_{\theta'=\theta-\pi}^{\theta'=\theta} \\
&\quad = \frac{\mu_0 r_R \ell_1 N_S}{2g} \Big(i_{Sa}\big(\sin(\theta) - \sin(\theta-\pi)\big) - i_{Sb}\big(\cos(\theta) - \cos(\theta-\pi)\big) \Big) \\
&\quad = \frac{\mu_0 r_R \ell_1 N_S}{g} \Big(i_{Sa}\sin(\theta) - i_{Sb}\cos(\theta) \Big). \qquad (6.8)
\end{aligned}$$

Remark Note that by taking the flux surface normal to be $d\vec{\mathbf{S}} = r_S d\theta' dz\hat{\mathbf{r}}$, the positive direction of travel around the turns of stator phase a at the angular position θ is out of the page on side a and into the page on side a'; that is, it is the same direction as positive current flow in stator phase a. If $-d\phi_{Sa}/dt > 0$, then this emf will push current in the same direction as that chosen for the direction of positive current.

Between the angular positions θ and $\theta + d\theta$ of stator phase a there are $(N_S/2)\sin(\theta)d\theta$ turns each having the flux (6.8) in them. The flux *linkage* in the turns of phase a between θ and $\theta + d\theta$ is then

$$\begin{aligned} d\lambda_{Sa} &\triangleq \phi_{Sa}(i_{Sa}, i_{Sb}, \theta)(N_S/2)\sin(\theta)d\theta \\ &= \frac{\mu_0 r_R \ell_1 N_S}{g}\left(i_{Sa}\sin(\theta) - i_{Sb}\cos(\theta)\right)\frac{N_S}{2}\sin(\theta)d\theta. \end{aligned} \tag{6.9}$$

To obtain the flux linkage in all the turns of stator phase a, simply integrate (6.9) as θ varies from 0 to π, that is,

$$\begin{aligned} \lambda_{Sa}(0,0,i_{Sa},i_{Sb},\theta_R) &= \int_{\substack{\text{All loops of} \\ \text{stator phase } a}} \vec{\mathbf{B}}_S \cdot d\vec{\mathbf{S}} \\ &= \int_{\theta=0}^{\theta=\pi} \frac{\mu_0 r_R \ell_1 N_S}{g}\Big(i_{Sa}\sin(\theta) - i_{Sb}\cos(\theta)\Big)\frac{N_S}{2}\sin(\theta)d\theta \\ &= \frac{\mu_0 r_R \ell_1 N_S^2}{2g}\int_{\theta=0}^{\theta=\pi}\Big(i_{Sa}\sin^2(\theta) - i_{Sb}\cos(\theta)\sin(\theta)\Big)d\theta \\ &= \frac{\mu_0 r_R \ell_1 N_S^2}{2g} i_{Sa}\frac{\pi}{2} \\ &= L_S i_{Sa} \end{aligned} \tag{6.10}$$

where

$$L_S \triangleq \frac{\mu_0 r_R \ell_1 \pi N_S^2}{4g} = \frac{\mu_0 \ell_1 \ell_2 \pi}{8g} N_S^2.$$

Note that for this two-phase machine, the magnetic field produced by the current i_{Sb} in phase b does not produce any net flux linkage in phase a. Similarly,

$$\begin{aligned} \lambda_{Sb}(0,0,i_{Sa},i_{Sb},\theta_R) &= \int_{\substack{\text{All loops of} \\ \text{stator phase } b}} \vec{\mathbf{B}}_S \cdot d\vec{\mathbf{S}} \\ &= \frac{\mu_0 r_R \ell_1 \pi N_S^2}{4g} i_{Sb} \\ &= L_S i_{Sb}. \end{aligned} \tag{6.11}$$

Stator Flux Linkage Produced by the Rotor Currents

The flux linkage in stator phase a due to magnetic field produced by the rotor currents is now computed from

$$\lambda_{Sa}(i_{Ra}, i_{Rb}, 0, 0, \theta_R) \triangleq \int_{\substack{\text{All loops of}\\ \text{stator phase } a}} \vec{\mathbf{B}}_R \cdot d\vec{\mathbf{S}}.$$

The flux surface of Figure 6.5 is also used for this computation. With $d\vec{\mathbf{S}} = r_S d\theta' dz \hat{\mathbf{r}}$ as the surface element vector, the flux $\phi_{Sa}(i_{Ra}, i_{Rb}, \theta - \theta_R)$ in each turn of stator phase a at the angular position θ of the stator due to the magnetic field $\vec{\mathbf{B}}_R$ is

$$\begin{aligned}
\phi_{Sa}(i_{Ra}, i_{Rb}, \theta - \theta_R) &\triangleq \int_{\theta'=\theta-\pi}^{\theta'=\theta} \int_{z=0}^{z=\ell_1} \kappa \frac{\mu_0 N_R r_R}{2g} \frac{1}{r_S} \Big(i_{Ra} \cos(\theta' - \theta_R) \\
&\qquad + i_{Rb} \sin(\theta' - \theta_R) \hat{\mathbf{r}} \cdot (r_S d\theta' dz \hat{\mathbf{r}}) \Big) \\
&= \kappa \frac{\mu_0 r_R \ell_1 N_R}{2g} \int_{\theta'=\theta-\pi}^{\theta'=\theta} \Big(i_{Ra} \cos(\theta' - \theta_R) + i_{Rb} \sin(\theta' - \theta_R) \Big) d\theta' \\
&= \kappa \frac{\mu_0 r_R \ell_1 N_R}{g} \Big(i_{Ra} \sin(\theta - \theta_R) - i_{Rb} \cos(\theta - \theta_R) \Big). \qquad (6.12)
\end{aligned}$$

Note that the factor κ has been included as the rotor's magnetic field is being evaluated on the stator side of the air gap. Between the angular positions θ and $\theta + d\theta$, there are $(N_S/2)\sin(\theta) d\theta$ turns, each having the flux (6.12) in them due to $\vec{\mathbf{B}}_R$. The flux linkage $d\lambda_{Sa}$ in the turns of stator phase a between θ and $\theta + d\theta$ produced by the rotor's magnetic field is then

$$\begin{aligned}
d\lambda_{Sa} &= \phi_{Sa}(i_{Ra}, i_{Rb}, \theta - \theta_R) \frac{N_S}{2} \sin(\theta) d\theta \\
&= \kappa \frac{\mu_0 r_R \ell_1 N_R N_S}{2g} \Big(i_{Ra} \sin(\theta - \theta_R) - i_{Rb} \cos(\theta - \theta_R) \Big) \sin(\theta) d\theta.
\end{aligned} \qquad (6.13)$$

To obtain the total flux linkage due to the rotor's magnetic field in all the turns of stator phase a, integrate (6.13) as θ varies from 0 to π to obtain

$$\lambda_{Sa}(i_{Ra}, i_{Rb}, 0, 0, \theta_R) = \int_{\substack{\text{All loops of}\\ \text{stator phase } a}} \vec{\mathbf{B}}_R \cdot d\vec{\mathbf{S}}$$

$$= \int_{\theta=0}^{\theta=\pi} \kappa \frac{\mu_0 r_R \ell_1 N_S N_R}{2g} \left(i_{Ra}\sin(\theta - \theta_R) - i_{Rb}\cos(\theta-\theta_R)\right)\sin(\theta) d\theta$$

$$= \kappa \frac{\mu_0 r_R \ell_1 N_S N_R}{2g}\Big(i_{Ra}\int_{\theta=0}^{\theta=\pi}\Big(\sin(\theta)\cos(\theta_R) - \cos(\theta)\sin(\theta_R)\Big)\sin(\theta)d\theta$$

$$-i_{Rb}\int_{\theta=0}^{\theta=\pi}(\cos(\theta)\cos(\theta_R) + \sin(\theta)\sin(\theta_R))\sin(\theta)d\theta\Big)$$

$$= \kappa \frac{\mu_0 r_R \ell_1 N_S N_R}{2g}\int_{\theta=0}^{\theta=\pi}\Big(i_{Ra}\sin^2(\theta)\cos(\theta_R) - i_{Rb}\sin^2(\theta)\sin\theta_R\Big)d\theta$$

$$= M\Big(i_{Ra}\cos(\theta_R) - i_{Rb}\sin(\theta_R)\Big) \tag{6.14}$$

where

$$M \triangleq \kappa \frac{\mu_0 \pi r_R \ell_1 N_S N_R}{4g} = \kappa \frac{\mu_0 \pi \ell_1 \ell_2 N_S N_R}{8g}.$$

Similarly,

$$\lambda_{Sb}(i_{Ra}, i_{Rb},, 0, 0, \theta_R) = \int_{\substack{\text{All loops of}\\ \text{stator phase } b}} \vec{\mathbf{B}}_R \cdot d\vec{\mathbf{S}}$$

$$= M\Big(i_{Ra}\sin(\theta_R) + i_{Rb}\cos(\theta_R)\Big). \tag{6.15}$$

Total Flux Linkage in the Stator Phases

Combining (6.10), (6.11), (6.14), and (6.15), the total flux linkage in each of the stator phases is then

$$\lambda_{Sa}(i_{Ra}, i_{Rb}, i_{Sa}, i_{Sb}, \theta_R) = \int_{\substack{\text{All loops of}\\ \text{stator phase } a}} (\vec{\mathbf{B}}_S + \vec{\mathbf{B}}_R) \cdot d\vec{\mathbf{S}}$$

$$= L_S i_{Sa} + M\Big(i_{Ra}\cos(\theta_R) - i_{Rb}\sin(\theta_R)\Big) \tag{6.16}$$

$$\lambda_{Sb}(i_{Ra}, i_{Rb}, i_{Sa}, i_{Sb}, \theta_R) = \int_{\substack{\text{All loops of}\\ \text{stator phase } b}} (\vec{\mathbf{B}}_S + \vec{\mathbf{B}}_R) \cdot d\vec{\mathbf{S}}$$

$$= L_S i_{Sb} + M\Big(i_{Ra}\sin(\theta_R) + i_{Rb}\cos(\theta_R)\Big).$$

6.3.2 *Flux Linkages in the Rotor Phases*

The flux linkages in the two sinusoidally wound phases of the rotor are computed according to

$$\lambda_{Ra}(i_{Ra}, i_{Rb}, i_{Sa}, i_{Sb}, \theta_R) \triangleq \int_{\substack{\text{All loops of}\\ \text{rotor phase } a}} (\vec{\mathbf{B}}_S + \vec{\mathbf{B}}_R) \cdot d\vec{\mathbf{S}}$$

$$\lambda_{Rb}(i_{Ra}, i_{Rb}, i_{Sa}, i_{Sb}, \theta_R) \triangleq \int_{\substack{\text{All loops of}\\ \text{rotor phase } b}} (\vec{\mathbf{B}}_S + \vec{\mathbf{B}}_R) \cdot d\vec{\mathbf{S}}.$$

Starting with rotor phase a and, as in the case of the stator flux linkages, this computation is done in two parts as

$$\lambda_{Ra}(0, 0, i_{Sa}, i_{Sb}, \theta_R) \triangleq \int_{\substack{\text{All loops of}\\ \text{rotor phase } a}} \vec{\mathbf{B}}_S \cdot d\vec{\mathbf{S}}$$

and

$$\lambda_{Ra}(i_{Ra}, i_{Rb}, 0, 0, \theta_R) \triangleq \int_{\substack{\text{All loops of}\\ \text{rotor phase } a}} \vec{\mathbf{B}}_R \cdot d\vec{\mathbf{S}}.$$

The total flux linkage is then

$$\lambda_{Ra}(i_{Ra}, i_{Rb}, i_{Sa}, i_{Sb}, \theta_R) = \lambda_{Ra}(0, 0, i_{Sa}, i_{Sb}, \theta_R) + \lambda_{Ra}(i_{Ra}, i_{Rb}, 0, 0, \theta_R).$$

Rotor Flux Linkage Produced by the Stator Currents

The flux in the rotor phase due to the magnetic field established by the stator currents is given by

$$\lambda_{Ra}(0, 0, i_{Sa}, i_{Sb}, \theta_R) \triangleq \int_{\substack{\text{All loops of}\\ \text{rotor phase } a}} \vec{\mathbf{B}}_S \cdot d\vec{\mathbf{S}}.$$

To do this computation, consider a single turn (loop) of the rotor phase at the angular position θ with $0 \leq \theta - \theta_R \leq \pi$ so that one side of the rotor loop is at θ and the other side is at $\theta - \pi$. The flux surface S is chosen to lie just inside the air gap as shown in Figure 6.6.

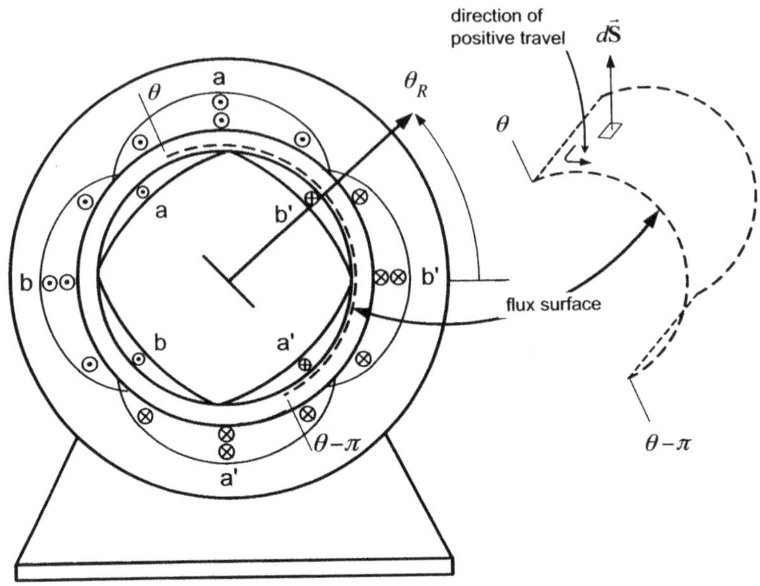

FIGURE 6.6. Surface to compute the flux in a single turn of rotor phase a produced by $\vec{\mathbf{B}}_S$.

On the flux surface, the surface element vector is chosen as $d\vec{\mathbf{S}} = r_R d\theta' dz \hat{\mathbf{r}}$. The flux $\phi_{Ra}(i_{Sa}, i_{Sb}, \theta)$ in each turn of rotor phase a at the angular position θ of the rotor is given by

$$\begin{aligned}
&\phi_{Ra}(i_{Sa}, i_{Sb}, \theta)\\
&\quad= \int_{\theta'=\theta-\pi}^{\theta'=\theta} \int_{z=0}^{z=\ell_1} \kappa \frac{\mu_0 r_R N_S}{2g} \frac{1}{r_R} \Big(i_{Sa}\cos(\theta') + i_{Sb}\sin(\theta') \Big) \hat{\mathbf{r}} \cdot (r_R d\theta' dz \hat{\mathbf{r}})\\
&\quad= \kappa \frac{\mu_0 r_R \ell_1 N_S}{2g} \int_{\theta'=\theta-\pi}^{\theta'=\theta} \Big(i_{Sa}\cos(\theta') + i_{Sb}\sin(\theta') \Big) d\theta'\\
&\quad= \kappa \frac{\mu_0 r_R \ell_1 N_S}{2g} \Big(i_{Sa}\sin(\theta') - i_{Sb}\cos(\theta') \Big)\Big|_{\theta'=\theta-\pi}^{\theta'=\theta}\\
&\quad= \kappa \frac{\mu_0 r_R \ell_1 N_S}{2g} \Big(i_{Sa}(\sin(\theta) - \sin(\theta-\pi)) - i_{Sb}(\cos(\theta) - \cos(\theta-\pi) \Big)\\
&\quad= \kappa \frac{\mu_0 r_R \ell_1 N_S}{g} \Big(i_{Sa}\sin(\theta) - i_{Sb}\cos(\theta) \Big).
\end{aligned} \tag{6.17}$$

Remarks Note that the factor κ has been included as the rotor's magnetic field is being evaluated on the stator side of the air gap. Also, note that by taking the flux surface normal to be $d\vec{\mathbf{S}} = r_R d\theta' dz \hat{\mathbf{r}}$, the positive direction of travel around the turns of rotor phase a at the angular position θ is out of the page on side a and into the page on side a'; that is, it coincides with the direction chosen for positive current flow in rotor phase a.

Between the angular positions θ and $\theta + d\theta$, there are $(N_R/2)\sin(\theta - \theta_R)d\theta$ turns each having the flux (6.17) in them. The flux linkage $d\lambda_{Ra}$ in the turns of phase a between θ and $\theta + d\theta$ produced by the stator's magnetic field is then

$$\begin{aligned} d\lambda_{Ra} &\triangleq \phi_{Ra}(i_{Sa}, i_{Sb}, \theta)\frac{N_R}{2}\sin(\theta - \theta_R)d\theta \\ &= \kappa\frac{\mu_0 r_R \ell_1 N_S N_R}{2g}\left(i_{Sa}\sin(\theta) - i_{Sb}\cos(\theta)\right)\sin(\theta - \theta_R)d\theta. \end{aligned} \tag{6.18}$$

To obtain the flux linkage in all the turns of rotor phase a, (6.18) is integrated between $\theta = \theta_R$ and $\theta = \theta_R + \pi$, that is,

$$\begin{aligned} &\lambda_{Ra}(0,0,i_{Sa},i_{Sb},\theta_R) = \int_{\substack{\text{All loops of}\\ \text{rotor phase } a}} \vec{\mathbf{B}}_S \cdot d\vec{\mathbf{S}} \\ &- \int_{\theta=\theta_R}^{\theta=\theta_R+\pi} \kappa\frac{\mu_0 r_R \ell_1 N_S N_R}{2g}\left(i_{Sa}\sin(\theta) - i_{Sb}\cos(\theta)\right)\sin(\theta - \theta_R)d\theta \\ &= \kappa\frac{\mu_0 r_R \ell_1 N_S N_R}{2g}\int_{\theta=\theta_R}^{\theta=\theta_R+\pi}\Big(i_{Sa}\sin(\theta - \theta_R + \theta_R)\sin(\theta - \theta_R) \\ &\quad - i_{Sb}\cos(\theta - \theta_R + \theta_R)\sin(\theta - \theta_R)\Big)\,d\theta \\ &= \kappa\frac{\mu_0 r_R \ell_1 N_S N_R}{2g} \\ &\quad\times \int_{\theta=\theta_R}^{\theta=\theta_R+\pi}\Big(i_{Sa}\sin^2(\theta - \theta_R)\cos(\theta_R) + i_{Sa}\cos(\theta - \theta_R)\sin(\theta_R)\sin(\theta - \theta_R) \\ &\quad - i_{Sb}\cos(\theta - \theta_R)\cos(\theta_R)\sin(\theta - \theta_R) + i_{Sb}\sin^2(\theta - \theta_R)\sin(\theta_R)\Big)d\theta \\ &= \kappa\frac{\mu_0 r_R \ell_1 N_S N_R}{2g}\Big(i_{Sa}\frac{\pi}{2}\cos(\theta_R) + \frac{\pi}{2}i_{Sb}\sin(\theta_R)\Big) \\ &= M\Big(i_{Sa}\cos(\theta_R) + i_{Sb}\sin(\theta_R)\Big) \end{aligned} \tag{6.19}$$

where, as before,

$$M \triangleq \kappa\frac{\mu_0 \pi r_R \ell_1 N_S N_R}{4g} = \kappa\frac{\mu_0 \pi \ell_1 \ell_2 N_S N_R}{8g}.$$

Similarly,

$$\begin{aligned} \lambda_{Rb}(0,0,i_{Sa},i_{Sb},\theta_R) &\triangleq \int_{\substack{\text{All loops of}\\ \text{rotor phase } b}} \vec{\mathbf{B}}_S \cdot d\vec{\mathbf{S}} \\ &= M\Big(-i_{Sa}\sin(\theta_R) + i_{Sb}\cos(\theta_R)\Big). \end{aligned} \tag{6.20}$$

Rotor Flux Linkage Produced by the Rotor Currents

The flux linkage in the rotor phases due to the magnetic field produced by the rotor currents is now computed. Consider a single turn (loop) of rotor phase a at the angular position θ with $0 \le \theta - \theta_R \le \pi$ and the same flux surface drawn in Figure 6.6.

To compute flux through the surface, $d\vec{\mathbf{S}} = r_R d\theta' dz\hat{\mathbf{r}}$ is chosen as the surface element vector. The flux $\phi_{Ra}(i_{Ra}, i_{Rb}, \theta - \theta_R)$ in each turn of rotor phase a at the angular position θ of the rotor is given by

$$\begin{aligned}
&\phi_{Ra}(i_{Ra}, i_{Rb}, \theta - \theta_R)\\
&\quad \triangleq \int_{\theta'=\theta-\pi}^{\theta'=\theta}\int_{z=0}^{z=\ell_1} \frac{\mu_0 r_R N_R}{2g}\frac{1}{r_R}\Big(i_{Ra}\cos(\theta'-\theta_R) + i_{Rb}\sin(\theta'-\theta_R)\Big)\hat{\mathbf{r}}\\
&\qquad \cdot \Big(r_R d\theta' dz\hat{\mathbf{r}}\Big)\\
&\quad = \frac{\mu_0 r_R \ell_1 N_R}{2g}\int_{\theta'=\theta-\pi}^{\theta'=\theta}\Big(i_{Ra}\cos(\theta'-\theta_R) + i_{Rb}\sin(\theta'-\theta_R)\Big)\,d\theta'\\
&\quad = \frac{\mu_0 r_R \ell_1 N_R}{2g}\Big(i_{Ra}\sin(\theta'-\theta_R) - i_{Rb}\cos(\theta'-\theta_R)\Big)\Big|_{\theta'=\theta-\pi}^{\theta'=\theta}\\
&\quad = \frac{\mu_0 r_R \ell_1 N_R}{2g}\big(i_{Ra}(\sin(\theta-\theta_R) - \sin(\theta-\theta_R-\pi))\\
&\qquad - i_{Rb}(\cos(\theta-\theta_R) - \cos(\theta-\theta_R-\pi))\big)\\
&\quad = \frac{\mu_0 r_R \ell_1 N_R}{g}\Big(i_{Ra}\sin(\theta-\theta_R) - i_{Rb}\cos(\theta-\theta_R)\Big).
\end{aligned} \tag{6.21}$$

Remark Again, by taking the flux surface normal to be $d\vec{\mathbf{S}} = r_R d\theta' dz\hat{\mathbf{r}}$, the positive direction of travel around the turns of rotor phase a at the angular position θ is out of the page on side a of the rotor phase and into the page on side a' of the rotor phase coinciding with the direction chosen for positive current flow in rotor phase a.

Between the angular positions θ and $\theta + d\theta$, there are $(N_R/2)\sin(\theta - \theta_R)d\theta$ turns, each having the flux (6.21) in them. The flux linkage in the turns of phase a between θ and $\theta + d\theta$ is then

$$\begin{aligned}
d\lambda_{Ra} &\triangleq \phi_{Ra}(i_{Ra}, i_{Rb}, \theta - \theta_R)\frac{N_R}{2}\sin(\theta-\theta_R)d\theta\\
&= \frac{\mu_0 r_R \ell_1 N_R^2}{2g}(i_{Ra}\sin(\theta-\theta_R) - i_{Rb}\cos(\theta-\theta_R))\sin(\theta-\theta_R)d\theta.
\end{aligned} \tag{6.22}$$

To obtain the flux linkage in all the turns of stator phase a, (6.22) is integrated from $\theta = \theta_R$ to $\theta = \theta_R + \pi$, that is,

$$\begin{aligned}
\lambda_{Ra}(i_{Ra}, i_{Rb}, 0, 0, \theta_R) &= \int_{\substack{\text{All loops of}\\ \text{rotor phase } a}} \vec{\mathbf{B}}_R \cdot d\vec{\mathbf{S}} \\
&= \int_{\theta=\theta_R}^{\theta=\theta_R+\pi} \frac{\mu_0 r_R \ell_1 N_R^2}{2g} \left(i_{Ra}\sin(\theta-\theta_R) - i_{Rb}\cos(\theta-\theta_R)\right)\sin(\theta-\theta_R)d\theta \\
&= \frac{\mu_0 r_R \ell_1 N_R^2}{2g} \int_{\theta=\theta_R}^{\theta=\theta_R+\pi} \left(i_{Ra}\sin^2(\theta-\theta_R) - i_{Rb}\cos(\theta-\theta_R)\sin(\theta-\theta_R)\right) d\theta \\
&= \frac{\mu_0 r_R \ell_1 N_R^2}{2g} i_{Ra} \frac{\pi}{2} \\
&= L_R i_{Ra}
\end{aligned} \tag{6.23}$$

where

$$L_R \triangleq \frac{\mu_0 r_R \ell_1 \pi N_R^2}{4g} = \frac{\mu_0 \ell_1 \ell_2 \pi}{8g} N_R^2.$$

Note that for this particular two-phase system, the magnetic field produced by the current i_{Rb} in rotor phase b does not produce any net flux linkage in rotor phase a. Similarly,

$$\lambda_{Rb}(i_{Ra}, i_{Rb}, 0, 0, \theta_R) = \int_{\substack{\text{All loops of}\\ \text{rotor phase } b}} \vec{\mathbf{B}}_R \cdot d\vec{\mathbf{S}} = L_R i_{Rb}. \tag{6.24}$$

Total Flux Linkage in the Rotor Phases

Combining the expressions (6.19), (6.20), (6.23), and (6.24), the total flux in the sinusoidally wound rotor phases due to both the stator and rotor magnetic fields is then

$$\begin{aligned}
\lambda_{Ra}(i_{Ra}, i_{Rb}, i_{Sa}, i_{Sb}, \theta_R) &= \int_{\substack{\text{All loops of}\\ \text{rotor phase } a}} (\vec{\mathbf{B}}_S + \vec{\mathbf{B}}_R) \cdot d\vec{\mathbf{S}} \\
&= L_R i_{Ra} + M\Big(i_{Sa}\cos(\theta_R) + i_{Sb}\sin(\theta_R)\Big) \\
\lambda_{Rb}(i_{Ra}, i_{Rb}, i_{Sa}, i_{Sb}, \theta_R) &= \int_{\substack{\text{All loops of}\\ \text{rotor phase } b}} (\vec{\mathbf{B}}_S + \vec{\mathbf{B}}_R) \cdot d\vec{\mathbf{S}} \\
&= L_R i_{Rb} + M\Big(-i_{Sa}\sin(\theta_R) + i_{Sb}\cos(\theta_R)\Big).
\end{aligned} \tag{6.25}$$

6.4 Torque Production in AC Machines

The stator currents establish a radial magnetic field in the air gap. The torque produced by this magnetic field on the currents in the sinusoidally wound rotor phases is now computed. In order to carry out this computation, a more convenient way of representing the stator and rotor magnetic fields is now used following Refs. [1] and [2].

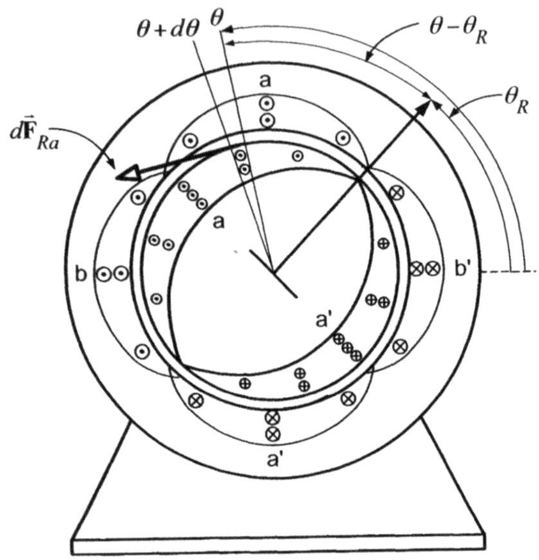

FIGURE 6.7. Computation of the torque on the current i_{Ra} in rotor phase a produced by $\vec{\mathbf{B}}_S$. Rotor phase b not shown for clarity.

The magnetic field at an arbitrary angle θ in the air-gap due to the stator currents is

$$\vec{\mathbf{B}}_S(i_{Sa}, i_{Sb}, \theta) = \frac{\mu_0 r_R N_S}{2g r_R}\Big(i_{Sa}(t)\cos(\theta) + i_{Sb}(t)\sin(\theta)\Big)\hat{\mathbf{r}}.$$

Let $i_S(t) \triangleq \sqrt{i_{Sa}^2(t) + i_{Sb}^2(t)}$, $\xi(t) \triangleq \tan^{-1}\left(i_{Sb}(t)/i_{Sa}(t)\right)$ so that $\vec{\mathbf{B}}_S$ may be rewritten as

$$\begin{aligned}\vec{\mathbf{B}}_S &= \frac{\mu_0 \ell_2 N_S}{4gr} i_S(t)\Big(\cos(\xi)\cos(\theta) + \sin(\xi)\sin(\theta)\Big)\hat{\mathbf{r}} \\ &= \frac{\mu_0 \ell_2 N_S}{4gr} i_S(t)\cos(\theta - \xi)\hat{\mathbf{r}}. \qquad (6.26)\end{aligned}$$

Similarly, the magnetic field at the angle θ in the air gap due to the rotor currents is

$$\vec{\mathbf{B}}_R(i_{Ra}, i_{Rb}, \theta - \theta_R) = \frac{\mu_0 \ell_2 N_R}{4gr}\Big(i_{Ra}(t)\cos(\theta - \theta_R) + i_{Rb}(t)\sin(\theta - \theta_R)\Big)\hat{\mathbf{r}}.$$

Let $i_R(t) \triangleq \sqrt{i_{Ra}^2(t) + i_{Rb}^2(t)}$, $\zeta(t) \triangleq \tan^{-1}(i_{Rb}(t)/i_{Ra}(t))$ so that $\vec{\mathbf{B}}_R$ may be rewritten as

$$\begin{aligned}\vec{\mathbf{B}}_R &= \frac{\mu_0 \ell_2 N_R}{4gr} i_R(t)\Big(\cos(\zeta)\cos(\theta-\theta_R) + \sin(\zeta)\sin(\theta-\theta_R)\Big)\hat{\mathbf{r}} \\ &= \frac{\mu_0 \ell_2 N_R}{4gr} i_R(t)\cos(\theta-\theta_R-\zeta)\hat{\mathbf{r}}. \qquad (6.27)\end{aligned}$$

The torque $\vec{\tau}_R$ produced by a two-phase machine is now computed. This is done by computing the force on the rotor current loops due to the magnetic field established by the stator currents. Referring to Figure 6.7, the differential magnetic force $d\vec{\mathbf{F}}_{Ra}$ produced on the loops of rotor phase a between θ and $\theta + d\theta$ by $\vec{\mathbf{B}}_S = B_S\hat{\mathbf{r}} = \kappa\dfrac{\mu_0 r_R N_S}{2gr_R} i_S(t)\cos(\theta-\xi)\hat{\mathbf{r}}$ is[4]

$$d\vec{\mathbf{F}}_{Ra} = \begin{cases} i_{Ra}(t)\dfrac{N_R}{2}\sin(\theta-\theta_R)d\theta(+\ell_1\hat{\mathbf{z}}) \times B_S\hat{\mathbf{r}}, & \theta_R \le \theta \le \theta_R + \pi \\ i_{Ra}(t)\dfrac{N_R}{2}|\sin(\theta-\theta_R)|d\theta(-\ell_1\hat{\mathbf{z}}) \times B_S\hat{\mathbf{r}}, & \theta_R + \pi \le \theta \le \theta_R + 2\pi \end{cases}$$

where $(N_R/2)|\sin(\theta-\theta_R)|d\theta$ is the number of axial sides of rotor phase a between θ and $\theta + d\theta$, each carrying the current $i_{Ra}(t)$. As $\sin(\theta-\theta_R) \le 0$ for $\theta_R + \pi \le \theta \le \theta_R + 2\pi$, it follows that $|\sin(\theta-\theta_R)| = -\sin(\theta-\theta_R)$ for $\theta_R + \pi \le \theta \le \theta_R + 2\pi$ so that $d\vec{\mathbf{F}}_{Ra}$ can be written more compactly as

$$\begin{aligned}d\vec{\mathbf{F}}_{Ra} &= i_{Ra}(t)\frac{N_R}{2}\sin(\theta-\theta_R)d\theta\ell_1\hat{\mathbf{z}} \times B_S\hat{\mathbf{r}} \\ &= i_{Ra}(t)\frac{N_R}{2}\sin(\theta-\theta_R)\ell_1 B_S d\theta\hat{\boldsymbol{\theta}} \\ &= i_{Ra}(t)\frac{N_R}{2}\sin(\theta-\theta_R)\ell_1\kappa\frac{\mu_0 r_R N_S}{2gr_R} i_S(t)\cos(\theta-\xi)d\theta\hat{\boldsymbol{\theta}} \\ &= \kappa\frac{\mu_0 \ell_1 N_S N_R}{4g} i_{Ra}(t) i_S(t)\sin(\theta-\theta_R)\cos(\theta-\xi)d\theta\hat{\boldsymbol{\theta}}\end{aligned}$$

for $\theta_R \le \theta \le \theta_R + 2\pi$. The differential torque $d\vec{\tau}_{Ra}$ is then given by

$$\begin{aligned}d\vec{\tau}_{Ra} &= (\ell_2/2)\hat{\mathbf{r}} \times d\vec{\mathbf{F}}_{Ra} \\ &= (\ell_2/2)\frac{\mu_0 \ell_1 N_S N_R}{4g} i_{Ra}(t) i_S(t)\sin(\theta-\theta_R)\cos(\theta-\xi)d\theta\hat{\mathbf{r}} \times \hat{\boldsymbol{\theta}} \\ &= \kappa\frac{\mu_0 \ell_1 \ell_2 N_S N_R}{8g} i_{Ra}(t) i_S(t)\sin(\theta-\theta_R)\cos(\theta-\xi)d\theta\hat{\mathbf{z}} \\ &= \frac{M}{\pi} i_{Ra}(t) i_S(t)\sin(\theta-\theta_R)\cos(\theta-\xi)d\theta\hat{\mathbf{z}} \\ &= \frac{M}{\pi} i_{Ra}(t) i_S(t)\frac{1}{2}\left(\sin(2\theta-\theta_R-\xi) + \sin(\xi-\theta_R)\right)d\theta\hat{\mathbf{z}}\end{aligned}$$

[4]Note that the leakage factor κ has been included as this is the stator magnetic field on the rotor side of the air gap.

where $M = \kappa\mu_0\pi\ell_1\ell_2 N_S N_R/(8g)$ is the coefficient of mutual inductance as previously defined. The torque on rotor phase a is then

$$\begin{aligned}
\vec{\boldsymbol{\tau}}_{Ra} &= \int_0^{2\pi} d\vec{\boldsymbol{\tau}}_{Ra} \\
&= \int_0^{2\pi} \frac{M}{2\pi} i_{Ra}(t) i_S(t) \Big(\sin(2\theta - \theta_R - \xi) + \sin(\xi - \theta_R)\Big) d\theta \hat{\mathbf{z}} \\
&= M i_{Ra}(t) i_S(t) \sin(\xi - \theta_R)\hat{\mathbf{z}} \\
&= M i_{Ra}(t) i_S(t) \Big(\sin(\xi)\cos(\theta_R) - \sin(\theta_R)\cos(\xi)\Big)\hat{\mathbf{z}} \\
&= M i_{Ra}(t) \Big(i_{Sb}(t)\cos(\theta_R) - i_{Sa}(t)\sin(\theta_R)\Big)\hat{\mathbf{z}}.
\end{aligned}$$

Similarly,

$$d\vec{\mathbf{F}}_{Rb} = \begin{cases} i_{Rb}(t)\dfrac{N_R}{2}\sin(\theta - \pi/2 - \theta_R) d\theta(\ell_1\hat{\mathbf{z}}) \times B_S\hat{\mathbf{r}}, & \theta_R + \dfrac{\pi}{2} \le \theta \le \theta_R + \dfrac{3\pi}{2} \\ i_{Rb}(t)\dfrac{N_R}{2}\left|\sin(\theta - \pi/2 - \theta_R)\right| d\theta(-\ell_1\hat{\mathbf{z}}) \times B_S\hat{\mathbf{r}}, & \theta_R - \dfrac{\pi}{2} \le \theta \le \theta_R + \dfrac{\pi}{2} \end{cases}$$

Again, as $\sin(\theta - \pi/2 - \theta_R) < 0$ for $\theta_R - \pi/2 \le \theta \le \theta_R + \pi/2$, $\left|\sin(\theta - \frac{\pi}{2} - \theta_R)\right| = -\sin(\theta - \pi/2 - \theta_R)$ for $\theta_R - \pi/2 \le \theta \le \theta_R + \pi/2 < 0$, the expression for the force $d\vec{\mathbf{F}}_{Rb}$ may be written more compactly for $\theta_R \le \theta \le \theta_R + 2\pi$ as

$$\begin{aligned}
d\vec{\mathbf{F}}_{Rb} &= i_{Rb}(t)\frac{N_R}{2}\sin(\theta - \pi/2 - \theta_R) d\theta(\ell_1\hat{\mathbf{z}}) \times (B_S\hat{\mathbf{r}}) \\
&= i_{Rb}(t)\frac{N_R}{2}\sin(\theta - \pi/2 - \theta_R)\ell_1 B_S d\theta\hat{\boldsymbol{\theta}} \\
&= -i_{Rb}(t)\frac{N_R}{2}\cos(\theta - \theta_R)\ell_1\kappa\frac{\mu_0 r_R N_S}{2g r_R} i_S(t)\cos(\theta - \xi) d\theta\hat{\boldsymbol{\theta}} \\
&= -\kappa\frac{\mu_0\ell_1 N_S N_R}{4g} i_{Rb}(t) i_S(t)\cos(\theta - \theta_R)\cos(\theta - \xi) d\theta\hat{\boldsymbol{\theta}}.
\end{aligned}$$

The differential torque $d\vec{\boldsymbol{\tau}}_{Rb}$ is then given by

$$\begin{aligned}
d\vec{\boldsymbol{\tau}}_{Rb} &= (\ell_2/2)\hat{\mathbf{r}} \times d\vec{\mathbf{F}}_{Rb} \\
&= -(\ell_2/2)\kappa\frac{\mu_0\ell_1 N_S N_R}{4g} i_{Rb}(t) i_S(t)\cos(\theta - \theta_R)\cos(\theta - \xi) d\theta\hat{\mathbf{r}} \times \hat{\boldsymbol{\theta}} \\
&= -\kappa\frac{\mu_0\ell_1\ell_2 N_S N_R}{8g} i_{Rb}(t) i_S(t)\cos(\theta - \theta_R)\cos(\theta - \xi) d\theta\hat{\mathbf{z}} \\
&= -\frac{M}{\pi} i_{Rb}(t) i_S(t)\cos(\theta - \theta_R)\cos(\theta - \xi) d\theta\hat{\mathbf{z}} \\
&= -\frac{M}{\pi} i_{Rb}(t) i_S(t)\frac{1}{2}\left(\cos(2\theta - \theta_R - \xi) + \cos(\xi - \theta_R)\right) d\theta\hat{\mathbf{z}}.
\end{aligned}$$

The torque on rotor phase b is then

$$\begin{aligned}
\vec{\tau}_{Rb} &= \int_0^{2\pi} d\vec{\tau}_{Rb} \\
&= -\int_0^{2\pi} \frac{M}{2\pi} i_{Rb}(t) i_S(t) \left(\cos(2\theta - \theta_R - \xi) + \cos(\xi - \theta_R)\right) d\theta \hat{\mathbf{z}} \\
&= -M i_{Rb}(t) i_S(t) \cos(\xi - \theta_R) \hat{\mathbf{z}} \\
&= -M i_{Rb}(t) i_S(t) \Big(\cos(\xi)\cos(\theta_R) + \sin(\xi)\sin(\theta_R)\Big) \hat{\mathbf{z}} \\
&= -M i_{Rb}(t) \Big(i_{Sa}(t)\cos(\theta_R) + i_{Sb}(t)\sin(\theta_R)\Big) \hat{\mathbf{z}}.
\end{aligned}$$

The total torque on the rotor is then

$$\begin{aligned}
\tau_R &= \tau_{Ra} + \tau_{Rb} \\
&= M\Big(-i_{Ra}(t) i_{Sa}(t)\sin(\theta_R) + i_{Ra}(t) i_{Sb}(t)\cos(\theta_R) \\
&\qquad - i_{Rb}(t) i_{Sa}(t)\cos(\theta_R) - i_{Rb}(t) i_{Sb}(t)\sin(\theta_R)\Big). \qquad (6.28)
\end{aligned}$$

Remark The assumption that the stator and rotor turns have negligible cross-sectional area was still in effect in the above derivation of the torque on the rotor. However, in a real motor, the windings are only approximately sinusoidally wound and this is accomplished by putting the turns in *slots* as explained in Chapter 4. In this case, it is pointed out by G. R. Slemon (see Section 3.4 of Ref. [32]) as well as by W. Leonhard (see p. 173 of Ref. [2]) that the magnetic forces are actually acting on the slides of rotor slots where the magnetic field enters the rotor iron. In the idealized model here, the slots do not exist and the forces/torques can be computed as described above. Further, the resulting mathematical model of the machine derived under these idealized assumptions will be shown to be a very good dynamic model of the machine (see Chapter 8).

6.5 Mathematical Model of a Sinusoidally Wound Induction Machine

The hard work of computing the expressions for the flux linkages in the phases of an AC machine as well as for the rotor torque has been completed. The analysis is now specialized to an induction motor which is illustrated in Figure 6.8. This figure is the same as Figure 6.1 except now it is explicitly assumed that the two ends of each rotor phase are shorted together, that is, there are no voltage sources for the rotor phases. With the rotor phases shorted, the expressions for the phase flux linkages and rotor torque are used to derive the nonlinear differential equation model of a two-phase sinusoidally wound induction motor.

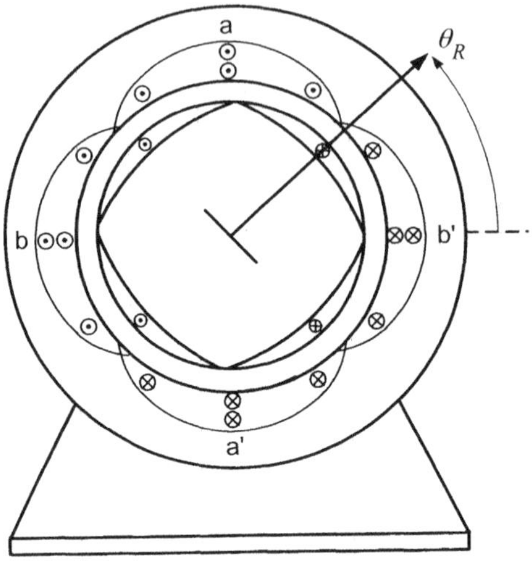

FIGURE 6.8. A symmetric sinusoidally wound two-phase induction machine.

By (6.16) and (6.25), the flux linkages in the stator and rotor phases in terms of the currents and rotor position are

$$\begin{aligned}
\lambda_{Sa}(i_{Ra}, i_{Rb}, i_{Sa}, i_{Sb}, \theta_R) &= L_S i_{Sa} + M\Big(+i_{Ra}\cos(\theta_R) - i_{Rb}\sin(\theta_R)\Big) \\
\lambda_{Sb}(i_{Ra}, i_{Rb}, i_{Sa}, i_{Sb}, \theta_R) &= L_S i_{Sb} + M\Big(+i_{Ra}\sin(\theta_R) + i_{Rb}\cos(\theta_R)\Big) \\
& \qquad\qquad\qquad\qquad\qquad\qquad\qquad\qquad (6.29) \\
\lambda_{Ra}(i_{Ra}, i_{Rb}, i_{Sa}, i_{Sb}, \theta_R) &= L_R i_{Ra} + M\Big(+i_{Sa}\cos(\theta_R) + i_{Sb}\sin(\theta_R)\Big) \\
\lambda_{Rb}(i_{Ra}, i_{Rb}, i_{Sa}, i_{Sb}, \theta_R) &= L_R i_{Rb} + M\Big(-i_{Sa}\sin(\theta_R) + i_{Sb}\cos(\theta_R)\Big).
\end{aligned}$$

By Faraday's and Ohm's laws, the equations describing the electrical dynamics of this system are

$$\begin{aligned}
-\frac{d\lambda_{Sa}}{dt} - R_S i_{Sa} + u_{Sa} &= 0 \\
-\frac{d\lambda_{Sb}}{dt} - R_S i_{Sb} + u_{Sb} &= 0 \\
-\frac{d\lambda_{Ra}}{dt} - R_R i_{Ra} &= 0 \\
-\frac{d\lambda_{Rb}}{dt} - R_R i_{Rb} &= 0.
\end{aligned}$$

Explicitly, this is

$$\begin{aligned}
L_S\frac{d}{dt}i_{Sa} + M\frac{d}{dt}\Big(+i_{Ra}\cos(\theta_R) - i_{Rb}\sin(\theta_R)\Big) + R_S i_{Sa} &= u_{Sa}\\
L_S\frac{d}{dt}i_{Sb} + M\frac{d}{dt}\Big(+i_{Ra}\sin(\theta_R) + i_{Rb}\cos(\theta_R)\Big) + R_S i_{Sb} &= u_{Sb}\\
L_R\frac{d}{dt}i_{Ra} + M\frac{d}{dt}\Big(+i_{Sa}\cos(\theta_R) + i_{Sb}\sin(\theta_R)\Big) + R_R i_{Ra} &= 0\\
L_R\frac{d}{dt}i_{Rb} + M\frac{d}{dt}\Big(-i_{Sa}\sin(\theta_R) + i_{Sb}\cos(\theta_R)\Big) + R_R i_{Rb} &= 0.
\end{aligned} \tag{6.30}$$

By equation (6.28), the torque produced by this machine is given by

$$\begin{aligned}
\tau_R = {} & M\Big(-i_{Ra}(t)i_{Sa}(t)\sin(\theta_R) + i_{Ra}(t)i_{Sb}(t)\cos(\theta_R)\\
& - i_{Rb}(t)i_{Sa}(t)\cos(\theta_R) - i_{Rb}(t)i_{Sb}(t)\sin(\theta_R)\Big)
\end{aligned} \tag{6.31}$$

so that, with τ_L the load torque, the mechanical equations are simply

$$\begin{aligned}
J\frac{d\omega_R}{dt} &= \tau_R - \tau_L\\
\frac{d\theta_R}{dt} &= \omega_R.
\end{aligned} \tag{6.32}$$

Equations (6.30), (6.31), and (6.32) are the set of nonlinear differential equations characterizing a two-phase sinusoidally wound induction motor.

A *space vector* representation of the equations of the induction motor is given in problem 5 (see also problem 6).

6.6 Total Leakage Factor

The leakage factor is defined as

$$\sigma = 1 - \frac{M^2}{L_S L_R}$$

Substituting the expressions

$$\begin{aligned}
L_R &= \frac{\mu_0 \ell_1 \ell_2 \pi}{8g} N_R^2\\
L_S &= \frac{\mu_0 \ell_1 \ell_2 \pi}{8g} N_S^2\\
M &= \frac{\kappa \mu_0 \pi \ell_1 \ell_2 N_S N_R}{8g}
\end{aligned}$$

results in

$$\sigma = 1 - \kappa^2.$$

Note that if $\kappa = 1$, corresponding to no spreading of the magnetic field as one goes across the air gap, then $\sigma = 0$. However, in a practical motor $\sigma > 0$. As pointed out in Ref. [2], σ typically varies from 0.05 for low leakage machines up 0.20.

6.7 The Squirrel Cage Rotor

The above mathematical model of the induction motor was derived assuming sinusoidally wound rotor phases. However, the most commonly used induction motor has a squirrel cage rotor which is illustrated in Figure 6.9.

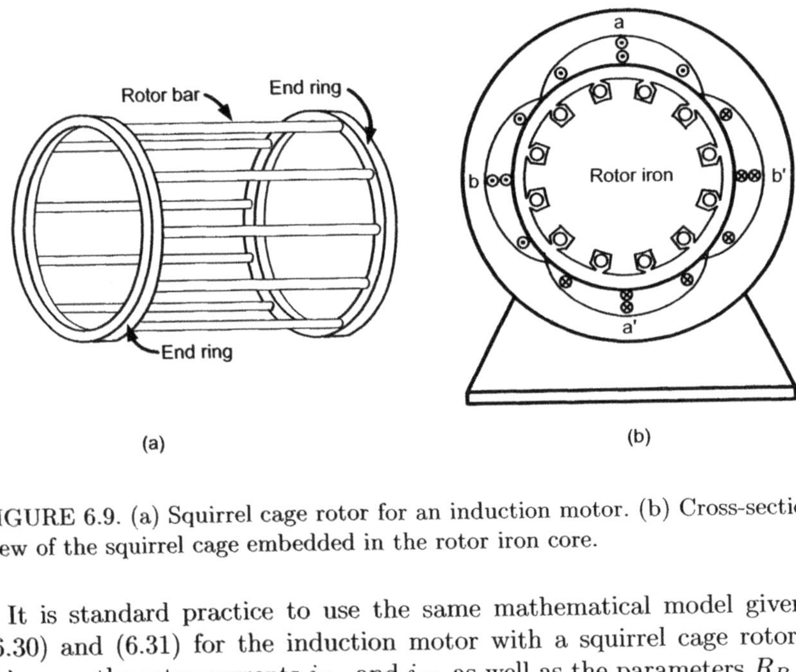

FIGURE 6.9. (a) Squirrel cage rotor for an induction motor. (b) Cross-sectional view of the squirrel cage embedded in the rotor iron core.

It is standard practice to use the same mathematical model given in (6.30) and (6.31) for the induction motor with a squirrel cage rotor. In this case, the rotor currents i_{Ra} and i_{Rb} as well as the parameters R_R and L_R must be given a different interpretation. There is actually a different current in each rotor bar of the squirrel cage and these together make up a current distribution around the periphery of the rotor (see the discussion of the squirrel cage rotor in Chapter 5). The parameters R_R and L_R in the model can be interpreted as those values that best fit the mathematical model (6.30), (6.31), and (6.32) to that of an actual squirrel cage motor. For example, in Stephan et al. [42] and Wang et al. [59] the parameters of the model (6.30), (6.31), and (6.32) were identified based on data from a squirrel cage motor using a least-squares algorithm. The simulated response of the system $(\omega_R, i_{Sa}, i_{Sb})$ based on the model (6.30), (6.31), and (6.32) using these parameters values was shown to be quite close to the actual measured response of the squirrel cage motor [42][59]. The parameter values determined in this manner are then interpreted as those values that best fit the model (6.30), (6.31), and (6.32) to the squirrel cage motor data in the least-squares sense. See Chapter 8 where it is shown how to use the least-squares method to identify the induction motor parameters.

6.8 Induction Machine With Multiple Pole Pairs

It is left to problem 10 at the end of this chapter to show that if the machine has n_p pole pairs, then the expressions for the flux linkages are modified to

$$\begin{aligned}
\lambda_{Sa} &= L_S i_{Sa} + M\Big(i_{Ra}\cos(n_p\theta_R) - i_{Rb}\sin(n_p\theta_R)\Big)\\
\lambda_{Sb} &= L_S i_{Sb} + M\Big(i_{Ra}\sin(n_p\theta_R) + i_{Rb}\cos(n_p\theta_R)\Big)\\
\lambda_{Ra} &= L_R i_{Ra} + M\Big(+i_{Sa}\cos(n_p\theta_R) + i_{Sb}\sin(n_p\theta_R)\Big)\\
\lambda_{Rb} &= L_R i_{Rb} + M\Big(-i_{Sa}\sin(n_p\theta_R) + i_{Sb}\cos(n_p\theta_R)\Big)
\end{aligned} \tag{6.33}$$

where

$$L_S \triangleq \frac{\mu_0\pi\ell_1\ell_2N_S^2}{8g},\quad L_R \triangleq \frac{\mu_0\pi\ell_1\ell_2N_R^2}{8g},\quad M \triangleq \frac{\kappa\mu_0\pi\ell_1\ell_2N_SN_R}{8g}.$$

Here N_S, and N_R are the number of stator and rotor windings, respectively, *per pole pair.* The total number of windings in a stator phase is n_pN_S and n_pN_R is the total number of windings in a rotor phase. The dynamic equations are

$$\begin{aligned}
u_{Sa} &= R_S i_{Sa} + \frac{d}{dt}\Big(L_S i_{Sa} + M\Big(i_{Ra}\cos(n_p\theta_R) - i_{Rb}\sin(n_p\theta_R)\Big)\Big)\\
u_{Sb} &= R_S i_{Sb} + \frac{d}{dt}\Big(L_S i_{Sb} + M\Big(i_{Ra}\sin(n_p\theta_R) + i_{Rb}\cos(n_p\theta_R)\Big)\Big)\\
0 &= R_R i_{Ra} + \frac{d}{dt}\Big(L_R i_{Ra} + M\Big(+i_{Sa}\cos(n_p\theta_R) + i_{Sb}\sin(n_p\theta_R)\Big)\Big)\\
0 &= R_R i_{Rb} + \frac{d}{dt}\Big(L_R i_{Rb} + M\Big(-i_{Sa}\sin(n_p\theta_R) + i_{Sb}\cos(n_p\theta_R)\Big)\Big)\\
J\frac{d\omega_R}{dt} &= n_pM\Big(i_{Sb}(i_{Ra}\cos(n_p\theta_R) - i_{Rb}\sin(n_p\theta_R)\\
&\quad - i_{Sa}(i_{Ra}\sin(n_p\theta_R) + i_{Rb}\cos(n_p\theta_R))\Big) - \tau_L\\
\frac{d\theta_R}{dt} &= \omega_R.
\end{aligned} \tag{6.34}$$

Consider two induction machines which are identical in all aspects except that one has only a single pole pair (i.e., $n_p = 1$) while the other has $n_p > 1$. That is, each has the same geometrical construction (ℓ_1, ℓ_2, and g are the same), the same number N_S of stator windings *per pole pair*, the same number of rotor windings N_R *per pole pair*, and the same current ratings. Then the torque output of the machine with n_p pole pairs will have a torque output that is a factor n_p *greater* than the single pole pair machine. This is clear from considering the coefficient n_pM in the torque expression (6.34). However, the n_p pole pair machine has n_p more turns/windings in each phase than a single pole pair machine.

6.9 Mathematical Model of a Wound Rotor Synchronous Machine

The mathematical model of a wound rotor synchronous machine is straightforwardly derived from the expressions for the phase flux linkages and rotor torque derived above. Figure 6.10 shows the synchronous motor to be modeled which differs from Figure 6.1 in that there is only one rotor phase and this rotor phase is explicitly assumed to have a voltage source.

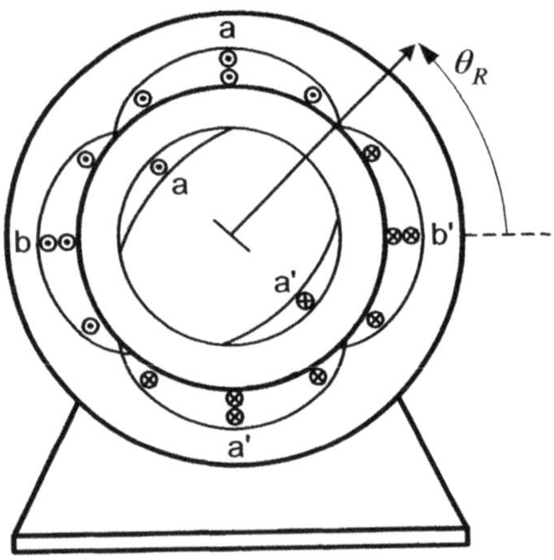

FIGURE 6.10. A symmetric two-phase synchronous machine with a sinusoidally wound rotor phase. The two stator phases have voltage sources u_{Sa} and u_{Sb}, respectively, and the single rotor phase has voltage source u_F.

The terminology for the synchronous machine changes to the rotor phase being referred to as the *field winding*; its current is called the *field current* i_F, the flux linkage in its windings is called the *field flux (linkage)* λ_F, and the voltage source u_F applied to the phase winding is called the *field* source.

Recall that the term "field" refers to the fact that during normal operation, i_F and λ_F are constant, which is the case for a synchronous machine. The flux linkages in this machine are found by simply setting $i_{Ra} = i_F$ and $i_{Rb} \equiv 0$ in (6.16) and (6.25) to obtain

$$\begin{aligned}\lambda_{Sa}(i_F, i_{Sa}, i_{Sb}, \theta_R) &= L_S i_{Sa} + M i_F \cos(\theta_R)\\ \lambda_{Sb}(i_f, i_{Sa}, i_{Sb}, \theta_R) &= L_S i_{Sb} + M i_F \sin(\theta_R)\\ \lambda_F(i_F, i_{Sa}, i_{Sb}, \theta_R) &= L_F i_F + M\left(i_{Sa}\cos(\theta_R) + i_{Sb}\sin(\theta_R)\right)\end{aligned} \tag{6.35}$$

where $L_F \triangleq L_R$. With R_S the resistance of each stator phase and R_F the resistance of the field winding (rotor phase), Faraday's and Ohm's laws give

$$\begin{aligned}-\frac{d\lambda_{Sa}}{dt} - R_S i_{Sa} + u_{Sa} &= 0\\ -\frac{d\lambda_{Sb}}{dt} - R_S i_{Sb} + u_{Sb} &= 0\\ -\frac{d\lambda_F}{dt} - R_F i_F + u_F &= 0\end{aligned} \tag{6.36}$$

where u_F is the voltage source for the field (rotor) winding.

Setting $i_{Ra} = i_F$ and $i_{Rb} \equiv 0$ in equation (6.28), the torque produced by this machine is

$$\tau_R = M i_F\Big(-i_{Sa}\sin(\theta_R) + i_{Sb}\cos(\theta_R)\Big). \tag{6.37}$$

Combining these equations together, a nonlinear differential equation model of a two-phase synchronous motor with a sinusoidally wound rotor is

$$\begin{aligned}-\frac{d}{dt}\Big(L_S i_{Sa} + M i_F\cos(\theta_R)\Big) - R_S i_{Sa} + u_{Sa} &= 0\\ -\frac{d}{dt}\Big(L_S i_{Sb} + M i_F\sin(\theta_R)\Big) - R_S i_{Sb} + u_{Sb} &= 0\\ -\frac{d}{dt}\Big(L_F i_F + M\left(i_{Sa}\cos(\theta_R) + i_{Sb}\sin(\theta_R)\right)\Big) - R_F i_F + u_F &= 0\\ M i_F\Big(-i_{Sa}\sin(\theta_R) + i_{Sb}\cos(\theta_R)\Big) - \tau_L &= J\frac{d\omega_R}{dt}\\ \frac{d\theta_R}{dt} &= \omega_R\end{aligned} \tag{6.38}$$

where τ_L is the load torque.

In a synchronous machine, the rotor current is often chosen be constant, that is,

$$i_F = I_F.$$

This can be done by choosing the voltage u_F in the field winding according to

$$u_F = K_P(I_F - i_F) + K_I\int_0^t (I_F - i_F)dt.$$

The mathematical model of a two-phase synchronous motor with a sinusoidally wound rotor with constant field current then reduces to

$$\begin{aligned}
L_S\frac{di_{Sa}}{dt} &= -R_S i_{Sa} + MI_F\sin(\theta_R)\omega_R + u_{Sa} \\
L_S\frac{di_{Sb}}{dt} &= -R_S i_{Sb} - MI_F\cos(\theta_R)\omega_R + u_{Sb} \\
J\frac{d\omega_R}{dt} &= MI_F\Big(-i_{Sa}\sin(\theta_R) + i_{Sb}\cos(\theta_R)\Big) - \tau_L \\
\frac{d\theta_R}{dt} &= \omega_R.
\end{aligned} \tag{6.39}$$

6.10 Mathematical Model of a PM Synchronous Machine

Consider the case of a two-phase permanent magnet machine with the permanent magnet of the rotor designed to produce a sinusoidally distributed radially directed magnetic field in the air gap (see Figure 6.11). That is,

$$\vec{\mathbf{B}}_R(r,\theta-\theta_R) = B_m\frac{r_R}{r}\cos(\theta-\theta_R)\hat{\mathbf{r}}.$$

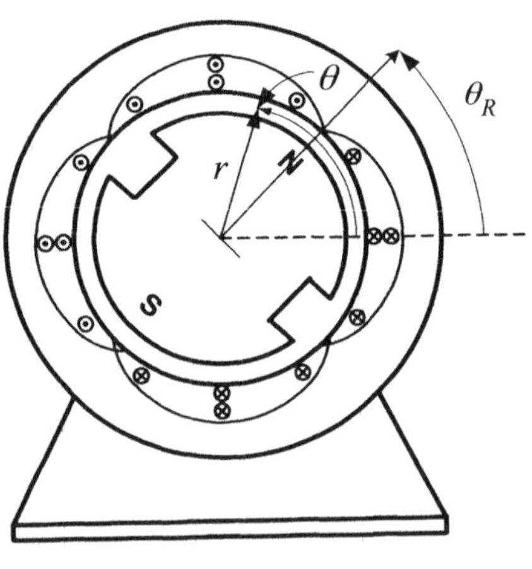

FIGURE 6.11. Two-phase permanent magnet synchronous machine.

This is essentially the same situation as a wound rotor synchronous machine with a constant field current. It is straightforward to show that its model has the same form as (6.39) given by (see problem 11)

$$\begin{aligned}
L_S\frac{di_{Sa}}{dt} &= -R_S i_{Sa} + K_m \sin(\theta_R)\omega_R + u_{Sa} \\
L_S\frac{di_{Sb}}{dt} &= -R_S i_{Sb} - K_m \cos(\theta_R)\omega_R + u_{Sb} \\
J\frac{d\omega_R}{dt} &= K_m\Big(-i_{Sa}\sin(\theta_R) + i_{Sb}\cos(\theta_R)\Big) - \tau_L \\
\frac{d\theta_R}{dt} &= \omega_R.
\end{aligned} \tag{6.40}$$

6.11 The Stator and Rotor Magnetic Fields of an Induction Machine Rotate Synchronously*

Recall the radial magnetic field in the air gap produced by the stator currents is given by

$$\vec{\mathbf{B}}_S(i_{Sa}, i_{Sb}, r, \theta) = \frac{\mu_0 r_R N_S}{2g}\frac{1}{r}\left(i_{Sa}\cos(\theta) + i_{Sb}\sin(\theta)\right)\hat{\mathbf{r}} \tag{6.41}$$

while the radial magnetic field in the air gap produced by the rotor currents is given by

$$\vec{\mathbf{B}}_R(i_{Ra}, i_{Rb}, r, \theta - \theta_R) = \frac{\mu_0 r_R N_R}{2g}\frac{1}{r}\left(i_{Ra}\cos(\theta - \theta_R) + i_{Rb}\sin(\theta - \theta_R)\right)\hat{\mathbf{r}}. \tag{6.42}$$

Applying the voltages

$$\begin{aligned}
u_{Sa} &= U_S\cos(\omega_S t) \\
u_{Sb} &= U_S\sin(\omega_S t)
\end{aligned}$$

to the stator phases, a steady-state solution for the currents and rotor position has the form

$$\begin{aligned}
i_{Sa} &= I_S\cos(\omega_S t + \phi_S) \\
i_{Sb} &= I_S\sin(\omega_S t + \phi_S) \\
i_{Ra} &= I_R\cos((\omega_S - \omega_R)t + \phi_R) \\
i_{Rb} &= I_R\sin((\omega_S - \omega_R)t + \phi_R) \\
\theta_R &= \omega_R t.
\end{aligned} \tag{6.43}$$

Problem 9 outlines the derivation of these expressions to show that the phase angles ϕ_S, ϕ_R are functions of the stator frequency ω_S and the normalized slip $(\omega_S - \omega_R)/\omega_S$; thus they are *constant* in steady state. Substituting the expressions (6.43) into (6.41), the steady-state stator magnetic field may be written as

$$\vec{\mathbf{B}}_S(r,\theta,t) = \frac{\mu_0 r_R N_S I_S}{2g}\frac{1}{r}\cos(\theta - (\omega_S t + \phi_S))\hat{\mathbf{r}}$$

where at the angular position $\theta_{B_S}(t) \triangleq \omega_S t + \phi_S$, $\vec{\mathbf{B}}_S$ has its maximum magnitude. Denote $\theta_{B_S}(t)$ as the *magnetic axis* of $\vec{\mathbf{B}}_S$ and note that the magnetic field distribution of $\vec{\mathbf{B}}_S$ is fixed relative to $\theta_{B_S}(t)$; that is, $\vec{\mathbf{B}}_S$ is a "magnet" that rotates at the angular rate ω_S. Similarly, substituting the expressions (6.43) into (6.42), the steady-state rotor magnetic field may then be written as

$$\begin{aligned}
\vec{\mathbf{B}}_R(r,\theta,t) &= \frac{\mu_0 r_R N_R I_R}{2g}\frac{1}{r}\Big(\cos((\omega_S - \omega_R)t + \phi_R)\cos(\theta - \omega_R t) \\
&\quad + \sin((\omega_S - \omega_R)t + \phi_R)\sin(\theta - \omega_R t)\Big)\hat{\mathbf{r}} \\
&= \frac{\mu_0 r_R N_R I_R}{2g}\frac{1}{r}\cos\Big(\theta - \omega_R t - \Big((\omega_S - \omega_R)t + \phi_R\Big)\Big)\hat{\mathbf{r}} \\
&= \frac{\mu_0 r_R N_R I_R}{2g}\frac{1}{r}\cos\Big(\theta - (\omega_S t + \phi_R)\Big)\hat{\mathbf{r}}
\end{aligned}$$

where at the angular position $\theta_{B_R}(t) \triangleq \omega_S t + \phi_R$, $\vec{\mathbf{B}}_R$ has its maximum magnitude. Denote $\theta_{B_R}(t)$ as the *magnetic axis* of $\vec{\mathbf{B}}_R$ and note that the magnetic field distribution of $\vec{\mathbf{B}}_R$ is fixed relative to $\theta_{B_R}(t)$; that is, $\vec{\mathbf{B}}_R$ is also a "magnet" that rotates at the angular rate ω_S.

Both of these magnetic fields rotate at the angular rate ω_S, with the angle between the two "magnets" being $\theta_{B_S}(t) - \theta_{B_R}(t) = \phi_S - \phi_R$ which is *constant*; that is, the two *magnetic fields* rotate synchronously together! Another way to view this is to note that by (6.43) the currents i_{Sa} and i_{Sb} have angular frequency ω_S and produce a magnetic field rotating at the angular speed ω_S. Also, equations (6.43) show that rotor currents i_{Ra} and i_{Rb} have angular frequency $\omega_S - \omega_R$ and thus produce a magnetic field that rotates at angular speed $\omega_S - \omega_R$ *with respect to the rotor*. However, the rotor is moving at angular speed ω_R so that the magnetic field produced by the rotor currents is rotating at angular speed $\omega_S - \omega_R + \omega_R = \omega_S$ *with respect to the stator*. Again, both fields rotate at the same angular speed ω_S.

Substituting the expressions (6.43) for the steady-state currents and $\theta_R(t)$ into equation (6.31) results in the steady-state torque being given by (see problem 9)

$$\begin{aligned}
\tau_R &= M I_S I_R \sin(\phi_S - \phi_R) \\
&= g2\pi(\ell_1/2)\ell_2\frac{1}{2\mu_0}\left(\kappa\frac{\mu_0 N_S I_S}{2g}\right)\left(\frac{\mu_0 N_R I_R}{2g}\right)\sin(\phi_S - \phi_R) \\
&= V_{\text{airgap}}\frac{1}{2\mu_0}B_{S\max}B_{R\max}\sin(\phi_S - \phi_R)
\end{aligned}$$

where $M = \kappa\mu_0\pi\ell_1\ell_2 N_S N_R/(8g)$, $B_{S\max} \triangleq \kappa\mu_0 N_S I_S/(2g)$ and $B_{R\max} \triangleq \mu_0 N_R I_R/(2g)$ are the maximum values of the stator and rotor magnetic fields, respectively, at the surface of the rotor, and $V_{\text{airgap}} = g2\pi(\ell_1/2)\ell_2$ is the approximate volume of the air gap for g small. This expression shows that the torque is proportional to the maximum strength of the stator and rotor magnetic fields and to the sine of the angle between them.[5]

Finally, the steady-state condition requires $J d\omega_R/dt \equiv 0$ so that the load-torque τ_L must be such that it satisfies

$$MI_S I_R \sin(\phi_S - \phi_R) - \tau_L = 0.$$

6.12 Torque, Energy, and Co-energy*

In this section it is shown how the torque of an AC machine can be derived from conservation of energy. The analysis is carried out for the two-phase induction motor while the synchronous machine is considered in problem 19.

6.12.1 Magnetic Field Energy

The equations describing the electrical dynamics of a two-phase induction motor are

$$\begin{aligned}
-\frac{d\lambda_{Sa}}{dt} - R_S i_{Sa} + u_{Sa} &= 0\\
-\frac{d\lambda_{Sb}}{dt} - R_S i_{Sb} + u_{Sb} &= 0\\
-\frac{d\lambda_{Ra}}{dt} - R_R i_{Ra} + u_{Ra} &= 0\\
-\frac{d\lambda_{Rb}}{dt} - R_R i_{Rb} + u_{Rb} &= 0
\end{aligned} \qquad (6.44)$$

where the flux linkages are given by ($n_p = 1$)

$$\begin{aligned}
\lambda_{Sa}(i_{Ra}, i_{Rb}, i_{Sa}, i_{Sb}, \theta_R) &= L_S i_{Sa} + M(+i_{Ra}\cos(\theta_R) - i_{Rb}\sin(\theta_R))\\
\lambda_{Sb}(i_{Ra}, i_{Rb}, i_{Sa}, i_{Sb}, \theta_R) &= L_S i_{Sb} + M(+i_{Ra}\sin(\theta_R) + i_{Rb}\cos(\theta_R))\\
\lambda_{Ra}(i_{Ra}, i_{Rb}, i_{Sa}, i_{Sb}, \theta_R) &= L_R i_{Ra} + M(+i_{Sa}\cos(\theta_R) + i_{Sb}\sin(\theta_R))\\
\lambda_{Rb}(i_{Ra}, i_{Rb}, i_{Sa}, i_{Sb}, \theta_R) &= L_R i_{Rb} + M(-i_{Sa}\sin(\theta_R) + i_{Sb}\cos(\theta_R)).
\end{aligned} \qquad (6.45)$$

---

[5] There is an expression identical in form for the torque of a synchronous motor!

Of course, normally $u_{Ra} \equiv 0, u_{Rb} \equiv 0$. Substituting these expressions for the flux linkages into the system equations (6.44), it follows that

$$\begin{aligned}
L_S\frac{d}{dt}i_{Sa} + M\frac{d}{dt}\left(+i_{Ra}\cos(\theta_R) - i_{Rb}\sin(\theta_R)\right) + R_S i_{Sa} &= u_{Sa} \\
L_S\frac{d}{dt}i_{Sb} + M\frac{d}{dt}\left(+i_{Ra}\sin(\theta_R) + i_{Rb}\cos(\theta_R)\right) + R_S i_{Sb} &= u_{Sb} \\
L_R\frac{d}{dt}i_{Ra} + M\frac{d}{dt}\left(+i_{Sa}\cos(\theta_R) + i_{Sb}\sin(\theta_R)\right) + R_R i_{Ra} &= u_{Ra} \\
L_R\frac{d}{dt}i_{Rb} + M\frac{d}{dt}\left(-i_{Sa}\sin(\theta_R) + i_{Sb}\cos(\theta_R)\right) + R_R i_{Rb} &= u_{Rb}.
\end{aligned} \tag{6.46}$$

In order to calculate the energy stored in the magnetic field of the induction machine (that is, the magnetic field in the stator/rotor iron as well as in the air gap), the rotor is locked and voltage is applied to the phases. Using the system (6.44), the power delivered to each of the phases is given by

$$\begin{aligned}
i_{Sa}u_{Sa} &= R_S i_{Sa}^2 + i_{Sa}\frac{d\lambda_{Sa}}{dt} \\
i_{Sb}u_{Sb} &= R_S i_{Sb}^2 + i_{Sb}\frac{d\lambda_{Sb}}{dt} \\
i_{Ra}u_{Ra} &= R_R i_{Ra}^2 + i_{Ra}\frac{d\lambda_{Ra}}{dt} \\
i_{Rb}u_{Rb} &= R_R i_{Rb}^2 + i_{Rb}\frac{d\lambda_{Rb}}{dt}.
\end{aligned} \tag{6.47}$$

Substituting the expressions (6.45) for the flux linkages into (6.47) and adding up the resulting four equations one obtains (see problem 17)

$$\begin{aligned}
i_{Sa}u_{Sa}+i_{Sb}u_{Sb} + i_{Ra}u_{Ra} + i_{Rb}u_{Rb} &= R_S(i_{Sa}^2 + i_{Sb}^2) + R_R(i_{Ra}^2 + i_{Rb}^2) \\
&+ \frac{d}{dt}\left(\frac{1}{2}L_S(i_{Sa}^2 + i_{Sb}^2) + \frac{1}{2}L_R(i_{Ra}^2 + i_{Rb}^2)\right) \\
&+ \frac{d}{dt}\left(M i_{Sa}\left(+i_{Ra}\cos(\theta_R) - i_{Rb}\sin(\theta_R)\right)\right)\Big|_{\theta_R \text{ constant}} \\
&+ \frac{d}{dt}\left(M i_{Sb}\left(+i_{Ra}\sin(\theta_R) + i_{Rb}\cos(\theta_R)\right)\right)\Big|_{\theta_R \text{ constant}}.
\end{aligned} \tag{6.48}$$

Defining the magnetic field energy W_{field} as

$$\begin{aligned}
W_{\text{field}}(i_{Sa}, i_{Sb}, i_{Ra}, i_{Rb}, \theta_R) \triangleq\ & \frac{1}{2}L_S(i_{Sa}^2 + i_{Sb}^2) + \frac{1}{2}L_R(i_{Ra}^2 + i_{Rb}^2) \\
&+ M i_{Sa}\Big(+i_{Ra}\cos(\theta_R) - i_{Rb}\sin(\theta_R)\Big) \\
&+ M i_{Sb}\Big(+i_{Ra}\sin(\theta_R) + i_{Rb}\cos(\theta_R)\Big)
\end{aligned} \tag{6.49}$$

equation (6.48) may be rewritten as

$$
\begin{aligned}
i_{Sa}u_{Sa} + i_{Sb}u_{Sb} + i_{Ra}u_{Ra} + i_{Rb}u_{Rb} &= R_S(i_{Sa}^2 + i_{Sb}^2) + R_R(i_{Ra}^2 + i_{Rb}^2) \\
&\quad + \frac{d}{dt}W_{\text{field}}\bigg|_{\theta_R \text{ constant}}.
\end{aligned}
$$

Here

$$
\begin{aligned}
\frac{d}{dt}W_{\text{field}} &= \frac{\partial W_{\text{field}}}{\partial i_{Sa}}\frac{di_{Sa}}{dt} + \frac{\partial W_{\text{field}}}{\partial i_{Sb}}\frac{di_{Sb}}{dt} + \frac{\partial W_{\text{field}}}{\partial i_{Ra}}\frac{di_{Ra}}{dt} + \frac{\partial W_{\text{field}}}{\partial i_{Rb}}\frac{di_{Rb}}{dt} \\
&\quad + \frac{\partial W_{\text{field}}}{\partial \theta_R}\frac{d\theta_R}{dt}
\end{aligned}
$$

$$
\begin{aligned}
\frac{d}{dt}W_{\text{field}}\bigg|_{\theta_R \text{ constant}} &= \frac{\partial W_{\text{field}}}{\partial i_{Sa}}\frac{di_{Sa}}{dt} + \frac{\partial W_{\text{field}}}{\partial i_{Sb}}\frac{di_{Sb}}{dt} + \frac{\partial W_{\text{field}}}{\partial i_{Ra}}\frac{di_{Ra}}{dt} \\
&\quad + \frac{\partial W_{\text{field}}}{\partial i_{Rb}}\frac{di_{Rb}}{dt}.
\end{aligned}
$$

The quantity $W_{\text{field}}(i_{Sa}, i_{Sb}, i_{Ra}, i_{Rb}, \theta_R)$ is the energy stored in the magnetic field established in the stator/rotor iron[6] and in the air gap by the phase currents. (This was derived in the same way that one derives the expression $Li^2/2$ for the energy in a standard inductor.) Defining

$$
\begin{aligned}
W_{\text{ohmic loss}} &= \int_0^t \left(R_S(i_{Sa}^2 + i_{Sb}^2) + R_R(i_{Ra}^2 + i_{Rb}^2)\right) dt \\
W_{\text{electrical}} &= \int_0^t (i_{Sa}u_{Sa} + i_{Sb}u_{Sb} + i_{Ra}u_{Ra} + i_{Rb}u_{Rb})\, dt
\end{aligned}
$$

it then follows that

$$W_{\text{electrical}} = W_{\text{field}} + W_{\text{ohmic loss}}.$$

This is simply an energy conservation equation. It says that, with the rotor locked, the energy supplied by the voltage sources u_{Sa}, u_{Sb}, u_{Ra}, and u_{Rb} goes into magnetic field energy W_{field} and into the heat energy $W_{\text{ohmic loss}}$ dissipated in the resistance of the windings.

The expression (6.49) for the field energy is valid whether or not the rotor is moving. This is because if the rotor is at the angle θ_R and the currents in the phases are i_{Sa}, i_{Sb}, i_{Ra}, and i_{Rb}, then the magnetic field at each point of the motor (i.e., in the stator/rotor iron and in the air

[6] Actually, as the iron is assumed to have infinite permeability ($\mu_r = \infty$), the stored energy is all in the air gap. This is simply because electromagnetic theory allows one to write the stored field energy as $W_{\text{field}} = (1/2)\int_{\substack{\text{rotor and}\\ \text{stator iron}}} (B^2/(\mu_0\mu_r))\, dV + (1/2)\int_{\text{air gap}} (B^2/\mu_0)\, dV = (1/2)\int_{\text{air gap}} (B^2/\mu_0)\, dV$ as $\mu_r = \infty$.

gap) is the same whether or not the rotor is moving. In other words, the stored magnetic energy depends only on $(i_{Sa}, i_{Sb}, i_{Ra}, i_{Rb}, \theta_R)$ and not on ω_R. The derivation was carried out with the rotor locked so that it could be concluded that the applied power went only into the field energy and ohmic losses (i.e., not into mechanical power) in deriving the expression (6.49) for the magnetic field energy.

6.12.2 Computing Torque From the Field Energy

In order to derive an expression for the torque in terms of the stored energy, the rotor is unlocked and allowed to move under the torque produced by the magnetic forces. Multiplying the first equation of (6.46) by i_{Sa}, the second by i_{Sb}, the third by i_{Ra}, and the fourth by i_{Rb} and adding them together give $(\omega_R = d\theta_R/dt \neq 0)$

$$\begin{aligned} i_{Sa}u_{Sa} + i_{Sb}u_{Sb} + i_{Ra}u_{Ra} + i_{Rb}u_{Rb} = {} & \frac{d}{dt}W_{\text{field}}\Big|_{\theta_R \text{ constant}} + 2\frac{\partial W_{\text{field}}}{\partial \theta_R}\frac{d\theta_R}{dt} \\ & + R_S(i_{Sa}^2 + i_{Sb}^2) + R_R(i_{Ra}^2 + i_{Rb}^2). \end{aligned} \tag{6.50}$$

The reader is asked to verify this expression in problem 18. The term $2(\partial W_{\text{field}}/\partial \theta_R)d\theta_R/dt$ is an additional term describing power being absorbed from the sources when there is rotor movement. As

$$dW_{\text{field}}/dt = dW_{\text{field}}/dt|_{\theta_R \text{ constant}} + (\partial W_{\text{field}}/\partial \theta_R)(d\theta_R/dt),$$

equation (6.50) may be rewritten as

$$\begin{aligned} i_{Sa}u_{Sa} + i_{Sb}u_{Sb} + i_{Ra}u_{Ra} + i_{Rb}u_{Rb} = {} & \frac{d}{dt}W_{\text{field}} + \frac{\partial W_{\text{field}}}{\partial \theta_R}\frac{d\theta_R}{dt} \\ & + R_S(i_{Sa}^2 + i_{Sb}^2) + R_R(i_{Ra}^2 + i_{Rb}^2). \end{aligned} \tag{6.51}$$

Integrating both sides with respect to time t gives the conservation of energy equation

$$W_{\text{electrical}} = W_{\text{field}} + \int_0^t \frac{\partial W_{\text{field}}}{\partial \theta_R}\omega_R dt + W_{\text{ohmic loss}}.$$

The electrical power supplied by the source goes into the magnetic field energy, heat losses in the phase windings and another term given by

$$\int_0^t \frac{\partial W_{\text{field}}}{\partial \theta_R}\omega_R dt.$$

The only other energy conversion is mechanical energy. Consequently, this term represents the energy from the supply converted into mechanical work.

The mechanical power $\tau_R\omega_R$ is therefore given by $\tau_R\omega_R = (\partial W_{\text{field}}/\partial\theta_R)\,\omega_R$ so that the torque produced by the machine is[7]

$$\tau_R = \frac{\partial W_{\text{field}}}{\partial \theta_R}. \tag{6.52}$$

Using the expression (6.49) for W_{field} results in

$$\begin{aligned}\tau_R &= M\Big(-i_{Ra}(t)i_{Sa}(t)\sin(\theta_R) + i_{Ra}(t)i_{Sb}(t)\cos(\theta_R)\\ &\quad -i_{Rb}(t)i_{Sa}(t)\cos(\theta_R) - i_{Rb}(t)i_{Sb}(t)\sin(\theta_R)\Big)\end{aligned}$$

which is the same expression as in equation (6.31) derived previously.

6.12.3 Computing Torque From the Co-energy

The above analysis was done assuming linear magnetics for the two-phase induction motor. A more general approach is now given that does not assume linear magnetics.

Recall the electrical dynamics of the two-phase induction motor can be written as

$$\begin{aligned}u_{Sa} - R_S i_{Sa} &= \frac{d\lambda_{Sa}}{dt}\\ u_{Sb} - R_S i_{Sb} &= \frac{d\lambda_{Sb}}{dt}\\ u_{Ra} - R_R i_{Ra} &= \frac{d\lambda_{Ra}}{dt}\\ u_{Rb} - R_R i_{Rb} &= \frac{d\lambda_{Rb}}{dt}.\end{aligned} \tag{6.53}$$

These equations are very general in the sense that they hold even in the case of magnetic saturation in the rotor or stator iron.[8] Define a new set of voltages $e_{Sa} \triangleq u_{Sa} - R_S i_{Sa}$, $e_{Sb} \triangleq u_{Sb} - R_S i_{Sb}$, $e_{Ra} \triangleq u_{Ra} - R_R i_{Ra}$, and $e_{Rb} \triangleq u_{Rb} - R_R i_{Rb}$ and consider these voltages as the *inputs* to the phase windings. In this way, the power loss to ohmic heating of the phase windings can be considered as part of the power supply and not the machine. The power delivered to the phases is then $i_{Sa}e_{Sa} + i_{Sb}e_{Sb} + i_{Ra}e_{Ra} + i_{Rb}e_{Rb}$ and, using (6.53), the basic equation for conservation of power may be written

[7] This derivation assumed that (6.45) held, that is, that magnetic saturation was not present. This result does not hold if magnetic saturation is present. A more general result that holds in the presence of magnetic saturation is given in the next subsection.

[8] However, the expressions (6.45) are valid only for the linear magnetics case, that is, no magnetic saturation.

as

$$\begin{aligned} i_{Sa}e_{Sa} + i_{Sb}e_{Sb} + i_{Ra}e_{Ra} + i_{Rb}e_{Rb} &= i_{Sa}\frac{d\lambda_{Sa}}{dt} + i_{Sb}\frac{d\lambda_{Sb}}{dt} + i_{Ra}\frac{d\lambda_{Ra}}{dt} \\ &\quad + i_{Rb}\frac{d\lambda_{Rb}}{dt} \\ &= \frac{d}{dt}W_{\text{field}}(i_{Sa}, i_{Sb}, i_{Ra}, i_{Rb}, \theta_R) \\ &\quad + \tau_R\frac{d\theta_R}{dt}. \end{aligned} \tag{6.54}$$

This equation just says that the power provided to the phases by the voltage sources equals the rate of change of the field energy and the mechanical power produced. This expression is perfectly general in that it holds in the case of magnetic saturation. However, it does assume that an expression for the field energy $W_{\text{field}}(i_{Sa}, i_{Sb}, i_{Ra}, i_{Rb}, \theta_R)$ in terms of the phase currents and rotor position has been found.[9]

Now,

$$\begin{aligned} \frac{d}{dt}W_{\text{field}} &= \frac{\partial W_{\text{field}}}{\partial i_{Sa}}\frac{di_{Sa}}{dt} + \frac{\partial W_{\text{field}}}{\partial i_{Sb}}\frac{di_{Sb}}{dt} + \frac{\partial W_{\text{field}}}{\partial i_{Ra}}\frac{di_{Ra}}{dt} + \frac{\partial W_{\text{field}}}{\partial i_{Rb}}\frac{di_{Rb}}{dt} \\ &\quad + \frac{\partial W_{\text{field}}}{\partial \theta_R}\frac{d\theta_R}{dt} \end{aligned} \tag{6.55}$$

and, with $\lambda = \lambda(i_{Sa}, i_{Sb}, i_{Ra}, i_{Rb}, \theta_R)$,

$$\frac{d\lambda}{dt} = \frac{\partial\lambda}{\partial i_{Sa}}\frac{di_{Sa}}{dt} + \frac{\partial\lambda}{\partial i_{Sb}}\frac{di_{Sb}}{dt} + \frac{\partial\lambda}{\partial i_{Ra}}\frac{di_{Ra}}{dt} + \frac{\partial\lambda}{\partial i_{Rb}}\frac{di_{Rb}}{dt} + \frac{\partial\lambda}{\partial\theta_R}\frac{d\theta_R}{dt}$$

where λ stands for $\lambda_{Sa}, \lambda_{Sb}, \lambda_{Ra}$, and λ_{Rb} (see the footnote[10]). Equation (6.54) holds for all $i_{Sa}, i_{Sb}, i_{Ra}, i_{Rb}$, and θ_R and their derivatives di_{Sa}/dt, di_{Sb}/dt, di_{Ra}/dt, di_{Rb}/dt, and $d\theta_R/dt$. The next step is to substitute the derivative of the flux linkages into left-hand side of (6.54) and the derivative

[9] The expression (6.49) is valid only for the linear magnetics case. To determine the field energy in the general case, one would lock the rotor so that $i_{Sa}e_{Sa} + i_{Sb}e_{Sb} + i_{Ra}e_{Ra} + i_{Rb}e_{Rb} = \frac{d}{dt}W_{\text{field}}(i_{Sa}, i_{Sb}, i_{Ra}, i_{Rb}, \theta_R)$ holds. Then one would try to integrate this expression to obtain W_{field}. Though the left-hand side is known, such a calculation would typically require numerical techniques in the presence of magnetic saturation. The "Achilles heel" in this approach is trying to find the field energy W_{field}.

[10] In the case of linear magnetics, the expressions (6.45) would be used. However, in the case of magnetic saturation, one proceeds as follows: For each phase, integrate $\lambda(t) = \int_0^t (u - Ri)dt$ to compute the phase flux as the currents i_{Sa}, i_{Sb}, i_{Ra}, and i_{Rb} are varied between $-I_{\max}$ and $I_{\max}$, and the rotor position is varied for $0 \le \theta_R \le \frac{2\pi}{n_p}$. This then gives the phase flux as a function of $i_{Sa}, i_{Sb}, i_{Ra}, i_{Rb}$, and θ_R.

of the field energy (6.55) into the right-hand side of (6.54). This results in

$$\begin{aligned}
&\left(i_{Sa}\frac{\partial\lambda_{Sa}}{\partial i_{Sa}}+i_{Sb}\frac{\partial\lambda_{Sb}}{\partial i_{Sa}}+i_{Ra}\frac{\partial\lambda_{Ra}}{\partial i_{Sa}}+i_{Rb}\frac{\partial\lambda_{Rb}}{\partial i_{Sa}}\right)\frac{di_{Sa}}{dt}\\
&+\left(i_{Sa}\frac{\partial\lambda_{Sb}}{\partial i_{Sb}}+i_{Sb}\frac{\partial\lambda_{Sb}}{\partial i_{Sb}}+i_{Ra}\frac{\partial\lambda_{Ra}}{\partial i_{Sb}}+i_{Rb}\frac{\partial\lambda_{Rb}}{\partial i_{Sb}}\right)\frac{di_{Sb}}{dt}\\
&+\left(i_{Sa}\frac{\partial\lambda_{Ra}}{\partial i_{Ra}}+i_{Sb}\frac{\partial\lambda_{Sb}}{\partial i_{Ra}}+i_{Ra}\frac{\partial\lambda_{Ra}}{\partial i_{Ra}}+i_{Rb}\frac{\partial\lambda_{Rb}}{\partial i_{Ra}}\right)\frac{di_{Ra}}{dt}\\
&+\left(i_{Sa}\frac{\partial\lambda_{Rb}}{\partial i_{Rb}}+i_{Sb}\frac{\partial\lambda_{Sb}}{\partial i_{Rb}}+i_{Ra}\frac{\partial\lambda_{Ra}}{\partial i_{Rb}}+i_{Rb}\frac{\partial\lambda_{Rb}}{\partial i_{Rb}}\right)\frac{di_{Rb}}{dt}\\
&+\left(i_{Sa}\frac{\partial\lambda_{Sa}}{\partial\theta_R}+i_{Sb}\frac{\partial\lambda_{Sb}}{\partial\theta_R}+i_{Ra}\frac{\partial\lambda_{Ra}}{\partial\theta_R}+i_{Rb}\frac{\partial\lambda_{Rb}}{\partial\theta_R}\right)\frac{d\theta_R}{dt}\\
&=\frac{\partial W_{\text{field}}}{\partial i_{Sa}}\frac{di_{Sa}}{dt}+\frac{\partial W_{\text{field}}}{\partial i_{Sb}}\frac{di_{Sb}}{dt}+\frac{\partial W_{\text{field}}}{\partial i_{Ra}}\frac{di_{Ra}}{dt}+\frac{\partial W_{\text{field}}}{\partial i_{Rb}}\frac{di_{Rb}}{dt}\\
&+\frac{\partial W_{\text{field}}}{\partial\theta_R}\frac{d\theta_R}{dt}+\tau_R\frac{d\theta_R}{dt}.
\end{aligned}\tag{6.56}$$

Given any set of values for i_{Sa}, i_{Sb}, i_{Ra}, i_{Rb}, and θ_R, by proper choice of u_{Sa},u_{Sb}, u_{Ra}, u_{Rb}, and τ_L, the derivatives di_{Sa}/dt, di_{Sb}/dt, di_{Ra}/dt, di_{Rb}/dt, and $d\theta_R/dt$ can be arbitrarily chosen (see problem 20). Therefore, the only way that (6.56) can be satisfied for any such values of di_{Sa}/dt, di_{Sb}/dt, di_{Ra}/dt, di_{Rb}/dt, and $d\theta_R/dt$ is that the coefficients of these derivatives in (6.56) must be equal for all time, that is,

$$\begin{aligned}
i_{Sa}\frac{\partial\lambda_{Sa}}{\partial i_{Sa}}+i_{Sb}\frac{\partial\lambda_{Sb}}{\partial i_{Sa}}+i_{Ra}\frac{\partial\lambda_{Ra}}{\partial i_{Sa}}+i_{Rb}\frac{\partial\lambda_{Rb}}{\partial i_{Sa}}&=\frac{\partial W_{\text{field}}}{\partial i_{Sa}}\\
i_{Sa}\frac{\partial\lambda_{Sb}}{\partial i_{Sb}}+i_{Sb}\frac{\partial\lambda_{Sb}}{\partial i_{Sb}}+i_{Ra}\frac{\partial\lambda_{Ra}}{\partial i_{Sb}}+i_{Rb}\frac{\partial\lambda_{Rb}}{\partial i_{Sb}}&=\frac{\partial W_{\text{field}}}{\partial i_{Sb}}\\
i_{Sa}\frac{\partial\lambda_{Ra}}{\partial i_{Ra}}+i_{Sb}\frac{\partial\lambda_{Sb}}{\partial i_{Ra}}+i_{Ra}\frac{\partial\lambda_{Ra}}{\partial i_{Ra}}+i_{Rb}\frac{\partial\lambda_{Rb}}{\partial i_{Ra}}&=\frac{\partial W_{\text{field}}}{\partial i_{Ra}}\\
i_{Sa}\frac{\partial\lambda_{Rb}}{\partial i_{Rb}}+i_{Sb}\frac{\partial\lambda_{Sb}}{\partial i_{Rb}}+i_{Ra}\frac{\partial\lambda_{Ra}}{\partial i_{Rb}}+i_{Rb}\frac{\partial\lambda_{Rb}}{\partial i_{Rb}}&=\frac{\partial W_{\text{field}}}{\partial i_{Rb}}\\
i_{Sa}\frac{\partial\lambda_{Sa}}{\partial\theta_R}+i_{Sb}\frac{\partial\lambda_{Sb}}{\partial\theta_R}+i_{Ra}\frac{\partial\lambda_{Ra}}{\partial\theta_R}+i_{Rb}\frac{\partial\lambda_{Rb}}{\partial\theta_R}&=\frac{\partial W_{\text{field}}}{\partial\theta_R}+\tau_R.
\end{aligned}\tag{6.57}$$

In particular, the last expression shows that

$$\tau_R=i_{Sa}\frac{\partial\lambda_{Sa}}{\partial\theta_R}+i_{Sb}\frac{\partial\lambda_{Sb}}{\partial\theta_R}+i_{Ra}\frac{\partial\lambda_{Ra}}{\partial\theta_R}+i_{Rb}\frac{\partial\lambda_{Rb}}{\partial\theta_R}-\frac{\partial W_{\text{field}}}{\partial\theta_R}.$$

Finally, making the definition of *co-energy* as

$$W_{\text{co-energy}}\triangleq i_{Sa}\lambda_{Sa}+i_{Sb}\lambda_{Sb}+i_{Ra}\lambda_{Ra}+i_{Rb}\lambda_{Rb}-W_{\text{field}},\tag{6.58}$$

it follows that the torque may be written as

$$\tau_R=\frac{\partial W_{\text{co-energy}}}{\partial\theta_R}.\tag{6.59}$$

This derivation assumed that all of the phases had voltage inputs. However, the expression for torque (6.59) still holds even if the rotor phases are shorted. That is, the expression (6.59) is true for any set of inputs, so it must be true for $u_{Ra} = u_{Rb} \equiv 0$.

The co-energy (6.58) does not have any physical significance and only represents a convenient mathematical quantity for computing the torque.

The reader is asked in problem 15 to show that, in the case of linear magnetics, the expression for W_{field} given in (6.49) equals the expression for $W_{\text{co-energy}}$ given in (6.58).

Problems

Torque Production in AC Machines

Problem 1 Torque on the Stator Winding

Use the expression (6.27) for radial magnetic field in the air gap due to the rotor currents and compute the total torque $\vec{\tau}_S$ *it produces on the currents in the stator windings. In particular, show that* $\vec{\tau}_S = -\vec{\tau}_R$.

Problem 2 $\vec{\mathbf{B}}_R$ Cannot Produce Torque on the Rotor

(a) Using the expression (6.6) [or (6.27)] for the radial magnetic field in the air gap produced by the rotor currents, show by direct computation that the total torque this magnetic field produces on the rotor *currents is zero.*

(b) Similarly, using the expression (6.5) [or (6.26)] for the radial magnetic field in the air gap produced by the stator currents, show by direct computation that the torque this magnetic field produces on the stator *currents is zero.*

(c) Explain why these two results must be true from basic principles of Physics.

Problem 3 The Induction Motor as a Transformer

Using the flux linkage equations (6.45), show that $[\sigma = 1 - M^2/(L_S L_R)]$

$$\begin{aligned}
\sigma L_S L_R i_{Sa} &= L_R \lambda_{Sa} - M(+\lambda_{Ra}\cos(\theta_R) - \lambda_{Rb}\sin(\theta_R)) \\
\sigma L_S L_R i_{Sb} &= L_R \lambda_{Sb} - M(+\lambda_{Ra}\sin(\theta_R) + \lambda_{Rb}\cos(\theta_R)) \\
\sigma L_S L_R i_{Ra} &= L_S \lambda_{Ra} - M(+\lambda_{Sa}\cos(\theta_R) + \lambda_{Sb}\sin(\theta_R)) \\
\sigma L_S L_R i_{Rb} &= L_S \lambda_{Rb} - M(-\lambda_{Sa}\sin(\theta_R) + \lambda_{Sb}\cos(\theta_R)).
\end{aligned}$$

If $\sigma = 0$, *find an expression for the rotor flux linkages in terms of the stator flux linkages. Use this expression to show that, with the rotor locked at some angle* θ_R, *the induction motor (with a sinusoidally wound rotor) is a (two-input two-output) transformer with a turns ratio of either* $(M/L_S)\cos(\theta_R) = (N_R/N_S)\cos(\theta_R)$ *or* $(M/L_S)\sin(\theta_R) = (N_R/N_S)\sin(\theta_R)$ *from a stator winding to a rotor winding. In particular, find* $\lambda_{Ra}/\lambda_{Sa}$ *and* $\lambda_{Rb}/\lambda_{Sb}$ *for* $\theta_R = 0$.

What if one had a squirrel cage induction motor rather than a sinusoidally wound rotor induction motor?

Mathematical Models of an Induction Machine

Problem 4 State-Space Model of the Induction Motor

Starting from the mathematical model (6.30) and (6.31), derive a state space model for the two-phase induction motor where the state variables are $i_{Sa}, i_{Sb}, i_{Ra}, i_{Rb}, \omega$, *and* θ. *(See problem 6 for a simpler, but equivalent, model.)*

Problem 5 Space Vector Model of the Induction Motor (Leonhard [1])

Let $\underline{i}_S \triangleq i_{Sa} + j i_{Sb}$, $\underline{i}_R \triangleq i_{Ra} + j i_{Rb}$, $\underline{u}_S \triangleq u_{Sa} + j u_{Sb}$ *and show that the equations of the induction motor (6.34) may be written succinctly as*

$$
\begin{aligned}
R_S \underline{i}_S + L_S \frac{d}{dt}\underline{i}_S + M\frac{d}{dt}\left(\underline{i}_R e^{j n_p \theta_R}\right) &= \underline{u}_S \\
R_R \underline{i}_R + L_R \frac{d}{dt}\underline{i}_R + M\frac{d}{dt}\left(\underline{i}_S e^{-j n_p \theta_R}\right) &= 0 \\
n_p M \mathrm{Im}\left\{\underline{i}_S(\underline{i}_R e^{j n_p \theta_R})^*\right\} - \tau_L &= J\frac{d\omega_R}{dt}
\end{aligned}
$$

where Im$\{\cdot\}$ *denotes the* imaginary *part and* $*$ *denotes complex conjugate. This is known as the* space vector *representation.*

Problem 6 A Standard Model of the Induction Motor

The model (6.34) of the induction motor (or its equivalent space vector representation given in problem 5) is quite complicated. By doing a change of variables, a simpler, but equivalent model can be found. In particular, a model can found where the explicit dependency on θ_R *[i.e., the* $\sin(n_p\theta_R)$ *and* $\cos(n_p\theta_R)$ *terms] can be eliminated. To do so, define new (fictitious) flux linkages as*

$$
\underline{\psi}_R \triangleq \psi_{Ra} + j\psi_{Rb} \triangleq \underline{\lambda}_R e^{j n_p \theta_R} = L_R \underline{i}_R e^{j n_p \theta_R} + M \underline{i}_S
$$

(a) Make this substitution into the representation given in problem 5 to show that the system of equations for the induction motor may be rewritten in the space vector form as

$$
\begin{aligned}
\frac{d}{dt}\underline{\psi}_R &= \left(-\frac{1}{T_R} + j n_p \omega_R\right)\underline{\psi}_R + \frac{M}{T_R}\underline{i}_S \\
\underline{u}_S &= \frac{M}{L_R}\frac{d}{dt}\underline{\psi}_R + \sigma L_S \frac{d}{dt}\underline{i}_S + R_S \underline{i}_S \\
\frac{d\omega_R}{dt} &= \mu\, \mathrm{Im}\left\{\underline{i}_S(\underline{\psi}_R)^*\right\} - \frac{\tau_L}{J}
\end{aligned}
$$

where $\mu \triangleq n_p M/(J L_R)$, $\sigma \triangleq 1 - M^2/L_S L_R$ *is the leakage factor, and* $T_R \triangleq L_R/R_R$ *is the rotor time constant. This model is often seen in the literature.*

(b) Equate real and imaginary parts in part (a) to write the model as five differential equations.

(c) Use the answer in part (b) to find a state space model of the motor with state variables $i_{Sa}, i_{Sb}, \psi_{Ra}, \psi_{Rb}$, *and* ω_R.

Remark *This problem illustrates the power of the space vector representation as the manipulations of the equations are all done with (complex) scalar quantities rather than dealing with vectors and matrices. The representations in parts (a) and (b) are standard forms seen in the literature.*

Often the real and imaginary parts of $\psi_{Ra} + j\psi_{Rb}$ *are referred to as the "rotor fluxes", but the actual rotor flux linkages are the real and imaginary parts of* $\underline{\lambda}_R = (\psi_{Ra} + j\psi_{Rb})\, e^{-jn_p\theta_R}$.

Problem 7 Simulation of an Induction Motor

(a) Using the state-space model derived in problem 6(c), simulate a two-phase induction motor. Set the parameters as $R_R = 3.9\ \Omega$, $R_S = 1.7\ \Omega$, $L_R = 0.014\ H$, $L_S = 0.014\ H$, $M = 0.0117\ H$, $J = 0.00011\ kg\text{-}m^2$, $f = 0\ N\text{-}m/rad/sec$, $n_p = 3$, $\omega_S = (2\pi)60\ rad/sec$, $V = V_{max} = 60\ V$ *(peak)*, $I_{max} = 12\ A$, $\tau_L = 0$. *Let* $u_{Sa} = V\cos(\omega_S t)$, $u_{Sb} = V\sin(\omega_S t)$ *and plot (a)* i_{Sa} *and* i_{Sb}, *(b)* i_{Ra} *and* i_{Rb}, *and (c)* ω. *Run the simulation long enough (about 0.5 sec) so that the speed* ω_R *goes to a constant steady-state value. Notice that at steady-state (constant) speed, the rotor currents are zero. Give a* physical *explanation as to why this happens.*

(b) Rerun the simulation with a load torque of $\tau_L = 0.2$ *N-m acting on the system* after *it has reached steady-state (synchronous) speed. Be sure to run the simulation until the motor speed has reached its new (lower) steady-state speed. Note that the steady-state rotor currents are no longer zero. Why?*

(c) What happens if $\tau_L = 0.5$ *N-m is used? Note that the frequency of the rotor currents increases as the load-torque increases. Give a* physical *explanation to why this happens.*

Problem 8 Induction Motor Model in Terms of the Flux Linkages

With λ_{Sa} *and* λ_{Sb} *the total flux linkage in stator phases a and b, respectively, define*

$$\underline{\lambda}_S \triangleq \lambda_{Sa} + j\lambda_{Sb} \triangleq L_S\underline{i}_S + M\underline{i}_R e^{+jn_p\theta_R}.$$

Similarly, let λ_{Ra} *and* λ_{Rb} *be the total flux linkage in rotor phases a and b, respectively, and define*

$$\underline{\lambda}_R \triangleq \lambda_{Ra} + j\lambda_{Rb} \triangleq L_R\underline{i}_R + M\underline{i}_S e^{-jn_p\theta_R}.$$

(a) Show that the electrical equations of the induction motor may be written as

$$\begin{aligned} R_S\underline{i}_S + \frac{d}{dt}\underline{\lambda}_S &= \underline{u}_S \\ R_R\underline{i}_R + \frac{d}{dt}\underline{\lambda}_R &= 0. \end{aligned}$$

(b) Using the fact that

$$\begin{bmatrix} \underline{i}_S \\ \underline{i}_R e^{+jn_p\theta_R} \end{bmatrix} = \frac{1}{L_S L_R - M^2}\begin{bmatrix} +L_R & -M \\ -M & +L_S \end{bmatrix}\begin{bmatrix} \underline{\lambda}_S \\ \underline{\lambda}_R e^{+jn_p\theta_R} \end{bmatrix}$$

show that

$$\begin{aligned}
\frac{R_S}{L_S L_R - M^2}\left(L_R \underline{\lambda}_S - M\underline{\lambda}_R e^{+jn_p\theta_R}\right) + \frac{d}{dt}\underline{\lambda}_S &= \underline{u}_S \\
\frac{R_R}{L_S L_R - M^2}\left(-M\underline{\lambda}_S e^{-jn_p\theta_R} + L_S\underline{\lambda}_R\right) + \frac{d}{dt}\underline{\lambda}_R &= 0 \\
\frac{M}{L_S L_R - M^2}\operatorname{Im}\{\underline{\lambda}_S(\underline{\lambda}_R e^{+jn_p\theta_R})^*\} - \tau_L &= J\frac{d\omega_R}{dt} \\
\frac{d\theta_R}{dt} &= \omega_R
\end{aligned}$$

is a representation of the induction motor in terms of the variables $\underline{\lambda}_S \triangleq \lambda_{Sa} + j\lambda_{Sb}, \underline{\lambda}_R \triangleq \lambda_{Ra} + j\lambda_{Rb}, \omega_R$, *and* θ_R.

(c) By equating real and imaginary parts of the answer to part (b), find a state space representation of the induction motor in terms of the state variables $\lambda_{Sa}, \lambda_{Sb}, \lambda_{Ra}, \lambda_{Rb}, \omega_R$, *and* θ_R.

Problem 9 Steady-State Solution of the Induction Motor Equations [1]

Use the space vector formulation given in problem 5 to find a sinusoidal steady-state solution to the induction motor equations. In particular, try a solution of the form

$$\begin{aligned}
\underline{u}_S &= u_{Sa} + ju_{Sb} = U_S\cos(\omega_S t + \phi_{Us}) + jU_S\sin(\omega_S t + \phi_{Us}) \\
&= U_S e^{j\phi_{Us}} e^{j\omega_S t} = \underline{U}_S e^{j\omega_S t}
\end{aligned}$$

$$\begin{aligned}
\underline{i}_S &= i_{Sa} + ji_{Sb} = I_S\cos(\omega_S t + \phi_S) + jI_S\sin(\omega_S t + \phi_S) \\
&= I_S e^{j\phi_S} e^{j\omega_S t} = \underline{I}_S e^{j\omega_S t}
\end{aligned}$$

$$\begin{aligned}
\underline{i}_R &= i_{Ra} + ji_{Rb} = I_R\cos\left((\omega_S - \omega_R)t + \phi_R\right) + jI_R\sin\left((\omega_S - \omega_R)t + \phi_R\right) \\
&= I_R e^{j\phi_R} e^{j(\omega_S - \omega_R)t} = \underline{I}_R e^{j(\omega_S - \omega_R)t}
\end{aligned}$$

$$\theta_R(t) = \omega_R t$$

where the stator electrical frequency ω_S, *rotor speed* ω_R *and* $\underline{U}_S = U_S e^{j\phi_{Us}}$ *are assumed to be known and constant. The stator and rotor current phasors given by* $\underline{I}_S \triangleq I_S e^{j\phi_S}$ *and* $\underline{I}_R \triangleq I_R e^{j\phi_R}$ *are to be determined; that is, the unknowns are* I_S, ϕ_S, I_R, *and* ϕ_R.

(a) Substitute these expressions into the space vector representation of the induction motor equations to obtain

$$\begin{aligned}
(R_S + j\omega_S L_S)\underline{I}_S + j\omega_S M\underline{I}_R &= \underline{U}_S \\
(R_R + j(\omega_S - \omega_R)L_R)\underline{I}_R + j(\omega_S - \omega_R)M\underline{I}_S &= 0 \\
M\operatorname{Im}\{\underline{I}_S(\underline{I}_R)^*\} - \tau_L &= 0.
\end{aligned}$$

With the normalized slip speed defined by $S \triangleq (\omega_S - \omega_R)/\omega_S$, *it follows that*

$$\begin{bmatrix} \underline{I}_S \\ \underline{I}_R \end{bmatrix} = \frac{1}{\det}\begin{bmatrix} R_R/S + j\omega_S L_R & -j\omega_S M \\ -j\omega_S M & R_S + j\omega_S L_S \end{bmatrix}\begin{bmatrix} \underline{U}_S \\ 0 \end{bmatrix}$$

where

$$\det \triangleq (R_S + j\omega_S L_S)(\frac{R_R}{S} + j\omega_S L_R) - (j\omega_S M)^2.$$

(b) Use these expressions to compute I_S, ϕ_S, I_R, ϕ_R, *and* $\tau = M\,\mathrm{Im}\{\underline{I}_S(\underline{I}_R)^*\} = MI_SI_R\sin(\phi_S - \phi_R)$ *as functions of the stator frequency* ω_S *and the normalized slip* S.

(c) Using the parameter values given in problem 7, write a program to plot $\tau = MI_SI_R\sin(\phi_S - \phi_R)$ *for a fixed* $\omega_S = 2\pi * 60$ *as* S *varies from* 0 *to* 20.

Problem 10 Model of an Induction Motor with Multiple Pole Pairs

Consider a two-phase motor with n_p *pole pairs. With* $n_p = 2$, *Figure 6.12(a) shows stator phase a and Figure 6.12(b) shows stator phase b.*

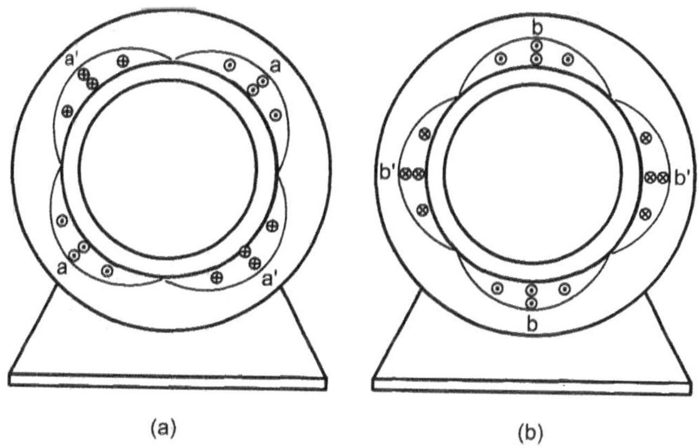

FIGURE 6.12. (a) Sinusoidally wound stator phase a with $n_p = 2$ and turns density $N_{Sa}(\theta) = (n_pN_S/2)\,|\sin(n_p\theta)|$. (b) Sinusoidally wound stator phase b with $n_p = 2$ and turns density $N_{Sb}(\theta) = (n_pN_S/2)\,|\sin(n_p(\theta - \pi/2))|$.

The sinusoidal turns density *for the windings of each phase are given by*

$$\begin{aligned} N_{Sa}(\theta) &= \frac{n_pN_S}{2}|\sin(n_p\theta)| \\ N_{Sb}(\theta) &= \frac{n_pN_S}{2}|\sin(n_p(\theta - \pi/2))| \\ N_{Ra}(\theta) &= \frac{n_pN_R}{2}|\sin(n_p(\theta - \theta_R))| \\ N_{Rb}(\theta) &= \frac{n_pN_R}{2}|\sin(n_p(\theta - \theta_R - \pi/2))|. \end{aligned}$$

For example, the first pole pair of stator phase a is made up of the turns for $-\pi/n_p < \theta < \pi/n_p$*, the second pole pair of stator phase a is made up of the turns for* $\pi/n_p < \theta < 3\pi/n_p$*, and so on. The total number of turns making up the first pole pair is* $\int_0^{\pi/n_p} N_{Sa}(\theta)d\theta = N_S$*. As there are* n_p *pole pairs, the total number of turns in phase a is* $n_p N_S$*; that is,* N_S *is the number of turns in phase a per pole pair.*

(a) Explain why Ampère's law $\oint \vec{\mathbf{H}} \cdot d\vec{\ell} = i_{enclosed}$ *applied to the closed path 1–2–3–4–1 indicated in Figure 6.13 is*

$$\oint_{1-2-3-4-1} \vec{\mathbf{H}}_{Sa} \cdot d\vec{\ell} = \int_0^{\theta} i_{Sa} \frac{n_p N_S}{2} \sin(n_p\theta')d\theta' \qquad (6.60)$$

for any *angle* θ.

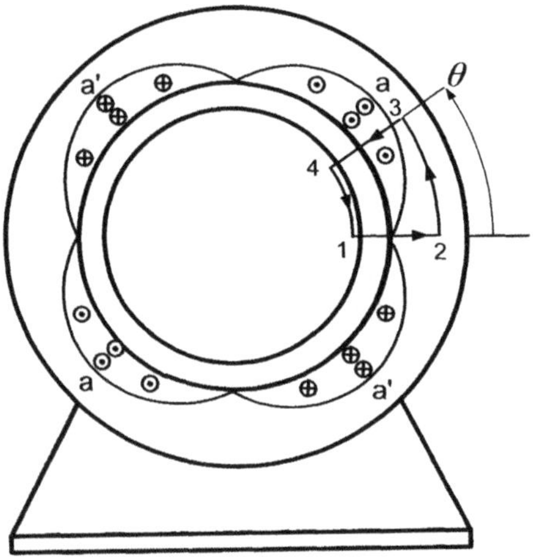

FIGURE 6.13. Use of Ampère's law to determine the air gap radial magnetic field produced by i_{Sa} in a n_p pole-pair machine.

Show that equation (6.60) leads to

$$B_{Sa}(i_{Sa}, \theta) = \frac{\mu_0 N_S}{2g} i_{Sa} \cos(n_p\theta). \qquad (6.61)$$

(b) Show that conservation of flux will hold in the air gap if this expression is modified by including the factor r_R/r *to obtain*

$$\vec{\mathbf{B}}_{Sa}(i_{Sa}, r, \theta) = \frac{\mu_0 N_S}{2g} \frac{r_R}{r} i_{Sa} \cos(n_p\theta)\hat{\mathbf{r}}.$$

(c) Show that

$$
\begin{aligned}
\vec{\mathbf{B}}_S(i_{Sa}, i_{Sb}, r, \theta) &= \vec{\mathbf{B}}_{Sa}(i_{Sa}, r, \theta) + \vec{\mathbf{B}}_{Sb}(i_{Sb}, r, \theta) \\
&= \frac{\mu_0 N_S}{2g}\frac{r_R}{r}\left(i_{Sa}\cos(n_p\theta) + i_{Sb}\sin(n_p\theta)\right)\hat{\mathbf{r}} \\
\vec{\mathbf{B}}_R(i_{Ra}, i_{Rb}, r, \theta - \theta_R) &= \vec{\mathbf{B}}_{Ra}(i_{Ra}, r, \theta - \theta_R) + \vec{\mathbf{B}}_{Rb}(i_{Rb}, r, \theta - \theta_R) \\
&= \frac{\mu_0 N_R}{2g}\frac{r_R}{r} \\
&\quad \times \Big(i_{Ra}\cos\left(n_p(\theta - \theta_R)\right) + i_{Rb}\sin\left(n_p(\theta - \theta_R)\right)\Big)\hat{\mathbf{r}}.
\end{aligned}
$$

(d) Show that

$$
\begin{aligned}
\lambda_{Sa}(i_{Ra}, i_{Rb}, i_{Sa}, i_{Sb}, \theta_R) &\triangleq \int_{\substack{\text{All loops of} \\ \text{stator phase } a}} (\vec{\mathbf{B}}_S + \vec{\mathbf{B}}_R)\cdot d\vec{\mathbf{S}} \\
&= L_S i_{Sa} + M\Big(i_{Ra}\cos(n_p\theta_R) - i_{Rb}\sin(n_p\theta_R)\Big) \\
\lambda_{Sb}(i_{Ra}, i_{Rb}, i_{Sa}, i_{Sb}, \theta_R) &= \int_{\substack{\text{All loops of} \\ \text{stator phase } b}} (\vec{\mathbf{B}}_S + \vec{\mathbf{B}}_R)\cdot d\vec{\mathbf{S}} \\
&= L_S i_{Sb} + M\Big(i_{Ra}\sin(n_p\theta_R) + i_{Rb}\cos(n_p\theta_R)\Big)
\end{aligned}
$$

$$
\begin{aligned}
\lambda_{Ra}(i_{Ra}, i_{Rb}, i_{Sa}, i_{Sb}, \theta_R) &= \int_{\substack{\text{All loops of} \\ \text{rotor phase } a}} (\vec{\mathbf{B}}_S + \vec{\mathbf{B}}_R)\cdot d\vec{\mathbf{S}} \\
&= L_R i_{Ra} + M\Big(i_{Sa}\cos(n_p\theta_R) + i_{Sb}\sin(n_p\theta_R)\Big) \\
\lambda_{Rb}(i_{Ra}, i_{Rb}, i_{Sa}, i_{Sb}, \theta_R) &= \int_{\substack{\text{All loops of} \\ \text{rotor phase } b}} (\vec{\mathbf{B}}_S + \vec{\mathbf{B}}_R)\cdot d\vec{\mathbf{S}} \\
&= L_R i_{Rb} + M\Big(-i_{Sa}\sin(n_p\theta_R) + i_{Sb}\cos(n_p\theta_R)\Big)
\end{aligned}
$$

$$
\begin{aligned}
\tau_R &= n_p M\Big(-i_{Ra}(t)i_{Sa}(t)\sin(n_p\theta_R) + i_{Ra}(t)i_{Sb}(t)\cos(n_p\theta_R) \\
&\quad - i_{Rb}(t)i_{Sa}(t)\cos(n_p\theta_R) - i_{Rb}(t)i_{Sb}(t)\sin(n_p\theta_R)\Big)
\end{aligned}
$$

where $L_S = \dfrac{\mu_0 \ell_1 \ell_2 \pi}{8g} N_S^2$, $L_R = \dfrac{\mu_0 \ell_1 \ell_2 \pi}{8g} N_R^2$, $M = \kappa \dfrac{\mu_0 \pi \ell_1 \ell_2}{8g} N_S N_R$.

Mathematical Model of a Two-Phase Synchronous Machine

Problem 11 Mathematical Model of a PM Synchronous Machine

Consider a two-phase synchronous machine in which the rotor consists of a permanent magnet which produces a sinusoidally distributed radial magnetic field in the air gap given by

$$\vec{\mathbf{B}}_R(r,\theta-\theta_R) = B_m\frac{r_R}{r}\cos(\theta-\theta_R)\hat{\mathbf{r}}.$$

At $r = r_S$ *this is written as*

$$\vec{\mathbf{B}}_R(r_S,\theta-\theta_R) = \kappa B_m\frac{r_R}{r_S}\cos(\theta-\theta_R)\hat{\mathbf{r}}$$

where the coupling factor κ *is included to account for leakage.*

The stator is identical to the wound rotor synchronous motor, that is, it has two sinusoidally wound phases 90° apart with voltage sources u_{Sa} *and* u_{Sb}. *The radial magnetic field in the air gap due to the stator currents is then given by (6.5).*

(a) Show that the flux linkages in the stator phases have the form

$$\begin{aligned}\lambda_{Sa}(t) &= L_S i_{Sa}(t) + K_m\cos(\theta_R)\\ \lambda_{Sb}(t) &= L_S i_{Sb}(t) + K_m\sin(\theta_R)\end{aligned}$$

and compute explicit expressions for L_S *and* K_m *in terms of* $\mu_0, \kappa, \ell_1, \ell_2, N_S$, B_m, *and* g.

(b) Find the torque $\vec{\tau}_S$ *on the currents in the stator windings produced by* $\vec{\mathbf{B}}_R(r_S,\theta-\theta_R)$. *Then compute the torque on the rotor as* $\vec{\tau}_R = -\vec{\tau}_S$.

(c) Give the complete set of equations characterizing a two-phase permanent magnet synchronous machine and show that it has the form (6.40).

Problem 12 Wound Rotor Synchronous Machine with Multiple Pole Pairs

Using the multiple-pole-pair model of a two-phase sinusoidally wound induction motor (see problem 10), show that it can be used directly to derive the multiple-pole-pair model of a two-phase sinusoidally wound synchronous motor. Specifically, show

$$\begin{aligned}\lambda_{Sa}(i_F,i_{Sa},i_{Sb},\theta_R) &= L_S i_{Sa} + M i_F\cos(n_p\theta_R)\\ \lambda_{Sb}(i_F,i_{Sa},i_{Sb},\theta_R) &= L_S i_{Sb} + M i_F\sin(n_p\theta_R)\\ \lambda_F(i_F,i_{Sa},i_{Sb},\theta_R) &= L_F i_{Fa} + M(i_{Sa}\cos(n_p\theta_R) + i_{Sb}\sin(n_p\theta_R))\\ \tau_R &= n_p M i_F\left(-i_{Sa}\sin(n_p\theta_R) + i_{Sb}\cos(n_p\theta_R)\right)\end{aligned}$$

where $L_S = \dfrac{\mu_0\ell_1\ell_2\pi}{8g}N_S^2, L_F = \dfrac{\mu_0\ell_1\ell_2\pi}{8g}N_F^2$, *and* $M = \kappa\dfrac{\mu_0\pi\ell_1\ell_2}{8g}N_S N_F$. N_S *is the number of winding per pole pair in each of the stator phases and* N_F *is the number of turns in the field winding. Use these expressions to deduce the dynamic equations of this machine.*

Problem 13 State-Space Model of a Wound Rotor Synchronous Machine

For the synchronous machine model of problem 12 [see also (6.38)], note that the relationship between the currents and the flux linkages may be written as

$$\begin{bmatrix} \lambda_{Sa} \\ \lambda_{Sb} \\ \lambda_F \end{bmatrix} \triangleq \begin{bmatrix} L_S & 0 & M\cos(n_p\theta_R) \\ 0 & L_S & M\sin(n_p\theta_R) \\ M\cos(n_p\theta_R) & M\sin(n_p\theta_R) & L_F \end{bmatrix} \begin{bmatrix} i_{Sa} \\ i_{Sb} \\ i_F \end{bmatrix}.$$

With $\Lambda(n_p\theta_R)$ denoting the 3×3 matrix on the right-hand side, it follows that

$$\begin{bmatrix} i_{Sa} \\ i_{Sb} \\ i_F \end{bmatrix} = \Lambda^{-1}(n_p\theta_R) \begin{bmatrix} \lambda_{Sa} \\ \lambda_{Sb} \\ \lambda_F \end{bmatrix}$$

where, with $\sigma \triangleq 1 - M^2/L_SL_F$,

$$\Lambda^{-1}(n_p\theta_R) = \frac{1}{\sigma L_S^2 L_F} \times$$

$$\begin{bmatrix} L_SL_F - M^2\sin^2(n_p\theta_R) & M^2\sin(n_p\theta_R)\cos(n_p\theta_R) & -L_SM\cos(n_p\theta_R) \\ M^2\sin(n_p\theta_R)\cos(n_p\theta_R) & L_SL_F - M^2\cos^2(n_p\theta_R) & -L_SM\sin(n_p\theta_R) \\ -L_SM\cos(n_p\theta_R) & -L_SM\sin(n_p\theta_R) & L_S^2 \end{bmatrix}.$$

(a) Use these relationships to find a state-space representation of the synchronous motor in terms of the flux linkages $\lambda_{Sa}, \lambda_{Sb}$, and λ_F. Be sure to include the torque equation.

(b) Use these relationships to find a state-space representation of the synchronous motor in terms of the currents i_{Sa}, i_{Sb}, and i_F. Be sure to include the torque equation.

Problem 14 DQ State-Space Model of a Synchronous Machine

For the synchronous machine model of Problem 12 [see also (6.38)], consider a change of coordinates to the so-called direct-quadrature *(dq) reference frame defined by*

$$\begin{bmatrix} u_d \\ u_q \\ u_F \end{bmatrix} \triangleq \begin{bmatrix} \cos(n_p\theta_R) & \sin(n_p\theta_R) & 0 \\ -\sin(n_p\theta_R) & \cos(n_p\theta_R) & 0 \\ 0 & 0 & 1 \end{bmatrix} \begin{bmatrix} u_{Sa} \\ u_{Sb} \\ u_F \end{bmatrix}$$

$$\begin{bmatrix} i_d \\ i_q \\ i_F \end{bmatrix} \triangleq \begin{bmatrix} \cos(n_p\theta_R) & \sin(n_p\theta_R) & 0 \\ -\sin(n_p\theta_R) & \cos(n_p\theta_R) & 0 \\ 0 & 0 & 1 \end{bmatrix} \begin{bmatrix} i_{Sa} \\ i_{Sb} \\ i_F \end{bmatrix}$$

$$\begin{bmatrix} \lambda_d \\ \lambda_q \\ \lambda_F \end{bmatrix} \triangleq \begin{bmatrix} \cos(n_p\theta_R) & \sin(n_p\theta_R) & 0 \\ -\sin(n_p\theta_R) & \cos(n_p\theta_R) & 0 \\ 0 & 0 & 1 \end{bmatrix} \begin{bmatrix} \lambda_{Sa} \\ \lambda_{Sb} \\ \lambda_F \end{bmatrix}.$$

(a) *Show that*

$$\begin{bmatrix} \lambda_d \\ \lambda_q \\ \lambda_F \end{bmatrix} = \begin{bmatrix} L_S & 0 & M \\ 0 & L_S & 0 \\ M & 0 & L_F \end{bmatrix} \begin{bmatrix} i_d \\ i_q \\ i_F \end{bmatrix}.$$

(b) *Show that*

$$\begin{bmatrix} i_d \\ i_q \\ i_F \end{bmatrix} = \frac{1}{\sigma L_S L_F} \begin{bmatrix} L_F & 0 & -M \\ 0 & \sigma L_F & 0 \\ -M & 0 & L_S \end{bmatrix} \begin{bmatrix} \lambda_d \\ \lambda_q \\ \lambda_F \end{bmatrix}.$$

(c) *Use the expressions from parts (a) and (b) to show that*

$$\begin{aligned} \frac{d\lambda_d}{dt} &= n_p\omega\lambda_q - R_S i_d + u_d \\ \frac{d\lambda_q}{dt} &= -n_p\omega\lambda_d - R_S i_q + u_q \\ \frac{d\lambda_F}{dt} &= -R_F i_F + u_F \\ J\frac{d\omega_R}{dt} &= M i_F i_q - \tau_L \end{aligned}$$

and so

$$\frac{d}{dt}\begin{bmatrix} L_S & 0 & M \\ 0 & L_S & 0 \\ M & 0 & L_F \end{bmatrix} \begin{bmatrix} i_d \\ i_q \\ i_F \end{bmatrix} = \begin{bmatrix} -R_S & n_p\omega L_S & 0 \\ -n_p\omega L_S & -R_S & -n_p\omega M \\ 0 & 0 & R_F \end{bmatrix} \begin{bmatrix} i_d \\ i_q \\ i_F \end{bmatrix} + \begin{bmatrix} u_d \\ u_q \\ u_F \end{bmatrix}.$$

(d) *Show that*

$$\begin{aligned} \frac{d}{dt}\begin{bmatrix} i_d \\ i_q \\ i_F \end{bmatrix} = \frac{1}{\sigma}\begin{bmatrix} -R_S/L_S & n_p\omega & -M/L_S \\ -\sigma n_p\omega & -\sigma R_S/L_S & -n_p\omega\sigma M/L_S \\ -MR_S/L_S L_F & -n_p\omega M/L_F & R_F/L_F \end{bmatrix} \begin{bmatrix} i_d \\ i_q \\ i_F \end{bmatrix} \\ + \frac{1}{\sigma L_S L_F} \begin{bmatrix} L_F & 0 & -M \\ 0 & \sigma L_F & 0 \\ -M & 0 & L_S \end{bmatrix} \begin{bmatrix} u_d \\ u_q \\ u_F \end{bmatrix} \end{aligned}$$

$$J\frac{d\omega_R}{dt} = M i_F i_q - \tau_L$$

where

$$\sigma \triangleq 1 - \frac{M^2}{L_S L_F}.$$

Magnetic Field Energy, Co-energy and Torque

Problem 15 Co-energy

The co-energy is defined by

$$W_{co\text{-}energy} \triangleq i_{Sa}\lambda_{Sa} + i_{Sb}\lambda_{Sb} + i_{Ra}\lambda_{Ra} + i_{Rb}\lambda_{Rb} - W_{field}.$$

The expression for the field energy given in equation (6.49) is valid if magnetic saturation is not *present. In this case, show that $W_{co\text{-}energy} = W_{field}$ and consequently, either (6.52) or (6.59) is valid for computing the torque. (The co-energy is* not *equal to the magnetic field energy when magnetic saturation is present in the iron.)*

Problem 16 Co-energy

Inverting the flux linkage equations (6.45) gives $\left(\sigma = 1 - \dfrac{M^2}{L_S L_R}\right)$

$$\begin{aligned}
i_{Sa} &= \frac{1}{\sigma L_S L_R}\Big(L_R\lambda_{Sa} - M\big(+\lambda_{Ra}\cos(\theta_R) - \lambda_{Rb}\sin(\theta_R)\big)\Big)\\
i_{Sb} &= \frac{1}{\sigma L_S L_R}\Big(L_R\lambda_{Sb} - M\big(+\lambda_{Ra}\sin(\theta_R) + \lambda_{Rb}\cos(\theta_R)\big)\Big)\\
i_{Ra} &= \frac{1}{\sigma L_S L_R}\Big(L_S\lambda_{Ra} - M\big(+\lambda_{Sa}\cos(\theta_R) + \lambda_{Sb}\sin(\theta_R)\big)\Big)\\
i_{Rb} &= \frac{1}{\sigma L_S L_R}\Big(L_S\lambda_{Rb} - M\big(-\lambda_{Sa}\sin(\theta_R) + \lambda_{Sb}\cos(\theta_R)\big)\Big).
\end{aligned}$$

Use this to find an expression for the field energy and torque in terms of $\lambda_{Sa}, \lambda_{Sb}, \lambda_{Ra}, \lambda_{Rb}$, and θ_R.

Problem 17 Co-energy

Verify equation (6.48).

Problem 18 Co-energy

Verify equation (6.50).

Problem 19 Energy, Co-energy, and Torque in Synchronous Machines

Use the equations (6.35) and (6.36) to compute the field energy and the co-energy in a two-phase wound rotor synchronous machine. Use $\tau = \partial W_{field}/\partial\theta_R$ to obtain an expression for the torque put out by this machine and compare it with the expression (6.37).

Problem 20 Specifying the Derivatives of the Induction Motor Model

Prove that given any state $i_{Sa}, i_{Sb}, i_{Ra}, i_{Rb}$, and θ_R of the motor, the derivatives di_{Sa}/dt, di_{Sb}/dt, di_{Ra}/dt, di_{Rb}/dt, and $d\theta_R/dt$ can be arbitrarily chosen by appropriately choosing $u_{Sa}, u_{Sb}, u_{Ra}, u_{Rb}$, and τ_L.

7

Symmetric Balanced Three-Phase AC Machines

The mathematical models of two-phase induction and synchronous machines have been derived. In this chapter, the mathematical models of *three-phase* induction and permanent magnet synchronous machines are developed. From the three-phase models, two-phase equivalent models are derived. A classical steady-state analysis of the induction motor is also presented including the development of an equivalent circuit from the nonlinear differential equation model. Finally, the chapter concludes with a discussion on why power systems are three-phase systems with sinusoidal voltages operating at 60 Hz is presented.

7.1 Mathematical Model of a Three-Phase Induction Motor

A three-phase induction motor model is now developed and its two-phase equivalent model is then derived. Figure 7.1 illustrates how the stator and rotor windings are laid out in a three-phase machine.

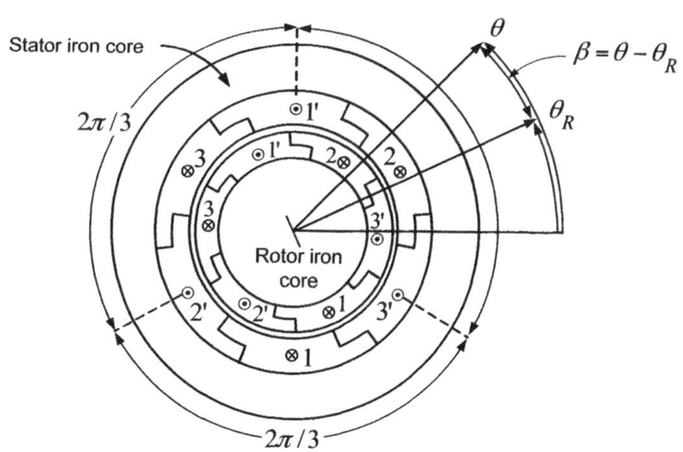

FIGURE 7.1. Cross-sectional view of a three-phase induction motor. The stator phases are $2\pi/3$ radians from each other as are the rotor phases.

To explain the winding layout in more detail, Figure 7.2 shows stator phase 1 where it is seen that one side of a loop is wound at $\pi/3$ (the other side at $4\pi/3$), two loop sides are wound at $\pi/2$ (the other two sides at $3\pi/2$), and, finally, one loop side is wound at $2\pi/3$ (the other side at $5\pi/3$).

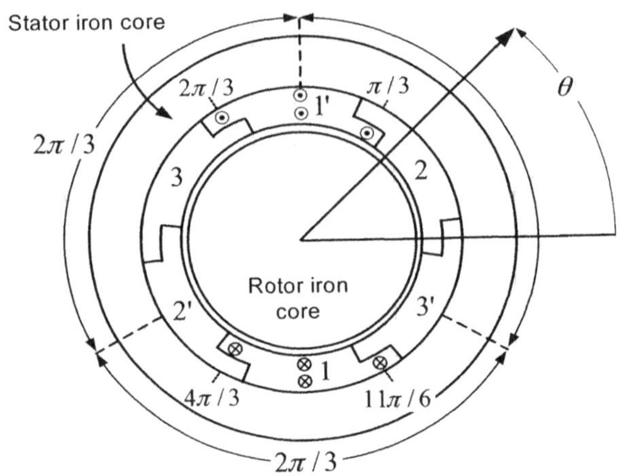

FIGURE 7.2. Cross-sectional view of stator phase 1.

Figures 7.1 and 7.2 show that the loops of the phases are wound in two layers which is a common layout in AC machines. Using Ampère's law,[1] one can straightforwardly compute the radial magnetic field in the air gap. The developed view in Figure 7.3 shows that the resulting radial magnetic field B_{S1} in the air gap due to the current i_{S1} in stator phase 1 is a staircase function that is approximately sinusoidal.

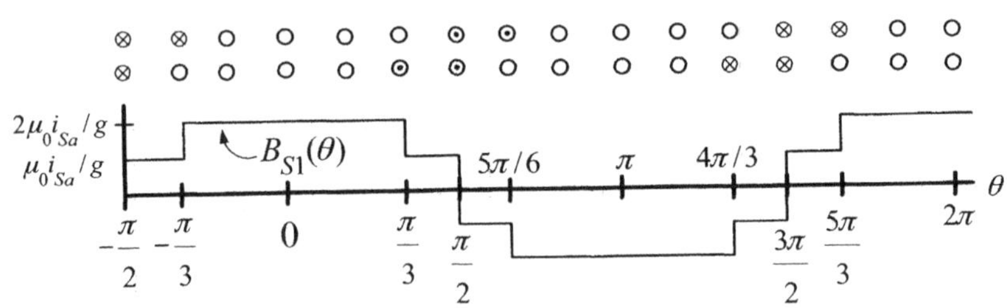

FIGURE 7.3. Developed view of the windings of stator phase 1.

The machine of Figure 7.1 (or Figure 7.2) is referred to as a *symmetric*

[1]Of course, in the use of Ampère's law, it is still assumed that the cross section of the winding is negligible in order to obtain a relatively simple analytical expression for the radial magnetic field in the air gap.

machine because the stator phases are identical in structure only rotated $2\pi/3$ radians from each other and similarly for the rotor phases.

In order to obtain a sinusoidally (spatially) distributed magnetic field in the air gap, it will be necessary to assume that the phases are sinusoidally wound. This is now described.

Sinusoidal Windings

Let θ_R denote the rotor position taken to coincide with the magnetic axis of rotor phase 1 as shown in Figure 7.1, θ is an arbitrary angular position in the air gap so that $\beta \triangleq \theta - \theta_R$ represents the same angle with respect to the rotor position.

Now consider the three stator phases to be sinusoidally wound with their stator turns densities given by

$$\begin{aligned} N_{S1}(\theta) &= \frac{N_S}{2}\left|\sin(\theta)\right| \\ N_{S2}(\theta) &= \frac{N_S}{2}\left|\sin(\theta - 2\pi/3)\right| \\ N_{S3}(\theta) &= \frac{N_S}{2}\left|\sin(\theta - 4\pi/3)\right|. \end{aligned}$$

Stator phases 1 and 2 are illustrated in Figures 7.4(a) and 7.4(b), respectively.

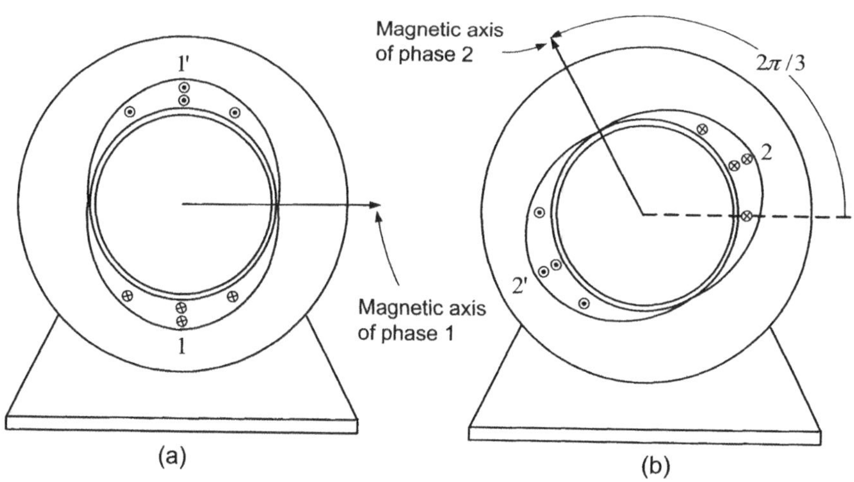

FIGURE 7.4. Sinusoidally wound stator phases. (a) Stator phase 1 windings. (b) Stator phase 2 windings.

Similarly, the rotor turns densities are given by

$$\begin{aligned} N_{R1}(\theta - \theta_R) &= \frac{N_R}{2}\left|\sin(\theta - \theta_R)\right| \\ N_{R2}(\theta - \theta_R) &= \frac{N_R}{2}\left|\sin(\theta - \theta_R - 2\pi/3)\right| \\ N_{R3}(\theta - \theta_R) &= \frac{N_R}{2}\left|\sin(\theta - \theta_R - 4\pi/3)\right|. \end{aligned}$$

This is referred to as a symmetric machine as the windings are all identical in construction being shifted by $2\pi/3$ with respect to each other and, similarly, the rotor windings are identical in construction also being shifted by $2\pi/3$ with respect to each other. As illustrated in Figure 7.5, the stator end windings $1', 2'$, and $3'$ are tied together to form the *motor neutral* $\hat{N}$, and the stator end windings $1, 2$, and 3 are connected to the source voltages u_{S1}, u_{S2}, and u_{S3}, respectively. The other end of the source voltages are tied together to form the *source neutral* N. This is called a *wye-connected* motor. Similarly, the rotor end windings $1, 2$, and 3 are shorted (connected) together and the other rotor end windings $1', 2'$, and $3'$ are also shorted together as shown in Figure 7.6. Due to the wye connection of the stator and rotor windings, the neutrals N and $\hat{N}$ in Figures 7.5 and 7.6, respectively, are isolated so that $i_{S1} + i_{S2} + i_{S2} \equiv 0$ and $i_{R1} + i_{R2} + i_{R2} \equiv 0$.

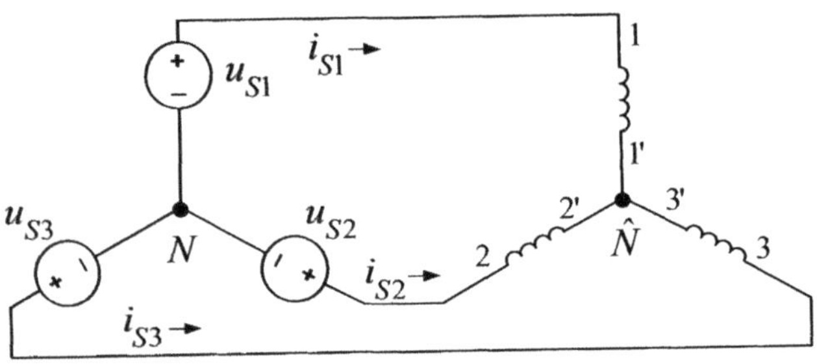

FIGURE 7.5. Wye-connected stator windings.

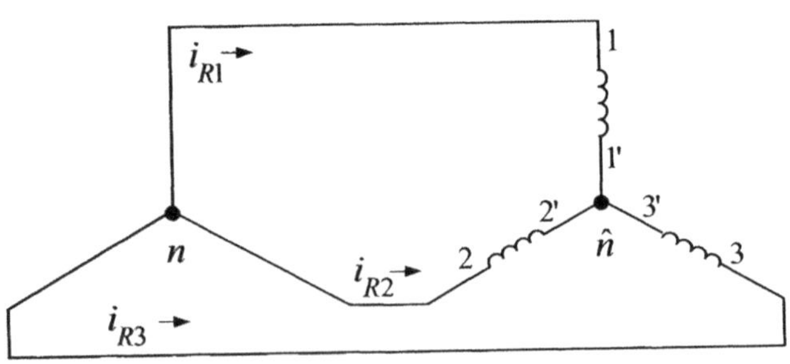

FIGURE 7.6. Wye-connected rotor windings.

The stator magnetic field is the vector sum of the fields due to each phase. These fields are

$$\begin{aligned}
\vec{\mathbf{B}}_{S1}(i_{S1}, r, \theta) &= \frac{\mu_0 N_S}{2g}\frac{r_R}{r} i_{S1}\cos(\theta)\hat{\mathbf{r}} \\
\vec{\mathbf{B}}_{S2}(i_{S2}, r, \theta) &= \frac{\mu_0 N_S}{2g}\frac{r_R}{r} i_{S2}\cos(\theta - 2\pi/3)\hat{\mathbf{r}} \\
\vec{\mathbf{B}}_{S3}(i_{S3}, r, \theta) &= \frac{\mu_0 N_S}{2g}\frac{r_R}{r} i_{S3}\cos(\theta - 4\pi/3)\hat{\mathbf{r}}
\end{aligned}$$

so that

$$\begin{aligned}
\vec{\mathbf{B}}_{S}(i_{S1}, i_{S2}, i_{S3}, r, \theta) &= \frac{\mu_0 N_S}{2g}\frac{r_R}{r}\Big(i_{S1}\cos(\theta) + i_{S2}\cos(\theta - 2\pi/3) \\
&\quad + i_{S3}\cos(\theta - 4\pi/3)\Big)\hat{\mathbf{r}}. \qquad (7.1)
\end{aligned}$$

If the balanced[2] three-phase set of currents

$$\begin{aligned}
i_{S1} &= I_S\cos(\omega_S t) \\
i_{S2} &= I_S\cos(\omega_S t - 2\pi/3) \\
i_{S3} &= I_S\cos(\omega_S t - 4\pi/3)
\end{aligned}$$

are applied to the stator phases, the corresponding stator magnetic field can be written as

$$\begin{aligned}
\vec{\mathbf{B}}_{S}(I_S, r, \theta, t) &= \frac{\mu_0 N_S I_S}{2g}\frac{r_R}{r}\Big(\cos(\omega_S t)\cos(\theta) + \cos(\omega_S t - \frac{2\pi}{3})\cos(\theta - \frac{2\pi}{3}) \\
&\quad + \cos(\omega_S t - \frac{4\pi}{3})\cos(\theta - 4\pi/3)\Big)\hat{\mathbf{r}} \\
&= \frac{\mu_0 N_S I_S}{2g}\frac{r_R}{r}\frac{3}{2}\cos(\theta - \omega_S t)\hat{\mathbf{r}}.
\end{aligned}$$

That is, with balanced three-phase currents, a radial rotating magnetic field is established in the air gap.

Similarly, the magnetic fields due to the rotor currents are given by

$$\begin{aligned}
\vec{\mathbf{B}}_{R1}(i_{R1}, r, \theta) &= \frac{\mu_0 N_R}{2g}\frac{r_R}{r} i_{R1}\cos(\theta - \theta_R)\hat{\mathbf{r}} \\
\vec{\mathbf{B}}_{R2}(i_{R2}, r, \theta) &= \frac{\mu_0 N_R}{2g}\frac{r_R}{r} i_{R2}\cos(\theta - \theta_R - 2\pi/3)\hat{\mathbf{r}} \\
\vec{\mathbf{B}}_{R3}(i_{R3}, r, \theta) &= \frac{\mu_0 N_R}{2g}\frac{r_R}{r} i_{R3}\cos(\theta - \theta_R - 4\pi/3)\hat{\mathbf{r}}
\end{aligned}$$

[2] Balanced just means $i_{S1} + i_{S2} + i_{S3} \equiv 0$.

so that, with $\beta = \theta - \theta_R$,

$$\begin{aligned}
&\vec{\mathbf{B}}_R(i_{R1}, i_{R2}, i_{R3}, r, \beta) \\
&= \frac{\mu_0 N_R}{2g}\frac{r_R}{r}\Big(i_{R1}\cos(\theta-\theta_R) + i_{R2}\cos(\theta-\theta_R-2\pi/3) \\
&\quad + i_{R3}\cos(\theta-\theta_R-4\pi/3)\Big)\hat{\mathbf{r}} \\
&= \frac{\mu_0 N_R}{2g}\frac{r_R}{r}\Big(i_{R1}\cos(\beta) + i_{R2}\cos(\beta-2\pi/3) + i_{R3}\cos(\beta-4\pi/3)\Big)\hat{\mathbf{r}}.
\end{aligned} \tag{7.2}$$

As explained in the previous chapter, the coupling factor κ is included in the expression for $\vec{\mathbf{B}}_R$ on the *stator side* of the air gap and is included in the expression for $\vec{\mathbf{B}}_S$ on the *rotor side* of the air gap to account for the spreading (leakage) of the magnetic field from the radial direction as one goes across the air gap. That is, with $\vec{\mathbf{B}}_S$ and $\vec{\mathbf{B}}_R$ given by (7.1) and (7.2) respectively, the total radial magnetic field $\vec{\mathbf{B}}$ on the stator side of the air gap is taken as

$$\begin{aligned}
\vec{\mathbf{B}}(i_{S1}, i_{S2}, i_{S3}, i_{R1}, i_{R2}, i_{R3}, r_S, \theta, \theta_R) \triangleq\; & \vec{\mathbf{B}}_S(i_{S1}, i_{S2}, i_{S3}, r_S, \theta) \\
& + \kappa\vec{\mathbf{B}}_R(i_{R1}, i_{R2}, i_{R3}, r_S, \theta-\theta_R).
\end{aligned} \tag{7.3}$$

On the rotor side of the air gap, the total radial magnetic field $\vec{\mathbf{B}}$ is taken as

$$\begin{aligned}
\vec{\mathbf{B}}(i_{S1}, i_{S2}, i_{S3}, i_{R1}, i_{R2}, i_{R3}, r_R, \theta, \theta_R) \triangleq\; & \kappa\vec{\mathbf{B}}_S(i_{S1}, i_{S2}, i_{S3}, r_R, \theta) \\
& + \vec{\mathbf{B}}_R(i_{R1}, i_{R2}, i_{R3}, r_R, \theta-\theta_R)
\end{aligned}$$

or, in terms of $\beta = \theta - \theta_R$ and an abuse of notation,

$$\begin{aligned}
\vec{\mathbf{B}}(i_{S1}, i_{S2}, i_{S3}, i_{R1}, i_{R2}, i_{R3}, r_R, \beta, \theta_R) \triangleq\; & \kappa\vec{\mathbf{B}}_S(i_{S1}, i_{S2}, i_{S3}, r_R, \beta+\theta_R) \\
& + \vec{\mathbf{B}}_R(i_{R1}, i_{R2}, i_{R3}, r_R, \beta).
\end{aligned} \tag{7.4}$$

Stator Flux Linkages

Using (7.3) for $\vec{\mathbf{B}}$ on the stator side of the air gap, the stator flux linkages are given by (see problem 1)

$$\begin{aligned}\psi_{S1}(t) &= \int_0^{\pi} \frac{N_S}{2}\sin(\theta) \\ &\quad \times \left(\int_{\theta-\pi}^{\theta} \ell_1 r_S B(i_{S1}, i_{S2}, i_{S3}, i_{R1}, i_{R2}, i_{R3}, r_S, \theta', \theta_R)d\theta'\right)d\theta \\ &= \frac{2}{3}L_S(i_{S1} + i_{S2}\cos(2\pi/3) + i_{S3}\cos(4\pi/3)) \\ &\quad + \frac{2}{3}M\Big(i_{R1}\cos(\theta_R) + i_{R2}\cos(\theta_R + 2\pi/3) + i_{R3}\cos(\theta_R + 4\pi/3)\Big)\end{aligned} \tag{7.5}$$

$$\begin{aligned}\psi_{S2}(t) &= \int_{2\pi/3}^{2\pi/3+\pi} \frac{N_S}{2}\sin(\theta - 2\pi/3) \\ &\quad \times \left(\int_{\theta-\pi}^{\theta} \ell_1 r_S B(i_{S1}, i_{S2}, i_{S3}, i_{R1}, i_{R2}, i_{R3}, r_S, \theta', \theta_R)d\theta'\right)d\theta \\ &= \frac{2}{3}L_S(i_{S1}\cos(2\pi/3) + i_{S2} + i_{S3}\cos(2\pi/3)) \\ &\quad + \frac{2}{3}M\Big(i_{R1}\cos(\theta_R - 2\pi/3) + i_{R2}\cos(\theta_R) + i_{R3}\cos(\theta_R + 2\pi/3)\Big)\end{aligned} \tag{7.6}$$

$$\begin{aligned}\psi_{S3}(t) &= \int_{4\pi/3}^{4\pi/3+\pi} \frac{N_S}{2}\sin(\theta - 4\pi/3) \\ &\quad \times \left(\int_{\theta-\pi}^{\theta} \ell_1 r_S B(i_{S1}, i_{S2}, i_{S3}, i_{R1}, i_{R2}, i_{R3}, r_S, \theta', \theta_R)d\theta'\right)d\theta \\ &= \frac{2}{3}L_S(i_{S1}\cos(4\pi/3) + i_{S2}\cos(2\pi/3) + i_{S3}) \\ &\quad + \frac{2}{3}M\Big(i_{R1}\cos(\theta_R - 4\pi/3) + i_{R2}\cos(\theta_R - 2\pi/3) + i_{R3}\cos(\theta_R)\Big)\end{aligned} \tag{7.7}$$

where

$$\begin{aligned}L_S &\triangleq \frac{3}{2}\frac{\pi\mu_0\ell_1\ell_2 N_S^2}{8g} \\ M &\triangleq \frac{3}{2}\kappa\frac{\pi\mu_0\ell_1\ell_2 N_S N_R}{8g} \\ L_R &\triangleq \frac{3}{2}\frac{\pi\mu_0\ell_1\ell_2 N_R^2}{8g}.\end{aligned} \tag{7.8}$$

The reason for the factor 2/3 in equations (7.5), (7.6), and (7.7), and the factor 3/2 in the expressions for $L_S, M,$ and L_R in (7.8) are so that the expressions (7.8) represent the *two-phase equivalent* coefficients of inductances in the two-phase equivalent model as will be shown below. In matrix form, the flux linkages may be written as

$$\begin{bmatrix} \psi_{S1}(t) \\ \psi_{S2}(t) \\ \psi_{S3}(t) \end{bmatrix} = \frac{2}{3}L_S \begin{bmatrix} 1 & \cos(2\pi/3) & \cos(4\pi/3) \\ \cos(2\pi/3) & 1 & \cos(2\pi/3) \\ \cos(4\pi/3) & \cos(-2\pi/3) & 1 \end{bmatrix} \begin{bmatrix} i_{S1}(t) \\ i_{S2}(t) \\ i_{S3}(t) \end{bmatrix}$$
$$+ \frac{2}{3}M \begin{bmatrix} \cos(\theta_R) & \cos(\theta_R + 2\pi/3) & \cos(\theta_R + 4\pi/3) \\ \cos(\theta_R - 2\pi/3) & \cos(\theta_R) & \cos(\theta_R + 2\pi/3) \\ \cos(\theta_R - 4\pi/3) & \cos(\theta_R - 2\pi/3) & \cos(\theta_R) \end{bmatrix} \begin{bmatrix} i_{R1}(t) \\ i_{R2}(t) \\ i_{R3}(t) \end{bmatrix}. \tag{7.9}$$

Note that $\theta_R - 4\pi/3 = \theta_R - 2\pi + 2\pi/3$ so that $\cos(\theta_R - 4\pi/3) = \cos(\theta_R + 2\pi/3)$. With the obvious definitions for the matrices C_1 and $C_2(\theta_R)$, the flux linkages are written more compactly as

$$\begin{bmatrix} \psi_{S1}(t) \\ \psi_{S2}(t) \\ \psi_{S3}(t) \end{bmatrix} = C_1 \begin{bmatrix} i_{S1}(t) \\ i_{S2}(t) \\ i_{S3}(t) \end{bmatrix} + C_2(\theta_R) \begin{bmatrix} i_{R1}(t) \\ i_{R2}(t) \\ i_{R3}(t) \end{bmatrix}. \tag{7.10}$$

Rotor Flux Linkages

Using (7.4) for $\vec{\mathbf{B}}$ on the rotor side of the air gap, the rotor flux linkages are given by (see problem 2)

$$\begin{aligned} \psi_{R1}(t) = & \int_0^{\pi} \frac{N_R}{2} \sin(\beta) \\ & \times \left(\int_{\beta-\pi}^{\beta} \ell_1 r_R B(i_{S1}, i_{S2}, i_{S3}, i_{R1}, i_{R2}, i_{R3}, r_R, \beta', \theta_R) d\beta' \right) d\beta \\ = & \frac{2}{3} L_R (i_{R1} + i_{SR2}\cos(2\pi/3) + i_{R1}\cos(4\pi/3)) \\ & + \frac{2}{3} M \Big(i_{S1}\cos(\theta_R) + i_{S2}\cos(\theta_R - 2\pi/3) + i_{S3}\cos(\theta_R - 4\pi/3) \Big) \end{aligned} \tag{7.11}$$

$$
\begin{aligned}
\psi_{R2}(t) &= \int_{2\pi/3}^{2\pi/3+\pi} \frac{N_R}{2} \sin(\beta - 2\pi/3) \\
&\quad \times \left(\int_{\beta-\pi}^{\beta} \ell_1 r_R B(i_{S1}, i_{S2}, i_{S3}, i_{R1}, i_{R2}, i_{R3}, r_R, \beta', \theta_R) d\beta' \right) d\beta \\
&= \frac{2}{3} L_R (i_{R1} \cos(2\pi/3) + i_{R2} + i_{R3} \cos(2\pi/3)) \\
&\quad + \frac{2}{3} M \Big(i_{S1} \cos(\theta_R + 2\pi/3) + i_{S2} \cos(\theta_R) + i_{S3} \cos(\theta_R - 2\pi/3) \Big)
\end{aligned}
\tag{7.12}
$$

$$
\begin{aligned}
\psi_{R3}(t) &= \int_{4\pi/3}^{4\pi/3+\pi} \frac{N_R}{2} \sin(\beta - 4\pi/3) \\
&\quad \times \left(\int_{\beta-\pi}^{\beta} \ell_1 r_R B(i_{S1}, i_{S2}, i_{S3}, i_{R1}, i_{R2}, i_{R3}, r_R, \beta', \theta_R) d\beta' \right) d\beta \\
&= \frac{2}{3} L_R (i_{R1} \cos(4\pi/3) + i_{R2} \cos(2\pi/3) + i_{R3}) \\
&\quad + \frac{2}{3} M \Big(i_{S1} \cos(\theta_R + 4\pi/3) + i_{S2} \cos(\theta_R + 2\pi/3) + i_{S3} \cos(\theta_R) \Big).
\end{aligned}
\tag{7.13}
$$

In matrix form, the rotor flux linkages may be written as

$$
\begin{aligned}
\begin{bmatrix} \psi_{R1}(t) \\ \psi_{R2}(t) \\ \psi_{R3}(t) \end{bmatrix} &= \frac{2}{3} L_R \begin{bmatrix} 1 & \cos(2\pi/3) & \cos(4\pi/3) \\ \cos(2\pi/3) & 1 & \cos(2\pi/3) \\ \cos(4\pi/3) & \cos(2\pi/3) & 1 \end{bmatrix} \begin{bmatrix} i_{R1}(t) \\ i_{R2}(t) \\ i_{R3}(t) \end{bmatrix} \\
&+ \frac{2}{3} M \begin{bmatrix} \cos(\theta_R) & \cos(\theta_R - 2\pi/3) & \cos(\theta_R - 4\pi/3) \\ \cos(\theta_R + 2\pi/3) & \cos(\theta_R) & \cos(\theta_R - 2\pi/3) \\ \cos(\theta_R + 4\pi/3) & \cos(\theta_R + 2\pi/3) & \cos(\theta_R) \end{bmatrix} \begin{bmatrix} i_{S1}(t) \\ i_{S2}(t) \\ i_{S3}(t) \end{bmatrix}
\end{aligned}
\tag{7.14}
$$

or, more compactly,

$$
\begin{bmatrix} \psi_{R1}(t) \\ \psi_{R2}(t) \\ \psi_{R3}(t) \end{bmatrix} = C_1 \begin{bmatrix} i_{R1}(t) \\ i_{R2}(t) \\ i_{R3}(t) \end{bmatrix} + C_2(-\theta_R) \begin{bmatrix} i_{S1}(t) \\ i_{S2}(t) \\ i_{S3}(t) \end{bmatrix} \tag{7.15}
$$

where C_1 and C_2 are defined as in (7.10).

Balanced Conditions

As described above, the source voltages $u_{S1}(t), u_{S2}(t)$, and $u_{S3}(t)$ are wye connected as are the stator and rotor phases (see Figures 7.5 and 7.6). Consequently, it is always true that

$$
i_{S1}(t) + i_{S2}(t) + i_{S3}(t) \equiv 0
$$

$$i_{R1}(t) + i_{R2}(t) + i_{R3}(t) \equiv 0.$$

The currents are then said to be *balanced* as they sum up to zero.

Using (7.10) and (7.15) it is straightforward to show that the stator and rotor fluxes are always balanced, that is,

$$\begin{aligned}\psi_{S1}(t) + \psi_{S2}(t) + \psi_{S3}(t) &\equiv 0\\ \psi_{R1}(t) + \psi_{R2}(t) + \psi_{R3}(t) &\equiv 0.\end{aligned}$$

(Note that these two equations hold whether or not the currents are balanced. See problem 4.)

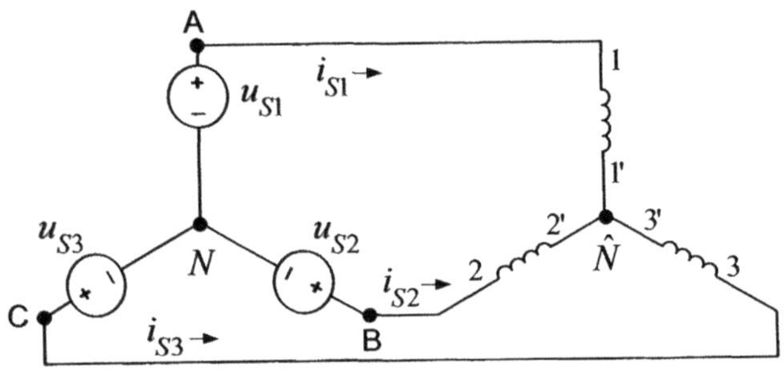

FIGURE 7.7. Wye-connected stator windings.

Referring to Figure 7.7, let $v_{A\hat{N}} = v_A - v_{\hat{N}}$, $v_{B\hat{N}}(t) = v_B - v_{\hat{N}}$, and $v_{C\hat{N}}(t) = v_C - v_{\hat{N}}$ denote the phase to *motor neutral* voltages. Faraday's and Ohm's laws give

$$\begin{aligned}v_{A\hat{N}} &= R_S i_{S1} + \frac{d\psi_{S1}(t)}{dt}\\ v_{B\hat{N}} &= R_S i_{S2} + \frac{d\psi_{S2}(t)}{dt}\\ v_{C\hat{N}} &= R_S i_{S3} + \frac{d\psi_{S3}(t)}{dt}.\end{aligned}$$

Adding these three equations results in

$$\begin{aligned}v_{A\hat{N}} + v_{B\hat{N}} + v_{C\hat{N}} &= R_S\left(i_{S1} + i_{S2} + i_{S3}\right) + \frac{d}{dt}\left(\psi_{S1}(t) + \psi_{S2}(t) + \psi_{S3}(t)\right)\\ &\equiv 0.\end{aligned}$$

That is, the phase to motor neutral voltages are always balanced. However, the voltages that are applied to the motor are the phase to *source neutral* voltages $u_{S1}(t), u_{S2}(t)$, and $u_{S3}(t)$. The next lemma relates the motor voltages $v_{A\hat{N}}, v_{B\hat{N}}$, and $v_{C\hat{N}}$ to the source voltages $u_{S1}(t), u_{S2}(t)$, and $u_{S3}(t)$ in terms of the voltage drop $v_{\hat{N}N}$ from the motor neutral to the source neutral.

Lemma
Let $v_{\hat{N}N} \triangleq v_{\hat{N}} - v_N$, then

$$v_{\hat{N}N} = \frac{u_{S1}(t) + u_{S2}(t) + u_{S3}(t)}{3}.$$

Proof[3]
Referring to Figure 7.7, write

$$\begin{aligned} u_{S1}(t) &= v_A - v_N = v_{A\hat{N}} + v_{\hat{N}N} \\ u_{S2}(t) &= v_B - v_N = v_{B\hat{N}} + v_{\hat{N}N} \\ u_{S3}(t) &= v_C - v_N = v_{C\hat{N}} + v_{\hat{N}N} \end{aligned}$$

and adding up these three equations, one obtains

$$\begin{aligned} u_{S1}(t) + u_{S2}(t) + u_{S3}(t) &= v_{A\hat{N}} + v_{B\hat{N}}(t) + v_{C\hat{N}}(t) + 3v_{\hat{N}N} \\ &= 3v_{\hat{N}N} \end{aligned}$$

or

$$v_{\hat{N}N} = \frac{u_{S1}(t) + u_{S2}(t) + u_{S3}(t)}{3}.$$

From this result, the following corollary is easily seen to be true.
Corollary
If the source voltages are balanced, that is,

$$u_{S1}(t) + u_{S2}(t) + u_{S3}(t) \equiv 0$$

then

$$v_{\hat{N}N} \equiv 0$$

so that

$$\begin{aligned} u_{S1}(t) &= v_{A\hat{N}} \\ u_{S2}(t) &= v_{B\hat{N}} \\ u_{S3}(t) &= v_{C\hat{N}}. \end{aligned}$$

By the above lemma, it follows that for a wye-connected motor,

$$\begin{aligned} v_{A\hat{N}} &= u_{S1}(t) - v_{\hat{N}N} = R_S i_{S1} + \frac{d\psi_{S1}(t)}{dt} \\ v_{B\hat{N}} &= u_{S2}(t) - v_{\hat{N}N} = R_S i_{S2} + \frac{d\psi_{S2}(t)}{dt} \\ v_{C\hat{N}} &= u_{S3}(t) - v_{\hat{N}N} = R_S i_{S3} + \frac{d\psi_{S3}(t)}{dt} \end{aligned} \tag{7.16}$$

where $v_{\hat{N}N} = 0$ if the source voltages are balanced.

---

[3] See, for example, Appendix 4.1 of Murphy and Turnbull [9].

Three-Phase to Two-Phase Transformation

Define a *three-phase to two-phase* transformation of the voltages by

$$\begin{bmatrix} u_{Sa}(t) \\ u_{Sb}(t) \\ u_{SO}(t) \end{bmatrix} \triangleq \sqrt{\frac{2}{3}} \begin{bmatrix} 1 & \cos(2\pi/3) & \cos(4\pi/3) \\ 0 & \sin(2\pi/3) & \sin(4\pi/3) \\ 1/\sqrt{2} & 1/\sqrt{2} & 1/\sqrt{2} \end{bmatrix} \begin{bmatrix} u_{S1}(t) \\ u_{S2}(t) \\ u_{S3}(t) \end{bmatrix}$$

$$= \sqrt{\frac{2}{3}} \begin{bmatrix} 1 & -1/2 & -1/2 \\ 0 & \sqrt{3}/2 & -\sqrt{3}/2 \\ 1/\sqrt{2} & 1/\sqrt{2} & 1/\sqrt{2} \end{bmatrix} \begin{bmatrix} u_{S1}(t) \\ u_{S2}(t) \\ u_{S3}(t) \end{bmatrix}.$$

If the source voltages are balanced,

$$u_{SO}(t) \triangleq \frac{1}{\sqrt{3}} \left(u_{S1}(t) + u_{S2}(t) + u_{S3}(t) \right) \equiv 0.$$

The inverse transformation is given by

$$\begin{bmatrix} u_{S1}(t) \\ u_{S2}(t) \\ u_{S3}(t) \end{bmatrix} = \sqrt{\frac{3}{2}} \begin{bmatrix} 2/3 & 0 & \sqrt{2}/3 \\ -1/3 & 1/\sqrt{3} & \sqrt{2}/3 \\ -1/3 & -1/\sqrt{3} & \sqrt{2}/3 \end{bmatrix} \begin{bmatrix} u_{Sa}(t) \\ u_{Sb}(t) \\ u_{SO}(t) \end{bmatrix}.$$

Denote the *3-2* transformation matrix as Q, that is,

$$Q \triangleq \sqrt{\frac{2}{3}} \begin{bmatrix} 1 & -1/2 & -1/2 \\ 0 & \sqrt{3}/2 & -\sqrt{3}/2 \\ 1/\sqrt{2} & 1/\sqrt{2} & 1/\sqrt{2} \end{bmatrix} \tag{7.17}$$

with inverse

$$Q^{-1} = \sqrt{\frac{3}{2}} \begin{bmatrix} 2/3 & 0 & \sqrt{2}/3 \\ -1/3 & +1/\sqrt{3} & \sqrt{2}/3 \\ -1/3 & -1/\sqrt{3} & \sqrt{2}/3 \end{bmatrix}. \tag{7.18}$$

Q is an *orthogonal* matrix, that is, $Q^{-1} = Q^T$. Define

$$\begin{bmatrix} i_{Sa}(t) \\ i_{Sb}(t) \\ i_{SO}(t) \end{bmatrix} \triangleq Q \begin{bmatrix} i_{S1}(t) \\ i_{S2}(t) \\ i_{S3}(t) \end{bmatrix}, \quad \begin{bmatrix} i_{Ra}(t) \\ i_{Rb}(t) \\ i_{RO}(t) \end{bmatrix} \triangleq Q \begin{bmatrix} i_{R1}(t) \\ i_{R2}(t) \\ i_{R3}(t) \end{bmatrix}$$

$$\begin{bmatrix} \lambda_{Sa}(t) \\ \lambda_{Sb}(t) \\ \lambda_{SO}(t) \end{bmatrix} \triangleq Q \begin{bmatrix} \psi_{S1}(t) \\ \psi_{S2}(t) \\ \psi_{S3}(t) \end{bmatrix}, \quad \begin{bmatrix} \lambda_{Ra}(t) \\ \lambda_{Rb}(t) \\ \lambda_{RO}(t) \end{bmatrix} \triangleq Q \begin{bmatrix} \psi_{R1}(t) \\ \psi_{R2}(t) \\ \psi_{R3}(t) \end{bmatrix}.$$

As shown in problem 4,

$$\begin{aligned} \lambda_{SO}(t) &= \frac{1}{\sqrt{3}} \left(\psi_{S1}(t) + \psi_{S2}(t) + \psi_{S3}(t) \right) \equiv 0 \\ \lambda_{RO}(t) &= \frac{1}{\sqrt{3}} \left(\psi_{R1}(t) + \psi_{R2}(t) + \psi_{R3}(t) \right) \equiv 0. \end{aligned}$$

As the phases are wye connected, it follows that

$$\begin{aligned} i_{S0}(t) &= \frac{1}{\sqrt{3}}\left(i_{S1}(t)+i_{S2}(t)+i_{S3}(t)\right) \equiv 0 \\ i_{R0}(t) &= \frac{1}{\sqrt{3}}\left(i_{R1}(t)+i_{R2}(t)+i_{R3}(t)\right) \equiv 0. \end{aligned}$$

Remark If in addition the source voltages are balanced, it also follows that

$$u_{S0}(t) = \frac{1}{\sqrt{3}}\left(u_{S1}(t)+u_{S2}(t)+u_{S3}(t)\right) \equiv 0.$$

That is, when the source voltages are balanced, all of the so-called *zero sequence* quantities $i_{S0}(t), i_{R0}(t), \lambda_{S0}(t), \lambda_{R0}(t)$, and $u_{S0}(t)$ are identically zero under the transformation Q and the original three-phase model consisting of the three phases $1, 2$, and 3 is now reduced to an *equivalent two-phase model* consisting of the phases a and b. This is not surprising as the balanced conditions imply that there are only two independent variables (e.g., if $u_{S1}(t)$ and $u_{S2}(t)$ are given, then $u_{S3}(t) = -\left(u_{S1}(t)+u_{S2}(t)\right)$ is determined).

In the case when the source voltages are not balanced, the phase to neutral voltages of the machine $v_{A\hat{N}}, v_{B\hat{N}}$, and $v_{C\hat{N}}$ transform as

$$\begin{bmatrix} u_{Sa}(t) \\ u_{Sb}(t) \\ u_{S0}(t) - \sqrt{3}v_{\hat{N}N} \end{bmatrix} = \sqrt{\frac{2}{3}}\begin{bmatrix} 1 & -1/2 & -1/2 \\ 0 & \sqrt{3}/2 & -\sqrt{3}/2 \\ 1/\sqrt{2} & 1/\sqrt{2} & 1/\sqrt{2} \end{bmatrix}\begin{bmatrix} u_{S1}(t) - v_{\hat{N}N} \\ u_{S2}(t) - v_{\hat{N}N} \\ u_{S3}(t) - v_{\hat{N}N} \end{bmatrix}.$$

and the dynamic equations for the stator flux linkages (7.16) then transform to

$$\begin{aligned} u_{Sa}(t) &= R_S i_{Sa} + \frac{d\lambda_{Sa}(t)}{dt} \\ u_{Sb}(t) &= R_S i_{Sb} + \frac{d\lambda_{Sb}(t)}{dt} \\ u_{S0}(t) - \sqrt{3}v_{\hat{N}N} &= R_S i_{S0} + \frac{d\lambda_{S0}(t)}{dt}. \end{aligned} \tag{7.19}$$

To find the expressions for the two-phase equivalent flux linkages in terms of the two-phase equivalent currents, recall that

$$\begin{bmatrix} \psi_{S1}(t) \\ \psi_{S2}(t) \\ \psi_{S3}(t) \end{bmatrix} = C_1\begin{bmatrix} i_{S1}(t) \\ i_{S2}(t) \\ i_{S3}(t) \end{bmatrix} + C_2(\theta_R)\begin{bmatrix} i_{R1}(t) \\ i_{R2}(t) \\ i_{R3}(t) \end{bmatrix}$$

so that

$$\begin{bmatrix} \lambda_{Sa}(t) \\ \lambda_{Sb}(t) \\ \lambda_{S0}(t) \end{bmatrix} = QC_1Q^{-1}\begin{bmatrix} i_{Sa}(t) \\ i_{Sb}(t) \\ i_{S0}(t) \end{bmatrix} + QC_2(\theta_R)Q^{-1}\begin{bmatrix} i_{Ra}(t) \\ i_{Rb}(t) \\ i_{R0}(t) \end{bmatrix}$$

$$= \begin{bmatrix} L_S & 0 & 0 \\ 0 & L_S & 0 \\ 0 & 0 & 0 \end{bmatrix} \begin{bmatrix} i_{Sa}(t) \\ i_{Sb}(t) \\ i_{S0}(t) \end{bmatrix} + \begin{bmatrix} M\cos(\theta_R) & -M\sin(\theta_R) & 0 \\ M\sin(\theta_R) & M\cos(\theta_R) & 0 \\ 0 & 0 & 0 \end{bmatrix} \begin{bmatrix} i_{Ra}(t) \\ i_{Rb}(t) \\ i_{R0}(t) \end{bmatrix} \tag{7.20}$$

or

$$\begin{aligned} \lambda_{Sa}(t) &= L_S i_{Sa}(t) + M\Big(i_{Ra}(t)\cos(\theta_R) - i_{Rb}(t)\sin(\theta_R)\Big) \\ \lambda_{Sb}(t) &= L_S i_{Sb}(t) + M\Big(i_{Ra}(t)\sin(\theta_R) + i_{Rb}(t)\cos(\theta_R)\Big) \\ \lambda_{S0} &\equiv 0. \end{aligned} \tag{7.21}$$

Similarly, the rotor flux linkages satisfy the system of equations

$$\begin{aligned} 0 &= R_R i_{R1} + \frac{d\psi_{R1}(t)}{dt} \\ 0 &= R_R i_{R2} + \frac{d\psi_{R2}(t)}{dt} \\ 0 &= R_R i_{R3} + \frac{d\psi_{R3}(t)}{dt} \end{aligned}$$

so that multiplying this system of equations through by Q results in

$$\begin{aligned} 0 &= R_R i_{Ra} + \frac{d\lambda_{Ra}(t)}{dt} \\ 0 &= R_R i_{Rb} + \frac{d\lambda_{Rb}(t)}{dt} \\ 0 &= R_R i_{R0} + \frac{d\lambda_{R0}(t)}{dt} \end{aligned} \tag{7.22}$$

where $i_{R0} \equiv 0, \lambda_{R0}(t) \equiv 0$. Also, as

$$\begin{bmatrix} \psi_{R1}(t) \\ \psi_{R2}(t) \\ \psi_{R3}(t) \end{bmatrix} = C_1 \begin{bmatrix} i_{R1}(t) \\ i_{R2}(t) \\ i_{R3}(t) \end{bmatrix} + C_2(-\theta_R) \begin{bmatrix} i_{S1}(t) \\ i_{S2}(t) \\ i_{S3}(t) \end{bmatrix}$$

it follows that

$$\begin{aligned} \begin{bmatrix} \lambda_{Ra}(t) \\ \lambda_{Rb}(t) \\ \lambda_{R0}(t) \end{bmatrix} &= QC_1Q^{-1} \begin{bmatrix} i_{Ra}(t) \\ i_{Rb}(t) \\ i_{R0}(t) \end{bmatrix} + QC_2(-\theta_R)Q^{-1} \begin{bmatrix} i_{Sa}(t) \\ i_{Sb}(t) \\ i_{S0}(t) \end{bmatrix} \\ &= \begin{bmatrix} L_R & 0 & 0 \\ 0 & L_R & 0 \\ 0 & 0 & 0 \end{bmatrix} \begin{bmatrix} i_{Ra}(t) \\ i_{Rb}(t) \\ i_{R0}(t) \end{bmatrix} \\ &\quad + \begin{bmatrix} M\cos(\theta_R) & M\sin(\theta_R) & 0 \\ -M\sin(\theta_R) & M\cos(\theta_R) & 0 \\ 0 & 0 & 0 \end{bmatrix} \begin{bmatrix} i_{Sa}(t) \\ i_{Sb}(t) \\ i_{S0}(t) \end{bmatrix} \end{aligned} \tag{7.23}$$

or

$$
\begin{aligned}
\lambda_{Ra}(t) &= L_R i_{Ra}(t) + M\Big(+i_{Sa}(t)\cos(\theta_R) + i_{Sb}(t)\sin(\theta_R)\Big)\\
\lambda_{Rb}(t) &= L_R i_{Rb}(t) + M\Big(-i_{Sa}(t)\sin(\theta_R) + i_{Sb}(t)\cos(\theta_R)\Big) \quad (7.24)\\
\lambda_{R0} &\equiv 0.
\end{aligned}
$$

Collecting together equations (7.19), (7.21) and (7.24), the two-phase equivalent equations for the stator and rotor currents of a *wye connected* three-phase motor become

$$
\begin{aligned}
u_{Sa} &= L_S\frac{d}{dt}i_{Sa} + M\frac{d}{dt}\Big(+i_{Ra}\cos(\theta_R) - i_{Rb}\sin(\theta_R)\Big) + R_S i_{Sa}\\
u_{Sb} &= L_S\frac{d}{dt}i_{Sb} + M\frac{d}{dt}\Big(+i_{Ra}\sin(\theta_R) + i_{Rb}\cos(\theta_R)\Big) + R_S i_{Sb}\\
u_{S0} &= \sqrt{3}v_{\hat{N}N} \qquad (7.25)\\
0 &= L_R\frac{d}{dt}i_{Ra} + M\frac{d}{dt}\Big(+i_{Sa}\cos(\theta_R) + i_{Sb}\sin(\theta_R)\Big) + R_R i_{Ra}\\
0 &= L_R\frac{d}{dt}i_{Rb} + M\frac{d}{dt}\Big(-i_{Sa}\sin(\theta_R) + i_{Sb}\cos(\theta_R)\Big) + R_R i_{Rb}
\end{aligned}
$$

where $i_{S0} \equiv 0, i_{R0} \equiv 0$. In what follows, it is now assumed that the voltages are balanced so that $u_{S0} = \sqrt{3}v_{\hat{N}N} \equiv 0$. In this case, there is a one-to-one correspondence between the two-phase variables i_{Sa}, i_{Sb}, i_{Ra}, i_{Rb}, u_{Sa}, and u_{Sb} and the three-phase variables i_{S1}, i_{S2}, i_{S3}, i_{R1}, i_{R2}, i_{R3}, u_{S1}, u_{S2}, and u_{S3}.

Torque

It was shown above that the stator currents established a radial magnetic field in the air gap. The torque produced by this magnetic field on the currents in the sinusoidally wound rotor phases is now computed. Recall that the stator magnetic field in terms of $\beta = \theta - \theta_R$ at $r = r_R$ is given by

$$
\vec{\mathbf{B}}_S(i_{S1}, i_{S2}, i_{S3}, r, \beta, \theta_R)\Big|_{r=r_R} = \kappa\frac{\mu_0 N_S}{2g}\frac{r_R}{r_R}\times
$$
$$
\Big(i_{S1}\cos(\beta + \theta_R) + i_{S2}\cos(\beta + \theta_R - 2\pi/3) + i_{S3}\cos(\beta + \theta_R - 4\pi/3)\Big)\hat{\mathbf{r}}.
$$

The torque is then (see problem 8)

$$\begin{aligned}
\vec{\boldsymbol{\tau}}_{R1} &= \int_{\beta=0}^{2\pi} (r_R\hat{\mathbf{r}}) \times \left(i_{R1}(t)\frac{N_R}{2}\sin(\beta)d\beta(+\ell_1\hat{\mathbf{z}}) \times \left(B_{S|_{r=r_R}}\hat{\mathbf{r}}\right)\right) \\
&= \int_{\beta=0}^{2\pi} r_R i_{R1}(t)\frac{\ell_1 N_R}{2}\sin(\beta)\left(\kappa\frac{\mu_0 N_S}{2g}\right)\Big(i_{S1}\cos(\beta+\theta_R) \\
&\quad + (i_{S2}\cos(\beta+\theta_R-2\pi/3) + i_{S3}\cos(\beta+\theta_R-4\pi/3))\,d\beta\Big)\hat{\mathbf{z}} \\
&= \kappa\frac{r_R\ell_1\mu_0 N_R N_S}{4g} i_{R1}(t)\Big(i_{S1}(-\pi\sin(\theta_R)) + i_{S2}\pi\cos(\theta_R-\frac{\pi}{6}) \\
&\quad + i_{S3}(-\pi\cos(\theta_R+\frac{\pi}{6}))\Big)\hat{\mathbf{z}} \\
&= \frac{2}{3}Mi_{R1}(t)\left(-i_{S1}\sin(\theta_R) + i_{S2}\cos(\theta_R-\frac{\pi}{6}) - i_{S3}\cos(\theta_R+\frac{\pi}{6})\right)\hat{\mathbf{z}}
\end{aligned} \tag{7.26}$$

where $M \triangleq (3/2)\kappa\pi\mu_0\ell_1\ell_2 N_S N_R/(8g)$. The torque on phase 2 of the rotor is then

$$\begin{aligned}
\vec{\boldsymbol{\tau}}_{R2} &= \int_{\beta=0}^{2\pi} (r_R\hat{\mathbf{r}}) \times \left(i_{R2}(t)\frac{N_R}{2}\sin(\beta-2\pi/3)d\beta(+\ell_1\hat{\mathbf{z}}) \times \left(B_{S|_{r=r_R}}\hat{\mathbf{r}}\right)\right) \\
&= \int_{\beta=0}^{2\pi} r_R i_{R2}(t)\frac{\ell_1 N_R}{2}\sin(\beta-2\pi/3)\left(\kappa\frac{\mu_0 N_S}{2g}\right) \times \Big(i_{S1}\cos(\beta+\theta_R) \\
&\quad + i_{S2}\cos(\beta+\theta_R-2\pi/3) + i_{S3}\cos(\beta+\theta_R-4\pi/3)\Big)\,d\beta\hat{\mathbf{z}} \\
&= \kappa\frac{r_R\ell_1\mu_0 N_R N_S}{4g} i_{R2}(t)\Big(i_{S1}(-\pi\cos(\theta_R+\frac{\pi}{6})) + i_{S2}(-\pi\sin(\theta_R)) \\
&\quad + i_{S3}\pi\sin(\theta_R+\frac{\pi}{3})\Big)\hat{\mathbf{z}} \\
&= \frac{2}{3}Mi_{R2}(t)\left(-i_{S1}\cos(\theta_R+\frac{\pi}{6}) - i_{S2}\sin(\theta_R) + i_{S3}\sin(\theta_R+\frac{\pi}{3})\right)\hat{\mathbf{z}}.
\end{aligned} \tag{7.27}$$

Finally, the torque on phase 3 is computed as

$$\begin{aligned}
\vec{\boldsymbol{\tau}}_{R3} &= \int_{\beta=0}^{2\pi} (r_R\hat{\mathbf{r}}) \times \left(i_{R3}(t)\frac{N_R}{2}\sin(\beta-4\pi/3)d\beta(+\ell_1\hat{\mathbf{z}}) \times \left(B_{S|_{r=r_R}}\hat{\mathbf{r}}\right)\right) \\
&= \int_{\beta=0}^{2\pi} r_R i_{R3}(t)\frac{\ell_1 N_R}{2}\sin(\beta-4\pi/3)\left(\kappa\frac{\mu_0 N_S}{2g}\right) \times \Big(i_{S1}\cos(\beta+\theta_R) \\
&\quad + i_{S2}\cos(\beta+\theta_R-2\pi/3) + i_{S3}\cos(\beta+\theta_R-4\pi/3)\Big)\,d\beta\hat{\mathbf{z}}
\end{aligned}$$

$$= \kappa\frac{r_R\ell_1\mu_0 N_R N_S}{4g} i_{R3}(t)\Big(i_{S1}\pi\cos(\theta_R - \frac{\pi}{6}) + i_{S2}\pi\sin(\theta_R - \frac{\pi}{3})$$
$$+ i_{S3}\left(-\pi\sin(\theta_R)\right)\Big)\hat{\mathbf{z}}$$
$$= \frac{2}{3}M i_{R3}(t)\Big(i_{S1}\cos(\theta_R - \frac{\pi}{6}) + i_{S2}\sin(\theta_R - \frac{\pi}{3}) - i_{S3}\sin(\theta_R)\Big)\hat{\mathbf{z}}. \tag{7.28}$$

The total torque is then

$$\begin{aligned}
\tau_R &= \tau_{R1} + \tau_{R2} + \tau_{R3} \\
&= \frac{2}{3}M\Big(i_{R1}(t)\Big(-i_{S1}\sin(\theta_R) + i_{S2}\cos(\theta_R - \frac{\pi}{6}) - i_{S3}\cos(\theta_R + \frac{\pi}{6})\Big) \\
&\quad + i_{R2}(t)\Big(-i_{S1}\cos(\theta_R + \frac{\pi}{6}) - i_{S2}\sin(\theta_R) + i_{S3}\sin(\theta_R + \frac{\pi}{3})\Big) \\
&\quad + i_{R3}(t)\Big(i_{S1}\cos(\theta_R - \frac{\pi}{6}) + i_{S2}\sin(\theta_R - \frac{\pi}{3}) - i_{S3}\sin(\theta_R)\Big)\Big).
\end{aligned} \tag{7.29}$$

With $i_{S0}(t) \equiv 0, i_{R0}(t) \equiv 0$, substitute

$$\begin{bmatrix} i_{S1}(t) \\ i_{S2}(t) \\ i_{S3}(t) \end{bmatrix} \triangleq Q^{-1}\begin{bmatrix} i_{Sa}(t) \\ i_{Sb}(t) \\ i_{S0}(t) \end{bmatrix} = \sqrt{\frac{3}{2}}\begin{bmatrix} 2/3 & 0 & \sqrt{2}/3 \\ -1/3 & 1/\sqrt{3} & \sqrt{2}/3 \\ -1/3 & -1/\sqrt{3} & \sqrt{2}/3 \end{bmatrix}\begin{bmatrix} i_{Sa}(t) \\ i_{Sb}(t) \\ i_{S0}(t) \end{bmatrix}$$
$$\begin{bmatrix} i_{R1}(t) \\ i_{R2}(t) \\ i_{R3}(t) \end{bmatrix} \triangleq Q^{-1}\begin{bmatrix} i_{Ra}(t) \\ i_{Rb}(t) \\ i_{R0}(t) \end{bmatrix} = \sqrt{\frac{3}{2}}\begin{bmatrix} 2/3 & 0 & \sqrt{2}/3 \\ -1/3 & 1/\sqrt{3} & \sqrt{2}/3 \\ -1/3 & -1/\sqrt{3} & \sqrt{2}/3 \end{bmatrix}\begin{bmatrix} i_{Ra}(t) \\ i_{Rb}(t) \\ i_{R0}(t) \end{bmatrix}$$

into (7.29) to obtain (see problem 10)

$$\begin{aligned}
\tau_R &= M\Big(-i_{Ra}(t)i_{Sa}(t)\sin(\theta_R) + i_{Ra}(t)i_{Sb}(t)\cos(\theta_R) \\
&\quad - i_{Rb}(t)i_{Sa}(t)\cos(\theta_R) - i_{Rb}(t)i_{Sb}(t)\sin(\theta_R)\Big).
\end{aligned} \tag{7.30}$$

Substituting the two-phase equivalent stator flux linkages (7.21) into (7.19) and the two-phase equivalent rotor flux linkages (7.24) into (7.22) along with torque equation (7.30) gives the following two-phase equivalent mathematical model of a three-phase wye-connected induction motor:

$$
\begin{aligned}
u_{Sa} &= L_S \frac{d}{dt} i_{Sa} + M \frac{d}{dt}\Big(+i_{Ra}\cos(\theta_R) - i_{Rb}\sin(\theta_R)\Big) + R_S i_{Sa} \\
u_{Sb} &= L_S \frac{d}{dt} i_{Sb} + M \frac{d}{dt}\Big(+i_{Ra}\sin(\theta_R) + i_{Rb}\cos(\theta_R)\Big) + R_S i_{Sb} \\
0 &= L_R \frac{d}{dt} i_{Ra} + M \frac{d}{dt}\Big(+i_{Sa}\cos(\theta_R) + i_{Sb}\sin(\theta_R)\Big) + R_R i_{Ra} \\
0 &= L_R \frac{d}{dt} i_{Rb} + M \frac{d}{dt}\Big(-i_{Sa}\sin(\theta_R) + i_{Sb}\cos(\theta_R)\Big) + R_R i_{Rb} \\
J\frac{d\omega_R}{dt} &= M\Big(-i_{Ra}(t)i_{Sa}(t)\sin(\theta_R) + i_{Ra}(t)i_{Sb}(t)\cos(\theta_R) \\
&\quad - i_{Rb}(t)i_{Sa}(t)\cos(\theta_R) - i_{Rb}(t)i_{Sb}(t)\sin(\theta_R)\Big) - \tau_L \\
\frac{d\theta_R}{dt} &= \omega_R.
\end{aligned}
\tag{7.31}
$$

The system (7.31) may be rewritten using a *space vector* representation as

$$R_S\underline{i}_S + L_S\frac{d}{dt}\underline{i}_S + M\frac{d}{dt}\left(\underline{i}_R e^{j\theta_R}\right) = \underline{u}_S \tag{7.32}$$

$$R_R\underline{i}_R + L_R\frac{d}{dt}\underline{i}_R + M\frac{d}{dt}\left(\underline{i}_S e^{-j\theta_R}\right) = 0 \tag{7.33}$$

$$M\,\mathrm{Im}\{\underline{i}_S(\underline{i}_R e^{j\theta_R})^*\} - \tau_L = J\frac{d\omega_R}{dt} \tag{7.34}$$

$$\frac{d\theta_R}{dt} = \omega_R \tag{7.35}$$

where $\underline{u}_S \triangleq u_{Sa} + ju_{Sb}$, $\underline{i}_S \triangleq i_{Sa} + ji_{Sb}$, and $\underline{i}_R \triangleq i_{Ra} + ji_{Rb}$ are the space vector form of the two-phase equivalent voltages and currents.

Remarks

The model (7.31) [or, equivalently, (7.32), (7.33), and (7.34)] is identical in form to the model derived for a two-phase motor in Chapter 6.

The parameters L_S, L_R, and M are the two-phase equivalent inductance values for the three-phase machine. As seen from equations (7.5), (7.6), (7.7), (7.11), (7.12), and (7.13), the actual stator, rotor, and mutual coefficients of inductance in the three-phase machine are $2L_S/3$, $2L_R/3$, and $2M/3$, respectively. However, the phase resistances R_S and R_R in the two-phase model are the same as in the three-phase model.

Simulation of the Three-Phase Machine

Figure 7.8 illustrates how the two-phase model is used to simulate a balanced three-phase machine.

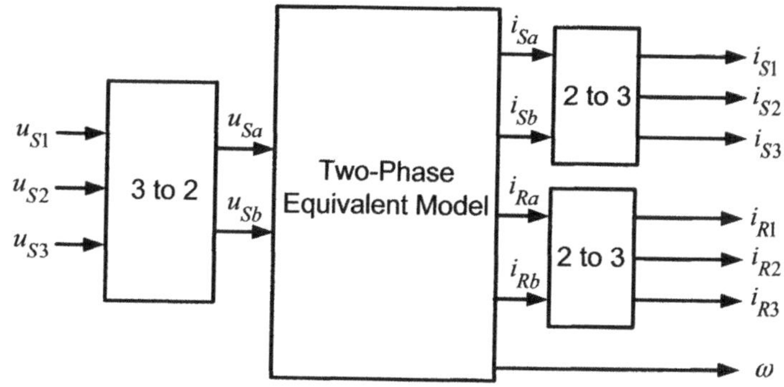

FIGURE 7.8. Block diagram for simulating a three-phase machine using the two-phase equivalent model.

One simply uses the transformation

$$\begin{bmatrix} u_{Sa}(t) \\ u_{Sb}(t) \end{bmatrix} \triangleq \sqrt{\frac{2}{3}} \begin{bmatrix} 1 & -1/2 & -1/2 \\ 0 & \sqrt{3}/2 & -\sqrt{3}/2 \end{bmatrix} \begin{bmatrix} u_{S1}(t) \\ u_{S2}(t) \\ u_{S3}(t) \end{bmatrix}$$

to obtain the voltages $u_{Sa}(t)$ and $u_{Sb}(t)$ which are then applied to the two-phase equivalent simulation model. The currents i_{Sa}, i_{Sb}, i_{Ra}, and i_{Rb} from the simulation are then put through the two-phase to three-phase (2-3) transformation

$$\begin{bmatrix} i_{S1}(t) \\ i_{S2}(t) \\ i_{S3}(t) \end{bmatrix} \triangleq \sqrt{\frac{3}{2}} \begin{bmatrix} 2/3 & 0 \\ -1/3 & 1/\sqrt{3} \\ -1/3 & -1/\sqrt{3} \end{bmatrix} \begin{bmatrix} i_{Sa}(t) \\ i_{Sb}(t) \end{bmatrix}$$

$$\begin{bmatrix} i_{R1}(t) \\ i_{R2}(t) \\ i_{R3}(t) \end{bmatrix} \triangleq \sqrt{\frac{3}{2}} \begin{bmatrix} 2/3 & 0 \\ -1/3 & 1/\sqrt{3} \\ -1/3 & -1/\sqrt{3} \end{bmatrix} \begin{bmatrix} i_{Ra}(t) \\ i_{Rb}(t) \end{bmatrix}$$

to obtain the (balanced) three-phase stator and rotor currents.

Zero Component

The three-phase to two-phase (3 to 2) transformation is given by

$$\begin{bmatrix} u_{Sa}(t) \\ u_{Sb}(t) \\ u_{S0}(t) \end{bmatrix} \triangleq \sqrt{\frac{2}{3}} \begin{bmatrix} 1 & \cos(2\pi/3) & \cos(4\pi/3) \\ 0 & \sin(2\pi/3) & \sin(4\pi/3) \\ 1/\sqrt{2} & 1/\sqrt{2} & 1/\sqrt{2} \end{bmatrix} \begin{bmatrix} u_{S1}(t) \\ u_{S2}(t) \\ u_{S3}(t) \end{bmatrix}. \quad (7.36)$$

Figure 7.9 shows the three magnetic axes s_1, s_2, and s_3 for the stator phases and two orthogonal axes which are denoted a and b, respectively. Note that the s_1, s_2, and s_3 axes are not orthogonal.

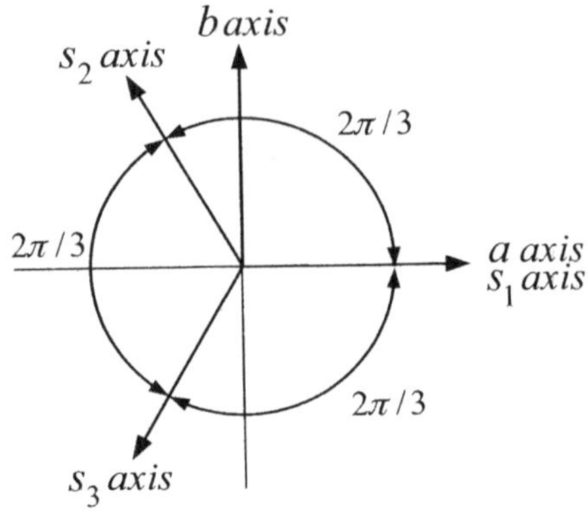

FIGURE 7.9. Magnetic axes for the three-phase and two-phase models.

An interpretation of the 3 to 2 transformation (7.36) is that the elements of the 3-tuple

$$\begin{bmatrix} u_{S1}(t) \\ u_{S2}(t) \\ u_{S3}(t) \end{bmatrix}$$

are the *components* of the vector

$$\vec{\mathbf{u}}_S \triangleq u_{S1}(t)\hat{\mathbf{e}}_{S1} + u_{S2}(t)\hat{\mathbf{e}}_{S2} + u_{S3}(t)\hat{\mathbf{e}}_{S3}$$

with respect to the basis of orthogonal unit vectors

$$\hat{\mathbf{e}}_{S1} = \sqrt{\frac{2}{3}}\begin{bmatrix} 1 \\ 0 \\ 1/\sqrt{2} \end{bmatrix},\ \hat{\mathbf{e}}_{S2} = \sqrt{\frac{2}{3}}\begin{bmatrix} \cos(2\pi/3) \\ \sin(2\pi/3) \\ 1/\sqrt{2} \end{bmatrix},\ \hat{\mathbf{e}}_{S3} = \sqrt{\frac{2}{3}}\begin{bmatrix} \cos(4\pi/3) \\ \sin(4\pi/3) \\ 1/\sqrt{2} \end{bmatrix}$$

while the 3-tuple

$$\begin{bmatrix} u_{Sa}(t) \\ u_{Sb}(t) \\ u_{S0}(t) \end{bmatrix}$$

are the *components* of this *same* vector

$$\vec{\mathbf{u}}_S = u_{Sa}(t)\hat{\mathbf{e}}_{Sa} + u_{Sb}(t)\hat{\mathbf{e}}_{Sb} + u_{S0}\hat{\mathbf{e}}_{S0}$$

with respect to the basis of orthogonal unit vectors

$$\hat{\mathbf{e}}_{Sa} = \begin{bmatrix} 1 \\ 0 \\ 0 \end{bmatrix},\ \hat{\mathbf{e}}_{Sb} = \begin{bmatrix} 0 \\ 1 \\ 0 \end{bmatrix},\ \hat{\mathbf{e}}_{S0} = \begin{bmatrix} 0 \\ 0 \\ 1 \end{bmatrix}.$$

The third component u_{S0} is referred to as the *zero component* because in a balanced three-phase system it is zero.

Note that the projections of $\hat{\mathbf{e}}_{S1}, \hat{\mathbf{e}}_{S2}$, and $\hat{\mathbf{e}}_{S3}$ onto the ab plane of Figure 7.9 are in the directions of the s_1, s_2, and s_3 axes, respectively, which are not orthogonal.

Remarks The 3 to 2 transformation (7.36) is also referred to as *Clarke's transformation* [6]. The three-phase windings whose magnetic axes are shown in Figure 7.9 are magnetically coupled. For example, a current i_{S1} in phase 1 will produce a net (nonzero) flux linkage in the other two phases. As pointed out by Holtz [60], the 3 to 2 transformation has (mathematically) replaced the magnetically coupled three-phase windings with an equivalent two-phase set of windings whose magnetic axes are orthogonal (in quadrature) and are *not* magnetically coupled. That is, a current i_{Sa} in phase a will not produce a net flux linkage in phase b. Often phase a is referred to as the *direct* or d axis and phase b is referred to as the *quadrature* or q axis. For example, in problem 13, a standard model of the induction is given in space vector form. In that formulation, the notation $(u_{Sd}^S, u_{Sq}^S, i_{Sd}^S, i_{Sq}^S, i_{Rd}^S, i_{Rq}^S)$ is often used for the equivalent two-phase variables (the superscript "S" refers to the "stationary frame"). However, in this book, the dq notation is reserved for the *field-oriented* coordinate system (see Chapters 8 and 9) and so the reader must be wary when reading different authors. Further, one also sees the notation $(u_{S\alpha}, u_{S\beta}, i_{S\alpha}, i_{S\beta}, i_{R\alpha}, i_{R\beta})$ and the notation $(u_{Sd}, u_{Sq}, i_{Sd}, i_{Sq}, i_{Rd}, i_{Rq})$ for these same variables.

7.2 Steady-State Analysis of the Induction Motor

The objective here is to find steady-state solutions to the mathematical model (7.31) of the induction motor. The space vector representation of the induction motor given by (7.32), (7.33), and (7.34) will be used because it is especially convenient for this analysis. This presentation is from the development in Chapter 10 of Professor Leonhard's book [1][2].

With $\underline{U}_S \triangleq |\underline{U}_S| e^{j\angle \underline{U}_S} = U_S e^{j\angle \underline{U}_S}$ a root mean square (rms) phasor[4] for the voltage, a balanced set of three-phase steady-state voltages is specified by

$$\begin{aligned}
u_{S1} &= \sqrt{2}U_S \cos(\omega_S t + \angle \underline{U}_S) \\
&= \frac{\sqrt{2}}{2}\left(\underline{U}_S e^{j\omega_S t} + \underline{U}_S^* e^{-j\omega_S t}\right) \\
u_{S2} &= \sqrt{2}U_S \cos(\omega_S t + \angle \underline{U}_S - 2\pi/3) \\
&= \frac{\sqrt{2}}{2}\left(\underline{U}_S e^{j(\omega_S t - 2\pi/3)} + \underline{U}_S^* e^{-j(\omega_S t - 2\pi/3)}\right) \\
u_{S3} &= \sqrt{2}U_S \cos(\omega_S t + \angle \underline{U}_S - 4\pi/3) \\
&= \frac{\sqrt{2}}{2}\left(\underline{U}_S e^{j(\omega_S t - 4\pi/3)} + \underline{U}_S^* e^{-j(\omega_S t - 4\pi/3)}\right).
\end{aligned} \tag{7.37}$$

The objective here is to apply these voltages to the motor and find the resulting steady-state currents and torque.

7.2.1 *Steady-State Currents and Voltages*

The three-phase to two-phase transformation (7.17) can be represented in the space vector form as (see problem 11)

$$\underline{u}_S \triangleq \sqrt{\frac{2}{3}}\left(u_{S1} + u_{S2}e^{j2\pi/3} + u_{S3}e^{j4\pi/3}\right). \tag{7.38}$$

Expanded out, this becomes

$$\begin{aligned}
\underline{u}_S &= \sqrt{\frac{2}{3}}\left(u_{S1} + u_{S2}\cos(2\pi/3) + u_{S3}\cos(4\pi/3)\right) \\
&\quad + j\sqrt{\frac{2}{3}}\left(u_{S2}\sin(2\pi/3) + u_{S3}\sin(4\pi/3)\right) \\
&= u_{Sa} + j u_{Sb}.
\end{aligned}$$

---

[4] In this section, $U_S \triangleq |\underline{U}_S|$ is an rms voltage rather than a peak voltage. This is to keep the notation in this section consistent with standard practice.

Substituting the expressions (7.37) into (7.38) and using the fact that $1 + e^{j4\pi/3} + e^{j8\pi/3} = 0$ results in

$$\begin{aligned}\underline{u}_S &= \frac{1}{\sqrt{3}}\left(\underline{U}_S e^{j\omega_S t} + \underline{U}_S^* e^{-j\omega_S t}\right)\\ &+\frac{1}{\sqrt{3}}\left(\underline{U}_S e^{j(\omega_S t - 2\pi/3)} + \underline{U}_S^* e^{-j(\omega_S t - 2\pi/3)}\right) e^{j2\pi/3}\\ &+\frac{1}{\sqrt{3}}\left(\underline{U}_S e^{j(\omega_S t - 4\pi/3)} + \underline{U}_S^* e^{-j(\omega_S t - 4\pi/3)}\right) e^{j4\pi/3}\\ &= \sqrt{3}\underline{U}_S e^{j\omega_S t}.\end{aligned}$$

This is the voltage to be applied to the induction motor model (7.32), (7.33), and (7.34).

The approach taken here is to look for a solution consisting of a balanced set of three-phase stator currents. With $\underline{I}_S \triangleq |\underline{I}_S|\, e^{j\angle \underline{I}_S} = I_S e^{j\angle \underline{I}_S}$ an rms current phasor, the stator currents take the form

$$\begin{aligned}i_{S1} &= \sqrt{2} I_S \cos(\omega_S t + \angle \underline{I}_S)\\ &= \frac{\sqrt{2}}{2}\left(\underline{I}_S e^{j\omega_S t} + \underline{I}_S^* e^{-j\omega_S t}\right)\\ i_{S2} &= \sqrt{2} I_S \cos(\omega_S t + \angle \underline{I}_S - 2\pi/3)\\ &= \frac{\sqrt{2}}{2}\left(\underline{I}_S e^{j(\omega_S t - 2\pi/3)} + \underline{I}_S^* e^{-j(\omega_S t - 2\pi/3)}\right)\\ i_{S3} &= \sqrt{2} I_S \cos(\omega_S t + \angle \underline{I}_S - 4\pi/3)\\ &= \frac{\sqrt{2}}{2}\left(\underline{I}_S e^{j(\omega_S t - 4\pi/3)} + \underline{I}_S^* e^{-j(\omega_S t - 4\pi/3)}\right)\end{aligned}$$

where the three-phase to two-phase transformation results in

$$\underline{i}_S = \sqrt{\frac{2}{3}}\left(i_{S1} + i_{S2} e^{j2\pi/3} + i_{S3} e^{j4\pi/3}\right) = \sqrt{3}\underline{I}_S e^{j\omega_S t}.$$

For the rotor currents, a solution consisting of a balanced set of three-phase rotor currents with angular frequency $\omega_S - \omega_R$ is sought. With $\underline{I}_R = |\underline{I}_R|\, e^{j\angle \underline{I}_R} = I_R e^{j\angle \underline{I}_R}$, the rotor currents take the form

$$\begin{aligned}i_{R1} &= \frac{\sqrt{2}}{2}\left(\underline{I}_R e^{j(\omega_S - \omega_R)t} + \underline{I}_R^* e^{-j(\omega_S - \omega_R)t}\right)\\ i_{R2} &= \frac{\sqrt{2}}{2}\left(\underline{I}_R e^{j((\omega_S - \omega_R)t - 2\pi/3)} + \underline{I}_R^* e^{-j((\omega_S - \omega_R)t - 2\pi/3)}\right)\\ i_{R3} &= \frac{\sqrt{2}}{2}\left(\underline{I}_R e^{j((\omega_S - \omega_R)t - 4\pi/3)} + \underline{I}_R^* e^{-j((\omega_S - \omega_R)t - 4\pi/3)}\right).\end{aligned}$$

The three-phase to two-phase transformation of the rotor currents results in

$$\underline{i}_R = \sqrt{\frac{2}{3}}\left(i_{R1} + i_{R2} e^{j2\pi/3} + i_{R3} e^{j4\pi/3}\right) = \sqrt{3}\underline{I}_R e^{j(\omega_S - \omega_R)t}.$$

Finally, with $\theta_R(t) = \omega_R t$,

$$\underline{i}_R(t)e^{j\theta_R(t)} = \sqrt{3}\underline{I}_R e^{j\omega_S t}.$$

7.2.2 Steady-State Equivalent Circuit Model

The above expressions for the steady-state currents are convenient for developing an equivalent circuit model of the induction motor in steady state. To do so, the expressions $\underline{u}_S = \sqrt{3}\underline{U}_S e^{j\omega_S t}$, $\underline{i}_S = \sqrt{3}\underline{I}_S e^{j\omega_S t}$, $\underline{i}_R = \sqrt{3}\underline{I}_R e^{j(\omega_S - \omega_R)t}$ and $\underline{i}_R e^{j\theta_R(t)} = \sqrt{3}\underline{I}_R e^{j\omega_S t}$ are substituted into (7.32) and (7.33). After canceling out the terms $\sqrt{3}e^{j\omega_S t}$ and $\sqrt{3}e^{j(\omega_S - \omega_R)t}$, one obtains

$$\begin{aligned} (R_S + j\omega_S L_S)\underline{I}_S + j\omega_S M\underline{I}_R &= \underline{U}_S && (7.39)\\ (R_R + j(\omega_S - \omega_R)L_R)\underline{I}_R + j(\omega_S - \omega_R)M\underline{I}_S &= 0. && (7.40)\end{aligned}$$

With $\omega_{slip} \triangleq \omega_S - \omega_R$, define the normalized slip

$$S = \frac{\omega_{slip}}{\omega_S} = \frac{\omega_S - \omega_R}{\omega_S}.$$

Replace $\omega_{slip} = \omega_S - \omega_R$ by $S\omega_S$ into (7.40) and rearrange to obtain

$$\begin{aligned} (R_S + j\omega_S L_S)\underline{I}_S + j\omega_S M\underline{I}_R &= \underline{U}_S && (7.41)\\ (R_R/S + j\omega_S L_R)\underline{I}_R + j\omega_S M\underline{I}_S &= 0. && (7.42)\end{aligned}$$

With $L_S = (1+\sigma_S)M, L_R = (1+\sigma_R)M$, equations (7.41) and (7.42) can be rearranged to obtain[5]

$$\begin{aligned} (R_S + j\omega_S\sigma_S M)\underline{I}_S + j\omega_S M(\underline{I}_S + \underline{I}_R) &= \underline{U}_S && (7.43)\\ (R_R/S + j\omega_S\sigma_R M)\underline{I}_R + j\omega_S M(\underline{I}_S + \underline{I}_R) &= 0. && (7.44)\end{aligned}$$

In this analysis, it was assumed that ω_S, ω_R, and therefore $\omega_{slip} = \omega_S - \omega_R$ are all *constant.* The two algebraic relationships (7.43) and (7.44) may be viewed as the result of applying Kirchhoff's voltage law to the two loops indicated in the *equivalent circuit* of Figure 7.10.

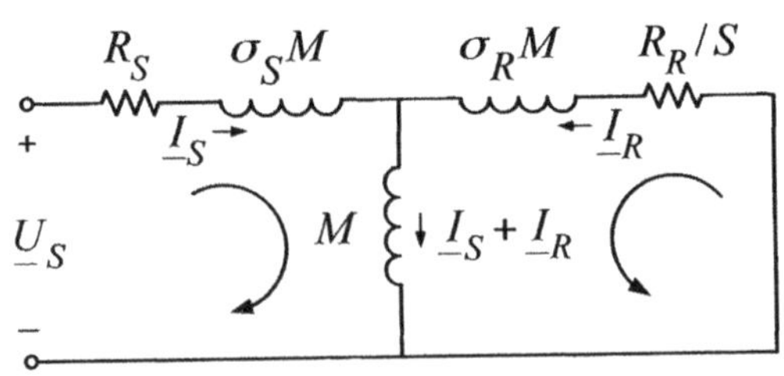

FIGURE 7.10. An equivalent circuit of the induction motor.

[5] Setting $L_S = (1+\sigma_S)M, L_R = (1+\sigma_R)M$ is just saying that L_S and L_R are slightly greater than M. Due to leakage, this is always true as $\sigma = 1 - M^2/(L_S L_R) > 0$.

The system (7.41) and (7.42) may be rewritten as

$$\begin{bmatrix} R_S + j\omega_S L_S & j\omega_S M \\ j\omega_S M & R_R/S + j\omega_S L_R \end{bmatrix} \begin{bmatrix} \underline{I}_S \\ \underline{I}_R \end{bmatrix} = \begin{bmatrix} \underline{U}_S \\ 0 \end{bmatrix}. \tag{7.45}$$

Solving for $\underline{I}_S$ and $\underline{I}_R$ gives

$$\begin{aligned} \begin{bmatrix} \underline{I}_S \\ \underline{I}_R \end{bmatrix} &= \frac{1}{(R_S + j\omega_S L_S)(R_R/S + j\omega_S L_R) - (j\omega_S M)^2} \\ &\quad \times \begin{bmatrix} R_R/S + j\omega_S L_R & -j\omega_S M \\ -j\omega_S M & R_S + j\omega_S L_S \end{bmatrix} \begin{bmatrix} \underline{U}_S \\ 0 \end{bmatrix}. \end{aligned}$$

The input impedance is then

$$\begin{aligned} Z_S &= \frac{\underline{U}_S}{\underline{I}_S} = \frac{(R_S + j\omega_S L_S)(R_R/S + j\omega_S L_R) - (j\omega_S M)^2}{R_R/S + j\omega_S L_R} \\ &= R_S + j\omega_S L_S - \frac{(j\omega_S M)^2}{R_R/S + j\omega_S L_R}. \end{aligned}$$

Using the fact that $M = L_S/(1+\sigma_S) = L_R/(1+\sigma_R)$ and $\sigma = 1 - \dfrac{1}{(1+\sigma_S)(1+\sigma_R)}$, this becomes

$$\begin{aligned} Z_S &= R_S + j\omega_S L_S \left(1 - \frac{\dfrac{S}{R_R} j\omega_S L_R \dfrac{1}{(1+\sigma_S)(1+\sigma_R)}}{1 + \dfrac{j\omega_S L_R S}{R_R}} \right) \\ &= R_S + j\omega_S L_S \left(\frac{1 + \dfrac{j\omega_S L_R S}{R_R}\left(1 - \dfrac{1}{(1+\sigma_S)(1+\sigma_R)}\right)}{1 + \dfrac{j\omega_S L_R S}{R_R}} \right) \\ &= R_S + j\omega_S L_S \left(\frac{1 + \dfrac{j\sigma\omega_S L_R S}{R_R}}{1 + \dfrac{j\omega_S L_R S}{R_R}} \right). \end{aligned}$$

Finally, defining S_p as

$$S_p \triangleq \frac{R_R}{\sigma\omega_S L_R}, \tag{7.46}$$

the input impedance may be written as

$$Z_S = R_S + j\omega_S L_S \left(\frac{1 + \dfrac{jS}{S_p}}{1 + \dfrac{jS}{\sigma S_p}} \right).$$

With $R_S = 0$[6], the stator current phasor $\underline{I}_S = \underline{U}_S/Z_S$ becomes

$$\underline{I}_S = \frac{\underline{U}_S}{j\omega_S L_S}\left(\frac{1+\dfrac{jS}{\sigma S_p}}{1+\dfrac{jS}{S_p}}\right) \tag{7.47}$$

$$= \underline{I}_{S0}\left(\frac{1+\sigma}{2\sigma} - \frac{1-\sigma}{2\sigma}\frac{1-\dfrac{jS}{S_p}}{1+\dfrac{jS}{S_p}}\right)$$

$$= \underline{I}_{S0}\left(\frac{1+\sigma}{2\sigma} - \frac{1-\sigma}{2\sigma}e^{-j2\tan^{-1}(S/S_p)}\right). \tag{7.48}$$

Here $\underline{I}_{S0} \triangleq \underline{U}_S/(j\omega_S L_S)$ is the no-load current phasor, that is, the current phasor when the slip is zero. For convenience, choose $\underline{U}_S = jU_S$ so that $\underline{I}_{S0} = I_{S0} = U_S/(\omega_S L_S)$ is real. Figure 7.11 shows a plot of phasor $\underline{I}_S$ versus S/S_p.

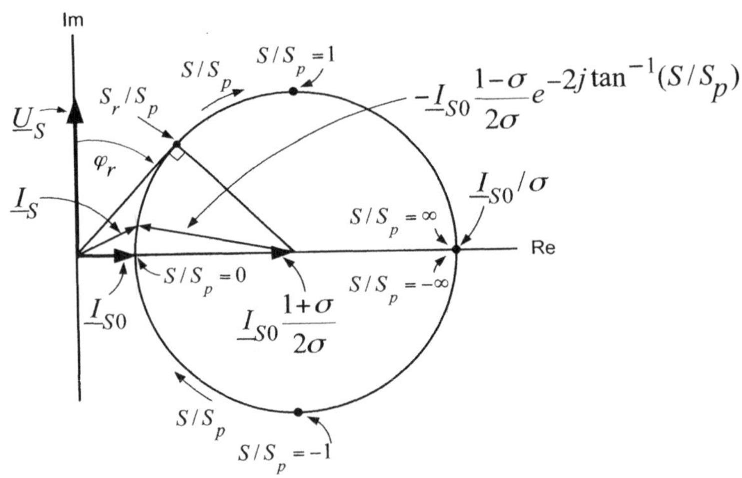

FIGURE 7.11. Circle diagram of the stator current phasor as a function of S/S_p with $R_S = 0$.

As S/S_p varies in the interval $-\infty \leq S/S_p \leqslant \infty$, the tip of the phasor $\underline{I}_S$ traces out a circle so that Figure 7.11 is referred to as the *circle diagram*. As illustrated in the figure, the tip of the phasor $\underline{I}_S$ lies on a

[6] This is primarily done so that the expressions that are now derived for the stator input impedance, power factor, and torque have a nice form.

circle whose center is at $\underline{I}_{S0}\frac{1+\sigma}{2\sigma}$ on the real axis and whose radial arm is $-\underline{I}_{S0}\frac{1+\sigma}{2\sigma}e^{-j2\tan^{-1}(S/S_p)}$.

The average power P_{stator} into the stator from the source is

$$P_{\text{stator}} = \frac{1}{2\pi/\omega_S}\int_0^{2\pi/\omega_S} (u_{S1}i_{S1} + u_{S2}i_{S2} + u_{S3}i_{S3})\,dt.$$

Then

$$\frac{1}{2\pi/\omega_S}\int_0^{2\pi/\omega_S} u_{S1}i_{S1}dt = \frac{1}{2\pi/\omega_S}\int_0^{2\pi/\omega_S}\left(\frac{\sqrt{2}}{2}\left(\underline{U}_S e^{j\omega_S t} + \underline{U}_S^* e^{-j\omega_S t}\right)\right) \times \left(\frac{\sqrt{2}}{2}\left(\underline{I}_S e^{j\omega_S t} + \underline{I}_S^* e^{-j\omega_S t}\right)\right) dt$$

$$= \frac{1}{2\pi/\omega_S}\int_0^{2\pi/\omega_S}\frac{1}{2}\left[\underline{U}_S\underline{I}_S^* + \underline{U}_S^*\underline{I}_S + \underline{U}_S\underline{I}_S e^{j2\omega_S t} + \underline{U}_S^*\underline{I}_S^* e^{-j2\omega_S t}\right] dt$$

$$= \frac{1}{2}\left[\underline{U}_S\underline{I}_S^* + \underline{U}_S^*\underline{I}_S\right] = \text{Re}\{\underline{U}_S\underline{I}_S^*\} = U_S I_S\cos(\angle\underline{U}_S - \angle\underline{I}_S)$$

as

$$\text{Re}\{\underline{U}_S\underline{I}_S^*\} = \text{Re}\left\{|\underline{U}_S|\,e^{j\angle\underline{U}_S}\,|\underline{I}_S|\,e^{-j\angle\underline{I}_S}\right\} = |\underline{U}_S|\,|\underline{I}_S|\cos(\angle\underline{U}_S - \angle\underline{I}_S).$$

The angle $\varphi \triangleq \angle\underline{U}_S - \angle\underline{I}_S$ is the angle between the phasors $\underline{U}_S$ and $\underline{I}_S$, and is called the *power factor angle*. Each phase contributes the same average power so that the total electrical power into the stator is

$$P_{\text{stator}} = 3U_S I_S\cos(\varphi)$$

where $\cos(\varphi)$ is the *power factor*. Using equation (7.47), the power factor angle φ is written as a function of S/S_p by

$$\varphi = \angle\underline{U}_S - \angle\underline{I}_S = \frac{\pi}{2} - \left(\tan^{-1}\left(\frac{S}{\sigma S_p}\right) - \tan^{-1}\left(\frac{S}{S_p}\right)\right).$$

The *rated* slip S_r is defined as the value of S that minimizes φ or equivalently, maximizes $\cos(\varphi)$. Solving $d\varphi/dS = 0$ gives

$$S_r = \sqrt{\sigma}S_p. \tag{7.49}$$

Again using equation (7.47), the ratio I_S/I_{S0} may be written as

$$\left(\frac{I_S}{I_{S0}}\right)^2 = \frac{1 + \left(\dfrac{S}{\sigma S_p}\right)^2}{1 + \left(\dfrac{S}{S_p}\right)^2}, \quad I_{S0} = \frac{U_S}{\omega_S L_S} \tag{7.50}$$

so that at rated slip, the *rated current* is

$$\frac{I_{S_r}}{I_{S0}} = \frac{1}{\sqrt{\sigma}}. \tag{7.51}$$

Also note that it follows from equation (7.50) that

$$\lim_{S\to\infty} \frac{I_S}{I_{S0}} = \frac{1}{\sigma}.$$

A plot of I_S/I_{S0} versus S/S_p is shown in Figure 7.12. At rated slip, the power factor angle is

$$\varphi_r \triangleq \varphi_{|S=S_r} = \pi/2 - \left(\tan^{-1}(1/\sqrt{\sigma}) - \tan^{-1}(\sqrt{\sigma})\right)$$

so that power factor is [see (7.57) below]

$$\cos(\varphi_r) = \frac{1-\sigma}{1+\sigma}. \tag{7.52}$$

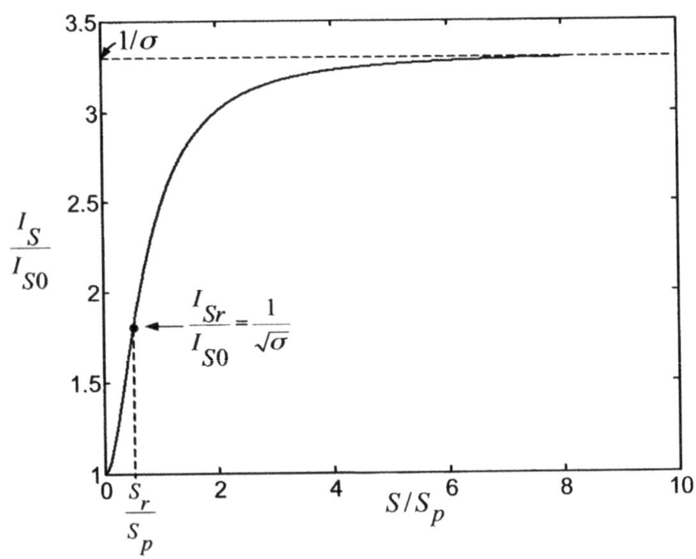

FIGURE 7.12. I_S/I_{S0} versus S/S_p, with $\sigma = 0.3$ and $R_S = 0$.

7.2.3 Rated Conditions

An interpretation to the definition of rated slip and the other corresponding rated variables is as follows: The balanced three-phase sinusoidal input voltages are constant in both magnitude and frequency. If the motor is operating at *rated* slip (done by choosing the appropriate load on the motor), then the power into the motor is

$$3U_S I_{S_r}\cos(\varphi_r) = 3U_S\frac{I_{S0}}{\sqrt{\sigma}}\frac{1-\sigma}{1+\sigma} \tag{7.53}$$

where $I_{S_r} = I_{S0}/\sqrt{\sigma}, \cos(\varphi_r) = (1-\sigma)/(1+\sigma)$ as previously shown. At rated slip, the power factor $\cos(\varphi_r)$ is at its maximum value so that the amount of stator current I_S required to achieve the particular input power level (7.53) is at a minimum. This power goes into mechanical work and losses given by[7]

$$3U_S I_{S_r}\cos(\varphi_r) = \tau_r\omega_R + 3R_R I_R^2.$$

The torque τ_r and speed ω_R under these operating conditions are called the *rated* torque and *rated* speed, respectively. A real motor has the stator current losses $3R_S I_S^2$ so that if motor is operating at rated slip, the amount of stator current required to produce the rated torque is minimized, making this an efficient operating point.

7.2.4 Steady-State Torque

Substituting $\underline{i}_S = \sqrt{3}\underline{I}_S e^{j\omega_S t}$ and $\underline{i}_R(t)e^{j\theta_R(t)} = \sqrt{3}\underline{I}_R e^{j\omega_S t}$ into (7.34) gives

$$\tau = M\,\mathrm{Im}\{\underline{i}_S(\underline{i}_R e^{j\theta_R})^*\} = 3M\,\mathrm{Im}\{\underline{I}_S\underline{I}_R^*\}. \tag{7.54}$$

From (7.45), the rotor current phasor is related to the stator current phasor by

$$\underline{I}_R = -\frac{j\omega_S M}{R_R/S + j\omega_S L_R}\underline{I}_S. \tag{7.55}$$

Substituting this into the steady-state torque expression (7.54) gives

$$\begin{aligned}\tau &= 3M\,\mathrm{Im}\{\underline{I}_S\underline{I}_R^*\} = 3M\,\mathrm{Im}\left\{\underline{I}_S\left(-\frac{j\omega_S M}{R_R/S + j\omega S L_R}\underline{I}_S\right)^*\right\}\\ &= 3M\,\mathrm{Im}\left\{\underline{I}_S\frac{j\omega_S M}{R_R/S - j\omega_S L_R}\underline{I}_S^*\right\} = 3MI_S^2\,\mathrm{Im}\left\{\frac{\dfrac{j\omega_S MS}{R_R}}{1-\dfrac{j\omega_S L_R S}{R_R}}\right\}.\end{aligned}$$

[7] Recall that this analysis assumes $R_S = 0$. The mechanical losses (friction, windage, etc.) have also been assumed to be zero in this analysis.

Recalling that $S_p = \dfrac{R_R}{\sigma\omega_S L_R}$, this becomes

$$\tau = 3MI_S^2\,\mathrm{Im}\left\{\frac{\dfrac{j\omega_S MS}{R_R}}{1-\dfrac{jS}{\sigma S_p}}\,\frac{1+\dfrac{jS}{\sigma S_p}}{1+\dfrac{jS}{\sigma S_p}}\right\} = 3MI_S^2\,\mathrm{Im}\left\{\frac{\dfrac{j\omega_S MS}{R_R}\left(1+\dfrac{jS}{\sigma S_p}\right)}{1+\left(\dfrac{S}{\sigma S_p}\right)^2}\right\}$$

$$= 3MI_S^2\frac{\omega_S MS/R_R}{1+\left(\dfrac{S}{\sigma S_p}\right)^2} = 3M\left(\frac{I_S}{I_{S0}}\right)^2 (I_{S0})^2\,\frac{\omega_S MS/R_R}{1+\left(\dfrac{S}{\sigma S_p}\right)^2}.$$

Using (7.50), this simplifies to

$$\begin{aligned}
\tau &= 3M\frac{1+\left(\dfrac{S}{\sigma S_p}\right)^2}{1+\left(\dfrac{S}{S_p}\right)^2}\left(\frac{U_S}{\omega_S L_S}\right)^2\frac{\omega_S MS/R_R}{1+\left(\dfrac{S}{\sigma S_p}\right)^2}\\
&= 3\frac{U_S^2}{\omega_S^2 L_S^2}\frac{\omega_S M^2 S/R_R}{1+\left(\dfrac{S}{S_p}\right)^2} = 3\frac{U_S^2}{\omega_S^2 L_S^2}\frac{S_p/S}{S_p/S+S/S_p}\frac{\omega_S M^2 S}{R_R}.
\end{aligned}$$

Rewriting $M^2 = \dfrac{L_S}{(1+\sigma_S)}\dfrac{L_R}{(1+\sigma_R)} = L_S L_R(1-\sigma)$, one obtains

$$\begin{aligned}
\tau &= 3\frac{U_S^2}{\omega_S^2 L_S^2}\frac{S_p/S}{S_p/S+S/S_p}L_S(1-\sigma)\frac{\omega_S L_R}{R_R}S\\
&= 3\frac{U_S^2}{\omega_S^2 L_S}\frac{S_p/S}{S_p/S+S/S_p}\frac{(1-\sigma)}{\sigma}\frac{S}{S_p}\\
&= 3\frac{U_S^2}{\omega_S^2 L_S}\frac{1}{S_p/S+S/S_p}\frac{(1-\sigma)}{\sigma}.
\end{aligned}$$

Finally,

$$\tau = \frac{3}{2}\frac{(1-\sigma)}{\sigma}\frac{U_S^2}{\omega_S^2 L_S}\frac{2}{S_p/S+S/S_p} = \tau_p\frac{2}{S_p/S+S/S_p} \tag{7.56}$$

where

$$\tau_p \triangleq \frac{3}{2}\frac{(1-\sigma)}{\sigma}\frac{U_S^2}{\omega_S^2 L_S}.$$

This is the well-known *torque-slip* curve for an induction motor and is plotted in Figure 7.13.

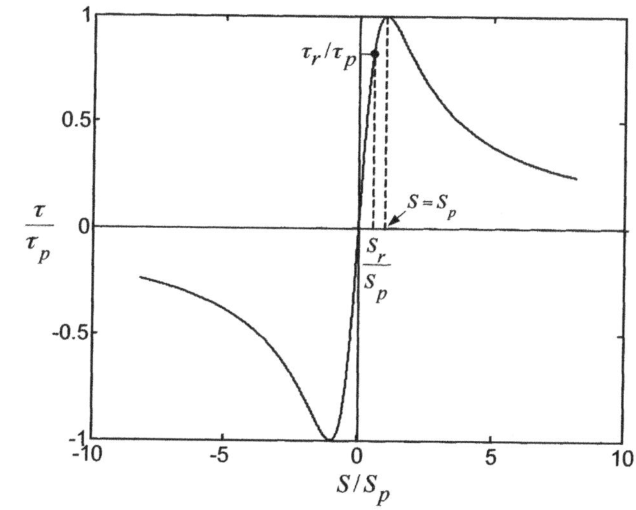

FIGURE 7.13. τ/τ_p versus S/S_p, with $R_S = 0$.

For a fixed U_S and ω_S, the torque is a maximum for $S = S_p$. S_p is called the *pull-out slip* and the corresponding maximum torque τ_p is called the *pull-out torque.* This terminology comes from the fact that for $S > S_p$ the torque *decreases* as the slip increases. Consequently, if the motor is operating at $S = S_p$ (putting out maximum torque) and then more load (load torque) is added, the rotor speed ω_R decreases and the (normalized) slip $S = (\omega_S - \omega_R)/\omega_S$ increases. But in this situation, even less torque is produced so the motor's speed continues to decrease and the motor ultimately comes to a stop. That is, for $S \geq S_p$ the motor is not at a stable operating point. On the other hand, if the motor is operating at a slip $S < S_p$, then putting more load on the motor will decrease the speed, thus increasing the slip, but in this case the torque increases. Consequently, the motor keeps running at a little lower speed, but with the additional load on it. At the rated slip $S_r = \sqrt{\sigma} S_p$, the corresponding rated torque τ_r is defined as

$$\tau_r \triangleq \tau_p \frac{2}{S/S_p + S_p/S}\,|_{S=S_r} = \tau_p \frac{2}{\sqrt{\sigma} + 1/\sqrt{\sigma}} = \tau_p \frac{2\sqrt{\sigma}}{1+\sigma}.$$

The ratio of the pull-out (or peak) torque τ_p to the rated torque τ_r is then

$$\frac{\tau_r}{\tau_p} = \frac{2\sqrt{\sigma}}{1+\sigma}.$$

This is a measure of the *overload capacity* in that, with the motor normally operating at rated torque, the factor $(1+\sigma)\,/\,(2\sqrt{\sigma})$ is how much the torque can be increased before pull-out [2]. It is normally desired to operate the motor at rated slip S_r. That is, one picks ("sizes") the motor so that its rated torque is close to the torque load expected on the motor. Then the motor will put out the required torque while minimizing the stator current and, therefore, the $R_S I_S^2$ losses.

7.2.5 Steady-State Power Transfer in the Induction Motor

Recall that the power into the stator is given by

$$P_{\text{stator}} = 3U_S I_S \cos(\varphi)$$

where

$$\begin{aligned} \frac{I_S}{I_{S0}} &= \sqrt{\frac{1+\left(\dfrac{S}{\sigma S_p}\right)^2}{1+\left(\dfrac{S}{S_p}\right)^2}} \\ I_{S0} &= \frac{U_S}{\omega_S L_S} \\ \varphi &= \angle \underline{U}_S - \angle \underline{I}_S = \frac{\pi}{2} - \left(\tan^{-1}\left(\frac{S}{\sigma S_p}\right) - \tan^{-1}\left(\frac{S}{S_p}\right)\right). \end{aligned}$$

The power factor may then be written as

$$\begin{aligned} \cos(\varphi) &= \cos\left(\frac{\pi}{2} - \tan^{-1}\left(\frac{S}{\sigma S_p}\right) + \tan^{-1}\left(\frac{S}{S_p}\right)\right) \\ &= \sin\left(\tan^{-1}\left(\frac{S}{\sigma S_p}\right) - \tan^{-1}\left(\frac{S}{S_p}\right)\right) \\ &= \sin\left(\tan^{-1}\left(\frac{S}{\sigma S_p}\right)\right)\cos\left(\tan^{-1}\left(\frac{S}{S_p}\right)\right) \\ &\quad - \sin\left(\tan^{-1}\left(\frac{S}{S_p}\right)\right)\cos\left(\tan^{-1}\left(\frac{S}{\sigma S_p}\right)\right) \\ &= \frac{\dfrac{S}{\sigma S_p}}{\sqrt{1+\left(\dfrac{S}{\sigma S_p}\right)^2}}\frac{1}{\sqrt{1+\left(\dfrac{S}{S_p}\right)^2}} - \frac{\dfrac{S}{S_p}}{\sqrt{1+\left(\dfrac{S}{S_p}\right)^2}}\frac{1}{\sqrt{1+\left(\dfrac{S}{\sigma S_p}\right)^2}} \end{aligned}$$

or

$$\cos(\varphi) = \frac{\dfrac{S}{\sigma S_p} - \dfrac{S}{S_p}}{\sqrt{1+\left(\dfrac{S}{\sigma S_p}\right)^2}\sqrt{1+\left(\dfrac{S}{S_p}\right)^2}}. \tag{7.57}$$

The power into the stator can now be written as

$$\begin{aligned}
P_{\text{stator}} &= 3U_S I_S \cos(\varphi) \\
&= 3U_S \sqrt{\frac{1+\left(\dfrac{S}{\sigma S_p}\right)^2}{1+\left(\dfrac{S}{S_p}\right)^2}} \frac{U_S}{\omega_S L_S} \frac{\dfrac{S}{\sigma S_p} - \dfrac{S}{S_p}}{\sqrt{1+\left(\dfrac{S}{\sigma S_p}\right)^2}\sqrt{1+\left(\dfrac{S}{S_p}\right)^2}} \\
&= 3\frac{U_S^2}{\omega_S L_S} \frac{\dfrac{S}{\sigma S_p} - \dfrac{S}{S_p}}{1+\left(\dfrac{S}{S_p}\right)^2} \\
&= 3\frac{U_S^2}{\omega_S L_S} \frac{1/\sigma - 1}{\dfrac{S_p}{S} + \dfrac{S}{S_p}} \\
&= \frac{3}{2}\frac{U_S^2}{\omega_S L_S}\frac{1-\sigma}{\sigma}\frac{2}{S_p/S + S/S_p} \\
&= \omega_S \tau.
\end{aligned}$$

The mechanical power produced is

$$P_{\text{mech}} = \omega_R \tau$$

and the difference between the power into the stator and the mechanical output power is

$$P_{\text{stator}} - P_{\text{mech}} = (\omega_S - \omega_R)\tau = \omega_S S \tau.$$

Where does this power go? It is now shown that this power is dissipated as heat in the rotor windings, that is, $\omega_S S \tau = 3I_R^2 R_R$. To do so, equation (7.45) is used to obtain

$$\begin{aligned}
I_R^2 &= \frac{\omega_S^2 M^2}{(R_R/S)^2 + \omega_S^2 L_R^2} I_S^2 = \frac{\left(\dfrac{\omega_S M S}{R_R}\right)^2}{1+\left(\dfrac{\omega_S L_R S}{R_R}\right)^2} I_S^2 \\
&= \frac{\dfrac{1}{(1+\sigma_R)(1+\sigma_S)}\left(\dfrac{\omega_S L_R S}{R_R}\right)\left(\dfrac{\omega_S L_S S}{R_R}\right)}{1+\left(\dfrac{S}{\sigma S_p}\right)^2} I_S^2 \\
&= \frac{(1-\sigma)\left(\dfrac{S}{\sigma S_p}\right)^2 \dfrac{L_S}{L_R}}{1+\left(\dfrac{S}{\sigma S_p}\right)^2} I_S^2
\end{aligned}$$

Using (7.50) to eliminate I_S^2, this further simplifies to

$$\begin{aligned} I_R^2 &= \frac{(1-\sigma)\left(\dfrac{S}{\sigma S_p}\right)^2 \dfrac{L_S}{L_R}}{1+\left(\dfrac{S}{\sigma S_p}\right)^2} \frac{1+\left(\dfrac{S}{\sigma S_p}\right)^2}{1+\left(\dfrac{S}{S_p}\right)^2}\left(\frac{U_S}{\omega_S L_S}\right)^2 \\ &= \frac{(1-\sigma)\left(\dfrac{S}{\sigma S_p}\right)^2 \dfrac{L_S}{L_R}}{1+\left(\dfrac{S}{S_p}\right)^2}\left(\frac{U_S}{\omega_S L_S}\right)^2 \\ &= \frac{U_S^2}{\omega_S^2 L_S}\frac{(1-\sigma)}{\sigma}\frac{\dfrac{S}{\sigma S_p}}{\dfrac{S_p}{S}+\dfrac{S}{S_p}}\frac{1}{L_R} \\ &= \frac{1}{2}\frac{U_S^2}{\omega_S^2 L_S}\frac{(1-\sigma)}{\sigma}\frac{2}{S_p/S+S/S_p}S\frac{\omega_S}{R_R} \end{aligned}$$

where

$$\frac{1}{\sigma S_p L_R} = \frac{1}{\sigma \dfrac{R_R}{\sigma \omega_S L_R} L_R} = \frac{\omega_S}{R_R}$$

was used. Then

$$3I_R^2 R_R = \left(\frac{3}{2}\frac{U_S^2}{\omega_S^2 L_S}\frac{(1-\sigma)}{\sigma}\frac{2}{S_p/S+S/S_p}\right) S\omega_S = \omega_S S\tau.$$

In summary, it has just been shown that

$$P_{\text{stator}} - P_{\text{mech}} = (\omega_S - \omega_R)\tau = \omega_S S\tau = 3I_R^2 R_R.$$

The mechanical power may be rewritten as

$$P_{\text{mech}} = \omega_R \tau = \frac{\omega_R}{\omega_S S}\omega_S S\tau = \frac{(1-S)\,\omega_S}{\omega_S S}3I_R^2 R_R = \frac{1-S}{S}3I_R^2 R_R$$

so that

$$P_{\text{stator}} = P_{\text{stator}} - P_{\text{mech}} + P_{\text{mech}} = 3I_R^2 R_R + \frac{1-S}{S}3I_R^2 R_R = 3I_R^2\frac{R_R}{S}.$$

In words, one can view all of the energy put into the rotor as being dissipated in an "equivalent" resistance of $R_R/S = R_R + (1-S)\,R_R/S$ where the energy dissipated in R_R are ohmic losses due to the rotor resistance while the energy "dissipated" in $(1-S)\,R_R/S$ is the mechanical energy produced by the machine (see Figure 7.14).

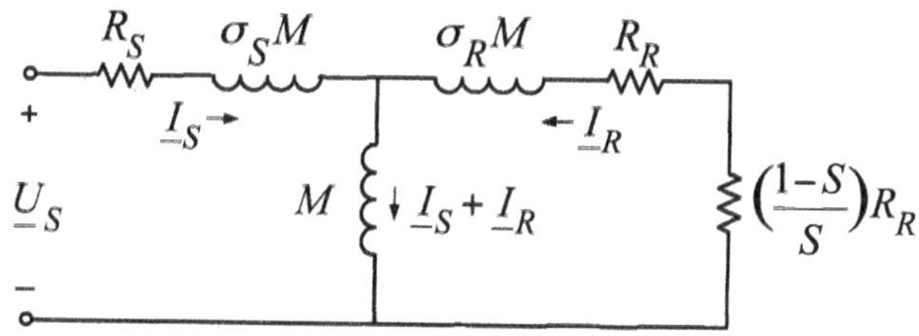

FIGURE 7.14. Mechanical energy represented as a loss in the resistor $(\frac{1-S}{S})R_R$.

Using the expressions (7.50), (7.51), (7.56), and (7.57), the normalized stator current I_S/I_{S_r}, the normalized torque τ/τ_p, and the power factor $\cos(\varphi)$ are plotted versus S/S_p in Figure 7.15.

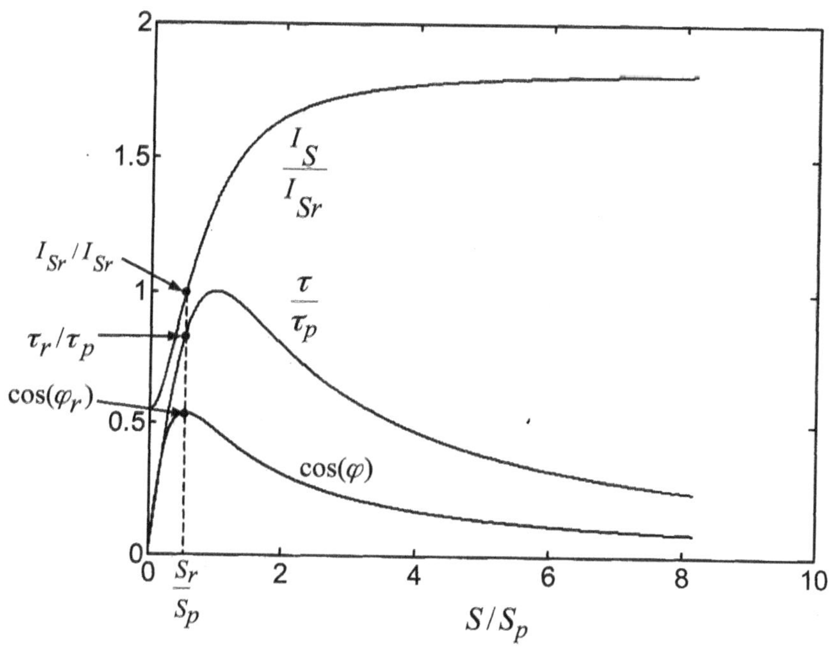

FIGURE 7.15. Plots of $\cos(\varphi)$, τ/τ_r, and I_S/I_{S_r} versus S/S_p with $\sigma = 0.3$, $S_r/S_p = \sqrt{\sigma} = 0.549$, and $\cos(\varphi_r) = (1-\sigma)/(1+\sigma) = 0.537$.

Efficiency

The efficiency[8] is defined as

$$\text{Efficiency} \triangleq \frac{P_{\text{mech}}}{P_{\text{stator}}} = \frac{\dfrac{1-S}{S} 3I_R^2 R_R}{3I_R^2 \dfrac{R_R}{S}} = 1 - S = \frac{\omega_R}{\omega_S} < 1.$$

The power input to the rotor that is not converted to mechanical power is lost as heat in the rotor windings so it is obviously important to have high efficiency. Consequently, the slip must not be too large and this must be traded off with the amount of torque required since the torque output goes down as the slip is decreased.

Theory Versus Experiment [1][2]

A comparison of the predicted values for the rated slip (7.49), the ratio of the no-load current to the rated current (7.51), the maximum power factor (7.52) and the steady-state torque (7.56) versus their measured values for two motors are reproduced in the tables below from Professor Leonhard's book [1][2]. In the tables, two different 50-Hz motors are used. The first one has two poles ($n_p = 1$) and a corresponding synchronous rotor speed[9] of 50 $\frac{\text{rev}}{\text{sec}} \times 60 \, \frac{\text{sec}}{\text{min}} \frac{1}{n_p} = 3000$ rev/min. The second motor has eight poles ($n_p = 4$) with a synchronous rotor speed of 50 $\frac{\text{rev}}{\text{sec}} \times 60 \, \frac{\text{sec}}{\text{min}} \frac{1}{n_p} = 750$ rev/min. The comparison for the two-pole machine (3000 rev/min) with $\sigma = 0.05$ is shown in the table given in Figure 7.16.

Quantity	Predicted Value $\sigma = 0.05$	Actual Value
S_r/S_p	$\sqrt{\sigma} = 0.22$	0.20
I_{S0}/I_{Sr}	$\sqrt{\sigma} = 0.22$	0.30
$\cos(\varphi_r)$	$\frac{1-\sigma}{1+\sigma} = 0.90$	0.90
τ_r/τ_p	$\frac{1+\sigma}{2\sqrt{\sigma}} = 2.35$	2.30

FIGURE 7.16. A table comparing the predicted values of S_r, I_r, φ_r, and τ_r with their measured values for a two-pole machine with $\sigma = 0.05$. From Table 10.2 of *Control of Electrical Drives,* 3rd edition by W. Leonhard, Springer-Verlag, 2001. Reprinted with permission.

---

[8] Remember that $R_S = 0$ is still assumed.

[9] The synchronous motor speed is the rotor speed with no load and rated frequency applied.

The comparison for the eight pole machine (750 rev/min) with $\sigma = 0.10$ is shown in the table given in Figure 7.17.

Quantity	Predicted Value $\sigma = 0.1$	Actual Value
S_r/S_p	$\sqrt{\sigma} = 0.32$	0.30
I_{S0}/I_{Sr}	$\sqrt{\sigma} = 0.32$	0.40
$\cos(\varphi_r)$	$\frac{1-\sigma}{1+\sigma} = 0.82$	0.84
τ_r/τ_p	$\frac{1+\sigma}{2\sqrt{\sigma}} = 1.82$	2.0

FIGURE 7.17. A table comparing the predicted values of S_r, I_r, φ_r, and τ_r with their measured values for an eight pole machine with $\sigma = 0.10$. From Table 10.2 of *Control of Electrical Drives*, 3rd edition by W. Leonhard, Springer-Verlag, 2001. Reprinted with permission.

The predicted values and the measured values are in very good agreement. It is interesting to note that these open-loop characteristics depend only on the leakage parameter σ. This is particularly remarkable as one recalls that in the derivation of the model (7.32)–(7.35), the leakage parameter σ is introduced through the parameter κ ($\sigma = 1 - \kappa^2$) in a pretty much ad hoc manner to account for fact that as the air gap is crossed the magnetic field spreads out in the axial and azimuthal directions. The nonlinear differential equation model (7.31) will be seen to be invaluable for designing feedback controllers for the induction motor, but a more detailed model is needed by machine designers to be able to specify the leakage and therefore the open loop operating characteristics of the machine.

7.3 Mathematical Model of a Three-Phase PM Synchronous Motor

Figure 7.18 shows a sinusoidally wound three-phase synchronous machine with a permanent magnet rotor. The derivation of its mathematical model is now presented. The stator magnetic field is the same as in the case of the induction motor so that it is given by (7.1) which is repeated below.

$$\vec{\mathbf{B}}_S(i_{S1}, i_{S2}, i_{S3}, r, \theta) = \frac{\mu_0 N_S}{2g}\frac{r_R}{r}\Big(i_{S1}\cos(\theta) + i_{S2}\cos(\theta - 2\pi/3) + i_{S3}\cos(\theta - 4\pi/3)\Big)\hat{\mathbf{r}}. \tag{7.58}$$

The magnetic field due to the permanent magnet is of the form

$$\vec{\mathbf{B}}_R(r, \theta - \theta_R) = B_m \frac{r_R}{r} \cos(\theta - \theta_R)\hat{\mathbf{r}}.$$

At $r = r_S$ this is written as

$$\vec{\mathbf{B}}_R(r_S, \theta - \theta_R) = \kappa B_m \frac{r_R}{r_S} \cos(\theta - \theta_R)\hat{\mathbf{r}} \tag{7.59}$$

where the coupling factor κ is included to account for leakage. The quantities that are to be computed are the stator flux linkages and the rotor torque. As the rotor has no windings, the torque is computed by finding the torque exerted on the stator windings and then letting $\vec{\tau}_R = -\vec{\tau}_S$. Consequently, the computations can all be done with the value of the total magnetic field at the inside surface of the stator.

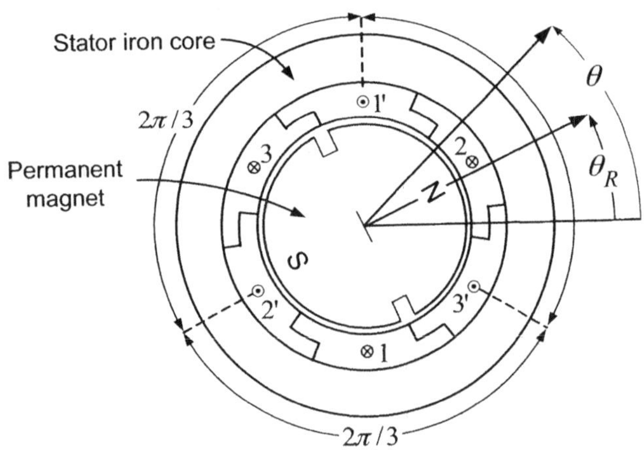

FIGURE 7.18. Three-phase synchronous machine with a permanent magnet rotor.

With $\vec{\mathbf{B}}_S$ and $\vec{\mathbf{B}}_R$ given by (7.58) and (7.59), respectively, the total radial magnetic field $\vec{\mathbf{B}}$ on the stator side of the air gap is

$$\vec{\mathbf{B}}(i_{S1}, i_{S2}, i_{S3}, r_S, \theta, \theta_R) \triangleq \vec{\mathbf{B}}_S(i_{S1}, i_{S2}, i_{S3}, r_S, \theta) + \vec{\mathbf{B}}_R(r_S, \theta - \theta_R). \tag{7.60}$$

Stator Flux Linkages

Using (7.60) for $\vec{\mathbf{B}}$ on the stator side of the air gap, the stator flux linkage in stator phase 1 is computed as

$$\begin{aligned}
\psi_{S1}(t) &= \int_0^\pi \frac{N_S}{2}\sin(\theta) \times \left(\int_{\theta-\pi}^{\theta} \ell_1 r_S B(i_{S1}, i_{S2}, i_{S3}, r_S, \theta', \theta_R)d\theta'\right)d\theta \\
&= \int_0^\pi \frac{N_S}{2}\sin(\theta) \times \left(\int_{\theta-\pi}^{\theta} \ell_1 r_S B_S(i_{S1}, i_{S2}, i_{S3}, r_S, \theta')d\theta'\right)d\theta \\
&\quad + \int_0^\pi \frac{N_S}{2}\sin(\theta) \times \left(\int_{\theta-\pi}^{\theta} \ell_1 r_S B_R(r_S, \theta' - \theta_R)d\theta'\right)d\theta.
\end{aligned}$$

The first integral is the same as in the wound rotor case and the induction motor so that by (7.5) this evaluates to

$$\begin{aligned}
&\int_0^\pi \frac{N_S}{2}\sin(\theta) \times \left(\int_{\theta-\pi}^{\theta} \ell_1 r_S B_S(i_{S1}, i_{S2}, i_{S3}, r_S, \theta')d\theta'\right)d\theta \\
&\quad = \frac{2}{3}L_S\Big(i_{S1} + i_{S2}\cos(2\pi/3) + i_{S3}\cos(4\pi/3)\Big)
\end{aligned}$$

where

$$L_S = \frac{3}{2}\frac{\pi\mu_0\ell_1\ell_2 N_S^2}{8g}. \tag{7.61}$$

The second integral evaluates as

$$\begin{aligned}
&\int_0^\pi \frac{N_S}{2}\sin(\theta)\left(\int_{\theta-\pi}^{\theta} \kappa B_m \frac{r_R}{r_S}\cos(\theta' - \theta_R)\ell_1 r_S d\theta'\right)d\theta \\
&= \int_0^\pi \frac{N_S}{2}\sin(\theta) 2\kappa B_m \frac{r_R}{r_S}\sin(\theta - \theta_R)\ell_1 r_S d\theta \\
&= \kappa\ell_1 r_R B_m N_S\cos(\theta_R)\int_0^\pi \sin^2(\theta)d\theta = \kappa\ell_1 r_R B_m N_S\cos(\theta_R)\frac{\pi}{2} \\
&= \kappa\ell_1 r_R B_m N_S\frac{\pi}{2}\cos(\theta_R) \\
&= \sqrt{\frac{2}{3}}K_m\cos(\theta_R)
\end{aligned}$$

where

$$K_m \triangleq \sqrt{\frac{3}{2}}\frac{\kappa\pi\ell_1\ell_1 B_m N_S}{4}. \tag{7.62}$$

That is,

$$\psi_{S1}(t) = \frac{2}{3}L_S\Big(i_{S1} + i_{S2}\cos(2\pi/3) + i_{S3}\cos(4\pi/3)\Big) + \sqrt{\frac{2}{3}}K_m\cos(\theta_R). \tag{7.63}$$

Similarly,

$$\begin{aligned}
&\psi_{S2}(t)\\
&= \int_{2\pi/3}^{2\pi/3+\pi} \frac{N_S}{2}\sin(\theta - 2\pi/3)\left(\int_{\theta-\pi}^{\theta} \ell_1 r_S B(i_{S1}, i_{S2}, i_{S3}, r_S, \theta', \theta_R)d\theta'\right)d\theta\\
&= \frac{2}{3}L_S(i_{S1}\cos(2\pi/3) + i_{S2} + i_{S3}\cos(2\pi/3)) + \sqrt{\frac{2}{3}}K_m\cos(\theta_R - 2\pi/3)
\end{aligned} \tag{7.64}$$

and

$$\begin{aligned}
&\psi_{S3}(t)\\
&= \int_{4\pi/3}^{4\pi/3+\pi} \frac{N_S}{2}\sin(\theta - 4\pi/3)\left(\int_{\theta-\pi}^{\theta} \ell_1 r_S B(i_{S1}, i_{S2}, i_{S3}, r_S, \theta', \theta_R)d\theta'\right)d\theta\\
&= \frac{2}{3}L_S(i_{S1}\cos(4\pi/3) + i_{S2}\cos(2\pi/3) + i_{S3}) + \sqrt{\frac{2}{3}}K_m\cos(\theta_R - 4\pi/3).
\end{aligned} \tag{7.65}$$

The factor 2/3 in front of L_S and the factor $\sqrt{2/3}$ in front of K_m in equations (7.63), (7.64), and (7.65) are so that the expressions for L_S and K_m are the *two-phase equivalent* coefficient of inductance and back-emf constant, respectively, in the two-phase equivalent model.

In matrix form, the flux linkages may be written as

$$\begin{aligned}
\begin{bmatrix} \psi_{S1}(t) \\ \psi_{S2}(t) \\ \psi_{S3}(t) \end{bmatrix} &= \frac{2}{3}L_S\begin{bmatrix} 1 & \cos(2\pi/3) & \cos(4\pi/3) \\ \cos(2\pi/3) & 1 & \cos(2\pi/3) \\ \cos(4\pi/3) & \cos(-2\pi/3) & 1 \end{bmatrix}\begin{bmatrix} i_{S1}(t) \\ i_{S2}(t) \\ i_{S3}(t) \end{bmatrix}\\
&\quad + \sqrt{\frac{2}{3}}K_m\begin{bmatrix} \cos(\theta_R) \\ \cos(\theta_R - 2\pi/3) \\ \cos(\theta_R - 4\pi/3) \end{bmatrix}.
\end{aligned} \tag{7.66}$$

With the obvious definition for matrix C_1, the flux linkages are written more compactly as

$$\begin{bmatrix} \psi_{S1}(t) \\ \psi_{S2}(t) \\ \psi_{S3}(t) \end{bmatrix} = C_1\begin{bmatrix} i_{S1}(t) \\ i_{S2}(t) \\ i_{S3}(t) \end{bmatrix} + \sqrt{\frac{2}{3}}K_m\begin{bmatrix} \cos(\theta_R) \\ \cos(\theta_R - 2\pi/3) \\ \cos(\theta_R - 4\pi/3) \end{bmatrix}. \tag{7.67}$$

With the stator voltages $u_{S1}(t), u_{S2}(t)$, and $u_{S3}(t)$ assumed to be balanced and R_S the resistance in each stator phase, Faraday's law gives

$$\begin{aligned}
u_{S1}(t) &= R_S i_{S1} + \frac{d\psi_{S1}(t)}{dt}\\
u_{S2}(t) &= R_S i_{S2} + \frac{d\psi_{S2}(t)}{dt}\\
u_{S3}(t) &= R_S i_{S3} + \frac{d\psi_{S3}(t)}{dt}.
\end{aligned} \tag{7.68}$$

Three-Phase to Two-Phase Transformation

The stator voltages are (assumed) balanced and the machine is assumed to be wye-connected so that the stator currents are also balanced. The *three-phase to two-phase* transformation defined by (7.17) is again used to simplify the flux equations by defining

$$\begin{bmatrix} i_{Sa}(t) \\ i_{Sb}(t) \\ i_{S0}(t) \end{bmatrix} \triangleq Q \begin{bmatrix} i_{S1}(t) \\ i_{S2}(t) \\ i_{S3}(t) \end{bmatrix}, \quad \begin{bmatrix} \lambda_{Sa}(t) \\ \lambda_{Sb}(t) \\ \lambda_{S0}(t) \end{bmatrix} \triangleq Q \begin{bmatrix} \psi_{S1}(t) \\ \psi_{S2}(t) \\ \psi_{S3}(t) \end{bmatrix}.$$

where (see problem 4)

$$\begin{aligned} \lambda_{S0}(t) &= \frac{1}{\sqrt{3}}(\psi_{S1}(t) + \psi_{S2}(t) + \psi_{S3}(t)) \equiv 0 \\ i_{S0}(t) &= \frac{1}{\sqrt{3}}(i_{S1}(t) + i_{S2}(t) + i_{S3}(t)) \equiv 0. \end{aligned}$$

The dynamic equations for the stator flux linkages (7.68) then transform to

$$\begin{aligned} u_{Sa}(t) &= R_S i_{Sa} + \frac{d\lambda_{Sa}(t)}{dt} \\ u_{Sb}(t) &= R_S i_{Sb} + \frac{d\lambda_{Sb}(t)}{dt} \\ u_{S0}(t) &= R_S i_{S0} + \frac{d\lambda_{S0}(t)}{dt}. \end{aligned} \tag{7.69}$$

To find the expressions for the two-phase equivalent flux linkages in terms of the two-phase equivalent currents, recall that

$$\begin{bmatrix} \psi_{S1}(t) \\ \psi_{S2}(t) \\ \psi_{S3}(t) \end{bmatrix} = C_1 \begin{bmatrix} i_{S1}(t) \\ i_{S2}(t) \\ i_{S3}(t) \end{bmatrix} + \sqrt{\frac{2}{3}} K_m \begin{bmatrix} \cos(\theta_R) \\ \cos(\theta_R - 2\pi/3) \\ \cos(\theta_R - 4\pi/3) \end{bmatrix}. \tag{7.70}$$

Then

$$\begin{aligned} \begin{bmatrix} \lambda_{Sa}(t) \\ \lambda_{Sb}(t) \\ \lambda_{S0}(t) \end{bmatrix} &= QC_1Q^{-1} \begin{bmatrix} i_{Sa}(t) \\ i_{Sb}(t) \\ i_{S0}(t) \end{bmatrix} + Q\sqrt{\frac{2}{3}} K_m \begin{bmatrix} \cos(\theta_R) \\ \cos(\theta_R - 2\pi/3) \\ \cos(\theta_R - 4\pi/3) \end{bmatrix} \\ &= \begin{bmatrix} L_S & 0 & 0 \\ 0 & L_S & 0 \\ 0 & 0 & 0 \end{bmatrix} \begin{bmatrix} i_{Sa}(t) \\ i_{Sb}(t) \\ i_{S0}(t) \end{bmatrix} + K_m \begin{bmatrix} \cos\theta_R \\ \sin\theta_R \\ 0 \end{bmatrix} \end{aligned} \tag{7.71}$$

or

$$\begin{aligned} \lambda_{Sa}(t) &= L_S i_{Sa}(t) + K_m \cos(\theta_R) \\ \lambda_{Sb}(t) &= L_S i_{Sb}(t) + K_m \sin(\theta_R) \\ \lambda_{S0} &\equiv 0. \end{aligned} \tag{7.72}$$

Collecting together equations (7.69) and (7.72), the two-phase equivalent equations for the stator and rotor currents of a *wye connected* three-phase motor become

$$\begin{aligned}
u_{Sa} &= L_S\frac{d}{dt}i_{Sa} + K_m\frac{d}{dt}\cos(\theta_R) + R_S i_{Sa} \\
u_{Sb} &= L_S\frac{d}{dt}i_{Sb} + K_m\frac{d}{dt}\sin(\theta_R) + R_S i_{Sb} \\
u_{S0} &= 0
\end{aligned} \tag{7.73}$$

where $i_{S0} \equiv 0$.

Torque

The strategy is to compute the rotor torque $\vec{\tau}_R$ by first determining the torque $\vec{\tau}_S$ that the magnetic field of the rotor's permanent magnet produces on the stator windings and then use $\vec{\tau}_R = -\vec{\tau}_S$. The rotor's magnetic field at the inside surface of the stator is

$$\vec{\mathbf{B}}_R(r_S, \theta - \theta_R) = \kappa B_m \frac{r_R}{r_S}\cos(\theta - \theta_R)\hat{\mathbf{r}}.$$

The torque on stator phase 1 is then computed by

$$\begin{aligned}
\vec{\tau}_{S1} &= \int_{\theta=0}^{2\pi} r_S\hat{\mathbf{r}} \times \left(i_{S1}(t)\frac{N_S}{2}\sin(\theta)d\theta(+\ell_1\hat{\mathbf{z}}) \times \left(B_{R|_{r=r_S}}\hat{\mathbf{r}}\right)\right) \\
&= \int_{\theta=0}^{2\pi} r_S i_{S1}(t)\frac{\ell_1 N_S}{2}\sin(\theta)\left(\kappa B_m\frac{r_R}{r_S}\right)\cos(\theta-\theta_R)d\theta\hat{\mathbf{z}} \\
&= i_{S1}(t)\frac{\kappa B_m \ell_1 r_R N_S}{2}\int_{\theta=0}^{2\pi}\sin(\theta)\cos(\theta-\theta_R)d\theta\hat{\mathbf{z}} \\
&= \kappa\ell_1 r_R B_m N_S\frac{\pi}{2}i_{S1}(t)\sin(\theta_R)\hat{\mathbf{z}} \\
&= \sqrt{\frac{2}{3}}K_m i_{S1}(t)\sin(\theta_R)\hat{\mathbf{z}}
\end{aligned} \tag{7.74}$$

where K_m is as given in (7.62).

The torque on stator phase 2 of the rotor is then

$$\begin{aligned}
\vec{\tau}_{S2} &= \int_{\theta=0}^{2\pi} r_S\hat{\mathbf{r}} \times \left(i_{S2}(t)\frac{N_S}{2}\sin(\theta-2\pi/3)d\theta(+\ell_1\hat{\mathbf{z}}) \times \left(B_{R|_{r=r_S}}\hat{\mathbf{r}}\right)\right)\\
&= \int_{\theta=0}^{2\pi} r_S i_{S2}(t)\frac{\ell_1 N_S}{2}\sin(\theta-2\pi/3)\left(\kappa B_m\frac{r_R}{r_S}\right)\cos(\theta-\theta_R)d\theta\hat{\mathbf{z}}\\
&= i_{S2}(t)\frac{\kappa B_m\ell_1 r_R N_S}{2}\int_{\theta=0}^{2\pi}\sin(\theta-2\pi/3)\cos(\theta-\theta_R)d\theta\hat{\mathbf{z}}\\
&= i_{S2}(t)\frac{\kappa B_m\ell_1 r_R N_S}{2}\int_{\theta=0}^{2\pi}\frac{1}{2}\left(\sin(2\theta-\frac{2\pi}{3}-\theta_R)+\sin(\theta_R-\frac{2\pi}{3})\right)d\theta\hat{\mathbf{z}}\\
&= \kappa\ell_1 r_R B_m N_S\frac{\pi}{2}i_{S2}(t)\sin(\theta_R-2\pi/3)\hat{\mathbf{z}}\\
&= \sqrt{\frac{2}{3}}K_m i_{S2}(t)\sin(\theta_R-2\pi/3)\hat{\mathbf{z}}. \qquad (7.75)
\end{aligned}$$

Finally, the torque on stator phase 3 is computed as

$$\begin{aligned}
\vec{\tau}_{S3} &= \int_{\theta=0}^{2\pi} r_S\hat{\mathbf{r}} \times \left(i_{S3}(t)\frac{N_S}{2}\sin(\theta-4\pi/3)d\theta(+\ell_1\hat{\mathbf{z}}) \times \left(B_{R|_{r=r_S}}\hat{\mathbf{r}}\right)\right)\\
&= \int_{\theta=0}^{2\pi} r_S i_{S3}(t)\frac{\ell_1 N_S}{2}\sin(\theta-4\pi/3)\left(\kappa B_m\frac{r_R}{r_S}\right)\cos(\theta-\theta_R)d\theta\hat{\mathbf{z}}\\
&= i_{S3}(t)\frac{\kappa B_m\ell_1 r_R N_S}{2}\int_{\theta=0}^{2\pi}\sin(\theta-4\pi/3)\cos(\theta-\theta_R)d\theta\hat{\mathbf{z}}\\
&= i_{S3}(t)\frac{\kappa B_m\ell_1 r_R N_S}{2}\int_{\theta=0}^{2\pi}\frac{1}{2}\left(\sin(2\theta-\frac{4\pi}{3}-\theta_R)+\sin(\theta_R-\frac{4\pi}{3})\right)d\theta\hat{\mathbf{z}}\\
&= \kappa\ell_1 r_R B_m N_S\frac{\pi}{2}i_{S3}(t)\sin(\theta_R-4\pi/3)\hat{\mathbf{z}}\\
&= \sqrt{\frac{2}{3}}K_m i_{S3}(t)\sin(\theta_R-4\pi/3)\hat{\mathbf{z}}. \qquad (7.76)
\end{aligned}$$

The total torque is then $\tau_S = \tau_{S1} + \tau_{S2} + \tau_{S3}$ or

$$\tau_S = \sqrt{\frac{2}{3}}K_m\Big(i_{S1}\sin(\theta_R) + i_{S2}\sin(\theta_R-2\pi/3) + i_{S3}\sin(\theta_R-4\pi/3)\Big). \qquad (7.77)$$

Substitute

$$\begin{bmatrix} i_{S1}(t)\\ i_{S2}(t)\\ i_{S3}(t)\end{bmatrix} \triangleq Q^{-1}\begin{bmatrix} i_{Sa}(t)\\ i_{Sb}(t)\\ i_{S0}(t)\end{bmatrix} = \sqrt{\frac{3}{2}}\begin{bmatrix} 2/3 & 0 & \sqrt{2}/3\\ -1/3 & 1/\sqrt{3} & \sqrt{2}/3\\ -1/3 & -1/\sqrt{3} & \sqrt{2}/3\end{bmatrix}\begin{bmatrix} i_{Sa}(t)\\ i_{Sb}(t)\\ i_{S0}(t)\end{bmatrix}$$

into (7.77) to obtain ($i_{S0}(t) \equiv 0$)

$$\tau_S = K_m\Big(i_{Sa}\sin(\theta_R) - i_{Sb}\cos(\theta_R)\Big). \qquad (7.78)$$

The torque on the rotor is then

$$\tau_R = -K_m\Big(i_{Sa}\sin(\theta_R) - i_{Sb}\cos(\theta_R)\Big). \tag{7.79}$$

Substituting the two-phase equivalent stator flux linkages (7.72) into (7.69) along with torque equation (7.79) gives the following two-phase equivalent mathematical model of a three-phase wye-connected permanent magnet synchronous motor

$$\begin{aligned}
u_{Sa} &= L_S\frac{di_{Sa}}{dt} + K_m\frac{d}{dt}\cos(\theta_R) + R_S i_{Sa}\\
u_{Sb} &= L_S\frac{di_{Sb}}{dt} + K_m\frac{d}{dt}\sin(\theta_R) + R_S i_{Sb}\\
J\frac{d\omega_R}{dt} &= K_m\Big(i_{Sb}\cos(\theta_R) - i_{Sa}\sin(\theta_R)\Big) - \tau_L\\
\frac{d\theta_R}{dt} &= \omega_R
\end{aligned} \tag{7.80}$$

which simplifies to

$$\begin{aligned}
L_S\frac{di_{Sa}}{dt} &= -R_S i_{Sa} - K_m\omega_R\sin(\theta_R) + u_{Sa}\\
L_S\frac{di_{Sb}}{dt} &= -R_S i_{Sb} + K_m\omega_R\cos(\theta_R) + u_{Sb}\\
J\frac{d\omega_R}{dt} &= K_m\Big(i_{Sb}\cos(\theta_R) - i_{Sa}\sin(\theta_R)\Big) - \tau_L\\
\frac{d\theta_R}{dt} &= \omega_R.
\end{aligned} \tag{7.81}$$

Figure 7.19 is a block diagram for using the two-phase equivalent model to simulate the three-phase model.

Remark

With $L = (2/3)L_S$, and $K = \sqrt{2/3}K_m$ the three-phase flux linkages can be written in a standard three-phase form as

$$\begin{bmatrix}\psi_{S1}\\ \psi_{S2}\\ \psi_{S3}\end{bmatrix} = \begin{bmatrix} L & -L/2 & -L/2\\ -L/2 & L & -L/2\\ -L/2 & -L/2 & L\end{bmatrix}\begin{bmatrix}i_{S1}\\ i_{S2}\\ i_{S3}\end{bmatrix} + K\begin{bmatrix}\cos(\theta_R)\\ \cos(\theta_R - 2\pi/3)\\ \cos(\theta_R - 4\pi/3)\end{bmatrix} \tag{7.82}$$

so that

$$\begin{aligned}
\begin{bmatrix} L & -L/2 & -L/2\\ -L/2 & L & -L/2\\ -L/2 & -L/2 & L\end{bmatrix}\begin{bmatrix}di_{S1}/dt\\ di_{S2}/dt\\ di_{S3}/dt\end{bmatrix} &= K\omega_R\begin{bmatrix}\sin(\theta_R)\\ \sin(\theta_R - 2\pi/3)\\ \sin(\theta_R - 4\pi/3)\end{bmatrix}\\
&\quad - R_S\begin{bmatrix}i_{S1}\\ i_{S2}\\ i_{S3}\end{bmatrix} + \begin{bmatrix}u_{S1}\\ u_{S2}\\ u_{S3}\end{bmatrix}.
\end{aligned} \tag{7.83}$$

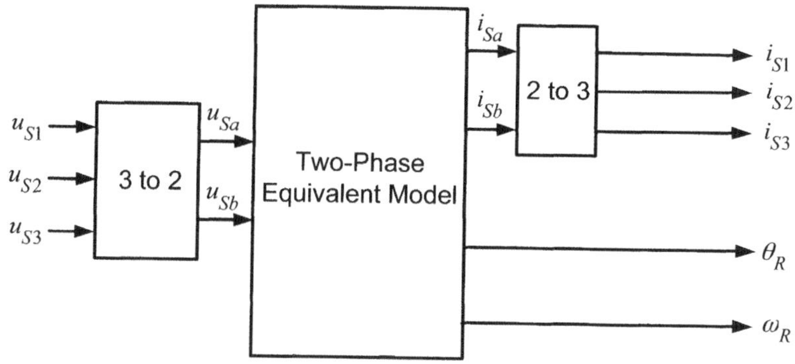

FIGURE 7.19. Three-phase PM motor simulated using a two-phase equivalent model.

The torque is then written as

$$\tau_R = -K\Big(i_{S1}\sin(\theta_R) + i_{S2}\sin(\theta_R - 2\pi/3) + i_{S3}\sin(\theta_R - 4\pi/3)\Big). \tag{7.84}$$

As the currents are balanced ($i_{S1} + i_{S2} + i_{S3} \equiv 0$), equation (7.82) may be rewritten as

$$\begin{bmatrix} \psi_{S1} \\ \psi_{S2} \\ \psi_{S3} \end{bmatrix} = \begin{bmatrix} 3L/2 & 0 & 0 \\ 0 & 3L/2 & 0 \\ 0 & 0 & 3L/2 \end{bmatrix} \begin{bmatrix} i_{S1} \\ i_{S2} \\ i_{S3} \end{bmatrix} + K \begin{bmatrix} \cos(\theta_R) \\ \cos(\theta_R - 2\pi/3) \\ \cos(\theta_R - 4\pi/3) \end{bmatrix} \tag{7.85}$$

and equation (7.83) becomes

$$\begin{aligned} \begin{bmatrix} di_{S1}/dt \\ di_{S2}/dt \\ di_{S3}/dt \end{bmatrix} &= -\frac{2}{3}\frac{R_S}{L} \begin{bmatrix} i_{S1} \\ i_{S2} \\ i_{S3} \end{bmatrix} + \frac{2}{3}\frac{K}{L}\omega_R \begin{bmatrix} \sin(\theta_R) \\ \sin(\theta_R - 2\pi/3) \\ \sin(\theta_R - 4\pi/3) \end{bmatrix} \\ &\quad + \frac{2}{3}\frac{1}{L} \begin{bmatrix} u_{S1} \\ u_{S2} \\ u_{S3} \end{bmatrix}. \end{aligned} \tag{7.86}$$

7.4 Three-Phase, Sinusoidal, 60-Hz Voltages*

Electric machines and power systems have been around for over 100 years. Basic choices in construction and operation of these systems were made along the way. In North America, this has resulted in a power system providing three-phase sinusoidal voltages operating at 60 Hz passing through three-phase transformers to power three-phase induction motors, light buildings, and so on. Why these specific choices? The answer to these questions are given in the book [61] by Laithewaite and Freris and are summarized below.

7.4.1 Why Three-Phase?

Modeling has been done of both two-phase and three-phase AC machines. One could consider a four-phase machine as well. Why are most electric machines three-phase? To answer this question, consider a simple generator shown in Figure 7.20 where a single-phase winding consists of three loops with one of the loops having a side in the slot at $\theta = \pi/3$ (the other side of this loop in the slot at $\theta = 4\pi/3$), another one at $\theta = \pi/2$, and the third one at $\theta = 2\pi/3$.

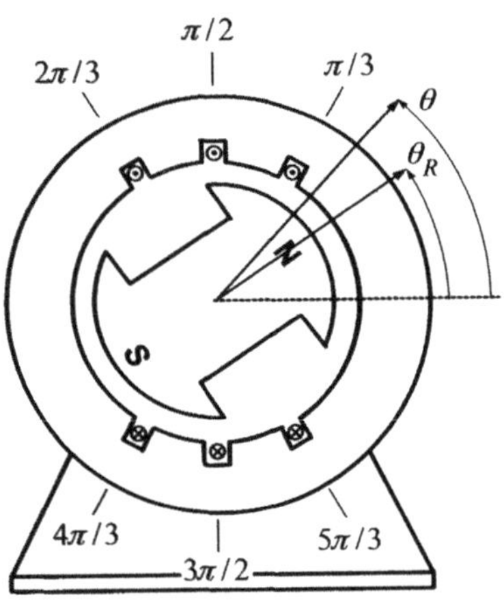

FIGURE 7.20. Computation of the emf in a distributed winding.

Consider the rotor to be a permanent magnet with the magnetic field in the air gap given by

$$\vec{\mathbf{B}}_R(\theta - \theta_R) = B_{\max}\frac{r_R}{r}\cos(\theta - \theta_R)\hat{\mathbf{r}}.$$

With $d\vec{\mathbf{S}} = r_S d\theta d\ell\hat{\mathbf{r}}$ and the rotor moving at a constant angular speed so that $\theta_R = \omega_R t$, the flux in the loop at θ_i is

$$\begin{aligned}\phi_{\theta_i} &= \int_{\substack{\text{Loop from}\\ \theta_i-\pi \text{ to } \theta_i}} \vec{\mathbf{B}}_R \cdot d\vec{\mathbf{S}} = \int_0^{\ell_1}\int_{\theta=\theta_i-\pi}^{\theta=\theta_i} B_{\max}\frac{r_R}{r_S}\cos(\theta-\theta_R)\hat{\mathbf{r}}\cdot(r_S d\theta d\ell\hat{\mathbf{r}})\\ &= 2\ell_1 r_R B_{\max}\sin(\theta_i-\theta_R).\end{aligned}$$

With $\theta_R = \omega_R t$, the corresponding emf is

$$\xi_{\theta_i} = -\frac{d\phi_{\theta_i}}{dt} = V_{\max}\cos(\theta_R-\theta_i) = \operatorname{Re}\{V_{\max}e^{j(\omega_R t-\theta_i)}\}$$

where $V_{\max} \triangleq 2\ell_1 r_R B_{\max}\omega_R$. Denoting the three loops shown in Figure 7.20 as phase 1, the total emf generated in this phase is

$$\begin{aligned}\xi_1 &= V_{\max}\operatorname{Re}\Big\{e^{j(\omega_R t-\pi/3)} + e^{j(\omega_R t-\pi/2)} + e^{j(\omega_R t-2\pi/3)}\Big\}\\ &= V_{\max}\operatorname{Re}\Big\{e^{j\omega_R t}\left(e^{-j\pi/3}+e^{-j\pi/2}+e^{-j2\pi/3}\right)\Big\}\\ &= V_{\max}(1+\sqrt{3})\cos(\omega_R t-\pi/2)\end{aligned}$$

as $e^{-j\pi/3}+e^{-j\pi/2}+e^{-j2\pi/3} = (1+\sqrt{3})e^{-j\pi/2}$.

On the other hand, if all three loops were located in the single pair of slots at $\theta = \pi/2$ and $-\pi/2$, then the generated voltage ζ_1 would be

$$\begin{aligned}\zeta_1 &= 3V_{\max}\operatorname{Re}\Big\{e^{j\omega_R t}e^{-j\pi/2}\Big\}\\ &= 3V_{\max}\cos(\omega_R t-\pi/2).\end{aligned}$$

The ratio of these two voltages is then

$$\frac{\xi_1}{\zeta_1} = \frac{1+\sqrt{3}}{3} = 0.91.$$

That is, by spreading the loops along the inside periphery of the stator, the voltage generated is about 91% of the voltage that would result if all the loops were put in the same slot. Why spread out the windings? A fundamental problem that limits the output power of an electric machine is the amount of heat it generates. So as to not damage the winding insulation by overheating, the amount of heat generated must be limited, and this is directly related to the amount of current in the windings. The addition of a fan on the rotor of the machine greatly enhances the heat transfer (removal) so that much larger currents could be used in the phase windings. (The payoff of being able to have larger currents in the windings greatly outweighs the loss of power due to the drag of the fan.) However, even with a fan, the amount of current in a winding at any location of the stator is still limited. With a current i_S in a phase winding, putting all three loops at one location will generate three times the heat of putting just one loop at the one location. So, if more windings (loops) are desired to obtain a greater generated voltage, they must be distributed around the periphery for cooling (heat transfer) to avoid overheating the insulation.

The above analysis is now generalized to distributing the loops around the complete inner surface of the stator. Reiterating, to obtain more generated voltage out of the machine without overheating it, the loops must be distributed along the inner periphery of the stator.

Single Phase

First consider the case of a *single-phase* machine where N_S stator loops are equally spread around the complete 2π radians of the inner stator surface as shown in Figure 7.21(a).

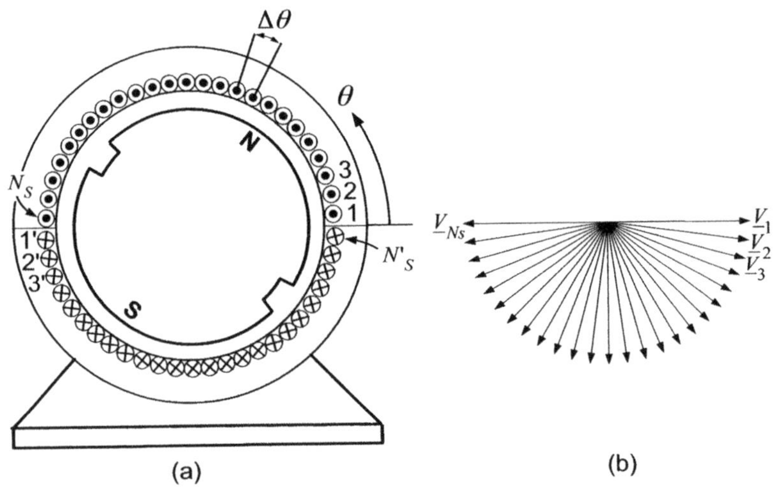

FIGURE 7.21. A single phase machine with a total of N_S turns/loops. (a) The first loop has sides 1, the second loop has sides 2, and so on. with the final loop having sides $N_S - N_S'$. These loops are all electrically connected in series to make up the single-phase winding. (b) The phasors for each loop.

The top sides of the N_S loops are placed in slots on the inside surface of the stator core at

$$\theta_i = (i-1)\Delta\theta \quad \text{for } i = 1, ..., N_S \tag{7.87}$$

where $\Delta\theta = \pi/N_S$ and the other side of each loop is at the diametrically opposite side of the stator. Then the total voltage $\xi_{1\text{ph}}$ generated in this single phase is

$$\begin{aligned}\xi_{1\text{ph}}(t) &= \text{Re}\left\{V_{\max}\left(\sum_{i=1}^{N_S} e^{j(\omega_R t-\theta_i)}\right)\right\} = \text{Re}\left\{\left(\sum_{i=1}^{N_S} \underline{V}_i\right) e^{j\omega_R t}\right\} \\ &= \text{Re}\left\{\underline{\xi}_{1\text{ph}} e^{j\omega_R t}\right\} = \text{Re}\left\{\left|\underline{\xi}_{1\text{ph}}\right| e^{j\angle\underline{\xi}_{1\text{ph}}} e^{j\omega_R t}\right\} \end{aligned} \qquad (7.88)$$

where

$$\underline{V}_i \triangleq V_{\max} e^{-j\theta_i} = V_{\max} e^{-j((i-1)\Delta\theta\,)}$$

is the voltage phasor for loop $i - i'$ and

$$\underline{\xi}_{1\text{ph}} \triangleq \sum_{i=1}^{N_S} \underline{V}_i.$$

is the phasor for the total emf generated in the phase.

With N_S large (so $\Delta\theta = \pi/N_S$ is small), the phasors $\underline{V}_i$ in Figure 7.21 put end to end trace out (to a good approximation) a semicircle as illustrated in Figure 7.22 with their sum $\underline{\xi}_{1\text{ph}} = \sum_{i=1}^{N_S} \underline{V}_i$ being the diameter of the semicircle.

On the other hand, if all N_S loops are in the single pair of slots at $\pi/2$ and $3\pi/2$, the generated voltage is ($\underline{V}_1 = V_{\max}$ for the single-phase machine)

$$\zeta_{1\text{ph}}(t) = \text{Re}\left\{\left(N_S \underline{V}_1 e^{-j\pi/2}\right) e^{j\omega_R t}\right\} = \text{Re}\left\{\underline{\zeta}_{1\text{ph}} e^{j\omega_R t}\right\}$$

with the magnitude of this voltage phasor given by

$$\left|\underline{\zeta}_{1\text{ph}}\right| = |N_S \underline{V}_1| = \sum_{i=1}^{N_S} |\underline{V}_i|.$$

With r the radius of the semicircle, $\left|\underline{\xi}_{1\text{ph}}\right|$ is just the circumferential length πr of the semicircle. It then follows that

$$\begin{aligned}\left|\underline{\xi}_{1\text{ph}}\right| &= \left|\sum_{i=1}^{N_S} \underline{V}_i\right| = 2r \\ \left|\underline{\zeta}_{1\text{ph}}\right| &= \sum_{i=1}^{N_S} |\underline{V}_i| = \pi r\end{aligned}$$

or

$$\frac{\left|\underline{\xi}_{1\text{ph}}\right|}{\left|\underline{\zeta}_{1\text{ph}}\right|} = \frac{2}{\pi} = 0.636.$$

That is, for a *single-phase* machine, only 63.6% of the voltage is obtained distributing the N_S loops over the inner stator surface compared to being able to put all N_S windings at a single location. However, the distributed winding can carry a much larger current without overheating and damaging the conductor insulation compared to having all the windings concentrated in one pair of slots.

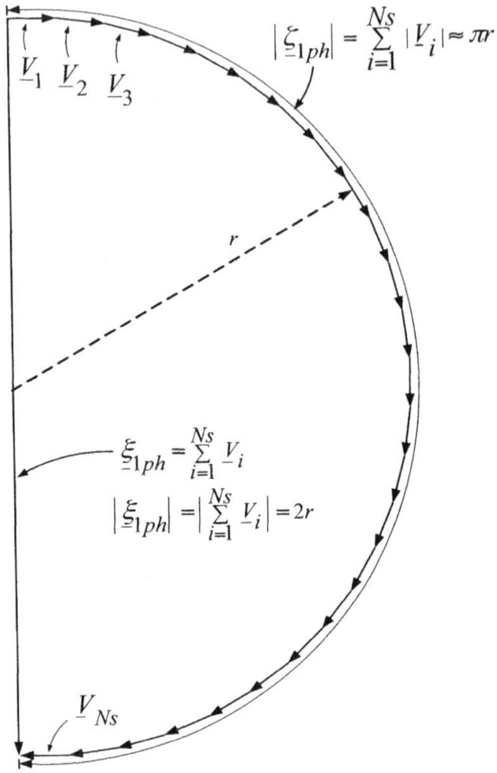

FIGURE 7.22. A geometric representation of the phasor sum $\underline{\boldsymbol{\xi}}_{1\text{ph}} = \sum_{i=1}^{N_S} \underline{\mathbf{V}}_i$. The diameter of the semicircle is $\left|\sum_{i=1}^{N_S} \underline{\mathbf{V}}_i\right| = 2r$ while its circumferential length is $\sum_{i=1}^{N_S} |\underline{\mathbf{V}}_i| = \pi r$.

Two Phases

For a two-phase machine with phases 1 and 2, the N_S loops of *each* phase are distributed over only π radians; that is, with $\Delta\theta = \pi/(2N_S)$, the top side of the loops of phase 1 are in the slots at

$$\theta_i = \frac{\pi}{4} + (i-1)\Delta\theta \quad \text{for } i = 1, ..., N_S$$

as shown in Figure 7.23(a).

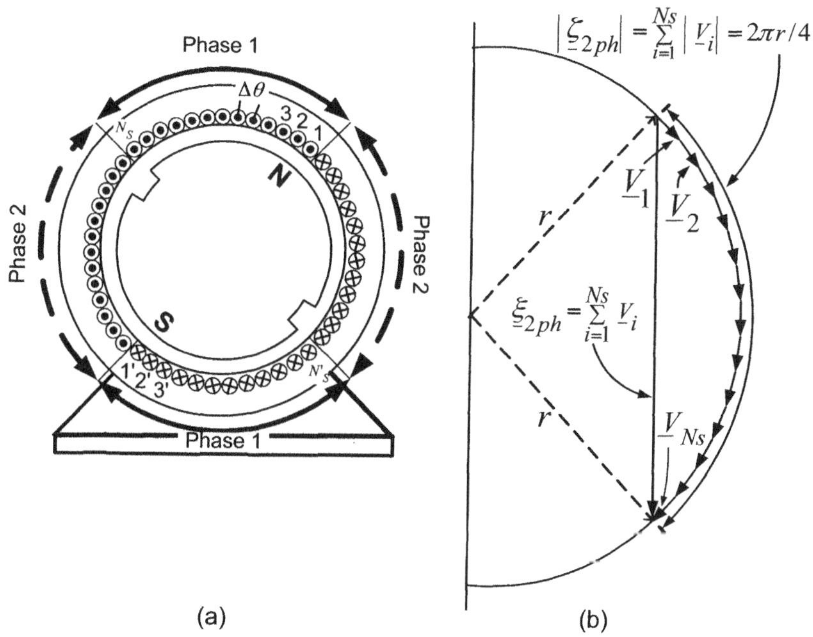

FIGURE 7.23. (a) Distributed two-phase windings. (b) Phasor diagram of the voltages in the loops of phase 1.

In this case, the voltage $\xi_{2\text{ph}}$ generated in this distributed winding is

$$\begin{aligned}\xi_{2\text{ph}}(t) &= \text{Re}\left\{V_{\max}\left(\sum_{i=1}^{N_S} e^{j(\omega_R t-\theta_i)}\right)\right\} = \text{Re}\left\{\left(\sum_{i=1}^{N_S} \underline{V}_i\right)e^{j\omega_R t}\right\} \\ &= \text{Re}\left\{\underline{\boldsymbol{\xi}}_{2\text{ph}} e^{j\omega_R t}\right\} = \text{Re}\left\{\left|\underline{\boldsymbol{\xi}}_{2\text{ph}}\right| e^{j\angle\underline{\boldsymbol{\xi}}_{2\text{ph}}} e^{j\omega_R t}\right\} \qquad (7.89)\end{aligned}$$

where

$$\underline{V}_i \triangleq V_{\max}e^{-j\theta_i} = V_{\max}e^{-j\left((i-1)\Delta\theta\right)}$$

is the voltage phasor for loop i–i' and

$$\underline{\boldsymbol{\xi}}_{2\text{ph}} \triangleq \sum_{i=1}^{N_S} \underline{V}_i$$

is the phasor of the total emf generated in the phase.

With N_S large, the phasors $\underline{V}_i$ put end-to-end trace out a quarter of a circle and their sum $\underline{\boldsymbol{\xi}}_{2\text{ph}} = \sum_{i=1}^{N_S} \underline{V}_i$ is a chord of the circle as illustrated in Figure 7.23(b). The magnitude of this phasor is

$$\left|\underline{\boldsymbol{\xi}}_{2\text{ph}}\right| = \left|\sum_{i=1}^{N_S} \underline{V}_i\right| = \sqrt{2}r$$

where r is the radius of the circle.

If all N_S loops are in a single-slot pair at $\pi/2$ and $3\pi/2$, the generated voltage is ($\underline{V}_1 = V_{\max}e^{-j\pi/4}$ for the two-phase machine so that $\underline{V}_1 e^{-j\pi/4} = V_{\max}e^{-j\pi/2}$)

$$\zeta_{2\text{ph}}(t) = \text{Re}\left\{\left(N_S\underline{V}_1 e^{-j\pi/4}\right)e^{j\omega_R t}\right\} = \text{Re}\left\{\underline{\zeta}_{2\text{ph}}e^{j\omega_R t}\right\}$$

where the magnitude of this voltage phasor given by

$$\left|\underline{\zeta}_{2\text{ph}}\right| = \left|N_S\underline{V}_1\right| = \sum_{i=0}^{N_S}\left|\underline{V}_i\right| = \frac{2\pi r}{4}$$

as it is just the circumferential length of the quarter circle. The ratio of the magnitude of these two generated emfs is

$$\frac{\left|\underline{\xi}_{2\text{ph}}\right|}{\left|\underline{\zeta}_{2\text{ph}}\right|} = \frac{2\sqrt{2}}{\pi} = 0.9.$$

Three Phases

For a three-phase motor with phases 1, 2 and 3, the N_S loops of *each* phase are distributed over $2\pi/3$ radians. With $\Delta\theta = \pi/(3N_S)$, the top side of the loops of phase 1 are in the slots at

$$\theta_i = \frac{\pi}{3} + (i-1)\Delta\theta \quad \text{for } i = 1, ..., N_S$$

as illustrated in Figure 7.24. The total voltage $\xi_{3\text{ph}}$ generated in phase 1 is

$$\begin{aligned}\xi_{3\text{ph}}(t) &= \text{Re}\left\{V_{\max}\left(\sum_{i=1}^{N_S} e^{j(\omega_R t-\theta_i)}\right)\right\} = \text{Re}\left\{\left(\sum_{i=1}^{N_S}\underline{V}_i\right)e^{j\omega_R t}\right\} \\ &= \text{Re}\left\{\underline{\xi}_{3\text{ph}}e^{j\omega_R t}\right\} = \text{Re}\left\{\left|\underline{\xi}_{3\text{ph}}\right|e^{j\angle\underline{\xi}_{3\text{ph}}}e^{j\omega_R t}\right\} \end{aligned} \quad (7.90)$$

where

$$\underline{V}_i \triangleq V_{\max}e^{-j\theta_i} = V_{\max}e^{-j\left((i-1)\Delta\theta\right)}$$

is the voltage phasor for loop i–i' and

$$\underline{\xi}_{3\text{ph}} \triangleq \sum_{i=1}^{N_S}\underline{V}_i$$

is the phasor of the total emf generated in the phase.

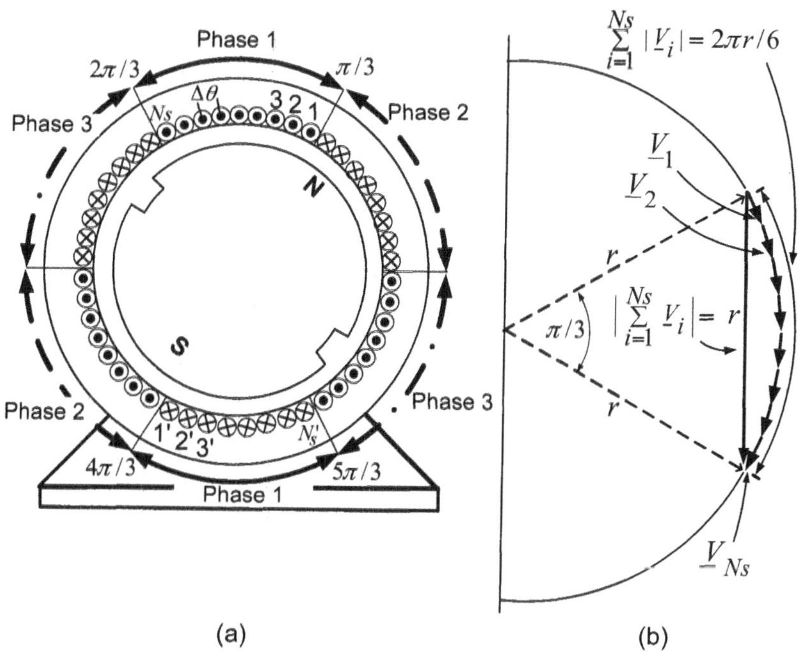

FIGURE 7.24. (a) Distributed three-phase windings. (b) Phasor diagram of the voltages in each loop of phase 1.

With N_S large, the phasors $\underline{V}_i$ put end to end trace out a sixth of a circle and the sum $\underline{\xi}_{3\text{ph}} = \sum_{i=1}^{N_S} \underline{V}_i$ is a chord of this circle as shown in Figure 7.24(b). The magnitude of the phasor $\underline{\xi}_{3\text{ph}}$ is given by

$$\left|\underline{\xi}_{3\text{ph}}\right| = \left|\sum_{i=1}^{N_S} \underline{V}_i\right| = 2r\sin(\pi/6) = r$$

where r is the radius of the circle.

If all N_S loops are in a single-slot pair at $\pi/2$ and $3\pi/2$, the generated voltage $\zeta_{3\text{ph}}(t)$ is given by ($\underline{V}_1 = V_{\max}e^{-j\pi/3}$ for the three-phase machine so that $\underline{V}_1 e^{-j\pi/6} = V_{\max}e^{-j\pi/2}$)

$$\zeta_{3\text{ph}}(t) = \text{Re}\left\{\left(N_S\underline{V}_1 e^{-j\pi/6}\right)e^{j\omega_R t}\right\} = \text{Re}\left\{\underline{\zeta}_{3\text{ph}}e^{j\omega_R t}\right\}$$

where the magnitude of this voltage phasor given by

$$\left|\underline{\zeta}_{3\text{ph}}\right| = |N_S\underline{V}_1| = \sum_{i=0}^{N_S}|\underline{V}_i| = \frac{2\pi r}{6}$$

as it is just the circumferential length of the sixth of a circle. The ratio of these generated emfs is

$$\frac{\left|\underline{\xi}_{3\mathrm{ph}}\right|}{\left|\underline{\zeta}_{3\mathrm{ph}}\right|} = \frac{3}{\pi} = 0.955.$$

Four or More Phases

Consider a four-phase machine where the N_S loops of *each* phase are distributed over $\pi/2$ radians, that is, with $\Delta\theta = \pi/(8N_S)$, the top side of the loops of phase 1 are in the slots at

$$\theta_i = \frac{\pi}{2} + (i-1)\Delta\theta \quad \text{for } i = 1, ..., N_S.$$

The ratio of the voltage $\left|\underline{\xi}_{4\mathrm{ph}}\right|$ generated in the distributed windings to the voltage $\left|\underline{\zeta}_{4\mathrm{ph}}\right|$ generated if all the windings are in a single slot is

$$\frac{\left|\underline{\xi}_{4\mathrm{ph}}\right|}{\left|\underline{\zeta}_{4\mathrm{ph}}\right|} \approx \frac{2r\sin(\pi/8)}{\pi r/(4)} = \frac{\sin(\pi/8)}{\pi/8} = 0.975.$$

In general, for a machine with n_{ph} phases, the ratio is given by

$$\frac{\left|\underline{\xi}_{\mathrm{nph}}\right|}{\left|\underline{\zeta}_{\mathrm{nph}}\right|} \approx \frac{2r\sin\left(\dfrac{\pi}{2n_{ph}}\right)}{2r\dfrac{\pi}{2n_{ph}}} = \frac{\sin\left(\dfrac{\pi}{2n_{ph}}\right)}{\dfrac{\pi}{2n_{ph}}} \to 1 \text{ as } n_{ph} \to \infty.$$

Finally, for a machine with n_{ph} phases and N_S large, the sum $\sum_{i=1}^{N_S}\underline{V}_i$ from each phase spans a circumferential length of $(1/2)\,(2\pi r/n_{ph})$, which is approximately $N_S V_{\max}$, that is,

$$\frac{1}{2}\frac{2\pi r}{n_{ph}} \approx N_S V_{\max} \implies 2r \approx \frac{N_S V_{\max}}{\pi/(2n_{ph})}.$$

Comparison

The table below summarizes the results of computing the ratio of the magnitude of the voltage in a distributed winding to that of a concentrated winding when the number of phases is 1, 2, 3, and 4.

n_{ph}	$\left\|\underline{\xi}_{\mathrm{nph}}\right\| / \left\|\underline{\zeta}_{\mathrm{nph}}\right\|$
1	0.636
2	0.900
3	0.955
4	0.975

The quantity N_S denotes the total number of loops per phase so that *total* number of loops in a machine with n_{ph} phases is $N_{\text{total}} = n_{ph}N_S$. To make a comparison of the voltage output of machines with different number of phases, it is assumed that the total number of windings N_{total} is the same in each machine. For example, in a three-phase machine, the N_S loops of *each* phase are distributed over $2\pi/3$ radians making full use of the inside stator surface with $N_{total} = 3N_S$. In a three-phase generator, all the available surface area of the inside stator surface is used and the three independent phase voltages coming out of the machine produce 95.5% of the voltage compared to putting all the windings of each phase in a single pair of slots.

A comparison of the machines is done on the basis of the power that can be produced. That is, all things being equal other than the number of phases, how do the number of phases impact the amount of electric power that can be extracted from the machine? The three-phase, two-phase and single-phase machines are now considered in this context.

Let

$$\begin{aligned}
\xi_{S1}(t) &= \text{Re}\Big\{\underline{\boldsymbol{\xi}}_{3\text{ph}} e^{j\omega_R t}\Big\} \\
\xi_{S2}(t) &= \text{Re}\Big\{\underline{\boldsymbol{\xi}}_{3\text{ph}} e^{-j2\pi/3} e^{j\omega_R t}\Big\} \\
\xi_{S3}(t) &= \text{Re}\Big\{\underline{\boldsymbol{\xi}}_{3\text{ph}} e^{-j4\pi/3} e^{j\omega_R t}\Big\}
\end{aligned}$$

be the generated voltages with $\underline{\boldsymbol{\xi}}_{3\text{ph}} = \sum_{i=1}^{N_S} \underline{V}_i$ the sum of the phasors in phase 1.

In a three-phase machine, the generated output power is

$$p_{3\text{phase}} = i_{S1}\xi_{S1} + i_{S2}\xi_{S2} + i_{S3}\xi_{S3}$$

and, in steady state with $\theta_R = \omega_R t$ and a balanced load, the currents constitute a three-phase balanced set of the form

$$\begin{aligned}
i_{S1} &= I\cos(\omega_R t + \angle\underline{\boldsymbol{\xi}}_{3\text{ph}} - \phi_{3\text{ph}}) \\
i_{S2} &= I\cos(\omega_R t - 2\pi/3 + \angle\underline{\boldsymbol{\xi}}_{3\text{ph}} - \phi_{3\text{ph}}) \\
i_{S3} &= I\cos(\omega_R t - 4\pi/3 + \angle\underline{\boldsymbol{\xi}}_{3\text{ph}} - \phi_{3\text{ph}}).
\end{aligned}$$

Under these conditions, the output power is constant and equal to

$$\begin{aligned}
p_{3\text{phase}} &= i_{S1}\xi_{S1} + i_{S2}\xi_{S2} + i_{S3}\xi_{S3} \\
&= \frac{\sin(\pi/6)}{\pi/6} N_S V_{\max} I \Big(\cos(\omega_R t + \angle\underline{\boldsymbol{\xi}}_{3\text{ph}}) \cos(\omega_R t + \angle\underline{\boldsymbol{\xi}}_{3\text{ph}} - \phi_{3\text{ph}}) \\
&\quad + \cos(\omega_R t + \angle\underline{\boldsymbol{\xi}}_{3\text{ph}} - 2\pi/3) \cos(\omega_R t - 2\pi/3 + \angle\underline{\boldsymbol{\xi}}_{3\text{ph}} - \phi_{3\text{ph}}) \\
&\quad + \cos(\omega_R t + \angle\underline{\boldsymbol{\xi}}_{3\text{ph}} - 4\pi/3) \cos(\omega_R t - 4\pi/3 + \angle\underline{\boldsymbol{\xi}}_{3\text{ph}} - \phi_{3\text{ph}})\Big) \\
&= \frac{3}{2}\frac{\sin(\pi/6)}{\pi/6} N_S V_{\max} I \cos(\phi_{3\text{ph}}) \\
&= (0.955)\,\frac{1}{2} N_{\text{total}} V_{\max} I \cos(\phi_{3\text{ph}})
\end{aligned}$$

where $N_{\text{total}} = 3N_S$ in a three-phase machine.

In a similar fashion, let

$$\begin{aligned}
\xi_{S1}(t) &= \text{Re}\left\{\underline{\boldsymbol{\xi}}_{2\text{ph}} e^{j\omega_R t}\right\} \\
\xi_{S2}(t) &= \text{Re}\left\{\underline{\boldsymbol{\xi}}_{2\text{ph}} e^{-j\pi/2} e^{j\omega_R t}\right\}
\end{aligned}$$

be the generated voltages in the two-phase machine where $\underline{\boldsymbol{\xi}}_{2\text{ph}} = \sum_{i=1}^{N_S} \underline{V}_i$ is the sum of the voltage phasors of phase 1. It is straightforward to show that the power in a two-phase machine is constant and given by ($N_{\text{total}} = 2N_S$ in a two-phase machine)

$$\begin{aligned}
p_{2\text{phase}} &= i_{S1}\xi_{S1} + i_{S2}\xi_{S2} \\
&= N_S \frac{\sin(\pi/4)}{\pi/4} V_{\max} I \Big(\cos(\omega_R t + \angle\underline{\boldsymbol{\xi}}_{2\text{ph}}) \cos(\omega_R t + \angle\underline{\boldsymbol{\xi}}_{2\text{ph}} - \phi_{2\text{ph}}) \\
&\qquad + \cos(\omega_R t + \angle\underline{\boldsymbol{\xi}}_{2\text{ph}} - \frac{\pi}{2}) \cos(\omega_R t + \angle\underline{\boldsymbol{\xi}}_{2\text{ph}} - \frac{\pi}{2} - \phi_{2\text{ph}})\Big) \\
&= \frac{1}{2}\frac{\sin(\pi/4)}{\pi/4} N_{\text{total}} V_{\max} I \cos(\phi_{2\text{ph}}) \\
&= (0.9)\,\frac{1}{2} N_{\text{total}} V_{\max} I \cos(\phi_{2\text{ph}}).
\end{aligned}$$

In a *single-phase* generator, there would be N_{total} windings distributed around the inner stator surface. The power is not constant in this case as

$$\begin{aligned}
\xi_{S1}(t) &= \text{Re}\left\{\underline{\boldsymbol{\xi}}_{1\text{ph}} e^{j\omega_R t}\right\} \\
i_{S1} &= I \cos(\omega_R t + \angle\underline{\boldsymbol{\xi}}_{1\text{ph}} - \phi_{1\text{ph}})
\end{aligned}$$

and

$$\begin{aligned} p_{\text{single-phase}} &= \xi_{S1} i_{S1} \\ &= \left(N_{\text{total}} V_{\max} \cos(\omega_R t + \angle\underline{\boldsymbol{\xi}}_{1\text{ph}})\right) \left(I \cos(\omega_R t + \angle\underline{\boldsymbol{\xi}}_{1\text{ph}} - \phi_{1\text{ph}})\right) \\ &= N_{\text{total}} V_{\max} I \frac{\sin(\pi/2)}{\pi/2} \frac{1}{2} \left(\cos(\phi_{1\text{ph}}) + \cos(2\omega_R t + 2\angle\underline{\boldsymbol{\xi}}_{1\text{ph}} - \phi_{1\text{ph}})\right). \end{aligned}$$

The average power is

$$\begin{aligned} \langle p_{\text{single-phase}} \rangle &\triangleq \frac{1}{2\pi} \int_0^{2\pi} p_{\text{single-phase}}(\omega_R t) d(\omega_R t) \\ &= (0.64) \frac{1}{2} N_{\text{total}} V_{\max} I \cos(\phi_{1\text{ph}}). \end{aligned}$$

For the same-size machine (that is, air gap length, rotor radius, stator radius, axial length, etc.) with the same total number of windings, the same current in the windings and operating at the same power factor (so that $\phi_{1\text{ph}} = \phi_{2\text{ph}} = \phi_{3\text{ph}}$), the single-phase machine can (on average) only generate 0.64/0.955 or 67% of the power of the three-phase system. Similarly, the two-phase machine can only generate 0.9/0.955 (94.2%) of a three-phase system. A three-phase machine can generate 95.5% (0.955) of the power that an ideal machine with an infinite number of phases could produce. Going to a four-phase machine results in a gain of only 2% compared to a three-phase machine (that is, $p_{4\text{phase}} = (0.975/0.955)p_{3\text{phase}} = (1.02)p_{3\text{phase}}$). This analysis shows that a three-phase generator is a good trade-off between complexity (that is, number of phases) and the achievable output power of the machine.

Sinusoidally Wound Stator

The windings in a practical machine can be approximately sinusoidally wound in two layers as illustrated in Figure 7.25 (see also Figures 7.1 and 7.2). As shown previously, the windings are distributed this way so that the radial magnetic field in the air gap produced by the stator currents is (approximately) sinusoidally distributed in space (as is the magnetic field due to the rotor currents). Note that the periphery of the inside surface of the stator has the loops evenly distributed (in two layers) over the complete circumference.

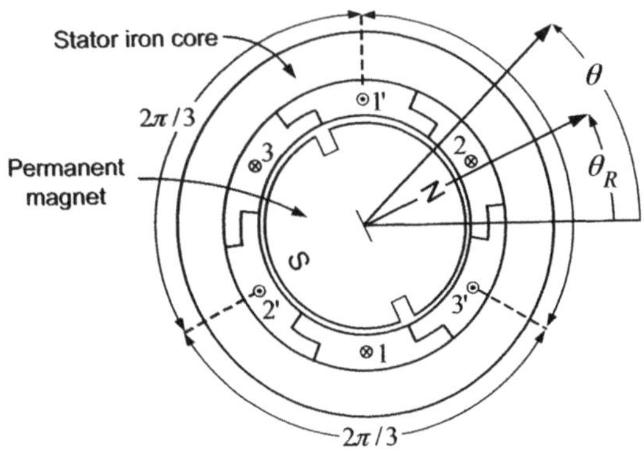

FIGURE 7.25. Three-phase sinusoidally wound PM synchronous motor.

In this case, the voltage produced by phase 1 is (see Figure 7.2)

$$\begin{aligned}\xi_1 &= V_{\max}\,\mathrm{Re}\Big\{e^{j(\omega_R t-\pi/3)} + 2e^{j(\omega_R t-\pi/2)} + e^{j(\omega_R t-2\pi/3)}\Big\}\\ &= V_{\max}\,\mathrm{Re}\Big\{e^{j\omega_R t}\left(e^{-j\pi/3} + 2e^{-j\pi/2} + e^{-j2\pi/3}\right)\Big\}\\ &= V_{\max}(2+\sqrt{3})\cos(\omega_R t - \pi/2)\end{aligned}$$

where $e^{-j\pi/3} + 2e^{-j\pi/2} + e^{-j2\pi/3} = (2+\sqrt{3})e^{-j\pi/2}, \omega_R \triangleq d\theta_R/dt$, and $V_{\max} \triangleq 2\ell_1 r_R B_{\max}\omega_R$.

This can be compared to the voltage produced by putting all four windings in a single slot where the ratio of the magnitude of these two voltages is $(2+\sqrt{3})/4 = 0.933$.

On the other hand, if only one layer was used, the four loops would be evenly spread over $2\pi/3$ radians. The resulting voltage would then be ($\Delta\theta = \frac{\pi/3}{4} = \frac{\pi}{12}, \theta_i = \frac{9\pi}{24} + (i-1)\frac{\pi}{12}$ for $i = 1, 2, 3, 4$.)

$$\begin{aligned}\xi_1 = V_{\max}\,\mathrm{Re}\Big\{&e^{j(\omega_R t-9\pi/24)} + e^{j(\omega_R t-11\pi/24)} + e^{j(\omega_R t-13\pi/24)}\\ &+ e^{j(\omega_R t-15\pi/24)}\Big\}\\ = V_{\max}\,\mathrm{Re}\Big\{&e^{j\omega_R t}\left(e^{-j9\pi/24} + e^{-j11\pi/24} + e^{-j13\pi/24} + e^{-j15\pi/24}\right)\Big\}\\ = V_{\max}&\Big(\sin(9\pi/24) + \sin(11\pi/24) + \sin(13\pi/24) + \sin(15\pi/24)\Big)\\ &\times\cos(\omega_R t - \pi/2)\\ = 3.83&V_{\max}\cos(\omega_R t - \pi/2)\end{aligned}$$

and the ratio would be $3.83/4 = 0.958$. Though this gives a higher voltage output, this configuration requires 24 slots (one side of a loop in each slot) compared to the one of Figure 7.25 which requires only 12 slots (two loops in each slot). The mechanical integrity of the machine usually favors going with 12 slots as there is more stator iron between the slots in this case.

7.4.2 Why AC?

Why is the power system AC and not DC? A three-phase generator supplies constant power by producing three balanced *sinusoidal* voltages to a balanced three-phase load. Historically, the use of AC power was important for efficient distribution of electric power over long distances. To explain, consider a three-phase generator connected to a balanced three-phase power line which is then connected to a balanced three-phase load. Let the output voltages of the generator be of the form

$$\begin{aligned} v_{S1} &= V\cos(\omega t - \pi/2) \\ v_{S2} &= V\cos(\omega t - \pi/2 - 2\pi/3) \\ v_{S3} &= V\cos(\omega t - \pi/2 - 4\pi/3) \end{aligned}$$

and the corresponding currents are of the form

$$\begin{aligned} i_{S1} &= I\cos(\omega t - \pi/2 - \phi) \\ i_{S2} &= I\cos(\omega t - \pi/2 - 2\pi/3 - \phi) \\ i_{S3} &= I\cos(\omega t - \pi/2 - 4\pi/3 - \phi). \end{aligned}$$

The frequency $\omega = 2\pi f$ is constant at 60 Hz (50 Hz in Europe). Here the quantities I and ϕ depend on the values of the inductances and resistances of the generator, the transmission line, and the load as well as the operating frequency. The power delivered by the generator is then

$$p_{\text{generator}} = v_{S1}i_{S1} + v_{S2}i_{S2} + v_{S3}i_{S3} = \frac{3}{2}VI\cos(\phi).$$

Let the power line have length ℓ and let R denote the resistance per unit length in each phase of the power line so that the average electric power lost in the line is then

$$p_{\text{lost}} = \frac{3}{2}\ell RI^2.$$

The power delivered to the load is then

$$p_{\text{load}} = p_{\text{generator}} - p_{\text{lost}} = \frac{3}{2}VI\cos(\phi) - \frac{3}{2}\ell RI^2$$

For a given total resistance ℓR in the line, one keeps the loss down by delivering the power using as little current as possible. For a given power factor $\cos(\phi)$ (fixed by the impedances of the load and power line), the power delivered $\frac{3}{2}VI\cos(\phi)$ depends on the product VI. The generator is designed to put out a large voltage V (accomplished by putting many turns in each stator slot and having a strong rotor magnetic field[10]) so that the current is (relatively) small and yet the generator still delivers the requisite power needed by the load.[11] However, though the power is efficiently delivered to the load, the voltage is too large to be used by the consumer (industrial or home). As this power is AC, it can be stepped down by a transformer to a level that can be used by the consumer for lighting, running motors, and so on.[12] The induction motor was (and still is) so important because it could be started and run open-loop off of a three-phase AC power supply.[13] (The synchronous motor would not work because it cannot be started by direct connection to a three-phase 60-Hz AC source and can stall out of control quite easily under changing loads when run in open loop.) For these reasons, AC power won out over DC power many years ago. An interesting article on this history is Ref. [62].

7.4.3 Why Sinusoidal Voltages?

The question still remains as to why sinusoids for the voltages. The voltage out of the generator will be periodic as long as the generator is run at constant speed. Why is the rotor iron designed to produce a sinusoidally distributed magnetic field in the air gap resulting in sinusoidal voltages induced in the sinusoidally distributed stator windings? Again, the answer to this question is provided in Laithwaite and Freris [61]. It comes down to the fact that sinusoids are the only signals that can be passed through a stable linear time-invariant system and still be sinusoids. For example, consider the transmission of the power through a transformer. Let the three-phase output voltages of the generator be of the form

$$\begin{aligned}
v_{S1} &= V\cos(\omega t)\\
v_{S2} &= V\cos(\omega t - 2\pi/3)\\
v_{S3} &= V\cos(\omega t - 4\pi/3).
\end{aligned}$$

---

[10] $V \propto N_{\text{total}} V_{\max} = N_{\text{total}} 2\ell_1 r_R B_{\max}\omega_R$

[11] As pointed out in Ref. [61], it is still a tradeoff in the sense that the higher the voltage adopted, the higher costs in the insulation for the power lines, the higher the transmission towers, the wider the right-of-way for the transmission lines, and so on. This cost must be traded off with the cost due to energy losses in the power line.

[12] Of course, transformers don't work with DC voltages and it was only in the latter part of the 20th century that power elecronics has provided the means to step down DC voltages.

[13] The argument for the DC generation was that the DC motor could be employed by the end user. However, as the voltage could not be large in order to run the DC motor, the transmitted voltages from the DC generator must be (relatively) small. The losses in the power lines delivering this DC power (low voltage and high current) over large distances are so great that a DC power grid would have required building a power station every 2 kilometers or so!

Let these voltages be input to a transformer in the power system so that with N_2/N_1 the transformer ratio, the output voltages are then

$$\begin{aligned} v'_{S1} &= \frac{N_2}{N_1} v_{S1} \\ v'_{S2} &= \frac{N_2}{N_1} v_{S2} \\ v'_{S2} &= \frac{N_2}{N_1} v_{S2} \end{aligned}$$

for an *ideal* transformer. The corresponding transfer function of the ideal transformer is then [see Figure 7.26(a)]

$$G_{\text{ideal}}(j\omega) \triangleq \frac{V'_{S1}(j\omega)}{V_{S1}(j\omega)} = \frac{N_2}{N_1}.$$

However, real transformers are not ideal. Figure 7.26(b) shows an equivalent circuit model for a nonideal transformer connected to a resistive load.

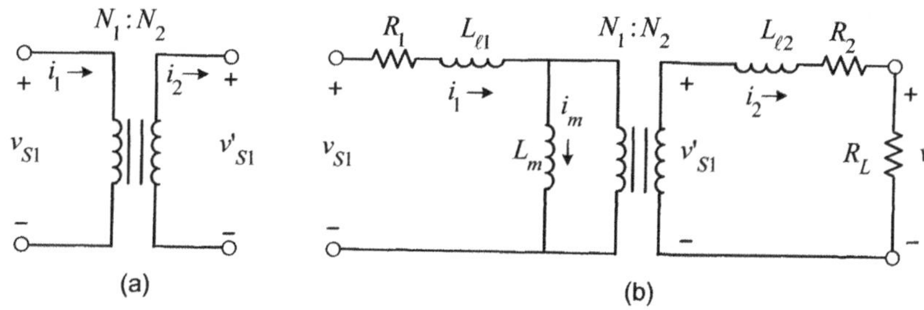

FIGURE 7.26. (a) Ideal transformer. (b) Nonideal transformer whose primary is connected to one phase of a three-phase source and whose secondary is connected to a load.

With

$$\begin{aligned} a_0 &= R_1 R_2 + R_1 R_L \\ a_1 &= L_{\ell 2} R_1 + L_m R_1 n^2 + L_{\ell 1} R_2 + L_m R_2 + L_{\ell 1} R_{\text{load}} + L_m R_L \\ a_2 &= L_{\ell 1} L_{\ell 2} + L_{\ell 2} L_m + L_{\ell 1} L_m n^2 \\ b_1 &= n R_L L_m \\ n &= N_2/N_1 \end{aligned}$$

the transfer function of the nonideal transformer is

$$G(s) \triangleq \frac{V_{L1}(s)}{V_{S1}(s)} = \frac{b_1 s}{a_2 s^2 + a_1 s + a_0}.$$

The output voltages are then

$$\begin{aligned}
v_{L1}(\omega t) &= |G(j\omega)|\, V \cos\Big(\omega t + \angle G(j\omega)\Big) \\
v_{L2}(\omega t) &= |G(j\omega)|\, V \cos\Big(\omega t - 2\pi/3 + \angle G(j\omega)\Big) \\
v_{L2}(\omega t) &= |G(j\omega)|\, V \cos\Big(\omega t - 4\pi/3 + \angle G(j\omega)\Big)
\end{aligned}$$

which is still a balanced three-phase set of sinusoidal voltages! That is, they are sinusoidal voltages of the same form as the input only shifted in phase by $\angle G(j\omega)$ and scaled in amplitude by the factor $|G(j\omega)|$. In particular, a three-phase induction motor is designed to run efficiently off any three-phase set of *sinusoidal* voltages.

Suppose the generator produced a periodic waveform $w_{S1}(\omega t)$ that is not sinusoidal. The waveform $w_{S1}(\omega t)$ can be expanded in a Fourier series as

$$w_{S1}(\omega t) = \sum_{n=1}^{\infty} A_n \cos(n\omega t + \Phi_i)$$

where $A_1 \neq 0$ and $A_n \neq 0$ for some $n \geq 2$ (in the sinusoidal case, $A_n = 0$ for *all* $n \geq 2$). When passed through the (nonideal) transformer, the output voltage will then be

$$w'_{S1}(\omega t) = \sum_{n=1}^{\infty} A_n \,|G(jn\omega)| \cos\Big(n\omega t + \Phi_i + \angle G(jn\omega)\Big).$$

The output waveform is distorted when compared to the input waveform as the different harmonic components ($n = 1, 2, ...$) of the waveform are shifted in phase by different amounts $\angle G(jn\omega)$ and are scaled in amplitude by different amounts $|G(jn\omega)|$. As a result, the output waveform does not have the same shape as the input voltage.

Of course, in addition to transformers, the generated power goes through transmission lines that are effectively modeled by RLC circuits and are also stable linear time-invariant systems. As long as the output voltage of the generator is a pure sinusoid, it may pass through any number of such systems and remain a pure sinusoid.

By designing the generator to produce a balanced set of three-phase *sinusoidal* voltages and designing the motors to run off of a balanced set of three-phase *sinusoidal* voltages, the transmission line and the transformer between the two will still output a (stepped-down) balanced three-phase set of *sinusoidal* voltages to provide the right match between the two.

7.4.4 Why 60 Hz?

The use of 60 Hz for the frequency of a power system is trade-off just as all the other choices in a power system [61]. A 10-Hz source would provide too much flicker in electric lights as perceived by the human eye to be of use. However, 50 Hz or 60 Hz is of high enough frequency so that no flicker is perceived. As far as an upper limit in frequency, note that the angle ϕ between the voltage and current increases as the operating frequency increases so that the power factor $\cos(\phi)$ goes down at higher frequency. Further, the upper limit in frequency is also limited by losses and resonance problems. For example, in the iron of electric machines and transformers connected to a power system, the hysteresis losses go up in proportion to the frequency ($p_{\text{hysteresis\_loss}}/\text{volume} = \frac{\omega}{2\pi}\oint_{cycle} HdB$) while the eddy current losses go up in proportion to the *square* of the frequency. Further, the distributed inductances and capacitances in the power line result in a transfer function for the power network that can have resonances at high frequencies, and thus the operating frequency of the power system must be well below any such resonant frequencies. Airplanes typically have a 400-Hz power system, but as the power system is spatially contained, such resonances can be avoided by design. In the case of aircraft, using a power system with higher frequency reduces the amount of iron, and therefore weight, required in transformers, which is obviously important.

Problems

Calculations

Problem 1 Stator Flux Linkages

Carry out the integrations to compute the expressions for the stator flux linkages given in equations (7.5), (7.6), and (7.7).

Problem 2 Rotor Flux Linkages

Carry out the integrations to compute the expressions for the stator flux linkages given in equations (7.11), (7.12), and (7.13).

Problem 3 Balanced Conditions

Show that

$$\begin{aligned}
1 + \cos(2\pi/3) + \cos(4\pi/3) &= 0 \\
\sin(2\pi/3) + \sin(4\pi/3) &= 0 \\
1 + e^{j2\pi/3} + e^{j4\pi/3} &= 0 \\
\cos(\theta_R) + \cos(\theta_R - 2\pi/3) + \cos(\theta_R - 4\pi/3) &\equiv 0 \\
\cos(\theta_R) + \cos(\theta_R + 2\pi/3) + \cos(\theta_R + 4\pi/3) &\equiv 0
\end{aligned}$$

Problem 4 Balanced Conditions

Using the identities from problem 3, and equations (7.10) and (7.15),

show that

$$\begin{aligned} \psi_{S1}(t)+\psi_{S2}(t)+\psi_{S3}(t) &\equiv 0 \\ \psi_{R1}(t)+\psi_{R2}(t)+\psi_{R3}(t) &\equiv 0. \end{aligned}$$

Problem 5 Three-Phase to Two-Phase Transformation

Verify equation (7.20).

Problem 6 Three-Phase to Two-Phase Transformation

Verify equation (7.23).

Problem 7 Induction Motor with Unbalanced Voltages

The system of equations (7.25) are valid even if the voltages are not balanced. Explain why the two-phase model (7.31) is no longer an equivalent *model, that is, knowledge of the two-phase variables $i_{Sa}, i_{Sb}, i_{Ra}, i_{Rb}, u_{Sa}, u_{Sb}$, ω, and θ is not enough to determine the three-phase variables $i_{S1}, i_{S2}, i_{S3}, i_{R1}$, $i_{R2}, i_{R3}, u_{S1}, u_{S2}, u_{S3}, \omega$, and θ. What if $u_{S0} = \sqrt{3}v_{\hat{N}N} = (u_{S1}+u_{S2}+u_{S3})/\sqrt{3}$ is known?*

Do the three-phase variables uniquely determine the two-phase variables if the voltages are unbalanced?

Problem 8 Induction Motor Torque

Verify the calculations used to derive the expressions for the phase torques given in equations (7.26), (7.27), and (7.28).

Problem 9 Three-Phase to Two-Phase Transformation

Show that with the three-phase to two-phase transformation given by (7.17), it follows that

$$u_{Sa}(t)i_{Sa}(t)+u_{Sb}(t)i_{Sb}(t) = u_{S1}(t)i_{S1}(t)+u_{S2}(t)i_{S2}(t)+u_{S3}(t)i_{S3}(t).$$

Problem 10 Three-Phase to Two-Phase Transformation

Do the calculations to verify that (7.30) is the correct expression for the two-phase equivalent torque.

Space Vector Induction Motor Models

Problem 11 Space Vector Form of the Three-Phase to Two-Phase Transformation

Defining $\underline{u}_S \triangleq u_{Sa} + ju_{Sb}$, show that the three-phase to two-phase transformation

$$\begin{bmatrix} u_{Sa}(t) \\ u_{Sb}(t) \end{bmatrix} \triangleq \sqrt{\frac{2}{3}} \begin{bmatrix} 1 & -1/2 & -1/2 \\ 0 & \sqrt{3}/2 & -\sqrt{3}/2 \end{bmatrix} \begin{bmatrix} u_{S1}(t) \\ u_{S2}(t) \\ u_{S3}(t) \end{bmatrix}$$

may be written as

$$\underline{u}_S = u_{Sa} + ju_{Sb} = \sqrt{\frac{2}{3}}\left(u_{S1} + u_{S2}e^{j2\pi/3} + u_{S3}e^{j4\pi/3}\right)$$

and similarly for $\underline{i}_S, \underline{i}_R$, *and so on.*

Problem 12 Alternative Definition of the Three-Phase to Two-Phase Transformation [1]

(a) Show that if the three-phase to two-phase transformation is defined by

$$\begin{aligned} \underline{u}_S &= u_{S1} + u_{S2}e^{j2\pi/3} + u_{S3}e^{j4\pi/3} \\ \underline{i}_S &= i_{S1} + i_{S2}e^{j2\pi/3} + i_{S3}e^{j4\pi/3} \\ \vdots &= \vdots \end{aligned}$$

that is, without the factor $\sqrt{2/3}$, *then the equations of the induction motor are given by*

$$\begin{aligned} R_S\underline{i}_S + L_S\frac{d}{dt}\underline{i}_S + M\frac{d}{dt}\left(\underline{i}_R e^{jn_p\theta_R}\right) &= \underline{u}_S \\ R_R\underline{i}_R + L_R\frac{d}{dt}\underline{i}_R + M\frac{d}{dt}\left(\underline{i}_S e^{-jn_p\theta_R}\right) &= 0 \\ \frac{2}{3}n_pM\,\mathrm{Im}\left\{\underline{i}_S\left(\underline{i}_R e^{jn_p\theta_R}\right)^*\right\} - \tau_L &= J\frac{d\omega_R}{dt}. \end{aligned}$$

(b) With the three-phase to two-phase transformation defined as in part (a), show that

$$u_{Sa}(t)i_{Sa}(t) + u_{Sb}(t)i_{Sb}(t) = \frac{3}{2}\Big(u_{S1}(t)i_{S1}(t) + u_{S2}(t)i_{S2}(t) + u_{S3}(t)i_{S3}(t)\Big).$$

Problem 13 Space Vector Representation of the Induction Motor

(a) Define

$$\underline{i}_r = \underline{i}_{ra} + j\underline{i}_{rb} \triangleq \underline{i}_R e^{jn_p\theta_R}.$$

Show that the system of equations

$$\begin{aligned} R_S\underline{i}_S + L_S\frac{d}{dt}\underline{i}_S + M\frac{d}{dt}\left(\underline{i}_R e^{jn_p\theta_R}\right) &= \underline{u}_S \\ R_R\underline{i}_R + L_R\frac{d}{dt}\underline{i}_R + M\frac{d}{dt}\left(\underline{i}_S e^{-jn_p\theta_R}\right) &= 0 \\ n_pM\,\mathrm{Im}\left\{\underline{i}_S\left(\underline{i}_R e^{jn_p\theta_R}\right)^*\right\} - \tau_L &= J\frac{d\omega_R}{dt} \end{aligned}$$

may be rewritten as

$$\begin{aligned} R_S\underline{i}_S + L_S\frac{d}{dt}\underline{i}_S + M\frac{d}{dt}\underline{i}_r &= \underline{u}_S \\ R_R\underline{i}_r + \sigma_R M\frac{d}{dt}\underline{i}_r + M\frac{d}{dt}\left(\underline{i}_S + \underline{i}_r\right) - jn_p\omega_R\underline{\psi}_R &= 0 \qquad (7.91) \\ n_pM\,\mathrm{Im}\left\{\underline{i}_S\left(\underline{i}_r\right)^*\right\} - \tau_L &= J\frac{d\omega_R}{dt} \end{aligned}$$

in terms of $\underline{i}_S$ *and* $\underline{i}_r$, *where*

$$\underline{\psi}_R \triangleq L_R \underline{i}_r + M \underline{i}_S.$$

[Note that the model (7.91) has no explicit dependence on θ_R*.]*

(b) With $L_S = (1+\sigma_S)M, L_R = (1+\sigma_R)M$, *and*

$$\begin{aligned}\underline{\lambda}_R &= L_R \underline{i}_R + M \underline{i}_S e^{-jn_p\theta_R} \\ \underline{\psi}_R &= \underline{\lambda}_R e^{jn_p\theta_R} = L_R \underline{i}_r + M \underline{i}_S\end{aligned}$$

show that the electrical equations of part (a) may be illustrated by the "equivalent circuit" of Figure 7.27.

(c) Put the result from part (a) in state space form.

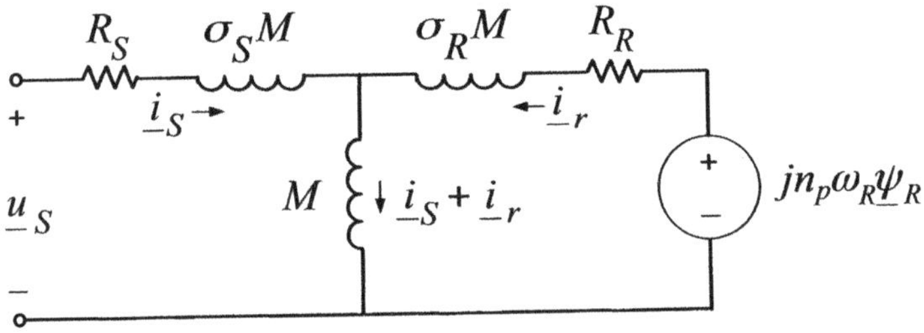

FIGURE 7.27. Circuit representation of the electrical equations of an induction motor.

Remark 1 *The quantity* $\underline{\lambda}_{ag} \triangleq M(\underline{i}_S + \underline{i}_r)$ *is called the* air gap (or magnetizing) flux linkage *(see Lipo and Novotny [7], page 75). The* air-gap voltage *is then defined as*

$$\underline{v}_{ag} = -\frac{d}{dt}\underline{\lambda}_{ag} = -\frac{d}{dt}M(\underline{i}_S + \underline{i}_r).$$

Remark 2 *The notation* $\underline{u}_S^S = u_{Sd} + ju_{Sq} \triangleq u_{Sa} + ju_{Sb}, \underline{i}_S^S = i_{Sd} + ji_{Sq} \triangleq i_{Sa} + ji_{Sb}$, *and* $\underline{i}_R^S = i_{Rd} + ji_{Rq} \triangleq \underline{i}_{ra} + j\underline{i}_{rb} = \underline{i}_r$ *is often used for the equivalent two-phase model developed in this problem. The superscript "S" refers to the "stationary frame" while the d and q subscripts refer to the direct and quadrature axis, respectively. However, in this book, the dq notation is reserved for the* field oriented *coordinate system (see Chapters 8 and 9) so the reader must be wary when reading different authors.*

Problem 14 Arbitrary Rotating Reference Frame

(a) Let

$$\begin{aligned}\underline{i}_{S\varphi} &\triangleq \underline{i}_S e^{-j\varphi} \\ \underline{i}_{R\varphi} &\triangleq i_{R\varphi a} + ji_{R\varphi b} \triangleq \underline{i}_R e^{-j(\varphi - n_p\theta_R)} \\ \underline{u}_{S\varphi} &\triangleq \underline{u}_S e^{-j\varphi}\end{aligned}$$

where $\varphi(t)$ is an arbitrary *function of time. With $\omega_\varphi \triangleq d\varphi/dt$, show that a space vector representation of the induction motor model*

$$\begin{aligned}
R_S\underline{i}_S + L_S\frac{d}{dt}\underline{i}_S + M\frac{d}{dt}\left(\underline{i}_R e^{jn_p\theta_R}\right) &= \underline{u}_S \\
R_R\underline{i}_R + L_R\frac{d}{dt}\underline{i}_R + M\frac{d}{dt}\left(\underline{i}_S e^{-jn_p\theta_R}\right) &= 0 \\
n_p M \,\mathrm{Im}\left\{\underline{i}_S\left(\underline{i}_R e^{jn_p\theta_R}\right)^*\right\} - \tau_L &= J\frac{d\omega_R}{dt}
\end{aligned}$$

in terms of $\underline{i}_{S\varphi}$, $\underline{i}_{R\varphi}$ and $\underline{u}_{S\varphi}$ is given by

$$\begin{aligned}
\underline{u}_{S\varphi} &= R_S\underline{i}_{S\varphi} + L_S\frac{d}{dt}\underline{i}_{S\varphi} + j\omega_\varphi L_S\underline{i}_{S\varphi} + M\frac{d}{dt}\underline{i}_{R\varphi} + j\omega_\varphi M\underline{i}_{R\varphi} \\
0 &= R_R\underline{i}_{R\varphi} + L_R\frac{d}{dt}\underline{i}_{R\varphi} + j\left(\omega_\varphi - n_p\omega_R\right)L_R\underline{i}_{R\varphi} + M\frac{d}{dt}\underline{i}_{S\varphi} \\
&\quad + j\left(\omega_\varphi - n_p\omega_R\right)M\underline{i}_{S\varphi} \\
J\frac{d\omega_R}{dt} &= n_p M\,\mathrm{Im}\left\{\underline{i}_{S\varphi}\left(\underline{i}_{R\varphi}\right)^*\right\} - \tau_L.
\end{aligned}$$

Common choices for φ are $\varphi = 0$ (the stationary coordinate system; see problem 13), $\varphi = n_p\omega_R$ (the rotor coordinate system), $\varphi = \angle\underline{u}_S$ (the synchronous coordinate system), and $\varphi = \angle\left(\underline{i}_S + (L_R/M)\underline{i}_R e^{jn_p\theta_R}\right) = \angle\underline{\psi}_R$ (the field-oriented coordinate system; see problem 6 of Chapter 6 *and problem 4 of Chapter 8).*

(b) Put the result from part (a) in state-space form.

Problem 15 Space Vector Model with Leakage on the Rotor Side

(a) Starting from the model (7.91) derived in problem 13, make the change of variables

$$\begin{aligned}
\underline{i}_r' &\triangleq \frac{M}{L_S}\underline{i}_r \\
\underline{\psi}_R' &\triangleq \frac{L_S}{M}\underline{\psi}_R = (L_S/M)\left(L_R\underline{i}_r + M\underline{i}_S\right) = L_S\underline{i}_S + (L_S/M)^2 L_R\underline{i}_r' \\
&= L_S\left(\underline{i}_S + \underline{i}_r'\right) + \frac{\sigma L_S}{1-\sigma}\underline{i}_r' \\
\underline{\lambda}_S &\triangleq L_S\underline{i}_S + M\underline{i}_r = L_S(\underline{i}_S + \underline{i}_r').
\end{aligned}$$

and show that the model may be rewritten as

$$\begin{aligned}
R_S\underline{i}_S + \frac{d}{dt}\underline{\lambda}_S &= \underline{u}_S \\
R_R'\underline{i}_r' + \frac{d}{dt}\underline{\lambda}_S - jn_p\omega\underline{\psi}_R' + L_\sigma\frac{d}{dt}\underline{i}_r' &= 0 \\
\frac{n_p M^2}{L_S L_R}\mathrm{Im}\left\{\underline{i}_S(\underline{\psi}_R')^*\right\} = \frac{n_p}{L_\sigma}\mathrm{Im}\left\{\underline{\lambda}_S(\underline{\psi}_R')^*\right\} &= \tau
\end{aligned}$$

where

$$\begin{aligned}\sigma &\triangleq 1 - M^2/L_S L_R\\ R'_R &\triangleq (L_S/M)^2 R_R\\ L_\sigma &\triangleq \sigma L_S/(1-\sigma).\end{aligned}$$

(b) Show that an equivalent circuit representation of this model is

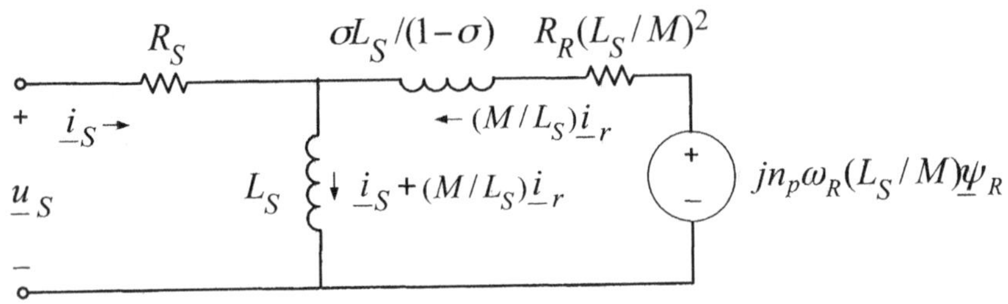

FIGURE 7.28.

where it is noted that there is leakage inductance only on the rotor side.

Problem 16 Space Vector Model with Leakage on the Stator Side
(a) Starting from the model (7.91) derived in problem 13 make the substitutions

$$\begin{aligned}\underline{i}''_r &\triangleq \frac{L_R}{M}\underline{i}_r\\ \underline{\psi}''_R &\triangleq \frac{M}{L_R}\underline{\psi}_R = (M/L_R)(L_R\underline{i}_r + M\underline{i}_S) = (M^2/L_R)\,\underline{i}_S + (M^2/L_R)\,L_R\underline{i}''_r\\ &= (M^2/L_R)(\underline{i}_S + \underline{i}''_r)\\ \underline{\lambda}_S &\triangleq L_S\underline{i}_S + M\underline{i}_r = L_S\underline{i}_S + (M^2/L_R)\,\underline{i}''_r = \sigma L_S\underline{i}_S + (M^2/L_R)(\underline{i}_S + \underline{i}''_r)\end{aligned}$$

and show that the model (7.91) may be rewritten as

$$\begin{aligned}R_S\underline{i}_S + \sigma L_S\frac{d}{dt}\underline{i}_S + \frac{d}{dt}(M^2/L_R)(\underline{i}_S + \underline{i}''_r) &= \underline{u}_S\\ (M/L_R)^2 R_R\underline{i}''_r + \frac{d}{dt}(M^2/L_R)(\underline{i}_S + \underline{i}''_r) - jn_p\omega\,(M/L_R)\,\underline{\psi}_R &= 0\\ \frac{n_p}{\sigma L_S}\,\mathrm{Im}\Big\{\underline{\lambda}_S(\underline{\psi}''_R)^*\Big\} &= \tau.\end{aligned}$$

(b) Show that an equivalent circuit representation of this model is

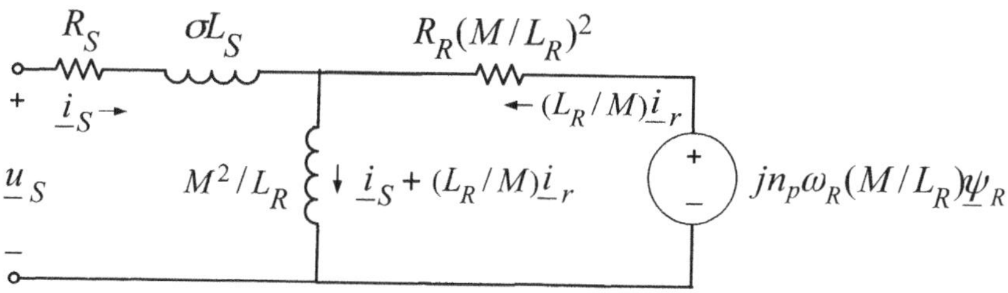

FIGURE 7.29.

where it is noted that there is leakage inductance only on the stator side.

Steady-State Induction Motor Models

Problem 17 Steady-State Solutions of the Induction Motor Equations

(a) Write a program to plot the circle diagram of Figure 7.11.

(b) Write a program to reproduce the plots of Figures 7.12, 7.13, and 7.15.

For the motor parameter values, use $R_R = 3.9\ \Omega$, $R_S = 1.7\ \Omega$, $L_R = 0.014\ H$, $L_S = 0.014\ H$, $M = 0.0117\ H$, $J = 0.00011\ kg\text{-}m^2$, $f = 0\ N\text{-}m/(rad/s)$, $n_p = 3$, $\omega_S = (2\pi)60\ rad/sec$, *and* $U_S = 80\ V$. *However, the analysis assumes the stator resistance is zero so set* $R_S = 0$.

Problem 18 Open-Loop Simulation of a Three-Phase Induction Machine

Let the parameter values of an induction motor be $R_R = 2.1\ \Omega$, $R_S = 1.85\ \Omega$, $L_R = 0.244\ H$, $L_S = 0.244\ H$, $M = 0.225\ H$, $J = 0.0185\ kg\text{-}m^2$, $f = 0\ N\text{-}m/(rad/s)$, $n_p = 3$, $\omega_S = (2\pi)60\ rad/sec$. *With* $V_{max} = 200$ *(peak),* $I_{max} = 10$ *(peak) simulate the open-loop machine based on the two-phase equivalent model as indicated in Figure 7.19 with a 208-V (rms) line-to-line voltage applied. (That is, the phase to neutral peak voltage is* $V_{3ph} = \sqrt{2}\,(208)\,/\sqrt{3} = 169.8\ V$.*)*

Problem 19 No-Load and Blocked-Rotor Tests

A standard method of determining the (two-phase equivalent) machine parameters is to use a couple of special tests called the no-load *test and the* blocked-rotor *test. Both of these tests involve applying a balanced three-phase set of sinusoidal voltages to the motor and measuring the resulting steady-state stator currents and stator power. In the no-load test, nothing is attached to the machine and so, after the voltages are applied, the machine quickly goes to steady state in which* $n_p\omega_R = \omega_S$ *(assuming no frictional losses). In this case,* $S = 0$ *and the equivalent circuit of Figure 7.10 reduces to that of Figure 7.30.*

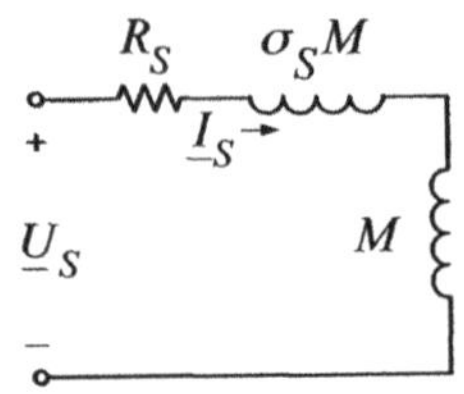

FIGURE 7.30. Equivalent circuit with no load so that $n_p\omega_R = \omega_S \Rightarrow S = 0$.

The blocked-rotor test refers to having the rotor held immobile so that $\omega_R \equiv 0$. *In this case,* $S = 1$ *and the equivalent circuit of Figure 7.10 reduces to that of Figure 7.31(a). Usually, the impedances in the two branches satisfy*

$$j\omega_S M >> j\omega_S \sigma_R M + R_R$$

so that the equivalent circuit reduces to that shown in Figure 7.31(b).

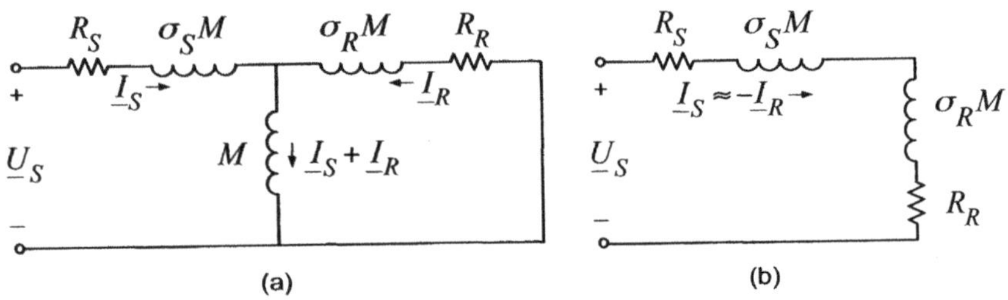

FIGURE 7.31. (a) Equivalent circuit with a locked rotor so that $\omega_R = 0 \Rightarrow S = 1$. (b) Equivalent circuit if $j\omega_S M >> j\omega_S \sigma_R M + R_R$.

(a) Run the simulation of problem 18 with no load. Assume that R_S *is known (perhaps measured using an ohmmeter). Use the steady-state stator voltages and currents to compute* $L_S = (1 + \sigma_S)M$. *(The amplitudes of the voltages and the currents along with the angle between them is needed.) How does this value of* L_S *compare with the actual value in the simulation? That is, what is the percent error?*

(b) Run the simulation of problem 18 with a blocked rotor (if the currents exceed I_{max}, *then reduce the input voltage) and assume that* $\sigma_S = \sigma_R$. *Using the known value of* R_S, *the value of* L_S *from part (a), and the measured stator voltages and currents, compute* $\sigma_S M$ *and* R_R *based on the equivalent circuit of Figure 7.31(b).*

(c) Use the information from parts (a) and (b) to compute M, L_R, *and* $\sigma = 1 - M^2/(L_S L_R)$. *How do they compare with the actual values in the simulation? That is, what is the percent error?*

(d) Repeat parts (a), (b), and (c) with the viscous friction coefficient set to $f = 0.00005$ *rather than* $f = 0$.

Problem 20 Induction Generator (M. Bodson [63])

Consider the equivalent circuit of a three-phase induction machine in Figure 7.32.

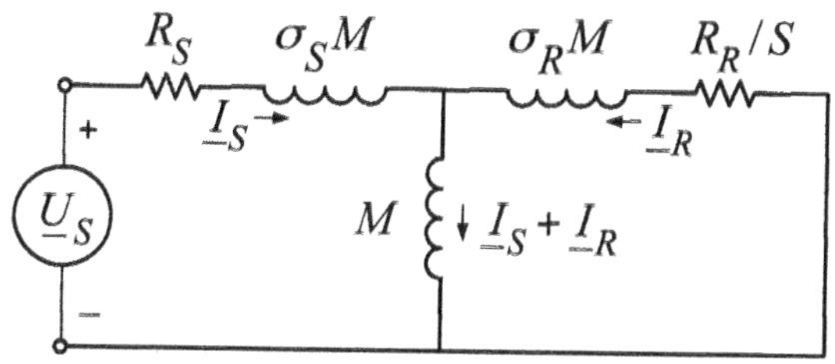

FIGURE 7.32. Equivalent circuit of an induction motor.

(a) Suppose the motor shaft (rotor) is connected to a prime mover (e.g., a windmill) that turns at a constant angular speed $\omega_R > \omega_S$, where ω_S is the angular frequency of the balanced three-phase voltage set applied to the stator terminals (e.g., connect the stator terminals to a three-phase transmission line). Show that the power to the motor is negative; that is, electrical power is being delivered to the source voltage converted from the mechanical power supplied to shaft. (Hint: note that $S < 0$.)

(b) Replace the three-phase power source by a three-phase impedance $Z_S = R + j\omega L$ as indicated schematically in Figure 7.33. The corresponding equivalent circuit is shown in Figure 7.34. Suppose the motor shaft is still connected to a prime mover (e.g., a windmill) that turns at a constant angular speed ω_R. Can there be a steady-state electrical power out of the motor into the impedance Z_S? Explain.

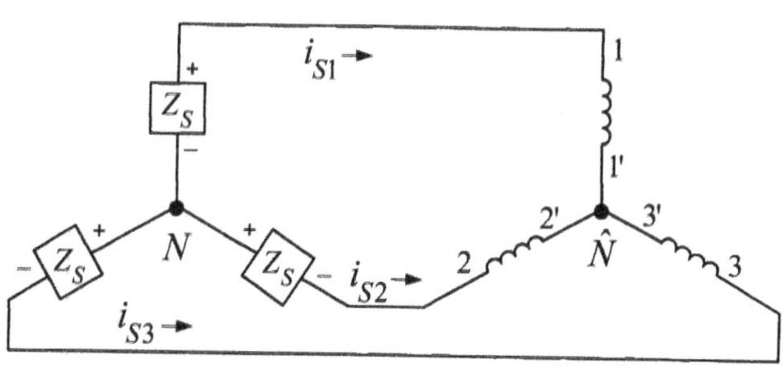

FIGURE 7.33. Stator windings of an induction motor connected to a symmetric three-phase load.

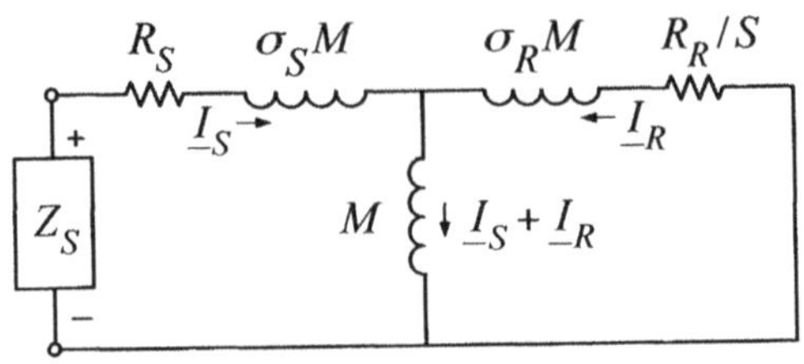

FIGURE 7.34. An equivalent circuit of the induction motor where the source voltage is replaced by a passive load.

Problem 21 Equivalent Circuit With Leakage on the Stator Side [2]
Show that the circuit in Figure 7.35 where the leakage inductance is completely on the stator *side is equivalent to the circuit given in Figure 7.10.*

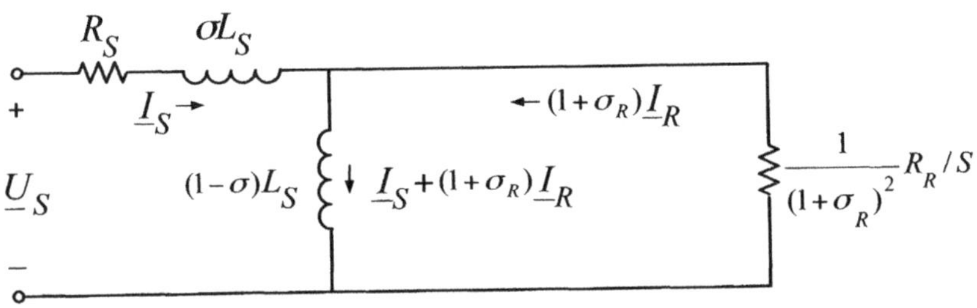

FIGURE 7.35. Equivalent circuit of induction motor with the leakage inductance completely on the stator side.

Problem 22 Equivalent Circuit With Leakage on the Rotor Side [2]
Show that the circuit in Figure 7.36 where the leakage inductance is completely on the rotor *side is equivalent to the circuit given in Figure 7.10.*

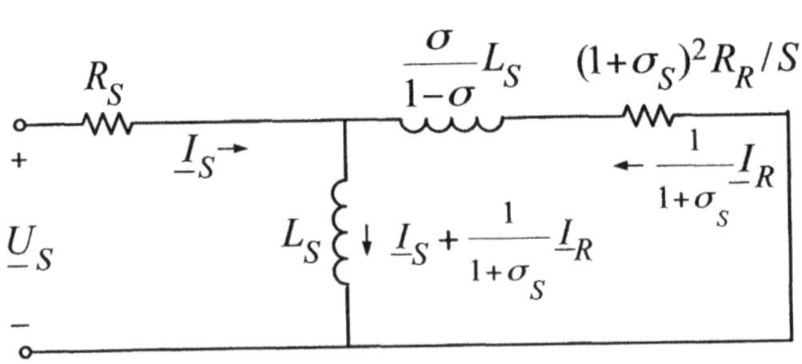

FIGURE 7.36. Equivalent circuit of induction motor with the leakage inductance completely on the rotor side.

Problem 23 Constant Speed Induction Machine as a Linear System

(a) Let $\left\{\underline{u}_S^{(1)}, \underline{i}_S^{(1)}, \underline{i}_R^{(1)}\right\}$ *and* $\left\{\underline{u}_S^{(2)}, \underline{i}_S^{(2)}, \underline{i}_R^{(2)}\right\}$ *each be a solution set to*

$$\begin{aligned} R_S\underline{i}_S + L_S\frac{d}{dt}\underline{i}_S + M\frac{d}{dt}\left(\underline{i}_R e^{j\theta_R(t)}\right) &= \underline{u}_S \\ R_R\underline{i}_R + L_R\frac{d}{dt}\underline{i}_R + M\frac{d}{dt}\left(\underline{i}_S e^{-j\theta_R(t)}\right) &= 0 \end{aligned}$$

where $\theta_R(t)$ *is the* same *function of time for both inputs [e.g.,* $\theta_R(t) = \omega_R t$ *with* ω_R *the* same *constant value for both inputs]. Under these conditions, show that the input*

$$\underline{u}_S = \underline{u}_S^{(1)} + \underline{u}_S^{(2)}$$

results in the stator and rotor currents being

$$\begin{aligned} \underline{i}_S &= \underline{i}_S^{(1)} + \underline{i}_S^{(2)} \\ \underline{i}_R &= \underline{i}_R^{(1)} + \underline{i}_R^{(2)}. \end{aligned}$$

(b) Assuming ω_R *is constant and that the inputs are sinusoids so that*

$$\underline{u}_S^{(1)} = \sqrt{2}\underline{U}_S^{(1)} e^{j\omega_{S1}t} = \sqrt{2}\left|\underline{U}_S^{(1)}\right| e^{j\angle\underline{U}_S^{(1)}} e^{j\omega_{S1}t}$$

then the phasor equations (7.39) and (7.40) can be solved for $\underline{I}_S^{(1)}$ *and* $\underline{I}_R^{(1)}$*, resulting in currents of the form*

$$\begin{aligned} \underline{i}_S^{(1)} &= \sqrt{2}\underline{I}_S^{(1)} e^{j\omega_{S1}t} = \sqrt{2}\left|\underline{I}_S^{(1)}\right| e^{j\angle\underline{I}_S^{(1)}} e^{j\omega_{S1}t} \\ \underline{i}_R^{(1)} &= \sqrt{2}\underline{I}_R^{(1)} e^{j(\omega_{S1}-\omega_R)t} = \sqrt{2}\left|\underline{I}_R^{(1)}\right| e^{j\angle\underline{I}_R^{(1)}} e^{j(\omega_{S1}-\omega_R)t}. \end{aligned}$$

Similarly, the input

$$\underline{u}_S^{(2)} = \sqrt{2}\underline{U}_S^{(2)} e^{j\omega_{S2}t} = \sqrt{2}\left|\underline{U}_S^{(2)}\right| e^{j\angle\underline{U}_S^{(2)}} e^{j\omega_{S2}t}$$

results in the stator and rotor currents being given by

$$\begin{aligned} \underline{i}_S^{(2)} &= \sqrt{2}\underline{I}_S^{(2)} e^{j\omega_{S2}t} = \sqrt{2}\left|\underline{I}_S^{(2)}\right| e^{j\angle\underline{I}_S^{(2)}} e^{j\omega_{S2}t} \\ \underline{i}_R^{(2)} &= \sqrt{2}\underline{I}_R^{(2)} e^{j(\omega_{S2}-\omega_R)t} = \sqrt{2}\left|\underline{I}_R^{(2)}\right| e^{j\angle\underline{I}_R^{(2)}} e^{j(\omega_{S2}-\omega_R)t}. \end{aligned}$$

Use the results of part (a) to compute the torque produced by the motor due to the input

$$\underline{u}_S = \underline{u}_S^{(1)} + \underline{u}_S^{(2)} = \sqrt{2}\underline{U}_S^{(1)} e^{j\omega_{S1}t} + \sqrt{2}\underline{U}_S^{(2)} e^{j\omega_{S2}t}. \tag{7.92}$$

(c) Consider the motor to be two-phase. Assuming $\theta_R = \omega_R t$ *with* ω_R *constant, use*

$$\tau = M\,\mathrm{Im}\{\underline{i}_S(\underline{i}_R e^{j\theta_R})^*\}$$

to find an expression for the torque produced by the input (7.92) in terms of the phasors $\left|\underline{I}_S^{(1)}\right| e^{j\angle \underline{I}_S^{(1)}}$, $\left|\underline{I}_R^{(1)}\right| e^{j\angle \underline{I}_R^{(1)}}$, $\left|\underline{I}_S^{(2)}\right| e^{j\angle \underline{I}_S^{(2)}}$, *and* $\left|\underline{I}_R^{(2)}\right| e^{j\angle \underline{I}_R^{(2)}}$. *What is the average torque* τ_{avg} *where* $\tau_{avg} \triangleq \lim_{T\to\infty}(1/T)\int_0^T \tau(t)dt$? *Of course, for* $\omega_{S2} = \omega_{S1}$, *the torque will be constant and equal to its average. Under what other conditions is the torque essentially (approximately) constant?*

(d) Assuming $R_S = 0$, *show that the average torque computed in part (c) may be written as*

$$\tau_{avg} = \frac{(1-\sigma)}{\sigma}\frac{U_{S1}^2}{\omega_{S1}^2 L_S}\frac{2}{S_p/S_1 + S_1/S_p} + \frac{(1-\sigma)}{\sigma}\frac{U_{S2}^2}{\omega_{S2}^2 L_S}\frac{2}{S_p/S_2 + S_2/S_p}$$

where

$$S_1 \triangleq \frac{\omega_{S1} - \omega_R}{\omega_{S1}}, \quad S_2 \triangleq \frac{\omega_{S2} - \omega_R}{\omega_{S2}}.$$

(e) Consider the special case with $\omega_{S2} = -\omega_{S1}$ *and*

$$\begin{aligned} \underline{U}_S^{(1)} &= \left|\underline{U}_S^{(1)}\right| e^{j\angle \underline{U}_S^{(1)}} = \frac{U_S}{2} e^{j\angle \underline{U}_S^{(1)}} \\ \underline{U}_S^{(2)} &= \left(\underline{U}_S^{(1)}\right)^* = \frac{U_S}{2} e^{-j\angle \underline{U}_S^{(1)}}. \end{aligned}$$

Show that the inputs to the two-phase motor are of the form

$$\begin{aligned} u_{Sa}(t) &= \sqrt{2}U_S \cos(\omega_S t + \angle \underline{U}_S) \\ u_{Sb}(t) &= 0. \end{aligned}$$

Show that the average torque is given by $\left(S \triangleq \dfrac{\omega_S - \omega_R}{\omega_S}\right)$

$$\begin{aligned} \tau_{avg} &= \tau_p\left(\frac{2}{S_p/S + S/S_p} - \frac{2}{S_p/(2-S) + (2-S)/S_p}\right) \\ \tau_p &\triangleq \frac{1}{4}\frac{(1-\sigma)}{\sigma}\frac{U_S^2}{\omega_{S1}^2 L_S}. \end{aligned}$$

Plot τ_{avg}/τ_p *as a function of* S/S_p. *What is the average torque if* $\omega_R = 0$?

Problem 24 Single-Phase Induction Motor with a Shorted Phase

Background: Motors are needed in homes for the compressors used in refrigerators and air conditioners as well as for washing machines, dryers, and so on. A typical home has only a single-phase source available, for example, a 120-V (rms) 60-Hz wall outlet and the (two-phase) induction motor requires two voltage sources 90 degrees out of phase in time. Though the DC machine can run off of a single-phase source, the induction motor is the motor of choice because it is much cheaper. Based on problem 23, this problem outlines how an induction motor can be run using only a single-phase source $u_{Sa} = \sqrt{2}U_S\cos(\omega_S t + \angle \underline{U}_S)$ *with the other stator phase shorted so that* $u_{Sb} = 0$. *(However, the standard way of running a single-phase induction motor is to* open *the second stator phase rather than short it.) Consider a two-phase induction motor with inputs* u_{Sa}, u_{Sb}. *Figure 7.37 illustrates a single-phase induction motor where the same voltage* u_{Sa} *is applied to both phases and phase b has a resistor* R_b *and capacitor* C *added in series in the winding. By part (e) of problem 23, if* $u_{Sa}(t) = \sqrt{2}\,|\underline{U}_S|\cos(\omega_S t + \angle\underline{U}_S)$ *and* $u_{Sb}(t) \equiv 0$ *with* $\omega_R = 0$, *then the average torque is zero. To start the motor, the switch in Figure 7.37 is opened. Denote by* $Z_a(j\omega) = R_S + j\omega L_S$ *and* $Z_b(j\omega) = R_S + R_b + j\omega L_S + 1/(j\omega C)$ *the impedances of the stator windings a and b, respectively, where the effect of the rotor has been neglected. Rewrite the stator phase b impedance as*

$$Z_b(j\omega) = R_S + R_b + j\omega L_S + 1/(j\omega C) = R_S + R_b + j\left(\omega^2 L_S C - 1\right)/\left(\omega C\right).$$

Choose $C = \dfrac{1}{(2\pi 60)^2 L_S}$ *so that* $Z_b(j\omega) = R_S + R_b$ *for* $\omega = \omega_S = 60$ *Hz. With* $\omega L_S >> R_S$, *the impedances* $Z_a(j\omega_S)$ *and* $Z_b(j\omega_S)$ *are*

$$\begin{aligned} Z_a(j\omega_S) &\approx j\omega_S L_S \\ Z_b(j\omega_S) &= R_S + R_b. \end{aligned}$$

The corresponding currents in the windings of phases a and b are then

$$\begin{aligned} i_{Sa}(t) &\approx \frac{\sqrt{2}U_S}{\omega_S L_S}\sin(\omega_S t + \angle\underline{U}_S) \\ i_{Sb}(t) &\approx \frac{\sqrt{2}U_S}{R_S + R_b}\cos(\omega_S t + \angle\underline{U}_S). \end{aligned}$$

Finally, design the phase b resistance R_b *such that* $R_S + R_b = \omega_S L_S$ *so that the phase currents are equal amplitude sinusoidal currents and* $\pi/2$ *radians out of phase in time. These currents establish a rotating sinusoidally distributed magnetic field in the air gap just as in the case of the symmetric motor. In reality, the inductance, resistance, and capacitance in the phase windings are not set precisely to these ideal values, but the motor will still start.*[14] *This is called a* capacitor-start *single-phase induction motor.*

[14]Even if the parameter values are set exactly as described, this is still approximate because the effect of the rotor on the stator impedance has been ignored (compare with the equivalent circuit model of the symmetric induction motor).

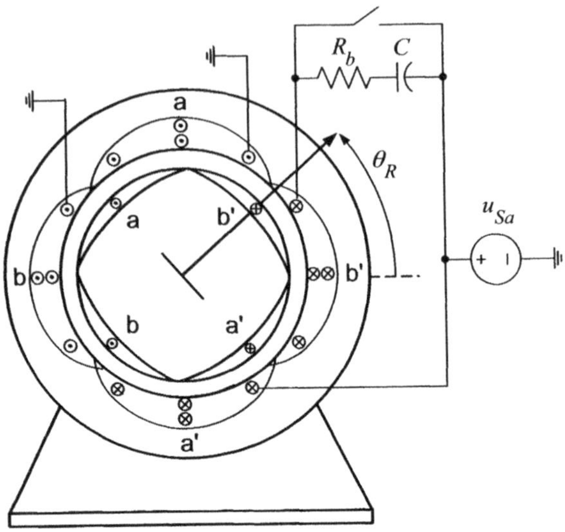

FIGURE 7.37. Capacitor-start single-phase induction motor.

(a) Is open circuiting phase b the same as applying zero voltage to it?

(b) Consider a capacitor-start single-phase induction motor as described above. Suppose that after the motor gets up to speed $\omega_R \approx \omega_S$, the switch in Figure 7.37 is closed so

$$\begin{aligned} u_{Sa}(t) &= \sqrt{2}U_S\cos(\omega_S t + \angle \underline{U}_S) \\ u_{Sb}(t) &= 0. \end{aligned}$$

Assuming the rotor speed is essentially constant, it was shown in problem 23(d) that the average torque produced is given by

$$\begin{aligned} \tau_{avg} &= \tau_p\left(\frac{2}{S_p/S + S/S_p} - \frac{2}{S_p/\left(2-S\right) + \left(2-S\right)/S_p}\right) \\ \tau_p &\triangleq \frac{1}{4}\frac{(1-\sigma)}{\sigma}\frac{U_S^2}{\omega_S^2 L_S} \\ S &\triangleq \frac{\omega_S - \omega_R}{\omega_S}. \end{aligned}$$

Plot τ_{avg}/τ_p for $S_p = 0.1, 0.2$ and 1. Based on these plots, explain why $0 < S_p \ll 1$ for the single-phase induction motor to work.

(c) Using the two-phase induction motor model, simulate a single-phase induction motor by applying a 60-Hz, 120-V (rms) source voltage to one phase and 0 V to the other. Set the initial speed to be $\omega_R(0) = \omega_S/n_p$ where $\omega_S = 2\pi 60$. Use the parameter values of $L_S = 0.157$ H, $L_R = 0.157$ H, $M = 0.148$ H, $R_R = 0.6\ \Omega$, $R_S = 2.9\ \Omega$, $n_P = 2$, $J = 0.01$ kg-m^2, $f = 0$ N-m/rad/sec, and $\tau_L = 0.8$ N-m. Set $R_S = 0$ for the calculations, but not

in the simulation. What is the value of S_p and τ_p for this motor? What is the value of S that the machine is operating at?

Problem 25 Torque-Slip Curve with Impressed Stator Currents

Suppose that the motor phases are fed by a three-phase current source rather than a three-phase voltage source. Use (7.47) to obtain U_S^2 in terms of I_S^2 and substitute this expression into (7.56) to show that the expression for the torque-slip curve in terms of the rms stator current I_S is

$$\tau = \frac{3}{2}(1-\sigma)I_S^2\frac{2}{s_p/S + S/s_p}$$

where $s_p \triangleq \sigma S_p = R_R/(\omega L_R)$.

Problem 26 Maximum Mechanical Power

The steady-state mechanical power is

$$P_{mech} = \tau\omega_R = \tau_p\frac{2}{S_p/S + S/S_p}(1-S)\omega_S.$$

(a) Compute the normalized slip $S_{maxpower}$ corresponding to maximum power *transfer for both motoring and generating.*

(b) For both motoring and generating, compute the power and torque achieved when the normalized slip is at the values found in part (a).

(c) Let σ = 0.05 *and compute the rated slip S_r and maximum power transfer slip $S_{maxpower}$. Which is larger? By approximately how much?*

(d) Compute the corresponding power factors when the machine is operating at S_r and $S_{maxpower}$. Which is larger and by how much?

(e) Compute the corresponding efficiencies when the machine is operating at S_r and $S_{maxpower}$. Which is larger and by how much?

(f) Compare operating the motor at S_r and $S_{maxpower}$ in terms of advantages and disadvantages.

Wound Rotor Synchronous Machines

These problems explore the mathematical modeling of wound rotor synchronous machines. For more detailed models of large (wound rotor) synchronous machines, see Krause et al. [4] and Ong [6].

Problem 27 Model of a Wound Rotor Synchronous Machine [1]

A mathematical model of a balanced three-phase wound rotor synchronous motor can be found directly from the induction motor expressions for the flux linkages and torque. The difference in this case is that the rotor currents are held constant in a synchronous machine by a voltage source u_F applied to the rotor windings. In order to keep the rotor currents balanced let

$$\begin{aligned} i_{R1} &= +i_f \\ i_{R2} &= -i_f/2 \\ i_{R3} &= -i_f/2 \end{aligned}$$

in equation (7.2) where i_f is nominally constant.

(a) Show that the rotor torque τ_R is given by

$$\tau_R = Mi_f\Big(-i_{S1}\sin(\theta_R) - i_{S2}\sin(\theta_R - \frac{2\pi}{3}) - i_{S3}\sin(\theta_R - \frac{4\pi}{3})\Big).$$

(b) Show that the flux linkages in the stator phases are given by

$$\begin{bmatrix} \psi_{S1}(t) \\ \psi_{S2}(t) \\ \psi_{S3}(t) \end{bmatrix} = \frac{2}{3}L_S \begin{bmatrix} 1 & \cos(2\pi/3) & \cos(4\pi/3) \\ \cos(-2\pi/3) & 1 & \cos(2\pi/3) \\ \cos(-4\pi/3) & \cos(-2\pi/3) & 1 \end{bmatrix} \begin{bmatrix} i_{S1}(t) \\ i_{S2}(t) \\ i_{S3}(t) \end{bmatrix}$$
$$+ Mi_f \begin{bmatrix} \cos(\theta_R) \\ \cos(\theta_R - 2\pi/3) \\ \cos(\theta_R - 4\pi/3) \end{bmatrix}.$$

(c) Show that an equivalent two-phase model of the wound rotor synchronous machine is given by

$$\begin{aligned} u_{Sa} &= L_S\frac{di_{Sa}}{dt} + M\frac{d}{dt}\Big(i_F\cos(\theta_R)\Big) + R_S i_{Sa} \\ u_{Sb} &= L_S\frac{di_{Sb}}{dt} + M\frac{d}{dt}\Big(i_F\sin(\theta_R)\Big) + R_S i_{Sb} \\ u_F &= L_R\frac{di_F}{dt} + M\frac{d}{dt}\Big(i_{Sa}\cos(\theta_R) + i_{Sb}\sin(\theta_R)\Big) + R_R i_F \\ J\frac{d\omega_R}{dt} &= Mi_F\Big(-i_{Sa}\sin(\theta_R) + i_{Sb}\cos(\theta_R)\Big) - \tau_L \end{aligned} \quad (7.93)$$

where $u_F \triangleq \sqrt{3/2}u_{R1}, i_F(t) \triangleq \sqrt{3/2}i_f(t)$. A block diagram for simulating a three-phase machine using a two-phase equivalent model is shown in Figure 7.38.

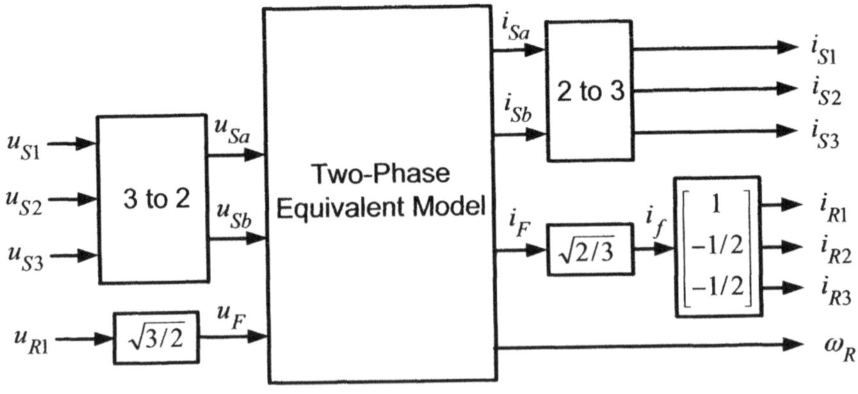

FIGURE 7.38. Using a two-phase equivalent model to simulate a three-phase machine.

Problem 28 State-Space Model of a Wound Rotor Synchronous Machine
Develop a state space representation of the wound rotor synchronous motor model (7.93) given in problem (27) using $i_{Sa}, i_{Sb}, i_F, \omega_R$, *and* θ_R *as the state variables and* u_{Sa}, u_{Sb}, *and* u_F *as the inputs.*

Problem 29 DQ model of a Wound Rotor Synchronous Machine
Define

$$\begin{bmatrix} i_d \\ i_q \end{bmatrix} = \begin{bmatrix} \cos(\theta_R) & \sin(\theta_R) \\ -\sin(\theta_R) & \cos(\theta_R) \end{bmatrix} \begin{bmatrix} i_{Sa} \\ i_{Sb} \end{bmatrix}$$

$$\begin{bmatrix} u_d \\ u_q \end{bmatrix} = \begin{bmatrix} \cos(\theta_R) & \sin(\theta_R) \\ -\sin(\theta_R) & \cos(\theta_R) \end{bmatrix} \begin{bmatrix} u_{Sa} \\ u_{Sb} \end{bmatrix}$$

(a) Rewrite the wound rotor synchronous motor model (7.93) given in problem (27) in terms of $i_d, i_q, i_F, \omega_R, \theta_R, u_d, u_q$, *and* u_F.

(b) Develop a state space dq model in terms of the state variables i_d, i_q, i_F, ω_R, *and* θ_R *and the inputs* u_d, u_q, *and* u_F.

Problem 30 Model of a Wound Rotor Synchronous Machine
Modify the approach in problem 27 to rederive the mathematical model of the wound rotor synchronous motor assuming that $i_{R1} = i_f, i_{R2} = 0$, *and* $i_{R3} = 0$. *That is, assume there is only one winding on the rotor.*

8

Induction Motor Control

In this chapter, the methods of field-oriented control and a modification of it known as input-output linearization control are described. These techniques allow one to achieve fast, precise tracking of demanding trajectories using an induction motor. These controllers are based on the nonlinear differential equation model of the induction motor and require (full) state feedback. As measurements of the rotor flux linkages (or, equivalently, the rotor currents) are not usually available, they must be estimated and techniques to do this are also described. Experimental results are presented to demonstrate the effectiveness of these control schemes. The chapter ends with two optional sections. The first describes a systematic procedure to obtain the maximum torque from the machine without violating the voltage or current constraints. The final section presents a nonlinear least-squares approach to identification of the induction motor parameters.

8.1 Dynamic Equations of the Induction Motor

The starting point for the control of the induction motor is the system of nonlinear differential equations which characterize its behavior. As derived previously, the dynamics of a n_p pole-pair two-phase induction motor[1] are given by the system of differential equations

$$
\begin{aligned}
u_{Sa} &= R_S i_{Sa} + L_S \frac{d}{dt} i_{Sa} + M \frac{d}{dt}\Big(i_{Ra}\cos(n_p\theta) - i_{Rb}\sin(n_p\theta) \Big) \\
u_{Sb} &= R_S i_{Sb} + L_S \frac{d}{dt} i_{Sb} + M \frac{d}{dt}\Big(i_{Ra}\sin(n_p\theta) + i_{Rb}\cos(n_p\theta) \Big) \\
0 &= R_R i_{Ra} + L_R \frac{d}{dt} i_{Ra} + M \frac{d}{dt}\Big(+i_{Sa}\cos(n_p\theta) + i_{Sb}\sin(n_p\theta) \Big) \\
0 &= R_R i_{Rb} + L_R \frac{d}{dt} i_{Rb} + M \frac{d}{dt}\Big(-i_{Sa}\sin(n_p\theta) + i_{Sb}\cos(n_p\theta) \Big) \\
J\frac{d\omega}{dt} &= n_p M \Big(i_{Sb}\Big(i_{Ra}\cos(n_p\theta) - i_{Rb}\sin(n_p\theta) \Big) \\
&\quad - i_{Sa}\Big(i_{Ra}\sin(n_p\theta) + i_{Rb}\cos(n_p\theta) \Big)\Big) - f\omega - \tau_L \\
\frac{d\theta}{dt} &= \omega
\end{aligned}
\tag{8.1}
$$

[1](Or, the two-phase equivalent model of a three phase motor.)

with the flux linkages of the motor phases given by

$$\begin{aligned} \lambda_{Sa} &= L_S i_{Sa} + M\Big(i_{Ra}\cos(n_p\theta) - i_{Rb}\sin(n_p\theta)\Big) \\ \lambda_{Sb} &= L_S i_{Sb} + M\Big(i_{Ra}\sin(n_p\theta) + i_{Rb}\cos(n_p\theta)\Big) \\ \lambda_{Ra} &= L_R i_{Ra} + M\Big(+i_{Sa}\cos(n_p\theta) + i_{Sb}\sin(n_p\theta)\Big) \\ \lambda_{Rb} &= L_R i_{Rb} + M\Big(-i_{Sa}\sin(n_p\theta) + i_{Sb}\cos(n_p\theta)\Big) \end{aligned} \quad (8.2)$$

where

$$L_S \triangleq \frac{\mu_0 \pi \ell_1 \ell_2 N_S^2}{8g}, \; L_R \triangleq \frac{\mu_0 \pi \ell_1 \ell_2 N_R^2}{8g}, \; M \triangleq \frac{\kappa \mu_0 \pi \ell_1 \ell_2 N_S N_R}{8g}.$$

Here N_S and N_R are the number of windings per pole-pair of the stator and rotor phases, respectively. The retarding torque produced by the friction in the ball bearings of the machine is modeled here by $-f\omega$, where f is called the *coefficient of viscous friction.* Of course, one could just consider this part of the load torque τ_L. However, as one can typically estimate f at the same time as estimating the motor parameters, $-f\omega$ then represents the "known" load torque and τ_L is then the "unknown" load torque. Also, to simplify the notation, the subscript "R" on θ_R and ω_R has now been dropped as there should be no confusion. As stated above, the mathematical model derived above is that of a sinusoidally wound, two-phase, n_p pole-pair induction motor. Nevertheless, it is standard practice to also use this model for induction motors with *squirrel cage* rotors (see Refs. [42][59] for experimental validation).

8.1.1 The Control Problem

The control problem is to choose u_{Sa} and u_{Sb} in such a way as to force ω and/or θ to track a given reference trajectory. Measurements of i_{Sa}, i_{Sb}, and θ are usually available for feedback control. However, measurements of i_{Ra} and i_{Rb} are typically not available. In fact, the most common type of induction motor is the squirrel cage motor, where rotor currents are distributed on the surface of the rotor making it very impractical to measure the current in each rotor bar. The resulting flux can be measured through the use of Hall sensors placed in the air gap, but such sensors are not economical and diminish the overall reliability of the control system.

The particular set of nonlinear differential equations (8.1) describing the induction motor is complicated and a control strategy is by no means self-evident. The solution lies in finding an *equivalent* set of equations which are simpler in form for which the control design becomes apparent. As a first step toward simplifying the above equations, the dynamic equations are rewritten in terms of some equivalent rotor flux linkages. This results in an equivalent model in which the $\cos(n_p\theta)$ and $\sin(n_p\theta)$ expressions are eliminated. To proceed, define an equivalent set of rotor flux linkages as

$$\begin{bmatrix} \psi_{Ra} \\ \psi_{Rb} \end{bmatrix} = \begin{bmatrix} \cos(n_p\theta) & -\sin(n_p\theta) \\ \sin(n_p\theta) & \cos(n_p\theta) \end{bmatrix} \begin{bmatrix} \lambda_{Ra} \\ \lambda_{Rb} \end{bmatrix}$$

where upon substituting the expressions for the rotor fluxes in (8.2), one obtains

$$\begin{aligned} \psi_{Ra} &= L_R\left(i_{Ra}\cos(n_p\theta) - i_{Rb}\sin(n_p\theta)\right) + Mi_{Sa} \\ \psi_{Rb} &= L_R\left(i_{Ra}\sin(n_p\theta) + i_{Rb}\cos(n_p\theta)\right) + Mi_{Sb}. \end{aligned} \tag{8.3}$$

The model of the induction motor in terms of the state variables $\psi_{Ra}, \psi_{Rb}, i_{Sa}, i_{Sb}$, and ω is now derived.[2] Using (8.3) to eliminate the rotor currents, the first two equations of (8.1) may be rewritten as

$$\begin{aligned} u_{Sa} &= R_S i_{Sa} + L_S\frac{d}{dt}i_{Sa} + M\frac{d}{dt}(\psi_{Ra} - Mi_{Sa})/L_R \\ u_{Sb} &= R_S i_{Sb} + L_S\frac{d}{dt}i_{Sb} + M\frac{d}{dt}(\psi_{Rb} - Mi_{Sb})/L_R \end{aligned}$$

or

$$\begin{aligned} u_{Sa} &= R_S i_{Sa} + L_S\left(1 - \frac{M^2}{L_R L_S}\right)\frac{d}{dt}i_{Sa} + \frac{M}{L_R}\frac{d}{dt}\psi_{Ra} \\ u_{Sb} &= R_S i_{Sb} + L_S\left(1 - \frac{M^2}{L_R L_S}\right)\frac{d}{dt}i_{Sb} + \frac{M}{L_R}\frac{d}{dt}\psi_{Rb}. \end{aligned}$$

The third and fourth equations of (8.1) can be rewritten as

$$\begin{aligned} \begin{bmatrix} 0 \\ 0 \end{bmatrix} &= \begin{bmatrix} R_R i_{Ra} \\ R_R i_{Rb} \end{bmatrix} + \frac{d}{dt}\left(\begin{bmatrix} \cos(n_p\theta) & \sin(n_p\theta) \\ -\sin(n_p\theta) & \cos(n_p\theta) \end{bmatrix}\right. \\ &\quad \times \left.\begin{bmatrix} L_R\left(\cos(n_p\theta)i_{Ra} - \sin(n_p\theta)i_{Rb}\right) + Mi_{Sa} \\ L_R\left(\sin(n_p\theta)i_{Ra} + \cos(n_p\theta)i_{Rb}\right) + Mi_{Sb} \end{bmatrix}\right) \end{aligned}$$

or, simply

$$\begin{bmatrix} 0 \\ 0 \end{bmatrix} = \begin{bmatrix} R_R i_{Ra} \\ R_R i_{Rb} \end{bmatrix} + \frac{d}{dt}\left(\begin{bmatrix} \cos(n_p\theta) & \sin(n_p\theta) \\ -\sin(n_p\theta) & \cos(n_p\theta) \end{bmatrix}\begin{bmatrix} \psi_{Ra} \\ \psi_{Rb} \end{bmatrix}\right).$$

Expanding,

$$\begin{aligned} \begin{bmatrix} 0 \\ 0 \end{bmatrix} &= \begin{bmatrix} R_R i_{Ra} \\ R_R i_{Rb} \end{bmatrix} + \frac{d}{dt}\left(\begin{bmatrix} \cos(n_p\theta) & \sin(n_p\theta) \\ -\sin(n_p\theta) & \cos(n_p\theta) \end{bmatrix}\right)\begin{bmatrix} \psi_{Ra} \\ \psi_{Rb} \end{bmatrix} \\ &\quad + \begin{bmatrix} \cos(n_p\theta) & \sin(n_p\theta) \\ -\sin(n_p\theta) & \cos(n_p\theta) \end{bmatrix}\frac{d}{dt}\begin{bmatrix} \psi_{Ra} \\ \psi_{Rb} \end{bmatrix} \end{aligned}$$

[2]See problem 2 where this set of equations is rederived in a simpler fashion using the space vector representation.

Multiplying both sides on the left by $\begin{bmatrix} \cos(n_p\theta) & -\sin(n_p\theta) \\ \sin(n_p\theta) & \cos(n_p\theta) \end{bmatrix}$ gives

$$\begin{bmatrix} 0 \\ 0 \end{bmatrix} = \begin{bmatrix} \cos(n_p\theta) & -\sin(n_p\theta) \\ \sin(n_p\theta) & \cos(n_p\theta) \end{bmatrix} \begin{bmatrix} R_R i_{Ra} \\ R_R i_{Rb} \end{bmatrix} + \begin{bmatrix} \cos(n_p\theta) & -\sin(n_p\theta) \\ \sin(n_p\theta) & \cos(n_p\theta) \end{bmatrix}$$

$$\times\, n_p\omega \begin{bmatrix} -\sin(n_p\theta) & \cos(n_p\theta) \\ -\cos(n_p\theta) & -\sin(n_p\theta) \end{bmatrix} \begin{bmatrix} \psi_{Ra} \\ \psi_{Rb} \end{bmatrix} + \frac{d}{dt}\begin{bmatrix} \psi_{Ra} \\ \psi_{Rb} \end{bmatrix}$$

which simplifies to

$$\begin{bmatrix} 0 \\ 0 \end{bmatrix} = \begin{bmatrix} (R_R/L_R)(\psi_{Ra} - Mi_{Sa}) \\ (R_R/L_R)(\psi_{Rb} - Mi_{Sb}) \end{bmatrix} - n_p\omega \begin{bmatrix} 0 & -1 \\ 1 & 0 \end{bmatrix} \begin{bmatrix} \psi_{Ra} \\ \psi_{Rb} \end{bmatrix} + \frac{d}{dt}\begin{bmatrix} \psi_{Ra} \\ \psi_{Rb} \end{bmatrix}.$$

Finally, the torque equation becomes

$$\begin{aligned} J\frac{d\omega}{dt} &= n_p M \left(i_{Sb}\frac{\psi_{Ra} - Mi_{Sa}}{L_R} - i_{Sa}\frac{\psi_{Rb} - Mi_{Sb}}{L_R} \right) - f\omega - \tau_L \\ &= n_p \frac{M}{L_R}(i_{Sb}\psi_{Ra} - i_{Sa}\psi_{Rb}) - f\omega - \tau_L. \end{aligned}$$

Collecting these equations together, the dynamic model of the induction motor in terms of the state variables θ, ω, ψ_{Ra}, ψ_{Rb}, i_{Sa}, and i_{Sb} is then

$$\begin{aligned} \frac{d\theta}{dt} &= \omega \\ \frac{d\omega}{dt} &= \frac{n_p M}{J L_R}(i_{Sb}\psi_{Ra} - i_{Sa}\psi_{Rb}) - \frac{f}{J}\omega - \frac{\tau_L}{J} \\ \frac{d\psi_{Ra}}{dt} &= -\frac{R_R}{L_R}\psi_{Ra} - n_p\omega\psi_{Rb} + \frac{MR_R}{L_R}i_{Sa} \\ \frac{d\psi_{Rb}}{dt} &= -\frac{R_R}{L_R}\psi_{Rb} + n_p\omega\psi_{Ra} + \frac{MR_R}{L_R}i_{Sb} \\ u_{Sa} &= R_S i_{Sa} + \sigma L_S \frac{di_{Sa}}{dt} + \frac{M}{L_R}\frac{d\psi_{Ra}}{dt} \\ u_{Sb} &= R_S i_{Sb} + \sigma L_S \frac{di_{Sb}}{dt} + \frac{M}{L_R}\frac{d\psi_{Rb}}{dt} \end{aligned} \tag{8.4}$$

where

$$\sigma \triangleq 1 - \frac{M^2}{L_R L_S}$$

is the *leakage* parameter. Substituting the expressions for $d\psi_{Ra}/dt$ and $d\psi_{Rb}/dt$ from the third and fourth equations of (8.4) into the fifth and sixth equations of (8.4) and rearranging, one obtains the state-space form of the system given by (see Refs. [64][65])

$$\begin{aligned}\frac{d\theta}{dt} &= \omega\\ \frac{d\omega}{dt} &= \mu\left(i_{Sb}\psi_{Ra} - i_{Sa}\psi_{Rb}\right) - (f/J)\omega - \tau_L/J\\ \frac{d\psi_{Ra}}{dt} &= -\eta\psi_{Ra} - n_p\omega\psi_{Rb} + \eta M i_{Sa}\\ \frac{d\psi_{Rb}}{dt} &= -\eta\psi_{Rb} + n_p\omega\psi_{Ra} + \eta M i_{Sb}\\ \frac{di_{Sa}}{dt} &= \eta\beta\psi_{Ra} + \beta n_p\omega\psi_{Rb} - \gamma i_{Sa} + u_{Sa}/\sigma L_S\\ \frac{di_{Sb}}{dt} &= \eta\beta\psi_{Rb} - \beta n_p\omega\psi_{Ra} - \gamma i_{Sb} + u_{Sb}/\sigma L_S\end{aligned} \tag{8.5}$$

with

$$\eta \triangleq \frac{R_R}{L_R}, \beta \triangleq \frac{M}{\sigma L_R L_S}, \mu \triangleq \frac{n_p M}{J L_R}, \gamma \triangleq \frac{M^2 R_R}{\sigma L_R^2 L_S} + \frac{R_S}{\sigma L_S}. \tag{8.6}$$

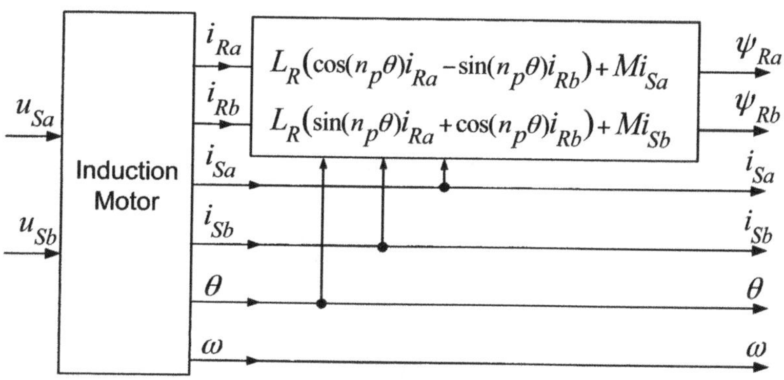

FIGURE 8.1. Block diagram of an equivalent induction motor model.

8.2 Field-Oriented and Input-Output Linearization Control of an Induction Motor

Recall that the fictitious rotor flux linkages ψ_{Ra} and ψ_{Rb} were used in the state-space formulation rather than the actual rotor flux linkages λ_{Ra} and λ_{Rb} in order to eliminate $\cos(\theta_R)$ and $\sin(\theta_R)$ from the mathematical model. However, even with this simplification, it is still not clear how to design a controller. That is, how does one choose u_{Sa} and u_{Sb} in order to control the torque, speed and/or position of the motor? The key idea of field-oriented control is to go to another *coordinate system* or *state variable representation* in which it becomes clear how to design a controller. There is an interesting analogy from mankind's understanding of the motion of the planets around the sun. Before the Polish astronomer Mikołaj Kopernik[3] came along in the 1500*s*, the view of the solar system was earth-centered; that is, the trajectories of the planets and the sun were described in terms of their motion with respect to the earth. The trajectories were quite complicated from this reference frame since the planets moved back and forth across the night sky while the sun appeared to go completely around the earth. Copernicus pointed out that if one viewed these motions from the sun's point of view, the planetary motions are quite simple. That is, one would jump into a new coordinate system where the sun is the stationary origin of the new coordinates. Then the erratic motions of the planets, as viewed from the earth, are now simple elliptical trajectories around the sun with the sun at one focus of the ellipse. The idea of field-oriented control is the same: Find a new coordinate system where the equations are simple enough that the way to control the system is self-evident.

For field-oriented control, the new coordinate system is a rotating system whose angular position is defined by

$$\rho \triangleq \tan^{-1}\left(\frac{\psi_{Rb}}{\psi_{Ra}}\right).$$

That is, instead of working with (ψ_{Ra}, ψ_{Rb}), one uses the polar coordinate representation (ρ, ψ_d) given by (see Ref. [64])

$$\begin{aligned}\rho &= \tan^{-1}\left(\frac{\psi_{Rb}}{\psi_{Ra}}\right)\\ \psi_d &= \sqrt{\psi_{Ra}^2 + \psi_{Rb}^2}.\end{aligned}$$

The phase currents and voltages are then transformed into the new coordinate system as

$$\begin{bmatrix} i_d \\ i_q \end{bmatrix} \triangleq \begin{bmatrix} \cos(\rho) & \sin(\rho) \\ -\sin(\rho) & \cos(\rho) \end{bmatrix} \begin{bmatrix} i_{Sa} \\ i_{Sb} \end{bmatrix} \tag{8.7}$$

$$\begin{bmatrix} u_d \\ u_q \end{bmatrix} \triangleq \begin{bmatrix} \cos(\rho) & \sin(\rho) \\ -\sin(\rho) & \cos(\rho) \end{bmatrix} \begin{bmatrix} u_{Sa} \\ u_{Sb} \end{bmatrix}. \tag{8.8}$$

The quantity ψ_d is referred to as the magnitude of the *rotor field flux* while ρ is the angle of the rotor field flux. This coordinate system is one that is moving (oriented) with this field flux and thus is called the *rotor-flux field-oriented* coordinate system. The currents i_d and i_q are called the direct and quadrature currents, respectively. Similarly, the voltages u_d and u_q are

[3] Better known as Copernicus.

called the direct and quadrature voltages, respectively. The rotation matrix used in (8.7) and (8.8) is called the direct quadrature or *dq* transformation.

A mathematical model of the induction motor in which ω, i_d, i_q, ψ_d, and ρ are the state variables and the inputs are u_d and u_q is now developed. The torque equation is $(\cos(\rho) = \psi_{Ra}/\psi_d, \sin(\rho) = \psi_{Rb}/\psi_d)$

$$
\begin{aligned}
\frac{d\omega}{dt} &= \mu(\psi_{Ra} i_{Sb} - \psi_{Rb} i_{Sa}) - (f/J)\omega - \tau_L/J \\
&= \mu\psi_d \left(\psi_{Ra}/\psi_d) i_{Sb} - (\psi_{Rb}/\psi_d) i_{Sa}\right) - (f/J)\omega - \tau_L/J \\
&= \mu\psi_d \left(-\sin(\rho) i_{Sa} + \cos(\rho) i_{Sb}\right) - (f/J)\omega - \tau_L/J \\
&= \mu\psi_d i_q - (f/J)\omega - \tau_L/J.
\end{aligned}
$$

The equation describing the dynamics of the magnitude of the rotor flux is

$$
\begin{aligned}
\frac{d\psi_d}{dt} &= \frac{d}{dt}\sqrt{\psi_{Ra}^2 + \psi_{Rb}^2} \\
&= \frac{1/2}{\sqrt{\psi_{Ra}^2 + \psi_{Rb}^2}}\left(2\psi_{Ra}\frac{d}{dt}\psi_{Ra} + 2\psi_{Rb}\frac{d}{dt}\psi_{Rb}\right) \\
&= \cos(\rho)\left(-\eta\psi_{Ra} - n_p\omega\psi_{Rb} + \eta M i_{Sa}\right) \\
&\quad + \sin(\rho)\left(-\eta\psi_{Rb} + n_p\omega\psi_{Ra} + \eta M i_{Sb}\right) \\
&= -\eta\left(\cos(\rho)\psi_{Ra} + \sin(\rho)\psi_{Rb}\right) + n_p\omega\left(-\cos(\rho)\psi_{Rb} + \sin(\rho)\psi_{Ra}\right) \\
&\quad + \eta M\left(\cos(\rho) i_{Sa} + \sin(\rho) i_{Sb}\right) \\
&= -\eta\psi_d + \eta M i_d.
\end{aligned}
$$

The rotor flux angle dynamics are then

$$
\begin{aligned}
\frac{d\rho}{dt} &= \frac{d}{dt}\tan^{-1}\left(\frac{\psi_{Rb}}{\psi_{Ra}}\right) \\
&= \frac{1}{1 + (\psi_{Rb}/\psi_{Ra})^2}\,\frac{\dot{\psi}_{Rb}\psi_{Ra} - \psi_{Rb}\dot{\psi}_{Ra}}{\psi_{Ra}^2} \\
&= \frac{1}{\psi_{Ra}^2 + \psi_{Rb}^2}\left(\psi_{Ra}(-\eta\psi_{Rb} + n_p\omega\psi_{Ra} + \eta M i_{Sb})\right. \\
&\quad \left. -\psi_{Rb}(-\eta\psi_{Ra} - n_p\omega\psi_{Rb} + \eta M i_{Sa})\right) \\
&= \frac{1}{\psi_d^2}\left(n_p\omega(\psi_{Ra}^2 + \psi_{Rb}^2)\right) + \eta M\frac{1}{\psi_d}\left((\psi_{Ra}/\psi_d) i_{Sb} - (\psi_{Rb}/\psi_d) i_{Sa}\right) \\
&= n_p\omega + \eta M\frac{1}{\psi_d}\left(\cos(\rho) i_{Sb} - \sin(\rho) i_{Sa}\right) \\
&= n_p\omega + \eta M\frac{i_q}{\psi_d}.
\end{aligned}
$$

The dynamic equation describing the direct current is

$$
\begin{aligned}
\frac{di_d}{dt} &= \frac{d}{dt}\left(\cos(\rho)i_{Sa} + \sin(\rho)i_{Sb}\right)\\
&= \left(-\sin(\rho)i_{Sa} + \cos(\rho)i_{Sb}\right)\frac{d\rho}{dt} + \cos(\rho)\frac{di_{Sa}}{dt} + \sin(\rho)\frac{di_{Sb}}{dt}\\
&= i_q\frac{d\rho}{dt} + \cos(\rho)\left(\eta\beta\psi_{Ra} + \beta n_p\omega\psi_{Rb} - \gamma i_{Sa} + \frac{u_{Sa}}{\sigma L_S}\right)\\
&\quad + \sin(\rho)\left(\eta\beta\psi_{Rb} - \beta n_p\omega\psi_{Ra} - \gamma i_{Sb} + \frac{u_{Sb}}{\sigma L_S}\right)\\
&= i_q\left(n_p\omega + \eta M\frac{i_q}{\psi_d}\right) + \eta\beta\left(\cos(\rho)\psi_{Ra} + \sin(\rho)\psi_{Rb}\right)\\
&\quad + n_p\omega\beta\left(\cos(\rho)\psi_{Rb} - \sin(\rho)\psi_{Ra}\right) - \gamma\left(\cos(\rho)i_{Sa} + \sin(\rho)i_{Sb}\right)\\
&\quad + \left(\cos(\rho)u_{Sa} + \sin(\rho)u_{Sb}\right)/\sigma L_S\\
&= i_q n_p\omega + \eta M\frac{i_q^2}{\psi_d} + \eta\beta\psi_d - \gamma i_d + \frac{u_d}{\sigma L_S}.
\end{aligned}
$$

The dynamic equation describing the quadrature current is

$$
\begin{aligned}
\frac{di_q}{dt} &= \frac{d}{dt}\left(-\sin(\rho)i_{Sa} + \cos(\rho)i_{Sb}\right)\\
&= -\left(\cos(\rho)i_{Sa} + \sin(\rho)i_{Sb}\right)\frac{d\rho}{dt} - \sin(\rho)\frac{di_{Sa}}{dt} + \cos(\rho)\frac{di_{Sb}}{dt}\\
&= -i_d\frac{d\rho}{dt} - \sin(\rho)\left(\eta\beta\psi_{Ra} + \beta n_p\omega\psi_{Rb} - \gamma i_{Sa} + \frac{u_{Sa}}{\sigma L_S}\right)\\
&\quad + \cos(\rho)\left(\eta\beta\psi_{Rb} - \beta n_p\omega\psi_{Ra} - \gamma i_{Sb} + \frac{u_{Sb}}{\sigma L_S}\right)\\
&= -i_d\left(n_p\omega + \eta M i_q/\psi_d\right) + \eta\beta\left(-\sin(\rho)\psi_{Ra} + \cos(\rho)\psi_{Rb}\right)\\
&\quad - n_p\omega\beta\left(\sin(\rho)\psi_{Rb} + \cos(\rho)\psi_{Ra}\right) - \gamma\left(-\sin(\rho)i_{Sa} + \cos(\rho)i_{Sb}\right)\\
&\quad + \left(-\sin(\rho)u_{Sa} + \cos(\rho)u_{Sb}\right)/\sigma L_S\\
&= -i_d n_p\omega - \eta M i_d i_q/\psi_d - n_p\omega\beta\psi_d - \gamma i_q + \frac{u_q}{\sigma L_S}.
\end{aligned}
$$

Collecting the above equations together, a state-space model in the field-oriented coordinate system is given by

$$\begin{aligned}
\frac{d\theta}{dt} &= \omega \\
\frac{d\omega}{dt} &= \mu\psi_d i_q - (f/J)\omega - \tau_L/J \\
\frac{d\psi_d}{dt} &= -\eta\psi_d + \eta M i_d \\
\frac{di_d}{dt} &= -\gamma i_d + (\eta M/\sigma L_R L_S)\psi_d + n_p\omega i_q + \eta M i_q^2/\psi_d + u_d/\sigma L_S \\
\frac{di_q}{dt} &= -\gamma i_q - (M/\sigma L_R L_S)n_p\omega\psi_d - n_p\omega i_d - \eta M i_q i_d/\psi_d + u_q/\sigma L_S \\
\frac{d\rho}{dt} &= n_p\omega + \eta M i_q/\psi_d.
\end{aligned} \tag{8.9}$$

This model is now used to develop a very effective method of control. It is first assumed that all of the state-variables θ, ω, ψ_d, i_d, i_q, and ρ are available for feedback. This would mean that one physically measures θ, ω, i_{Sa}, i_{Sb}, i_{Ra}, and i_{Rb} and then uses these measurements to compute ψ_d, i_d, i_q, and ρ in *real time*. In practice, one is rarely able to measure the rotor currents i_{Ra} and i_{Rb}. This problem can be resolved using an *observer* to estimate ρ and ψ_d as will be shown later. Then i_d and i_q can be computed using the estimated value for ρ and the measured stator currents i_{Sa} and i_{Sb}.

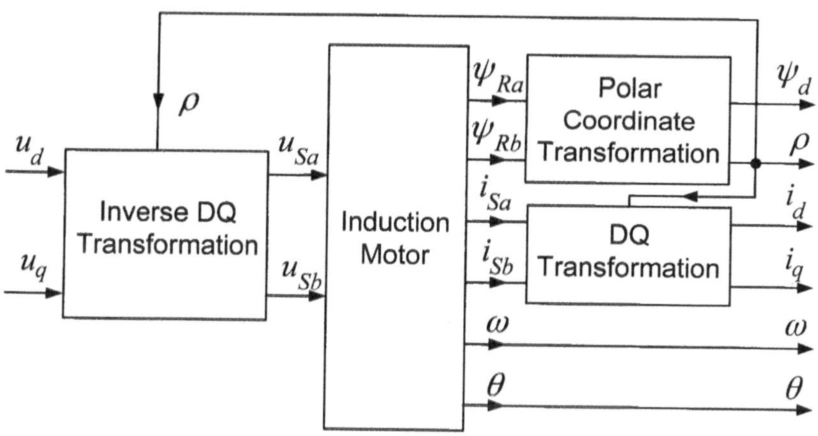

FIGURE 8.2. Transformation to the field-oriented (dq) coordinate system.

8.2.1 *Current-Command Field-Oriented Control*

Note that the electromagnetic torque $\tau = J\mu\psi_d i_q$ is now just proportional to the product of two state variables. However, the differential equations for i_d and i_q still contain quite complicated nonlinearities. One possibility to simplify these current dynamics is to use feedback cancellation; that is, let

$$\begin{aligned} u_d &= -(\sigma L_S)\left(-\gamma i_d + \eta\beta\psi_d + n_p\omega i_q + \eta M i_q^2/\psi_d\right) + (\sigma L_S)v_d \\ u_q &= -(\sigma L_S)\left(-\gamma i_q - \beta n_p\omega\psi_d - n_p\omega i_d - \eta M i_q i_d/\psi_d\right) + (\sigma L_S)v_q \end{aligned} \tag{8.10}$$

so that (8.9) simplifies to

$$\begin{aligned} d\omega/dt &= \mu\psi_d i_q - (f/J)\omega - \tau_L/J \\ d\psi_d/dt &= -\eta\psi_d + \eta M i_d \\ di_d/dt &= v_d \\ di_q/dt &= v_q \\ d\rho/dt &= n_p\omega + \eta M i_q/\psi_d \end{aligned} \tag{8.11}$$

For current-command control, one may choose

$$\begin{aligned} v_d &= K_{dI}\int_0^t (i_{dr} - i_d)dt + K_{dP}(i_{dr} - i_d) \\ v_q &= K_{qI}\int_0^t (i_{qr} - i_q)dt + K_{qP}(i_{qr} - i_q) \end{aligned}$$

and by appropriate choice of the PI gains, $i_d \to i_{dr}$ and $i_q \to i_{qr}$. However, in practice, this is *not* done. To explain, note that in applying the feedback (8.10), there is some uncertainty in the knowledge of the motor parameters and the state variables. The motor parameters R_R and R_S can vary significantly due to ohmic heating while L_R and L_S can also vary due to magnetic saturation. Furthermore, the state variables ρ and ψ_d must be estimated leaving uncertainty not only in their values, but in the currents i_d and i_q as the value of ρ is also needed to estimate them.

Experience has shown that the effect of the nonlinear terms in (8.9) can be eliminated by forcing the system into *current-command* mode using high-gain feedback [66]. That is, one applies the PI current loops

$$\begin{aligned} u_d &= K_{dI}\int_0^t (i_{dr} - i_d)dt + K_{dP}(i_{dr} - i_d) \\ u_q &= K_{qI}\int_0^t (i_{qr} - i_q)dt + K_{qP}(i_{qr} - i_q) \end{aligned} \tag{8.12}$$

directly to (8.9) to force i_d and i_q to track their corresponding references i_{dr} and i_{qr}, respectively. Values of the gains in (8.12) can be found that result in fast responses $i_d \to i_{dr}$ and $i_q \to i_{qr}$ for the currents. Consequently, i_{dr}

and i_{qr} can then be considered as the new inputs for the control design where the system equations simplify to (i.e., are approximated as)

$$\frac{d\theta}{dt} = \omega \tag{8.13}$$

$$\frac{d\omega}{dt} = \mu\psi_d i_{qr} - (f/J)\omega - \tau_L/J \tag{8.14}$$

$$\frac{d\psi_d}{dt} = -\eta\psi_d + \eta M i_{dr} \tag{8.15}$$

$$\frac{d\rho}{dt} = n_p\omega + \eta M i_q/\psi_d. \tag{8.16}$$

Remark Note the the subscript "r" is not used on i_q in (8.16). The reason for this is that one is only interested in controlling θ, ω, and ψ_d, but not ρ. However, one must estimate ρ (as well as ψ_d), and this equation will be used for that purpose using the estimated value i_q rather than the reference value i_{qr}.

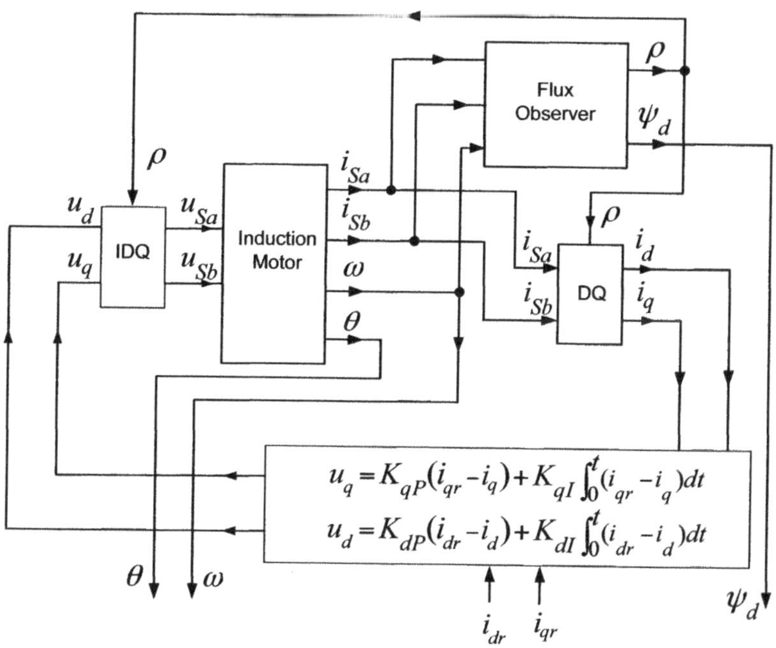

FIGURE 8.3. Block diagram of a current-command field-oriented system model.

A block diagram for the current-command field-oriented system model is given in Figure 8.3. The "Flux Observer" block will be described in Section 8.3.1. For now, the reader should just understand this block to represent an algorithm to compute ρ and ψ_d from measurements of i_{Sa}, i_{Sb}, and ω.

Field-oriented control consists of using i_{dr} to force ψ_d to track the *constant* flux reference $\psi_{d0} = Mi_{d0}$. As a consequence, equations (8.13) and (8.14) become

$$\begin{aligned} \frac{d\theta}{dt} &= \omega & (8.17) \\ \frac{d\omega}{dt} &= \mu\psi_{d0}i_{qr} - (f/J)\omega - \tau_L/J & (8.18) \end{aligned}$$

so that there is now a *linear* relationship between the input i_{qr} and the speed. To complete the control design, let

$$\begin{bmatrix} \theta_{ref}(t) \\ \omega_{ref}(t) \\ \alpha_{ref}(t) \end{bmatrix} = \begin{bmatrix} \theta_{ref}(t) \\ d\theta_{ref}/dt \\ d^2\theta_{ref}/dt^2 \end{bmatrix}$$

be the mechanical reference trajectory the motor is required to track. The control of speed/position is then done through the input i_{qr}. Specifically, choose

$$\begin{aligned} i_{qr} &= \left(K_0 \int_0^t (\theta_{ref} - \theta)dt + K_1(\theta_{ref} - \theta) + K_2(\omega_{ref} - \omega)\right. \\ &\quad \left. + (f/J)\omega + \alpha_{ref}\right) / \mu\psi_{d0}. \qquad (8.19) \end{aligned}$$

With the proper choice of the gains K_0, K_1, and K_2 (see problem 8), $\theta \rightarrow \theta_{ref}$ and $\omega \rightarrow \omega_{ref}$ even with a constant load torque τ_L acting on the motor.

The flux ψ_d is regulated to a constant value ψ_{d0} by

$$i_{dr} = K_{\psi I} \int_0^t (\psi_{d0} - \psi_d)dt + K_{\psi P}(\psi_{d0} - \psi_d) + i_{d0}. \qquad (8.20)$$

That is, with proper choice of the feedback gains (see problem 7), $\psi_d \rightarrow \psi_{d0} = Mi_{d0}$.

A block diagram of this trajectory tracking controller is given in Figure 8.4.

Remarks

The roles of ψ_d and i_q (i_{dr} and i_{qr}) could, in principle, be reversed, but this is not done because the input i_d's (i_{dr}'s) ability to rapidly change ψ_d is limited by the rotor time constant $T_R = 1/\eta = L_R/R_R$, which is not the case for $i_q(i_{qr})$.

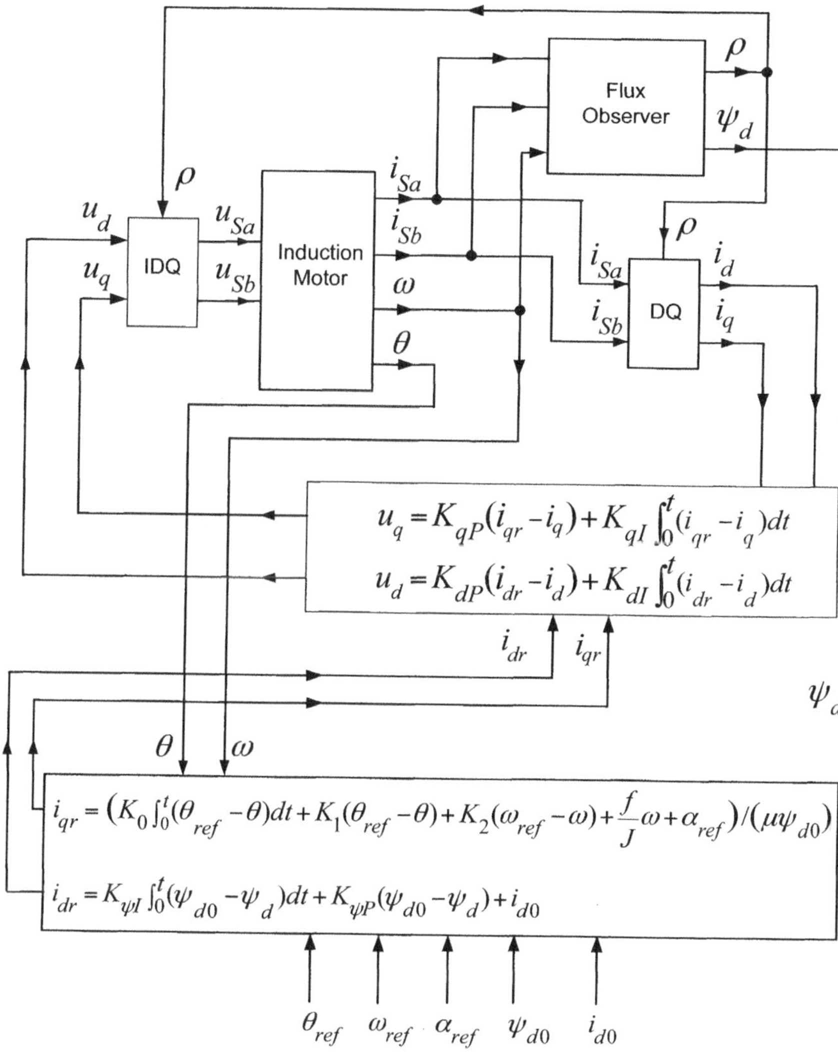

FIGURE 8.4. Block diagram of a trajectory tracking controller using a field-oriented controller.

As ψ_{d0} is constant, one usually incorporates $\mu\psi_{d0}$ directly into the controller gains as

$$i_{qr} = \bar{K}_0 \int_0^t (\theta_{ref} - \theta)dt + \bar{K}_1(\theta_{ref} - \theta) + \bar{K}_2(\omega_{ref} - \omega) - \bar{f}\omega + \bar{\alpha}_{ref} \quad (8.21)$$

where $\bar{K}_0 = K_0/(\mu\psi_{d0})$, $\bar{K}_1 = K_1/(\mu\psi_{d0})$, $\bar{K}_2 = K_2/(\mu\psi_{d0})$, $\bar{f} = f/(J\mu\psi_{d0})$, and $\bar{\alpha}_{ref} = \alpha_{ref}/(\mu\psi_{d0})$.

Also, in practice, one typically leaves out the term $\bar{\alpha}_r$ in (8.21) and the term i_{d0} in (8.20) and finds that there is no significant deterioration in tracking if the gains can be chosen large enough.

8.2.2 *Experimental Results Using a Field-Oriented Controller*

Some experimental results showing the type of servo performance that can be achieved are now presented [67]. The motor is a 6-pole ($n_p = 3$) 1/12-horsepower two-phase induction motor with a squirrel cage rotor that is rated for 2.4 A (continuous) and 60 V. The rest of the hardware (see Figure 8.5) consists of a Motorola DSP56001 digital signal processing system for implementation of the control algorithm, two PWM amplifiers (± 80 V and ± 6 A), two analog-to-digital (A/D) converters to sample the stator currents, two digital-to-analog (D/A) converters to command voltage to the amplifiers, an electronic interface boards for the A/Ds and D/As, and a 2000 pulse/rev encoder (resolution of $360°/2000 = 0.18°$) to sense the angular position of the rotor. The PC indicated in Figure 8.5 is used to write, edit, and download the assembly language programs to the DSP which execute the controller algorithms. The parameters were identified using the technique given in Stephan et al. [42] and are $M = 0.0117$ H, $R_R = 3.9$ Ω, $R_S = 1.7$ $\Omega, L_R = 0.014$ H$, L_S = 0.014$ $H, f = 0.00014$ N-m/rad/sec and $J = 0.00011$ kg-m^2. The constant flux reference ψ_{d0} was chosen to be $\psi_{d0} = Mi_{d0} = 0.0117(5.5/\sqrt{2}) = 0.0455$ Wb. (See problem 13 for an explanation of the choice $i_{d0} = 5.5/\sqrt{2}$.)

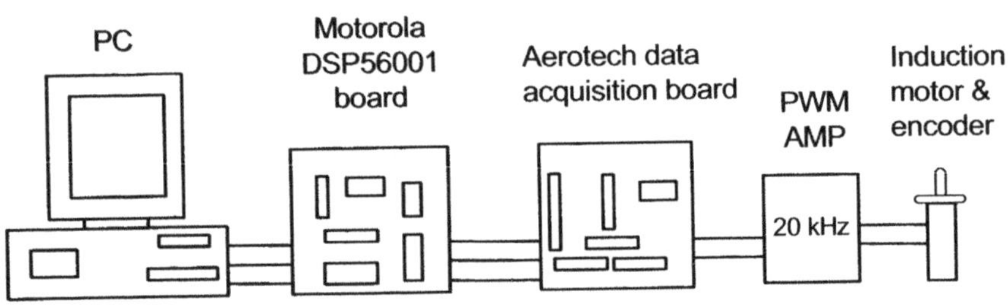

FIGURE 8.5. Hardware setup.

Figure 8.6 is a plot of the measure motor position and its reference trajectory used to carry out the turn. This figure shows the controller was able to turn the motor 180° in 73 msec. The tracking of the motor position to its reference is close enough that they are indistinguishable in Figure 8.6. Their difference in encoder counts is plotted in Figure 8.7 where it is seen that the measured error is less than two counts [$= 2(2\pi/2000) = 0.00314$ radians] for the entire move.

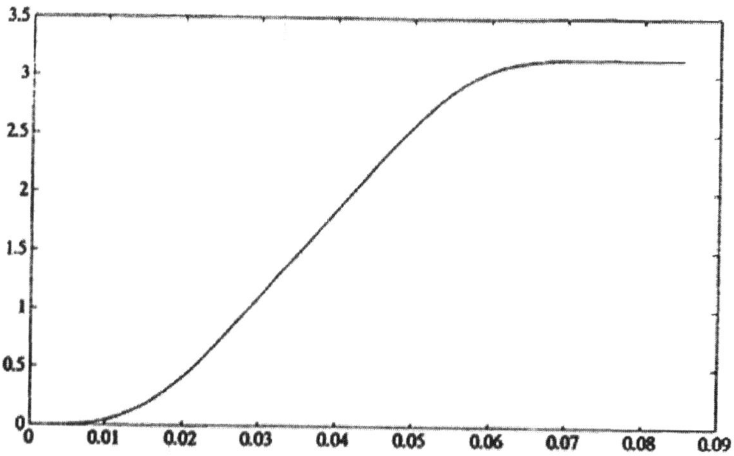

FIGURE 8.6. θ and θ_{ref} in radians versus time in seconds. From Bodson et al. [67], © 2004 IEEE.

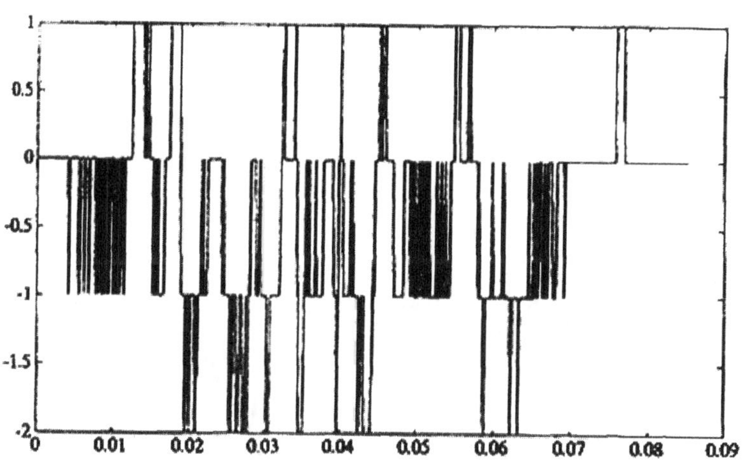

FIGURE 8.7. $\theta_{ref} - \theta$ in encoder counts versus time in seconds. From Bodson et al. [67], © 2004 IEEE.

The corresponding estimated speed and the speed reference are shown in Figure 8.8.

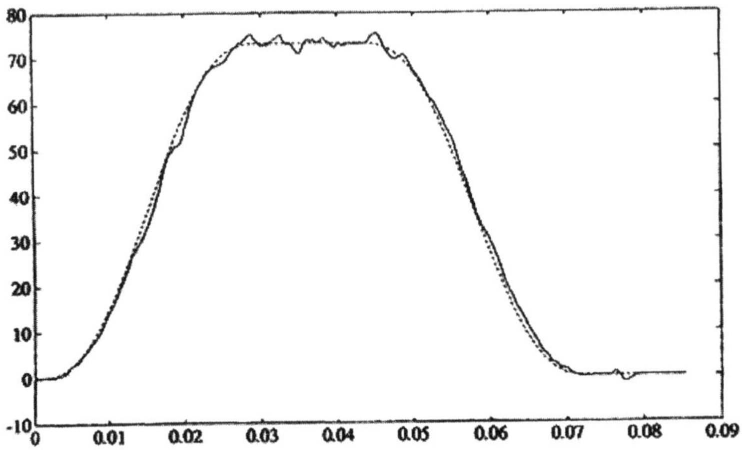

FIGURE 8.8. ω and ω_{ref} in radians/sec versus time in seconds. From Bodson et al. [67], © 2004 IEEE.

Finally, the quadrature current i_q and $i_{qref} = \alpha_{ref}/\psi_{d0}$ are given in Figure 8.9 while the flux ψ_d and its reference $\psi_{dref} = \psi_{d0}$ are given in Figure 8.10.

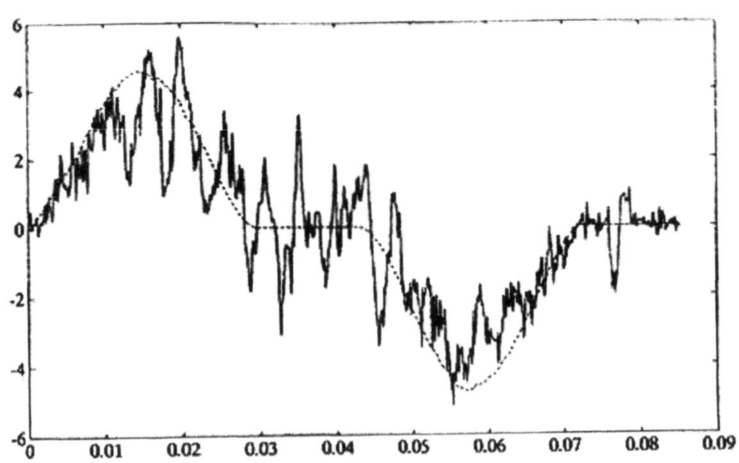

FIGURE 8.9. i_q and i_{qref} in amperes versus time in seconds. From Bodson et al. [67], © 2004 IEEE.

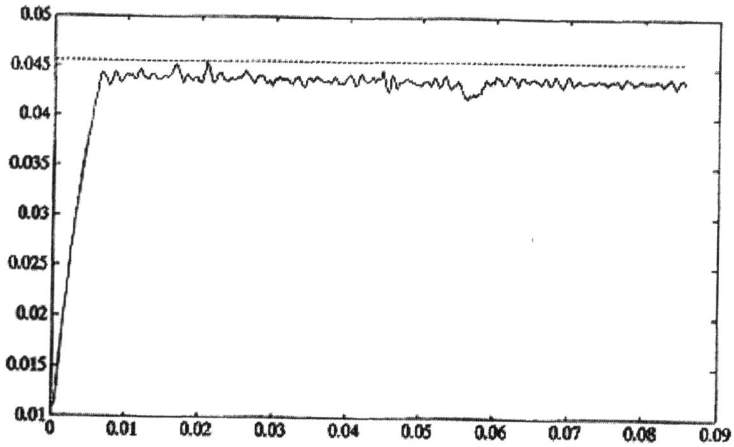

FIGURE 8.10. ψ_d and ψ_{dref} in webers versus time in seconds. From Bodson et al. [67], © 2004 IEEE.

The PI current gains for the tracking error in (8.12) where chosen as $K_{dI} = 740, K_{dP} = 6.4, K_{qI} = 740$, and $K_{qP} = 6.4$. The gains for the trajectory tracking controller in (8.19) were set as

$$\begin{aligned} K_0 &= 1.07 \times 10^6 \\ K_1 &= 7.04 \times 10^5 \\ K_2 &= 1.07 \times 10^3 \end{aligned} \tag{8.22}$$

and the gains for the flux regulator (8.20) were set to

$$\begin{aligned} K_{\psi P} &= 1600 \\ K_{\psi I} &= 23000. \end{aligned} \tag{8.23}$$

8.2.3 Field Weakening

The torque is controlled through the quadrature current i_q. Inspection of the fifth equation in (8.9) reveals that in order to maintain or increase i_q, the voltage u_q must be chosen such that $di_q/dt \geq 0$, that is,

$$\begin{aligned} u_q &\geq \sigma L_S \gamma i_q + M n_p \omega \psi_d / L_R + \sigma L_S n_p \omega i_d + \sigma L_S \eta M i_q i_d / \psi_d \\ &= \left(\frac{M^2}{L_R^2} R_R + R_S \right) i_q + \frac{M}{L_R} n_p \omega \psi_d + \sigma L_S n_p \omega i_d + \sigma L_S \frac{R_R}{L_R} i_q M i_d / \psi_d \end{aligned}$$

where (8.6) was used. Consider the machine in steady state so that $\psi_d = Mi_d$ and use the approximation $L_S \approx L_R \approx M$ so that this inequality reduces to

$$u_q \geq (R_R + R_S)\, i_q + n_p\omega\psi_d + \sigma n_p\omega\psi_d + \sigma R_R i_q.$$

At higher speeds, the dominant term on the right-hand side of this inequality is $n_p\omega\psi_d$. When the motor speed reaches a high enough speed, called the *base speed* ω_{base}, the required voltage to maintain the current i_q equals the maximum voltage $V_{\max}$ the amplifier can produce. To get around this problem, the flux ψ_d is decreased at higher speeds. The standard way to choose the flux reference is given by

$$\psi_{dref} = \begin{cases} \psi_{d0} & \text{for} \quad |\omega| < \omega_{base} \\ \psi_{d0}\omega_{base}/\,|\omega| & \text{for} \quad |\omega| \geq \omega_{base} \end{cases} \tag{8.24}$$

which is plotted versus ω in Figure 8.11.

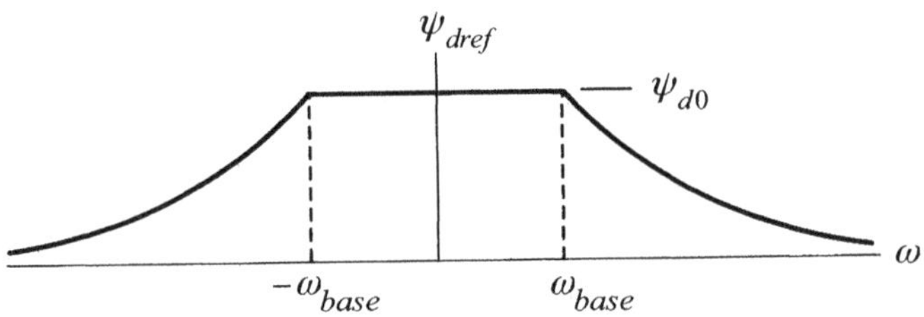

FIGURE 8.11. Flux reference for field weakening.

The direct current i_{dr} is chosen to force ψ_d to track ψ_{dref} specified by (8.24) so that above the base speed ω_{base} the term $n_p\omega\psi_d = n_p\omega_{base}\psi_{d0}$ is constant. This reduction of ψ_d above ω_{base} is referred to as *field weakening.* One way to choose the base speed ω_{base} is to set $n_p\omega_{base}\psi_{d0} = V_{\max}$ or $\omega_{base} = V_{\max}/(n_p\psi_{d0})$, where $V_{\max}$ is the (equivalent two-phase) stator voltage maximum. The trade-off here is that one can have the motor achieve speeds above the base speed, but less torque is produced as $\tau = J\mu\psi_d i_q$ so that for the same i_q, decreasing ψ_d means that less torque is produced.

Another way to formulate the problem of selecting the flux reference is as a static optimization problem. That is, at each constant rotor speed ω, maximize

$$\tau = J\mu\psi_d i_q \tag{8.25}$$

subject to the constraints

$$u_d^2 + u_q^2 \leq V_{\max}^2,\, i_d^2 + i_q^2 \leq I_{\max}^2. \tag{8.26}$$

Such an approach is given in Section 8.4 based on Refs. [68] and [69].

8.2.4 Input-Output Linearization

In the above field-oriented control scheme, the speed ω and flux ψ_d are only *asymptotically* decoupled [64]. In other words, the speed equation (8.14) is linear only *after* ψ_d is constant. For high speeds, field weakening (i.e., decreasing ψ_d), is necessary so as not to exceed the stator voltage limit. Since in the field-weakening region, the flux reference varies as a function of the speed, the dynamics of ψ_d then interfere with the dynamics of ω. This coupling of the flux and speed dynamics can be eliminated by considering an *input-output linearizing* controller [64][65][70][71]. Recall the system (8.13)-(8.16) repeated below:

$$\begin{aligned}\frac{d\theta}{dt} &= \omega\\ \frac{d\omega}{dt} &= \mu\psi_d i_{qr} - (f/J)\omega - \tau_L/J\\ \frac{d\psi_d}{dt} &= -\eta\psi_d + \eta M i_{dr}\\ \frac{d\rho}{dt} &= n_p\omega + \eta M i_q/\psi_d.\end{aligned}$$

In this model, i_{dr} and i_{qr} are the inputs. Let them now be specified by

$$\begin{aligned}i_{dr} &= u_1\\ i_{qr} &= \frac{u_2}{\mu\psi_d}\end{aligned}\qquad(8.27)$$

so that the system becomes

$$\begin{aligned}d\theta/dt &= \omega\\ d\omega/dt &= u_2 - (f/J)\omega - \tau_L/J\\ d\psi_d/dt &= -\eta\psi_d + \eta M u_1\\ d\rho/dt &= n_p\omega + \eta M i_q/\psi_d.\end{aligned}\qquad(8.28)$$

The flux dynamics are now decoupled from the speed dynamics. That is, the first three equations of (8.28) may be written as two decoupled linear subsystems

$$d\psi_d/dt = -\eta\psi_d + \eta M u_1\qquad(8.29)$$

and

$$\begin{aligned}d\theta/dt &= \omega\\ d\omega/dt &= u_2 - (f/J)\omega - \tau_L/J.\end{aligned}\qquad(8.30)$$

The input u_1 is chosen to force the *linear* system (8.29) to track a given flux reference trajectory ψ_{dr}, and the input u_2 is used to force the *linear* system (8.30) to track a given mechanical reference trajectory $(\theta_{ref}(t), \omega_{ref}(t), \alpha_{ref}(t))$. After the input u_2 is specified in order to accomplish the control task, the current reference i_{qr} is then simply chosen to be $i_{qr} = u_2/(\mu\psi_d)$.

As system (8.28) is *linear* from the inputs u_1, u_2 to the outputs ω, ψ, this controller is referred to as an *input-output linearization* controller. However, the overall system (8.28) is still *nonlinear* as the dynamics of ρ are nonlinear. The boundedness (stability) of ρ is guaranteed since it is an angle and reset to 0 every 2π radians.

The new inputs u_1 and u_2 may be chosen as follows:

Let $\psi_{dref}(t)$ be a desired flux reference and choose

$$u_1 = i_{dr} = K_{\psi P}(\psi_{dref} - \psi_d) + K_{\psi I}\int_0^t(\psi_{dref} - \psi_d)dt + i_{dref} \qquad (8.31)$$

where i_{dref} satisfies

$$d\psi_{dref}/dt = -\eta\psi_{dref} + \eta M i_{dref}.$$

With proper choice of the feedback gains $K_{\psi P}$ and $K_{\psi I}$ (see problem 10), $\psi_d \rightarrow \psi_{dref}$.

Let $(\theta_{ref}(t), \omega_{ref}(t), \alpha_{ref}(t))$ be the mechanical reference trajectory the motor is required to track. Choosing

$$\begin{aligned} u_2 &= K_0\int_0^t(\theta_{ref} - \theta)dt + K_1(\theta_{ref} - \theta) + K_2(\omega_{ref} - \omega) \\ &\quad + \alpha_{ref} + (f/J)\omega \end{aligned} \qquad (8.32)$$

and along with a proper choice of the gains K_0, K_1, and K_2 (see problem 11) results in $\theta \rightarrow \theta_{ref}$ and $\omega \rightarrow \omega_{ref}$ even with a constant load torque τ_L acting on the motor. Figure 8.12 is a block diagram of the input–output linearization controller. The "speed observer" block will be described in Section 8.3.2. For now, the reader should understand it to be an algorithm to obtain an estimate $\hat{\omega}$ of the speed ω from knowledge of ψ_d, i_q, and θ.

Voltage Command Input-Output Control

An input-output linearization controller may also be designed for the full-order (voltage command) system model (8.11). Specifically, the transformation

$$\begin{aligned} \omega &\triangleq \omega \\ \alpha &\triangleq \mu\psi_d i_q \\ \psi_d &\triangleq \psi_d \\ \psi_d' &\triangleq -\eta\psi_d + \eta M i_d \\ \rho &\triangleq \rho \end{aligned}$$

is applied to (8.11), resulting in

$$\begin{aligned}
d\omega/dt &= \alpha - (f/J)\omega - \tau_L/J \\
d\alpha/dt &= \mu\psi_d' i_q + (\mu\psi_d)v_q \\
d\psi_d/dt &= \psi_d' \\
d\psi_d'/dt &= -\eta\psi_d' + (\eta M)v_d \\
d\rho/dt &= n_p\omega + \eta M\alpha/(\mu\psi_d^2).
\end{aligned}$$

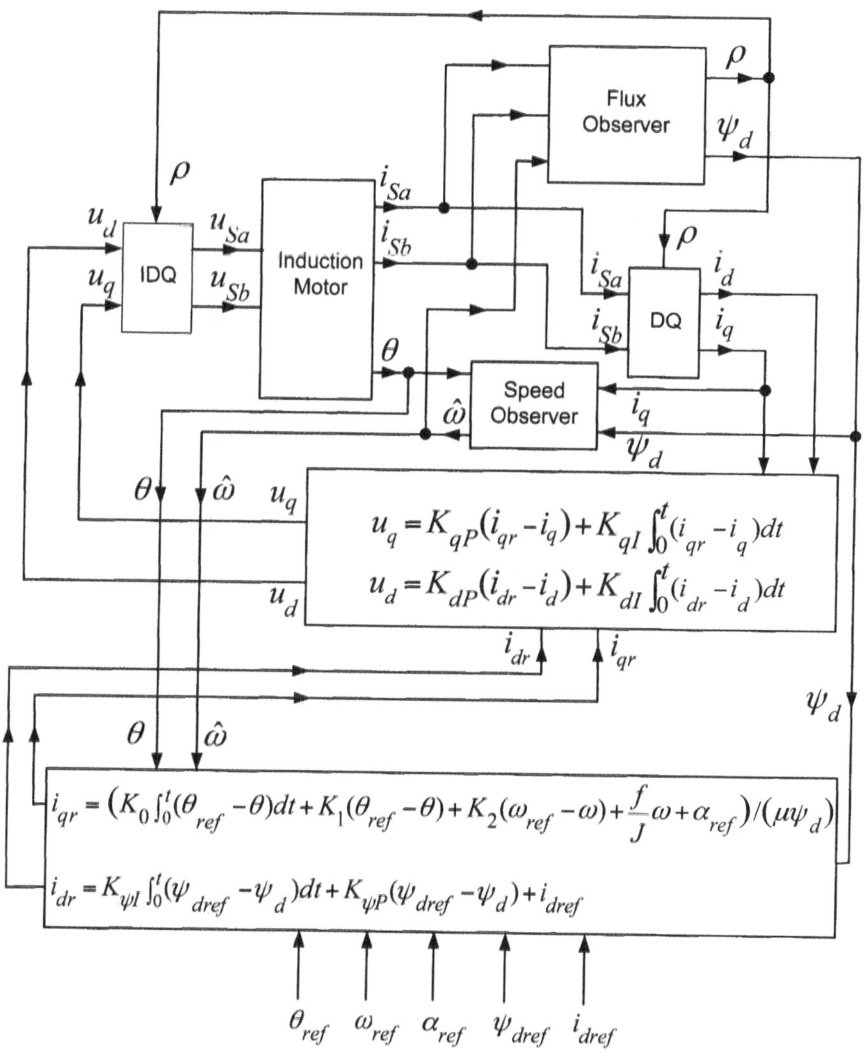

FIGURE 8.12. Block diagram of an input-output linearization controller.

With w_d and w_q being two new inputs, application of the feedback

$$\begin{aligned} v_d &= \psi'_d/M + w_d/(\eta M) \\ v_q &= -\mu\psi'_d i_q/(\mu\psi_d) + w_q/(\mu\psi_d) \end{aligned} \tag{8.33}$$

results in the input-output linearized system

$$\begin{aligned} d\omega/dt &= \alpha - (f/J)\omega - \tau_L/J \\ d\alpha/dt &= w_q \\ d\psi_d/dt &= \psi'_d \\ d\psi'_d/dt &= w_d \\ d\rho/dt &= n_p\omega + \eta M\mu\alpha/\psi_d^2. \end{aligned}$$

8.2.5 Experimental Results Using an Input-Output Controller

Some experimental results are now presented to show the increased performance that can be extracted from the motor using an input–output linearization controller [65][69]. The motor and experimental setup is the same as that used for the field-oriented controller.

An experiment requiring a point-to-point position move was carried out in which the motor was brought up to a speed of 8000 rev/min in 0.38 seconds and brought down from 8000 rev/min to 0 rev/min in 0.265 sec (see Figure 8.15). The PI current gains in (8.12) were set at $K_{dI} = 9000, K_{dP} = 15, K_{qI} = 9000$, and $K_{qP} = 15$. The PI gains for the flux tracking controller (8.31) were chosen as $K_{\psi P} = 10,000, K_{\psi I} = 420,000$. The PID gains for the tracking error in (8.32) were chosen as $K_0 = 3.0 \times 10^5, K_1 = 5.5 \times 10^4$, and $K_2 = 125$. The sample rate was 4 kHz.

Figure 8.13 is a plot of the position and its reference where the tracking is close enough that they are indistinguishable in this figure. The difference $\theta_{ref} - \theta$ in encoder counts is given in Figure 8.14. The maximum error is 34 encoder counts.

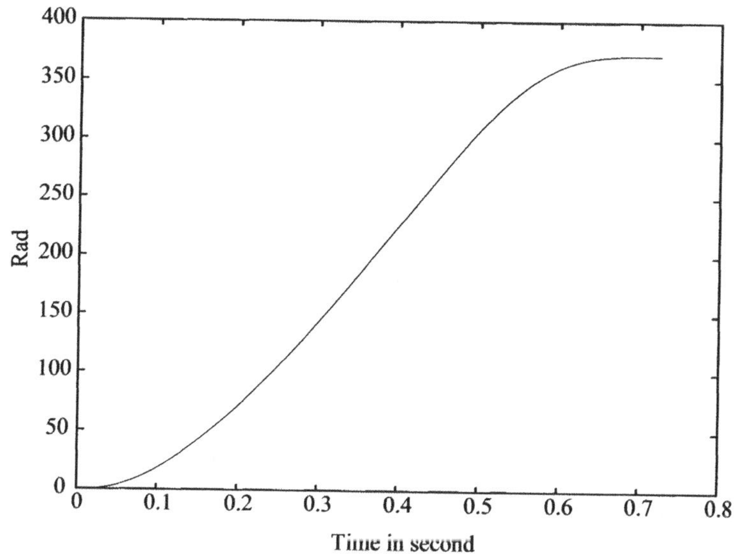

FIGURE 8.13. θ and θ_{ref} in radians versus time in seconds. From Bodson et al. [65], © 2004 IEEE.

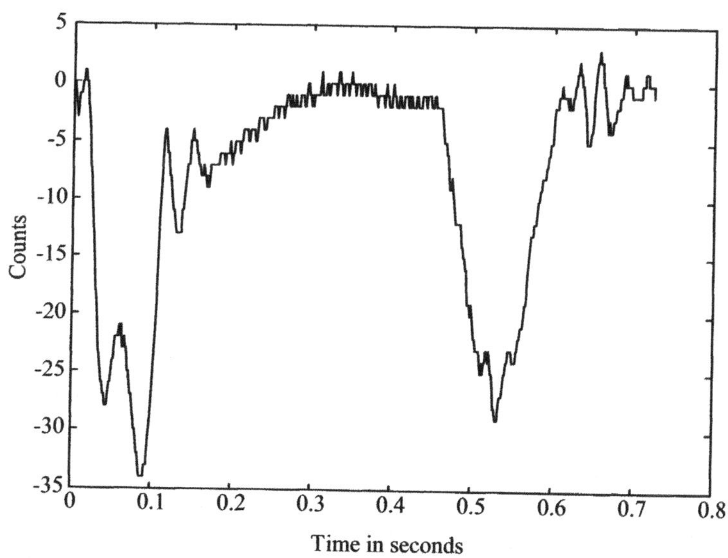

FIGURE 8.14. $\theta_{ref} - \theta$ in encoder counts versus time in seconds. From Bodson et al. [65], © 2004 IEEE.

As Figure 8.14 shows, the final position error is zero at the end of the run which is a requirement of a point-to-point move. Comparing with Figure 8.18, it is seen that the maximum position errors occur when either the acceleration or deceleration is at a maximum. This is especially true at about 0.1 sec where, in addition, the jerk ($\triangleq d\alpha/dt$) is discontinuous.

Figure 8.15 is a plot of the estimated speed and the reference speed. Note the excellent speed tracking despite the time-varying flux (see Figure 8.17).

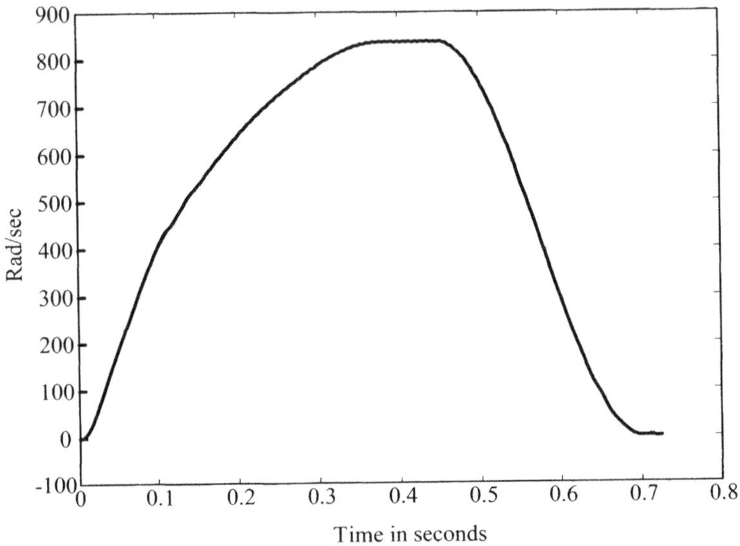

FIGURE 8.15. ω and ω_{ref} in radians/sec versus time in seconds. From Bodson et al. [65], © 2004 IEEE.

Figure 8.16 is a plot of i_q along with its reference $i_{qref} \triangleq \alpha_{ref}/(\mu\psi_{dref})$. Figure 8.17 shows a plot of ψ_{dref} and ψ_d, where ψ_{dref} is the result of solving the static optimization problem of equations (8.25) and (8.26) (see Section 8.4 and Refs. [68][69]). Figures 8.16 and 8.17 indicate that ψ_d and i_q vary significantly, yet the input–output controller forces the resultant torque $\tau = J\mu\psi_d i_q$ to provide smooth tracking of the position and speed as shown in Figures 8.13 and 8.15, respectively.

The dashed line in Figure 8.18 is a plot of the reference torque τ_{ref} ($\alpha_{ref} = \tau_{ref}/J$) corresponding to the position and speed references given in Figures 8.13 and 8.15. The solid line is a plot of the motors' optimum achievable torque by the motor given the voltage and current constraints [i.e., the solution $\tau_{optimum}$ for τ in equations (8.25) and (8.26)]. Note the closeness of the reference to the optimum.

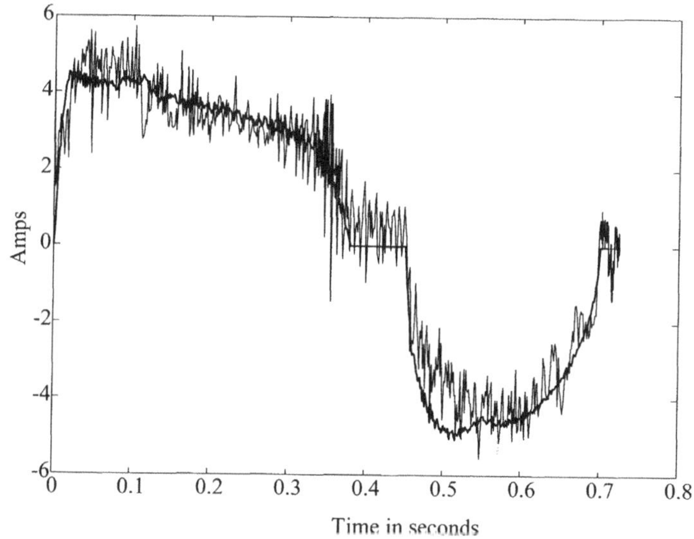

FIGURE 8.16. i_q and i_{qref} in amperes versus time in seconds. From Bodson et al. [65], © 2004 IEEE.

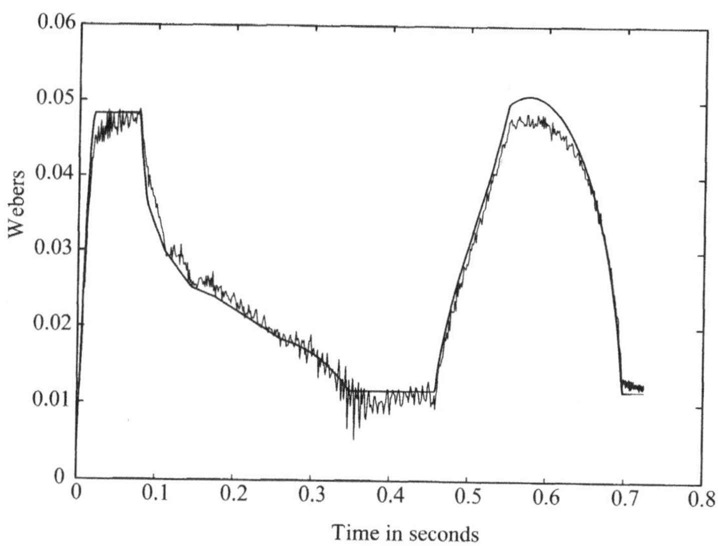

FIGURE 8.17. ψ_d and ψ_{dref} in webers versus time in seconds. From Bodson et al. [65], © 2004 IEEE.

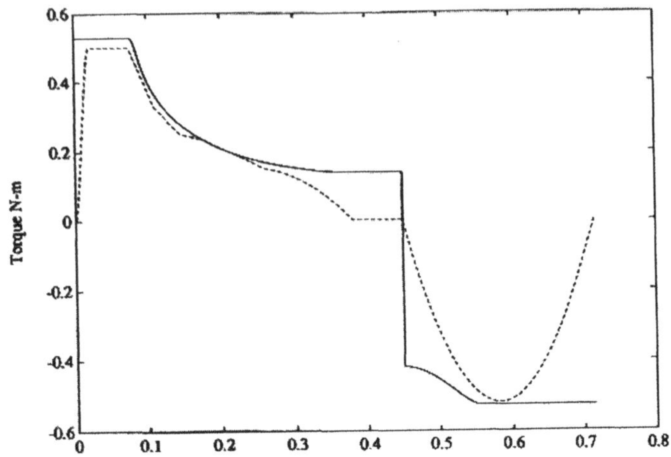

FIGURE 8.18. $\tau_{optimum}$ and τ_{ref} in N-m versus time in seconds. From Bodson et al. [65], © 2004 IEEE.

This high-speed experiment (i.e., the trajectories of Figures 8.13, 8.15, and 8.17) was attempted using a field-oriented controller and was not successful as the phase voltages exceed their limits. That is, the input-output linearization controller was essential for obtaining such high performance.

Remarks This control algorithm depends on the inner PI current control loops working satisfactorily. That is, they must be given sufficient time, as well as sufficient source voltage, to track the current references. As pointed out above, the dominant term in the equation for di_q/dt [fifth equation of (8.9)] is the *back emf* term given by $Mn_p\omega\psi_d/L_R \approx n_p\omega\psi_d$ (see problem 9). When n_p is large, a small change in ω results in a correspondingly large change in the back emf, making it difficult (i.e., takes time) for the PI current controller to overcome this disturbance in order to track the reference i_{qr}. Typically, induction motors do not have a large number of pole pairs; and even if they do (say $n_p = 12$), they are not usually run at high speeds. Consequently, the PI current loops usually perform satisfactorily. On the other hand, permanent magnet stepping (synchronous) motors typically have 50 pole pairs. In that case, the current dynamics must be taken into account (rather than swamped out by high-gain feedback) in order to achieve high performance. That is, for the same trajectory, less source voltage may be needed compared to the current-command input-output controller. (See Ref. [72] where voltage command was essential in achieving high-performance motion control of a 50-pole-pair PM stepper motor.)

8.3 Observers

The field-oriented and input-output controllers require the rotor flux linkages in order to implement them. The simplest type of algorithm to estimate them based on observer theory is now presented. Also, when a position sensor is used, a speed sensor is typically *not* available. Instead, one usually numerically differentiates the position measurement to get the speed. However, it is shown how a smoother estimate can be found using a speed observer.

8.3.1 Flux Observer

Recall from (8.4) the dynamic equations of the flux linkages are given by ($\eta = 1/T_R$)

$$\begin{aligned} \frac{d}{dt}\psi_{Ra} &= -\eta\psi_{Ra} - n_p\omega\psi_{Rb} + \eta M i_{Sa} \\ \frac{d}{dt}\psi_{Rb} &= -\eta\psi_{Rb} + n_p\omega\psi_{Ra} + \eta M i_{Sb}. \end{aligned} \tag{8.34}$$

A straightforward way to estimate the fluxes ψ_{Ra} and ψ_{Rb} is to simply implement a *real-time* simulation of the equations (8.34) on the controller processor. That is, the currents i_{Sa} and i_{Sb} are sampled from the motor through analog to digital (A/D) converters, the speed ω is known through a sensor and these quantities are then used to run the following *real-time* simulation of the flux linkage equations

$$\begin{aligned} \frac{d}{dt}\hat{\psi}_{Ra} &= -\eta\hat{\psi}_{Ra} - n_p\omega\hat{\psi}_{Rb} + \eta M i_{Sa} \\ \frac{d}{dt}\hat{\psi}_{Rb} &= -\eta\hat{\psi}_{Rb} + n_p\omega\hat{\psi}_{Ra} + \eta M i_{Sb} \end{aligned} \tag{8.35}$$

on the controller processor. The solutions to these equations are then used as the *estimates* of the fluxes for use in the feedback control algorithm.

Note that for this to work, it is implicitly assumed that (i) the model (8.34) is the correct dynamic model of the flux linkages, (ii) the parameters η and M are known, (iii) the currents and speed are measured precisely, and (iv) the numerical integration of (8.35) is done accurately. Under these conditions, it is now shown that the flux linkage estimates given by the solution to (8.35) will converge to the correct values even if the initial conditions $\psi_{Ra}(0)$ and $\psi_{Rb}(0)$ for (8.35) are *unknown*. This is important because it also implies the observer will work even in the presence of disturbances. For example, suppose at time t_1 there is noise on the measurement of $\omega(t_1)$. Further, at time $t_1 + \delta$, say, the noise is no longer present on the speed measurement. Then, for $t_1 < t < t_1 + \delta$, the speed is measured as $\omega(t) + n(t)$ rather than the correct value of $\omega(t)$. This incorrect measurement is then being used in (8.35) to calculate the estimates $\hat{\psi}_{Ra}(t)$ and $\hat{\psi}_{Rb}(t)$. Consequently, these estimates are now incorrect. However, for $t > t_1 + \delta$, the measured speed is again $\omega(t)$, that is, the correct value. Thus, one can model the situation at time $t = t_1 + \delta$ as (8.35) being the *correct* equations with the initial conditions $\hat{\psi}_{Ra}(t_1 + \delta)$ and $\hat{\psi}_{Rb}(t_1 + \delta)$ being *unknown.* If the estimates from (8.35) converge to the true values irrespective of the initial conditions, the estimator (8.35) is able to recover from disturbances in the measurements.

To show the convergence, subtract (8.35) from (8.34) to obtain error system

$$\begin{aligned}
\dot{\varepsilon}_{Ra} &= -\eta\varepsilon_{Ra} - n_p\omega\varepsilon_{Rb} \\
\dot{\varepsilon}_{Rb} &= -\eta\varepsilon_{Rb} + n_p\omega\varepsilon_{Ra}
\end{aligned} \tag{8.36}$$

where $\varepsilon_{Ra} \triangleq \psi_{Ra} - \hat{\psi}_{Ra}$, $\varepsilon_{Rb} \triangleq \psi_{Rb} - \hat{\psi}_{Rb}$ are the errors in the estimates.

Consider a (Lyapunov) function defined by

$$\begin{aligned}
V(t) &\triangleq \left(\psi_{Ra}(t) - \hat{\psi}_{Ra}(t)\right)^2 + \left(\psi_{Rb}(t) - \hat{\psi}_{Rb}(t)\right)^2 \\
&= \varepsilon_{Ra}^2(t) + \varepsilon_{Rb}^2(t).
\end{aligned}$$

If it can be shown $V(t) \to 0$ as $t \to \infty$, then $\hat{\psi}_{Ra}(t) \to \psi_{Ra}(t)$, $\hat{\psi}_{Rb}(t) \to \psi_{Rb}(t)$ as $t \to \infty$. To do this, compute

$$\begin{aligned}
dV/dt &= 2\varepsilon_{Ra}\dot{\varepsilon}_{Ra} + 2\varepsilon_{Rb}\dot{\varepsilon}_{Rb} \\
&= 2\varepsilon_{Ra}\left(-\eta\varepsilon_{Ra} - n_p\omega\varepsilon_{Rb}\right) + 2\varepsilon_{Rb}\left(-\eta\varepsilon_{Rb} + n_p\omega\varepsilon_{Ra}\right) \\
&= -2\eta\left(\varepsilon_{Ra}^2 + \varepsilon_{Rb}^2\right) \\
&= -2\eta V.
\end{aligned}$$

That is, $dV/dt = -2\eta V$ with solution $V(t) = V(0)e^{-2\eta t}$. Now, $V(0) = \left(\psi_{Ra}(0) - \hat{\psi}_{Ra}(0)\right)^2 + \left(\psi_{Rb}(0) - \hat{\psi}_{Rb}(0)\right)^2$ is unknown, but $V(t) \to 0$ regardless of the value of $V(0)$ and thus $\hat{\psi}_{Ra}(t) \to \psi_{Ra}(t)$, $\hat{\psi}_{Rb}(t) \to \psi_{Rb}(t)$ as $t \to \infty$ independent of the initial conditions used for (8.35).

The nice thing about the development of this observer based on the model (8.34) is that one is able to show that the estimate converges to the actual flux. An equivalent representation of the system (8.34) is its polar coordinate form. That is, defining

$$\begin{aligned}
\rho &\triangleq \tan^{-1}\left(\psi_{Rb}/\psi_{Ra}\right) \\
\psi_d &\triangleq \sqrt{\psi_{Ra}^2 + \psi_{Rb}^2}
\end{aligned}$$

one then defines a flux estimator by

$$\begin{aligned}
\frac{d\hat{\rho}}{dt} &= n_p\omega + \eta M\hat{i}_q/\hat{\psi}_d \\
&= n_p\omega + \eta M\Big(-i_{Sa}\sin(\hat{\rho}) + i_{Sb}\cos(\hat{\rho})\Big)/\hat{\psi}_d \\
\frac{d\hat{\psi}_d}{dt} &= -\eta\hat{\psi}_d + \eta M\hat{i}_d \\
&= -\eta\hat{\psi}_d + \eta M\Big(i_{Sa}\cos(\hat{\rho}) + i_{Sb}\sin(\hat{\rho})\Big).
\end{aligned} \tag{8.37}$$

As the estimator (8.37) is just the polar coordinate form of (8.35), it follows that its solution will converge to the true values for ρ and ψ_d. The usefulness of the representation (8.37) is that the variables ψ_d, i_d, and i_q vary much slower than variables $\psi_{Ra}(t)$ and $\psi_{Rb}(t)$. For example, if the motor is running at a constant speed, the variables ψ_d, i_d, and i_q are constant no matter how high the speed. In contrast, $\psi_{Ra}(t)$ and $\psi_{Rb}(t)$ vary at the stator frequency, which at high speeds is quite large. This high frequency makes the numerical integration of (8.35) difficult in that a small time step is required.

Remarks

To implement either observer (8.35) or (8.37), η and M must be known. The parameter $\eta = 1/T_R = R_R/L_R$ and the resistance R_R can vary by 100% due to ohmic heating of the rotor windings [2].

8.3.2 Speed Observer

In a position (servo) control system, typically one has a position measurement (e.g., from an optical encoder), but a speed measurement is not available. It is standard practice in industry to compute the speed by a discrete differentiation of the position output (backward difference) from the optical encoder as

$$\hat{\omega}_{bd}(kT) = \frac{\theta(kT) - \theta((k-1)T)}{T} \tag{8.38}$$

where T is the sample period. The difference $\theta(kT) - \theta((k-1)T)$ can be in error by no more than one count so that, using a 2000 pulse/rev encoder, the error in this estimate of the speed is tightly bounded by $(2\pi/2000)/T$. This noise is particularly significant at high sample rates and moderate to low speeds as less encoder counts are detected per sample period than at higher speeds. For example, in the experiment presented in Section 8.2.2 where the motor is turned 180° in 73 msec, the top speed of the motor was less than 75 rad/sec and a sample rate of 8 kHz was required. In this circumstance, the error bound on the speed computed by differentiation is $(2\pi/2000)8000 = 25.13$ rad/sec. This is shown in Figure 8.19. However, this noise (the difference $\omega - \hat{\omega}_{bd}$) has a high frequency, and for the most part the machine responds to the average value of $\hat{\omega}_{bd}$ over several time steps, which is close to the actual value.

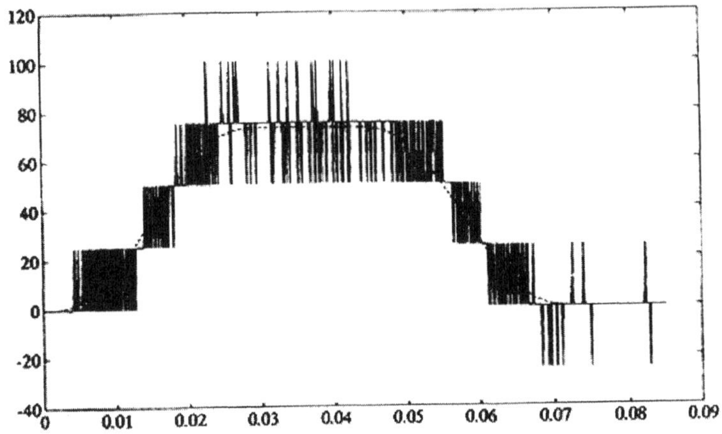

FIGURE 8.19. $\hat{\omega}_{bd}$ and ω_{ref} in radians/sec versus time in seconds. From Bodson et al. [67], © 2004 IEEE.

The speed estimate used to obtain Figure 8.8 (repeated in Figure 8.20) of Section 8.2.2 is much smoother and was computed using an observer.

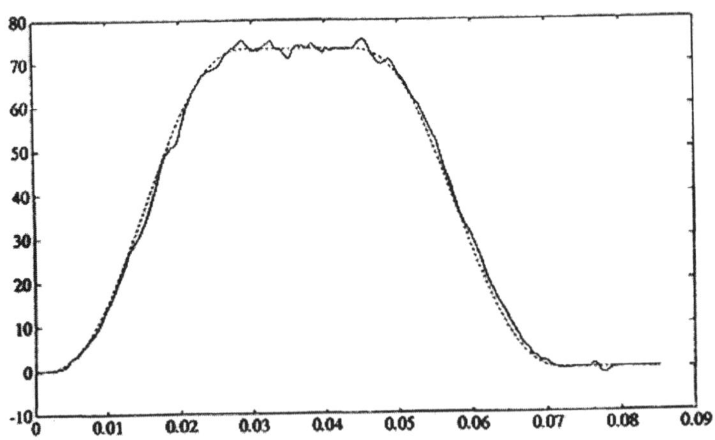

FIGURE 8.20. $\hat{\omega}$ and ω_{ref} in radians/sec versus time in seconds. From Bodson et al. [67], © 2004 IEEE.

This approach is now described. In the field-oriented coordinates, the equation governing the speed in an induction motor is given by

$$\begin{aligned} \frac{d\theta}{dt} &= \omega \\ \frac{d\omega}{dt} &= \mu\psi_d i_q - (f/J)\omega \end{aligned} \tag{8.39}$$

The quantities ψ_d and i_q are not known, but can be estimated as shown above. Thus, consider an observer defined by

$$\begin{aligned} \frac{d\hat{\theta}}{dt} &= \hat{\omega} + \ell_1(\theta - \hat{\theta}) \\ \frac{d\hat{\omega}}{dt} &= \mu\hat{\psi}_d\hat{i}_q - (f/J)\hat{\omega} + \ell_2(\theta - \hat{\theta}). \end{aligned} \tag{8.40}$$

Assuming that $\hat{\psi}_d\hat{i}_q \to \psi_d i_q$ fast enough so that there are essentially equal, one subtracts (8.40) from (8.39) to obtain the error system

$$\begin{aligned} \frac{de_1}{dt} &= e_2 - \ell_1 e_1 \\ \frac{de_2}{dt} &= -(f/J)e_2 - \ell_2 e_1 \end{aligned} \tag{8.41}$$

where $e_1 = \theta - \hat{\theta}$ and $e_2 = \omega - \hat{\omega}$. The characteristic polynomial for the error system is then

$$\det\left(s\begin{bmatrix} 1 & 0 \\ 0 & 1 \end{bmatrix} - \begin{bmatrix} -\ell_1 & 1 \\ -\ell_2 & -f/J \end{bmatrix}\right) = s^2 + (\ell_1 + f/J)s + \ell_2 + \ell_1(f/J).$$

Choosing

$$\begin{aligned} \ell_1 &= r_1 + r_2 - f/J \\ \ell_2 &= r_1 r_2 - \ell_1(f/J) \end{aligned}$$

puts the roots of the characteristic polynomial at $-r_1$ and $-r_2$. It turns out in practice that this estimate of ω is much better than plain differentiation of the position [67]. This estimate is then used for the speed feedback and in the flux observer (8.37).

With the gains set as $\ell_1 = 1.8 \times 10^3$ and $\ell_2 = 8 \times 10^5$, the result of using the speed observer (8.40) is given in Figure 8.20. This same observer (with the same gain settings) was also used to obtain the speed estimate for the input output linearization controller in Figure 8.15. This approach to estimating ω can be modified to account for a load torque on the motor (see problem 25). A disadvantage of a speed observer is that the machine parameters must be known accurately [μ, f/J for (8.40) and $\eta = 1/T_R, M$ for the flux observer] while the backward difference estimator (8.38) is independent of the machine parameters.

Remark The flux estimator (8.37) and speed estimator (8.40) are coupled; that is, to estimate the flux and speed, it is required to integrate the fourth-order system given by

$$
\begin{aligned}
\frac{d\hat{\rho}}{dt} &= n_p\hat{\omega} + \eta M\left(-i_{Sa}\sin(\hat{\rho}) + i_{Sb}\cos(\hat{\rho})\right)/\hat{\psi}_d \\
\frac{d\hat{\psi}_d}{dt} &= -\eta\hat{\psi}_d + \eta M\left(i_{Sa}\cos(\hat{\rho}) + i_{Sb}\sin(\hat{\rho})\right) \\
\frac{d\hat{\theta}}{dt} &= \hat{\omega} + \ell_1(\theta - \hat{\theta}) \\
\frac{d\hat{\omega}}{dt} &= \mu\hat{\psi}_d\left(-i_{Sa}\sin(\hat{\rho}) + i_{Sb}\cos(\hat{\rho})\right) - (f/J)\hat{\omega} + \ell_2(\theta - \hat{\theta})
\end{aligned}
$$

where i_{Sa}, i_{Sb} and θ are the measured "inputs" to this observer.

*8.3.3 Verghese–Sanders Flux Observer**

Using the observer (8.35), it was shown that the magnitude of the error $V(t) = (\psi_{Ra}(t) - \hat{\psi}_{Ra}(t))^2 + (\psi_{Rb}(t) - \hat{\psi}_{Rb}(t))^2$ went to zero exponentially as $V(t) = V(0)e^{-2\eta t}$. For example, with $L_R = 0.085$ Henries and $R_R = 0.8$ ohms, $\eta = 9.4$ so that $V(t) = V(0)e^{-2(9.4)t} = V(0)e^{-t/.05}$ and the error dies out after four time constants or $4 \times 0.05 = 0.2$ sec. It is usually preferable to force the error to go to zero *arbitrarily* fast rather than just as $e^{-2\eta t}$.

An ingenious approach has been given by Verghese and Sanders [73] that allows one to specify an *arbitrary* rate of convergence for the flux estimator. This was done by first observing that the four electrical equations in (8.4) may be broken up into two subsystems consisting of the flux *system model*

$$
\begin{aligned}
\frac{d}{dt}\psi_{Ra} &= -\frac{R_R}{L_R}\psi_{Ra} - n_p\omega\psi_{Rb} + \frac{MR_R}{L_R}i_{Sa} \\
\frac{d}{dt}\psi_{Rb} &= -\frac{R_R}{L_R}\psi_{Rb} + n_p\omega\psi_{Ra} + \frac{MR_R}{L_R}i_{Sb}
\end{aligned}
\qquad (8.42)
$$

and the voltage *measurement model*

$$
\begin{aligned}
u_{Sa} &= R_S i_{Sa} + \sigma L_S\frac{di_{Sa}}{dt} + \frac{M}{L_R}\frac{d\psi_{Ra}}{dt} \\
u_{Sb} &= R_S i_{Sb} + \sigma L_S\frac{di_{Sb}}{dt} + \frac{M}{L_R}\frac{d\psi_{Rb}}{dt}.
\end{aligned}
\qquad (8.43)
$$

The insight given by this decomposition is that a measurement of the voltages (assuming the currents and their derivatives are known) provides an estimate of the *derivative* of the flux linkages by equation (8.43), which in turn can then be used to force the flux error to zero arbitrarily fast. Specifically, consider the estimator based on (8.42) defined by

$$\begin{aligned}\frac{d}{dt}\hat{\psi}_{Ra} &= -\frac{R_R}{L_R}\hat{\psi}_{Ra} - n_p\omega\hat{\psi}_{Rb} + \frac{MR_R}{L_R}i_{Sa} + k(\hat{u}_{Sa} - u_{Sa})\\ \frac{d}{dt}\hat{\psi}_{Rb} &= -\frac{R_R}{L_R}\hat{\psi}_{Rb} + n_p\omega\hat{\psi}_{Ra} + \frac{MR_R}{L_R}i_{Sb} + k(\hat{u}_{Sb} - u_{Sb})\end{aligned} \tag{8.44}$$

with the *predicted measurements* $\hat{u}_{Sa}$ and $\hat{u}_{Sb}$ given by

$$\begin{aligned}\hat{u}_{Sa} &= R_S i_{Sa} + \sigma L_S \frac{di_{Sa}}{dt} + \frac{M}{L_R}\frac{d\hat{\psi}_{Ra}}{dt}\\ \hat{u}_{Sb} &= R_S i_{Sb} + \sigma L_S \frac{di_{Sb}}{dt} + \frac{M}{L_R}\frac{d\hat{\psi}_{Rb}}{dt}.\end{aligned} \tag{8.45}$$

These expressions for $\hat{u}_{Sa}$ and $\hat{u}_{Sb}$ (8.45) are then substituted into (8.44) and, after some rearrangement, result in the flux estimator given by

$$\begin{aligned}\frac{d}{dt}\hat{\psi}_{Ra} &= \frac{1}{1-k\dfrac{M}{L_R}}\Bigg(-\frac{R_R}{L_R}\hat{\psi}_{Ra} - n_p\omega\hat{\psi}_{Rb} + \frac{MR_R}{L_R}i_{Sa}\\ &\qquad + k\Big(R_S i_{Sa} + \sigma L_S\frac{di_{Sa}}{dt} - u_{Sa}\Big)\Bigg)\\ \frac{d}{dt}\hat{\psi}_{Rb} &= \frac{1}{1-k\dfrac{M}{L_R}}\Bigg(-\frac{R_R}{L_R}\hat{\psi}_{Rb} + n_p\omega\hat{\psi}_{Ra} + \frac{MR_R}{L_R}i_{Sb}\\ &\qquad + k\Big(R_S i_{Sb} + \sigma L_S\frac{di_{Sb}}{dt} - u_{Sb}\Big)\Bigg).\end{aligned} \tag{8.46}$$

The idea here is that the system (8.46) is implemented in real time using the measured voltages u_{Sa} and u_{Sb} and currents i_{Sa} and i_{Sb}, the measured speed ω, and the motor parameter values. The solutions $\hat{\psi}_{Ra}$ and $\hat{\psi}_{Rb}$ are then used as the estimate of the state variables ψ_{Ra} and ψ_{Rb}.

To show that the estimator (8.46) works (i.e., $\hat{\psi}_{Ra} \to \psi_{Ra}, \hat{\psi}_{Rb} \to \psi_{Rb}$ if the gain k is chosen correctly), subtract (8.42) from (8.44) to obtain the error system

$$\begin{aligned}\frac{de_{Ra}}{dt} &= -\eta e_{Ra} - n_p\omega e_{Rb} + k(\hat{u}_{Sa} - u_{Sa})\\ \frac{de_{Rb}}{dt} &= -\eta e_{Rb} + n_p\omega e_{Ra} + k(\hat{u}_{Sb} - u_{Sb}).\end{aligned} \tag{8.47}$$

where $e_{Ra} \triangleq \hat{\psi}_{Ra} - \psi_{Ra}, e_{Rb} \triangleq \hat{\psi}_{Rb} - \psi_{Rb}$.

Using (8.43) and (8.45), this is rewritten as

$$\begin{aligned} \frac{de_{Ra}}{dt} &= -\eta e_{Ra} - n_p\omega e_{Rb} + \frac{kM}{L_R}\frac{de_{Ra}}{dt} \\ \frac{de_{Rb}}{dt} &= -\eta e_{Rb} + n_p\omega e_{Ra} + \frac{kM}{L_R}\frac{de_{Rb}}{dt} \end{aligned} \qquad (8.48)$$

or

$$\begin{aligned} \frac{de_{Ra}}{dt} &= \frac{1}{1-k(M/L_R)}\left(-\eta e_{Ra} - n_p\omega e_{Rb}\right) \\ \frac{de_{Rb}}{dt} &= \frac{1}{1-k(M/L_R)}\left(-\eta e_{Rb} + n_p\omega e_{Ra}\right). \end{aligned} \qquad (8.49)$$

Use of the (Lyapunov) function

$$V(e_{Ra}, e_{Rb}) \triangleq \left(e_{Ra}^2 + e_{Rb}^2\right)/2 \qquad (8.50)$$

with the error system (8.49) results in

$$\begin{aligned} \frac{dV}{dt} &= e_{Ra}\frac{de_{Ra}}{dt} + e_{Rb}\frac{de_{Rb}}{dt} \\ &= \frac{1}{1-k(M/L_R)}\Big(e_{Ra}(-\eta e_{Ra} - n_p\omega e_{Rb}) + e_{Rb}(-\eta e_{Rb} + n_p\omega e_{Ra})\Big) \\ &= -\frac{\eta}{1-k(M/L_R)}\left(e_{Ra}^2 + e_{Rb}^2\right) \\ &= -\frac{2\eta}{1-k(M/L_R)}V. \end{aligned}$$

Clearly then,

$$V(t) = V(0)e^{-\left(\frac{2\eta}{1-kM/L_R}\right)t}$$

which can be made to go to zero arbitrarily fast by choosing the gain k close to (but less than) L_R/M. Note that by choosing $k > L_R/M$, the estimation error actually diverges. However, as $L_R = (1+\sigma_R)M$ where $\sigma_R > 0$ is a small leakage constant [1], it is always true that $L_R/M > 1$. Consequently, by choosing $k \leq 1$, one is assured of avoiding this problem.

The estimator (8.46) requires the derivative of the current to implement it. This can be avoided using the method as described in Ref. [73]. Specifically, (8.45) is substituted into (8.44) and rearranged by bringing all of the terms with derivatives to the left-hand side, resulting in

$$\begin{aligned}\left(1-k\frac{M}{L_R}\right)\frac{d\hat{\psi}_{Ra}}{dt}-k\sigma L_S\frac{di_{Sa}}{dt} &= -\frac{R_R}{L_R}\hat{\psi}_{Ra}-n_p\omega\hat{\psi}_{Rb}+\frac{MR_R}{L_R}i_{Sa}\\ &\quad +k\Big(R_S i_{Sa}-u_{Sa}\Big)\\ \left(1-k\frac{M}{L_R}\right)\frac{d\hat{\psi}_{Rb}}{dt}-k\sigma L_S\frac{di_{Sb}}{dt} &= -\frac{R_R}{L_R}\hat{\psi}_{Rb}+n_p\omega\hat{\psi}_{Ra}+\frac{MR_R}{L_R}i_{Sb}\\ &\quad +k\Big(R_S i_{Sb}-u_{Sb}\Big).\end{aligned} \tag{8.51}$$

Defining the auxiliary variables

$$\begin{aligned}\hat{z}_1 &= \left(1-k\frac{M}{L_R}\right)\hat{\psi}_{Ra}-k\sigma L_S i_{Sa}\\ \hat{z}_2 &= \left(1-k\frac{M}{L_R}\right)\hat{\psi}_{Rb}-k\sigma L_S i_{Sb}\end{aligned}$$

with inverse

$$\begin{aligned}\hat{\psi}_{Ra} &= \Big(\hat{z}_1+k\sigma L_S i_{Sa}\Big)\Big/\left(1-k\frac{M}{L_R}\right)\\ \hat{\psi}_{Rb} &= \Big(\hat{z}_2+k\sigma L_S i_{Sb}\Big)\Big/\left(1-k\frac{M}{L_R}\right)\end{aligned} \tag{8.52}$$

the system (8.51) has the form

$$\begin{aligned}\frac{d\hat{z}_1}{dt} &= f_1(\hat{z}_1,\hat{z}_2,\omega,i_{Sa},i_{Sb},u_{Sa})\\ \frac{d\hat{z}_2}{dt} &= f_2(\hat{z}_1,\hat{z}_2,\omega,i_{Sa},i_{Sb},u_{Sb})\end{aligned} \tag{8.53}$$

where f_1 and f_2 are simply the right-hand sides of (8.51) after substituting the expressions for $\hat{\psi}_{Ra}$ and $\hat{\psi}_{Rb}$ in terms of $\hat{z}_1$ and $\hat{z}_2$ from (8.52). The real-time solutions to (8.53) are then substituted into (8.52) to obtain the flux estimates.

Remarks As before, there are numerical issues to consider. In this case, the system (8.53) is driven by the stator currents which vary at the electrical frequency $\omega_S = \omega_{slip} + n_p\omega$ which can be quite large. This necessitates a small step size to accurately carry out the numerical integration of (8.53). An appropriate way to handle this issue is given in Ref. [73] (see problem 22).

As in any model-based scheme, these estimates are sensitive to parameter error. In this case, the significant parameter is the rotor time constant $T_R = 1/\eta$ as its value can vary as the rotor windings heat up or cool down. An analysis of this sensitivity is also given in [73]. A generalization of this observer is the Martin–Rouchon observer Refs. [74][75] outlined in problem 27.

8.4 Optimal Field Weakening*

In Section 8.2.3, field weakening was introduced as means to achieve higher speeds without violating the stator voltage constraints with the trade off being that the available torque is less. In that approach, the reference for the (magnitude of the) flux linkage was specified by (8.24). In this section, the general problem of extracting the maximum torque from an induction motor without violating the voltage or current constraints is addressed systematically. There has been considerable work in this area including the approaches in Refs. [71][76][77][78][79][80]. Here the approach in Refs. [68][69] is presented. The field-oriented model of the motor as given in (8.9) is used as a starting point and, in order to obtain a tractable solution, the problem is formulated with the machine in steady state. Specifically, consider the following formulation:

Maximize the steady-state torque ($J\mu = n_p M/L_R$)

$$\frac{n_p M}{L_R}\psi_d i_q$$

subject to the constraints

$$\begin{aligned} u_d^2 + u_q^2 &\leq V_{\max}^2 \\ i_d^2 + i_q^2 &\leq I_{\max}^2. \end{aligned}$$

Remark The actual physical constraints are

$$\begin{aligned} |u_{Sa}| &\leq V_{\max}, \quad |u_{Sb}| \leq V_{\max} \\ |i_{Sa}| &\leq V_{\max}, \quad |i_{Sb}| \leq V_{\max}. \end{aligned}$$

However, at constant rotor speed in steady state, the voltages and currents have the form

$$\begin{aligned} u_{Sa} &= V\cos(\omega_S t + \phi) \\ u_{Sb} &= V\cos(\omega_S t + \phi) \\ i_{Sa} &= I\cos(\omega_S t + \varphi) \\ i_{Sb} &= I\cos(\omega_S t + \varphi) \end{aligned}$$

where ω_S is the stator electrical frequency. As a consequence,

$$\begin{aligned} V^2 &= u_{Sa}^2 + u_{Sb}^2 = u_d^2 + u_q^2 \leq V_{\max}^2 \\ I^2 &= i_{Sa}^2 + i_{Sb}^2 = i_d^2 + i_q^2 \leq I_{\max}^2 \end{aligned}$$

so that in steady state the conditions are equivalent.

To proceed with the solution of the torque optimization problem, the system of equations (8.9) are written in steady state as

$$0 = \mu\psi_d i_q - (f/J)\omega - \tau_L/J \tag{8.54}$$

$$0 = -\psi_d + M i_d \tag{8.55}$$

$$0 = -\left(\frac{M^2 R_R}{L_R^2} + R_S\right) i_d + \frac{\eta M}{L_R}\psi_d + \sigma L_S n_p \omega i_q + \sigma L_S M \frac{i_q^2}{T_R \psi_d} + u_d \tag{8.56}$$

$$0 = -\left(\frac{M^2 R_R}{L_R^2} + R_S\right) i_q - n_p \omega \psi_d \frac{M}{L_R} - \sigma L_S n_p \omega i_d - \sigma L_S M \frac{i_q i_d}{T_R \psi_d} + u_q \tag{8.57}$$

$$\frac{d\rho}{dt} = n_p\omega + M i_q / \left(T_R \psi_d\right) = \text{ constant.} \tag{8.58}$$

Using $T_S \triangleq L_S/R_S$, $\psi_d = M i_d$, and making the substitution $\sigma = 1 - M^2/(L_S L_R)$ into (8.57) results in (8.56) and (8.57) simplifying to

$$\frac{u_d}{R_S} = i_d - \sigma T_S i_q \left(n_p\omega + \frac{i_q}{T_R i_d}\right) \tag{8.59}$$

$$\frac{u_q}{R_S} = i_q + T_S i_d \left(n_p\omega + \frac{i_q}{T_R i_d}\right). \tag{8.60}$$

8.4.1 Torque Optimization Under Current Constraints

In the case of only the current constraint being active, the problem reduces to maximizing

$$\tau = \frac{n_p M}{L_R}\psi_d i_q = \frac{n_p M^2}{L_R} i_d i_q \tag{8.61}$$

subject to the constraint

$$i_d^2 + i_q^2 = I^2 \leq I_{\max}^2.$$

Defining

$$\delta = i_q / i_d$$

the expression for the torque may be rewritten as

$$\tau = \frac{n_p M^2}{L_R}\delta i_d^2 = \frac{n_p M^2}{L_R} I^2 \frac{\delta}{1+\delta^2}. \tag{8.62}$$

Optimizing (max or min) the torque requires $I = I_{\max}$. Also, solving $\partial\tau/\partial\delta = 0$ gives $\delta = \pm 1$ and checking $\partial^2\tau/\partial\delta^2$ at $\delta = \pm 1$ results in the maximum torque for $I = I_{\max}$ and $\delta = 1$ ($i_d = I_{\max}/\sqrt{2}$) and the minimum torque for $I = I_{\max}$ and $\delta = -1$ ($i_d = -I_{\max}/\sqrt{2}$).

8.4.2 Torque Optimization Under Voltage Constraints

In the case of only the voltage constraint being active, the problem reduces to maximizing

$$\tau = \frac{n_p M}{L_R}\psi_d i_q = \frac{n_p M^2}{L_R} i_d i_q \tag{8.63}$$

subject to the constraint

$$u_d^2 + u_q^2 = V^2 \leq V_{\max}^2.$$

Again letting $\delta = i_q/i_d$ and using (8.59) and (8.60), the square of the voltage V^2 may be written as

$$\begin{aligned}\frac{V^2}{R_S^2} &= \frac{u_d^2 + u_q^2}{R_S^2}\\ &= \left(i_d - \sigma T_S i_q\left(n_p\omega + \frac{i_q}{T_R i_d}\right)\right)^2 + \left(i_q + T_S i_d\left(n_p\omega + \frac{i_q}{T_R i_d}\right)\right)^2\\ &= i_d^2\left(\left(1 - \sigma T_S\delta(n_p\omega + \delta/T_R)\right)^2 + \left(\delta + T_S\left(n_p\omega + \delta/T_R\right)\right)^2\right)\end{aligned} \tag{8.64}$$

and the torque is then rewritten as

$$\begin{aligned}\tau &= \frac{n_p M^2}{L_R}\delta i_d^2\\ &= \frac{n_p M^2}{L_R R_S^2}V^2\frac{\delta}{\left(1 - \sigma T_S\delta\left(n_p\omega + \delta/T_R\right)\right)^2 + \left(\delta + T_S\left(n_p\omega + \delta/T_R\right)\right)^2}.\end{aligned} \tag{8.65}$$

Remark

This expression for the torque is comparable to the standard torque-slip equation for the torque. Specifically, with $\omega_S = n_p\omega + \delta/T_R$ the electrical frequency, $S_p = 1/(\sigma\omega_S T_R)$ the normalized pull-out slip, and $S = (\omega_S - n_p\omega)/\omega_S = (\delta/T_R)/\omega_S$ the normalized slip, this last expression for the torque reduces to (letting $R_S \to 0$)

$$\tau = n_p \frac{3}{2}\frac{1-\sigma}{\sigma}\frac{U_S^2}{\omega_S^2 L_S}\frac{2}{S/S_p + S_p/S} \tag{8.66}$$

where $U_S \triangleq \sqrt{2/3}\,(V/\sqrt{2})$ (see problem 31). ($V/\sqrt{2}$ is the two-phase equivalent rms voltage so that $\sqrt{2/3}\,(V/\sqrt{2})$ is the line-to-neutral rms voltage in the three-phase machine.) The expression (8.65) gives the torque as a function of the slip ($\delta/T_R = \omega_S - n_p\omega$) with the *rotor speed* ω held constant while the expression (8.66) gives the torque as function of the (normalized) slip with the *electrical frequency* ω_S held constant.

■

The denominator of (8.65) is always positive so that $\tau > 0$ for $\delta > 0$ and $\tau < 0$ for $\delta < 0$. As the torque (8.65) is proportional to V^2, the optimum torque is achieved choosing δ to optimize (8.65) with $V = V_{\max}$. To compute the maxima and minima of τ with respect to δ, one sets

$$\begin{aligned}\frac{\partial \tau}{\partial \delta} &= \frac{n_p M^2}{L_R R_S^2} V^2 \frac{p(\delta)}{\left((1 - \sigma T_S \delta\,(n_p\omega + \delta/T_R))^2 + (\delta + T_S\,(n_p\omega + \delta/T_R))^2\right)^2} \\ &= 0 \end{aligned} \tag{8.67}$$

where

$$\begin{aligned} p(\delta) &= -3\sigma^2 \left(\frac{T_S}{T_R}\right)^2 \delta^4 - 4\sigma^2 \left(\frac{T_S}{T_R}\right) (T_S n_p \omega)\,\delta^3 \\ &\quad + \left(2\sigma\left(\frac{T_S}{T_R}\right) - \sigma^2 (T_S n_p\omega)^2 - \left(1 + \left(\frac{T_S}{T_R}\right)\right)^2\right)\delta^2 \\ &\quad + 0\cdot\delta + 1 + (T_S n_p \omega)^2 . \end{aligned} \tag{8.68}$$

The solutions to $p(\delta)$ give the values of δ corresponding to local maxima and minima. Let δ_0 be a solution to $p(\delta) = 0$. The second derivative at such a δ_0 is given by

$$\begin{aligned}\left.\frac{\partial^2 \tau}{\partial \delta^2}\right|_{\delta_0} &= \frac{n_p M^2}{L_R R_S^2} V^2 \\ &\quad \times \frac{p'(\delta_0)}{\left((1 - \sigma T_S \delta\,(n_p\omega + \delta_0/T_R))^2 + (\delta_0 + T_S\,(n_p\omega + \delta_0/T_R))^2\right)^2}\end{aligned} \tag{8.69}$$

where

$$\begin{aligned}
p'(\delta) &= -12\sigma^2\left(\frac{T_S}{T_R}\right)^2\delta^3 - 12\sigma^2\left(\frac{T_S}{T_R}\right)(T_S n_p\omega)\,\delta^2 \\
&\quad + 2\left(2\sigma\left(\frac{T_S}{T_R}\right) - \sigma^2(T_S n_p\omega)^2 - \left(1+\left(\frac{T_S}{T_R}\right)\right)^2\right)\delta \\
&= -\delta\left(12\sigma^2\left(\frac{T_S}{T_R}\right)^2\delta^2 + 12\sigma^2\left(\frac{T_S}{T_R}\right)^2(T_R n_p\omega)\,\delta\right. \\
&\quad \left. + 2 + 4(1-\sigma)\left(\frac{T_S}{T_R}\right) + 2\left(\frac{T_S}{T_R}\right)^2 + 2\sigma^2\left(\frac{T_S}{T_R}\right)^2(T_R n_p\omega)^2\right).
\end{aligned} \tag{8.70}$$

The sign of $p'(\delta_0)$ determines whether or not δ_0 corresponds to a local minima or a local maxima.

Example 1 To study equation (8.68), consider an induction machine with parameter values $n_p = 3$, $R_S = 1.7$ ohms, $L_S = 0.014$ H, $R_R = 3.9$ ohms, $L_R = 0.014$ H, $M = 0.0117$ H, $V_{\max} = 80$ V, $I_{\max} = 6$ A, $J = 1.1 \times 10^{-4}$ kg-m^2 and $f = 1.4 \times 10^{-4}$ N-m/rad/sec. Figure 8.21 is a plot of the four roots of $p(\delta) = 0$ as the speed ω varies from 0 to 8000 rpm. In this example motor, only two of the roots are real in this speed range.

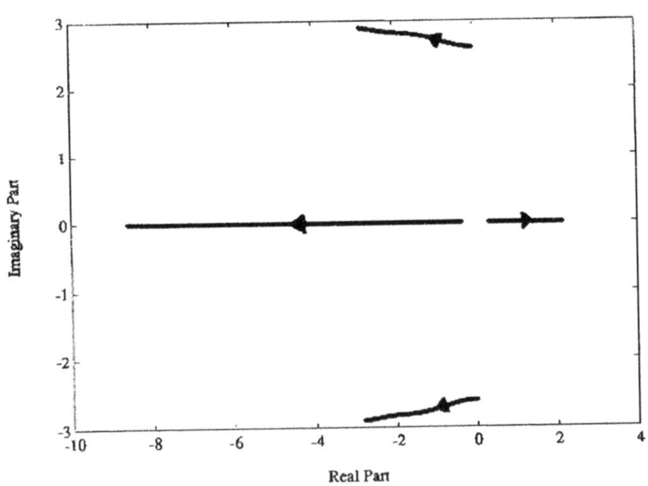

FIGURE 8.21. The four roots of $p(\delta) = 0$ for $0 \le \omega \le 8000$ rpm. From Bodson et al. [69], © 2004 IEEE.

Solutions of $p(\delta) = 0$

The polynomial $p(\delta)$ is fourth order in δ and our interest here is to determine which of the roots optimizes the torque. For $\omega \geq 0$ and constant, there are two cases to consider: $\tau > 0$ and $\tau < 0$.

$\boldsymbol{\tau > 0}$: The first case is $\tau > 0$ so that the machine is motoring. In this case $\delta > 0$ as both i_d and i_q are positive. The Routh–Hurwitz test [37][81][46] (see problem 32) implies that $p(\delta) = 0$ has exactly one positive root for all $\omega \geq 0$. Furthermore, for any $\delta > 0$, it follows by equation (8.70) that $\partial^2\tau/\partial\delta^2 < 0$ so this unique positive root corresponds to a global maximum. In Figure 8.21, this unique positive root starts out at approximately 0.323 for $\omega = 0$ and ends at approximately 2.2 for $\omega = 800$ rpm. The asymptotic value of this root is found by dividing (8.68) by $\sigma^2(T_S/T_R)^2\,(T_S n_p\omega)^2$ and letting $\omega \to \infty$ where it is seen that $p(\delta)/\left(\sigma^2(T_S/T_R)^2\,(T_S n_p\omega)^2\right) \to (T_R/T_S)^2\left(-\delta^2 + 1/\sigma^2\right)$ and the positive root goes to $1/\sigma$ ($= 3.33$ in Figure 8.21). Further, as $\omega \to \infty$,

$$\tau(\delta) \to \tau(1/\sigma) = \left(n_p M^2/L_R R_S^2\right) V^2/\left(2\sigma(T_S n_p \omega_R)^2\right).$$

$\boldsymbol{\tau < 0}$: The second case is $\tau < 0$ so that the induction machine is generating. In this case, the minimum of τ is sought (i.e., maximizing $|\tau|$ with $\tau < 0$) so that δ is negative with $i_q > 0$ and $i_d < 0$. As $p(\delta)$ is fourth order, it has four roots of which exactly one root is in the right-half plane. The remaining three roots all have negative real parts of which there are two possibilities. Either there are three real negative roots (two local minima and a local maximum) or there is one real negative root (single global minimum) and a complex conjugate pair with negative real parts.

First consider the solutions of (8.68) for ω small. In general, equation (8.68) has a unique negative root for sufficiently small speeds. To see this, let $\omega = 0$ in (8.68) to obtain

$$p(\delta) = -3\sigma^2\left(\frac{T_S}{T_R}\right)^2\delta^4 + \left(2\sigma\left(\frac{T_S}{T_R}\right) - \left(1 + \left(\frac{T_S}{T_R}\right)\right)^2\right)\delta^2 + 1 = 0.$$

Then with

$$a \triangleq 3\sigma^2\left(\frac{T_S}{T_R}\right)^2,\; b \triangleq \left(\frac{T_S}{T_R}\right)^2 + 2(1-\sigma)\left(\frac{T_S}{T_R}\right) + 1,\; c \triangleq -1$$

the quadratic formula gives

$$\delta^2 = \frac{-b \pm \sqrt{b^2 - 4ac}}{2a} \tag{8.71}$$

where $-4ac > 0$ so that $\sqrt{b^2 - 4ac} > b$. Consequently, one must choose the

+ sign in (8.71) for δ^2 to be positive. Doing so, let

$$\delta_{pos} \triangleq \sqrt{\frac{-b+\sqrt{b^2-4ac}}{2a}} \tag{8.72}$$

$$\delta_{neg} \triangleq -\sqrt{\frac{-b+\sqrt{b^2-4ac}}{2a}} \tag{8.73}$$

be the (unique) positive and negative roots, respectively, of (8.68) for $\omega = 0$. Note that $\delta_{pos} = 0.323$ in Figure 8.21.

At high enough speeds, there are always two local minima. To see this, rewrite (8.68) as

$$\begin{aligned} p(\delta) = &-3\sigma^2\left(\frac{T_S}{T_R}\right)^2\delta^4 - 4\sigma^2\left(\frac{T_S}{T_R}\right)^2(T_R n_p\omega)\,\delta^3 \\ &+\left(1+2(1-\sigma)\left(\frac{T_S}{T_R}\right)+\left(\frac{T_S}{T_R}\right)^2+\sigma^2\left(\frac{T_S}{T_R}\right)^2(T_R n_p\omega)^2\right)\delta^2 \\ &+0\cdot\delta+1+\left(\frac{T_S}{T_R}\right)^2(T_R n_p\omega)^2. \end{aligned} \tag{8.74}$$

It is now shown that one of the roots $\delta \to -T_R n_p\omega$ as $\omega \to \infty$. To do so, let $x \triangleq \delta/(T_R n_p\omega)$ in (8.74) and divide through by $-3\sigma^2\left(\frac{T_S}{T_R}\right)^2(T_R n_p\omega)^4$ to obtain

$$x^4+\frac{4}{3}x^3+\frac{1}{3}x^2 = -\frac{1+2(1-\sigma)\left(\frac{T_S}{T_R}\right)+\left(\frac{T_S}{T_R}\right)^2-1-\left(\frac{T_S}{T_R}\right)^2(T_R n_p\omega)^2}{3\sigma^2\left(\frac{T_S}{T_R}\right)^2(T_R n_p\omega)^4}. \tag{8.75}$$

Equation (8.75) shows that $x = -1$ is a solution as $\omega \to \infty$; that is, $-T_R n_p\omega$ is an asymptotic solution to $p(\delta) = 0$ for large ω. Substitution of $\delta = -T_R n_p\omega$ into (8.69) shows that $\partial^2\tau/\partial\delta^2|_{\delta=-T_R n_p\omega} > 0$ and thus corresponds to a minimum. Also, $\tau(-T_R n_p\omega) \to \left(n_p M^2/L_R R_S^2\right)V^2/\left(T_R n_p\omega_R\right)$ as $\omega \to \infty$. The other negative root corresponding to a local minimum is found by dividing (8.68) through by $\sigma^2\left(T_S/T_R\right)^2(T_R n_p\omega)^2$ and letting $\omega \to \infty$ to find $\delta \to -1/\sigma$. Again using (8.69), it is seen that this root does indeed correspond to a local minimum where

$$\tau(-1/\sigma) \to -\frac{\left(n_p M^2/L_R R_S^2\right)V^2}{2\sigma\left(T_S n_p\omega_R\right)^2}$$

as $\omega \to \infty$. Comparing the expressions for the torque at the two minima, it is seen that for ω sufficiently large, $\tau(-T_R n_p\omega) < \tau(-1/\sigma)$. As a consequence, $-1/\sigma$ corresponds to an asymptotic local minimum while $-T_R n_p\omega$

corresponds to the asymptotic global minimum for the torque. In general, at sufficiently high speeds, the complex conjugate pair of roots must come together so that there are three negative real roots corresponding to the two minima and one local maximum. (From specific examples, it appears that the δ corresponding to the global minimum is the negative root that starts δ_{neg} for $\omega = 0$ and remains real and negative throughout the speed range.)

For the specific motor example used in Figure 8.21, there is a unique negative root for the speed range $0 \leq \omega \leq 8000$ rpm, that is, throughout the speed rating of the motor. However, it was shown above that if the speed is large enough, the complex conjugate pair of roots eventually come together.

Example 2 Consider a motor with the parameter values $n_p = 3, R_S =$ 1.85 ohms, $L_S = 0.244$ H, $R_R = 2.1$ ohms, $L_R = 0.244$ H, $M = 0.224$ H, $V_{\max} = 225$ V, and $I_{\max} = 15$ A. Using these parameter values, Figure 8.22 is a plot of $\tau(\delta)$ versus δ for the fixed motor speed $\omega = 200$ radians/sec. The torque is a maximum (motoring) at $\delta_{\max} = 5.7$, but for generating, there is a global minimum at $\delta_a = -69$, a local minimum at $\delta_b - -8.8$, and a local maximum at $\delta_c = -20.8$.

Remark For future reference, the values of δ_1^{vc} and δ_2^{vc} in Figure 8.22 are where $V = V_{\max}, I = I_{\max}$. In particular, for $\delta_1^{vc} < \delta < \delta_2^{vc}$, it turns out that $I > I_{\max}$; for δ outside of this interval, it happens that $I \leq I_{\max}$. This will be shown below.

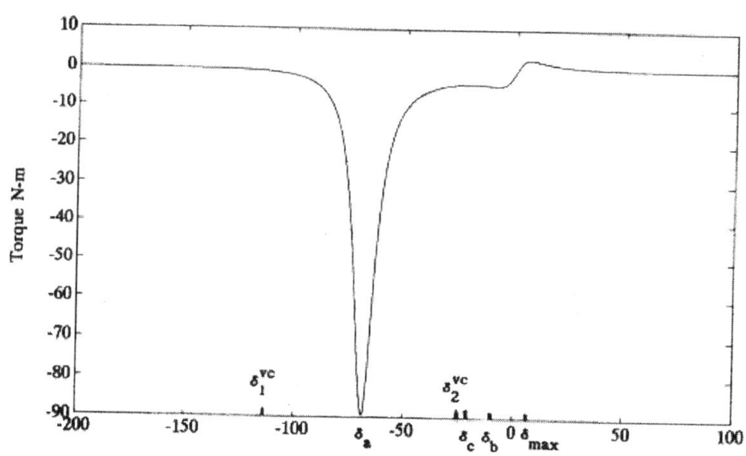

FIGURE 8.22. $\tau(\delta)$ in N-m versus δ for $\omega = 200$ rad/sec. At δ_1^{vc} and δ_2^{vc}, both constraints are encountered (i.e., $V = V_{\max}$ and $I = I_{\max}$). The torque $\tau(\delta)$ has a global minimum at δ_a, a relative minimum at δ_b, and a relative maximum at δ_c. From Bodson et al. [69], © 2004 IEEE.

Numerical Solution of $p(\delta) = 0$

The complete set of solutions to (8.68) can be found with a Newton–Raphson technique provided it is initialized properly. A closed-form approximate solution for δ can be found if the slip $\omega_{slip} = i_q/(T_R i_d) = \delta/T_R$ is small (in magnitude) compared to the stator frequency $\omega_S = n_p\omega + i_q/(T_R i_d)$; that is, $\omega_S \approx n_p\omega$ so that (8.59) and (8.60) become

$$\begin{aligned}\frac{u_d}{R_S} &= i_d - \sigma T_S i_q n_p \omega && (8.76)\\ \frac{u_q}{R_S} &= i_q + T_S i_d n_p \omega && (8.77)\\ \tau(\delta) &= \frac{n_p M^2}{L_R R_S^2} V^2 \frac{\delta}{(1-\sigma T_S \delta n_p \omega)^2 + (\delta + T_S n_p \omega)^2}. && (8.78)\end{aligned}$$

Setting $\partial\tau/\partial\delta = 0$ gives

$$\delta^2 = \frac{1 + (T_S \delta n_p \omega)^2}{1 + (\sigma T_S \delta n_p \omega)^2}$$

or

$$\delta = \pm\sqrt{\frac{1 + (T_S \delta n_p \omega)^2}{1 + (\sigma T_S \delta n_p \omega)^2}}. \tag{8.79}$$

At higher speeds for which $\sigma T_S \delta n_p \omega >> 1$, this reduces further to $\delta^2 = 1/\sigma^2$. As shown above, $\delta = 1/\sigma$ corresponds to the global maximum for the torque $\tau > 0$ while the solution $\delta = -1/\sigma$ corresponds to a local minimum for the torque $\tau < 0$. Also, as shown above, $\delta = -T_R n_p \omega$ is the asymptotic solution for the global minimum. [Note that the slip in this case is $\omega_{slip} = \delta/T_R = -n_p\omega$ which is not small in magnitude and, therefore, the equations (8.76)–(8.78) do not hold.]

To proceed with the numerical solution of (8.68), the global maximum is found by initializing the Newton–Raphson algorithm with the positive square root of (8.79). In the numerical examples given here, five iterations were sufficient to obtain a solution in the speed range $0 \le \omega \le 8000$ rpm. To obtain the global minimum, the Newton–Raphson algorithm is initialized with δ_{neg} [see equation (8.73)] for speeds below 1 rad/sec and by $-T_R n_p \omega$ for all higher speeds. In the numerical examples here, this initialization resulted in a solution after five iterations in the speed range $0 \le \omega \le 8000$ rpm.

Remark It is straightforward to show that the problem just considered of choosing $\delta = i_q/i_d$ to maximize the torque $\tau = (n_p M^2/L_R) i_d i_q$ subject to the voltage constraint $u_d^2 + u_q^2 = V^2 \le V_{\max}^2$ is equivalent to minimizing the voltage $u_d^2 + u_q^2 = V^2$ subject to the constant torque constraint $\tau = (n_p M^2/L_R) i_d i_q =$ constant.

8.4.3 Torque Optimization Under Voltage and Current Constraints

The previous two subsections considered the problem of obtaining the optimum torque from an induction machine without violating the current and voltage constraints, respectively. At low speeds the back emf of the machine is small and the inrush of current results encountering the current limits. At high speeds, the back emf is dominates and one encounters the voltage limits. Between these limiting cases, it is not surprising that one would encounter both voltage and current limits simultaneously. This is now considered.

To proceed, rewrite (8.65) for the torque as

$$\begin{aligned}\tau(\delta) &= \frac{n_p M^2}{L_R} V^2 Q(\delta,\omega)\\ Q(\delta,\omega) &\triangleq \frac{\delta/R_S^2}{\left(1-\sigma T_S\delta\left(n_p\omega+\delta/T_R\right)\right)^2+\left(\delta+T_S\left(n_p\omega+\delta/T_R\right)\right)^2}.\end{aligned} \quad (8.80)$$

Recall by equation (8.62) that the torque can also be written as

$$\tau(\delta) = \frac{n_p M^2}{L_R} I^2 \frac{\delta}{1+\delta^2}. \quad (8.81)$$

Equation (8.80) was convenient when operating at the voltage constraint while equation (8.81) was convenient when operating at the current constraint. Under both constraints, the optimization problem is formulated as follows: Given

$$\tau(\delta) = \frac{n_p M^2}{L_R} I^2 \frac{\delta}{1+\delta^2} = \frac{n_p M^2}{L_R} V^2 Q(\delta,\omega), \quad (8.82)$$

for each fixed speed ω, find the value of δ that optimizes the torque (maximum for motoring or minimum for generating) with $V^2 \leq V_{\max}^2$ and $I^2 \leq I_{\max}^2$. As V^2 and I^2 appear linearly in (8.82), the optimal torque must occur with $V^2 = V_{\max}^2$, or $I^2 = I_{\max}^2$, or both. It has been shown in Section 8.4.2 that for a fixed voltage magnitude V the torque is optimized for some δ satisfying $\partial Q(\delta,\omega)/\partial\delta = 0$, while for a fixed current magnitude I the torque is optimized for $\delta = \pm 1$. For each fixed ω, let $\delta_{opt}(\omega)$ denote the value of δ that optimizes the torque under both voltage and current constraints. Then at each such fixed speed, one of the following three cases must hold:

Case I $I^2 = I_{\max}^2$ and $V^2 < V_{\max}^2 \implies \delta_{opt} = \pm 1$

Case II $V^2 = V_{\max}^2$ and $I^2 < I_{\max}^2 \implies \partial Q(\delta_{opt},\omega)/\partial\delta = 0$

Case III $I^2 = I_{\max}^2$ and $V^2 = V_{\max}^2 \implies I_{\max}^2 \dfrac{\delta_{opt}}{1+\delta_{opt}^2} = V_{\max}^2 Q(\delta_{opt},\omega)$

Generating Mode

The following algorithm is then used to obtain δ_{opt} and $\tau(\delta_{opt})$ for generator mode.

1. Let $\delta_1 = -1$ and compute τ_1, V_1 according to

$$\begin{aligned}\tau_1 &\triangleq \frac{n_p M^2}{L_R} I_{\max}^2 \frac{\delta_1}{1+\delta_1^2}\\ V_1 &\triangleq \frac{\tau_1}{\dfrac{n_p M^2}{L_R} Q(\delta_1, \omega)}.\end{aligned}$$

   If $V_1 > V_{\max}$, then τ_1 is not the minimum torque and proceed to step 2. Otherwise, store the pair (δ_1, τ_1) as a possible optimum pair $(\delta_{opt}, \tau_{opt})$ and then proceed to step 2.

2. Solve $\partial Q(\delta, \omega)/\partial \delta = 0$ for its roots as explained in Section 8.4.2. There are then two possible subcases (a) and (b):

   (a) One real negative root δ_a (the other two complex conjugate roots). In this case, compute

$$\begin{aligned}\tau_2 &\triangleq \frac{n_p M^2}{L_R} V_{\max}^2 Q(\delta_a, \omega)\\ I_a^2 &\triangleq \frac{\tau_2}{(n_p M^2/L_R)} \frac{1+\delta_a^2}{\delta_a}.\end{aligned}$$

   If $I_a > I_{\max}$, then go to step 3 as this solution violates the current constraint. Otherwise, store the pair (δ_a, τ_2) as a possible candidate for $(\delta_{opt}, \tau_{opt})$ and proceed to step 3.

   (b) Three real negative roots denoted as δ_a, δ_b, and δ_c. In this case, let δ_a correspond to the global minimum, δ_b correspond to the local minimum, and δ_c correspond to the local maximum (see Figure 8.22). Next compute

$$\begin{aligned}\tau_a &\triangleq \frac{n_p M^2}{L_R} V_{\max}^2 Q(\delta_a, \omega), I_a^2 \triangleq \frac{\tau_a}{(n_p M^2/L_R)} \frac{1+\delta_a^2}{\delta_a}\\ \tau_b &\triangleq \frac{n_p M^2}{L_R} V_{\max}^2 Q(\delta_b, \omega), I_b^2 \triangleq \frac{\tau_b}{(n_p M^2/L_R)} \frac{1+\delta_b^2}{\delta_b}\end{aligned}$$

   If $I_a < I_{\max}$, then store (δ_a, τ_a) as a possible candidate for $(\delta_{opt}, \tau_{opt})$ and proceed to step 3. Otherwise, check if $I_b < I_{\max}$ is satisfied. If so, then store (δ_b, τ_b) as a possible candidate for $(\delta_{opt}, \tau_{opt})$ and proceed to step 3. If not, just proceed to step 3.

3. Solve

$$\frac{n_p M^2}{L_R} I_{\max}^2 \frac{\delta}{1+\delta^2} = \frac{n_p M^2}{L_R} V_{\max}^2 Q(\delta, \omega)$$

and let δ_3 denote the value corresponding to the minimum torque and denote the minimum torque as τ_3. Store (δ_3, τ_3) as a possible candidate for $(\delta_{opt}, \tau_{opt})$.

4. Choose τ_{opt} as the minimum value of the torque from steps 1, 2, and 3 and δ_{opt} as the corresponding value of δ.

Motoring Mode

The algorithm is similar, but simpler, in the motoring mode.

1. Let $\delta_1 = 1$ and compute τ_1, V_1 according to

$$\begin{aligned} \tau_1 &\triangleq \frac{n_p M^2}{L_R} I_{\max}^2 \frac{\delta_1}{1+\delta_1^2} \\ V_1 &\triangleq \frac{\tau_1}{\frac{n_p M^2}{L_R} Q(\delta_1, \omega)}. \end{aligned}$$

If $V_1 > V_{\max}$, then τ_1 is not the maximum torque and proceed to step 2. Otherwise, store the pair (δ_1, τ_1) as a possible optimum pair $(\delta_{opt}, \tau_{opt})$ and also proceed to step 2.

2. Solve $\partial Q(\delta, \omega)/\partial \delta = 0$ for its roots as explained in Section 8.4.2. Let the unique positive root be denoted as δ_a and compute

$$\begin{aligned} \tau_2 &\triangleq \frac{n_p M^2}{L_R} V_{\max}^2 Q(\delta_a, \omega) \\ I_a^2 &\triangleq \frac{\tau_2}{(n_p M^2/L_R)} \frac{1+\delta_a^2}{\delta_a}. \end{aligned}$$

If $I_a > I_{\max}$, then go to step 3 as this solution violates the constraint. Otherwise, store the pair (δ_a, τ_2) as a possible candidate for $(\delta_{opt}, \tau_{opt})$ and then proceed to step 3.

3. Solve

$$\frac{n_p M^2}{L_R} I_{\max}^2 \frac{\delta}{1+\delta^2} = \frac{n_p M^2}{L_R} V_{\max}^2 Q(\delta, \omega)$$

and let δ_3 denote the value corresponding to the maximum torque and denote the maximum torque as τ_3. Store (δ_3, τ_3) as a possible candidate for $(\delta_{opt}, \tau_{opt})$.

4. Choose τ_{opt} as the maximum value of the torque from steps 1, 2, and 3 and δ_{opt} as the corresponding value of δ.

Remark If the induction machine is operating in the generating mode, step 2 above can lead to interesting solutions when the machine is at both the current and voltage limits. For example, in step 2 of the generating mode, for either subcase (a) or (b), the global minimum at δ_a will typically result in $I^2 > I^2_{\max}$. However, in subcase (b), the local minimum at δ_b (found under the voltage constraint only) can turn out to be the global minimum when the current constraint is also taken into account. This is in fact the situation depicted in Figure 8.22. Specifically, at $\omega = 200$ rads/sec, step 1 violates the voltage constraint while for step 2 the *global* minimum at $\delta_a = -69$ violates the current constraint (it can be shown that for $\delta_1^{vc} < \delta < \delta_2^{vc}$, $I > I_{\max}$). The local minimum at $\delta_b = -8.8$ producing the torque $\tau(\delta_b) = -4.4$ N-m does not violate the current constraint, that is,

$$I_b^2(\delta_b) \triangleq \left(\tau(\delta_b)/\left(n_p M^2/L_R\right)\right)\left(1+\delta_b^2\right)/\delta_b < I^2_{\max}.$$

Going on to step 3 of the algorithm, one solves $I^2_{\max}\delta/\left(1+\delta^2\right) = V^2_{\max}Q(\delta,\omega)$ for its real roots to obtain $\delta_1^{vc} = -22.9, \tau\left(\delta_1^{vc}\right) = -4.06$ N-m and $\delta_2^{vc} = -116.4, \tau(\delta_2^{vc}) = -0.8$ N-m which are indicated in Figure 8.22 (the other two roots are complex conjugates). Consequently, $\delta_{opt} = \delta_b = -8.8$ produces the global minimum torque of $\tau_{opt} = -4.4$ N-m.

Numerical Solution of $I^2_{\max}\delta/\left(1+\delta^2\right) = V^2_{\max}Q(\delta,\omega)$

Step 3 of the algorithm requires solving $I^2_{\max}\delta/\left(1+\delta^2\right) = V^2_{\max}Q(\delta,\omega)$ for δ with ω fixed. This is rewritten as a polynomial in δ by

$$\begin{aligned}
\frac{V^2_{\max}}{R_S^2 I^2_{\max}}\left(1+\delta^2\right) &= \sigma^2\left(\frac{T_S}{T_R}\right)^2\delta^4 + 2\sigma^2\left(\frac{T_S}{T_R}\right)^2 T_R n_p\omega\delta^3 \\
&+ \left(1 + 2(1-\sigma)\left(\frac{T_S}{T_R}\right) + \left(\frac{T_S}{T_R}\right)^2 + \sigma^2\left(\frac{T_S}{T_R}\right)^2\left(T_R n_p\omega\right)^2\right)\delta^2 \\
&+ \left(2(1-\sigma)\left(\frac{T_S}{T_R}\right) + 2\left(\frac{T_S}{T_R}\right)^2\right)T_R n_p\omega\delta + 1 + \left(\frac{T_S}{T_R}\right)^2\left(T_R n_p\omega\right)^2. \quad (8.83)
\end{aligned}$$

This equation does not lend itself to as complete an analysis as (8.68) because the coefficient of δ in (8.83) is not zero as was the case for (8.68). The four roots of (8.83) are plotted in Figure 8.23. Upon computing the values of the torque corresponding to these values of δ, it turns out the positive real root between 1 and 1.5 in Figure 8.23 is the root corresponding to the (global) maximum torque for speeds between 2940 rpm and 3330 rpm. Similarly, the negative real root which is approximately between -1 and -2 in Figure 8.23 corresponds to the (global) minimum torque (i.e., $\max|\tau|, \tau < 0$) for speeds greater than 5410 rpm. These two sets of roots are easily found using the Newton–Raphson method and initializing it with $\delta = 1$ and $\delta = -1$ for the maximum and the minimum torques, respectively.

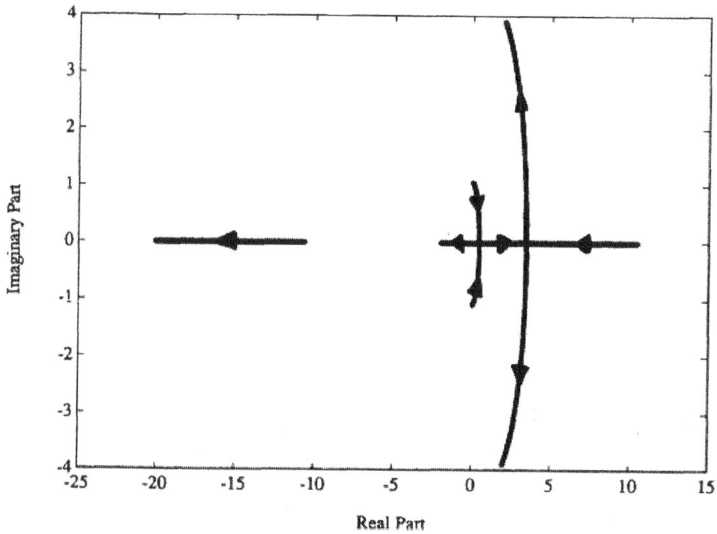

FIGURE 8.23. The four roots of $I_{\max}^2\delta/\left(1+\delta^2\right) = V_{\max}^2 Q(\delta,\omega)$ for $0 \leq \omega \leq 8000$ rpm. From Bodson et al. [69], © 2004 IEEE.

Base Speed

For low enough speeds, the machine is operating at the current limit to achieve optimum torque. The *base speed* ω_{base} can be defined as the speed where the flux linkage magnitude ψ_d is first decreased to avoid the voltage limit. Using the above analysis, a natural procedure to determine this base speed is possible as follows. Below the base speed, where $\delta = \delta_1$ with $\delta_1 = 1$ for motoring and $\delta_1 = -1$ for generating, the torque is given by

$$\tau\left(\delta_1\right) = \frac{n_p M^2}{L_R} I_{\max}^2 \frac{\delta_1}{1+\delta_1^2} = \frac{n_p M^2}{L_R}\frac{I_{\max}^2}{2}\delta_1.$$

The voltage required to achieve this torque is then

$$\begin{aligned} V^2 &= \frac{\tau\left(\delta_1\right)}{\dfrac{n_p M^2}{L_R} Q\left(\delta_1,\omega\right)} \\ &= \frac{n_p M^2}{L_R}\frac{I_{\max}^2}{2}\delta_1 \frac{\left(1-\sigma T_S\delta_1(n_p\omega+\delta_1/T_R)\right)^2 + \left(\delta_1 + T_S\left(n_p\omega+\delta_1/T_R\right)\right)^2}{\dfrac{n_p M^2}{L_R}\dfrac{\delta_1}{R_S^2}} \\ &= \frac{R_S^2 I_{\max}^2}{2}\left(\left(1-\sigma T_S\delta_1\left(n_p\omega+\delta_1/T_R\right)\right)^2 + \left(\delta_1 + T_S\left(n_p\omega+\delta_1/T_R\right)\right)^2\right). \end{aligned} \tag{8.84}$$

To satisfy the voltage constraint, the condition $V^2 \le V_{\max}^2$ must hold. To compute the maximum speed for which this holds, set $V^2 = V_{\max}^2$ in (8.84) and rearrange to obtain

$$\begin{aligned}2\frac{V_{\max}^2}{R_S^2 I_{\max}^2} &= \delta_1^2 + 2(1-\sigma)\frac{T_S}{T_R}\delta_1^2 + \frac{T_S^2}{T_R^2}\delta_1^2\left(1+\sigma^2\delta_1^2\right) + \omega^2 T_S^2 n_p^2\left(1+\sigma^2\delta_1^2\right)\\ &\quad + \omega 2 T_S \delta_1 n_p\left(1-\sigma+\frac{T_S}{T_R}\left(1+\sigma^2\delta_1^2\right)\right)+1.\end{aligned}$$

Using the fact that $\delta_1^2 = 1$, this simplifies to

$$\begin{aligned}0 = {} & \omega^2 T_S^2 n_p^2\left(1+\sigma^2\right) + \omega 2 T_S \delta_1 n_p\left(1-\sigma+\frac{T_S}{T_R}\left(1+\sigma^2\right)\right)\\ & +1+2(1-\sigma)\frac{T_S}{T_R}+\frac{T_S^2}{T_R^2}\left(1+\sigma^2\right)+1-2\frac{V_{\max}^2}{R_S^2 I_{\max}^2}.\end{aligned} \tag{8.85}$$

In a practical motor, $V_{\max}^2 >> R_S^2 I_{\max}^2$ so that

$$1+2(1-\sigma)\frac{T_S}{T_R}+\frac{T_S^2}{T_R^2}\left(1+\sigma^2\right)+1-2\frac{V_{\max}^2}{R_S^2 I_{\max}^2} \approx 1+\left(1+\frac{T_S}{T_R}\right)^2-2\frac{V_{\max}^2}{R_S^2 I_{\max}^2} < 0. \tag{8.86}$$

That is, the constant term of (8.85) is negative. The coefficient of ω is positive for $\delta_1 = 1$ and negative for $\delta_1 = -1$. Using these facts, it follows using the Routh–Hurwitz stability test (see problem 33) that (8.86) has exactly one positive root each for $\delta_1 = 1$ and for $\delta_1 = -1$. For the motor parameter values used in Figure 8.21, the positive solution of (8.86) for $\delta_1 = 1$ is $\omega_{base} = 2990$ rpm (motoring) and the positive solution of (8.86) for $\delta_1 = -1$ is $\omega_{base} = 5410$ rpm (generating). These speeds can be seen in Figure 8.24.

Numerical Results

Here the motor used in example 1 (page 532 above) is analyzed in terms of the optimum torque it can produce. Its parameter values are $n_p = 3$, $R_S = 1.7$ ohms, $L_S = 0.014$ H, $R_R = 3.9$ ohms, $L_R = 0.014$ H, $M = 0.0117$ H, $V_{\max} = 80$ V, $I_{\max} = 6$ A, $J = 1.1 \times 10^{-4}$ kg-m^2 and $f = 1.4 \times 10^{-4}$ N-m/rad/sec. Using the above algorithm, the maximum and minimum torques were computed and are shown in Figure 8.24 versus ω with the corresponding δ versus ω shown in Figure 8.25.

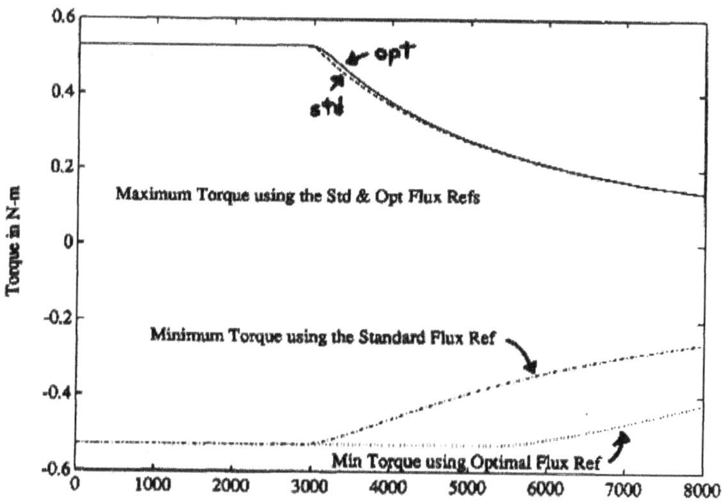

FIGURE 8.24. Maximum and minimum torque achievable versus ω (in rpm) for both optimal and standard field weakening. For the optimal field weakening, ω_{base} = 2990 rpm for motoring and ω_{base} = 5410 rpm for generating. From Bodson et al. [69], © 2004 IEEE.

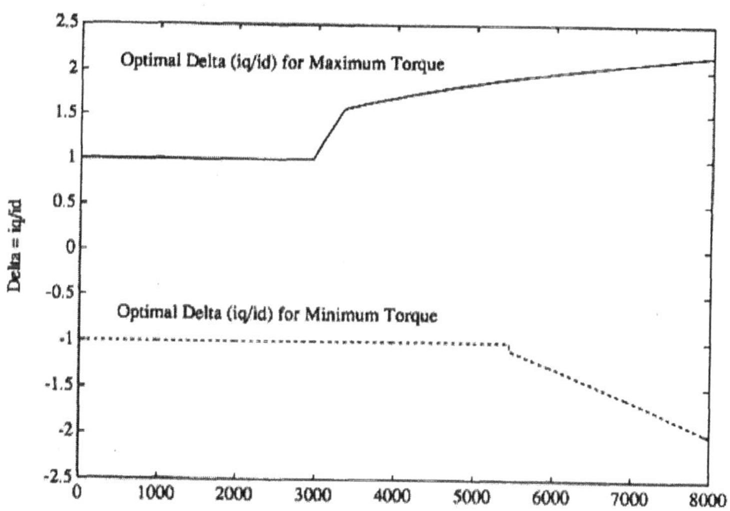

FIGURE 8.25. δ versus ω in rpm corresponding to the maximum and minimum torque. From Bodson et al. [69], © 2004 IEEE.

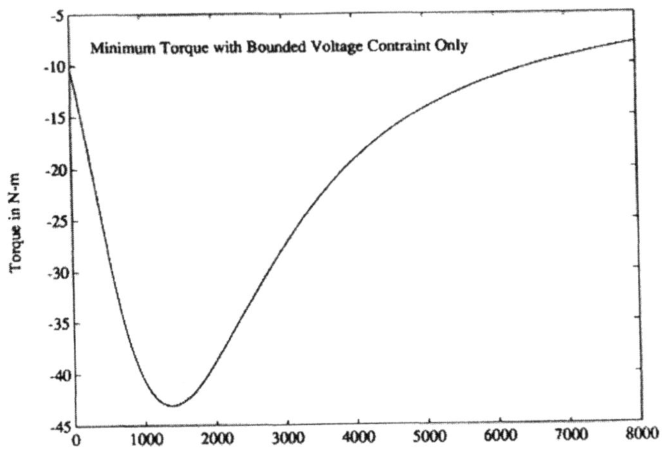

FIGURE 8.26. Minimum torque versus ω in rpm under just the voltage contraint. From Bodson et al. [69], © 2004 IEEE.

The maximum torque is limited by the current constraint up to a speed of 2990 rpm and by the voltage constraint for speeds above 3330 rpm. There is a small speed range from 2990 rpm to 3330 rpm where both the voltage and current constraints are encountered. This last observation is more clearly illustrated in Figure 8.25. For the minimum torque, Figure 8.24 shows it is determined by the current constraint up to 5410 rpm and then by both the voltage and current constraints for all higher speeds. In other words, the minimum torque under just the voltage constraint alone is never achieved because it requires too much current. This is demonstrated vividly by Figure 8.26, which is a plot of the minimum torque achievable with only the voltage constraint and no current constraint. Comparing with Figure 8.24, it is clear that the current limits prevent these large negative torques from ever being achieved.

Comparison with Standard Field Weakening

Recall that the standard approach to field weakening is to choose the flux as

$$\psi_{dref}(\omega) = \begin{cases} \psi_{d0} & \text{for } \ |\omega| < \omega_{base} \\ \psi_{d0}\omega_{base}/|\omega| & \text{for } \ |\omega| \geq \omega_{base} \end{cases} \tag{8.87}$$

where $\psi_{d0} = Mi_{d0}$. For the machine of example 1 (page 532 above), the direct current reference is chosen as $i_{d0} = 6/\sqrt{2}$ A to make the torque using the flux reference (8.87) equal to that of the optimal torque (i.e., choosing $\delta = 1$ so that $i_d = i_q = I_{\max}/\sqrt{2}$) below the base speed. The base speed is chosen as $\omega_{base} = 2990$ rpm (≈ 300 rads/sec) so as to coincide with the speed at which the maximum possible torque achievable starts to decrease (see Figure 8.24).

To provide a fair comparison of the two approaches to field weakening, it is necessary to determine the maximum torque that can be delivered under the standard field weakening approach without violating the constraints. The motor torque is $J\mu\psi_d i_q$ and, as the flux is determined by (8.87) at any constant speed, the optimum torque in this situation is determined by optimizing i_q subject to the voltage and current constraints. To proceed, in (8.64) set $V = V_{\max}, i_d(\omega) = \psi_{dref}(\omega)/M$ with $\psi_{dref}(\omega)$ given by (8.87) and compute its roots. With $\delta_{\max}$ the maximum real root of (8.64), the maximum i_q subject to the voltage constraint is simply $\delta_{\max} i_d$. Similarly, with $\delta_{\min}$ the minimum real root of (8.64), the minimum i_q subject to the voltage constraint is $\delta_{\min} i_d$. To account for the current constraint, one just takes the values of i_q as just explained and tests if $|i_q| \leq I_{q\max} \triangleq \sqrt{I_{\max}^2 - i_d(\omega)}$. If $|i_q| \leq I_{q\max}$ then i_q is left unchanged and it is at its optimum value. Otherwise, i_q is set to $\pm I_{q\max}$ where the $+$ is taken if $i_q > 0$ and the $-$ sign is taken if $i_q < 0$. Figure 8.24 shows the maximum and minimum torques computed using the standard flux reference. It is interesting to note that this maximum torque is essentially the same as that produced by the optimum torque *provided* that i_{d0} and ω_{base} are chosen as in the optimum torque case.

The Optimal Flux Reference

The algorithm above for computing $\delta = i_q/i_d$ to achieve the optimum torque under both voltage and current constraints was derived assuming constant speed. However, achieving the optimum torque is of particular interest in high-performance applications where the speed varies greatly. Here it is shown that the algorithm is indeed applicable to such applications due to the fact that the electrical dynamics of a machine are typically much faster than the mechanical dynamics. Figure 8.24 shows that the torque optimization will have a significant advantage during deceleration. In fact, the flux and current references used for the input output controller of Section 8.2.4 were developed using the algorithm of this section. The speed trajectory of Figure 8.15 shows the motor goes from 0 rpm to 8000 rpm in 0.38 seconds and from 8000 rpm to 0 rpm in only 0.265 seconds. The corresponding position reference trajectory is given in Figure 8.13. For the speed reference trajectory of Figure 8.15 the corresponding optimum achievable torque τ_{opt} is shown as the solid curve in Figure 8.18 while the dashed curve is the torque reference τ_{ref}. Note that the torque reference is close to the optimum reference throughout most of the speed range. The quadrature current reference is chosen as

$$i_{qref} \triangleq \frac{\tau_{ref}}{J\mu\psi_{dref}}$$

where, as stated above, τ_{ref} is the dashed curve in Figure 8.18. The direct current reference is then specified as

$$i_{ref}(\omega) = \frac{i_{qref}}{\delta(\omega_{ref})} = \frac{\tau_{ref}}{J\mu\delta(\omega_{ref})\psi_{dref}}$$

where $\delta(\omega_{ref})$ is the optimal i_q/i_d to optimize the torque and is shown in Figure 8.27.

This derivation has been done under the assumption of constant speed so that choosing $\psi_{dref} = Mi_{ref}$ would be consistent with this assumption. However, while the currents can be changed relatively fast, the flux may take time to build up. As a consequence, the flux reference is chosen as the solution to

$$\begin{aligned}\frac{d\psi_{dref}}{dt} &= -\frac{1}{T_R}\psi_{dref} + \frac{M}{T_R}i_{dref}\\ &= -\frac{1}{T_R}\psi_{dref} + \frac{M}{J\mu T_R}\frac{\tau_{ref}}{\delta(\omega_{ref})\psi_{dref}}.\end{aligned}$$

Figure 8.28 is a plot of the commanded voltage to the amplifier where it is noted that the it stays quite close to the 80-V limit as the motor goes from 0 rpm to 8000 rpm.

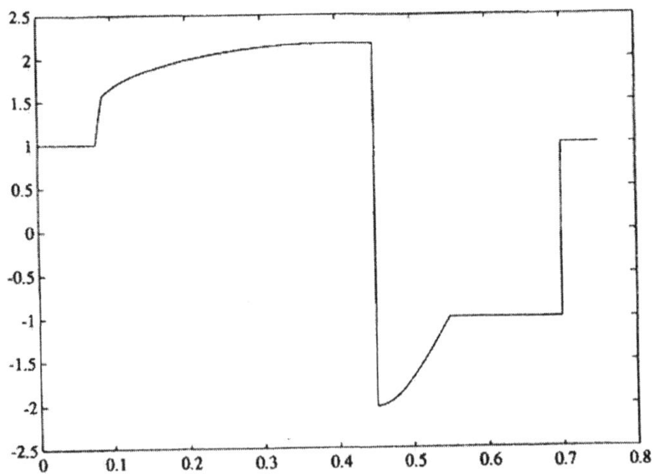

FIGURE 8.27. Optimal $\delta(\omega_{ref}(t)) = i_q/i_d$ versus time for the speed trajectory of Figure 8.15. From Bodson et al. [69], © 2004 IEEE.

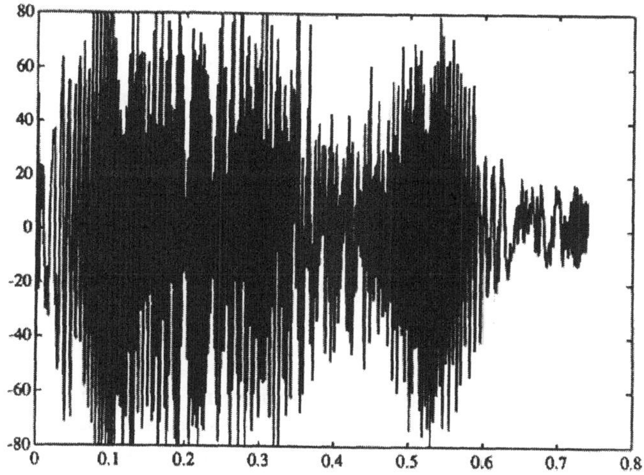

FIGURE 8.28. u_{Sa} versus time. From Bodson et al. [69], © 2004 IEEE.

Figure 8.29 shows the phase current i_{Sa} versus time.

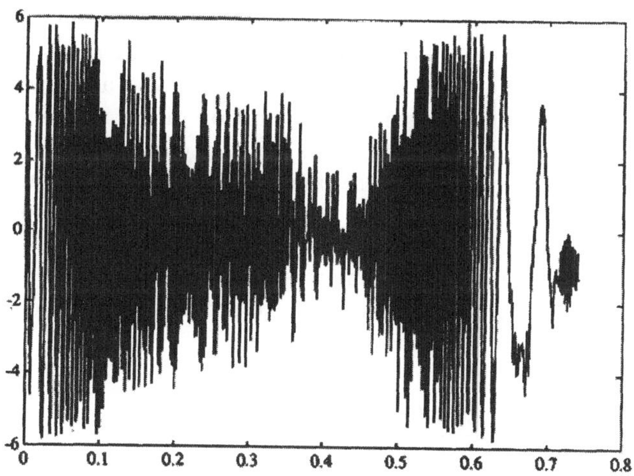

FIGURE 8.29. i_{Sa} versus time. From Bodson et al. [69], © 2004 IEEE.

The flux reference ψ_{dref} and the (estimated) flux ψ_d are shown in Figure 8.17. The position reference and measured position are shown in Figure 8.13 with the position error given in Figure 8.14 in encoder counts. This shows that the position error for the last 100 samples is less than one encoder count verifying that the motor comes into the final position to a complete stop.

The generalization of this approach to the situation in which the main magnetic path of the machine undergoes significant saturation has been considered in [82][83][84].

8.5 Identification of the Induction Motor Parameters*

The induction motor parameters are the mutual inductance M, the stator inductance L_S, the rotor inductance L_R, the stator resistance R_S, the rotor resistance R_R, the inertia of the rotor J and the load torque τ_L. Standard methods for the estimation of induction motor parameters include the locked rotor test, the no-load test, and the standstill frequency response test (see problem 19 of Chapter 7). In Ref. [85], an automatic procedure is described in which a sequence of such tests is performed, each designed to isolate and measure a specific parameter. The method is applied for the automatic tuning (self-commissioning) of an induction motor drive. In Ref. [86], another procedure is described, based primarily on the identification of the motor transfer function at standstill. The model is then refined to account for magnetic saturation and adaptation is included to compensate for the effects of heating. However, these approaches cannot be used online, that is, during normal operation of the machine. This is a disadvantage as some of the parameters do vary during operation. For example, field-oriented control requires knowledge of the rotor time constant $T_R = L_R/R_R$ in order to estimate the rotor flux linkages and R_R varies significantly due to ohmic heating. Based on the work in Refs. [42][59][87], a method for identifying the parameters of the induction motor that can be implemented *online* is now described. To proceed, recall the state-space model (8.5) of the induction motor given by

$$\begin{aligned}
\frac{d\omega}{dt} &= \frac{n_p M}{J L_R}\left(i_{Sb}\psi_{Ra} - i_{Sa}\psi_{Rb}\right) - \frac{\tau_L}{J} \\
\frac{d\psi_{Ra}}{dt} &= -\frac{1}{T_R}\psi_{Ra} - n_p\omega\psi_{Rb} + \frac{M}{T_R} i_{Sa} \\
\frac{d\psi_{Rb}}{dt} &= -\frac{1}{T_R}\psi_{Rb} + n_p\omega\psi_{Ra} + \frac{M}{T_R} i_{Sb} \\
\frac{di_{Sa}}{dt} &= \frac{\beta}{T_R}\psi_{Ra} + \beta n_p\omega\psi_{Rb} - \gamma i_{Sa} + \frac{1}{\sigma L_S} u_{Sa} \\
\frac{di_{Sb}}{dt} &= \frac{\beta}{T_R}\psi_{Rb} - \beta n_p\omega\psi_{Ra} - \gamma i_{Sb} + \frac{1}{\sigma L_S} u_{Sb}
\end{aligned} \tag{8.88}$$

where again

$$T_R = L_R/R_R, \qquad \sigma = 1 - M^2/(L_S L_R)$$
$$\beta = M/(\sigma L_S L_R), \qquad \gamma = R_S/(\sigma L_S) + M^2 R_R/(\sigma L_S L_R^2)$$

have been used to simplify the expressions. This model is then transformed into a coordinate system attached to the rotor where, for example, the current variables are transformed according to

$$\begin{bmatrix} i_{Sx} \\ i_{Sy} \end{bmatrix} = \begin{bmatrix} \cos(n_p\theta) & \sin(n_p\theta) \\ -\sin(n_p\theta) & \cos(n_p\theta) \end{bmatrix} \begin{bmatrix} i_{Sa} \\ i_{Sb} \end{bmatrix}. \tag{8.89}$$

An advantage of this transformation is that the signals in the moving frame (i.e., the (x, y) frame) typically vary slower than those in the (a, b) frame (they vary at the slip frequency rather than at the stator frequency). At the same time, the transformation does not depend on any unknown parameter in contrast to the field-oriented dq transformation. The stator voltages and the rotor flux linkages are transformed as the currents resulting in the following model [87]:

$$\frac{di_{Sx}}{dt} = \frac{1}{\sigma L_S}u_{Sx} - \gamma i_{Sx} + \frac{\beta}{T_R}\psi_{Rx} + n_p\beta\omega\psi_{Ry} + n_p\omega i_{Sy} \tag{8.90}$$
$$\frac{di_{Sy}}{dt} = \frac{1}{\sigma L_S}u_{Sy} - \gamma i_{Sy} + \frac{\beta}{T_R}\psi_{Ry} - n_p\beta\omega\psi_{Rx} - n_p\omega i_{Sx} \tag{8.91}$$
$$\frac{d\psi_{Rx}}{dt} = \frac{M}{T_R}i_{Sx} - \frac{1}{T_R}\psi_{Rx} \tag{8.92}$$
$$\frac{d\psi_{Ry}}{dt} = \frac{M}{T_R}i_{Sy} - \frac{1}{T_R}\psi_{Ry} \tag{8.93}$$
$$\frac{d\omega}{dt} = \frac{n_p M}{J L_R}(i_{Sy}\psi_{Rx} - i_{Sx}\psi_{Ry}) - \frac{\tau_L}{J}. \tag{8.94}$$

8.5.1 *Linear Overparameterized Model*

Measurements of the stator currents i_{Sa} and i_{Sb} and voltages u_{Sa} and u_{Sb} as well as the position θ of the rotor are assumed to be available (velocity may then be reconstructed from position measurements). However, the rotor flux linkages ψ_{Rx} and ψ_{Ry} are not assumed to be measured. Standard methods for parameter estimation are based on equalities where known signals depend *linearly* on unknown parameters. However, the induction motor model described above does not fit in this category unless the rotor flux linkages are measured. The first step is to eliminate the fluxes ψ_{Rx} and ψ_{Ry} and their derivatives $d\psi_{Rx}/dt$ and $d\psi_{Ry}/dt$. The four equations (8.90), (8.91), (8.92), (8.93) can be used to solve for ψ_{Rx}, ψ_{Ry}, $d\psi_{Rx}/dt$, and $d\psi_{Ry}/dt$, but one is left without another independent equation(s) to set up a regressor system for the identification algorithm (see Chapter 2).

A new set of independent equations are found by differentiating equations (8.90) and (8.91) to obtain

$$\begin{aligned}\frac{1}{\sigma L_s}\frac{du_{Sx}}{dt} &= \frac{d^2 i_{Sx}}{dt^2}+\gamma\frac{di_{Sx}}{dt}-\frac{\beta}{T_R}\frac{d\psi_{Rx}}{dt}-n_p\beta\omega\frac{d\psi_{Ry}}{dt}-n_p\beta\psi_{Ry}\frac{d\omega}{dt}\\ &\quad -n_p\omega\frac{di_{Sy}}{dt}-n_p i_{Sy}\frac{d\omega}{dt} \qquad (8.95)\end{aligned}$$

$$\begin{aligned}\frac{1}{\sigma L_s}\frac{du_{Sy}}{dt} &= \frac{d^2 i_{Sy}}{dt^2}+\gamma\frac{di_{Sy}}{dt}-\frac{\beta}{T_R}\frac{d\psi_{Ry}}{dt}+n_p\beta\omega\frac{d\psi_{Rx}}{dt}+n_p\beta\psi_{Rx}\frac{d\omega}{dt}\\ &\quad +n_p\omega\frac{di_{Sx}}{dt}+n_p i_{Sx}\frac{d\omega}{dt}. \qquad (8.96)\end{aligned}$$

Next, equations (8.90), (8.91), (8.92), and (8.93) are solved for ψ_{Rx}, ψ_{Ry}, $d\psi_{Rx}/dt$, and $d\psi_{Ry}/dt$ and substituted into equations (8.95) and (8.96) to obtain

$$\begin{aligned}0 = &-\frac{d^2 i_{Sx}}{dt^2}+\frac{di_{Sy}}{dt}n_p\omega+\frac{1}{\sigma L_S}\frac{du_{Sx}}{dt}-(\gamma+\frac{1}{T_R})\frac{di_{Sx}}{dt}-i_{Sx}(-\frac{\beta M}{T_R^2}+\frac{\gamma}{T_R})\\ &+i_{Sy}n_p\omega(\frac{1}{T_R}+\frac{\beta M}{T_R})+\frac{u_{Sx}}{\sigma L_S T_R}+n_p\frac{d\omega}{dt}i_{Sy}-n_p\frac{d\omega}{dt}\frac{1}{\sigma L_S(1+n_p^2\omega^2 T_R^2)}\\ &\times\Big(-\sigma L_S T_R\frac{di_{Sy}}{dt}-\gamma i_{Sy}\sigma L_S T_R-i_{Sx}n_p\omega\sigma L_S T_R-\frac{di_{Sx}}{dt}n_p\omega\sigma L_S T_R^2\\ &-\gamma i_{Sx}n_p\omega\sigma L_S T_R^2+i_{Sy}n_p^2\omega^2\sigma L_S T_R^2+n_p\omega T_R^2 u_{Sx}+T_R u_{Sy}\Big) \qquad (8.97)\end{aligned}$$

and

$$\begin{aligned}0 = &-\frac{d^2 i_{Sy}}{dt^2}-\frac{di_{Sx}}{dt}n_p\omega+\frac{1}{\sigma L_S}\frac{du_{Sy}}{dt}-(\gamma+\frac{1}{T_R})\frac{di_{Sy}}{dt}-i_{Sy}(-\frac{\beta M}{T_R^2}+\frac{\gamma}{T_R})\\ &-i_{Sx}n_p\omega(\frac{1}{T_R}+\frac{\beta M}{T_R})+\frac{u_{Sy}}{\sigma L_S T_R}-n_p\frac{d\omega}{dt}i_{Sx}+n_p\frac{d\omega}{dt}\frac{1}{\sigma L_S(1+n_p^2\omega^2 T_R^2)}\\ &\times\Big(-\sigma L_S T_R\frac{di_{Sx}}{dt}-\gamma i_{Sx}\sigma L_S T_R+i_{Sy}n_p\omega\sigma L_S T_R+\frac{di_{Sy}}{dt}n_p\omega\sigma L_S T_R^2\\ &+\gamma i_{Sy}n_p\omega\sigma L_S T_R^2+i_{Sx}n_p^2\omega^2\sigma L_S T_R^2-n_p\omega T_R^2 u_{Sy}+T_R u_{Sx}\Big). \qquad (8.98)\end{aligned}$$

The set of equations (8.97) and (8.98) may be rewritten in regressor form as

$$y(t) = W(t)K \qquad (8.99)$$

where $y \in \mathbb{R}^2, W \in \mathbb{R}^{2\times 15}$, and $K \in \mathbb{R}^{15}$ are given by

$$\begin{aligned}K \triangleq \Big[&\gamma \quad \beta M \quad \frac{1}{\sigma L_S} \quad \frac{\beta M}{T_R^2} \quad \frac{1}{T_R} \quad \frac{\gamma}{T_R} \quad \frac{\beta M}{T_R} \quad T_R \quad \gamma T_R \quad \beta M T_R\\ &T_R^2 \quad \gamma T_R^2 \quad \frac{T_R^2}{\sigma L_S} \quad \frac{1}{\sigma L_S T_R} \quad \frac{T_R}{\sigma L_S}\Big]^T\end{aligned}$$

$$W(t) \triangleq \left[\begin{matrix} -\dfrac{di_{Sx}}{dt} & n_p^2\omega^2 i_{Sx} & \dfrac{du_{Sx}}{dt} & i_{Sx} & n_p\omega i_{Sy} - \dfrac{di_{Sx}}{dt} & -i_{Sx} & n_p\omega i_{Sy} \\ -\dfrac{di_{Sy}}{dt} & n_p^2\omega^2 i_{Sy} & \dfrac{du_{Sy}}{dt} & i_{Sy} & -n_p\omega i_{Sx} - \dfrac{di_{Sy}}{dt} & -i_{Sy} & -n_p\omega i_{Sx} \end{matrix} \right.$$

$$\begin{matrix} -n_p^2\omega^2\dfrac{di_{Sx}}{dt} + n_p^3\omega^3 i_{Sy} + \dfrac{d\omega}{dt}\left(n_p\dfrac{di_{Sy}}{dt} + n_p^2\omega i_{Sx}\right) & n_p i_{Sy}\dfrac{d\omega}{dt} - n_p^2\omega^2 i_{Sx} \\ -n_p^2\omega^2\dfrac{di_{Sy}}{dt} - n_p^3\omega^3 i_{Sx} + \dfrac{d\omega}{dt}\left(-n_p\dfrac{di_{Sx}}{dt} + n_p^2\omega i_{Sy}\right) & -n_p i_{Sx}\dfrac{d\omega}{dt} - n_p^2\omega^2 i_{Sy} \end{matrix}$$

$$\begin{matrix} n_p^3\omega^3 i_{Sy} & n_p^2(\omega\dfrac{di_{Sx}}{dt}\dfrac{d\omega}{dt} - \omega^2\dfrac{d^2 i_{Sx}}{dt^2}) + n_p^3\omega^3\dfrac{di_{Sy}}{dt} & n_p^2(\omega i_{Sx}\dfrac{d\omega}{dt} - \omega^2\dfrac{di_{Sx}}{dt}) \\ -n_p^3\omega^3 i_{Sx} & n_p^2(\omega\dfrac{di_{Sy}}{dt}\dfrac{d\omega}{dt} - \omega^2\dfrac{d^2 i_{Sy}}{dt^2}) - n_p^3\omega^3\dfrac{di_{Sx}}{dt} & n_p^2(\omega i_{Sy}\dfrac{d\omega}{dt} - \omega^2\dfrac{di_{Sy}}{dt}) \end{matrix}$$

$$\left. \begin{matrix} n_p^2(\omega^2\dfrac{du_{Sx}}{dt} - u_{Sx}\omega\dfrac{d\omega}{dt}) & u_{Sx} & n_p^2\omega^2 u_{Sx} - n_p u_{Sy}\dfrac{d\omega}{dt} \\ n_p^2(\omega^2\dfrac{du_{Sy}}{dt} - u_{Sy}\omega\dfrac{d\omega}{dt}) & u_{Sy} & n_p^2\omega^2 u_{Sy} + n_p u_{Sx}\dfrac{d\omega}{dt} \end{matrix} \right]$$

and

$$y(t) \triangleq \begin{bmatrix} \dfrac{d^2 i_{Sx}}{dt^2} - n_p i_{Sy}\dfrac{d\omega}{dt} - n_p\omega\dfrac{di_{Sy}}{dt} \\ \dfrac{d^2 i_{Sy}}{dt^2} + n_p i_{Sx}\dfrac{d\omega}{dt} + n_p\omega\dfrac{di_{Sx}}{dt} \end{bmatrix}.$$

The system (8.99) is linear in the parameters. However, this system is *over-parameterized*, meaning that the components K_i of the parameter vector K are *not* independent of each other. The relationship between the parameters is

$$\begin{aligned} K_1 &= K_6K_8, K_2 = K_4K_8^2, K_3 = K_8K_{14}, K_5 = 1/K_8, \\ K_7 &= K_4K_8, K_9 = K_6K_8^2, K_{10} = K_4K_8^3, K_{11} = K_8^2, \\ K_{12} &= K_6K_8^3, K_{13} = K_{14}K_8^3, K_{15} = K_{14}K_8^2 \end{aligned} \tag{8.100}$$

so that only the four parameters K_4, K_6, K_8, and K_{14} are independent. If the parameter vector is computed according to $K = R_W^{-1}R_{Wy} \in \mathbb{R}^{15}$ as in Chapter 2, one would find out that the problem is numerically *ill-conditioned.* This means that small changes in the data matrices R_W and R_{Wy} will lead to large changes in the computed value for $K \in \mathbb{R}^{15}$ and therefore the value of K could not be considered valid. Furthermore, there is no guarantee that the components of K computed in this way would satisfy the constraints (8.100). Note also that not all five electrical parameters R_S, L_S, R_R, L_R, and M can be retrieved from the K_i's. The four parameters K_4, K_6, K_8, and K_{14} determine only the four independent parameters R_S, L_S, σ, and T_R by

$$R_S = \frac{K_6 - K_4}{K_{14}}, \; T_R = K_8, \; L_S = \frac{1 + K_4 K_8^2}{K_{14} K_8}, \; \sigma = \frac{1}{1 + K_4 K_8^2}. \tag{8.101}$$

As $T_R = L_R/R_R$ and $\sigma = 1 - M^2/(L_S L_R)$, only L_R/R_R and M^2/L_R can be obtained and not M, L_R and R_R independently. This situation is inherent to the identification problem when some of the state variables (in this case the rotor flux linkages) are unknown and is not specific to the method. If the rotor flux linkages are not measured, machines with different R_R, L_R, and M, but identical L_R/R_R and M^2/L_R, will have the same input/output (i.e., voltage to stator currents and speed) characteristics. However, machines with different K_i parameters, yet satisfying the nonlinear relationships (8.100), will be distinguishable. (For a related discussion of this issue see Bellini et al. [88], where parameter identification is performed using torque–speed and stator current–speed characteristics.)

8.5.2 *Nonlinear Least-Squares Identification*

In this section, it is shown how the linear least-squares method described in Chapter 2 is modified to account for the constraint (8.100). Recall from Chapter 2 that the least-squares problem is to minimize

$$E^2(K) = \sum_{n=1}^{N} \left| y(nT) - W(nT)K \right|^2 = R_y - 2R_{Wy}^T K + K^T R_W K \tag{8.102}$$

subject to the constraints

$$\begin{aligned} K_1 &= K_6 K_8, K_2 = K_4 K_8^2, K_3 = K_8 K_{14}, K_5 = 1/K_8, \\ K_7 &= K_4 K_8, K_9 = K_6 K_8^2, K_{10} = K_4 K_8^3, K_{11} = K_8^2, \\ K_{12} &= K_6 K_8^3, K_{13} = K_{14} K_8^3, K_{15} = K_{14} K_8^2. \end{aligned} \tag{8.103}$$

On physical grounds, the parameters K_4, K_6, K_8, and K_{14} are constrained to

$$0 < K_i < \infty \quad \text{for } i = 4, 6, 8, 14. \tag{8.104}$$

Also, based on physical grounds, the squared error $E^2(K)$ will be minimized in the interior of this region. Let

$$\begin{aligned} E^2(K_p) &\triangleq \sum_{n=1}^{N} \left| y(nT) - W(nT)K \right|^2_{\substack{K_1=K_6K_8 \\ K_2=K_4K_8^2 \\ \vdots}} \\ &= R_y - 2R_{Wy}^T K\Big|_{\substack{K_1=K_6K_8 \\ K_2=K_4K_8^2 \\ \vdots}} + (K^T R_W K)\Big|_{\substack{K_1=K_6K_8 \\ K_2=K_4K_8^2 \\ \vdots}} \end{aligned} \tag{8.105}$$

where

$$K_p \triangleq \begin{bmatrix} K_4 & K_6 & K_8 & K_{14} \end{bmatrix}^T.$$

As just explained, the minimum of (8.105) must occur in the interior of the region and therefore at an extremum point. This then entails solving the four equations

$$r_1(K_p) \triangleq \frac{\partial E^2(K_p)}{\partial K_4} = 0 \tag{8.106}$$

$$r_2(K_p) \triangleq \frac{\partial E^2(K_p)}{\partial K_6} = 0 \tag{8.107}$$

$$r_3(K_p) \triangleq \frac{\partial E^2(K_p)}{\partial K_8} = 0 \tag{8.108}$$

$$r_4(K_p) \triangleq \frac{\partial E^2(K_p)}{\partial K_{14}} = 0. \tag{8.109}$$

The partial derivatives in (8.106)–(8.109) are *rational* functions in the parameters K_4, K_6, K_8, and K_{14}. Defining

$$p_1(K_p) \triangleq K_8 r_1(K_p) = K_8 \frac{\partial E^2(K_p)}{\partial K_4} \tag{8.110}$$

$$p_2(K_p) \triangleq K_8 r_2(K_p) = K_8 \frac{\partial E^2(K_p)}{\partial K_6} \tag{8.111}$$

$$p_3(K_p) \triangleq K_8^3 r_3(K_p) = K_8^3 \frac{\partial E^2(K_p)}{\partial K_8} \tag{8.112}$$

$$p_4(K_p) \triangleq K_8 r_4(K_p) = K_8 \frac{\partial E^2(K_p)}{\partial K_{14}} \tag{8.113}$$

results in the $p_i(K_p)$ being *polynomials* in the parameters K_4, K_6, K_8, and K_{14} and having the same positive zero set (i.e., the same roots satisfying $K_i > 0$) as the system (8.106)–(8.109). The degrees of the polynomials p_i

are given in the following table.

	deg K_4	deg K_6	deg K_8	deg K_{14}
$p_1(K_p)$	1	1	7	1
$p_2(K_p)$	1	1	7	1
$p_3(K_p)$	2	2	8	2
$p_4(K_p)$	1	1	7	1

All possible solutions to this set may be found using elimination theory as is now summarized.

Solving Systems of Polynomial Equations [89][90]

The question at hand is "Given two polynomial equations $a(K_1, K_2) = 0$ and $b(K_1, K_2) = 0$, how does one solve them simultaneously to eliminate (say) K_2?". A systematic procedure to do this is known as *elimination theory* and uses the notion of *resultants* (see the appendix at the end of this chapter). Briefly, one considers $a(K_1, K_2)$ and $b(K_1, K_2)$ as polynomials in K_2 whose coefficients are polynomials in K_1. Then, for example, letting $a(K_1, K_2)$ and $b(K_1, K_2)$ have degrees 3 and 2, respectively, in K_2, they may be written in the form

$$\begin{aligned} a(K_1, K_2) &= a_3(K_1)K_2^3 + a_2(K_1)K_2^2 + a_1(K_1)K_2 + a_0(K_1) \\ b(K_1, K_2) &= b_2(K_1)K_2^2 + b_1(K_1)K_2 + b_0(K_1). \end{aligned}$$

The $n \times n$ *Sylvester* matrix, with

$$n \triangleq \deg_{K_2}\{a(K_1, K_2)\} + \deg_{K_2}\{b(K_1, K_2)\} = 3 + 2 = 5$$

is defined by

$$S_{a,b}(K_1) \triangleq \begin{bmatrix} a_0(K_1) & 0 & b_0(K_1) & 0 & 0 \\ a_1(K_1) & a_0(K_1) & b_1(K_1) & b_0(K_1) & 0 \\ a_2(K_1) & a_1(K_1) & b_2(K_1) & b_1(K_1) & b_0(K_1) \\ a_3(K_1) & a_2(K_1) & 0 & b_2(K_1) & b_1(K_1) \\ 0 & a_3(K_1) & 0 & 0 & b_2(K_1) \end{bmatrix}.$$

The *resultant* polynomial is then defined by

$$r(K_1) = \text{Res}\Big(a(K_1, K_2), b(K_1, K_2), K_2\Big) \triangleq \det S_{a,b}(K_1) \tag{8.114}$$

and is the result of *eliminating* the variable K_2 from $a(K_1, K_2) = 0$ and $b(K_1, K_2) = 0$. In fact, the following is true (see Refs. [89] and [90]).

Theorem Any solution (K_1^0, K_2^0) of $a(K_1, K_2) = 0$ and $b(K_1, K_2) = 0$ must satisfy $r(K_1^0) = 0$.

Though the converse of this theorem is not necessarily true, the finite number of solutions of $r(K_1) = 0$ are the *only* possible candidates for the first coordinate (partial solutions) of the common zeros of $a(K_1, K_2)$ and $b(K_1, K_2)$. Whether or not such a partial solution results in a full solution is easily determined by back solving and checking the solution (see the Appendix).

Using the polynomials (8.110)–(8.113) and a computer algebra software program (e.g., MATHEMATICA [91]), the variable K_4 is eliminated first to obtain three polynomials in three unknowns as

$$r_{p1p2}(K_6, K_8, K_{14}) \triangleq \text{Res}\Big(p_1(K_4, K_6, K_8, K_{14}), p_2(K_4, K_6, K_8, K_{14}), K_4\Big)$$
$$r_{p1p3}(K_6, K_8, K_{14}) \triangleq \text{Res}\Big(p_1(K_4, K_6, K_8, K_{14}), p_3(K_4, K_6, K_8, K_{14}), K_4\Big)$$
$$r_{p1p4}(K_6, K_8, K_{14}) \triangleq \text{Res}\Big(p_1(K_4, K_6, K_8, K_{14}), p_4(K_4, K_6, K_8, K_{14}), K_4\Big)$$

where

	deg K_6	deg K_8	deg K_{14}
r_{p1p2}	1	14	1
r_{p1p3}	2	22	2
r_{p1p4}	1	14	1

Next K_6 is eliminated to obtain two polynomials in two unknowns as

$$r_{p1p2p3}(K_8, K_{14}) \triangleq \text{Res}\Big(r_{p1p2}(K_6, K_8, K_{14}), r_{p1p3}(K_6, K_8, K_{14}), K_6\Big)$$
$$r_{p1p2p4}(K_8, K_{14}) \triangleq \text{Res}\Big(r_{p1p2}(K_6, K_8, K_{14}), r_{p1p4}(K_6, K_8, K_{14}), K_6\Big)$$

where

	deg K_8	deg K_{14}
r_{p1p2p3}	50	2
r_{p1p2p4}	28	1

Finally K_{14} is eliminated to obtain a single polynomial in K_8 as

$$r(K_8) \triangleq \text{Res}\Big(r_{p1p2p3}(K_8, K_{14}), r_{p1p2p4}(K_8, K_{14}), K_{14}\Big)$$

where

$$\deg_{K_8}\{r(K_8)\} = 104.$$

The parameter K_8 was chosen as the variable *not* eliminated because its degree was the highest at each step, meaning it would have a larger (in dimension) Sylvester matrix than using any other variable. The positive roots of $r(K_8) = 0$ are found, which are then substituted into $r_{p1p2p3} = 0$ (or $r_{p1p2p4} = 0$) and which in turn are solved to obtain the partial solutions (K_8, K_{14}). The partial solutions (K_8, K_{14}) are then substituted into $r_{p1p2} = 0$ (or $r_{p1p3} = 0$ or $r_{p1p4} = 0$), which are solved to obtain the partial solutions (K_6, K_8, K_{14}) so that they in turn may be substituted into $p_1 = 0$ (or $p_2 = 0$, or $p_3 = 0$, or $p_4 = 0$) which is then solved to obtain the candidate solutions (K_4, K_6, K_8, K_{14}). These candidate solutions are then checked to see which ones satisfy the complete system of polynomials equations (8.110)–(8.113) and those that do constitute the *candidate* solutions for the minimization. Based on physical considerations, the set of candidate solutions is nonempty. From the set of candidate solutions, the one that gives the smallest squared error is chosen.

Computational Issues

Due to the high degrees of the resultant polynomials, care must be taken to compute their roots. The data set is collected and brought into MATHEMATICA [91]. The matrices R_y, R_W, and R_{Wy} are then computed and their entries converted to *rational* form. Finally the roots of the resultant polynomials are computed in MATHEMATICA using rational arithmetic with 16 digits of precision.

Numerical Conditioning of the Nonlinear Least-Squares Solution

After finding the solution that gives the minimal value for $E^2(K_p)$, one needs to know if the solution makes sense. For example, in the *linear* least-squares problem, there is a unique well-defined solution provided that the regressor matrix R_W is non singular (or in practical terms, its condition number is not too large). In the nonlinear case here, a Taylor series expansion about the computed minimum point $K_p^* = [K_4^*, K_6^*, K_8^*, K_{14}^*]^T$ to obtain $(i, j = 4, 6, 8, 14)$

$$E^2(K_p) = E^2(K_p^*) + \frac{1}{2}\left[K_p - K_p^*\right]^T \frac{\partial^2 E^2(K_p^*)}{\partial K_i \partial K_j}\left[K_p - K_p^*\right] + \cdots \qquad (8.115)$$

One then checks that the Hessian matrix $\dfrac{\partial^2 E^2(K_p^*)}{\partial K_i \partial K_j}$ is positive definite to ensure that the data set is sufficiently rich to identify the parameters. Its condition number is also computed to obtain a measure of the numerically reliability of the computations [92].

8.5.3 *Calculating the Parametric Error Indices*

The expression for the error index in Chapter 2 is still applicable for the nonlinear case. However, the method for computing the parametric error indices must be modified for the nonlinear least-squares estimation problem. Here, the parametric error index is defined as the *maximum* value of δK_i for $i = 4, 6, 8, 14$ such that

$$E^2(K_p^* + \delta K_p) = 1.25E^2(K_p^*) \qquad (8.116)$$

where $\delta K_p \triangleq \begin{bmatrix} \delta K_4 & \delta K_6 & \delta K_8 & \delta K_{14} \end{bmatrix}^T$. Recall that in Chapter 2, $E^2(K_p^* + \delta K_p) = 2E^2(K_p^*)$ was used rather than (8.116). The choice is somewhat arbitrary as explained in Chapter 2.

In words, for all δK_p that result in a 25% increase in the residual error, the parametric index is the maximum possible value of δK_i. Mathematically, for $i = 4, 6, 8$, and 14, one maximizes

$$\delta K_i$$

subject to (8.116). This is straightforwardly setup as an unconstrained optimization using Lagrange multipliers by maximizing

$$\delta K_i + \lambda \Big(E^2(K_p^* + \delta K_p) - 1.25 E^2(K_p^*) \Big) \tag{8.117}$$

over all possible $\delta K_p \triangleq \left[\begin{array}{cccc} \delta K_4 & \delta K_6 & \delta K_8 & \delta K_{14} \end{array} \right]^T$ and λ. For example, with $i = 4$, the extrema are solutions to

$$1 + \lambda \frac{\partial \left(E^2(K_p^* + \delta K_p) - 1.25 E^2(K_p^*) \right)}{\partial \delta K_4} = 0 \tag{8.118}$$

$$\lambda \frac{\partial \left(E^2(K_p^* + \delta K_p) - 1.25 E^2(K_p^*) \right)}{\partial \delta K_6} = 0 \tag{8.119}$$

$$\lambda \frac{\partial \left(E^2(K_p^* + \delta K_p) - 1.25 E^2(K_p^*) \right)}{\partial \delta K_8} = 0 \tag{8.120}$$

$$\lambda \frac{\partial \left(E^2(K_p^* + \delta K_p) - 1.25 E^2(K_p^*) \right)}{\partial \delta K_{14}} = 0 \tag{8.121}$$

$$E^2(K_p^* + \delta K_p) - 1.25 E^2(K_p^*) = 0. \tag{8.122}$$

The equations (8.118) through (8.122) are transformed to five polynomial equations in the five unknowns $\delta K_4, \delta K_6, \delta K_8, \delta K_{14}$, and λ, and elimination theory is then used to solve this set.

8.5.4 Mechanical Parameters

Once the electrical parameters have been found, the two mechanical parameters J, f ($\tau_L = -f\omega$) can be found using a linear least-squares algorithm. To do so, equations (8.95) and (8.96) are solved for $M\psi_{Rx}/L_R$ and $M\psi_{Ry}/L_R$, resulting in

$$\begin{aligned} \left[\begin{array}{c} M\psi_{Rx}/L_R \\ M\psi_{Ry}/L_R \end{array} \right] &= \sigma L_S \frac{1}{\left(1/T_R\right)^2 + n_p^2 \omega^2} \left[\begin{array}{cc} 1/T_R & -\omega \\ \omega & 1/T_R \end{array} \right] \\ &\times \left[\begin{array}{c} di_{Sx}/dt - u_{Sx}/\left(\sigma L_S\right) + \gamma i_{Sx} - n_p \omega i_{Sy} \\ di_{Sy}/dt - u_{Sy}/\left(\sigma L_S\right) + \gamma i_{Sy} + n_p \omega i_{Sx} \end{array} \right]. \end{aligned} \tag{8.123}$$

Noting that

$$\gamma = \frac{R_S}{\sigma L_S} + \frac{1}{\sigma L_S} \frac{1}{T_R} \frac{M^2}{L_R} = \frac{R_S}{\sigma L_S} + \frac{1}{\sigma L_S} \frac{1}{T_R} \left(1 - \sigma\right) L_S, \tag{8.124}$$

it is seen that the quantities on the right-hand side of (8.123) are all known once the electrical parameters have been computed. With $K_{16} \triangleq n_p/J, K_{17} \triangleq f/J$, equation (8.94) may be rewritten as

$$\frac{d\omega}{dt} = \begin{bmatrix} (M/L_R)(\psi_{Rx} i_{Sy} - \psi_{Ry} i_{Sx}) & -\omega \end{bmatrix} \begin{bmatrix} K_{16} \\ K_{17} \end{bmatrix}$$

so that the standard linear squares approach of Chapter 2 is directly applicable. Then

$$J \triangleq n_p/K_{16}, f \triangleq n_p K_{17}/K_{16}. \tag{8.125}$$

8.5.5 *Simulation Results*

The algorithm is first applied to simulated "data" to see how it performs under ideal conditions. In particular, the simulations are helpful in evaluating the usefulness of the parametric error indices. Here, a two pole-pair ($n_p = 2$), three-phase induction motor model was simulated using SIMULINK with parameter values chosen to be

$$\begin{aligned} R_S &= 9.7\ \Omega, R_R = 8.6\ \Omega, L_S = L_R = 0.67\ \text{H} \\ M &= 0.64\ \text{H}, \sigma = 0.088, J = 0.011\ \text{kg-m}^2, \tau_L = 3.7\ \text{N-m}. \end{aligned}$$

A 2048-line position encoder and 12-bit A/D converters were included in the simulation model in order to make the simulation data closer to actual experimental measurements. These values correspond to a 1/2-kW machine with a synchronous frequency of 50 Hz. These machine parameter values correspond to the following K values:

$$\begin{aligned} K_1 &= 299.15, K_2 = 10.42, K_3 = 17.05, K_4 = 1717.18, K_5 = 12.84 \\ K_6 &= 3839.8, K_7 = 133.78, K_8 = 0.0779, K_9 = 23.31, K_{10} = 0.8120 \\ K_{11} &= 0.0061, K_{12} = 1.82, K_{13} = 0.1035, K_{14} = 218.83, K_{15} = 1.328 \end{aligned}$$

for the electrical parameters and to

$$K_{16} = 181.82, \quad K_{17} = 336.36$$

for the mechanical parameters.

Figure 8.30 shows the simulated speed response (from standstill) of the induction motor after a balanced (open-loop) set of three-phase voltages of amplitude 466.7 V and frequency 50 Hz was applied to the machine.

The "data" $\{u_{Sa}, u_{Sb}, i_{Sa}, i_{Sb}, \theta\}$ were collected between 0 sec and 0.2 sec. The quantities $u_{Sx}, u_{Sy}, du_{Sx}/dt, du_{Sy}/dt, i_{Sx}, i_{Sy}$ $di_{Sx}/dt, di_{Sy}/dt, d^2 i_{Sx}/dt^2$, $d^2 i_{Sy}/dt^2$, $\omega = d\theta/dt$, and $d\omega/dt$ were calculated and the regressor matrices R_W, R_y, and R_{Wy} were computed. The procedure explained in Section 8.5.2 was then carried out to compute K_4, K_6, K_8, and K_{14}. In this case, there was only one extremum point that had *positive* values for all the K_i. The table

Parameter	True Value	Estimated Value	Parametric Error Index
K_4	1717.2	1743.4	13.23
K_6	3839.8	3917.5	24.59
K_8	0.0779	0.0780	0.00031
K_{14}	218.8	222.2	8.27

compares the electrical parameter values determined from the nonlinear least-squares procedure to their actual values (which are known only because this is simulation data). Also given in the table are the corresponding parametric error indices. The residual error index for the parameters K_4, K_6, K_8, and K_{14} was computed to be 3.15%.

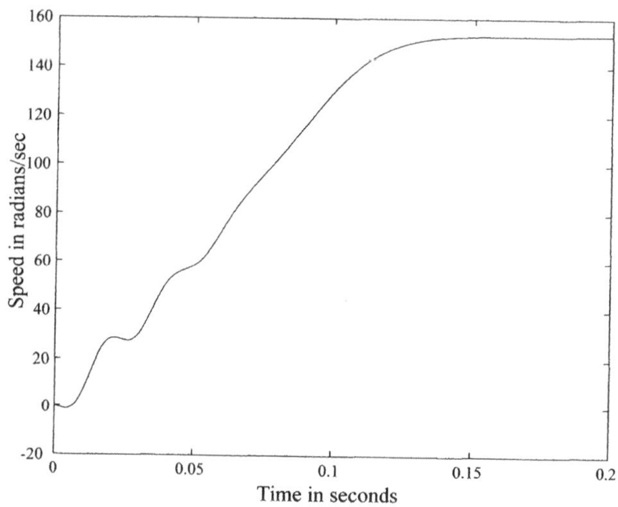

FIGURE 8.30. Rotor speed versus time.

Using these values, the scaled flux linkages $M\psi_{Rx}/L_R$ and $M\psi_{Ry}/L_R$ were reconstructed according to (8.123) to estimate the mechanical parameters K_{16} and K_{17}. The table

Parameter	True Value	Estimated Value	Parametric Error Index
K_{16}	181.8	195.3	4.29
K_{17}	336.4	359.3	9.78

compares the mechanical parameter values determined from the least-squares procedure to their actual values along with the corresponding parametric error indices. The residual error index for the mechanical parameters K_{16} and K_{17} was computed to be 4.26%.

The Hessian matrix for the identification of the parameters K_4, K_6, K_8, and K_{14} was calculated at the minimum point according to (8.115), resulting in

$$\left\{\frac{\partial^2 E^2(K_p^*)}{\partial K_i \partial K_j}\right\} = \begin{bmatrix} 7.534 & 0.2175 & 40.33 & -89.39 \\ 0.2175 & 18.04 & 10.17 & -44.35 \\ 40.33 & 10.17 & 5.308 \times 10^4 & 1.112 \times 10^3 \\ -89.39 & -44.35 & 1.112 \times 10^3 & 1.876 \times 10^3 \end{bmatrix}$$

which is positive definite and has a condition number of 1.88×10^4. The table below compares the estimated machine parameters to their actual values.

Parameter	True Value	Estimated Value
T_R	0.0779 sec	0.0780 sec
σ	0.088	0.086
L_S	0.67 H	0.6698 H
R_S	9.7 ohms	9.8 ohms
J	0.011 kg-m^2	0.010 kg-m^2
τ_L	3.7 N-m	3.68 N-m

8.5.6 *Experimental Results*

A three-phase, 0.5-Hp, 1735-rpm ($n_p = 2$ pole-pair) induction machine was used for the experiments. A 4096-pulse/rev optical encoder was attached to the motor for position measurements. The motor was connected to a three-phase, 60-Hz, 230-V source through a switch with no load on the machine. When the switch was closed, the stator currents and voltages along with the rotor position were sampled at 4 kHz. Filtered differentiation (using digital filters) was used for calculating the acceleration and the derivatives of the voltages and currents. Specifically, the signals were filtered with a low-pass digital Butterworth filter followed by reconstruction of the derivatives using $dx(t)/dt = (x(t) - x(t-T))/T$, where T is the sampling interval. The voltages and currents were put through a 3–2 transformation to obtain the two-phase equivalent voltages u_{Sa} and u_{Sb} which are plotted in Figure 8.31.

The sampled two-phase equivalent current i_{Sa} and it simulated response i_{Sa_sim} are shown in Figure 8.32 (the simulated current i_{Sa_sim} is discussed below). The phase b current i_{Sb} is similar, but shifted by $\pi/(2n_p)$. The calculated speed ω (from the position measurements) and the simulated speed ω_{sim} are shown in Figure 8.33 (the simulated speed ω_{sim} will also be discussed below).

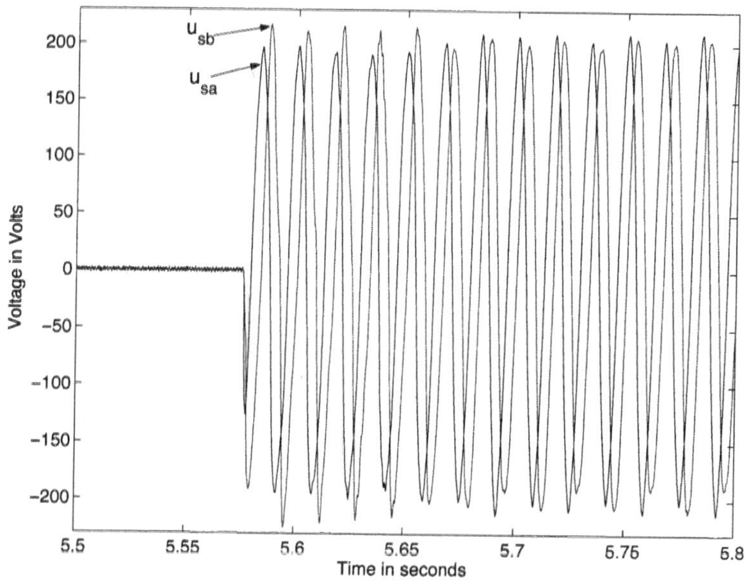

FIGURE 8.31. Sampled two-phase equivalent voltages u_{Sa} and u_{Sb}.

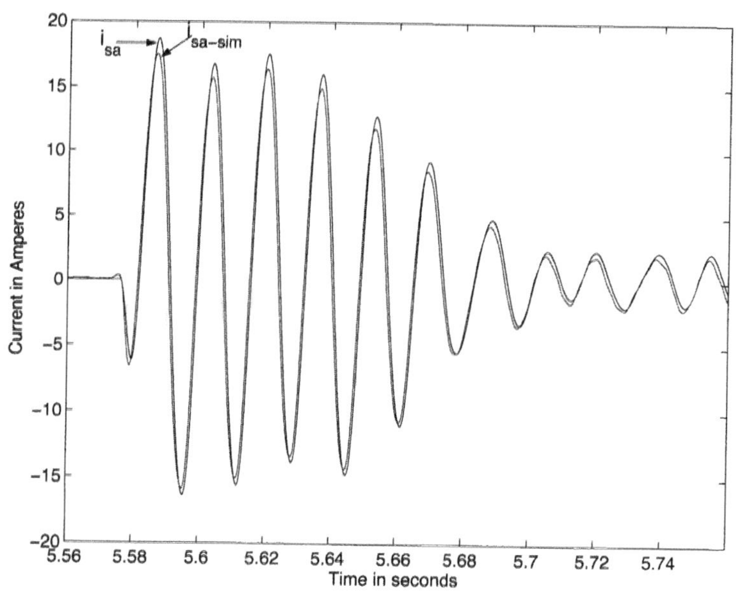

FIGURE 8.32. Phase a current i_{Sa} and its simulated response i_{Sa_sim}.

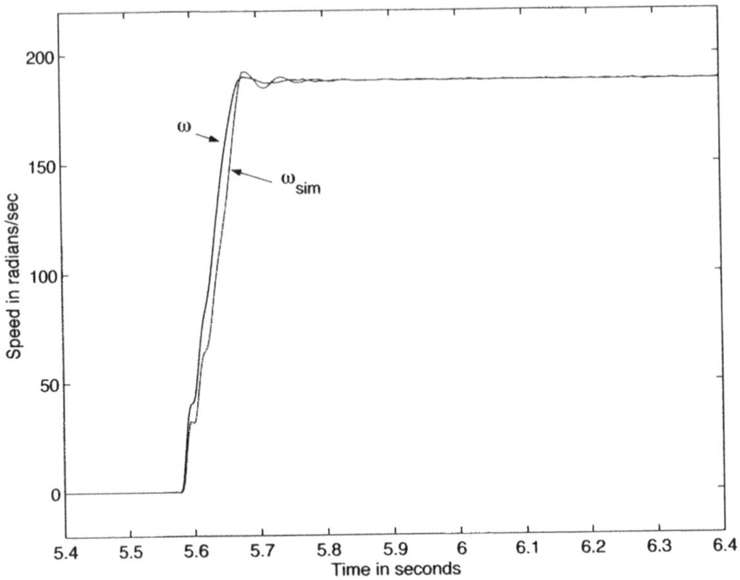

FIGURE 8.33. Calculated speed ω and simulated speed ω_{sim}.

Using the data $\{u_{Sa}, u_{Sb}, i_{Sa}, i_{Sb}, \theta\}$ collected between 5.57 sec to 5.8 sec, the quantities $u_{Sx}, u_{Sy}, du_{Sx}/dt, du_{Sy}/dt, i_{Sx}, i_{Sy}\ di_{Sx}/dt, di_{Sy}/dt, d^2i_{Sx}/dt^2$, d^2i_{Sy}/dt^2, and $\omega = d\theta/dt$, $d\omega/dt$ were calculated and the regressor matrices R_W, R_y, and R_{Wy} were computed. The procedure explained in Section 8.5.2 was then carried out to compute K_4, K_6, K_8, and K_{14}. In this case, there was only one extremum point that had *positive* values for all the K_i. The table below presents the parameter values determined using the nonlinear least-squares methodology along with their corresponding parametric error indices.

Parameter	Estimated Value	Parametric Error Index
$K4$	519.7	185.8
$K6$	1848.3	796.4
$K8$	0.1311	0.0103
$K14$	259.5	59.4

The residual error index was calculated to be 13.43%. The motor's electrical parameters are computed using (8.101) to obtain

$$R_S = 5.12 \text{ ohms} \quad (8.126)$$
$$T_R = 0.1311 \text{ sec} \quad (8.127)$$
$$L_S = 0.2919 \text{ H} \quad (8.128)$$
$$\sigma = 0.1007 \quad (8.129)$$

By way of comparison, the stator resistance was measured using an ohmmeter giving the value of 4.9 ohms and a no-load test was also run to compute the value of L_S resulting in 0.33 H.

The Hessian matrix for the identification of the parameters K_4, K_6, K_8, and K_{14} was calculated at the minimum point according to (8.115), resulting in

$$\left\{ \frac{\partial^2 E^2(K_p^*)}{\partial K_i \partial K_j} \right\} = \begin{bmatrix} 0.0574 & 0.1943 & -0.0034 & -0.7655 \\ 0.1943 & 2.584 & 10.17 & -44.35 \\ -0.0034 & 10.17 & 631.4 & 193.8 \\ -0.7655 & -44.35 & 193.8 & 3012 \end{bmatrix}$$

which is positive definite and has a condition number of 8.24×10^4.

Using the electrical parameters, the rotor flux linkages $(M/L_R)\,\psi_{Rx}$ and $(M/L_R)\,\psi_{Ry}$ were reconstructed and used to identify the mechanical parameters. The table below gives the estimated values and the parametric error indices.

Parameter	Estimated Value	Parametric Error Index
K_{16}	952.38	126.92
K_{17}	0.5714	0.1528

The residual error index was calculated to be 18.6%. The corresponding values for the motor parameters J and f are then computed using (8.125) to obtain

$$J = n_p/K_{16} = 0.0021 \text{ kg-m}^2 \quad (8.130)$$

$$f = n_p K_{17}/K_{16} = 0.0012 \text{ N-m/rad/sec.} \quad (8.131)$$

Simulation of the Experimental Motor

Another useful way to evaluate the identified parameters (8.126)–(8.129) and (8.130)–(8.131) is to simulate the motor using these values with the measured voltages as input. One then compares the simulation's output (stator currents) with the measured outputs. To proceed in this manner, recall that only R_S, T_R, L_S, and σ can be identified, but this is all that is needed for the simulation. Specifically, defining

$$\phi_{Ra} \triangleq \frac{M}{L_R}\psi_{Ra}, \quad \phi_{Rb} \triangleq \frac{M}{L_R}\psi_{Rb}$$

the model (8.88) may be rewritten as

$$\begin{aligned}
\frac{d\omega}{dt} &= \frac{n_p}{J}\left(i_{Sb}\phi_{Ra} - i_{Sa}\phi_{Rb}\right) - \frac{\tau_L}{J} \\
\frac{d\phi_{Ra}}{dt} &= -\frac{1}{T_R}\phi_{Ra} - n_p\omega\phi_{Rb} + \frac{1}{T_R}\frac{M^2}{L_R}i_{Sa} \\
\frac{d\phi_{Rb}}{dt} &= -\frac{1}{T_R}\phi_{Rb} + n_p\omega\phi_{Ra} + \frac{1}{T_R}\frac{M^2}{L_R}i_{Sb} \\
\frac{di_{Sa}}{dt} &= \frac{1}{\sigma L_S T_R}\phi_{Ra} + \frac{1}{\sigma L_S}n_p\omega\phi_{Rb} - \gamma i_{Sa} + \frac{1}{\sigma L_S}u_{Sa} \\
\frac{di_{Sb}}{dt} &= \frac{1}{\sigma L_S T_R}\phi_{Rb} - \frac{1}{\sigma L_S}n_p\omega\phi_{Ra} - \gamma i_{Sb} + \frac{1}{\sigma L_S}u_{Sb}
\end{aligned} \tag{8.132}$$

where

$$\frac{M^2}{L_R} = (1-\sigma)L_S, \gamma = \frac{R_S}{\sigma L_S} + \frac{1}{\sigma L_S}\frac{1}{T_R}\frac{M^2}{L_R}.$$

The model (8.132) uses only parameters that can be estimated. The experimental voltages shown in Figure 8.31 were then used as input to a simulation of the model (8.132) using the parameter values from (8.126)–(8.129) and (8.130)–(8.131). The resulting phase a current i_{Sa_sim} from the simulation is shown in Figure 8.32, and corresponds well with the actual measured current i_{Sa}. Similarly, the resulting speed ω_{sim} from the simulation is shown in Figure 8.33, where it is seen that the simulated speed is somewhat more oscillatory than the measured speed ω.

Remarks As explained above, M is not identifiable by the above method. However, it can be (and often is) estimated by assuming that $L_R = L_S$ (rotor and stator leakages are equal) so that $M = \sqrt{L_S L_R (1-\sigma)} = L_S\sqrt{1-\sigma}$. For the above experimental motor, this gives $M = 0.2768$ H. Under this same assumption, an estimate for the rotor resistance is given by $R_R = L_R/T_R = L_S/T_R = 2.23$ ohms.

The combined parameter identification and velocity estimation problem is considered in Refs. [93][94][95][96] while in Ref. [97] the sensitivity of measurable outputs to the machine parameters are discussed.

During operation of the machine, the values of the rotor time constant $T_R = L_R/R_R$ and the stator resistance R_S change due to ohmic heating. If one is only concerned with the online identification (tracking) of these two parameter values, then the problem simplifies considerably. This particular case has been developed in [98] (see also [99]).

Appendix

Elimination Theory and Resultants

Given two polynomials $a(x_1, x_2)$ and $b(x_1, x_2)$, how does one find their common zeros? That is, the values (x_{10}, x_{20}) such that

$$a(x_{10}, x_{20}) = b(x_{10}, x_{20}) = 0.$$

Consider $a(x_1, x_2)$ and $b(x_1, x_2)$ as polynomials in x_2 whose coefficients are polynomials in x_1. For example, let $a(x_1, x_2)$ and $b(x_1, x_2)$ have degrees 3 and 2, respectively, in x_2 so that they may be written in the form

$$\begin{aligned} a(x_1, x_2) &= a_3(x_1)x_2^3 + a_2(x_1)x_2^2 + a_1(x_1)x_2 + a_0(x_1) \\ b(x_1, x_2) &= b_2(x_1)x_2^2 + b_1(x_1)x_2 + b_0(x_1). \end{aligned}$$

Then there exist polynomials $\alpha(x_1, x_2)$ and $\beta(x_1, x_2)$ of the form

$$\begin{aligned} \alpha(x_1, x_2) &= \alpha_1(x_1)x_2 + \alpha_0(x_1) \\ \beta(x_1, x_2) &= \beta_2(x_1)x_2^2 + \beta_1(x_1)x_2 + \beta_0(x_1) \end{aligned}$$

that is, satisfying

$$\begin{aligned} \deg_{x2}\{\alpha(x_1, x_2)\} &= \deg_{x2}\{b(x_1, x_2)\} - 1 \\ \deg_{x2}\{\beta(x_1, x_2)\} &= \deg_{x2}\{a(x_1, x_2)\} - 1 \end{aligned}$$

and a polynomial $r(x_1)$ in one variable such that

$$\alpha(x_1, x_2)a(x_1, x_2) + \beta(x_1, x_2)b(x_1, x_2) = r(x_1).$$

The polynomial $r(x_1)$ is called the *resultant polynomial.* So if $a(x_{10}, x_{20}) = b(x_{10}, x_{20}) = 0$, then $r(x_{10}) = 0$. That is, if (x_{10}, x_{20}) is a *common zero* of the pair $\{a(x_1, x_2), b(x_1, x_2)\}$, then the first coordinate x_{10} is a zero of $r(x_1) = 0$. The roots of $r(x_1)$ are easy to find (numerically) as it is a polynomial in one variable. To find the common zeros of $\{a(x_1, x_2), b(x_1, x_2)\}$, one computes all roots x_{1i}, $i = 1, ..., n_1$, of $r(x_1)$. Next, for each such x_{1i}, one (numerically) computes the roots of

$$a(x_{1i}, x_2) = 0 \tag{8.133}$$

and the roots of

$$b(x_{1i}, x_2) = 0. \tag{8.134}$$

Any root x_{2j} that is in the solution set of both (8.133) and (8.134) for a given x_{1i} results in the pair (x_{1i}, x_{2j}) being a common zero of $a(x_1, x_2)$ and $b(x_1, x_2)$. Thus, this gives a method of solving polynomials in one variable to compute the common zeros of $\{a(x_1, x_2), b(x_1, x_2)\}$.

To see how one obtains $r(x_1)$, let

$$\begin{aligned} a(x_1,x_2) &= a_3(x_1)x_2^3 + a_2(x_1)x_2^2 + a_1(x_1)x_2 + a_0(x_1) \\ b(x_1,x_2) &= b_2(x_1)x_2^2 + b_1(x_1)x_2 + b_0(x_1). \end{aligned} \tag{8.135}$$

Next, one sees if polynomials of the form

$$\begin{aligned} \alpha(x_1,x_2) &= \alpha_1(x_1)x_2 + \alpha_0(x_1) \\ \beta(x_1,x_2) &= \beta_2(x_1)x_2^2 + \beta_1(x_1)x_2 + \beta_0(x_1) \end{aligned} \tag{8.136}$$

can be found such that

$$\alpha(x_1,x_2)a(x_1,x_2) + \beta(x_1,x_2)b(x_1,x_2) = r(x_1). \tag{8.137}$$

Proceeding, the expressions (8.135) for $a(x_1,x_2)$ and $b(x_1,x_2)$ as well as the expressions (8.136) for $\alpha(x_1,x_2)$ and $\beta(x_1,x_2)$ are substituted into (8.137). Equating powers of x_2, equation (8.137) may then be represented in matrix form as

$$\begin{bmatrix} a_0(x_1) & 0 & b_0(x_1) & 0 & 0 \\ a_1(x_1) & a_0(x_1) & b_1(x_1) & b_0(x_1) & 0 \\ a_2(x_1) & a_1(x_1) & b_2(x_1) & b_1(x_1) & b_0(x_1) \\ a_3(x_1) & a_2(x_1) & 0 & b_2(x_1) & b_1(x_1) \\ 0 & a_3(x_1) & 0 & 0 & b_2(x_1) \end{bmatrix} \begin{bmatrix} \alpha_0(x_1) \\ \alpha_1(x_1) \\ \beta_0(x_1) \\ \beta_1(x_1) \\ \beta_2(x_1) \end{bmatrix} = \begin{bmatrix} r(x_1) \\ 0 \\ 0 \\ 0 \\ 0 \end{bmatrix}.$$

The 5×5 matrix on the left-hand side is called the *Sylvester* matrix and is denoted here by $S_{a,b}(x_1)$. The inverse of $S_{a,b}(x_1)$ has the form

$$S_{a,b}^{-1}(x_1) = \frac{1}{\det S_{a,b}(x_1)} \operatorname{adj}\Big(S_{a,b}(x_1)\Big)$$

where $\operatorname{adj}(S_{a,b}(x_1))$ is the adjoint matrix and is a 5×5 *polynomial* matrix in x_1. Solving for $\alpha_i(x_1), \beta_i(x_1)$ gives

$$\begin{bmatrix} \alpha_0(x_1) \\ \alpha_1(x_1) \\ \beta_0(x_1) \\ \beta_1(x_1) \\ \beta_2(x_1) \end{bmatrix} = \frac{\operatorname{adj} S_{a,b}(x_1)}{\det S_{a,b}(x_1)} \begin{bmatrix} r(x_1) \\ 0 \\ 0 \\ 0 \\ 0 \end{bmatrix}.$$

If $r(x_1)$ is chosen as $r(x_1) = \det S_{a,b}(x_1)$, then

$$\begin{bmatrix} \alpha_0(x_1) \\ \alpha_1(x_1) \\ \beta_0(x_1) \\ \beta_1(x_1) \\ \beta_2(x_1) \end{bmatrix} = \operatorname{adj} S_{a,b}(x_1) \begin{bmatrix} 1 \\ 0 \\ 0 \\ 0 \\ 0 \end{bmatrix}.$$

This then guarantees that

$$\alpha_0(x_1), \alpha_1(x_1), \beta_0(x_1), \beta_1(x_1), \beta_2(x_1)$$

are polynomials in x_1. That is, the *resultant polynomial* is defined by $r(x_1) \triangleq \det S_{a,b}(x_1)$ and is the polynomial required for (8.137) to hold.

In summary, the polynomials $\{a(x_1, x_2), b(x_1, x_2)\}$ have a common zero at (x_{10}, x_{20}) only if $r(x_{10}) \triangleq \det S_{a,b}(x_{10}) = 0$. For an arbitrary pair of polynomials $\{a(x), b(x)\}$ of degrees n_a and n_b in x, respectively, the *Sylvester* matrix $S_{a,b}$ is of dimension $(n_a + n_b) \times (n_a + n_b)$ (see Refs. [89][90][100]).

Remark It was just shown that if $a(x_{10}, x_{20}) = b(x_{10}, x_{20}) = 0$, then $r(x_{10}) \triangleq \det S_{a,b}(x_{10}) = 0$ as a simple consequence of (8.137). Does

$$r(x_{10}) \triangleq \det S_{a,b}(x_{10}) = 0$$

imply that there exists x_{20} such that

$$a(x_{10}, x_{20}) = b(x_{10}, x_{20}) = 0?$$

Not necessarily. However, the answer is yes if either of the leading coefficients in x_2 of $a(x_1, x_2), b(x_1, x_2)$ are not zero at x_{10}, that is, $a_3(x_{10}) \neq 0$ *or* $b_2(x_{10}) \neq 0$ (see [89][90][100] for detailed explanations).

Example Let

$$\begin{aligned} a(x_1, x_2) &= x_2^2 + x_2 + x_1 \\ b(x_1, x_2) &= -x_1 x_2^2 + x_2 + 1. \end{aligned}$$

The resultant matrix is

$$S_{a,b}(x_1) = \begin{bmatrix} a_0(x_1) & 0 & b_0(x_1) & 0 \\ a_1(x_1) & a_0(x_1) & b_1(x_1) & b_0(x_1) \\ a_2(x_1) & a_1(x_1) & b_2(x_1) & b_1(x_1) \\ 0 & a_2(x_1) & 0 & b_2(x_1) \end{bmatrix} = \begin{bmatrix} x_1 & 0 & 1 & 0 \\ 1 & x_1 & 1 & 1 \\ 1 & 1 & -x_1 & 1 \\ 0 & 1 & 0 & -x_1 \end{bmatrix}$$

and the resultant polynomial is

$$r(x_1) = \det S_{a,b}(x_1) = (3 + x_1^2) x_1^2$$

with roots

$$\left\{0, 0, j\sqrt{3}, -j\sqrt{3}\right\}.$$

The root at $x_1 = 0$ gives

$$\begin{aligned} a(0, x_2) &= x_2^2 + x_2 = 0 \\ b(0, x_2) &= x_2 + 1 = 0 \end{aligned}$$

making

$$\{(0, -1)\}$$

a common zero.

The root at $x_1 = j\sqrt{3}$ gives

$$a(j\sqrt{3}, x_2) = x_2^2 + x_2 + j\sqrt{3} = 0$$

with roots

$$\left\{ \frac{1}{2} - j\frac{\sqrt{3}}{2}, -\frac{3}{2} + j\frac{\sqrt{3}}{2} \right\}$$

and

$$b(j\sqrt{3}, x_2) = -i\sqrt{3}x_2^2 + x_2 + 1 = 0$$

with roots

$$\left\{ \frac{1}{2} - j\frac{\sqrt{3}}{2}, -\frac{1}{2} + j0.2887 \right\}.$$

The only common zero is then

$$\left(j\sqrt{3}, \frac{1}{2} - j\frac{\sqrt{3}}{2} \right).$$

Similarly, the root at $x_1 = -j\sqrt{3}$ results in the common zero

$$\left(-j\sqrt{3}, \frac{1}{2} + j\frac{\sqrt{3}}{2} \right).$$

The set of common zeros of $a(x_1, x_2)$ and $b(x_1, x_2)$ is

$$\left\{ (0, -1), \left(j\sqrt{3}, \frac{1}{2} - j\frac{\sqrt{3}}{2} \right), \left(-j\sqrt{3}, \frac{1}{2} + j\frac{\sqrt{3}}{2} \right) \right\}.$$

Problems

Mathematical Models of the Induction Motor

Problem 1 State Space Form

Derive the state-space representation (8.5) from the representation (8.4).

Problem 2 Induction Motor Model in the Space Vector Formulation

The derivation of the model (8.4) is easier using the space vector representation

$$\begin{aligned}
R_S \underline{i}_S + L_S \frac{d}{dt} \underline{i}_S + M \frac{d}{dt} \left(\underline{i}_R e^{j n_p \theta} \right) &= \underline{u}_S \\
R_R \underline{i}_R + L_R \frac{d}{dt} \underline{i}_R + M \frac{d}{dt} \left(\underline{i}_S e^{-j n_p \theta} \right) &= 0 \\
n_p M \, \mathrm{Im}\{ \underline{i}_S (\underline{i}_R e^{j n_p \theta})^* \} - \tau_L &= J \frac{d\omega}{dt}.
\end{aligned}$$

To see this, define new (fictitious) flux linkages as

$$\underline{\psi}_R \triangleq \psi_{Ra} + j\psi_{Rb} \triangleq \underline{\lambda}_R e^{jn_p\theta} = L_R \underline{i}_R e^{jn_p\theta} + M\underline{i}_S$$

and show that system of equations for the induction motor can rewritten in the space vector form as

$$\begin{aligned}
\frac{d}{dt}\underline{\psi}_R &= \left(-\frac{1}{T_R} + jn_p\omega\right)\underline{\psi}_R + \frac{M}{T_R}\underline{i}_S \\
\underline{u}_S &= \frac{M}{L_R}\frac{d}{dt}\underline{\psi}_R + \sigma L_S \frac{d}{dt}\underline{i}_S + R_S\underline{i}_S \\
\frac{d\omega}{dt} &= \mu\,\mathrm{Im}\{\underline{i}_S(\underline{\psi}_R)^*\} - \frac{\tau_L}{J}
\end{aligned}$$

where $\mu \triangleq n_pM/(JL_R)$, $\sigma \triangleq 1 - M^2/L_SL_R$ *is the leakage factor, and* $T_R \triangleq L_R/R_R$ *is the rotor time constant. This problem illustrates the power of the space vector representation as the manipulations of the equations are carried out using (complex) scalar variables rather than dealing with matrix multiplications, matrix inverses, and so on.*

Problem 3 Induction Motor Model in Terms of the Flux Linkages

With $\lambda_{Sa}, \lambda_{Sb}$ *the flux linkage in stator phases a and b respectively, define*

$$\underline{\lambda}_S \triangleq \lambda_{Sa} + j\lambda_{Sb} \triangleq L_S\underline{i}_S + M\underline{i}_R e^{+jn_p\theta}.$$

Similarly, let λ_{Ra} *and* λ_{Rb} *be the flux linkage in rotor phases a and b, respectively, and define*

$$\underline{\lambda}_R \triangleq \lambda_{Ra} + j\lambda_{Rb} \triangleq L_R\underline{i}_R + M\underline{i}_S e^{-jn_p\theta}.$$

(a) Show that the electrical equations of the induction motor may be written as

$$\begin{aligned}
R_S\underline{i}_S + \frac{d}{dt}\underline{\lambda}_S &= \underline{u}_S \\
R_R\underline{i}_R + \frac{d}{dt}\underline{\lambda}_R &= 0.
\end{aligned}$$

(b) Using the fact that

$$\begin{bmatrix} \underline{i}_S \\ \underline{i}_R e^{+jn_p\theta} \end{bmatrix} = \frac{1}{L_SL_R - M^2}\begin{bmatrix} +L_R & -M \\ -M & +L_S \end{bmatrix}\begin{bmatrix} \underline{\lambda}_S \\ \underline{\lambda}_R e^{+jn_p\theta} \end{bmatrix}$$

show that

$$\begin{aligned}
\frac{R_S}{L_SL_R - M^2}\left(L_R\underline{\lambda}_S - M\underline{\lambda}_R e^{+jn_p\theta}\right) + \frac{d}{dt}\underline{\lambda}_S &= \underline{u}_S \\
\frac{R_R}{L_SL_R - M^2}\left(-M\underline{\lambda}_S e^{-jn_p\theta} + L_S\underline{\lambda}_R\right) + \frac{d}{dt}\underline{\lambda}_R &= 0 \\
\frac{M}{L_SL_R - M^2}\mathrm{Im}\{\underline{\lambda}_S(\underline{\lambda}_R e^{+jn_p\theta})^*\} - \tau_L &= J\frac{d\omega}{dt}
\end{aligned}$$

is a representation of the induction motor in terms of $\underline{\lambda}_S \triangleq \lambda_{Sa} + j\lambda_{Sb}$, $\underline{\lambda}_R \triangleq \lambda_{Ra} + j\lambda_{Rb}$, *and* ω.

Problem 4 Field-Oriented Induction Motor Model

The space vector model of the induction motor

$$\begin{aligned}\frac{d}{dt}\underline{\psi}_R &= \left(-\frac{1}{T_R} + jn_p\omega\right)\underline{\psi}_R + \frac{M}{T_R}\underline{i}_S \\ \underline{u}_S &= \frac{M}{L_R}\frac{d}{dt}\underline{\psi}_R + \sigma L_S\frac{d}{dt}\underline{i}_S + R_S\underline{i}_S \\ \frac{d\omega}{dt} &= \mu\,\mathrm{Im}\{\underline{i}_S(\underline{\psi}_R)^*\} - \frac{\tau_L}{J}\end{aligned}$$

is derived in problem 2. Let $\underline{\psi}_R \triangleq \psi_{Ra} + j\psi_{Rb} = \psi_d e^{j\rho}$ *where* $\psi_d \triangleq \sqrt{\psi_{Ra}^2 + \psi_{Rb}^2}$, $\rho \triangleq \tan^{-1}(\psi_{Rb}/\psi_{Ra})$ *and further let* $\underline{i}_{dq} \triangleq i_d + ji_q = \underline{i}_S e^{-j\rho}$, *and* $\underline{u}_{dq} \triangleq u_d + ju_q = \underline{u}_S e^{-j\rho}$.

(a) Find the field-oriented space vector representation of the induction motor, that is, in terms of $\psi_d, \rho, \underline{i}_{dq} \triangleq i_d + ji_q$, *and* $\underline{u}_{dq} \triangleq u_d + ju_q$

(b) Use the answer from part (a) to find a state space representation of the field-oriented induction motor model, that is, in terms of the state variables $i_d, i_q, \psi_d, \rho, \omega$, *and* θ *and the inputs* u_d *and* u_q.

Problem 5 Stator Flux Induction Motor Model

(a) The stator flux linkage is given by

$$\underline{\lambda}_S = L_S\underline{i}_S + M\left(\underline{i}_R e^{jn_p\theta}\right) = \lambda_S e^{j\rho_S}.$$

Rewrite the system model

$$\begin{aligned}R_S\underline{i}_S + L_S\frac{d}{dt}\underline{i}_S + M\frac{d}{dt}\left(\underline{i}_R e^{jn_p\theta}\right) &= \underline{u}_S \\ R_R\underline{i}_R + L_R\frac{d}{dt}\underline{i}_R + M\frac{d}{dt}\left(\underline{i}_S e^{-jn_p\theta}\right) &= 0 \\ n_p M\,\mathrm{Im}\left\{\underline{i}_S\left(\underline{i}_R e^{jn_p\theta}\right)^*\right\} - \tau_L &= J\frac{d\omega}{dt}\end{aligned}$$

in terms of $\underline{\lambda}_S, \underline{i}_S$, *and* ω.

(b) Put the result from part (a) in state space form.

Problem 6 ψ_d and ψ_q

Show that $\cos(\rho) = \psi_{Ra}/\psi_d$, $\sin(\rho) = \psi_{Rb}/\psi_d$, *and*

$$\begin{bmatrix}\psi_d \\ \psi_q\end{bmatrix} \triangleq \begin{bmatrix}\cos(\rho) & \sin(\rho) \\ -\sin(\rho) & \cos(\rho)\end{bmatrix}\begin{bmatrix}\psi_{Ra} \\ \psi_{Rb}\end{bmatrix}$$

where $\psi_q \equiv 0$.

Feedback Control and Reference Trajectories

Problem 7 Tracking a Constant Flux Reference

Let $e_0 = \int_0^t (\psi_{d0} - \psi_d)dt$ *and* $e_1 = \psi_{d0} - \psi_d$ *and assuming that* $\psi_{d0} = Mi_{d0}$, *show how to choose the gains* $K_{\psi P}$ *and* $K_{\psi I}$ *in (8.20) so that* $e_0(t) \to 0, e_1(t) \to 0$ *as* $t \to \infty$.

Problem 8 Trajectory Tracking via Field-Oriented Control

Let $e_0 = \int_0^t (\theta_{ref} - \theta)dt$, $e_1 = \theta_{ref} - \theta$, $e_2 = \omega_{ref} - \omega$, *and, assuming that* $\psi_d = \psi_{d0}$, *show how to choose the gains* K_0, K_1, *and* K_2 *in (8.19) so that the closed-loop poles are in the open left-half plane located at* $-r_1, -r_2, -r_3$ $(r_1 > 0, r_2 > 0, r_3 > 0)$. *If* τ_L *is constant, show that* $e_1(t) \to 0, e_2(t) \to 0$ *as* $t \to \infty$. *(Hint: Review Chapter 2.)*

Problem 9 Back Emf

With reference to the field-oriented model (8.9) of the induction motor, explain why $Mn_p\omega\psi_d/L_R$ *is the* back emf. *(The term "back emf" refers to the voltage* $v_{backemf}$ *that results in energy conversion from electrical to mechanical, that is,* $i_q v_{backemf} = \tau\omega$.*)*

Problem 10 Tracking a Time Varying Flux Reference

Consider the input-output linearization controller presented in Section 8.2.4. Let $\psi_{dref}(t)$ *be a desired flux reference with* i_{dref} *chosen to satisfy*

$$d\psi_{dref}/dt = -\eta\psi_{dref} + \eta M i_{dref}.$$

Choose

$$u_1 = i_{dr} = K_{\psi P}(\psi_{dref} - \psi_d) + K_{\psi I}\int_0^t (\psi_{dref} - \psi_d)dt + i_{dref}$$

and with $e_0 = \int_0^t (\psi_{dref} - \psi_d)dt, e_1 = \psi_{dref} - \psi_d$, *show how to choose the gains* $K_{\psi P}$ *and* $K_{\psi I}$ *such that* $e_0(t) \to 0, e_1(t) \to 0$ *as* $t \to \infty$. *(In practice,* i_{dref} *is usually set to zero with no significant deterioration in tracking.)*

Problem 11 Trajectory Tracking Via Input-Output Linearization

With $e_0 = \int_0^t (\theta_{ref} - \theta)dt$, $e_1 = \theta_{ref} - \theta$, $e_2 = \omega_{ref} - \omega$ *show how to choose the gains* K_0, K_1, *and* K_2 *in (8.32) so that the closed-loop poles are in the open left-half plane located at* $-r_1, -r_2, -r_3$ $(r_1 > 0, r_2 > 0, r_3 > 0)$. *Show that if* τ_L *is constant, then* $e_1(t) \to 0, e_2(t) \to 0$ *as* $t \to \infty$.

Problem 12 Field Weakening

Explain why field weakening is only needed when the induction machine is motoring (accelerating). That is, even if the motor is running at high speed, but one wants to decelerate *it down to zero (so it is generating rather than motoring), then the voltage limit* V_{max} *is not a constraint in bringing the flux* ψ_d *back up to* ψ_{d0}. *(However, most amplifiers are switching amplifiers and the blocking voltage rating of the switches must be respected.)*

Problem 13 Maximum Torque

In steady state the currents in a two-phase motor are of the form $i_{Sa} = I\cos(\omega_S t+\phi_S), i_{Sb} = I\cos(\omega_S t+\phi_S-\pi/2)$ *so that* $i_{d0}^2+i_{q0}^2 = i_{Sa}^2+i_{Sb}^2 = I^2$. *Show that the steady-state torque* $\tau = J\mu\psi_{d0}i_{q0} = J\mu M i_{d0}i_{q0}$ *subject to* $i_{d0}^2 + i_{q0}^2 \leq I_{max}^2$ *is maximized by choosing* $i_{d0} = i_{q0} = I_{max}/\sqrt{2}$. *This is used only at lower speeds where the voltage constraint* $u_{Sa}^2+u_{Sb}^2 \leq V_{max}$ *will not be violated. (Note that if, for example,* $I_{max} = 6.0$ *A, then one might choose* $i_{d0} = 5.5/\sqrt{2}$ *so as to not operate at the current limit.)*

Problem 14 Closed-Loop Poles of the Field-Oriented and Input-Output Controller

(a) Using the motor parameters given in Section 8.2.2 and the gains of the field-oriented controller given by (8.22) and (8.23), compute the closed loop poles of the mechanical trajectory tracking error system as well as the closed-loop poles of the flux tracking error system.

(b) Repeat part (a) using the gain values of the input-output controller of Section 8.2.5.

Problem 15 Simulation of an Input-Output Linearization Controller

Let the parameters of the system be $n_p = 3$, $M = 0.0117$ *H*, $R_R = 3.9$ Ω, $R_S = 1.7$ Ω, $L_R = 0.014$ *H*, $L_S = 0.014$ *H*, $f = 0$ *N-m/rad/sec, and* $J = 0.00011$ *kg-m*^{2}, $i_{d0} = 5.5/\sqrt{2}$, $\psi_{d0} = M i_{d0}$, $\omega_{base} = 250$ *rad/sec*, $I_{max} = 6$ *A*, $V_{max} = 60$ *V [42].*

Simulate the complete input–output linearization controller for the induction motor. To simplify the matter, do the complete simulation in continuous time and assume that both speed and position measurements are available. Specifically,

Add the flux observer using equations (8.37). Check that this is working by comparing ρ *and* ψ_d *from the estimator with* $\tan^{-1}(\psi_{Rb}/\psi_{Ra})$ *and* $\sqrt{\psi_{Rb}^2+\psi_{Ra}^2}$ *from the simulation of the motor. Note from (8.37) that the direct and quadrature currents* i_d *and* i_q *are computed in the flux estimator.*

Add the PI current controllers described by (8.12) to the simulation so that the setup of Figure 8.3 results. Check that current controllers are working by commanding constant currents $i_{dr} = i_{d0}$ *and* $i_{qr} = i_{q0}$ *to see if* $i_d \to i_{d0}, i_q \to i_{q0}$, *respectively.*

Choose i_{dr} *and* i_{qr} *as given in (8.27) where* u_1 *is chosen as given in (8.31) and* u_2 *is chosen as given in (8.32). Let the flux reference be given by (8.24) where* $\omega_{base} = 250$ *rads/sec and* $\psi_{d0} = M i_{d0}$ *with* $i_{d0} = 5.5/\sqrt{2}$. *Also, just take* $i_{dref} = 0$.

Add the trajectory generator to the simulation.

(a) Try a trajectory specified by $\omega_{\max} = 75$ *rads/sec,* $t_1 = 0.025$ *sec and* $\theta_f = \pi$. *Set the gains of the system to obtain tracking of the mechanical trajectory* $(\theta_{ref}, \omega_{ref})$ *and the flux reference* ψ_{d0}, *and run the simulation. Do you need to let the flux* ψ_d *build up before starting the* $180°$ *turn?*

(b) Try a high-speed trajectory specified by $\omega_{\max} = (2\pi/60)8000 = 838$

rads/sec, $t_1 = 1.5$ *seconds,* $t_2 = t_1 + 0.1$ *sec* $(\theta_f = \omega_{\max} t_2)$. *Set the gains of the system to obtain tracking of the mechanical trajectory* $(\theta_{ref}, \omega_{ref})$ *and the flux reference* ψ_{dref}, *and run the simulation.*

Problem 16 i_{dref}
Given a mechanical trajectory mechanical trajectory $(\theta_{ref}, \omega_{ref}, \alpha_{ref})$, *show how to obtain the reference current* i_{dref} *corresponding to the flux reference* ψ_{dref} *specified by (8.24).*

Problem 17 Nested Loop Control Structure [2]
Consider the (nested loop) control structure of Figure 8.34 following Figures 12.14 *and* 12.27 *of Ref.* [2]. *The block labeled "induction motor" is considered to be the model (8.9) with the decoupling feedback (8.10) applied, that is, the system (8.11). Using the model (8.11), and the controller given in Figure 8.34 with* ψ_{dref} *given by (8.24), do an analysis of this controller. In particular, with* ψ_{dref} *constant, can you show how to specify the controller gains so that the closed-loop system is stable? What if* ψ_{dref} *is not constant?*

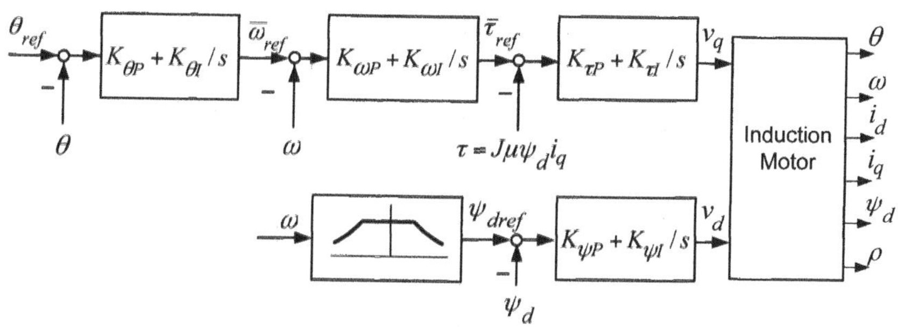

FIGURE 8.34. Nested loop control structure for the induction motor.

Field Energy and Torque

Problem 18 Torque from Conservation of Energy
(a) Multiply the fourth equation of (8.9) by $\sigma L_S i_d$, *the fifth equation of (8.9) by* $\sigma L_S i_q$ *and add to obtain*

$$\begin{aligned} i_d u_d + i_q u_q &= \frac{1}{2}\sigma L_S \frac{d}{dt}\left(i_d^2 + i_q^2\right) + \sigma L_S \gamma \left(i_d^2 + i_q^2\right) - (\eta M/L_R)\psi_d i_d \\ &+ (n_p M/L_R)\omega \psi_d i_q. \end{aligned}$$

(b) In Chapter 6 *it was shown that the field energy (the energy stored in the magnetic field of the air gap) is given by*

$$\begin{aligned} W_{field}(i_{Sa}, i_{Sb}, i_{Ra}, i_{Rb}, \theta) &\triangleq \frac{1}{2}L_S(i_{Sa}^2 + i_{Sb}^2) + \frac{1}{2}L_R(i_{Ra}^2 + i_{Rb}^2) \\ &+ M i_{Sa}\Big(i_{Ra}\cos(\theta) - i_{Rb}\sin(\theta)\Big) + M i_{Sb}\Big(i_{Ra}\sin(\theta) + i_{Rb}\cos(\theta)\Big). \end{aligned}$$

Show that in the $(i_{Sa}, i_{Sb}, \psi_{Ra}, \psi_{Rb}, \theta, \omega)$ *coordinate system, the field energy is given by*

$$W_{field} = \frac{1}{2}L_S(i_{Sa}^2 + i_{Sb}^2) + \frac{1}{2}\frac{1}{L_R}\left(\psi_{Ra}^2 + \psi_{Rb}^2\right) - \frac{1}{2}\frac{M^2}{L_R}\left(i_{Sa}^2 + i_{Sb}^2\right)$$

and therefore in the dq coordinate system it is given by

$$W_{field} = \frac{1}{2}\sigma L_S(i_d^2 + i_q^2) + \frac{1}{2}\frac{1}{L_R}\psi_d^2.$$

(c) Conservation of energy (power) requires that

$$\frac{d}{dt}W_{field} + R_S(i_{Sa}^2 + i_{Sb}^2) + R_R(i_{Ra}^2 + i_{Rb}^2) + \tau\omega = i_{Sa}u_{Sa} + i_{Sb}u_{Sb}$$

where $\tau\omega$ *is the mechanical power produced. Rewrite this in the field-oriented coordinate system and compare with the answer to part (a) to conclude that* $\tau = (n_pM/L_R)\psi_d i_q$.

Observers

Problem 19 Discretization of the Flux Observer

Using Euler's approximation for the derivatives, that is,

$$\begin{aligned}\dot{\psi}_{Ra}(kT) &\approx \left(\psi_{Ra}((k+1)T) - \psi_{Ra}(kT)\right)/T\\ \dot{\psi}_{Rb}(kT) &\approx \left(\psi_{Rb}((k+1)/T) - \psi_{Rb}(kT)\right)/T\end{aligned}$$

a discretized version of (8.34) is given by

$$\begin{aligned}\psi_{Ra}((k+1)T) &= (1-\eta T)\psi_{Ra}(kT) - n_pT\omega(kT)\psi_{Rb}(kT) + \eta MTi_{Sa}(kT)\\ \psi_{Rb}((k+1)T) &= n_pT\omega(kT)\psi_{Ra}(kT) + (1-\eta T)\psi_{Rb}(kT) + \eta MTi_{Sb}(kT).\end{aligned}$$

An observer for this discretized system is then a discretized version of (8.35) given by

$$\begin{aligned}\hat{\psi}_{Ra}((k+1)T) &= (1-\eta T)\hat{\psi}_{Ra}(kT) - n_pT\omega(kT)\hat{\psi}_{Rb}(kT) + \eta MTi_{Sa}(kT)\\ \hat{\psi}_{Rb}((k+1)T) &= n_pT\omega(kT)\hat{\psi}_{Ra}(kT) + (1-\eta T)\hat{\psi}_{Rb}(kT) + \eta MTi_{Sb}(kT).\end{aligned}$$

Simulate this discrete *version of the observer (8.35) with the inputs to the motor given by* $u_{Sa} = V\cos(\omega_S t)$, $u_{Sb} = V\sin(\omega_S t)$, $\omega_S = (2\pi)60$ *rad/sec,* $V = 80$ *V, and* $\tau_L = 0$. *Plot out the discrete estimates* $\hat{\psi}_{Ra}$ *and* $\hat{\psi}_{Rb}$ *and compare with the actual values* ψ_{Ra} *and* ψ_{Rb}, *respectively. Try sample rates of* 2 *kHz,* 10 *kHz, and* 20 *kHz.*

Problem 20 Stability of a Discrete Flux Observer

Consider the stability of the discrete-time flux estimator of problem 19, define

$$\begin{aligned}e_{Ra}((k+1)T) &\triangleq \psi_{Ra}((k+1)T) - \hat{\psi}_{Ra}((k+1)T)\\ e_{Rb}((k+1)T) &\triangleq \psi_{Rb}((k+1)T) - \hat{\psi}_{Rb}((k+1)T)\end{aligned}$$

and show that these error variables satisfy

$$\begin{bmatrix} e_{Ra}((k+1)T) \\ e_{Rb}((k+1)T) \end{bmatrix} = \begin{bmatrix} 1-\eta T & -n_p\omega(k)T \\ n_p\omega(k)T & 1-\eta T \end{bmatrix} \begin{bmatrix} e_{Ra}(kT) \\ e_{Rb}(kT) \end{bmatrix}.$$

Next, define the discrete-time function $V(kT)$ as

$$V(kT) \triangleq e_{Ra}^2(kT) + e_{Rb}^2(kT)$$

and show that

$$V((k+1)T) = \left((1-\eta T)^2 + (n_p\omega(kT)T)^2\right) V(kT).$$

Use this to find the conditions on the sample period T and speed ω such that $V(kT) \to 0$ as $t \to \infty$. In particular, find the value of T that minimizes $(1-\eta T)^2 + (n_p\omega_{\max}T)^2$, where $\omega_{\max}$ is the maximum speed the motor is to be run. With $\omega_{\max} = 2\pi(60)$ rads/sec and sample rates of 330 Hz, 1 kHz, and 10 kHz, is the system (8.35) stable? Are your results consistent with your simulation?

Problem 21 Discretization of the Flux Observer in the Field-Oriented Coordinate System

Simulate a discrete-time version of the equations (8.37). Plot out the discrete estimates $\hat{\rho}$ and $\hat{\psi}_d$, compute $\hat{\psi}_{Ra} = \hat{\psi}_d \cos(\hat{\rho})$ and $\hat{\psi}_{Rb} = \hat{\psi}_d \sin(\hat{\rho})$ and compare with ψ_{Ra} and ψ_{Rb}, respectively. Try sample rates of 330 Hz, 1 kHz, and 10 kHz.

Problem 22 Discretization of the Flux Observer Equations [73]

Consider another discretization of the flux equations (8.34) by writing them in the form

$$\frac{d}{dt}\begin{bmatrix} \psi_{Ra} \\ \psi_{Rb} \end{bmatrix} = \begin{bmatrix} -\eta & -n_p\omega \\ n_p\omega & -\eta \end{bmatrix} \begin{bmatrix} \psi_{Ra} \\ \psi_{Rb} \end{bmatrix} + \eta M \begin{bmatrix} i_{Sa} \\ i_{Sb} \end{bmatrix}$$

(a) Assuming that $\omega(t), i_{Sa}(t)$, and $i_{Sb}(t)$ are constant over the sample period T, show that

$$\begin{aligned} \begin{bmatrix} \psi_{Ra}((k+1)T) \\ \psi_{Rb}((k+1)T) \end{bmatrix} &= e^{(-\eta\mathbf{I}+n_p\omega(kT)\mathbf{J})T} \begin{bmatrix} \psi_{Ra}(kT) \\ \psi_{Rb}(kT) \end{bmatrix} \\ &+ \left(\int_0^T e^{(-\eta\mathbf{I}+n_p\omega(kT)\mathbf{J})(T-t)}\eta M\,dt\right) \begin{bmatrix} i_{Sa}(kT) \\ i_{Sb}(kT) \end{bmatrix} \end{aligned}$$

where $\mathbf{I} \triangleq \begin{bmatrix} 1 & 0 \\ 0 & 1 \end{bmatrix}, \mathbf{J} \triangleq \begin{bmatrix} 0 & -1 \\ 1 & 0 \end{bmatrix}.$

(b) Using the expressions

$$\begin{aligned}
\int e^{at}\cos(bt)dt &= \frac{e^{at}}{a^2+b^2}(a\cos(bt)+b\sin(bt))\\
\int e^{a(T-t)}\cos(b(T-t))dt &= -\frac{e^{a(T-t)}}{a^2+b^2}(a\cos(b(T-t))+b\sin(b(T-t)))\\
\int e^{at}\sin(bt)dt &= \frac{e^{at}}{a^2+b^2}(a\sin(bt)-b\cos(bt))\\
\int e^{a(T-t)}\sin(b(T-t))dt &= -\frac{e^{a(T-t)}}{a^2+b^2}(a\sin(b(T-t))-b\cos(b(T-t)))
\end{aligned}$$

show that

$$e^{(-\eta\mathbf{I}+n_p\omega(kT)\mathbf{J})T} = \begin{bmatrix} e^{-\eta T} & 0\\ 0 & e^{-\eta T}\end{bmatrix}\begin{bmatrix}\cos(n_p\omega(kT)) & -\sin(n_p\omega(kT)\\ \sin(n_p\omega(kT)) & \cos(n_p\omega(kT))\end{bmatrix}$$

and

$$\int_0^T e^{(-\eta\mathbf{I}+n_p\omega(kT)\mathbf{J})(T-t)}dt = \frac{1}{\eta^2+(n_p\omega(kT))^2} \times\begin{bmatrix} e^{-\eta T}d_1(kT)+\eta & -e^{-\eta T}d_2(kT)-n_p\omega(kT)\\ e^{-\eta T}d_2(kT)+n_p\omega(kT) & e^{-\eta T}d_1(kT)+\eta\end{bmatrix}$$

where

$$\begin{aligned}
d_1(kT) &\triangleq -\eta\cos(n_p\omega(kT)T)+n_p\omega(kT)\sin(n_p\omega(kT)T)\\
d_2(kT) &\triangleq -\eta\sin(n_p\omega(kT)T)-n_p\omega(kT)\cos(n_p\omega(kT)T).
\end{aligned}$$

(c) Use this discretization of the flux linkage equations to define an observer.

(d) With the flux estimate error defined by

$$\begin{aligned}
e_{Ra}((k+1)T) &\triangleq \psi_{Ra}((k+1)T)-\hat{\psi}_{Ra}((k+1)T)\\
e_{Rb}((k+1)T) &\triangleq \psi_{Rb}((k+1)T)-\hat{\psi}_{Rb}((k+1)T)
\end{aligned}$$

use the (Lyapunov) function

$$V(kT)\triangleq e_{Ra}^2(kT)+e_{Rb}^2(kT)$$

to show that

$$V((k+1)T)=e^{-2\eta T}V(kT).$$

(e) Simulate this discrete-time flux observer with the inputs to the motor given by $u_{Sa}=V\cos(\omega_S t)$, $u_{Sb}=V\sin(\omega_S t)$ *where* $\omega_S=(2\pi)60$ *rad/sec and* $V=80$ V, $\tau_L=0$. *Plot out the discrete estimates* $\hat{\psi}_{Ra}$ *and* $\hat{\psi}_{Rb}$, *and compare with their actual values* ψ_{Ra} *and* ψ_{Rb}. *Try sample rates of* 330 *Hz,* 1 *kHz, and* 10 *kHz.*

Problem 23 Flux Observer Based on Position Measurements [67]

Consider the flux observer (8.37). Make a change of variables from (ρ, ψ_d) *to* (s, ψ_d) *where* $s(t) \triangleq \rho(t) - n_p\theta(t)$ *and show that*

$$\frac{d\hat{s}}{dt} = \eta M\left(-i_{Sa}\sin(\hat{\rho}) + i_{Sb}\cos(\hat{\rho})\right)/\hat{\psi}_d$$

$$\frac{d\hat{\psi}_d}{dt} = -\eta\hat{\psi}_d + \eta M\left(i_{Sa}\cos(\hat{\rho}) + i_{Sb}\sin(\hat{\rho})\right)$$

where

$$\hat{\rho}(t) \triangleq n_p\theta(t) + \hat{s}(t)$$

is an estimator for ρ, ψ_d. *Note that this flux estimator requires the stator currents and rotor position, but not the rotor speed.*

Problem 24 Speed Observer

(a) Implement a discretized version of the speed observer

$$\frac{d\hat{\theta}}{dt} = \hat{\omega} + \ell_1(\theta_c - \hat{\theta})$$
$$\frac{d\hat{\omega}}{dt} = \mu\hat{\psi}_d\hat{\imath}_q + \ell_2(\theta_c - \hat{\theta})$$

using the Euler approximation for the derivatives where $\theta_c(t) \triangleq \frac{2\pi}{2000}N(t)$ *in which* $N(t)$ *is the position of the motor in encoder counts. Set the inputs to the motor as* $u_{Sa} = V\cos(\omega_S t)$, $u_{Sb} = V\sin(\omega_S t)$ ($\omega_S = (2\pi)60$ *rad/sec, and* $V = 80$ *V). To simulate the encoder count* N, *simply set* $N =$ `floor`$(\frac{2000}{2\pi}\theta)$, *where* `floor(x)` *is the largest integer less than or equal to* x.

(b) Compare the speed estimate $\hat{\omega}$ *from the observer with that obtained from numerically differentiating the output of the encoder as* $\hat{\omega}_1(kT) = (\theta_c(kT) - \theta_c((k-1)T))/T$. *Also, compare with the actual speed* ω. *Try sample rates of* 1 *kHz,* 2 *kHz, and* 10 *kHz.*

(c) Show that a (Euler) discretized version of the error *equations for the above speed observer are given by*

$$\begin{bmatrix} e_1(k+1) \\ e_2(k+1) \end{bmatrix} = \begin{bmatrix} 1-\ell_1 T & T \\ -\ell_2 & 1 \end{bmatrix}\begin{bmatrix} e_1(k) \\ e_2(k) \end{bmatrix}$$

where T *is the sample period,* ℓ_1 *and* ℓ_2 *are the observer gains,* $e_1(k) = \theta_c(kT) - \hat{\theta}(kT)$ *and* $e_2(k) = \omega(kT) - \hat{\omega}(kT)$. *Compute the eigenvalues of the error system in terms of* T, ℓ_1, *and* ℓ_2. *Is the discretized speed observer stable, that is, are the magnitudes of the eigenvalues less than* 1 *for the gains and sample rates chosen in your simulation?*

Problem 25 Speed and Load-Torque Observer

Consider a modification of the speed observer to estimate any constant (or, slowly varying) load torque. Specifically, with the load torque τ_L *assumed constant, consider a speed and load-torque estimator given by*

$$\begin{aligned}\frac{d\hat{\theta}}{dt} &= \hat{\omega} + \ell_1(\theta - \hat{\theta})\\ \frac{d\hat{\omega}}{dt} &= \mu\hat{\psi}_d\hat{i}_q - \hat{\tau}_L/J + \ell_2(\theta - \hat{\theta})\\ \frac{d(\hat{\tau}_L/J)}{dt} &= 0 + \ell_3(\theta - \hat{\theta}).\end{aligned}$$

(a) With $e_1 = \theta - \hat{\theta}, e_2 = \omega - \hat{\omega}$, *and* $e_3 = \tau_L/J - \hat{\tau}_L/J$, *find the equations describing the error system assuming* $\hat{\psi}_d\hat{i}_q = \psi_d i_q$. *For an appropriate choice of the observer gains* ℓ_1, ℓ_2, *and* ℓ_3, *show that* $e(t) = (e_1(t), e_2(t), e_3(t)) \to 0$ *as* $t \to \infty$.

(b) Assume that $\hat{\psi}_d\hat{i}_q \neq \psi_d i_q$, *but that their difference is constant. That is, let* $\varepsilon_{\tau/J} \triangleq \mu\hat{\psi}_d\hat{i}_q - \mu\psi_d i_q$ *be constant. Show that it still true that* $e_1(t) = \theta(t) - \hat{\theta}(t) \to 0$ *and* $e_2(t) = \omega(t) - \hat{\omega}(t) \to 0$ *as* $t \to \infty$.

Problem 26 Flux and Speed Observer

Consider the following approach to estimating the flux linkages and speed simultaneously. The stator currents i_{Sa} *and* i_{Sb} *and flux linkages* ψ_{Ra} *and* ψ_{Rb} *are rotated by the angle* $n_p\theta$ *to obtain*

$$\begin{bmatrix} i_{Sx} \\ i_{Sy} \end{bmatrix} \triangleq \begin{bmatrix} \cos(n_p\theta) & \sin(n_p\theta) \\ -\sin(n_p\theta) & \cos(n_p\theta) \end{bmatrix} \begin{bmatrix} i_{Sa} \\ i_{Sb} \end{bmatrix}$$

$$\begin{bmatrix} \psi_{Rx} \\ \psi_{Ry} \end{bmatrix} \triangleq \begin{bmatrix} \cos(n_p\theta) & \sin(n_p\theta) \\ -\sin(n_p\theta) & \cos(n_p\theta) \end{bmatrix} \begin{bmatrix} \psi_{Ra} \\ \psi_{Rb} \end{bmatrix}.$$

The currents i_{Sx} *and* i_{Sy} *are* known *as the angle* θ *is measured. Note also that* $\psi_{Rx} = \lambda_{Ra}$ *and* $\psi_{Ry} = \lambda_{Rb}$, *that is, they are the actual flux linkages in the rotor windings as given in (8.2). In order to estimate the load torque, it is modeled as a constant so that its dynamic equation is taken to be* $d(\tau_L/J)/dt = 0$.

(a) In terms of the new state variables $i_{Sx}, i_{Sy}, \psi_{Rx} = \lambda_{Ra}$, *and* $\psi_{Ry} =$

λ_{Rb}, *show that*

$$
\begin{aligned}
\frac{d\psi_{Rx}}{dt} &= -\frac{R_R}{L_R}\psi_{Rx} + \frac{MR_R}{L_R}i_{Sx} \\
\frac{d\psi_{Ry}}{dt} &= -\frac{R_R}{L_R}\psi_{Ry} + \frac{MR_R}{L_R}i_{Sy} \\
\frac{d\theta}{dt} &= \omega \\
\frac{d\omega}{dt} &= \mu(i_{Sy}\psi_{Rx} - i_{Sx}\psi_{Ry}) - \frac{f}{J}\omega - \tau_L/J \\
\frac{d(\tau_L/J)}{dt} &= 0.
\end{aligned}
$$

The key point here is that in this state-space representation, the (unmeasured) speed ω *has been* eliminated *from the flux equations.*

(b) Define an estimator for the flux linkages, speed, and load torque by

$$
\begin{aligned}
\frac{d\hat{\psi}_{Rx}}{dt} &= -\frac{R_R}{L_R}\hat{\psi}_{Rx} + \frac{MR_R}{L_R}i_{Sx} \\
\frac{d\hat{\psi}_{Rx}}{dt} &= -\frac{R_R}{L_R}\hat{\psi}_{Ry} + \frac{MR_R}{L_R}i_{Sy} \\
\frac{d\theta}{dt} &= \hat{\omega} + \ell_1(\theta - \hat{\theta}) \\
\frac{d\omega}{dt} &= \mu(i_{Sy}\hat{\psi}_{Rx} - i_{Sx}\hat{\psi}_{Ry}) - \frac{f}{J}\omega - \frac{\hat{\tau}_L}{J} + \ell_2(\theta - \hat{\theta}) \\
\frac{d(\hat{\tau}_L/J)}{dt} &= 0 + \ell_3(\theta - \hat{\theta}).
\end{aligned}
$$

With $e_{Rx} \triangleq \psi_{Rx} - \hat{\psi}_{Rx}, e_{Ry} \triangleq \psi_{Ry} - \hat{\psi}_{Ry}$, $e_\theta \triangleq \theta - \hat{\theta}$, $e_\omega \triangleq \omega - \hat{\omega}$, *and* $e_{\tau_L} \triangleq \tau_L/J - \hat{\tau}_L$, *show that the error system is given by*

$$
\begin{aligned}
\frac{de_{Rx}}{dt} &= -\frac{R_R}{L_R}e_{Rx} \\
\frac{de_{Ry}}{dt} &= -\frac{R_R}{L_R}e_{Ry} \\
\frac{de_\theta}{dt} &= e_\omega - \ell_1(\theta - \hat{\theta}) \\
\frac{de_\omega}{dt} &= \mu(i_{Sy}e_{Rx} - i_{Sx}e_{Ry}) - \frac{f}{J}e_\omega - e_{\tau_L} - \ell_2(\theta - \hat{\theta}) \\
\frac{de_{\tau_L}}{dt} &= 0 - \ell_3(\theta - \hat{\theta}).
\end{aligned}
$$

(c) As long as the currents i_{Sx} *and* i_{Sy} *are bounded (consistent with the assumed current-command operation), show that the error system is stable. Note that the rate of convergence of the error dynamics of this observer is still limited by the rotor time constant* $T_R = L_R/R_R = 1/\eta$ *[73].*

Flux Observers with an Arbitrary Rate of Convergence

Martin and Rouchon [74][75] have developed an observer that generalizes the results of Verghese and Sanders [73] presented in Section 8.3.3. Specifically, their method results in an error system with complex eigenvalues whose rate of convergence can be arbitrarily specified. This is outlined in the next problem.

Problem 27 Martin-Rouchon Observer [74][75]
The equations in space vector form may be written as

$$\begin{aligned} R_S\underline{i}_S + \frac{d}{dt}\underline{\lambda}_S &= \underline{u}_S \\ R_R\underline{i}_R + \frac{d}{dt}\underline{\lambda}_R &= 0 \end{aligned} \tag{8.138}$$

where $\underline{i}_S = i_{Sa} + ji_{Sb}$, *and so on, and the flux linkages and currents are related by*

$$\begin{aligned} \underline{\lambda}_S &\triangleq \lambda_{Sa} + j\lambda_{Sb} \triangleq L_S\underline{i}_S + M\underline{i}_R e^{+j\theta_R} \\ \underline{\lambda}_R &\triangleq \lambda_{Ra} + j\lambda_{Rb} \triangleq L_R\underline{i}_R + M\underline{i}_S e^{-j\theta_R}. \end{aligned} \tag{8.139}$$

(a) Rearrange the first equation of (8.139) to obtain

$$\underline{i}_R e^{j\theta_R} = \frac{1}{M}\left(\underline{\lambda}_S - L_S\underline{i}_S\right)$$

and show that substituting this into the second equation of (8.139) results in

$$\underline{\lambda}_R e^{j\theta_R} = \frac{L_R}{M}\underline{\lambda}_S + \sigma M\underline{i}_S$$

where $\sigma = 1 - M^2/(L_SL_R)$.

(b) Multiply the second equation of (8.138) by $e^{-j\theta_R}$ *and substitute the expressions for* $\underline{i}_R e^{j\theta_R}$ *and* $\underline{\lambda}_R e^{j\theta_R}$ *from part (a) into (8.138) to show that the electrical equations have the equivalent form*

$$\begin{aligned} \frac{d}{dt}\underline{\lambda}_S &= -R_S\underline{i}_S + \underline{u}_S \\ \frac{d}{dt}\underline{i}_S &= \frac{(R_R - j\omega L_R)\underline{\lambda}_S - (R_RL_S + R_SL_R - j\omega\sigma L_RL_S)\underline{i}_S + L_R\underline{u}_S}{\sigma L_RL_S}. \end{aligned}$$

(c) Consider the observer proposed in Refs. [74][75] given by

$$\begin{aligned} \frac{d}{dt}\hat{\underline{\lambda}}_S &= -R_S\underline{i}_S + \underline{u}_S + c_1(\underline{i}_S - \hat{\underline{i}}_S)(R_R - j\omega L_R) \\ \frac{d}{dt}\hat{\underline{i}}_S &= \frac{(R_R - j\omega L_R)\hat{\underline{\lambda}}_S - (R_RL_S + R_SL_R - j\omega\sigma L_RL_S)\underline{i}_S + L_R\underline{u}_S}{\sigma L_RL_S} \\ &\quad + \frac{c_2(\underline{i}_S - \hat{\underline{i}}_S)(R_R - j\omega L_R)}{\sigma L_RL_S} \end{aligned}$$

where c_1 *and* c_2 *are* complex *gains to be chosen.*

(d) Set $\underline{\varepsilon}_\lambda = \underline{\lambda}_S - \hat{\underline{\lambda}}_S$ *and* $\underline{\varepsilon}_i = \underline{i}_S - \hat{\underline{i}}_S$ *and show the error system is given by*

$$\begin{aligned}\frac{d}{dt}\underline{\varepsilon}_\lambda &= -(R_R - j\omega L_R)c_1\underline{\varepsilon}_i \\ \frac{d}{dt}\underline{\varepsilon}_i &= \frac{(R_R - j\omega L_R)}{\sigma L_R L_S}\left(\underline{\varepsilon}_\lambda - c_2\underline{\varepsilon}_i\right).\end{aligned}$$

More compactly, one may write

$$\begin{aligned}\frac{d}{dt}\begin{bmatrix}\underline{\varepsilon}_\lambda \\ \underline{\varepsilon}_i\end{bmatrix} &= (R_R - j\omega L_R)\begin{bmatrix} 0 & -c_1 \\ \dfrac{1}{\sigma L_R L_S} & -\dfrac{1}{\sigma L_R L_S}c_2\end{bmatrix}\begin{bmatrix}\underline{\varepsilon}_\lambda \\ \underline{\varepsilon}_i\end{bmatrix} \\ \frac{d}{dt}\underline{\varepsilon} &= (R_R - j\omega L_R)A\underline{\varepsilon} \end{aligned} \qquad (8.140)$$

with the obvious definitions for $\underline{\varepsilon}$ *and* A.

(e) Show the eigenvalues of A *can be arbitrarily assigned by choosing* c_1 *and* c_2 *appropriately.*

(f) Show that the solution to (8.140) is given by

$$\underline{\varepsilon}(t) = e^{(R_R t - j\theta(t)L_R)A}\underline{\varepsilon}(0).$$

(g) Show that c_1 *and* c_2 *can be chosen so that* $s_1 = s_{1a} + js_{1b}sign(\theta(t))$ *and* $s_2 = s_{2a} + js_{2b}sign(\theta(t))$ *are the eigenvalues of* A *with* $s_{1a} < 0$, $s_{1b} \le 0$, $s_{2a} < 0$, $s_{2b} \le 0$. *That is, they are the roots of the polynomial*

$$s^2 + \frac{c_2}{\sigma L_R L_S}s + \frac{c_1}{\sigma L_R L_S}.$$

(As c_1 *and* c_2 *are complex gains, this is possible to do with* s_1 *and* s_2 not *being complex conjugates.)*

(h) Assuming $s_1 \ne s_2$, *show that the error* $\varepsilon(t)$ *is a linear combination of the complex exponentials*

$$e^{s_{1a}R_R t + s_{1b}sign(\theta(t))L_R\theta(t) + j\left(s_{1b}sign(\theta(t))\,R_R t - s_{1a}L_R\theta(t)\right)}$$

$$e^{s_{2a}R_R t + s_{2b}sign(\theta(t))L_R\theta(t) + j\left(s_{2b}sign(\theta(t))\,R_R t - s_{2b}L_R\theta(t)\right)}$$

which decay to zero.

Remark *One may choose* $s_{1b} = s_{2b} = 0$ *to simplify the expressions in part (h). However, as explained in [74][75], there is an advantage in choosing* c_1 *and* c_2 *such that* s_1 *and* s_2 *are not real. After the rotor position increases,* $L_R\theta(t)$ *typically is much larger than* $R_R t$ *so that a faster convergence rate results by having* s_{1b} *and* s_{2b} *both negative.*

Dynamic Feedback Linearization

The concept behind feedback linearization is to find a coordinate system transformation such that the nonlinear dynamics represented in the new coordinate system can be canceled out by state feedback. This is a quite appealing approach when such a transformation exists. This theory has been worked out in that (checkable) necessary and sufficient conditions exist to determine the existence of such a transformation (see the excellent books by A. Isidori [52], H. Nijmeijer and A. J. van der Schaft [101], and R. Marino and P. Tomei [102] that described this theory is detail). However, it turns out that the system model (8.5) [or equivalently, the system model (8.9)] of the induction motor is *not* feedback linearizable [64]. Furthermore, the reduced-order model (8.13)–(8.16) with inputs i_{dr}, i_{qr} and state variables ω, ψ_d, ρ is also *not* feedback linearizable. However, a quite interesting result is that for some dynamic systems which are not feedback linearizable, the addition of integrators in some of the inputs can result in the (higher-order) system being feedback linearizable [103][104][105]. This is a phenomenon of multi-input systems, in that a single-input system is dynamically feedback linearizable if and only if it is (statically) feedback linearizable [105]. This approach is explored in the next two problems.

Problem 28 Addition of an Integrator to the d-axis Input [106][107]

Consider the current command induction motor model as given in the system model (8.13)–(8.16) whose inputs are taken to be i_{dr} and i_{qr}. A so-called dynamic feedback linearization controller can be found by the addition of an integrator to the d axis of field-oriented model. Specifically, define $x_1 \triangleq \omega$, $x_2 \triangleq \psi_d$, $x_3 \triangleq \rho$ and $x_4 \triangleq i_{dr}$, $dx_4/dt \triangleq v_1$, $v_2 \triangleq i_{qr}$ so that the equations are now

$$\begin{aligned} dx_1/dt &= \mu x_2 v_2 - \tau_L/J \\ dx_2/dt &= -\eta x_2 + \eta M x_4 \\ dx_3/dt &= n_p x_1 + \eta M v_2/x_2 \\ dx_4/dt &= v_1. \end{aligned}$$

More compactly, this may be written as

$$\frac{dx}{dt} = f(x) + g_1 v_1 + g_1 v_2$$

with

$$f(x) = \begin{bmatrix} -\tau_L/J \\ -\eta x_2 + \eta M x_4 \\ n_p x_1 \\ 0 \end{bmatrix}, \; g_1 = \begin{bmatrix} 0 \\ 0 \\ 0 \\ 1 \end{bmatrix}, \; g_2 = \begin{bmatrix} \mu x_2 \\ 0 \\ \eta M/x_2 \\ 0 \end{bmatrix} \in \mathbb{R}^4.$$

(a) Consider the nonlinear state-space transformation given by

$$\begin{aligned} z_1 &= x_2 \\ z_2 &= -\eta x_2 + \eta M x_4 \\ z_3 &= x_1 - \mu x_2^2 x_3/\eta M \\ z_4 &= \frac{2\mu}{M} x_2 x_3 (x_2 - M x_4) - \frac{\mu n_p}{\eta M} x_2^2 x_1 - \tau_L/J. \end{aligned}$$

Show that in these new coordinates, the system equations become

$$\begin{aligned} \frac{dz_1}{dt} &= z_2 \\ \frac{dz_2}{dt} &= a_1(x) + v_1 b_{11} + v_2 b_{12}(x) \\ \frac{dz_3}{dt} &= z_4 \\ \frac{dz_4}{dt} &= a_2(x) + v_1 b_{21} + v_2 b_{22}(x) \end{aligned}$$

where

$$\begin{aligned} a_1(x) &= -\eta(-\eta x_2 + \eta M x_4) \\ a_2(x) &= -2\eta M \mu x_3 x_4^2 + 6\eta\mu x_2 x_3 x_4 - 4\mu n_p x_1 x_2 x_4 + \frac{\mu n_p}{\eta M J}\tau_L x_2^2 \\ &\quad + \frac{1}{M}(-4\eta\mu x_2^2 x_3 + 4\mu n_p x_1 x_2^2) \end{aligned}$$

and $b_{11} = \eta M, b_{12}(x) = -2\mu x_2 x_3, b_{21} = 0, b_{22}(x) = -2\mu(-\eta x_2 + \eta M x_4) - \frac{\mu^2 n_p}{\eta M} x_2^3.$

(b) Show that application of the feedback

$$\begin{bmatrix} v_1 \\ v_2 \end{bmatrix} = \begin{bmatrix} b_{11} & b_{12} \\ b_{21} & b_{22} \end{bmatrix}^{-1} \begin{bmatrix} u_1 - a_1(x) \\ u_2 - a_2(x) \end{bmatrix}$$

results in two decoupled second-order linear *systems*

$$\begin{aligned} dz_1/dt &= z_2 \\ dz_2/dt &= u_1 \\ dz_3/dt &= z_4 \\ dz_4/dt &= u_2. \end{aligned}$$

(c) The controller is singular when

$$\det\begin{bmatrix} b_{11} & b_{12} \\ b_{21} & b_{22} \end{bmatrix} = -2\eta M\mu(-\eta x_2 + \eta M x_4) - \mu^2 n_p x_2^3 = 0.$$

Rewrite this singularity condition in the dq coordinates and show that it becomes

$$-(2\eta M\mu)d\psi_d/dt - \mu^2 n_p\psi_d^3 = 0.$$

Remark 1 *This singularity can be avoided by choosing, for example, the input u_1 to regulate ψ_d to a positive constant so that $d\psi_d/dt = 0$. On the other hand, during field weakening when ψ_d is reduced, this condition just restricts how fast ψ_d can be decreased, that is,*

$$0 \geq d\psi_d/dt > -(\frac{\mu n_p}{2\eta M})\psi_d^3 = -(\frac{n_p^2}{2JR_R})\psi_d^3.$$

At start-up of the motor, when the flux is brought up from zero to its nominal value, the singularity is not encountered as $d\psi_d/dt > 0$.

Remark 2 *Note that none of the state variables z_i are linearly related to the speed ω. Consequently, $d\theta/dt = \omega$ is* not *a linear function of the z_i so that the position cannot be appended to the coordinate system defining the z_i's without losing linearity of the system. That is, if $d\theta/dt = \omega = x_1$ is appended to the z_i coordinate transformation, then it is straightforward to show that the resulting system is* not *feedback linearizable (a similar situation arises in the case of a shunt-connected DC motor [55]). This situation also complicates the speed control design, as the speed is indirectly controlled through the input u_2 vis-a-vis $z_3 = x_1 - \mu x_2^2 x_3/(\eta M) = \omega - \mu\psi_d^2\rho/(\eta M)$. For example, if a constant speed ω_0 is desired, $z_3 = \omega_0 - \mu\psi_d^2\rho/(\eta M)$ must be tracked using the input u_2. This is sensitive to errors in the parameter values, as well as errors in the estimation of the state variables ψ_d and ρ.*

Problem 29 Addition of an Integrator to the q-axis Input [106][107]

It is quite natural to consider adding an integrator in the q axis of the system model (8.13)–(8.16) rather than the d axis for comparison. To proceed, let $x_1 = \omega$, $x_2 = \psi_d$, $x_3 = \rho$ and $x_4 = i_{qr}$, $dx_4/dt = v_2, v_1 = i_{dr}$ so that the system (8.13)–(8.16) becomes

$$\begin{aligned}dx_1/dt &= \mu x_2 x_4 - \tau_L/J\\ dx_2/dt &= -\eta x_2 + \eta M v_1\\ dx_3/dt &= n_p x_1 + \eta M x_4/x_2\\ dx_4/dt &= v_2.\end{aligned}$$

More compactly, this may be written as

$$\frac{dx}{dt} = f(x) + g_1 v_1 + g_1 v_2$$

with

$$f(x) = \begin{bmatrix}\mu x_2 x_4 - \tau_L/J\\ -\eta x_2\\ n_p x_1 + \eta M x_4/x_2\\ 0\end{bmatrix},\ g_1 = \begin{bmatrix}0\\ \eta M\\ 0\\ 0\end{bmatrix},\ g_2 = \begin{bmatrix}0\\ 0\\ 0\\ 1\end{bmatrix} \in \mathbb{R}^4.$$

A feedback linearizing transformation is then

$$\begin{aligned} z_1 &= x_1 \\ z_2 &= \mu x_2 x_4 - \tau_L/J \\ z_3 &= x_3 \\ z_4 &= n_p x_1 + \eta M x_4/x_2. \end{aligned}$$

(a) Show that in these new coordinates, the system equations become

$$\begin{aligned} \frac{dz_1}{dt} &= z_2 \\ \frac{dz_2}{dt} &= a_1(x) + v_1 b_{11} + v_2 b_{12} \\ \frac{dz_3}{dt} &= z_4 \\ \frac{dz_4}{dt} &= a_2(x) + v_1 b_{21} + v_2 b_{22} \end{aligned}$$

where

$$\begin{aligned} a_1(x) &= -\eta\mu x_2 x_4 \\ a_2(x) &= \mu n_p x_2 x_4 + \eta^2 M \frac{x_4}{x_2} - \frac{n_p \tau_L}{J} \end{aligned}$$

and $b_{11} = \eta M \mu x_4, b_{12} = \mu x_2, b_{21} = -\eta^2 M^2 x_4/x_2^2, b_{22} = \eta M/x_2.$

(b) Show that application of the feedback

$$\begin{bmatrix} v_1 \\ v_2 \end{bmatrix} = \begin{bmatrix} b_{11} & b_{12} \\ b_{21} & b_{22} \end{bmatrix}^{-1} \begin{bmatrix} u_1 - a_1(x) \\ u_2 - a_2(x) \end{bmatrix}$$

results in two decoupled second-order linear *systems*

$$\begin{aligned} dz_1/dt &= z_2 \\ dz_2/dt &= u_1 \\ dz_3/dt &= z_4 \\ dz_4/dt &= u_2. \end{aligned}$$

results in two decoupled second-order systems as in the d-axis case.

Remark *This controller is singular when*

$$\det \begin{bmatrix} b_{11} & b_{12} \\ b_{21} & b_{22} \end{bmatrix} = 2\eta^2 M^2 \mu \frac{x_4}{x_2} = 2\eta^2 M^2 \mu \frac{i_{qr}}{\psi_d} = 0.$$

This singularity condition is comparable to the condition that the torque $\tau = J\mu\psi_d i_q = J\mu x_2 x_4 \neq 0$ *given in Ref.* [108]. *However, in contrast to Ref.* [108], *the approach here requires only a single transformation, and the resulting controller requires relatively little feedback computation. A drawback* *here is the requirement that the quadrature current be nonzero ($\psi_d > 0$ is straightforward to maintain and is common to other controllers* [1][64][70]*). Consequently, to use this controller would require switching to another control strategy when it is desired to change the sign of the torque. Note that, in contrast to the case where the integrator was in the d axis, one of the transformed state variables is the speed, which is easily controlled using the input u_1. By appending the state variable $z_0 = \theta$ to the z-coordinate transformation gives a feedback linearizing controller for* position *control. However, to avoid the singularity of the controller, it is still required that $i_q \neq 0$.*

Problem 30 The Induction Motor is a Flat System [74][109]

Another way of viewing dynamic feedback linearization is through the notion of flatness *of a system. The concept of a system being* flat *is due to Michel Fliess* [109] *and has been applied to the induction motor by Martin and Rouchon* [74]*. The book* [110] *by H. Sira-Ramírez and S. K. Agrawal presents a nice introduction to the theory of flat systems in terms of applications and the book* [111] *by G. Conte, C. H. Moog and A. M. Perdon presents a nice introduction to the underlying mathematics. To motivate the concept of flatness, consider the Chua circuit example discussed by Sira-Ramírez and Agrawal* ([110], *p. 196), which is modeled by the third-order nonlinear system*

$$\frac{dx_1}{dt} = p\left(-x_1 + x_2 - x_1\left(1 - x_1^2\right)\right) \tag{8.141}$$

$$\frac{dx_2}{dt} = x_1 - x_2 + x_3 \tag{8.142}$$

$$\frac{dx_3}{dt} = -qx_2 + u \tag{8.143}$$

where p and q are known constant parameters. Let the (flat) output be $y = x_1$. The state can be written as function of y and its derivatives as follows:

$$\begin{aligned}
x_1 &= y \\
x_2 &= \dot{y}/p + y + y(1 - y^2) && \textit{(rearranging (8.141))} \\
x_3 &= \frac{d}{dt}\left(\dot{y}/p + y + y(1 - y^2)\right) + \left(\dot{y}/p + y + y(1 - y^2)\right) - y \\
&= \ddot{y}/p + (2 + 1/p)\,\dot{y} - 3y^2\dot{y} + y(1 - y^2) && \textit{(rearranging (8.142))}
\end{aligned}$$

or, more compactly,

$$x = \varphi(y, \dot{y}, \ddot{y}) \in \mathbb{R}^3.$$

Also, the input u can be written as function of y and its derivatives by rearranging (8.143) to obtain

$$\begin{aligned} u &= \frac{d}{dt}\left(\ddot{y}/p + (2+1/p)\,\dot{y} - 3y^2\dot{y} + y(1-y^2)\right) + q\left(\dot{y}/p + y + y(1-y^2)\right) \\ &= \dddot{y}/p + (2+1/p)\ddot{y} - 6y\dot{y} - 3y^2\ddot{y} + \dot{y} - 3y^2\dot{y} + (q/p)\dot{y} + qy + qy(1-y^2) \qquad (8.144) \\ &= \phi(y,\dot{y},\ddot{y},\dddot{y}) \in \mathbb{R}. \end{aligned}$$

The notion of flatness refers to defining an output variable y such that both the state and the input can be written as a function of this output and its derivatives. Thus, the Chua circuit model is a flat system. To see why the notion of flatness is useful, in the Chua circuit example define the new state variables $z_1 = y, z_2 = \dot{y}$, and $z_3 = \ddot{y}$ so that

$$\begin{aligned} \frac{dz_1}{dt} &= z_2 \\ \frac{dz_2}{dt} &= z_3 \\ \frac{dz_3}{dt} &= \dddot{y}. \end{aligned}$$

However, by equation (8.144),

$$\dddot{y} = pu - p\left((2+1/p)\ddot{y} - 6y\dot{y} - 3y^2\ddot{y} + \dot{y} - 3y^2\dot{y} + (q/p)\dot{y} + qy + qy(1-y^2)\right)$$

so that setting

$$u = (2+1/p)\ddot{y} - 6y\dot{y} - 3y^2\ddot{y} + \dot{y} - 3y^2\dot{y} + (q/p)\dot{y} + qy + qy(1-y^2) + v/p$$

the system becomes

$$\begin{aligned} \frac{dz_1}{dt} &= z_2 \\ \frac{dz_2}{dt} &= z_3 \\ \frac{dz_3}{dt} &= v \end{aligned}$$

which is a linear system. Furthermore, the flatness property allows a straightforward design of the reference trajectory and reference input. Specifically, let $y_{ref}(t)$ be the desired reference output so that

$$\begin{aligned} x_{ref} &= \varphi\left(y_{ref}, \dot{y}_{ref}, \ddot{y}_{ref}\right) \in \mathbb{R}^3 \\ u_{ref} &= \phi\left(y_{ref}, \dot{y}_{ref}, \ddot{y}_{ref}, \dddot{y}_{ref}\right) \in \mathbb{R} \\ z_{ref} &= \left[y_{ref}, \dot{y}_{ref}, \ddot{y}_{ref}\right]^T \in \mathbb{R}^3 \end{aligned}$$

are the reference signals.

Remark *For single-input systems, flatness is equivalent to feedback linearizability of which the Chua circuit is an example. For multi-input systems, flatness implies the system is dynamically feedback linearizable* [110].

(a) Consider the space vector formulation of the induction motor model, that is,

$$\begin{aligned} R_S\underline{i}_S + L_S\frac{d}{dt}\underline{i}_S + M\frac{d}{dt}\left(\underline{i}_R e^{jn_p\theta_R}\right) &= \underline{u}_S \\ R_R\underline{i}_R + L_R\frac{d}{dt}\underline{i}_R + M\frac{d}{dt}\left(\underline{i}_S e^{-jn_p\theta_R}\right) &= 0 \\ n_p M\,\mathrm{Im}\left\{\underline{i}_S(\underline{i}_R e^{jn_p\theta_R})^*\right\} - \tau_L &= J\frac{d\omega_R}{dt} \end{aligned}$$

where $\underline{i}_S \triangleq i_{Sa} + ji_{Sb}$, $\underline{i}_R \triangleq i_{Ra} + ji_{Rb}$, $\underline{u}_S \triangleq u_{Sa} + ju_{Sb}$. *With*

$$\begin{aligned} \underline{\lambda}_S &\triangleq \lambda_{Sa} + j\lambda_{Sb} \triangleq L_S\underline{i}_S + M\underline{i}_R e^{+jn_p\theta_R} \\ \underline{\lambda}_R &\triangleq \lambda_{Ra} + j\lambda_{Rb} \triangleq L_R\underline{i}_R + M\underline{i}_S e^{-jn_p\theta_R} \end{aligned}$$

the electrical equations of the induction motor may be written as

$$\begin{aligned} R_S\underline{i}_S + \frac{d}{dt}\underline{\lambda}_S &= \underline{u}_S \\ R_R\underline{i}_R + \frac{d}{dt}\underline{\lambda}_R &= 0. \end{aligned}$$

In problem 8 of Chapter 6, it was shown that using

$$\begin{bmatrix} \underline{i}_S \\ \underline{i}_R e^{+jn_p\theta_R} \end{bmatrix} = \frac{1}{L_S L_R - M^2}\begin{bmatrix} +L_R & -M \\ -M & +L_S \end{bmatrix}\begin{bmatrix} \underline{\lambda}_S \\ \underline{\lambda}_R e^{+jn_p\theta_R} \end{bmatrix},$$

a model for the induction motor in terms of the flux linkages is

$$\begin{aligned} \frac{R_S}{\sigma L_S L_R}\left(L_R\underline{\lambda}_S - M\underline{\lambda}_R e^{+jn_p\theta_R}\right) + \frac{d}{dt}\underline{\lambda}_S &= \underline{u}_S \\ \frac{R_R}{\sigma L_S L_R}\left(-M\underline{\lambda}_S e^{-jn_p\theta_R} + L_S\underline{\lambda}_R\right) + \frac{d}{dt}\underline{\lambda}_R &= 0 \\ \frac{M}{\sigma L_S L_R}\mathrm{Im}\{\underline{\lambda}_S(\underline{\lambda}_R e^{+jn_p\theta_R})^*\} - \tau_L &= J\frac{d\omega_R}{dt} \\ \frac{d\theta_R}{dt} &= \omega_R. \end{aligned}$$

Show that

$$\frac{M}{\sigma L_S L_R}\mathrm{Im}\{\underline{\lambda}_S(\underline{\lambda}_R e^{+jn_p\theta_R})^*\} = -\,\mathrm{Im}\left\{\underline{i}_R\underline{\lambda}_R^*\right\}.$$

(b) Show that

$$\begin{aligned} J\frac{d\omega}{dt} &= -\operatorname{Im}\{\underline{i}_R\underline{\lambda}_R^*\} - \tau_L \\ &= -\frac{1}{R_R}\operatorname{Im}\left\{\frac{d\underline{\lambda}_R}{dt}\underline{\lambda}_R^*\right\} - \tau_L \\ &= -\frac{1}{R_R}\lambda_R^2\frac{d}{dt}\rho_R - \tau_L \end{aligned}$$

where $\underline{\lambda}_R \triangleq \lambda_R e^{j\rho_R}$.

(c) Show that

$$\begin{bmatrix} y_1 \\ y_2 \end{bmatrix} \triangleq \begin{bmatrix} \theta \\ \rho_R \end{bmatrix}$$

is a flat output. *That is, show that*

$$\begin{aligned} \underline{\lambda}_R &= \underline{\varphi}_1(y, \dot{y}, \ddot{y}) \\ \underline{\lambda}_R &= \underline{\varphi}_2(y, \dot{y}, \ddot{y}, \dddot{y}) \\ \underline{u}_S &= \underline{\phi}(y, \dot{y}, \ddot{y}, \dddot{y}, \ddddot{y}). \end{aligned}$$

Explicitly compute the functions $\underline{\varphi}_1, \underline{\varphi}_2$, *and* $\underline{\phi}$.

(d) When is the transformation singular?

Optimal Field Weakening

Problem 31 Torque Expressions

Show that with $\omega_S = n_p\omega + \delta/T_R$ *the electrical frequency,* $S_p = 1/(\sigma\omega_S T_R)$ *the normalized pull out slip,* $S = (\omega_S - n_p\omega)/\omega_S = (\delta/T_R)/\omega_S$ *the normalized slip, and letting* $R_S \to 0$, *the torque expression*

$$\tau = \frac{n_p M^2}{L_R R_S^2} V^2 \frac{\delta}{\left(1 - \sigma T_S \delta\left(n_p\omega + \delta/T_R\right)\right)^2 + \left(\delta + T_S\left(n_p\omega + \delta/T_R\right)\right)^2}$$

reduces to

$$\tau = n_p \frac{3}{2}\frac{1-\sigma}{\sigma}\frac{U_S^2}{\omega_S^2 L_S}\frac{2}{S/S_p + S_p/S}$$

where $U_S \triangleq \sqrt{\dfrac{2}{3}}\dfrac{V}{\sqrt{2}}$. *(*$V/\sqrt{2}$ *is the two-phase equivalent rms voltage so that* $\sqrt{2/3}\,(V/\sqrt{2})$ *is the line-to-neutral rms voltage in the three-phase machine.)*

Problem 32 Positive root of $p(\delta) = 0$

Use the Routh-Hurwitz test and the expression (8.68) to show that $p(\delta) = 0$ *has exactly one root in the right-half plane for all* $\omega \geq 0$.

Problem 33 Base Speed

Use the Routh-Hurwitz test to show that equation (8.85) has only one root satisfying $\omega > 0$ *for* $\delta_1 = \pm 1$.

9

PM Synchronous Motor Control

Starting with a mathematical model of the PM synchronous machine, a field-oriented controller is developed for high-performance control. Experimental results are presented to illustrate the methodology. Following this, a systematic approach to field weakening is developed based on maximizing the torque obtainable from the machine without violating the voltage or the current constraints. Next, it is shown how the least-squares identification procedure developed in Chapter 2 can be used to determine the parameters of a PM synchronous machine. Finally, the structure and operation of PM stepper motors are described.

9.1 Field-Oriented Control

The approach presented here for high-performance control of a PM synchronous machine is based on the development in [72][112]. To proceed, the two-phase model of a PM synchronous machine with sinusoidally wound stator phases and n_p pole pairs is given by (see equation (6.40) of Chapter 6 where $n_p = 1$)

$$\begin{aligned} L_S\frac{di_{Sa}}{dt} &= -R_S i_{Sa} + K_m\sin(n_p\theta)\omega + u_{Sa} \\ L_S\frac{di_{Sb}}{dt} &= -R_S i_{Sb} - K_m\cos(n_p\theta)\omega + u_{Sb} \\ J\frac{d\omega}{dt} &= K_m(-i_{Sa}\sin(n_p\theta) + i_{Sb}\cos(n_p\theta)) - \tau_L \\ \frac{d\theta}{dt} &= \omega. \end{aligned} \tag{9.1}$$

This is also the two-phase equivalent model of a three-phase PM synchronous machines [see equation (7.81) in Chapter 7]. Equation (9.1) is the starting point for the design of a field-oriented controller for the PM synchronous machine. For synchronous machines, the *dq* transformations of the phase voltages and currents are given by

$$\begin{bmatrix} u_d \\ u_q \end{bmatrix} \triangleq \begin{bmatrix} \cos(n_p\theta) & \sin(n_p\theta) \\ -\sin(n_p\theta) & \cos(n_p\theta) \end{bmatrix} \begin{bmatrix} u_{Sa} \\ u_{Sb} \end{bmatrix} \tag{9.2}$$

$$\begin{bmatrix} i_d \\ i_q \end{bmatrix} \triangleq \begin{bmatrix} \cos(n_p\theta) & \sin(n_p\theta) \\ -\sin(n_p\theta) & \cos(n_p\theta) \end{bmatrix} \begin{bmatrix} i_{Sa} \\ i_{Sb} \end{bmatrix}. \tag{9.3}$$

The direct current i_d corresponds to the component of the stator magnetic field along the axis of the rotor magnetic field, while the quadrature current i_q corresponds to the orthogonal component. The application of the *dq* transformation to the original system (9.1) yields the system of equations

$$L_S \frac{di_d}{dt} = -R_S i_d + n_p \omega L_S i_q + u_d \tag{9.4}$$

$$L_S \frac{di_q}{dt} = -R_S i_q - n_p \omega L_S i_d - K_m \omega + u_q \tag{9.5}$$

$$J \frac{d\omega}{dt} = K_m i_q - \tau_L \tag{9.6}$$

$$\frac{d\theta}{dt} = \omega \tag{9.7}$$

where u_d is the direct voltage, u_q is the quadrature voltage, i_d is the direct current, i_q is the quadrature current, ω is the angular velocity, and θ is the angular position. In this coordinate system, the transformed currents i_d and i_q vary approximately at the *mechanical* frequency of the motor. These variables typically have bandwidths in the range of 0–100 Hz compared with 0–5 kHz bandwidth for u_{Sa}, u_{Sb}, i_{Sa} and i_{Sb}.

The resulting *dq* system model (9.4)–(9.7) is still nonlinear, however the nonlinear terms can now be canceled by state feedback. Specifically, choosing u_d and u_q to be of the form

$$u_d = R_S i_d - n_p \omega L_S i_q + L_S v_d \tag{9.8}$$

$$u_q = R_S i_q + n_p \omega L_S i_d + K_m \omega + L_S v_q \tag{9.9}$$

results in the *feedback linearized* system

$$\frac{di_d}{dt} = v_d \tag{9.10}$$

$$\frac{di_q}{dt} = v_q \tag{9.11}$$

$$\frac{d\omega}{dt} = (K_m/J) i_q - \tau_L/J \tag{9.12}$$

$$\frac{d\theta}{dt} = \omega. \tag{9.13}$$

Note that the original fourth-order system has been transformed into a first-order linear system (9.10) and a third-order linear system (equations (9.11)–(9.13)) which are decoupled from each other. As a consequence, *linear* control techniques can be used for the system (9.10)–(9.13) in these new variables.

9.1.1 Design of the Reference Trajectory and Inputs

The linearized system (9.10)–(9.13) is decoupled. However, constraints on the magnitude of the voltages introduce couplings. To achieve a desired (reference) speed ω_{ref} and acceleration α_{ref}, the required quadrature current is found from equation (9.6) to be

$$i_{qref} = \frac{J}{K_m}\alpha_{ref} + \frac{f}{K_m}\omega_{ref}$$

where $\tau_L = -f\omega$ is now assumed. The quadrature component i_q of the current produces torque while the direct component i_d does not produce any torque. However, in order to attain higher speeds, it is necessary to apply a *negative* direct current to cancel the effect of the back emf of the motor. Specifically, note from equation (9.5), the back-emf term in the dq coordinates is $K_m\omega$ and a decoupling control law as in (9.9) requires that u_q *cancel* this back-emf term. For example, in the experiments to be presented below, $K_m = 0.19$ V/rad/sec so that the back-emf voltage at a speed of 2000 rpm (209 radians/sec) is about 41 V. However, the source voltage for this motor is only 40 V, so that cancellation of the back emf would lead to saturation at the input. However, if i_d were forced to be negative (field-weakening) by the correct choice of v_d, the term $-n_p\omega L_S i_d$ would help to cancel the back-emf term $-K_m\omega$ [see equation (9.5) or, equivalently, equation (9.9)]. The design of an appropriate reference for i_d is thus essential to avoid saturation of the phase voltages at high rotor speeds.

The desired (reference) direct current i_{dref} is found by maximizing the torque, that is, $K_m i_q$ at constant speed subject to the constraint $u_d^2 + u_q^2 = u_{Sa}^2 + u_{Sb}^2 \leq V_{\max}^2$ where $V_{\max}$ is fixed. In order to obtain a tractable solution, the optimization is carried out at constant speed under steady-state conditions. Under steady-state conditions, the relations (9.4)–(9.7) become

$$\begin{aligned} u_d &= R_S i_d - n_p\omega L_S i_q \\ u_q &= R_S i_q + n_p\omega L_S i_d + K_m\omega \\ K_m i_q &= \tau_L. \end{aligned}$$

The actual physical constraints for u_{Sa}, u_{Sb} are of the form $|u_{Sa}| \leq V_{max}$, $|u_{Sb}| \leq V_{max}$ rather than $u_{Sa}^2 + u_{Sb}^2 \leq V_{\max}^2$. However, in steady-state operation, these conditions are equivalent (see problem 1). Using standard optimization techniques, the maximum is found to correspond to

$$u_{qref}(\omega) = \frac{R_S}{\sqrt{R_S^2 + (n_p\omega L_S)^2}} V_{\max} \tag{9.14}$$

$$\frac{u_{dref}(\omega)}{u_{qref}(\omega)} = -\frac{n_p\omega L_S}{R_S} \tag{9.15}$$

$$i_{dref}(\omega) = -\frac{n_p L_S K_m \omega^2}{R_S^2 + (n_p\omega L_S)^2}. \tag{9.16}$$

The inverse tangent of (9.15) is often referred to as the optimal lead angle [113][114][115]. It is the angle advance that the phase voltages u_{Sa} and

u_{Sb} need to be commanded relative to $n_p\theta$ (corresponding to the position of the rotor) to achieve the maximum torque at any given speed ω without violating the voltage constraints. The relationship (9.16) gives the corresponding direct current as a function of speed required to obtain the maximum torque. Although (9.14) through (9.16) were derived assuming constant speed, it will be seen below that they are very useful even under dynamic conditions.

Choosing the direct current to be negative as in (9.16) is referred to as *field weakening*. This comes from considering the terms

$$-n_p\omega L_S i_d - K_m\omega = -\left(n_p L_S i_d + K_m\right)\omega$$

in (9.5) where, as the value of K_m is proportional to the strength of rotor's magnetic field, making $i_d < 0$ is viewed as weakening this field. A more general approach to obtaining the current references that also take into account the current constraints is given in Section 9.2.

After specifying a desired mechanical reference trajectory $\theta_{ref}, \omega_{ref}, \alpha_{ref}$, the corresponding state trajectory $i_{dref}, i_{qref}, \omega_{ref}, \theta_{ref}$ and reference input voltages u_{dref}, u_{qref} are computed. They are chosen to satisfy the system equations (9.4)–(9.7), that is,

$$L_S\frac{di_{dref}}{dt} = -R_S i_{dref} + n_p\omega_{ref}L_S i_{qref} + u_{dref} \tag{9.17}$$

$$L_S\frac{di_{qref}}{dt} = -R_S i_{qref} - n_p\omega_{ref}L_S i_{dref} - K_m\omega_{ref} + u_{qref} \tag{9.18}$$

$$J\frac{d\omega_{ref}}{dt} = K_m i_{qref} - f\omega_{ref} \tag{9.19}$$

$$\frac{d\theta_{ref}}{dt} = \omega_{ref} \tag{9.20}$$

The desired quadrature current is found by solving (9.19) for i_{qref} and its derivative di_{qref}/dt is then found by differentiating this expression. The desired direct current i_{dref} is given by (9.16) with ω replaced by ω_{ref}, and its derivative is found by simply differentiating this expression. Doing these calculations gives

$$i_{dref} = \frac{-n_p L_S K_m \omega_{ref}^2}{R_S^2 + (n_p\omega_{ref}L_S)^2} \tag{9.21}$$

$$i_{qref} = \frac{J}{K_m}\alpha_{ref} + \frac{f}{K_m}\omega_{ref} \tag{9.22}$$

$$\frac{di_{dref}}{dt} = \frac{-2K_m n_p \omega_{ref} L_S R_S^2}{(R_S^2 + (n_p\omega_{ref}L_S)^2)^2}\alpha_{ref} \tag{9.23}$$

$$\frac{di_{qref}}{dt} = \frac{J}{K_m}\frac{d\alpha_{ref}}{dt} + \frac{f}{K_m}\alpha_{ref}. \tag{9.24}$$

Solving (9.17) and (9.18) for v_{dref} and v_{qref} results in

$$u_{dref} = L_S \frac{di_{dref}}{dt} + R_S i_{dref} - n_p \omega_{ref} L_S i_{qref} \tag{9.25}$$

$$u_{qref} = L_S \frac{di_{qref}}{dt} + R_S i_{qref} + n_p \omega_{ref} L_S i_{dref} + K_m \omega_{ref} \tag{9.26}$$

where the expressions in (9.21) through (9.24) are substituted into their respective places on the right-hand sides of (9.25) and (9.26). An important point here is that these references are chosen so as to not violate the current and voltage constraints, which typically requires some trial and error. For example, if i_{qref} is too large, then the mechanical trajectory must be modified so that α_{ref} is decreased [see equation (9.22)].

Remark Note that u_{dref}, u_{qref} are *not* given by (9.14) and (9.15) because these are constant speed values for the voltages. Rather, one *sets* (9.21) as the desired direct current and (9.21) as the desired quadrature current, and then chooses the reference input voltages according to (9.25) and (9.26).

9.1.2 State Feedback Controller

Subtracting the system model (9.4)–(9.7) from the reference model (9.17)–(9.20) gives

$$\begin{aligned} L_S \frac{d}{dt}(i_{dref} - i_d) &= -R_S(i_{dref} - i_d) + n_p \omega_{ref} L_S i_{qref} - n_p \omega L_S i_q \\ &\quad + u_{dref} - u_d \end{aligned} \tag{9.27}$$

$$\begin{aligned} L_S \frac{d}{dt}(i_{qref} - i_q) &= -R_S(i_{qref} - i_q) - n_p \omega_{ref} L_S i_{dref} + n_p \omega L_S i_d \\ &\quad - K_m(\omega_{ref} - \omega) + u_{qref} - u_q \end{aligned} \tag{9.28}$$

$$J \frac{d}{dt}(\omega_{ref} - \omega) = K_m(i_{qref} - i_q) - f(\omega_{ref} - \omega) \tag{9.29}$$

$$\frac{d}{dt}(\theta_{ref} - \theta) = \omega_{ref} - \omega. \tag{9.30}$$

The first step in specifying the state feedback control law is to use a *feedback linearizing* controller given by

$$u_d = -n_p \omega L_S i_q + u_{dref} + n_p \omega_{ref} L_S i_{qref} - L_S v_d \tag{9.31}$$

$$u_q = +n_p \omega L_S i_d + u_{qref} - n_p \omega_{ref} L_S i_{dref} - L_S v_q \tag{9.32}$$

where v_d and v_q are new inputs yet to be defined. Substituting equations (9.31) and (9.32) for u_d and u_q into (9.27)–(9.30) results in the *linear* system of equations

$$L_S\frac{d}{dt}(i_{dref}-i_d) = -R_S(i_{dref}-i_d)+v_d \tag{9.33}$$

$$L_S\frac{d}{dt}(i_{qref}-i_q) = -R_S(i_{qref}-i_q)-K_m(\omega_{ref}-\omega)+v_q \tag{9.34}$$

$$J\frac{d}{dt}(\omega_{ref}-\omega) = K_m(i_{qref}-i_q)-f(\omega_{ref}-\omega) \tag{9.35}$$

$$\frac{d}{dt}(\theta_{ref}-\theta) = \omega_{ref}-\omega. \tag{9.36}$$

Define the tracking error as

$$e = \begin{bmatrix} e_1 \\ e_2 \\ e_3 \\ e_4 \\ e_5 \end{bmatrix} \triangleq \begin{bmatrix} i_{dref}-i_d \\ i_{qref}-i_q \\ \omega_{ref}-\omega \\ \theta_{ref}-\theta \\ \int_0^t(\theta_{ref}(\tau)-\theta(\tau))d\tau \end{bmatrix} \tag{9.37}$$

where the integrator is added to the controller to eliminate any steady-state error due to constant disturbances (see Chapter 2). Using (9.33)–(9.36), the error (9.37) satisfies

$$\frac{de}{dt} = \begin{bmatrix} -R_S/L_S & 0 & 0 & 0 & 0 \\ 0 & -R_S/L_S & -K_m/L_S & 0 & 0 \\ 0 & K_m/J & -f/J & 0 & 0 \\ 0 & 0 & 1 & 0 & 0 \\ 0 & 0 & 0 & 1 & 0 \end{bmatrix} e + \begin{bmatrix} 1/L_S & 0 \\ 0 & 1/L_S \\ 0 & 0 \\ 0 & 0 \\ 0 & 0 \end{bmatrix} v. \tag{9.38}$$

where $v \triangleq \begin{bmatrix} v_d & v_q \end{bmatrix}^T$. More compactly,

$$\frac{de}{dt} = Ae + Bv$$

with the obvious definitions for A and B. Through the use of a nonlinear state transformation, input transformation, and nonlinear feedback, a *linear time-invariant* error system has been obtained. The input v is chosen as the linear state feedback

$$v = -Ke \tag{9.39}$$

where K is taken to be of the form

$$K = \begin{bmatrix} k_{11} & 0 & 0 & 0 & 0 \\ 0 & k_{22} & k_{23} & k_{24} & k_{25} \end{bmatrix}. \tag{9.40}$$

The closed-loop error system is then

$$\frac{de}{dt} = (A - BK)\,e.$$

Using the techniques of Chapter 2, one can choose K such that the closed-loop poles of $A - BK$ are placed at any desired location (see problem 2).

9.1.3 Speed Observer

The above controller design assumed full state measurements. However, a typical hardware setup provides only the current measurements and the position measurements. In order to estimate the speed from the position measurement (optical encoder) and current measurements, a reduced-order observer can be implemented according to

$$\frac{d\hat{\theta}}{dt} = \hat{\omega} + \ell_1(\theta - \hat{\theta}) \tag{9.41}$$

$$\frac{d\hat{\omega}}{dt} = (K_m/J)i_q - (f/J)\hat{\omega} + \ell_2(\theta - \hat{\theta}) \tag{9.42}$$

where ℓ_1 and ℓ_2 are the observer gains and i_q is a known input to the observer as i_{Sa}, i_{Sb}, and θ are measured. Subtracting (9.41) from (9.7) and (9.42) from (9.6) results in

$$\frac{d\varepsilon_1}{dt} = \varepsilon_2 - \ell_1\varepsilon_1 \tag{9.43}$$

$$\frac{de_2}{dt} = -(f/J)\varepsilon_2 - \ell_2\varepsilon_1 \tag{9.44}$$

where $\varepsilon_1 \triangleq \theta - \hat{\theta}$, $\varepsilon_2 \triangleq \omega - \hat{\omega}$. One can then choose the observer gains ℓ_1, ℓ_2 such that $\varepsilon_1(t) \to 0, \varepsilon_2(t) \to 0$ at any prescribed rate (see problem 3 and Chapter 2).

A block diagram illustrating the state feedback controller for a PM synchronous motor is given in Figure 9.1.

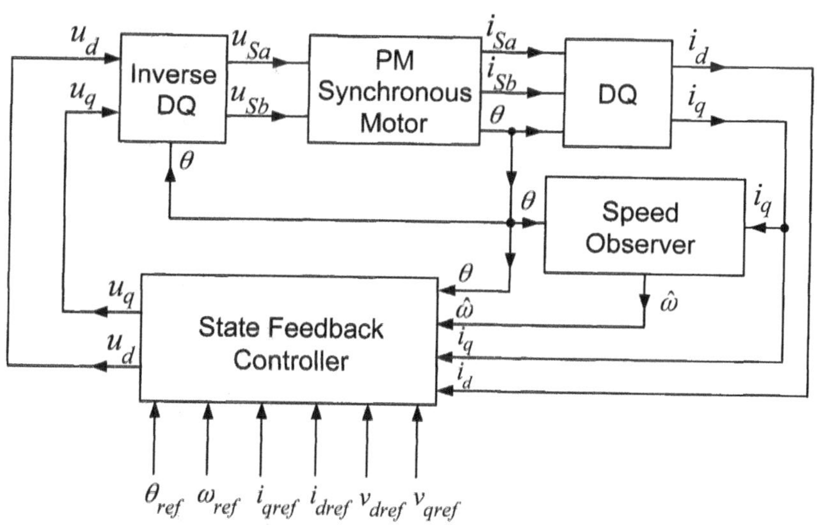

FIGURE 9.1. Controller block diagram. The state feedback controller is the composition of (9.31) and (9.32) with (9.39).

9.1.4 Experimental Results

The PM motor used in the experiments has $n_p = 50$ pole pairs, two stator phases, and the following parameter values $I_{max} = 6.0$ A, $V_{max} = 40$ V, $R_S = 0.55\ \Omega$, $L_S = 1.5$ mH, $K_m = 0.19$ N-m/A, $J = 4.5 \times 10^{-5}$ kg-m^2, and $f = 0.0008$ N-m/rads/sec. The hardware setup used to implement the feedback controller consisted Motorola's Advanced Development System (ADS) including a Motorola DSP56001 digital signal processor, a DSP extension board, two Aerotech 20-kHz PWM amplifier, and the above PM motor with a 2000-counts/rev optical encoder.

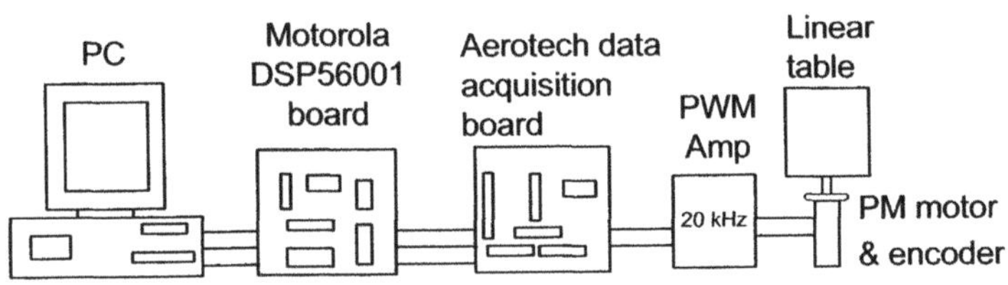

FIGURE 9.2. Experimental setup. The motor is connected to a linear positioning table through a ball screw.

The data acquisition board was built by Aerotech, Inc. and has four 8-bit A/D converters for sampling the voltages and currents, two 12-bit D/A converters to output the control voltages and a timer chip that is used to set the sample rate and to count the optical encoder pulses.

The control algorithm described above was tested on two different trajectories. The first one was a fast point-to-point move in which the motor turned 0.9π radians in 30 msec. The second move forces the motor up to a speed of 3000 rpm and brings it back down into position. This move was chosen to study high-speed control of the PM motor where the use of the reference trajectory (9.21) is required.

Sample Period and Observer Gains

A sample rate of 10 kHz ($T = 100\ \mu$sec) is required due to the high number of pole pairs. Specifically, the motor reaches speeds up to 3000 rpm, which corresponds to an electrical frequency of $(2\pi/60) \times 3000 \times n_p = 1.57$ kHz for $n_p = 50$.

The observer gains for both moves were chosen as $\ell_1 = 5272$ and $\ell_2 = 7.0 \times 10^6$, resulting in both closed-loop poles at -2646. The observer computation, that is, the integration of the system (9.41), (9.42) as well as the control algorithm is carried out every $T = 0.1$ msec (10 kHz).

As shown below (see Figure 9.5), the observer performed well and was essential to get the controller to track the particular demanding trajectories used (see Figures 9.3 and 9.5). It is reported in Ref. [72] that a speed estimator based on the backward difference estimate $\omega(kT) = [\theta(kT) - \theta((k-1)T)]/T$ did not result in closed-loop tracking for these same trajectories. As explained in Ref. [72], this is due to the fact when the velocity is computed by differentiating the position, the resolution of the velocity depends on the sample rate as well as the resolution of the encoder. Note that a 2000-counts/rev encoder gives a minimum velocity resolution of $(2\pi/2000)/T$ (see Chapter 2), which for $T = 1/10^4$ sec, gives a resolution of 31.4 rads/sec or 300 rpm! As the reference trajectory for speed (Figure 9.5) requires the motor to go from zero to 141 rads/sec (1350 rpm) in 10 msec, a sample rate of 10 kHz results in 100 iterations through the control loop in this time. On the other hand, a sample rate of 1 kHz would give a speed resolution of 3.14 rads/sec, but only 10 control loop iterations! The simple difference algorithm will work if a 50,000-counts/rev encoder is used (see Ref. [112]). However, this is not a practical choice for an industrial setup due to the cost and the fact that such an encoder is typically limited to lower motor speeds to work properly. (See Chapter 2 for more details on how an optical encoder works.)

Control of a 1.8-mm Move

A typical high-performance application for a PM synchronous motor is in a positioning system. Specifically, consider the requirement to have the motor move (point-to-point) a small linear positioning table 1.8 mm under 30 msec using a smooth mechanical trajectory so as to not excite vibrations in the table. The specified position θ_{ref}, speed $\omega_{ref} = d\theta_{ref}/dt$ and acceleration $\alpha_{ref} = d\omega_{ref}/dt$ are chosen to make the table move 1.8 mm. This turns out to be a 0.9π radian turn of the motor as the motor is attached to the linear stage through a ball screw. Figures 9.3 and 9.5 show the desired position and speed references θ_{ref} and ω_{ref}, respectively. That this move represents a very stringent performance requirement can be seen in Figure 9.6, which shows that the corresponding the reference input v_{dr} goes below -25 V leaving less than 15 V for the feedback controller. Furthermore, the current i_{qref} is shown in Figure 9.8 and is over 4 A (limit of 6 A) in order to achieve the required torque $K_m i_q$.

In the implementation, the gains for the 1.8-mm move were set as

$$K = \begin{bmatrix} 1.06 \times 10^4 & 0 & 0 & 0 & 0 \\ 0 & 1.78 \times 10^4 & 3.40 \times 10^2 & 3.80 \times 10^6 & 1.07 \times 10^7 \end{bmatrix}$$

to place the closed-loop poles at -18178 for the first-order subsystem and at $-11047, -28.3, -54.89 \pm j1190.7$ for the fourth-order subsystem. Figure 9.3 gives the actual and desired position response for the 1.8-mm move of the positioning table (corresponding to a 0.9π radian rotation of the motor) using the state feedback controller of Figure 9.1. The last part of the position response is shown in more detail in Figure 9.4, showing that the motor achieves the required final position within one encoder count.

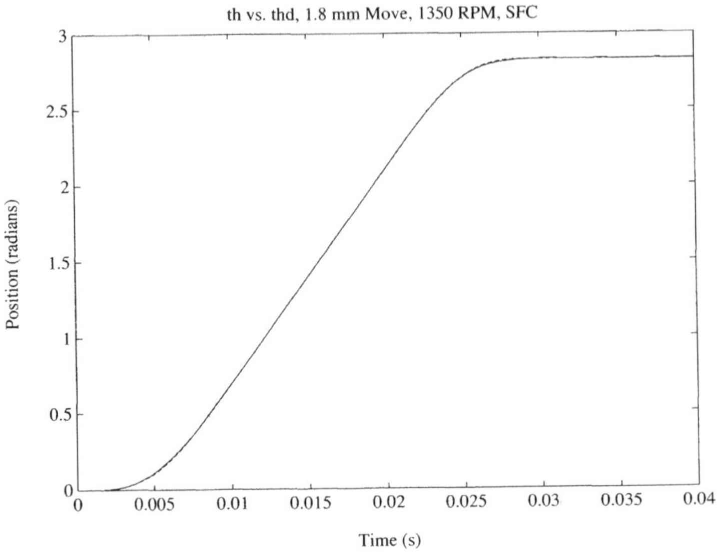

FIGURE 9.3. θ and θ_{ref} for the 1.8 mm move. From Bodson et al. [72], © 2004 IEEE.

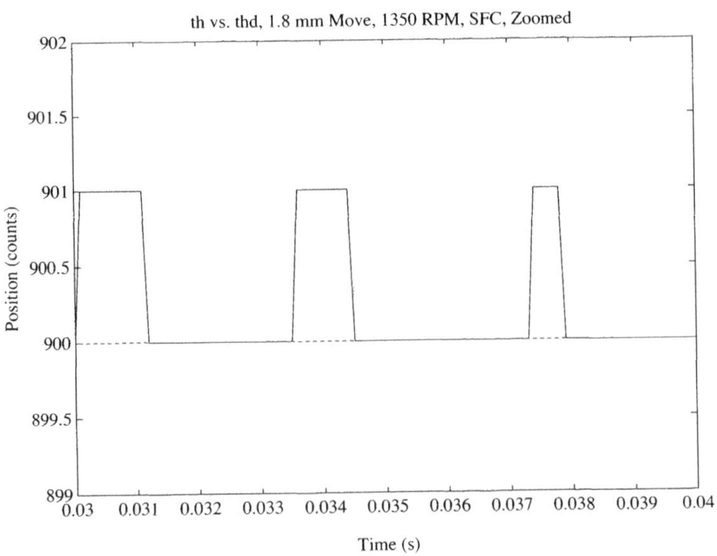

FIGURE 9.4. θ in counts after 30 msecs. Desired position is 900 counts ($= 2\pi \times 900/2000 = 0.9\pi$). From Bodson et al. [72], © 2004 IEEE.

The desired and estimated (observer) speeds are shown in Figure 9.5.

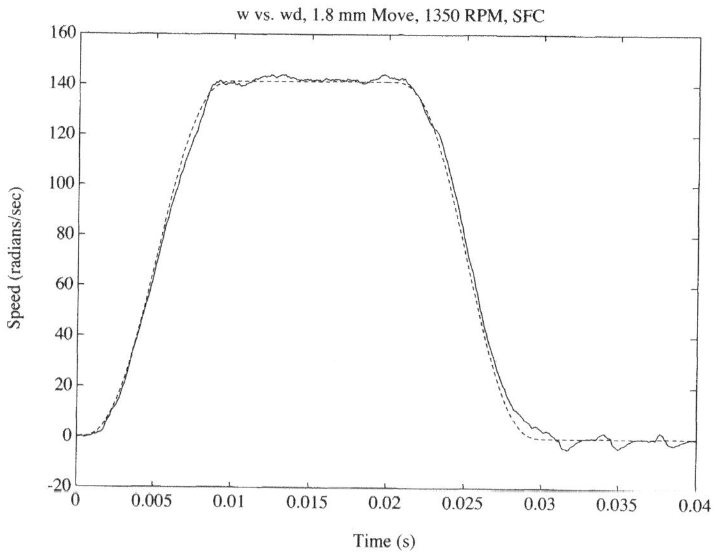

FIGURE 9.5. $\hat{\omega}$ and ω_{ref} for the 1.8-mm move. From Bodson et al. [72], © 2004 IEEE.

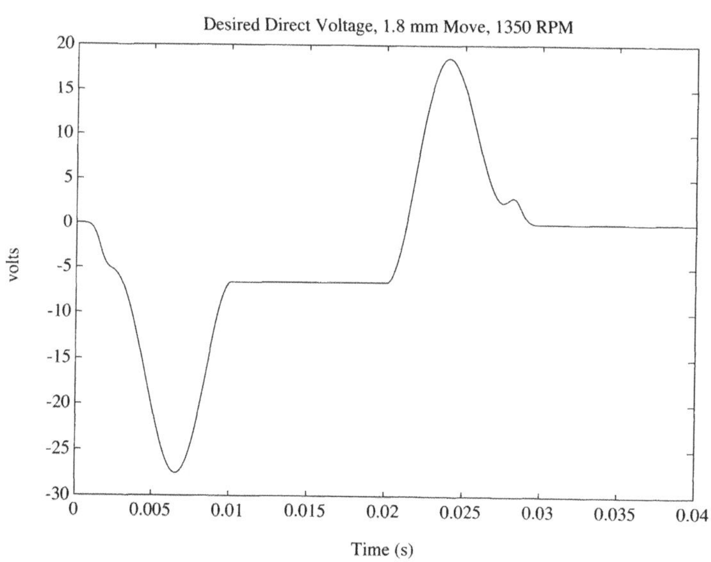

FIGURE 9.6. v_{dref} for the 1.8-mm move. From Bodson et al. [72], © 2004 IEEE.

Figures 9.7 and 9.8 show the tracking of the currents i_d and i_q.

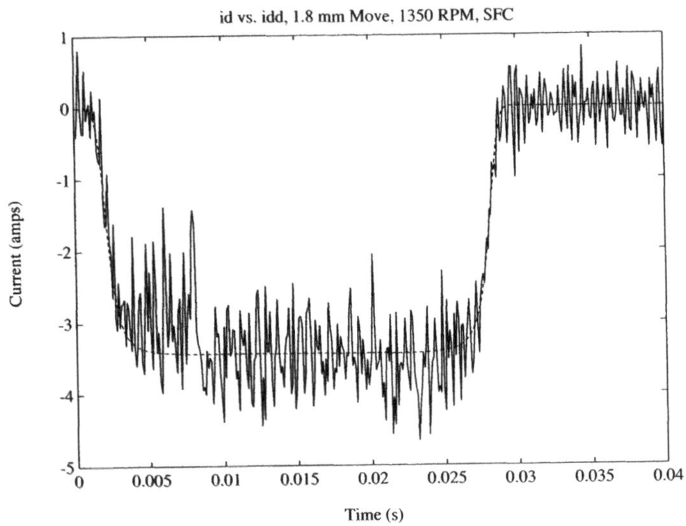

FIGURE 9.7. i_d and i_{dref} for the 1.8-mm move. From Bodson et al. [72], © 2004 IEEE.

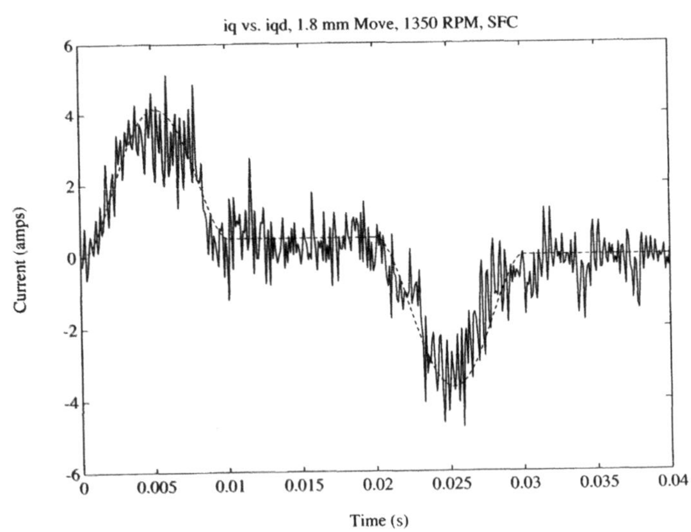

FIGURE 9.8. i_q and i_{qref} for the 1.8-mm move. From Bodson et al. [72], © 2004 IEEE.

The position and speed tracking are very good despite large fluctuations in the measured/calculated currents i_d and i_q.

Current Fluctuations in i_d and i_q

Simulations show that the fluctuations in the currents shown in Figures 9.7 and 9.8 are due to the limited resolution of the position measurement (2000-counts/rev encoder) and not to the PWM noise [72]. Specifically, the oscillatory behavior is present in the simulation when a 2000 counts/rev encoder was modeled, but not when a 50,000-counts/rev encoder was modeled. As pointed out above, Figure 9.4 (note the units are in *counts* not radians) shows that the motor stays within one encoder count of the final position after 30 msec, which is the highest accuracy the controller can achieve with any given encoder. The oscillatory behavior of $\hat{\omega}$ at the end of the run in Figure 9.5 is a direct result of integrating the oscillating quadrature current i_q, that is, the noise due to quantization of the position θ. Consequently, the estimate of speed $\hat{\omega}$ is responding to the noisy oscillations in i_q and does not indicate that the controller is not maintaining the correct final position.

Remark It is straightforward to show that a permanent magnet DC brush motor with a torque constant of $K_T = 0.2$ N-m/A ($K_m = 0.19$ N-m/A for the stepper motor used here), a moment of inertia $J = 3.6 \times 10^{-4}$ kg-m^2 (the inertia is much larger than the stepper motor due to windings on the rotor), voltage ceiling limits of ± 80 V, and a peak current limit of 22 A is unable to make such a move under 30 msec. This is due to the fact that field weakening is not possible. However, the above PM synchronous motor with voltage limits of ± 40 V and current limits of 6 A can indeed make such a move.

Control of a 3000-rpm Run

Figure 9.9 shows the actual and desired (reference) position using the controller to bring the motor up to 3000 rpm. The actual and estimated speeds are shown in Figure 9.10. Again, the tracking of the mechanical trajectory is quite good despite oscillatory behavior in i_d and i_q (see Figures 9.12 and 9.13). These oscillations are again attributed to the lack of resolution in the 2000-counts/rev encoder. Note also that the reference input v_{dref} (Figure 9.11) reaches -30 V, leaving less than 10 V for the feedback controller.

In the implementation, the gains for the 3000-rpm run were set as

$$K =$$

$$\begin{bmatrix} 2.587 \times 10^4 & 0 & 0 & 0 & 0 \\ 0 & 2.446 \times 10^4 & 6.031 \times 10^2 & 2.935 \times 10^6 & 1.004 \times 10^7 \end{bmatrix}$$

to place the closed-loop poles at -26314 for the first-order subsystem and at $-24727, -3.4, -101 \pm j893$ for the fourth-order subsystem.

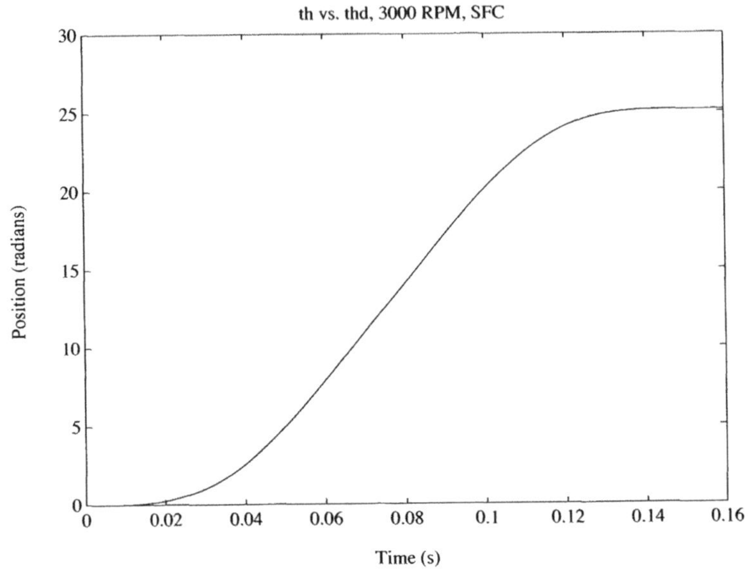

FIGURE 9.9. θ and θ_{ref} for the 3000-rpm move. From Bodson et al. [72], © 2004 IEEE.

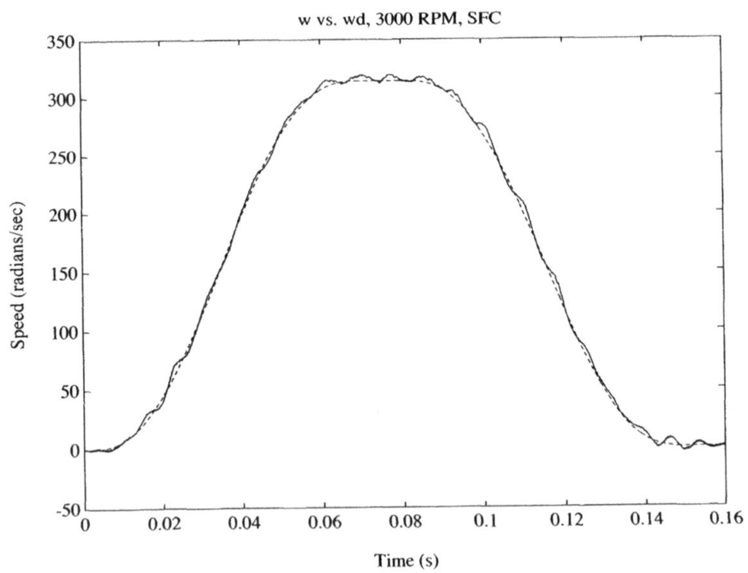

FIGURE 9.10. $\hat{\omega}$ and ω_{ref} for the 3000-rpm move. From Bodson et al. [72], © 2004 IEEE.

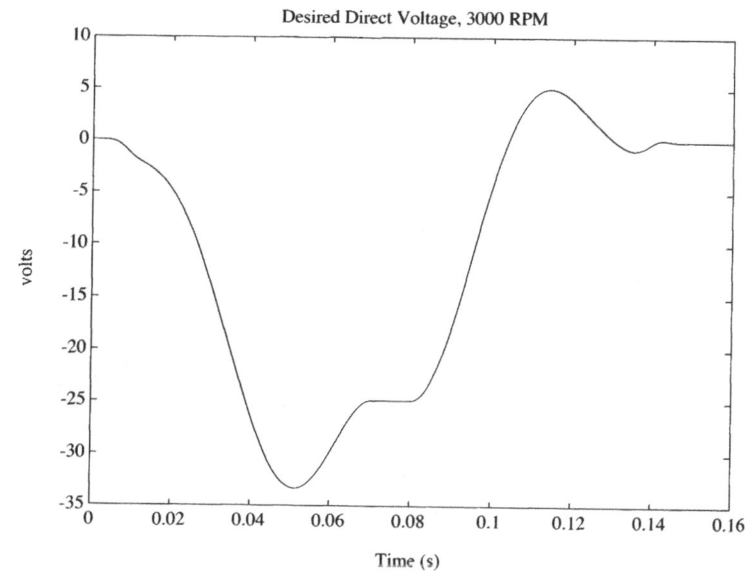

FIGURE 9.11. v_{dref} for the 3000-rpm move. From Bodson et al. [72], © 2004 IEEE.

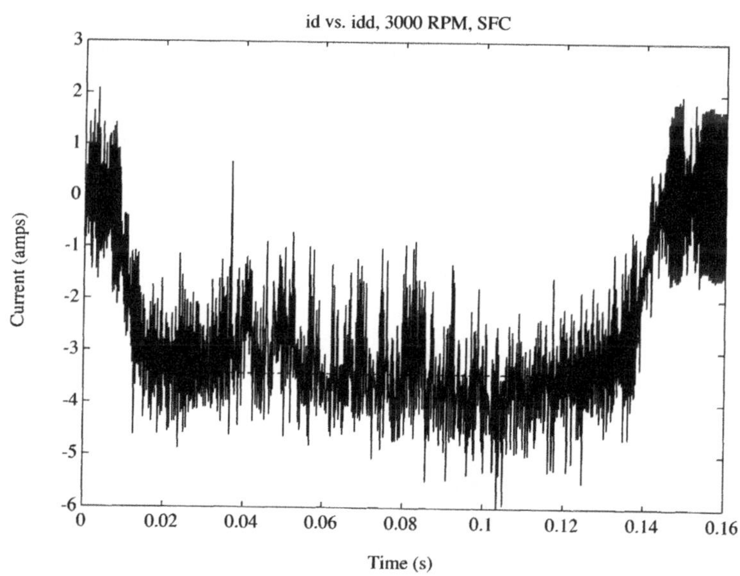

FIGURE 9.12. i_d and i_{dref} for the 3000-rpm move. From Bodson et al. [72], © 2004 IEEE.

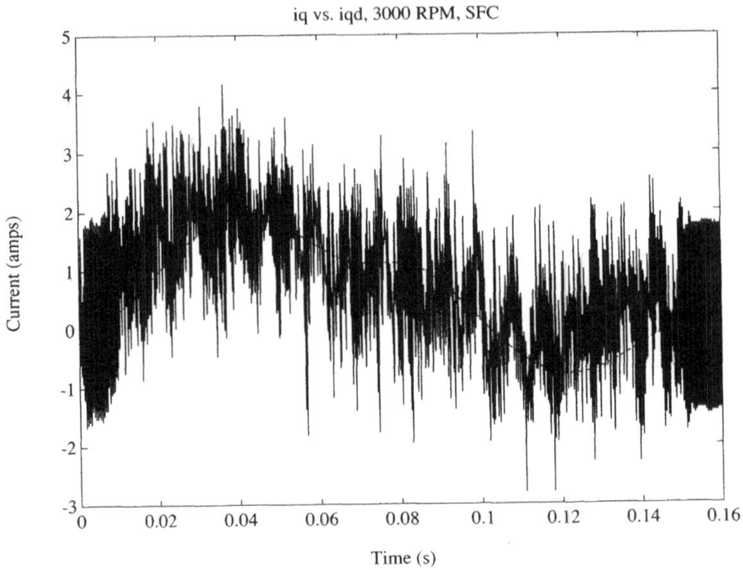

FIGURE 9.13. i_q and i_{qref} for the 3000-mm move. From Bodson et al. [72], © 2004 IEEE.

9.1.5 Current Command Control

In the previous section the motor was voltage-controlled. The experiments presented there used a PM synchronous motor with $n_p = 50$ pole pairs, making voltage control essential to obtain high performance. To explain, the equations of the PM machine are (9.4)–(9.7) which are repeated below.

$$L_S \frac{di_d}{dt} = -R_S i_d + n_p \omega L_S i_q + u_d \quad (9.45)$$

$$L_S \frac{di_q}{dt} = -R_S i_q - n_p \omega L_S i_d - K_m \omega + u_q \quad (9.46)$$

$$J \frac{d\omega}{dt} = K_m i_q - \tau_L \quad (9.47)$$

$$\frac{d\theta}{dt} = \omega \quad (9.48)$$

Equations (9.45) and (9.46) show that $n_p \omega L_S$ is an impedance to changing i_d, i_q through the input voltages u_d, u_q. With n_p large, the time it takes to change a current through an input voltage can be of the same order of time it takes to change the speed through i_q. This made it essential in the previous section (where $n_p = 50$) to design a voltage controller that accounts for the impedances $n_p \omega L_S$ in the current equations. However, it is also common to find PM synchronous motors with $n_p = 2$ rather than

$n_p = 50$. In this case, the impedance is 25 times less and one does not usually need to use voltage control, but, rather, current control methods can be employed. That is, using a PI current controller of the form

$$u_q = K_p(i_{qref} - i_q) + K_I \int_0^t (i_{qref} - i_q)dt \tag{9.49}$$

$$u_d = K_p(i_{dref} - i_d) + K_I \int_0^t (i_{dref} - i_q)dt \tag{9.50}$$

and adjusting the gains appropriately, $i_{qref} \to i_q, i_{dref} \to i_d$ fast enough that one can consider $i_q \approx i_{qref}, i_d \approx i_{dref}$. Then a (mechanical) trajectory tracking controller can be designed based on the reduced-order system model

$$\frac{d\omega}{dt} = (K_m/J)\, i_{qref} - \tau_L/J \tag{9.51}$$

$$\frac{d\theta}{dt} = \omega \tag{9.52}$$

where i_{qref} is now considered as the input. Figure 9.14 is a block diagram illustrating a current command configuration. Experience shows that this works quite well if the number of pole pairs n_p is small. Equations (9.51) and (9.52) are identical in form to a DC motor so that, as in Chapter 2, one can choose a trajectory tracking controller in the form

$$i_{qref} = \frac{J}{K_m}\left(K_0 \int_0^t (\theta_{ref} - \theta)dt + K_1(\theta_{ref} - \theta) + K_2(\omega_{ref} - \omega) + \alpha_{ref}\right)$$

and choose the gains K_0, K_1, and K_2 such that $\theta \to \theta_{ref}, \omega \to \omega_{ref}$ despite any constant load torque τ_L acting on the motor.

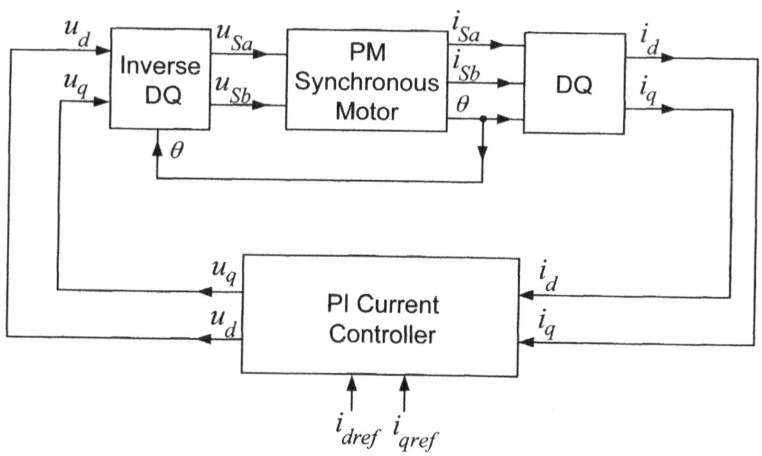

FIGURE 9.14. Current command control of a PM synchronous motor.

One often chooses $i_{dref} \equiv 0$ in (9.49) if the speeds are not too high. However, field weakening may be required if one desires to run the machine at high speeds. In this case, the general approach to field weakening given in the next section would be a useful way to specify i_{dref}.

9.2 Optimal Field Weakening*

Maximization of the torque produced by electric motors is an important practical consideration, as optimization may yield the use of a smaller motor for a given application, or higher performance for a given motor. In this section, the problem of extracting all the available torque from a PM synchronous motor is considered. More specifically, given any fixed speed ω as well as the current and voltage constraints $I_{\max}$ and $V_{\max}$, respectively, it is shown how the voltages and currents in the machine must be chosen to obtain the optimum torque without violating the constraints. The approach here is from the development in Refs. [116][117].

As shown in the previous section, one of the reasons the available torque of electric motors decreases at high speed is due to the voltage limits. This is because the source voltage must overcome the increasing back-emf voltage to the detriment of building up the stator currents. Maximization of the torque at high speeds requires the use of field-weakening which, for synchronous motors, may be achieved optimally using a relatively simple strategy. At low speeds, the current limits restrict how much torque can be produced and the voltage limits play no role. At intermediate speeds as well as higher speeds, both the voltage and the current limits are active or just the voltage limits. This section provides algorithms to compute the speeds at which transitions occur between these modes, and describes control strategies required in each case.

9.2.1 Formulation of the Torque Maximization Problem

The state-space model in the *dq* coordinates is

$$L_S\frac{di_d}{dt} = -R_S i_d + n_p\omega L_S i_q + u_d \tag{9.53}$$

$$L_S\frac{di_q}{dt} = -R_S i_q - n_p\omega L_m i_d - K_m\omega + u_q \tag{9.54}$$

$$J\frac{d\omega}{dt} = K_m i_q - \tau_L \tag{9.55}$$

$$\frac{d\theta}{dt} = \omega. \tag{9.56}$$

For a constant speed ω, equations (9.53)–(9.55) yield

$$\begin{aligned} u_d &= R_S i_d - n_p \omega L_S i_q && (9.57) \\ u_q &= R_S i_q + n_p \omega L_S i_d + K_m \omega && (9.58) \\ K_m i_q &= \tau_L. && (9.59) \end{aligned}$$

The objective here is the maximization of the electrical torque $K_m i_q$ within the voltage and current constraints. The voltage and current constraints are incorporated into the problem statement by using the fact that, at constant speed, the phase voltages and the phase currents are sinusoidal, with peak magnitudes given by $\sqrt{u_d^2 + u_q^2}$ and $\sqrt{i_d^2 + i_q^2}$, respectively. Therefore, the phase voltages bounded by V_{max} and the phase currents bounded by I_{max} yield constraints in the dq variables given by

$$\begin{aligned} V &\triangleq \sqrt{u_d^2 + u_q^2} \le V_{max} && (9.60) \\ I &\triangleq \sqrt{i_d^2 + i_q^2} \le I_{max}. && (9.61) \end{aligned}$$

9.2.2 *Speed Ranges and Transition Speeds*

Because the torque may, in theory, be made arbitrarily large by increasing the voltage and current levels, optimal operation under constraints is always achieved with the limits on either the currents, the voltages or both being reached. A typical motor is characterized by three ranges of speed: a low-speed range where operation is constrained by the current limit only, an intermediate speed range where operation is constrained by both the voltage and the current limit, and a high-speed range where operation is constrained by the voltage limit only.

At a specific speed, torque maximization will be achieved for a specific choice of i_d and i_q for a current controlled drive, or a specific choice of u_d and u_q for a voltage controlled drive. The optimal values are denoted by i_d^*, i_q^*, u_d^*, and u_q^*. At each such speed, the torque optimization will result in the machine being at one of the three constraint conditions: (i) $V = V_{\max}$, (ii) $I = I_{\max}$, or (iii) $V = V_{\max}, I = I_{\max}$. Knowing what the solution is for each of these three possible constraint conditions, it remains to determine the speeds at which the system transitions from one constraint to another. The speed at which the system transitions between $V = V_{\max}$ and $V = V_{\max}, I = I_{\max}$ is referred to as the *first transition speed* and is denoted by ω_1. Similarly, the speed at which the system transitions between $I = I_{\max}$ and $V = V_{\max}, I = I_{\max}$ is referred to as the *second transition speed* and is denoted by ω_2. These speeds can be calculated after having obtained the form of the optimal solution in each speed range. In the analysis that follows, only the case $\omega \ge 0$ is considered because straightforward symmetry arguments easily yield the case $\omega \le 0$.

Case 1: Optimization under just Current Constraints

First consider the case where only the current constraint is active. In this case, the optimum i_d^* must be zero and $i_q^* = \pm I_{\max}$ to optimize the torque. The overall solution is simply

$$
\begin{aligned}
i_d^*(\omega) &= 0 && (9.62)\\
i_q^*(\omega) &= \pm I_{max} && (9.63)\\
u_d^*(\omega) &= \pm(-n_p\omega L_S I_{max}) && (9.64)\\
u_q^*(\omega) &= \pm R_S I_{max} + K_m\omega. && (9.65)
\end{aligned}
$$

where the positive sign is chosen for positive torque and the negative sign is chosen for negative torque.

Case 2: Maximization under just Voltage Constraints

Torque maximization under the voltage constraint alone yields a solution which is referred to as *optimal field-weakening* and is well known (cf. Ref. [1], p. 266, and Ref. [114]). To obtain the solution, one solves (9.57) and (9.58) for i_d and i_q, so that

$$
\begin{bmatrix} i_d \\ i_q \end{bmatrix} = \frac{1}{Z_S^2(\omega)} \begin{bmatrix} R_S u_d + n_p\omega L_S(u_q - K_m\omega) \\ -n_p\omega L_S u_d + R_S(u_q - K_m\omega) \end{bmatrix}, \qquad (9.66)
$$

where $Z_S^2(\omega) \triangleq R_S^2 + (n_p\omega L_S)^2$. Choosing ϕ so that $u_d = V\cos(\phi), u_q = V\sin(\phi)$, the current i_q is given by

$$
i_q = \frac{-n_p\omega L_S V\cos(\phi) + R_S V\sin(\phi) - R_S K_m\omega}{Z_S^2(\omega)}. \qquad (9.67)
$$

Equation (9.67) shows that the maximum current i_q, and therefore the maximum electrical torque, is obtained for

$$
\begin{aligned}
V &= \pm V_{max}\\
\frac{d}{d\phi}(-n_p\omega L_S\cos(\phi) + R_S\sin(\phi)) &= 0
\end{aligned}
$$

where the positive sign is chosen for positive torque and the negative sign is chosen for negative torque. Solving the second of these equations gives $\tan(\phi^*) = -R_S/(n_p\omega L_S)$. The optimal u_d^* and u_q^* therefore satisfy

$$
\frac{u_d^*}{u_q^*} = -\frac{n_p\omega L_S}{R_S} = \frac{1}{\tan(\phi^*)}. \qquad (9.68)
$$

This equation gives the value of the *optimal lead angle* for voltage-controlled drives. Equation (9.68) yields

$$
\begin{aligned}
\sin(\phi^*) &= R_S/Z_S(\omega)\\
\cos(\phi^*) &= -n_p\omega L_S/Z_S(\omega)
\end{aligned}
$$

and therefore

$$u_d^*(\omega) = \pm\frac{-n_p\omega L_S}{Z_S(\omega)}V_{max} \tag{9.69}$$

$$u_q^*(\omega) = \pm\frac{R_S}{Z_S(\omega)}V_{max}. \tag{9.70}$$

where the + sign is for positive torque and the − sign is for negative torque. Using these values for the voltages, one finds the resulting direct current i_d^* is given by [see (9.66) and (9.68)]

$$i_d^*(\omega) = -\frac{(n_p\omega L_S)(K_m\omega)}{Z_S^2(\omega)} = \frac{-n_pL_SK_m\omega^2}{R_S^2 + (n_p\omega L_S)^2}, \tag{9.71}$$

which can be viewed as the *optimal field weakening current* (compare with (9.21)). Similarly, the quadrature current is given by

$$i_q^*(\omega) = \frac{\pm V_{max}Z_S(\omega) - K_m\omega R_S}{Z_S^2(\omega)}. \tag{9.72}$$

Note that if $R_S = 0$, then the motoring and braking quadrature currents are the negative of each other.

Case 3: Maximization under Voltage and Current Constraints

In the case where both the voltage and the current constraints are active, the optimization reduces to the solution of a system of algebraic equations, because there are as many constraints as degrees of freedom. The steady-state equations (9.57) and (9.58) are substituted into $u_d^2 + u_q^2 = V_{max}^2$ and after some rearranging, results in

$$Z_S^2(\omega)(i_d^2 + i_q^2) + K_m^2\omega^2 + 2R_SK_m\omega i_q + 2n_pL_SK_m\omega^2 i_d = V_{max}^2. \tag{9.73}$$

Eliminating i_d using $i_d^2 + i_q^2 = I_{max}^2$ leads to a quadratic equation for i_q^* of the form

$$\alpha(\omega)\, i_q^{*2} + \beta(\omega)\, i_q^* + \gamma(\omega) = 0, \tag{9.74}$$

where

$$\alpha(\omega) \triangleq 4R_S^2\omega^2K_m^2 + 4n_p^2\omega^4L_S^2K_m^2 \tag{9.75}$$

$$\beta(\omega) \triangleq -4R_SK_m\omega(V_{max}^2 - K_m^2\omega^2 - Z_S^2(\omega)I_{max}^2) \tag{9.76}$$

$$\gamma(\omega) \triangleq (V_{max}^2 - K_m^2\omega^2 - Z_S^2(\omega)I_{max}^2)^2 - 4I_{max}^2n_p^2\omega^4L_S^2K_m^2. \tag{9.77}$$

In this case, where both the voltage and the current constraints are active, $\beta^2(\omega) - 4\alpha(\omega)\gamma(\omega) \geq 0$ (see problem 4) and equation (9.74) has two real roots

$$i_q^*(\omega) = \frac{-\beta(\omega) \pm \sqrt{\beta^2(\omega) - 4\alpha(\omega)\gamma(\omega)}}{2\alpha(\omega)} \tag{9.78}$$

with the positive sign for motoring and the negative sign for braking. It then follows that

$$
\begin{aligned}
i_d^*(\omega) &= -\sqrt{I_{max}^2 - i_q^{*2}(\omega)} & (9.79)\\
u_d^* &= R_S i_d^* - n_p \omega L_S i_q^* & (9.80)\\
u_q^* &= R_S i_q^* + n_p \omega L_S i_d^* + K_m \omega. & (9.81)
\end{aligned}
$$

Again note that if $R_S = 0$ ($\Rightarrow \beta(\omega) = 0$), the motoring and braking quadrature currents are just the negative of each other.

First Transition Speed ω_1

As previously explained, the speed at which the system transitions between $V = V_{\max}$ and $V = V_{\max}, I = I_{\max}$ is referred to as the *first transition speed* and is denoted by ω_1. To calculate ω_1, consider the machine to be operating under the current constraint only so that $i_q = \pm I_{\max}$ and find the value of the speed for which the voltage limit is first encountered. Specifically, substitute (9.64) and (9.65) into $u_d^2(\omega) + u_q^2(\omega) = V_{max}^2$ and rearrange to obtain the following equation for the first transition speed:

$$
(K_m^2 + (n_p L_S I_{max})^2)\omega_1^2 + 2K_m K_m I_{max}\omega_1 = V_{max}^2. \tag{9.82}
$$

The equation has two real roots $\omega_{1pos} > 0$ and $\omega_{1neg} < 0$ (see problem 5). The positive root ω_{1pos} corresponds to the case of positive torque. The negative root corresponds to the braking mode, that is, $|\omega_{1neg}|$ is the transition speed when $i_q = -I_{\max}$ (see problem 5). If $R_S = 0$, $\omega_{1neg} = -\omega_{1pos}$.

Second Transition Speed ω_2

Similarly, the speed at which the system transitions between $I = I_{\max}$ and $V = V_{\max}, I = I_{\max}$ is referred to as the *second transition speed* and is denoted by ω_2. To calculate ω_2, consider the machine to be operating under the voltage constraint only so that u_d^*, u_q^* are given by (9.69) and (9.70) resulting in the currents (9.71) and (9.72), then find the value of the speed for which the current limit is first encountered. Specifically, substitute the currents (9.71) and (9.72) into $i_d^2(\omega) + i_q^2(\omega) = I_{max}^2$ and rearrange to obtain

$$
Z_S^2(\omega_2)\left(V_{max}^2 + (K_m\omega_2)^2 - Z_S^2(\omega_2)\, I_{max}^2\right) = \pm 2K_m\omega_2 R_S V_{max} Z_S(\omega_2). \tag{9.83}
$$

Upon canceling $Z_S(\omega_2)$ and squaring both sides, one obtains

$$
Z_S^2(\omega_2)\left((V_{max}^2 - R_S^2 I_{max}^2) - (n_p^2 L_S^2 I_{max}^2 - K_m^2)\omega_2^2\right)^2 = (2K_m R_S V_{max})^2\omega_2^2. \tag{9.84}
$$

This is a cubic equation in ω_2^2 of the form

$$
(x - a)^2(x + b) = cx, \tag{9.85}
$$

where

$$x \triangleq \omega_2^2 \tag{9.86}$$

$$a \triangleq \frac{V_{max}^2 - R_S^2 I_{max}^2}{n_p^2 L_S^2 I_{max}^2 - K_m^2} \tag{9.87}$$

$$b \triangleq \frac{R_S^2}{(n_p L_S)^2} \tag{9.88}$$

$$c \triangleq \frac{(2K_m R_S V_{max})^2}{(n_p L_S)^2 (n_p^2 L_S^2 I_{max}^2 - K_m^2)^2}. \tag{9.89}$$

To study the roots of equation (9.85), rearrange it as

$$1 - c\frac{x}{(x-a)^2(x+b)} = 0. \tag{9.90}$$

Its root locations can then be found using the properties of the *complementary root locus* [37][46][81], that is, the root locus corresponding to a zero at $x = 0$, a pole at $x = -b$, a double pole at $x = a$, and a *negative* gain $k = -c$. This analysis is now broken down into two cases.

Case 1: $a > 0$

The most common situation is the one in which $a > 0$. The quantity $V_{max}^2 - R_S^2 I_{max}^2 > 0$ for any practical motor, so that the case $a > 0$ actually means that $n_p^2 L_S^2 I_{max}^2 - K_m^2 > 0$. In this case, the complementary root locus is shown in Fig. 9.15. For $c > 0$, there is one negative root $x < -b$, a positive root $x < a$, and a positive root $x > a$. As $x = \omega_2^2$, the root $x < -b$ is extraneous and therefore ignored. To understand which of the other two roots may apply, rearrange equation (9.83) to obtain

$$Z_S^2(\omega)\left(n_p^2 L_S^2 I_{max}^2 - K_m^2\right)\left(\frac{V_{max}^2 - R_S^2 I_{max}^2}{n_p^2 L_S^2 I_{max}^2 - K_m^2} - \omega^2\right) = \pm 2K_m \omega R_S V_{max} Z_S(\omega). \tag{9.91}$$

This expression shows that the root satisfying $x < a$ corresponds to positive torque ($+V_{max}$) while the root satisfying $x > a$ corresponds to the negative torque ($-V_{max}$).

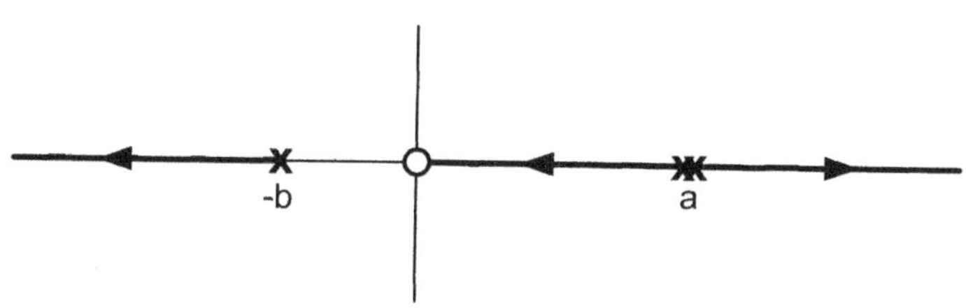

FIGURE 9.15. Locus of roots for $x = \omega_2^2$ of (9.90) with $a > 0$.

Case 2: $a < 0$

The solutions of (9.90), or equivalently (9.85), when $a < 0$ are now considered. This case requires $K_m > n_p L_S I_{max}$. In this case, there is never a second transition speed in brake mode, while in the motor mode, there is either *no* second transition speed, or *two* second transition speeds (that is, *two* speeds $\omega_{21}$and ω_{22} where the system transitions between the $V = V_{\max}$ constraint region and the $V = V_{\max}, I = I_{\max}$ constraint region.

In the case $a < 0$, there are two possibilities for the complementary root locus as shown in Figure 9.16. These two possibilities are determined by the break-away points which are the roots of

$$\frac{dc}{dx} = \frac{d}{dx}\frac{(x-a)^2(b+x)}{x} = \frac{(x-a)(2x^2+bx+ab)}{x^2} = 0. \tag{9.92}$$

The solutions of this equation that belong to the complementary root locus are

$$\begin{aligned} x_{b1} &= a \\ x_{b2} &= \frac{-b+\sqrt{b^2-8ab}}{4} \\ x_{b3} &= \frac{-b-\sqrt{b^2-8ab}}{4} \quad \text{if } -a > b. \end{aligned} \tag{9.93}$$

The second root x_{b2} is the only break-away point on the positive real axis and therefore the only one that has physical significance. For roots of (9.90) to be real and positive, the given value of c must be greater than or equal to the value corresponding to the break-away point x_{b2}. In other words, there are two positive roots x_{21}, x_{22} of (9.90) satisfying $x_{21} < x_{b2}$ and $x_{22} > x_{b2}$ if and only if

$$1 - c\frac{x_{b2}}{(x_{b2}-a)^2(b+x_{b2})} > 0. \tag{9.94}$$

Note that, as $a < 0$ (that is, $n_p^2 L_S^2 I_{max}^2 - K_m^2 < 0$), both x_{21} and x_{22} satisfy (9.91) [equivalently (9.83)], with $+$ sign chosen ($+V_{\max}$) rather than the $-$ sign ($-V_{\max}$). In other words, *both* transition speeds $\omega_{21} = \sqrt{x_{21}}, \omega_{22} = \sqrt{x_{22}}$ are associated with the motoring mode in contrast to the case $a > 0$ where they were split between the two modes. As a result, there is no second transition speed in the brake mode. In motor mode, there are actually *four* speed ranges in this case. They are, in order of increasing speed, constrained by (1) current limits (2) current and voltage limits (3) voltage limits and (4) current and voltage limits.

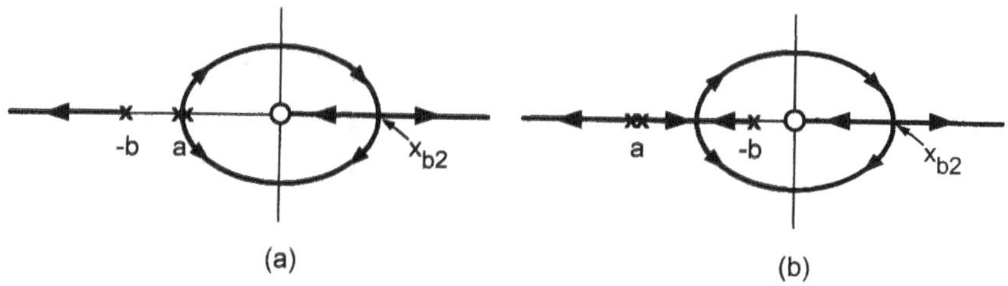

FIGURE 9.16. Locus of roots for $x = \omega_2^2$ of (9.90) with $a < 0$. (a) $-a < b$. (b) $-a > b$.

Although hypothetical motor parameters can be selected to obtain the case with $a < 0$ and two second transition speeds, they do not correspond to realistic motors. The typical case is the one in which (9.94) is *not* satisfied, and there is no second transition speed in either motor or brake mode. Then, there are only two speed ranges for torque optimization: a low-speed range constrained by current limits, and a high-speed range constrained by voltage and current limits. Examples of both the typical case (Figure 9.15) and the atypical case (Figure 9.16) are given next.

9.2.3 *Two Examples*

The motor chosen for consideration is a three-phase PM synchronous motor (BM 500 from Aerotech Inc., Pittsburgh, PA) whose parameter values are given by $L = 9.3 \times 10^{-4}$ H, $R_S = 0.25$ ohms, $J = 13.9 \times 10^{-5}$ kg-m^2, $K = 0.198$ N-m/A and $n_p = 4$. The limits are given by $I_{\max}$ (continuous) = 18 A, $I_{\max}$ (peak) = 55 A, and V_{bus} (bus voltage) = 160 V. The motor bearings are rated for 10,000 rpm or about 1000 rads/sec, which is taken to be the maximum speed of the motor. The corresponding equivalent two-phase parameters are (see the first appendix at the end of this chapter) $L_S = L + L/2 = 14 \times 10^{-4}$ H, $K_m = (\sqrt{3/2})K = 0.243$ N-m/A, $I_{\max}$(continuous) $= (\sqrt{3/2})i_{\max}$ (continuous) = 22 A, $I_{\max}$ (peak) = $\sqrt{3/2}i_{\max}$ (peak) = 67.4 A, $V_{\max} = \sqrt{3/2}(2/\pi)V_{bus} = 124.8$ V.

Continuous Current Limit

With $I_{\max} = 22$ A and $V_{\max} = 124.8$ V, Figure 9.17 is a plot of the maximum torque for $V = V_{\max}$, Figure 9.18 is a plot of the maximum torque for $I = I_{\max}$, and Figure 9.19 is a plot of the maximum torque for $V = V_{\max}$ and $I = I_{\max}$, all as a function of the speed ω (see the second appendix at the end of this chapter). Both positive and negative torques are given. The curves give the torques obtained by the three strategies, accounting for the voltage and current limits.

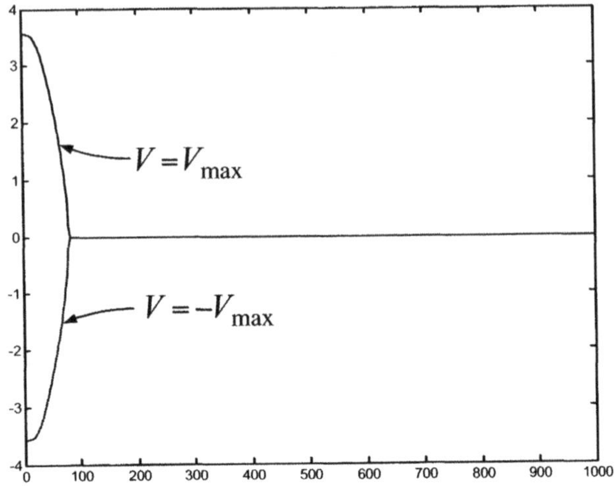

FIGURE 9.17. Maximum torque in N-m versus speed in rad/sec with $V = V_{\mathrm{max}}$.

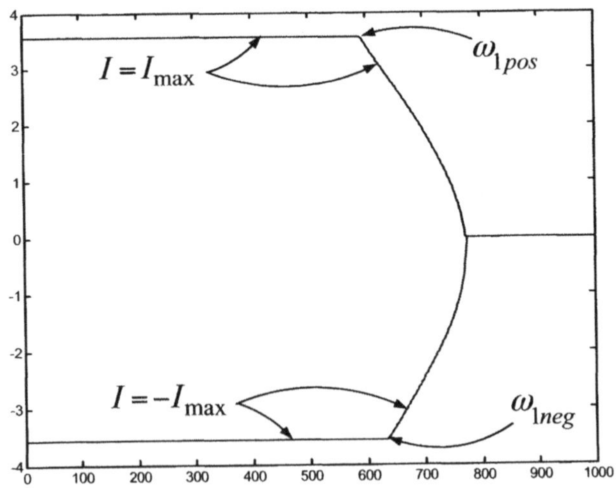

FIGURE 9.18. Maximum torque in N-m versus speed in rad/sec with $I = I_{\mathrm{max}}$.

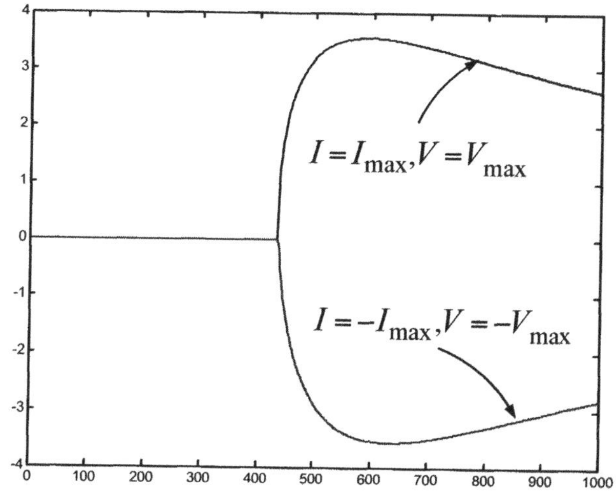

FIGURE 9.19. Maximum torque in N-m versus speed in rad/sec with $I = I_{max}$ and $V = V_{max}$.

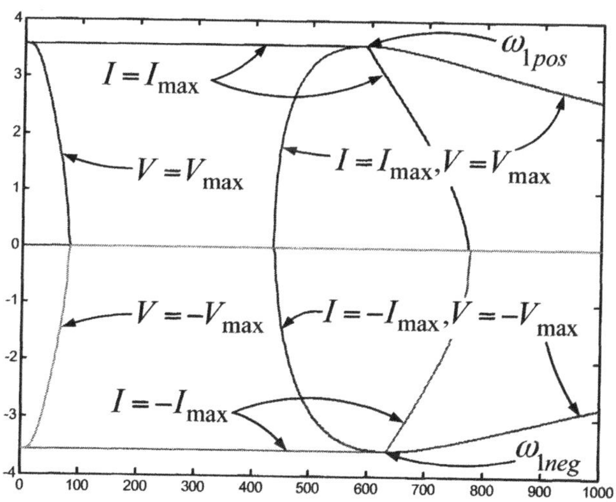

FIGURE 9.20. All three torque components in N-m versus speed in rad/sec.

Figure 9.20 is a plot of all three torque components where it is seen that the actual maximum (minimum) torque attainable is the largest (smallest) of the three curves at each speed (the outer envelope of the curves).

In this case, the first transition speed for the motor mode is $\omega_{1pos} = 592$ rads/sec (see Figures 9.18 and 9.20), while the first transition speed for the brake mode is $\omega_{1neg} = 635$ rads/sec (see Figures 9.18 and 9.20). Note that there are no second transition speeds. As one can see from Figure 9.17, the reason for this situation is that the torque curve obtained with optimal field-weakening (that is, operating at the voltage limit $V = V_{\max}$) falls rapidly due to the current limit. The torque is zero at the speed where the optimal field-weakening current is equal to the maximum current, leaving no torque-producing current. For this reason, a solution accounting for both current and voltage limits must be used at high speed.

Peak Current Limit

Now consider the peak current limit so that $I_{\max} = 67.4$ A and $V_{\max} = 124.8$ V. Figure 9.21 is a plot of the maximum torque for $V = V_{\max}$, Figure 9.22 is a plot of the maximum torque for $I = I_{\max}$, and Figure 9.23 is a plot of the maximum torque for $V = V_{\max}$ and $I = I_{\max}$, all as a function of the speed ω. Both positive and negative torques are given. The curves give the torques obtained by the three strategies, accounting for the voltage and current limits.

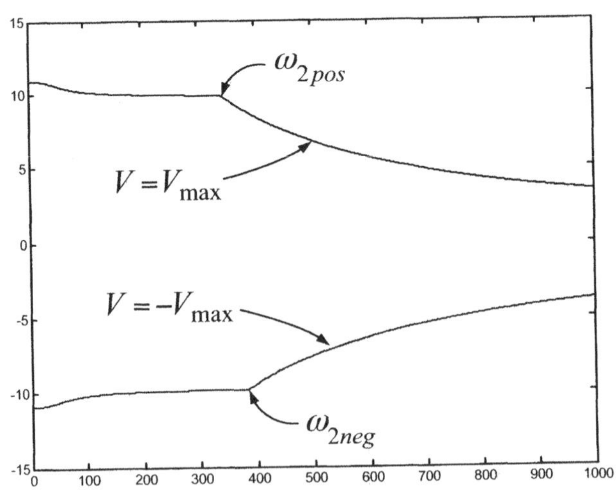

FIGURE 9.21. Maximum torque in N-m versus speed in rad/sec with $V = V_{\max}$.

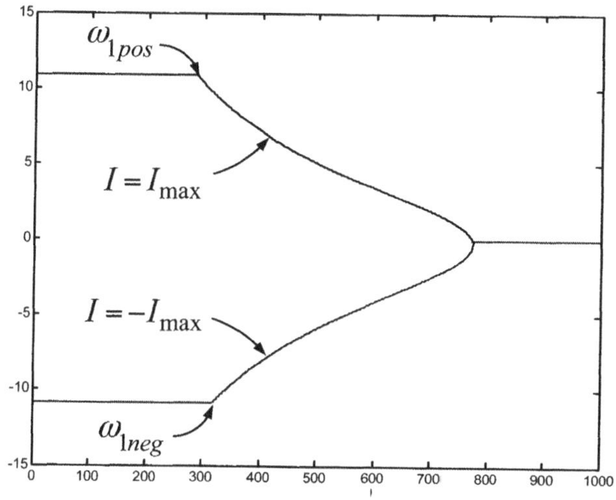

FIGURE 9.22. Maximum torque in N-m versus speed in rad/sec with $I = I_{max}$.

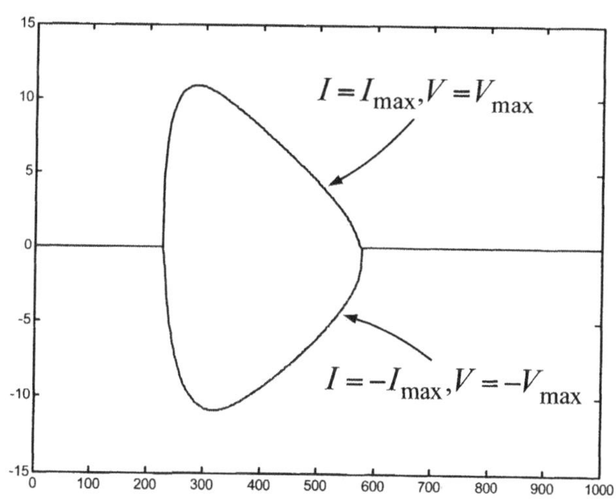

FIGURE 9.23. Maximum torque in N-m versus speed in rad/sec with $I = I_{max}$ and $V = V_{max}$.

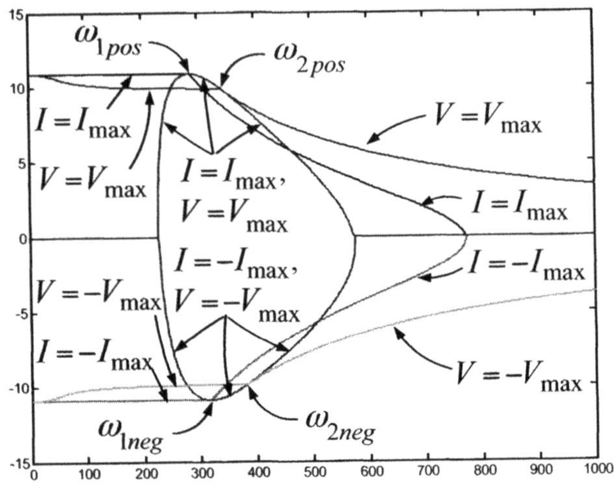

FIGURE 9.24. All three torque components in N-m versus speed in rad/sec.

Figure 9.24 is a plot of all three torque components. Again, the actual maximum (minimum) torque is simply the maximum (minimum) of the three curves at each speed. In this case, the first transition speed for the motor mode is $\omega_{1pos} = 288$ rads/sec (see Figures 9.22 and 9.24) while the second transition speed for the motor mode is $\omega_{1pos} = 320.6$ rads/sec (see Figures 9.21 and 9.24). The first transition speed for the brake mode is $\omega_{1neg} = 340.8$ rads/sec (see Figures 9.22 and 9.24) while the second transition speed for the brake mode is $\omega_{2neg} = 383$ rads/sec (see Figures 9.21 and 9.24).

Simulation

The above analysis assumed constant speed, steady-state conditions. To see how this approach works under non-constant speed conditions, a time-domain simulation can be performed (see problem 8). In this simulation, the continuous current limit of $I_{\max} = 22$ A and voltage limit of $V_{\max} = 124.8$ V are used with the motor driven by the maximum possible torque until the speed reaches approximately the maximum speed 1000 rads/sec (10,000 rpm or the rated speed of the motor bearings). As shown above, this motor with the continuous current limit of 22 A has only the two modes

$$V = V_{\max} \tag{9.95}$$

$$i_d^*(\omega) = 0 \tag{9.96}$$

$$i_q^*(\omega) = \pm I_{\max} \tag{9.97}$$

and

$$\begin{aligned} V &= V_{\max}, \quad I = I_{\max} && (9.98)\\ i_q^*(\omega) &= \frac{-\beta(\omega) \pm \sqrt{\beta^2(\omega) - 4\alpha(\omega)\gamma(\omega)}}{2\alpha(\omega)} && (9.99)\\ i_d^*(\omega) &= -\sqrt{I_{max}^2 - i_q^{*2}} && (9.100) \end{aligned}$$

with

$$\begin{aligned} \alpha(\omega) &\triangleq 4R_S^2\omega^2K_m^2 + 4n_p^2\omega^4L_S^2K_m^2\\ \beta(\omega) &\triangleq -4R_SK_m\omega(V_{max}^2 - K_m^2\omega^2 - Z_S^2(\omega)I_{max}^2)\\ \gamma(\omega) &\triangleq (V_{max}^2 - K_m^2\omega^2 - Z_S^2(\omega)I_{max}^2)^2 - 4I_{max}^2n_p^2\omega^4L_S^2K_m^2 \end{aligned}$$

where the positive sign is for motoring and the negative sign is for braking. Consequently, for maximum positive torque, the current commands are (9.96) and (9.97) with the + sign up to $\omega_{1pos} = 592$ rads/sec and then (9.99) and (9.100) are used as the current commands with the + sign for higher speeds. For negative torque, the current commands are (9.96) and (9.97) (with the − sign) up to $\omega_{1neg} = 635$ rads/sec and then (9.99) and (9.100) are used as the current commands (with the − sign) for higher speeds. A basic block diagram for running the motor with maximum torque is given in Figure 9.25.

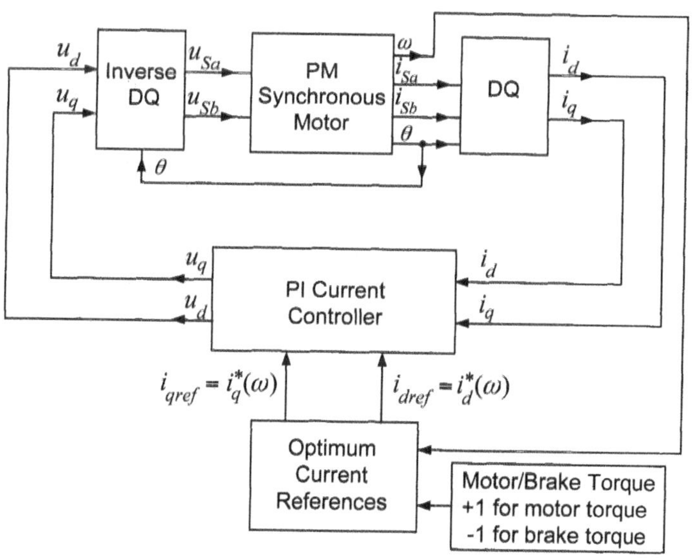

FIGURE 9.25. Block diagram for commanding maximum torque.

The normalized optimal values $i_d^*/I_{\max}$ and $i_q^*/I_{\max}$ are graphed in Figure 9.26 corresponding to the normalized speed $\omega_{\text{normalized}} \triangleq \omega/\left(2\pi 10000/60\right))$ given in Figure 9.27.

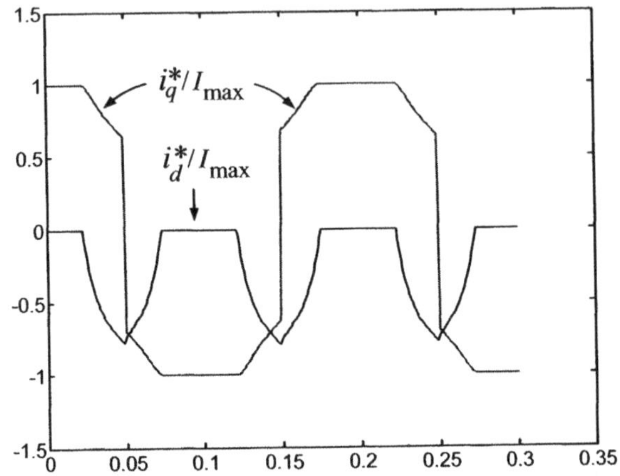

FIGURE 9.26. Optimal direct and quadrature currents versus time in seconds.

Figure 9.27 shows that the motor reaches the speed of 10,000 rpm in about 40 msec.

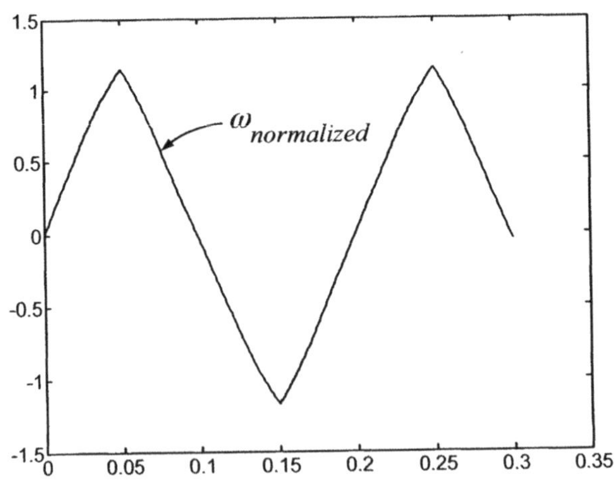

FIGURE 9.27. Normalized speed versus time in seconds.

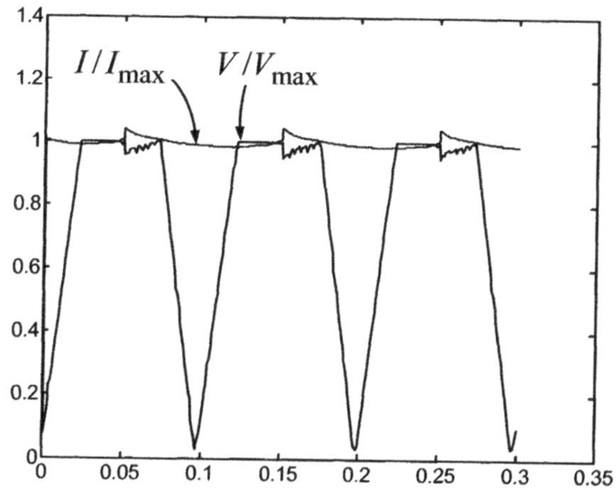

FIGURE 9.28. Normalized current and voltage versus time in seconds.

The normalized magnitude of the voltage $V/V_{\max}$ where $V = \sqrt{u_d^2 + u_q^2}$ and the normalized current $I/I_{\max}$ where $I = \sqrt{i_d^2 + i_q^2}$ are shown in Figure 9.28. Note that the voltage has peaks which are greater than the normalized limit of 1 at those times when the commanded current i_q^* undergoes step changes from the maximum positive value to the minimum negative value (or vice versa). The PI current controller given by $u_q = K_p(i_q^* - i_q) + K_I \int_0^t (i_q^* - i_q)dt$ [see equation (9.49)] is commanding a large voltage due to the large step change in the error $i_q^* - i_q$ at these points in time. However, the voltage actually applied to the motor in the simulation was limited to $V_{\max}$, indicating that these commanded overvoltage peaks have no effect on the performance.

Remarks Of course, the general situation is that one is not always requesting maximum torque from the machine. One possibility is to set the direct current $i_d = i_d^*(\omega)$ so that the maximum torque is always readily available.

Note that the most complicated calculations are square roots and cubic roots so that the whole set of computations could be performed in real-time if the motor parameters were determined online, using an identification procedure such as described in the next section.

In general, there is an intermediate speed range where both the voltage and the current limits are reached. Consideration of this speed range not only provides the complete solution to the problem, but also provides a smooth connection between the solutions for the low and high speed regimes. Further, there are cases where the optimal operation at high speed is determined solely by the joint voltage/current limit solution. In such cases, the conventional optimal field weakening formula never provides the optimum torque.

9.3 Identification of the PM Synchronous Motor Parameters*

This section requires an understanding of the material in Chapter 2 on least-squares identification. The identification of the parameters of a PM synchronous machine are conveniently carried out with the model in the dq coordinate system given by

$$\begin{aligned} L_S\frac{di_d}{dt} &= -R_S i_d + n_p\omega L_S i_q + u_d && (9.101)\\ L_S\frac{di_q}{dt} &= -R_S i_q - n_p\omega L_S i_d - K_m\omega + u_q && (9.102)\\ K\frac{d\omega}{dt} &= K_m i_q - \tau_L. && (9.103)\end{aligned}$$

For the identification of the motor parameters, the load torque is now taken to be a linear combination of viscous friction and coulomb friction given by

$$\tau_L = -f\omega - f_c\text{sgn}(\omega)$$

where

$$\text{sgn}(\omega) = \begin{cases} 1, & \omega > 0\\ 0, & \omega = 0\\ -1, & \omega < 0.\end{cases}$$

The model (9.101), (9.102), (9.103) is rewritten in regressor form as

$$y(nT) = W(nT)K \tag{9.104}$$

where the *output vector* y, *regressor matrix* W, and *parameter vector* K are defined as

$$y(nT) \triangleq \begin{bmatrix} u_d(nT)\\ u_q(nT)\\ 0\end{bmatrix} \in \mathbb{R}^3,$$

$$W(nT) \triangleq \begin{bmatrix} i_d(nT) & \dfrac{di_d(nT)}{dt} - n_p\omega(nT)i_q(nT) & 0 & 0 & 0 & 0\\ i_q(nT) & \dfrac{di_q(nT)}{dt} + n_p\omega(nT)i_d(nT) & \omega(nT) & 0 & 0 & 0\\ 0 & 0 & -i_q(nT) & \dfrac{d\omega(nT)}{dt} & \omega(nT) & -\text{sgn}(\omega(nT))\end{bmatrix} \in \mathbb{R}^{3\times 6}$$

and

$$K \triangleq \begin{bmatrix} R_S & L_S & K_m & J & f & f_c \end{bmatrix}^T \in \mathbb{R}^6.$$

As the model (9.101), (9.102), (9.103), or equivalently (9.104), is *linear* in the parameters, the linear least-squares technique of Chapter 2 applies directly. The *error equation* (9.105) is formed by subtracting the actual (measured) output $y(nT)$ from the estimated output $W(nT)K$ to obtain

$$e(nT) \triangleq W(nT)K - y(nT) \in \mathbb{R}^2. \qquad (9.105)$$

If the model (9.104) of the motor were exact, and the voltages, currents, position, and speed could be measured exactly, then there is a parameter vector K such that $e(nT) = 0$ for all time. However, (9.104) is not exactly satisfied due to modeling errors, nor can the voltages, currents, and so on. be measured exactly. In other words, the error $e(nT)$ is never zero for all time. By definition, for any value of the parameter vector K, the *residual error* $E^2(K)$ is equal to the sum of the norm squared error $e^T(nT)e(nT)$ over an interval $[n_0T, n_1T]$, that is,

$$E^2(K) = \sum_{n=n_0}^{n_1} e^T(nT)e(nT) \qquad (9.106)$$

$$= \sum_{n=n_0}^{n_1} \Big(W(nT)K - y(nT)\Big)^T\Big(W(nT)K - y(nT)\Big). \qquad (9.107)$$

The *least-squares estimate* K^* is the one that minimizes the residual error. Expanding (9.107), the residual error becomes

$$\begin{aligned}E^2(K) &= \sum_{n=n_0}^{n_1} \Big(y^T(nT)y(nT) - y^T(nT)W(nT)K - K^TW^T(nT)y(nT) \\ &\quad + K^TW^T(nT)W(nT)K\Big) \\ &= \sum_{n=1}^{N} y^T(nT)y(nT) - \left(\sum_{n=1}^{N} y^T(nT)W(nT)\right)K \\ &\quad - K^T\left(\sum_{n=1}^{N} W^T(nT)y(nT)\right) + K^T\left(\sum_{n=1}^{N} W^T(nT)W(nT)\right)K.\end{aligned}$$

Making the definitions

$$R_W \triangleq \sum_{n=n_0}^{n_1} W^T(nT)W(nT) \in \mathbb{R}^{6\times 6}, \quad R_{Wy} \triangleq \sum_{n=n_0}^{n_1} W^T(nT)y(nT) \in \mathbb{R}^{6\times 1}$$

$$R_{yW} \triangleq \sum_{n=n_0}^{n_1} y^T(nT)W(nT) \in \mathbb{R}^{1\times 6}, \quad R_y \triangleq \sum_{n=n_0}^{n_1} y^T(nT)y(nT) \in \mathbb{R}$$

it is written compactly as

$$\begin{aligned} E^2(K) &= R_y - R_{yW}K - K^T R_{Wy} + K^T R_W K \\ &= R_y - R_{yW}R_W^{-1}R_{Wy} + \left(K - R_W^{-1}R_{Wy}\right)^T R_W\left(K - R_W^{-1}R_{Wy}\right) \end{aligned} \tag{9.108}$$

where it is assumed R_W is invertible (The machine is sufficiently excited). The residual error is then minimized by choosing

$$K = K^* \triangleq R_W^{-1}R_{Wy}. \tag{9.109}$$

This approach is the standard least-squares approach presented in Chapter 2.

Error Index

The residual error indicates how well the measured output $y(nT)$ fits the linear model $W(nT)\hat{K}$. A residual error of zero would indicate a perfect match for every point n. However, as noted above, due to noise, quantization errors, and unmodeled dynamics, the actual residual error is greater than zero. To quantify this fact, recall that the notion of an error index was developed in Chapter 2. Specifically, if all the parameters were set to zero, the residual error would be equal to $E^2(0) = R_y$. As $E^2(K^*)$ is the minimum residual error, it must be less than or equal to R_y. Using these two facts, an indication of how well the least-squares solution fits the data is found by defining a residual error index as

$$\text{Error Index} \triangleq \sqrt{\frac{E^2(K^*)}{E^2(0)}} = \sqrt{\frac{R_y - R_{yW}R_W^{-1}R_{Wy}}{R_y}} \le 1. \tag{9.110}$$

Multiplying this by 100, the error index can be viewed as a percentage of how well the data fits the linear relationship (9.104).

Parametric Error Index

The residual error index indicates how well the data set fits the model, but gives no indication as to how well the parameters are determined from the data. In Chapter 2, the idea of a parametric error index was developed to address this issue. Recall that to define the parametric error index, variations in the parameters that cause the residual error to double were considered. The residual error due to a variation of $\delta K = K - K^* = K - R_W^{-1}R_{Wy}$ in the parameters is given by $E^2(K) = E^2(K^*) + \delta K^T R_W \delta K$, which follows directly from (9.108). Setting $E^2(K)$ equal to $2E^2(K^*)$ yields

$$E^2(K^*) = \delta K^T R_W \delta K. \tag{9.111}$$

If there were just a single parameter (so that R_W is a scalar), this would give an estimate of the parametric sensitivity as

$$\delta K = \sqrt{\frac{E^2(K^*)}{R_W}}.$$

The general solution to (9.111) for $K \in \mathbb{R}^n$ was derived in Chapter 2 and is given by

$$\delta K_i = \sqrt{E^2(K^*)(R_W^{-1})_{ii}}$$

and is taken to be the *parametric error* index. The percent parametric error index is then defined by

$$PE_i \triangleq \frac{\delta K_i}{K_i^*} \times 100. \quad (9.112)$$

This index is equal to the percentage change in the parameter K_i corresponding to a doubling of the residual error and provides an indication of how sensitive the residual error is to each parameter estimate. For example, if $PE_i = 10\%$, then the residual error $E^2(K)$ is doubled for only a 10% change in the ith parameter relative to the optimal estimate K_i^* and therefore $E^2(K)$ is relatively sensitive to the estimated value of K_i. PE_i can be interpreted as a measure of the order of magnitude of the errors on the parameters.

The parametric error indices were chosen to correspond to a doubling of the residual error. As pointed out in Chapter 2, this is an arbitrary as well as a conservative choice; it means that one looks for a parametric variation that would lead to an increase of error equal to 100% of the minimum residual error. If another level had been chosen, all the parametric error indices would have been scaled by the same constant factor. The parametric error indices are especially useful to compare relative sensitivities of different parameters and to give some idea of the values of the errors to be anticipated.

9.3.1 Experimental Results

Experimental results from Refs. [118][119] are now presented to illustrate the identification procedure. A 50-pole-pair ($n_p = 50$) PM synchronous machine is used with a 2000-pulses/rev encoder. Input voltages were commanded to the motor so that it would accelerate reasonably fast while not going over their limits. The sampling frequency is $T = 50$ kHz (see the subsection below on signal processing). Figure 9.29 shows both the current i_q resulting from the input voltages and the simulated current $i_{q\text{sim}}$ resulting from a simulation using the identified parameters and the same input as the experiment. Similarly, Figure 9.30 shows the corresponding speed response ω and the simulated speed ω_{sim}.

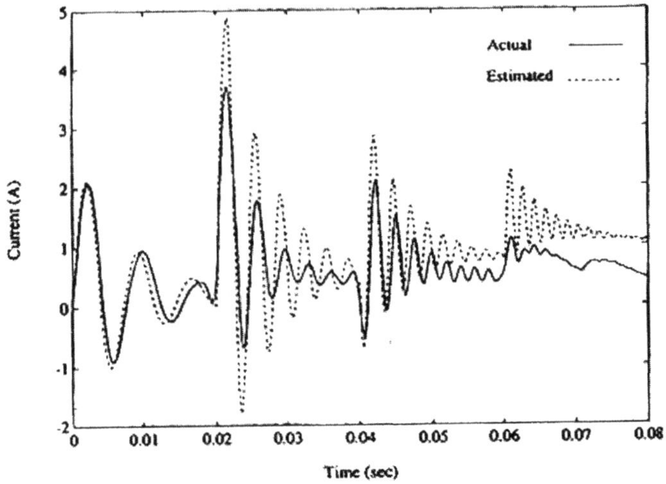

FIGURE 9.29. i_q and $i_{q\text{sim}}$ for the identification experiment. From Blauch et al. [119], © 2004 IEEE.

Using the data collected between 0.01 sec and 0.03 sec, the parameters were computed according to (9.109) and their percent parametric error according to (9.112) and are given in the table below. The residual error computed using (9.110) was found to be 12%.

Parameter	Estimate	Parametric Error
R_S ohms	0.269	399%
L_S H	0.0027	47.6%
K_m N-m/A	0.515	19.8%
J kg-m^2/ sec^2	0.000187	312%
f N-m/rad/sec	0.0032	5040%
f_c N-m	0.0693	4500%

For comparison, the resistance R_S was measured using an ohmmeter and was found to be 0.6 ohm. By locking the rotor at $\theta = 0$, phase a is just a simple RL circuit [see equation (9.1)] with a time constant of L_S/R_S. Commanding step voltages to phase a and computing the decay rate (time constant) in the current response, L_S/R_S, was found to be $0.0024/0.6 = 0.004$, giving a value of $L_S = 0.0024$ H for the inductance. The value of the resistance computed using the static rotor tests is further from its least-squares value than the value of the inductance. This is simply due to the fact that the value of R_S is not very significant compared to the value of $n_p\omega L_S$ if the speed is just a little above zero. Note that in Figure 9.30 the speed is

10 rad/sec or higher after 0.01 sec so that $n_p \omega L_S \geq 50 \times 10 \times 0.0024 = 1.2$ while $R_S = 0.27$ ohm. That is, the value of R_S does not impact the value of the residual error compared to the value of L_S, and this is reflected in the relative values of their parametric errors.

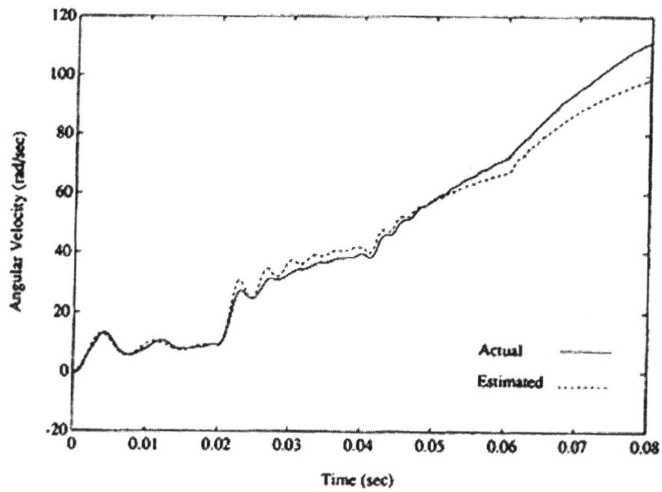

FIGURE 9.30. ω and ω_{sim} for the identification experiment. From Blauch et al. [119], © 2004 IEEE.

Note that the corresponding simulations in Figures 9.29 and 9.30 show a good correspondence with the actual measured data.

Two-Stage Identification

Note from the table above that values of the parametric indices of the friction parameters f and f_c indicate that they are much more uncertain than the stator resistance value, and that the moment of inertia J appears to be almost as uncertain as R_S. However, recall that the parametric indices are not absolute indicators of error. In fact, by doing the identification in two stages (i.e., first identifying the electrical parameters R_S, L_S, and K_m and then using these values to identify the mechanical parameters J, f, and f_c), the values of the parametric indices are significantly reduced though the computed values of J, f, and f_c remain the same. In more detail, let

$$W_{elec}(nT) \triangleq \begin{bmatrix} i_d(nT) & \dfrac{di_d(nT)}{dt} - n_p\omega(nT)i_q(nT) & 0 \\ i_q(nT) & \dfrac{di_q(nT)}{dt} + n_p\omega(nT)i_d(nT) & \omega(nT) \end{bmatrix}$$

$$y_{elec}(nT) = \begin{bmatrix} u_d(nT) \\ u_q(nT) \end{bmatrix}, \quad K_{elec} \triangleq \begin{bmatrix} R_S \\ L_S \\ K_m \end{bmatrix}$$

so that

$$y_{elec}(nT) = W_{elec}(nT)K_{elec}.$$

Defining

$$\begin{aligned} R_{W_{elec}} &\triangleq \sum_{n=n_0}^{n_1} W_{elec}^T(nT)W_{elec}(nT) \in \mathbb{R}^{2\times 2} \\ R_{W_{elec}y} &\triangleq \sum_{n=n_0}^{n_1} W_{elec}^T(nT)y_{elec}(nT) \in \mathbb{R}^{3\times 1} \\ R_{yW_{elec}} &\triangleq \sum_{n=n_0}^{n_1} y_{elec}^T(nT)W_{elec}(nT) \in \mathbb{R}^{1\times 3} \\ R_{y_{elec}} &\triangleq \sum_{n=n_0}^{n_1} y_{elec}^T(nT)y_{elec}(nT) \in \mathbb{R}, \end{aligned}$$

the optimal estimate in the least-squares sense is then given by

$$K_{elec} = R_{W_{elec}}^{-1} R_{W_{elec}y}.$$

Using the same data as above, the parameters and parametric errors were computed and are given in the table below. The values of the parameters are essentially the same as before.

Parameter	Estimate	Parametric Error
R_S ohms	0.268	399%
L_S H	0.0027	47.6%
K_m N-m/A	0.515	19.8%

Next, using the value of K_m just found, the mechanical parameters are identified. To proceed, let

$$W_{mech}(nT) \triangleq \begin{bmatrix} \dfrac{d\omega(nT)}{dt} & \omega(nT) & \text{sgn}(\omega(nT)) \end{bmatrix}$$

$$y_{mech}(nT) = K_m i_q(nT), K_{mech} \triangleq \begin{bmatrix} J \\ f \\ f_c \end{bmatrix}$$

so that

$$y_{mech}(nT) = W_{mech}(nT)K_{mech}.$$

Also, let

$$\begin{aligned} R_{W_{mech}} &\triangleq \sum_{n=n_0}^{n_1} W_{mech}^T(nT)W_{mech}(nT) \in \mathbb{R}^{2\times 2} \\ R_{W_{mech}y} &\triangleq \sum_{n=n_0}^{n_1} W_{mech}^T(nT)y_{mech}(nT) \in \mathbb{R}^{3\times 1} \\ R_{yW_{mech}} &\triangleq \sum_{n=n_0}^{n_1} y_{mech}^T(nT)W_{mech}(nT) \in \mathbb{R}^{1\times 3} \\ R_{y_{elec}} &\triangleq \sum_{n=n_0}^{n_1} y_{mech}^T(nT)y_{mech}(nT) \in \mathbb{R}, \end{aligned}$$

so that optimal estimate in the least-squares sense is given by

$$K_{mech} = R_{W_{mech}}^{-1} R_{W_{mech}y}.$$

The parameter estimates and their parametric errors are

Parameter	Estimate	Parametric Error
J kg-m^2/sec^2	0.000187	13.2%
f N-m/rad/sec	0.0032	214%
f_c N-m	0.0693	191%

The error index was 10%. Note that parameter estimates are the same as before, but the parametric errors are significantly reduced and the error index is reduced slightly. However, the parametric errors still show the same *relative* uncertainty in that the estimate of J is much more certain than the values of f and f_c.

Signal Processing

A few words about processing the signals i_d, i_q, and θ when n_p is large are in order. Recall that the currents i_{Sa} and i_{Sb} are sampled and i_d and i_q computed according to

$$\begin{bmatrix} i_d \\ i_q \end{bmatrix} = \begin{bmatrix} \cos(n_p\theta) & \sin(n_p\theta) \\ \sin(n_p\theta) & \cos(n_p\theta) \end{bmatrix} \begin{bmatrix} i_{Sa} \\ i_{Sb} \end{bmatrix}. \tag{9.113}$$

As the number of pole pairs is $n_p = 50$, a mechanical speed of 3000 rpm corresponds to an electrical frequency of 1.57 kHz. So, even if one samples at 50 kHz, a delay of one sampling period corresponds to a rotor movement of $n_pT \times \left(3000\text{ rpm}/60\frac{\text{sec}}{\text{min}}\right) \times 360° = 18°$ for $n_p = 50, T = 1/\left(50 \times 10^3\right)$, and, therefore, a phase lag in i_d and i_q of this same amount. A phase shift in the measurement of the phase currents leads to a rotation by the same angle in the dq coordinates. Consequently, as the electrical torque is proportional to i_q only, such a shift could significantly affect the torque. For this reason, the delays present in the system must be accounted for precisely [119]. Due to the zero-order holds, the D/A converters create a delay between the commanded and actual voltages equal to half the sampling period. A delay is also caused by the time required to measure the state variables, calculate the dq transformation, and send the new commanded voltages. The PWM circuit also contains a low-pass filter to reduce circulation in the voltage regulator of the high-frequency noise. The filter has a DC gain of 1 and a cutoff frequency of 10 kHz. The delays due to the filter, PWM inverter, and zero-order hold were combined into a single delay of approximately $1.6T$. If the commanded voltages are used (rather than the measured voltages), they must be delayed by this amount to reflect the voltages effectively applied to the motor. However, if n_p is small, these delays are typically of no significance.

Another issue of significance in the experiments is the filtering of the noise before processing by the least squares algorithm. The collected currents and voltages were transformed into the dq coordinate system according to (9.113) and were filtered in order to remove as much of the measurement (PWM) noise as possible. A third-order Butterworth filter was used with a cutoff frequency of 500 Hz for the dq currents and 100 Hz for the dq voltages. This filtering was performed twice, forward and backward, to avoid the introduction of delays using the MATLAB function *filtfilt* (see Chapter 2). To obtain all the necessary signals for the identification procedure, the derivatives were reconstructed using the backward difference calculation $dx(kT)/dt \triangleq (x(kT) - x((k-1)T))\,/T$. As such a reconstruction enhances high-frequency noise, the derivatives are filtered again using a third-order Butterworth with a cutoff frequency of 500 Hz.

9.4 PM Stepper Motors*

Motion (servo) control systems often use a particular type of permanent magnet machine called the stepper motor. Permanent magnet stepper motors are synchronous electric machines, but have a quite different construction than the PM synchronous machine described in Chapters 6 and 7. A typical PM stepper motor consists of a stator made of soft iron equipped with windings/coils and a permanent magnet rotor. As shown in Figure 9.31, the rotor has two sets of teeth (left and right in Figure 9.31) which are out of alignment with each other by a tooth width. Typically, there are 50 teeth in each set; one set is magnetized as south poles and the other set is magnetized as north poles, resulting in $n_p = 50$ pole pairs.

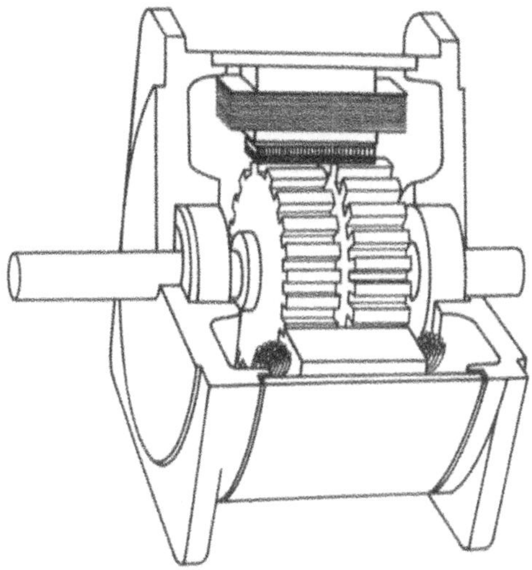

FIGURE 9.31. Cutaway view of a PM stepper motor. Drawn by Sharon Katz. From Section 9.7 of Ref. [120] and reproduced with permision of Marcel Dekker.

A photograph of a PM stepper motor is shown in Figure 9.32 where the rotor has been pulled out for viewing purposes.

FIGURE 9.32. Photograph of a PM stepper motor with the rotor pulled out from the stator. Courtesy of Prof. J. Douglas Birdwell of The University of Tennessee.

The stator by itself is shown in Figure 9.33 and has eight windings. Four of these windings make up phase a and the other four windings make up phase b. The four windings of phase a are each separated by a winding of phase b and vice versa.

FIGURE 9.33. Stator windings of a PM stepper motor. Four of the windings make up phase a and the other four windings make up phase b. Courtesy of Prof. J. Douglas Birdwell of The University of Tennessee.

Figure 9.34 shows a cross-sectional view of both the left and right sides of a simplified PM stepper motor with only 5 rotor teeth on each side and a stator with four windings. Here the left side of the PM rotor is magnetized as a south pole and the right side of the PM rotor is magnetized as a north pole. The rotor is made of a cylindrical permanent magnet core with the sets of teeth made of soft iron pressed onto either end.

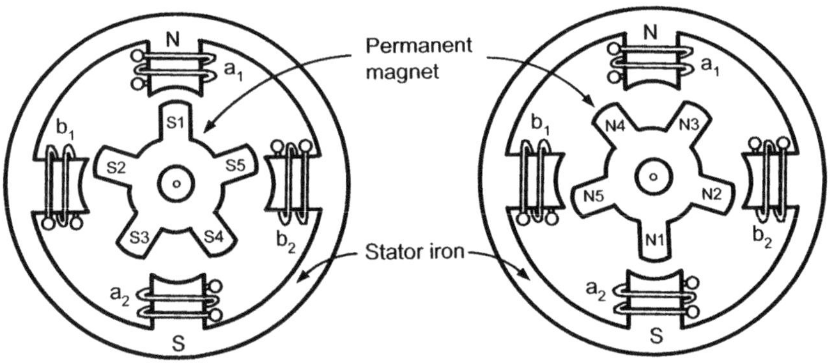

FIGURE 9.34. Cross-sectional views of the left and right sides of a PM stepper motor. Adapted from Deltoro [29] and drawn by Sharon Katz. From Section 9.7 of Ref. [120] and reproduced with permision of Marcel Dekker.

The permanent magnet magnetizes the soft iron of the rotor teeth, making one set a south pole and the other a north pole. The motor is operated as a two-phase machine by connecting windings (coils) a_1 and a_2 in series to make phase a and windings b_1 and b_2 in series to make phase b. This connection as a two-phase machine is now assumed is all that follows.

9.4.1 Open-Loop Operation of the Stepper Motor

Stepper motors were originally designed to be used in open-loop [113][114][29]. Their inherent *stepping* ability allows for accurate positioning to a set of discrete positions without feedback. To explain this, consider Figure 9.35(a) with the motor at $\theta = 0$. The direction for positive current for phase a is as shown in the figure. As indicated there, when positive current is in phase a, winding a_1 of the stator is magnetized as a north pole and winding a_2 is magnetized as a south pole. Conversely, when negative current is in phase a, winding a_1 of the stator is magnetized as a south pole and winding a_2 is magnetized as a north pole. Similar remarks hold for phase b. Following the development in Ref. [29], the stepping motion is illustrated in Figures 9.35(a)–(e), which are adapted from Figure 9–23 of Deltoro [29]. Figure 9.35(a) illustrates the starting point with phase a having positive current for a north–south (NS) orientation for windings $a_1 - a_2$ and phase b having no current. In steady-state with $u_{Sa} = u_0$, it follows that $i_{Sa} = i_0$ ($u_0 = R_S i_0$, where R_S is the resistance of phase a) and the motor is held by magnetic attraction at this location. For the first step, i_{Sa} is set to zero and $i_{Sb} = i_0$ so that rotor tooth $N2$ of (the right side) Figure 9.35(a) is attracted to stator winding b_2 [see the right side of Figure 9.35(b)]. Similarly, the rotor tooth $S2$ on the left side of Figure 9.35(a) is attracted to stator winding b_1 [see the left side of Figure 9.35(b)].

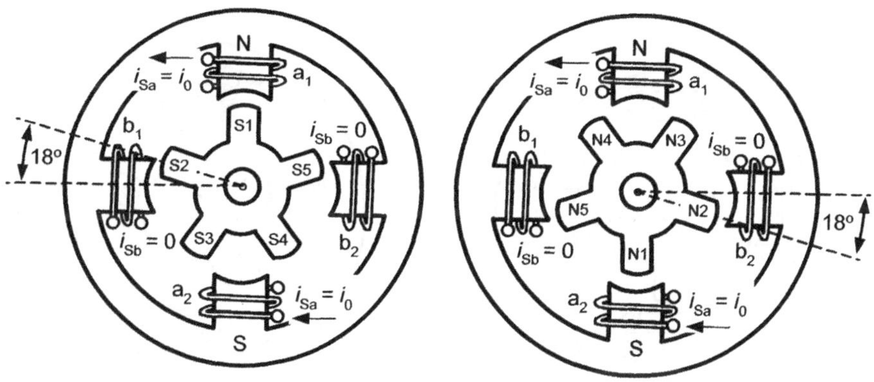

FIGURE 9.35. (a) Starting point with $\theta = 0$ where $u_{Sa} = u_0$, $u_{Sb} = 0$, $i_{Sa} = i_0$, and $i_{Sb} = 0$. Adapted from Deltoro [29] and drawn by Sharon Katz. From Section 9.7 of Ref. [120] and reproduced with permision of Marcel Dekker.

There are $360°/5 = 72°$ between each rotor tooth and, by the geometry of the right side of Figure 9.35(a), $N2$ is 18° from being aligned with winding b_2. The rotor then rotates counterclockwise 18° and, after a short transient period, ends up in the position shown in Figure 9.35(b).

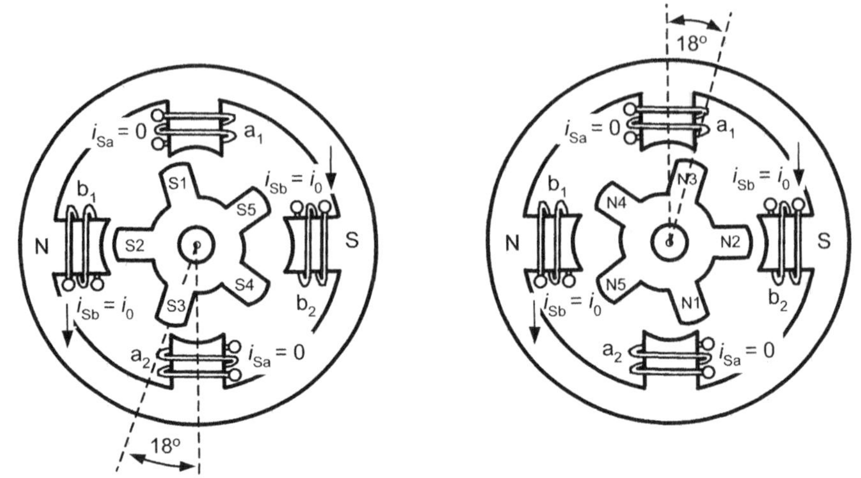

Figure 9.35(b) First step where $\theta = 18°$ with $u_{Sa} = 0$, $u_{Sb} = u_0$, $i_{Sa} = 0$, $i_{Sb} = i_0$.

For the second step, one sets $u_{Sa} = -u_0$, $u_{Sb} = 0$, $i_{Sa} = -i_0$, and $i_{Sb} = 0$ so that winding a_1 is now a south pole and winding a_2 is a north pole [Figure 9.35(c)]. Rotor tooth $N3$ is then attracted to stator winding a_1 and rotor tooth $S3$ is attracted to stator winding a_2. The rotor then rotates another 18°; after a transient period, θ is 36°, ending up in the position shown in 9.35(c).

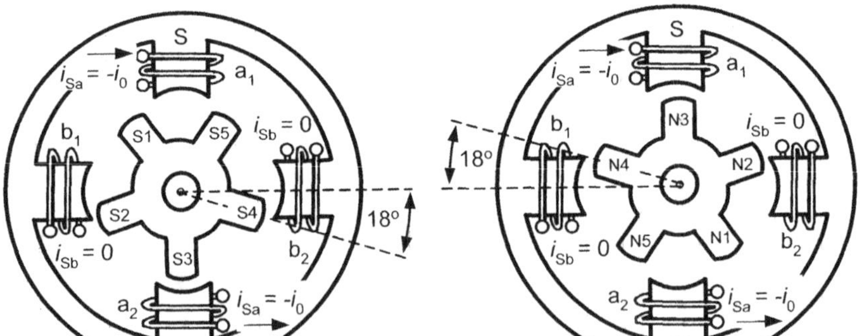

Figure 9.35(c) Second step where $\theta = 36°$ with $u_{Sa} = -u_0$, $u_{Sb} = 0$, $i_{Sa} = -i_0$, $i_{Sb} = 0$.

For the third step, one sets $u_{Sa} = 0$, $u_{Sb} = -u_0$, $i_{Sa} = 0$, and $i_{Sb} = -i_0$ so that winding b_1 is now a south pole and winding b_2 is a north pole [Figure 9.35(d)]. Rotor tooth $N4$ is then attracted to stator winding b_1 and rotor tooth $S4$ is attracted to stator winding b_2. The rotor then rotates another 18°; after a transient period, θ is 54° as shown in 9.35(d).

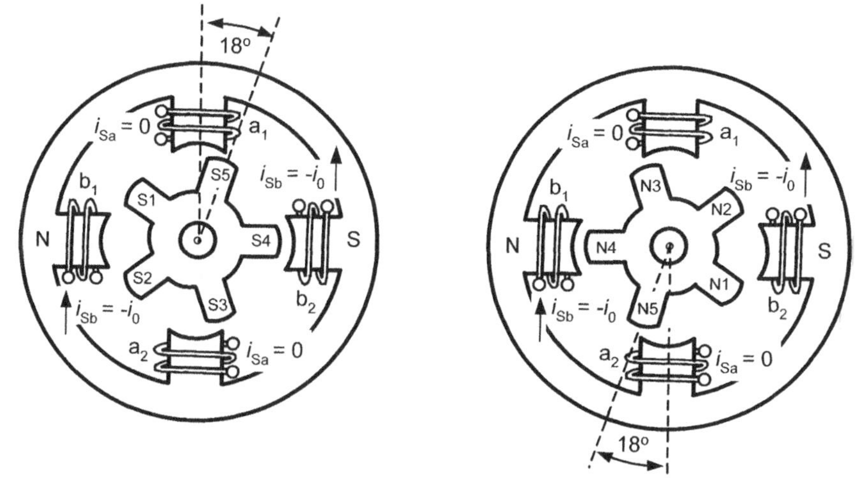

Figure 9.35(d) Third step where $\theta = 54°$ with
$u_{Sa} = 0$, $u_{Sb} = -u_0$, $i_{Sa} = 0$, $i_{Sb} = -i_0$.

For the fourth step, one sets $u_{Sa} = u_0$, $u_{Sb} = 0$, $i_{Sa} = i_0$, and $i_{Sb} = 0$ so that winding a_1 is now a north pole and winding a_2 is a south pole [Figure 9.35(e)]. Rotor tooth $S5$ is then attracted to stator winding a_1 and rotor tooth $N5$ is attracted to stator winding a_2. The rotor then rotates another 18°; and after a transient period, θ is 72° as shown in 9.35(e).

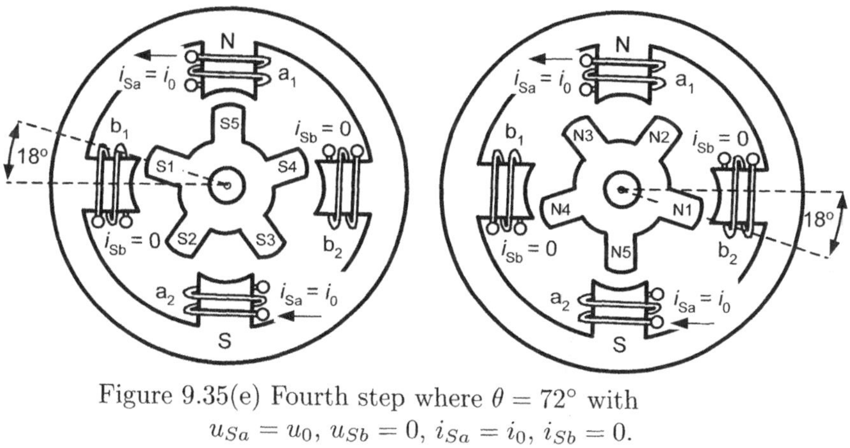

Figure 9.35(e) Fourth step where $\theta = 72°$ with
$u_{Sa} = u_0$, $u_{Sb} = 0$, $i_{Sa} = i_0$, $i_{Sb} = 0$.

Note that Figure 9.35(e) is identical to Figure 9.35(a) except for the labeling of the rotor teeth. That is, after four steps, the rotor has moved by an angular distance equal to the angle between two successive rotor teeth. In general, the step size is given by $360°/(4n_p)$. In this particular example, $n_p = 5$ so that the step size is $360°/20 = 18°$ and, with $n_p = 50$, it follows that the step size is 1.8°.

A typical step response of a PM stepper motor with $n_p = 50$ is shown in Figure 9.36, in which the motor rotates 1.8°. For example, starting with the motor as shown in Figure 9.35(a), set $u_{Sa} = 0$ and $u_{Sb} = u_0$, and the response shown in Figure 9.36 results. This oscillatory behavior is due to the fact that the rotor motion (for small steps) is approximately the same as that of a pendulum with the only damping due to the viscous friction of the bearings (see below). To run the motor open loop at constant speed requires the stepping motion illustrated in Figures 9.35(a) through 9.35(e). If a step is initiated before the transient from the previous step have settled out sufficiently, the motor can lose synchronism, resulting in the position being off by a step or more, from which it cannot recover [114]. In fact, the motor can even reverse its direction [121]. As a result, the open-loop operation is traded off with a relatively long settling time and low-speed performance. Fast and precise positioning requires closed-loop control.

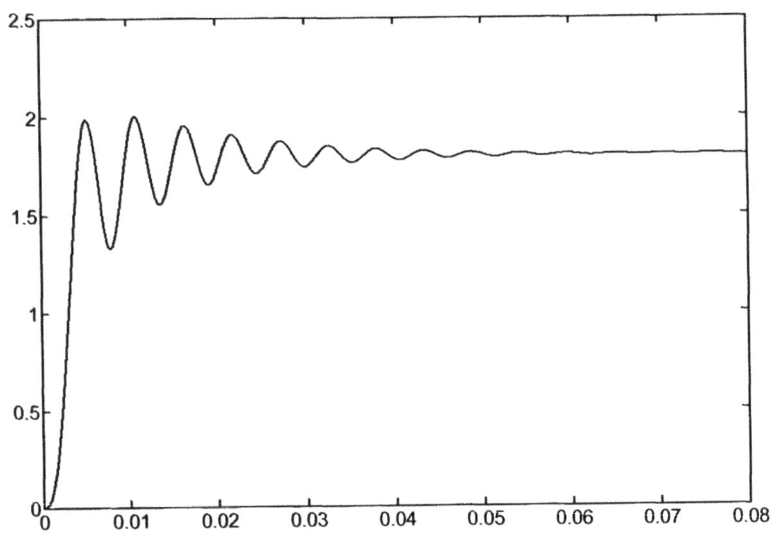

FIGURE 9.36. Step response showing rotor position going from the position in Figure 9.35(a) to the position in Figure 9.35(b). (θ in degrees versus time in seconds.)

9.4.2 Mathematical Model of a PM Stepper Motor

Following the developments in Refs. [113][114][115], the mathematical model is now derived using conservation of energy. With L_S the self inductance of each phase winding, the total flux linkage in the windings of the stator phases is given by

$$\begin{aligned}\lambda_{Sa} &= L_S i_{Sa} + \phi_a(\theta)\\ \lambda_{Sb} &= L_S i_{Sb} + \phi_b(\theta)\end{aligned}$$

where $\phi_a(\theta)$ and $\phi_b(\theta)$ are the flux linkages in stator phases a and b, respectively, produced by the rotor's permanent magnet. The surfaces of the rotor and stator teeth are shaped so that the flux linkages $\phi_a(\theta)$ and $\phi_b(\theta)$ are given by (see Refs. [113][115][114])

$$\begin{aligned}\phi_a(\theta) &= \lambda_M \cos(n_p\theta)\\ \phi_b(\theta) &= \lambda_M \sin(n_p\theta)\end{aligned}$$

where λ_M is a constant and $\theta = 0$ corresponds to the rotor as shown in Figure 9.35. By Faraday's law, Ohm's law, and Kirchhoff's voltage law,

$$\begin{aligned}u_{Sa} - R_S i_{Sa} - \frac{d}{dt}\lambda_{Sa} &= 0\\ u_{Sb} - R_S i_{Sb} - \frac{d}{dt}\lambda_{Sb} &= 0\end{aligned}$$

or

$$\begin{aligned}L_S\frac{di_{Sa}}{dt} &= -R_S i_{Sa} + K_m\omega\sin(n_p\theta) + u_{Sa}\\ L_S\frac{di_{Sb}}{dt} &= -R_S i_{Sb} - K_m\omega\cos(n_p\theta) + u_{Sb}\end{aligned}$$

where $K_m \triangleq n_p\lambda_M$ and $\omega = d\theta/dt$ is the rotor's angular speed. The power supplied by the source voltages u_{Sa} and u_{Sb} is $i_{Sa}u_{Sa} + i_{Sb}u_{Sb}$ and thus

$$\begin{aligned}i_{Sa}u_{Sa} + i_{Sb}u_{Sb} &= R_S i_{Sa}^2 + R_S i_{Sb}^2 + L_S i_{Sa}\frac{di_{Sa}}{dt} + L_S i_{Sb}\frac{di_{Sb}}{dt}\\ &\quad - (i_{Sa}e_{Sa} + i_{Sb}e_{Sb})\\ &= R_S\left(i_{Sa}^2 + i_{Sb}^2\right) + \frac{d}{dt}\frac{1}{2}L_S\left(i_{Sa}^2 + i_{Sb}^2\right) - (i_{Sa}e_{Sa} + i_{Sb}e_{Sb})\end{aligned}$$

where the back-emf voltages e_{Sa} and e_{Sb} are defined by

$$\begin{aligned}e_{Sa} &\triangleq -\frac{d}{dt}\phi_a(\theta) = K_m\omega\sin(n_p\theta)\\ e_{Sb} &\triangleq -\frac{d}{dt}\phi_b(\theta) = -K_m\omega\cos(n_p\theta).\end{aligned}$$

The term $R_S\left(i_{Sa}^2+i_{Sb}^2\right)$ represents ohmic losses in the resistance of the phases ultimately appearing as heat while the term $\dfrac{d}{dt}\dfrac{1}{2}L_S\left(i_{Sa}^2+i_{Sb}^2\right)$ represents the power stored in the inductance of the phases. The only other power in the system is the mechanical power $\tau\omega$ and thus the term $-\left(i_{Sa}e_{Sa}+i_{Sb}e_{Sb}\right)$ must be the mechanical power delivered to the rotor. That is,

$$\tau\omega=-\left(i_{Sa}e_{Sa}+i_{Sb}e_{Sb}\right)$$

so that the torque is given by

$$\begin{aligned}\tau &= -\left(i_{Sa}e_{Sa}+i_{Sb}e_{Sb}\right)/\omega\\ &= -i_{Sa}K_m\sin(n_p\theta)+i_{Sb}K_m\cos(n_p\theta).\end{aligned}\tag{9.114}$$

With J the rotor's moment of inertia and τ_L the load torque, the equations of a PM stepper motor are given by

$$\begin{aligned}L_S\frac{di_{Sa}}{dt} &= -R_Si_{Sa}+K_m\omega\sin(n_p\theta)+u_{Sa}\\ L_S\frac{di_{Sb}}{dt} &= -R_Si_{Sb}-K_m\omega\cos(n_p\theta)+u_{Sb}\\ J\frac{d\omega}{dt} &= -i_{Sa}K_m\sin(n_p\theta)+i_{Sb}K_m\cos(n_p\theta)-\tau_L\\ \frac{d\theta}{dt} &= \omega.\end{aligned}$$

Oscillatory Behavior of the Open-Loop PM Stepper Motor

Figure 9.36 shows the oscillatory behavior of a PM stepper motor about it final equilibrium position. To understand the nature of this oscillation, consider the torque expression (9.114) around $\theta=0$. Specifically, let the stepper motor be in the position shown in Figure 9.35(a), where $\theta=0$ and with $u_{Sa}=u_0$, $u_{Sb}=0$, $i_{Sa}=i_0$, $i_{Sb}=0$. Then

$$\tau\approx -i_{Sa}K_m\sin(n_p\theta)\approx -i_0n_pK_m\theta\tag{9.115}$$

for θ close to zero. The slope of the torque vs. θ curve at $\theta=0$ is $-i_0n_pK_m$ which is large for n_p large (e.g., $n_p=50$). That is, a slight disturbance of the rotor position θ from 0 results in a corresponding large increase in the restoring torque. Also, with $\tau_L=-f\omega$ a viscous friction load torque, the equation describing the rotor position for $|\theta|$ small is given by

$$J\frac{d^2\theta}{dt^2}=-\left(i_0n_pK_m\right)\theta-f\frac{d\theta}{dt}$$

or, with $\omega_n^2\triangleq i_0n_pK_m/J, 2\zeta\omega_n\triangleq f/J$, this may be rewritten as

$$\frac{d^2\theta}{dt^2}+2\zeta\omega_n\frac{d\theta}{dt}+\omega_n^2\theta=0.$$

This is simply the equation of a pendulum (simple harmonic motion) with viscous damping and, as such, the response is oscillatory.

9.4.3 High-Performance Control of a PM Stepper Motor

As the mathematical model of the PM stepper motor is identical to that of a standard PM synchronous machine, the control is identical. In fact, the experimental motor used in Section 9.1.4 is a PM stepper motor.

Appendices

Two-Phase Equivalent Parameters

In this Appendix, the calculation of the motor parameters from the manufacturer's data sheet is given. The manufacturer's parameters (Aerotech, Inc., Pgh PA) for their $BM500$ PM synchronous motor are given as follows: line-to-line inductance $L_{\ell-\ell} = 0.028$ H, line-to-line resistance $R_{\ell-\ell} = 0.5$ ohms, line-to-line back-emf constant $K_b^{\ell-\ell} = 23.6$ V(peak)/krpm, torque constant $K_T = 0.28$ N-m/A(rms).

From Chapter 7, equation (7.83), the electrical equations of a three-phase PM synchronous machine are given by ($n_p = 1$)

$$L\frac{di_{S1}}{dt} - \frac{L}{2}\frac{di_{S2}}{dt} - \frac{L}{2}\frac{di_{S3}}{dt} = K\omega\sin(\theta) - R_S i_{S1} + u_{S1} \tag{9.116}$$

$$-\frac{L}{2}\frac{di_{S1}}{dt} + L\frac{di_{S2}}{dt} - \frac{L}{2}\frac{di_{S3}}{dt} = K\omega\sin(\theta - 2\pi/3) - R_S i_{S2} + u_{S2} \tag{9.117}$$

$$-\frac{L}{2}\frac{di_{S1}}{dt} - \frac{L}{2}\frac{di_{S2}}{dt} + L\frac{di_{S3}}{dt} = K\omega\sin(\theta - 4\pi/3) - R_S i_{S3} + u_{S3} \tag{9.118}$$

The line-to-line inductance and resistance measurements are done with the rotor held fixed and applying voltage between phase 1 and 2 with phase 3 open-circuited. Subtracting equation (9.117) from equation (9.116), with $i_{S2} = -i_{S1}, i_{S3} = 0, \omega = 0$ gives

$$3L\frac{di_{S1}}{dt} = u_{S1} - u_{S2} - 2R_S i_{S1}.$$

Consequently, $3L = 0.028$ H and as $L = (2/3)L_S$ one obtains

$$L_S = \frac{3}{2}\frac{0.028}{3}\text{ H } = 0.014\text{ H}$$

while $2R_S = 0.5$ ohm gives

$$R_S = 0.25\text{ ohm}.$$

The line-to-line back-emf voltage is measured by externally driving the motor shaft at constant speed with the phases open circuited and measuring the voltage between two-phases. In this setup, the voltages are given by

$$
\begin{aligned}
u_{S1} &= -K\omega\sin(\theta) \\
u_{S2} &= -K\omega\sin(\theta - 2\pi/3) = -K\omega\Big(-0.5\sin(\theta) - \sqrt{3}/2\cos(\theta)\Big) \\
u_{S3} &= -K\omega\sin(\theta - 4\pi/3) = -K\omega\Big(-0.5\sin(\theta) + \sqrt{3}/2\cos(\theta)\Big)
\end{aligned}
$$

where $\theta = \omega t$. Then

$$
\begin{aligned}
u_{S2} - u_{S3} &= \sqrt{3}K\omega\cos(\theta) \\
\frac{(u_{S2} - u_{S3})_{\text{peak}}}{\omega} &= \sqrt{3}K\,\frac{\text{V}}{\text{rads/sec}}
\end{aligned}
$$

and

$$
K_b^{\ell-\ell} = \sqrt{3}K\frac{2\pi/60 \text{ rads/sec}}{\text{rpm}}\frac{1000 \text{ rpm}}{\text{krpm}}. \tag{9.119}
$$

From Chapter 7 (see the remark on page 456), $K = \sqrt{2/3}K_m$ so that

$$
\begin{aligned}
K_m &= \sqrt{\frac{3}{2}}K = \frac{1}{\sqrt{2}}K_b^{\ell-\ell}\frac{\text{V(peak)}}{\text{krpm}}\frac{\text{krpm}}{1000 \text{ rpm}}\frac{\text{rpm}}{2\pi/60 \text{ rads/sec}} \\
&= \frac{1}{\sqrt{2}}23.6\frac{1}{1000}\frac{60}{2\pi} = 0.159\,\frac{\text{V(peak)}}{\text{rads/sec}}
\end{aligned} \tag{9.120}
$$

To compute the torque constant (which must equal the back-emf constant), consider the steady-state situation in which the currents are given by

$$
\begin{aligned}
i_{S1} &= I_{3ph}\cos(\omega t) \\
i_{S2} &= I_{3ph}\cos(\omega t - 2\pi/3) \\
i_{S3} &= I_{3ph}\cos(\omega t - 4\pi/3)
\end{aligned}
$$

where I_{3ph} is the peak current. The $3-2$ phase transformation results in

$$
\begin{bmatrix} i_{Sa} \\ i_{Sb} \end{bmatrix} = \begin{bmatrix} \sqrt{3/2}I_{3ph}\cos(\omega t) \\ \sqrt{3/2}I_{3ph}\sin(\omega t) \end{bmatrix}
$$

so that the torque is given by

$$
\begin{aligned}
\tau &= -\sqrt{\frac{3}{2}}Ki_{Sa}\sin(n_p\theta) + \sqrt{\frac{3}{2}}Ki_{Sb}\cos(n_p\theta) \\
&= \sqrt{\frac{3}{2}}K\sqrt{\frac{3}{2}}I_{3ph}\sin(\omega t - n_p\theta) \\
&= \sqrt{2}\frac{3}{2}KI_{3ph\_rms}\sin(\omega t - n_p\theta) \\
&= K_T I_{3ph\_rms}\sin(\omega t - n_p\theta)
\end{aligned}
$$

where I_{3ph_rms} is the rms current. The manufacturer's value for the torque constant K_T is the torque per unit of rms phase current which this expression shows is equal to $K_T = \sqrt{2}\frac{3}{2}K$ N-m/A(rms) or

$$K = \frac{1}{\sqrt{2}}\frac{2}{3}K_T.$$

K_T is given in units of N-m/A(rms) while K_m is in N-m/A(peak). Finally, putting in the numerical values gives

$$K_m = \sqrt{\frac{3}{2}}K = \frac{K_T}{\sqrt{3}} = \frac{0.28}{\sqrt{3}}\frac{\text{N-m}}{\text{A(peak)}} = 0.162\frac{\text{N-m}}{\text{A(peak)}}. \tag{9.121}$$

Consequently, computing K_m using the back-emf constant (9.120) gives the same value (within numerical accuracy) as computing K_m using the torque constant (9.121) as it must be.

Current Plots

In this appendix, the procedure used to compute the dq currents given in Figures 9.17–9.24 is described in detail.

Case 1

Here $i_d^* = 0, i_q^* = \pm I_{max}$ and the magnitude of the voltage required to achieve the maximum current I_{max} must satisfy

$$\begin{aligned} u_d^{*2} + u_q^{*2} &= (-n_p\omega L I_{max})^2 + (R_S I_{max} + K_m\omega)^2 \\ &= Z_S^2(\omega) I_{max}^2 + 2R_S K_m \omega I_{\max} + K_m^2\omega^2 \leq V_{\max}^2. \end{aligned}$$

For higher speeds, this constraint cannot be satisfied and the current i_q^* must be reduced below $I_{\max}$. Setting $i_q^* = I$, the maximum current available for i_q^* is found by solving

$$Z_S^2(\omega) I^2 + 2R_S K_m \omega I + K_m^2\omega^2 = V_{\max}^2.$$

Solving,

$$I_\pm(\omega) = \frac{-R_S K_m\omega \pm \sqrt{(R_S K_m\omega)^2 + V_{\max}^2 - Z_S^2(\omega)(K_m\omega)^2}}{Z_S^2(\omega)}$$

where the + sign is for motoring and the − sign is for braking. In summary, $i_d^* \equiv 0$ is the direct current reference and

$$i_q^* = \begin{cases} \pm I_{\max}, & u_d^{*2} + u_q^{*2} \le V_{\max}^2 \\ I_\pm(\omega), & u_d^{*2} + u_q^{*2} \ge V_{\max}^2 \text{ and } \\ & (R_S K_m \omega)^2 + V_{\max}^2 - Z_S^2(\omega)(K_m\omega)^2 > 0 \\ 0 & \text{otherwise} \end{cases}$$

with + for motoring and − braking.

Case 2

Operating under only the voltage constraint, the optimal field weakening current is given by

$$i_d^* = -\frac{(n_p \omega L_S)(K_m \omega)}{Z_S^2(\omega)}$$

and the corresponding quadrature current is

$$i_q^* = \frac{\pm V_{max} Z_S(\omega) - K_m \omega R_S}{Z_S^2(\omega)}.$$

This is the overall optimum as long as

$$i_d^{*2} + i_q^{*2} \le I_{\max}^2.$$

If this constraint is violated for $-I_{\max} < i_d^* < 0$, then the quadrature current is limited to

$$i_q^* = \pm\sqrt{I_{\max}^2 - i_d^{*2}}$$

otherwise $i_q^* = 0$. In summary, with $i_d^* = -(n_p \omega L)(K_m \omega)/Z_S^2(\omega)$ the direct current reference, the corresponding quadrature current is

$$i_q^* = \begin{cases} \dfrac{\pm V_{max} Z_S(\omega) - K_m \omega R_S}{Z_S^2(\omega)}, & i_d^{*2} + i_q^{*2} \le I_{\max}^2 \\ \pm\sqrt{I_{\max}^2 - i_d^{*2}}, & i_d^{*2} + i_q^{*2} \ge I_{\max}^2 \text{ and } -I_{\max} < i_d^* < 0 \\ 0 & \text{otherwise} \end{cases}$$

with + for motoring and − braking.

Case 3

This case gives a direct current reference as $i_d^* = -\sqrt{I_{\max}^2 - i_q^{*2}}$ and the corresponding quadrature current is

$$i_q^* = \begin{cases} \dfrac{-\beta(\omega) \pm \sqrt{\beta^2(\omega) - 4\alpha(\omega)\gamma(\omega)}}{2\alpha(\omega)}, & \beta^2(\omega) - 4\alpha(\omega)\gamma(\omega) \geq 0 \\ 0 & \text{otherwise} \end{cases}$$

with + for motoring and − braking.

Problems

Problem 1 Maximum Torque Under the Voltage Constraint

(a) With the PM synchronous machine at constant speed in steady state so that

$$\begin{aligned} u_d &= R_S i_d - n_p \omega L_S i_q \\ u_q &= R_S i_q + n_p \omega L_S i_d + K_m \omega \\ K_m i_q &= \tau_L \end{aligned}$$

hold, show that the torque is maximized under the constraint $u_d^2 + u_q^2 \leq V_{max}^2$ *by choosing the voltages and currents as in (9.14)–(9.16).*

(b) Explain why in steady state the physical constraints $|u_{Sa}| \leq V_{max}$, $|u_{Sb}| \leq V_{max}$ *are equivalent to* $u_{Sa}^2 + u_{Sb}^2 \leq V_{max}^2$.

Problem 2 State Feedback

Show how to choose the state feedback gain (9.40) such that the state feedback specified by (9.39) places the poles of the system (9.38) at any arbitrary location.

Problem 3 Observer

Show how to choose the gains ℓ_1, ℓ_2 *in (9.43) and (9.44) so that the error state* $e(t) = \begin{bmatrix} e_1(t) & e_2(t) \end{bmatrix}^T \to 0$ *at any given prescribed rate.*

Problem 4 Maximum Torque Reference with $V = V_{max}, I = I_{max}$

Show that $\beta^2(\omega) - 4\alpha(\omega)\gamma(\omega) \geq 0$ *with* $\alpha(\omega)$, $\beta(\omega)$, *and* $\gamma(\omega)$ *given by (9.75), (9.76), and (9.77).*

Problem 5 First Transition Speed

(a) Show that equation (9.82) has one positive root and one negative root.

(b) Show that $|\omega_{1neg}|$, *where* ω_{1neg} *is the real negative root of (9.82), is the transition speed for* $i_q = -I_{max}$.

Problem 6 Open-Loop Simulation of a PM Synchronous Machine

Let a two-phase PM machine have parameter values $n_p = 50$ *rotor teeth,* $I_{max} = 6.0\ A$, $I_{max} = 6.0\ A$, $V_{max} = 40\ V$, $R_S = 0.55\ \Omega$, $L_S = 1.5\ mH$, $K_m = 0.19\ N\text{-}m/A$, $J = 4.5 \times 10^{-5}\ kg\text{-}m^2$, *and* $f = 0.0008\ N\text{-}m/rad/sec$. *Simulate the machine based on equations (9.1). Can you get it to run open loop?*

Problem 7 Closed-Loop Control of a PM Synchronous Machine

Using the parameter values of problem 6, simulate the complete closed-loop controller of Figure 9.1. Do this for two trajectories.

(a) Do a fast turn of the motor from $\theta = 0$ *to* $\theta = 0.9\pi$ *radians using the trajectories of Figures 9.3–9.8.*

(b) Do a high speed move from 0 *to* 25π *radians using the trajectories of Figures 9.9–9.13.*

Problem 8 Simulation of a PM Synchronous Machine Producing Maximum Torque

Let a three-phase PM synchronous motor have parameters values given by $L = 9.3 \times 10^{-4}\ H$, $R = 0.25\ ohms$, $J = 13.9 \times 10^{-5}\ kg\text{-}m^2$, $K = 0.198\ N\text{-}m/A$, $n_p = 4$. *The limits are given by* i_{max} *(continuous)* $= 18\ A$, i_{max} *(peak)* $= 55\ A$, *and* V_{bus} *(bus voltage)* $= 160\ V$. *The motor bearings are rated for* $10,000$ *rpm or about* 1000 *rads/sec which is taken to be the maximum speed of the motor. The corresponding equivalent two-phase parameters are* $L_S = L + L/2 = 14 \times 10^{-4}\ H$, $K_m = (\sqrt{3/2})K = 0.243\ N\text{-}m/A$, I_{max}*(continuous)* $= (\sqrt{3/2})i_{max}$ *(continuous)* $= 22\ A$, I_{max} *(peak)* $= \sqrt{3/2}i_{max}$ *(peak)* $= 67.4\ A$, $V_{max} = \sqrt{3/2}(2/\pi)V_{bus} = 124.8\ V$.

Do a simulation that implements the block diagram of Figure 9.25. Run the simulation for both the continuous current limit and the peak current limit.

Problem 9 Controlling Two PM Motors with One Amplifier [122]

Consider a situation in which two identical PM synchronous machines are connected in parallel to the same amplifier. This means that the same voltages u_{Sa}, u_{Sb} *are simultaneously applied to both motors. The dq voltages of each motor are given by*

$$\begin{bmatrix} u_{d1} \\ u_{q1} \end{bmatrix} = \begin{bmatrix} \cos(n_p\theta_1) & \sin(n_p\theta_1) \\ -\sin(n_p\theta_1) & \cos(n_p\theta_1) \end{bmatrix} \begin{bmatrix} u_{Sa} \\ u_{Sb} \end{bmatrix}$$

$$\begin{bmatrix} u_{d2} \\ u_{q2} \end{bmatrix} = \begin{bmatrix} \cos(n_p\theta_2) & \sin(n_p\theta_2) \\ -\sin(n_p\theta_2) & \cos(n_p\theta_2) \end{bmatrix} \begin{bmatrix} u_{Sa} \\ u_{Sb} \end{bmatrix}$$

and note that these dq voltages are not the same as the angle of the two motors may not be equal. The model of the two motors in the dq-coordinate system are then

$$\begin{aligned} L\frac{di_{d1}}{dt} &= -R_S i_{d1} + n_p\omega_1 L i_{q1} + v_{d1} \\ L\frac{di_{q1}}{dt} &= -R_S i_{q1} - n_p\omega_1 L i_{d1} - \omega_1 K_{eq} + v_{q1} \\ J\frac{d\omega_1}{dt} &= K_{eq} i_{q1} - \tau_{L1} \\ \frac{d\theta_1}{dt} &= \omega_1 \end{aligned}$$

and

$$\begin{aligned} L\frac{di_{d2}}{dt} &= -R_S i_{d2} + n_p\omega_2 L i_{q2} + v_{d2} \\ L\frac{di_{q2}}{dt} &= -R_S i_{q2} - n_p\omega_2 L i_{d2} - \omega_2 K_{eq} + v_{q2} \\ J\frac{d\omega_2}{dt} &= K_{eq} i_{q2} - \tau_{L2} \\ \frac{d\theta_2}{dt} &= \omega_2. \end{aligned}$$

Show that using the feedback

$$\begin{bmatrix} v_{q1} \\ v_{q2} \end{bmatrix} = \begin{bmatrix} -\sin(n_p\theta_1) & \cos(n_p\theta_1) \\ -\sin(n_p\theta_2) & \cos(n_p\theta_2) \end{bmatrix} \begin{bmatrix} u_{Sa} \\ u_{Sb} \end{bmatrix}$$

and current command control

$$\begin{aligned} v_{q1} &= K_p(i_{q1ref} - i_{q1}) + K_I \int_0^t (i_{q1ref} - i_{q1})dt \\ v_{q2} &= K_p(i_{q2ref} - i_{q2}) + K_I \int_0^t (i_{q2ref} - i_{q2})dt \end{aligned}$$

that the position and speed of each motor can be forced to track a trajectory. Is there any singularity in the controller? How should i_{q1ref}, i_{q2ref} be specified?

Problem 10 Half-Stepping of a PM Stepper Motor

Assume that the stepper motor starts out in the position shown in Figure

9.35. Fill in the table below to show the stepping of the motor.

u_{Sa}	u_{Sb}	θ *(degrees)*
u_0	0	0
u_0	u_0	
0	u_0	
$-u_0$	u_0	
$-u_0$	0	
$-u_0$	$-u_0$	
0	$-u_0$	
u_0	$-u_0$	
u_0	0	

Problem 11 PM Synchronous Motor with a Salient Rotor [51]

Some PM synchronous motors are constructed in such a way that the rotor exhibits significant saliency due to a nonuniform air gap. In this case, the flux linkages in the stator phases due to the stator currents depends on the rotor position. Specifically, the stator flux linkages are given by

$$\begin{aligned}\lambda_{Sa} &= L_S i_{Sa} + K_m \cos(n_p\theta) + L_g\left(i_{Sa}\cos(2n_p\theta) + i_{Sb}\sin(2n_p\theta)\right)\\ \lambda_{Sb} &= L_S i_{Sb} + K_m \sin(n_p\theta) + L_g\left(i_{Sa}\sin(2n_p\theta) - i_{Sb}\cos(2n_p\theta)\right)\end{aligned}$$

where the third term in each expression is due to the nonuniform air gap (saliency) of the rotor. Notice that these terms have a period of π/n_p while the flux linkage due to the permanent magnet rotor has a period of $2\pi/n_p$.

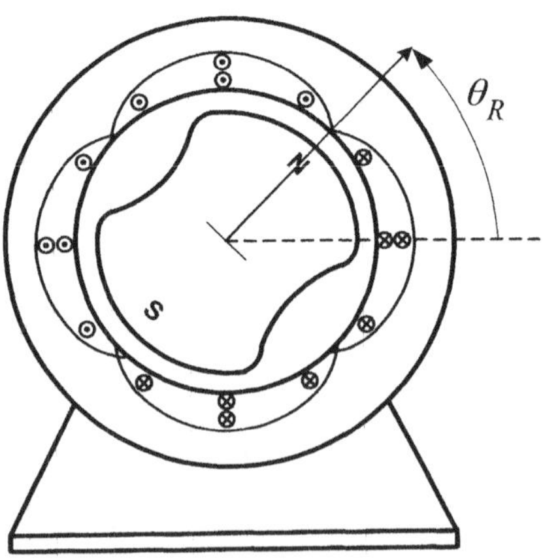

FIGURE 9.37. PM sychronous motor with a salient rotor ($n_p = 1$).

(a) Use a conservation of energy (power) argument to show that the

motor torque is given by

$$\begin{aligned}\tau &= 2n_pL_g\Big((i_{Sa}^2 - i_{Sb}^2)\sin(2n_p\theta) + 2i_{Sa}i_{Sb}\cos(2n_p\theta)\Big)\\ &\quad + n_pK_m\Big(-i_{Sa}\sin(n_p\theta) + i_{Sb}\cos(n_p\theta)\Big).\end{aligned}$$

(b) Show that using the dq transformation given by

$$\begin{aligned}\begin{bmatrix} i_d \\ i_q \end{bmatrix} &= \begin{bmatrix} \cos(n_p\theta) & \sin(n_p\theta) \\ -\sin(n_p\theta) & \cos(n_p\theta) \end{bmatrix}\begin{bmatrix} i_{Sa} \\ i_{Sb} \end{bmatrix}\\ \begin{bmatrix} u_d \\ u_q \end{bmatrix} &= \begin{bmatrix} \cos(n_p\theta) & \sin(n_p\theta) \\ -\sin(n_p\theta) & \cos(n_p\theta) \end{bmatrix}\begin{bmatrix} u_{Sa} \\ u_{Sb} \end{bmatrix}\end{aligned}$$

that the equations of the salient PM synchronous motor are given by

$$\begin{aligned}L_d\frac{di_d}{dt} &= -Ri_d - n_p\omega L_qi_q + u_d\\ L_q\frac{di_q}{dt} &= -Ri_q - n_p\omega L_di_d - n_p\omega K_m + u_q\\ J\frac{d\omega}{dt} &= n_p\left(K_mi_q + (L_d - L_q)i_qi_d\right)\\ \frac{d\theta}{dt} &= \omega\end{aligned}$$

where

$$\begin{aligned}L_d &\triangleq L_S + L_g\\ L_q &\triangleq L_S - L_g.\end{aligned}$$

(c) Consider the change of coordinates

$$\begin{aligned}\theta &= \theta\\ \omega &= \omega\\ \alpha &= \frac{n_p}{J}\left(K_mi_q + (L_d - L_q)i_qi_d\right)\\ i_d &= i_d\end{aligned}$$

and show that the motor model is then of the form

$$\begin{aligned}\frac{d\theta}{dt} &= \omega\\ \frac{d\omega}{dt} &= \alpha\\ \frac{d\alpha}{dt} &= f_1(\omega, i_d, i_q) + 2L_gn_pu_d/L_d + n_p\left(K_m + 2L_gi_d\right)u_q/L_q\\ \frac{di_d}{dt} &= f_2(\omega, i_d, i_q) + u_d/L_d.\end{aligned}$$

Give the explicit expressions for f_1 *and* f_2.

(d) Show that the nonlinear feedback

$$\begin{bmatrix} u_d \\ u_q \end{bmatrix} = \begin{bmatrix} \dfrac{2L_g n_p}{L_d} & \dfrac{n_p\left(K_m + 2L_g i_d\right)}{L_q} \\ \dfrac{n_p\left(K_m + 2L_g i_d\right)}{L_q} & 0 \end{bmatrix}^{-1} \begin{bmatrix} -f_1 + v_d \\ -f_2 + v_q \end{bmatrix}$$

results in a linear *system.*

(e) Under what conditions is the nonlinear feedback of part (d) singular? Is it a practical problem? Explain.

10

Trapezoidal Back-Emf PM Synchronous Motors (BLDC)

In this chapter a mathematical model of a trapezoidal back-emf synchronous motor is derived and a control scheme for it is developed. This machine is also known as the brushless DC (BLDC) motor, and this terminology will be explained in Section 10.8.1. Despite the name "brushless DC motor", this is a PM *synchronous* machine where the magnetic fields are *uniformly* distributed in the air gap (rather than sinusoidally distributed). With the motor running at constant speed, this results in a back emf that has a trapezoidal shape in time, and hence the name *trapezoidal back emf* motor.

10.1 Construction

Figure 10.1 shows the basic construction of a three-phase BLDC motor [25]. Figure 10.2 is a redrawing of Figure 10.1 without the rotor so as to show the layout of the three stator phases more clearly.

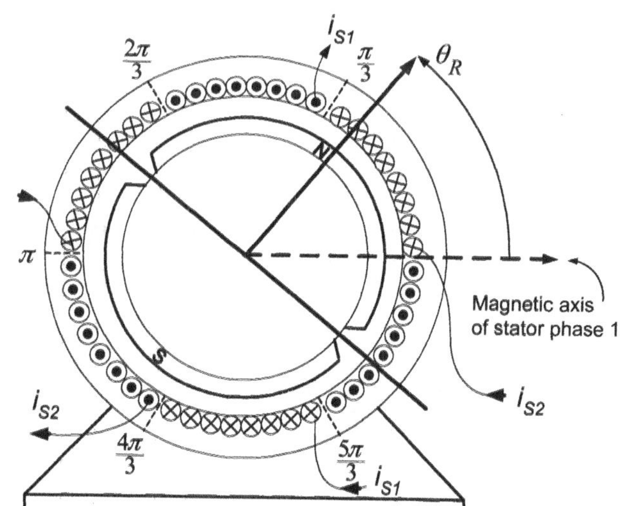

FIGURE 10.1. Construction of a PM trapezoidal back-emf machine. Adapted from Mohan [25].

Figure 10.2 shows that there are three sets of windings which are uniformly wound[1] on the inside surface of the stator. Phase 1 with current i_{S1} has its turns uniformly wound between $\pi/3$ and $2\pi/3$ on one side of the stator's inside surface and between $4\pi/3$ and $5\pi/3$ on the other side of the stator. Similarly, phase 2 with current i_{S2} has its turns uniformly wound between π and $4\pi/3$ and between 0 and $\pi/3$ while phase 3 with current i_{S3} has its turns uniformly wound between $2\pi/3$ and π and between $5\pi/3$ and 2π.

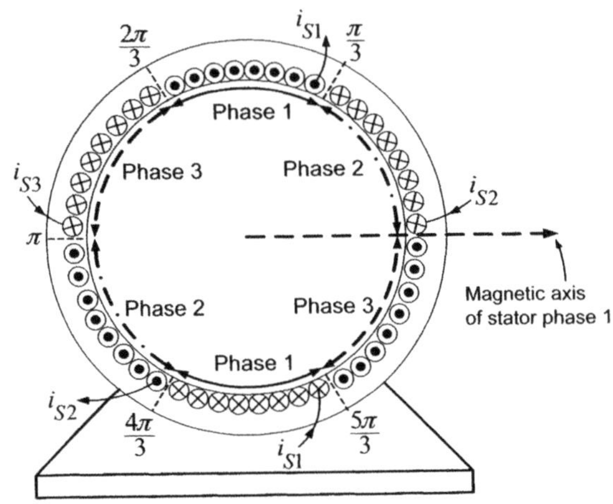

FIGURE 10.2. Three uniformly wound stator phases in a BLDC machine.

Figure 10.3 shows the PM rotor of Figure 10.1 by itself [25].

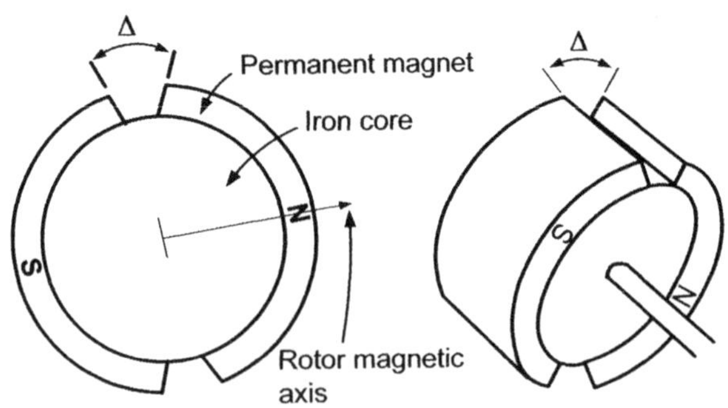

FIGURE 10.3. Rotor of a single pole-pair PM magnet rotor for a BLDC motor. Adapted from Mohan [25].

---

[1] By *uniformly wound*, it is meant that the number of turns between θ and $\theta + d\theta$ is constant independent of the angle θ.

The magnetic axis of the rotor as well as the angle Δ between the north and south poles are indicated. The permanent magnet rotor is constructed by bonding permanent magnets on the surface of a cylindrical core of laminated soft iron. As usual, the rotor position is taken to be along the magnetic axis of the rotor (see Figure 10.3). The permanent magnets on the rotor set up a uniformly (in θ) distributed magnetic field in the air gap given by

$$\vec{\mathbf{B}}_R(r,\theta-\theta_R) = \begin{cases} B_{R0}\dfrac{r_R}{r}\hat{\mathbf{r}} & \text{for} \quad -\dfrac{\pi}{2}+\dfrac{\Delta}{2} \le \theta-\theta_R \le \dfrac{\pi}{2}-\dfrac{\Delta}{2} \\ -B_{R0}\dfrac{r_R}{r}\dfrac{(\theta-\theta_R-\pi/2)}{\Delta/2}\hat{\mathbf{r}} & \text{for} \quad \dfrac{\pi}{2}-\dfrac{\Delta}{2} \le \theta-\theta_R \le \dfrac{\pi}{2}+\dfrac{\Delta}{2} \\ -B_{R0}\dfrac{r_R}{r}\hat{\mathbf{r}} & \text{for} \quad \dfrac{\pi}{2}+\dfrac{\Delta}{2} \le \theta-\theta_R \le \dfrac{3\pi}{2}-\dfrac{\Delta}{2} \\ B_{R0}\dfrac{r_R}{r}\dfrac{(\theta-\theta_R-3\pi/2)}{\Delta/2}\hat{\mathbf{r}} & \text{for} \quad \dfrac{3\pi}{2}-\dfrac{\Delta}{2} \le \theta-\theta_R \le \dfrac{3\pi}{2}+\dfrac{\Delta}{2} \end{cases} \tag{10.1}$$

where the parameter Δ represents the angular distance between the north and south pole magnets on the rotor. Also, the factor r_R/r ensures that conservation of flux holds in the air gap. The radial magnetic field distribution in the air gap due to the rotor's permanent magnet is illustrated in Figure 10.4.

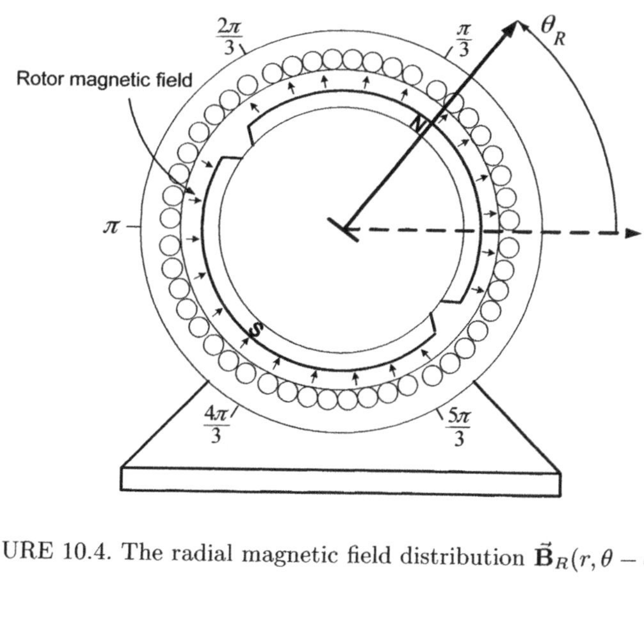

FIGURE 10.4. The radial magnetic field distribution $\vec{\mathbf{B}}_R(r,\theta-\theta_R)$.

In Sections 10.2 through 10.7, Δ will be taken to be zero to simplify the analysis with no significant impact on the resulting mathematical model. In this case, (10.1) simplifies to

$$\vec{\mathbf{B}}_R(r,\theta-\theta_R)=\begin{cases} B_{R0}\dfrac{r_R}{r}\hat{\mathbf{r}} & \text{for} \quad -\pi/2 \le \theta-\theta_R \le \pi/2 \\ -B_{R0}\dfrac{r_R}{r}\hat{\mathbf{r}} & \text{for} \quad +\pi/2 \le \theta-\theta_R \le 3\pi/2. \end{cases} \tag{10.2}$$

10.2 Stator Magnetic Field $\vec{\mathbf{B}}_S$

The magnetic field $\vec{\mathbf{B}}_S$ produced stator currents is now derived. As illustrated in Figure 10.1, the stator winding density of a BLDC is uniform over inner periphery of the stator. Specifically, stator phase 1 has a winding density given by

$$N_{S1}(\theta) = \begin{cases} \dfrac{N_S}{\pi/3} & \text{for } \dfrac{\pi}{3} \le \theta \le \dfrac{2\pi}{3} \text{ and } \dfrac{4\pi}{3} \le \theta \le \dfrac{5\pi}{3} \\ 0 & \text{elsewhere.} \end{cases}$$

The total number of windings (turns or loops) making up phase 1 is then

$$\int_{\pi/3}^{2\pi/3} \frac{N_S}{\pi/3} d\theta = N_S.$$

Similarly, $N_{S2}(\theta) = N_{S1}(\theta - 2\pi/3)$ and $N_{S3}(\theta) = N_{S1}(\theta - 4\pi/3)$. To determine the radial magnetic field in the air gap produced by the current in stator phase 1, Ampère's law with $\vec{\mathbf{H}} \equiv 0$ in the iron is applied to the path 1–2–3–1 of Figure 10.5 to obtain [by symmetry, $H_{S1}(i_{S1}, \pi+\theta) = -H_{S1}(i_{S1},\theta)$]

$$2gH_{S1}(i_{S1},\theta) = \begin{cases} N_S i_{S1}, & -\pi/3 \le \theta \le \pi/3 \\ \displaystyle\int_{\theta}^{2\pi/3} \frac{N_S i_{S1}}{\pi/3} d\theta - \int_{4\pi/3}^{\theta+\pi} \frac{N_S i_{S1}}{\pi/3} d\theta, & \pi/3 \le \theta \le 2\pi/3 \\ -N_S i_{S1}, & 2\pi/3 \le \theta \le 4\pi/3 \\ \displaystyle\int_{\theta}^{5\pi/3} -\frac{N_S i_{S1}}{\pi/3} d\theta + \int_{7\pi/3}^{\theta+\pi} \frac{N_S i_{S1}}{\pi/3} d\theta, & 4\pi/3 \le \theta \le 5\pi/3 \end{cases}$$

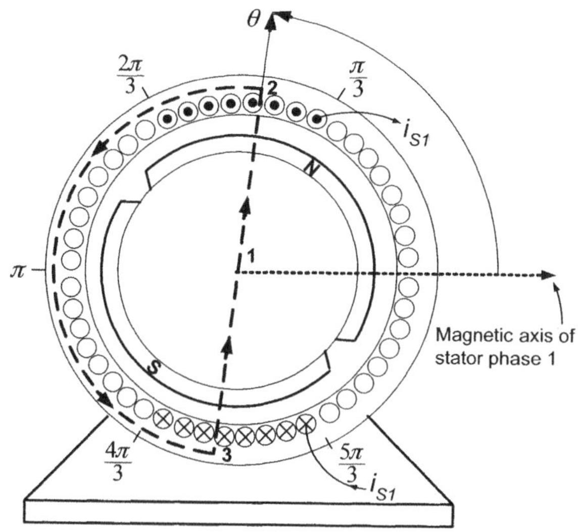

FIGURE 10.5. Path 1–2–3–1 for applying Ampère's law. Note that for the path drawn, $\pi/3 \le \theta \le 2\pi/3$.

or

$$\vec{\mathbf{H}}_{S1}(i_{S1},\theta) = \begin{cases} \dfrac{N_S i_{S1}}{2g}\hat{\mathbf{r}} & \text{for} \quad -\pi/3 \le \theta \le \pi/3 \\[2ex] \dfrac{N_S i_{S1}}{2g}\dfrac{6}{\pi}\left(\dfrac{\pi}{2}-\theta\right)\hat{\mathbf{r}} & \text{for} \quad \pi/3 \le \theta \le 2\pi/3 \\[2ex] -\dfrac{N_S i_{S1}}{2g}\hat{\mathbf{r}} & \text{for} \quad 2\pi/3 \le \theta \le 4\pi/3 \\[2ex] \dfrac{N_S i_{S1}}{2g}\dfrac{6}{\pi}\left(\theta-\dfrac{3\pi}{2}\right)\hat{\mathbf{r}} & \text{for} \quad 4\pi/3 \le \theta \le 5\pi/3 \end{cases}$$

where it was assumed that $\vec{\mathbf{H}}_{S1}$ is constant across the air gap. Then $\vec{\mathbf{B}}_{S1} = \mu_0\vec{\mathbf{H}}_{S1}$ in the air gap. A factor of r_R/r is now included to ensure that $\vec{\mathbf{B}}_{S1}$ satisfies conservation of flux in the air gap. Consequently, the magnetic field $\vec{\mathbf{B}}_{S1} = B_{S1}(i_{S1}, r, \theta)\hat{\mathbf{r}}$ at any point (r,θ) in the air gap due to the current

i_{S1} in phase 1 is given by

$$B_{S1}(i_{S1},r,\theta)\hat{\mathbf{r}} = \begin{cases} \dfrac{\mu_0 N_S i_{S1}}{2g}\dfrac{r_R}{r}\hat{\mathbf{r}} & \text{for} \quad -\pi/3 \le \theta \le \pi/3 \\ \dfrac{\mu_0 N_S i_{S1}}{2g}\dfrac{r_R}{r}\dfrac{6}{\pi}\left(\dfrac{\pi}{2}-\theta\right)\hat{\mathbf{r}} & \text{for} \quad \pi/3 \le \theta \le 2\pi/3 \\ -\dfrac{\mu_0 N_S i_{S1}}{2g}\dfrac{r_R}{r}\hat{\mathbf{r}} & \text{for} \quad 2\pi/3 \le \theta \le 4\pi/3 \\ \dfrac{\mu_0 N_S i_{S1}}{2g}\dfrac{r_R}{r}\dfrac{6}{\pi}\left(\theta-\dfrac{3\pi}{2}\right)\hat{\mathbf{r}} & \text{for} \quad 4\pi/3 \le \theta \le 5\pi/3. \end{cases} \tag{10.3}$$

Similarly,

$$\begin{aligned} B_{S2}(i_{S2},r,\theta) &= B_{S1}(i_{S2},r,\theta-2\pi/3) \\ B_{S3}(i_{S3},r,\theta) &= B_{S1}(i_{S3},r,\theta-4\pi/3). \end{aligned}$$

A plot of the normalized radial magnetic field $B_{S1}(i_{S1},r,\theta)/\left(\frac{\mu_0 N_S i_{S1}}{2g}\frac{r_R}{r}\right)$ due to the current in stator phase 1 for $-\pi/3 \le \theta \le 2\pi$ is in Figure 10.6.

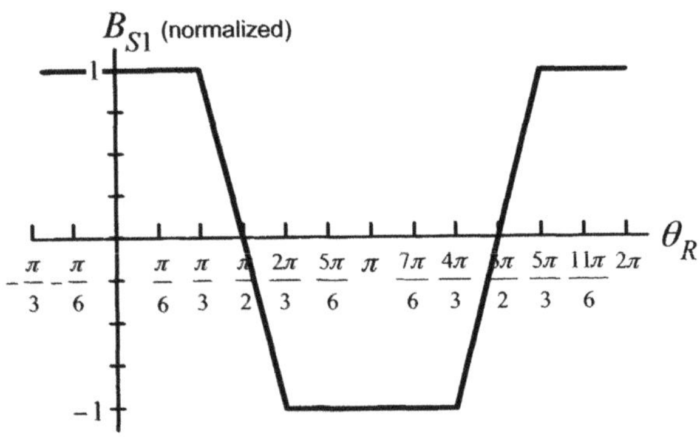

FIGURE 10.6. Normalized magnetic field $B_{S1}(i_{S1},r,\theta)/\left(\frac{\mu_0 N_S i_{S1}}{2g}\frac{r_R}{r}\right)$.

This is also illustrated in Figure 10.7 which shows the direction of radial magnetic field distribution $B_{S1}(i_{S1},r,\theta)$ in the air gap at each angular position θ. In particular, the magnetic field due to i_{S1} is radially out to the right of the dashed vertical line and radially in to the left of the vertical line.

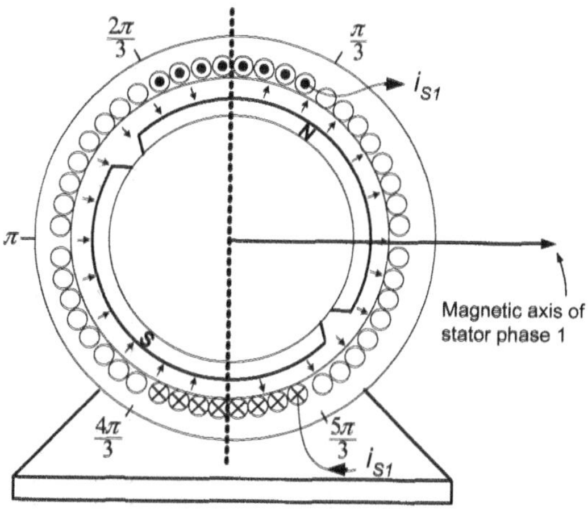

FIGURE 10.7. The radial magnetic field distribution $\vec{\mathbf{B}}_{S1}(i_{S1}, r, \theta)$ in the air gap due to the current i_{S1} in stator phase 1.

10.3 Stator Flux Linkage Produced by $\vec{\mathbf{B}}_S$

The flux $\phi_{11}(i_{S1}, \theta)$ in a winding of stator phase 1 at the angle θ where $\pi/3 \le \theta \le 2\pi/3$, due to the current in stator phase 1, is now computed using the flux surface shown in Figure 10.8.

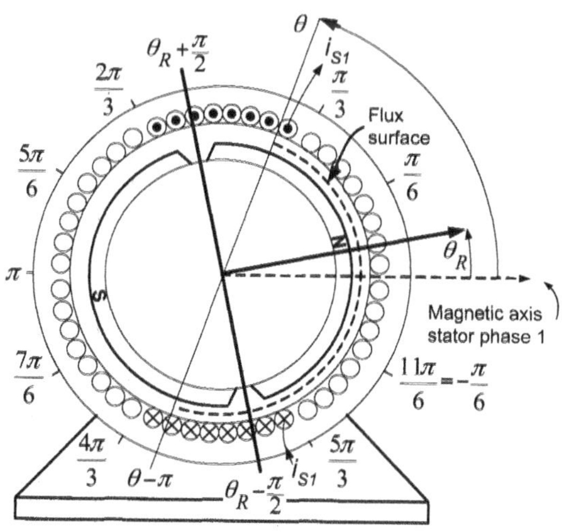

FIGURE 10.8. Flux surface for a winding of stator phase 1 to calculate the flux produced by the magnetic field of i_{S1}.

Using the expression (10.3) for $\vec{\mathbf{B}}_{S1}$, the flux produced by $\vec{\mathbf{B}}_{S1}$ in a winding of stator phase 1 with one side at θ and the other at $\theta-\pi$ is given by

$$\begin{aligned}
\phi_{11}(i_{S1},\theta) &= \int_0^{\ell_1}\int_{\theta-\pi}^{\theta} \vec{\mathbf{B}}_{S1}(i_{S1},r_S,\theta')\cdot(r_S d\theta' d\ell\hat{\mathbf{r}})\\
&= r_S\ell_1\int_{-\pi/3}^{\pi/3}\frac{\mu_0 N_S i_{S1}}{2g}\frac{r_R}{r_S}d\theta' + r_S\ell_1\int_{\pi/3}^{\theta}\frac{\mu_0 N_S i_{S1}}{g\pi/3}\frac{r_R}{r_S}\left(\frac{\pi}{2}-\theta'\right)d\theta'\\
&\quad + r_S\ell_1\int_{\theta+\pi}^{5\pi/3}\frac{\mu_0 N_S i_{S1}}{g\pi/3}\frac{r_R}{r_S}\left(\theta'-\frac{3\pi}{2}\right)d\theta'\\
&= \frac{\mu_0 r_R\ell_1 N_S i_{S1}}{2g}\left(\frac{2\pi}{3}-\frac{1}{2}\frac{6}{\pi}\left(\frac{\pi}{2}-\theta'\right)^2\Big|_{\pi/3}^{\theta}+\frac{1}{2}\frac{6}{\pi}\left(\theta'-\frac{3\pi}{2}\right)^2\Big|_{\theta+\pi}^{5\pi/3}\right)\\
&= \frac{\mu_0 r_R\ell_1 N_S i_{S1}}{2g}\left(\frac{2\pi}{3}-\frac{3}{\pi}\left(\left(\frac{\pi}{2}-\theta\right)^2-\left(\frac{\pi}{6}\right)^2\right)\right.\\
&\quad\left.+\frac{3}{\pi}\left(\left(\frac{\pi}{6}\right)^2-\left(\theta-\frac{\pi}{2}\right)^2\right)\right)\\
&= \frac{\mu_0 r_R\ell_1 N_S i_{S1}}{2g}\left(-\frac{6}{\pi}\left(\theta-\frac{\pi}{2}\right)^2+\frac{5\pi}{6}\right).
\end{aligned}$$

As the outward normal was used to compute the flux, $-d\phi_{11}/dt>0$ means that this induced emf in stator phase 1 will push current in the same direction as that chosen for positive current flow in stator phase 1.

The total flux linkage $\lambda_{S1}(i_{S1},0,0)$ in stator phase 1 due to the current in phase 1, is then found by adding up the fluxes in each winding of phase 1. That is,

$$\begin{aligned}
\lambda_{S1}(i_{S1},0,0) &= \int_{\pi/3}^{2\pi/3}\phi_{11}(i_{S1},\theta)\frac{N_S}{\pi/3}d\theta\\
&= \frac{\mu_0 r_R\ell_1 N_S^2 i_{S1}}{2g}\frac{3}{\pi}\int_{\pi/3}^{2\pi/3}\left(-\frac{6}{\pi}\left(\theta-\frac{\pi}{2}\right)^2+\frac{5\pi}{6}\right)d\theta\\
&= \frac{\mu_0 r_R\ell_1 N_S^2 i_{S1}}{2g}\left(-\frac{18}{\pi^2}\frac{1}{3}\left(\theta-\frac{\pi}{2}\right)^3\Big|_{\pi/3}^{2\pi/3}+\frac{5}{2}\left(\frac{2\pi}{3}-\frac{\pi}{3}\right)\right)\\
&= \frac{\mu_0 r_R\ell_1 N_S^2 i_{S1}}{2g}\left(-\frac{\pi}{18}+\frac{5}{2}\frac{\pi}{3}\right)=\frac{\mu_0 r_R\ell_1\pi N_S^2 i_{S1}}{2g}\left(\frac{14}{18}\right)\\
&= L_S i_{S1}
\end{aligned}$$

where ($r_R=\ell_2/2$)

$$L_S \triangleq \left(\frac{7}{3}\right)\frac{\mu_0\ell_1\ell_2\pi}{12g}N_S^2.$$

The flux $\phi_{21}(i_{S1},\theta)$ in a winding of stator phase 2 at an angle θ ($\pi\le\theta\le 4\pi/3$), produced by the magnetic field of stator phase 1, is computed using the flux surface shown in Figure 10.9. With $\pi \le \theta \le 4\pi/3$, this flux is computed as

$$\begin{aligned}
\phi_{21}(i_{S1},\theta) &= \int_0^{\ell_1}\int_{\theta-\pi}^{\theta} \vec{\mathbf{B}}_{S1}(i_{S1},r_S,\theta')\cdot\left(r_S d\theta' d\ell\hat{\mathbf{r}}\right)\\
&= r_S\ell_1\int_{\theta-\pi}^{\pi/3}\frac{\mu_0 N_S i_{S1}}{2g}\frac{r_R}{r_S}d\theta' + r_S\ell_1\int_{\pi/3}^{2\pi/3}\frac{\mu_0 N_S i_{S1}}{g\pi/3}\frac{r_R}{r_S}\left(\frac{\pi}{2}-\theta'\right)d\theta'\\
&\quad + r_S\ell_1\int_{2\pi/3}^{\theta}-\frac{\mu_0 N_S i_{S1}}{2g}\frac{r_R}{r_S}d\theta'\\
&= \frac{\mu_0 r_R\ell_1 N_S i_{S1}}{2g}\left(\frac{4\pi}{3}-\theta-\frac{6}{\pi}\frac{1}{2}\left(\theta'-\frac{\pi}{2}\right)^2\Big|_{\pi/3}^{2\pi/3}-\left(\theta-\frac{2\pi}{3}\right)\right)\\
&= \frac{\mu_0 r_R\ell_1 N_S i_{S1}}{g}(\pi-\theta).
\end{aligned} \tag{10.4}$$

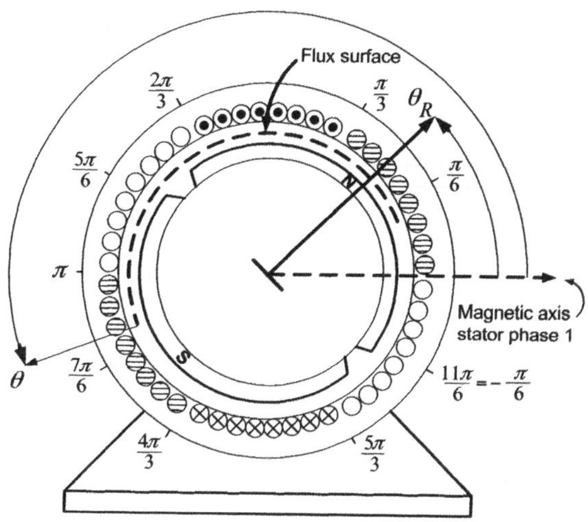

FIGURE 10.9. Flux surface for a winding of stator phase 2 to calculate the flux produced by the magnetic field of i_{S1}. The windings of phase 2 are shown shaded.

The total flux linkage $\lambda_{S2}(i_{S1},0,0)$ in stator phase 2 produced by the current in stator phase 1 is then

$$\begin{aligned}
\lambda_{S2}(i_{S1},0,0) &= \int_{\pi}^{4\pi/3}\phi_{21}(i_{S1},\theta)\frac{N_S}{\pi/3}d\theta = -\frac{3}{\pi}\frac{\mu_0 r_R\ell_1 N_S^2}{g}i_{S1}\frac{1}{2}(\pi-\theta)^2\Big|_{\pi}^{4\pi/3}\\
&= -\frac{\mu_0 r_R\ell_1\pi N_S^2}{6g}i_{S1} = -Mi_{S1}
\end{aligned}$$

where $(r_R = \ell_2/2)$

$$M \triangleq \frac{\mu_0 \ell_1 \ell_2 \pi}{12g} N_S^2.$$

The flux $\phi_{31}(i_{S1}, \theta)$ in a winding of stator phase 3 at an angle θ where $5\pi/3 \le \theta \le 2\pi$ produced by the magnetic field of stator phase 1 is computed using the flux surface shown in Figure 10.10.

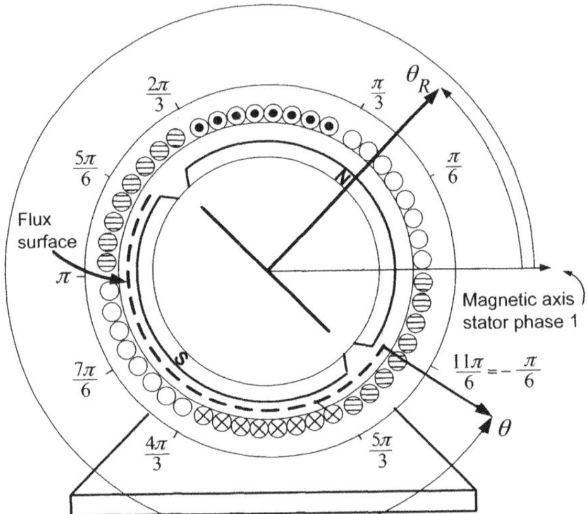

FIGURE 10.10. Flux surface for a winding of stator phase 3 to calculate the flux produced by the magnetic field of i_{S1}. The windings of phase 3 are shown shaded.

Proceeding, with $5\pi/3 \le \theta \le 2\pi$,

$$\begin{aligned}
\phi_{31}(i_{S1}, \theta) &= \int_0^{\ell_1} \int_{\theta-\pi}^{\theta} \vec{\mathbf{B}}_{S1}(i_{S1}, r_S, \theta') \cdot (r_S d\theta' d\ell \hat{\mathbf{r}}) \\
&= r_S \ell_1 \int_{\theta-\pi}^{4\pi/3} -\frac{\mu_0 N_S i_{S1}}{2g} \frac{r_R}{r_S} d\theta' + r_S \ell_1 \int_{4\pi/3}^{5\pi/3} \frac{\mu_0 N_S i_{S1}}{g\pi/3} \frac{r_R}{r} \left(\theta' - \frac{3\pi}{2}\right) d\theta' \\
&\quad + r_S \ell_1 \int_{5\pi/3}^{\theta} \frac{\mu_0 N_S i_{S1}}{2g} \frac{r_R}{r_S} d\theta' \\
&= \frac{\mu_0 r_R \ell_1 N_S i_{S1}}{2g} \left(-\left(\frac{7\pi}{3} - \theta\right) + \frac{6}{\pi} \frac{1}{2} \left(\theta' - \frac{3\pi}{2}\right)^2 \Bigg|_{4\pi/3}^{5\pi/3} + \left(\theta - \frac{5\pi}{3}\right) \right) \\
&= \frac{\mu_0 r_R \ell_1 N_S i_{S1}}{g} (\theta - 2\pi).
\end{aligned} \tag{10.5}$$

The total flux linkage in stator phase 3 produced by the current in stator phase 1 is then

$$\begin{aligned}\lambda_{S3}(i_{S1},0,0) &= \int_{5\pi/3}^{2\pi}\phi_{31}(i_{S1},\theta)\frac{N_S}{\pi/3}d\theta = \frac{3}{\pi}\frac{\mu_0 r_R \ell_1 N_S^2}{g}i_{S1}\frac{1}{2}\left(\theta-2\pi\right)^2\Big|_{5\pi/3}^{2\pi}\\ &= -\frac{\mu_0 r_R \ell_1 \pi N_S^2}{6g}i_{S1} = -Mi_{S1}\end{aligned}$$

with

$$M \triangleq \frac{\mu_0 \ell_1 \ell_2 \pi}{12g}N_S^2.$$

The other phases are computed similarly. In summary, the flux linkages in the stator phases due to the stator currents are given by

$$\begin{aligned}\lambda_{S1}(i_{S1},i_{S2},i_{S3}) &= +L_S i_{S1} - Mi_{S2} - Mi_{S3}\\ \lambda_{S2}(i_{S1},i_{S2},i_{S3}) &= -Mi_{S1} + L_S i_{S2} - Mi_{S3}\\ \lambda_{S3}(i_{S1},i_{S2},i_{S3}) &= -Mi_{S1} - Mi_{S2} + L_S i_{S3}\end{aligned}$$

where

$$L_S = \frac{7}{3}\frac{\mu_0 \ell_1 \ell_2 \pi}{12g}N_S^2, \quad M = \frac{\mu_0 \ell_1 \ell_2 \pi}{12g}N_S^2. \tag{10.6}$$

In matrix form, the flux linkages are given by

$$\begin{bmatrix}\lambda_{S1}\\ \lambda_{S2}\\ \lambda_{S3}\end{bmatrix} = \begin{bmatrix}L_S & -M & -M\\ -M & L_S & -M\\ -M & -M & L_S\end{bmatrix}\begin{bmatrix}i_{S1}\\ i_{S2}\\ i_{S3}\end{bmatrix}.$$

The inverse of the inductance matrix on the right is

$$\frac{1}{(L_S-2M)(M+L_S)}\begin{bmatrix}L_S-M & M & M\\ M & L_S-M & M\\ M & M & L_S-M\end{bmatrix}$$

where by (10.6), $L_S = (7/3)M$ results in $(L_S-2M)(M+L_S) > 0$.

10.4 Stator Flux Linkage Produced by $\vec{\mathbf{B}}_R$

The flux linkage $\lambda_{S1\_R}(\theta_R)$ in the windings of phase 1 due to the magnetic field of the PM rotor is now computed. The first step is to compute the flux $\phi_{S1\_R}(i_{S1},\theta)$ in each of the windings of stator phase 1 produced by the rotor's magnetic field. Recall that the top sides of the windings of phase 1 are located in the interval $\pi/3 \le \theta \le 2\pi/3$.

Case 1. As Figure 10.11 illustrates, in this case the computation of the flux in a loop of stator phase 1 produced by the rotor's magnetic field is done with rotor angle in the interval $\pi/6 \le \theta_R \le 5\pi/6$.

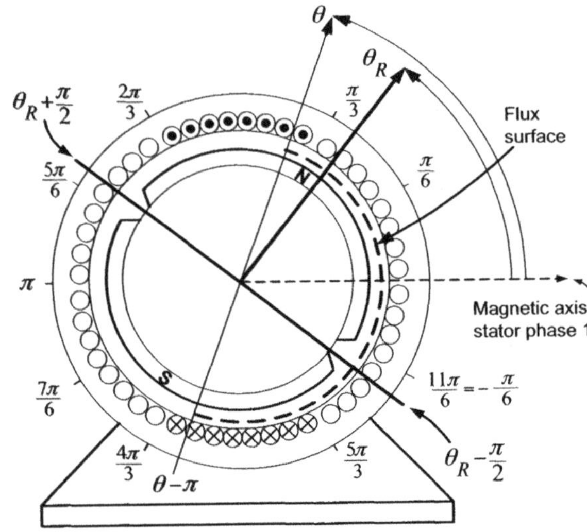

FIGURE 10.11. Computation of the flux in a loop of stator phase 1 produced by the rotor's magnetic field when $\pi/6 \leq \theta_R \leq 5\pi/6$.

Case 2. Figure 10.12 illustrates the computation of the flux in a loop of stator phase 1 produced by the rotor's magnetic field when $-\pi/6 \leq \theta_R \leq \pi/6$ and $\theta_R + \pi/2 \leq \theta \leq 2\pi/3$.

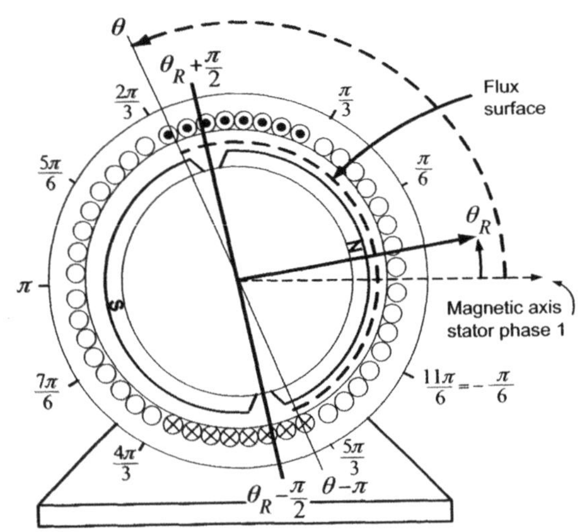

FIGURE 10.12. Computation of the flux in a loop of stator phase 1 produced by the rotor's magnetic field when $-\pi/6 \leq \theta_R \leq \pi/6$ and $\theta_R + \pi/2 \leq \theta \leq 2\pi/3$.

Case 3. Figure 10.13 illustrates the computation of the flux in a loop of stator phase 1 produced by the rotor's magnetic field when $-\pi/6 \le \theta_R \le \pi/6$ and $\pi/3 \le \theta \le \theta_R + \pi/2$.

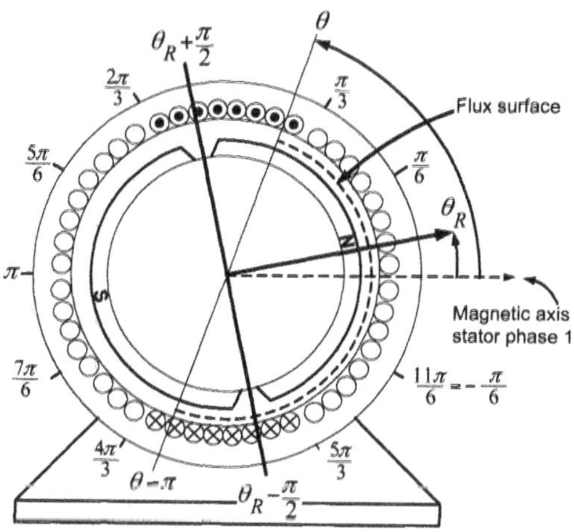

FIGURE 10.13. Computation of the flux in a loop of stator phase 1 produced by the rotor's magnetic field when $-\pi/6 \le \theta_R \le \pi/6$ and $\pi/3 \le \theta \le \theta_R + \pi/2$.

With reference to Figures 10.11, 10.12 and 10.13, and using the expression (10.2) for the magnetic field produced by the rotor, it follows that

$$\phi_{S1\_R}(i_{S1}, \theta) = \int_0^{\ell_1} \int_{\theta-\pi}^{\theta} \vec{\mathbf{B}}_R(r_S, \theta' - \theta_R) \cdot (r_S d\theta' d\ell \hat{\mathbf{r}})$$

$$= \begin{cases} \displaystyle\int_0^{\ell_1} \int_{\theta_R-\pi/2}^{\theta} B_{R0} r_R d\theta' d\ell + \int_0^{\ell_1} \int_{\theta-\pi}^{\theta_R-\pi/2} -B_{R0} r_R d\theta' d\ell \\ \qquad \text{for } \pi/6 \le \theta_R \le 5\pi/6 \\ \displaystyle\int_0^{\ell_1} \int_{\theta_R+\pi/2}^{\theta} -B_{R0} r_R d\theta' d\ell + \int_0^{\ell_1} \int_{\theta-\pi}^{\theta_R+\pi/2} B_{R0} r_R d\theta' d\ell \\ \qquad \text{for } \theta_R + \pi/2 \le \theta \le 2\pi/3 \\ \qquad \text{and } -\pi/6 \le \theta_R \le \pi/6 \\ \displaystyle\int_0^{\ell_1} \int_{\theta_R-\pi/2}^{\theta} B_{R0} r_R d\theta' d\ell + \int_0^{\ell_1} \int_{\theta-\pi}^{\theta_R-\pi/2} -B_{R0} r_R d\theta' d\ell \\ \qquad \text{for } \pi/3 \le \theta \le \theta_R + \pi/2 \\ \qquad \text{and } -\pi/6 \le \theta_R \le \pi/6. \end{cases}$$

Simplifying, this becomes

$$\phi_{S1\_R}(i_{S1},\theta) = \begin{cases} 2r_R\ell_1 B_{R0}\left(\theta-\theta_R\right) & \text{for} \quad \pi/6 \le \theta_R \le 5\pi/6 \\ -2r_R\ell_1 B_{R0}\left(\theta-\theta_R-\pi\right) & \text{for} \quad \theta_R+\pi/2 \le \theta \le 2\pi/3 \\ & \text{and} \quad -\pi/6 \le \theta_R \le \pi/6 \\ 2r_R\ell_1 B_{R0}\left(\theta-\theta_R\right) & \text{for} \quad \pi/3 \le \theta \le \theta_R+\pi/2 \\ & \text{and} \quad -\pi/6 \le \theta_R \le \pi/6. \end{cases}$$

The total flux linkage $\lambda_{S1\_R}(\theta_R)$ in stator phase 1 produced by the rotor's magnetic field is then

$$\lambda_{S1\_R}(\theta_R) = \int_{\pi/3}^{2\pi/3} \phi_{S1\_R}(i_{S1},\theta)\frac{N_S}{\pi/3}d\theta.$$

Evaluating, this becomes

$$\lambda_{S1\_R}(\theta_R)$$

$$= \begin{cases} \displaystyle\int_{\pi/3}^{2\pi/3} 2r_R\ell_1 B_{R0}\left(\theta-\theta_R\right)\frac{N_S}{\pi/3}d\theta & \text{for } +\pi/6 \le \theta_R \le 5\pi/6 \\ \displaystyle\int_{\pi/3}^{\theta_R+\pi/2} 2r_R\ell_1 B_{R0}\left(\theta-\theta_R\right)\frac{N_S}{\pi/3}d\theta & \\ \displaystyle\quad+\int_{\theta_R+\pi/2}^{2\pi/3} -2r_R\ell_1 B_{R0}\left(\theta-\theta_R-\pi\right)\frac{N_S}{\pi/3}d\theta & \text{for } -\pi/6 \le \theta_R \le \pi/6. \end{cases}$$

Simplifying, this becomes

$$\lambda_{S1\_R}(\theta_R) = \begin{cases} \dfrac{3}{\pi}r_R\ell_1 N_S B_{R0}\left(\theta-\theta_R\right)^2\Big|_{\pi/3}^{2\pi/3} & \text{for } +\pi/6 \le \theta_R \le 5\pi/6 \\ r_R\ell_1 B_{R0}\dfrac{N_S}{\pi/3}\left(\left(\theta-\theta_R\right)^2\Big|_{\pi/3}^{\theta_R+\pi/2} - \left(\theta-\theta_R-\pi\right)^2\Big|_{\theta_R+\pi/2}^{2\pi/3}\right) & \\ & \text{for } -\pi/6 \le \theta_R \le \pi/6 \end{cases}$$

which reduces to

$$\lambda_{S1\_R}(\theta_R) = \begin{cases} r_R\ell_1 N_S B_{R0}\Big(2(\dfrac{\pi}{3}-\theta_R)+\dfrac{\pi}{3}\Big) & \text{for} \quad +\pi/6 \le \theta_R \le 5\pi/6 \\ r_R\ell_1 N_S B_{R0}\Big(-\dfrac{6}{\pi}\theta_R^2+\dfrac{5\pi}{6}\Big) & \text{for} \quad -\pi/6 \le \theta_R \le \pi/6. \end{cases}$$

By symmetry, $\lambda_{S1\_R}(\theta_R \pm \pi) = -\lambda_{S1\_R}(\theta_R)$ so that the total flux linkage in stator phase 1 due to the rotor's magnetic field may be written as

$$\lambda_{S1\_R}(\theta_R) = \begin{cases} +r_R\ell_1 N_S B_{R0}\left(-\frac{6}{\pi}\theta_R^2 + \frac{5\pi}{6}\right), & -\frac{\pi}{6} \le \theta_R \le \frac{\pi}{6} \\ +r_R\ell_1 N_S B_{R0}\left(2(\frac{\pi}{3} - \theta_R) + \frac{\pi}{3}\right), & \frac{\pi}{6} \le \theta_R \le \frac{5\pi}{6} \\ -r_R\ell_1 N_S B_{R0}\left(-\frac{6}{\pi}(\theta_R - \pi)^2 + \frac{5\pi}{6}\right), & \frac{5\pi}{6} \le \theta_R \le \frac{7\pi}{6} \\ -r_R\ell_1 N_S B_{R0}\left(2\left(\frac{\pi}{3} - (\theta_R - \pi)\right) + \frac{\pi}{3}\right), & \frac{7\pi}{6} \le \theta_R \le \frac{11\pi}{6}. \end{cases}$$

Define $\lambda_R(\theta_R)$ as

$$\lambda_R(\theta_R) \triangleq \lambda_{S1\_R}(\theta_R)/M_{SR}$$

where $M_{SR} \triangleq 2r_R\ell_1 N_S B_{R0}$ is the *coefficient of mutual inductance* between the stator and the rotor. A plot of $\lambda_R(\theta_R)$ versus θ_R is given in Figure 10.14.

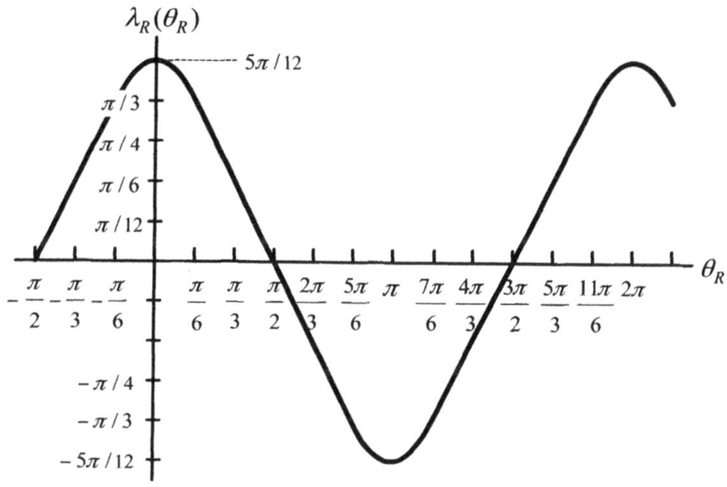

FIGURE 10.14. $\lambda_R(\theta_R)$ versus θ_R.

A simple computation shows that

$$-1 \le \frac{\partial \lambda_R(\theta_R)}{\partial \theta_R} \le 1.$$

The total flux linkage in each stator phase may now be written in terms of $\lambda_R(\theta_R)$ as

$$\begin{aligned}
\lambda_{S1_R}(\theta_R) &= M_{SR}\lambda_R(\theta_R)\\
\lambda_{S2_R}(\theta_R) &= M_{SR}\lambda_R(\theta_R - 2\pi/3)\\
\lambda_{S3_R}(\theta_R) &= M_{SR}\lambda_R(\theta_R - 4\pi/3).
\end{aligned}$$

10.5 Emf in the Stator Windings Produced by $\vec{\mathbf{B}}_R$

By Faraday's law, the back emf induced in the windings of phase 1 by rotor's magnetic field is given by

$$e_{S1} \triangleq -\frac{d}{dt}M_{SR}\lambda_R(\theta_R) = \begin{cases} +M_{SR}\dfrac{6\theta_R}{\pi}\omega_R, & -\pi/6 \le \theta_R \le \pi/6 \\ +M_{SR}\omega_R, & \pi/6 \le \theta_R \le 5\pi/6 \\ -M_{SR}\dfrac{6\,(\theta_R - \pi)}{\pi}\omega_R, & 5\pi/6 \le \theta_R \le 7\pi/6 \\ -M_{SR}\omega_R, & 7\pi/6 \le \theta_R \le 11\pi/6 \end{cases} \tag{10.7}$$

where $\omega_R = d\theta_R/dt$. With $e_p \triangleq M_{SR} = 2r_R\ell_1 N_S B_{R0}$ and

$$e(\theta_R) \triangleq \frac{e_{S1}}{M_{SR}\omega_R} = -\frac{\partial\lambda_R(\theta_R)}{\partial\theta_R}, \tag{10.8}$$

the back emf in each stator phase may now be written succinctly as

$$\begin{aligned}
e_{S1} &= e_p e(\theta_R)\omega_R\\
e_{S2} &= e_p e(\theta_R - 2\pi/3)\omega_R\\
e_{S2} &= e_p e(\theta_R - 4\pi/3)\omega_R.
\end{aligned}$$

As $-1 \le e(\theta_R) \le 1$, $e_p\omega_R$ is the peak value of the back emf. A plot of the back emfs $e_{S1}(\theta_R), e_{S2}(\theta_R)$, and $e_{S3}(\theta_R)$ at constant angular speed ω_R are given in Figures 10.15, 10.16, and 10.17, respectively. Note that each one has a trapezoidal shape and is the reason this machine is referred to as a *trapezoidal back-emf* synchronous motor.

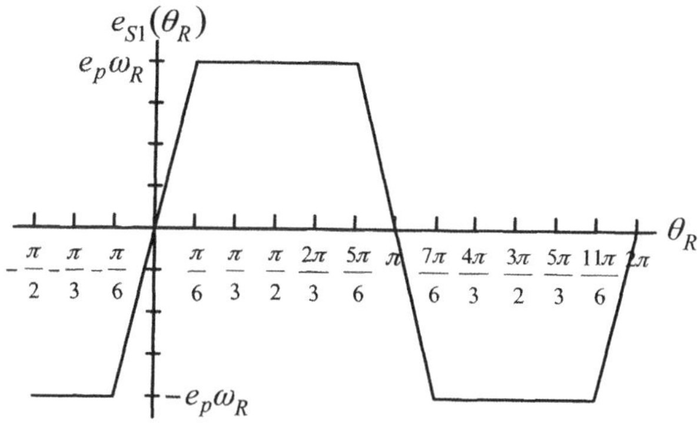

FIGURE 10.15. $e_{S1}(\theta_R)$ versus θ_R.

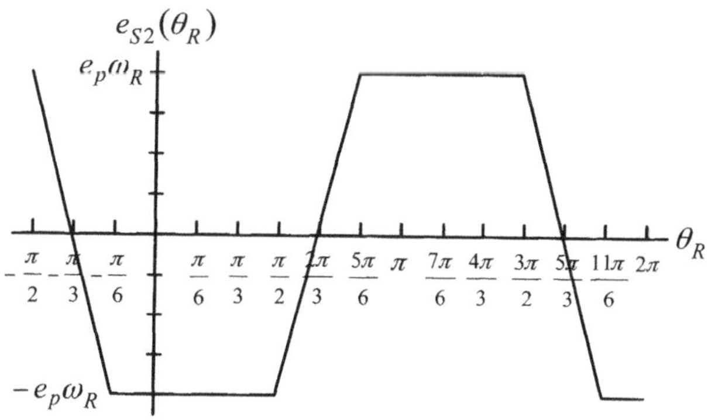

FIGURE 10.16. $e_{S2}(\theta_R)$ versus θ_R.

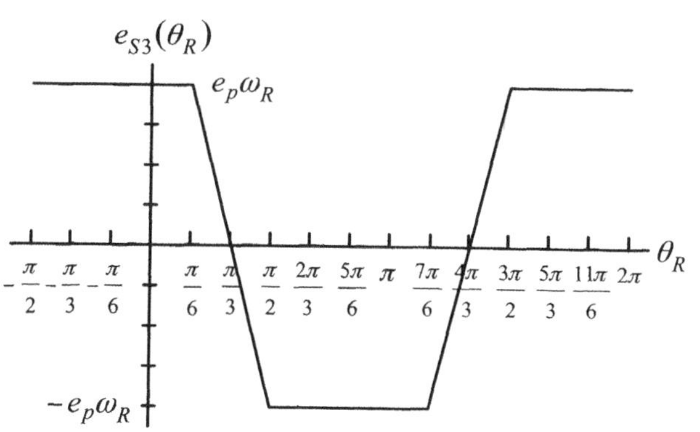

FIGURE 10.17. $e_{S3}(\theta_R)$ versus θ_R.

10.6 Torque

The magnetic force exerted by the rotor's magnetic field $\vec{\mathbf{B}}_R(r_S, \theta - \theta_R)$ on the top axial side (i.e., $\pi/3 \le \theta \le 2\pi/3$) of a winding of phase 1 is (see Figure 10.18)

$$\begin{aligned}\vec{\mathbf{F}}_{S1}(r, \theta - \theta_R) &= i_{S1}\,(\ell_1 \hat{\mathbf{z}}) \times \vec{\mathbf{B}}_R(r_S, \theta - \theta_R)\\ &= i_{S1}\ell_1 B_R(r_S, \theta - \theta_R)\hat{\mathbf{z}} \times \hat{\mathbf{r}}\\ &= i_{S1}\ell_1 B_R(r_S, \theta - \theta_R)\hat{\boldsymbol{\theta}}.\end{aligned}$$

On the bottom side (i.e., $4\pi/3 \le \theta \le 5\pi/3$) the force is

$$\begin{aligned}\vec{\mathbf{F}}_{S1}(r, \theta - \theta_R) &= i_{S1}\,(-\ell_1 \hat{\mathbf{z}}) \times \vec{\mathbf{B}}_R(r_S, \theta - \theta_R)\\ &= -i_{S1}\ell_1 B_R(r_S, \theta - \theta_R)\hat{\mathbf{z}} \times \hat{\mathbf{r}}\\ &= -i_{S1}\ell_1 B_R(r_S, \theta - \theta_R)\hat{\boldsymbol{\theta}}.\end{aligned}$$

The total torque on the axial sides of the windings of phase 1 produced by the rotor's magnetic field is then

$$\begin{aligned}\tau_{S1} &= \int_{\pi/3}^{2\pi/3} \vec{\mathbf{r}} \times \vec{\mathbf{F}}_{S1} \frac{N_S}{\pi/3} d\theta + \int_{4\pi/3}^{5\pi/3} \vec{\mathbf{r}} \times \vec{\mathbf{F}}_{S1} \frac{N_S}{\pi/3} d\theta\\ &= \int_{\pi/3}^{2\pi/3} i_{S1}\ell_1 B_R(r_S, \theta - \theta_R) r_S \frac{N_S}{\pi/3} d\theta \hat{\mathbf{r}} \times \hat{\boldsymbol{\theta}}\\ &\quad - \int_{4\pi/3}^{5\pi/3} i_{S1}\ell_1 B_R(r_S, \theta - \theta_R) r_S \frac{N_S}{\pi/3} d\theta \hat{\mathbf{r}} \times \hat{\boldsymbol{\theta}}\\ &= i_{S1}\ell_1 \frac{N_S}{\pi/3}\left(\int_{\pi/3}^{2\pi/3} r_S B_R(r_S, \theta - \theta_R) d\theta\right.\\ &\quad \left. - \int_{4\pi/3}^{5\pi/3} r_S B_R(r_S, \theta - \theta_R) d\theta\right)\hat{\mathbf{z}}\\ &= 2 i_{S1}\ell_1 \frac{N_S}{\pi/3} \int_{\pi/3}^{2\pi/3} r_S B_R(r_S, \theta - \theta_R) d\theta \hat{\mathbf{z}}\end{aligned}$$

where the fact that $B_R(r_S, \theta - \theta_R) = -B_R(r_S, \theta - \theta_R \pm \pi)$ was used to obtain the last expression.

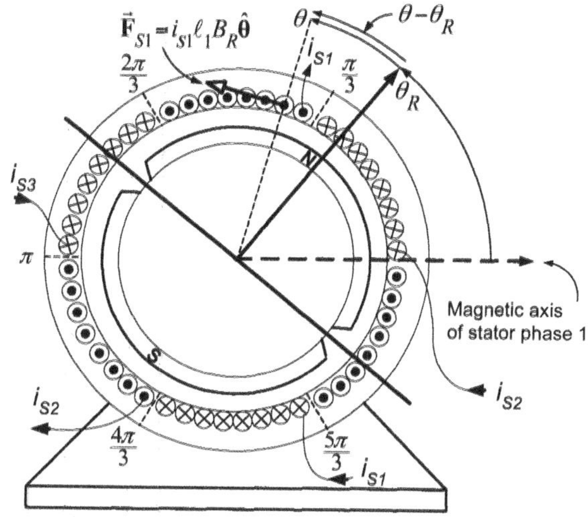

FIGURE 10.18. The force $\vec{\mathbf{F}}_{S1}(r, \theta - \theta_R) = -i_{S1}\ell_1 B_R(r_S, \theta - \theta_R)\hat{\boldsymbol{\theta}}$ exerted on a winding on the top side of stator phase 1 by the magnetic field of the rotor's permanent magnet.

Using the expression for the rotor's magnetic field given in (10.2), the integral is broken up as

$$\int_{\pi/3}^{2\pi/3} r_S B_R(r_S, \theta - \theta_R) d\theta$$

$$= \begin{cases} \displaystyle\int_{\pi/3}^{\theta_R+\pi/2} r_R B_{R0} d\theta + \int_{\theta_R+\pi/2}^{2\pi/3} -r_R B_{R0} d\theta & \text{for} \quad -\dfrac{\pi}{6} \le \theta_R \le \dfrac{\pi}{6} \\ \displaystyle\int_{\pi/3}^{2\pi/3} r_R B_{R0} d\theta & \text{for} \quad \dfrac{\pi}{6} \le \theta_R \le \dfrac{5\pi}{6} \\ \displaystyle\int_{\pi/3}^{\theta_R-\pi/2} -r_R B_{R0} d\theta + \int_{\theta_R-\pi/2}^{2\pi/3} r_R B_{R0} d\theta & \text{for} \quad \dfrac{5\pi}{6} \le \theta_R \le \dfrac{7\pi}{6} \\ \displaystyle\int_{\pi/3}^{2\pi/3} -r_R B_{R0} d\theta & \text{for} \quad \dfrac{7\pi}{6} \le \theta_R \le \dfrac{11\pi}{6} \end{cases}$$

which simplifies to

$$\int_{\pi/3}^{2\pi/3} r_S B_R(r_S, \theta - \theta_R) d\theta$$

$$= \begin{cases} +r_R B_{R0}(\theta_R + \pi/6) - r_R B_{R0}(\pi/6 - \theta_R), & -\dfrac{\pi}{6} \leq \theta_R \leq \dfrac{\pi}{6} \\ +r_R B_{R0}\pi/3, & \dfrac{\pi}{6} \leq \theta_R \leq \dfrac{5\pi}{6} \\ -r_R B_{R0}(\theta_R - 5\pi/6) + r_R B_{R0}(7\pi/6 - \theta_R), & \dfrac{5\pi}{6} \leq \theta_R \leq \dfrac{7\pi}{6} \\ -r_R B_{R0}\pi/3, & \dfrac{7\pi}{6} \leq \theta_R \leq \dfrac{11\pi}{6} \end{cases}$$

Combining terms results in

$$\int_{\pi/3}^{2\pi/3} r_S B_R(r_S, \theta - \theta_R) d\theta = \begin{cases} +2r_R B_{R0}\theta_R, & -\pi/6 \leq \theta_R \leq \pi/6 \\ +r_R B_{R0}\pi/3, & \pi/6 \leq \theta_R \leq 5\pi/6 \\ -2r_R B_{R0}(\theta_R - \pi), & 5\pi/6 \leq \theta_R \leq 7\pi/6 \\ -r_R B_{R0}\pi/3, & 7\pi/6 \leq \theta_R \leq 11\pi/6. \end{cases}$$

Thus, the total torque exerted on the axial windings of phase 1 is then

$$\tau_{S1} = \begin{cases} +2\ell_1 r_R N_S B_{R0}\dfrac{6\theta_R}{\pi} i_{S1} & \text{for } -\pi/6 \leq \theta_R \leq \pi/6 \\ +2\ell_1 r_R N_S B_{R0} i_{S1} & \text{for } \pi/6 \leq \theta_R \leq 5\pi/6 \\ -2\ell_1 r_R N_S B_{R0}\dfrac{6(\theta_R - \pi)}{\pi} i_{S1} & \text{for } 5\pi/6 \leq \theta_R \leq 7\pi/6 \\ -2\ell_1 r_R N_S B_{R0} i_{S1} & \text{for } 7\pi/6 \leq \theta_R \leq 11\pi/6. \end{cases}$$

The torque exerted on the rotor by the magnetic field produced by the stator currents of phase 1 is then $\tau_{R1} = -\tau_{S1}$, that is,

$$\tau_{R1} = \begin{cases} -2\ell_1 r_R N_S B_{R0}\dfrac{6\theta_R}{\pi} i_{S1} & \text{for } -\pi/6 \leq \theta_R \leq \pi/6 \\ -2\ell_1 r_R N_S B_{R0} i_{S1} & \text{for } \pi/6 \leq \theta_R \leq 5\pi/6 \\ +2\ell_1 r_R N_S B_{R0}\dfrac{6(\theta_R - \pi)}{\pi} i_{S1} & \text{for } 5\pi/6 \leq \theta_R \leq 7\pi/6 \\ +2\ell_1 r_R N_S B_{R0} i_{S1} & \text{for } 7\pi/6 \leq \theta_R \leq 11\pi/6. \end{cases}$$

With $\tau_p = 2\ell_1 r_R N_S B_{R0}$ (note that $\tau_p = M_{SR} = e_p$), this may be rewritten compactly as

$$\tau_{R1}(\theta_R, i_{S1}) = \begin{cases} -\tau_p \dfrac{6\theta_R}{\pi} i_{S1} & \text{for} \quad -\pi/6 \leq \theta_R \leq \pi/6 \\ -\tau_p i_{S1} & \text{for} \quad \pi/6 \leq \theta_R \leq 5\pi/6 \\ +\tau_p \dfrac{6(\theta_R - \pi)}{\pi} i_{S1} & \text{for} \quad 5\pi/6 \leq \theta_R \leq 7\pi/6 \\ +\tau_p i_{S1} & \text{for} \quad 7\pi/6 \leq \theta_R \leq 11\pi/6. \end{cases}$$

However, by (10.7) and (10.8), it follows that

$$\tau_{R1}(\theta_R, i_{S1}) = -\tau_p e(\theta_R) i_{S1}$$

and similarly

$$\begin{aligned} \tau_{R2}(\theta_R, i_{S2}) &= -\tau_p e(\theta_R - 2\pi/3) i_{S2} \\ \tau_{R3}(\theta_R, i_{S3}) &= -\tau_p e(\theta_R - 4\pi/3) i_{S2}. \end{aligned}$$

10.7 Mathematical Model

The stator flux linkages are given by

$$\begin{aligned} \lambda_1(i_{S1}, i_{S2}, i_{S3}) &= +L_S i_{S1} - M i_{S2} - M i_{S3} + e_p \lambda_R(\theta_R) \\ \lambda_2(i_{S1}, i_{S2}, i_{S3}) &= -M i_{S1} + L_S i_{S2} - M i_{S3} + e_p \lambda_R(\theta_R - 2\pi/3) \\ \lambda_3(i_{S1}, i_{S2}, i_{S3}) &= -M i_{S1} - M i_{S2} + L_S i_{S3} + e_p \lambda_R(\theta_R - 4\pi/3) \end{aligned}$$

and the phase torques are

$$\begin{aligned} \tau_{R1}(\theta_R, i_{S1}) &= -\tau_p e(\theta_R)\, i_{S1} \\ \tau_{R2}(\theta_R, i_{S2}) &= -\tau_p e(\theta_R - 2\pi/3) i_{S2} \\ \tau_{R3}(\theta_R, i_{S3}) &= -\tau_p e(\theta_R - 4\pi/3) i_{S3}. \end{aligned}$$

With R_S the resistance in each stator phase and u_{S1}, u_{S2}, and u_{S3} the voltages applied to each stator phase, the mathematical model is then

$$\begin{aligned} \frac{d}{dt}\lambda_1(i_{S1}, i_{S2}, i_{S3}) &= -R_S i_{S1} + u_{S1} \\ \frac{d}{dt}\lambda_2(i_{S1}, i_{S2}, i_{S3}) &= -R_S i_{S2} + u_{S2} \\ \frac{d}{dt}\lambda_3(i_{S1}, i_{S2}, i_{S3}) &= -R_S i_{S3} + u_{S3} \\ J\frac{d\omega}{dt} &= \tau_{R1}(\theta_R, i_{S1}) + \tau_{R2}(\theta_R, i_{S2}) + \tau_{R3}(\theta_R, i_{S3}) - \tau_L \\ \frac{d\theta_R}{dt} &= \omega_R \end{aligned}$$

or

$$
\begin{aligned}
\frac{d}{dt}\Big(L_S i_{S1} - M i_{S2} - M i_{S3} + e_p \lambda_R(\theta_R)\Big) &= -R_S i_{S1} + u_{S1} \\
\frac{d}{dt}\Big(-M i_{S1} + L_S i_{S2} - M i_{S3} + e_p \lambda_R(\theta_R - 2\pi/3)\Big) &= -R_S i_{S2} + u_{S2} \\
\frac{d}{dt}\Big(-M i_{S1} - M i_{S2} + L_S i_{S3} + e_p \lambda_R(\theta_R - 4\pi/3)\Big) &= -R_S i_{S3} + u_{S3} \\
J\frac{d\omega_R}{dt} &= \tau - \tau_L \\
\frac{d\theta_R}{dt} &= \omega_R
\end{aligned}
$$

with

$$\tau \triangleq -\tau_p e\,(\theta_R)\, i_{S1} - \tau_p e\,(\theta_R - 2\pi/3)\, i_{S2} - \tau_p e\,(\theta_R - 4\pi/3)\, i_{S3}.$$

This system is often written in the form $\left(e\,(\theta_R) = -\partial\lambda_R(\theta_R)/\partial\theta_R\right)$

$$
\begin{aligned}
\begin{bmatrix} u_{S1} \\ u_{S2} \\ u_{S3} \end{bmatrix} &= \begin{bmatrix} L_S & -M & -M \\ -M & L_S & -M \\ -M & -M & L_S \end{bmatrix} \frac{d}{dt} \begin{bmatrix} i_{S1} \\ i_{S2} \\ i_{S3} \end{bmatrix} + R_S \begin{bmatrix} i_{S1} \\ i_{S2} \\ i_{S3} \end{bmatrix} \\
&\quad - e_p \begin{bmatrix} e\,(\theta_R) \\ e\,(\theta_R - 2\pi/3) \\ e\,(\theta_R - 4\pi/3) \end{bmatrix} \omega_R \\
J\frac{d\omega_R}{dt} &= -\tau_p e\,(\theta_R)\, i_{S1} - \tau_p e\,(\theta_R - 2\pi/3)\, i_{S2} - \tau_p e\,(\theta_R - 4\pi/3)\, i_{S3} - \tau_L \\
\frac{d\theta_R}{dt} &= \omega_R.
\end{aligned}
$$

With *balanced* stator currents, this may be rewritten as (see problem 1)

$$
\begin{aligned}
\begin{bmatrix} u_{S1} \\ u_{S2} \\ u_{S3} \end{bmatrix} &= \begin{bmatrix} L_S + M & 0 & 0 \\ 0 & L_S + M & 0 \\ 0 & 0 & L_S + M \end{bmatrix} \frac{d}{dt} \begin{bmatrix} i_{S1} \\ i_{S2} \\ i_{S3} \end{bmatrix} + R_S \begin{bmatrix} i_{S1} \\ i_{S2} \\ i_{S3} \end{bmatrix} \\
&\quad - e_p \begin{bmatrix} e\,(\theta_R) \\ e\,(\theta_R - 2\pi/3) \\ e\,(\theta_R - 4\pi/3) \end{bmatrix} \omega_R \\
J\frac{d\omega_R}{dt} &= -\tau_p e\,(\theta_R)\, i_{S1} - \tau_p e\,(\theta_R - 2\pi/3)\, i_{S2} - \tau_p e\,(\theta_R - 4\pi/3)\, i_{S3} - \tau_L \\
\frac{d\theta_R}{dt} &= \omega_R.
\end{aligned}
$$

In state-space form this becomes

$$
\begin{aligned}
\frac{di_{S1}}{dt} &= \frac{e_p}{L_S+M}\omega_R e(\theta_R) - \frac{R_S}{L_S+M}i_{S1} + \frac{1}{L_S+M}u_{S1} \\
\frac{di_{S2}}{dt} &= \frac{e_p}{L_S+M}\omega_R e(\theta_R - 2\pi/3) - \frac{R_S}{L_S+M}i_{S2} + \frac{1}{L_S+M}u_{S2} \\
\frac{di_{S3}}{dt} &= \frac{e_p}{L_S+M}\omega_R e(\theta_R - 4\pi/3) - \frac{R_S}{L_S+M}i_{S3} + \frac{1}{L_S+M}u_{S3} \\
\frac{d\omega_R}{dt} &= -(\tau_p/J)e\left(\theta_R\right)i_{S1} - (\tau_p/J)e\left(\theta_R - 2\pi/3\right)i_{S2} \\
&\quad -(\tau_p/J)e\left(\theta_R - 4\pi/3\right)i_{S3} - \tau_L/J \\
\frac{d\theta_R}{dt} &= \omega_R
\end{aligned}
$$

where $e_p = \tau_p$.

Remark

In this formulation of the mathematical model, $e\left(\theta_R\right) = -\partial\lambda_R(\theta_R)/\partial\theta_R$ was used [see equation (10.8)] so that the back emfs in each phase $e_{S1} = e_p e\left(\theta_R\right), e_{S2} = e_p e\left(\theta_R - 2\pi/3\right)$, and $e_{S3} = e_p e\left(\theta_R - 4\pi/3\right)$ have the *same* sign convention as the corresponding applied stator voltages u_{S1}, u_{S2}, and u_{S3}, respectively. This is consistent with Faraday's law and with the sign conventions chosen for the back emf in the induction and synchronous machines. However, in the literature, it is common to use $\varepsilon(\theta_R) \triangleq \partial\lambda_R(\theta_R)/\partial\theta_R$ for the back emf [15]; that is, the sign convention is opposite to that used here. This formulation is explored in problem 4. (In Chapter 1, both sign conventions were considered for the back emf of the DC motor.)

10.8 Operation and Control

At constant angular speed, Figure 10.19 is a graphical representation of the relationships between the back emf voltages (note their trapezoidal shape)

$$
\begin{aligned}
e_{S1} &= e_p e(\theta_R)\omega_R \\
e_{S2} &= e_p e(\theta_R - 2\pi/3)\omega_R \\
e_{S3} &= e_p e(\theta_R - 4\pi/3)\omega_R,
\end{aligned}
\tag{10.9}
$$

the phase currents i_{S1}, i_{S2}, and i_{S3}, the power absorbed in each phase by the back emfs $p_{S1} \triangleq e_{S1}i_{S1}, p_{S2} \triangleq e_{S2}i_{S2}, p_{S3} \triangleq e_{S3}i_{S3}$, and the total power absorbed by the back emfs $p \triangleq e_{S1}i_{S1} + e_{S2}i_{S2} + e_{S3}i_{S3}$. In order to obtain the current profile i_{S1} shown in Figure 10.19, the current reference i_{S1r} for the current in phase 1 is chosen as

$$
i_{S1r} = \begin{cases} -I_p & \text{if} \quad e_{S1} = +E_p = +e_p\omega_R \\ +I_p & \text{if} \quad e_{S1} = -E_p = -e_p\omega_R \\ 0 & \text{otherwise} \end{cases}
$$

or, from the expressions (10.7) and (10.8) for e_{S1}, it is seen that i_{S1r} may be written as

$$i_{S1r}(\theta_R) = I_p i_S(\theta_R)$$

where

$$i_S(\theta_R) \triangleq \begin{cases} 0 & \text{for} \quad -\pi/6 \leq \theta_R \leq \pi/6 \\ -1 & \text{for} \quad \pi/6 \leq \theta_R \leq 5\pi/6 \\ 0 & \text{for} \quad 5\pi/6 \leq \theta_R \leq 7\pi/6 \\ 1 & \text{for} \quad 7\pi/6 \leq \theta_R \leq 11\pi/6. \end{cases} \tag{10.10}$$

Choosing the gains in the PI current controller

$$u_{S1} = K_P(i_{S1r} - i_{S1}) + K_I \int_0^t (i_{S1r}(\tau) - i_{S1}(\tau))d\tau \tag{10.11}$$

appropriately results in $i_{S1} \to i_{S1r}$. Similarly, one chooses $i_{S2r} = I_p i_S(\theta_R - 2\pi/3)$ and $i_{S3r} = I_p i_S(\theta_R - 4\pi/3)$ and then employs PI current controllers of the form (10.11) to force $i_{S2} \to i_{S2r}, i_{S3} \to i_{S3r}$.

Figure 10.19 shows that the current is commutated (i.e., is switched from one phase to another) every $\pi/3$ radians or 60°. For example, at $\theta_R = \pi/6$ the current $-I_p$ in phase 3 goes (is commutated) to phase 1. After another $\pi/3$ radians so that $\theta_R = \pi/2$, the current I_p in phase 2 is commutated to phase 3. Note that one need only know the rotor position at the discrete locations $30°, 60°, 90°, 150°, 210°, 270°$, and $330°$ to carry out the commutation of the phases, i.e., to implement $i_{S1r}, i_{S2r}, i_{S3r}$. Usually these motors are equipped with Hall effect sensors that detect these discrete rotor positions for current commutation.

To determine torque produced using these current references, consider the electrical power absorbed by the back emfs which is given by

$$\begin{aligned} e_{S1}i_{S1} + e_{S2}i_{S2} + e_{S3}i_{S3} &= e_p e\,(\theta_R)\omega_R i_{S1} + e_p e\,(\theta_R - 2\pi/3)\omega_R i_{S2} \\ &\quad + e_p e\,(\theta_R - 4\pi/3)\omega_R i_{S3}. \end{aligned}$$

Choosing the current references as just described, it is seen from the plots in Figure (10.19) that this last expression reduces to

$$\begin{aligned} e_{S1}i_{S1} + e_{S2}i_{S2} + e_{S3}i_{S3} &= e_p e\,(\theta_R)\omega_R i_{S1r} + e_p e\,(\theta_R - 2\pi/3)\omega_R i_{S2r} \\ &\quad + e_p e\,(\theta_R - 4\pi/3)\omega_R i_{S3r} \\ &= -2e_p\omega_R I_p \end{aligned}$$

which is negative for $\omega_R > 0, I_p > 0$. Where does this power go? It is converted into mechanical power as is now shown (see also problem 2).

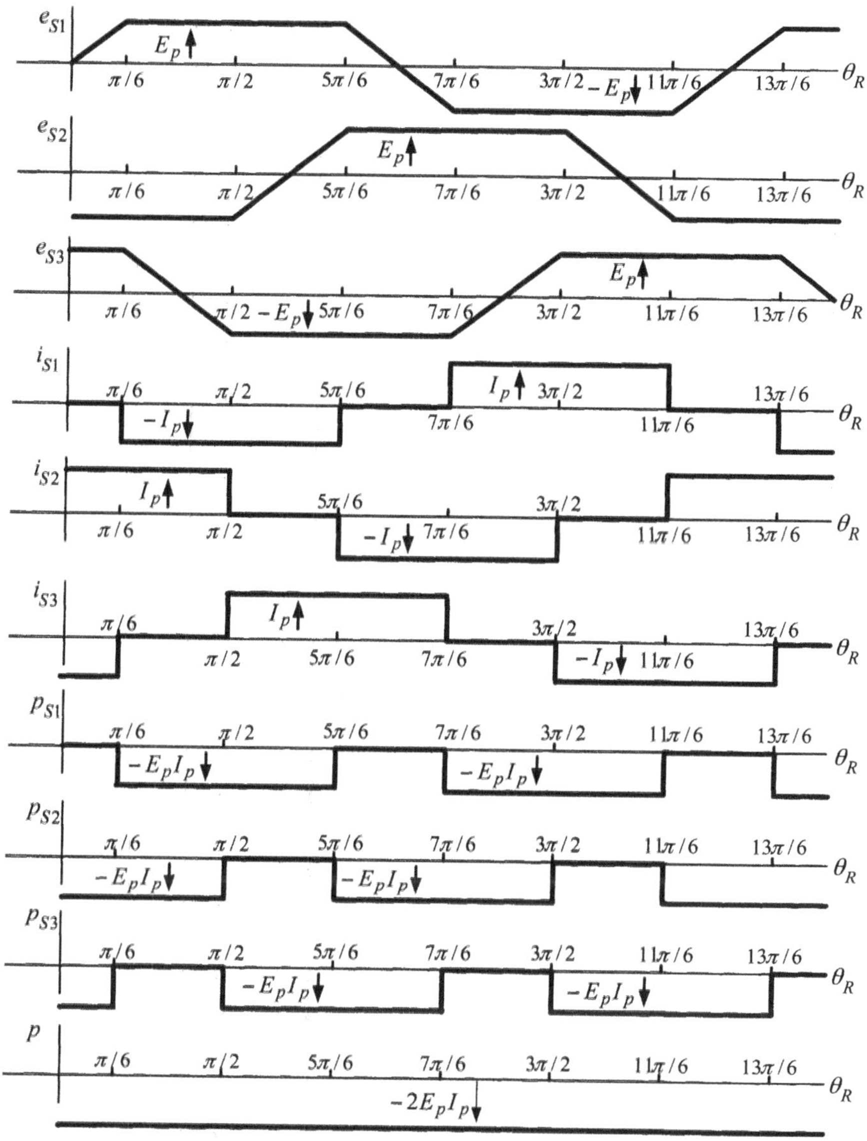

FIGURE 10.19. Back emf, currents, and power waveforms ($E_p \triangleq e_p\omega_R$). Note that the commutation of the phase currents occurs every $\pi/3$ radians at $\pi/6, \pi/2, 5\pi/6, 7\pi/6, 9\pi/6$, and $11\pi/6$. Adapted from Krishnan [15].

Specifically, with the currents chosen as above, the mechanical power is

$$\begin{aligned}\tau\omega_R &= -\tau_p e\,(\theta_R)\omega_R i_{S1} - \tau_p e\left(\theta_R - \frac{2\pi}{3}\right)\omega_R i_{S2} - \tau_p e\left(\theta_R - \frac{4\pi}{3}\right)\omega_R i_{S3} \\ &= -e_p e\,(\theta_R)\omega_R i_{S1r} - e_p e\left(\theta_R - \frac{2\pi}{3}\right)\omega_R i_{S2r} - e_p e\left(\theta_R - \frac{4\pi}{3}\right)\omega_R i_{S3r} \\ &= 2e_p\omega_R I_p \\ &= 2\tau_p I_p\omega_R.\end{aligned}$$

That is, the torque is simply given by

$$\tau = 2\tau_p I_p.$$

This suggests a control scheme in which the current references are chosen as

$$\begin{aligned}i_{S1r} &= I_p i_S(\theta_R) \\ i_{S2r} &= I_p i_S(\theta_R - 2\pi/3) \\ i_{S3r} &= I_p i_S(\theta_R - 4\pi/3)\end{aligned} \tag{10.12}$$

and

$$\begin{aligned}u_{S1} &= K_P(i_{S1r} - i_{S1}) + K_I\int_0^t (i_{S1r}(\tau) - i_{S1}(\tau))d\tau \\ u_{S2} &= K_P(i_{S2r} - i_{S2}) + K_I\int_0^t (i_{S2r}(\tau) - i_{S2}(\tau))d\tau \\ u_{S1} &= K_P(i_{S3r} - i_{S3}) + K_I\int_0^t (i_{S3r}(\tau) - i_{S3}(\tau))d\tau.\end{aligned} \tag{10.13}$$

If the PI current controller performs well enough to assume $i_{S1} = i_{S1r}$, $i_{S2} = i_{S2r}$ and $i_{S3} = i_{S3r}$, then

$$\frac{d\omega}{dt} = \frac{\tau}{J} - \frac{\tau_L}{J} = \frac{2\tau_p}{J} I_p - \frac{\tau_L}{J}.$$

One then simply chooses

$$I_p = \frac{J}{2\tau_p}\alpha_r \tag{10.14}$$

$$\begin{aligned}\alpha_r &= K_I\int_0^t \Big(\theta_{ref}(\tau) - \theta(\tau)\Big)\, d\tau + K_P\Big(\theta_{ref}(t) - \theta(t)\Big) \\ &\quad + K_D\Big(\omega_{ref}(t) - \omega(t)\Big) + \alpha_{ref}.\end{aligned} \tag{10.15}$$

With $e_0(t) \triangleq \int_0^t(\theta_{ref}(\tau) - \theta(\tau)\,)d\tau, e_1(t) \triangleq \theta_{ref}(t) - \theta(t)$, and $e_2(t) \triangleq \omega_{ref}(t) - \omega(t)$, the closed-loop system for the mechanical variables is

$$\frac{d}{dt}\begin{bmatrix} e_0 \\ e_1 \\ e_2 \end{bmatrix} = \begin{bmatrix} 0 & 1 & 0 \\ 0 & 0 & 1 \\ -K_I & -K_P & -K_D \end{bmatrix}\begin{bmatrix} e_0 \\ e_1 \\ e_2 \end{bmatrix} + \begin{bmatrix} 0 \\ 0 \\ 1/J \end{bmatrix}\tau_L.$$

With $r_1 > 0, r_2 > 0, r_3 > 0$, set the gains as

$$\begin{aligned} K_D &= r_1 + r_2 + r_3 \\ K_P &= r_1 r_2 + r_1 r_3 + r_2 r_3 \\ K_I &= r_1 r_2 r_3 \end{aligned}$$

so that the closed-loop characteristic polynomial is

$$\begin{aligned} a(s) &= \det\left(sI - \begin{bmatrix} 0 & 1 & 0 \\ 0 & 0 & 1 \\ -K_I & -K_P & -K_D \end{bmatrix}\right) \\ &= s^3 + K_I s^2 + K_P s + K_I \\ &= s^3 + (r_1 + r_2 + r_3)s^2 + (r_1 r_2 + r_1 r_3 + r_2 r_3)s + r_1 r_2 r_3 \\ &= (s + r_1)(s + r_2)(s + r_3). \end{aligned}$$

With this choice of gains, the closed-loop poles are $p_1 = -r_1$, $p_2 = -r_2$, and $p_3 = -r_3$. As shown in Chapter 2, if the load torque is constant, then $e_1(t) \to 0, e_2(t) \to 0$ as $t \to \infty$. Of course, this controller assumes that $i_{S1} \to i_{S1r}, i_{S2} \to i_{S2r}$, and $i_{S3} \to i_{S3r}$ fast enough that it can be assumed that currents are *equal* to their reference values.

Remark

Note that the set of currents $i_{S1r} = I_p i_S(\theta_R), i_{S2r} = I_p i_S(\theta_R - 2\pi/3)$, and $i_{S3r} = I_p i_S(\theta_R - 4\pi/3) i_{S1r}$ are balanced, that is,

$$I_p i_S(\theta_R) + I_p i_S(\theta_R - 2\pi/3) + I_p i_S(\theta_R - 4\pi/3) \equiv 0.$$

However, the back emf voltages $e_{S1}(\theta_R)$, $e_{S2}(\theta_R)$ and $e_{S3}(\theta_R)$ are not balanced as

$$\begin{aligned} e_{S1}(\theta_R) + e_{S2}(\theta_R) + e_{S3}(\theta_R) &= -e_p \omega_R \Big(e(\theta_R) + e\left(\theta_R - 2\pi/3\right) \\ &\qquad + e\left(\theta_R - 4\pi/3\right) \Big) \end{aligned}$$

is not identically zero (see Figure 10.19). This is in contrast to a three-phase PM synchronous machine with sinusoidally distributed windings in which the back emf voltages are balanced.

10.8.1 *The Terminology "Brushless DC Motor"*

The terminology *brushless DC motor* or *BLDC* is used for this machine because usually the motor is combined with an optical encoder, current measurements, Hall sensors for current commutation, an amplifier and feedback controller so that it behaves like a DC motor. That is, as indicated in Figure 10.20, the currents and motor position are fed back to the controller;

the controller then uses PI current loops of the form (10.11) to force the currents to track the references

$$\begin{aligned} i_{S1r}(\theta_R) &= I_p i_S(\theta_R) \\ i_{S2r}(\theta_R) &= I_p i_S(\theta_R - 2\pi/3) \\ i_{S3r}(\theta_R) &= I_p i_S(\theta_R - 4\pi/3) \end{aligned} \tag{10.16}$$

where $i_S(\theta_R)$ is given by (10.10). The input to the controller is simply I_p so that with the inner current control loops working properly, the equations of the motor become

$$\begin{aligned} J\frac{d\omega}{dt} &= 2\tau_p I_p - \tau_L \\ \frac{d\theta}{dt} &= \omega \end{aligned}$$

which is the same form as current command DC motor with torque constant $K_T = 2\tau_p$.

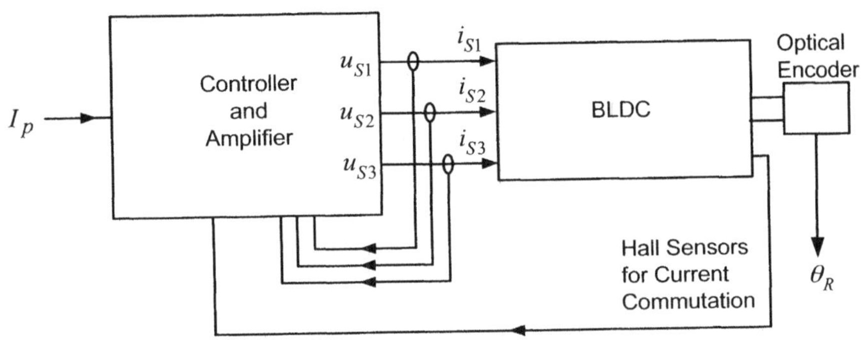

FIGURE 10.20. Block diagram for a brushless DC motor.

The system of Figure 10.20 is what one refers to as a "brushless DC motor" which the user obtains as a complete system from the manufacturer. Then, in the same way as for a DC motor, one simply chooses the "DC current" I_p as in equations (10.14) and (10.15) to use the BLDC as a servo (positioning) system.

Remark Typically, the position sensor for the current commutation, that is, for tracking the current references (10.16), is done with Hall effect sensors rather than an encoder. This is because to track the current references (10.16), the phase current plots in Figure 10.19 show that one only needs to determine the position of the rotor at multiples of $\pi/3$ or $60°$ as the current in any particular phase changes only at some multiple of $60°$. Hall effect sensors provide a simple inexpensive way to detect the rotor position at multiples of $60°$.

10.9 Microscopic Viewpoint of BLDC Machines*

The microscopic viewpoint of the Physics of BLDC machine is described in this section. In this analysis, the system will be assumed to be in steady-state so that $\theta_R = \omega_R t - \delta, i_{S1} = I_S\cos(\omega_S t), i_{S2} = I_S\cos(\omega_S t - 2\pi/3), i_{S3} = I_S\cos(\omega_S t - 4\pi/3)$, and $\omega_S = \omega_R$.

10.9.1 Axial Electric Field $\vec{\mathbf{E}}_R$

It is now shown that the rotating radial magnetic field of the PM rotor produces an axial electric field in the air gap. Recall that the magnetic field produced by the rotor magnet is

$$\vec{\mathbf{B}}_R(r,\theta-\theta_R)$$
$$= \begin{cases} B_{R0}\dfrac{r_R}{r}\hat{\mathbf{r}} & \text{for} \quad -\dfrac{\pi}{2}+\dfrac{\Delta}{2} \le \theta-\theta_R \le \dfrac{\pi}{2}-\dfrac{\Delta}{2} \\ -B_{R0}\dfrac{r_R}{r}\dfrac{(\theta-\theta_R-\pi/2)}{\Delta/2}\hat{\mathbf{r}} & \text{for} \quad \dfrac{\pi}{2}-\dfrac{\Delta}{2} \le \theta-\theta_R \le \dfrac{\pi}{2}+\dfrac{\Delta}{2} \\ -B_{R0}\dfrac{r_R}{r}\hat{\mathbf{r}} & \text{for} \quad \dfrac{\pi}{2}+\dfrac{\Delta}{2} \le \theta-\theta_R \le \dfrac{3\pi}{2}-\dfrac{\Delta}{2} \\ B_{R0}\dfrac{r_R}{r}\dfrac{(\theta-\theta_R-3\pi/2)}{\Delta/2}\hat{\mathbf{r}} & \text{for} \quad \dfrac{3\pi}{2}-\dfrac{\Delta}{2} \le \theta-\theta_R \le \dfrac{3\pi}{2}+\dfrac{\Delta}{2} \end{cases} \tag{10.17}$$

where the parameter $\Delta > 0$ is now taken to be small, but nonzero. Maxwell's equations require that the divergence of the magnetic induction field $\vec{\mathbf{B}}_R$ be identically zero. In cylindrical coordinates with $\vec{\mathbf{B}} = B_r\hat{\mathbf{r}} + B_\theta\hat{\boldsymbol{\theta}} + B_z\hat{\mathbf{z}}$, the divergence is written as

$$\nabla\cdot\vec{\mathbf{B}} = \frac{1}{r}\frac{\partial}{\partial r}(rB_r) + \frac{1}{r}\frac{\partial B_\theta}{\partial\theta} + \frac{\partial B_z}{\partial z}.$$

A straightforward calculation (note that $B_{R\theta} = 0$, $B_{Rz} = 0$) shows that $\nabla\cdot\vec{\mathbf{B}}_R = 0$.

The rotor's magnetic field in the air gap induces an *axial* electric field in the air gap. This electric field can be computed using Faraday's law in differential form, that is,

$$\nabla\times\vec{\mathbf{E}}_R = -\frac{\partial\vec{\mathbf{B}}_R}{\partial t}.$$

In cylindrical coordinates, the curl of $\vec{\mathbf{E}}_R$ is given by $\nabla\times\vec{\mathbf{E}}_R =$

$$\left(\frac{1}{r}\frac{\partial E_{Rz}}{\partial\theta} - \frac{\partial E_{R\theta}}{\partial z}\right)\hat{\mathbf{r}} + \left(\frac{\partial E_{Rr}}{\partial z} - \frac{\partial E_{Rz}}{\partial r}\right)\hat{\boldsymbol{\theta}} + \left(\frac{1}{r}\frac{\partial(rE_{R\theta})}{\partial r} - \frac{1}{r}\frac{\partial E_{Rr}}{\partial\theta}\right)\hat{\mathbf{z}}.$$

Then

$$-\frac{\partial \vec{\mathbf{B}}_R(r,\theta-\theta_R)}{\partial t}=\begin{cases} 0 & \text{for} \quad -\frac{\pi}{2}+\frac{\Delta}{2}\leq\theta-\theta_R\leq\frac{\pi}{2}-\frac{\Delta}{2} \\ -B_{R0}\frac{r_R}{r}\frac{2}{\Delta}\omega_R\hat{\mathbf{r}} & \text{for} \quad \frac{\pi}{2}-\frac{\Delta}{2}\leq\theta-\theta_R\leq\frac{\pi}{2}+\frac{\Delta}{2} \\ 0 & \text{for} \quad \frac{\pi}{2}+\frac{\Delta}{2}\leq\theta-\theta_R\leq\frac{3\pi}{2}-\frac{\Delta}{2} \\ B_{R0}\frac{r_R}{r}\frac{2}{\Delta}\omega_R\hat{\mathbf{r}} & \text{for} \quad \frac{3\pi}{2}-\frac{\Delta}{2}\leq\theta-\theta_R\leq\frac{3\pi}{2}+\frac{\Delta}{2}. \end{cases}$$

Now,

$$\frac{\partial E_{R\theta}}{\partial z}=0 \quad \text{and} \quad \frac{\partial E_{Rr}}{\partial z}=0$$

as it is assumed there is no dependence in z. Thus,

$$\nabla\times\vec{\mathbf{E}}_R=\frac{1}{r}\frac{\partial E_{Rz}}{\partial\theta}\hat{\mathbf{r}}$$

and Faraday's law in differential form reduces to

$$\frac{1}{r}\frac{\partial E_{Rz}}{\partial\theta}=-\frac{\partial B_R(r,\theta-\theta_R)}{\partial t}.$$

The solution is

$$E_{Rz}(\theta-\theta_R,t)=-\int_{\theta_R+\pi/2}^{\theta}r\frac{\partial B_R(r,\theta-\theta_R)}{\partial t}d\theta$$

where $\theta_R+\pi/2$ was chosen as the reference position for the integration as $E_{Rz}(\theta-\theta_R,t)|_{\theta=\theta_R+\pi/2}=0$ (see Figure 10.21 and problem 7). Evaluating the integral gives

$$E_{Rz}(\theta-\theta_R,t)$$
$$=\begin{cases} -\frac{2r_RB_{R0}}{\Delta}\omega_R(\theta-\theta_R-\pi/2) & \text{for} \quad \frac{\pi}{2}-\frac{\Delta}{2}\leq\theta-\theta_R\leq\frac{\pi}{2}+\frac{\Delta}{2} \\ -r_RB_{R0}\omega_R & \text{for} \quad \frac{\pi}{2}+\frac{\Delta}{2}\leq\theta-\theta_R\leq\frac{3\pi}{2}-\frac{\Delta}{2} \\ \frac{2r_RB_{R0}}{\Delta}\omega_R(\theta-\theta_R-3\pi/2) & \text{for} \quad \frac{3\pi}{2}-\frac{\Delta}{2}\leq\theta-\theta_R\leq\frac{3\pi}{2}+\frac{\Delta}{2} \\ r_RB_{R0}\omega_R & \text{for} \quad -\frac{\pi}{2}+\frac{\Delta}{2}\leq\theta-\theta_R\leq\frac{\pi}{2}-\frac{\Delta}{2}. \end{cases}$$

Letting $\Delta \to 0$, this becomes

$$E_{Rz}(\theta - \theta_R, t)\hat{\mathbf{z}} = \begin{cases} +r_R B_{R0}\omega_R \hat{\mathbf{z}} & \text{for} \quad -\pi/2 \le \theta - \theta_R \le \pi/2 \\ -r_R B_{R0}\omega_R \hat{\mathbf{z}} & \text{for} \quad \pi/2 \le \theta - \theta_R \le 3\pi/2 \end{cases} \tag{10.18}$$

which is illustrated in Figure 10.21.

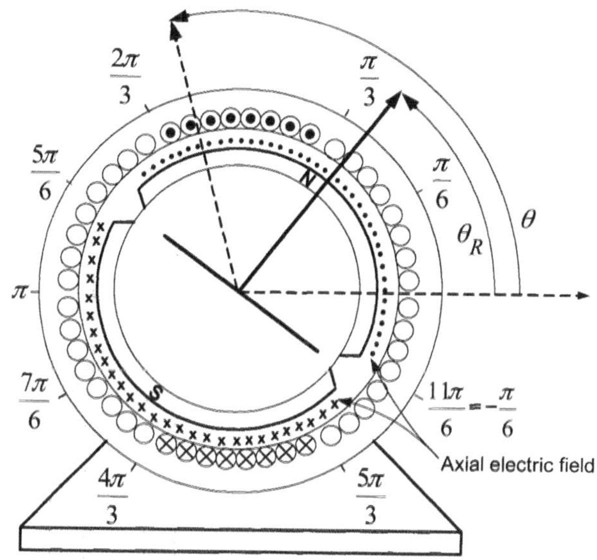

FIGURE 10.21. Axial electric field $\vec{\mathbf{E}}_R$ produced by $\vec{\mathbf{B}}_R$.

10.9.2 Emf Induced in the Stator Phases

Consider a winding of phase 1 where the axial side 1 is at the location θ with $2\pi/3 \le \theta \le \pi/3$ and the other axial side $1'$ of the winding is at $\theta + \pi$. To compute the emf $\zeta_{S1}(\theta)$ in this winding, let

$$d\vec{\ell} = \begin{cases} d\ell\hat{\mathbf{z}} & \text{for side } 1 \\ -d\ell\hat{\mathbf{z}} & \text{for side } 1'. \end{cases}$$

For $\pi/3 \le \theta \le 2\pi/3$ and using (10.18), it follows that

$$\zeta_{S1}(\theta) = \int_{\substack{\text{stator winding whose} \\ \text{sides are at } \theta \text{ and } \theta+\pi}} \vec{\mathbf{E}}_R(\theta) \cdot d\vec{\ell}$$

$$= \begin{cases} +2r_R\ell B_{R0}\omega_R & \text{for} \quad \pi/6 \le \theta_R \le 5\pi/6 \\ -2r_R\ell B_{R0}\omega_R & \text{for} \quad \begin{array}{c} \theta_R + \pi/2 \le \theta \le 2\pi/3 \\ -\pi/6 \le \theta_R \le \pi/6 \end{array} \\ +2r_R\ell B_{R0}\omega_R & \text{for} \quad \begin{array}{c} \pi/3 \le \theta \le \theta_R + \pi/2 \\ -\pi/6 \le \theta_R \le \pi/6 \end{array} \\ -2r_R\ell B_{R0}\omega_R & \text{for} \quad 7\pi/6 \le \theta_R \le 11\pi/6 \\ -2r_R\ell B_{R0}\omega_R & \text{for} \quad \begin{array}{c} \pi/3 \le \theta \le \theta_R - \pi/2 \\ 5\pi/6 \le \theta_R \le 7\pi/6 \end{array} \\ +2r_R\ell B_{R0}\omega_R & \text{for} \quad \begin{array}{c} \theta_R - \pi/2 \le \theta \le 2\pi/3 \\ 5\pi/6 \le \theta_R \le 7\pi/6. \end{array} \end{cases}$$

The total emf ξ_{S1} induced in stator phase 1 due to the rotor's magnetic field is then

$$\xi_{S1} = \int_{\substack{\text{All windings of} \\ \text{stator phase 1}}} \vec{\mathbf{E}}_R \cdot d\vec{\ell} = \int_{\pi/3}^{2\pi/3} \zeta_{S1}(\theta)\frac{N_S}{\pi/3}d\theta.$$

Evaluating the integral gives

$$\xi_{S1} = \begin{cases} \displaystyle\int_{\pi/3}^{2\pi/3} 2r_R\ell B_{R0}\omega_R\frac{N_S}{\pi/3}d\theta & \text{for} \quad \pi/6 \le \theta_R \le 5\pi/6 \\ \displaystyle\int_{\theta_R+\pi/2}^{2\pi/3} -2r_R\ell B_{R0}\omega_R\frac{N_S}{\pi/3}d\theta & \\ \displaystyle\quad + \int_{\pi/3}^{\theta_R+\pi/2} 2r_R\ell B_{R0}\omega_R\frac{N_S}{\pi/3}d\theta & \text{for} \quad -\pi/6 \le \theta_R \le \pi/6 \\ \displaystyle\int_{\pi/3}^{2\pi/3} -2r_R\ell B_{R0}\omega_R\frac{N_S}{\pi/3}d\theta & \text{for} \quad 7\pi/6 \le \theta_R \le 11\pi/6 \\ \displaystyle\int_{\pi/3}^{\theta_R-\pi/2} -2r_R\ell B_{R0}\omega_R\frac{N_S}{\pi/3}d\theta & \\ \displaystyle\quad + \int_{\theta_R-\pi/2}^{2\pi/3} 2r_R\ell B_{R0}\omega_R\frac{N_S}{\pi/3}d\theta & \text{for} \quad 5\pi/6 \le \theta_R \le 7\pi/6. \end{cases}$$

This simplifies to

$$\xi_{S1} = \begin{cases} 2r_R\ell N_S B_{R0}\omega_R & \text{for} \quad \dfrac{\pi}{6} \le \theta_R \le \dfrac{5\pi}{6} \\ -2r_R\ell B_{R0}\omega_R\dfrac{N_S}{\pi/3}\left(\dfrac{2\pi}{3} - \theta_R - \dfrac{\pi}{2}\right) & \\ +2r_R\ell B_{R0}\omega_R\dfrac{N_S}{\pi/3}\left(\theta_R + \dfrac{\pi}{2} - \dfrac{\pi}{3}\right) & \text{for} \quad -\dfrac{\pi}{6} \le \theta_R \le \dfrac{\pi}{6} \\ -2r_R\ell N_S B_{R0}\omega_R & \text{for} \quad \dfrac{7\pi}{6} \le \theta_R \le \dfrac{11\pi}{6} \\ -2r_R\ell B_{R0}\omega_R\dfrac{N_S}{\pi/3}\left(\theta_R - \dfrac{\pi}{2} - \dfrac{\pi}{3}\right) & \\ +2r_R\ell B_{R0}\omega_R\dfrac{N_S}{\pi/3}\left(\dfrac{2\pi}{3} - \theta_R + \dfrac{\pi}{2}\right) & \text{for} \quad \dfrac{5\pi}{6} \le \theta_R \le \dfrac{7\pi}{6} \end{cases}$$

or, finally

$$\xi_{S1} = \int_{\substack{\text{All windings of} \\ \text{stator phase 1}}} \vec{\mathbf{E}}_R \cdot d\vec{\ell}$$

$$= \begin{cases} +2r_R\ell N_S B_{R0}\omega_R & \text{for} \quad \pi/6 \le \theta_R \le 5\pi/6 \\ +2r_R\ell N_S B_{R0}\omega_R\dfrac{6\theta_R}{\pi} & \text{for} \quad -\pi/6 \le \theta_R \le \pi/6 \\ -2r_R\ell N_S B_{R0}\omega_R & \text{for} \quad 7\pi/6 \le \theta_R \le 11\pi/6 \\ -2r_R\ell N_S B_{R0}\omega_R\dfrac{6\left(\theta_R - \pi\right)}{\pi} & \text{for} \quad 5\pi/6 \le \theta_R \le 7\pi/6 \end{cases}$$

which is the same expression as (10.7) derived in the macroscopic case.

Problems

Problem 1 BLDC Motor Model

The electrical equations for the BLDC were derived as

$$\begin{bmatrix} u_{S1} \\ u_{S2} \\ u_{S3} \end{bmatrix} = R_S \begin{bmatrix} i_{S1} \\ i_{S2} \\ i_{S3} \end{bmatrix} + \frac{d}{dt}\begin{bmatrix} L_S & -M & -M \\ -M & L_S & -M \\ -M & -M & L_S \end{bmatrix}\begin{bmatrix} i_{S1} \\ i_{S2} \\ i_{S3} \end{bmatrix} + e_p \begin{bmatrix} e(\theta_R) \\ e(\theta_R - 2\pi/3) \\ e(\theta_R - 4\pi/3) \end{bmatrix}\omega_R.$$

Assuming the stator currents are balanced, show that

$$\begin{bmatrix} u_{S1} \\ u_{S2} \\ u_{S3} \end{bmatrix} = R_S \begin{bmatrix} i_{S1} \\ i_{S2} \\ i_{S3} \end{bmatrix} + \frac{d}{dt}\begin{bmatrix} L_S + M & 0 & 0 \\ 0 & L_S + M & 0 \\ 0 & 0 & L_S + M \end{bmatrix}\begin{bmatrix} i_{S1} \\ i_{S2} \\ i_{S3} \end{bmatrix} - e_p \begin{bmatrix} e(\theta_R) \\ e(\theta_R - 2\pi/3) \\ e(\theta_R - 4\pi/3) \end{bmatrix}\omega_R.$$

Problem 2 Conservation of Energy

The model of the BLDC may be written as

$$\frac{d}{dt}\Lambda \begin{bmatrix} i_{S1} \\ i_{S2} \\ i_{S3} \end{bmatrix} - e_p \begin{bmatrix} e(\theta_R) \\ e(\theta_R - 2\pi/3) \\ e(\theta_R - 4\pi/3) \end{bmatrix}\omega_R + R_S \begin{bmatrix} i_{S1} \\ i_{S2} \\ i_{S3} \end{bmatrix} = \begin{bmatrix} u_{S1} \\ u_{S2} \\ u_{S3} \end{bmatrix}$$

$$-\tau_p e(\theta_R) i_{S1} - \tau_p e(\theta_R - 2\pi/3) i_{S1} - \tau_p e(\theta_R - 4\pi/3) i_{S1} - \tau_L = J\frac{d\omega_R}{dt}.$$

Do an energy balance based on these equations to account for the energy stored in the magnetic fields in the air gap/iron due to the stator currents, the ohmic losses in the stator windings, the energy stored in the rotational motion of the rotor and the energy converted to mechanical power. In particular, show that $e_p = \tau_p$ must hold for energy conservation to hold.

Problem 3 Power Density in PM Synchronous Machines [15]

Consider two PM motors, one is a three-phase BLDC motor and the second one is a three-phase PM synchronous (sinusoidally wound stator) motor. Let both motors have the same stator resistance R_S and the same torque/back-emf constants ($\tau_p = K_T$ or, equivalently, $e_p = K_m$), and rotate at the same constant speed. Let I_{bldc} and I_{synch} be the peak current limit in the BLDC and the PM synchronous motor, respectively.

(a) Show that the rms current in each phase of the BLDC is $\sqrt{2/3}I_{bldc}$ while in the PM synchronous motor it is $I_{synch}/\sqrt{2}$.

(b) Let the peak currents in the motors be chosen so they both have the same ohmic losses, that is,

$$3R_S\left(\sqrt{\frac{2}{3}}I_{bldc}\right)^2 = 3R_S\left(\frac{I_{synch}}{\sqrt{2}}\right)^2$$

or

$$I_{bldc} = \frac{\sqrt{3}}{2}I_{synch}.$$

Choose the current references in the BLDC according to (10.12) in order to obtain the maximum torque out of the machine for a given current. In this case, the steady-state power produced by the BLDC is constant and equal to

$$2e_pI_p.$$

Assume in the PM synchronous motor that it is possible to have the stator currents in phase with their corresponding stator voltages (unity power factor) so that the steady-state power produced by the PM motor is

$$3\frac{K_m}{\sqrt{2}}\frac{I_{synch}}{\sqrt{2}}$$

where $K_m = K_T$ is the back emf constant. Show that, under these conditions, the ratio of the output power of the BLDC machine to the PM synchronous machine is 1.155.

Problem 4 Mathematical Model of the BLDC

In the literature, the normalized back emf is usually taken to be $\varepsilon(\theta_R) \triangleq \partial\lambda_R(\theta_R)/\partial\theta_R$ rather than $e(\theta_R) \triangleq -\partial\lambda_R(\theta_R)/\partial\theta_R$ as in equation (10.8). That is, the back emfs are then taken to be

$$\begin{aligned}\varepsilon_{S1} &= e_p\varepsilon(\theta_R)\omega_R\\ \varepsilon_{S2} &= e_p\varepsilon(\theta_R - 2\pi/3)\omega_R\\ \varepsilon_{S3} &= e_p\varepsilon(\theta_R - 4\pi/3)\omega_R\end{aligned}$$

(compare with 10.9). This is just a different sign convention for the back emfs; that is, these back emfs have a sign convention that is opposite to that of the applied phase voltages u_{S1}, u_{S2}, and u_{S3}.

(a) Rewrite the mathematical model of the motor in terms of $\varepsilon(\theta_R)$.

(b) How do the waveforms in Figure 10.19 change as a result of using $\varepsilon(\theta_R)$?

Problem 5 Simulation of a BLDC

Simulate an open-loop three-phase trapezoidal back-emf synchronous motor. Set the parameters as $R_S = 0.7\ \Omega$, $L_S = 2.72$ mH, $M = 0.15$ mH, $\tau_p = e_p = 0.5128$ N-m/A, $J = 0.0002$ kg-m^2, $f = 0.002$ N-m/rad/sec, $n_p =$

1, $\omega_S = (2\pi)10$ *rad/sec,* $V = V_{max} = 200$ *V (peak),* $I_{max} = 10$ *A, and* $\tau_L = 0$ *[15]. Apply a three-phase set of trapezoidal voltages to the motor with a peak of* V_{max} *and plot (a)* i_{Sa} *and* i_{Sb}*, (b)* ω*, and (c)* θ*. Does the speed* ω *need to be initialized to a value close to the synchronous speed value* ω_S *in order to run open loop?*

Problem 6 Closed-Loop Control of a BLDC

Add the closed-loop controller (10.13), (10.14), and (10.15) to the simulation of the BLDC in problem 5. Use only proportional control in the current controller. Have the motor do a point-to-point move in which it goes from 0 *to* 240 *radians and achieves a maximum speed of* 400 *rads/sec in* 0.5 *sec.*

Problem 7 Rotor Electric Field

Show that at $\theta = \theta_R + \pi/2$ *the axial electric field due to motion of the rotor magnet is zero, that is,*

$$E_{Rz}(\theta - \theta_R, t)|_{\theta=\theta_R+\pi/2} = 0.$$

Hint: Consider a half-cylindrical-shaped loop containing the rotor with one axial side fixed at the angular position θ_0 *in the air gap and, consequently, the other side fixed at* $\theta_0 + \pi$ *(the loop does not rotate). Use Faraday's law to show that the emf generated in this loop by the rotor's magnetic field is zero when the position of the rotor is such that* $\theta_0 = \theta_R + \pi/2 \pm n\pi$*.*

Problem 8 Permeability of a Permanent Magnet

Explain why, as far as the stator magnetic field $\vec{\mathbf{B}}_S$ *is concerned, the permeability of the permanent magnets on the rotor is* $\mu = \mu_0\mu_r = \mu_0$*. Show that this implies that the "effective" air gap (i.e., the value for* g*) is the sum of the actual air-gap length and the thickness of the PM magnets on the rotor surface.*

Trigonometric Table and Identities

Trigonometric Table

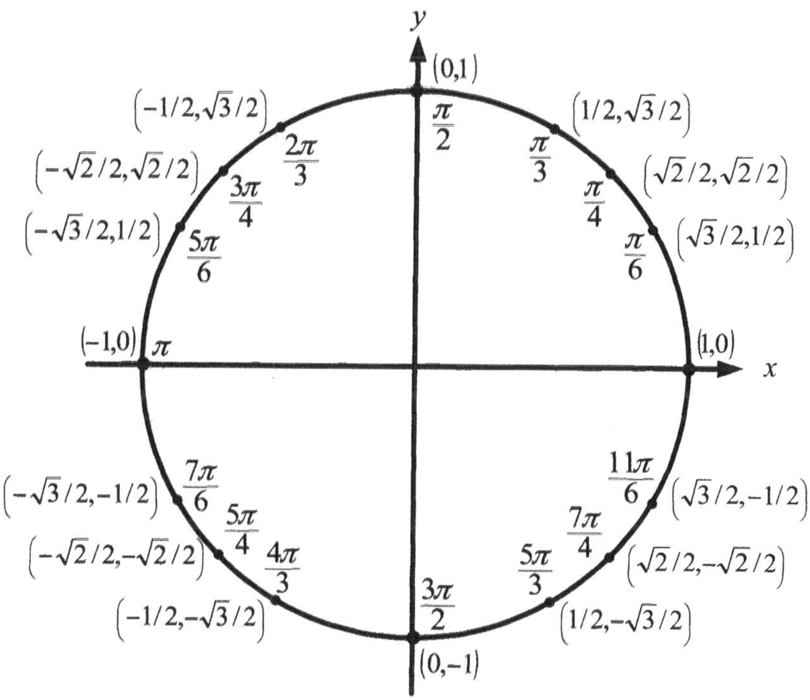

Trigonometric Identities

$$\begin{aligned}
e^{j\theta} &= \cos(\theta) + j\sin(\theta) \\
a\cos(\theta) + b\sin(\theta) &= \sqrt{a^2+b^2}\cos\left(\theta + \tan^{-1}(b/a)\right) \\
\sin(2\theta) &= 2\sin(\theta)\cos(\theta) \\
\cos(2\theta) &= \cos^2(\theta) - \sin^2(\theta) = 2\cos^2(\theta) - 1 \\
&= 1 - 2\sin^2(\theta) \\
\cos^2(\theta) &= \frac{1+\cos(2\theta)}{2} \\
\sin^2(\theta) &= \frac{1-\cos(2\theta)}{2} \\
\cos(\theta_1 \pm \theta_2) &= \cos(\theta_1)\cos(\theta_2) \mp \sin(\theta_1)\sin(\theta_2) \\
\sin(\theta_1 \pm \theta_2) &= \sin(\theta_1)\cos(\theta_2) \pm \cos(\theta_1)\sin(\theta_2)
\end{aligned}$$

$$\begin{aligned}
\cos(\theta_1)\cos(\theta_2) &= \frac{1}{2}\cos(\theta_1+\theta_2) + \frac{1}{2}\cos(\theta_1-\theta_2) \\
\sin(\theta_1)\sin(\theta_2) &= \frac{1}{2}\cos(\theta_1-\theta_2) - \frac{1}{2}\cos(\theta_1+\theta_2) \\
\sin(\theta_1)\cos(\theta_2) &= \frac{1}{2}\sin(\theta_1+\theta_2) + \frac{1}{2}\sin(\theta_1-\theta_2) \\
\cos(\theta_1)\sin(\theta_2) &= \frac{1}{2}\sin(\theta_1+\theta_2) - \frac{1}{2}\sin(\theta_1-\theta_2)
\end{aligned}$$

$$\begin{aligned}
\frac{3}{2} &= \cos^2(\theta) + \cos^2(\theta - 2\pi/3) + \cos^2(\theta - 4\pi/3) \\
\frac{3}{2} &= \sin^2(\theta) + \sin^2(\theta - 2\pi/3) + \sin^2(\theta - 4\pi/3) \\
0 &= \sin(\theta)\cos(\theta) + \sin(\theta - 2\pi/3)\cos(\theta - 2\pi/3) \\
&\quad + \sin(\theta - 4\pi/3)\cos(\theta - 4\pi/3) \\
0 &= \cos(\theta) + \cos(\theta - 2\pi/3) + \cos(\theta - 4\pi/3) \\
0 &= \sin(\theta) + \sin(\theta - 2\pi/3) + \sin(\theta - 4\pi/3)
\end{aligned}$$

$$\begin{aligned}\frac{3}{2}\cos(5\theta_1+\theta_2) &= \sin(5(\theta_1+\pi/2))\cos(\theta_2)\\ &\quad + \sin\left(5(\theta_1+\pi/2-2\pi/3)\right)\cos(\theta_2-2\pi/3)\\ &\quad + \sin\left(5(\theta_1+\pi/2-4\pi/3)\right)\cos(\theta_2-4\pi/3)\end{aligned}$$

$$\begin{aligned}-\frac{3}{2}\cos(7\theta_1-\theta_2) &= \sin(7(\theta_1+\pi/2))\cos(\theta_2)\\ &\quad + \sin\left(7(\theta_1+\pi/2-2\pi/3)\right)\cos(\theta_2-2\pi/3)\\ &\quad + \sin\left(7(\theta_1+\pi/2-4\pi/3)\right)\cos(\theta_2-4\pi/3)\end{aligned}$$

$$\begin{aligned}\frac{3}{2}\sin(\theta_1-\theta_2) &= \sin(\theta_1)\cos(\theta_2)+\sin(\theta_1-2\pi/3)\cos(\theta_2-2\pi/3)\\ &\quad + \sin(\theta_1-4\pi/3)\cos(\theta_2-4\pi/3)\end{aligned}$$

$$\begin{aligned}\frac{3}{2}\cos(\theta_1-\theta_2) &= \sin(\theta_1)\sin(\theta_2)+\sin(\theta_1-2\pi/3)\sin(\theta_2-2\pi/3)\\ &\quad + \sin(\theta_1-4\pi/3)\sin(\theta_2-4\pi/3)\end{aligned}$$

$$\begin{aligned}-\frac{3}{2}\sin(\theta_1-\theta_2) &= \cos(\theta_1)\sin(\theta_2)+\cos(\theta_1-2\pi/3)\sin(\theta_2-2\pi/3)\\ &\quad + \cos(\theta_1-4\pi/3)\sin(\theta_2-4\pi/3)\end{aligned}$$

$$\begin{aligned}\frac{3}{2}\cos(\theta_1-\theta_2) &= \cos(\theta_1)\cos(\theta_2)+\cos(\theta_1-2\pi/3)\cos(\theta_2-2\pi/3)\\ &\quad + \cos(\theta_1-4\pi/3)\cos(\theta_2-4\pi/3)\end{aligned}$$

$$\begin{aligned}\frac{3}{2}\sin(\theta_1+\theta_2) &= \sin(\theta_1)\cos(\theta_2)+\sin(\theta_1-4\pi/3)\cos(\theta_2-2\pi/3)\\ &\quad + \sin(\theta_1-2\pi/3)\cos(\theta_2-4\pi/3)\end{aligned}$$

$$\begin{aligned}-\frac{3}{2}\cos(\theta_1+\theta_2) &= \sin(\theta_1)\sin(\theta_2)+\sin(\theta_1-4\pi/3)\sin(\theta_2-2\pi/3)\\ &\quad + \sin(\theta_1-2\pi/3)\sin(\theta_2-4\pi/3)\end{aligned}$$

$$\begin{aligned}\frac{3}{2}\sin(\theta_1+\theta_2) &= \cos(\theta_1)\sin(\theta_2)+\cos(\theta_1-4\pi/3)\sin(\theta_2-2\pi/3)\\ &\quad + \cos(\theta_1-2\pi/3)\sin(\theta_2-4\pi/3)\end{aligned}$$

References

[1] Werner Leonhard, *Control of Electrical Drives*, Springer-Verlag, Berlin, 1985.

[2] Werner Leonhard, *Control of Electrical Drives*, 3rd Edition, Springer-Verlag, Berlin, 2001.

[3] Paul C. Krause, *Analysis of Electric Machinery*, McGraw-Hill, New York, 1986.

[4] Paul C. Krause, Oleg Wasynczuk, and Scott D. Sudhoff, *Analysis of Electric Machinery and Drive Systems*, 2nd Edition, John Wiley & Sons, New York, 2002.

[5] C. B. Gray, *Electrical Machines and Drive Systems*, Longman Scientific & Technical, New York, 1989.

[6] Chee-Mun Ong, *Dynamic Simulation of Electric Machinery using Matlab/Simulink*, Prentice-Hall, Upper Saddle River, NJ, 1998.

[7] D. W. Novotny and T. A. Lipo, *Vector Control and Dynamics of AC Drives*, Oxford University Press, New York, 1996.

[8] Peter Vas, *Vector Control of AC Machines*, Clarendon Press, Oxford, 1990.

[9] J. M. D. Murphy and F. G. Turnbull, *Power Electronic Control of AC Motors*, Pergamon, Elmsford, NY, 1988.

[10] I. Boldea and S. A. Nasar, *Vector Control of AC Drives*, CRC Press, Boca Raton, FL, 1992.

[11] B. Adkins and R. G. Harley, *The General Theory of Alternating Current Machines*, Chapman and Hall, New York, 1975.

[12] A. M. Trzynadlowski, *The Field Orientation Principle in Control of Induction Motors*, Kluwer Academic Publishers, Hingham, MA, 1994.

[13] M. P. Kazmierkowski and H. Tunia, *Automatic Control of Converter Fed Drives*, Elsevier, Amsterdam, 1994.

[14] B. K. Bose, *Modern Power Electronics and AC Drives*, Prentice-Hall, Upper Saddle River, NJ, 2002.

[15] R. Krishnan, *Electric Motor Drives: Modeling, Analysis, and Control*, Prentice-Hall, Upper Saddle River, NJ, 2001.

[16] Romeo Ortega, Antonio Loria, Per Johan Nicklasson, and Hebert Sira-Ramírez, *Passivity-Based Control of Euler–Lagrange Systems: Mechanical, Electrical and Electromechanical Applications*, Springer-Verlag, Berlin, 1998.

[17] Darren M. Dawson, Jun Hu, and Timothy C. Burg, *Nonlinear Control of Electric Machinery*, Marcel Dekker, New York, 1998.

[18] F. Khorrami, P. Krishnamurthy, and H. Melkote, *Modeling and Nonlinear Adaptive Control of Electric Motors*, Springer-Verlag, Berlin, 2003.

[19] Stephen J. Chapman, *Electric Machinery Fundamentals*, McGraw-Hill, New York, 1985.

[20] H. H. Woodson and J. Melcher, *Electromechanical Dynamics: Part 1 Discrete Systems*, John Wiley & Sons, New York, 1968.

[21] L. W. Matsch and J. Derald Morgan, *Electromagnetic and Electromechanical Machines*, 3rd edition, John Wiley & Sons, New York, 1986.

[22] George McPherson and Robert D. Laramore, *An Introduction to Electrical Machines and Transformers*, 2nd edition, John Wiley & Sons, New York, 1990.

[23] D. V. Richardson, *Rotating Electric Machinery and Transformer Technology*, Reston Publishing Co., Reston, VA, 1978.

[24] Paul C. Krause and O. Wasynczuk, *ELECTROMECHANICAL Motion Devices*, McGraw-Hill, New York, 1989.

[25] Ned Mohan, *ELECTRIC DRIVES: An Integrative Approach*, MNPERE, Minneapolis, 2001.

[26] G. R. Slemon and A. Straughen, *Electric Machines*, Addison-Wesley, Reading, MA, 1980.

[27] J. Sokira and W. Jaffe, *Brushless DC Motors Electronic Commutation and Controls*, TAB Books, Blue Ridge Summit, PA, 1990.

[28] George J. Thaler and Milton L. Wilcox, *Electric Machines: Dynamics and Steady State*, John Wiley & Sons, New York, 1966.

[29] Vincent Deltoro, *Electric Machines and Power Systems*, Prentice-Hall, Englewood Cliffs, NJ, 1985.

[30] Mohamed E. El-Hawary, *Principles of Electric Machines with Power Electronic Applications*, John Wiley & Sons, New York, 2002.

[31] P.C. Sen, *Principles of Electric Machines and Power Electronics*, John Wiley & Sons, New York, 1989.

[32] G. R. Slemon, *Electric Machines and Drives*, Addison-Wesley, Reading, MA, 1992.

[33] Uri Haber-Schaim, John H. Dodge, Robert Gardner, and Edward Shore, *PSSC Physics*, 7th Edition, Kendall/Hunt, Dubuque, IA, 1991.

[34] D. Halliday and R. Resnick, *Physics Volume II*, John Wiley & Sons, New York, 1962.

[35] Melvin Schwartz, *Principles of Electrodynamics*, Dover Publications, New York, 1972.

[36] Roald K. Wangsness, *Electromagnetic Fields*, John Wiley & Sons, New York, 1979.

[37] K. Ogata, *Modern Control Engineering*, Prentice-Hall, Englewood Cliffs, NJ, 1970.

[38] A. R. Bergen and V. Vittal, *Power Systems Analysis*, 2nd Edition, Prentice-Hall, Upper Saddle River, NJ, 2000.

[39] C. W. deSilva, *Control Sensors and Actuators*, Prentice-Hall, Englewood Cliffs, NJ, 1989.

[40] C. W. deSilva, *Mechatronics: An Integrative Approach*, CRC Press, Boca Raton, FL, 2004.

[41] T. Kailath, *Linear Systems*, Prentice-Hall, Englewood Cliffs, NJ, 1980.

[42] J. Stephan, M. Bodson, and J. Chiasson, "Real-time estimation of induction motor parameters", *IEEE Transactions on Industry Applications*, vol. 30, no. 3, pp. 746–759, May/June 1994.

[43] Alan V. Oppenheim and Alan S. Willsky, *Signals & Systems*, 2nd Edition, Prentice-Hall, Englewood Cliffs, NJ, 1997.

[44] E. W. Kamen and B. S. Heck, *Fundamentals of Signals and Systems using the Web and Matlab*, 2nd Edition, Prentice-Hall, Upper Saddler River, NJ, 2000.

[45] Michael J. Roberts, *Signals and Systems – Analysis using Transform Methods and MATLAB*, McGraw-Hill, New York, 2003.

[46] G. F. Franklin, J. D. Powell, and A. Emami-Naeini, *Feedback Control of Dynamic Systems*, Addison-Wesley, Reading, MA, 1986.

[47] William A. Wolovich, *Automatic Control Systems*, Saunders College Publishing, Fort Worth, TX, 1994.

[48] Peter Dorato, *ANALYTICAL FEEDBACK SYSTEM DESIGN An Interpolation Approach*, Brooks/Cole, Pacific Grove, CA, 1999.

[49] Graham C. Goodwin, Stefan F. Graebe, and Mario E. Salgado, *Control System Design*, Prentice-Hall, Upper Saddle River, NJ, 2001.

[50] Vladimir Kučera, *Discrete Linear Control: The Polynomial Approach*, John Wiley & Sons, New York, 1979.

[51] M. Bodson and J. Chiasson, "Differential-geometric methods for control of electric motors", *International Journal of Robust and Nonlinear Control*, vol. 8, pp. 923–954, 1998.

[52] A. Isidori, *Nonlinear Control Systems*, Springer-Verlag, Berlin, 1989.

[53] J. Chiasson, "Nonlinear differential-geometric techniques for control of a series DC motor", *IEEE Transactions on Control Systems Technology*, vol. 2, no. 1, pp. 35–42, March 1994.

[54] Samir Mehta and John Chiasson, "Nonlinear control of a series DC motor: Theory and experiment", *IEEE Transactions on Industrial Electronics*, vol. 45, no. 1, pp. 134–141, February 1998.

[55] J. Chiasson and M. Bodson, "Nonlinear control of a shunt DC motor", *IEEE Transactions on Automatic Control*, vol. 38, no. 11, pp. 1662–1665, November 1993.

[56] Bodine Electric Co., *HANDBOOK: Small Motor, Gearmotor and Control*, Bodine Electric Company, Chicago, IL, 1993.

[57] John Kassakian, Martin Schlecht, and George Verghese, *Principles of Power Electronics*, Addison-Wesley, Reading, MA, 1991.

[58] A. E. Fitzgerald, C. Kingsley Jr., and S. D. Umans, *Electric Machinery*, McGraw-Hill, New York, 1983.

[59] Kaiyu Wang, John Chiasson, Marc Bodson, and Leon M. Tolbert, "A nonlinear least-squares approach for estimation of the induction motor parameters", in *Proceedings of the IEEE Conference on Decision and Control*, December 2004, pp. 3856–3861.

[60] Joachim Holtz, "On the spatial propagation of transient magnetic fields in ac machines", *IEEE Transactions on Industry Applications*, vol. 32, no. 4, pp. 927–937, July/August 1996.

[61] E. R. Laithwaite and L. L. Freris, *Electric Energy: Its Generation, Transmission and Use*, McGraw-Hill, New York, 1980.

[62] Frank Wicks, "How George Westinghouse changed the world", *Mechanical Engineering*, pp. 74–79, October 1996.

[63] Marc Bodson, "Induction machines as generators", Private communication, May 2003, Denali National Park, Alaska.

[64] R. Marino, S. Peresada, and P. Valigi, "Adaptive input-output linearizing control of induction motors", *IEEE Transactions on Automatic Control*, vol. 38, no. 2, pp. 208–221, February 1993.

[65] M. Bodson, J. Chiasson, and R. Novotnak, "High performance induction motor control via input-output linearization", *IEEE Control Systems Magazine*, vol. 14, no. 4, pp. 25–33, August 1994.

[66] R. Lessmeier, W. Shumacher, and W. Leonhard, "Microprocessor-controlled ac-servo drives with synchronous or induction motors: Which is preferable?", *IEEE Transactions Industry Applications*, vol. IA-22, no. 5, pp. 812–819, September/October 1986.

[67] M. Bodson, J. Chiasson, and R. Novotnak, "Nonlinear speed observer for high-performance induction motor control", *IEEE Transactions on Industrial Electronics*, vol. 42, no. 4, pp. 337–343, August 1995.

[68] Marc Bodson and John Chiasson, "A systematic approach to selecting optimal flux references in induction motors", in *Record of the 27th Annual Meeting of the Industry Applications Society*, October 1992, Houston, TX.

[69] M. Bodson, J. Chiasson, and R. Novotnak, "A systematic approach to selecting flux references for torque maximization in induction motors", *IEEE Transactions on Control Systems Technology*, vol. 3, no. 4, pp. 388–997, December 1995.

[70] Z. Krzeminski, "Nonlinear control of the induction motor", in *10th IFAC World Congress*, 1987, pp. 349–354.

[71] G. S. Kim, I. J. Ha, and M. S. Ko, "Control of induction motors for both high dynamic performance and high-power efficiency", *IEEE Transactions on Industrial Electronics*, vol. 39, no. 4, pp. 323–333, August 1992.

[72] M. Bodson, J. Chiasson, R. Novotnak, and R. Rekowski, "High performance nonlinear control of a permanent magnet stepper motor", *IEEE Transactions Control Systems Technology*, vol. 1, no. 1, pp. 5–14, March 1993.

[73] George C. Verghese and Seth Sanders, "Observers for flux estimation in induction machines", *IEEE Transactions on Industrial Electronics*, vol. 35, no. 1, pp. 85–84, February 1988.

[74] Phillipe Martin and Pierre Rouchon, "Two remarks on induction motors", in *CESA'96 IMACS Multiconference, Symposium on Control, Optimization and Supervision*, Lilles, 1996, pp. 76–79.

[75] Phillipe Martin and Pierre Rouchon, "Two simple flux observers for induction motors", *International Journal of Adaptive Control and Signal Processing*, vol. 14, pp. 171–175, 2000.

[76] X. Xu and D. W. Novotny, "Selection of the flux reference for induction machines in the field weakening region", *IEEE Transactions on Industry Applications*, vol. 28, no. 6, pp. 1353–1358, 1992.

[77] F. M. H. Khater, R. D. Lorenz, D. W. Novotny, and K. Tang, "Selection of the flux reference in field oriented induction machine controllers with consideration of magnetic saturation effects", *IEEE Transactions on Industry Applications*, vol. 23, no. 2, pp. 276–282, 1987.

[78] V. R. Stefanovic and T. H. Barton, "Static torque characteristics of an induction motor with variable frequency supply", in *Proceedings of the IEEE PES Winter Meeting*, 1972, pp. 1–3, New York.

[79] I. T. Wallace, D. W. Novotny, R. D. Lorenz, and D. M. Divan, "Increasing the dynamic torque per ampere capability of induction machines", *IEEE Transactions on Industry Applications*, vol. 30, no. 1, pp. 146–153, January/February 1994.

[80] S. H. Kim, S. K. Sul, and M. H. Park, "Maximum torque control of an induction machine in the field weakening region", in *Conference Record of the Industrial Applications Society*, 1993, pp. 401–407.

[81] B. C. Kuo, *Automatic Control Systems*, Prentice-Hall, Englewood Cliffs, NJ, 1987.

[82] R. Novotnak, J. Chiasson, and M. Bodson, "High-performance motion control of an induction motor with magnetic saturation", in *Proceedings of the 34th IEEE Conference on Decision and Control*, December 1995, pp. 2145–2150, New Orleans, LA.

[83] R. Novotnak, *Nonlinear Control of an Induction Motor with Saturation: Theory and Experiment*, PhD thesis, University of Pittsburgh, 1995.

[84] R. Novotnak, J. Chiasson, and M. Bodson, "High-performance motion control of an induction motor with magnetic saturation", *IEEE Transactions on Control Systems Technology*, vol. 7, no. 3, pp. 315–327, May 1999.

[85] A. Khambadkone and J. Holtz, "Vector-controlled induction motor drive with a self-commissioning scheme", *IEEE Transactions on Industrial Electronics*, vol. 38, no. 5, pp. 322–327, October 1991.

[86] G. Heinemann and W. Leonhard, "Self-tuning field-oriented control of an induction motor drive", in *Proceedings of the 1990 International Power Electronics Conference*, April 1990, Tokyo, Japan.

[87] J. Stephan, "Real-time estimation of the parameters and fluxes of induction motors", Master's thesis, Carnegie Mellon University, 1992.

[88] A. Bellini, A. De Carli, and M. La Cava, "Parameter identification for induction motor simulation", *Automatica*, vol. 12, pp. 383–386, 1976.

[89] David Cox, John Little, and Donal O'Shea, *IDEALS, VARIETIES, AND ALGORITHMS An Introduction to Computational Algebraic Geometry and Commutative Algebra*, 2nd Edition, Springer-Verlag, Berlin, 1996.

[90] Joachim von zur Gathen and Jürgen Gerhard, *Modern Computer Algebra*, Cambridge University Press, Cambridge, UK, 1999.

[91] Stephen Wolfram, *Mathematica, A System for Doing Mathematics by Computer*, 2nd Edition, Addison-Wesley, Reading, MA, 1992.

[92] David Kincaid and Ward Cheney, *Numerical Analysis: Mathematics of Scientific Computing*, 3rd edition, Brooks/Cole, Pacific Grove, CA, 2002.

[93] Miguel Vélez-Reyes, W. L. Fung, and J. E. Ramos-Torres, "Developing robust algorithms for speed and parameter estimation in induction machines", in *Proceedings of the IEEE Conference on Decision and Control*, 2001, pp. 2223–2228, Orlando, Florida.

[94] M. Vélez-Reyes, K. Minami, and G. Verghese, "Recursive speed and parameter estimation for induction machines", in *Proceedings of the IEEE Industry Applications Conference*, 1989, pp. 607–611, San Diego, California.

[95] Miguel Vélez-Reyes and George Verghese, "Decomposed algorithms for speed and parameter estimation in induction machines", in *Proceedings of the IFAC Nonlinear Control Systems Design Symposium*, 1992, pp. 156–161, Bordeaux, France.

[96] Miguel Vélez-Reyes, *Decomposed algorithms for parameter estimation*, PhD thesis, Massachusetts Institute of Technology, 1992.

[97] Miguel Vélez-Reyes, Milan Mijalković, Aleksandar M. Stanković, Silva Hiti, and James Nagashima, "Output selection for tuning of field-oriented controllers: Steady-state analysis", in *Conference Record of Industry Applications Society*, October 2003, pp. 2012–2016, Salt Lake City, UT.

[98] Kaiyu Wang, John Chiasson, Marc Bodson, and Leon M. Tolbert, "Tracking the rotor time constant of an induction motor traction drive for HEVs", in *Proceedings of the IEEE Workshop on Power Electronics in Transportation (WPET)*, October 2004, pp. 83–88.

[99] John Chiasson, Kaiyu Wang, Mengwei Li, Marc Bodson, and Leon M. Tolbert, "Algebraic methods for nonlinear systems: Parameter identification and state estimation", in *Current Trends in Nonlinear Systems and Control*. Laura Menini, Luca Zaccarian and Chaouki T. Abdallah, Eds., 2005, pp. 57–76, Birkhäuser.

[100] David Cox, John Little, and Donal O'Shea, *Using Algebraic Geometry*, Springer-Verlag, Berlin, 1998.

[101] Henk Nijmeijer and A. J. van der Schaft, *Nonlinear Dynamical Control Systems*, Springer-Verlag, 1990.

[102] R. Marino and P. Tomei, *Nonlinear Control Design - Geometric, Adaptive and Robust*, Prentice-Hall, Englewood Cliffs, NJ, 1995.

[103] B. Charlet, J. Levine, and R. Marino, "Two sufficient conditions for dynamic feedback linearization", in *Analysis and Optimization of Systems, Lecture Notes in Control and Information Sciences*. A. Bensoussan and J. L. Lions, Eds., 1988, vol. 111, pp. 181–192, Springer-Verlag, Berlin.

[104] B. Charlet, J. Levine, and R. Marino, "On dynamic feedback linearization", *Systems and Control Letters*, vol. 13, no. 2, pp. 143–151, August 1989.

[105] B. Charlet, J. Levine, and R. Marino, "Sufficient conditions for dynamic feedback linearization", *SIAM Journal on Control and Optimization*, vol. 29, no. 1, pp. 38–57, January 1991.

[106] J. Chiasson, "Nonlinear controllers for induction motors", in *Proceedings of the IFAC Conference on System Structure and Control*, July 1995, pp. 572–583, Nantes, France.

[107] J. Chiasson, "A new approach to dynamic feedback linearization control of an induction motor", *IEEE Transactions on Automatic Control*, pp. 391–397, March 1998.

[108] J. Chiasson, "Dynamic feedback linearization of the induction motor", *IEEE Transactions on Automatic Control*, vol. 28, no. 10, pp. 1588–1594, 1993.

[109] M. Fliess, J. Levine, P. Martin, and P. Rouchon, "Flatness and defect of nonlinear systems: introductory theory and applications", *International Journal of Control*, vol. 61, no. 6, pp. 1327–1361, 1995.

[110] Hebertt Sira-Ramírez and Sunil K. Agrawal, *Differentially Flat Systems*, Marcel-Dekker, New York, 2004.

[111] G. Conte, C. H. Moog, and A. M. Perdon, *Noninear Control Systems, Lecture Notes in Control and Information Sciences, Volume 242*, Springer-Verlag, 1999.

[112] R. Novotnak, "A state feedback tracking controller for a PM stepper motor", Master's thesis, University of Pittsburgh, 1991.

[113] B. Kuo and J. Tal, *Incremental Motion Control, Step Motors and Control Systems, Vol. II*, SRL Publishing, Champaign, IL, 1987.

[114] P. P. Acarnley, *Stepping Motors: A Guide to Modern Theory and Practice*, Peter Peregrinus, Ltd., London, 1982.

[115] T. Kenjo, *Stepping Motors and Their Microprocessor Control*, Clarendon, New York, 1984.

[116] A. Verl and Marc Bodson, "Torque maximization for permanent magnet synchronous motors", *IEEE Transactions on Control Systems Technology*, vol. 6, no. 6, pp. 740–745, November 1998.

[117] Marc Bodson, John Chiasson, and Leon M. Tolbert, "A complete characterization of torque maximization for permanent magnet non-salient synchronous motors", in *Proceedings of the American Control Conference*, June 2001, Arlington, VA.

[118] A. Blauch, "High-speed parameter estimation of step motors", Master's thesis, Carnegie Mellon University, 1992.

[119] A. Blauch, M. Bodson, and J. Chiasson, "High-speed parameter estimation of stepper motors", *IEEE Transactions on Control Systems Technology*, vol. 1, no. 4, pp. 270–279, April 1993.

[120] Marc Bodson and John Chiasson, "Open and closed loop control of a permanent magnet stepper motor", in *Handbook of Electric Motors.* Richard H. Engelmann and William H. Middendorf, Eds., 1995, pp. 540–552, Marcel Dekker, New York.

[121] Marc Bodson, Jeffrey S. Sato, and Stephen R. Silver, "Spontaneous speed reversals in stepper motors", in *Proceedings of the 2003 Conference on Decision and Control,* December 2003, Maui, Hawaii.

[122] J. Chiasson, D. Seto, F. Sun, A. Stanković, and S. Bortoff, "Independent control of two PM motors using a single inverter: Application to elevator doors", *International Journal of Mechatronics*, vol. 15, no. 1, pp. 95–110, February 2005.

Index

Printed and bound by CPI Group (UK) Ltd, Croydon, CR0 4YY

16/09/2026

14963443-0007